T0200651

33rd EUROPEAN SYMPOSIUM ON COMPUTER AIDED PROCESS ENGINEERING

VOLUME 1

COMPUTER-AIDED CHEMICAL ENGINEERING, 52

33rd EUROPEAN SYMPOSIUM ON COMPUTER AIDED PROCESS ENGINEERING

VOLUME 1

Edited by

Prof. Antonios C. Kokossis
National Technical University of Athens
School of Chemical Engineering
Athens, Greece

Prof. Michael C. Georgiadis
Aristotle University of Thessaloniki
School of Engineering
Department of Chemical Engineering
Greece

Prof. Efstratios Pistikopoulos
Chemical Engineering
Texas A&M Energy Institute
TX, USA

Amsterdam – Boston – Heidelberg – London – New York – Oxford
Paris – San Diego – San Francisco – Singapore – Sydney – Tokyo

Elsevier
Radarweg 29, PO Box 211, 1000 AE Amsterdam, Netherlands
The Boulevard, Langford Lane, Kidlington, Oxford OX5 1GB, UK
50 Hampshire Street, 5th Floor, Cambridge, MA 02139, USA

British Library Cataloguing in Publication Data
A catalogue record for this book is available from the British Library

Library of Congress Cataloging-in-Publication Data
A catalog record for this book is available from the Library of Congress

ISBN (Volume 1): 978-0-443-23553-5
ISBN (Set) : 978-0-443-15274-0
ISSN: 1570-7946

For information on all Elsevier publications visit our website at https://www.elsevier.com/

Publisher: Candice Janco
Acquisition Editor: Anita Koch
Editorial Project Manager: Lena Sparks
Production Project Manager: Paul Prasad Chandramohan
Designer: Greg Harris

Typeset by STRAIVE

Contents

Preface

The 3 volumes contain papers presented at the 33rd European Symposium of Computer Aided Process Engineering (ESCAPE-33) held in Athens, Greece from 18-21 June 2023. The ESCAPE Series started in 1992 and builds on a strong foundation of 32 previous events of the European Federation of Chemical Engineers (EFCE) Working Party on Computer Aided Process Engineering (CAPE). Hosting countries to the conference have been Austria (1993, 2018), Ireland (1994), Slovenia (1995, 2016), Greece (1996, 2011), Norway (1997), Belgium (1998), Hungary (1999, 2014), Italy (2000, 2010, 2020), Denmark (1992, 2001, 2015), The Netherlands (2002, 2019), Finland (2003, 2013), Portugal (2004), Spain (2005, 2017), Germany (2016), Romania (2007), France (1992, 2008, 2022), Poland (2009), United Kingdom (2012), and Turkey (2021).

ESCAPE-33 addresses emerging and significant challenges in **Process Systems Engineering as a driver to rebuild industry with a systemic and holistic approach.** Contributions relate to the sustainable development of chemical processes, the development and the systematic evaluation of processing technologies and process innovations, and research promoting **circular economy** paradigms including social aspects and social engagement networks to incentivize and involve citizens. The conference recognizes the exceptional importance of **industrial biotechnology** as a driver and enabler for new chemistries, alongside **Artificial Intelligence and data engineering** as a technological challenge with an apparent impact on many systems technologies. Overall, ESCAPE 33 covered up-to-date topics in Process Systems Engineering and attracted a significant and diversified number of people from academia, research institutions, and industrial organizations worldwide. The themes of the conference included:

Modelling and optimization for multi-scale integration
Coordinators: Antonio Espuña and Fani Boukouvala

Control, scheduling, and operability at the process and enterprise-level
Coordinators: Ana Barbosa-Povoa and Chrysanthos Gounaris

Safe and sustainable products by design
Coordinators: Andrè Bardow and Fengqi You

Green and sustainable processes for the circular economy
Coordinators: Seyed Mansouri and Bhavik Bakshi

Systems methods in industrial biotechnology and biomedical applications
Coordinators: Joern Viell and Christos Maravelias

Multi-scale energy systems engineering (organized by the EFCE energy section)
Coordinators: Francois Marèchal and Fabrizio Bezzo

Sustainable supply chains and ecosystems
Coordinators: Franjo Cecelja and Edwin Zondervan

Education and knowledge transfer
Coordinators: Ludovic Montastruc and Iqbal Mujtaba

The conference has been oversubscribed and we had to face a difficult task to review an excess of 800 submitted abstracts. The review process included reviews of abstracts, reviews of manuscripts, and the final selection of the revised manuscripts. Reviews involved 185 reviewers. They were completed primarily thanks to the generous support by members of the ESCAPE international scientific committee who contributed with evaluations, comments, and recommendations. The review process converged to 619 contributions from 57 countries. A total of 561 contributions by means of 6-page papers are included in these 3 volumes.

We hope that the books will serve as valuable reference documents to the scientific and industrial community and that they will contribute to the effective and innovative use of process systems engineering in the design, operation, and development of sustainable processes.

Antonis C. Kokossis
Michael C. Georgiadis
Stratos Pistikopoulos

LOCAL ORGANIZING COMMITTEE

Chair board
Antonis Kokossis (Chair) – National Technical University of Athens
Michael Georgiadis (Co-chair) – Aristotle University of Thessaloniki
Ioannis Kookos (Co-chair) – University of Patras
Konstantina Kosmidou(Treasurer) – National Technical University of Athens
Polyxeni Lazaropoulou (Secretary) – National Technical University of Athens

Members
Kostas Pyrgakis – National Technical University of Athens
Melina Psycha – National Technical University of Athens
Theodoros Damartzis – Aristotle University of Thessaloniki
Nikolaos A. Diangelakis – Technical University of Crete
Athanasios Papadopoulos – Centre for Research & Technology Hellas (CERTH)
Effie Marcoulaki – National Centre for Scientific Research 'Demokritos'
Chrysoula Kappatou – Imperial College London

Antonis Kokossis, Michael C. Georgiadis, Efstratios N. Pistikopoulos (Eds.)
PROCEEDINGS OF THE 33rd European Symposium on Computer Aided Process Engineering (ESCAPE33), June 18-21, 2023, Athens, Greece
 http://dx.doi.org/10.1016/B978-0-443-15274-0.50001-9

An exploratory model-based design of experiments technique to aid parameters identification and reduce prediction uncertainty

Francesca Cenci,[a] Arun Pankajakshan,[b] Solomon Gajere Bawa,[b] Asterios Gavriilidis, [b] Pierantonio Facco,[a] Federico Galvanin[b]

[a]*CAPE-Lab – Computer-Aided Process Engineering Laboratory, Department of Industrial Engineering, University of Padova, via Marzolo 9, 35131 Padova, Italy*

[b] *Department of Chemical Engineering, University College London, Torrington Place, London WC1E 7JE, United Kingdom*

Abstract

When developing mathematical models to describe reaction processes, model parameters require to be estimated from experimental data. Experiments are traditionally designed through techniques aiming at space exploration, like space-filling methods (e.g., Latin Hypercube sampling or LHS), or at information maximization, like model-based design of experiments (MBDoE). However, the former methods do not minimize parameters uncertainty, while the latter do not ensure a minimization of model prediction uncertainty in the entire experimental design space. In this work, we propose a novel exploratory MBDoE (*eMBDoE*) approach based on G-optimality calculation (*G-map* eMBDoE) to simultaneously enhance space exploration and minimize model prediction variance. The method is tested on a case study related to the identification of kinetic parameters of catalytic total methane oxidation in a flow microreactor. Results show that the method is more explorative than conventional MBDoE and more efficient than LHS and MBDoE in reducing model prediction uncertainty and parameters uncertainty.

Keywords: model-based design of experiments; model prediction uncertainty; parameters identification; design space exploration; total methane oxidation

1. Introduction

For a model to be representative of the system under study, its parameters must be estimated from experimental data. However, not all experiments are equally informative for estimating model parameters and scarce information may lead to unsatisfactory parameters precision even with a considerable amount of data. The information of each possible experiment can be estimated through the Fisher information matrix (FIM; Fisher, 1950), which depends on the sensitivity of the model response with respect to every model parameter. This is used by model-based design of experiments (MBDoE; Asprey and Macchietto, 2000), which is an optimization problem where the objective function is a scalar measure of the FIM and the result is a set of values for the manipulated variables that ensure a maximization of the information. Maximum information leads to the minimization of parameters uncertainty. However, optimal experimental design techniques are inherently exploitative and may suffer from scarce space exploration. Moreover, there is no guarantee that the minimization of parameters uncertainty leads to a minimization of model prediction uncertainty in the whole design space. In this work, we propose a modification of the MBDoE optimization framework in order to enhance space exploration and to minimize model prediction variance in the entire domain of model utilization. The proposed procedure is based on the G-optimality (Kiefer and Wolfowitz, 1959), which estimates model prediction variance starting from the FIM, thus it does not require the acquisition of additional data. Therefore, a mapping of G-optimality values (*G-maps*) can be built for the whole design space and MBDoE optimization is performed in the regions that satisfy a user-defined requirement on the G-optimality value. This explorative MBDoE procedure based on G-map (*G-map eMBDoE*) is applied to a case study on the identification of kinetic parameters of Mars-van Krevelen reaction mechanism for total oxidation of methane. Herein,an automated flow micropacked bed catalytic reactor platform is simulated to generate in silico data. Thanks to simulated experimental campaignsG-map eMBDoE is compared against a purely information-based MBDoE and a purely exploration-based technique, namely Latin Hypercube sampling (LHS; McKay et al., 1979).

2. Methods

The estimation of the N_{θ} model parameters $\hat{\boldsymbol{\theta}}$ is carried out through Maximum Likelihood Estimation (MLE; Bard, 1974). The reference techniques for design of experiments are: model-based design of experiments (MBDoE), which aims at maximizing the experiments information, and Latin Hypercube sampling (LHS), which aims at an explorative design. The former can be represented by the optimization problem:

$$\boldsymbol{\varphi}_{opt} = \arg\min_{\boldsymbol{\varphi}} \psi(\boldsymbol{H}_{\hat{\theta}}) \tag{1}$$

where $\boldsymbol{\varphi}$ is the design vector, i.e. the set of control variables that determine the experimental conditions, and $\psi(\mathbf{H}_{\hat{\theta}})$ is a scalar measure of the FIM $\mathbf{H}_{\hat{\theta}}$. Consider a general differential and algebraic model $\mathbf{f}$: $\mathbf{f}(\dot{\mathbf{x}}, \mathbf{x}, \mathbf{u}, t, \boldsymbol{\theta}) = \mathbf{0}$, $\mathbf{y} = \mathbf{h}(\mathbf{x})$, with N_y measurable responses ($\hat{\mathbf{y}}$), $N_{\hat{\theta}}$ parameters ($\hat{\boldsymbol{\theta}}$), N_u control variables ($\mathbf{u}$), N_x-dimensional vectors of state variables and their first derivatives ($\mathbf{x}$ and $\dot{\mathbf{x}}$, respectively) that change over time (t). Then, the $N_{\theta} \times N_{\theta}$ FIM is defined as:

$$\boldsymbol{H}_{\hat{\theta}}(\hat{\boldsymbol{\theta}}, \boldsymbol{\varphi}) = \left[\boldsymbol{V}_{\hat{\theta}}^{0}\right]^{-1} + \sum_{i=1}^{n_{sp}} \left(\frac{d\hat{\mathbf{y}}}{d\hat{\boldsymbol{\theta}}}\right)_i^T \boldsymbol{\Sigma}_y^{-1} \left(\frac{d\hat{\mathbf{y}}}{d\hat{\boldsymbol{\theta}}}\right)_i \quad \text{, with:} \quad \boldsymbol{H}_{\hat{\theta}}(\hat{\boldsymbol{\theta}}, \boldsymbol{\varphi}) \cong [\boldsymbol{V}_{\hat{\theta}}]^{-1} \tag{2}$$

Thus, the FIM depends on the prior information $\left[\mathbf{V}_{\hat{\theta}}^{0}\right]^{-1}$(related to parameters bounds; Bard, 1974), on the $N_y \times N_y$ response variance covariance matrix $\boldsymbol{\Sigma}_y$ and on the $N_y \times N_{\theta}$ matrix of sensitivities $\left(\frac{d\hat{\mathbf{y}}}{d\hat{\boldsymbol{\theta}}}\right)_i$ calculated at the i-th sampling point (n_{sp} is the total number of sampling points, given by the sum of the N_{sp} sampling points for every response). As shown in Eq. (2), the parameters variance covariance matrix $\mathbf{V}_{\hat{\theta}}$ can be approximated as the FIM inverse: therefore, the higher the experiments information content, the smaller the parameters uncertainty region. In this work, we minimize the direction of maximum variability of the uncertainty region by maximizing the minimum FIM eigenvalue $\psi(\mathbf{H}_{\hat{\theta}}) = -(\lambda_{\min}(\mathbf{H}_{\hat{\theta}}))$ (E-optimal criterion).

After parameters identification, the reliability of model predictions can be assessed through the G-optimality V_y, i.e. a $N_y N_{\mathrm{sp}} \times N_y N_{\mathrm{sp}}$ matrix where the ji-th diagonal element represents the prediction variance of the j-th response at the i-th time point. The ji-th element $V_y(\hat{\boldsymbol{\theta}}, \boldsymbol{\varphi})\big|_{j,i}$ is calculated as:

$$V_y(\hat{\boldsymbol{\theta}}, \boldsymbol{\varphi})\big|_{j,i} = \left(\frac{d\hat{y}_j}{d\hat{\boldsymbol{\theta}}}\right)_i^T [\boldsymbol{H}_{\hat{\theta}}]^{-1} \left(\frac{d\hat{y}_j}{d\hat{\boldsymbol{\theta}}}\right)_i \qquad \text{for } j = 1, \dots, N_y;\ i = 1, \dots, N_{sp} \tag{3}$$

where $\left(\frac{d\hat{y}_j}{d\boldsymbol{\theta}}\right)_i$ is the $N_\theta \times 1$ vector of sensitivities. V_y depends on the current parameter estimates $\hat{\boldsymbol{\theta}}$ and it is evaluated at each condition $\boldsymbol{\varphi}$ in the design space. For the same $\boldsymbol{\varphi}$, $N_y N_{sp}$ values of $V_y(\hat{\boldsymbol{\theta}}, \boldsymbol{\varphi})\big|_{j,i}$ are calculated; a scalar measure J_G summarizing them all can be defined as:

$$J_G = \sum_{j=1}^{N_y} \sum_{i=1}^{N_{sp}} V_y(\hat{\boldsymbol{\theta}}, \boldsymbol{\varphi})\big|_{j,i} \tag{4}$$

We define *G-map* as the mapping of J_G in the whole design space and we integrate it into the MBDoE optimization seeking enhancement of space exploration and minimization of model prediction uncertainty in the whole domain of model utilization. This is done through the steps illustrated in Fig.1:

- the design space is discretized into a grid of equally spaced points (black dots in (a)) representing experimental conditions $\boldsymbol{\varphi}$. In the figure, two inputs are shown for sake of simplicity, but it can be extended to any number N_u;
- every point is characterized in terms of information content ψ and total variance J_G;
- only a subset of points is retained for the optimization of Eq. (1). These are the $\boldsymbol{\varphi}$ satisfying the condition: $J_G \geq J_{G,\text{thr}} J_{G,\max}$ (blue circles in (b)), where $J_{G,\max}$ is the maximum J_G calculated in the grid, while $J_{G,\text{thr}} \in [0,1]$ is a user defined threshold that allows to balance space exploration and information maximization;
- finally, the $\boldsymbol{\varphi}_{\text{opt}}$ satisfying Eq. (1) is selected among the candidates (red star in (c)). It is used as initial guess for a subsequent constrained optimization in Python 3.9 with the Scipy optimization package.

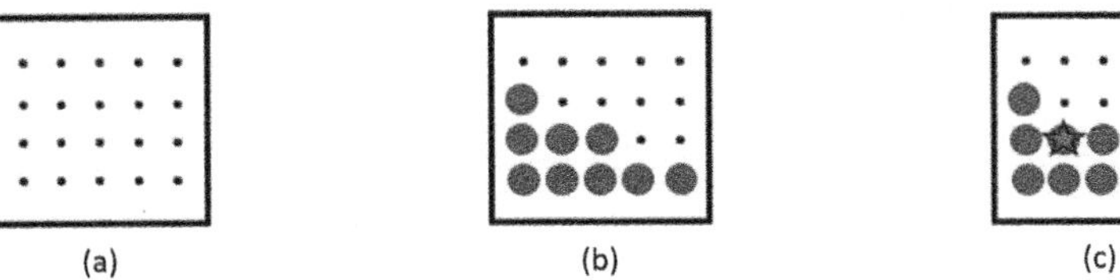

Figure 1. *Graphical representation of the explorative MBDoE based on G-maps.*

This procedure is iterated until the budget of experiments (N_e) is reached; model calibration is performed at every iteration as soon as the new experiment is designed.

3. Results and discussion

To test the performance of the G-map eMBDoE, a case study related to the identification of reaction models for catalytic total oxidation of methane over a 5% Pd/Al_2O_3- catalyst is considered. The mechanism is described through Mars-van Krevelen reaction kinetics in a flow microreactor system, which proved to be the most suitable kinetic model in Bawa et al. (2022). The control variables and their ranges are: temperature (u_1), 250-350 °C; total flow rate (u_2), 20-30 Nml min^{-1}; inlet methane concentration (u_3), 0.005-0.025 mol mol^{-1}; oxygen to methane mole ratio (u_4), 2-4 mol mol^{-1}. The three responses $\boldsymbol{y}$ are: unreacted methane, unreacted oxygen and carbon dioxide mole fractions, with measurement errors standard deviations $\boldsymbol{\sigma}_y = [\sigma_{CH_4}, \sigma_{O_2}, \sigma_{CO_2}] = [0.00043, 0.00202, 0.00051]$. Consider the kinetic rate of methane:

$$r_{CH_4} = k_1 k_2 P_{CH_4} P_{O_2} / (k_1 P_{O_2} + 2k_2 P_{CH_4} + (k_1 k_2 / k_3) P_{O_2} P_{CH_4}) \tag{5}$$

To aid parameters identification, Arrhenius and Van't Hoff equations are reparametrized:

$$k(T) = exp(logk(T_{ref}) - (E_a/R)(1/T - 1/T_{ref})) \quad (6)$$

$$K(T) = exp(logK(T_{ref}) - (\Delta H/R)(1/T - 1/T_{ref})) \quad (7)$$

Thus, six parameters must be estimated: $\widehat{\boldsymbol{\theta}} = [(-\log k_1(T_{\text{ref}})), \left(\frac{E_{a,1}}{10^4}\right), (-\log k_2(T_{\text{ref}})), \left(\frac{E_{a,2}}{10^4}\right), (-\log k_3(T_{\text{ref}})), \left(\frac{E_{a,3}}{10^4}\right)]$. Parameters estimation is carried out with initial guesses $\widehat{\boldsymbol{\theta}}_{\text{guess}} = [3, 8, 3, 10, 10, 8]$ and $[0,15]$ as parameter Bounds. Experiments are executed in silico by simulating the flow reactor model concentration responses assuming the following set of parameters to represent the system: $\widehat{\boldsymbol{\theta}}_{\text{true}} = [5.31, 6.96, 4.88, 10.49, 10.44, 7.95]$ and by adding random gaussian noise with zero mean and standard deviation $\boldsymbol{\sigma}_{\text{y}}$.

A G-map eMBDoE with thresholds $J_{G,thr} = 0.50$ and $J_{G,thr} = 0.70$ is compared against classical MBDoE and LHS. Since LHS can change randomly the allocation of design points in the design space, three different runs are considered. Finally, 8 LHS points are used as preliminary experiments to obtain initial parameters estimates as well as to gain minimum threshold information prior to applying MBDoE; they are the same in all simulations. The results are analyzed in terms of:

- space exploration, evaluated as deviation from the MBDoE design: $d_i = \text{abs}(u_{i,l} - u_{i,\text{ref}})/u_{i,\text{ref}}$, with $i = 1,2,3,4$; l refers to $J_G = \{0.50, 0.70\}$, while the reference ref is MBDoE;
- mean and maximum J_G in the whole space; the smaller these indices, the smaller the model prediction uncertainty in the whole space of model utilization;
- parameters precision tests satisfied when $t_{\text{value},i} \geq t_{\text{ref}}$ (Aprey and Naka, 1999).

As seen in Fig. 2, the higher the G-optimality threshold, the greater the deviation of the optimal input values from the ones calculated through a standard MBDoE, especially at the first iterations of the experimental campaign. For instance, variable u_1 with $J_G = 0.70$ has $d_i > 0$ starting from the 3rd iteration, while with $J_G = 0.50$ it has $d_i > 0$ after the 8th iteration. The results in terms of LHS are not shown since their departure from the MBDoE design is expected due to its explorative nature.

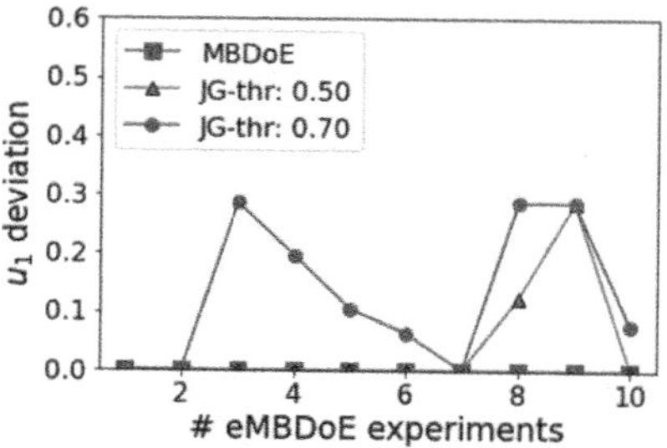

Figure 2. *Deviation of the Gmap eMBDoE design with* $J_G = \{\mathbf{0.50}, \mathbf{0.70}\}$ *from the MBDoE optimal design. Only the deviations of variable 1 is shown for sake of conciseness. The x-axis shows the number of eMBDoE experiments after 8 LHS preliminary experiments.*

The effect of this deviation from MBDoE conditions is assessed in terms of reduction of model prediction variance (Fig. 3 a-b) and increase of parameters precision (Fig. 3 c-h).

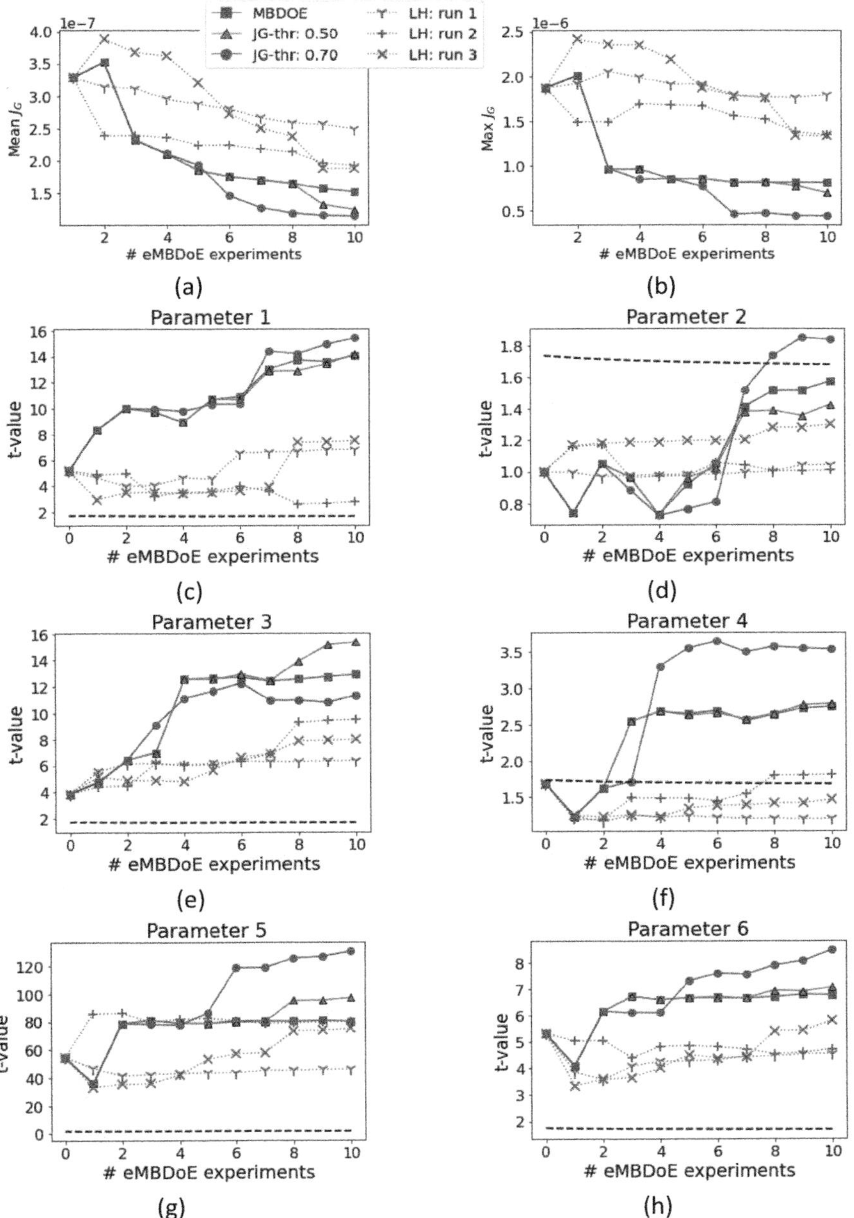

Figure 3. *Results of eMBDoE compared to LHS and MBDoE: (a)-(b) mean and maximum J_G in the whole design space; (c)-(h) t-values for the full set of model parameters. In (c)-(h), the black dotted line represents the reference t-value, while the 0 in the x-axes represents the results of model calibration with the 8 LHS preliminary experiments only.*

As shown in Fig.3 a-b, a purely explorative technique such as Latin Hypercube sampling has the worst performance in the reduction of model prediction uncertainty; this is true for all the three runs considered. MBDoE is able to reduce model prediction uncertainty, but the best performance is obtained by the explorative MBDoE, particularly when a stricter threshold (namely, $J_{G,thr} = 0.70$) is used.

Finally, assuming a maximum budget of 10 experiments, parameters identification is characterized by (Fig. 2 c-h):

- statistically satisfactory estimation of parameters $\hat{\theta}_1, \hat{\theta}_3, \hat{\theta}_5$ and $\hat{\theta}_6$ with the preliminary dataset in all scenarios;
- $\hat{\theta}_4$ estimated precisely from 3 optimal experiments for MBDoE and eMBDoE, while the estimate is more critical for LHS (estimated in one run over 3);
- the most critical parameter, e.g. $\hat{\theta}_2$, is sufficiently precise only when a Gmap eMBDoE is used with a strict threshold of 0.70.

This suggests that the proposed explorative MBDoE does not cause a loss of information content.

Conclusion

A novel MBDoE technique is proposed in this paper in order to favor experimental design space exploration towards regions characterized by higher prediction uncertainty. Such regions are detected by using G-optimality maps, while the trade-off between information maximization and space exploration can be handled through a G-optimality threshold. The stricter the threshold, the higher the deviation of eMBDoE design from a conventional MBDoE. The new explorative and optimal design has benefits in terms of reduction of prediction variance as it reduces the mean and maximum G-optimality values in the entire design space with respect to MBDoE. Furthermore, G-map eMBDoE is more effective on estimating the full set of model parameters, since all t-tests are passed within 8 eMBDoE experiments. However, results show how a LHS space filling design, which is inherently explorative, can perform poorly both in terms of prediction variance minimization and increase of parameters precision. This suggests that the best experimental design performance is obtained only if a balance between space exploration and information maximization can be achieved.

References

S. P. Asprey, S. Macchietto (2000). Statistical Tools for Optimal Dynamic Model Building. *Comput. Chem. Eng*, **24**, 1261-1267.

S. P. Asprey, Y. Naka, (1999). Mathematical problems in fitting kinetic models-some new perspectives. *J. Chem. Eng. Jpn*, **32(3)**, 328–337

Y. Bard (1974). Nonlinear Parameter Estimation. New York, NY: Academic Press.

S. G. Bawa, A.Pankajakshan, C. Waldron, E. Cao, F. Galvanin, A. Gavriilidis (2022). Rapid Screening of Kinetic Models for Methane Total Oxidation using an Automated Gas Phase Catalytic Microreactor Platform. *Chemistry—Methods*, e202200049.

R. A., Fisher (1950). Contributions to Mathematical Statistics. Papers 10, 11,38, JWS.

J. Kiefer, Wolfowitz (1960), The equivalence of two extremum problems, *Canad. J. Math.* **12**, 363-366.

M.D. McKay, R.J. Beckman, W.J. Conover (1979). "A Comparison of Three Methods for Selecting Values of Input Variables in the Analysis of Output from a Computer Code". *Technometrics*. American Statistical Association. **21** (2), 239–245.

Antonis Kokossis, Michael C. Georgiadis, Efstratios N. Pistikopoulos (Eds.)
PROCEEDINGS OF THE 33rd European Symposium on Computer Aided Process Engineering (ESCAPE33), June 18-21, 2023, Athens, Greece
 http://dx.doi.org/10.1016/B978-0-443-15274-0.50002-0

GHG emission reduction via multiple fuel options in carbon-constrained industrial clusters

Christelle Bechara[a], Sabla Y. Alnouri[b*]

[a]*Baha and Walid Bassatne Department of Chemical and Petroleum Engineering, American University of Beirut, P.O. Box 11-0236, Riyad El-Solh, Beirut, Lebanon*
[b]*Gas Processing Center, College of Engineering, Qatar University, Doha, Qatar*
sabla@qu.edu.qa

Abstract

This paper primarily investigates the effect of incorporating multiple fuel options onto carbon constrained industrial clusters. The purpose of doing so was to maximize the profit through the optimal allocation of multiple types of fuel. The cluster consists of a number of different production plants and a power plant for the processing of value-added products as well as the generation of electricity. The power plant is connected to the electric grid and constitutes the main source of electricity for the cluster. It also has the ability to produce more power and export it to the grid. The integration of additional plants that could generate new products, convert, or store carbon dioxide through CCUS technologies, and produce cleaner power through the use of renewable and/or multiple and blended fuels is assessed. It was found that the incorporation of multiple types of fuel within a single cluster resulted in the lowest post-capture emissions with 9469 t CO_2/d generating a high-end profit of $907.16 million/y, when compared to cases which relied on a single fuel option only.

Keywords: GHG Emissions, Industrial Clusters, Biomass, Energy, Fuel

1. Introduction

Fossil fuels, mainly oil, coal, and gas, have been dominating the world's energy market. In fact, they are considered to be the most commercially attractive, (Wang and Economides, 2009). Moreover, the versatile functions that are associated with such fossil fuels, such as the production of value-added products to the supply of electricity and power, has all led to a drastic increase in industrial dependence on such energy sources. The industrial sector's energy consumption has been forecasted to grow by 32% by the year 2050 (Wang, 2019), which inevitably instigates major environmental concerns. Fossil fuel utilization is associated with high CO_2 emissions. Recent studies have shown that 18% of global emissions are a result from the combustion of natural gas, 40% from the combustion of coal, and 31% from the combustion of oil and (Olivier and Peters, 2018). This has pushed many governments to start enforcing carbon taxes on industrial clusters and set CO_2 emission limits in compliance with the targets imposed by the Paris Agreement. The need to satisfy the rising energy demands, while conforming to the set emission bounds, has pushed mitigation techniques that could be implemented in industrial clusters. Bechara and Alnouri (2022) have previously assessed and compared different energy scenarios in industrial clusters. In this work, the incorporation of multiple and simultaneous fuel options within an industrial cluster is studied, and compared against single type fuel clusters.

2. Methodology

The generic plant in this model can have several active sinks and sources, as shown in Figure 1. The sink options include: a fuel sink, a power sink, and a CO_2 sink, while the source options include: a product source, a CO_2 source, and a power source. The plant can be of two types: a production plant or a power plant with one source of fuel allocated to it. In terms of inputs, the plant requires a fuel input either to produce a value-added product or to generate power, a power input to satisfy its electricity requirements, and/or a carbon input, as in the case of a carbon utilization plant. In terms of outputs, a plant can produce products to be exported out of the industrial cluster, generate power to either satisfy the cluster's need or to be exported as revenue to the electricity grid, and emit carbon dioxide, which in turn is either released into the atmosphere or captured, stored, or utilized by the carbon capture and utilization technologies. Moreover, each production plant contributes to the total CO_2 emissions of the cluster based on the number of products it produces, the quantity of fuel, and the amount of electricity it requires to meet the set production. The power plant, in turn, contributes based on the amount of electricity the cluster requires and the quantity exported to the grid as an additional revenue source. The integration of the additional plants, the option of utilizing several types of fuels as well as renewable power within the cluster serve the following functions:

- The reduced dependence of the cluster on one type of fossil fuel.
- The reduction of the total CO_2 footprint of the cluster through the use of carbon utilization technology with the possibility of monetizing the captured carbon.
- The production of a clean source of power to satisfy the cluster's requirement and reduce the fossil fuel requirement for power production.

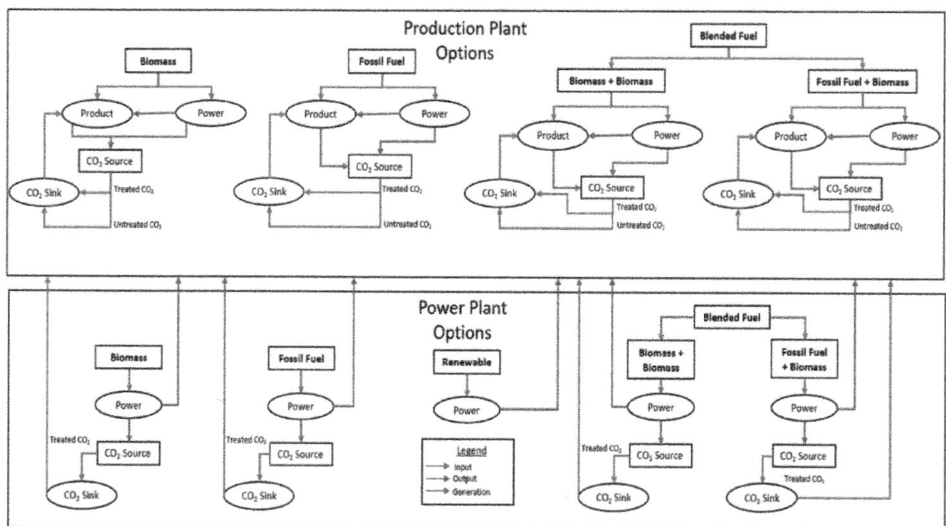

Figure 1: A generic illustration of sinks and sources within an industrial cluster

Because of the large number of variables involved in the optimization problem, from fuel allocation to CO_2 utilization, and power generation, a superstructure network approached is used to capture all the available configurations and determine the optimal selection. The cluster has an infrastructure suitable for the allocation and transportation of the different types of fuels that are introduced, from natural gas, to coal, to biomass, and blended fuel. All the product streams as well as the power exported to the grid are considered as revenue streams that serve the end profit of the cluster. All active power sources are connected to the cluster's electricity grid which is supplied by the existing power plant. The CO_2 streams, treated and untreated, are connected to the CO_2 active sinks of all plants and the uncaptured CO_2 sources are emitted into the atmosphere. Taking all the aforesaid connections and networks into account, the optimization model was primarily designed determine the optimal scenario that respects all constraints and results in the maximum profit for the cluster. To achieve the aforementioned, this work presents a superstructure optimization-based approach that aims at developing sustainable fuel and carbon utilization strategies to meet the environmental restrictions at a maximum end-profit. To summarise, the optimization model is able to determine:

- The selection the production plants and the type of power plant.
- The integration of renewable energy as a replacement to the fuel-derived energy with regard to the economic and environmental aspects.
- The optimal allocation of the fuel options to the production and power plants.
- The integration of carbon capture technologies onto the cluster.
- The capturing, treating, and monetizing the captured carbon.

Moreover, the following factors are assumed to be known:

- The existing production plants, their capacities, and their locations in the cluster.
- The products that are produced in every production plant in the cluster.
- The characteristics and costs of the fuels to be allocated.
- The power requirements of the production plants in the cluster.
- The maximum renewable percentage that can replace the fuel-derived power.
- The product-to-fuel and power-to-fuel requirements.
- The parameter that describes the carbon emitted per ton of product, kWh of power produced, during the transport of solid fuels
- The revenue/cost parameter of every revenue/cost stream.
- The capital and operating costs of every plant.
- The cost of the carbon integration network (transportation, capture, treatment, compression, and pumping parameters) and the carbon capture limit.

3. Case Study

The integration of different types of fuels was assessed based on both the economic and environmental feasibilities. The cluster investigated includes a set of existing and optional production plants. The existing plants were Methanol A, Aluminium, Cement and the natural gas power plant. Whereas the optional entities were Methanol B, Greenhouse, Saline storage and Enhanced Oil Recovery. Moreover, a renewable power plant with PV units has also been considered as an optional addition. Choosing between a natural gas, coal, biomass, and blended fuel power plants, and carbon utilization technologies were considered as part of the optimization problem. All process data has been obtained from Al-Mohannadi at al. (2017). The following additional assumptions were used:

- Natural gas, coal, biomass, and blended fuel are all integrated in the cluster.
- Only one type of fuel-derived power plant exists.
- Each production plant can have one type of fuel allocated to it.
- The net carbon capture rate is set to a minimum of 20%. The optimization model will select the optimal carbon capture percentage based on the set constraints.

The Mixed Integer Linear Program (MILP) involved a total of 634 variables, with a CPU time of 3 mins and 20s using What'sBest (Lindo systems, 2006) Lindo Branch-and-Bound solver for MS-Excel 2013 via a laptop PC with Intel Core i7 Duo processor, 8 GB RAM and a 64-bit operating System. It was found that with a natural gas flow of 444.25 t natural gas/d, a coal flow of 888.35 t coal/d, a biomass flow of 585.39 t biomass/d, and a blended fuel flow of 756.14 t blended fuel/d and a carbon capture rate constraint set to a minimum of 20%, the cluster produced 5000 tMeOH(A), 10000 t Portland-cement, 1118.44 tAluminum, and 2600 tMeOH(B). The optimization selected the natural gas plant as the operating power plant, which satisfied the cluster's power requirements. The option of exporting power to the grid was not reported in the optimal solution, instead, the option of operating the renewable energy plant at the highest capacity to replace 5% (1010526.32 kWh/d) of the power supplied by the fossil fuel plant. The cluster's profit turned out to be $907.16 million/y, slightly lower than that for the natural gas case study, but higher than the remaining case studies. However, the different fuel requirements are much lower, reducing the dependency on one type and showing the advantage of integrating more than one fuel into an industrial cluster. The activation of the renewable solar plant further shows the opportunity of gradually integrating more sources of green energy, such as green hydrogen, without being limited by capacity or by efficiency. To better understand the cluster's economics, with multiple fuels enabled, a cost distribution is presented in Figure 2. With the cluster's large product requirements, the highest costs resulted from the production sector, requiring $1010.12 million/y, followed by the power sector, consisting of the natural gas power plant and the solar photovoltaic plant costs with $151.53 million/y.

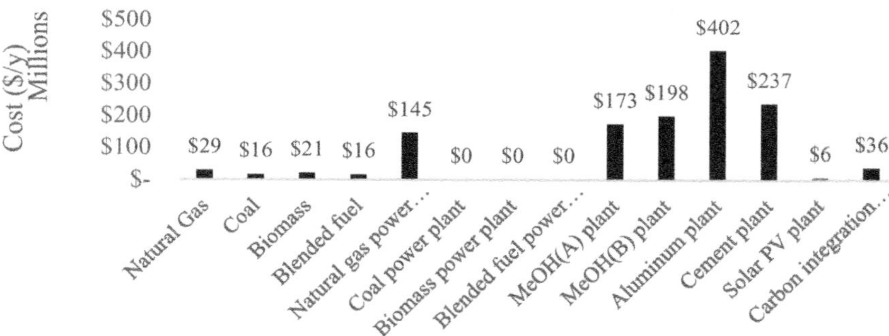

Figure 2: Optimal cost distribution attained

Analysing the carbon integration network's cost distribution, displayed in Figure 3, it can be seen that the highest cost is attributed to the treatment unit (Almohannadi and Alnouri, 2021), accounting for 65.28% of the total carbon integration costs. This high number is also associated with the fact that the optimization model selected to treat 99.91% of the carbon streams.

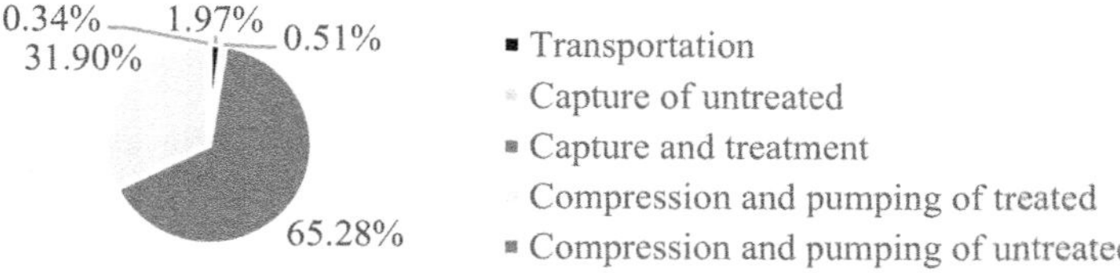

Figure 3: Carbon Network Cost Distribution

Regarding the carbon footprint, the total emissions of the cluster without capture was found to be 17640 t CO_2/d, lower than those for the natural gas, biomass, and blended fuel cases, but slightly higher than that for the coal case study. The latter can be attributed to the higher production capacity of this cluster in comparison to the coal cluster. After applying carbon capture and utilization technologies and selecting the optimal source to sink mapping (refer to Figure 4) the carbon captured amounted to 13796 t CO_2/d reducing the cluster's emissions by 47% resulting in a final footprint of 9345 t CO_2/d, and a capture rate of 79%. Compared to the standalone and blended fuel case studies, the resulting emissions post capture of the multiple fuel case study turned out to be the lowest showing the great economic and environmental potentials of integrating multiple fuels into a cluster to mitigate its carbon footprint. Furthermore, the optimization chose to operate the solar photovoltaic plant at full capacity. Therefore, adding more capacity or integrating more than one type of renewable energy could result in a higher replacement of the fossil-fuel derived energy. Figure 4 summarizes the optimal source to sink mapping obtained. A treated carbon stream was allocated from aluminum to methanol (B), treated carbon from cement and a mixture of treated and untreated carbon from the natural gas power plant to the enhanced oil recovery, treated carbon from methanol (B) to the greenhouse, and treated carbon from methanol (A) and the natural gas power plant to the saline storage.

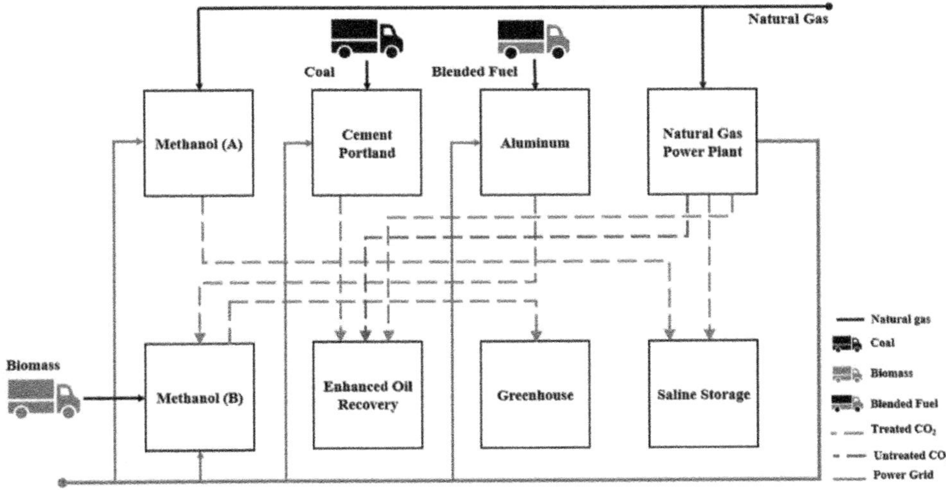

Figure 4: multiple fuel CO_2 network superstructure

Looking at the carbon emissions distribution in the cluster (illustrated in Figure 5) it can be seen that the production and power sector generated the highest emissions with 10780.58 t CO_2/d and 6302.69 t CO_2/d. Comparing the power emissions in every case study, natural gas resulted in 0.000691 tCO2/kWh, coal in 0.00077 tCO2/kWh, biomass in 0.000773 tCO2/kWh, blended fuel in 0.00077 tCO2/kWh, and finally multiple fuels in 0.000328 tCO2/kWh. The values were obtained by dividing the carbon emissions by the fuel derived power. Hence, the incorporation of multiple fuels yielded the lowest emissions per kWh, proving the advantage of adding renewable sources of energies into the power portfolio of a cluster. The flow rates of the different types of fuel are represented in Figure 6, showing the lower flow requirements needed in this analysis compared to those of the standalone and blended fuels. This further proves that integrating different types of fuel results in a reduction in the reliance of the industry on one type. The total attained revenue for this case was found to be $2170 million/y.

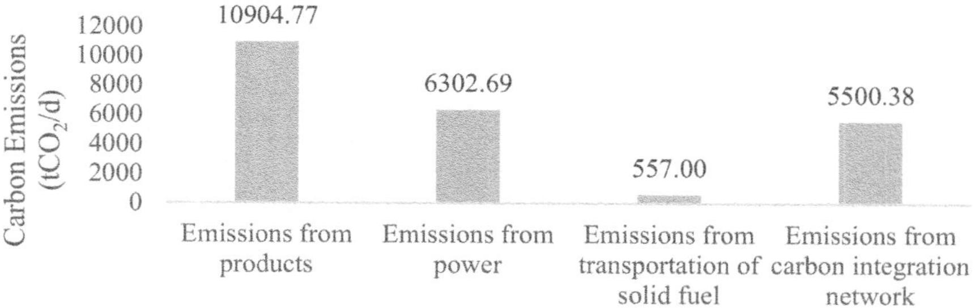

Figure 5: carbon emissions distribution in the cluster

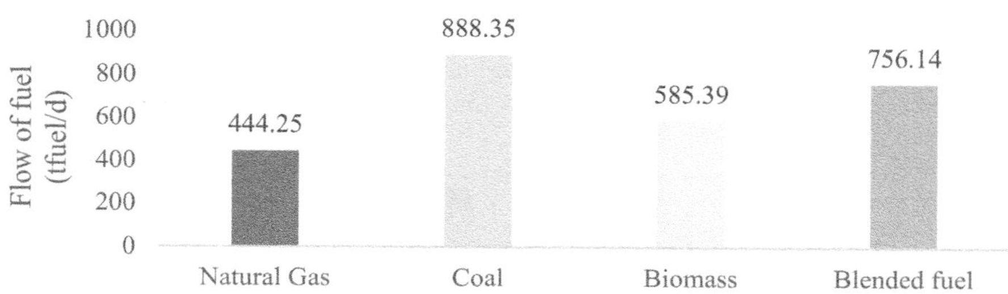

Figure 6: Total fuel flows fuel into the multiple fuel industrial cluster

4. Conclusion

A mathematical model for a systematic approach that integrates multiple types of fuels into industrial clusters was developed, by integrating the following multiple fuel options into the cluster: natural gas, coal, biomass, and blended fuel. The results obtained pertaining to fuel use, and its distribution amongst product and power sinks have been discussed. Moreover, the carbon emission footprint has analysed, and potential reduction strategies via CCUS utilization have been proposed, so as to meet the set emission target.

References

C. Bechara, S. Y. Alnouri, 2022, Energy assessment strategies in carbon-constrained industrial clusters. Energy Conversion and Management, 254, 38, 115204.

D. Al-Mohannadi, K. Abdulaziz, S. Y. Alnouri, P. Linke, 2017, On the synthesis of carbon constrained natural gas monetization networks, Journal of Cleaner Production, 168, 735-745.

D. Al-Mohannadi, S. Y. Alnouri, 2021, Studying the effect of solvent properties in treatment design within CO_2 Integration Networks, Computer Aided Chemical Engineering, 50, 295–300

R. E. Ooi, D. C. Foo, R. R. Tan, D. K. Ng, R. Smith, 2013, Carbon Constrained Energy Planning (CCEP) for Sustainable Power Generation Sector with Automated Targeting Model. Industrial & Engineering Chemistry Research, 52, 29, 9889-9896.

X. Wang, M. Economides, 2009, Natural Gas Supply, Alternative Energy Sources, and the Environment, Advanced Natural Gas Engineering, 303–331.

T. Wang, 2019, Forecast for worldwide industrial energy consumption from 2018 to 2050, by fuel, In. Statistia 2020.

J. G. Olivier, J. A. Peters, 2018, Trends in Global CO_2 and Total Greenhouse Gas Emissions, The Hague: www.pbl.nl/en

Antonis Kokossis, Michael C. Georgiadis, Efstratios N. Pistikopoulos (Eds.)
PROCEEDINGS OF THE 33rd European Symposium on Computer Aided Process Engineering (ESCAPE33), June 18-21, 2023, Athens, Greece
 http://dx.doi.org/10.1016/B978-0-443-15274-0.50003-2

An optimization-free Fisher information driven approach for online design of experiment

Andrea Friso,[a] Federico Galvanin,[a]

[a]*Department of Chemical Engineering University College London, Torrington Place, WC1E 7JE London , United Kingdom*

Abstract

Developing mathematical models used to describe reaction kinetics is pivotal for the design, control and optimization of chemical processes. One of the most challenging tasks in the model development procedure is the identification of the unknown parameters within the model. This problem can be addressed using Model-Based Design of Experiment (MBDoE) techniques that allow experiments to be designed in such a way that parameters can be precisely estimated with the minimum number of runs and analytical resources. However, MBDoE techniques rely on an optimization procedure that is affected by the uncertainty related to the identified parameters and can be computationally expensive and prone to local optimality issues. MBDoE techniques are also applied to online procedures for faster identification of the kinetic model in autonomous platforms, and for this reason it is necessary to ensure a fast convergence and avoid numerical convergence issues during the operation. In this paper, a new optimization-free technology is proposed to tackle the above-mentioned problems.

Keywords: design of experiment, parameter estimation, Fisher Information matrix

1. Introduction

In chemical engineering, the role of kinetic models is fundamental to describe in a quantitative way the progress of chemical reactions occurring in a reaction system. These models, in the form of mathematical equations, play a key role in the design, control and optimization of chemical processes. When conventional model-building procedures are used reaction kinetics are modelled by proposing a set of candidate models based on preliminary experimental observations and hypotheses on potential reaction mechanisms. The key subsequent step is to determine the best model among the candidates to describe the system under analysis. The standard sequential methodology proposed by Franceschini and Macchietto (2000) consists of three different stages: 1) Preliminary analysis based on the identifiability and distinguishability of the model structure; 2) Model-based Design of Experiments to discriminate among the rival models that passed the first stage (MBDoE-MD); 3) Model-based Design of Experiments to improve the precision of the identified parameters for the best model selected in step 2 (MBDoE-PE). If no model passes the identifiability test of step 1 it is necessary to propose a new set of candidates. When a new experiment must be performed (either in steps 1 and 2) the new experimental conditions are optimally computed according to the expected amount of information for model discrimination (step 2) or parameter precision (step 3) the new experiments will eventually return to the experimenter. Automated model identification platforms were recently used to estimate the parameters of kinetic models online and speed up the identification procedure to reduce the costs of experimentation (Waldron et al, 2020, Pankajakshan et al, 2019). However, the parameter identification problem is often ill-conditioned and consequently the objective function of the MBDoE problem may become difficult to compute (Quaglio et al. 2019). Moreover, optimization in these platforms can be difficult to handle in online model identification procedures for the following additional reasons: *i)* when the system under analysis is structurally very complex, and consequently requires model candidates containing a large set of variables and model parameters, numerical optimization algorithms may easily incur into infeasible solutions; *ii)* particularly for systems characterized by a high number of state variables and model parameters MBDoE techniques can become computationally very expensive, thus reducing the benefits of online identification of kinetic models. A new MBDoE approach to increase the precision of parameter estimation is proposed in this paper where a new optimization-free Fisher information-driven algorithm has been developed to overcome the above-mentioned problems. The method has been tested on a benchmark case study for the identification of the parameters of the Bakers yeast growth model (Asprey et al. 2000). Results show that the information-driven approach is comparable with MBDoE methods in terms of experiments required for the precise identification of the parameters, but the information-driven approach is less computationally expensive.

2. Proposed framework and methodologies

The set of initial candidates used in the MBDoE procedures are formulated as differential and algebraic equations in the form:

$$\begin{cases} \mathrm{f}(\dot{\mathbf{x}}(\mathbf{t}), \mathbf{x}(\mathbf{t}), \mathbf{u}(\mathbf{t}), \boldsymbol{\theta}, \mathbf{t}) = 0 \\ \hat{\mathrm{y}} = \mathrm{g}(\mathrm{x}(\mathrm{t})) \end{cases} \tag{1}$$

In Eq. 1 $\mathbf{x}$ is the vector of N_x states variables, $\dot{\mathbf{x}}$ is the vector of the derivative of the state variables, $\mathbf{u}$ is the vector of the input of the system under analysis, $\boldsymbol{\theta}$ is the vector of the model parameters of N_θ-dimensions, t is the time variable, $\hat{\mathbf{y}}$ is the vector of the measured output of the system. The objective of the proposed methodology is to reduce the number of experiments needed to estimate the parameters of the model with adequate precision, so the objective is to perform experiments that lead to estimate precisely the model parameters, i.e. with minimum uncertainty. The proposed procedure for the optimization-free Fisher Information Matrix (FIM)-driven MBDoE algorithm is divided in 7 sequential steps, as reported in Figure 1:

1. Definition of experimental design bounds: in this step, all the control variables that can be manipulated and optimized during the process are collected in the design vector: $\boldsymbol{\varphi} = \boldsymbol{\varphi}[\mathbf{u}(\mathbf{t}), \mathbf{y}^{\mathbf{0}}, \boldsymbol{\tau}, \overline{\mathbf{w}}, \mathbf{t}_{\mathbf{sp}}]$ where $\boldsymbol{u}(t)$ is the vector of the time-varying input controls, $\boldsymbol{y}^{\mathbf{0}}$ is the vector of the initial conditions, $\boldsymbol{\tau}$ is the experiment duration, $\overline{\boldsymbol{w}}$ is the vector of the time-invariant input controls and $\boldsymbol{t_{sp}}$ is the vector of the sampling times. In addition, the bounds on the experimental vector are defined to create the experimental space;
2. Definition of preliminary experiments using DoE techniques: N_{exp} sets of experimental conditions to test are computed using a space-filling DoE method, for example, Latin Hypercube Sampling (LHS).
3. Selection of the first experiment to run: since no information about the system is available at this stage the selection of the first experiment is random.
4. Estimation of the parameter: after collecting the experimental data, in step 4 the model parameters are estimated minimizing the loss function:

$$LOSS = \sum_{i=1}^{N_{exp}} \left(\frac{y_i - \hat{y}_i}{\sigma} \right)^2 \tag{2}$$

where y_i is the measurement of experiment *i*, $\hat{y}_i$ is the predicted output of experiment *i* obtained using the model and σ is the standard deviation. The initial guesses on model parameters are updated with the new estimation ($\hat{\theta}$).

5. Statistics on model parameters and FIM evaluation: the *t*-test is used to evaluate the relative precision of the estimates. At this point the FIM (Zullo 1991) is evaluated using the following equation:

$$\mathbf{M}(\hat{\boldsymbol{\theta}}, \boldsymbol{\varphi}) = \sum_{j=1}^{n_{exp}} \sum_{r=1}^{n_{resp}} \sum_{s=1}^{n_{resp}} \tilde{\sigma}_{rs,j} \mathbf{Q}_{r,j}^{T} \mathbf{Q}_{s,j} + \mathbf{M}^{0} = [\mathbf{V}(\boldsymbol{\theta}, \boldsymbol{\varphi})]^{-1} \tag{3}$$

In (3) n_{exp}, n_{resp} are respectively the number of experiments and the number of responses, **Q** is the sensitivity matrix and $\mathbf{M^0}$ is the prior information on the parametric system. By inverting the matrix **M**, it is possible to obtain the variance-covariance matrix (**V**).

6. The evaluation of **M** is done for each experimental design point generated in step 2 to compute the expected amount of information achievable from each experiment. Experiments are then ranked based on the computed information content using the concept of Relative Fisher information (RFI) (Galvanin et al. 2016).
7. At this point the experimental conditions with the highest expected information are selected (Eq. 4) and the corresponding experiment is performed.

$$\boldsymbol{\varphi}_{i+1} = \operatorname{argmax}\left(\mathbf{M}(\hat{\boldsymbol{\theta}}, \boldsymbol{\varphi})\right) \tag{4}$$

Steps 4, 5, 6 and 7 are repeated iteratively for the fixed number of experiments set in step 2. Within this framework, it is possible to select at each iteration the most informative experiment to run in the candidate set without performing an optimization. The optimization-free approach is less prone to local optimization issues affecting the standard MBDoE approaches because an ‘exploratory part’ in the selection of the experiments is forced in step 2 as the initial selection of the experiments to perform using the LHS as preliminary DoE. In fact, since the optimal experiment is only selected based on the maximum evaluation of information from Eq. 4 there is no risk of incurring in local optima. The absence of optimization can, however, also be a limitation of the methodology, because the solution found through this procedure will always be sub-optimal. In addition, the fact that the first experiment is selected randomly (as would be done in practice if no preliminary knowledge of the system is available) can affect the computational performance and efficiency in subsequent iterations.

3. Case study description

The proposed framework has been tested using an in-silico case study related to a fed-batch reactor in which a fermentation reaction is carried out using baker’s yeast (Asprey et al. 2000). The model used to simulate the experiments is presented in Eq. 3.

$$\begin{cases} \dfrac{dx_1}{dt} = \left(\dfrac{\theta_1 x_2}{\theta_2 x_1 + x_2} - u_1 - \theta_4\right) x_1 \\ \dfrac{dx_2}{dt} = -\dfrac{\theta_2 x_1 x_2}{(\theta_2 x_1 + x_2)\theta_3} + u_1(u_2 - x_2) \end{cases} \tag{5}$$

This system has been chosen to test the method because the governing equations and their parameters are very well known. The experimenter aims to compute the dynamics of yeast concentrations $x_1(t)$ $[gL^{-1}]$ and the substrate concentrations $x_2(t)$ $[gL^{-1}]$ as a function of two different inputs: the dilution factor, u_1 $[h^{-1}]$, and the substrate concentration in the feed u_2 $[gL^{-1}]$. In this model the vector of model parameters that need to be identified precisely is $\boldsymbol{\theta} = [\theta_1, \theta_2, \theta_3, \theta_4]^T$. Parameters and constant input variables used to perform in-silico experiments are reported in Table 1.

Table 1: Parameter and constant input variables values used for the in-silico experiments.

Parameter	θ_1	θ_2	θ_3	θ_4	$x_2(0)$	u_2
Values	0.31	0.18	0.55	0.05	0.01	5
Initial guess	5	5	5	5	-	-

The design vector is $\boldsymbol{\varphi} = [\mathbf{x_1}, \mathbf{u_1}]$, where x_2 and u_2 are kept constant. The bounds used to define the experimental space are the same used by Asprey et al. (2000) and are reported in Table 2.

Table 2: Bounds on the experimental conditions.

Parameter	Lower Bound	Upper Bound
$x_1(0)$ $[gL^{-1}]$	1	10
$u_1[h^{-1}]$	0.05	0.2

The proposed FIM-driven method is compared with a classical MBDoE approach and with a DOE LHS approach. In the MBDoE framework, the experimental conditions are selected iteratively after a sequential optimization procedure as in Asprey et al (2000). In the DOE procedure the same experiments generated for the optimization-free method in step 2 are used but randomly selected so without operating a ranking of the experiments.

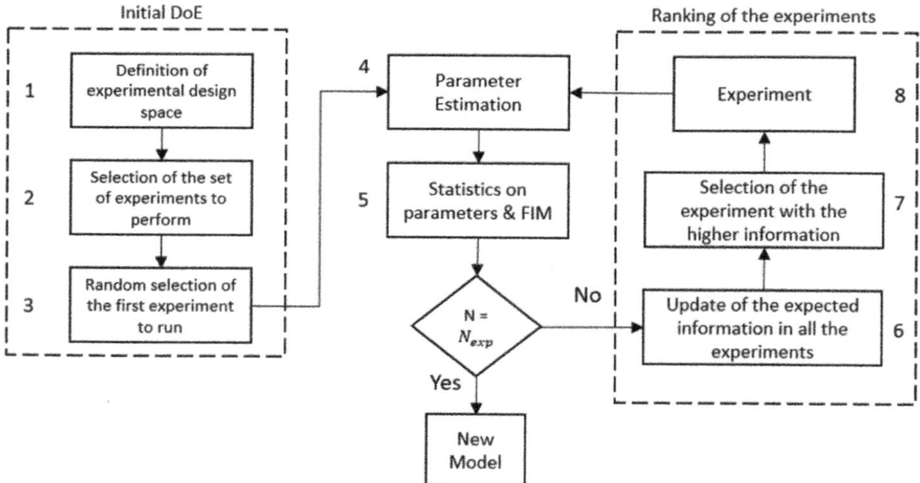

Figure 1: Block diagram representation of the proposed methodology for the design of experiment for parameter identification.

4. Results

In this section, three experimental design methodologies are compared: DOE (LHS), the proposed FIM-driven approach and a standard MBDoE. The comparison of the profiles of parameter estimates with the number of experiments is reported in Figure 2 for the full set of model parameters. The profile of estimated values and the variance of parameters are reported to quantify the precision of the estimates during the model identification procedure.

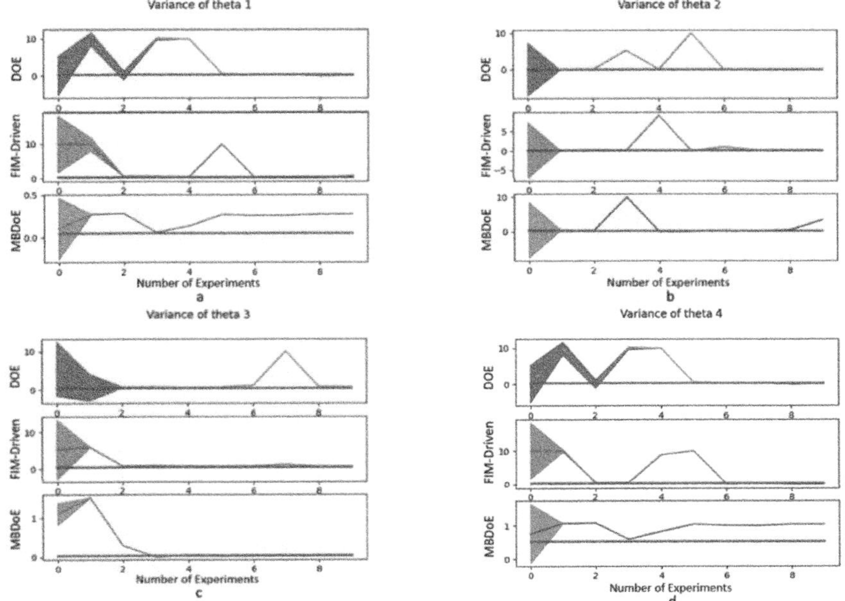

Figure 2 Profiles of θ_1 (a), θ_2 (b), θ_3 (c) and θ_4 (d) estimated values and relative variances.

From Figure 2 it is possible to observe that by using MBDoE it is possible to obtain the fastest reduction in parameter variance. Instead, the variance reduction obtained with the proposed FIM-driven approach is slower than the one obtained with MBDoE and faster than the one obtained with a conventional DOE methodology. As expected, in MBDoE the variance is generally lower and decreases faster than the variance obtained from the other two methods. This means that the precision of the estimations increases with every iteration faster than the other frameworks. The proposed methodology allows convergence to the assumed true value and negligible variance in 6 experiments. Instead, DoE requires more experiments (9) to obtain the same results. In Figure 3 the profiles of the *t*-values of the identified parameters are reported to evaluate and compare the relative precision of the estimates using the three methods. In Figure 2 the blue line represents the reference *t*-value, and the orange line represents the profile of the t-value of the parameter. If the computed *t*-values of the parameters are higher than the reference *t*-value the parameters are evaluated with good precision. From Figure 3 it is possible to notice that with a DoE method, the *t*-value for parameter θ_3 is lower than the reference until the sixth experiment and so the estimate is uncertain. Regarding the convergence speed of the three different procedures, the CPU time has been evaluated for all three methods. DoE is the fastest method (1.70 s) followed by FIM-driven (2.42 s) while MBDoE is the slowest method (10.47 s). These algorithms have been tested using an Intel® Core™ i9-10885H @ CPU 2.40 GHz with 16.0 GB RAM.

5. Conclusion

This study allows to define a new framework for the fast identification of parameters of kinetic models using an optimization-free Fisher Information Matrix-driven approach.

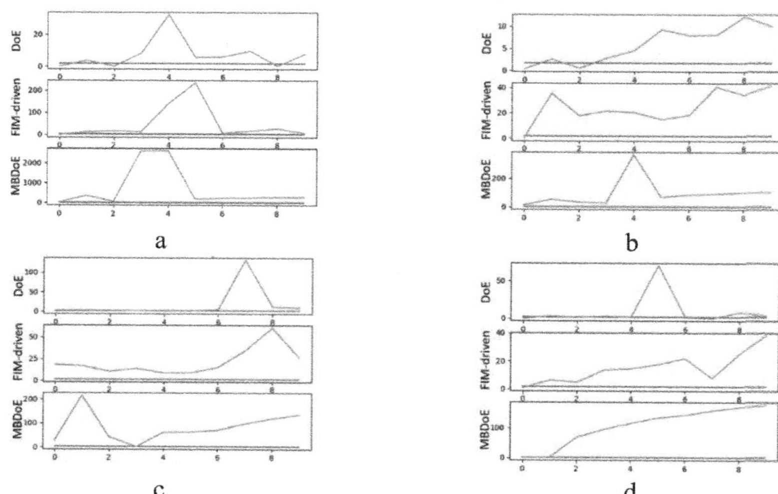

Figure 3 t-values of the parameters θ_1 (a), θ_2 (b), θ_3 (c) and θ_4 (d)

Preliminary results show how the proposed method can outperform standard space-filling design managing to identify the full set of model parameters precisely with a limited number of runs and providing results in parameter accuracy that are comparable to a standard sequential MBDoE but less computationally intensive. Future works will include re-sampling methods in the proposed procedure and the application to real chemical systems in automated platforms to test the robustness of the method and the comparison in terms of computational time required in practical lab settings.

Acknowledgment

This project has received funding from EPSRC (EP/R032807/1). The support is gratefully acknowledged.

References

Asprey, S.P., Macchietto, S., 2000. Statistical tools for optimal dynamic model building. Computers and Chemical Engineering 24, 1261–1267

Quaglio, M., Waldron, C., Pankajakshan, A., Cao, E., Gavriilidis, A., Fraga, E. A., Galvanin, F., 2019, An online reparametrisation approach for robust parameter estimation in automated model identification platforms, Comp. Chem. Eng., 124, 270.

Galvanin, F., Cao, E., Al-Rifai, N., Gavriilidis, A., Dua,V., 2016, A joint model-based experimental design approach for the identification of kinetic models in continuous flow laboratory reactors, Comp. Chem. Eng., 95, 202.

Waldron C., Pankajakshan A., Quaglio M., Cao E., Galvanin F., Gavriilidis A., Model-based design of transient flow experiments for the identification of kinetic parameters, React. Chem. Eng., 2020, 5, 1, 112-123, The Royal Society of Chemistry.

Pankajakshan A., Waldron C., Quaglio M., Gavriilidis A., Galvanin F., A Multi-Objective Optimal Experimental Design Framework for Enhancing the Efficiency of Online Model Identification Platforms, Engineering, Volume 5, Issue 6, 2019.

Zullo, L., 1991. Computer aided design of experiments. An engineering approach. Ph.D. Thesis, University of London, UK.

Antonis Kokossis, Michael C. Georgiadis, Efstratios N. Pistikopoulos (Eds.)
PROCEEDINGS OF THE 33rd European Symposium on Computer Aided Process Engineering (ESCAPE33), June 18-21, 2023, Athens, Greece
 http://dx.doi.org/10.1016/B978-0-443-15274-0.50004-4

A Python-based approach for thermodynamic consistency tests of binary VLE data

Pedro Cezareth,[a] Martina Costa Reis,[a]

[a]*University of São Paulo, School of Engineering, São Paulo 05508-000, Brazil*

Abstract

Although reliable experimental data are fundamental in process design and optimization, critical assessment of experimental vapor-liquid equilibrium data is usually time-consuming as it requires data mining and multiple thermodynamic and statistical analyses. Given this, any initiative towards the development of computational tools that helps to speed up the assessment of experimental vapor-liquid equilibrium data is essential. Therefore, in this work, a Python-based approach for thermodynamic consistency tests of vapor-liquid equilibrium data of binary mixtures is presented. While the consistency of experimental isobaric vapor-liquid equilibrium data was checked through the L-W and Redlich-Kister tests, the thermodynamic consistency of isothermal vapor-liquid equilibrium data was analyzed through the area, Redlich-Kister, and L-W tests. The objective of this work is to introduce an open-source, efficient computational tool implemented with Python that embeds different thermodynamic consistency tests which helps chemical engineers make rational use of experimental data in process simulation and model parameterization.
Keywords: Thermodynamic consistency tests, vapor-liquid equilibrium, binary mixtures.

1. Introduction

A fundamental requirement for the design and optimization of chemical processes is the use of a reliable set of experimental data of pure component and mixture properties. Since the design and optimization of chemical processes depend on thermodynamic models whose key adjustable parameters are determined by regression of experimental data, the use of a thermodynamically consistent data set is not only desirable but also imperative. To this end, it is fundamental that the chemical engineer is familiar with the different thermodynamic consistency tests available in the literature, as well as their pitfalls and limitations. In general, it is improbable that a particular data set succeeds in all thermodynamic consistency tests. As noted by Wisniak et al. (2017), there is no unique test capable of providing an absolute answer to the question "Are the experimental data thermodynamically consistent?". This is exactly why it is important to use a data-informed approach, where the chemical engineer can submit all data sets to several consistency tests, analyze the output, and then decide whether the experimental data set must be rejected or not.

Therefore, in this work, a Python-based approach for thermodynamic consistency tests of VLE (vapor-liquid equilibrium) data of binary mixtures is developed. For the sake of convenience, all consistency tests embedded in the Python-based approach only depend on tabulated physicochemical properties, which are easily found in compendia of physical properties. Unlike some well-known computerized systems for the retrieval, correlation, and prediction of phase equilibria data, this work presents an open-source, efficient computational tool that helps chemical engineers to make rational use of experimental data in process simulation and model parameterization.

2. Theoretical background of the thermodynamic consistency tests

Overall, the Gibbs-Duhem equation plays a very important role in the development of thermodynamic consistency tests. Since one usually neglects the effect of the pressure $\boldsymbol{p}$ on the activity coefficient $\boldsymbol{\gamma_i}$, for a binary mixture at constant temperature $\boldsymbol{T}$ the Gibbs-Duhem equation becomes

$$x_1 d\ln\gamma_1 + x_2 d\ln\gamma_2 = 0, \tag{1}$$

whose mathematical structure implies that the activity coefficients γ_i are coupled, that is, if one knows the value of γ_1 at composition x_1, then the value of γ_2 at composition x_2 is known too. In practical terms, if γ_1 and γ_2 are determined separately, Equation (1) may be used to check the consistency of the data. Hence, if both γ_1 and γ_2 satisfy Equation (1) over the range $0 \leq x_1 \leq 1$, the data are said to be thermodynamically consistent.

Equation (1) may be rewritten as

$$x_1\left(\frac{d\ln\gamma_1}{dx_1}\right) + x_2\left(\frac{d\ln\gamma_2}{dx_1}\right) = 0, \tag{2}$$

which is the main equation of the slope test. Likewise, Equation (1) may be integrated over the range $0 \leq \ln\gamma_i \leq \ln\gamma_i^\infty$ to give

$$\int_{\ln\gamma_1^\infty}^{0} x_1 d\ln\gamma_1 + \int_{0}^{\ln\gamma_2^\infty} x_2 d\ln\gamma_2 = \underbrace{\int_0^1 \ln\gamma_1\, dx_1}_{A_1} + \underbrace{\int_1^0 \ln\gamma_2\, dx_2}_{A_2} = 0, \tag{3}$$

where γ_i^∞ is the activity coefficient at infinite dilution. Equation (3) is the basis for the area test that asserts that in a plot of $\ln\gamma_i$ versus x_i the quantities $|A_1|$ and $|A_2|$ must be equal. Of course, because the measurements of isothermal VLE data are associated to an experimental error the quantities $|A_1|$ and $|A_2|$ are in practice not equal. Thus, it is common to assume a more lenient criterion for the thermodynamic consistency: the difference between $|A_1|$ and $|A_2|$ should be smaller than 2.0% of the sum $|A_1| + |A_2|$. Here, note that the requirement that $|A_1|$ and $|A_2|$ be equal is only a necessary condition for the thermodynamic consistency of the experimental data. In fact, it might be the case that some experimental data satisfy Equation (3), but not Equation (2).

Another test that may be applied to isothermal VLE data is the Redlich-Kister test. After some algebra, the Gibbs-Duhem equation gives

$$\ln\gamma_1\, dx_1 + \ln\gamma_2\, dx_2 = 0, \tag{4}$$

which can be integrated over the range $0 \leq x_1 \leq 1$ to give

$$\int_0^1 \ln\left(\frac{\gamma_1}{\gamma_2}\right) dx_1 = 0. \tag{5}$$

Equation (5) indicates that in a plot of $\ln\left(\gamma_1/\gamma_2\right)$ versus x_1 the areas above A_A and below A_B the axis x_1 must be equal. However, because of the experimental errors, this criterion is relaxed to

$$D = 100\frac{|A_A - A_B|}{A_A + A_B} < 10. \tag{6}$$

If, however, the Redlich-Kister test is applied to isobaric VLE data, then the contribution of the excess enthalpy on the excess Gibbs energy g^E is not negligible. In this case, the criterion of thermodynamic consistency becomes

$$D = \left|100\frac{|A_A - A_B|}{A_A + A_B} - 150\frac{|\Delta T^{max}|}{T^{min}}\right| < 10, \tag{7}$$

where ΔT^{max} is the largest difference of temperature observed and T^{min} is the lowest temperature value in the data set.

A more comprehensive consistency test for VLE data is the *L-W* test proposed by Wisniak (1993). The starting point of the *L-W* test is the equations $g^E = RT\sum_i x_i \ln \gamma_i$ and $\gamma_i = {y_i p}/{x_i p_i^*}$, where R is the gas constant, p_i^* is the vapor pressure, and y_i and x_i are the mole fraction of the species i in the vapor and liquid phase, respectively. By restricting the analysis to mixtures composed by species whose boiling points are not too different, so that the Clausius-Clapeyron equation may be used, one obtains after some algebra the following expression

$$\underbrace{\frac{\sum_i x_i \Delta s_i^* T_i^*}{\sum_i x_i \Delta s_i^*} - T}_{L_i} = \underbrace{\frac{g^E - RT\sum_i x_i \ln\left(y_i/x_i\right)}{\sum_i x_i \Delta s_i^*}}_{W_i}, \tag{8}$$

where Δs_i^* and T_i^* are the entropy of vaporization and the boiling point of the pure constituents at the pressure p of the solution. By integrating Equation (8) over $0 \le x_1 \le 1$, one has:

$$\underbrace{\int_0^1 L_i dx_1}_{L} - \underbrace{\int_0^1 W_i dx_1}_{W} = 0. \tag{9}$$

Equations (8) and (9) comprise the two equations that the *L-W* test relies on. While Equation (8) may be used to check the consistency of every single point of the experimental data set, Equation (9) may be employed to test the consistency of the full data set. According to the *L-W* test, a single experimental point is thermodynamically consistent, if the condition $0.92 < \left(L_i/W_i\right) < 1.08$ is fulfilled. In addition, an experimental data set is said to be consistent, if the difference between L and W is smaller than 3.0% of the sum $L + W$.

3. Implementation of the thermodynamic consistency tests

The thermodynamic consistency tests herein discussed were implemented in Python by using the libraries Numpy, Pandas, Scypy, Matplotlib, and Pyplot. Once experimental phase equilibrium data have been imported as a Python dataframe, the user must specify the type of data and provide some pure component parameters. Next, the thermodynamic consistency of the experimental data set is checked and the results are exported as a worksheet, which should be analyzed by the user. A sketch of the algorithm architecture is shown in Figure 1.

For the sake of illustration, Figure 2 shows the results of thermodynamic consistency tests for experimental VLE data retrieved from the Dortmund Data Bank (2022). In addition to the numerical results exported in the worksheet (Table 1), the results of some thermodynamic consistency tests (e.g. area test, Redlich-Kister test, and *L-W* test) may also be analyzed through plots. This approach is particularly interesting to identify possible outliers in the data set and help the user to decide if the data set should be totally or partially rejected. In addition, note that the results of the thermodynamic consistency tests for VLE data strongly depend on the quality of the pure component parameters, the applicable temperature range of the Antoine equation parameters, the amount of available experimental data, and how large the deviations from the ideality are. Regarding this last aspect, thermodynamic consistency tests based on the Gibbs-Duhem equation tend to face some issues when dealing with VLE data of very ideal mixtures. One of the reasons for this behavior is the fact that the measured pressure and temperature values are usually the same order of magnitude as the experimental error and, as a consequence, the output of the thermodynamic consistency tests is somehow imprecise with large values for the parameter *D*, as illustrated in Table 1. Thus, in such cases, the results of the thermodynamic consistency tests might be dubious and should be carefully analyzed.

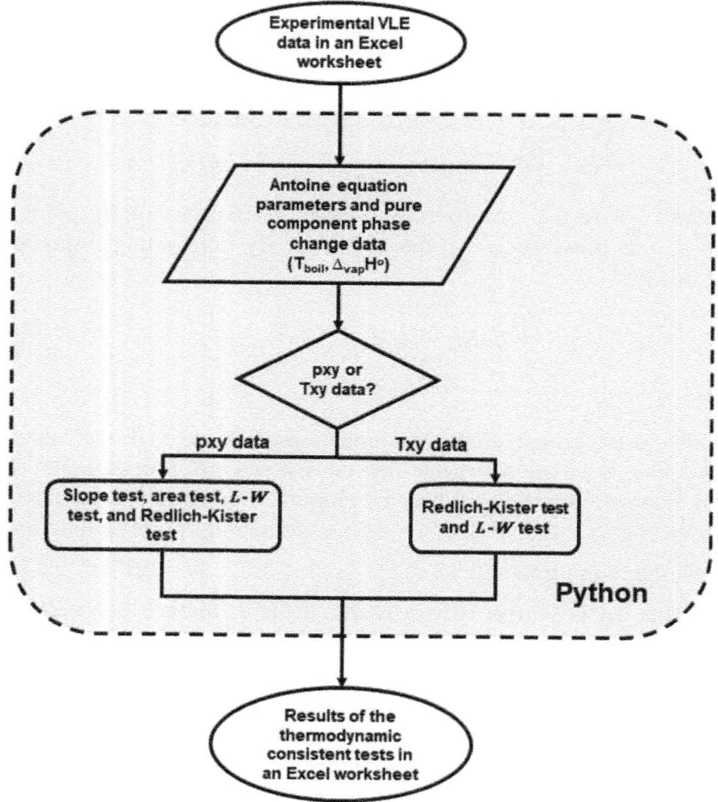

Figure 1. Structure of the proposed algorithm for thermodynamic consistency tests.

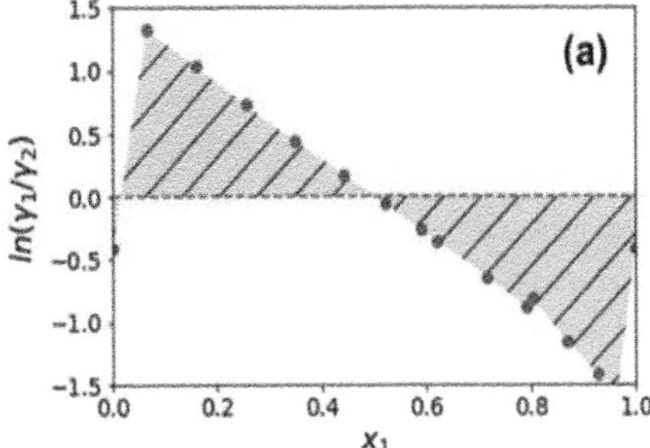

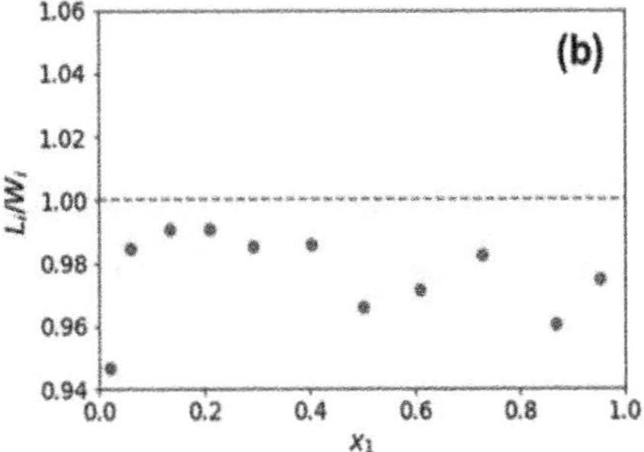

Figure 2. (a) Graphical visualization of the Redlich-Kister test (D= 1.48) for the isothermal VLE data of the mixture acetone-hexane at 318.15 K and (b) results of the L-W test for each point of the isobaric VLE data of the mixture acetone-ethanol at 101.33 kPa.

Table 1. Results of thermodynamic consistency tests for isobaric and isothermal VLE data.

			Thermodynamic Consistency Tests				
			Area	L-W			Redlich-Kister
			D	*L*	*W*	*D*	*D*
Butanol-Chlorobenzene	101.33 kPa	Txy	-	5.9	6.1	1.8	5.2
Methanol-Water	97.22 kPa	Txy	-	9.3	8.7	3.0	9.8
Acetone-Ethanol	101.33 kPa	Txy	-	4.4	4.7	3.2	7.1
Benzene-Toluene	333.15 K	pxy	1.6	35.9	2.4	87.7	14.9
Benzene-Toluene	353.15 K	pxy	13.1	16.0	2.3	75.0	11.1
1-Hexene-Hexane	328.15 K	pxy	77.9	11.2	0.3	95.2	11.3
1-Hexene-Hexane	333.15 K	pxy	38.3	4.7	0.2	91.1	5.8
Acetone-Hexane	318.15 K	pxy	10.6	15.4	10.1	21.1	1.5
Chloroform-Benzene	298.20 K	pxy	70.4	33.4	1.1	93.9	15.5
Diethyl ether-Acetonitrile	293.65 K	pxy	0.2	46.2	37.5	10.3	0.5

4. Conclusion and future perspectives

Although checking the reliability of experimental data is a long-established problem, the development of new thermodynamic consistency tests and computational tools for the assessment of VLE experimental data is still the motive of intensive research in chemical engineering thermodynamics. There are some commercial tools (Frenkel et al., (2005)) that were designed to dynamically evaluate the consistency of experimental data, but unfortunately, open-source codes are still non-existent. Then, since the assessment of experimental VLE data is a time-consuming and laborious task, any initiative towards the development of computational tools that help in the critical data-evaluation process is not only welcome but also necessary.

Therefore, in this work, an open-source algorithm written in Python was used to check the thermodynamic consistency of VLE data of binary mixtures. To this end, initially, experimental data are imported as a Python dataframe and then submitted to a series of thermodynamic consistency tests, according to its type. Next, the obtained results are exported as a worksheet that should be carefully analyzed by the user since passing a thermodynamic consistency test does not imply that the data set is accurate. In fact, as observed in this work, thermodynamic consistency tests based on the Gibbs-Duhem equation seem to be not so efficient when applied to very ideal solutions. One of the reasons for this trend might be related to the fact that the measured pressure and temperature values are usually the same order of magnitude as the experimental error, which leaves room for further research in chemical thermodynamics.

Even though the open-source tool presented in this work is still under development, it may be already used to test the thermodynamic consistency of VLE data. In the near term, other consistency tests for vapor-liquid and solid-liquid equilibria data should be implemented (Kojima test, Fredenslund test, McDermott-Ellis test, Kang test, and Cunico test), as well as tests of internal and external consistency.

References

J. Wisniak, J. Ortega, L. Fernández, 2017, A Fresh Look at the Thermodynamic Consistency of Vapour-Liquid Equilibria Data, The Journal of Chemical Thermodynamics, v. 105, p. 385-395.

J. Wisniak, 1993, A New Test for the Thermodynamic Consistency of Vapor-Liquid Equilibrium, Industrial & Engineering Chemistry Research, v. 32, p. 1531-1533.

Dortmund Data Bank, 2022, www.ddbst.com

NIST Chemistry WebBook, 2022, NIST Standard Reference Database Number 69, National Institute of Standards and Technology, https://doi.org/10.18434/T4D303

M. Frenkel, R. D. Chirico, V. Diky, X. Yan, Q. Dong, C. Muzny, 2005, ThermoData Engine (TDE): Software Implementation of the Dynamic Data Evaluation Concept, Journal of Chemical Information and Modeling, v. 45, p. 816-83.

Antonis Kokossis, Michael C. Georgiadis, Efstratios N. Pistikopoulos (Eds.)
PROCEEDINGS OF THE 33rd European Symposium on Computer Aided Process Engineering (ESCAPE33), June 18-21, 2023, Athens, Greece
 http://dx.doi.org/10.1016/B978-0-443-15274-0.50005-6

Modelling building's life-cycle decarbonization in China: A multi-level and multi-region optimization approach

Chenxi Li,[a] Pei Liu,[*] Zheng Li

[a]*State Key Lab of Power Systems, Department of Energy and Power Engineering, Tsinghua-BP Clean Energy Center, Tsinghua University, Beijing 100084, China*
liu_pei@tsinghua.edu.cn

Abstract

The decarbonization of life-cycle emissions is a huge challenge to the building sector because it involves multiple levels. However, previous studies rarely evaluated the potential emission reduction contribution of various technologies in the life cycle, especially China, whose life-cycle CO_2 emissions of the building sector account for over 40 percent of the total emissions. A multi-level and multi-region optimization model is proposed to quantify the life-cycle CO_2 emission reduction path of China's building sector in this paper. Considering both indirect and direct emissions, this model can find the lowest-cost path of CO_2 emission reduction in China's building sector under specified emission scenarios. Results show that extending the building life from 30 years to 50 years has a great impact on emission reduction, leading to a CO_2 emission reduction of more than 50 percent if low-carbon building materials are also used. Moreover, electrification and emission reduction of the power system also have the potential of reducing the CO_2 emissions of buildings by nearly 40 percent, with an end-use electrification rate of more than 70 percent by 2060.

Keywords: building sector, decarbonization, optimization, life-cycle, China

1. Introduction

The building sector is an important sector which connects the upstream industries and the user terminal, leading to its high life-cycle CO_2 emissions. To achieve the goal of CO_2 mitigation in the whole society, the life-cycle emission reduction of the building sector which connects the upstream and downstream is very important, while the multi-level structure of the building sector brings great challenges to the planning of the life-cycle decarbonization.

Previous studies assessing the CO_2 mitigation of the building sector most focused on one level, such as the material of buildings, the operation of buildings or the reconstruction of a specific house. Li et al. (2017) and Ma et al. (2016) quantified the impacts of decarbonization in China's cement sector and steel sector by using China TIMES model separately. In order to achieve the goal of climate change, the energy demand for heating and cooling of German residential buildings was assessed comprehensively (Olonscheck et al., 2011). A multi-objective optimization model was utilized to find a building retrofit strategy to improve energy efficiency (Asadi et al., 2012). Although the above studies described the CO_2 mitigation pathway of the building sector or one specific house detailedly, the research is still insufficient because of a lack of life-cycle analyzing. Each level of the building sector, material, construction, operation and demolition, affects each other, making it essential to carry out life-cycle CO_2 mitigation planning.

However, most life-cycle research on the building sector emphasized the accounting of historical CO_2 emissions instead of the mitigation potential of different low-carbon methods. Yang et al. (2018), Clabeaux et al. (2020), and Atmaca et al.(2022) counted the life-cycle emissions of a specific building or a building complex in China, USA and Turkey separately.

Overall, previous studies are rarely able to evaluate the emission reduction potential of buildings in the life cycle. Therefore, a multi-level and a multi-region optimization model is established to plan the life-cycle mitigation pathway of the building sector. The model is applied to the building sector of China, whose CO_2 emissions surpasses 4 billion tonnes, accounting for nearly 20 percent of the total CO_2 emissions all over the world.

2. Methodology

2.1. Model structure and Assumptions

The lifecycle of buildings includes material, construction, operation and demolition, which is shown in Figure 1. In this multi-level and multi-region optimization model, three main building materials and three main building structures are considered. Buildings are divided into three categories according to their use, and six energy demands are regarded. In this model, 40 technologies can be applied to the decarbonization of building sector, which are listed in Table 1.

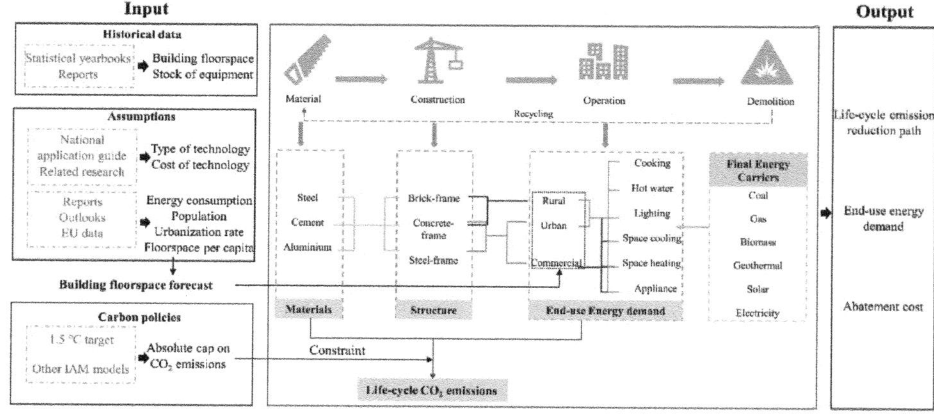

Figure 1. The structure of multi-level and multi-region optimization model of buildings

2.2. Mathematical formulation

Mathematical equations of the optimization model are presented in this model, including objective function and model constraints. Eight sets, t, mat, u, str, tech, r, op, ref stand for time, material, type of utilization, structure, technology, region, energy demand and insulation of buildings.

2.2.1. Objective function

The objective function of this model is to minimize the total cost of the building sector in its lifetime from 2020 to 2060. The total cost is composed of material costs, end-use device costs, central heating costs, insulation material costs and fuel costs, as Equations (1)-(2) shows.

$$objective = \sum_{t=2020}^{2060} \frac{sumcost_t}{(1+I)^{t-2020}} \quad (1)$$

$$sumcost_t = matc_t + devicec_t + heatc_t + refc_t + fuelc_t \quad (2)$$

Table 1. The technologies used in the model

Type of technology		Name of technology
Low carbonization of building materials		Far-infrared ceramic coating mixture, Electric arc furnace, Hydrogen steelmaking, New cement clinker, Solid regenerative fuel, Carbon capture and storage, Carbon capture and usage, 600kA potcell
Changing of building structure		Brick-frame, Concrete-frame, Steel-frame
Decarbonization of central heating		Coal-fired boiler, Gas-fired boiler, Coal-fired cogeneration, Biomass boiler, Geothermal, Solar heating, Heat pump
Insulation of buildings		Single glazing windows, 2bl glazing windows, Low-e windows, Stone wool insulation materials, Glass wool insulation materials, Expanded polystyrene, Sprayed polyurethane
End-use electrification	Cooking	Coal-fired cooking stove, Gas-fired cooking stove, Electric cooker, Biomass-fired cooking stove
	Hot water	Electric water heater, Solar thermal water heater, Gas-fired water heater
	Lighting	Incandescent lamp, Fluorescent lamp
	Space cooling	AC-I, AC-II, AC-III
	Decentralized space heating	Coal-burning fireplace, Solar heating, Heat pump

2.2.2. Material constrains

The demand of various materials depends on the newly-built area of different structures, which can be described by Equation (3). Although there may be demolition areas and new areas every year, it is still necessary to ensure that the total area is not less than the required area, listed as Equation (4). The CO_2 emissions caused by building materials are calculated by emission factors and the costs of materials are computed by unit price.

$$matd_{t,mat} = \sum_{str,u} narea_{t,u,str} * MATPRO_{str,mat} \quad (3)$$

$$\sum_{str}(area_{t-1,u,str} + narea_{t,u,str} - darea_{t,u,str}) \geq AREAD_{t,u} \quad (4)$$

$$matco2_t = \sum_{mat,tech} matp_{t,mat,tech} * TECHCO2_{t,mat,tech} \quad (5)$$

$$matc_t = \sum_{mat,tech} matp_{t,mat,tech} * MATPC_{t,mat,tech} \quad (6)$$

2.2.3. End-use demand constrains

The end-use demand can be divided into six categories: cooking, hot water, lighting, space cooling, space heating and electricity. Decentralized space heating and central heating constitute space heating. The decentralized equipment at the user end is supposed to meet the end-use demand, including cooking, hot water, lighting, space cooling and decentralized space heating, according to Equation (7). Due to the use of thermal insulation materials, the heating and cold demands of buildings will be reduced accordingly, as described in Equation (8). The coefficient $COEFC_{r,ref}$ and $COEFH_{r,ref}$ is related to the climate of the area where the building is located, and is estimated according to the average temperature. The heat provided by central heating should not be less than the demand, with the operation and heat supply of central heating boilers meeting Equation (9). The total power demand is the sum of all appliance electricity consumption, as shown in Equation (10). The consumption of fossil energy, biomass and other renewable energy is calculated according to the actual operation time of the equipment, as shown in Equation (11). The CO_2 emissions from fossil energy and electricity are computed by emission factors, which are described by Equation (12).

$$ENDUSED_{t,r,op} \leq \sum_{u,de,f} ndevice_{t,r,u,de} * OH_{t,r,u,de} * CONSUMPTION_{t,de,f} * CV_f \quad (7)$$

$$endused_{t,r,cooling} = ENDUSED_{t,r,op} - \sum_{ref} refarea_{t,ref} * COEFC_{r,ref} \quad (8a)$$

$$endused_{t,r,heating} = ENDUSED_{t,r,op} - \sum_{ref} refarea_{t,ref} * COEFH_{r,ref} \quad (8b)$$

$$\sum_{th} centralheat_{t,r,th} \leq CENTRALHEATD_{t,r} \quad (9a)$$

$$nbheat_{t,r,th} * OHMIN_{t,r,th} \leq heat_{t,r,th} \leq nbheat_{t,r,th} * OHMAX_{t,r,th} \quad (9b)$$

$$elecd_{t,r} = ELECD_{appliance,t,r} + devicefuel_{t,elec} \quad (10)$$

$$devicefuel_{t,f} = \sum_{u,de,r} ndevice_{t,r,u,de} * OH_{t,r,u,de} * DCON_{t,de,f} \quad (11)$$

$$endco2_t = \sum_f devicef_{t,f} * FCO2_f \quad (12)$$

2.2.4. CO_2 emissions constrains

The life-cycle CO_2 emissions of the building sector are composed of material emissions and end-use emissions, which can be described as Equation (13). The total emissions in the life cycle are supposed to meet the upper limit requirements of CO_2 emissions, as shown in Equation (14).

$$sumco2_t = matco2_t + endco2_t \quad (13)$$

$$sumco2_t \leq UPPERCO2_t \quad (14)$$

The model is implemented in General Algebraic Modelling System (GAMS) by using the CPLEX solver.

3. Case Study

The proposed model is implied in a case study of the building sector of China, which is divided into 9 regions according to the climate and data availability, as shown in Figure 2. The average life of buildings in China is about 30 years. It is assumed that the life of buildings are extended to 50 years to achieve the decarbonization goals. Several basic assumptions are listed in Table 2.

Table 2. Basic assumptions

Year		2030	2040	2050	2060
Floor area/100 million m^2 (ICCSD, 2021)		806	844	861	828
Population/100 million		14.0	13.4	12.6	11.3
Emission factor of electricity /kg·kWh^{-1}	BAU	0.36	0.24	0.21	0.21
(Song et al., 2022)	Decarbonization	0.34	0.13	0.06	0
Total energy demand/100 million tce (Chen et al., 2015)		7.1	8.0	8.8	8.5

3.1. Decarbonization of building materials

Selecting low-carbon materials enables the building sector to lower its life cycle CO_2 emissions. According to the proposed model, average carbon emission factor of materials given in Figure 3 can be referred to elect low-carbon materials when new buildings are constructed in the future. In 2060, the average CO_2 emissions factor of steel reduces to half of that of 2020, and that of cement reduces to one-third of that of 2020. Due to the carbon neutralization of the power system, the average CO_2 emissions factor of aluminium is close to 0.

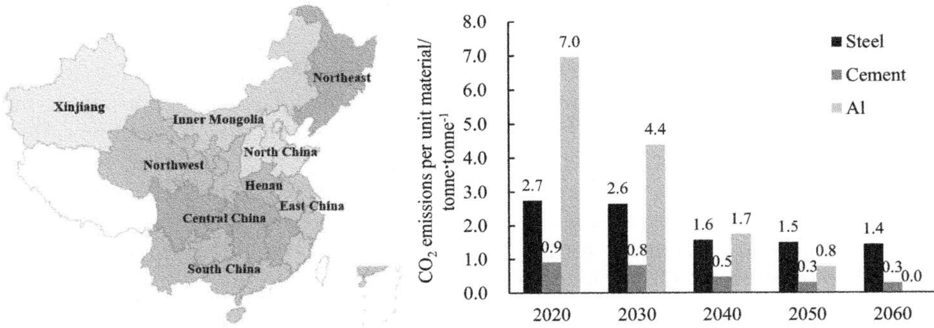

Figure 2 Geographic division of China Figure 3 Emission factor of building materials

3.2. Decarbonization of building operation

Owing to the application of thermal insulation materials, heat loss is reduced in each region, especially after 2040, which is depicted in Figure 4. Because of the low temperature and long heating period, the application of thermal insulation materials in the northeast is significantly earlier than that in other regions. Despite rare heating demand in South China, thermal insulation materials are supposed to be used after 2050, which can reduce the demand for refrigeration, beneficial to the decarbonization.

As can be seen from Figure 5, which shows the consumption of primary energy, there is a peak both in coal and natural gas. The maximum coal consumption occurs before 2025, which is not much higher than that in 2020. The maximum gas consumption is supposed to appear between 2040 and 2045, with a peak of 250bcm. The electricity demand rises steadily until 2050. Although the electricity demand decreases after 2050 because of the shrinking population and total energy demand, the electrification rate gradually increases from 2020 to 2060, and exceeds 70% in 2060 according to Table 3.

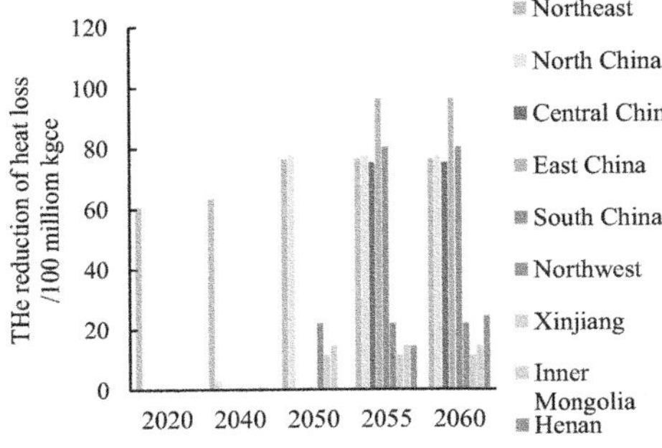

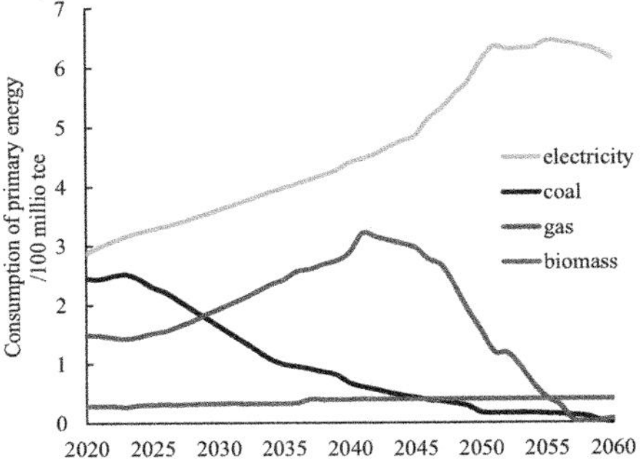

Figure 4 The reduction of heat loss Figure 5 Fuel consumption during operation

Table 3. Proportion of clean energy

Year	2020	2030	2040	2050	2060
End-use Penetration of renewable energy	4.6%	5.0%	5.3%	12.4%	26.7%
End-use electrification rate	40.2%	27.7%	52.4%	68.2%	72.5%

3.3. Comparision of life cycle decarbonization potential in the building sector

Table 4 illustrates the decarbonization potential of various emission reduction methods. It is apparent from this table that extending the average lifespan of Chinese buildings from 30 years to 50 years can significantly reduce the life-cycle emissions, due to the demand for building materials being greatly reduced. This mitigation measure can be achieved by government order. Combined with low-carbon building materials, the contribution rate of CO_2 reduction in non-operational processes of buildings is higher than 50%. What's more, although previous studies have implied the importance of building reconstruction (Motalebi et al., 2022), only the replacement of windows and the addition of building insulation layers contribute little to decarbonization pursuant to our study. In addition, even if there are no measures taken by the building sector, one fifth of the life-cycle CO_2 emissions of the building sector can be reduced by the power sector. The decarbonization of electricity and the electrification of the building contribute nearly 40% to the reduction of emissions in the building sector, meaning that support for electrification is very necessary for the Chinese government to the public to realize the low-carbon transition of the building sector at a lower cost.

4. Conclusions

This paper proposes a multi-level and multi-region optimization model to analyze the life-cycle decarbonization pathway of the building sector and China is taken as a case study. It can be found that life extension plays a significant role in buildings' life-cycle CO_2 emissions reduction by lowering the demand for building materials. During the operation stage, natural gas becomes one of the main fuels in the next 20 years and the consumption of natural gas declines rapidly, which is consistent with the current policy of China. Although biomass-related technologies are relatively mature, the decarbonization transition of China's building sector cannot be entirely dependent on biomass due to lack of biomass resources. The improvement of end-use electrification rate is essential for the emission mitigation of buildings, which is supposed to be increased from 40% in 2020 to 70% in 2060. Therefore, to achieve life-cycle CO_2 reduction in the building sector in China, attention to building life extension and electrification must be paid, meaning that reasonable urban planning and electrification reform of buildings should be the major work of China's building sector in the future.

Table 4. The decarbonization potential of various emission reduction methods

Methods	CO_2 reduction in 2060 /100 million tonnes	Proportion	Total CO_2 reduction in 40 years /100 million tonnes	Proportion
Life extension	3.4	14.5%	268.7	41.3%
The use of low-carbon materials	1.7	7.2%	70.6	10.8%
Decarbonization of central heating	4.0	17.0%	62.9	9.7%
Insulation	0.2	0.9%	3.4	0.5%
Electrification of distributed heating, cooking and hot water	6.1	26.0%	109.4	16.8%
Decarbonization of electricity	8.1	34.4%	135.8	20.9%

References

E. Asadi, M. G. da Silva, C. H. Antunes, L. Dias, 2012, Multi-objective optimization for building retrofit strategies: A model and an application, Energy and Buildings, 44, 81-87.

A. Atmaca, N. Atmaca, 2022, Carbon footprint assessment of residential buildings, a review and a case study in Turkey, Journal of Cleaner Production, 340, 130691.

P. Chen, Y. Da, S. Guo, S. Hu, J. Yi, 2015, Building energy use in China: Ceiling and scenario, Energy and Buildings, 102, 307-316.

R. Clabeaux, M. Carbajales-Dale, D. Ladner, T. Walker, 2020, Assessing the carbon footprint of a university campus using a life cycle assessment approach, Journal of Cleaner Production, 273, 122600.

ICCSD, 2021, The Comprehension of Carbon Neutral: The Low Carbon Development Action Roadmap of China during 2020–2050, China Citic Press, Beijing, China.

N. Li, D. Ma, W. Chen, 2017, Quantifying the impacts of decarbonisation in China's cement sector: A perspective from an integrated assessment approach, Applied Energy, 185, 1840-1848.

D. Ma, W. Chen, X. Yin, L. Wang, 2016, Quantifying the co-benefits of decarbonisation in China's steel sector: An integrated assessment approach, Applied Energy, 162, 1225-1237.

M. Motalebi, A. Rashidi, M. M. Nasiri, 2022, Optimization and BIM-based lifecycle assessment integration for energy efficiency retrofit of buildings, Journal of Building Engineering, 49, 104022.

M. Olonscheck, A. Holsten, J. P. Kropp, 2011, Heating and cooling energy demand and related emissions of the German residential building stock under climate change, Energy Policy, 39, 9, 4795-4806.

S. Song, T. Li, P. Liu, Z. Li, 2022, The transition pathway of energy supply systems towards carbon neutrality based on a multi-regional energy infrastructure planning approach: A case study of China, Energy, 238, 122037.

X. Yang, M. Hu, J. Wu, B. Zhao, 2018, Building-information-modeling enabled life cycle assessment, a case study on carbon footprint accounting for a residential building in China, Journal of Cleaner Production, 183, 729-743.

Antonis Kokossis, Michael C. Georgiadis, Efstratios N. Pistikopoulos (Eds.)
PROCEEDINGS OF THE 33rd European Symposium on Computer Aided Process Engineering (ESCAPE33), June 18-21, 2023, Athens, Greece
 http://dx.doi.org/10.1016/B978-0-443-15274-0.50006-8

Automated Kinetic Model Discovery – A Methodological Framework

Miguel Ángel de Carvalho Servia [a], Ilya Orson Sandoval [a], Dongda Zhang [b,*], Klaus Hellgardt [a], King Kuok (Mimi) Hii [c], Ehecatl Antonio del Rio Chanona [a,*]

[a] *Department of Chemical Engineering, Imperial College London, South Kensington, London SW7 2AZ, UK*

[b] *Department of Chemical Engineering, The University of Manchester, M13 9PL, UK*

[c] *Department of Chemistry, Imperial College London, 82, Wood Lane, London W12 0BZ, UK*

[*] *Corresponding authors: dongda.zhang@manchester.ac.uk; a.del-rio-chanona@imperial.ac.uk*

Abstract

The industrialization of catalytic processes benefits strongly from kinetic models for optimization and control purposes. Nevertheless, mechanistic models are difficult to construct; data-driven and hybrid models lack interpretability and the flexibility to leverage physical knowledge. Thus, a different approach called automated knowledge discovery has been recently popularized. Existing methods in literature suffer from important drawbacks: necessitating assumptions about model structures, a lack of model selection automation, and sensitivity to noise. To overcome these challenges, the present work constructs a methodological framework for the automated generation of catalytic kinetic models. We leverage symbolic regression for model generation, a hybrid optimization algorithm for parameter estimation, and a robust criterion for model selection. The framework is tested with an illustrative isomerization case study, where it showcases the ability to retrieve the underlying kinetic model with a limited amount of noisy data from the catalytic system.

Keywords: catalysis, kinetic model generation, automated knowledge discovery, information criteria, machine learning.

1. Introduction

The industrialization of catalytic processes requires kinetic models, which are a mathematical representation of the dynamical trajectories of a chemical system, typically presented in the form of ordinary differential equations. Kinetic models can be constructed using any of the three classical modelling paradigms: mechanistic (white-box), data-driven (black-box) or hybrid (grey-box). Mechanistic models are constructed by using existing fundamental laws, such as mass, energy, and momentum balances, making them interpretable and extrapolatory (Baker, 2018). However, the construction of these models can be time-consuming and experimentally expensive. Data-driven models do not use any fundamental laws for their construction. Typically, these models are constructed by solely using collected data of a system, making their construction time-efficient but experimentally expensive. Furthermore, these models generally have poor extrapolatory abilities and are uninterpretable, which may classify their usage for certain industrial scenarios as unsafe. In comparison, hybrid models aim to exploit the advantages of mechanistic and data-driven modelling. Hybrid models are composed of two parts: a mechanistic backbone and a data-driven block which aims to improve the backbone's fit. Generally, hybrid models retain the extrapolation capabilities of a mechanistic model and the flexibility and ease of construction of a data-driven model.

Hybrid modelling offers an elegant solution to the problems posed by the other classical modelling paradigms, but not an optimal one. A better solution would be to utilize existing data to automatically generate and select mechanistic models by exploiting state-of-the-art statistical and machine learning methods. In this way, the benefits of mechanistic models are maintained, whilst some of their drawbacks are eliminated. This modelling paradigm has been coined *automated knowledge discovery* and has recently gained popularity. In current literature, many automated knowledge discovery frameworks have been developed, such as: the ALAMO approach (Wilson, 2017), the SINDy algorithm (Brunton, 2016), and symbolic regression using genetic programming (Koza, 1994). These frameworks have three drawbacks that limit their ability to retrieve underlying ground-truth models, and consequently, their real-world applicability. Firstly, they necessitate substantial structural assumptions of the underlying data-generating model. This is particularly true of ALAMO and SINDy, as a design matrix needs to be constructed for their execution (Wilson, 2017; Brunton, 2016). Secondly, they lack a motivated and rigorous model selection routine (Wilson, 2017; Brunton, 2016; Koza, 1994). Lastly, they are sensitive to noisy data (Wilson, 2017; Brunton, 2016; Koza, 1994). Hence, this work aims to overcome these obstacles by constructing a generalizable and robust methodological framework that integrates a rigorous model selection routine for the automated kinetic model discovery of catalytic systems.

The rest of the paper will be structured as follows: in Section 2, the proposed automated knowledge discovery framework is motivated and described in detail; in Section 3, the illustrative case study used to showcase the capabilities of the proposed methodology is introduced; in Section 4, the results of the study are amply discussed; in Section 5, the key findings are presented, and the outlook of the research is explored.

2. Methodological Framework

Given the objective of automated knowledge discovery, the proposed methodological framework constitutes three main stages: model generation, model refinement, and model selection and discrimination.

Symbolic regression using genetic programming (SR-GP) is often considered the most generalizable and reliable model generation method found in literature, as this method does not require the construction of a design matrix. In other words, assumptions regarding an underlying ground-truth model are minimal. The basic concept of SR-GP is to specify a set of state variables and operators (e.g.: '+', '/') that may be present in the underlying mechanistic model. With this, an initial population of models can be constructed, and based on Darwin's theory of evolution, the best models – based on a specified performance metric – are evolved via genetic operations (e.g.: crossover and mutation), and the worst models are discarded. This process is iterated until convergence is achieved or a termination criterion is met. As such, for the model generation stage, we propose a weak reformulation of SR-GP, where models generated represent a mapping between input and output variables in the derivative hyperspace. In other words, this reformulation generates functions dependent on the state variables that can map the output via an integration step. In this way, weak SR-GP receives the dynamic trajectories of concentration as inputs, proposes rate models ($r = f(C_1, C_2, \dots)$ where C_i is the concentration of species $i \in \mathbb{Z}^+$) integrates the rate models with respect to time at each given time-step where concentration data is available ($\int r dt = \int \frac{1}{v_i}\frac{dC_i}{dt}dt = \frac{1}{v_i}C_i(t)$), and compares the results from the integration with the original dataset.

Once promising rate models are proposed by the weak reformulation of SR-GP, model refinement needs to take place. Specifically, the model refinement comes as a parameter estimation problem, where the error between the model's response and the data are minimized by finding the best set of kinetic parameters. This optimization problem is solved by deploying a hybrid optimization algorithm composed of an explorative phase, carried out by an artificial bee colony (ABC) algorithm, and an exploitative phase carried out by a limited memory Broyden Fletcher Goldfarb Shanno (LBFGS) algorithm. The former is a stochastic optimization algorithm shown to have excellent explorative characteristics (Cho, 2021), while the latter is a gradient-based optimization algorithm shown to have excellent exploitative characteristics (Malouf, 2002). The objective function used for the parameter estimation was the negative log-likelihood (NLL), presented below:

$$f(\boldsymbol{\theta}) = \sum_{i,j,k}\left[\frac{\left(C_{i,j,k} - y_i(\boldsymbol{\theta},\boldsymbol{x}_0)_{j,k}\right)^2}{2\hat{\sigma}_i^2} - \log\left(\frac{1}{\sqrt{2\pi\hat{\sigma}_i^2}}\right)\right] \quad (1)$$

where $C_{i,j,k}$ is the simulated concentration (i.e.: in-silico data) of species $i \in S$ for dataset $j \in D$ at time $k \in T$, where S, D, and T represent the species set, data set and time set, respectively; $y_i(\boldsymbol{\theta}, \boldsymbol{x}_0)_{j,k}$ is the concentration of species $i \in S$ for dataset $j \in D$ at time $k \in T$ proposed by an arbitrary model which is dependent on its parameters $\boldsymbol{\theta}$ and the initial conditions $\boldsymbol{x}_0$; $\hat{\sigma}_i^2$ is the variance of the noise that we assume the concentrations of species $i \in S$ has. Once all the promising models have been explicitly optimized, the models are ranked based on their Akaike information criterion (AIC) values. The AIC was selected after a thorough analysis of the performance of different criteria under several conditions (e.g.: different amounts of additive noise, different amounts of data). This analysis concluded that AIC has a statistically higher probability of selecting the correct data-generating rate model than the other criteria tested.

After the ranking of the optimized models has been established based on AIC values, the Akaike weights of each model, $w_i(\boldsymbol{\theta}_i, \boldsymbol{x}_0)$, (i.e.: a statistical measure that quantifies the probability of a model structure being the correct one based on AIC values) are also evaluated. Provided a user-defined tolerance $\epsilon \in \mathbb{R}^+$, if $|w_1(\theta_1, x_0) - w_2(\theta_2, x_0)| \leq \epsilon$, then the modeler cannot be confident enough of the AIC selection, extra discriminatory experiments should be performed, and the process repeated. The discriminatory experiment can be determined by solving the Hunter-Reiner criterion, which aims at finding the experimental conditions that will maximize the difference between the response of two competing models. If $|w_1(\theta_1, x_0) - w_2(\theta_2, x_0)| > \epsilon$, then the modeler can be confident of the AIC selection and should integrate the proposed rate model and compare it with the dynamic trajectories of the concentrations. If the results are satisfactory, the methodological framework is terminated. Otherwise, further data should be collected by applying (model-based) design of experiments – a strategy aimed at computing maximally informative experiments. For clarity and simplicity, the flowchart of the proposed methodological framework is presented in Figure 1.

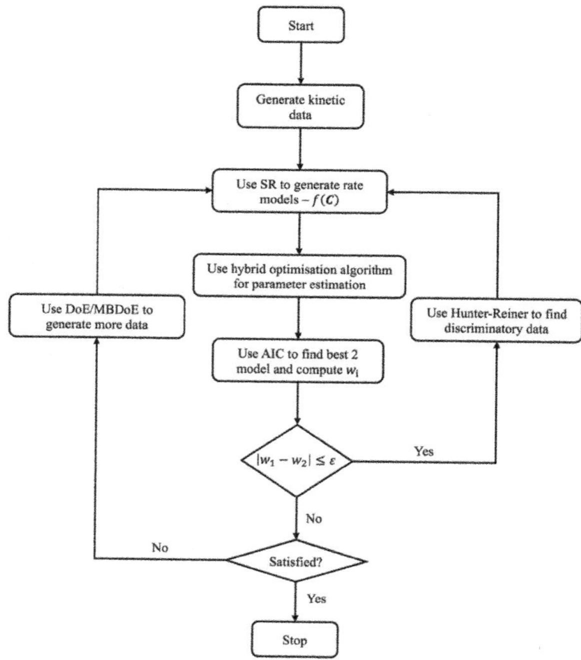

Figure 1: The flowchart of the proposed methodological framework featuring a weak reformulation of SR-GP

3. Kinetic Case Study

To showcase the performance of the proposed methodology, an illustrative case study of an isomerization reaction was chosen, where A is transformed to B reversibly over a catalytic active site. The reaction mechanism and the King-Altman graph of the case study are displayed in Figure 2(a).

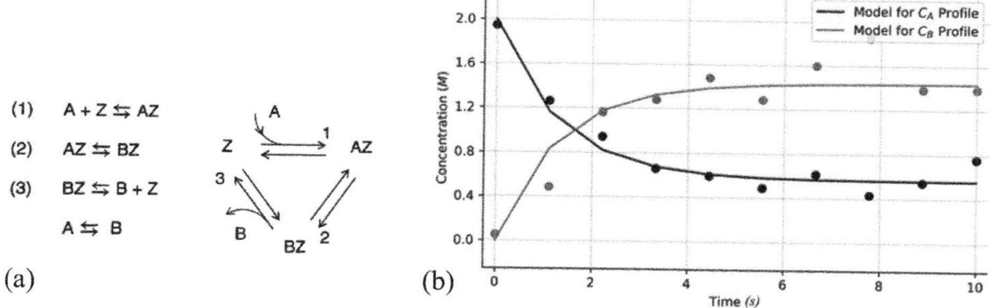

Figure 2: (a) Mechanism and King-Altman graph of illustrative isomerization reaction (Marin, 2019); (b) response of the chosen rate model for one of the three computational experiments

Given the presented mechanism, a rate model was derived and is shown in Equation 2. A detailed derivation of the rate model is found in (Marin, 2019).

$$r = -\frac{dC_A}{dt} = \frac{dC_B}{dt} = \frac{k_A C_A - k_B C_B}{k_C C_A + k_D C_B + k_E} \quad (2)$$

In Equation 2, C_A and C_B represent the concentration of reactant A and product B, respectively. The kinetic parameters of the rate model are represented by k_i where $i \in$

$\{A, B, \ldots, E\}$. To generate the necessary data to test the proposed framework, three computational experiments were carried out, each with different initial conditions. The simulations were run with the following initial conditions (in molar units): $(C_{A0}, C_{B0}) \in \{(2,0), (6,1), (10,2)\}$. For each simulation, the concentration of the reactant and product are recorded at evenly spaced intervals between time $t = 0$ s and $t = 10$ s. For the simulations, the kinetic parameters were defined as: $k_A = 7$ M s^{-2}, $k_B = 3$ M s^{-2}, $k_C = 4$ s^{-1}, $k_D = 2$ s^{-1} and $k_E = 6$ M s^{-1}. To ensure that the generated data is as realistic as possible, Gaussian noise was added to the simulation results. The defined Gaussian noise had zero mean and a standard deviation of 0.13 for the concentrations of A, and 0.21 for the concentrations of B (the standard deviations represent 5% of the mean of the simulated concentrations of A and B). For the parameter estimation task, it would be futile to assume that as modellers the exact variance of the noise would be known. Thus, a conservative assumption was made by setting $\hat{\sigma}_A = 0.26$ and $\hat{\sigma}_B = 0.42$, (i.e.: assuming a standard deviation of 10% of the mean of the simulated concentrations of A and B).

4. Results and Discussion

As explained in Section 2, the proposed methodological framework implements a weak reformulation of SR-GP which automatically evolves rate models by integrating and comparing them with the concentration data available. Note that the law of conservation of mass is satisfied by construction under this integrating scheme. Departing from the kinetic data generated, the SR algorithm was used to generate rate models. The expression construction rules exclusively included the arithmetic operators '+', '-', '×' and '/', since rates including other operators are less common. The rate models generated were a function of the species whose concentrations were measured, $r(C_A, C_B)$ in this case. The best expression proposed by the weak SR-GP, sorted by degree of complexity (i.e.: the number of operators and variables), are shown in Table 1.

The estimation of each kinetic parameter, k_i where $i \in \{1,2, \ldots, 6\}$, was carried out as explained previously: NLL was used as the objective function where the assumed variances for NLL are double of the real variances used to generate the Gaussian noise, and this is solved by using the hybrid algorithm comprised of the ABC and the LBFGS algorithms. The NLL, AIC and Akaike weight (AW) values for each of the weak SR-GP proposed and optimized models are also presented in Table 1 below. This shows that r_6 and r_3 are the best and second-best models for the given data, and by setting an Akaike weight tolerance of $\epsilon = 0.05$, we can be confident in the AIC's choice that r_6 is the best model to represent the dynamical catalytic reactive system under investigation, without requiring further discriminatory data. The response of the selected model is shown in Figure 2(b). For the sake of brevity, only one of the experiments is presented. The final rate model output by the proposed framework had a near identical structure to the data-generating one, differing only by a single parameter in the numerator. However, the hybrid optimization algorithm used for parameter estimation was able to determine that the extra parameter was practically zero, and therefore may be non-existent in the actual model. The discrepancy in other parameters may be attributed to the additive Gaussian noise in the data. Table 2 details the estimated kinetic parameter values and the original values of the true data-generating rate model.

Table 1: The NLL value, the AIC values and the Akaike weight (AW) values of all weak SR-GP proposed rate models

Model	NLL Value	AIC Value	AW Value
$r_1 = k_1$	540.410	1082.821	0.000

$r_2 = k_1 C_A$	53.857	109.715	0.000
$r_3 = \frac{k_1 C_A - k_2 C_B}{k_3 C_A}$	-0.831	4.337	0.244
$r_4 = \frac{k_1 C_A - k_2 C_B - k_3}{k_4 C_A}$	-0.988	6.025	0.105
$r_5 = \frac{k_1 C_A - k_2 C_B - k_3}{k_4 C_A + k_5}$	-2.724	4.551	0.219
$r_6 = \frac{k_1 C_A - k_2 C_B - k_3}{k_4 C_A + k_5 C_B + k_6}$	-4.275	3.450	0.380
$r_7 = \frac{k_1 C_A^2 C_B - k_2 C_A C_B^2 - k_3 C_A + k_4 C_B}{k_5 C_A^2 C_B - k_6 C_A}$	-2.289	7.422	0.052

Table 2: The estimated kinetic parameters from the chosen rate model proposed by proposed framework and the original kinetic parameter's values

Kinetic Parameter	Estimated Value	True Value
k_1, k_A	5.557 M s^{-2}	7.000 M s^{-2}
k_2, k_B	2.335 M s^{-2}	3.000 M s^{-2}
k_3	0.000 M^2 s^{-2}	N/a
k_4, k_C	3.114 s^{-1}	4.000 s^{-1}
k_5, k_D	1.379 s^{-1}	2.000 s^{-1}
k_6, k_E	6.033 M s^{-1}	6.000 M s^{-1}

5. Conclusions and Outlook

In this work, we proposed an automated knowledge discovery framework for the generation of catalytic kinetic models from data. A combination of a reformulation of SR-GP, parameter estimation through hybrid optimization, and a rigorous model selection and discrimination routine alleviate drawbacks of alternative automated knowledge discovery methods in literature. The presented case study showcases the framework's ability to retrieve the underlying kinetic model of a catalytic system with a limited amount of noisy data. Future work will pursue strategies to incorporate physical constraints into the model generation, to increase its efficiency through physical guidance.

References

Ruth E. Baker, Jose-Maria Peña, Jayaratnam Jayamohan, and Antoine Jérusalem, may 2018, Mechanistic models versus machine learning, a fight worth fighting for the biological community? *Biology Letters*, 14(5):20170660, pages 2

Zachary T. Wilson and Nikolaos V. Sahinidis, nov 2017, The ALAMO approach to machine learning. *Computers & Chemical Engineering*, 106:785–795, pages 3, 4, 7, 8, 9, 10

Steven L. Brunton, Joshua L. Proctor, and J. Nathan Kutz, mar 2016, Discovering governing equations from data by sparse identification of nonlinear dynamical systems. *Proceed- ings of the National Academy of Sciences*, 113(15):3932–3937, pages 3, 4, 10, 12, 13

JohnR. Koza, jun 1994, Genetic programming as a means for programming computers by natural selection. *Statistics and Computing*, 4(2), pages 14

Bovinille Anye Cho, Miguel Ángel Carvalho Servia, Ehecatl Antonio del Río Chanona, Robin Smith, and Dongda Zhang, mar 2021, Synergising biomass growth kinetics and trans- port mechanisms to simulate light/dark cycle effects on photo-production systems. *Biotechnology and Bioengineering*, 118(5):1932–1942, pages 31

Robert Malouf, 2002, A comparison of algorithms for maximum entropy parameter estima- tion. In *proceeding of the 6th conference on Natural language learning - COLING-02*. Association for Computational Linguistics, pages 31

Guy B Marin, Gregory S Yablonsky, and Denis Constales, 2019, *Kinetics of chemical reac- tions: decoding complexity*. John Wiley & Sons, pages 36

Antonis Kokossis, Michael C. Georgiadis, Efstratios N. Pistikopoulos (Eds.)
PROCEEDINGS OF THE 33rd European Symposium on Computer Aided Process Engineering (ESCAPE33), June 18-21, 2023, Athens, Greece
 http://dx.doi.org/10.1016/B978-0-443-15274-0.50007-X

Assessing process systems models for pharmaceutical development

Margherita Geremia[a], Samir Diab[b], Charalampos Christodoulou[c], Gabriele Bano[d], Massimiliano Barolo[a], Fabrizio Bezzo[a,*]

[a] *CAPE-Lab – Computer-Aided Process Engineering Laboratory. Department of Industrial Engineering, University of Padova, via Marzolo 9, 35131 Padova (PD), Italy*
[b] *GlaxoSmithKline, Park Road, Ware SG12 0DP, United Kingdom*
[c] *GlaxoSmithKline, Gunnels Wood Road, Stevenage SG1 2NY, United Kingdom*
[d] *GlaxoSmithKline, 1250 S Collegeville Rd, Collegeville (PA) 19426, United States*
[*] *Corresponding author e-mail: fabrizio.bezzo@unipd.it*

Abstract

Quantitative models are useful tools to accelerate the development of pharmaceutical processes. The assessment of the predictive capability of such models is fundamental to enhance their usage in a systematic way, particularly when the focus is on flowsheet models. In this study, we propose a systematic procedure that combines techniques that are typically used in different modeling contexts – namely, global sensitivity analysis, model-based design of experiments, and data reduction by means of multivariate statistical methods, with the advantage of enhancing readability and interpretation. The methodology is effective for the assessment of model fidelity and can support practitioners in the development of pharmaceutical processes. A direct compression systems model for manufacturing oral solid dosage products is used as a case study. Results show that just a subset of model parameters require precise estimation to meet the target critical quality attribute, and that calibrating flowsheet models on a unit-by-unit basis may be unnecessary when the focus is on one final quality attribute.

Keywords: systems model, pharmaceutical development, model identifiability, parameter estimation, model fidelity, pharmaceutical engineering.

1. Introduction

Quantitative models have been progressively adopted to accelerate the development of pharmaceutical manufacturing processes that traditionally requires time and resource-intensive experimental campaigns (Bano et al, 2022). Despite the potential benefits of modeling in pharmaceutical process development, the systematic use of these methods is not widespread, as stakeholders generally show a lack of confidence in the prediction capability of quantitative models with respect to key performance indicators (KPIs) or critical quality attributes (CQAs), which we will generically call key indicators (KIs) (Braakman et al, 2022). It is worth adding that pharmaceutical processes involve very complex phenomena, which may not be easily captured by first-principles models; therefore, high in-house expertise and resources are required (Boukouvala et al, 2011; Polak et al, 2023).

The prediction reliability of a quantitative model largely depends on the precision of model parameter estimates. A particular case is given by flowsheet models (or *systems models*, as they are often called in the pharmaceutical sector), where several unit operation models are connected, and the KIs are the outputs of some units (often the final one). It is evident that, in general, the prediction fidelity of a KI depends not only on the parameters of the specific unit operation model, but also on the parameters of all unit operation models impacting the unit of interest. In this context, one may wonder whether *all* parameters should be estimated in a statistically satisfactory way or just a subset of them, and what parametric precision is sufficient to satisfy the prediction requirements for the KI. In this study, we propose a model evaluation framework to support the critical usage of quantitative models within pharmaceutical manufacturing process development. The methodology combines traditional tools typically adopted for parameter estimation purposes – namely global sensitivity analysis (GSA) and model-based design of experiments (MBDoE) – with techniques based on data analytics, that allow to graphically represent both model parameters and the KI onto a common space. A direct compression systems model for manufacturing oral solid dosage (OSD) products is used as a case study.

2. Methodology

In this section we describe the methodology for the assessment and quantification of parameters impact on the prediction reliability of quantitative models.

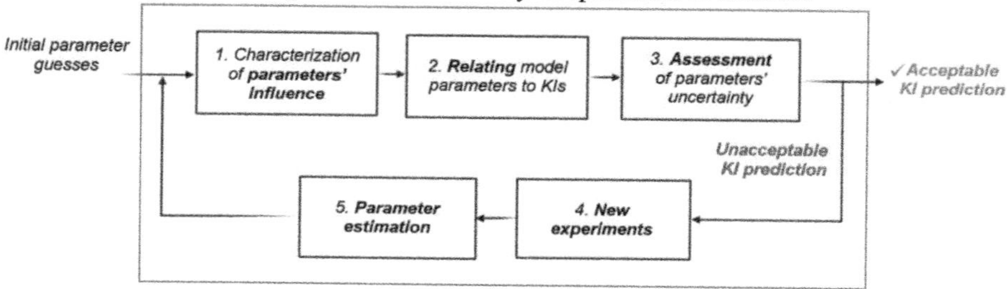

Figure 1: Schematic of the proposed methodology to quantify the model parameters' impact on the prediction reliability of quantitative models.

The workflow consists of multiple sequential steps (Fig. 1):

- Step 1. The objective is to characterize the parameter impacts on the prediction of the selected KI, so that the most influential parameters can be identified. We use Sobol's GSA to understand how the variability of the selected KI can be allocated to each model parameter.
- Step 2. The goal is quantifying the parameter impacts on the model prediction fidelity. In order to assist the interpretation of these results, here we suggest using a partial least-squares (PLS) regression model to relate the model parameters to the selected KI. Details are reported in § 2.1.
- Step 3. It is necessary to assess if the precision of the model parameters is sufficient. To this purpose, the estimated uncertainties of model parameters are projected onto the PLS model. If some projections fall outside the KI acceptability region, new experiments are needed to increase the parameters' precision. No improvement is required if all uncertainties fall inside the KI acceptability region (stop criterion).
- Step 4. If the parameters' precision requires improvement based on the output of Step 3, new experiments should be designed to collect new data. Experiments are designed through MBDoE techniques (Franceschini and Macchietto, 2008) to strengthen the precision of the most influential parameters identified at Step 1.
- Step 5. Once new data are available, all model parameters are estimated using a maximum likelihood estimator.

The procedure is iteratively repeated until satisfaction of the stop criterion.

2.1. Use of a PLS model

Let $\boldsymbol{X}$ $[N \times V]$ be the input (regressor) matrix of N observations of V variables, and $\boldsymbol{Y}$ $[N \times L]$ the response matrix of L variables. PLS (Geladi and Kowalski, 1986) is a multivariate regression technique that projects the regressor and response variables onto a common latent space, according to the model structure:

$$\boldsymbol{X} = \boldsymbol{T}\boldsymbol{P}^T + \boldsymbol{E}_X \;\; ; \quad \boldsymbol{Y} = \boldsymbol{T}\boldsymbol{Q}^T + \boldsymbol{E}_Y \quad . \tag{1}$$

First, we generate a set of parameters combinations using Monte Carlo simulations, given their initial uncertainties. These combinations of model parameters make the set of regressors $\boldsymbol{X}$, while the predicted KI is the response variable $\boldsymbol{y}$. $\boldsymbol{T}$ $[N \times V]$ is the score matrix, $\boldsymbol{P}$ $[V \times A]$ and $\boldsymbol{Q}$ $[L \times A]$ are arrays of loadings, A = number of latent variables (LVs), while $\boldsymbol{E}_X$ and $\boldsymbol{E}_Y$ are the residuals in the inputs and responses, respectively. We fix operating conditions so that the KI is equal to the target value, given the initial parameters guesses. Then, the PLS model is used in its inverse form (Jaeckle and MacGregor, 2000) to determine the set of input combinations that yield the desired tolerance on the target KI. The quality target is described by inequality constraints (bound values for the response prediction), which in turn defines the KI acceptability region. If the dimensionality of the latent space is greater than the y-space, multiple solutions of the model inversion problem exist, and are defined by the null space:

$$W = \{(\boldsymbol{t}_{new} + \boldsymbol{t}), \boldsymbol{t} \in \ker(\boldsymbol{Q})\} , \tag{2}$$

where $\boldsymbol{t}_{new} = (\boldsymbol{Q}^T\boldsymbol{Q})^{-1}\,\boldsymbol{Q}^T\,y^{*T}$ is the score vector for the target response variable y^*. The use of a PLS model to relate parameters combinations and KIs gives us the possibility to monitor the evolution of parameter uncertainties along each iteration of the methodology in Fig. 1. We obtain a clear graphical representation of both regressor and response variables onto a common latent space that can be easily interpreted.

3. Case Study

A direct compression systems model for manufacturing OSD products is used as a case study. The systems model comprises a tablet press unit operation and a tablet disintegration unit.

3.1. Tablet press unit operation

First, the variation in tablet solid fraction caused by the compaction pressure, P [MPa], exerted by the press is expressed according to Kawakita and Lüdde (1971):

$$sf = \frac{a_{sf}(1 + b_{sf}P)}{1 + a_{sf}b_{sf}P} \quad , \tag{3}$$

where sf [–] is the attained tablet solid fraction, while a_{sf} [–] and b_{sf} [MPa^{-1}] are model parameters to be estimated. The Kushner (2012) equation is used to relate the effect of the extent of lubrification K [dm] attained in the upstream powder blending on the tablet tensile strength:

$$TS = TS_0\big((1-\beta) + \beta\exp(-\gamma K)\big) \;, \tag{4}$$

where TS [MPa] is the tensile strength, TS_0 [MPa] is the tensile strength at zero porosity, γ [dm^{-1}] is the lubrication rate constant, and β [–] is the total fraction of tensile strength that can be lost due to lubrication. The empirical model by Nassar et al. (2021) accounts for the dependence of the Kushner parameters on the attained tablet solid fraction:

$$TS_0 = a_1\exp\big(b_1(1-sf)\big) \quad , \tag{5}$$

$$\beta = a_2(1-sf) + b_2 \quad . \tag{6}$$

The model parameters a_1 [MPa], b_1 [–], a_2 [–] and b_2 [–] need to be estimated.

3.2. Tablet disintegration

Both erosion and swelling (Markl et al., 2017) mechanisms are considered.

Erosion is described as:

$$V_c = (H_{coat} - \dot{\epsilon} t)A_t \,, \tag{7}$$

where V_c [m^3] is coating volume varying with time t [s], H_{coat} [m] is the thickness of the coating layer, A_t [m^2] is the tablet surface area, and $\dot{\epsilon}$ [m s^{-1}] is the constant erosion rate. The dynamic evolution of the penetration depth due to swelling is modelled as:

$$\frac{dP_d}{dt} = \left(\frac{P}{F_L/A_t}\right)^{n(T_{t/2}-P_d)/T_{t/2}} \left[\frac{d_h^2\, \varepsilon}{S_P\, \mathcal{T}_{or}^2 \mu P_d}\right] p_c \,, \tag{8}$$

where P_d [m] is the water penetration depth, F_L [MN] the tablet hardness, d_h [m] the tablet hydraulic diameter, $\mathcal{T}_{or}$ [–] the average tablet tortuosity, μ [Pa s] the liquid viscosity, p_c [Pa] the capillary pressure. S_p [–] and n [–] are formulation-dependent model parameters to be estimated. Parameter n can be estimated if dynamic penetration depth data are available, e.g., via terahertz (THz) spectroscopy. However, this is a relatively new technology if compared to a standard end-point disintegration time test. If we only have end-point disintegration data, the term $\left(\frac{P}{F_L/A_t}\right)^{n(T_{t/2}-P_d)/T_{t/2}}$ can be replaced with a lumped parameter Λ [–] which can be fitted. $T_{t/2}$ [m] is the time-dependent half tablet thickness, and ε [–] represents the average porosity of the swollen product. The stress due to tablet expansion from swelling is defined according to Peppas and Colombo (1989):

$$\tau = -TS + C_2 w_l + C_3\sqrt{w_l} \,, \tag{9}$$

where τ [MPa] is the total stress, w_l [–] is the liquid content in the tablet, and C_2 [MPa] and C_3 [MPa] are model parameters to be estimated. From τ, we can compute ε to be included in Eq. (8):

$$\tau = \frac{G_0 \exp\left(-\frac{B\varepsilon}{1-\varepsilon}\right) \lambda t}{T_{t/2}} \,, \tag{10}$$

with G_0 [MPa] and B [–] elastic constants from literature, and λ [s^{-1}] the swelling rate (Kuentz and Leuenberger, 1998). We define the disintegration time as the time for which the tablet stops disintegrating, i.e., $dP_d/dt = 0$. Five model parameters need to be estimated: C_2, C_3, $\dot{\varepsilon}$, n, S_p.

3.3. Product quality assessment

We consider an immediate release product, with a target disintegration time of 4 min. According to USP <701> (2011) disintegration test specifications, the time limit for the formulation to completely disintegrate is 5 min; therefore, we set ± 1 min as the admissible tolerance with respect to the target value of the disintegration time. Results from Sobol's GSA (Step 1) are collected in two different metrics: the first-order sensitivity index S_i, which accounts for the direct effect of each parameter on the KI, and the total sensitivity index $S_{i,TOT}$, which also accounts for interactions with the other model parameters (Table 1). The use of both indices allows us to identify possible parameter interactions. Uniform distributions for the model parameters were assumed, and bounds were chosen on typical values that can be found in the literature (Peppas and Colombo, 1989; Nassar et al., 2021). The initial parameters' uncertainties were assumed equal to $\pm 50\%$ of the nominal values, as reported in Table 2. The five most influential parameters, as characterized by the greatest total sensitivity indices, are: a_{sf}, b_1, b_{sf}, n and S_p. Note that a_{sf}, b_1 and b_{sf} are parameters related to the tablet press model, suggesting that considering only the second unit may cause a loss of information compared to the consideration of both the tablet press and the disintegration together.

Table 1. Sobol's sensitivity indices for the parameters of the direct compression systems model with respect to the disintegration time. The most influential model parameters are in boldface.

Parameter	S_i	$S_{i,TOT}$
a_1	4.767×10^{-4}	1.925×10^{-2}
a_2	3.762×10^{-4}	5.478×10^{-3}
$\boldsymbol{a_{sf}}$	$\mathbf{8.425 \times 10^{-2}}$	**0.286**
$\boldsymbol{b_1}$	$\mathbf{1.562 \times 10^{-3}}$	$\mathbf{5.431 \times 10^{-2}}$
b_2	1.273×10^{-4}	1.634×10^{-2}
$\boldsymbol{b_{sf}}$	$\mathbf{2.245 \times 10^{-2}}$	**0.139**
γ	2.332×10^{-4}	2.695×10^{-4}
C_2	2.815×10^{-4}	3.992×10^{-3}
C_3	2.209×10^{-4}	3.731×10^{-3}
$\dot{\varepsilon}$	3.603×10^{-4}	5.042×10^{-4}
$\boldsymbol{n}$	**0.641**	**0.881**
$\boldsymbol{S_p}$	$\mathbf{1.312 \times 10^{-2}}$	**0.101**

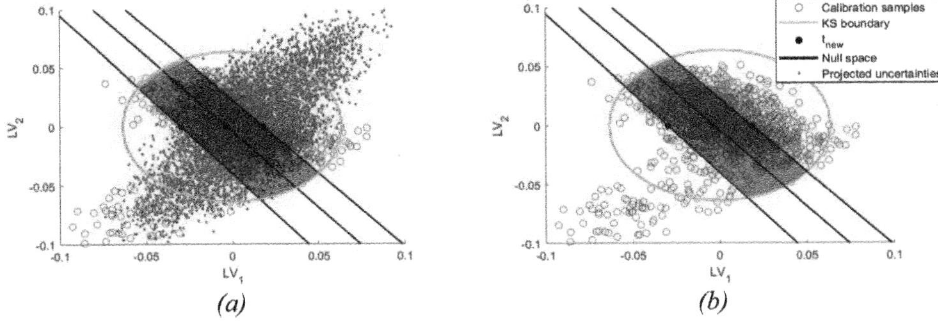

(a) (b)

Figure 2. Application of the proposed methodology to the direct compression systems model for the assessment of tablet disintegration time: (a) initial iteration, (b) final iteration where all projected uncertainties expressed using new estimates CIs (red points) fall inside the KI acceptability region (black area).

The relationship between model parameters and the KI is assessed by building a PLS model (Step 2). We verified that two LVs are sufficient to capture enough variance of the data. With reference to the first iteration of the proposed workflow in Fig. 1, LV1 explains ~44% the total variance, while LV2 ~28%. Confidence limits are considered in the latent space in the shape of an ellipse, whose semi-axis $s_a$along the ath direction is defined as $s_a = \sqrt{\lambda_a T_{lim}^2}$ (λ_a = eigenvalue of the matrix $\boldsymbol{X}^T\boldsymbol{y}\boldsymbol{y}^T\boldsymbol{X}$; T_{lim}^2 = 95% confidence limit); this region defines the knowledge space (KS) boundary. By assessing the effect of the initial model parameter uncertainty on the KI (Step 3), it can be observed that the prediction fidelity requirement for disintegration time cannot be satisfied, as many projections lie outside of the KI acceptability region (Fig. 2a). Thus, MBDoE techniques are applied to increase the precision of the five relevant parameters (Step 4); (in-silico) experimental data are used to re-estimate all parameters (Step 5). Seven iterations are required to reach the required model fidelity with respect to the KI prediction, i.e., seven experimental runs would need to be simultaneously performed for the two units. The results after the final iteration are shown in Fig. 2b: all projected uncertainties of parameter estimates fall inside the KI acceptability region. Finally, we report the estimated values of model parameters with their 95% confidence intervals (CIs) and *t*-values (Table 2). It can be observed that parameters C_2, C_3 and $\dot{\varepsilon}$ do not require precise estimation (*t*-values are lower than the reference) to meet the KI specification, and additional experimental effort can be saved.

Table 2. Estimated values of model parameters with their 95% CIs and *t*-values. † = 95% CI larger than ±50% the parameter guess value, * = precision is not statistically satisfactory.

Parameter	Units	Nominal	Estimated	95% CI	*t*-value
a_1	MPa	11.04	10.31	0.294	35.05
a_2	-	1.091	0.658	0.336	1.96
a_{sf}	-	0.463	0.439	2.021×10^{-3}	2.170×10^{2}
b_1	-	-8.202	-9.037	6.638×10^{-2}	1.361×10^{2}
b_2	-	0.326	0.150	6.670×10^{-2}	2.26
b_{sf}	MPa^{-1}	2.460×10^{-2}	2.984×10^{-2}	3.863×10^{-4}	1.036×10^{2}
γ	dm^{-1}	1.211×10^{-3}	1.148×10^{-3}	7.309×10^{-5}	15.71
					t_{ref} = 1.690
C_2	MPa	1.000×10^{2}	1.032×10^{2}	2.636×10^{4} †	3.942×10^{-3}*
C_3	MPa	1.000×10^{2}	82.46	7.231×10^{2} †	0.114*
$\dot{\varepsilon}$	m s^{-1}	1.000×10^{-3}	8.361×10^{-4}	7.493×10^{-2} †	0.112*
n	-	0.900	0.929	3.591×10^{-2}	1.037×10^{2}
S_p	-	0.524	0.488	8.451×10^{-2}	5.77
					t_{ref} = 1.646

4. Conclusions

In this study, a systematic workflow for the development and assessment of the predictive capability of quantitative models has been proposed and assessed. The methodology combines traditional tools with techniques based on data analytics, with the advantage of graphically represent both model parameters and the KI onto a common latent space. The methodology can be applied to any model, and be used to guide practitioners in model development and the subsequent usage for pharmaceutical process development.

Acknowledgements

This study was funded by a Digital Design capability project at GlaxoSmithKline (GSK).

References

G. Bano, R.M. Dhenge, S. Diab, D.J. Goodwin, L. Gorringe, M. Ahmed, R. Elkes, S. Zomer, 2022. Streamlining the development of an industrial dry granulation process for an immediate release tablet with systems modelling. Chem. Eng. Res. Des., 178, 421–37.

F. Boukouvala, F.J. Muzzio, M.G. Ierapetritou, 2011. Dynamic data-driven modeling of pharmaceutical processes. Ind. Eng. Chem. Res., 50, 6743–6754.

S. Braakman, P. Pathamanathan, H. Moore, 2022. Evaluation framework for systems models. CPT Pharmacometrics Syst. Pharmacol., 11, 264–289.

G. Franceschini, S. Macchietto, 2008. Model-based design of experiments for parameters precision: state of the art. Chem. Eng. Sci., 63, 4864–4872.

P. Geladi, B.R. Kowalski, 1996. Partial least-squares regression: A tutorial. Anal. Chim. Acta. 185,1–17.

C.M. Jaeckle, J.F. MacGregor, 2000. Industrial applications of product design through the inversion of latent variable models. Chemom. Intell. Lab Syst., 50, 199–210.

K. Kawakita, K.H. Lüdde, 1971. Some considerations on powder compression equations. Powder Technol., 4, 61–68.

M. Kuentz, H. Leuenberger, 1998. Modified Young's modulus of microcrystalline cellulose tablets and the directed continuum percolation model. Pharm. Dev. Technol., 3, 13–19.

J. Kushner, 2012. Incorporating Turbula mixers into a blending scale-up model for evaluating the effect of magnesium stereate on tablet tensile strength and bulk specific volume. Int. J. Pharm., 399, 19–30.

D. Markl, S. Yassin, D.I. Wilson, D.J. Goodwin, A. Anderson, J.A. Zeitler, 2017. Mathematical modelling of liquid transport in swelling immediate release tablets, Int. J. Pharm., 526, 1–10.

J. Nassar, B. Williams, C. Davies, K. Lief, R. Elkes, 2021. Lubrication empirical model to predict tensile strength of directly compressed powder blends. Int. J. Pharm., 592, 119980.

N.A. Peppas, P. Colombo, 1989, Development of disintegration forces during water penetration in porous pharmaceutical systems, J. Control. Release, 10, 245–250.

J. Polak, M. Von Stosch, M. Sokolov, L. Piccioni, A. Streit, B. Schenkel, B. Guelat, 2023. Hybrid modeling supported development of an industrial small-molecule flow chemistry process. Comput. Chem. Eng., 170, 108127.

USP <701> Disintegration. The United States Pharmacopeial Convention. 2011.

Antonis Kokossis, Michael C. Georgiadis, Efstratios N. Pistikopoulos (Eds.)
PROCEEDINGS OF THE 33rd European Symposium on Computer Aided Process Engineering (ESCAPE33), June 18-21, 2023, Athens, Greece
 http://dx.doi.org/10.1016/B978-0-443-15274-0.50008-1

Simulation and optimisation of a medium scale industrial reverse osmosis desalination system

Mudhar A. Al-Obaidi [a], Alanood A. Alsarayreh [b], and Iqbal M. Mujtaba [c]

[a] *Middle Technical University, Technical Institute of Baquba, Baquba, Dayala – Iraq*
[b] *Chemical Engineering Department, Mu'tah University, Al Karak – Jordan*
[c,*] *Chemical Engineering Department, University of Bradford, Bradford BD7 1DP, UK. Email: I.M.Mujtaba@bradford.ac.uk*

Abstract

Desalination is one of the techniques used for meeting increasing water demand around the world. However, as the technology is energy intensive, the increasing energy price around the globe will force the practitioners to optimise the process from the point of minimising water production cost. The motivation of this study is to carry out simulation and optimisation of a reverse osmosis (RO) system to estimate and minimise the freshwater production cost. For this purpose, a detailed mathematical model of the process is developed using a combination of an earlier model of RO developed by the authors and a set of cost model equations collected from the open literature. The medium-scale industrial brackish water RO desalination system of Arab Potash Company in Jordan is selected as a case study. The model is first used in simulation mode which provides the detailed insight of the process and feasible operation envelope. The model is then embedded in a single objective non-linear optimisation framework to determine the best operating parameters in order to minimise the freshwater production cost while optimising several operating parameters and meeting the desired water quality specification in terms of salinity. Sensitivity of energy cost on the optimum operating conditions will also be presented in detail.

Keywords: Brackish water desalination; Reverse Osmosis system; Simulation; Optimisation; Water production cost.

1. Introduction

One of the methods being utilised to fulfil the rising water demand of the world is desalination (Mujtaba and Sowgath, 2022). However, such processes require huge amount of energy, and the rising cost of energy globally compels practitioners to optimise the process from the standpoint of reducing the freshwater production cost. Brackish Water Reverse Osmosis (BWRO) process is commonly used as a successful desalination method in many countries including Jordan and Saudi Arabia using groundwater with salinity ranging between 1,000 ppm to 10,000 ppm (Afonso et al., 2004).

According to the latest study of Pearson et al. (2021), the freshwater production cost varies between 0.39$ to 0.66$ per cubic meter of freshwater for water productivities ranging between 10,000 to 70,000 m^3/d. However, there is still a scope of reducing the freshwater production cost further while enhancing the BWRO performance. Model based techniques can be applied to BWRO processes to identify such opportunity. Several researchers were successful in identifying optimal design and operating conditions of RO processes via simulation and optimisation leading to reduced freshwater production cost of water. For instance, Arroyo and Shirazi (2012) evaluated the total freshwater production cost of six brackish water RO desalination facilities in Texas with productivity ranging between 5455.3 to 125017.47 m^3/day and the cost ranging between 0.24 to 0.528$ per cubic meter of freshwater. Ghaffour et al. (2013) carried out a techno-economic evaluation of the freshwater production cost considering the main influential parameters of different water desalination facilities resulting in0.5 $/$m^3$ of freshwater. Using simulation and optimisation, an economic study was achieved by Atab et al. (2016) to assess the specific energy consumption and freshwater production cost of brackish water RO desalination system which utilised a pressure exchanger as an energy recovery device. A specific model of RO process was used to carry out. The study resulted in 24000 m^3/day of productivity at 0.11 $/$m^3$ and freshwater salinity of 400 ppm from feed water salinity 15,000 ppm with specific energy consumption of 2.8 to 0.8 kWh/m^3.

Recently, Pearson et al. (2021) attempted to obtain real cost information of seven BWRO plants in Florida, built between 2004 to 2013, for the period between 2017 and 2019. Considering the calculations of capital cost and operating cost of the plants with capacity ranging between 1364 to 56818 m^3/day and brackish water salinity ranging between 2,000 to 6,000 ppm, the freshwater production cost varied between 0.23 to 0.63 $/$m^3$ of freshwater. Using solar energy for different RO-PV designs, most recently Shalaby et al. (2022) reported the freshwater cost between 0.74-1.58 $/$m^3$.

The cost of desalination varies significantly depending on the location and is influenced by several factors, such as the feed water source, feed water quality, plant size, process type and design including energy recovery, intake type, pre- and post-treatment processes, concentrate disposal method, regulatory issues, land costs, and water conveyance to and from the plant. Because of these variations in water quality and quantity from site to site, and occasionally even within the same site, it is expected that the cost of RO brackish water desalination will be the same for all cases and condition. However, each case should be optimised to obtain the minimum freshwater production cost. To the best of our knowledge, there has not been any study to predict the freshwater production cost of a medium scale industrial BWRO system. This study estimates the freshwater production cost of the RO plant via simulation and optimisation using model-based technique. To carry out this task, an earlier RO model developed by the same authors and a set of cost model equations gathered from the open literature are combined to create a thorough mathematical model of the RO process comprising the estimation of freshwater production cost and associated specific energy consumption. The model is initially utilised in simulation mode, which offers an understanding of the actual freshwater production cost and a workable operational envelope. In order to reduce the cost of producing water while optimising a number of operating parameters and achieving the appropriate salinity specification, the model is then incorporated into a single objective non-linear optimisation framework.

2. Description of BWRO desalination system

Fig. 1 depicts the design of a brackish water RO for the APC plant with a 1200 m^3/day capacity. Permeate and retentate reprocessing designs are used in two passes. Two stages of pressure vessels are arranged in the first pass in 4:2 order. Three pressure vessel stages are present in the second pass and arranged in 2:1:1 order. The permeate from the first pass is transferred to the second pass for further purification. The brine line from the first pass is drained. A high-quality water of 2 ppm is produced from the second pass. The retentate of the second pass is returned and mixed with the raw feed water to maintain a high productivity.

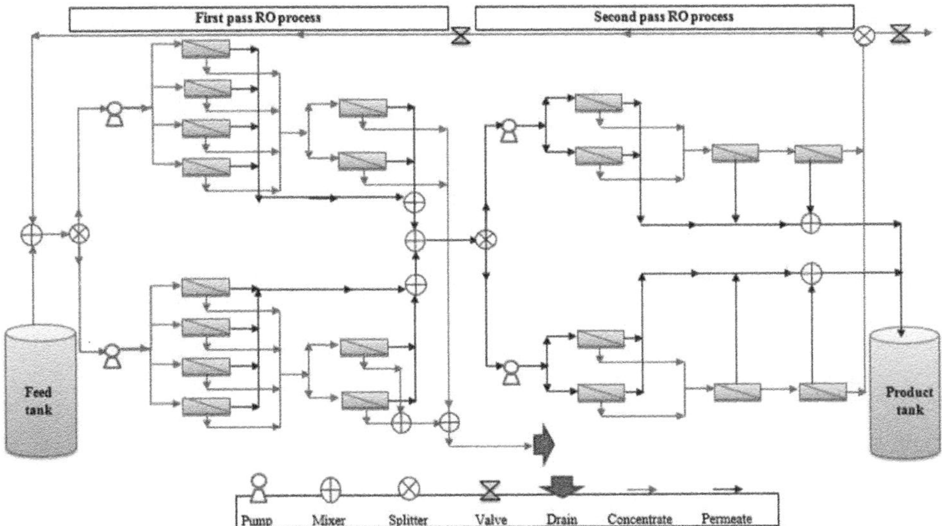

Fig. 1. A schematic representation of BWRO system of APC

3. Modelling of BWRO desalination system

A steady-state model was created by Al-Obaidi et al. (2018) for the APC medium-scale brackish water RO desalination plant. The model was constructed based on a number of assumptions. The most important ones are: 1) steady state operation, 2) fixed membrane features and channel geometries, 3) the film theory expresses the concentration polarisation phenomenon, 4) 1 atm in the permeate channel, 5) fixed temperature, and 6) The spacer properties are used to quantify the pressure drop in the feed channel. Before being utilised to examine the plant performance under various operating situations, the model was verified using APC plant data. In this work, this model has been improved by including a sub-model to calculate the cost of producing freshwater using data from several research (Filippini et al., 2018; Al-Obaidi et al., 2019; Malek et al., 1996). Table 1 presents a set of important equations of the RO model and the sub economic model used in this study. For further details of the model developed, parameters and variables and associated effects on the performance indicators can all be found in Al-Obaidi et al. (2018).

4. Simulation of BWRO desalination system

Simulation of the process is carried out to estimate the most important performance indicators including the freshwater production cost and specific energy consumption. Table 2 depicts the simulation results for the set of original inlet conditions of 988.93 ppm (salinity), 9.22 atm (pressure of 1st pass), 9.832 atm (pressure of 2nd pass), 2034.33 m^3/day (feed flowrate) and 25 °C (temperature).

Furthermore, the estimation of electrical cost of pumping the brackishwater into the 1st and 2nd passes of RO system is important to be addressed. The power price (\$/day) is calculated using the following correlation:

$$Power\ price\ per\ day\ \left(\frac{USD\$}{day}\right) = SEC\ \left(\frac{kWh}{m^3}\right) x\ productivity\ \left(\frac{m^3}{day}\right) x\ Factor\left(\frac{\$}{kWh}\right)$$

Factor is the specific price of each kWh in Jordan which is 0.123$/kWhr (Jordan electricty prices). Table 2 shows the energy price per day.

Table 1. RO model

Equation	Description	Unit
$Q_p = A_{w(T)}\left(P_f - \frac{\Delta P_{drop,E}}{2} - P_p - \pi_w - \pi_p\right) A_m$	Water flux	(m³/s)
$Q_s = B_{s(T)}(C_w - C_p)$	Solute flux	(kg/m² s)
$\Delta P_{drop,E} = \frac{9.8692x10^{-6} A^* \rho_b Q_b{}^2 L}{2 d_h Re_b^n (W t_f \epsilon)^2}$	Pressure drop per element	(atm, -)
$\frac{(C_w - C_p)}{(C_b - C_p)} = exp\left(\frac{Q_p/A_m}{k}\right)$	Membrane surface concentration	(-)
$k = 0.664\, k_{dc}\, Re_b^{0.5}\, Sc^{0.33}\left(\frac{D_b}{d_h}\right)\left(\frac{2d_h}{L_f}\right)^{0.5}$	Mass transfer coefficient	(m/s, -)
$C_p = \frac{B_s C_f\ e^{\frac{J_w}{k}}}{J_w + B_s\ e^{\frac{J_w}{k}}}$	Permeate concentration	(kg/m³)
$Rej = \frac{C_f - C_p}{C_f}$ $Rec = \frac{Q_p}{Q_f}$	Solute rejection and water recovery	(-)
$SEC = \frac{\frac{(P_{f(plant)} x101325)\ Q_{f(plant)})}{Q_{p(Total)}\ \varepsilon_{pump}}}{36x10^5}$	Specific energy consumption	(kWh/m³)
$TAC = TCC + AOC$	Total annual cost	($/yr)
$TCC = [(C_{wip} + C_{Pump} + C_{me})\ SD\ CC]$	Total capital cost	($/yr)
$AOC = OC_{Pu} + OC_{sc} + OC_{ch} + OC_{me} + OC_{lab} + OC_{maint} + OC_{bd}$	Total annual operating cost	($/yr)
$C_{wip} = 996\ (86400\ Q_{f(plant)})^{0.8}$	Water intake and pre-treatment cost	($)
$C_{Pump} = [52\ (3600\ Q_{f(plant)}\ (P_{f(plant)}\ 0.101325))^{0.96}]$	Capital cost of high-pressure pump	($)
$C_{me} = N_s\, N_{PV}(C_{ele}\, N_{ele} + C_{PV})$	Membrane module and pressure vessel capital cost	($)
$OC_{pu} = 365x24\left[\left(\frac{(3600\ (P_{f(plant)}\ 0.101325))\ Q_{f(plant)})}{3.6\ \varepsilon_{pump}\ \varepsilon_{motor}}\right)\right] E_c\ L_f$	Pumping operating cost	($/yr)
$OC_{sc} = 3600x24x365\ C_{cf}\ Q_{p(plant)}\ L_f$	Annual operating spares cost	($/yr)
$OC_{bd} = 3600x24x365\ C_{bd}\ Q_{p(plant)}\ L_f$	Effluents disposal cost	($/yr)
$OC_{ch} = 3600x24x365\ C_{ct}\ Q_{f(plant)}\ L_f$	Annual chemical treatment cost	($/yr)
$OC_{me} = 0.2\ C_{me}$	Annual membrane replacement cost	($/yr)
$OC_{lab} = C_{lab}\ 3600x24x365\ Q_{p(plant)}$	Annual labour cost	($/yr)
$FWC_{RO} = \frac{\left(\frac{TCC}{CCRF}\right) + AOC}{3600x\ 24x\ 365\ Q_{p(plant)}}$	Freshwater production cost	($/m³)
$CCRF = \left[\frac{(i+1)^n - 1}{i\ (i+1)^n}\right]$	Capital cost recovery factor	(yr)
$OC_{main} = 0.02\ PUC\ 3600x24x365\ Q_{p(plant)}$	Annual maintenance costs	($/yr)

Table 2. Simulation results

Water productivity (m³/day)	Product salinity (ppm)	Salt rejection (%)	Water recovery (%)	Specific energy consumption (kWh/m³)	Total capital cost ($)	Total operation cost ($)	Freshwater production cost ($/m³)	Energy price per day ($/day)
1164.51	2.0	99.79	57.24	0.834	64005.66	75753.3	0.192	119.45

5. Optimisation of BWRO desalination system

This section intends to optimise the operation of BWRO system of APC to guarantee the lowest freshwater production cost (objective function) via determining the optimal operating condition of feed pressure of the 1st and 2nd passes and feed flowrate. The other operating conditions of groundwater salinity and temperature will be taken as fixed values of 988.93 ppm and 25 °C, respectively. Furthermore, the optimisation framework has considered several inequality constraints including the upper and lower feed flowrate for each membrane. Also, the specific energy consumption has been constrained at its simulation value. The non-linear single-objective optimisation framework can be described as:

Given: Feed water pressure and flowrate, RO module specifications.
Optimise: The optimisation variables of feed pressure, and flow rate.
Minimise: the freshwater production cost (*FWPC*) of RO system.
Subject to: Equality and inequality constraints and limits of optimisation variables.
Thus, the optimisation problem can be mathematically represented as:

$$\min_{Q_{f(plant)},\,P_{p(plant)}} FWPC$$

Subject to: Equality constraints: RO process model and fixed specific energy consumption (simulation value) (SEC = 0.834 kWh/m^3)

Inequality constraints:

a) lower and upper limits of feed flow rate of RO system

$$(696.96\ \text{m}^3\text{/day})\ {Q_{f(plant)}}^L \leq Q_{f(plant)} \leq {Q_{f(plant)}}^U\ (3707.52\ \text{m}^3\text{/day})$$

a) lower and upper limits of feed pressure of the 1st and 2nd passes of RO system

$$(5\ \text{atm})\ {P_{f(1st\ pass)}}^L \leq P_{f(1st\ pass)} \leq {P_{f(1st\ pass)}}^U\ (12\ \text{atm})$$

$$(5\ \text{atm})\ {P_{f(2nd\ pass)}}^L \leq P_{f(2nd\ pass)} \leq {P_{f(2nd\ pass)}}^U\ (12\ \text{atm})$$

c) lower and upper limits of feed flow rate of each membrane module

$$(87.12\ \text{m}^3\text{/day})\ {Q_{f(membrane)}}^L \leq Q_{f(membrane)} \leq {Q_{f(membrane)}}^U\ (463.44\ \text{m}^3\text{/day})$$

Table 3 shows the optimisation results of several perforamcne indicators including the freshwater production cost and the optimal feed pressure and flowrate. Note that the freshwater production cost could be reduced by 6.7% by increasing the pressure of the 1st and 2nd passes from 9.22 and 9.832 atm to 9.93 and 10.59 atm, respectively. Also, it can be noticed that the feed flowrate has been decreased from 2034.33 to 1900.8 m^3/day to assure increasing the rate of permeation through the membranes due to increased residence time of the water inside the module. The productivity increases by 8% while maintaining a high-quality freshwater of 2.1 ppm (as desired). However, the electrcial cost is inevatibly increased (by 8%) due to increasing the productivity in cubic meter per day.

Table 3. Optimisation results

Optimised inlet conditions			Optimised results		
Parameter	Value	Unit	Parameter	Value	Unit
Feed pressure of 1st pass	9.93	atm	Water recovery	66.33	%
Feed pressure of 2nd pass	10.59	atm	Salt rejection	99.78	%
Feed flow rate	1900.8	m^3/day	Salinity of product water	2.1	ppm
			Specific energy consumption	0.834	kWh/m^3
			Water productivity	1260.81	m^3/day
			Freshwater production cost	0.179	\$/m^3
			Electricity cost	129.33	\$/day
Reduction of freshwater production cost (in comparison to simulation) = 6.7%					

6. Conclusions

This paper attempted to reduce the overall freshwater production cost of a medium scale brackish water RO desalination system of Arab Potash Company (Jordan) via process modelling, simulation and optimisation. The simulation of the studied RO system obtained the values of performance indicators while the optimisation has obtained the minimum freshwater production cost with optimal feed pressure of the 1st and 2nd passes and inlet feed flowrate for the same specific energy consumption (as in the real plant and simulation). The freshwater production cost is reduced by 6.7% in comparison to the original simulation value while achieving a high-quality of product freshwater. Note, the actual electrical cost increases by more than 8% due to increase in the water productivity of RO system (compared to actual plant productivity). In the context of model refinement, relaxing the model assumptions would certainly provide more accurate predictions of the performance metrics. Furthermore, it should be noted that the developed model did not thoroughly incorporate the influence of fouling on the performance metrics. The enhancement of the established model is expected to introduce actual estimation of the freshwater production cost, which should be addressed in future study.

References

A.A. Alsarayreh, M.A. Al-Obaidi, S.K. Farag, R. Patel, I.M. Mujtaba, 2021, Performance evaluation of a medium-scale industrial reverse osmosis brackish water desalination plant with different brands of membranes. A simulation study, Desalination, 503, 114927.

A. Malek, M.N.A. Hawlader, J.C. Ho, 1996, Design and economics of RO seawater desalination. Desalination, 105, 245–261.

G. Filippini, M.A. Al-Obaidi, F. Manenti, I.M. Mujtaba, 2018, Performance analysis of hybrid system of multi effect distillation and reverse osmosis for seawater desalination via modeling and simulation. Desalination, 448, 21–35.

I.M. Mujtaba, M.T. Sowgath, 2022, Desalination Technology: Design and Operation, Elsevier.

J. Arroyo, and S. Shirazi, 2012. Cost of brackish groundwater desalination in Texas. Texas Water Development Board, 1–7.

J.L. Pearson, P.R. Michael, N. Ghaffour, T.M. Missimer, 2021. Economics and Energy Consumption of Brackish Water Reverse Osmosis Desalination: Innovations and Impacts of Feedwater Quality. Membranes, 11(8), 616.

Jordan electricity prices. GlobalPrices.com. https://www.globalpetrolprices.com/Jordan/electricity_prices/ (Accessed at 3-11-2022).

M.A. Al-Obaidi, G. Filippini, F. Manenti, I.M. Mujtaba, 2019, Cost evaluation and optimisation of hybrid multi effect distillation and reverse osmosis system for seawater desalination. Desalination, 456, 136–149.

M.D. Afonso, J.O. Jaber, M.S. Mohsen, 2004. Brackish groundwater treatment by reverse osmosis in Jorden. Desalination, 164, 157–171.

M.S. Atab, A.J. Smallbone, A.P. Roskilly, 2016. An operational and economic study of a reverse osmosis desalination system for potable water and land irrigation. Desalination, 397, 174-184.

N. Ghaffour, T.M. Missimer, G.L. Amy, 2013. Technical review and evaluation of the economics of water desalination: current and future challenges for better water supply sustainability. Desalination, 309, 197–207.

S.M. Shalaby, M.K. Elfakharany, I.M. Mujtaba, B. M. Moharram, H. F. Abosheiasha, 2022. Development of an efficient nano-fluid cooling/preheating system for PV-RO water desalination pilot plant, Energy Conversion and Management, 268, 115960

Antonis Kokossis, Michael C. Georgiadis, Efstratios N. Pistikopoulos (Eds.)
PROCEEDINGS OF THE 33rd European Symposium on Computer Aided Process Engineering (ESCAPE33), June 18-21, 2023, Athens, Greece
 http://dx.doi.org/10.1016/B978-0-443-15274-0.50009-3

Modeling strategies in multi-scale food-energy-water nexus system optimization

Marcello Di Martino[a,b], Patrick Linke[a,c], Efstratios N. Pistikopoulos[a,b]

[a]*Artie McFerrin Department of Chemical Engineering, Texas A&M University, College Station, TX, USA*

[b]*Texas A&M Energy Institute, Texas A&M University, College Station, TX, USA*

[c]*Department of Chemical Engineering, Texas A&M University at Qatar, Education City, Doha, Qatar*

Abstract

The modeling and optimization of multi-scale process systems is based on several interconnected process sub-systems. Due to the complexity of each individual sub-model, the resulting integrated process framework optimization formulations are computationally challenging to solve. While the richness of the multi-scale model employed is desired to maintain in order to obtain a solution with some degree of accuracy, simpler surrogate models are typically more attractive as a means to tame the underlying complexity, albeit often leading to an increase of the problem size. Here, the food-energy-water nexus (FEWN) is selected as a representative multi-scale process system, with focus on the reverse osmosis (RO) water supply sub-system. Based on a RO desalination model, two models are developed and compared in terms of accuracy, model complexity and size as well as computational efficiency, (i) a mixed-integer non-linear programming (MINLP) surrogate model, and (ii) a mixed-integer linear programming (MILP) surrogate model of reduced complexity but larger size. The results indicate that improved computational times can be obtained for a valid (lower bound) solution based on the MILP modeling strategy within the same level of accuracy, further underlying the importance of the selection of an appropriate surrogate model.

Keywords: optimization, surrogate modeling, food-energy-water nexus, reverse osmosis.

1. Introduction

The optimization of multi-scale process systems, which are composed of several interconnected sub-systems, remains challenging. Encountered challenges besides the multi-scale nature, encompass the complexity, as well as the size of the resulting optimization models. One approach to resolve these challenges is the usage of surrogate models to approximate sub-systems of the overall process systems (Kakodkar et al. 2022, Guillén-Gosálbez et al. 2019). This work evaluates in detail how changes in the employed surrogate model can have significant impact on the solution generation of the integrated process system. One of these interconnected multi-scale systems is the food-energy-water nexus (FEWN), which postulates that sustainable solutions regarding food, energy and water systems can only be obtained when all resources are considered holistically. Recently, more and more interest has been placed on the FEWN due to depleting natural resources and increasing global demands (Di Martino et al. 2019, Garcia and You 2016). FEWN models result in complex optimization programs, where surrogate models can mitigate computational difficulties by reducing the complexity of the system. However, such systems usually increase in number of constraints and number of variables, which in turn can result in an intractable optimization model as well, especially if the surrogate is solved over a given time horizon multiple times integrated in an overarching process system (Allen et al. 2022). Thus, the scope of this work is to analyze and compare a mixed-integer linear programming (MILP) and a mixed-integer non-linear programming (MINLP) surrogate model which are representative of a sub-system in an interconnected multi-scale process model. As an example for such an interconnected system the FEWN is selected, with focus on a reverse osmosis (RO) desalination plant as the water supply system. The comparison is based on the accuracy of the obtained minimized solutions, the computational efficiency in the form of the necessary computational time to obtain solutions and the difference in the size and complexity of the optimization models. Next, the system under investigation is described. Then, the detailed modeling equations are discussed, before analyzing and discussing the results.

2. System Under Investigation

To analyze and compare the impact of MILP and MINLP surrogate models for the integration in a multi-scale process system the performance of RO desalination plant models is compared. As visualized in Figure 1, the goal is to integrate the RO model in a multi-scale FEWN model which in turn specifies restrictions regarding the design and operation of the RO plant. The goal is to solve this integrated model at an hourly time scale for a one-year time horizon, resulting in 8760 time points. Solutions specify the sustainable and interconnected water, energy and food supply system design and operation (Cook et al. 2022, Di Martino et al. 2022a). Thus, it is important to evaluate the model approximations in terms of accuracy and computational efficiency for the solution generation. All obtained surrogate models are based on one-and-a-half years of operational RO plant data, as presented in Di Martino et al. 2022b.

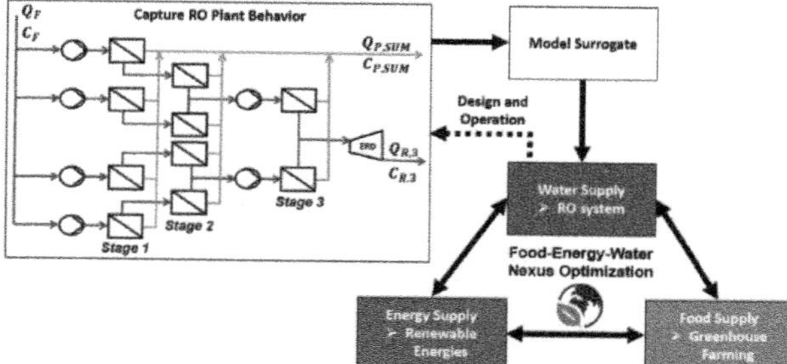

Figure 1: Schematic overview of the placement of the sub-system (RO) in the multi-scale process system (FEWN).

3. Modeling

The RO desalination plant is modeled and optimized according to Eq. 1 to 9 (Di Martino et al. 2022b).

$$min\ SEC^* = \frac{1}{4\cdot\eta_1}\sum_{j=1}^{4} P_{f,1,j} + \frac{1}{10\cdot\eta_1}\left(\sum_{j=1}^{2} P_{f,3,j} - \frac{1}{2}\sum_{j=1}^{4} P_{r,2,j}\right) - \frac{\eta_2}{20}\sum_{j=1}^{2} \Delta P_{ERD,j} \quad (1)$$

s.t.

$$WR_{sys} \geq WR^{res} \leftrightarrow WR_i \geq WR_i^{res}\ \forall i \in \{1,2,3\} \quad (2)$$

$$Q_f \geq Q_f^{res} \quad (3)$$

$$C_{p,t-1} = C_{p,t} \leq C_p^{res} \quad (4)$$

$$WR_i = WR_i^{max} \cdot \frac{\sum_j P_{i,j}}{\sum_j \max(P_{i,j})} \; \forall i \in \{1,2,3\} \tag{5}$$

$$P_{r,i,j} = f_1(P_{f,i,j}) \; \forall i = \{1,2\}, \forall j = \{1,2,3,4\} \tag{6}$$

$$P_{f,2,j} = P_{r,1,j} \; \forall j \in \{1,2,3,4\} \tag{7}$$

$$\Delta P_{ERD,j} = f_2(P_{f,3,j}) \; \forall j = \{1,2\} \tag{8}$$

$$C_{p,t} = f_3(Q_f, WR_1, WR_2, WR_3, C_{p,t-1}) \tag{9}$$

The presented objective function (Eq. 1) minimizes the linearized specific energy consumption of the system (SEC^*) based on the feed ($P_{f,i,j}$) and retentate pressures of each stage i and parallel flow j ($P_{r,i,j}$). η_1 and η_2 denote the efficiency of the utilized pumps and energy recovery device, respectively. Eq. 2 to 9 summarize all necessary linear constraints of the system to specify the operation and design of the RO plant. The overall water recovery ($WR_{sys} = Q_p/Q_f$), the feed flow (Q_f) and the permeate concentration at the given and previous time point ($C_{p,t}$, $C_{p,t-1}$) are restricted to satisfy given water demand scenarios (Eq. 2 to 4). It is important to note that WR_{sys} is implicitly restricted by restricting the water recovery of each stage. The water recovery of each stage ($WR_i \; \forall i \in \{1,2,3\}$) is estimated based on the observed maximum water recovery ($WR_i^{max} \; \forall i \in \{1,2,3\}$) and pressures, according to Eq. 5. Eq. 6 and 7 summarize the pressure modeling of the system, with the linear regression ($f_1(P_{f,i,j})$) used for calculating the retentate pressure of stages one and two for all parallel flows based on the respective feed pressure and the utilization of the retentate of stage one as the feed of stage 2. The pressure difference across the energy recovery device ($\Delta P_{ERD,j} \; \forall j = \{1,2\}$) as well as the overall permeate concentration ($C_{p,t}$) are calculated based on the MILP reformulation of a feedforward artificial neural network (ANN) with rectified linear units (ReLU) as activation functions, as stated in Eq. 8 and 9 ($f_2(P_{f,3,j}) \; \forall j = \{1,2\}$, $f_3(Q_f, WR_1, WR_2, WR_3, C_{p,t-1})$).

3.1. MINLP Surrogate Model

SEC^* is derived by applying process specific, non-generic assumptions as described in Di Martino et al. 2022b. Therefore, obtained solutions actually specify lower bounds of the true solution. Accordingly, the non-linearized formula to calculate the exact specific energy consumption (SEC) of the system is utilized in this study, as shown in Eq. 10. Furthermore, to obtain the energy consumption of the system (EC), SEC has to be multiplied by WR_{sys} and Q_f, resulting in a trilinear term (Eq. 11). In addition, WR_{sys} can be derived based on $WR_i \; \forall i \in \{1,2,3\}$ with the nonlinear equation presented in Eq. 12. Thus, the MINLP surrogate model is given by minimizing Eq. 10 subject to the constraints summarized from Eq. 2 to 9, 11 and 12. With this modified model formulation no non-generic process specific assumptions are required.

$$SEC = \frac{1}{WR_{sys}} \cdot \left(\frac{1}{4 \cdot \eta_1} \sum_{j=1}^{4} P_{f,1,j} + \frac{(1-WR_1)\cdot(1-WR_2)}{\eta_1} \left[\frac{1}{2} \sum_{j=1}^{2} P_{f,3,j} - \frac{1}{4} \sum_{j=1}^{4} P_{r,2,j} \right] - \frac{(1-WR_1)\cdot(1-WR_2)\cdot(1-WR_3)\cdot\eta_2}{2} \sum_{j=1}^{2} \Delta P_{ERD,j} \right) \tag{10}$$

$$EC = SEC \cdot WR_{sys} \cdot Q_f \tag{11}$$

$$WR_{sys} = WR_1 + (1 - WR_1) \cdot WR_2 + (1 - WR_1) \cdot (1 - WR_2) \cdot WR_3 \tag{12}$$

3.2. MILP Surrogate Model

Next, a linearized version of the nonlinear model is obtained by training ANNs with ReLUs as activation functions and reformulating them as MILPs for the calculation of SEC and EC, ultimately resulting in piecewise linear surrogate models. Here the results of the MILP approximations are specified as $\widehat{SEC}$ (see Eq. 13 to 17) and $\widehat{EC}$ (see Eq. 18 to 22). The obtained weights and biases together with the R^2 value of both surrogate models are summarized in Table 1. The auxiliary positive variables $(x_{SEC}^{(k_{SEC})}, s_{SEC}^{(k_{SEC})}, x_{EC}^{(k_{EC})}, s_{EC}^{(k_{EC})})$, binary variables $(z_{SEC}^{(k_{SEC})}, z_{EC}^{(k_{EC})})$, together with lower and upper bounds $(LB_{SEC}^{(k_{SEC})}, LB_{EC}^{(k_{EC})}, UB_{SEC}^{(k_{SEC})}, UB_{EC}^{(k_{EC})})$ are introduced for the MILP reformulation.

$$W_{SEC,1}^{(k_{SEC})} \cdot [WR_1\ WR_2\ WR_3\ Q_f]^T + b_{SEC,1}^{(k_{SEC})} = x_{SEC}^{(k_{SEC})} - s_{SEC}^{(k_{SEC})}, \forall k_{SEC} \in \{1,2\} \tag{13}$$

$$x_{SEC}^{(k_{SEC})} - z_{SEC}^{(k_{SEC})} \cdot UB_{SEC}^{(k_{SEC})} \leq 0, \forall k_{SEC} \in \{1,2\} \tag{14}$$

$$s_{SEC}^{(k_{SEC})} - \left(1 - z_{SEC}^{(k_{SEC})}\right) \cdot LB_{SEC}^{(k_{SEC})} \leq 0, \forall k_{SEC} \in \{1,2\} \tag{15}$$

$$x_{SEC}^{(k_{SEC})}, s_{SEC}^{(k_{SEC})} \geq 0, z_{SEC}^{(k_{SEC})} \in \{0,1\}, \forall k_{SEC} \in \{1,2\} \tag{16}$$

$$\widehat{SEC} = \sum_{k_{SEC}} W_{SEC,2}^{(k_{SEC})} \cdot x_{SEC}^{(k_{SEC})} + b_{SEC,2} \tag{17}$$

$$W_{EC,1}^{(k_{EC})} \cdot [SEC\ WR_{sys}\ Q_f]^T + b_{EC,1}^{(k_{EC})} = x_{EC}^{(k_{EC})} - s_{EC}^{(k_{EC})}, \forall k_{EC} \in \{1,2,3\} \tag{18}$$

$$x_{EC}^{(k_{EC})} - z_{EC}^{(k_{EC})} \cdot UB_{EC}^{(k_{EC})} \leq 0, \forall k_{EC} \in \{1,2,3\} \tag{19}$$

$$s_{EC}^{(k_{EC})} - \left(1 - z_{EC}^{(k_{EC})}\right) \cdot LB_{EC}^{(k_{EC})} \leq 0, \forall k_{EC} \in \{1,2,3\} \tag{20}$$

$$x_{EC}^{(k_{EC})}, s_{EC}^{(k_{EC})} \geq 0, z_{EC}^{(k_{EC})} \in \{0,1\}, \forall k_{EC} \in \{1,2,3\} \tag{21}$$

$$\widehat{EC} = \sum_{k_{EC}} W_{EC,2}^{(k_{EC})} \cdot x_{EC}^{(k_{EC})} + b_{EC,2} \tag{22}$$

Table 1: Weights ($W_{na,1}^{(k)}$, $W_{na,2}^{(k)}$), biases ($b_{na}^{(k)}$, $b_{na,2}$) and R^2 of the surrogate models for the approximation of the specific energy consumption (SEC) and energy consumption of the system (EC).

na	k_{na}	$W_{na,1}^{(k)}$	$W_{na,2}^{(k)}$	$b_{na,1}^{(k)}$	$b_{na,2}$	R^2
SEC	1	$[1.1556 - 0.5436\ 0.4116\ 0.4630]$	0.4638	1.0465	−0.9950	0.99
	2	$[0.2843\ 0.1137\ 0.1071\ 1.3843]$	0.5371	−0.3829		
EC	1	$[0.7492 - 0.3236\ 0.4624]$	0.8180	0.4966	−0.9443	0.99
	2	$[-1.0350\ 2.5363\ 1.3228]$	0.2926	−0.3701		
	3	$[-0.1111\ - 0.1307 - 0.7936]$	−0.2929	−1.1133		

Lastly, Eq. 12 is linearized with a first-order Taylor expansion around the nominal operating point of the RO plant, as shown in Eq. 23 ($\overline{WR_{sys}} = 0.6039, \overline{WR_1} = 0.3113, \overline{WR_2} = 0.2935, \overline{WR_3} = 0.1860$). For the relevant operating range of $0.4 \leq WR_{sys} \leq 0.85$, the approximation results in $R^2 = 0.98$. Overall, the MILP surrogate model is composed of minimizing $\widehat{SEC}$ subject to Eq. 2 to 9 and Eq. 13 to 23.

$$\begin{aligned}\widehat{WR}_{sys} \approx \overline{WR_{sys}} &+ (1 - WR_2 - WR_3 + WR_2 \cdot WR_3) \cdot (WR_1 - \overline{WR_1}) \\ &+ (1 - WR_1 - WR_3 + WR_1 \cdot WR_3) \cdot (WR_2 - \overline{WR_2}) \\ &+ (1 - WR_1 - WR_2 + WR_1 \cdot WR_2) \cdot (WR_3 - \overline{WR_3})\end{aligned} \tag{23}$$

The described nonlinearities are overcome by introducing piecewise linear approximations resulting in a less complex optimization model. However, the resulting system increases in size, with an increased number of constraints (60 vs. 80) and variables (continuous 47 vs. 57; binaries 9 vs. 14). The optimization models have been solved in GAMS with the CPLEX solver for the MILP surrogate model and the BARON solver for the MINLP surrogate model. All surrogate models have been derived in MATLAB.

4. Results and Discussion

To analyze the trade-offs between system complexity and system size of the presented MILP and MINLP surrogates, the monthly energy consumption of the year 2017 is minimized based on the obtained operational RO data. The results of the optimization are visualized in Figure 2. The results of both models are of the same order of magnitude, where for the same set of restrictions in terms of WR_{sys} and Q_f the MINLP surrogate results in higher energy consumptions than the MILP surrogate, as expected.

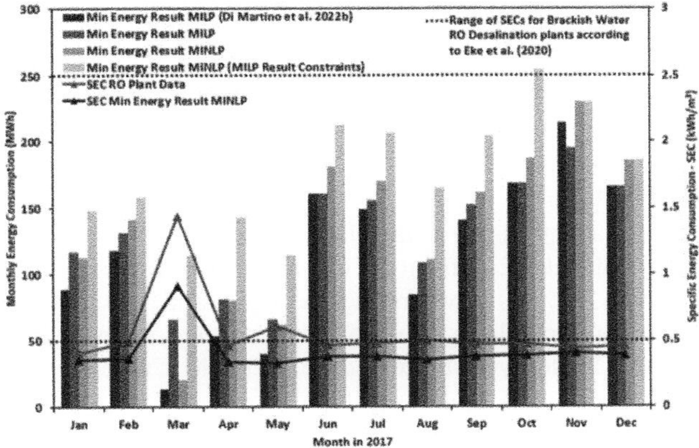

Figure 2: Comparison of the monthly energy consumption of the RO plant based on the optimization of the MILP, MINLP and MINLP with MILP result constraints.

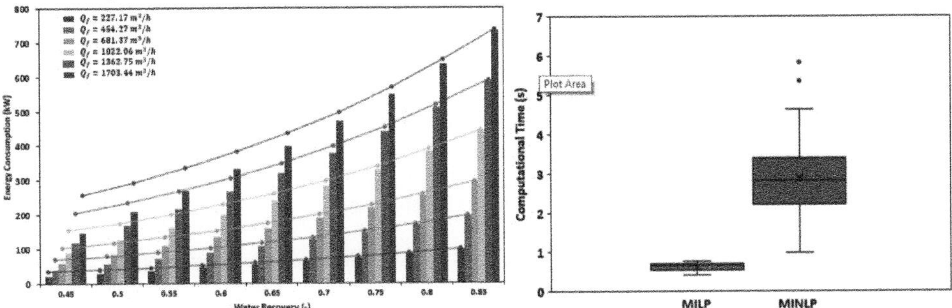

Figure 3: Left: Pareto front of the energy consumption of the RO system dependent on the system's water recovery and feed flow; bars – results of MILP, points connected with line – results of MINLP. Right: Comparison of the necessary computational time to solve the MILP and MINLP; boxplot: {min – 1st quartile – median – 3rd quartile – max}, x – mean, outliers – 1.5 times the interquartile range larger than the third quartile.

Next, the energy consumption of the RO plant is systematically minimized for different sets of WR_{sys} and Q_f resulting in the Pareto front shown in Figure 3. Again, for all observed results, the MILP denotes the lower bound of the solution. The difference between the MINLP and MILP solution decreases with increasing water recovery. This trend is observed for all feed flows under investigation. In addition, the computational efficiency of obtained solutions was recorded, as shown in Figure 3. In all observed cases, the computational time for the MILP solution generation is less than the MINLP one, even in extreme cases. On average, the difference between the observed computational times is 2.28*s*. Thus, utilizing the MILP indeed results in significant reduction of computational time (for an hourly time scale over one year, 5.55 hours) while the obtained lower bounds are good approximations of the true optimum for high water recoveries.

5. Conclusion

This work analyzed and compared a MILP and MINLP surrogate model for the integration in a multi-scale process system. For illustration, the surrogate modeling of a RO process for the integration in a FEWN system has been studied due to its importance for sustainable solution generation. The derived MILP surrogate provides lower bounds to the true optimum which are of the same level of accuracy as the MINLP surrogate solutions, while resulting in significant increase of computational efficiency, viz. the reduction of computational time. This trade-off between increasing the size of the optimization model to linearize it, is particularly important to consider in applications for interconnected multi-scale process systems.

6. Acknowledgments

This publication was made possible by the National Priorities Research Program (NPRP) grant No. NPRP11S-0107–180216 from the Qatar National Research Fund (a member of Qatar Foundation). The findings herein reflect the work, and are solely the responsibility, of the authors. The authors also gratefully acknowledge support from Texas A&M University, Texas A&M Energy Institute and Texas A&M University at Qatar.

References

R. C. Allen, S. G. Baratsas, R. Kakodkar, S. Avraamidou, C. D. Demirhan, C. F. Heuberger-Austin, M. Kolokkenburg, E. N. Pistikopoulos, 2022, A multi-period integrated planning and scheduling approach for developing energy systems, Optim. Control. Appl. Methods.

J. Cook, M. Di Martino, R. C. Allen, E. N. Pistikopoulos, S. Avraamidou, 2022, A decision-making framework for the optimal design of renewable energy systems under energy-water-land nexus considerations, Sci. Total Environ., 827, 154185.

J. Eke, A. Yusuf, A. Giwa, A. Sodiq, 2020, The global status of desalination: An assessment of current desalination technologies, plants and capacity, Desalination, 495, 114633.

M. Di Martino, S. Avraamidou, J. Cook, E. N. Pistikopoulos, 2021, An Optimization Framework for the Design of Reverse Osmosis Desalination Plants under Food-Energy-Water Nexus Considerations, Desalination, 503, 114937.

M. Di Martino, S. Namany, S. Avraamidou, T. Al-Ansari, P. Linke, E. N. Pistikopoulos, 2022a, Energy-water Scheduling Decisions for Agricultural Scenario Planning, 32nd European Symposium on Computer Aided Process Engineering (ESCAPE-32), Elsevier, 1573-1578.

M. Di Martino, S. Avraamidou, E. N. Pistikopoulos, 2022b, A Neural Network Based Superstructure Optimization Approach to Reverse Osmosis Desalination Plants, Membranes, 12 (2), 199.

D. J. Garcia, F. You, 2016, The water-energy-food nexus and process systems engineering: a new focus, Compu. Chem. Eng, 91, 49–67.

G. Guillén-Gosálbez, F. You, A. Galan-Martin, C. Pozo, I. E. Grossmann, 2019, Process systems engineering thinking and tools applied to sustainability problems: Current landscape and future opportunities, Current Opinion in Chemical Engineering 26, 170–179.

R. Kakodkar, G. He, C. D. Demirhan, M. Arbabzadeh, S. G. Baratsas, S. Avraamidou, D. Mallapragada, I. Miller, R. C. Allen, E. Gencer, E. N. Pistikopoulos, 2022, A review of analytical and optimization methodologies for transitions in multi-scale energy systems, Renewable and Sustainable Energy Reviews, 160, 112277.

Antonis Kokossis, Michael C. Georgiadis, Efstratios N. Pistikopoulos (Eds.)
PROCEEDINGS OF THE 33rd European Symposium on Computer Aided Process Engineering (ESCAPE33), June 18-21, 2023, Athens, Greece
 http://dx.doi.org/10.1016/B978-0-443-15274-0.50010-X

MULTI-FIDELITY SURROGATE MODELING FOR CHEMICAL PROCESSES WITH PHYSICS-INFORMED NEURAL NETWORKS

Yu-Ting Liu[a], Chuan-Yu Wu[b], Tao Chen[b], Yuan Yao[a]*

[a] *Department of Chemical Engineering, National Tsing Hua University, Hsinchu, 30013, Taiwan, ROC*

[b] *Department of Chemical and Process Engineering, University of Surrey, Guildford GU2 7XH, UK*

**Corresponding Author's E-mail: yyao@mx.nthu.edu.tw*

Abstract

In this study, a multi-fidelity surrogate modeling method based on physics-informed neural network (PINN) was proposed, which integrates high-fidelity simulation data and low-fidelity governing equations described by differential equations. By leveraging governing equations in the training of deep neural networks, the reliance on large amount of data has been relaxed. In the meantime, imposing physical laws ensures that the achieved surrogate models have clear physical meanings, which also improves the extrapolation performance of the models. Herein, the proposed multi-fidelity PINN surrogate modeling method was implemented to the simulation of the startup phase of a continuous stirred-tank reactor (CSTR) for illustrating its feasibility and advantages. From the computer experiment results, it is observed that the proposed method successfully reduced the sample size needed in model training and significantly improved the model extrapolation performance, facilitating its potential industrial applications.

Keywords: physics-informed neural network, multi-fidelity, surrogate modeling, deep learning.

1. Introduction

Numerical simulations, which use mathematical formulae to simulate complex physical systems, are becoming increasingly important in engineering tasks and decision-making processes (Moin and Mahesh, 1998), but the computational burden is a major practical problem limiting their applications. Surrogate models are a good solution to this problem. Nevertheless, data collection from high-fidelity simulations for training surrogate models is still computational expensive. Multi-fidelity surrogate modeling is an alternative approach, which utilizes both high-fidelity and low-fidelity data for model training (Fernández-Godino et al., 2019). A remaining problem is that, without a guidance of physical laws, the performance of surrogate models usually degrades significantly in the extrapolation applications.

In this study, we combined the strengths of data-driven and physics-based models to reduce the requirement on the amount of data for surrogate model training and improve the model extrapolation performance. In detail, we used the low-fidelity governing law described by a differential equation to constrain the training of the neural network model and obtain a multi-fidelity surrogate model based on physics-informed neuron network (PINN) (Raissi et al., 2019). With the aid of the physical law, the multi-fidelity PINN (MFPINN) surrogate model better approximates the simulated process especially in the situation of small data or extrapolation. These advantages facilitate the potential applications of MFPINN to industries.

2. Numerical Simulation

In this study, a non-isothermal continuous stirred-tanked reactor (CSTR) was simulated using ANSYS Fluent to demonstrate the feasibility of proposed method. Figure 1 show a full-scale three-dimensional (3D) model of the system and Table 1 lists the system parameters and design space of input variables (Chuang et al., 2018). Here, C_{Ai} is the inlet concentrations of species A, Ti is the inlet temperature, V is the volume of the reactor, F_i is the inlet and outlet mass flow rate, S is the stirring speed of impeller, and $\tau=V/F_i$ is the time constant of the CSTR. The physical characteristics, such as material density ρ, specific heat C_p, thermal conductivity κ, and dynamic viscosity η, are assumed to be constants with respect to temperature and compositions. This system consists of a first-order reaction, $A \rightarrow B$, where the activation energy is E_A and the pre-exponential Arrhenius constant is k_A. HR_A is the molar heats of the reaction. This 3D simulation provides information on product concentration and temperature as outputs, and we have selected the outlet concentration C_A as the predicted variable in presenting our results in this study due to its significance in determining product quality.

The mesh shape is tetrahedral with a minimum orthogonal quality of 0.238 and maximum skewness of 0.762 for accurate simulation results. It has 1867511 elements and 342112 nodes. The time consumption of each simulation run is about five to six hours. Consequently, this high-fidelity model is not suited for the engineering applications requiring multiple runs of simulations, such as optimization and sensitivity analysis. To deal with this problem, an MFPINN is proposed in this work, which can be implemented to build a surrogate model with a relatively small amount of simulation data.

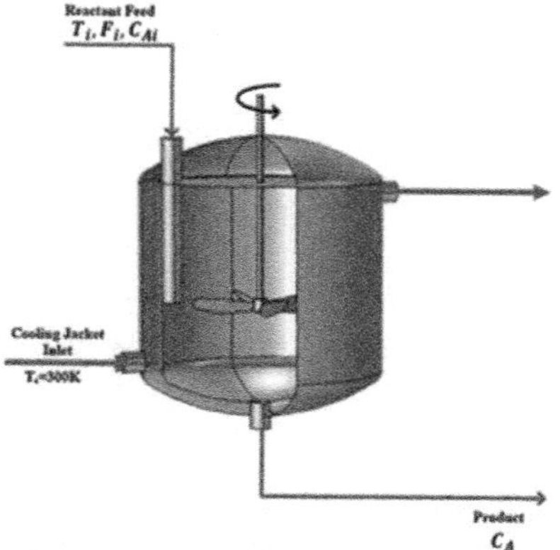

Figure 1 Schematic diagram of 3D model of CSTR process

Table 1. System parameters and their values used in the simulation

Parameters	Values	Units	Descriptions
k_A	1.4×10^4	1/sec	Physical constant
HR_A	-2.12×10^7	J/kgmol	Physical constant
E_A	3.64×10^7	J/kgmol	Physical constant

ρ	1180	Kg/m^3	Physical constant
C_p	3.2×10^3	J/kg/K	Physical constant
κ	0.61	W/m/K	Physical constant
η	0.0008	Kg/m/s	Physical constant
V	18.764	dm^3	Physical constant
S	120	rpm	Physical constant
R	8.314	J/mol/K	Physical constant
C_{Ai}	2-3	M	Design variable
F_i	0.0167-1.67	L/s	Design variable
T_i	300-370	K	Design variable
C_A		mol/dm^3	System output

3. Multi-Fidelity Surrogate Model base on PINN

In the course of surrogate modeling of high-fidelity simulations that are computationally expensive, the cost of simulation data acquisition for surrogate model training is prohibitive. Therefore, we are inevitably confronted with the challenge of small data. Most conventional machine learning techniques, including most deep neural networks, are not robust in such a situation, whose performance cannot be guaranteed. The state-of-the-art PINN technique proposed by Raissi et al. (2019), which aims to solve supervised learning problems while respecting physical laws described by differential equations, can be adopted to address this issue. Inspired with PINN, a multi-fidelity surrogate model named MFPINN is developed by integrating both high-fidelity simulation data and low-fidelity physical law.

The low-fidelity physical law used in the MFPINN captures the essential physics of the process by assuming that the CSTR is perfectly mixed and isothermal at 300K. The following ordinary differential equation (ODE) can be obtained according to the conservation law to the mass.

$$\tau\frac{dC_A}{dt} = C_{Ai} - C_A + r_A\tau \tag{1}$$

With a first-order reaction, - $r_A = k\,C_A$ and $\tau = \frac{V}{v}$, where k is defined by the Arrhenius equation.

$$k = k_A e^{\frac{-E_A}{RT}} \tag{2}$$

Consequently, eq. (1) is expressed as:

$$\frac{\mathrm{d}C_A}{\mathrm{d}t} + \frac{1+\tau k}{\tau}C_A - \frac{C_{Ai}}{\tau} = 0 \tag{3}$$

MFPINN can be understood as a feedforward neural network (FFNN) trained with high-fidelity data generated with the numerical simulations described in Section 2 and the low-fidelity ODE described in eq.(3), whose loss function is composed of both two terms, i.e. a prediction loss and an ODE loss as depicted in eq.(4) ~ (6).

$$Loss = Prediction\ loss + ODE\ loss \tag{4}$$

$$prediction\ loss = \frac{1}{N_{training}} \sum_{i=0}^{N_{training}} (\widehat{C_A}(T_i, F_i, C_{Ai}, t_i) - C_A)^2 \tag{5}$$

$$ODE\ loss = \frac{1}{N_{collocation} + N_{training}} \sum_{i=0}^{N_{collocation}+N_{training}} (\frac{d\widehat{C_A}(T_i, F_i, C_{Ai}, t_i)}{dt} + \frac{1+\tau k}{\tau}\widehat{C_A} - \frac{C_{Ai}}{\tau})^2 \tag{6}$$

where T_i, F_i, C_{Ai} are three process parameters, which represent inlet temperature, inlet volume flow rate, and inlet concentration, respectively, t_i is the time value, $\widehat{C_A}(T_i, F_i, C_{Ai}, t_i)$ is the prediction of the system output, i.e. the outlet concentration C_A, provided by the neural network, $N_{training}$ denotes the number of training data points, and $N_{collocation}$ denotes the number of collocation points which are chosen by Latin hypercube sampling (LHS) (Stein, 1987) within the design space of C_{Ai}, F_i, and T_i as listed in Table 1. The utilization of collocation points reduces the requirement on the amount of training data points. The automatic differentiation techniques (Baydin et al., 2018) was adopted for calculating the differentials. The training process of the MFPINN model is illustrated in Figure 2, where θ^* denotes the model parameters including all weights and biases.

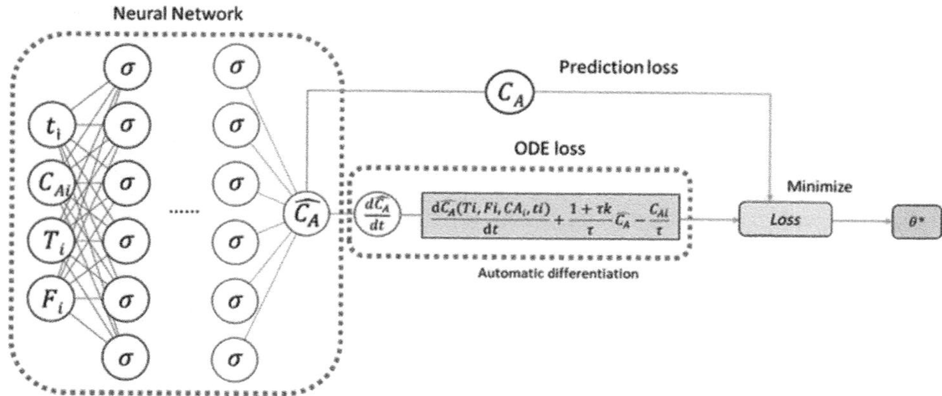

Figure 2. Training process of MFPINN

4. Case Study

For the training and testing of both the proposed MFPINN model and the conventional FFNN model used for comparison, LHS was adopted to generate eight different operating conditions within the design space of C_{Ai}, F_i, T_i as listed in Table 1. The process was simulated under each operating condition for 150 seconds and the sampling time interval was set as 0.1 seconds for data generation. The hyperparameters of the MFPINN model are given in Table 2. As shown in Table 2, the number of hidden layers and neurons per hidden layer of both the MFPINN and FFNN models are 8 and 100, respectively. Moreover, the number of collocation points collected for MFPINN is 100,000.

Table 2. Hyperparameter setup of both MFPINN and FFNN models

Number of hidden layers	8
Number of neurons per hidden layer	100
Number of collocation points	100,000

4.1 Modeling with Small Data

In the first scenario, we used the simulation data collected from the 100th to 400th time intervals under each operating condition as the test set for both models. The training and validation sets with different sizes were randomly selected from the rest of the data. The ratio between the sizes of the training and validation sets was 4:1. The model test results show that, when the size of training set is relatively large, e.g. over 300, the performance of the two surrogate models are comparable. However, when the size of training set is reduced to 200, the MFPINN model is significantly better than the FFNN model, especially when applied to operating conditions 5~7. Tables 3 and 4 show the values of the coefficient of determination (R^2) and mean squared error (MSE) of both models.

Table 3. Test results of MFPINN and FFNN when the size of training set is 300

	R^2 of MFPINN	R^2 of FFNN	MSE of MFPINN	MSE of FFNN
Condition 1	0.9964	**0.9996**	0.0002	**$1.87*10^{-5}$**
Condition 2	**0.9987**	0.9952	**$8.97*10^{-5}$**	0.0003
Condition 3	**0.9996**	**0.9996**	**$2.57*10^{-5}$**	**$2.57*10^{-5}$**
Condition 4	0.9962	**0.9990**	$6.49*10^{-5}$	**$1.79*10^{-5}$**
Condition 5	0.9777	**0.9817**	$6.68*10^{-5}$	**$5.46*10^{-5}$**
Condition 6	**0.9994**	0.9651	**$3.89*10^{-5}$**	0.0023
Condition 7	0.9955	**0.9997**	0.0003	**$1.90*10^{-5}$**
Condition 8	**0.9998**	0.9972	**$7.40*10^{-6}$**	0.0001

Table 4. Test results of MFPINN and FFNN when the size of training set is 200

	R^2 of MFPINN	R^2 of FFNN	MSE of MFPINN	MSE of FFNN
Condition 1	**0.9898**	0.9534	**0.0005**	0.0021

Condition 2	0.9983	**0.9999**	0.0001	**$8.09*10^{-6}$**
Condition 3	**0.9999**	0.9905	**$7.10*10^{-6}$**	0.0006
Condition 4	0.9981	**0.9993**	$3.22*10^{-5}$	**$1.13*10^{-5}$**
Condition 5	**0.9183**	0.4434	**0.0002**	0.0017
Condition 6	**0.9862**	-0.9912	**0.0009**	0.1316
Condition 7	**0.9982**	0.3650	**0.0001**	0.0073
Condition 8	**0.9865**	0.9568	**0.0006**	0.0020

4.2 Extrapolation

In order to evaluate the extrapolation capability of the proposed model, the second scenario was designed, where all data collected under operating conditions 2 and 6 were only used for test, while the data of other operating conditions were divided to the training and validation sets under a ratio of 4:1 The test results are shown in Table 5, while Figure 3 compares the ground truth and the prediction values of different models. Obviously, the proposed MFPINN model outperforms its competitor. Because the MFPINN model is guided by the physical law during the model training process, it provides a predicted profile of CA much closer to the high-fidelity simulation data, i.e. the ground truth. Its R2 values for operating condition 2 and 6 reach 0.967 and 0.977, respectively, although this is an extrapolation application. In contrast, although the training data is sufficient in this scenario, the FFNN model only achieves R2 values of 0.769 and 0.887 for operating condition 2 and 6, respectively. It is easy to understand that the extrapolation performance cannot be guaranteed unless the physics is taken into consideration.

Table 5 Regression analysis of MFPINN and FFNN models

	R^2 of MFPINN	R^2 of FFNN	MSE of MFPINN	MSE of FFNN
Condition 2	**0.9673**	0.7694	**$7.10*10^{-3}$**	$5.03*10^{-2}$
Condition 6	**0.9765**	0.8873	**$4.80*10^{-3}$**	$2.28*10^{-3}$

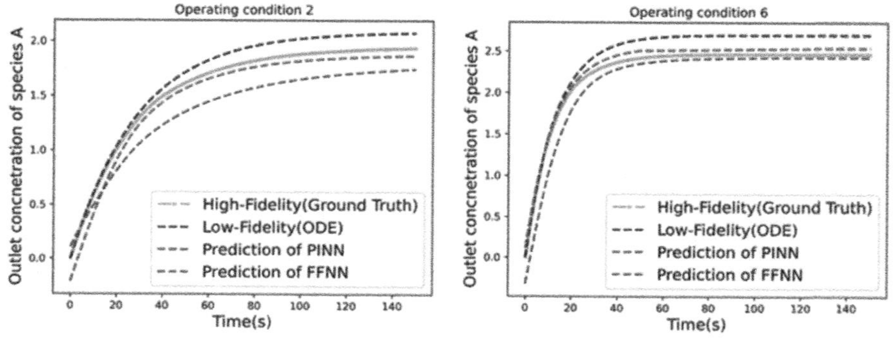

Figure 3. Predicted C_A of different models

5. Conclusions

In this study, we integrate the high-fidelity simulation data and the low-fidelity physical law to build an MFPINN model for surrogate modeling. The experimental results show that, thanks to the guidance of the physical law, the proposed MFPINN outperforms the conventional FFNN when the training data is insufficient. Furthermore, the physical law comes in handy in the extrapolation. As a result, the MFPINN model predicts the outlet concentration of the CSTR system more accurately than the FFNN model. The application to the CSTR system illustrates the feasibility and potentials of the proposed MFPINN model.

Acknowledgment

This work was supported in part by the National Science and Technology Council under project numbers NSTC 111-2622-8-007-011 and NSTC 111-2221-E-007-005.

References

A.G. Baydin, B.A. Pearlmutter, A.A. Radul and J.M. Siskind, (2018). Automatic Differentiation in Machine Learning: a Survey, Journal of Machine Learning Research, 18(1453), 1-43.

S. Chakraborty, (2021). Transfer Learning Based Multi-Fidelity Physics Informed Deep Neural Network, Journal of Computational Physics, 426, 109942.

M.G. Fernández-Godino, C. Park, N.H. Kim and R.T. Haftka (2019), Issues in Deciding Whether to Use Multifidelity Surrogates, AIAA Journal, 57(5), 2039-2054.

S. Markidis, (2021), The Old and the New: Can Physics-Informed Deep-Learning Replace Traditional Linear Solvers?, arXiv:2103.09655.

P. Moin and K. Mahesh, (1998), DIRECT NUMERICAL SIMULATION: A Tool in Turbulence Research, Annual Review of Fluid Mechanics, 30(1), 539-578.

M. Raissi, P. Perdikaris and G.E. Karniadakis, (2019), Physics-Informed Neural Networks: A Deep Learning Framework for Solving Forward and Inverse Problems Involving Nonlinear Partial Differential Equations, Journal of Computational Physics, 378, 686-707.

M. Stein, (1987), Large Sampling Properties of Simulations Using Latin Hypercube Sampling, Technometrics, 29(2), 143-151.

Antonis Kokossis, Michael C. Georgiadis, Efstratios N. Pistikopoulos (Eds.)
PROCEEDINGS OF THE 33rd European Symposium on Computer Aided Process Engineering (ESCAPE33), June 18-21, 2023, Athens, Greece
 http://dx.doi.org/10.1016/B978-0-443-15274-0.50011-1

A CFD-model-based approach to continuous freezing process design for human induced pluripotent stem cells

Yusuke Hayashi[a,*], Benedikt X. Scholz[a], Hirokazu Sugiyama[a]

[a]*Department of Chemical System Engineering, The University of Tokyo, Japan*

[*]*y-hayashi@pse.t.u-tokyo.ac.jp*

Abstract

This work presents a computer fluid dynamics (CFD)-model-based approach to continuous freezing process design for human induced pluripotent stem (hiPS) cells. A process model was developed that combined a hybrid single-cell freezing model covering the cell level with a CFD model covering the cryovial and freezer levels. Given an inlet coolant temperature, an inlet coolant velocity, and a residence time, the developed hybrid model can calculate the cell survival rate as a quality indicator. We applied the developed process model to three operating conditions of continuous freezing processes for hiPS cells. As a result, the survival rate in one of the operating conditions could achieve more than 0.90. Thus, it was demonstrated that designing the continuous freezing process could be performed with appropriate value selection of the decision parameters.

Keywords: Regenerative medicine, Cryopreservation, Cell therapy, Hybrid model, Optimization.

1. Introduction

Human induced pluripotent stem (hiPS) cells (Takahashi et al., 2007) are regarded as one of the most promising sources for regenerative medical products because of various advantages over the conventional sources, e.g., human embryonic stem cells (Narsinh et al., 2011). Along with recent successful clinical studies, e.g., Parkinson's disease (Morizane, 2019) and retinitis pigmentosa (Tagawa et al., 2021), the implementation of medical treatments using hiPS cells is in progress.

In hiPS cell manufacturing, freezing processes are one of the most critical steps because the process is needed to the transportation and preservation of the cells. Generally, two methods can be adopted for the freezing process of hiPS cells: slow freezing and vitrification. Slow freezing has been applied to many cell types, where cryovials filled with cell suspension are cooled in a freezer at a predetermined cool rate. In vitrification, the vials are immediately cooled using liquid nitrogen. At a commercial scale, vitrification is rarely adopted because of vitrification is rarely adopted in the cell therapy industry because of a scale limitation and process complexity (Vajta and Nagy, 2006). Therefore, in this work, we focus on the slow freezing option.

In general, the slow freezing of hiPS cells has been performed by a direct contact freezer that can accommodate only a limited number of cryovials. On the other hand, with the expected future commercialization, there is a need to design continuous slow freezing processes for hiPS cells that can handle many cryovials. Most recently, a computer fluid dynamics (CFD) model-based approach to scale-up batch slow freezing processes for hiPS cells using forced convection-based freezers was published by the authors' research group (Scholz et al., 2022). However, a process design of continuous slow cell freezing was yet to be performed.

This work presents a CFD-model-based approach to continuous freezing process design for hiPS cells. A process model was developed that combined a hybrid single-cell freezing model covering the cell level with a CFD model covering the cryovial and freezer levels. The cell level model was based on our previous publications (Hayashi et al., 2021, 2020), and the cryovial and freezer level models were newly constructed. Given an inlet coolant temperate, an inlet coolant velocity, and a residence time, the developed hybrid model can calculate the cell survival rate as a quality indicator. We applied the developed model to three operating conditions of continuous freezing processes for hiPS cells.

2. Continuous freezing processes for hiPS cells

The continuous freezing process assumed in this work is represented in Figure 1. The freezer adopted convection-based cooling and consisted of four stations according to the three-temperature zones proposed by Hayashi et al. (2021), i.e., dehydration (Station 1), nucleation (Stations 2 and 3), and further cooling (Station 4). The decision parameters in each station were defined as the inlet coolant temperate, T_i^{st} [K], the inlet coolant velocity, u_i^{st} [m s^{-1}], and the residence time, t_i^{st} [s] (i = 1, 2, 3, 4).

3. Process model

Figure 2 shows an overview of the developed process model consisting of the freezer, cryovial, and cell levels. The inputs were defined as the decision parameters, i.e., T_i^{st}, u_i^{st}, and t_i^{st}, and the output was defined as the cell survival rate, r_{surv} [–]. The freezer and cryovial level models constituted of the CFD model, and the cell model was based on Hayashi et al. (2021, 2020). The process model was defined, based on the following assumptions:

- The number of rows in the freezer is one.
- The values of the decision parameters are strictly controlled.
- The flow is laminar ($Re_{\mathrm{max}} = 2.0\times10^2$).

3.1. Freezer and cryovial level models

3.1.1. Flow calculation

For the fluid flow calculation, the continuity equation and the Navier-Stokes equation were adopted as follows:

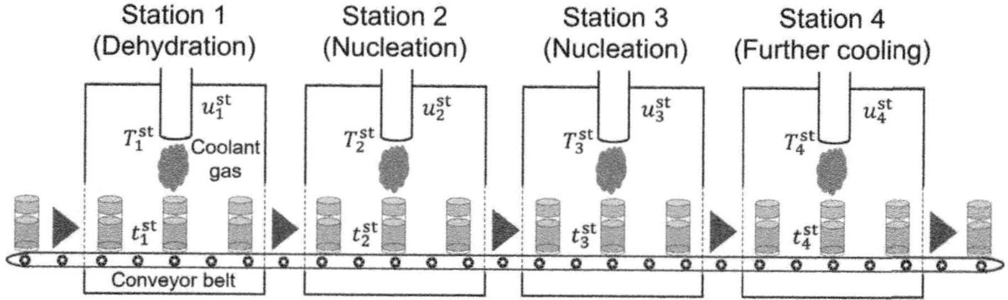

Figure 1. Continuous freezing processes for hiPS cells.

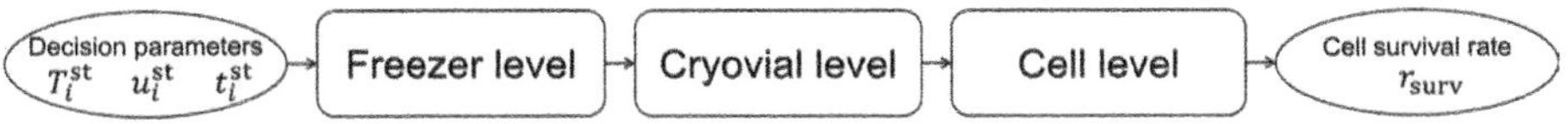

Figure 2. Overview of the developed process model.

$$\nabla \cdot v_f = 0 \tag{1}$$

$$\rho_f \frac{\partial v_f}{\partial t} + \rho_f (v_f \cdot \nabla) v_f = \nabla \cdot [-pI + \mu_f (\nabla v_f + (\nabla v_f)^T)] \tag{2}$$

where ρ_f [kg m^{-3}] is the density, μ_f [Pa s] is the dynamic viscosity, v_f [m s^{-1}] is the velocity, p [Pa] is the pressure, and t [s] is the time.

3.1.2. Heat transfer

The energy balance of the fluid in the freezer was represented using the following equation:

$$\rho_f c_{p,f} \left[\frac{\partial T}{\partial t} + (v_f \cdot \nabla) T \right] = -\nabla q_{cond} - \frac{T}{\rho_f} \frac{\partial \rho_f}{\partial T}\bigg|_p \left(\frac{\partial p}{\partial t} + (v_f \cdot \nabla) p \right) \tag{3}$$

where $c_{p,f}$ [J kg^{-1} K^{-1}] is the specific heat capacity of the fluid, T [K] is the temperature, q_{cond} [J s^{-1} m^{-3}] is the conductive heat flux. The energy balance of the solid was also modeled as follows:

$$\rho_s c_{p,s} \frac{\partial T}{\partial t} = -(\nabla \cdot q_{cond}) \tag{4}$$

3.2. Cell level model

3.2.1. Mass transfer

The following equation was adopted to model the mass transport of water across the cell membrane:

$$\frac{dV_{cell}}{dt} = L_p A_{cell} \Delta \Pi \tag{5}$$

where V [m^3] is the volume, L_p [m s^{-1} Pa^{-1}] is the water permeability, A [m^2] is the surface area, $\Delta\Pi$ [Pa] is the pressure difference. The normalized maximum cell volume change, $\overline{\Delta V_{cell}^{max}}$ [–], was calculated as follows:

$$\overline{\Delta V_{cell}^{max}} = \max \left\{ \frac{\left| V_{cell}^{fin} - V_{cell}^{init} \right|}{V_{cell}^{init}} \right\} \tag{6}$$

where the superscripts initial and final represent the initial and final state of freezing, respectively.

3.2.2. Crystallization

The radius at time t of an intracellular crystal that nucleated at time, τ [s], was estimated as follows:

$$r_{\text{ice},i} = \begin{cases} 0 \quad (0 \le N_{\text{ice}} < 1) \\ \sqrt{\int_{\tau_i}^{t} \gamma_{\text{ice}}^2 \bar{D} dt} \quad (N_{\text{ice}} \ge 1) \end{cases} \tag{7}$$

where r_{ice} [m] is the radius of an ice crystal, N_{ice} [–] is the number of intracellular ice crystals, γ_{ice} [–] is the nondimensional ice crystal growth parameter, and $\bar{D}$ [m^2 s^{-1}] is the average water diffusion coefficient. The number of intracellular ice crystals, N_{ice}, was calculated using the following equations:

$$N_{\text{ice}} = \text{int}[\bar{N}_{\text{ice}}] \tag{8}$$

$$\frac{d\bar{N}_{\text{ice}}}{dt} = J_{\text{ice}} \tag{9}$$

where $\bar{N}_{\text{ice}}$ [–] is the ensemble average of the number of intracellular ice crystals and J_{ice} [s^{-1}] is the ice nucleation rate. The total volume of intracellular ice crystals, V_{ice} [m^3], was modeled as follows:

$$V_{\text{ice}} = \sum_{i=1}^{N_{\text{ice}}} \frac{4}{3} \pi r_{\text{ice},i}^3 \tag{10}$$

The normalized maximum ice crystal volume, $\overline{V_{\text{ice}}^{\text{max}}}$ [–], was calculated using the following equation:

$$\overline{V_{\text{ice}}^{\text{max}}} = \max\left[\frac{V_{\text{ice}}^{\text{fin}}}{V_{\text{cell}}^{\text{fin}}}\right] \tag{11}$$

3.2.3. Cell survival rate

The cell survival rate was estimated as follows (Hayashi et al., 2021):

$$r_{\text{surv}} = \omega_1 + \omega_2 \overline{\Delta V_{\text{cell}}^{\text{max}}} + \omega_3 \overline{V_{\text{ice}}^{\text{max}}} + \omega_4 \overline{\Delta V_{\text{cell}}^{\text{max}}} \cdot \overline{V_{\text{ice}}^{\text{max}}} \tag{12}$$

where ω_1 [–], ω_2 [–], ω_3 [–], and ω_4 [–] are the fitting coefficients, for which the same values published in Hayashi et al. (2021) were used in the calculation.

4. Results and discussion

Three operating conditions of continuous freezing processes for hiPS cells were applied as defined in Table 1. The conditions were selected based on the results presented in Hayashi et al. (2021). Figure 3 shows (a) the temperature profiles of the vial center and (b) the cell survival rates for the three operating conditions. The cell survival rate in Condition 1 was highest, followed by Condition 2 and Condition 3. This result can be explained using the three-temperature zones proposed by Hayashi et al. (2021). Above 233 K, fast cooling would be preferred because cell dehydration mainly occurred in the temperature range. In contrast, the cooling rate between 233 K and 213 K should be slow because intracellular ice nucleation mainly happened in the temperature range. In addition, below 213 K, any cooling rates could be applied because the cooling rate would

Table 1. Detailed values of the inlet coolant temperature, T_i^{st}, the inlet coolant velocity, u_i^{st}, and the residence time, t_i^{st}, applied in this work

Station	Condition 1			Condition 2			Condition 3		
	T_i^{st} [K]	u_i^{st} [m s^{-1}]	t_i^{st} [s]	T_i^{st} [K]	u_i^{st} [m s^{-1}]	t_i^{st} [s]	T_i^{st} [K]	u_i^{st} [m s^{-1}]	t_i^{st} [s]
1	233	0.100	625	213	0.100	625	233	0.100	315
2	223	0.100	625	203	0.100	625	223	0.100	315
3	213	0.100	625	198	0.100	625	213	0.100	315
4	178	0.100	625	178	0.100	625	178	0.100	315

not affect the cell survival rate. In the range of above 233 K, Condition 2 would be the best for the cell survival rate because the cooling rate was the fastest. On the other hand, in the range of between 233 K and 213 K, Condition 1 would be the most desirable for the cell survival rate because the cooling rate was the slowest. According to Mazur et al. (1972), cell damage due to intracellular ice nucleation was more critical than due to cell dehydration. Therefore, the highest survival rate (more than 0.90) was observed in Condition 1. Thus, it was demonstrated that designing continuous freezing processes for hiPS cells could be performed with appropriate value selection of the decision parameters.

The flow and heat transfer calculations were performed using COMSOL Multiphysics® version 6.1, and the calculations of mass transfer, crystallization, and cell survival rate were conducted by Python 3.9. The total CPU time that the results shown in Figure 3 were obtained was about 60 hours, using Intel® Xeon® Gold 6230 CPU @ 2.10 GHz with 512 GB RAM memory.

5. Conclusions and outlook

This work presented a CFD-model-based approach to continuous freezing process design for hiPS cells. A process model was developed that combined a hybrid single-cell freezing model covering the cell level with a CFD model covering the cryovial and freezer levels. The developed process model enabled calculating the cell survival rate of hiPS cells, given an inlet coolant temperature, an inlet coolant velocity, and a residence time. We

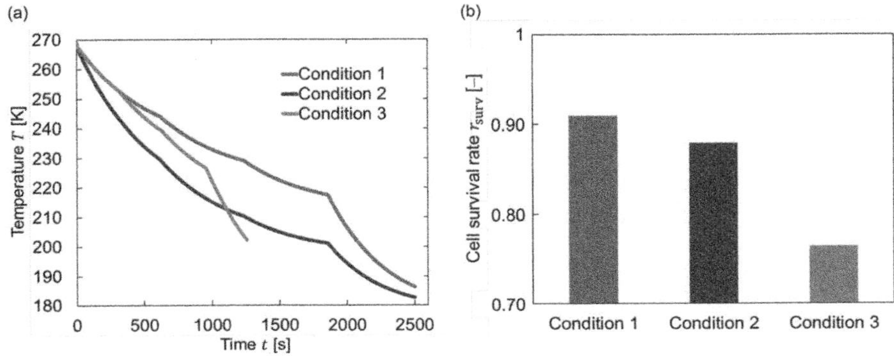

Figure 3. (a) Temperature profile of the vial center and (b) cell survival rate for the three operating conditions.

applied the developed process model to three operating conditions of continuous freezing processes for hiPS cells. As a result, the cell survival rate in one of the three operating conditions could achieve more than 0.90. Thus, it was demonstrated that designing the continuous freezing process could be performed with appropriate value selection of the decision parameters. In the field of computer-aided process engineering, cell therapy related studies are becoming relevant, e.g., Triantafyllou et al. (2022) and Hirono et al. (2022). Further model-based studies in this area are encouraged.

Acknowledgements

H. S. is thankful for financial support by a Grant-in-Aid for Challenging Research (Exploratory) No. 20K21102 from the Japan Society for the Promotion of Science.

References

Y. Hayashi, I. Horiguchi, M. Kino-oka, H. Sugiyama, 2021, Model-based assessment of temperature profiles in slow freezing for human induced pluripotent stem cells, Comput. Chem. Eng., 144, 107150.

Y. Hayashi, I. Horiguchi, M. Kino-oka, H. Sugiyama, 2020, Slow freezing process design for human induced pluripotent stem cells by modeling intracontainer variation, Comput. Chem. Eng., 132, 106597.

K. Hirono, I. Udugama, Y. Hayashi, M. Kino-oka, H. Sugiyama, 2022, Design space determination of mesenchymal stem cell cultivation by dynamic modeling under uncertainty, Comput. Aided Chem. Eng., 51, 721.

P. Mazur, S. Leibo, E. Chu, 1972, A two-factor hypothesis of freezing injury: Evidence from Chinese hamster tissue-culture cells, Exp. Cell Res., 71, 345.

A. Morizane, 2019, Cell therapy for Parkinson's disease with induced pluripotent stem cells, Rinsho Shinkeigaku, 59, 119.

K. Narsinh, J. Plews, J. Wu, 2011, Comparison of human induced pluripotent and embryonic stem cells: fraternal or identical twins?, Mol. Ther., 19, 635.

B. Scholz, Y. Hayashi, I. Udugama, M. Kino-oka, H. Sugiyama, 2022, A multilayered approach to scale-up forced convection-based freezing of human induced pluripotent stem cells, Comput. Chem. Eng., 163, 107851.

M. Tagawa, H. Ikeda, Y. Inoue, S. Iwai, Y. Iida, M. Hata, I. Asaka, A. Tsujikawa, 2021, Deterioration of phagocytosis in induced pluripotent stem cell-derived retinal pigment epithelial cells established from patients with retinitis pigmentosa carrying Mer tyrosine kinase mutations, Exp. Eye Res., 205, 108503.

K. Takahashi, K. Tanabe, M. Ohnuki, M. Narita, T. Ichisaka, K. Tomoda, S. Yamanaka, 2007, Induction of pluripotent stem cells from adult human fibroblasts by defined factors, Cell, 131, 861.

N, Triantafyllou, A. Bernardi, M. Lakelin, N. Shah, M. Papathanasiou, 2022, A bi-level decomposition approach for CAR-T cell therapies supply chain optimization, Comput. Aided Chem. Eng., 49, 2197.

G. Vajta, Z. Nagy, 2006, Are programmable freezers still needed in the embryo laboratory? Review on vitrification, Reprod. Biomed. Online, 12, 779.

Antonis Kokossis, Michael C. Georgiadis, Efstratios N. Pistikopoulos (Eds.)
PROCEEDINGS OF THE 33rd European Symposium on Computer Aided Process Engineering (ESCAPE33), June 18-21, 2023, Athens, Greece
 http://dx.doi.org/10.1016/B978-0-443-15274-0.50012-3

Multiscale Modeling of Spatial Area-Selective Thermal Atomic Layer Deposition

Matthew Tom[a], Sungil Yun[a], Henrik Wang[a], Feiyang Ou[a], Gerassimos Orkoulas[b], Panagiotis D. Christofides[a,c,*]

[a]*Department of Chemical and Biomolecular Engineering, University of California, Los Angeles, CA 90095-1592, USA*
[b]*Department of Chemical Engineering, Widener University, Chester, PA 19013, USA*
[c]*Department of Electrical and Computer Engineering, University of California, Los Angeles, CA 90095-1592, USA*
pdc@seas.ucla.edu

Abstract

Area-selective atomic layer deposition (ASALD) is an advancement of conventional thin-layer deposition processes that utilizes an additional chemoselective inhibition reaction to improve nanopatterning. However, ASALD is a challenging procedure that is difficult to characterize in the surface and fluid phases. A three-dimensional (3D) multiscale model that employs an atomistic-mesoscopic density functional theory (DFT) and kinetic Monte Carlo (kMC) simulation and macroscopic computational fluid dynamics (CFD) modeling are performed for a spatial rotary reactor configuration for the ASALD of a SiO_2/Al_2O_3 substrate *in silico*. Due to the complexity of the multiscale simulation, various reactor configurations are devised to minimize the computational requirements of the dynamic mesh of the reactor while reproducing results that are analogous to experimental data.

Keywords: Area-selective atomic layer deposition; multiscale modeling; computational fluid dynamics; kinetic Monte Carlo simulation; density functional theory.

1. Introduction

Area-selective atomic layer deposition is an attractive thin-layer deposition process that does not require subsequent processing steps (e.g., etching and lithography) to achieve effective nanopatterning and self-limiting deposition on growth areas of the substrate. In particular, semiconducting materials require thin-layer deposition processes to promote the self-aligned stacking of transistors, which is essential for producing high-performance wafers. In accordance with the predictions of Moore's Law, the stacking of the transistors is needed to improve the computational performance and efficiency of electronics; however, the lack of self-alignment introduces a fabrication challenge for wafers. Although conventional thin-layer deposition reactors are widely used, these reactors require extensive purging steps to maintain the self-limiting nature of the process. Spatial reactor configurations, including the rotary type, introduce reagent in physically isolated regions to prevent intermixing while maintaining the self-limiting tendency of the reactions (Poodt et al., 2012). The stringent dimensions of these transistors require much experimentation through a tedious procedure of optimizing the reactor design such that reagent intermixing is minimized and substrate exposure to the reagents is uniformly distributed. However, multiscale modeling has emerged as a cost-effective and timely solution for relating multiphase deposition processes for various time and length scales while reproducing experimental results with high fidelity. This work will employ an atomistic-mesoscopic density functional theory (DFT) and kinetic Monte Carlo model in conjunction with a three-dimensional (3D) macroscopic computational fluid dynamics (CFD) to simultaneously simulate surface-scale reaction kinetics comprising steric hindrance and repulsion effects and macroscopic fluid transport phenomena for a silicon wafer (SiO_2/Al_2O_3) substrate in accordance to the ABC-type reaction kinetics proposed by Merkx et al. (2020) and Mameli et al. (2017) composed of (A) inhibition, (B) adsorption, and (C) oxidation steps. Additionally, the relation of various constraints in the design of the rotary reactor (e.g., gap distance, geometries of reaction zones, and orientations of inlets) to the surface uniformity and deposition growth rate are studied for process optimization. Such constraints provide significant insight for application to current fabrication procedures.

2. Multiscale Simulation

2.1. Atomistic-mesoscopic modeling

Area-selective atomic layer deposition (ASALD) processes integrate a variety of synergistic, chemoselective, and self-limiting reactions that deposit monolayers of substrate material. Mameli et al. (2017) proposed an ABC-type mechanism composed of three steps (Steps A, B, and C). First, an inhibition preprocessing step (Step A) to deactivate subsequent reactions in the non-growth area using a small molecule inhibitor, acetylacetone (Hacac). Next, a precursor, bis(diethylamino)silane (BDEAS), adsorbs onto the growth area to produce a modified surface layer (Step B) composed of hydrogen-terminated ligands. Finally, an oxidation step (Step C) using ozone (O_3) oxidizes the hydrogen terminals to produce hydroxyl-ligands, thereby concluding a single monolayer of the deposition cycle. To ensure the self-limiting behavior of these reactions, each step is followed by a purging action by employing an inert gas such as nitrogen, N_2.

The abundance of reaction mechanisms and the lack of reference and experimental data would make the characterization of the kinetics difficult to quantify in the atomistic and mesoscopic phases. Thus, *ab initio* quantum mechanics that uses first principles density functional theory (DFT) is performed through the open-source software, Quantum ESPRESSO (QE) to compute the thermophysical and kinetics parameters of species and reactions, respectively. QE performs molecular structure and electronic optimization through pseudopotential data and nudged elastic band (NEB) methods to compute the minimal energy path to compute the activation energy of the reaction, to be substituted into the Arrhenius Equation. Additionally, a typical elementary reaction mechanism for rate-determining reaction steps is proposed for each step (A, B, and C) in the ASALD process to integrate transition state theory (TST) for non-adsorption reactions and collision theory (CT) for adsorption reactions.

With the kinetics mechanisms and reaction parameters defined, reaction parameters are defined into the kinetic Monte Carlo (kMC) algorithm that adopts a stochastic procedure to exemplify the behavior of realistic reactions. The kMC method, performed through the Python programming language, adopts a sequential procedure that sums probable reaction rates at a surface site to preferentially weight reactions of greater magnitude while devising a random number to arbitrarily select a reaction for the active site. Such procedure enables the computation of species generation and consumption terms from surface pressure and temperature data (from the CFD simulation), which will be integrated into the CFD simulation. Subsequently, the kMC method simulates the progression of the reaction path by calculating process time through a secondary random number described by Yun et al. (2022a). Additionally, steric hindrance effects are simulated by randomly selecting the angle of rotation for adsorption reactions encountered in Step A, and portraying the deflection of adsorption sites that are hindered by bulky molecules (Roh et al., 2022).

2.2. Macroscopic modeling

Computational fluid dynamics (CFD) is performed to simulate the transport phenomena in the fluid regime through the commercial multiphysics software, ANSYS Fluent. Prior to conducting the CFD simulation, a three-dimensional (3D) rotary reactor model, illustrated in Fig. 4 is constructed through ANSYS DesignModeler, which is then meshed through ANSYS Workbench's meshing software using tetrahedral cells, until an ideal mesh quality (per ANSYS standards) and minimal number of cells is obtained. An ideal mesh is composed of a minimal number of cells to lessen the dependence of robust computational power while maintaining accurate computations that are reflective of experimental results recorded in academia and ensuring convergence is obtainable. Additionally, the movement of the wafers between reaction zones are achieved through a dynamic mesh method that performs remeshing and diffusion-based smoothing methods to retain mesh quality and preserve solution convergence. In Fig. 4, the reactor design separates each wafer into 10 sections to define source generation and consumption flux terms (evaluated by the kMC simulation) on the wafer surface through user-defined functions (UDFs) in a "C++ like" script. The CFD simulation is performed using a fixed time step of 0.001 s, employing a pressure-based solver method, and adopting a first-order implicit numerical solver method that is defined through a journal file and executed through a Scheme script. Lastly, various thermophysical parameters are defined by integrating results from the atomistic *ab initio* first principles quantum mechanics simulations and through the ANSYS Fluent database of gaseous molecules.

In addition to the optimization of the reactor mesh, the chamber design of the rotary reactor and the operating conditions have a significant role in the quality and conformance of the substrate (De la Huerta et al., 2018). The reactor design must ensure that effective reagent separation is made to prevent species intermixing, which disrupts the perceived self-limiting behavior of the ASALD process. Reagent intermixing is prevented by specifying an optimal gap distance, the distance between the substrate surface and the ceiling of the reactor that separates the reaction zones, introducing output streams at precise locations to remove byproducts generated, and defining a large purge flow rate. Following the procedures to minimize reagent intermixing, the delivery of gases to the surface of the wafer and the exposure time are of paramount importance to ensure that a total surface deposition is achieved. Constraints including the reagent (Hacac, BDEAS, O_3) flow rates and rotation speed have a profound impact on the deposition rate onto the substrate surface (Pan et al., 2017). Although large reagent flow rates and low rotation speeds ensure complete deposition is obtained, such quantities are also nonideal from an economics and production perspective. Lastly, the uniformity of the deposition is a qualitative parameter that discusses the effectiveness of the nanopatterning performance. This modification provides better control of the film uniformity to ensure that all regions in the wafer have the same exposure time. The aforementioned quantities provide a meaningful study for optimizing the reactor design and process operation for practical applications.

2.3. Multiscale modeling

The multiscale simulation allows the calculation of various time and length scale in various domains including the atomistic, mesoscopic, and macroscopic regimes (Raimondeau and Vlachos, 2002) as described in Fig. 1. Such multiscale modeling allows the relating of various phases that cannot be discerned through *ex silico* methods. The multiscale simulation is performed autonomously through a Linux cluster system through the UCLA Hoffman2 Cluster and allows the interconnection of various programming languages and application software. This work will utilize 36-core compute nodes consisting of 192 GB of dynamic random-access memory (DRAM) that are standard for ANSYS Fluent to conduct the multiscale simulation by adopting both a parallel computation strategy in the macroscopic simulation and serial method for the microscopic simulation taking an average of 1 to 3 days, which depends on the rotation speed of the reactor. Such coding logic allows ANSYS Fluent to evaluate surface pressure (P) and temperature (T) data on the wafer surface to be read by the kMC coding script to evaluate the source generation and consumption terms, S_h and $S_{m,i}$, for the heat and mass of species i balances, which will then be defined through the UDF by ANSYS Fluent as summarized in the process diagram in Fig. 2. The simulation will conclude until the termination condition (full coverage) is obtained or if the wafer exits the reaction zone completely.

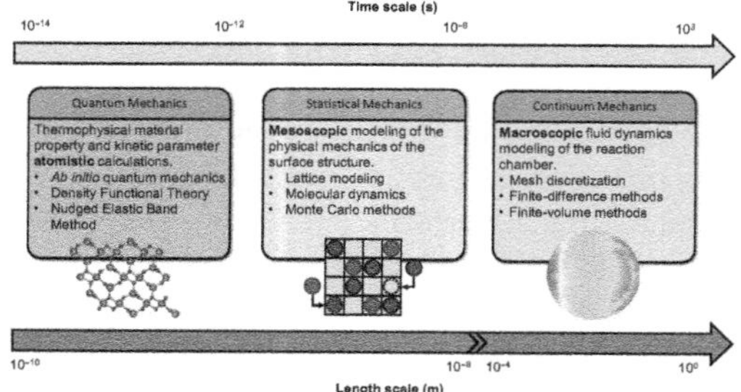

Figure 1: Multiscale model application for various phases and time and length domains.

Figure 2: Multiscale model simulation process diagram.

3. Results

3.1. Atomistic-mesoscopic simulation results

The atomistic-mesoscopic simulation is conducted for a pressure range of 10 to 500 Pa in 10 Pa intervals and a temperature range of 423 to 573 K in 10 K intervals for a total of 800 data points. The collection of a diverse data set is employed to determine the effects of the operating conditions on the total process time to obtain full coverage on the wafer surface for Steps A, B, and C on a 100 × 100 lattice. Graphical results for the process times as a function of the temperature and pressure are provided in Fig. 3. Results from

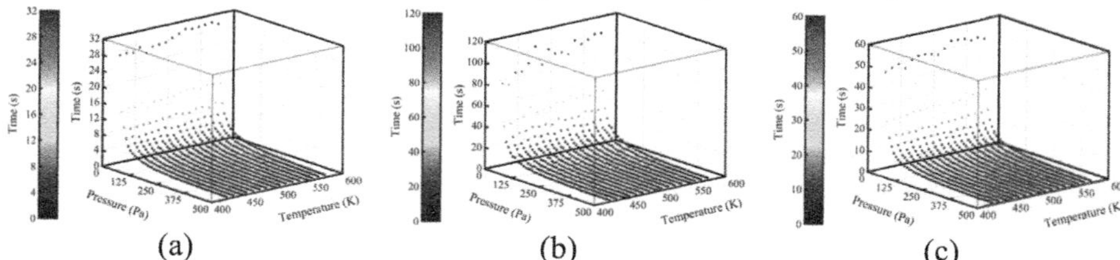

Figure 3: Kinetic Monte Carlo process times for (a) Hacac inhibition, (b) BDEAS adsorption, and (c) O_3 oxidation at various pressure and temperature conditions.

Step A in Fig. 3(a) demonstrates that the Hacac pulse time of 3.0 s or more above was observed for pressures below the 100 Pa, which was analogous to reported findings by Mameli et al. (2017) who reported a saturation dosage time of 5.0 s for such operating conditions. Additionally, the BDEAS adsorption in Step B displayed in Fig. 3(b) has a BDEAS dosage time of 2.3 s for a pressure of 400 Pa to reach full coverage, which is comparable to the reported findings of 2.0 s of BDEAS saturation times by Merkx et al. (2020). The aforementioned results indicate that the combined atomistic and mesoscopic simulations are reflective of experimental results; thus, the kMC simulation will have a minimal contribution to deviation of results in the multiscale simulation.

3.2. Multiscale computational fluid dynamics simulation results

The reactor mesh is fully optimized using less than 1.3 million tetrahedral cells and a gap distance of 5 mm by adopting a reactor geometry with reaction barrier zones conforming to the radial direction of the reactor as illustrated in Fig. 4. In addition to the radial separation of the vacuum ports, the wafers were also separated into 10 regions in the radial direction for the UDF specifications of the source terms. Results from the multiscale CFD simulation in Fig. 5 reveal that the dividers provide substantial reagent exposure in the radial direction when rotating with an angular velocity of 0.40 rad/s. The

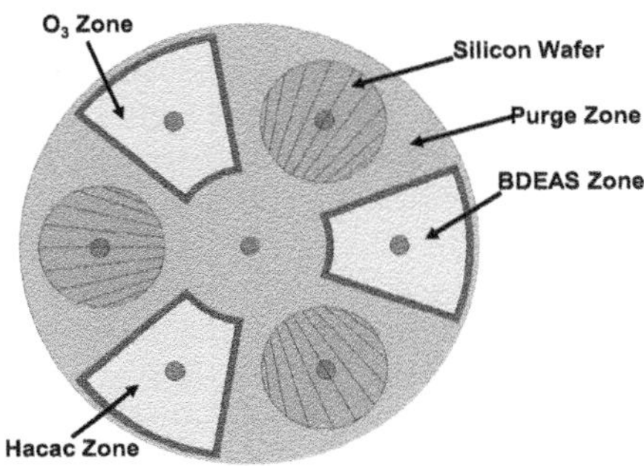

Figure 4: Top view of the rotary reactor design. Inlet and outlet streams are in blue and red, respectively.

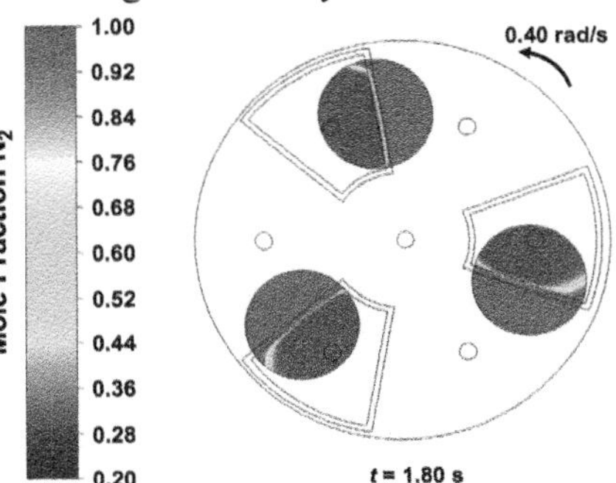

Figure 5: Multiscale CFD simulation results illustrated in N_2 mole fraction wafer surface contours.

contours of mole fraction of N_2 also illustrate that reagent intermixing is effectively minimized, thereby maintaining the self-limiting nature of the ASALD process. However, it is notable that intermixing is visible in reaction zone corners; thus, further study is needed to minimize the effects of the purge and reagents intermixing within the reaction zone by

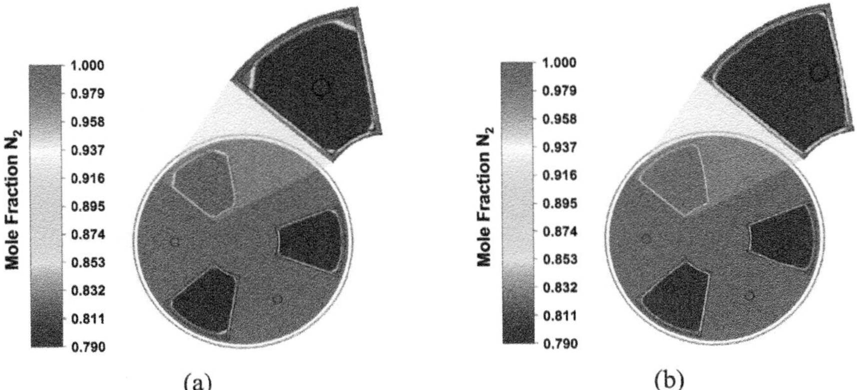

(a) (b)

Figure 6: Contours of N_2 mole fraction illustrating intermixing for a conventional rotary reactor design (a) and the mitigation of intermixing by the asymmetric design (b).

adjusting the orientation of the inlets, modifying the diameters of the outlets, and adjusting the reaction zone geometry. The optimization of the reactor configuration is illustrated in Fig. 6 to demonstrate the effects of adjusting the inlet location to an asymmetric model in Fig. 6b, which moderates the effects of purge transfer into the reaction zones for the conventional reactor configuration in Fig. 6a. Also, the process operation parameters can be considered for further study such as balancing the reagent flow rates with a proper composition of reagent. The collection of a diverse data set is still needed to further characterize various constraints (e.g., exposure uniformity and deposition growth rate).

4. Conclusion

A novel multiscale simulation of an area-selective ASALD process was performed to study the relation between various operating conditions and reactor geometry on the uniformity and deposition rate of a thin film on a SiO_2/Al_2O_3 substrate in various phases. First, an atomistic-mesoscopic model was developed in conjunction with *ab initio* first principles quantum mechanics simulation to calculate kinetic parameters of rate-determining elementary reactions and thermophysical parameters of species and with a kinetic Monte Carlo (kMC) algorithm to simulate the stochastic nature of surface reactions on the wafer. The resulting mesoscopic simulation was performed with computational fluid dynamics (CFD) to study the transport phenomena of the gases. It was determined that the atomistic-mesoscopic model, when compared to experimental results found in literature, demonstrated accurate process time results in exact operating conditions. The multiscale CFD simulation illustrated that the reactor design successfully minimized the effects of reagent intermixing while improving exposure uniformity to enhance nanopatterning. Although further study is needed to characterize the relation of various reactor design and operation constraints to fully understand their effects on the deposition growth rate and film uniformity, initial results reveal that the reactor design and operation conditions determined by the model lead to improved deposition performance.

References

De la Huerta, C.M., Nguyen, V.H., Dedulle, J.M., Bellet, D., Jiménez, C., Muñoz-Rojas, D., 2018. Influence of the geometric parameters on the deposition mode in spatial atomic layer deposition: a novel approach to area-selective deposition. Coatings 9, 5.

Mameli, A., Merkx, M.J.M., Karasulu, B., Roozeboom, F., Kessels, W.E.M.M., Mackus, A.J.M., 2017. Area-selective atomic layer deposition of SiO_2 using acetylacetone as a chemoselective inhibitor in an ABC-type cycle. ACS Nano 11, 9303–9311.

Merkx, M.J.M., Sandoval, T.E., Hausmann, D.M., Kessels, W.M.M., Mackus, A.J.M., 2020. Mechanism of precursor blocking by acetylacetone inhibitor molecules during areaselective atomic layer deposition of SiO_2. Chemistry of Materials 32, 3335–3345.

Pan, D., Jen, T.C., Yuan, C., 2016. Effects of gap size, temperature and pumping pressure on the fluid dynamics and chemical kinetics of in-line spatial atomic layer deposition of Al_2O_3. International Journal of Heat and Mass Transfer 96, 189–198.

Poodt, P., Cameron, D.C., Dickey, E., George, S.M., Kuznetsov, V., Parsons, G.N., Roozeboom, F., Sundaram, G., Vermeer, A., 2012. Spatial atomic layer deposition: A route towards further industrialization of atomic layer deposition. Journal of Vacuum Science & Technology A 30, 010802.

Raimondeau, S., Vlachos, D.G, 2002. Recent developments on multiscale, hierarchical modeling of chemical reactors. Chemical Engineering Journal 90, 3–23.

Roh, H., Kim, H.L., Khumaini, K., Son, H., Shin, D., Lee, W.J., 2022. Effect of deposition temperature and surface reactions in atomic layer deposition of silicon oxide using bis(diethylamino)silane and ozone. Applied Surface Science 571, 151231.

Yun, S., Tom, M., Luo, J., Orkoulas, G., Christofides, P.D., 2022a. Microscopic and data-driven modeling and operation of thermal atomic layer etching of aluminum oxide thin films. Chemical Engineering Research and Design 177, 96–107.

Yun, S., Tom, M., Orkoulas, G., Christofides, P.D., 2022b. Multiscale computational fluid dynamics modeling of spatial thermal atomic layer etching. Computers & Chemical Engineering 163, 107861.

Yun, S., F. Ou, H. Wang, M. Tom, G. Orkoulas and P. D. Christofides, 2022c. Atomistic-mesoscopic modeling of area-selective thermal atomic layer deposition. Chemical Engineering Research and Design 188, 271–286.

Antonis Kokossis, Michael C. Georgiadis, Efstratios N. Pistikopoulos (Eds.)
PROCEEDINGS OF THE 33rd European Symposium on Computer Aided Process Engineering (ESCAPE33), June 18-21, 2023, Athens, Greece
 http://dx.doi.org/10.1016/B978-0-443-15274-0.50013-5

Modeling the shrinkage effect during the drying process of a green ceramic material

*Achilleas L. Arvanitidis,[a] Margaritis Kostoglou,[b] Michael C. Georgiadis[a,]**
[a]Department of Chemical Engineering, Aristotle University of Thessaloniki, University Campus, Thessaloniki, 54124, Greece
[b]Department of Chemistry, Aristotle University of Thessaloniki, University Campus, Thessaloniki, 54124, Greece
Correspending author, e-mail: mgeorg@auth.gr

Abstract

In this work, a 1-D diffusion model is developed to describe the drying process of a ceramic slab with respect to shrinkage. The proposed mathematical framework consists of a set of partial differential and algebraic equations along with moving boundary conditions, making it suitable for predicting the drying behavior and the physical phenomena that accompany the drying process of a green ceramic slab, such as shrinkage and porosity increase. Initially, the developed model is validated against experimental data from the open literature. Then, the drying behavior of a clay material, produced at a local factory, is investigated. This requires the characterization of the material with respect to its desorption kinetics. The parameter values of a semi-empirical desorption isotherm equation are estimated based on drying tests conducted in the lab. Finally, the drying kinetics of the porous material is studied and the shrinkage effect is discussed.

Keywords: Ceramic drying, shrinkage, porosity, moving boundary problem

1. Introduction

The drying process constitutes an essential step in the ceramic sector, which aims at removing the water from the green ceramic material before firing takes place. This is a critical production step because if the material is inserted wet into the extreme environment conditions of the firing chamber, then the probability of moisture evaporation inside the pores of the material would be high. As a consequence, this could lead to the development of intense stresses inside the porous body that can potentially lead to cracks and fracture. Cracks could also occur during the drying process itself, because of the non-uniform moisture distribution along the body and hence, non-uniform shrinkage (Kowalski and Pawłowski, 2010).

The understanding of the shrinkage mechanism during the drying procedure of a ceramic green body plays a key role for the product quality assurance and the fabrication of new drying strategies. Mancuhan et al. (2016) studied the shrinkage behavior of ceramic green bricks through Bigot curves. They proposed that during the initial stages of drying, the volume variation of the material is equal to the volume of the removed water. However, as drying proceeds, a critical moisture content is reached. This indicates that most clay particles of the solid matrix are already in direct contact and cannot move any further. At that point, shrinkage is almost complete and drying proceeds without any volume variations taking place.

It is a common practice in the open literature, to calculate the drying-induced shrinkage using mechanical interaction models which use strain – stress relationships (Hammouda and Mihoubi, 2014). Other approaches use more simplistic methods to account for the volume variation of a porous material during drying. Barbosa de Lima et al. (2016) used polynomial expressions to correlate the moisture content of a clay material to its volumetric strain. However, this method of approaching the shrinkage calculation does not provide any physical insight for the drying process.

In this work, the drying-induced shrinkage calculation is approached by using a shrinkage velocity, which reflects the velocity at which a specific nodal point of the clay matrix recedes to the interior of the bulk body. The proposed modeling framework leads to high quality predictive results with low computational complexity and facilitates the numerical calculation of key physical quantities, such as the solid concentration and the body porosity, which are closely associated to qualitative characteristics of the green material.

2. Mathematical model

In this section, a mathematical model is developed to capture the drying behavior of a shrinking ceramic roof tile. Since the material aspect ratio is very small, its geometry is simplified to that of a semi-infinite tile. Consequently, the model considers that heat and mass transfer phenomena occur only on the thickness dimension of the tile.

The evolution of the water mass concentration, C_w and temperature, T, with time along the green material is given by equations. (1) and (2), respectively. Both equations also consider the shrinkage effect via a local shrinkage velocity term, v_{shr}, which determines the velocity at which a nodal point of the clay matrix moves towards the bulk body.

$$\frac{\partial C_w}{\partial t} = \frac{\partial}{\partial z}\left[D_{eff}\frac{\partial C_w}{\partial z} - v_{shr}C_w\right] \quad , \quad z \in (0, L) \tag{1}$$

$$\rho c_p \frac{\partial T}{\partial t} = \frac{\partial}{\partial z}\left[k_{eff}\frac{\partial T}{\partial z} - v_{shr}\rho c_p T\right] \quad , \quad z \in (0, L) \tag{2}$$

Where D_{eff} is the effective diffusivity of the moisture and $k_{eff}$is the effective thermal conductivity of the material. Both properties are constant for the present study.

Preheated air flows over the wet tile and drying is achieved through forced convection. This model assumes that moisture diffuses from the bulk body towards the air-solid interface, where it instantaneously evaporates and desorbs to the blowing air. As such the mass balance boundary equation is given by equation (3).

$$-D_{eff}\left.\frac{\partial C_w}{\partial z}\right|_{z=L} = h_m \frac{MW_w}{R_g}\left(\frac{P_{sat}(T_{z=L})}{T_{z=L}}RH_{eq} - \frac{P_{sat}(T_{air})}{T_{air}}RH\right) \tag{3}$$

Where, h_m, is the convective mass transfer coefficient, MW_w is the water molecular weight and R_g is the global gas constant. Also P_{sat} is the saturation pressure of water and RH is the relative humidity of the air, which is defined as $P_w/P_{w,sat}$.

It is assumed that at the air - solid interface, the moisture content is at chemical equilibrium with the air humidity. This is given by a desorption isotherm correlation.

$$RH_{eq} = f(X|_{z=L}, T|_{z=L}) \tag{4}$$

Where X is the moisture mass fraction (dry base) of the material and is defined as $X = C_w/C_s$.

As the preheated air flows over the colder wet tile surface, heat is transferred through convection. Additionally, the latent heat of vaporization, ΔH_{vap}, is abducted from the drying surface of the material to achieve the water evaporation.

$$-k_{eff}\left.\frac{\partial T}{\partial z}\right|_{z=L} = h_T(T|_{z=L} - T_{air}) - \Delta H_{vap}D_{eff}\left.\frac{\partial C_w}{\partial z}\right|_{z=L} \quad (5)$$

The bottom surface of the green body is not in direct contact with the air.

$$\left.\frac{\partial C_w}{\partial z}\right|_{z=0} = 0 \quad (6)$$

$$\left.\frac{\partial T}{\partial z}\right|_{z=0} = 0 \quad (7)$$

The only mechanism that affects the solid concentration is the shrinkage-induced convection. Hence, the temporal evolution of the solid concentration C_s, is given by (8).

$$\frac{\partial C_s}{\partial t} = -\frac{\partial[v_{shr}C_s]}{\partial z} \quad , \quad z \in (0, L) \quad (8)$$

The local porosity, ε, is calculated from the phase volume balance.

$$1 = \varepsilon + \frac{C_s}{\rho_{intrinsic}} + \frac{C_w}{\rho_w} \quad (9)$$

Where, $\rho_{intrinsic}$ and ρ_w are the intrinsic solid density and water density respectivelly. Since all heat and mass transfer phenomena are considered exclusively on the thickness direction, the tile thickness, L, evolution should be descibed by equation (10).

$$\frac{dL}{dt} = v_{shr}|_{z=L} \quad (10)$$

The shrinkage velocity is a local term, which shows the velocity at which a nodal point recedes to the interior of the body. More specifically, the shrinkage velocity is the velocity at which a volume of water diffuses towards the air – solid interface.

$$v_{shr} = \int_0^z \alpha \frac{\partial}{\partial z}\left[\frac{D_{eff}}{\rho_w}\frac{\partial C_w}{\partial z}\right] dz \quad , \quad z \in [0, L] \quad (11)$$

Where α is a proportionality factor. Its physical meaning is that when α equals to 1, the volume variation of the porous body is respective to the volume variation of the removed water (ideal shrinkage case). When $\alpha = 0$, no volume variation takes place as water leaves the body (no shrinkage case).

As explained earlier, during the initial stages of drying, the shrinkage is ideal (e.g. $\alpha = 1$). However, when the moisture content, X, reaches a critical value, X_{cr}, almost all shrinkage phenomena are complete because most of the clay particles are already in direct contact with each other (Mancuhan et al., 2016). At that point, shrinkage stops (e.g. $\alpha = 0$) and the porosity of the material starts to increase since air penetrates into the body to substitute the volume of the removed water.

$$\alpha = \begin{cases} 1, & X \geq X_{cr} \\ 0, & X < X_{cr} \end{cases} \quad , \quad z \in [0, L] \quad (12)$$

3. Solution Strategy

The integration of the above system of equations on the (z, t) plane is not trivial because the boundary conditions are located on a moving boundary. The position of the $z = L$ node must be determined as an integral part of the solution at any time. That for, a simple Lagrangean transformation is used, by introducing the spatial variable $y = z/L$ to immobilize the moving boundary. The y spatial variable lies between 0 and 1, and enables the normal discretization of the thickness dimension. The newly transformed model is implemented into gPROMS^{TM} modeling environment. The partial differential equations are approximated using 2nd order Central Finite Differences and the differential – algebraic solver DASOLV is used for the integration.

4. Model Validation

For the validation of the model, a suitable set of experimental data from the open literature was used. The experimental measurements that are presented in the work of Heydari et al. (2018) facilitate the corroboration of the model predicted shrinkage behavior of the green ceramic body. In Fig. 1, the model predicted moisture content and volumetric strain is plotted against the experimental measurements.

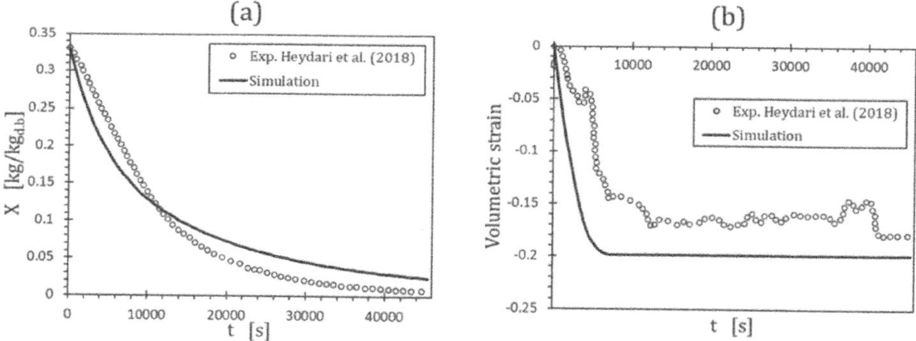

Fig. 1: Comparison of experimental data (Heydari et al., 2018) and model predictions for (a) moisture content and (b) volumetric strain.

The results indicate that a sufficiently good agreement is achieved. The deviations of the model predictions from the experimental points are ought to the fact that the studied material of Heydari et al. (2018) has much larger aspect ratio than the semi-infinite plate that is of interest in the present work. It is expected that the use of a higher order model could result to a better match between the model predictions and the experimental data.

5. Parameter Estimation

The material of interest in this work is a clay tile that is produced in a large-scale production plant in Greece. The simulation of the drying process of the material requires the establishment of its desorption kinetics. That for, equation (4) has to be substituted by an equation that correlates the relative humidity to the moisture content of the material and its temperature. For this reason, the GAB model is selected.

$$X_{eq} = \frac{Ckm_0RH_{eq}}{(1 - kRH_{eq})(1 - kRH_{eq} + CkRH_{eq})} \tag{13}$$

Where C, k, m_0 are temperature dependent parameters and are described by Arrhenious type equations :$C = C'e^{-\frac{\Delta H_C}{R_g T}}$, $k = k'e^{-\frac{\Delta H_k}{R_g T}}$ and $m_0 = m_0'e^{-\frac{\Delta H m_0}{R_g T}}$.
The parameters C', k', m_0' and $\Delta H_C, \Delta H_k, \Delta H_{m_0}$ are estimated based on drying tests that were conducted in the lab and the their values are illustrated in Table 1.

Table 1: GAB model parameters

GAB parameters					
C''	k''	m_0''	$\Delta H_{C'}$	$\Delta H_{k'}$	$\Delta H_{m_0'}$
0.195	0.0254	0.175	-1.62×10^7	-5.05×10^6	2.42×10^5

6. Results and Discussion

Simulation results of the drying process for the studied material are summarized here, to gain insight on how the shrinkage phenomemon affects the dynamics of various quantities of the process. The model inputs required for the simulation are summarized as follows. The blowing air temperature, T_{air}, is 100°C and the air relative humidity, RH, is 3%. The initial water concentration and initial solid concentration are $316\, kg/m^3$ and $1754\ kg/m^3$ respectively. The initial tile thickness is 1.87 cm and the critical moisture content, which determines when shrinkage stops, is $X_{cr} = 0.1\ kg_w/kg_{d.s.}$. As for the heat and mass transfer coefficients, their values are chosen arbitrarily as 28 W/m^2K and 0.028 m/s, respectively.

Fig. 2 illustrates the water concentration and body temperature time profiles. It is observed that the water concentration inside the body decreases with time as its temperature rises. Finally, a steady state is reached and the drying procedure is complete.

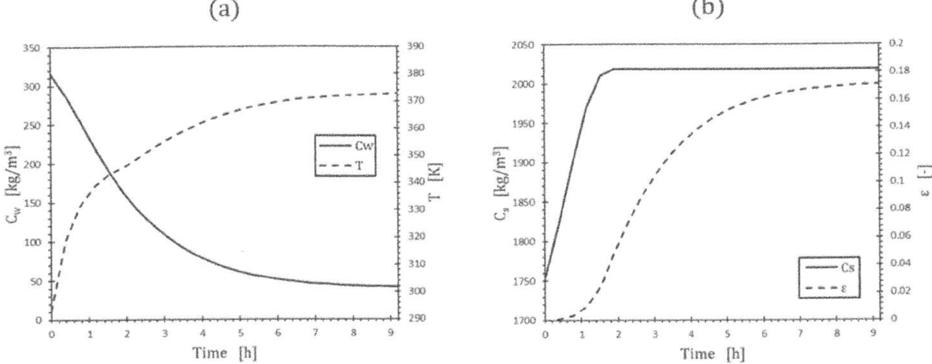

***Fig. 2**: (a) Mean water concentration and temperature vs time and (b) Mean solid concentration and body porosity vs time*

The solid concentration and porosity temporal evolutions are illustrated in Fig. 2(b). At the beginning of the process, the body porosity remains unchanged and the solid concentration rises abruptly. This is justified by the fact that during these times shrinkage is ideal and $\alpha = 1$. When the critical moisture content, X_{cr}, is achieved, a transition to the no-shrinkage state occurs and shrinkage stops. Consequently, the solid concentration stops to variate and air starts to replace the removed water volume, which leads to the porosity increase.

In Fig. 3, the shrinkage parameter on the moving boundary and the tile thickness are plotted against time. It is observed that the body shrinks for some time despite that the shrinkage parameter $\alpha|_{z=L}$ obtains the zero value. This is because even if $X < X_{cr}$ at the

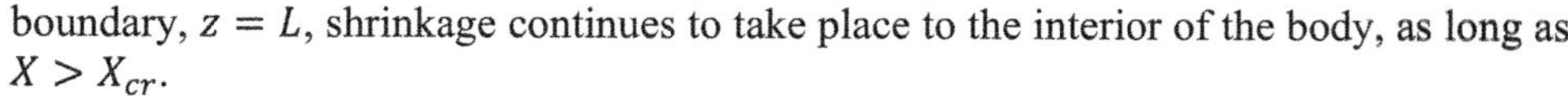
boundary, $z = L$, shrinkage continues to take place to the interior of the body, as long as $X > X_{cr}$.

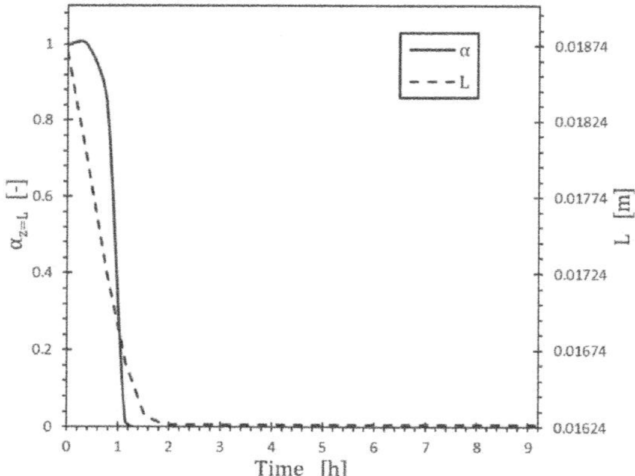

Fig. 3: α parameter on the boundary $z = L$ and slab thickness vs time

7. Conclussions

A 1-dimensional model is developed to predict the drying behavior of a green ceramic tile with respect to shrinkage. The proposed model facilitates the calculation of key physical quantities that have not been studied extensively in the literature. Such quantities are the local solid concentration and porosity. The prediction of the transient evolution of these quantities is crucial because they are severely associated to most qualitative characteristics of the green product.

Acknowledgments

This research has been co-financed by the European Union and Greek national funds through the Operational Program Competitiveness, Entrepreneurship and Innovation, under the specific call "Aquacultures"- "Industrial Materials"- "Open Innovation in Civilization" (project code: T6YBΠ-00251). Project title: Development of Computer-Aided Tools for Optimal Energy Consumption in Industrial Ceramics (CATOPEC-IC)

References

I. Hammouda, D. Mihoubi, 2014, Modeling of Thermo-Hydro-Viscoelastic Behavior of a Partially Saturated Ceramic Material During Drying. *Drying Technol.,32*,1219-1230.

M. Heydari M, K. Khalili,S.Y. Ahmadi-Brooghani, 2018, More comprehensive 3D modeling of clay-like material drying. *AIChE J.,64*,1469-1478.

E. Mancuhan, S. Özen, P. Sayan, S.T. Sargut, 2016, Experimental investigation of green brick shrinkage behavior with Bigot's curves. *Drying Technol.,34*,1535-1545.

S.J. Kowalski, A. Pawłowski, 2010, Modeling of kinetics in stationary and intermittent drying. *Drying Technol.,28*,1023-1031.

A.G. Barbosa de Lima, J. Barbosa da Silva, G.S. Almeida, J.J.S. Nascimento, F.V.S. Tavares, V.S. Silva, 2016, Clay products convective drying: Foundations, modeling and applications. *Adv Struct Mater.,63*,43-70.

Antonis Kokossis, Michael C. Georgiadis, Efstratios N. Pistikopoulos (Eds.)
PROCEEDINGS OF THE 33rd European Symposium on Computer Aided Process Engineering (ESCAPE33), June 18-21, 2023, Athens, Greece
 http://dx.doi.org/10.1016/B978-0-443-15274-0.50014-7

Investigating physics-informed neural networks for bioprocess hybrid model construction

Alexander William Rogers,[a] Ilya Orson Sandoval Cardenas,[b] Ehecatl Antonio Del Rio-Chanona,[b] Dongda Zhang,[a*]

[a] *Department of Chemical Engineering, University of Manchester, Oxford Road, Manchester, M1 3AL, UK.*

[b] *Department of Chemical Engineering, Imperial College London, South Kensington Campus, London, SW7 2AZ, UK.*

[*] *Corresponding author email: dongda.zhang@manchester.ac.uk*

Abstract

Integrating physical knowledge and machine learning is a cost-efficient solution to modelling complex biochemical processes when the underlying mechanisms are not fully understood. However, hybrid model structure identification is still time-consuming for new processes, requiring iteration over different hypotheses to explain the observed process dynamics while minimizing over-parameterization. Unfortunately, conventional approaches to automatic model structure identification do not always converge for highly nonlinear models and cannot estimate time-varying model parameters. To address this and accelerate the design of new biochemical processes, a Reinforcement Learning (RL) based framework recently reformulated synchronous hybrid model structure-parameter identification into a process optimal control problem. To further investigate other possible solutions, in this study, a novel Physics Informed Neural Network (PINN) based framework was proposed for the first time to infer time-varying kinetic parameters. This framework first combines possible kinetic structures from phenomenological knowledge, then simultaneously identifies the most likely hybrid model structure and time-varying parameter trajectories. To demonstrate the performance of the PINN based framework, several in-silico case studies were conducted using a known ground truth bioprocess. We thoroughly examined the advantages and limitations of the framework, elucidating its potential for high-fidelity hybrid model construction in biochemical engineering research.

Keywords: automatic model structure identification, time-varying parameter estimation, physics-informed neural network, hybrid modelling, machine learning.

1. Introduction

Mathematical modelling is pivotal to understanding and designing biochemical processes. A thoroughly validated dynamic model can predict biomass growth and product synthesis under different operating conditions, reducing the number of experiments required to characterize and optimize novel biochemical processes. Kinetic and data-driven models have been proposed to describe biological processes. However, identifying a suitable kinetic model structure is time-consuming, and they are often overparameterized (i.e., multiple structures and many parameters) in an effort to capture the complex dynamics, leading to high parameter and propagated state uncertainties. On the other hand, data-driven models risk overfitting without enough experimental data – which is time-consuming to generate – and poor generalization to new operating conditions.

Integrating physical knowledge and machine learning is a cost-efficient solution to modelling complex biochemical processes when the underlying mechanisms are not fully understood. Hybrid models either use a data-driven model to correct the discrepancy between a kinetic model and the observed process dynamics or update selected time-varying kinetic model parameters. A properly validated hybrid model can effectively resolve the issue of incomplete physical knowledge and low-quality-quantity data (Zhang et al., 2020) while improving prediction accuracy and confidence compared with pure kinetic models (Vega-Ramon et al., 2021). However, hybrid models inherit the risk of over-parameterizing the kinetics and overfitting the data-driven model, which can lead to high uncertainty and poor generalization. Consequently, it is essential to identify the kinetic model structure that best represents the underlying mechanisms to reduce the burden on the data-driven model to compensate and risk overfitting. This is no trivial task since the current bioprocess kinetics depend on the present culture conditions and microenvironmental history due to stochastic effects on the controlling mechanisms or metabolic stores and systematic intracellular metabolic regulation mechanisms. As a result, the kinetic model parameters and structure will evolve with time and history.

Despite this challenge, there are few attempts outside case-specific studies to resolve automatic model structure identification for complex combinatorial, history-dependent or time-varying kinetics. Conventional approaches such as mixed integer nonlinear programming (MINLP) do not always converge for highly nonlinear models and cannot estimate time-varying model parameters. To address this, we recently proposed a novel Reinforcement Learning (RL) based framework to reformulate synchronous hybrid model structure-parameter identification into a process optimal control problem (Wu et al., 2022). The RL-based framework proved promising for recovering the correct kinetic model structure or time-varying kinetic parameters, but the combined case with more combinatorial options and time-varying kinetic parameters remains unexplored. In recent years, Physics-Informed Neural Networks (PINNs) (Raissi et al., 2019) have emerged as a novel approach to discovering underlying governing equations; however, they have not been applied before to infer time-varying physical parameters. Therefore, in this work we aim to propose a novel PINN-based framework for hybrid bioprocess model construction. Several in-silico case studies will thoroughly examine the advantages and limitations of this technique for synchronous hybrid model structure-parameter identification.

2. Problem Statement

The PINN-based framework first combines several possible kinetic structures from phenomenological knowledge, then simultaneously identifies the most likely hybrid model structure and time-varying parameter trajectories. To demonstrate the performance of this framework, several in-silico case studies were conducted using a known ground truth model, shown in Equation 1, that we developed in our previous work (Rogers et al., 2022). This high-fidelity hybrid model can predict the temperature-dependent biomass growth, glucose consumption and γ-linolenic acid (GLA) accumulation rates during fermentation of the fungus *Cunninghamella echinulata*. In Equation 1, X, F, S and P are the total biomass, fat-free biomass, glucose and GLA concentration, respectively, while T is temperature. The half-saturation K_S and maintenance k_S coefficients, the total k_X and fat-free k_F biomass decay rates and specific GLA decay rate k_P are all constants, while the total Y_X and fat-free Y_F biomass growth rates, the glucose-to-biomass yield coefficient Y_S and growth-dependent GLA yield coefficient Y_P are all time-varying parameters. In our previous work, we estimated the constants and built the Gaussian Process (GP) model mapping the states and temperature to the time-varying parameters (Rogers et al., 2022).

$$\frac{\mathrm{d}X}{\mathrm{d}t} = Y_X(X,F,S,P,T) \cdot \frac{S}{S + K_S \cdot X} \cdot X - k_X \cdot X \tag{1a}$$

$$\frac{\mathrm{d}F}{\mathrm{d}t} = Y_F(X,F,S,P,T) \cdot \frac{S}{S + K_S \cdot X} \cdot X - k_F \cdot X \tag{1b}$$

$$\frac{\mathrm{d}S}{\mathrm{d}t} = Y_S(X,F,S,P,T) \cdot \frac{S}{S + K_S \cdot X} \cdot X - k_S \cdot X \cdot \frac{S}{S + 0.1} \tag{1c}$$

$$\frac{\mathrm{d}P}{\mathrm{d}t} = Y_P(X,F,S,P,T) \cdot \frac{S}{S + K_S \cdot X} \cdot X - k_P \cdot X \tag{1d}$$

Now the ground-truth model was used to generate six in-silico batches from low ($S_0 = 60\ \mathrm{g\,L^{-1}}$) and high ($S_0 = 100\ \mathrm{g\,L^{-1}}$) initial glucose concentrations under three different temperatures ($T = 14$°C, $T = 28$°C, $T = 37$°C). The four GP models updated the four time-varying parameters every 24 hours while the constants were fixed. Each batch was 'fermented' for 408 hours, and the four state variables 'measured' every 24 hours.

3. Methodology

3.1. Physics-Informed Neural Network Structure

The PINN aims to learn the solution to the system of ordinary differential equations (ODEs) presented in Equation 1 by reformulating the parameter estimation and numerical integration problem into a single nonlinear optimization problem. The ANN aims to learn the concentration profile and kinetic parameters that satisfy the measured states and the system of ODEs of the form: $dx/dt = \mathcal{N}(\boldsymbol{x}, \boldsymbol{\phi}, \boldsymbol{\theta})$, where $\boldsymbol{x} = [X, F, S, P]^T$, while $\boldsymbol{\phi}$ and $\boldsymbol{\theta}$ are vectors of time-varying and time-constant parameters, respectively. The unknown solution to the concentration profile $\boldsymbol{u}_x(\boldsymbol{\lambda}, t, \boldsymbol{x}_0)$ and kinetic parameters $\boldsymbol{u}_\phi(\boldsymbol{\lambda}, t, \boldsymbol{x}_0)$ was represented by a single ANN as a function of batch time $t \in [0, 408\ \text{hours}]$ and the initial state $\boldsymbol{x}_0 = [X_0, F_0, S_0, P_0]^T$, where $\boldsymbol{\lambda}$ denotes the tunable weights and biases.

$$\boldsymbol{\lambda}^* = \min_{\boldsymbol{\lambda}} \mathcal{L}(\boldsymbol{\lambda}) = \min_{\boldsymbol{\lambda}} [\omega_0 \mathcal{L}_0(\boldsymbol{\lambda}) + \omega_d \mathcal{L}_d(\boldsymbol{\lambda}) + \omega_c \mathcal{L}_c(\boldsymbol{\lambda})] \tag{2a}$$

$$\mathcal{L}_0(\boldsymbol{\lambda}) = \frac{1}{N_0} \sum_{i=1}^{N_0} \left| \widehat{\boldsymbol{x}}_0^i - \boldsymbol{x}_0^i \right|^2 \tag{2b}$$

$$\mathcal{L}_d(\boldsymbol{\lambda}) = \frac{1}{N_d} \sum_{i=1}^{N_d} \left| \widehat{\boldsymbol{x}}_d^i - \boldsymbol{x}_d^i \right|^2 \tag{2c}$$

$$\mathcal{L}_c(\boldsymbol{\lambda}) = \frac{1}{N_c} \sum_{i=1}^{N_c} \left| \frac{\partial \widehat{\boldsymbol{x}}_c^i}{\partial t} - \mathcal{N}\left(\widehat{\boldsymbol{x}}_c^i, \widehat{\boldsymbol{\phi}}_d^i, \widehat{\boldsymbol{\theta}}\right) \right|^2 \tag{2d}$$

The PINN was trained by minimizing the composite loss function in Equation 2, where ω_0, ω_d and ω_c balance the interplay between the different loss terms during training. Here $\{\boldsymbol{x}_0^i\}_{i=1}^{N_0}$, $\{t_d^i, \boldsymbol{x}_d^i\}_{i=1}^{N_d}$ and $\{t_c^i, \boldsymbol{x}_c^i\}_{i=1}^{N_c}$ are the time-state value pairs for the initial conditions, measured data points and collocation points, respectively. Each batch is divided into N intervals of $\Delta t = 24$ hours, where $t_d \in \{0,\ \Delta t,\ 2\Delta t,\ 3\Delta t \ldots N\Delta t\}$, while each interval is subdivided into $\Delta t/\delta t = 5$ collocation points, such that $t_c \in \{0,\ \delta t,\ 2\delta t,\ 3\delta t \ldots n\delta t\}$ is on the same timescale. The terms $\widehat{\boldsymbol{x}}_d = \boldsymbol{u}_x(\boldsymbol{\lambda}, t_d, \boldsymbol{x}_0)$ and $\boldsymbol{x}_d$ denote the predicted and measured states, respectively, at the known data points, while $\widehat{\boldsymbol{x}}_c = \boldsymbol{u}_x(\boldsymbol{\lambda}, t_c, \boldsymbol{x}_0)$ denotes the predicted state at the collocation points for which the states were not measured. Once $\widehat{\boldsymbol{x}}_c$ is evaluated, the gradients with respect to time $\partial\widehat{\boldsymbol{x}}_c/dt$ were computed by reverse automatic differentiation. The time-constant $\widehat{\boldsymbol{\theta}} = \boldsymbol{u}_{\phi}(\boldsymbol{\lambda}, 0,0)$ and time-varying $\widehat{\boldsymbol{\phi}}_d = \boldsymbol{u}_{\phi}(\boldsymbol{\lambda}, t_d, \boldsymbol{x}_0)$ kinetic parameters were both predicted by the ANN, implicitly coupling them to the same network parameters for more stable training. However, unlike the continuous state variables, $\widehat{\boldsymbol{\theta}}$ is a fixed constant, while $\widehat{\boldsymbol{\phi}}$ is a piecewise constant. To represent this behaviour, $\widehat{\boldsymbol{\theta}}$ was evaluated at a fixed input while $\widehat{\boldsymbol{\phi}}_d$ was evaluated at each data sampling point t_d and the result used over $t_d \leq t_c < t_d + \Delta t$ when computing $\mathcal{N}(\widehat{\boldsymbol{x}}_c, \widehat{\boldsymbol{\phi}}_d, \widehat{\boldsymbol{\theta}})$. The network was built from a single hidden layer of 40 neurons with a hyperbolic tangent activation function in the hidden layer and a linear output layer. This combination mirrors the typical bioprocesses sigmoid-like growth and parameter profiles.

3.2. Physics-Informed Neural Network Training and Simulation

The PINN was fitted simultaneously to all six in-silico batches. The PINN was trained in two stages: first (i) the network was trained for 6000 epochs with $\{\omega_0 = 5, \omega_d = 1, \omega_c = 0\}$ and a learning rate of 5×10^{-2} until $\mathcal{L}_d(\boldsymbol{\lambda})$ converged, then (ii) the network was trained for 10000 epochs with $\{\omega_0 = 5, \omega_d = 1, \omega_c = 5 \times 10^3\}$ and a learning rate of 5×10^{-4} until $\mathcal{L}_c(\boldsymbol{\lambda})$ converged. Fitting the state profile before the kinetic parameters approximates the initial gradients $\partial\widehat{\boldsymbol{x}}_c/dt$ and was found empirically to be more robust to becoming trapped within low-quality local optima than single-stage training.

Once the PINN was constructed, the multistep-ahead state trajectory was re-simulated using Equation 3 from $\boldsymbol{x}_0 = [X_0, F_0, S_0, P_0]^T$ using the predicted time-constant parameters $\widehat{\boldsymbol{\theta}}^* = \boldsymbol{u}_{\phi}(\boldsymbol{\lambda}^*, 0,0)$ and the predicted time-varying parameters $\widehat{\boldsymbol{\phi}}_d^* = \boldsymbol{u}_{\phi}(\boldsymbol{\lambda}^*, t_d, \boldsymbol{x}_0)$ updated once every 24 hours.

$$\widehat{\boldsymbol{x}}'_{d+1} = \widehat{\boldsymbol{x}}'_d + \int_{t_d}^{t_{d+1}} \mathcal{N}(\boldsymbol{x}', \widehat{\boldsymbol{\phi}}_d^*, \widehat{\boldsymbol{\theta}}^*)\, dt \tag{3}$$

4. Results and Discussion

Once the PINN was fitted to the six in-silico batches, the fitted state $\widehat{\boldsymbol{x}}_d^* = \boldsymbol{u}_x(\boldsymbol{\lambda}^*, t_d, \boldsymbol{x}_0)$ trajectory, the time-constant parameters $\widehat{\boldsymbol{\theta}}^* = \boldsymbol{u}_{\phi}(\boldsymbol{\lambda}^*, 0,0)$ and the time-varying kinetic parameter $\widehat{\boldsymbol{\phi}}_d^* = \boldsymbol{u}_{\phi}(\boldsymbol{\lambda}^*, t_d, \boldsymbol{x}_0)$ trajectory was retrieved. In addition, the fitted PINN was used to re-simulate the multistep-ahead state trajectory $\widehat{\boldsymbol{x}}'_d$ for the six in-silico batches from their different initial states and operating temperatures, as described in Section 3.2. Figure 1 compares the fitted (i.e., $\widehat{\boldsymbol{x}}_d^*$) and the re-simulated (i.e., $\widehat{\boldsymbol{x}}'_d$) state trajectories for one of the six in-silico batches, showing that the PINN could fit and re-simulate the state trajectory with very high accuracy, with a mean absolute percentage error (MAPE) of 1% and 5%, respectively. Since $\boldsymbol{x}_0$ is satisfied (MAPE of 1%) by strong regularization (i.e., $\omega_0 > \omega_d$) the mismatch between the fitted and re-simulated trajectories is due to the PINN not being strictly required to satisfy the system of ODEs during PINN training.

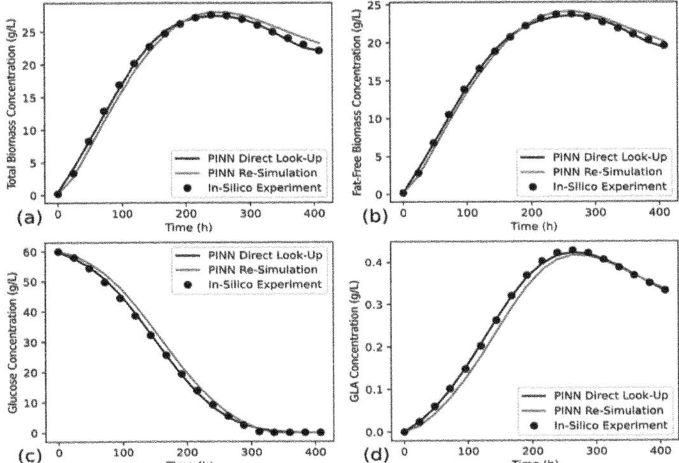

Figure 1: Fitted and re-simulated state trajectory for one of the six in-silico training batches for total biomass (a), fat-free biomass (b), glucose (c) and GLA (d) concentration.

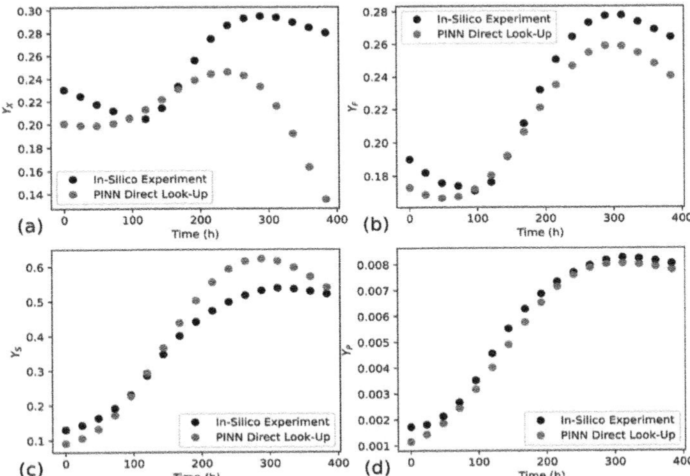

Figure 2: Ground truth and estimated time-varying parameters for one of the six in-silico batches for the specific growth rate Y_X (a) and Y_F (b), Y_S (c) and Y_P (d) yield coefficients.

Table 1 compares the PINN estimated and ground-truth constants, demonstrating that the PINN can accurately recover (MAPE of 8%) the correct constants when there is no measurement uncertainty. The residual between the ground truth and estimated values can likely be attributed to the practical identifiability of the parameters given only six in-silico batches. Figure 2 compares the PINN estimated and ground-truth time-varying parameter trajectories, demonstrating that the PINN can recover Y_F, Y_S and Y_P well (MAPE of 15%), particularly considering the time-varying parameters (i.e., Y_X, Y_F, Y_S and Y_P) become non-identifiable as $S \rightarrow 0$ at the end of the batch, as their influence becomes negligible compared to the specific decay constants (i.e., k_X, k_F, k_S and k_P). However, the PINN could not recover well Y_X (MAPE of 150%). Given the excellent fit in Figure 1, there could be two reasons for this: either (i) the PINN cannot converge to the globally optimum ground-truth parameters, and the model is insensitive to the exact parameter values, or (ii) K_S is partially non-identifiable, and extra regularization is required.

Table 1: Comparison of ground-truth and estimated time-constant parameters.

Parameter	K_S	k_X	k_F	k_S	k_P
Ground-Truth	48.64	1.70E-03	1.35E-03	3.10E-03	3.23E-05
Estimated	49.60	1.50E-03	1.32E-03	2.57E-03	2.96E-05
Absolute Error (%)	2.0	11.8	2.2	17.1	8.4

5. Conclusion

The PINN-based approach has potential for kinetic parameter estimation and for providing a way to understand real bioprocess underlying system dynamics. At present, embedded hybrid model construction follows a two-step procedure: (i) simultaneous time-constant and time-varying parameter estimation, then (ii) data-driven model construction to correlate the time-varying parameters with the state and operating conditions for predictive simulation. This novel PINN-based framework has the potential to accelerate hybrid model construction by directly building the predictive data-driven model. However, there remains the challenge of identifying a high-quality solution to the PINN network parameters. Therefore, future work will explore more advanced techniques, such as adaptive time-sampling, self-attenuation, time-marching and causal training (Wang et al., 2022). In addition, the PINN-based framework is currently only employed to estimate the kinetic parameters for a pre-defined kinetic model structure. Therefore, future work will also explore synchronous parameter-structure identification and compare the performance of the PINN-based framework against an extended RL-based framework (Wu et al., 2022) when multiple possible structures are possible.

References

Raissi, M., Perdikaris, P., & Karniadakis, G. E. (2019). Physics-informed neural networks: A deep learning framework for solving forward and inverse problems involving nonlinear partial differential equations. *Journal of Computational Physics*, *378*, 686–707.

Rogers, A. W., Song, Z., Vega-Ramon, F., Jing, K., & Zhang, D. (2022). Investigating 'greyness' of hybrid model for bioprocess predictive modelling. *Biochemical Engineering Journal*.

Vega-Ramon, F., Zhu, X., Savage, T. R., Petsagkourakis, P., Jing, K., & Zhang, D. (2021). Kinetic and hybrid modeling for yeast astaxanthin production under uncertainty. *Biotechnology and Bioengineering*, *118*(12), 4854–4866.

Wang, S., Sankaran, S., & Perdikaris, P. (2022). *Respecting causality is all you need for training physics-informed neural networks*.

Wu, C., Mowbray, M. R., Rogers, A. W., Rio-Chanona, E. A. Del, & Zhang, D. (2022). *A reinforcement learning based hybrid modelling framework for bioprocess kinetics identification*.

Zhang, D., Savage, T. R., & Cho, B. A. (2020). Combining model structure identification and hybrid modelling for photo-production process predictive simulation and optimisation. *Biotechnology and Bioengineering*, *117*(11), 3356–3367.

Antonis Kokossis, Michael C. Georgiadis, Efstratios N. Pistikopoulos (Eds.)
PROCEEDINGS OF THE 33rd European Symposium on Computer Aided Process Engineering (ESCAPE33), June 18-21, 2023, Athens, Greece
 http://dx.doi.org/10.1016/B978-0-443-15274-0.50015-9

Decision Making Approaches to Improve Resilience in Food Supply Chains and Enhance Food Security Against Climate Change Risks

Bashar Hassna, Sarah Namany, Mohammad Alherbawi, Adel Elomri, Tareq Al-Ansari*

College of Science and Engineering, Hamad Bin Khalifa University, Doha, Qatar
**talansari@hbku.edu.qa*

Abstract

As far as food security is concerned, the world is challenged by growing resource scarcity, changing diets, environmental challenges, volatile prices, and supply chain disruptions. Various technological, political, economic, and social factors drive food system dynamics that fluctuate in terms of their impact on food security. Many studies have attempted to alleviate risks governing food systems as means to achieve food security by suggesting numerous modelling approaches and decision-making tools. The purpose of this paper is to design a composite indicator analysing the risk status of a set of trade partners supplying Qatar with a predetermined food basket. The latter involves perishable fruits and vegetables. To design risk indicators, a set of individual factors reflecting the impact of climate change, inclusive of seasonality and sudden risks, are considered. The evaluation of these risks is performed based on their impact on Qatar's demand satisfaction. Whereas the assessment is conducted using the Analytical Hierarchy Process (AHP). The second stage of this work formulates a multi-objective optimization model in MATLAB that determines the optimal network of suppliers to satisfy the local need of Qatar, considering their climate risk profiles and their production capacities. The purpose of the optimization framework is to identify the contribution percentage of each trading partner to the total demand for the predetermined food basket while minimizing economic costs and risk composite factors. Results of the study assert that diversifying trade partners is the most resilient option to satisfy local demands yet with a high economic cost.

Keywords: food imports, climate change, risks, AHP, multi-objective optimisation.

1. Introduction

One of the most pressing issues in today's world is climate change, which has significantly impacted the earth's ecosystems. Although the world has always experienced some degree of climate change in the last 100 years, the rate of this variation has multiplied in recent decades. Since the nineteenth century, anthropogenic activities have caused an increase in average temperature of 0.9 °C, primarily as a result of greenhouse gas (GHG) releases into the atmosphere. This rise is anticipated to reach 1.5 °C by 2050 due to continuous deforestation and the contamination of air, soil, and water bodies (Roe et al., 2019).

Long-term variations in temperature and precipitation patterns contribute to changing the comparative advantage of food production around the world and may even slow the expansion of agricultural productivity in areas with high rates of hunger. As a result, global agriculture and food systems may be altered, causing an inter-annual volatility in food supplies. If the current climate change patterns continue to evolve, crop losses may rise in the very near future at an unprecedented rate; significantly reducing production, driving up food costs, and making it difficult to meet the rising demands of a growing population. Therefore, future food security must be strengthened by considering the inter-annual and inter-decadal effects of climate change.

Considering food systems are susceptible to climate change and natural disasters. The effect of these factors either can be sudden or can be slow, resulting in a positive or negative change (Gomez-Zavagliaa, 2020), noting that it mainly affects the production stages, which have been found to only contribute to around 20% of the value chain (Cucagna & Goldsmith, 2018). Whereas the complexity of the global agricultural markets has increased due to more focus on food standards that are related to food safety, food quality, and technological advances, which creates changes in the industry along the value chain. These considerations make the task of creating a robust and resilient food supply system even more challenging.

Therefore, it is imperative to adopt a holistic value chain to enhance resilience rather than considering the production stage alone (Ringsberg, 2014). Consequently, there is a need to increase the efforts related to identifying and classifying risks related to food systems considering data from different parts of the value chain.

Hence, the main aim of this study is to develop a clearer understanding of the food system, influencing factors, and respective risk management tools, with a focus on climate change as an external and unpredictable risk to food systems.

In this study, food system's associated risks are evaluated considering their impact on the demand satisfaction for the State of Qatar. While the Analytical Hierarchy Process (AHP) tool is utilised to evaluate the relative weight of these risk factors based on quantitative and qualitative measures.

A multi-objective optimization model is then developed and solved using MATLAB's Genetic Algorithm. The goal of the optimization framework is to select optimal suppliers and define their contribution percentage towards the food basket of Qatar, while ensuring the local demand is satisfied within the suppliers' capacities, with reduced costs and minimal associated risks. The model in hand may provide insights on possible enhancement to global food systems considering economic, social, and environmental aspects.

2. Methodology

2.1. The Analytical Hierarchy Process to design a climate risk indicator

The Analytical Hierarchy Process (AHP) was used to develop a composite risk index which involves five different climate change risk related indicators counting, surface temperature change, water stress, arable land, sea level rise and natural disasters management in addition to agriculture policy adoption. Four countries were selected which are USA, India, China, and Iran. The AHP is used to quantify the level of risk of each trade partner through determining its priority rank and following the decision tree described in figure 1. Priorities were selected based on the situation of the country in each of the indicators and was based on historical data and grey literature (Dahlman, (n.d); EEA(2021) Level of Water Stress (2023); The Economist Group (2018)).

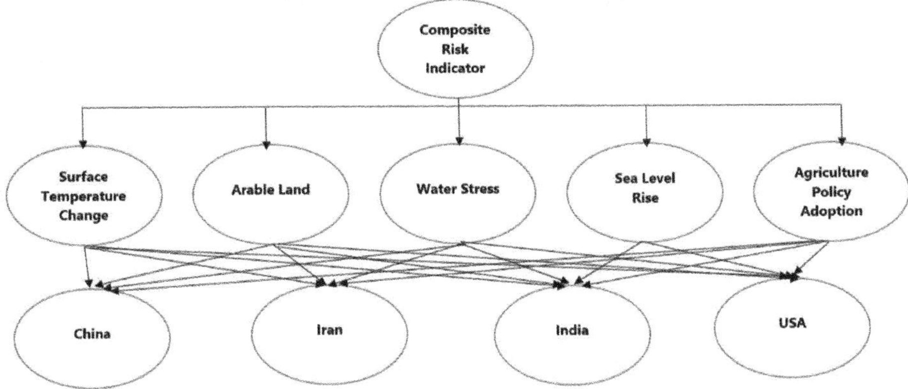

Figure 1. The hierarchical decision tree

2.2. Optimization model

The purpose of this part of the methodology is to design a multi-objective optimisation model that determines the optimal network of suppliers that can satisfy the local need of Qatar, considering their climate risk profiles and their production capacities. The purpose of the optimization framework is to identify the per cent contribution of each trading partner to the total demand for the predetermined food basket while minimizing economic costs, and risk composite indicators. The following table 1 presents the mathematical formulation of the suggested optimisation model. Considering the variable used, x_i presents the decision variable of the model which is the percent contribution of each country to the food basket. Q is the yearly local demand for food in kg. As for c_i, it represents the food basket average overall unit cost in $/kg. r_i is the overall risk indicator generated from the AHP model. As for the cap_i, it represents the allowable exportable quantities in kg that can be supplied by each trade partner.

Table 1. The mathematical formulation of the optimisation model

	Equation	Connotation
Objective function 1	$min \sum_{i=1}^{5} Qc_i x_i$	Identifies the optimal set of trade partners that supply Qatar's food basket while minimising the overall cost.
Objective Function 2	$min \sum_{i=1}^{5} r_i x_i$	Identifies the optimal set of trade partners that supply Qatar's food basket while minimising the overall composite risk indicators.
Constraint 1	$\sum_{i=1}^{5} x_i = 100\%$	Logical constraint implies that the sum of all contributions per cent of water technologies must be 100%.
Constraint 2	$Qx_i < cap_i$	The imported quantity from each trading partner should not exceed their individual exportable capacity.

3. Results and discussion

After conducting the AHP, 4 composite risks factors were generated for each country describing its overall risk performance (table 4). Those risks factors were derived from the composite indicators involving the set of climate risks sub-indicators as highlighted in table 3.

Composite risk factors were later used in the optimisation framework as parameters to select the optimal mix of exporting countries that can supply the market with food products.

Table 3. Results of the AHP priority score for each exporting country

	China	USA	India	Iran
Disaster Risk Management	0.0611	0.0614	0.0656	`0.0738
Surface Temperature Change	0.14065	0.1484	0.1322	0.1264
Arable land	0.1612	0.1738	0.3004	0.2546
Water Stress	0.1388	0.1921	0.2432	0.2726
Sea level rise	0.4522	0.3755	0.2116	0.2255
Agricultural Policy Adoption	0.0459	0.0482	0.0467	0.0466

Table 4. Composite risk factors for each exporting country ((Namany et al., 2019)

	China	USA	India	Iran
Unit cost ($/kg)	1.22	7.08	1.45	0.81
Composite Risk Factor	0.2222	0.2155	0.2114	0.2088

Results of the optimisation model are summarised in Figure 2 and 3. The optimal solution suggests a slightly diversified network of exporting countries with a large dominance of Iran with around 90%, while the remaining 10% is distributed between China and India. As for USA, it's contribution is below the 1%. This can be explained by the large unit costs of products originating from the USA and the relatively low risk factors and costs of the other trade partners. Taking the graphical average from the Pareto front, the average optimal solution, generates an average total cost of $672M.

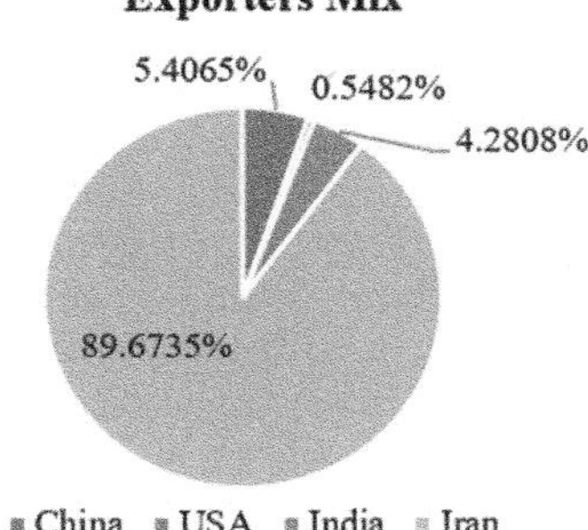

Figure 2. The average optimal trade partners' network

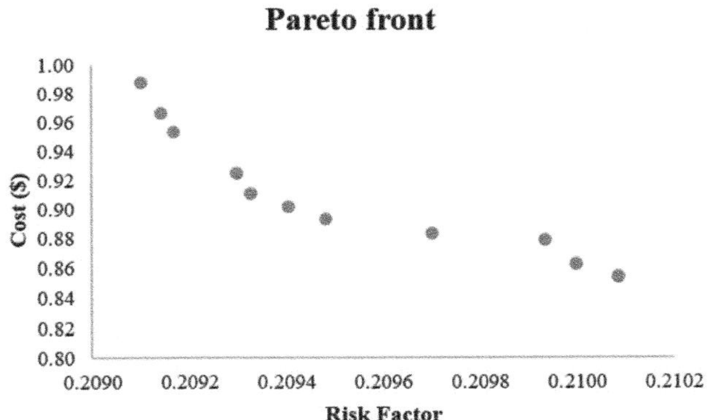

Figure 3. The Pareto front.

4. Conclusion

In this paper a multi-objective optimization model was developed to identify the ideal network of suppliers that can meet Qatar's local demand while taking into account their production capacities and risk profiles related to the climate. The optimization framework's goal was to calculate each trade partner's percentage contribution to the overall demand for the predefined food basket while minimizing composite risk factors and economic costs four different countries. According to study findings, diversifying trade partners is the most durable way to meet local wants, however it can come at a high financial cost.

5. Acknowledgments

This research was made possible by an Award (GSRA7-1-0521-20080) and supported by proposal number NPRP11S-0107-180216 from Qatar National Research Fund (a member of Qatar Foundation). The contents herein are solely the responsibility of the authors[s].

6. References

Cucagna, M. E., & Goldsmith, P. D. (2018). Value adding in the agri-food value chain. International Food and Agribusiness Management Review, 21(3), 293-316.

Dahlman, R. L. A. N. D. L. A. (n.d.). Climate change: Global temperature. NOAA Climate.gov. Retrieved February 5, 2023, from https://www.climate.gov/news-features/understanding-climate/climate-change-global-temperature

Global and European sea-level rise. European Environment Agency. (2021, May 11). Retrieved February 5, 2023, from https://www.eea.europa.eu/data-and-maps/indicators/sea-level-rise-6/assessment

Gomez-Zavaglia, A., Mejuto, J., & Simal-Gandara, J. (2020). Mitigation of emerging implications of climate change on food production systems. Food Research International, 134, 109256. https://doi.org/10.1016/j.foodres.2020.109256

Namany, S., Al-Ansari, T., & Govindan, R. (2019). Optimisation of the energy, water, and food nexus for food security scenarios. Computers & Chemical Engineering, 129, 106513. https://doi.org/10.1016/J.COMPCHEMENG.2019.106513

Level of Water Stress: Freshwater withdrawal as a proportion of available freshwater resources - sustainable development goals - united nations economic commission for Europe. (n.d.). Retrieved February 5, 2023, from https://w3.unece.org/SDG/en/Indicator?id=140

Ringsberg, H. (2014). Perspectives on food traceability: a systematic literature review. Supply Chain Management: An International Journal.

Roe, S., Streck, C., Obersteiner, M., Frank, S., Griscom, B., Drouet, L., ... & Lawrence, D. (2019). Contribution of the land sector to a 1.5 C world. Nature Climate Change, 9(11), 817-828.

The Economist Group. (2018). Global Food Security Index 2018: Build resilience in the face of rising food-security risks. The Economist Group.

Antonis Kokossis, Michael C. Georgiadis, Efstratios N. Pistikopoulos (Eds.)
PROCEEDINGS OF THE 33rd European Symposium on Computer Aided Process Engineering (ESCAPE33), June 18-21, 2023, Athens, Greece
 http://dx.doi.org/10.1016/B978-0-443-15274-0.50016-0

Lot-sizing and Production Scheduling of a Beverage Industry

Maria E. Samouilidou[a], Eleonora Diakoumi[a], Georgios P. Georgiadis[a], Antonios Dikaiakos[b], Michael C. Georgiadis[a]

[a]*Department of Chemical Engineering, Aristotle University of Thessaloniki, Thessaloniki 54124, Greece*
[b]*Green Cola Hellas, National Road Athens – Lamia & Kalamatas 2, New Kifisia Athens, 14564, Greece*
mgeorg@auth.gr

Abstract

In this work, the lot-sizing and production scheduling problem of a real-life beverage industry is addressed. In particular, a Big bucket-Small bucket Mixed-Integer Linear Programming (MILP) framework is proposed for the optimal scheduling of a Greek soft-drink production facility. The production process consists of two main stages, a batch syrup preparation stage and a continuous bottling stage. The synchronization between the syrup tanks and filling lines is of key importance for the feasibility and efficiency of the production schedules. The main goal is the generation of an optimal weekly production schedule which satisfies a given demand. A rigorous scheduling formulation for the lot-sizing and production scheduling of an industrial facility including both batch and continuous processes is presented. The derived optimal schedules lead to an increased plant productivity while reducing the utilization and cost of labor resources.

Keywords: production scheduling, optimization, beverage industries, MILP

1. Introduction

Nowadays, process industries operate in a challenging economic environment with low profit margins, where the demands they have to meet significantly increase and competition is rapidly intensifying. The ever-increasing number and diversity of products, as well as the complexity of production processes, especially in the food and beverage industry, render production scheduling an essential procedure for acquiring efficient production plans and ensuring the plant's viability and profitability. The main challenge regarding these industries is the integrated modelling of all the production stages (Toledo et al., 2015). In the beverage industry, a few scheduling techniques have been proposed in the literature, with even fewer addressing complex real-life industrial applications (Georgiadis et al., 2021). Baldo et al. (2014) proposed an MILP model, combined with MIP-based heuristics for the production lot-sizing and scheduling problem of a brewery industry with long lead times and unfixed bottleneck. Mediouni et al. (2022) studied the lot-sizing and scheduling problem in a dairy soft-drink production process, considering the limited shelf life of intermediate products and utilizing a relax-and-fix heuristic. Ferreira et al. (2009) presented an MILP model that integrates production lot-sizing and scheduling decisions of beverage plants with sequence-dependent setup costs and times and applied it to a real life instance generating improved results. The efficient lot-sizing and scheduling problem of a real-life soft-drink production facility is examined in this work, including the integrated modelling of all production stages. The proposed mathematical framework tackles the existing production restrictions and provides optimal results regarding the productivity and the production costs of the plant.

2. Problem Statement

The plant includes both batch and continuous processes. More than 60 final products are produced over a 5-day time horizon. A daily 8-hour shift is used, while overnight and weekend shifts can take place if needed. It is noted that more than one product can be produced from the same syrup. A brief layout of the production process is shown in Figure 1.

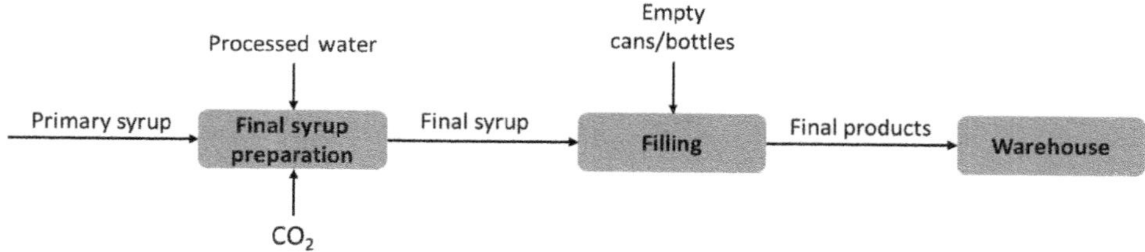

Figure 1. Production process layout

According to the production recipe, initially, a primary flavored syrup is prepared. There are two tanks available for the preparation of syrups. At the same time, water is treated and deaerated so as to be suitable for mixing with the primary syrup. Inline mixing takes place between the primary syrup, the treated water and CO_2 and the final mixture reaches the filling lines. In the meanwhile, empty cans/bottles are also transferred to the appropriate packing lines, where they are rinsed. There is a total of three packing lines in the facility, one for cans, one for PET bottles and one for glass bottles, but according to the company's policy only two of them can operate per day. Then, the cans/bottles are filled with the final mixture and before they finally form a multipack, labels and other messages are printed on them. Additionally, a brief quality check happens at this point. Finally, the final products are transferred to the warehouse, where they are stored until they are later distributed according to customer demand. It can be considered that the production process consists of two main stages, a batch syrup preparation stage and a continuous packing stage. The production of soft-drinks requires the synchronization between these two stages, in order to secure feasible schedules and to minimize overtime costs. Furthermore, time consuming changeovers take place in both stages depending on factors such as flavor, packing size and label. Taking into account all of these limitations while satisfying a given demand is quite challenging in real life industrial environments. Therefore, the production scheduling of a facility presenting the aforementioned characteristics is examined in this work.

3. Mathematical framework

This model constitutes an extension of the mathematical frameworks previously developed by Ferreira et al. (2009) and Mediouni et al. (2022). It is a Big bucket-Small bucket Mixed-Integer Linear Programming (MILP) framework extended in such a way that up to three products can be produced in the same Small Bucket (micro-period) from a single syrup and additionally the option of overtime production is considered. The planning horizon is divided in T macro-periods with the duration of a day and each macro-period is further divided into S micro-periods with flexible lengths. In each micro-period only one syrup can be prepared in each tank. A brief description of the mathematical model follows below.

1st stage constraints

$$\sum_{l\in LU_{l,u}} Yl_{u,l,s} \leq 1 \quad \forall\, u,s \tag{1}$$

$$\sum_{m} Yum_{u,m,l,s} = Yl_{u,l,s} \quad \forall\, u,s,l \in LU_{l,u} \tag{2}$$

$$\sum_{l,u\in UL_{u,l}} Yum_{u,m,l,s} \leq 1 \quad \forall\, m,s \tag{3}$$

$$Zl_{u,k,l,s} \geq Yl_{u,k,s-1} + Yl_{u,l,s} - 1 \quad \forall\, u,t,s\in St_{s,t}, k \in LU_{k,u}, l \in LU_{l,u}, k \neq l \tag{4}$$

$$\sum_{u\in UL_{u,l}} q_{l,u} \cdot Yum_{u,m,l,s} \leq \sum_{j\in JML_{j,l}\cap JM_{j,m}} r_{j,l} \cdot x_{m,j,s} \leq \sum_{u\in UL_{u,l}} Ku_u \cdot Yum_{u,m,l,s} \quad \forall\, m,s,l \tag{5}$$

2nd stage constraints

$$In_{j,t-1} + \sum_{m\,\in\, MJ_{m,j}, s\,\in\, St_{s,t}} x_{m,j,s} = In_{j,t} + d_{j,t} \quad \forall j,t \tag{6}$$

$$\sum_{s\,\in\, St_{s,t}} Ts_{m,s} \leq w_{m,t} + Ov_t \quad \forall\, m,t \tag{7}$$

$$Ts_{m,s} \geq \sum_{j\in JM_{j,m}} a_{m,j} \cdot x_{m,j,s} + \sum_{i\in JM_{i,m}, j\in JM_{j,m}, i\neq j} bb_{i,j} \cdot Zij_{m,i,j,s} + \sum_{u,k\,\in\, LU_{k,u}, l\,\in\, LU_{l,u}, k\neq l} b_{k,l} \cdot Zum_{u,m,k,l,s} + tr \cdot \sum_{u,l\,\in\, LU_{l,u}} Yum_{u,m,l,s} \quad \forall\, m,s \tag{8}$$

$$Ts_{m,s} \geq \sum_{j\in JM_{j,m}} a_{m,j} \cdot x_{m,j,s} + \sum_{i\in JM_{i,m}, j\in JM_{j,m}, i\neq j} bb_{i,j} \cdot Zij_{m,i,j,s} + \sum_{i\in JM_{i,m}, j\in JM_{j,m}, i\neq j} bb_{i,j} \cdot Zj_{m,i,j,s} \quad \forall\, m,s \tag{9}$$

$$\sum_{i\in JM_{i,m}, i\neq j} Zij_{m,i,j,s} + \sum_{ii\in JM_{ii,m}, ii\neq j} Zij_{m,j,ii,s} + 1 \geq Yj_{m,j,s} + Yj_{m,jj,s}, \quad \forall\, m, j\in JM_{j,m}, jj\in JM_{jj,m}, j \neq jj, s \tag{10}$$

$$\sum_{i\in JM_{i,m}, j\in JM_{j,m}} Zij_{m,i,j,s} + V_{m,s} = \sum_{j\in JM_{j,m}} Yj_{m,j,s} \quad \forall\, m,s \tag{11}$$

$$G \leq Zj_{m,i,j,s} \leq 0.5 + \frac{G}{2} \quad \forall\, m, i\in JM_{i,m}, j\in JM_{j,m}, i \neq j, t, s \;\;, \textit{ where} \tag{12}$$

$$G = Yj_{m,i,s-1} + Yj_{m,j,s} - 1 - \sum_{jj\in JM_{jj,m}, i\neq jj,} Zij_{m,i,jj,s-1} - \sum_{ii\in JM_{ii,m}, j\neq ii,} Zij_{m,ii,j,s} \leq Zj_{m,i,j,s}$$

Objective function

$$\min \sum_{m,t,s\,\in\, St_{s,t}} Ts_{m,s} \tag{13}$$

Constraint set (1) imposes that in each micro-period s, each tank u can prepare at most one syrup l, using the binary allocation variable $Yl_{u,l,s}$. Constraints (2) and (3) express that if a syrup l is prepared in tank u in micro-period s, then it must be transferred to a single packing line m and this packing line cannot receive another syrup from another tank. The binary variable $Yum_{u,m,l,s}$ indicates that the tank u prepares syrup l in micro-period s for the line m. Constraints (4) define the changeover of syrups in a tank using the immediate precedence binary variable $Zl_{u,k,l,s}$. With constraints (5) it is ensured that the amount of final products produced from the same syrup in a micro-period s (where $r_{j,l}$ is the required amount of syrup l to produce one unit of product j and $x_{m,j,s}$ represents the produced units of product j on packing line m in micro-period s), is bounded by the minimum quantity of filling $q_{l,u}$ and the maximum capacity Ku_u of the tank. Constraints (6) demonstrate that the inventory of product j in macro-period t, $In_{j,t}$, equals the inventory of the previous macro-period plus the produced amount in the current macro-period, minus the customer demand $d_{j,t}$. Constraint set (7) guarantees that the total production time in macro-period t, $\sum_{s \in St_{s,t}} Ts_{m,s}$, does not exceed the daily production time horizon $w_{m,t}$ including any possible overtime shifts Ov_t . Constraints (8) and (9) establish that the functioning time of line m in micro-period s, $Ts_{m,s}$, must be equal to or greater than the packing time (with $a_{m,j}$ being the required production time of one unit of product j on line m), plus the changeover time between products from the same syrup $bb_{i,j}$, plus the line's waiting time. The line's waiting time in a micro-period is either equal to the time needed for the tank's changeover $b_{k,l}$ and preparation of syrup tr or to the line's changeover $bb_{i,j}$ regarding the previous micro-period, depending on which one lasts longer. Constraints (10) and (11) monitor the changeover between products in the same micro-period $Zij_{m,i,j,s}$ while constraints (12) define the changeover between products in consecutive micro-periods $Zj_{m,i,j,s}$. The binary variable $Yj_{m,j,s}$ specifies whether product j is being packed in line m in micro-period s and the binary variable $V_{m,s}$ is activated if line m functions at micro-period s. The objective of the model, described by constraint (13), is the minimization of the total production time.

4. Results & Discussion

An industrial case study concerning the production plant of Green Cola Hellas is demonstrated. The MILP model was implemented in GAMS and solved using CPLEX 12.0. Optimality was reached in acceptable computational times by the company in all cases, as shown in Table 1. Figures 2 and 3 illustrate the generated schedules for two given weekly demands. According to the company's policy two of the three available packing lines operate per week. Figure 4 depicts the plant's weekly schedule if all three packing lines were to operate in a week. Moreover, Figure 5 shows the percentage of functioning time each machine devotes to different kinds of operations in this case. It is noted that in the following Gantt charts only the syrup preparation and changeover time are illustrated with colors for the tanks, whereas the time dedicated to feeding the packing lines is not. The different syrup flavors prepared are annotated as RX while the final products are annotated as PX.

Table 1. Solution report for the satisfaction of three weekly product demands

	CPU (s)	GAP (%)	Overtime (hrs)	Weekly functioning time (hrs)				
				Tank 1	Tank 2	Can line	PET line	Glass line
Demand 1	28	0	16.3	40.3	38.1	30.6	-	23.8
Demand 2	900	0.4	12.6	36.1	47.0	25.9	21.3	-
Demand 3	900	0.3	16.3	46.4	38.0	31.7	12.8	23.8

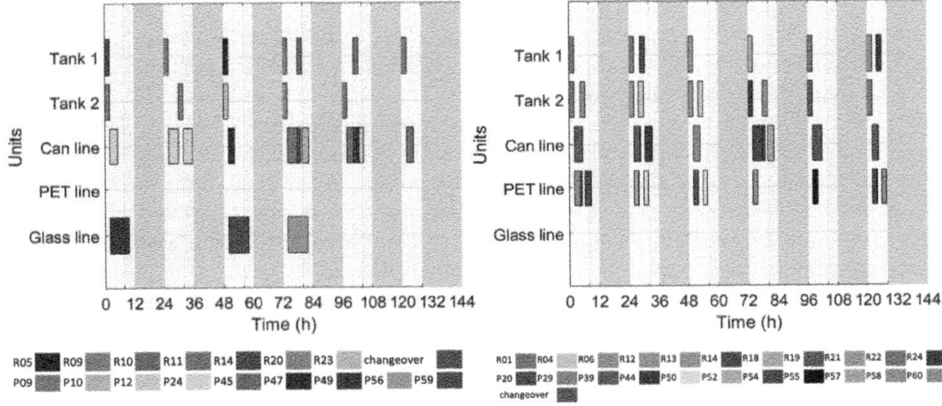

Figure 2. Weekly schedule for Demand 1

Figure 3. Weekly schedule for Demand 2

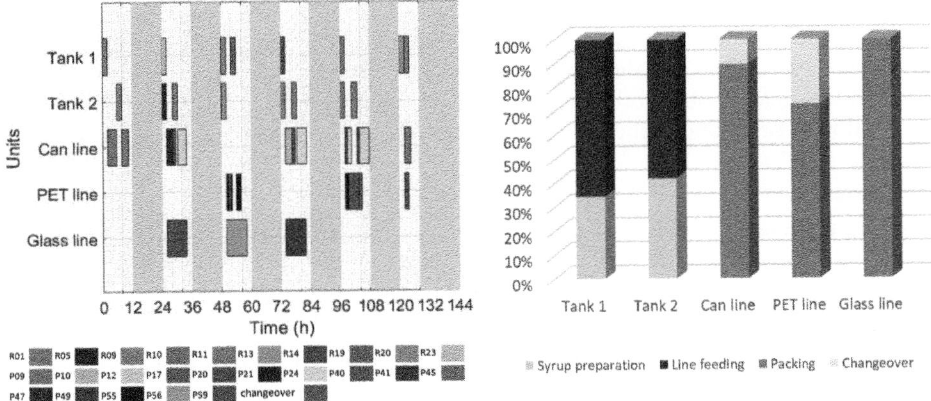

Figure 4. Weekly operation of all three lines

Figure 5. Functioning time allocation for each machine (Demand 3)

In all cases, overtime is required to satisfy the given demand. In Figures 2 and 4, it is observed that multiple products are produced from a single syrup, a flexibility given by the extended mathematical model presented in this work. Furthermore, the correct synchronization of the various units is achieved. The syrup tanks are identified as the production's bottleneck since they are utilized during the whole production horizon, either preparing syrups or feeding the packing lines. Syrup preparation requires only a little time, nevertheless a tank cannot start preparing a new syrup unless it is empty. In fact, significantly more time is dedicated to line feeding compared to the time dedicated to syrup preparation, as shown in Figure 5. The operation of all three available packing lines in a week gives flexibility to the production while larger product demands can be satisfied. However, this leads to an increased inventory.

5. Conclusions

The lot-sizing and production scheduling problem of a real-life beverage industry is presented in this work. The production facility under consideration can be identified as multiproduct and multistage, with each stage including both batch and continuous processes, making the problem particularly complex. Furthermore, the synchronization between these stages is important for the feasibility and efficiency of the production schedules, an issue often met in food and beverage industries. For the optimal scheduling of the facility, an MILP framework is proposed. To the best of our knowledge this study is the first to allow multiple products to be produced in the same Small Bucket from a single syrup, thus significantly improving the facility's productivity. This work's novelty is further extended by the inclusion of overtime production if necessary, for the timely satisfaction of the given weekly demand. Optimal production schedules are generated in short computational times leading to an increased plant productivity and reduced cost of labor resources. Moreover, the syrup tanks are identified as the production's bottleneck. Future work focuses on the short-term storage of the intermediate syrups in the preparation tanks while taking into account their shelf life and also on the introduction of buffer tanks between the preparation stage and the packing lines so that the syrup tanks can function at their full capacity.

Acknowledgements

This research is co-financed by the European Union and Greek National Funds through the Region of Central Macedonia, under the operational program "Region of Central Macedonia 2014-2020" and the specific action/call "Investment Plans on Innovation". (Project code: KMP6-0077560). Project title: Development of a software tool for the optimization of production scheduling in the manufacturing industries.

References

D. Ferreira, R. Morabito & S. Rangel, 2009, Solution approaches for the soft drink integrated production lot sizing and scheduling problem, European Journal of Operational Research, 196(2), 697-706

G. P. Georgiadis, A. P. Elekidis & M. C. Georgiadis, 2021, Optimal production planning and scheduling in breweries, Food and Bioproducts Processing, 125, 204-221

A. Mediouni, N. Zufferey, M. Rached & N. Cheikhrouhou, 2022, The multi-period multi-level capacitated lot-sizing and scheduling problem in the dairy soft-drink industry, Supply Chain Forum: An International Journal, 23(3), 272 - 284

C. F. Toledo, A. Kimms, P. M. França & R. Morabito, 2015, The synchronized and integrated two-level lot sizing and scheduling problem: evaluating the generalized mathematical model, Mathematical Problems in Engineering, 2015

T. A. Baldo, M. O. Santos, B. Almada-Lobo & R. Morabito, 2014, An optimization approach for the lot sizing and scheduling problem in the brewery industry, Computers & Industrial Engineering, 72, 58-71

Antonis Kokossis, Michael C. Georgiadis, Efstratios N. Pistikopoulos (Eds.)
PROCEEDINGS OF THE 33rd European Symposium on Computer Aided Process Engineering (ESCAPE33), June 18-21, 2023, Athens, Greece
 http://dx.doi.org/10.1016/B978-0-443-15274-0.50017-2

Modeling and Analysis of Clinical & Municipal Waste Incineration Process using Aspen Plus

Mohamad Rizza Othman,[a] Sivanesh Kumar Anpalagam,[a] Nur Fitriyanni Jafary,[a]

[a]*Dep. of Chem. Eng. Tech., Fac. of Chem. & Process Eng. Tech., Universiti Malaysia Pahang, Lebuhraya Tun Razak, 26300 Gambang, Pahang, Malaysia*

Abstract

Modeling and analyzing the solid waste incineration process is a salient study for future advanced waste disposal technology and optimizing process efficiency. However, the development of a mechanistic model that describes the process deal with several challenges and has yet to be extensively explored. Thus, this paper aims to develop a mechanistic model for clinical and municipal solid waste incineration in Aspen Plus. The waste was modeled using customized non-conventional (NC) solid properties, and a general coal enthalpy model with heat correlations to calculate the heating value. The model results accurately depict the waste heating values found in the literature. Further analysis reveals the ability of the model to describe the effect of waste feed quality on incinerator performance.

Keywords: Clinical and municipal waste, incineration, Aspen Plus.

1. Introduction

The generation of municipal solid waste (MSW) has been increasing globally owing, primarily due to an increase in global population, urbanization, and economic growth, as well as changes in production and consumption behavior (Lohri et al., 2017). On the other hand, the amount of clinical waste (CW) has increased over the past two years due to a high number of COVID-19 infections. Studies found that the weight of CW has increased by 27% due to higher cases of COVID-19 being recorded. In 2013 about 50 metric tons of CW was generated; however, for now, it is estimated that about 90 metric tons of CW are generated each day, and about 25 metric tons of it is due to COVID-19 (Agamuthu & Barasarathi, 2021). Various solid waste management treatments exist, including conventional sanitary landfill treatment, biological treatment, and thermal treatment. Predominantly, landfill is the primary method of waste disposal in most countries around the world. However, such a common method is ineffective in managing the increasing volume of solid waste apart from its drawback towards land, air, ground, and surface water pollution. Hence, this combustible waste may be incinerated or combusted as an alternative to landfill.

The advantages of waste incineration over landfilling are the reduction in waste volume, waste stabilization, waste-to-energy (WtE) conversion, and sanitization of waste, especially clinical waste. Besides, it also runs stably for a long period. In the incinerator, waste is burned at 850 to 1000°C, where CO2, SO2, NO, NO2, HCl, HF, CO, dioxins, and furans are released during incineration. The government law has mandated that waste incineration facilities use gas purification technologies to reduce volume and hazardous characteristics and capture or destroy potentially harmful substances that may be released during incineration. As a result, atmospheric gas emission limits are typically set to minimize health and environmental impacts. According to Williams (2005), the combustion properties and emissions produced by the combustion system will be influenced by the waste's composition and characteristics. The solid waste's fuel properties and the proximate and ultimate analyses play an important role in how the waste will burn in the incinerator and the emissions that are likely to result. The composition and heating value were the most important parameters governing thermal processing (Zhou et al., 2014). Heating values measure the chemical energy bound in a feedstock where this energy is released during combustion (Erol et al., 2010) and categorized into two, namely higher heating value (HHV) and lower heating value (LHV). The former involves the heat of condensation of water vapor produced in the heat of combustion, while the latter is related to the HHV via the heat of vaporization of water. In some instances, the HHV and LHV are associated with the heat of vaporization of both HCl and water if the waste contains chlorine and produces HCl as one of the products.

There are two methods for experimentally determining the heating values: a full-scale boiler as a calorimeter and a laboratory bomb calorimeter. Heating values can also be determined by calculation using empirical models, and several models have been developed by researchers (Ogwueleka, 2010). To this day, the most popular method for predicting heating value is linear regression. Using regression analysis to develop an empirical model to predict the heat value is thought to be easier and more cost-effective (Khuriati et al., 2017). A model that describes the oxidation, pyrolysis, and incineration process is useful to provide insights into the incineration performance and operations. However, the development of such a model deal with several challenges and has yet to be extensively explored. Therefore, the focus of this study was to develop a working model by employing Aspen Plus simulation software. The establishment of such a model is useful to describe the solid waste incineration process, determine the factors impacting its performance, and predict incineration performance under various operating conditions.

2. Methodology

2.1. Calculation of heating value

The solid waste was modeled in Aspen Plus as a non-conventional (NC) solid. The NC component models for enthalpy and density properties calculations were empirical correlations that require solid material characterization information derived from constituent ultimate and proximate analyses. The estimates were based on the general coal enthalpy model, including several correlations to compute the heat of combustion, heat of formation, and heat capacity. In this work, the heat of combustion in dry, mineral-matter-free basis, $\Delta_c h_i^{dm}$, was calculated using the Boie correlation according to the following equation,

$$\Delta_c h_i^{dm} = \left(a_{1i} w_{C,i}^{dm} + a_{2i} w_{H,i}^{dm} + a_{3i} w_{St,i}^{dm} + a_{4i} w_{O,i}^{dm} + a_{5i} w_{N,i}^{dm}\right) 10^2 + a_{6i} \tag{1}$$

Where w is the weight fraction and the constant terms, a_{1i} to a_{6i} were obtained from the Institute of Gas (IGT) study. Products were in the form of ash, liquid water, and gaseous CO2, SO2, and NO2. The heat of formation, $\Delta_f h_i^d$ was based on the standard heat of combustion-based correlation. The numerical coefficients were combinations of stoichiometric coefficients and heat of formation for CO2, H2O, HCl, and NO2 at 298.15K represented by,

$$\Delta_f h_i^d = \Delta_c h_i^d - (1.418 \times 10^6 w_{H,i}^d + 3.278 \times 10^5 w_{C,i}^d + 9.264 \times 10^4 w_{S,i}^d - 2.418 \times 10^4 w_{N,i}^d - 1.426 \times 10^4 w_{Cl,i}^d)10^2 \quad (2)$$

Whereas the heat capacity calculations were based on Kirov Correlations (Kirov, 1965), which treats the heat capacity as a weighted sum of the heat capacities of the constituents, which include moisture, fixed carbon, volatile matter, and ash.

2.2. Aspen Plus flowsheet for CW incineration

Modeling and simulation for solid waste incineration were performed in Aspen Plus V12.1. The solid waste was fed constantly, and the combustion chamber's process was adiabatic. Gaseous equilibrium is achieved between the primary and secondary chambers. Heat transfer occurs within the chamber via the mixing of hot gases as well as radiation. The flowsheet comprised two reactor blocks, RYIELD and RSTOIC. The former was used to convert non-conventional components into conventional components and the heat of formation required, while the latter was used to simulate the combustion process. In addition, two calculator blocks were added to convert the ultimate analysis element into a wet basis component for the RYIELD reactor block, while the second calculator block was used to calculate the heating value. The IDEAL base method was used as the property method. The thermophysical properties (enthalpy, density, and specific heat) of solid waste were calculated using the built-in coal property models HCOALGEN and DCOALIGT based on their proximate and ultimate analysis. For calculating heating values, all unit operations use 25°C and 1 bar as the reference state. Excess air with 78% nitrogen and 21% oxygen was used as the oxidizing agent.

2.3. Case study

Table 1 shows the proximate and ultimate analysis and its heating values of several municipal and clinical samples found in the literature. Each sample's proximate and ultimate values were inserted into the model to calculate the heating values. The calculated results were then compared. For all cases, the waste fed rate was assumed to be 500 kg/h. To ensure excess air between a stoichiometric ratio of 1.5 to 2.0, the combustion air fed rate used was 7000 kg/h.

Table 1: Proximate and ultimate analysis of municipal and clinical waste samples

	CW (Erdogan & Yilmazoglu, 2021)			MSW (Gebreslassie et al., 2020)		
	Sample 1	Sample 2	Sample 3	Sample 4	Sample 5	Sample 6
Proximate						
Moisture (%)	0.32	0.29	7.04	0.24	5.76	2.99
Ash (%)	0	2.3	1.89	0.12	6.23	13.73
Volatiles (%)	99.13	78.52	82.37	94.44	77.81	75.23
Carbon (%)	0.55	18.89	8.7	5.2	10.2	8.05
Ultimate						
C (%)	81.81	65.92	48.98	63.91	39.32	36.51
H (%)	12.17	10.03	7.2	5.331	7.2	5.31
O (%)	5.76	23.09	43.52	30.714	53.48	58.18
N (%)	0.15	0.74	0.22	0	0	0
S (%)	0.11	0.22	0.08	0.045	0	0
HV (kJ/kg)	42650[a]	29120[a]	15570[a]	23193[b]	16217[b]	12174[b]

[a] LHV, [b] HHV

2.4. Statistical analysis

To determine the accuracy of the proposed model, mean absolute percentage error (MAPE) was used to denote the closeness of the calculated heating value to the actual value. MAPE is mainly used to determine model accuracy and is calculated using the following equation,

$$MAPE = \left(\frac{1}{n}\sum_{i=1}^{n}\frac{|E_i - P_i|}{E_i}\right) \times 100\% \quad (3)$$

where n is the sample number, E_i is the actual value, and P_i is the calculated value. Based on Lewis (1982), a MAPE value of less than 10% gives a highly accurate prediction.

2.5. Model analysis

Analysis of the model was performed to illustrate the effect of process parameters on the incineration performance of CW and MSW. Two parameters were studied for each type: moisture and VM content. Moisture content, in CW, for example, could consist of up to 38% (Li & Jeng, 1993), and for some biomedical compositions, VM values were from 30% (Olanrewaju, 2019). These two values were varied, and the corresponding normalized proximate and ultimate analyses were then used in the model to determine the heating value and combustion temperature. Furthermore, some waste, i.e., CW, must be incinerated to around 1000 C due to regulatory requirements. Therefore, additional heat duty was calculated to determine the combustion heat duty required to achieve the desired decomposition temperature. Typically, the extra heat duty was supplied by auxiliary fuel i.e., diesel and natural gas.

3. Result & Discussion

3.1. Model validation

In Table 2, the model calculated Sample 1 to have the highest heating values, whereas Sample 6 had the lowest. Moreover, each sample's descending pattern of calculated heating values was consistent with the experimental heating values. In general, carbon elements have a direct impact on heating value and the results were coherent with the carbon composition in Table 1. Table 2 also shows the percentage difference between the calculated value and the experimental data. The percent MAPE range for all samples were between 3-12%. On average, the percentage error was 7%. This implies that the simulation model estimated the heating value of the CW and MSW samples with an excellent range of accuracy.

Table 2: Heating values comparison and simulated combustion temperature results.

Category	Sample #	Experiment, kJ/kg	Simulation, kJ/kg	Simulated comb. temp, C	MAPE, |%|
Clinical waste	1	42650[a]	39042	1075	8
	2	29120[a]	29868	857	3
	3	15570[a]	17471	543	12
Municipal solid waste	4	23193[b]	24824	719	7
	5	16217[b]	16533	444	2
	6	12174[b]	13693	398	12
Average MAPE, %					7

[a] LHV, [b] HHV

3.2. Model analysis

The results for variation in moisture and VM content of CW and MSW feed to heating value, and combustion temperature is shown in Figure 1. Figure 1a shows that when the moisture content is increased, the heating value decreases. These findings are intuitive and agree with Liang et al. (2008). Higher moisture content typically leads to lower calorific value as various moisture content levels in mass burning affect combustion performance. Furthermore, the higher moisture content also reduces reaction temperature and results in poor ignition (Johari et al., 2012). This, in return, lowers the combustion temperature, as shown by the decreasing combustion temperature. On the other hand, the increase in VM had an inverse effect on heating values. As shown in Figure 1b, the heating values increase upon increasing VM content. High VM content contributes to easy fuel ignition, where a lower minimum ignition temperature is needed. The impact, however was less significant compared to moisture content. Nevertheless, the result ties well with theoretical data wherein VM significantly impacts thermal decomposition.

Specifically for CW, an additional heat duty is required to increase the temperature to around 1000 C for complete disintegration. Figure 2 shows the additional heat duty required with variation in moisture content. This is typically achieved through burning of fuel i.e., diesel and natural gas. Obviously, a larger temperature difference needs more heat or fuel. Therefore, proper handling, segregation, sorting, waste volume, and size and feed time of the waste are detrimental to reducing the interference with reaction product combustion, increasing combustion quality, and consequently decreasing the fuel cost.

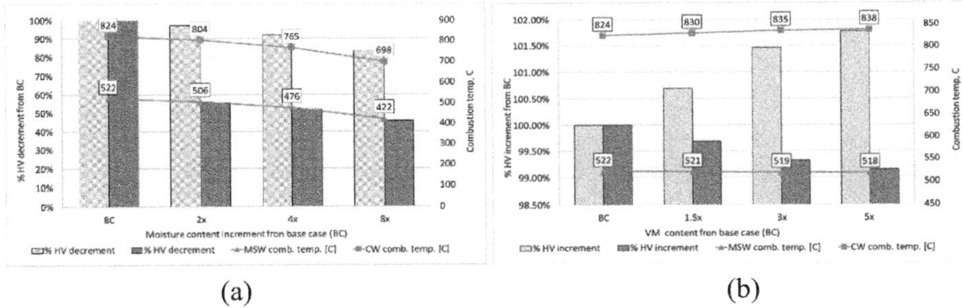

(a) (b)

Figure 1: Effect on LHV and combustion temperature of CW and MSW for variation in (a) moisture and (b) VM content.

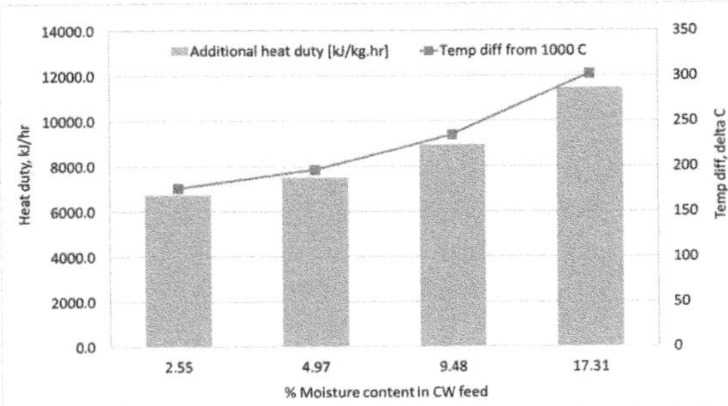

Figure 2: Additional heat required and temperature difference to achieve the CW decomposition temperature of 1000 C.

4. Conclusion

The objective of this study to develop a mechanistic model of the CW and MSW incineration process was achieved. Using customized NC solid properties, flowsheet, and a general coal enthalpy model with heat correlations in Aspen Plus, the model demonstrates excellent agreement with experimental data with a mean absolute percentage error of 7%. Model analysis of waste feed content agrees well with results found in the literature. The model shows that lower moisture and higher VM content contribute to higher heating values. However, the effect of moisture content is much more significant than VM content. The model was also extended to predict combustion temperature and the additional heat required to achieve decomposition temperature. The model reveals that waste feed quality is detrimental to the incinerator's performance. In future work, we will expand the model to other applications, such as optimization, air pollution control, and waste-to-energy systems.

5. Acknowledgements

The authors wish to acknowledge assistance and support from Universiti Malaysia Pahang through grant number PDU213003-1.

References

A. Johari, H. Hashim, R. Mat, H. Alias, M.H. Hassim, M. Rozainee, 2012, Generalization, formulation and heat contents of simulated MSW with high moisture content, Journal of Engineering Science and Technology, 7(6), pp. 701–710

A.A. Erdogan, M.Z. Yilmazoglu, 2021, Plasma gasification of the medical waste, International Journal of Hydrogen Energy, 46(57), pp. 29108-25

C.D. Lewis, 1982, Industrial and business forecasting methods: A practical guide to exponential smoothing and curve fitting', London; Boston: Butterworth Scientific

C.R. Lohri, S. Diener, I. Zabaleta, A. Mertenat, C. Zurbrügg, 2017, Treatment technologies for urban solid biowaste to create value products: a review with focus on low-and middle-income settings, Reviews in Environmental Science and Biotechnology, 16(1), pp.81-130

C.S. Li, F.T. Jeng, 1993. Physical and chemical composition of hospital waste. Infection Control & Hospital Epidemiology, 14(3), pp.145-150.

H. Zhou, A.H. Meng, Y.Q. Long, Q.H. Li, Y.G. Zhang, 2014, An Overview of Characteristics of Municipal Solid Waste Fuel in China. Physical, Chemical Composition and Heating Value. Renewable And Sustainable Energy Reviews, 36, pp. 107-122

M.G. Gebreslassie, H.B. Gebreyesus, M.T. Gebretsadik, S.T. Bahta, S.E. Birkie, 2020, Characterization of Municipal Solid waste's Potential for Power Generation at Mekelle City as a Waste Minimisation strategy, Int. Journal of Sustainable Engineering, 13(1), pp. 68-75

N.Y. Kirov, 1965, Specific Heats and Total Heat Contents of Coals and Related Materials are Elevated Temperatures, BCURA Monthly Bulletin, pp. 29-33

O. Olanrewaju, 2019. Quantification and characterization of medical waste in public health care facilities within Akure Metropolis, Ondo State, Nigeria. EPH-International Journal of Agriculture and Environmental Research, 55, pp.15-30.

P. Agamuthu, J. Barasarathi, 2021, Clinical waste management under COVID-19 scenario in Malaysia', Waste Management & Research, 39(1_suppl), pp. 18-26

P.T. Williams, 2005, Waste treatment and disposal, John Wiley & Sons

R.S. Liang, Jun Fei, Shaohua Wu, Xiang Liu, Kui Dai, Na Yao, 2008, Experimental study on effects of moisture content on combustion characteristics of simulated municipal solid wastes in a fixed bed. , 99(15), pp. 7238–7246

T.Ch. Ogwueleka, F.N. Ogwueleka, 2010, Modelling Energy Content Of Municipal Solid Waste Using Artificial Neural Network, Iran. J. Environ. Health. Sci. Eng., 7(3), pp. 259-266

Antonis Kokossis, Michael C. Georgiadis, Efstratios N. Pistikopoulos (Eds.)
PROCEEDINGS OF THE 33rd European Symposium on Computer Aided Process Engineering (ESCAPE33), June 18-21, 2023, Athens, Greece
 http://dx.doi.org/10.1016/B978-0-443-15274-0.50018-4

On the minimization of cycle time in periodic production scheduling

Alexandros Koulouris,[a] Georgios P. Georgiadis,[b]

[a]*International Hellenic University, P.O. Box 141, Sindos 57400, Greece*
[b]*Intelligen Europe, S. Kazantzidi 47, Thermi 57001,Greece*

Abstract

Periodic scheduling (commonly implemented in batch processing) is characterized by the process cycle time. Minimizing the cycle time maximizes throughput. This paper proposes two approaches for the calculation of the minimum cycle time for batch processes with resource sharing among tasks: an algorithmic and an MILP-based. The first approach implements an exact, exhaustive algorithm which also allows the calculation of all feasible cycle time ranges. The second approach is optimization-based, yields only the minimum cycle time but is more flexible than the algorithmic in the type of problems it can handle. Both approaches are demonstrated with illustrative examples.

Keywords: periodic scheduling, cycle time minimization, MILP.

1. Introduction

Cyclic, and more specifically, periodic scheduling is preferred by many batch industries because of the simplified shop floor control, the easily implementable schedules and easier shift decisions for the staff. Furthermore, studies show that cyclic schedules tend to be more robust compared to non-cyclic ones. Despite those facts, cyclic scheduling has not attracted much attention from the PSE research community.

Periodic scheduling is characterized by the process cycle time, i.e. the constant time interval between the start (or end) of two consecutive batches. The cycle time is directly related to throughput; minimizing the cycle time is essential in the operation of a plant and in capacity studies. Many researchers proposed exact algorithmic polynomial solvable methods for cyclic scheduling problems, where tasks do not share resources (Hanen, 1994). However, the methods available in the literature cannot optimize cases with sharable resources. Few researchers developed novel mathematical optimization approaches for cyclic scheduling (Pinto and Grossmann, 1994; Wu and Maravelias, 2019). In most of these formulations, each task can only be processed in exactly one unit throughout the cyclic schedule. Moreover, they do not address the cycle time minimization problem.

The objective of this paper is to present solutions to the minimization of cycle time in the presence of tasks sharing multiple resources. Two approaches are presented: an algorithmic and an MILP-based. In the algorithmic case, the non-overlapping constraint between tasks on the same resource yields bounds on the feasible cycle time between any pair of tasks under the assumption that all available resources are used periodically by all tasks. The developed MILP model is based on the general precedence framework (Kopanos et al., 2011) and takes into account the cyclic nature of batch scheduling through the introduction of a new set of constraints. In comparison to the solutions found in literature, the model relaxes the requirement of using a single resource for each task throughout the schedule allowing for a better utilization of the available resources. The proposed model is relatively general and can be easily extended to consider any specific operational constraint, thus providing a basis for tackling realistic large-scale problems. Two case studies are presented, demonstrating the applicability of the proposed solutions in real-life industrial scenarios and highlighting their differences. The case studies have been modeled and scheduled within the scheduling software SchedulePro (by Intelligen, Inc., Scotch Plains, NJ, USA, www.intelligen.com) where the algorithms have also been implemented. The MILP models have been solved using the HiGHS open-source solver (Huangfu and Hall, 2018).

2. Periodic Scheduling Set-up

Let $T = \{1,2, \dots n\}$ be a set of n tasks corresponding to a batch that is to be executed indefinitely often. Let $p_i > 0$ be the duration of task i. The batches must be executed periodically with a fixed cycle time, α. Let $t(i,k)$ denote the start of the execution of task i in batch k and t_i denote the start of task i for the (randomly selected) reference batch 0. Then,

$$t(i,k) = t_i + \alpha k \tag{1}$$

Within each iteration, the tasks are to be executed at a specific order with fixed time intervals between their starts. Let $D_{i,j}$ $\forall i,j \in T$ be the fixed time distance between the start of task j and the start of task i within the same batch. Then,

$$\begin{gathered} D_{ij} = t(j,k) - t(i,k) \Rightarrow D_{ij} = t_j - t_i \\ D_{ji} = -D_{ij}, \quad D_{ii} = 0 \end{gathered} \tag{2}$$

It is assumed that all tasks within T are ordered in ascending start times. Each task uses a pool of available resources to be used for its execution. Different tasks may use the same resources, therefore, the resource pools may overlap (partially or fully) between tasks. Feasible cycle times are those that do not create overlaps of tasks (in the same or different batches) on the same resource. Under this setting, a unique schedule can be generated with the assignment of the cycle time, a, and the allocation of resources for each task in every batch.

3. Algorithmic Approach

To obtain the minimum and feasible cycle time ranges in an algorithmic way, it is assumed that each task uses exclusively its own resource pool, or, if tasks are to share resources, the resource pools are identical. It is further assumed that if common resources are available, they are used in a rotation mode. As a result, all tasks will eventually "meet" on each resource but it is not known a priori to what batch these tasks will belong. Under these assumptions, a periodic schedule is completely determined by setting the value of the cycle time and the resource used for every task at the first batch. To find the minimum cycle time, it is possible to split the task set into subsets composed of either a single task (using its own resources) or of a group of tasks sharing a common resource pool. If a task uses exclusively one or multiple resources, then its feasible cycle time range is simply $[p_i/m_i, \infty)$ where m_i is the number of resources available for task i. The analysis below will deal with tasks sharing a pool of multiple resources.

Due to the periodicity and the cyclic rotation between all resources, each task i is executed on the same resource every m batches (where m is the number of common resources). The indices of repetitions of the same task i on the same resource can therefore be represented by $mk_i + s_i$, where $k_i, s_i \in Z$ and s_i $(0 \le s_i < m)$ is an integer representing the index of the batch (with respect to reference batch 0) for the first occurrence of task i in a specific resource. Note that s_i is not the same for all resources. The non-overlap constraints between two tasks i, j on a given resource can be written as:

$$\begin{aligned} t(i, mk_i + s_i) + p_i \le t(j, mk_j + s_j) \quad &or, \quad t(j, mk_j + s_j) + p_j \le t(i, mk_i + s_i) \\ t_i + \alpha mk_i + as_i + p_i \le t_j + \alpha mk_j + as_j \quad &or, \quad t_j + \alpha mk_j + as_j + p_j \le t_i + \alpha mk_i + as_i \\ t_j - t_i \ge p_i - \alpha(mK_{ij} + S_{ij}) \quad &or, \quad t_j - t_i \le -p_j - \alpha(mK_{ij} + S_{ij}) \\ D_{ij} \ge p_i - \alpha(mK_{ij} + S_{ij}) \quad &or, \quad D_{ij} \le -p_j - \alpha(mK_{ij} + S_{ij}) \end{aligned} \tag{3}$$

where $K_{ij} = k_j - k_i,\ and\ S_{ij} = s_j - s_i,\ K_{ij}, S_{ij} \in Z, -m < S_{ij} < m$. Note that because of the cyclical rotation over all resources, and despite the fact that s_i and s_j are resource-dependent, their difference S_{ij} is not.

For a feasible schedule to exist, for every pair of tasks i, j executed on the same resource with given value of S_{ij}, there must exist an integer value of K_{ij} so that D_{ij} belongs to one of the allowed intervals, namely $[p_i - a(mK_{ij} + S_{ij}), -p_j - a(m(K_{ij} - 1) + S_{ij})]$. In other words, the non-overlapping constraints can be re-written as:

$$p_i - \alpha(mK_{ij} + S_{ij}) \le D_{ij} \le -p_j - a(m(K_{ij} - 1) + S_{ij}) \tag{4}$$

or, if the two inequalities are separated,

$$\begin{aligned} &D_{ij} \ge p_i - \alpha(mK_{ij} + S_{ij}) \\ D_{ij} \le -p_j - \alpha(m(K_{ij} - 1) + S_{ij}) \Rightarrow \quad &D_{ji} \ge p_j - \alpha(m(1 - K_{ij}) - S_{ij}) \end{aligned} \tag{5}$$

These two constraints are, in fact, symmetrical provided that in the second inequality we use: $K_{ji} = 1 - K_{ij}\ and\ S_{ji} = -S_{ij}$. The feasible values of the cycle time, α, should therefore satisfy the following constraints:

$$\begin{aligned} D_{ij} \ge p_i - \alpha(mK_{ij} + S_{ij}) \quad \forall i, j \in T \\ K_{ij} + K_{ji} = 1 \\ S_{ji} = -S_{ij} \end{aligned} \tag{6}$$

Depending on the sign of the term $p_i - D_{ij}$, Eq. (6) provides upper or lower bounds on a for different values of K_{ij}, S_{ij}. After a few algebraic steps (outlined in Koulouris and Georgiadis, 2022), these bounds can be calculated as follows:

if $p_i - D_{ij} > 0$ then,

$$a \ge \frac{p_i + p_j}{m}$$

$$a \ge \max\left\{\frac{p_j + D_{ij}}{m - S_{ij}}, \frac{p_i - D_{ij}}{S_{ij}}\right\} \text{ for } S_{ij} \ne 0$$

if $p_i - D_{ij} = 0$ then, (7)

$$a \geq \frac{p_j + D_{ij}}{m - S_{ij}}$$

if $p_i - D_{ij} < 0$ then,

$$\frac{p_j + D_{ij}}{m(1 - K_{ij}) - S_{ij}} \leq a \leq \frac{p_i - D_{ij}}{mK_{ij} + S_{ij}} \text{ for } \frac{p_i - D_{ij}}{p_i + p_j} - \frac{S_{ij}}{m} \leq K_{ij} < -\frac{S_{ij}}{m} \text{ and } K_{ij} \neq 0$$

$$a \geq \frac{p_j + D_{ij}}{m - S_{ij}} \quad \text{for} \quad K_{ij} = 0$$

As seen in Eq. (7), the values of K_{ij} are either fixed or bounded on both sides and that means that the search of feasible ranges is also bounded. Using different values of K_{ij}, S_{ij} within their admissible ranges, bounds on cycle time are calculated. The union of all these feasible ranges provides the values of the cycle time that, for a given pair of tasks, guarantee repetitive execution of these tasks with no overlaps. This calculation process must be repeated for all task pairs; the intersection of all pair-wise feasible ranges will result in the feasible cycle time ranges for all tasks sharing common resources. The lower bound of these ranges is the minimum process cycle time.

4. MILP Approach

While computationally efficient, the previously described exact method is based on a heuristic of utilizing the resources in a cyclical manner that could affect the output resulting to very good but potentially suboptimal solutions. Therefore, an MILP-based approach, that can ensure the optimality of the generated solutions and can tackle problems with non-identical resource pools, is also developed. The model is based on the known general precedence framework, which has been extended to address periodic scheduling optimization problems. Given a specific number of batches B of tasks T to be processed in the available resources R, the optimization model generates a complete optimal schedule with minimum cycle time. A set of binary allocation variables ($Y_{i,b,r}$) and binary sequencing variables ($X_{i,b,i',b'}$) are employed. The first are enabled when a batch b of task i is processed by resource r, while the latter when a batch b of task i is processed prior to batch b' of task i' using the same resource. The proposed model consists of constraints (8)-(15) and the objective function (16), which are described below.

$$\sum_{r \in IR_{i,r}} Y_{i,b,r} = 1 \quad \forall i \in T, b \in IN_b \tag{8}$$

$$L_{i,b} + \sum_{r \in IR_{i,r}} (Y_{i,b,r} \cdot p_i) = C_{i,b} \qquad \forall i \in T, b \in B \tag{9}$$

$$L_{i,b} = t_i + (b-1) \cdot \alpha \qquad \forall i \in T, b \in B \tag{10}$$

$$L_{i',b'} \geq C_{i,b} - M \cdot (1 - X_{i,b,i',b'}) - M \cdot (2 - Y_{i,b,r} - Y_{i',b',r}) \ \forall i, i' \in T, i < i', b, b' \in B, r \in (IR_{i,r} \cap IR_{i',r}) \tag{11}$$

$$L_{i,b} \geq C_{i',b'} - M \cdot X_{i,b,i',b'} - M \cdot (2 - Y_{i,b,r} - Y_{i',b',r}) \ \forall i, i' \in T, i < i', b, b' \in B, r \in (IR_{i,r} \cap IR_{i',r}) \tag{12}$$

$$L_{i,b'} \geq C_{i,b} - M \cdot (2 - Y_{i,b,r} - Y_{i,b',r}) \ \forall i \in T, b, b' \in IN_b, b < b', r \in IR_{i,r} \tag{13}$$

$$Y_{i,b,r} = Y_{i,b+NB,r} \quad \forall i \in R, b \leq |B| - NB, r \in IR_{i,r} \tag{14}$$

$$L_{i,b+NB} \geq C_{i,b} \quad \forall i \in T, b \leq |B| - NB \tag{15}$$

$$Min\ \alpha \tag{16}$$

Constraint set (8) ensures that each batch b of a task i will be processed by exactly one resource r. Notice that in order to reduce the model's size, we only consider the batches in the set IN_b. This set denotes the batches for which we optimize the allocation decisions. Let us assume that we want to create an optimal schedule for a set of batches $B=\{b0, b1, \ldots, b10\}$ and decide that $IN_b=\{b0, b1\}$, then the allocation of the tasks in the available resources is repeated every two batches. For example, if task $i1$ is allocated to resource $r2$ and $r3$ in batches $b0$ and $b1$, then the allocation for the upcoming batches will be $b2 \rightarrow r2$, $b3 \rightarrow r3$ etc. The size of set IN_b is a user-defined parameter depending on the needs and wants of the production engineer using the model. A larger size means potentially lower cycle times but requires more CPU time. Next, constraint set (9) guarantees the proper connection between the starting and completion times of every batch of each task. In particular, the completion time $C_{i,b}$ must equal the starting time $L_{i,b}$ plus the required processing time p_i. Constraints (10) are essential since they define the problem as a cyclic scheduling problem. More specifically, we define that the starting time of a batch b of task i must be equal to the start time of the first batch t_i, plus the cycle time times the batch number minus 1. For example, let us assume a task with starting time 5 hours, if the optimal cycle time is calculated to be 2 hours, then the starting time of the first batch will be 5 hours, of the second 7 hours, the third 9 hours and so on. Constraints (11) and (12) are the precedence constraints that are necessary in order to respect the disjunctive constraints (any resource can only process one task at a time). Constraints (11) and (12) are complementary. If a batch b of a task i is processed prior to a batch b' of task i' ($X_{i,b,i',b'}=1$) and both of them are processed on the same resource ($Y_{i,b,r}=Y_{i',b',r}=1$), then the batch b' of task i' must start after the completion of batch b of task i. M is a large number (big-M parameter), which is necessary in order to relax the constraints, in case the previously mentioned conditions are not valid. Constraint set (13) states that a batch b' of task i must start after the completion of a batch $b<b'$ of the same task in case both are processed by the same resource. Again, to reduce the model's size we only consider this constraint for the subset IN_b. The next two constraints ensure that the generated schedule will be a cyclic one. NB is the number of batches in the subset IN_b. Constraints (14) guarantee that a batch $b+NB$ of a task i will be processed in the same resource with batch b, since the schedule must be cyclic. Similarly, for the same couple of batches, batch $b+NB$ must start after the completion of batch b. Goal is the minimization of the cycle time α.

5. Examples

To demonstrate the use of the above approaches, two examples are presented here. The first example refers to a process producing a small molecule API. It consists of 11 tasks: 4 tasks sharing 3 reactors, 5 tasks sharing 2 tanks and two tasks using exclusive equipment (1 filter and 3 dryers, respectively). Details about the start time and duration of these tasks can be extracted from Intelligen, Inc. (2022) where this process is presented as a SchedulePro case study. As an extension to this example, it was assumed that 2 transfer panels are used as auxiliary equipment for all material transfers (11 in total). Their duration and start times in hours are $\{4, 1, 1, 1, 1.5, 1, 1, 1, 1, 8, 2\}$ and $\{0, 11, 15, 16, 17, 19.5, 20.5, 24, 25, 43, 51\}$, respectively. Both the algorithmic and the MILP approaches yield instantaneously the same minimum cycle time of 17.67 hours. The bottleneck resources are the transfer panels. A Gantt chart showing the bottleneck equipment in a sequence of 15 batches scheduled at the minimum cycle time is shown in Figure 1. The algorithmic approach yields also all feasible cycle time ranges which are: $\{[17.67, 18.5], [19.5, 20], [27, 27.75], 29.25, [31.5, 32.5], [33.5, \infty)\}$.

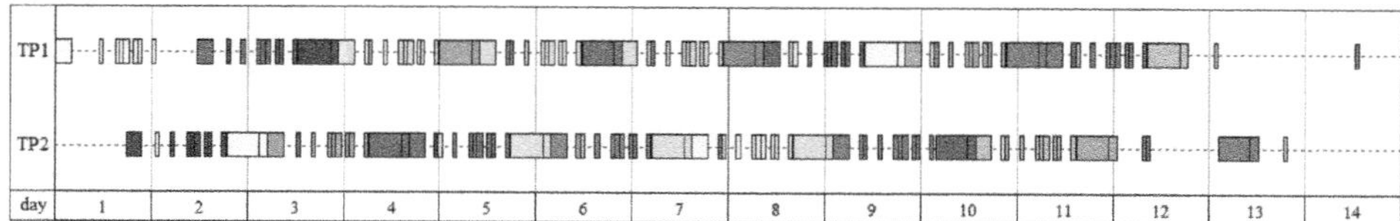

Figure 1. Gantt chart of 15 batches for the small molecule API case study.

Knowledge of the feasible ranges is useful if the plant wants to operate at specific cycle time which is not necessarily the minimum.

The second case is an example process where the algorithmic approach yields a suboptimal solution as a result of the resource rotation assumption. The process has 3 tasks with durations and start times $\{6, 2, 11\}$ and $\{0, 2, 5\}$ hours, respectively. It is assumed that 3 equipment resources (Eq1-3) are available. In this case, the algorithmic approach yields a minimum cycle time of 8 hours while the optimal value calculated by the MILP approach (IN_b=3) is 7 hours. As shown in Figure 2, unlike the algorithmic approach, the optimal policy assigns an exclusive resource (Eq1) to the first task.

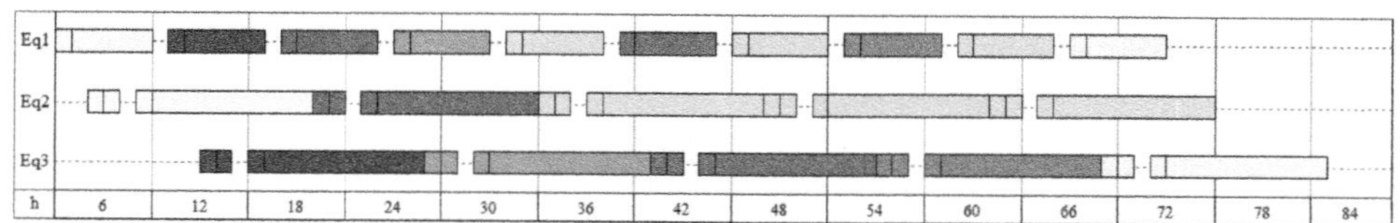

Figure 2. Gantt chart of 15 batches for the example process.

6. Conclusions

Two approaches are presented in this paper for the calculation of the minimum cycle time in periodic batch process scheduling with resource sharing. The algorithmic approach is an exact, exhaustive method that yields the minimum cycle time and all feasible cycle time ranges for an infinite scheduling horizon under the assumption that all common resources are used in rotation. The MILP approach is more general since it makes no a priori assumption on resource assignments and generates optimal solutions for predefined scheduling horizons in short computational times.

References

Hanen, C., 1994. Study of a NP-hard cyclic scheduling problem: The recurrent job-shop. *European journal of operational research*, *72*(1), pp.82-101.

Huangfu, Q. and Hall, J.J., 2018. Parallelizing the dual revised simplex method. *Mathematical Programming Computation*, *10*(1), pp.119-142.

Intelligen Inc. (2022) https://www.intelligen.com/products/schedulepro-examples/ (access November 2022)

Kopanos G., Puigjaner L., Georgiadis M.C., 2011. Production Scheduling in Multiproduct Multistage Semicontinuous Food Processes. *Industrial & Engineering Chemistry Research*, *50* (10), pp. 6316-6324.

Koulouris, A., Georgiadis, G., 2022. An exact algorithm for calculating the minimum and feasible ranges of cycle time in periodic scheduling with shared resources, *submitted to Computers and Chemical Engineering*.

Pinto, J.M. and Grossmann, I.E., 1994. Optimal cyclic scheduling of multistage continuous multiproduct plants. *Computers & Chemical Engineering*, *18*(9), pp.797-816.

Wu, Y. and Maravelias, C.T., 2019. A general model for periodic chemical production scheduling. *Industrial & Engineering Chemistry Research*, *59*(6), pp.2505-2515.

Antonis Kokossis, Michael C. Georgiadis, Efstratios N. Pistikopoulos (Eds.)
PROCEEDINGS OF THE 33rd European Symposium on Computer Aided Process Engineering (ESCAPE33), June 18-21, 2023, Athens, Greece
 http://dx.doi.org/10.1016/B978-0-443-15274-0.50019-6

Non-Pareto optimal solutions as enablers for versatile heat exchanger networks

David Huber[a], Felix Birkelbach[a], René Hofmann[a]

[a]*Institute for Energy Systems and Thermodynamics*, TU Wien, Getreidemarkt 9/BA, Wien, 1060, Austria
david.huber@tuwien.ac.at

Abstract

Assessing a heat exchanger network (HEN) based on total annual costs (TAC) alone is no longer appropriate. Climate-damaging emissions must also be considered. Thus, we use CO_2 emissions and TAC as objectives in the fully linearized heat exchanger network synthesis (HENS). Due to uncertainties concerning emissions and equipment costs, a refurbishment of the HEN may become necessary even before its life cycle has been reached. Since a significant share of the system costs is related to the piping, it is economical to change only the heat exchanger (HEX) areas. To assess the potential for the refurbishment of a given HEN in the future, we constrain the Pareto optimal HEN configurations and recompute sub-Pareto fronts. Our results show that there are obviously superior configurations regarding distance and coverage of the sub-Pareto front concerning the initial Pareto front. The hypervolume indicator (HVI) and the coverage were used as quantitative indicators to evaluate different configurations. However, these characteristic parameters do not correlate with empirically determined high-quality configurations. This contribution has laid an essential foundation towards developing versatile and optimal HEN.

Keywords: MILP, multi-objective optimization, non-dominated solutions, locked HEN configuration

1. Introduction

To save energy and costs, industrial energy systems can connect hot and cold process streams using a HEN. The optimal interconnection of the HEN can be calculated using HENS. If we look at various literature on HENS, the aim is almost entirely to achieve a cost minimum. The single-objective approach cannot consider environmental impacts. To produce a holistic set of choices for decision-makers, multi-objective optimization (MOO) with CO_2 emissions and costs is the way to go.

Many optimal HEN can provide a comprehensive basis for decision-making. However, determining which HEN to choose is a significant challenge for decision-makers. The selected HEN should meet the current conditions and be reasonable in case of unforeseen developments. For example, political or economic changes such as increased CO_2 certificate costs or energy costs can adversely influence the selected HEN's performance. Since such changes can never be fully considered, it is more effective to present decision-makers with fewer but more robust solutions.

According to Peters et al. (2003), the piping costs in fluid processing plants can be up to 80 % of equipment costs. This corresponds to about 25 % of the total investment costs. The costs associated with adjusting the operating point of the HEN can be reduced significantly when keeping the piping configuration as it is. With an unchanged HEN, the operating point, respectively the load on the utilities can be influenced by varying the HEX areas.

In this paper, we use HENS to propose a method to assess the versatility of Pareto-optimal configurations and to identify the ones that can be refurbished economically in the future if necessary. To determine the versatility of a configuration, we compute sub-Pareto fronts from the Pareto optimal configurations. Investigation of the Pareto front has shown that only a few unique configurations of HEN exist as non-dominated solutions. Sub-Pareto fronts can be generated using those unique configurations, which differ only in the utility load and HEX areas. The results show that the sub-Pareto fronts cover large areas of the initial Pareto front while containing a variety of only weakly dominated solutions.

2. Methods

The HENS formulation used in this paper is based on the stage-wise superstructure formulation according to Yee and Grossmann (1990). With this formulation, heat exchange can occur in counterflow between streams in k stages with stream splits. The hot and cold utilities are located at the ends of the streams, respectively.

2.1. Objectives

The first of the two objectives describes the TAC and was adopted from Yee and Grossmann (1990). The second objective represents the CO_2 emissions due to the operation and manufacturing of the HEN, according to Pintarič et al. (2019). This method evaluates CO_2 emissions based on hot utility consumption and economic activity level according to the economic input-output life cycle assessment (EIO-LCA). In contrast to Pintarič et al. (2019), we also consider emissions from the circulation pumps power consumption.

2.2. Linearization

To apply highly efficient mixed integer linear programming (MILP) solvers, all non-linear terms (reduced stream HEX area, reduced utility HEX area and logarithmic mean temperature difference (LMTD)) are linearized. The non-linear reduced stream HEX area is tightened to the feasible solution space according to Beck and Hofmann (2018) and approximated with super positioned planes. The convex approximation allows the efficient transfer to MILP without using binary variables. Analogously, the reduced utility HEX area is tightened. In contrast to Beck and Hofmann (2018), the non-linear relation is approximated with several straight lines. The function of LMTD is approximated using plane triangles on a regular grid. The transformation to MILP is done using a highly efficient logarithmic coding of the binary variables introduced by Vielma and Nemhauser (2011). With this method, T Simplices require only $\lceil log_2(T) \rceil$, instead of T binary variables.

2.3. Optimization Approach

The developed method can be divided into the following main steps:

1. Calculate Pareto optimal solutions
2. Constrain incidence matrix in Pareto optimal HEN configuration
3. Calculate Pareto optimal solutions of subproblems

The Pareto front is calculated using the adaptive weighted sum method from Kim and de Weck (2005). Compared to the weighted sum method, the Pareto front can also be solved in non-convex regions and with well-distributed solutions. In the remainder of this paper, we will refer to this Pareto front as an initial Pareto front.

In the second step, the unique configurations of the initial Pareto front are identified. From the resulting set of unique configurations, new subproblems are created. The incidence matrix representing the HEN configuration is blocked with constraints. Therefore, the degree of freedom and thus, the complexity can be reduced significantly, leading to short computation time.

In the final step, a new Pareto front is calculated for each subproblem with the adaptive weighted sum method. Note that the Pareto optimal solutions of the subproblems are not optimal regarding the initial problem. The sub-Pareto fronts always contain at least one non-dominated Pareto optimal solution concerning the initial problem.

3. Use Case

The method presented in the previous section was applied to a representative use case with four hot and five cold process streams, according to Linnhoff and Flower (1978). The stream data, HEX, and utility costs are summarized in Table 1. The CO_2 emissions factor for hot utilities ω_{HU} and investment costs ω_{INV} were taken from Pintarič et al. (2019). The emission factor for the cold utilities is calculated assuming that 10 % of the cooling load must be supplied as electrical energy. With 8000 h full load hours and an average electricity mix, ω_{CU} results in $0.0016\,{}^{\mathrm{t}}/_{\mathrm{kWy}}$. The minimum allowed approach temperature was set to $\Delta T_{\min} = 20°\mathrm{C}$ and heat transfer is possible in two stages.

Table 1: Case-study: stream and cost data according to Linnhoff and Flower (1978).

Stream	T_{in} / °C	T_{out} / °C	F / kW/K	H / kW/m²K
H1	327	40	100	0.50
H2	220	160	160	0.40
H3	220	60	60	0.14
H4	160	45	400	0.30
C1	100	300	100	0.35
C2	35	164	70	0.70
C3	85	138	350	0.50
C4	60	170	60	0.14
C5	140	300	200	0.60
HU	330	250	-	0.50
CU	15	30	-	0.50

HEX data: $c_{\mathrm{f}} = 2000\,{}^{\$}/_{\mathrm{y}}$, $c_{\mathrm{v}} = 70\,{}^{\$}/_{\mathrm{m}^{2\beta}\mathrm{y}}$, $\beta = 1$, $\omega_{INV} = 399 \cdot 10^{-6}\,{}^{\mathrm{t}}/_{\$\mathrm{y}}$

Utility data: $c_{\mathrm{HU}} = 60\,{}^{\$}/_{\mathrm{kWy}}$, $c_{\mathrm{CU}} = 6\,{}^{\$}/_{\mathrm{kWy}}$

CO_2 emissions: $\omega_{\mathrm{HU}} = 1.436\,{}^{\mathrm{t}}/_{\mathrm{kWy}}$, $\omega_{\mathrm{CU}} = 0.0016\,{}^{\mathrm{t}}/_{\mathrm{kWy}}$

3.1. Optimization Framework

The optimization problem has been modeled using *Yalmip* R20210331 Lofberg (2004) in *MATLAB* R2022a. The optimization problem was solved with the state-of-the-art solver *Gurobi* 9.5.2 on a 128-core system (AMD EPYC 7702P) with 128 GB RAM.

3.2. Linearization Settings

The reduced stream HEX areas were approximated with as many planes as necessary to obtain a root mean square error (RMSE) of less than 1.0 %. In most cases, this was achieved with four planes. The LMTD was approximated with 32 plane triangles resulting in an RMSE of 1.0 %. The reduced HEX areas of the utilities were approximated with two straight lines resulting in an RMSE below 2.0 %.

3.3. Pareto Front Settings

The problem was normalized between the nadir and utopia point to calculate the Pareto front. The solutions were estimated to have a minimum relative distance of 2.5 %. Below that, solutions are considered overlapping and excluded. Within the adapted weighted sum method, iteration was performed until a maximum relative point distance of 5.0 % was met. If larger distances occur in the Pareto front, no solution exists.

The normalization of the problem with locked HEN configurations is done based on the respective nadir and utopia points. Each subproblem was thus normalized with respect to its nadir and utopia point, different from the initial solution.

4. Results

Figure 1 shows the calculated Pareto fronts of the use case with 27 different solutions. Out of 2^{49} possible HEN configurations, only 13 are identified as a solution to the problem. Identical configurations are shown in the same color.

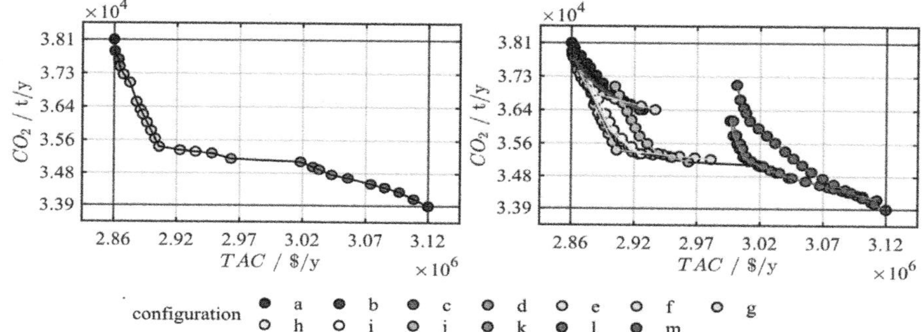

Figure 1: Left: initial Pareto front with 27 non-dominant solutions and 13 different HEN configurations. Right: initial Pareto front with 27 non-dominant, sub-Pareto fronts with 125 dominated solutions and 13 different HEN configurations.

The resulting 13 sub-Pareto fronts with locked HEN configuration are shown in Figure 1 (right). Obviously, some solutions are closer to the initial Pareto front and cover a broader range than others. The hypervolume indicator (HVI) according to Guerreiro et al. (2021) was calculated for each Pareto front to evaluate the quality of the different configurations. The reference point is always the nadir point of the initial Pareto front. The ratio of the HVI between a sub-Pareto front and the initial Pareto front can be interpreted as a dimensionless area ratio. The closer the HVI ratio is to 100 %, the closer the sub-Pareto front is to the initial Pareto front. The results are summarized in Table 2. Since the single-criterion evaluation of the different configurations considers versatility only to a limited extent, the coverage is also evaluated and listed in Table 2. The coverage quantifies the overlap of the sub-Pareto front concerning the initial Pareto front.

Since there is no direct correlation between coverage and HVI ratio, it is up to the decision-maker to find a trade-off between these two indicators.

Table 2: Characteristic values of subproblems with locked HEN configuration.

config.	# of solutions	HVI ratio	coverage annual CO_2 emissions	coverage TAC
/ -	/ -	/ %	/ %	/ %
a	7	32.42	9.90	23.95
b	9	52.23	22.49	34.91
c	8	52.64	21.12	31.14

d	2	33.46	3.67	8.16
e	7	52.51	23.90	21.39
f	8	85.60	16.92	40.99
g	12	88.57	37.13	32.21
h	9	85.56	15.94	41.01
i	9	84.74	16.20	43.33
j	11	54.16	23.39	45.11
k	9	55.93	18.67	34.49
l	15	55.93	45.67	48.00
m	19	48.46	47.44	75.22

Configurations f-i (light green to orange in Figure 1) have a high HVI ratio of over 80 % but cover only small areas of the Pareto front. The peripheral areas are not covered at all. Configuration g offers the best trade-off between HVI ratio and coverage. However, configurations l and m (red and dark red in Figure 1) cover at least the peripheral area with low CO_2 emissions well. Due to the higher HVI ratio, configuration l is preferable. The second configuration (blue in Figure 1) covers a substantial area in the low TAC region. Based on this empirical analysis, the obviously relevant solutions can be reduced to the three configurations b, g and l. The stream plots of the three configurations can be found in the Appendix. As can be seen in Figure 2, these solutions cover almost the entire Pareto front. Remarkably, all solutions have large areas with only weakly dominated solutions.

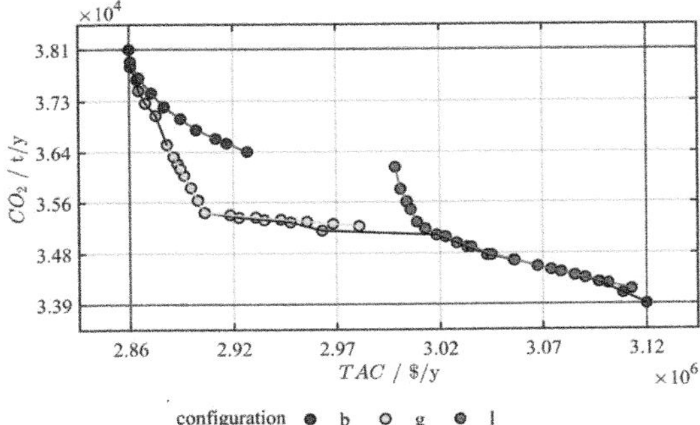

Figure 2: Pareto front with three non-dominant solutions and 33 weakly-dominated solutions of configuration b (blue), g (light green) and l (red).

5. Conclusion

In this paper, we used MOO to incorporate TAC and CO_2 emissions into the design optimization process of a HEN. Our results show that some sub-Pareto fronts cover the initial Pareto front to a large extent. Although each sub-Pareto front contains at least one Pareto optimal solution, some HEN configurations are obviously superior to others. We argue that a desirable configuration is one whose solutions are as close as possible to the initial Pareto front and at the same time provide high coverage and therefore versatility. In this paper, we attempted to quantify the trade-off between versatility and optimality of a given configuration. HVI and coverage were considered, but results were not satisfactory. The analysis of Pareto-optimal configurations, which we proposed in this paper, can be a useful tool to identify the most versatile HEN. More research on quantitative measures for the optimality of these configurations is still required.

6. Acknowledgment

This work was funded by the Austrian research promotion agency (FFG) under grant number 884340. The authors want to express their great acknowledgment.

References

M. S. Peters, K. D. Timmerhaus, R. E. West, 2003, *Plant Design and Economics for Chemical Engineers*, 5th ed. McGraw-Hill Chemical Engineering Series, New York: McGraw-Hill.

T. F. Yee, I. E. Grossmann, 1990, Simultaneous Optimization Models for Heat Integration - II. Heat Exchanger Network Synthesis, *Computers & Chemical Engineering*, 14 (10): 1165-84, https://doi.org/10.1016/0098-1354(90)85010-8.

A. Beck, R. Hofmann, 2018, A Novel Approach for Linearization of an MINLP Stage-Wise Superstructure Formulation, *Computers & Chemical Engineering*, 112 (April): 17-26, https://doi.org/10.1016/j.compchemeng.2018.01.010.

J. P. Vielma, G. L. Nemhauser, 2011, Modeling Disjunctive Constraints with a Logarithmic Number of Binary Variables and Constraints, *Mathematical Programming*, 128 (1): 49-72, https://doi.org/10.1007/s10107-009-0295-4.

I. Y. Kim, O. L. de Weck, 2005, Adaptive Weighted-Sum Method for Bi-Objective Optimization: Pareto Front Generation, *Structural and Multidisciplinary Optimization*, 29 (2): 149–58, https://doi.org/10.1007/s00158-004-0465-1.

B. Linnhoff, J. R. Flower, 1978, Synthesis of Heat Exchanger Networks: I. Systematic Generation of Energy Optimal Networks, *AIChE Journal*, 24 (4): 633-42, https://doi.org/10.1002/aic.690240411.

Z. N. Pintarič, P. S. Varbanov, J. J. Klemeš, Z. Kravanja, 2019, Multi-Objective Multi-Period Synthesis of Energy Efficient Processes under Variable Environmental Taxes, *Energy*, 189 (December): 116182, https://doi.org/10.1016/j.energy.2019.116182.

J. Lofberg, 2004, YALMIP: A toolbox for modeling and optimization in MATLAB, *IEEE International Conference on Robotics and Automation (IEEE Cat. No.04CH37508)*, 284-289, https://doi.org/10.1109/CACSD.2004.1393890.

A. P. Guerreiro, C. M. Fonseca, L. Paquete, 2021, The Hypervolume Indicator: Problems and Algorithms, *ACM Computing Surveys*, 54 (6): 1-42, https://doi.org/10.1145/3453474.

Appendix

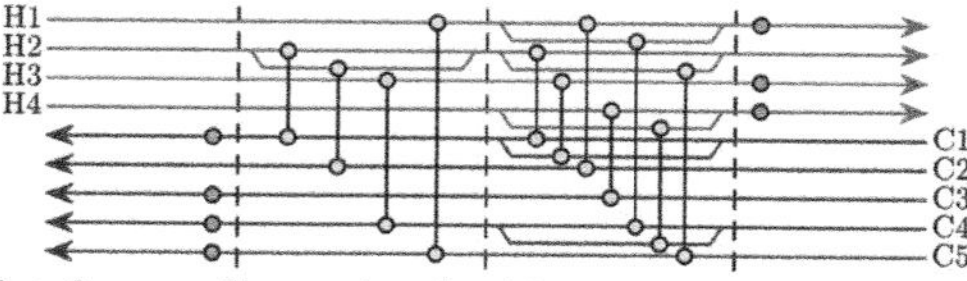

Figure 3: Stream plot for configuration b: 11 stream HEX, 4 hot and 3 cold utilities.

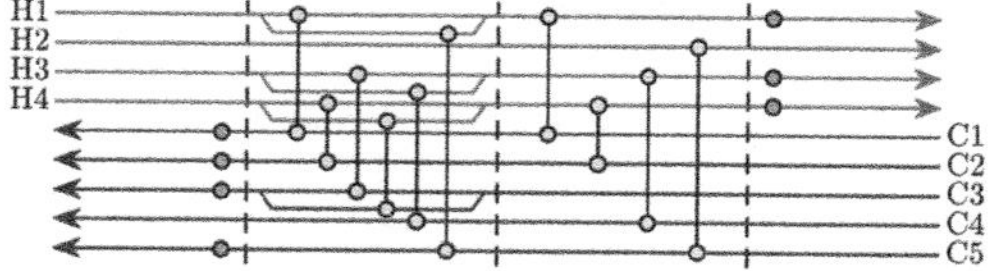

Figure 4: Stream plot for configuration g: 10 stream HEX, 4 hot and 3 cold utilities.

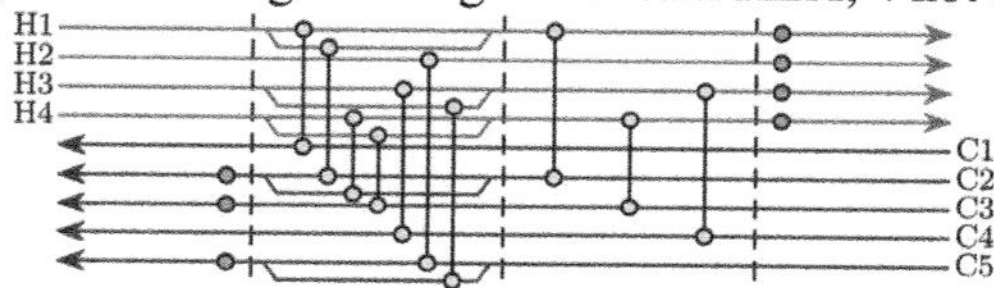

Figure 5: Stream plot for configuration l: 10 stream HEX, 3 hot and 4 cold utilities.

Antonis Kokossis, Michael C. Georgiadis, Efstratios N. Pistikopoulos (Eds.)
PROCEEDINGS OF THE 33rd European Symposium on Computer Aided Process Engineering (ESCAPE33), June 18-21, 2023, Athens, Greece
 http://dx.doi.org/10.1016/B978-0-443-15274-0.50020-2

Enhanced Automatic Initialization Method for Solving Large Nonlinear Algebraic Systems

Saskia Bublitz[a], Erik Esche*[a], Jens-Uwe Repke[a]

[a]*Technische Universität Berlin, Chair of Process Dynamics and Operation, Str. des 17. Juni 135, 10623 Berlin, Germany*
*erik.esche@tu-berlin.de

Abstract

New aspects of an interval-arithmetic (IA) based automatic initialization scheme for root-finding algorithms to efficiently solve nonlinear algebraic equation systems (NLEs) are presented. Using the model of a partial condenser, it is demonstrated how additional constraints may eliminate non-physical solutions by the help of interval arithmetic for initialization. Finally, the number of stages of a multicomponent distillation column is varied to investigate the required time of the approach to obtain initial values and solutions by root-finding as a function of the system size. The overall method finds a physically feasible solution for the distillation column with 20 stages without any initial values required.

Keywords: NLE, initialization, interval arithmetic, numerical iteration

1. Motivation

Chemical processes can become quite complex so that corresponding mathematical models are frequently hard to solve. Newton-based methods guarantee convergence in Lipschitz continuous areas, but fail, in case functions or derivatives are not defined at individual iteration points. In our study, we present further developments of an already introduced hybrid approach (Bublitz et al., 2021) that can automatically initialize nonlinear algebraic systems for root-finding algorithms using IA-based iteration methods. Its aim is to ease the challenging task of finding a converging initial point. The approach is designed to solve square problems of the form

$$\boldsymbol{f}(\boldsymbol{x}) = \boldsymbol{0} \ . \tag{1}$$

2. Notation

Throughout this work, multidimensional quantities are printed in bold. An interval $\underline{\overline{x}}$ is a convex set of real numerical values delimited by the lower bound $\underline{x}$ and the upper bound $\bar{x}$. Intervals of multiple variables span a convex variable space, which we term a box $\underline{\overline{\boldsymbol{x}}}$. The midpoint $m(\underline{\overline{x}})$, width $w(\underline{\overline{x}})$ and magnitude $|\underline{\overline{x}}|$ of an interval are defined as

$$m(\underline{\overline{x}}) := \frac{\underline{x} + \overline{x}}{2}; \quad w(\underline{\overline{x}}) := \bar{x} - \underline{x}; \quad |\underline{\overline{x}}| = \max\{|\underline{x}|;\ |\overline{x}|\} \tag{2}$$

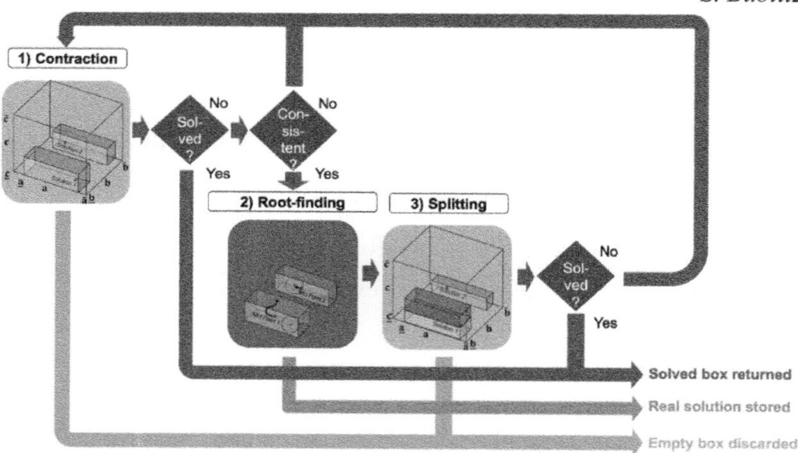

Figure 1. Three steps of the hybrid approach: 1) Contractions; 2)Root-finding; 3) Splitting

3. Hybrid Approach

The hybrid approach applied, consists of three steps, which are shown in Figure 1. As input, the user must provide initial intervals for all iteration variables. They span the initial box, which is reduced in the first step via IA-based contraction methods to one or more boxes (see section 3.1). A box $\underline{\overline{\boldsymbol{x}}}$ is termed solved when all its intervals $\underline{\overline{x}}$ satisfy the condition

$$\Delta_{Abs} + \Delta_{Rel} \cdot |\underline{\overline{x}}| \geq w(\underline{\overline{x}}) \quad . \tag{3}$$

The absolute and relative tolerances, Δ_{Abs} and Δ_{Rel}, must be specified. Unsolved boxes are reduced until they are consistent, i.e., their intervals no longer change from reduction step $k-1$ to k and both conditions,

$$\Delta_{Abs} + \Delta_{Rel} \cdot \max\{|\underline{x}^{(k-1)}|;\ |\underline{x}^{(k)}|\} \geq w(|\underline{x}^{(k)} - \underline{x}^{(k-1)}|), \tag{4}$$

$$\Delta_{Abs} + \Delta_{Rel} \cdot \max\left\{\left|\overline{x}^{(k-1)}\right|;\ \left|\overline{x}^{(k)}\right|\right\} \geq w\left(\left|\overline{x}^{(k)} - \overline{x}^{(k-1)}\right|\right) \tag{5}$$

hold for all of them. In the second step, a consistent, unresolved box is passed on to a Newton-based solver that starts from the center of the box to potentially find a root in its interior. A solution $\boldsymbol{x}^*$ to Eq. (1) in a set tolerance Δ_{FTOL} is found when

$$\|\boldsymbol{f}(\boldsymbol{x}^*)\|_2 \leq \Delta_{FTOL} \tag{6}$$

holds, and is immediately returned to the user. If only one numerical solution is of interest, the procedure can be configured so that it terminates after a successful numerical iteration. Otherwise, the consistent box is split in step 3 (see section 3.3), and then further processed in contraction.

3.1. Contraction

Contraction methods can reduce an initial variable space $\underline{\overline{\boldsymbol{x}}}$ based on the given constraints from Eq. (1). This can be generalized by the introduction of the contracting operator $\boldsymbol{\Gamma}(\boldsymbol{f}(\boldsymbol{x}), \underline{\overline{\boldsymbol{x}}})$ (Granvilliers, 2020):

$$\Gamma\big(f(x), \underline{\overline{x}}\big) \subseteq \underline{\overline{x}} \quad . \tag{7}$$

While removing inconsistent variable domains, contraction methods never lose any solution within the initial box. A general contraction step is defined as

$$\left\{ \underline{\overline{x}}^{(c+1)} \right\} := \underline{\overline{x}}^{(c)} \cap \Gamma\left(f(x), \underline{\overline{x}}^{(c)} \right) \quad . \tag{8}$$

Eq. (8) either breaks a box into one or more sub-boxes, verifies that the box cannot contain any solution, or proves that the box has a unique solution (Moore et al., 2009). For contraction, we use a combination of our own method Bnormal, presented in Bublitz et al. (2021), as well as the method HC4revise from Benhamou & Puget (1998). Both contracting operators have different requirements and capabilities. Details of which can be found in the respective publications. The latter method is particularly fast in reducing wide intervals. Our method on the other hand can find tighter enclosures for variable domains in equations with many instances of the same variable. For example, the equation

$$0 = x^3 - x^2 + x + y \tag{9}$$

can be used to reduce the initial interval $[-10;\ 10]$ of x to $[-1.37;\ 1.57]$ by HC4revise in 17 ms[i] through the given variable interval $[0;\ 10]$ of y. Bnormal needs 29 ms but the reduced interval of x is $[-0.55;\ 0]$. Besides, Bnormal can filter infeasible regions within variable intervals to pave the way for the upcoming root-finding step. However, to use the advantages of both methods, a combination of them is also tested, in which HC4revise is applied first due to its speed. It continues until a box becomes consistent and forwards the box subsequently to Bnormal. When Bnormal could further contract the box, it returns it to HC4revise, and the alternation continues until the box does not change anymore according to Eq. (4) and (5). This alternation is denoted as a box reduction step. One box reduction step usually encompasses many contraction steps to reach consistency.

3.2. Root-finding

The root-finding problem is reformulated to a minimization problem of the sum of the quadratic function residuals of Eq. (1)

$$\min_x \frac{0.5}{\Delta_{FTOL}} \cdot \sum f(x)^2 \tag{10}$$

and can be solved by any nonlinear optimization solver, e.g., ipopt (Wächter & Biegler, 2006). The reduced variable bounds from the contraction are directly considered here. To avoid an early termination of ipopt at a point with a low curvature of the objective, Eq. (10) is scaled by the hyperparameter Δ_{FTOL}. In addition, a Newton method is started on the root-finding problem at the local minimum in case it does not fulfill Eq. (6), to find the root of Eq (1) with the required tolerance.

3.3. Splitting

The applied splitting scheme bisects the variable interval, which has changed the least since the last split. If no split has occurred yet, it refers to the initial box. If there are several variable intervals that were reduced equally little, the interval is taken that causes

Table 1. Initial bounding of 36 iteration variables of the partial condenser.

Iteration Variables	Symbol	Initial Interval	Unit
Mole fractions	x_i; y_i	[0; 1]	-
Outlet flow rates	F^L; F^V	[0; 75]	kmol/h
Compressibility factors	Z^L; Z^V	[0; 10^2]	-
Cohesion pressure	a^L; a^V	[0; 10^1]	Pa
First derivative (EoS)	δ^L_{d1}; δ^V_{d1}	[0; 10^9]	-
Second derivative (EoS, Liquid)	δ^L_{d2}	[-10^9; 0]	-
Second derivative (EoS, Vapor)	δ^V_{d2}	[0; 10^9]	-
Other iteration variables		[-10^9; 10^9]	various

the greatest box reduction in a subsequent contraction. Ideally, one of the bisected boxes can be identified as empty and is directly discarded.

4. Computer Experiments

We apply the hybrid approach on two NLEs: a partial condenser and a distillation column. The hybrid procedure stops as soon as ipopt finds a solution. The tolerances of the box reduction Δ_{Abs} and Δ_{Rel} are set to 10^{-12} and 10^{-10}. These values ensure that no solution near an infeasibility is filtered accidentally. This could otherwise be the case due to low mole fractions in the order of 10^{-6} . For root-finding, Δ_{FTOL} is set to 10^{-8} to ensure that all state variables are sufficiently converged. A maximum of 10 box reduction steps and 50,000 iteration steps during root-finding are applied. Each test run is repeated three times and the averaged CPU time is presented in the results[i]. The success of the box reduction at reduction step k is measured by parameter ε:

$$\varepsilon^{(k)} := \sum_{b=1}^{n_{box}} \left(\sum_{v=1}^{n_{var}} \frac{w\left(\underline{\overline{x}}_v^{(k)}\right)}{w\left(\underline{\overline{x}}_v^{(0)}\right)} \cdot \frac{1}{n_{var}} \right) \cdot \frac{1}{n_{box}} \quad . \tag{11}$$

After reduction step k, the contraction methods might have split the initial variable space into n_{box} reduced boxes. A value of $\varepsilon = 1$ corresponds to no reduction of the initial box, while a value of $\varepsilon = 0$ equals tight enclosures of all real valued solution(s), which fulfill the required tolerances, i.e., all solutions have been found.

4.1. Partial Condenser

In this unit, a gas mixture of Toluene, Biphenyl, Benzene, Methane and Hydrogen is partially condensed to separate the low-boiling components Methane and Hydrogen at a given pressure and temperature of 35 bar and 298.15 K. The inlet flow rate is 75 kmol/h. Thermodynamic equilibrium between the ideally mixed liquid and vapor phases prevails in the unit. The chemical equilibrium according to the φ-φ approach as well as the enthalpies rates are calculated using the cubic equation of state of Soave-Redlich-Kwong (SRK) (Soave, 1972). The full model consists of 36 equations and can be found in MOSAICmodeling[ii]. Table 1 summarizes its coarse initialization.

For each phase, a dimensionless equation of state is implemented (Rao, 1997) to determine the compressibility factor Z

$$0 = Z^3 - Z^2 + (A - B - B^2) \cdot Z - A \cdot B \quad . \tag{12}$$

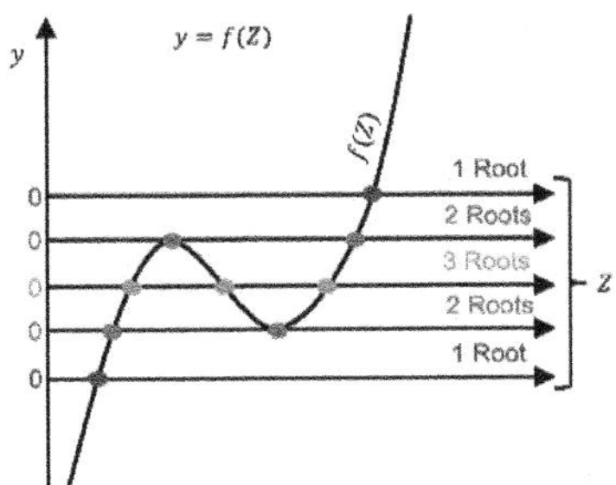

Figure 2. Number of real valued roots of a cubic equation $y = f(Z)$ depending on its distance to the origin.

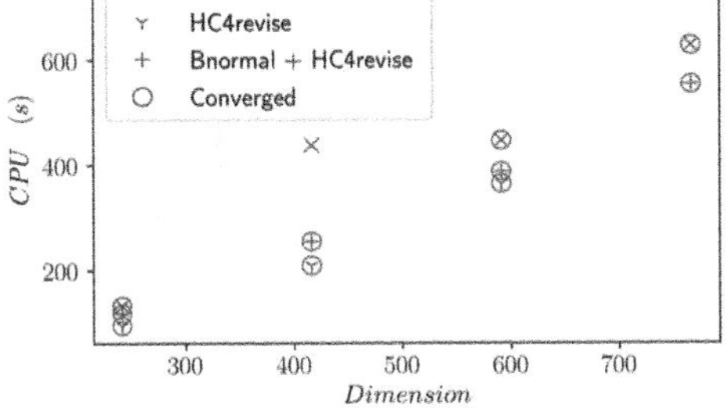

Figure 3. CPU time and convergence results after one box reduction step for different number of trays of the distillation column[i].

According to Figure 2, such a cubic equation has one up to three real roots for Z. Physically interesting, however, are only the smallest root for the liquid phase and the largest one for the vapor phase. A potential third solution in between can be filtered by requiring the first order derivative of Eq. (12) with respect to Z to be greater than or equal to zero

$$\delta_{d1} = 3 \cdot Z^2 - 2 \cdot Z + (A - B - B^2), \quad \delta_{d1} \geq 0 \ . \tag{13}$$

The compressibility factors of the liquid and the vapor phases differ in their second order derivatives. To obtain the compressibility factor of the liquid phase Z^L the condition

$$\delta_{d2}^L = 6 \cdot Z^L - 2, \quad \delta_{d2} \leq 0 \tag{14}$$

must be fulfilled, while for Z^V the condition

$$\delta_{d2}^V = 6 \cdot Z^V - 2, \quad \delta_{d2} \geq 0 \tag{15}$$

holds. To avoid interval dependency during contraction, Eq. (12) und Eq. (13) are reformulated to become

$$0 = Z^3 - Z^2 + (A - (B - 0.5)^2 + 0.25) \cdot Z - A \cdot B \ , \tag{16}$$

$$\delta_{d1} = \left(\sqrt{3} \cdot Z - \frac{1}{\sqrt{3}}\right)^2 - \frac{1}{3} + (A - (B - 0.5)^2 + 0.25) \ . \tag{17}$$

This ensures that, e.g., $\underline{\overline{\delta_{d1}}}$ can be further tightened. Let us assume $\underline{\overline{A}} = [0; 5]$, $\underline{\overline{B}} = [-1;\ 1]$, and $\underline{\overline{Z}} = [0;\ 1]$: Evaluating the righthand side of Eq. (13) by IA can only reduce $\underline{\overline{\delta_{d1}}}$ to $[-4; 9]$, while its reformulation according to Eq. (17) achieves $[-2.\bar{3};\ 6.25]$. Eq. (14) to (17) have been implemented to limit the variable space to the desired solution without having to make any case differentiation.

Table 2 shows the results of the tested contraction schemes. While Bnormal finds a physical solution after one box reduction step, HC4revise has not solved the system numerically even after 10 reduction steps. The combination of both finds a solution. It is slower than Bnormal but has a better box reduction performance in terms of ϵ.

Table 2. Results after one box reduction step applied on the partial condenser[i].

Method	Solved	CPU (s)	ϵ (-)
Bnormal	yes	4.63	0.60
HC4revise	no	2.13	0.59
HC4revise + Bnormal	yes	4.82	0.57

4.2. Distillation Column

A mixed liquid stream of the same components as in the previous example is thermally separated in a distillation column. The unit removes the byproduct Biphenyl from the hydrodealkylation process. Its model was originally implemented by Rajes (2020) and includes equilibrium trays, a total condenser and a total reboiler. The number of trays is varied from 5, 10, 15 to 20 which corresponds to 241, 416, 591, and 766 state variables. The processing time in dependence of the problem size is of interest in this investigation. The vapor liquid equilibrium and the enthalpies are also calculated via SRK. The initialized NLEs can be found in MOSAICmodeling[iii]. According to Figure 3 the combination of Bnormal and HC4revise outperforms the single methods in most cases regarding its run time. Unlike the other two methods, it always finds a solution for all NLEs after one box reduction step.

5. Conclusion

Without a laborious search for suitable initial values, just one box reduction step of our IA-based initialization approach suffices to reduce the initial variable space to such an extent that the applied root-finder determines feasible solutions for both test examples. Nevertheless, it should be noted that overly wide variable intervals, e.g., of $[-10^9;\ 10^9]$, may slow down the method due to extreme box splitting. Hence, process knowledge remains indispensable to narrow the operating range. The more precise this estimation is, the faster also the hybrid approach becomes.

References

A. Wächter, L. T. Biegler, 2006, On the implementation of an interior-point filter line-search algorithm for large-scale nonlinear programming, Math.Programming, 106, 1, p. 25 – 57

F. Benhamou, J.-F. Puget, 1998, Revising hull and box consistency, 16-th Conference Logic Programming, p. 230 – 244

G. Soave, 1972, Equilibrium constants from a modified Redlich-Kwong equation of state, Chem. Eng. Science, 27, 6, p. 1197 – 1203

L. Granvilliers, 2020, Filtering Domains of Factorable Functions Using Interval Contractors, Optimization of Complex Systems, p. 99 – 108

R. E. Moore, R. Baker Kearfott, M. J. Cloud, 2009, Introduction to interval analysis, SIAM

S. Bublitz, E. Esche, J.-U. Repke, 2021, Automatic initial value generation with interval arithmetic for nonlinear process models, Comput. Chem. Eng., 151, p. 107342

S. Rajes, 2020, Modellierung und Modellvalidierung des Hydrodealkylierungsprozesses, Master Thesis

Y. V. C. Rao, 1997, Chemical Engineering Thermodynamics, Sangam Books

---

[i] Tested on a Notebook Intel i7 Processor (8x 1.8 GHz, 8. Generation)

[ii] Internal ID (Partial condenser): 170665, free account from https://mosaic-modeling.de/

[iii] Internal IDs (Distillation column), 5 to 20 trays: 170907, 171127 171022 171020

Antonis Kokossis, Michael C. Georgiadis, Efstratios N. Pistikopoulos (Eds.)
PROCEEDINGS OF THE 33rd European Symposium on Computer Aided Process Engineering (ESCAPE33), June 18-21, 2023, Athens, Greece
 http://dx.doi.org/10.1016/B978-0-443-15274-0.50021-4

Electroplating scheduling: Closing a research gap from an automation vendor's perspective

Sophie Reimschüssel,[a] Uwe Fuchs,[b] Guido Sand,[a]

[a]*Pforzheim University of Applied Science, Tiefenbronner Straße 65, 75175 Pforzheim, Germany*

[b]*UF-automation GmbH, Am hohen Markstein 13, 75177 Pforzheim, Germany*

Abstract

The production scheduling of batch operated electroplating plants gives rise to hoist scheduling problems. A literature review and real-world case studies reveal that a broad spectrum of electroplating scheduling problems can be modeled as mixed integer programs. Since problems of real-world size cannot be solved by state-of-the-art standard solvers in reasonable times, problem-specific simplification heuristics can be manually applied. An automation vendor strives for an automated solution for a wide spectrum of hoist scheduling problems to reduce the engineering effort. Therefore, the selection and parameterization of models and heuristics need to be automated. This paper argues for an approach which combines mixed integer programming with hyper-heuristics to close this research gap.

Keywords: Hoist Scheduling, Electroplating Plants, Mixed Integer Programming, Hyper-Heuristics

1. Electroplating scheduling

In batch-operated electroplating plants, piece or bulk material is coated with thin layers of metal using electrolysis. Such a plant, schematically shown in Figure 1, consists of several tanks arranged in one or more coupled rows, which are filled with liquids for coating as well as for pre- and post-treatment of the products.

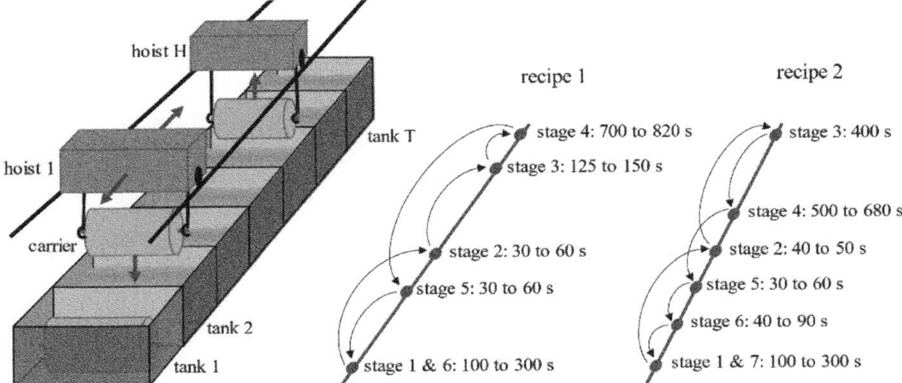

Figure 1 Schematic representation of a batch operated electroplating plant (left) and different recipes with their production sequence (right)

Items to be coated are mounted on racks or filled into barrels (in general carriers) and dipped into specific tanks in a sequence given by recipes. The processing times may vary for some stages within a given range (see Figure 1). The carriers are automatically picked up, transported and lowered by hoists, moving above the tanks on a rail. Multiple hoists operate on one rail, so they cannot pass each other and must not collide. The plant operation constitutes Hoist Scheduling Problems (HSPs), which can be thought of as Flexible Job Shop Problems (FJSP) with additional consideration of hoists. The dominating degrees of freedom are the movements of the hoists, which also determine the occupancy of the tanks. The carriers usually move back and forth and must be returned to the start of the line, as shown in Figure 1. The most common scheduling objectives are the minimization of the makespan or cycle time. In the HSPs of the electroplating processes at hand, no-wait conditions as well as time intervals for process durations must be satisfied, so that the hoists arrive at the tanks in time. The reachability combined with the collision avoidance of the hoists increase the complexity of the HSPs compared to FJSPs and simplification strategies become necessary to solve real-world problems.

Multiple aspects like the combination of products, their geometrical shape, base and coating material, layer-thickness or the required space for the plant determines the plant structure and the recipes of the products. Virtually each electroplating plant is designed individually for a specific application according to these requirements. Consequently, a broad spectrum of scheduling problems exists, which differ from application to application and require different simplification strategies.

2. State of research and technology

2.1. *MIP-based methods*

For the HSP, different Mixed Integer Programming (MIP) models for different plant and process structures exist. The HSP belongs to the class of NP-hard problems (Lei et al., 1989) and, depending on the size of the problem instance, cannot be solved in reasonable time with a standard solver like CPLEX (ibm.com/de-de/analytics/cplex-optimizer). Therefore, often simplification strategies are applied to compute a feasible schedule for one or a few days. A first distinction of simplification strategies is made between cyclic and acyclic scheduling. Cyclic scheduling is suitable for plants processing few product types in large quantities. Each cyclic schedule covers one or a few products and is executed several times in succession (Feng et al., 2018). There also exist MIP-models to switch from one cyclic schedule to another in an optimal way (Steneberg, 2013). Acyclic scheduling is suitable for plants processing many product types in small quantities. To compute an acyclic schedule in reasonable times, a rolling horizon approach can be applied, where the entire scheduling horizon is decomposed into smaller time segments (Ramin et al., 2022). If cyclic or acyclic scheduling with rolling horizons does not reduce the solution time sufficiently additional simplification heuristics become necessary.

The simplification heuristics can be classified as follows: (1) decomposition, (2) fixing variables, (3) adding constraints, (4) aggregation, (5) relaxation of constraints and (6) algorithm tuning. Cyclic scheduling (decomposition by products) and rolling horizon (decomposition by time) can be assigned to class (1). Furthermore, there are different zone partitioning approaches, in which the electroplating line is decomposed into zones which are assigned to the hoists (Li et al., 2015). A further decomposition can be done by the choice of the cycle degree (decomposition by batches). A higher degree leads to an equal or shorter cycle time but to a longer scheduling horizon and thus to a longer solution time (Feng et al., 2018). Simplification heuristics from class (2) are the assignment of certain transport operations to hoists (Fröhlich et al., 2009) or the fixed assignment of stages to identical tanks (Reimschüssel et al., 2022). Simplification strategies from class (3) include fixed processing times (Che et al., 2010), restriction of the transport direction (Chtourou et al., 2013), and prevention of a cycle transition during a transport operation (Leung et al., 2004). The aggregation of picking up, transporting, and lowering a carrier is a simplification from class (4) (Mao et al., 2018). Different approaches exist for relaxing constraints (5) and dealing with infeasible schedules caused by this simplification. For instance, Yan et al. (2018) check whether the solution violates relaxed constraints and reinserts them into the model. Algorithm tuning (6) can be done by tighter model formulations (Leung et al., 2004) or passing a warmstart solution (Ramin et al., 2022) or lower/upper bounds on variables to the solver (Che et al., 2007).

2.2. Metaheuristics, constructive heuristics and hyper-heuristics

In addition to MIP, some metaheuristics are applied to solve the HSP, such as evolutionary algorithms (Amraoui et al., 2013), variable neighborhood search (Laajili et al., 2021) or tabu search (Wang et al., 2019). There are also various constructive heuristics like variations of the earliest starting time heuristic (Amraoui et al., 2016) or the adaptive time window heuristic (Paul et al., 2007). Hyper-heuristics are used to automate the selection and parameterization of metaheuristics or constructive heuristics. For a given problem instance, a hyper-heuristic (often a metaheuristic itself or a neural network) selects a suitable metaheuristic/constructive heuristic from a pool of low-level heuristics and determines an appropriate parameterization (Burke et al., 2019).

2.3. Commercial solutions

Generic production planning systems are usually not applicable for the electroplating scheduling since they neglect the hoists. Scheduling without hoists and adding them to an existing schedule later is usually infeasible due to collisions of hoists or hoists not reaching transport operations in time. Specific software solutions for scheduling batch-operated electroplating plants exist, such as GALVATEK (galvatek.eu), DiTEC (ditec-gmbh.com) or ICOM Automation (icom-automation.de), but their market penetration is limited due to complex manual adjustments required for the specific problem instances.

3. Automation vendor's perspective

In practice the scheduling is typically done individually for each plant through the manual application of different types of heuristics by experienced experts. Sometimes the scheduling procedure is split between the automation vendor and the plant operator. However, this manual heuristic procedure suffers from long response times, a lack of experienced experts and a considerable loss of optimization potential – in particular for plants processing many product types in small quantities.

To overcome the mentioned drawbacks of the manual heuristic procedure, automated MIP-based scheduling services are proposed. In order to exploit economy of scale effects, they should be offered by automation vendors. However, a key requirement for an automated scheduling procedure for a broad spectrum of problem instances is an efficient and widely automated engineering process. The engineering process is meant to adapt the generic scheduling procedure to individual problems.

4. Research gap

This chapter outlines the gap between an automation vendor's requirements for an automated solution to schedule individual electroplating plants as described in chapter 3 and existing approaches.

4.1. MIP-based solution

Mathematical programming methods offer a high modeling flexibility and standard solvers that are independent of the model. The applicability of MIP methods to real-world electroplating scheduling problems depends on two prerequisites: The models need to represent the real-world problems sufficiently well and the computational limits of the standard solvers need to be overcome.

4.1.1. Case studies

Three real-world use cases were examined: (a) a medium-scale manually operated plant, (b) a small- and (c) a large-scale automatically operated plant. Their key properties can be adequately reflected by MIP models. To solve them by CPLEX, (a) was simplified by an aggregation approach and (b) and (c) by a product-decomposition approach. For (b), the optimized schedule was computed within 2 min CPU-time and with a 14,9 % shorter cycle time than the manually created schedule. Further it was shown that varying problem instances require different models and simplification heuristics. The model for (a) cannot be used for (b) and (c) because the hoists are not covered and depending on the problem size, certain combinations of simplifications are no longer sufficient to compute a schedule in reasonable time. For (c) for example, a schedule could be computed within 48 hours only for small recipes. The selection of a suitable model and suitable heuristics for a given problem instance is based on an educated guess.

4.1.2. Literature analysis

A complementary literature analysis revealed, that according to the classification scheme by Manier (2003), nearly the full spectrum of HSPs can be modelled as MIPs. However, an analysis of the papers also showed that different problems require different models and heuristics, as well as different parameterizations of the heuristics. A typical approach in the literature for selecting models and heuristics and parameterizing heuristics for MIP is also an educated guess.

4.2. Automated model and heuristic selection

The models existing in literature collectively represent almost the entire spectrum of the HSPs. Considered individually, the models as well as the heuristics are suitable for a small subset of problem instances and are usually applied to individual selected or benchmark instances. However, an automation vendor needs a solution that can be applied to a spectrum of real-world problem instances automatically. Which model and heuristics should be applied for a given problem instance to compute a solution in a short solution time with high quality is still not considered in the state of research and not yet automated. Manually selecting the models and the simplification heuristics for each instance takes much time and requires a lot of knowledge. From an automation vendor's perspective an automated engineering process i.e. an automated selection, parameterization and solution of models and heuristics for specific instances is crucial.

5. Approach

Studies on real-world instances show that MIP can be used in practice for hoist scheduling in electroplating plants if simplification heuristics are applied. As described in section 4.2, an automated selection and parameterization is needed. According to literature the application of hyper-heuristics is limited to metaheuristics and constructive heuristics. Methods to select and parameterize simplification heuristics used to solve MIP models are not known to the authors. The authors propose to combine the principles of hyper-heuristics with MIP as follows. The real-world problem instances as well as the models are classified according to the scheme of Manier. Based on the classification, a preselection of models suitable for the problem is possible. The suitable models are parameterized for the problem instance and a pool of suitable heuristics is chosen. The preselection of the models already decides which low-level heuristics are to be added to the pool, since only a subset of heuristics is applicable to each model. In the research phase, a parameterized model with a suitable heuristic and its parameterizations is to be determined with an evolutionary algorithm. A possible criterion to evaluate the fitness of individuals could be the objective value after a given solution time. This process is performed for multiple problem instances, resulting in pairs of scheduling problems with model-heuristic combinations. These will be used to train a neural network to avoid selecting a model and heuristic each time again with the evolutionary algorithm during the application phase.

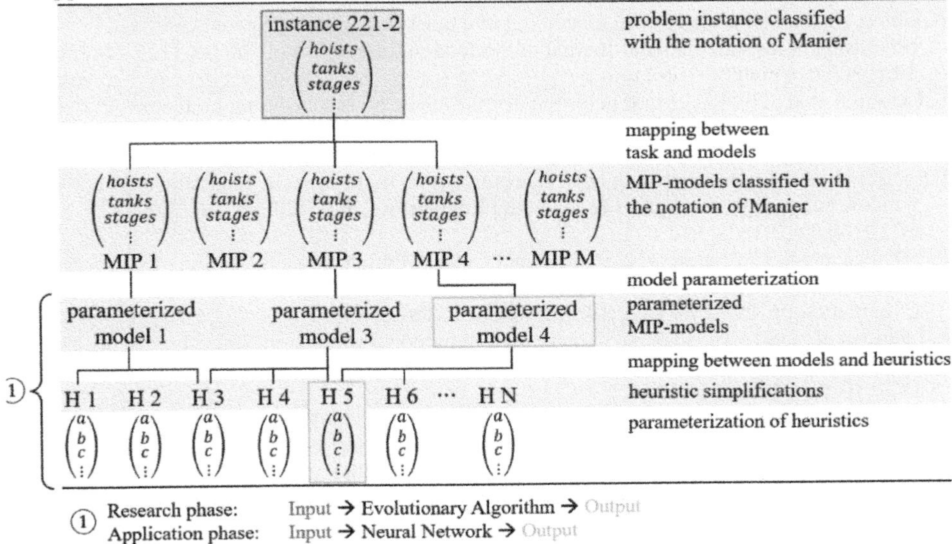

Figure 2 Procedure to select and parameterize a model and a heuristic for a scheduling problem

6. Outlook

Young research areas like hyper-heuristics deal with automated algorithm selection and parameterization but have not yet been applied to simplification heuristics for solving MIP models. The approach presented in this paper combines hyper-heuristics with MIP to compute good schedules in reasonable time for new, unknown problem instances. The study of this approach shall provide a deeper insight into the problem as well as knowledge about the relationships between the scheduling problem and suitable models-heuristic combinations. It is expected that the research will reveal the potentials of an intelligent, automated model and heuristic selection, regarding solution time and scheduling quality, compared to the educated guesses applied in literature and at the automation vendor. Furthermore, information on suitable structures for hyper-heuristics for selecting MIP models and heuristics as well as their essential differences compared to classical hyper-heuristics shall be obtained. The results can not only be transferred to further applications in the field of HSPs such as robotic cells or cluster tools, but also to optimization problems for which many models and heuristics exist which are differently suited for different problem instances. The expected overall goal is to bring the theoretical results in the form of MIP models and simplification heuristics into practice through the approach presented here.

References

A. E. Amraoui et al., 2013, A genetic algorithm approach for a single hoist scheduling problem with time windows constraints, Engineering Applications of Artificial Intelligence, Vol. 26, pp. 1761-1771, DOI: 10.1016/j.engappai.2013.02.004

A. E. Amraoui et al., 2016, An efficient new heuristic for the hoist scheduling problem, Computers & Operations Research, Vol. 67, pp. 184-192, DOI: 10.1016/j.cor.2015.10.006

E.K. Burke et al., 2019, A Classification of Hyper-Heuristic Approaches: Revisited, International Series in Operations Research & Management Science, Vol 272, pp. 453-477, DOI: 10.1007/978-3-319-91086-4_14

A. Che et al., 2007, Cyclic hoist scheduling in large real-life electroplating lines, OR Spectrum, Vol. 29, pp. 445-470, DOI: 10.1007/s00291-006-0040-9

A. Che et al., 2010, Optimal cyclic scheduling of a hoist and multi-type parts with fixed processing times, International Journal of Production Research, Vol. 48, pp. 1225-1243, DOI: 10.1080/00207540802552659

S. Chtourou et al., 2013, A hybrid algorithm for the cyclic hoist scheduling problem with two transportation resources, Computers & Industrial Engineering, Vol. 65, pp. 426-437, DOI: 10.1016/j.cie.2013.03.013

J. Feng et al., 2018, Cyclic jobshop hoist scheduling with multi-capacity reentrant tanks and time-window constraints, Computers & Industrial Engineering, Vol. 120, pp. 382-391, DOI: 10.1016/j.cie.2018.04.046

R. Fröhlich and S. C. Steneberg, 2009, Optimal Cyclic Multiple Hoist Scheduling for Processes with Loops and Parallel Resources, IEEE International Conference on Systems, Man, and Cybernetics, pp. 293-298, DOI: 10.1109/ICSMC.2009.5346151

E. Laajili et al., 2021, An Adapted Variable Neighborhood Search based algorithm for the cyclic multi-hoist design and scheduling problem, Computers & Industrial Engineering, Vol. 157, DOI: 10.1016/j.cie.2021.107225

L. Lei and T. Wang, 1989, A Proof: The Cyclic Hoist Scheduling Problem is NP-Complete, Technical Report 89-0016, Graduate School of Management, Rutgers University

J. M. Leung et al., 2004, Optimal Cyclic Multi-Hoist Scheduling: A Mixed Integer Programming Approach, Operations Research, Vol. 52, pp. 965-976, DOI: 10.1287/opre.1040.0144

X. Li et al., 2015, Optimal multi-degree cyclic scheduling of multiple robots without overlapping in robotic flowshops with parallel machines, Journal of Manufacturing Systems, Vol. 36, pp. 62-75, DOI: 10.1016/j.jmsy.2015.03.003

M. Manier, 2003, A Classification for Hoist Scheduling Problems, International Journal of Flexible Manufacturing Systems, Vol. 15, pp. 37-55, DOI: 10.1023/A:1023952906934

Y. Mao et al., 2018, Mixed-integer linear programming method for multi-degree and multi-hoist cyclic scheduling with time windows, Engineering Optimization, Vol. 50, pp. 1978-1995, DOI: 10.1080/0305215X.2017.1418865

H. J. Paul et al., 2007, A heuristic scheduling procedure for multi-item hoist production lines, International Journal of Production Economics, Vol. 105, pp. 54–69, DOI: 10.1016/j.ijpe.2005.11.008

R. Reimschüssel et al., 2022, AALE Conference 2022, DOI: 10.33968/2022.06

D. Ramin et al., 2022, Dynamic Hoist Scheduling for Multi-Recipe and Multi-Stage Production Lines: A Logical Framework, SSRN Electronic Journal, DOI: 10.2139/ssrn.4105670

S.C. Steneberg, 2013, MILP model for multi-product, multi-part and multi-hoist cycle shops, 2013 IEEE International Conference on Industrial Technology (ICIT), pp. 1339-1346, DOI: 10.1109/ICIT.2013.6505867

H. Wang et al., 2019, The Printed-Circuit-Board Electroplating Parallel-Tank Scheduling With Hoist and Group Constraints Using a Hybrid Guided Tabu Search Algorithm, IEEE Access, Vol. 7, DOI: 10.1109/ACCESS.2019.2915587

P. Yan et al., 2018, A dynamic scheduling approach for optimizing the material handling operations in a robotic cell, Computers & Operations Research, Vol. 99, pp. 166-177, DOI: 10.1016/j.cor.2018.05.009

Antonis Kokossis, Michael C. Georgiadis, Efstratios N. Pistikopoulos (Eds.)
PROCEEDINGS OF THE 33rd European Symposium on Computer Aided Process Engineering (ESCAPE33), June 18-21, 2023, Athens, Greece
 http://dx.doi.org/10.1016/B978-0-443-15274-0.50022-6

Real time prediction of ozone pollution using data-enabled deep learning modeling

Fugui Hong, Cheng Ji, Fangyuan Ma, Chang Chen, Wei Sun*
College of Chemical Engineering, Beijing University of Chemical Technology, North Third Ring Road 15, Chaoyang District, Beijing, 100029, China
**Corresponding Author's E-mail: sunwei@mail.buct.edu.cn*

Abstract

Tropospheric ozone pollution is one of the most harmful secondary air pollutants that could has brought challenges to air pollution prevention. Therefore, early prediction and warning of high-level tropospheric ozone concentration plays an important role in the management of ambient air pollution. Deep learning methods have been widely applied for their powerful data fitting ability, but nonstationary characteristic is exhibited in ozone concentration data owing to the instability of the atmosphere, which pose a challenge to the prediction of data with high-level ozone concentration. Considering the effectiveness of difference in handling nonstationary sequence, a nonstationary modeling method called single difference-embedded long short-term neural network (SDELSTM) is applied to realize the prediction of high-level tropospheric ozone concentration. Case study on an ozone dataset between 2013 and 2017 in Los Angeles is used to verify the effectiveness of the SDELSTM.

Keywords: SDELSTM, high-level tropospheric ozone concentration, nonstationary characteristics

1. Introduction

Tropospheric ozone pollution is a secondary pollutant produced by nitrogen oxides (NO_x) and volatile organic compounds (VOCs) under solar radiation (Jenkin and Clemitshaw, 2000). The long-term exposure to tropospheric ozone contributes to the risk of respiratory and circulatory mortality, and is potentially associated with cardiovascular disease risk and premature death (Jerrett et al., 2009, Turner et al., 2016). The real time prediction of tropospheric ozone pollution can realize the early alarm and prevention of ozone pollution and reduce the risk of human health. Generally, data-driven methods are widely used to achieve this task. Among them, LSTM is widely used in ozone prediction since the it could model long-term time dependence (Cabaneros et al., 2019). However, the air quality data show short-term nonstationary characteristics due to the unstable atmospheric composition and the complexity in the formation mechanism of pollutants. Under this circumstance, traditional data-driven models show poor performance in the field of ozone prediction. Zakaria et al. (Zakaria et al., 2021) introduced stationary and nonstationary model to forecast high ground-level concentrations and found that nonstationary models showed better performance in ozone prediction. Therefore, the consideration of nonstationary characteristics of data can enhance the performance of ozone prediction model. Difference is widely used in statistics to deal with nonstationary time series, which could eliminate the impact of random fluctuations in data, so as to improve the performance of the model. It is reported that several difference methods have been combined with deep learning methods for time series prediction. Zhou et al. (Zhou et al., 2022) proposed difference-LSTM (DLSTM) in which the difference of input variables was seen as a dynamic input, and the method was verified by the prediction of key variables in complex industrial process. Wang et al. (Wang et al., 2019)proposed the memory in memory (MIM) networks. Then the MIM module was utilized to replace the forgotten gate and retain more nonstationary characteristics. In this study, a model called SDELSTM is introduced to perform ozone prediction from the perspective of nonstationary characteristics of data. In this model, an additional network layer including a differential module is embedded into LSTM neural network to capture the nonstationary characteristics of data under the premise of avoiding the loss of original data information(Ji et al., 2023). The effectiveness of the SDELSTM model is validated by performing ozone prediction using the historical tropospheric data from Los Angeles.

2. Description of the model

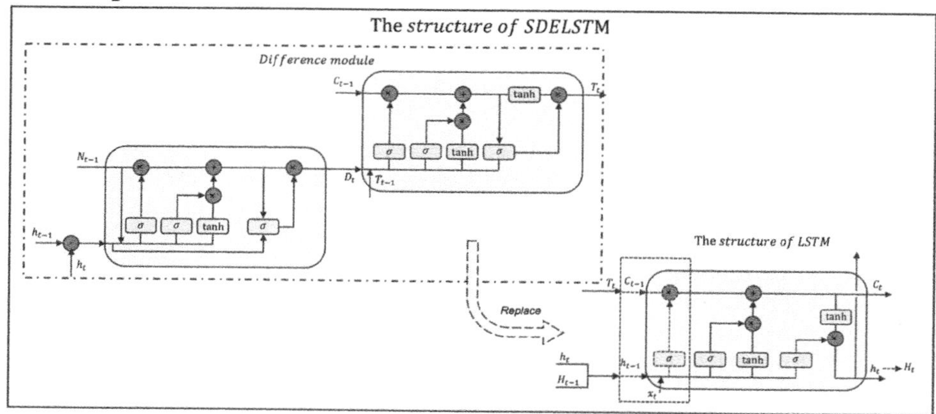

Figure 1 The structure of LSTM and SDELSTM

2.1. long short-term memory neural network

The LSTM neural network is a special recurrent neural network which contains many neurons in the hidden layer. The structure of LSTM is shown in Figure 1. The initial input h_{t-1} and x_t represent the state of the previous hidden layer and input of current time respectively. C_t and C_{t-1} record the cell state which is the key value to LSTM. h_t represents the output of the current layer. The neurons in hidden layer contains three gates including a forget gate, an input gate and an output gate which could control the transmission of information. The unique structure helps LSTM learn the long-term characteristics of the data. The forget gate control which part of information should be retained. Then the input gate determines the information to be added to cell state and the output gate control the output of the cell state. However, the short-term nonstationary characteristics cannot be obtained by a single forget gate and some measures have been adopted to make LSTM become nonstationary model.

2.2. The structure of single difference-embedded long short-term neural network

To address this problem, the SDELSTM model is improved from LSTM where an additional layer including a differential module is embedded into the hidden layer. The dashed box indicates the replacement of the forget gate by the differential module. The input of difference module is states of hidden layers of the current and previous moment (h_t and h_{t-1}) which is regarded as difference information. N_{t-1} represents cell state of the previous difference layer. The output of difference layer is a new cell state D_t which contains difference information. The next layer combines D_t and C_{t-1} whose purpose is selective retain difference characteristics and long-term characteristics of the original sequence. T_t represents cell state which records both short-term and long-term information. Eventually, in order to avoid the loss of original information, h_t and H_{t-1} which represents the state of SDELSTM are utilized as the input of LSTM neural network and H_t represents the output of SDELSTM. Then, the short-term nonstationary characteristics can be obtained by the unique forget gate module. In summary, the proposed SDELSTM model could learn short-term nonstationary characteristics and long-term dependence at the same time by a difference module, and therefore the performance of ozone concentration prediction can be improved.

2.3. Implementation procedure of SDELSTM

The implementation procedure of SDELSTM is shown in Figure 2 which contains three parts: data processing, stationary analysis and modelling. The procedures can be summarized as follows,

Step 1: The missing values are filled by K-Nearest Neighbor (KNN).

Step 2: The datasets are divided into training data and test data in a ratio of 4:1 and 20 percent of training data are grouped into validation data.

Step 3: The stationarity of time series is tested by the ADF test.

Step 4: The SDELSTM model is set up by training data and the hyperparameters of the model are determined by validation data.

Step 5: The performance of the model is verified by test data.

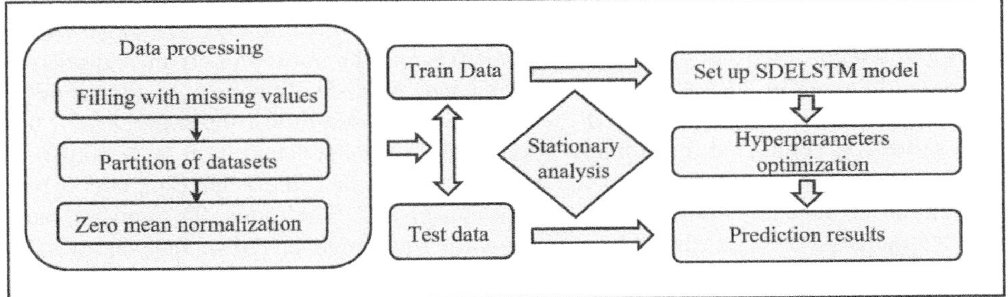

Figure 2. The training process of SDELSTM model

3. Results and discussions

3.1. Data preparation

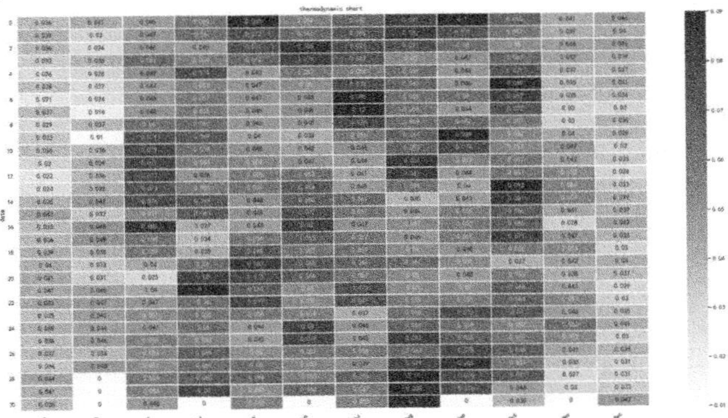

Figure 3 Daily Maximum ozone distribution thermal map of Los Angeles in 2017

Due to the special geographical location of Los Angeles and the year-round high temperature and little rain characteristics, ozone pollution is serious. In 2017, more than 100 million people in the United States were exposed to tropospheric ozone pollution, especially in Los Angeles (Zhang et al., 2019). Figure 3 shows that the thermal map of daily maximum ozone of Los Angeles in 2017. The national ambient air quality standards from EPA for ozone concentration is 0.070 parts per million (ppm). In addition, according to world health organization (WHO) air quality guidelines, the standard for ozone concentration is 0.05ppm which suggests more people are exposed to tropospheric ozone pollution. It is found that the darker the color, the more ozone pollution there is and most ozone pollution is concentrated during the warmer months. Besides the dataset mentioned above, the Los Angeles air quality datasets between 2013 and 2017 are obtained from the air quality system of environment protection agency (EPA) which were collected from the monitoring site at N main St in Los Angeles. Datasets from 2013 to 2016 are used as training sets and the data of 2017 is selected as the test set. Considering that there are missing values in the data, KNN is utilized to fill missing data. After data preprocessing, twelve air monitoring variables are selected, including SO_2, NO_2, $PM_{2.5}$, PM_{10}, RH_DP, TEMP, CO_0, CO_1, NO_x, NO_y, NO_0, and NO_1.

3.2. Case study

3.2.1. Stationary analysis

Since the complexity of atmospheric environment, the air quality monitoring data obtained from monitoring stations would fluctuate randomly within a certain range, which is easy to be ignored from a long-term perspective, and the nonstationary characteristics generated by random fluctuations will affect the performance of ozone prediction model. The Augmented Dickey-Fuller (ADF) test was widely used to test the stationary of the time series and conducted for monthly tropospheric ozone concentration to demonstrate the non-stationary characteristics of the data. It is shown in Table 1 that 8 months ozone concentration data are nonstationary at 1% confidence. Therefore, the nonstationary characteristics of the data cannot be ignored while traditional LSTM models do not take these characteristics into account. To address this issue, the SDELSTM model is introduced to learn the nonstationary characteristics within the air quality data, which could enhance the performance of ozone prediction.

Table 1 Los Angeles monthly data ADF statistics in 2017

Months	Test statistics	1%	5%	10%
January	**-2.97**	-3.71	-2.98	-2.63
February	**-2.44**	-3.66	-2.96	-2.62
March	-4.79	-3.67	-2.96	-2.62
April	-4.97	-3.78	-3.01	-2.64
May	**-3.34**	-3.66	-2.96	-2.62
June	**-2.53**	-3.66	-2.96	-2.62
July	-4.07	-3.75	-2.99	-2.63
August	-4.92	-3.67	-2.96	-2.62
September	**-3.27**	-3.66	-2.96	-2.62
October	**-3.43**	-3.66	-2.96	-2.62
November	**-1.75**	-3.71	-2.98	-2.63
December	**-2.99**	-3.66	-2.96	-2.62

3.2.2. Performance of SDELSTM on ozone concentration

With the current air quality data fed in, the model is able to predict ozone concentration for the next hour. It is shown in Figure 4 that the prediction results of the model for different months. The model makes accurate predictions of the concentration data where there is a health risk while the model performed poorly in the prediction of low concentration data. It is found that ozone data show successive near-zero values at regular intervals after the analysis of actual hourly ozone data. Then we speculate that it is caused by artificial manipulation which leads to the deviation in the predicted result graph. However, only high concentrations of ozone (>0.05ppm) pose a huge threaten to human health. Therefore, when evaluating the predictive performance of the model, the prediction results of low concentration values will not greatly affect the predictive performance of the model.

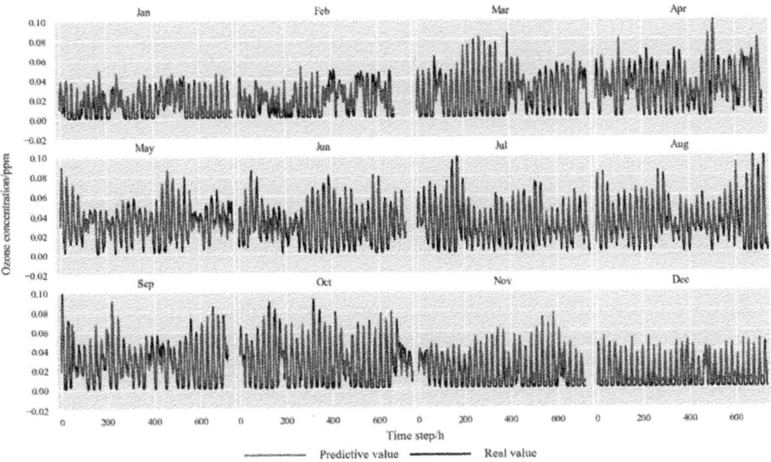

Figure 4 Comparison of true and predicted values for different months

In order to evaluate the prediction effect of the model more scientifically and objectively, it is necessary to use quantitative indicators to further evaluate the model, coefficient (R^2), root mean square error (RMSE) are widely used to evaluate the performance of prediction models. The indicators are defined as follows,

$$RMSE = \sqrt{\frac{1}{N}\sum_{i=1}^{N}(y_j - \hat{y}_j)^2} \tag{1}$$

$$R^2 = 1 - \frac{\sum_{i=1}^{N}(y_j - \hat{y}_j)^2}{\sum_{i=1}^{N}(y_j - \bar{y}_j)^2} \tag{2}$$

where N is the number of test samples. y_j and $\hat{y}_j$ indicate the actual and predicted values at time j. $\bar{y}_j$ indicates the mean of the predicted values over all times. The specific calculation results are presented in the next section along with the comparison of other models.

3.2.3. Model comparison

In order to demonstrated the performance of the SDELSTM model in ozone prediction, the related models including artificial neural network (ANN), LSTM, gate recurrent unit

(GRU) are utilized to compare the results with the SDELSTM model. The specific model comparison results are shown in Table 3. It is shown in Table 2 that the SDELSTM model performs better than other models in all indicators. The R^2 of the model is 13.4% higher than that of the traditional LSTM and RMSE is 34.6% lower than other models. The SDELSTM model shows more accuracy on the tropospheric ozone concentration because nonstationary characteristics of the ozone data cannot be considered by traditional predictive models.

Table 2 Comparison of different models

	RMSE	R^2
ANN	0.0120	0.6119
GRU	0.0110	0.6869
LSTM	0.0104	0.7265
SDELSTM	0.0068	0.8237

4. Conclusions

In this work, a model called SDELSTM is introduced to perform ozone prediction from the perspective of nonstationary characteristics of data. An additional network layer including a differential module is embedded into LSTM neural network to capture the nonstationary characteristics of data under the premise of avoiding the loss of original data information. Based on literature research, the ADF stability test was conducted on the ozone data of Los Angeles in 2017, which confirmed the nonstationary characteristics of the ozone data. Then the effectiveness of SDELSTM model is validated by performing ozone prediction using the historical tropospheric data from Los Angeles. The results demonstrates that the SDELSTM model shows better performance in hourly troposphere ozone prediction than other related models, indicating its application prospect in preventing ozone pollution.

References

Cabaneros, S. M., Calautit, J. K. and Hughes, B. R., 2019. A review of artificial neural network models for ambient air pollution prediction. Environmental Modelling & Software, 119 285-304.

Jenkin, M. E. and Clemitshaw, K. C., 2000. Ozone and other secondary photochemical pollutants: chemical processes governing their formation in the planetary boundary layer. Atmospheric Environment, 34 (16), 2499-2527.

Jerrett, M., Burnett, R. T., Pope III, C. A., Ito, K., Thurston, G., Krewski, D., Shi, Y., Calle, E. and Thun, M., 2009. Long-term ozone exposure and mortality. New England Journal of Medicine, 360 (11), 1085-1095.

Ji, C., Ma, F., Wang, J. and Sun, W., 2023. Profitability related industrial-scale batch processes monitoring via deep learning based soft sensor development. Computers & Chemical Engineering, 170 108125.

Turner, M. C., Jerrett, M., Pope, C. A., 3rd, Krewski, D., Gapstur, S. M., Diver, W. R., Beckerman, B. S., Marshall, J. D., Su, J., Crouse, D. L. and Burnett, R. T., 2016. Long-Term Ozone Exposure and Mortality in a Large Prospective Study. Am J Respir Crit Care Med, 193 (10), 1134-1142. 10.1164/rccm.201508-1633OC.

Wang, Y., Zhang, J., Zhu, H., Long, M., Wang, J. and Yu, P. S., 2019. Memory in memory: A predictive neural network for learning higher-order non-stationarity from spatiotemporal dynamics. Proceedings of the IEEE/CVF Conference on Computer Vision and Pattern Recognition,

Zakaria, S. A., Amin, N. A. M., Radi, N. F. A. and Hamidin, N., 2021. Stationary and Non-Stationary Models of Extreme Ground-Level Ozone in Peninsular Malaysia. Mathematics and Statistics, 9 (3), 357-370. 10.13189/ms.2021.090318.

Zhang, J., Wei, Y. and Fang, Z., 2019. Ozone pollution: a major health hazard worldwide. Frontiers in immunology, 10 2518.

Zhou, J., Wang, X., Yang, C. and Xiong, W., 2022. A Novel Soft Sensor Modeling Approach Based on Difference-LSTM for Complex Industrial Process. IEEE Transactions on Industrial Informatics, 18 (5), 2955-2964. 10.1109/tii.2021.3110507.

Antonis Kokossis, Michael C. Georgiadis, Efstratios N. Pistikopoulos (Eds.)
PROCEEDINGS OF THE 33rd European Symposium on Computer Aided Process Engineering (ESCAPE33), June 18-21, 2023, Athens, Greece
 http://dx.doi.org/10.1016/B978-0-443-15274-0.50023-8

Numerical Study of a Single-Spout and Z-shaped Fluidized Bed Biomass Gasifier

Fatima Ez-zahra El Hamra,[a] and Radouan Boukharfane[a,*]

[a]*Mohammed VI Polytechnic University (UM6P), MSDA Group, Benguerir, Morocco*
radouan.boukharfane@um6p.ma

Abstract

Gas-solid multiphase systems are ubiquitous in several applications such as pharmaceutical, mining, and petrochemical industry (Geldart, 1986). The vigorous interaction between gas and solid particles involves complex dynamics in the determination of mass and heat transfer. Most simulations and experiments use fluidized beds with simple geometries, but industrial multiphase reactors generally have complex configurations (Zhao et al., 2020). The aim of this work is to study the gas–solid flow characteristics of a Z-shaped gas-solid fluidized bed by means of an in-house solver DEM-LES solver. The fluid phase is computed through solving the volume-averaged four-way coupling Navier–Stokes equations in which the Smagorinsky sub-grid scale tensor model is used. Results show that particles and fluid phases are not uniformly distributed, and a core-annulus structure is observed. These findings provide valuable insight into the development and optimization of particle purification appliances in multiphase reactors.

Keywords: Fluidized bed, Computational fluid dynamics, DEM-LES, Biomass particles, Particle sorting.

1. Introduction

Finding sustainable, clean, and low-carbon alternatives to traditional fossil fuels is becoming crucial given the world's rising energy needs and the severity of global warming. The many inherent benefits of biomass for bioenergy, such as its global availability, renewability, and carbon neutrality, make it one of the most attractive possibilities. Torrefaction, pyrolysis, gasification, and catalytic conversion are among the thermochemical conversion processes that may turn raw biomass feedstocks into high-quality energy products. Because it can transform biomass into three different products-biogas, bio-oil, and biochar, pyrolysis has drawn the most attention among these processes. The fluidized-bed pyrolyzer is one of the most popular reactors for biomass pyrolysis because of its excellent particle mixing capability, high heat transfer efficiency, and operational flexibility. The fluidized-bed reactor, however, typically presents a complicated system in which the hydrodynamics of dense gas-solid flow, heat/mass transfer, chemical reactions, and turbulence are intimately inter-coupled, making it challenging to fully comprehend the mechanics of biomass pyrolysis.

In order to develop and scale-up fluidized systems, it is necessary to understand the flow characteristics between particles and gas but doing so through experiments is challenging. The usage of numerical simulation as an alternate tool for engineering analysis has increased with the advancement of numerical methods and processing power.

Particle-resolved direct numerical simulation (PR-DNS), two-fluid model (TFM), and CFD-DEM are the three categories into which numerical methods of particle fluidization can be categorized. PR-DNS is only useful for flow systems with no more than 1000 particles due to its high computing requirements. The TFM technique treats fluid and solid as two distinct continuous phases. The CFD-DEM model uses Newton's second law to monitor particles, which is why a good match was obtained between simulations and experiments.

Numerous studies have concentrated on how operating parameters affect the particle flow characteristics. Ejim et al., (2017) investigated how the inner geometry of mesoscale tubes with baffles affected the axial dispersion and suspension of materials. Hu et al. (2019) assessed how collisional characteristics affected the hydrodynamics and heat transport in a spouted bed. In a full-loop circulating fluidized bed at varied surface gas velocities, Xu et al. (2018) computationally and empirically examined the particle movements, proving that solid behaviors (such as residence duration, dispersion, and mixing) are connected to the fluidization regions. Particle size and gas velocity effects on tar yield in rapid pyrolysis reactors were discussed by Xue et al. (2012). Wang et al. (2018) also thoroughly investigated the particle behaviors, including the particle force and velocity. The effectiveness of chemical reactions is impacted by the fact that biomass is made up of heterogeneous particles of various densities, sizes, and forms. Consequently, a number of separation and purification techniques were suggested.

Konrath et al. (2014) investigated the circumstances of fine particle separation in centrifugal classification. The influence of particle organization and spigot widths on the enhancement of separation of the hydrocyclone was numerically modeled by Vega-Garcia et al. (2020). Further discussion was made regarding the impact of dipleg shape on the functionality of gas-solid cyclones.

Lv et al. (2018) investigated the motion and separation behavior of coal in a fluidized bed for gas-solid separation.

Since most previous research places emphasis on the effects of operating conditions but gives the reactor structure relatively little attention, the current work aims to investigate the influence of the reactor structure on biomass fluidized-bed pyrolysis behavior. The separation of two-component suspension has typically been shown to be improved by the fluidized bed structure's inclination (Masliyah et al., 1989). In the present study, multiple inclination cross-sections that make up a Z--shaped channel, which has considerable potential for the fluidization and separation of mixed biomass particles, is investigated.

2. Mathematical modeling

The single-spout fluidized bed is simulated in the current study using a coupled Computational Fluid Dynamics and Discrete Element Method (CFD-DEM). The Newtonian laws of motion control the dynamics of the particles suspended in the bed. To describe the fluid phase, the locally volume-filtered Navier-Stokes equations are solved. The whole momentum exchange between the solid and fluid phases is then included in these sets of equations (two-way coupling).

$$\begin{cases} \partial_t\left(\theta_f\, \overline{\varrho_f}\right) + \partial_i\left(\theta_f \overline{\varrho_f}\widetilde{u_{f,i}}\right) = 0, & (1a) \\ \partial_t\left(\theta_f \overline{\varrho_f}\widetilde{u_{f,i}}\right) + \partial_j\left(\theta_f \overline{\varrho_f}\widetilde{u_{f,i}}\widetilde{u_{f,j}}\right) = -\partial_j \overline{p} + \partial_j\left(\overline{\tau_{ij}} + \overline{\tau_{ij}}^{SGS}\right) + \theta_f \overline{\varrho_f} g_i + \mathcal{F}_i^{inter}, & (1b) \end{cases}$$

Where θ_f, ϱ_f, p, and u_f are the fluid-phase volume fraction, density, dynamic pressure, and velocity, respectively. In Eq.(1b), g stands for the acceleration due to gravity and F^{inter} for the inter-phase exchange term that arises from filtering the divergence of the stress tensor. The volume-filtered stress tensor, τ, is expressed as $\overline{\tau_{ij}} = \mu\left[\partial_i \overline{u_{f,j}} + \partial \overline{u_{f,i}} - \frac{2}{3}\partial_i \overline{u_{f,i}}\delta_{ij}\right] + \mathcal{R}_{\mu,ij}$, where μ is the dynamic viscosity and $\delta_{ij}$is the Kronecker function. The SGS stress tensor $\overline{\tau_{ij}}^{SGS} = \overline{\varrho_f u_{f,i} u_{f,j}} - \overline{\varrho}\widetilde{u_{f,i}}\widetilde{u_{f,j}}$ is modeled within the Boussinesq's framework evaluated using the classical Smagorinsky (Smagorinsky, 1963) model.

A Lagrangian particle-tracking approach is used for the solid phase. The displacement of an individual solid particle indicated by the subscript p is calculated using Newton-Euler equations for rigid body motion as follows:

$$\begin{cases} m_p \dfrac{du_{p,i}}{dt} = f_{p,i}^{inter} + f_{p,i}^{col} + m_p g_i, \ with \dfrac{dx_{p,i}}{dt} = u_{p,i} \ ) & (2a) \\ \mathcal{I}_p \dfrac{d\omega_{p,i}}{dt} = \mathcal{M}_{p,i}^{drag} + \mathcal{M}_{p,i}^{col} & (2b) \end{cases}$$

where the particle mass is defined by $m_p = \pi\varrho_p d_p^3/6$ and u_p is the centroid particle velocity. The force f_p^{col} accounts for the particle-particle and particle-wall repulsion modeled using a soft-sphere model (Cundall & Strack, 1979). The force f_p^{inter} exerted on a single particle p by the surrounding fluid is related to the inter-phase exchange term in Eq.(1b) by $\mathcal{F}^{inter} = \sum_{p=1}^{\mathcal{N}_p} \xi(|x - x_p|) f_p^{inter}$, where $\mathcal{N}_p$ is the total number of particles, ξ is the filtering kernel used to volume filter the Navier-Stokes equations, x_p is the position of the pth particle, and f_p^{inter} is approximated by $f_{p,i}^{inter} \approx \mathcal{V}_p \partial_j \tau_{ij} + f_{p,i}^{drag} \approx -\mathcal{V}_p \partial_i p^{©} + f_{p,i}^{drag}$, where $\mathcal{V}_p$ is the volume of the pth particle given by $\pi d_p^3/6$ and $\partial_i p^{©}$ is the local pressure gradient interpolated at the center of the particle. The drag force is computed using the drag force coefficient of Tenneti et al. (2011). Particle-particle and particle-wall collisions are modeled using the adaptive collision time model **ACTM** proposed by Kempe & Fröhlich (2012), which is a variation of the classical soft-sphere model. In this work, a fourth-order central scheme is used for the spatial integration, and a third-order accurate semi-implicit Crank-Nicolson scheme is employed for time integration. The reader is referred to (el Hamra et al., 2022) for further information on the current CFD-DEM solver as well as to review the numerical validation.

3. Single-spout and Z-shaped fluidized bed system

In our study, three spouted-fluidized beds with a single orifice with the same cross-section are considered as schematically depicted in Fig. 1. The difference between the second and third configuration is the angle between the two sections, which is equal to 155° in the Case 2, while it is equal to 132° In all base case simulations carried out for this work, the spouted-fluidized bed height (BH) and depth are 400 mm and 14 mm, respectively. There are two portions

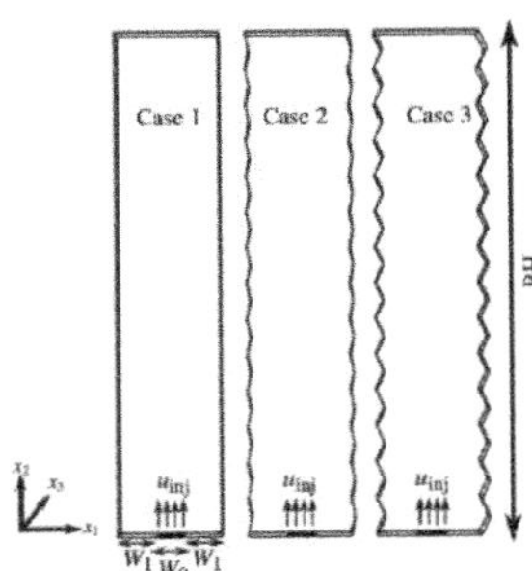

Figure 1: Different geometric models of Z-shaped particle fluidized system.

across the bed's width (shown as $W_0 = 0.007$ mm and $W_1 = 0.007$ mm in Fig. 1).

The computational domain is divided into 82x12x370 quad grids. In the spouted region, 12 grids are considered. Gas flows upwards through W_0, which induces three distinct regions: a dilute central core, where the solid phase is entrained in the concurrent jetting gas flow, a fountain region in the uppermost part, and a dense annular region with a counter-current flow between both phases. Simulation settings of the validation cases are summarized in Table1.

Fluid phase	
Fluid density ϱ_f[kg/m3]	1.165
Fluid viscosity $\mu[Pa \cdot s]$	2.0 x 10^{-5}
Superficial mass flow rate at the injection [kg/s]	2.0 x 10^{-3}
Solid phase	
Number of particles	93808
Bed mass m_{bed}[kg]	0.122
Particle diameter d_p[mm]	1.0
Particle density ϱ_p[kg/m3]	2500
Inter-particle & Particle-wall restitution coefficient	0.97
Inter-particle & Particle-wall friction coefficient	0.1

Table 1:Simulation settings of the validation case for the three spouted bubbling fluidized configurations.

A typical illustration of particle velocity distribution for the three cases at t=25 s can be seen in Fig. 2, where the mesoscopic bubble formed in the mixture with different sizes is observed. During this process, the particles are impulsed into the bubbles and scattered to the freeboard and fall back to bed due to the gravitational force. As the angle between the two sections in the Z-shaped fluidized bed, the bed height is higher. A higher bed expansion rate and an intense particle motion is therefore caused by applying the change of lateral wall's shape compared to the flat wall case.

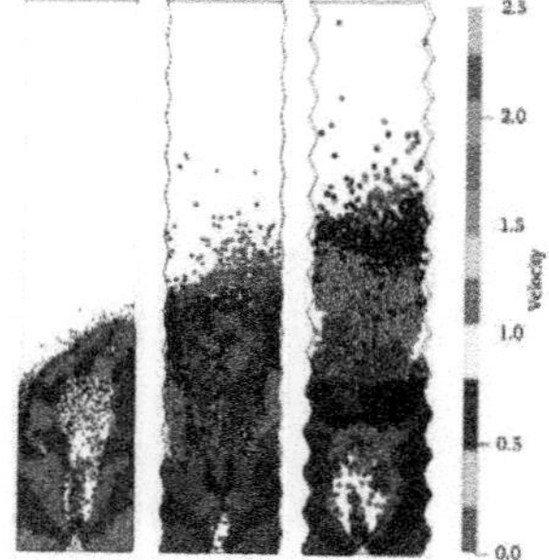

Figure 2: Instantaneous distribution characteristic of particles colored by particle velocity in the three cases.

The time-averaged fluid volume fraction and gas velocity at the midplane of the fluidized bed is portrayed in Fig. 3 for three configurations.

A large velocity gradient is observed around the irregular wall. The Z-shaped fluidized bed breaks the flow symmetry, thus allowing the mixed particles to achieve effective screening in each section. Furthermore, in the corners, heaps of unfluidized particles accumulate in much larger quantities in the case of flat walls. while less unfluidized zones are obtained when the angle between the two sections of the Z-shaped fluidized bed gets smaller. Note that due to these corners actually act as dead zones and create more resistance to bubble movement.

Because gas and solids transfer mass at slower rates in dead zones than they do in a fully fluidized bed, mixing rates may be lowered as a result. Dead zones in corners thereby reduce the cross section available for bubble passage, which in turn reduces bubble size and/or speed.

The mixing rate in case 1 bed is lower than in case 3 bed because dead zone particles move into the active zone of the bed extremely slowly. In other words, Z-shaped fluidized beds remove dead zones and enhance the movement, homogeneous mixing, and subsequent fluidization of the solids.

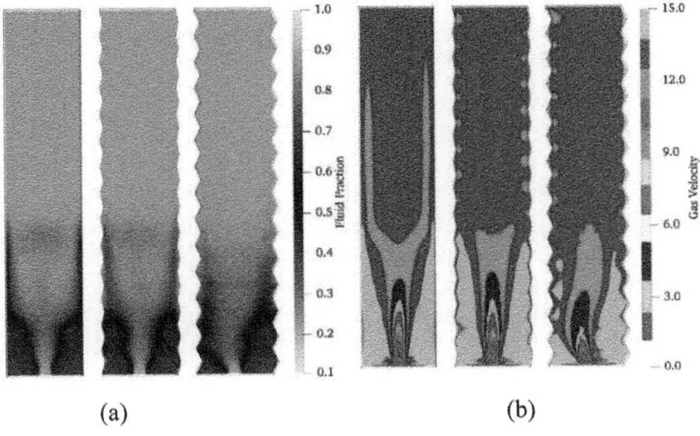

Figure 3: Time-averaged (a) fluid volume fraction and (b) gas velocity at the midplane for the three studied cases.

Solid flow rate is a key parameter for the operation and optimization of fluidized beds. To quantify this parameter, the x_2-direction of the bed, and mainly the interval x_2∈[0,0.35] m, is subdivided into seven equal sections, and the mean number of particles within each section is computed for t∈[5,25]s as can be seen in Fig. 4. The section #1 is the lowest one (at the injection) and the section #7 is the highest one. It is confirmed that the Z-shaped beds tend to provide sufficient space compared to flat walls for the separation and purification of particles as they occupy more space in the x_2-direction.

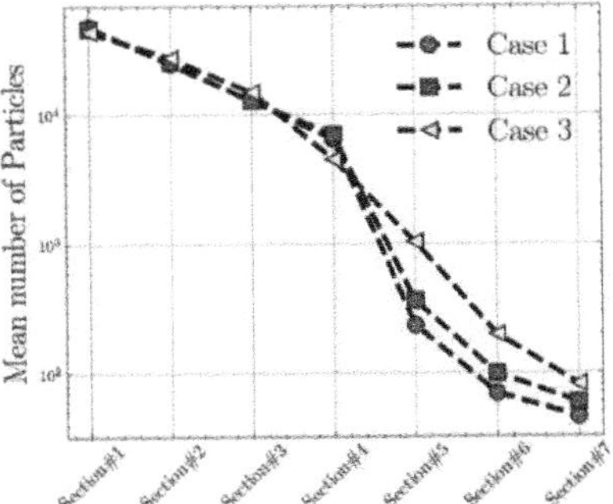

Figure 4: Mean number of particles in different sections for three studied cases.

Due to collisions and energy loss, many particles congregate near the corners of flat walls. However, in cases 2 and 3, where zigzag patterns are present, larger floccules form, break, and then slide down the wall.

4. Conclusion

This work presented a preliminary attempt to use single-spout and Z-shaped fluidized beds as fluidized bed gasifiers of mixed biomass particles. A numerical study based on a large-eddy Simulation (LES) solver coupled with Discrete Element Method (DEM) was employed to study the effect of zigzagging the lateral walls with two angles. The Z-shaped fluidized bed was found to modify the uniformity of the flow domain and to allow an enhanced particle's mixing. These results offer an important direction for designing clean particle products and can be expanded to include research into environmental chemical and process engineering.

Acknowledgement: This work was supported by OCP Group (Morocco). The authors gratefully acknowledge the support and computing resources from the African Supercomputing Center (ASCC) and SIMLAB HPC center at UM6P (Morocco).

References

Cundall, P. A., & Strack, O. D. L. (1979). A discrete numerical model for granular assemblies. *Geotechnique*, *29*(1), 47–65.

Ejim, L. N., Yerdelen, S., McGlone, T., Onyemelukwe, I., Johnston, B., Florence, A. J., & Reis, N. M. (2017). A factorial approach to understanding the effect of inner geometry of baffled meso-scale tubes on solids suspension and axial dispersion in continuous, oscillatory liquid–solid plug flows. *Chemical Engineering Journal*, *308*, 669–682.

el Hamra, F., Boukharfane, R., Er-raiy, A., & Chakraborty, N. (2022). Development and assessment of algorithms for DEM-LES simulations of fluidized bed. *Submitted.*

Geldart, D. (1986). *Gas fluidization technology*.

Hu, C., Luo, K., Wang, S., Junjie, L., & Fan, J. (2019). The effects of collisional parameters on the hydrodynamics and heat transfer in spouted bed: A CFD-DEM study. *Powder Technology*, *353*, 132–144.

Kempe, T., & Fröhlich, J. (2012). Collision modelling for the interface-resolved simulation of spherical particles in viscous fluids. *Journal of Fluid Mechanics*, *709*, 445–489.

Konrath, M., Hackbarth, M., & Nirschl, H. (2014). Process monitoring and control for constant separation conditions in centrifugal classification of fine particles. *Advanced Powder Technology*, *25*(3), 991–998.

Lv, B., Luo, Z., Zhang, B., Qin, X., & Zhu, C. (2018). Particle motion and separation behavior of coal in gas–solid separation fluidized bed. *Powder Technology*, *339*, 344–353.

Masliyah, J. H., Nasr-El-Din, H., & Nandakumar, K. (1989). Continuous separation of bidisperse suspensions in inclined channels. *International Journal of Multiphase Flow*, *15*(5), 815–829.

Smagorinsky, J. (1963). General circulation experiments with the primitive equations: I. The basic experiment. *Monthly Weather Review*, *91*(3), 99–164.

Tenneti, S., Garg, R., & Subramaniam, S. (2011). Drag law for monodisperse gas–solid systems using particle-resolved direct numerical simulation of flow past fixed assemblies of spheres. *International Journal of Multiphase Flow*, *37*(9), 1072–1092.

Vega-Garcia, D., Cilliers, J. J., & Brito-Parada, P. R. (2020). CFD modelling of particle classification in mini-hydrocyclones. *Separation and Purification Technology*, *251*, 117253.

Wang, S., Luo, K., Hu, C., Sun, L., & Fan, J. (2018). Effect of superficial gas velocity on solid behaviors in a full-loop CFB. *Powder Technology*, *333*, 91–105.

Xu, J., Lu, X., Zhang, W., Chen, J., Wang, Q., Chen, Y., & Guo, Q. (2018). Effects of superficial gas velocity and static bed height on gas-solid flow characteristics in a 60-meter-high transparent CFB riser. *Chemical Engineering Journal*, *334*, 545–557.

Xue, Q., Dalluge, D., Heindel, T. J., Fox, R. O., & Brown, R. C. (2012). Experimental validation and CFD modeling study of biomass fast pyrolysis in fluidized-bed reactors. *Fuel*, *97*, 757–769.

Zhao, P., Xu, J., Ge, W., & Wang, J. (2020). A CFD-DEM-IBM method for Cartesian grid simulation of gas-solid flow in complex geometries. *Chemical Engineering Journal*, *389*, 124343.

Antonis Kokossis, Michael C. Georgiadis, Efstratios N. Pistikopoulos (Eds.)
PROCEEDINGS OF THE 33rd European Symposium on Computer Aided Process Engineering (ESCAPE33), June 18-21, 2023, Athens, Greece
 http://dx.doi.org/10.1016/B978-0-443-15274-0.50024-X

Intensification of ATJ Process Using Catalytic Distillation, Optimization Considering Control and Economic Issues

Gabriel Contreras-Zarazúa1,[a,b] Juan José Quiroz Ramírez,[a] Eduardo Sánchez-Ramirez,[b] Juan Gabriel Segovia-Hernández[b]

[a]*CONACyT-CIATEC A.C Center of Applied Innovation in Competitive Technologies, Omega 201, Industrial Delta, 37545 León, Gto., Mexico.*
[b]*Department of Chemical Engineering University of Guanajuato, Noria Alta S/N, Noria Alta,36000, Guanajuato, Gto., Mexico.*

Abstract

In this work the design and simultaneous optimization of an intensified process for jet fuel production by the ATJ process was developed. The intensified process consists of a catalytic reactive distillation column. The process integrates the oligomerization, hydrogenation and purification stages in a single unit. The design and simultaneous optimization process were performed using the differential evolution with tabu list (DETL) algorithm. Total annual cost and condition number were considered as objective functions to evaluate cost issues and control properties, respectively. The optimized process was contrasted against the original intensified process. The results indicate that the optimized process reduces energy consumption by 30% and total annual cost by up to 80%, due to reductions in energy consumption and equipment size. Regarding the control properties, it is observed that the optimized process has a condition number of 82.52 against 915,000 of the original process. This means that the optimized process withstands disturbances better, which facilitates control and ensures the quality of the bioturbosine in a more reliable way.

Keywords: Biojet-fuel, reactive distillation, process intensification.

1. Introduction

he world is currently facing increasing energy demands. This growth in energy is associated with an increase in the production of goods and services, increase in the quality of life, among others. Unfortunately, most of the energy produced by humans derives from the burning of fossil fuels such as coal, natural gas, gasoline, etc., which generates severe environmental problems by the emission of greenhouse gases such as CO2. Carbon dioxide is the main greenhouse gas, recent studies reveal that by 2018 its concentration in the atmosphere had gone from 280 to 408ppm, taking the year 1900 as a reference. This increase means a 0.8°C increase in the earth's global temperature and it is expected that by the year 2100 the temperature increase will reach 5°C(Bains et al., 2017).
In order to reduce their emissions, various industries such as the aviation sector have started to develop technologies to reduce their CO2 emissions. The aircraft industry is responsible for 2.6% of CO2 emissions and it is expected that by mid-century this industry will emit 20.2% of CO2 emissions. An interesting alternative to reduce polluting emissions is to replace jet fuel (also called biojet-fuel), which comes from fossil sources,

with biojet-fuel, which is a renewable fuel derived from biomass residues, vegetable oils, etc. Several processes and biomasses have been proposed to produce biojet-fuel, however, a little explored alternative is the ATJ (Alcohol to Jet fuel) process, which consists of converting lignocellulosic residues into sugars, which are transformed into hydrocarbons (Wang and Tao, 2016). As with other alternatives to produce biojet-fuel or even other biofuels, economic viability remains the main challenge for this type of technology. In the case of the ATJ process, a critical and energy-intensive stage is the oligomerization and hydrogenation stage, since it is at this stage that hydrocarbons are produced and separated. In this sense, process intensification coupled with rigorous optimization techniques can help to reduce energy consumption and process operating costs, in order to increase the profitability and adoption of these processes. In this work, an intensified process based on catalytic reactive distillation is proposed to replace the current oligomerization and hydrogenation steps in the ATJ process. This column is simultaneously designed and optimized using the differential evolution stochastic optimization method with tabu list (DETL). The minimization of the total annual cost is proposed as the economic objective function. On the other hand, because the catalytic column is notably more complex and in order to ensure a viable operation of this equipment, the minimization of the condition number is proposed as the second objective function in order to evaluate the control properties.

2. Methodology

First, the intensified catalytic reactive distillation process is generated from the conventional process. As shown in Figure 1, the conventional process consists of 2 reactive zones, the first one is the oligomerization stage, where ethylene is fed to a fixed bed reactor, then the reactor products which are light compounds (C2- C5) and oligomers (C6-C17) are fed to a distillation column, where the light compounds are recirculated, while the heavy oligomers are sent to the hydrogenation zone. In this second reaction zone, the oligomers are transformed to alkanes by hydrogenation reactions, which are carried out in a second fixed-bed reactor. Finally, the products of the hydrogenation reactor are separated into three main fractions, light, jet-fuel and diesel(Wang and Tao, 2016).

The catalytic column integrates these separation and reaction stages in a single equipment, the oligomerization zone is located at the top of the column, it is in this zone where the proper distribution of hydrocarbons is obtained. The compounds that descend from this zone are transformed into olefins (alkanes), which at the same time as they are produced are separated into the 3 main fractions. The mathematical modeling of this intensified process was performed in Aspen Plus software, considering an ethylene input flow of 2100kg/hr. The kinetic models were taken from (Goortani et al., 2015). NRTL-RK (Interan, 2021) was selected as thermodynamic model. It is important to mention that the original design of the intensified column was obtained through a sensitivity analysis of some important design parameters, such as hold ups, number of hold ups, feed stages, etc.

The preliminary design of the catalytic column will be used as the starting point for the optimization algorithm. As mentioned earlier, the total annual cost (TAC) and the condition number were the objective functions to be minimized. The total annual cost consists of the sum of the annualized capital costs, plus the operating costs. As shown below:

$$TAC = \frac{\text{Capital cost}}{\text{Payback period}} + \text{Operating cost} \quad (1)$$

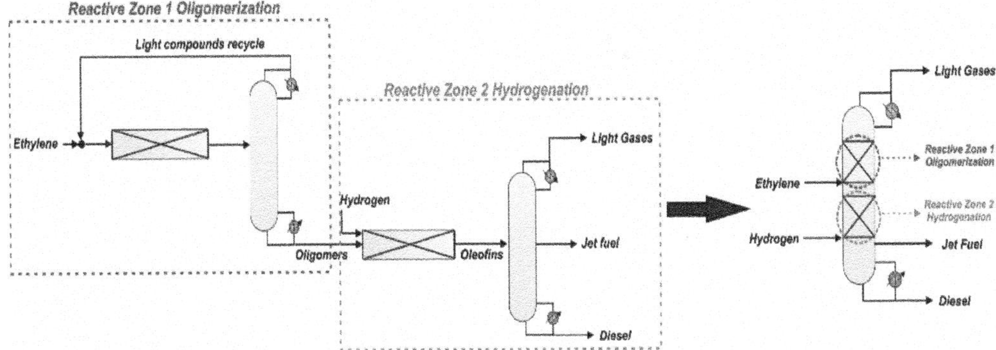

Figure 1. Intensification of ATJ process using catalytic distillation

The capital cost is associated to the construction cost of process equipment, such as condensers, reboilers, distillation columns, trays, process vessels and compressors; whereas the operating cost corresponds to the cost of use of electricity, cooling water, steam and solvents. The Guthrie method is used to calculate the total annual cost. The parameters and equations required to estimate the cost of equipment and utilities were taken from Turton et al., 2018. Carbon steel is the construction material considered, and a payback period of ten years is used, which corresponds to an average payback time of a chemical plant according to (Turton et al., 2018). Trays type Sieve with 0.61m spacing were considered for distillation columns. The utilities costs are calculated considering 8,500 hours of operation per year.Cooling water (0.355USD/GJ), electricity (16.8 USD/kwh) and Fired heat (20.92 USD/GJ) were the utilities considered (Turton et al., 2018).

The condition number is a common index used to determine in a qualitative way the controllability of a specific process. It has proven be a powerful tool as it analyzes the control properties of a process in order to detect potential operational problems. Mathematically, the condition number is calculated through a singular value decomposition of the gain matrix according to equation 2:

$$K = W \cdot S \cdot V^T \quad (2)$$

Where K is the relative gains matrix, W and V are unitary matrices and S is the singular value matrix. From the matrix S we take the maximum singular value (σ*) and the minimum singular value, to calculate the condition number as shown below:

$$\gamma = \frac{\sigma^*}{\sigma_*} \quad (3)$$

The condition number represents the sensitivity of a system to assimilate a disturbance without process destabilization. indicate that the processes could be susceptible to destabilization by small perturbations. Furthermore, low condition numbers represent a robust process that is resistant to perturbations

Based on the above, the optimization problem can be stated mathematically as equation 4.

$$\min\ Z=\left[TAC,\gamma\right]$$

$$\text{Subject to} \begin{array}{l} y_i \geq TE_i \\ w_i \geq u_i \end{array} \tag{4}$$

The objective functions are constrained to meet the minimum boiling point specifications (TE_i) for each of the products which are 220°C for jet fuel and 300°C for diesel. On the other hand, the optimization problem is also subject to meet minimum production quantities (u_i) which were 1000kg/hr for jet fuel and 100kg for diesel. As decision variables, the number of stages, feed stages, number of reactive stages, hold up values, reflux ratio and thermal load were taken.

Differential evolution with tabu list (DETL), the mathematical method used to solve the optimization problem. DETL is a direct population-based search method, specially designed for non-continuous and highly non-linear functions. This method consists of the 4 stages: generation of the initial population, mutation, crossover and selection. Its main feature is its ability to avoid or revisit areas with bad values of objective functions, by tracking previous searches using the tabu list, improving its computational efficiency (Srinivas and Rangaiah, 2007). The parameters used for the algorithm are: population size: 120, number of generations: 1250, Tabu list size 60, crossover factor: 0.8, mutation factor: 0.3 and tabu radius of 0.01. The algorithm parameters were taken from (Contreras-Zarazúa et al., 2017). Finally, the implementation of the optimization strategy involves a hybrid platform, where, the optimization algorithm is programmed in Excel and the mathematical model of the process and equation solving is in Aspen Plus.

3. Results

This section presents the results obtained during the design and simultaneous optimization stage of the catalytic column. In order to determine the solution with the best balance between both objective functions, the utopia point methodology was used to determine the solution, which requires the Pareto front analysis. The Pareto front of the solution is shown in Figure 2.

As can be seen in Figure 2, the range of condition numbers for the different solutions is quite wide, ranging from 11.34 to 2,500 000, indicating that the design of the column design parameters has a strong impact on the process accounting. Note the presence of a red dot in the Pareto, this dot corresponds to the solution with the best trade-off between the two objectives. Table 1 shows the comparison of the original intensified process versus the process obtained by the optimization technique.

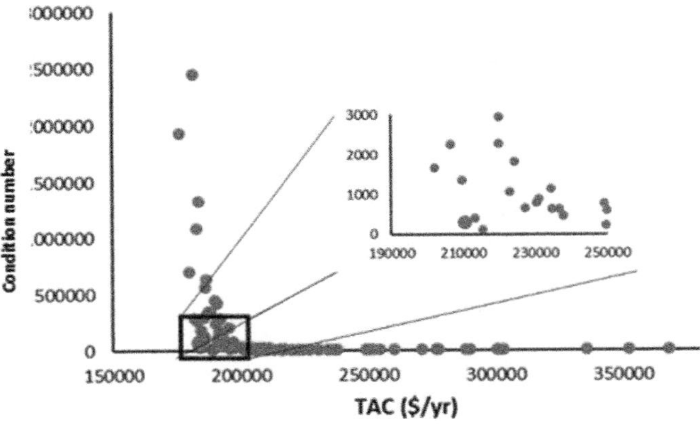

Figure2. Pareto front for catalytic distillation

Table 1. Design parameters for the original process and intensified process

Process variable	Original process	Optimized process
Number of stages	80	29
Ethylene feed stage	15	21
Hydrogen feed stage	59	26
Reactive stages (oligomerization)	2-29	12-16
Reactive stages (Hydrogenation)	30-59	17-25
Reflux ratio	100	51.55
Reboiler duty (kW)	170	117.642
Hydrogen mass flowrate (kg/hr)	22	20.6417
Diameter(m)	4	1.39
Condition number	915, 000	82.58
Total annual cost (USD/yr)	1,028,914	215,553

As can be seen in Table 1, the intensified process shows remarkable improvements over its original counterpart. It can be seen that the optimized process presents energy savings of 30% while the cost savings are almost 80%. As for the control properties, the optimized process has a condition number of 82.58 against 915,000 of the original process. This means that the optimized process is less sensitive to changes and disturbances. This is due to the reactive stages and column diameters. The oligomerization and hydrogenation reactions are very sensitive to temperature changes, so having many reactive stages for these areas means that it is easier for the process to have a temperature perturbation, which dramatically affects hydrocarbon conversion. In contrast, the optimized process has determined that it is not necessary to have many reactive stages for these zones, and the optimization process has identified the optimal location of the column temperature reactive zones, thus avoiding the process to be sensitive to disturbances.

4. Conclusions

In this work, the design and simultaneous optimization of a reactive distillation column for jet fuel production by the ATJ process was developed. The objective functions, total annual cost and condition number were used to evaluate the cost and control properties. The results indicate that the optimization algorithm is able to generate a design that has 30% energy savings and 80% cost savings over the original process. Also, the optimized process has a condition number of 82.58 which is a notably lower value than the original process. The results indicate that the design of a process using robust optimization techniques generates a significantly cheaper and more efficient process. Future work proposes the optimization of the conventional process and a rigorous control analysis in order to have a broader picture of the improvements of the optimized process.

References

Combust. Sci. 63, 146–172. https://doi.org/10.1016/j.pecs.2017.07.001

Contreras-Zarazúa, G., Vázquez-Castillo, J.A., Ramírez-Márquez, C., Segovia-Hernández, J.G., Alcántara-Ávila, J.R., 2017. Multi-objective optimization involving cost and control properties in reactive distillation processes to produce diphenyl carbonate. Comput. Chem. Eng. 105. https://doi.org/10.1016/j.compchemeng.2016.11.022

Goortani, B.M., Gaurav, A., Deshpande, A., Ng, F.T.T., Rempel, G.L., 2015. Production of isooctane from isobutene: Energy integration and carbon dioxide abatement via catalytic distillation. Ind. Eng. Chem. Res. 54, 3570–3581. https://doi.org/10.1021/ie5032056

Interan, R.R., 2021. Optimización De Sistemas Intensificados Para La Producción De Bioturbosina A Través De Alcoholes De Forma Sustentable. Universidad de Guanajuato.

Srinivas, M., Rangaiah, G.P., 2007. Differential evolution with tabu list for solving nonlinear and mixed-integer nonlinear programming problems. Ind. Eng. Chem. Res. 46, 7126–7135. https://doi.org/10.1021/ie070007q

Turton, R., Bailie, R.C., Whiting, W.B., Shaeiwitz, J.A., 2018. Analysis, synthesis and design of chemical processes. Pearson Education., 1-1200.

Wang, W.C., Tao, L., 2016. Bio-jet fuel conversion technologies. Renew. Sustain. Energy Rev. 53, 801–822. https://doi.org/10.1016/j.rser.2015.09.016

Antonis Kokossis, Michael C. Georgiadis, Efstratios N. Pistikopoulos (Eds.)
PROCEEDINGS OF THE 33rd European Symposium on Computer Aided Process Engineering (ESCAPE33), June 18-21, 2023, Athens, Greece
 http://dx.doi.org/10.1016/B978-0-443-15274-0.50025-1

Efficient physical model building algorithm using equations extracted from documents

Shota Kato and Manabu Kano

Department of Systems Science, Kyoto University, Yoshida-honmachi, Sakyo-ku, Kyoto 606-8501, Japan

Abstract

To facilitate arduous physical model building tasks, we aim to develop automated physical model builder, AutoPMoB, which automatically builds physical models from multiple documents. This study focuses on a method for building desired models from equations, a fundamental technology for realizing AutoPMoB. We defined two requirements desired models must fulfill and proposed a gradual method. The gradual method obtains combinations of input equations satisfying one requirement and then modifies the combinations to fulfill the other. For comparison, we used an exhaustive method, which obtains all possible combinations of the equations and checks whether each combination fulfills the requirements. We compared the models built by the two methods and their computational time in four cases. In all cases, both methods built models including all correct models. The gradual method and exhaustive method took 1.0×10^{-4} s and 30 s, respectively, to build models from 23 equations.

Keywords: Physical model building, Modeling algorithm, Process systems engineering.

1. Introduction

In the process industry, a digital twin, which is a model that mimics the behavior of a real process like a twin, is expected to yield significant benefits. Digital twins should be based on physical models derived from scientific principles rather than statistical models derived only from data since data collection through experiments under various conditions, especially abnormal ones, in the manufacturing process is difficult from cost and safety perspectives. However, the physical model building requires a deep understanding of the process and trial-and-error improvements in the model's accuracy. To facilitate the arduous task, we aim to develop automated physical model builder, AutoPMoB, which automatically builds physical models from multiple documents [Kato and Kano (2022)]. AutoPMoB 1) retrieves documents related to the target process from literature databases, 2) unifies the documents' formats, 3) extracts information vital for model building from the documents, 4) unifies the expression of the extracted information, and 5) combines the information to build physical models and presents the models with their rankings. Users of AutoPMoB decide which model to use based on the rankings.

This study focuses on how to build models by combining equations extracted from documents to perform task 5. We assume that users give input variables (IVs) and output variables (OVs) of the models. For example, when a user builds a model to predict a variable, the variable is an OV, and the other measurable variables are IVs.

The present study proposes a gradual method and uses an exhaustive method as a baseline. We compare the models built by the two methods and the required computational time.

2. Methods

The number of degrees of freedom (DoF) N_{DF} of a model calculating the OVs from the IVs must be equal to the number of the IVs N_{IV}. We propose a gradual method for building such desired models from a set of N_{e} equations $S_{\mathrm{e}} = \{e_1, \ldots, e_{N_e}\}$. Each desired model is an equation group (EG) that meets the following requirements: 1) the EG includes the IVs and OVs, and 2) N_{DF} of the EG is equal to N_{IV}. Here, the given set of IVs and that of OVs are denoted by $S_{\mathrm{IV}} = \{v_1, \ldots, v_{N_{\mathrm{IV}}}\}$ and $S_{\mathrm{OV}} = \{v'_1, \ldots, v'_{N_{\mathrm{OV}}}\}$, respectively. Since there has not existed any method for comparison, we use an exhaustive method, which obtains all combinations of N_{e} equations and checks whether each combination fulfills the requirements, as a baseline.

2.1. Exhaustive method

The exhaustive method is shown in Algorithm 1. The method obtains all possible combinations of N_{e} equations (ll. 3–6) and checks whether each combination fulfills the requirements above (ll. 7–13). The total number of the combinations is

$$\sum_{n=1}^{N_{\mathrm{e}}} {}_{N_{\mathrm{e}}}C_n = 2^{N_{\mathrm{e}}} - 1. \tag{1}$$

This method checks whether each of the $2^{N_{\mathrm{e}}} - 1$ EGs satisfies the requirements and outputs EGs that meet the requirements. The time complexity of this exhaustive method is $O(2^{N_{\mathrm{e}}})$.

Algorithm 1: Exhaustive method for building model

Input: set of equations $S_{\mathrm{e}} = \{e_1, \ldots, e_{N_{\mathrm{e}}}\}$
set of input variables $S_{\mathrm{IV}} = \{v_1, \ldots, v_{N_{\mathrm{IV}}}\}$
set of output variables $S_{\mathrm{OV}} = \{v'_1, \ldots, v'_{N_{\mathrm{OV}}}\}$
Output: set of equation groups S_{EG}
1: $S_{\mathrm{EG}} \leftarrow \emptyset$
2: $S'_{\mathrm{EG}} \leftarrow \emptyset$
3: **for** $n_{\mathrm{e}} = 1$ to N_{e} **do**
4: $S''_{\mathrm{EG}} \leftarrow n_{\mathrm{e}}$ combinations of S_{e}
5: add S''_{EG} to S'_{EG}
6: **end for**
7: **for** G' in S'_{EG} **do**
8: $S_{\mathrm{v},G'} \leftarrow$ set of variables in G'
9: $N_{\mathrm{DF},G'} \leftarrow$ the number of degrees of freedom of G'
10: **if** $(S_{\mathrm{IV}} \cup S_{\mathrm{OV}}) \subseteq S_{\mathrm{v},G'}$ **and** $N_{\mathrm{DF},G'}$ is equal to N_{IV} **then**
11: add G' to S_{EG}
12: **end if**
13: **end for**
14: **return** S_{EG}

2.2. Gradual Method

Algorithm 2 shows the pseudocode of the gradual method. The method first obtains sets of variables contained in each formula (ll. 2–4), sets of equations containing each of the IVs and OVs (ll. 5–8), and sets of EGs that include both IVs and OVs (l. 9). Then, the method checks each of the obtained EGs, G'. If the number of DoF of G', $N_{\mathrm{DF},G'}$, is equal to N_{IV}, G' is judged desired (ll. 10–13). If not, to change $N_{\mathrm{DF},G'}$, the method searches the remaining equations for the equations containing the variables that appear in G' (ll. 14–17). If such equations exist, one or more of the equations are combined with G', and the combined EGs whose number of DoF is equal to N_{IV} are decided desired (ll. 18–24). Finally, the method outputs all the desired EGs.

Algorithm 2: Gradual method for building model

Input: set of equations $S_{\mathrm{e}} = \{e_1, \ldots, e_{N_{\mathrm{e}}}\}$
set of input variables $S_{\mathrm{IV}} = \{v_1, \ldots, v_{N_{\mathrm{IV}}}\}$
set of output variables $S_{\mathrm{OV}} = \{v'_1, \ldots, v'_{N_{\mathrm{OV}}}\}$
Output: set of equation groups S_{EG}
1: $S_{\mathrm{EG}} \leftarrow \emptyset$
2: **for** $n = 1$ to N_{e} **do**
3: $S_{\mathrm{v},e_n} \leftarrow$ set of variables in e_n
4: **end for**
5: $N_{\mathrm{IOV}} = N_{\mathrm{IV}} + N_{\mathrm{OV}}$
6: **for** $n = 1$ to N_{IOV} **do**
7: $S_{\mathrm{e},v''_n} \leftarrow$ set of equations including v''_n in S_{e}, where $v''_n \in (S_{\mathrm{IV}} \cup S_{\mathrm{OV}})$
8: **end for**
9: $S'_{\mathrm{EG}} \leftarrow \{\{e'_1, \ldots, e'_{N_{\mathrm{IOV}}}\} \mid e'_n \in S_{\mathrm{e},v''_n}$ for every $n \in \{1, \ldots, N_{\mathrm{IOV}}\}\}$
10: **for** G' in S'_{EG} **do**
11: $N_{\mathrm{DF},G'} \leftarrow$ the number of degrees of freedom of G'
12: **if** $N_{\mathrm{DF},G'}$ is equal to N_{IV} **then**
13: add G' to S_{EG}
14: **else**
15: $S_{\mathrm{v},G'} \leftarrow \bigcup_{e \in G'} S_{\mathrm{v},e}$
16: $\tilde{S}_{\mathrm{v},G'} \leftarrow S_{\mathrm{v},G'} \setminus (S_{\mathrm{IV}} \cup S_{\mathrm{OV}})$
17: $S''_{\mathrm{EG}} \leftarrow \{\{e''_1, \ldots, e''_{|\tilde{S}_{\mathrm{v},G'}|}\} \mid e''_n \in S_{\mathrm{e},\tilde{v}_n}$ for every $\tilde{v}_n \in \tilde{S}_{\mathrm{v},G'}$
18: **For** G'' in S''_{EG} **do**
19: $\tilde{G} = G'' \cup G'$
20: $N_{\mathrm{DF},\tilde{G}} \leftarrow$ the number of degrees of freedom of $\tilde{G}$
21: **if** $N_{\mathrm{DF},\tilde{G}}$ is equal to N_{IV} **then**
22: add $\tilde{G}$ to S_{EG}
23: **end if**
24: **end for**
25: **end if**
26: **end for**
27: **return** S_{EG}

The time complexity of the gradual method is

$$O(N_{\mathrm{e}}N_{\mathrm{IOV}}) + O\left(\prod_{n=1}^{N_{\mathrm{IOV}}} |S_{\mathrm{e},v_n''}|\right) + \sum_{i=1}^{|S_{\mathrm{EG}}'|} O\left(\prod_{n=1}^{|\tilde{S}_{\mathrm{v},G'}(G_i')|} |S_{\mathrm{e},\tilde{v}_n}(G_i')| + |S_{\mathrm{EG}}''(G_i')|\right), \quad (2)$$

where G_i' denotes the ith element of S_{EG}' and $S(G_i')$ indicates that S depends on G_i'. The first, second, and third terms indicate the time complexities at lines 1–8, 9, and 10–26, respectively. The gradual method obtains combinations focusing on IVs and OVs and judges whether each combination satisfies the requirements; hence, the time complexity is smaller than the exhaustive method.

3. Experiments

3.1. Datasets

We generated datasets consisting of S_{e}, S_{IV}, and S_{OV} shown in Table 1 based on the textbook about process control [Seborg et al. (2010)] and defined correct models, which can calculate the values of the OVs by substituting values into IVs. Then, we used these S_{e}, S_{IV}, and S_{OV} as the input of the model building methods and checked whether the correct models could be obtained.

3.2. Results and Discussion

The model building results of the exhausted method and the gradual method are summarized in Table 2. In Case 1, the two methods obtained the same correct models. In Case 2, where equations with only variables not included in IVs and OVs exist, the gradual method built the same models as in Case 1, while the exhaustive method built 2,097,152 models, most of which were incorrect. Since adding unnecessary equations does not change the number of DoF, the exhaustive method outputs much more models than the gradual method. In Case 3, similar to Case 1, the same models were obtained by the two methods. In Case 4, similar to Case 2, the models obtained by the two methods comprised all the correct models, and the number of models built by the gradual method is less than that by the exhaustive method.

All the built models in Table 2 meet the predefined requirements, but some models, such as those built by only the exhaustive method in Case 2 and the second model in Case 3, are impractical because the values of the OVs cannot be obtained. This is because the two methods judged only the necessary conditions. To remove such impractical models and obtain only practical ones, we need to judge whether the values of OVs can be obtained. If all the equations in one model are linear, the additional check can be done using the coefficient matrix. However, built models usually include nonlinear equations, as shown in Case 3. We must develop a method for selecting practical models.

Equations (1) and (2) indicate that the gradual method can build models within a shorter time than the exhaustive method, even when the number of equations is large. In Case 2, which took the longest computational time among the four cases, the computational time of the exhaustive method was 30 s, and that of the gradual method was 1.0×10^{-4} s. We used MacBook Pro with an Apple M1 Max processor, 10 Cores, 64GB RAM, and macOS 13.0.1.

Table 1 Equation set, input variable (IV) set, and output variable (OV) set used in experiments.

Case	Equation set S_e	IV set S_{IV}	OV set S_{OV}
1	$y = x_1 + x_2$ $x_2 = 0.1$ $x_2 = 1$	x_1	y
2	$y = x_1 + x_2$ $x_2 = 0.1$ $x_2 = 1$ $a_i = i\ (i = 1, \dots, 20)$	x_1	y
3	$V\frac{dC_A}{dt} = v_0 C_{A0} - v_0 C_A + r_A V$ $-r_A = kC_A^n\ (n = 0, 1, 2)$ $k = k_0 \exp(a/T)$ $k_0 = 0.1$ $a = 10$	v_0 C_{A0} T V C_A	$\frac{dC_A}{dt}$
4	$V\frac{dC_A}{dt} = v_0 C_{A0} - v_0 C_A + r_A V$ $-r_A = kC_A^n\ (n = 0, 1, 2)$ $k = k_0 \exp(a/T)$ $k_0 = 0.1$ $a = 10$ $Q = UA(T_c - T)$ $V\rho\frac{dT}{dt} = wC(T_i - T) + H_r V r_A + Q$	v_0 C_{A0} T V C_A T_c U A ρ w C T_i H_r	$\frac{dC_A}{dt}$ $\frac{dT}{dt}$

4. Conclusion

We proposed a gradual method to build desired models from equations extracted from documents. We used an exhaustive method for comparison and compared the model building results of the two methods in four cases where the equations, IVs, and OVs were generated based on the textbook about process control [Seborg et al. (2010)]. Both methods built models including all correct models in all cases. In terms of the computational time, the gradual method took 1.0×10^{-4} s while the exhaustive method took 30 s when the number of equations was 23. However, the models built by the two methods sometimes had unnecessary ones. We will develop a method for filtering the obtained models to only correct models in our future work.

Table 2 Model building results of exhaustive method and gradual method.

Case	Method	#built models	Examples of built models
1	Exhaustive	2	$\begin{cases} y = x_1 + x_2 \\ x_2 = 1 \end{cases}$ $\begin{cases} y = x_1 + x_2 \\ x_2 = 0.1 \end{cases}$
	Gradual	2	
2	Exhaustive	2,097,152	$\begin{cases} y = x_1 + x_2 \\ x_2 = 1 \\ a_1 = 1 \\ a_2 = 2 \end{cases}$
	Gradual	2	
3	Exhaustive	10	$\begin{cases} V\frac{\mathrm{d}C_\mathrm{A}}{\mathrm{d}t} = v_0 C_{\mathrm{A}0} - v_0 C_\mathrm{A} + r_\mathrm{A} V \\ -r_\mathrm{A} = k \\ k = k_0 \exp(a/T) \\ k_0 = 0.1 \\ a = 10 \end{cases}$
	Gradual	10	$\begin{cases} V\frac{\mathrm{d}C_\mathrm{A}}{\mathrm{d}t} = v_0 C_{\mathrm{A}0} - v_0 C_\mathrm{A} + r_\mathrm{A} V \\ -r_\mathrm{A} = k \\ -r_\mathrm{A} = kC_\mathrm{A} \\ -r_\mathrm{A} = kC_\mathrm{A}^2 \\ k = k_0 \exp(a/T) \end{cases}$
4	Exhaustive	22	$\begin{cases} V\frac{\mathrm{d}C_\mathrm{A}}{\mathrm{d}t} = v_0 C_{\mathrm{A}0} - v_0 C_\mathrm{A} + r_\mathrm{A} V \\ -r_\mathrm{A} = k \\ k = k_0 \exp(a/T) \\ k_0 = 0.1 \\ a = 10 \\ Q = UA(T_\mathrm{c} - T) \\ V\rho\frac{\mathrm{d}T}{\mathrm{d}t} = wC(T_\mathrm{i} - T) + H_\mathrm{r} V r_\mathrm{A} + Q \end{cases}$
	Gradual	13	

References

S. Kato and M. Kano, 2022, Towards An Automated Physical Model Builder: CSTR Case Study, Computer Aided Chemical Engineering, Volume 49, Pages 1669–1674.

D. E. Seborg, *et al.*, 2010, Process dynamics and control, John Wiley & Sons.

Acknowledgments

This work was supported by JSPS KAKENHI Grant Number JP21K18849.

Antonis Kokossis, Michael C. Georgiadis, Efstratios N. Pistikopoulos (Eds.)
PROCEEDINGS OF THE 33rd European Symposium on Computer Aided Process Engineering (ESCAPE33), June 18-21, 2023, Athens, Greece
 http://dx.doi.org/10.1016/B978-0-443-15274-0.50026-3

Developing a rigorous chemical reaction network reduction strategy for *n*-hexadecane hydroisomerisation

Fernando Vega-Ramon,[a] Wei Wang,[b] Wei Wu,[b] Dongda Zhang,*[a]

[a] *Department of Chemical Engineering and Analytical Science, the University of Manchester, The Mill, Sackville Street, Manchester, M1 3AL, UK.*

[b] *School of Chemistry and Material Sciences, Heilongjiang University, Harbin, 150080, China*

* *Corresponding author. Email: dongda.zhang@manchester.ac.uk*

Abstract

The hydroisomerisation of long-chain alkanes on bifunctional zeolite catalysts is a promising synthetic route for the sustainable production of diesel fuel derived from vegetable oils. Significant efforts have been made in literature to propose potential reaction mechanisms and materials with improved catalytic activity, but the development of a kinetic model remains a challenge due to the complexity of the reaction network. In this study a microkinetic model was initially constructed to simulate both observable and non-observable states in the reaction mechanism. Upon fitting the model against packed bed reactor experimental data, two rigorous model simplification strategies were proposed and adopted to guide reaction network reduction. In the physics-based approach the ratios of the fitted microkinetic rate constants are evaluated against a threshold to identify irreversible reactions, whereas in the mathematics-based approach an additional penalty term is introduced during microkinetic parameter estimation to penalise the number of active chemical pathways. The reaction network reduction results were similar under both strategies and provided valuable insight on the irreversibility of isomerisation, cracking and desorption reactions, as also supported by observations in literature. Finally, simplified kinetic models that are only capable of simulating observable states based on the steady-state assumption were also constructed using the two reduced reaction networks. Through comparison, it is concluded that both simplified kinetic models can yield satisfactory fitting result of the process; however, their model structures and parameter values are highly sensitive to the mechanism reduction strategy adopted. This strongly indicates the impact of different reaction network reduction strategies on kinetic model construction, and that directly building a steady-state kinetic model may not the best approach to investigate intrinsic reaction mechanisms. The kinetic modelling and reaction network reduction frameworks proposed in this work, therefore, provide a new avenue to infer mechanistic knowledge from kinetic data in a more efficient manner than traditional steady-state approaches.

Keywords: Hydroisomerisation, kinetic model, reaction network reduction, catalytic reaction, intrinsic mechanisms.

1. Introduction

1.1. Background and motivation

The hydroisomerisation of *n*-alkanes found in waxy feedstock is considered a promising sustainable route for diesel synthesis. Recent research in this area has mainly focused on novel catalyst development in an attempt to minimize the yield toward cracking by-products and improve isomerization selectivites. This sets the motivation behind microkinetic modelling approaches, which can help elucidate potential reaction mechanisms, streamline the design of catalysts with improved performance, and provide guidelines for model-based design of experiments and process optimisation (Motagamwala & Dumesic, 2021). In this work, a microkinetic modelling and reaction network simplification framework was proposed to analyse experimental data of *n*-hexadecane hydrosiomerisation and cracking over a Pd/ZSM-12 catalyst.

2. Methodology

2.1. Kinetic model construction

2.1.1. Base-case reaction mechanism

To construct a microkinetic model comprising equations for the rate of change of reactants, products and intermediate species, the reaction network presented in Fig. 1 was proposed by considering the plausible elementary steps occurring at the catalyst surface.

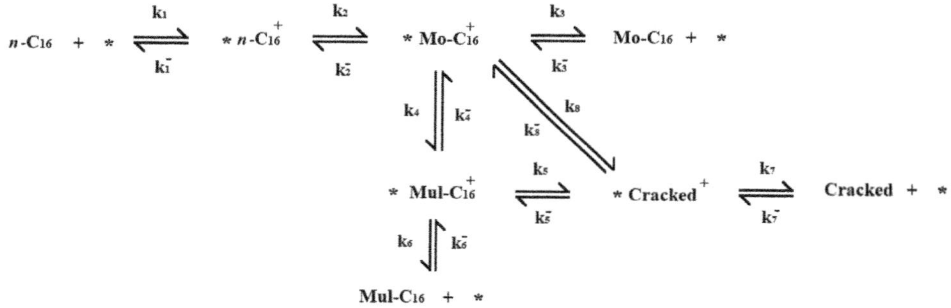

$$n\text{-}C_{16} + * \underset{k_1^-}{\overset{k_1}{\rightleftharpoons}} *\,n\text{-}C_{16}^+ \underset{k_2^-}{\overset{k_2}{\rightleftharpoons}} *\,\text{Mo-}C_{16}^+ \underset{k_3^-}{\overset{k_3}{\rightleftharpoons}} \text{Mo-}C_{16} + *$$

$$*\,\text{Mo-}C_{16}^+ \underset{k_4^-}{\overset{k_4}{\rightleftharpoons}} *\,\text{Mul-}C_{16}^+ \qquad *\,\text{Mo-}C_{16}^+ \underset{k_8^-}{\overset{k_8}{\rightleftharpoons}} *\,\text{Cracked}^+$$

$$*\,\text{Mul-}C_{16}^+ \underset{k_5^-}{\overset{k_5}{\rightleftharpoons}} *\,\text{Cracked}^+ \underset{k_7^-}{\overset{k_7}{\rightleftharpoons}} \text{Cracked} + *$$

$$*\,\text{Mul-}C_{16}^+ \underset{k_6^-}{\overset{k_6}{\rightleftharpoons}} \text{Mul-}C_{16} + *$$

Figure 1: Proposed reaction mechanism. n-C_{16}, Mo-C_{16} and Mul- C_{16} refer to unbranched, monobranched and multibranched hexadecane isomers, Cracked refers to cracking by-products. Their respective carbenium ion intermediates are denoted by the subscript $^+$, while * represents an active site. k_i and k_i^- are the forward and reverse reaction rates of elementary step i, respectively.

In the classical mechanism of *n*-alkane hydroisomerisation described by Weitkamp (2012), the unbranched alkane reactant is first adsorbed at the metallic sites of the bifunctional catalyst, where it is dehydrogenated into its corresponding *n*-alkene. The *n*-alkene intermediate is then readily protonated at the acid catalytic sites, forming a primary alkylcarbenium ion intermediate. This intermediate then undergoes successive skeletal rearrangement reactions into more stable monobranched, dibranched and tribranched isomers. The carbenium isomers desorb from the acid sites as their corresponding branched alkenes, which are then hydrogenated at the metallic sites to yield branched isomers of the starting *n*-alkane. The branched isomers may also crack due to β-scission reactions, resulting in shorter-chain alkenes and carbenium ions which can also participate in further cracking reactions.

In this study, the packed-bed reactor (PBR) experimental data lumped dibranched and tribranched isomer products into a single multibranched species. Similarly, the by-products of the cracking side-reactions were lumped into a single species for convenience. The mechanism proposed in this study was therefore modified accordingly such that it was in line with the measured states, as seen in Fig. 1. Furthermore, the fractional site coverage of surface species is not observable during practical applications, and so including all the reaction intermediates and active site balances in the kinetic analysis would lead to non-identifiable model structures. Coonradt and Garwood (1964) suggested that if there are sufficiently high concentrations of metallic sites for hydrogenation/dehydrogenation, the alkene intermediates assume their equilibrium concentrations. This, in turn, means that the acid site-catalysed isomerisation and cracking reactions become selectivity- and rate-determining for the overall reaction (Vandegehuchte et al., 2014). Thus, the metal-catalysed surface reactions were neglected in the present study, and it was assumed that the gas-phase species directly adsorb to (or desorb from) their corresponding alkylcarbenium ion intermediates at the acidic sites. For this reason, the alkene intermediates and the metallic active site balance were not considered during kinetic model construction.

2.1.2. Microkinetic model construction

The microkinetic model developed in this work consists of balance equations for both the gas-phase species as well as the non-observable intermediate states in the proposed reaction network (Fig. 1). These are presented in Eq. 1(a) – Eq 1(i), where x_i refers to the mass fraction of gas-phase species i, τ is the weight-based spacetime (h), θ_i is the fractional coverage of the alkylcarbenium intermediate i, the subscripts "r", "mo", "mul", and "cr" refer to unbranched, monobranched, multibranched, and cracked species, respectively. Forward and backward reaction constants (g_c mol^{-1} h^{-1}) of the jth elementary step are denoted by k_j and k_j^-, respectively. c_{θ_T} is the total active site concentration of the catalyst (mol g_c^{-1}), ρ_b is the catalyst bulk density (g_c m^{-3}) and $c_{r,0}$ is the initial concentration of the reactant (mol m^{-3}). In addition, the model assumes 1:1 overall reaction stoichiometry (this is exact for the isomerisation reactions but clearly an approximation for the cracking reactions) and isothermal reactor operation, such that the total pressure of the reacting mixture can be assumed constant.

$$\frac{dx_r}{d\tau} = c_{\theta_T} \cdot (-k_1 \cdot x_r \cdot \theta_V + k_1^- \cdot \theta_r) \quad \text{Eq. 1(a)}$$

$$\frac{d\theta_r}{d\tau} = \frac{c_{r0}}{\rho_b} \cdot [k_1 \cdot x_r \cdot \theta_V + k_2^- \cdot \theta_{Mo} - (k_1^- + k_2) \cdot \theta_r] \quad \text{Eq. 1(b)}$$

$$\frac{dx_{Mo}}{d\tau} = c_{\theta_T} \cdot (k_3 \cdot \theta_{Mo} - k_3^- \cdot x_{Mo} \cdot \theta_V) \quad \text{Eq. 1(c)}$$

$$\frac{d\theta_{Mo}}{d\tau} = \frac{c_{r0}}{\rho_b} \cdot [k_3^- \cdot x_{Mo} \cdot \theta_V + k_2 \cdot \theta_r + k_4^- \cdot \theta_{Mul} + k_8^- \cdot \theta_{Cr} - (k_2^- + k_3 + k_4 + k_8) \cdot \theta_{Mo}] \quad \text{Eq. 1(d)}$$

$$\frac{dx_{Mul}}{d\tau} = c_{\theta_T} \cdot (k_6 \cdot \theta_{Mul} - k_6^- \cdot x_{Mul} \cdot \theta_V) \quad \text{Eq. 1(e)}$$

$$\frac{d\theta_{Mul}}{d\tau} = \frac{c_{r0}}{\rho_b} \cdot [k_6^- \cdot x_{Mul} \cdot \theta_V + k_4 \cdot \theta_{Mo} + k_5^- \cdot \theta_{cr} - (k_4^- + k_5 + k_6) \cdot \theta_{Mul}] \quad \text{Eq. 1(f)}$$

$$\frac{dx_{cr}}{d\tau} = c_{\theta_T} \cdot (k_7 \cdot \theta_{cr} - k_7^- \cdot x_{cr} \cdot \theta_V) \quad \text{Eq. 1(g)}$$

$$\frac{d\theta_{cr}}{d\tau} = \frac{c_{r0}}{\rho_b} \cdot [k_7^- \cdot x_{Cr} \cdot \theta_V + k_5 \cdot \theta_{Mul} + k_8 \cdot \theta_{Mo} - (k_5^- + k_7 + k_8^-) \cdot \theta_{Cr}] \quad \text{Eq. 1(h)}$$

$$\theta_V = 1 - \theta_r - \theta_{Mo} - \theta_{Mul} - \theta_{Cr} \quad \text{Eq. 1(i)}$$

2.1.3. Kinetic parameter estimation

The kinetic models were fitted against the PBR data; the parameter estimation problem was formulated as a weighted least squares objective function (Eq. 4(a)), subject to nonlinear process constraints and parameter bound constraints (Eq. 4(b)):

$$\min \sum_j (\boldsymbol{x}_{j,E} - \boldsymbol{x}_{j,M})^T \boldsymbol{\Lambda} (\boldsymbol{x}_{j,E} - \boldsymbol{x}_{j,M}) \quad \text{Eq. 4(a)}$$

$$\text{st.}\quad \frac{d\boldsymbol{x}}{d\tau} = \boldsymbol{f}(\boldsymbol{x},\boldsymbol{\theta},\boldsymbol{k}),\qquad \frac{d\boldsymbol{\theta}}{d\tau} = \boldsymbol{g}(\boldsymbol{x},\boldsymbol{\theta},\boldsymbol{k}),\qquad \boldsymbol{k}_{lb} < \boldsymbol{k} < \boldsymbol{k}_{ub} \qquad \text{Eq. 4(b)}$$

where $\boldsymbol{x}_{j,E}$ and $\boldsymbol{x}_{j,M}$ are the experimental and model estimates of the state variables, respectively. $\boldsymbol{f}(\boldsymbol{x},\boldsymbol{\theta},\boldsymbol{k})$ and $\boldsymbol{g}(\boldsymbol{x},\boldsymbol{\theta},\boldsymbol{k})$ refer to the kinetic model differential equations for the gas-phase and surface species, respectively. $\boldsymbol{k}_{lb}$ and $\boldsymbol{k}_{ub}$ are the lower and upper bounds of the optimization variables $\boldsymbol{k}$. The weighting matrix $\boldsymbol{\Lambda}$ is used to normalise the residuals. The differential process constraints were fully discretised into algebraic profiles by applying orthogonal collocation on finite elements. The arising nonlinear programming problem was then solved via interior-point optimization (IPOPT) algorithms. Both the numerical discretization and IPOPT procedures were implemented via the *Pyomo* modelling environment in *Python* programming language.

2.2. Reaction network simplification

2.2.1. Physics-based network reduction

The physics-based reduction approach aims to identify kinetically negligible reactions by comparison of the forward and backward reaction rates of each elementary step. Similar rate-based procedures have been applied in literature to automatically discriminate elementary steps in complex reaction mechanisms (Goldsmith & West, 2017). In this study, the estimated equilibrium constants were compared against an upper and lower threshold:

- If $\frac{k_i}{k_i^-} > 10$, then the backward reaction is negligible ($k_i^- \approx 0$)
- If $\frac{k_i}{k_i^-} < 0.1$, then the forward reaction is neglibible ($k_i \approx 0$)

These thresholds have been chosen on the basis that most of the estimated rate constants in the proposed mechanism were 1 order of magnitude apart (10^5- 10^6), and that both observable and non-observable states in the kinetic model were expressed in normalised units (mass fractions and fractional surface coverages, respectively). Upon identifying negligible steps, the microkinetic model was reconstructed in accordance with the physics-reduced mechanism and subsequently fitted against experimental data.

2.2.2. Mathematics-based network reduction

In the mathematics-based network reduction strategy, an additional penalty term is included in the parameter estimation objective function to penalise the number of non-zero reaction rate constants:

$$\min \sum_j \left(\boldsymbol{x}_{j,E} - \boldsymbol{x}_{j,M}\right)^T \boldsymbol{\Lambda}\left(\boldsymbol{x}_{j,E} - \boldsymbol{x}_{j,M}\right) + \omega \sum_i \left(\frac{k_i}{k_i + 1} + \frac{k_i^-}{k_i^- + 1}\right) \qquad \text{Eq. 5}$$

For non-negligible microkinetic parameters, we have $k_i \gg 1$ and the corresponding penalty term $\frac{k_i}{k_i+1} \approx 1$, whereas for negligible reactions we obtain $\frac{k_i}{k_i+1} \approx 0$. This, in turn, allows for the identification of kinetically redundant reactions without the introduction of binary variables and mixed-integer programming. The penalty weight ω must be manually tuned so the sparsity of the model does not compromise its fitting accuracy.

3. Results

3.1. Microkinetic model results

The microkinetic model for the proposed mechanism exhibited good fitting performance, with an overall mean percentage error of 6.94%; the simulated process trajectories for the Pd/ZSM-12 catalyst with 0.5 Pd wt% composition are presented in Fig. 2 below. The estimated microkinetic constants not only allow for the simulation of the process, but also provide mechanistic insight on the underlying reaction network. For example, comparison of the estimated rate constants $k_3 = 8.02 \times 10^5, k_6 = 5.99 \times 10^5$ and $k_7 = 1.45 \times 10^6$ g_c mol^{-1} h^{-1} indicate that the monobranched and cracking intermediates are more readily desorbed to the gas-phase than their multibranched counterpart. This trend can be justified with reference to the higher diffusional energy barriers of multibranched paraffins in zeolite catalysts (Oenema et al., 2020), recalling that the desorption step involves the diffusion of an alkene intermediate from the acid sites to the metallic sites. Similarly, the estimated multibranched cracking rate constant ($k_5 = 3.44 \times 10^6$ g_c mol^{-1} h^{-1}) is much larger than its desorption rate constant ($k_6 = 5.99 \times 10^5$ g_c mol^{-1} h^{-1}), thus providing a mechanistic justification for the high yield towards cracking by-products and the comparatively low yield towards the multibranched hexadecane isomers. Previous microkinetic studies of n-alkane hydroconversion have reached similar conclusions regarding the cracking affinity of multibranched alkanes and its effect on the product distribution (Vandegehuchte et al., 2014).

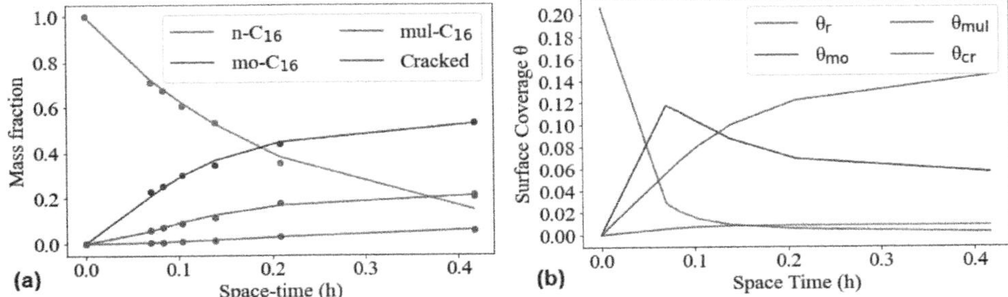

Figure 2: (a) Microkinetic model fitting results for gas-phase species profiles. Dots indicate experimental data while solid lines are model predictions; (b) Simulated coverages of surface species.

3.2. Reaction network simplification results

Physics- and mathematics-based network reduction analysis were conducted upon fitting the microkinetic model for the proposed mechanism; in either case five elementary steps were deemed to be irreversible, as seen in the simplified reaction networks presented in Fig. 3(a) and 3(b). The microkinetic model was also reconstructed in accordance with the simplified mechanisms, yielding fitting results (6.22% and 6.98% overall MAPE) of similar accuracy as the original model (6.15%) despite having 5 less parameters. This indicates that the neglected reactions were indeed kinetically insignificant.

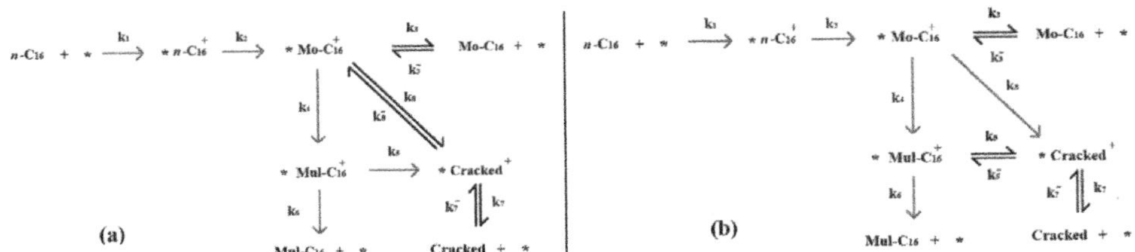

Figure 3: (a) Physics-reduced reaction network, (b) mathematics-reduced network. Red arrows indicate elementary steps that have been identified as irreversible.

Both simplification approaches identified the isomerization reactions (steps 2 and 4 in Fig. 3(a) and Fig. 3(b)) as irreversible. The irreversibility of these skeletal rearrangements has been previously analysed in literature and may be attributed to operating conditions and reactant conversions far away from thermodynamic equilibrium, such that the rates of the reverse isomerization reaction are negligible (Steijns & Froment, 1981). The main dissimilarity between the two reduced networks concerns the reversibility of the two cracking reactions in the mechanism (steps 5 and 8). The relative stabilities of the alkylcarbenium intermediates suggest that the cracking products would re-alkylate more readily into the multibranched species (i.e. $k_5^- > k_8^-$). This is indeed what is observed in the mathematics-reduced network ($k_8^- = 0$), but not in the physics-reduced network ($k_5^- = 0$). This suggests that the reduced microkinetic mechanisms identified by different model reduction strategies should be further evaluated before finally applying them.

4. Conclusion

In this work, a microkinetic modelling approach was adopted to simulate observable gas-phase species and non-observable intermediate states in the proposed *n*-hexadecane hydroisomerisation network. Upon parameter estimation, the model provided accurate fitting of the packed bed reactor data and mechanistic insight was gained through evaluation of the estimated microkinetic constants. Mathematics and physics-based network reduction strategies were then applied to identify kinetically insignificant reactions in the mechanism. The microkinetic model was modified in accordance with the network reduction results, also yielding satisfactory fitting of the experimental data. This study suggests that the two proposed reaction network reduction strategies have great potentials to be applied to generic processes for reaction mechanism investigation.

References

Coonradt, H. L., & Garwood, W. E. (1964). Mechanism of Hydrocracking. Reactions of Paraffins and Olefins. *Industrial & Engineering Chemistry Process Design and Development*, *3*(1), 38–45. https://doi.org/10.1021/i260009a010

Goldsmith, C. F., & West, R. H. (2017). Automatic Generation of Microkinetic Mechanisms for Heterogeneous Catalysis. *The Journal of Physical Chemistry C*, *121*(18), 9970–9981. https://doi.org/10.1021/acs.jpcc.7b02133

Motagamwala, A. H., & Dumesic, J. A. (2021). Microkinetic Modeling: A Tool for Rational Catalyst Design. *Chemical Reviews*, *121*(2), 1049–1076. https://doi.org/10.1021/acs.chemrev.0c00394

Oenema, J., Harmel, J., Vélez, R. P., Meijerink, M. J., Eijsvogel, W., Poursaeidesfahani, A., Vlugt, T. J. H., Zečević, J., & de Jong, K. P. (2020). Influence of Nanoscale Intimacy and Zeolite Micropore Size on the Performance of Bifunctional Catalysts for n -Heptane Hydroisomerization. *ACS Catalysis*, *10*(23), 14245–14257. https://doi.org/10.1021/acscatal.0c03138

Steijns, M., & Froment, G. F. (1981). Hydroisomerization and hydrocracking. 3. Kinetic analysis of rate data for n-decane and n-dodecane. *Industrial & Engineering Chemistry Product Research and Development*, *20*(4), 660–668. https://doi.org/10.1021/i300004a014

Vandegehuchte, B. D., Thybaut, J. W., & Marin, G. B. (2014). Unraveling Diffusion and Other Shape Selectivity Effects in ZSM5 Using n -Hexane Hydroconversion Single-Event Microkinetics. *Industrial & Engineering Chemistry Research*, *53*(40), 15333–15347. https://doi.org/10.1021/ie500164q

Weitkamp, J. (2012). Catalytic Hydrocracking—Mechanisms and Versatility of the Process. *ChemCatChem*, *4*(3), 292–306. https://doi.org/10.1002/cctc.201100315

Antonis Kokossis, Michael C. Georgiadis, Efstratios N. Pistikopoulos (Eds.)
PROCEEDINGS OF THE 33rd European Symposium on Computer Aided Process Engineering (ESCAPE33), June 18-21, 2023, Athens, Greece
 http://dx.doi.org/10.1016/B978-0-443-15274-0.50027-5

Troubleshooting high-pressure issues in an industrial biorefinery process by feature-oriented modeling

Elia Arnese-Feffin[a], Pierantonio Facco[a], Daniele Turati[b], Fabrizio Bezzo[a], Massimiliano Barolo[a,*]

[a]*CAPE-Lab – Computer-Aided Process Engineering Laboratory, Department of Industrial Engineering, University of Padova, via Marzolo 9, 35131 Padova, Italy*
[b]*Novamont S.p.A., via G. Fauser 8, 28100 Novara, Italy*
[*]*Corresponding author: max.barolo@unipd.it*

Abstract

Biorefinery plants manufacture chemicals by conversion of sustainable raw materials. Membrane separation processes are commonly used to separate cells from fermentation broths, and account for most of the operating costs of a biorefinery process. Membrane fouling can disrupt normal operation, increasing cleaning costs and process downtime (hence, overall manufacturing costs). Whereas investigating the causes of fouling through mechanistic models can be challenging in an industrial environment, information in process data historians can be leveraged through data-driven modeling. In this study, principal component analysis and feature-oriented modeling are combined to identify potential causes of fouling in a semi-continuous membrane separation process of an industrial biorefinery. This approach can effectively address batch duration variability issues, while exploiting process knowledge to enhance information on effects of fouling. Membrane age and operating temperature were found to be the major variables related to fouling.

Keywords: biorefinery; membrane separation processes; process understanding; feature-oriented models; data analytics

1. Introduction

Biorefineries are facilities that integrate biomass conversion processes and equipment to sustainably produce fuels, power, and chemicals from biomass (Martín *et al.*, 2013). Process operations typically include: media preparation, production of microorganisms, and large-scale fermentation in the upstream section; broth sterilization, cell separation, and product recovery and purification in the downstream section (Bähner *et al.*, 2021).
Operating costs of downstream in biorefineries usually range between 40% and 60% of the total processing cost. Energy is the main cost, as typical downstream operations are evaporation, distillation, and, recently, membrane separation processes (Bähner *et al.*, 2021). Membranes are becoming common in biorefineries to separate cells and/or large molecules from the fermentation broth (Ables *et al.*, 2013), for examples by ultrafiltration and nanofiltration operated (semi-)continuously. Such membrane separation processes can determine the major operating cost of a biorefinery (Satam *et al.*, 2019).
Pressure-driven membranes separation processes (i.e., membrane filtration) are the most

widely used, especially to remove cells from the fermentation broth. However, this operation can greatly suffer from membrane fouling, as highlighted in many studies. Ables *et al.* (2013) highlighted a permeate flux decrease in constant pressure filtration, while Klimkiewicz *et al.* (2016) recorded remarkable pressure build-up in constant flow-rate separation. The latter case is particularly relevant as, besides obvious disruption of operation and increase in cleaning costs, it also implies an increase in energy expenditure.
Mathematical modelling is a valuable tool to analyze this problem. Membrane fouling mechanisms are well known for some membrane separation processes, and models are available (Meindersma *et al.*, 1997) to help diagnosing the root-causes of fouling and support decision making to mitigate its effect. However, reliable fouling models often feature remarkable complexity and require high quality data for parameter estimation (Bolton *et al.*, 2016). Although modern biorefineries are heavily sensorized and provide a wealth of data recorded online, data usually regard the history of process operation, and might not be adequate for estimation of parameters of fouling models.
Data historians contain valuable information on process operation and can be leveraged to enhance performance (Cuellar *et al.*, 2020). To this end, data-driven models such as principal component analysis (PCA; Wold *et al.*, 1987), proved valuable when applied to membrane separation processes, as shown by Klimkiewicz *et al.* (2016) and Naessens *et al.* (2017). Applying such methods to membrane separation processes requires some extra care, mostly because of the semi-continuous nature of the process. This implies frequent process downtimes for membrane cleaning and strong variability in the duration of operating periods (runs), in the process variables within runs, and in the profiles of process variables between runs. Variable batch duration entails a lack-of-synchronization issue when process datasets are explored by data analytics.
Feature-oriented modeling (Yoon *et al.*, 2001; Rendall, 2019) offers an elegant way to address the lack-of-synchronization issue, while also emphasizing the phenomena one wants to model by properly defining features (Rendall *et al.*, 2017). In this study, knowledge-driven feature-oriented PCA (Wold *et al.*, 2009) is used to investigate the observable effects of fouling, i.e., high-pressure issues, in the membrane separation section of an industrial biorefinery process. The analysis of features summarizing time-profiles of process variables can highlight potential causes of fouling, thus paving the way for targeted experimental investigations.

2. Materials and methods

2.1. The membrane separation process

The focus of this study is membrane filtration of a fermentation broth, aimed at separating cells and large molecules from the stream containing the product. The operation is performed as a part of the downstream processing of an industrial biorefinery process where a biopolymer is manufactured. Sensible details on process and data are not disclosed for confidentiality.
The sterilized fermentation broth is accumulated in two parallel tanks, and, during operation, is fed to the first membrane module of the sequence at a constant flow-rate. Part of the retentate of each module gets discharged to the retentate manifold, while the remainder is fed to the following module. The first modules perform a standard filtration, while last ones operate in diafiltration: part of the permeate of each module is sent to the permeate manifold, and the remainder is fed back to the preceding module; water is fed to the very last one. A simplified scheme of the process and sensors is shown in Figure 1.
The process runs in semi-continuous mode. The membrane modules are initially filled with water, which is displaced by the fermentation broth during the startup phase. In the

steady-state phase, the pressure on the feed/retentate side of the membranes is manipulated to keep flow-rates constant. When the feed tanks are almost empty, the flushing phase begins, and the modules are filled again with water before cleaning-in-place takes place. However, premature interruption of the process occurs frequently due to excessive pressure build-up following membrane fouling. Investigating the potential causes of this issue is the objective of this study.

Production data over a seven-month timespan are available for this investigation. Namely, online measurements from all sensors are available; offline measurements from the upstream process are available as well, to characterize the broth being processed.

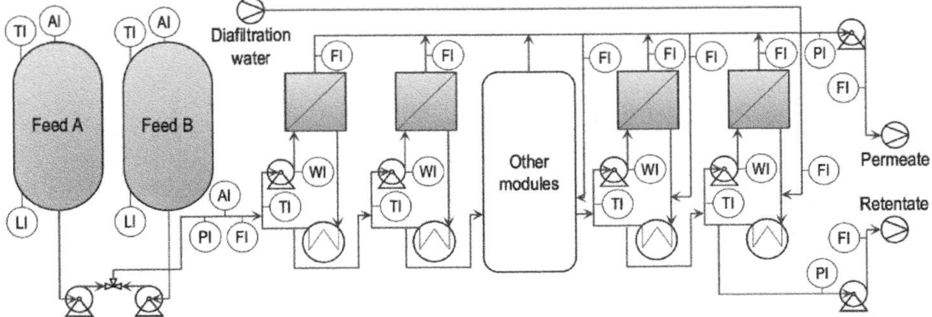

Figure 1. Simplified diagram of the membrane separation process under analysis. Analyzers on feed tanks measure pH, while the one on feed manifold measures conductivity.

2.2. Principal component analysis

PCA (Wold *et al.*, 1987) is a multivariate statistical method that allows extracting a sequence of orthogonal variables, called principal components (PCs), from a data matrix $\mathbf{X} \in \mathbb{R}^N \times \mathbb{R}^V$ collecting N observations of V variables. Assuming that $\mathbf{X}$ is mean-centered (and possibly scaled to unit variance), the PCA model is provided as a matrix decomposition in the form $\mathbf{X} = \mathbf{T} \cdot \mathbf{P}^{\mathrm{T}} + \mathbf{E}$, where $\mathbf{T} \in \mathbb{R}^N \times \mathbb{R}^A$ is the score matrix, $\mathbf{P} \in \mathbb{R}^V \times \mathbb{R}^A$ is the loading matrix, A is the number of PCs of the model (to be set prior to model calibration), and $\mathbf{E} \in \mathbb{R}^N \times \mathbb{R}^V$ is the residual matrix. The scores are projections of observations in $\mathbf{X}$ onto the PC space and describe the relation among observations. The loadings describe the correlation among variables in $\mathbf{X}$, and the relation between the original space and the PC space. The columns of $\mathbf{P}$ are set as to maximize the variance of data explained by the PCs (while respecting orthogonality constraints) and can be obtained by truncated singular value decomposition of the sample covariance matrix of data.

2.3. Feature-oriented modelling of membrane separation processes

PCA requires operating onto two-dimensional data arrays. However, the available data on the membrane separation process under investigation are a sequence $\{\mathbf{X}_1, \dots, \mathbf{X}_B\}$, where $\mathbf{X}_b \in \mathbb{R}^{N_b} \times \mathbb{R}^V$ is a matrix collecting N_b observations of the V process variables during the b-th run of the process, with $b = 1, 2, \cdots, B$, and B is the total number of runs (i.e., batches). To apply PCA to this process, one would need to concatenate all $\mathbf{X}_b$ matrices as a single matrix. However, if the aim is to investigate on runs that suffered from high-pressure issues, this approach may be inadequate due to observations representing single time instants rather than entire runs. In this study, we tackle the problem by means of feature-oriented modeling (Yoon *et al.*, 2001; Rendall *et al.*, 2019). The rationale is to summarize profiles of variables into numerical indexes called features. This approach offers an elegant solution to the lack-of-synchronization issue, while also allowing to emphasize the phenomena one is interested in by properly defining features. To this end, knowledge-driven features (Wold *et al.*, 2009; Rendall *et al.*, 2017) can be derived from time profiles, such as integrals, averages, slopes, minima, and maxima of measured variables in a run (note that all these operations are independent of N_b). Furthermore, the available process knowledge can be used to improve such "standard" features. An example is splitting time-profiles into phases to better characterize the process (e.g., startup, steady operation, and washing). Process knowledge can also be used to design special features.

Feature synthesis can be interpreted as an operator $\mathcal{F}: \mathbb{R}^{N_b} \times \mathbb{R}^{V} \rightarrow \mathbb{R}^{F} \mid \mathbf{X}_b \mapsto \mathbf{h}_b$, where $\mathbf{h}_b \in \mathbb{R}^F$ is a vector containing F features. As one vector of features is obtained from each run, a matrix $\mathbf{H} \in \mathbb{R}^B \times \mathbb{R}^F$ is obtained, whose rows characterize single runs. Such a matrix can be analyzed by PCA for process understanding and troubleshooting purposes.

3. Results

After preliminary screening, data for $B = 176$ runs were selected for analysis. Concerning data from process sensors, only the steady-state phase was used, as the startup and flushing phases usually feature excessive/unstructured variability and significant nonlinearities. Sensor measurements were first augmented with additional variables (e.g., pressure profile slope) to detect occurrence of fouling. Some engineering variables were added as performance indexes, for instance average energy consumption and volume conversion ratio (VCR, ratio of retentate flow-rate and feed flow-rate). A total of $V = 61$ online variables was obtained. These variables, together with data from upstream process (characteristics of the broth being processed), were used to compute $F = 179$ features for PCA analysis. Some of the most important features are collected in Table 1. Note that some features were also computed on sub-phases of the steady-state phase, whenever that seemed meaningful.

Table 1. Examples of features extracted from profiles of online variables. Features marked with a star are encoded as binary variables.

Average flow-rates in manifolds	Average VCR
Average feed conductivity	Energy consumed
Slope of feed conductivity	Processed volume
Max/min of pressures	Run duration
Average slope of pressures	Feed tanks average temperature
Max/min of pressure slopes	Feed tanks average pH
Average trans-membrane pressure	Concentration of cells in the feed
Average flow-rates of module permeates	Concentration of organic acids in the feed
Average temperatures of modules	Concentration of ions in the feed
Durations of pump steps	Concentration of product in the feed
Average pump power over steps	Contamination of upstream fermenter*
Pressure features computed over pump steps	Chemicals used in cleaning*

A preliminary PCA model including all variables featured limited interpretability. Therefore, a stepwise approach was adopted. Features were first grouped in subsets according to their "origin", such as upstream, cleaning, single modules. The most important subset is the one of global features characterizing the membrane filtration, which are basically the ones regarding manifolds: flow-rates, pressures, and performance indexes. A first PCA model is developed on global features alone. Pressure-related features were identified as the main drivers of variability (first PC), as expected; the second source of variability (second PC) regards average flow-rates and some of the performance indexes, among which VCR. The explained variances of the other PCs were far lower than the ones of the first two, and loadings appeared to model random variability. Therefore, the features modelled by the first two PCs were selected as the reference set of features; cell concentration in the feed was added to this set, as it was expected to be an important factor for fouling. A PCA model developed on these selected features proved to properly identify runs that suffered from high pressure issues, which are separated from the bulk of the runs in the score plot reported in Figure 2 (triangles).

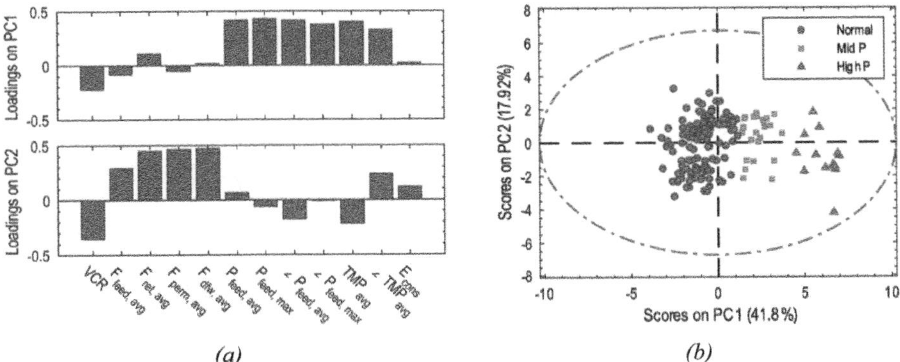

Figure 2. (a) Loadings and (b) scores of the PCA model developed on the set of global features. Runs suffering from high pressure clearly group along PC1.

More features were then added to the data matrix for PCA modelling, in a group-by-group fashion, to assess the occurrence of correlation with the pressure-related features. When such a correlation was found and was physically meaningful, newly added features correlating with pressure were regarded as potential causes of fouling.

The most important factor for high pressure issues was found to be membrane age. This is highlighted by features regarding permeates of single modules, which are proxies for membrane age (i.e., permeate flux decreases run after run). The second most important factor was found to be temperature: more precisely, runs featuring a high temperature of flows entering membrane modules suffered from high-pressures more frequently than runs with low inlet temperature. Such conclusions were verified by targeted analysis of raw data (see Figure 3) and make engineering sense. Therefore, they were deemed as potential causes of fouling, and are currently being verified by designed experiments.

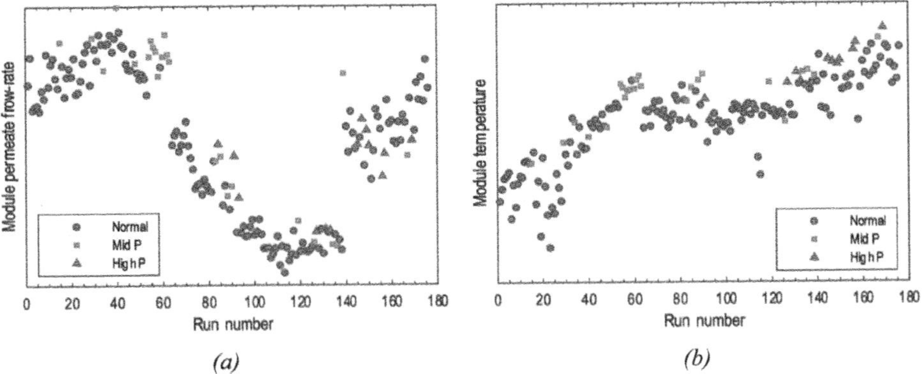

Figure 3. (a) Average permeate flow-rate and (b) temperature of a selected module. Pressure issues are more frequent for old membranes (flow-rate is a proxy for membrane age) and when modules operate at high temperature.

4. Conclusions

Interpretable data-driven models are valuable tools to investigate issues in membrane separation processes, such as membrane fouling, relying solely on data collected during process operation. This study proved how principal component analysis can identify potential causes of fouling by analysis of data concerning the observable effect of this complex phenomenon, namely, pressure increase in membrane filtration. A feature-oriented approach was adopted: instead of using process data directly, numerical values characterizing each operating period were obtained so as to summarize time profiles into time-independent numerical features. This allowed addressing issues (e.g, lack of synchronization and uneven batch duration) that can complicate the use of other approaches. Process knowledge was exploited to design features that enhanced the phenomenon under investigation, thus maximizing the dataset information content and enhancing interpretability with respect to other methods. Incorporation of process knowledge into the data analysis workflow proved essential to identify potential causes of fouling, which are currently being verified by targeted experimental studies.

References

C. Ables, F. Carstensen, M. Wessling, 2013, Membrane processes in biorefinery applications, Journal of Membrane Science, 444, 285–317

F. Bähner, O. A. Prado-Rubio, J. K. Huusom, 2021, Challenges in Optimization and Control of Biobased Process Systems: An Industrial-Academic Perspective, Industrial and Engineering Chemistry Research, 60, 42, 14985–15003

G. Bolton, D. LaCasse, R. Kuriyel, 2006, Combined models of membrane fouling: Development and application to microfiltration and ultrafiltration of biological fluids, Journal of Membrane Science, 277, 1–2, 75–84

M. Cuellar, A. JJ Straathof, 2020, Downstream of the bioreactor: advancements in recovering fuels and commodity chemicals, Current Opinion in Biotechnology, 62, 189–195

A. Klimkiewicz, A. E. Cervera-Padrell, F. W. J. van den Berg, 2016, Multilevel Modeling for Data Mining of Downstream Bio-Industrial Processes, Chemometrics and Intelligent Laboratory Systems, 154, 62–71

M. Martín, I. E. Grossmann, 2013, On the systematic synthesis of sustainable biorefineries, Industrial and Engineering Chemistry Research, 52, 9, 3044–3064

G. Meindersma, J. Augeraud, F. H. P. Vergossen, 1997, Separation of a biocatalyst with ultrafiltration or filtration after bioconversion, Journal of Membrane Science, 125, 2, 333–349

W. Naessens, T. Maere, G. Gilabert-Oriol, V. Garcia-Molina, I. Nopens, 2017, PCA as tool for intelligent ultrafiltration for reverse osmosis seawater desalination pretreatment, Desalination, 419, June, 188–196

R. Rendall, B. Lu, I. Castillo, S. T. Chin, L. H. Chiang, M. S. Reis, 2017, A Unifying and Integrated Framework for Feature Oriented Analysis of Batch Processes, Industrial & Engineering Chemistry Research, 56, 30, 8590–8605

R. Rendall, L. H. Chiang, M. S. Reis, 2019, Data-driven methods for batch data analysis – A critical overview and mapping on the complexity scale, Computers & Chemical Engineering, 124, 1–13

C. Satam, M. Daub, M. J. Realff, 2019, Techno-economic analysis of 1,4-butanediol production by a single-step bioconversion process, Biofuels, Bioproducts and Biorefining, 13, 5, 1261–1273

S. Wold, K. Esbensen, P. Geladi, 1987, Principal Component Analysis, Chemometrics and Intelligent Laboratory Systems, 8, 1–3, 37–52

S. Wold, N. Kettaneh-Wold, J. F. MacGregor, K. G. Dunn, 2009, Batch Process Modeling and MSPC, In: Comprehesinve Chemometrics, 2, 163–197

S. Yoon, J. F. MacGregor, 2001, Incorporation of External Information into Multivariate PCA/PLS Models, IFAC Proceedings Volumes, 34, 27, 105–110

Antonis Kokossis, Michael C. Georgiadis, Efstratios N. Pistikopoulos (Eds.)
PROCEEDINGS OF THE 33rd European Symposium on Computer Aided Process Engineering (ESCAPE33), June 18-21, 2023, Athens, Greece
 http://dx.doi.org/10.1016/B978-0-443-15274-0.50028-7

Optimal Determination of the Binary Interaction Parameters for the Vapor-Liquid-Liquid Equilibrium on the Vinyl Acetate/Acetic Acid/Water System

Jose Alfredo Paredes-Ortiz [a], Úrsula Fabiola Rodríguez-Zúñiga [b], Fernando Israel Gómez-Castro [a*].

[a] *Departamento de Ingeniería Química, División de Ciencias Naturales y Exactas, Campus Guanajuato, Universidad de Guanajuato, Noria Alta S/N, Guanajuato, Gto. 36050, México. fgomez@ugto.mx*

[b] *Universidad de Ingeniería y Tecnología, Departamento de Ingeniería Química, Jr. Medrano Silva 165, Barranco, Lima, Perú, 15063.*

Abstract

The design of the vinyl acetate purification process requires modelling the vapor-liquid-liquid equilibrium (VLLE) of the mixture vinyl acetate/acetic acid/water. In this work, the equilibrium of the system is represented through the gamma-phi formulation using the Hayden O'Connell (HOC) model for the vapor phase and the universal quasi-chemical (UNIQUAC) model for the liquid phases. The parameters for the model are estimated through optimization tools, using the BARON global optimization algorithm. The objective function to be minimized was the least-squares function. The results show a good correlation for the LLE and VLE of vinyl acetate with water with errors less than 2%. In the case of the VLE equilibrium of water with acetic acid the error is close to 5% with a larger deviation at acid mole fractions above 0.4. The parameters determined allow a proper modeling of the VLLE for the analyzed mixture.

Keywords: Binary interaction parameters, vapor-liquid-liquid equilibrium, deterministic optimization

1. Introduction

Vinyl acetate is one of the most widely produced chemicals in the world. It is used in the production of polyvinyl acetate and polyvinyl alcohol. At present, the main route for vinyl acetate synthesis is via the acetoxylation reaction of ethylene in the presence of oxygen and palladium catalysts. During the vinyl acetate synthesis process, the mixture vinyl acetate/acetic acid/water is generated; this mixture is commonly separated by distillation. Over the last few years, several authors have conducted research related to the improvement of the vinyl acetate process focusing on both the design and control areas, with particular emphasis on the purification of the mixture. Nevertheless, the mixture presents two liquid phases, which may difficult the separation. On the last years, authors such as Li et al. (2022) and Xie et al. (2020) point to the use of intensified separation operations such as extractive distillation or azeotropic distillation to achieve the purification task.

The vinyl acetate purification by intensified processes involves modelling the VLLE of the mixture. However, this implies that the binary interaction parameters of the equilibrium models must properly predict both vapor-liquid equilibrium (VLE) and liquid-liquid equilibrium (LLE). Unfortunately, the currently available binary interaction parameters accurately describing the VLE show inaccurate predictions for the LLE, and vice versa. The data fitting of these models for the representation of vapor-liquid-liquid equilibrium is a complex task with a high probability for the solution to get stuck in a local optimum. Some authors propose global optimization strategies based on minimization of squares to solve this problem (e.g., Wyczesany, 2014). Therefore, in the present work, a least-squares optimization was performed using the BARON global optimization algorithm to obtain the binary interaction parameters that properly represent VLE and LLE for the mixture vinyl acetate/acetic acid/water.

2. Methodology

For mixtures with carboxylic acids, as is the case due to the presence of acetic acid, the HOC model describes the behavior quite well since the carboxylic acid interacts with itself or with other compounds in vapor phase due to dimerization. In addition to this model, the NRTL or UNIQUAC models proposed by Hsieh et al. (2008) are often used for this type of mixtures due to the presence of non-ideal solutions. Thus, in this work, the UNIQUAC-HOC combination has been used.

The Gamma-Phi approach is used to model the vapor-liquid equilibrium, as represented in equation 1.

$$\phi_i^V y_i P = x_i \gamma_i \phi_i^S P_i^S \tag{1}$$

Where the left-hand terms are the vapor phase fugacity of the component, the mole fraction of the component in vapor phase, and the pressure. On the right-hand side the terms appearing are the liquid phase mole fraction, the activity coefficient, the saturation fugacity, and the saturation pressure, all for the respective component *i*.

For the calculations of the vapor phase, the relationship between the apparent fugacity coefficients and the real fugacity coefficients is used, as well as the true mole fraction of the vapor phase, based on the chemical theory that also complements and uses the equation of state HOC, equation 2.

$$\phi_i^V y_i = z_i \phi_i^{\#} \tag{2}$$

Equation 2 corrects for the deviation generated due to dimer formation. In this case it is used because acetic acid has a predominance in dimer formation.

The Gamma-Gamma representation has been used for the liquid-liquid equilibrium, this is shown in equation 3, where the upper indices refer to each of the liquid phases.

$$x_i^I \gamma_i^I = x_i^{II} \gamma_i^{II} \tag{3}$$

The binary interaction parameters of the UNIQUAC model are represented with an independent and a temperature-dependent parameter (equation 4).

$$\tau_{ij} = \exp\left(a_{ij} + \frac{b_{ij}}{T}\right) \tag{4}$$

For the correction of the binary interaction parameters, the experimental data reported by Shanghai College Chemical Engineering (1976) and Zhang et al. (2011) has been used for the liquid-vapor equilibrium modeling. Complementarily, the data reported by Gao et al. (2017) has been employed for the liquid-liquid equilibrium correlation.

The least squares ratio for the experimental and the estimated data is established as the objective function, equation 5.

$$F = \sum_{i=1}^{N_{LLE}} \sum_{j=1}^{N_C} \sum_{k=1}^{2} \left(x_{e_{k,j,i}} - x_{t_{k,j,i}}\right)^2 + \sum_{i=1}^{N_{VLE}} \sum_{j=1}^{N_C} \left(y_{e_{j,i}} - x_{t_{j,i}}\right)^2 + \sum_{i=1}^{N_{VLE}} \left(T_{e_i} - x_{t_i}\right)^2 \quad (5)$$

The objective function refers to the number of experimental points reported (*i*), the number of components (*j*) and the number of phases (*k*), organic and aqueous.

The model was run through the GAMS interface using the Global BARON optimization algorithm created by Sahinidis (1996) through the NEOS server, Hosted by the Wisconsin Institute for Discovery at the University of Wisconsin in Madison.

3. Results and Discussion

Based on the experimental data reported by Shanghai College Chemical Engineering (1976), Zhang et al. (2011) and Gao et al. (2017), and the adjustment of the interaction parameters, the following results have been obtained.

3.1. LLE Vinyl Acetate/Acetic Acid/Water

As shown in Figure 1, favorable results have been obtained with respect to the liquid-liquid equilibrium of the mixture. The curve fit with the model is very close to the data reported by Gao et al. (2017), as observed from the deviation data shown in Table 1.

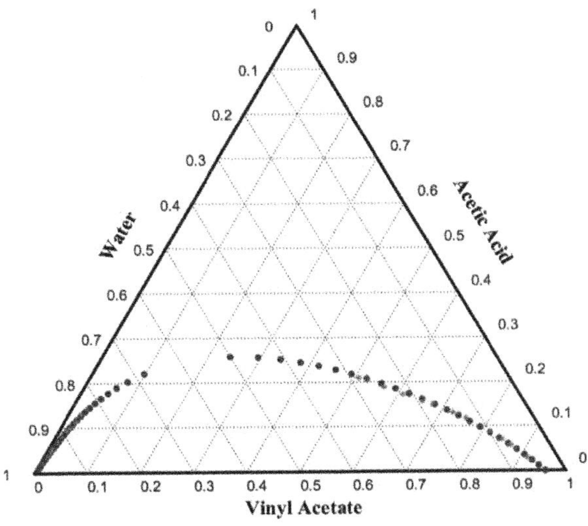

Fig 1. LLE for the ternary system vinyl acetate/ acetic acid / water: (red) experimental data at 298.15 K, (blue) estimated values.

Table 1. Deviation of the liquid-liquid equilibrium of the Vinyl Acetate (1) / Acetic Acid (2)/ Water (3) mixture at 298 K y 1 atm.

	Δx_i					
ΔT	$(1)^I$	$(2)^I$	$(3)^I$	$(1)^{II}$	$(2)^{II}$	$(3)^{II}$
5.29E-01	-3.16E-04	-2.26E-04	5.42E-04	-6.85E-04	1.58E-03	-8.95E-04
-2.34E-01	2.23E-04	-6.12E-04	3.89E-04	3.83E-04	-1.35E-03	9.72E-04
1.69E-01	-5.22E-05	-9.17E-04	9.69E-04	-2.87E-03	4.46E-05	2.83E-03
-1.27E-01	1.20E-04	5.66E-04	-6.86E-04	1.92E-03	-2.63E-03	7.04E-04

-4.17E-01	4.82E-04	1.08E-03	-1.56E-03	8.62E-03	-6.67E-03	-1.95E-03
-1.52E-01	2.26E-04	3.35E-04	-5.61E-04	3.77E-03	-2.87E-03	-9.05E-04
1.01E-01	-2.27E-04	-2.84E-03	3.06E-03	-6.30E-03	6.76E-03	-4.61E-04
-1.94E-02	-4.08E-07	2.89E-04	-2.88E-04	4.33E-04	1.32E-03	-1.75E-03
9.04E-02	-3.10E-04	1.16E-03	-8.49E-04	-2.88E-03	3.86E-03	-9.80E-04
9.03E-02	-2.33E-04	-1.30E-03	1.53E-03	-5.96E-03	7.06E-03	-1.10E-03
-3.15E-02	3.84E-04	-4.17E-03	3.78E-03	2.69E-03	2.39E-03	-5.08E-03

3.2. VLE Vinyl Acetate/Acetic Acid

A good correlation has been obtained between the model and the experimental data for the mixture of vinyl acetate with acetic acid (Figure 2), obtaining an adequate prediction, as shown in Table 2 in terms of the deviations.

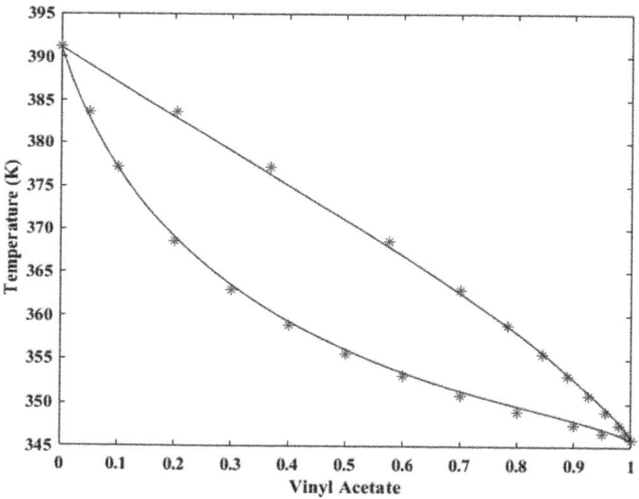

Fig 2. VLE of vinyl acetate/ acetic acid at 1 atm; (red) experimental data, (blue) estimated values.

Table 2. Deviation of the vapor-liquid equilibrium of the Vinyl Acetate (1) / Acetic Acid (2) at 1 atm.

ΔT	$\Delta x_{(1)}$	$\Delta x_{(2)}$	$\Delta y_{(1)}$	$\Delta y_{(2)}$
-5.78E-02	0.00E+00	0.00E+00	0.00E+00	0.00E+00
-6.60E-01	-2.02E-05	2.02E-05	-1.03E-03	1.03E-03
-1.13E-01	2.69E-05	-2.69E-05	-1.99E-02	1.99E-02
4.80E-01	1.78E-04	-1.78E-04	-2.80E-02	2.80E-02
4.96E-01	4.17E-04	-4.17E-04	-1.97E-02	1.97E-02
4.26E-01	5.37E-04	-5.37E-04	-1.15E-02	1.15E-02
3.85E-01	6.37E-04	-6.37E-04	-6.53E-03	6.53E-03

3.33E-01	2.61E-04	-2.61E-04	-2.67E-03	2.67E-03
4.51E-01	2.05E-04	-2.05E-04	-2.67E-03	2.67E-03
5.79E-01	1.75E-04	-1.75E-04	-2.98E-03	2.98E-03
3.91E-01	1.23E-04	-1.23E-04	-2.38E-03	2.38E-03
3.52E-01	5.14E-05	-5.14E-05	-9.35E-04	9.35E-04
7.39E-02	0.00E+00	0.00E+00	0.00E+00	0.00E+00

3.3. VLE Acetic Acid/Water

The fit between the model and the experimental data in Figure 3 is good. However, the results presented in Figures 1 and 2 present a better fit to the data. The higher the acetic acid composition, the larger the deviation. The error with respect to the experimental points is detailed in Table 3. Finally, the computed parameters are presented in Table 4.

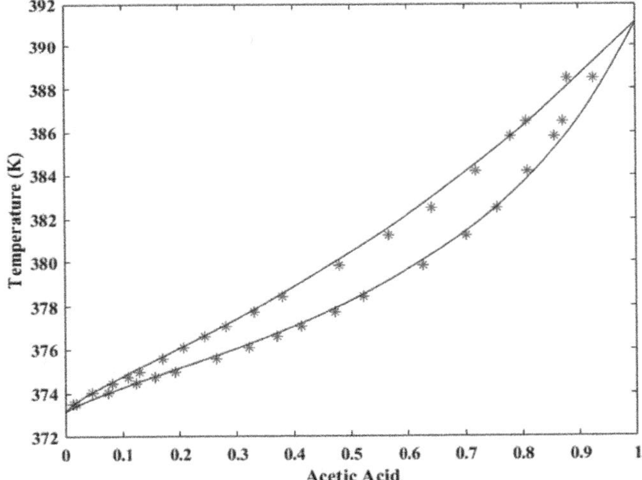

Fig 3. VLE of acetic acid/water at 1 atm; (red) experimental data, (blue) estimated values.

Table 3. Deviation of the vapor-liquid equilibrium of the Acetic Acid (1) / Water (2) at 1 atm.

ΔT	$\Delta x_{(1)}$	$\Delta x_{(2)}$	$\Delta y_{(1)}$	$\Delta y_{(2)}$
-5.00E-02	6.85E-05	-6.85E-05	-4.29E-03	4.29E-03
5.43E-03	3.32E-05	-3.32E-05	-1.92E-03	1.92E-03
5.93E-02	4.98E-05	-4.98E-05	-3.09E-03	3.09E-03
4.25E-02	1.37E-04	-1.37E-04	-8.86E-03	8.86E-03
9.35E-02	9.54E-05	-9.54E-05	-6.27E-03	6.27E-03
1.13E-01	-4.11E-05	4.11E-05	2.07E-03	-2.07E-03
1.54E-01	-1.38E-04	1.38E-04	7.22E-03	-7.22E-03
1.43E-01	-1.54E-04	1.54E-04	7.80E-03	-7.80E-03
1.39E-01	-1.14E-04	1.14E-04	5.15E-03	-5.15E-03
1.75E-01	-1.04E-04	1.04E-04	3.26E-03	-3.26E-03

1.04E-01	1.53E-05	-1.53E-05	-3.13E-03	3.13E-03
2.20E-01	-3.31E-05	3.31E-05	-2.83E-03	2.83E-03
1.91E-01	-4.26E-06	4.26E-06	-9.44E-03	9.44E-03
5.49E-02	1.18E-04	-1.18E-04	-2.30E-02	2.30E-02
-2.79E-01	2.01E-04	-2.01E-04	-3.13E-02	3.13E-02
-5.33E-01	1.71E-04	-1.71E-04	-2.53E-02	2.53E-02
-7.34E-01	2.12E-04	-2.12E-04	-3.11E-02	3.11E-02
-7.49E-01	1.12E-04	-1.12E-04	-1.78E-02	1.78E-02

Table 4. UNIQUAC binary interaction parameters of the Vinyl Acetate (1) / Acetic Acid (2)/ Water (3) mixture.

		A_{IJ}	B_{IJ}	A_{JI}	B_{JI}
(1)	(2)	-4.118	936.916	-0.554	416.761
(1)	(3)	-10.000	2469.161	7.513	-2500.000
(2)	(3)	5.724	-2500.000	-3.855	1675.908

4. Conclusions

The results of the methodology employed show that the UNIQUAC equation, combined with the HOC equation of state, can properly describe the LLE and VLE of the ternary mixture vinyl acetate/acetic acid/water with adequate accuracy if the parameters are simultaneously fitted for both equilibria. The VLE of acetic acid with water has a higher deviation, up to 3.13×10^{-2} for the vapor composition of acetic acid. Nevertheless, the parameters obtained allow to adequately represent the ELLV of the evaluated mixture. This data will be useful for the simulation and analysis of intensified processes for the separation of the mixture vinyl acetate/acetic acid/water in the vinyl acetate production process.

References

J. Gao, D. Guan, D. Xu, L. Zhang, Z. Zhang, 2017, Measurement and modeling of liquid-liquid equilibrium for the systems vinyl Acetate + acetic acid/ethanol + water at 298.15 and 308.15 K, Journal of Chemical and Engineering Data, 62, 4, 1240–1246.

C.-T. Hsieh, W.-Y. Ji, H.-m Lin, M.- J. Lee, 2008, Multiphase equilibria for mixtures containing water, acetic acid, propionic acid, methyl acetate and methyl propionate, Fluid Phase Equilibria, 271, 1–2, 69–75.

Z. Li, K. Wang, X. Luo, X. Xu, 2022. Optimization and simulation of vinyl acetate process based on Aspen Plus, Proceedings of the International Conference on Optoelectronic Materials and Devices, 121641U.

Shanghai College Chemical Engineering, 1976, Study on the vapor liquid equilibrium data for acetic acid-water-vinyl acetate ternary system- I.The mutually miscible liquid phase region, Acta Chimica Sinica, 34, 79–93.

Wyczesany, A. 2014. Calculation of vapor-liquid-liquid equilibria at atmospheric and high pressures. Industrial and Engineering Chemistry Research, 53(6), 2509–2519.

Y.N. Xie, Z. Meng, F. Li, 2020, Purification process design of vinyl acetate based on Aspen Plus, IOP Conference Series: Earth and Environmental Science, 545.

C. Zhang, H. Wan, L. Xue, G. Guan, 2011, Investigation on isobaric vapor liquid equilibrium for acetic acid + water + (n-propyl acetate or iso-butyl acetate), Fluid Phase Equilibria, 305, 1, 68–75.

Antonis Kokossis, Michael C. Georgiadis, Efstratios N. Pistikopoulos (Eds.)
PROCEEDINGS OF THE 33rd European Symposium on Computer Aided Process Engineering (ESCAPE33), June 18-21, 2023, Athens, Greece
 http://dx.doi.org/10.1016/B978-0-443-15274-0.50029-9

Assessment of subgrid-scale models for large-eddy simulation of a gas-liquid stirred reactor

Safae Elmisaoui,[a,b] Radouan Boukharfane,[a] Lhachmi Khamar,[a,c] and Jean-Michel Ghidaglia [a,d]

[a] *Mohammed VI Polytechnic University, MSDA Program, Benguerir 43150, Maroc*
[b] *Mohamed V University, Laboratoire du génie civil et environnement, 11060, Maroc*
[c] *Sultan Moulay Slimane University, LIPIM, ENSA Khouribga, 25000, Maroc*
[d] *Centre Borrelli, ENS Paris-Saclay, CNRS, Université Paris-Saclay, 94235 Cachan, France*
Radouan.boukharfane@um6p.ma

Abstract

The LES-SGS simulations of a three-dimensional multiphase flow in an industrial preneutralizer reactor is performed, and the Multiple Reference Frame (MRF) approach is used. The effectiveness of the three SGS turbulent closures (Smagorinsky–Lilly model, Wall Adapting Local Eddy Viscosity (WALE) model, and the Algebraic Wall Modeled LES (WMLES) model) to predict the flow hydrodynamics and to capture the eddies near to high turbulent zones (in liquid injection point and in the turbine rotation zone) is assessed. The results show that the SGS model has a significant impact not only on the instantaneous flow field, but also on the time-average velocity magnitude, therefore the hydrodynamics of unsteady behaviour in turbulent chemical reactors is found to be significantly influenced by the choice of the SGS closure model.

Keywords: CFD, LES- SGS, turbulence model, hydrodynamics, multiphase flow

1. Introduction

In the phosphate industry, Di–Ammonium Phosphate (DAP) is considered to be a very efficient chemical fertilizers that is used to provide the required nutrients to the plants and soil. The DAP manufacturing processes begin with a preneutralizer reactor that consists of the production of homogeneous slurry produced by mixing the Ammonia gas and Phosphoric acid (ACP), under specific conditions. The preneutralizer reactor is a non-standard stirred tank reactor, equipped with a pitched blades turbine agitator that (i) promotes mass and heat transfer, (ii) reduces foaming, and (iii) improves ammonia dispersion in the liquid phase. This unit operation represents the masterpiece that conducts the physicochemical phenomenon called "preneutralization" chemical reaction (Elmisaoui et al., 2020).

The reaction between ammonia and ACP in the reactor leads to MAP or DAP depending on the ratio of N (Nitrogen) and P (Phosphorus) ratio N/P. Other species are formed but in small quantities in such a way that they do not significantly affect the total nitrogen or P_2O_5 in the final product (Campbell et al., 2006). When the ratio N/P is greater than unity, the produced MAP in the first reaction is converted to DAP in the second one (Campbell et al., 2006). The overall reactions are listed as follows

$$\begin{cases} NH_3 + H_3PO_4 \longrightarrow NH_4H_2PO_4 + \text{ Heat} & (1a) \\ NH_3 + NH_4H_2PO_4 \longrightarrow (NH_4)_2HPO_4 + \text{ Heat} & (1b) \end{cases}$$

These reactions occur immediately after Ammonia gas NH_3 and $ACP(H_3PO_4)$ are mixed. Produced slurry feeds the granulator, to be sprayed onto the granular fertilizer bed existing in the granulator, and granule growth occurs. Granules leaving the granulator are first dried and then screened to separate out the product size.

For the CFD modeling studies of multiphase flows in stirred reactors, different turbulence models have been developed to predict the hydrodynamics of the flow, each one of them demonstrating its effectiveness and robustness in each particular application (Derksen, 2001). The age-old CFD modeling method based on the Reynolds Averaged Navier-Stokes (RANS) approach is still considered the most employed turbulence modeling approach due to its acceptable accuracy and affordable computational cost for predicting the hydrodynamics involving complex geometries, but with the development of the HPC capabilities, the employment of the hybrid Detached Eddy Simulation (DES) model, which blends the RANS approach with LES founds its applications. The DES methodology shows its increasing potential and accuracy in flow prediction more than the RANS models. Especially for its informative ability to predict the flow regimes in the moving zone characterizing the impeller rotation. This work aims at evaluating the performance of three main Large-Eddy Simulation (LES) subgrid-scale (SGS) turbulence models in a stirred reactor.

2. Physical model of the preneutralizer

In this study, numerical simulations have been carried out using the workbench of the CFD commercial software ANSYS Fluent 2021-R2. The physical model tailored in this work is based on the actual size of a preneutralization chemical reactor (PN) with a working volume of 47 m^3 to ensure throughput of around 120 m^3/hr. The ACP is injected from the top, and eight uniform Ammonia inlet nozzles are arranged in the bottom, which are 3m away from it. The rotational agitator is equipped with a Pitch Blade Turbine (PBT), consisting of four impellers slanted with an angle of 45° from the horizontal plane. Detailed characteristics of the CAD file (cf. Fig. 1) are based on the original preneutralizer geometry and are previously described in Elmisaoui et al. (2022).

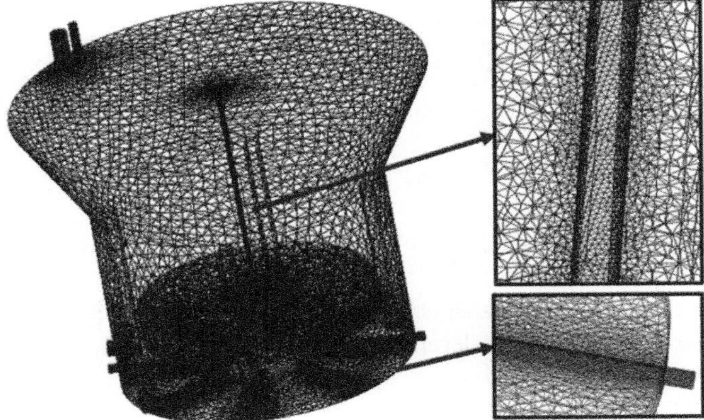

Figure 1: Mesh grid of the simulation domain, with zoom in shaft axis and gas spargers

3. Mathematical modeling and numerical schemes

LES comes in between the Direct Numerical Simulation (DNS) and the Reynolds Averaged Navier-Stokes (RANS) approaches in terms of turbulent scale resolution. Large eddies are directly resolved in LES, whereas small eddies are modeled. The time-dependent Navier-Stokes equations are filtered. The space-filtered equations for the conservation of mass and momentum of an incompressible Newtonian fluid can be written as:

$$\begin{cases} \frac{\partial \bar{u}_i}{\partial x_i} = 0 & \text{(2a)} \\ \frac{\partial \bar{u}_i}{\partial t} + \bar{u}_i \frac{\partial \bar{u}_j}{\partial x_i} = -\frac{\partial \bar{P}}{\partial x_i} + v \frac{\partial^2 \bar{u}_i}{\partial x_j^2} - \frac{\partial \tau_{ij}}{\partial x_j} & \text{(2b)} \\ \tau_{ij} = \bar{u}_i \bar{u}_j - \bar{u}_i \bar{u}_j & \text{(2c)} \end{cases}$$

where τ_{ij} is the sub-grid scale stress tensor, which reflects the effect of the unresolved scales on the resolved scales. Subgrid-scale stresses resulting from the filtering operation are unknown, and require modeling. The subgrid-scale turbulence models employ the Boussinesq hypothesis as in the RANS models, computing subgrid-scale turbulent stresses from:

$$\tau_{ij} = \frac{1}{3}\tau_{kk}\delta_{ij} - 2\mu_t \bar{S}_{ij} \quad (3)$$

Three SGS turbulence models are considered in this work:

- Smagorinsky-Lilly model

$$\mu_t = \rho L_s^2 |\bar{S}| \quad (4)$$

- Wall Adapting Local Eddy Viscosity (WALE) model

$$\mu_t = \rho L_s^2 \frac{\left(s_{ij}^d s_{ij}^d\right)^{3/2}}{\left(\bar{S}_{ij}\bar{S}_{ij}\right)^{5/2} + \left(s_{ij}^d s_{ij}^d\right)^{5/4}} \tag{5}$$

- Algebraic Wall-Modeled LES Model

$$\mu_t = \rho \, \mathrm{m} \; [(\kappa d_{\text{wall}})^2, (C_k \Delta)^2] \cdot S \cdot \{1 - \exp(-(y^+/25))\} \tag{6}$$

where d_{wall} is the wall distance, S is the strain rate, $\kappa = 0.41$, and $C_k = 0.2$ are constants.
To consider chemical species transport, the previous system of equations is coupled to the species transport equation, as described in Elmisaoui et al. (2022).
3.1 Numerical simulation conditions
As previously stated in many CFD simulations of stirred reactors, a variety approaches may be used to cope with the movement of the impeller blades. The Multiple Reference Frame (MRF) approach was employed in this investigation Elmisaoui et al. (2022), and the SIMPLE algorithm was employed for pressure-velocity coupling.

3.1 Computational domain
The computational domain grid is composed of two parts: an inner spinning cylindrical volume holding the turbine and an outer, stationary volume containing the remainder of the tank. In both portions, structured grids of non-uniformly distributed hexahedral cells are utilized. To provide a more precise description of the impeller, the grid employed in the impeller region is refined. In the tank, there are a total of 970997 grid nodes.
It is worth noting that a grid-dependency study is carried out using ANSYS Mesher to evaluate mesh suitability for the present configuration. A three-dimensional mesh grid of the computational domain is discretized into unstructured cells with specific refined zones (cf. Fig. 1). Three mesh featuring grids are generated and defined as coarse, medium, and fine meshes, respectively. Refinements are carried out in the shaft region, near to the nozzles, at the inlets, and at the outlet of the domain, allowing for a minimum and maximum mesh size of 25 mm and 15 mm for the medium mesh grid. The last refinement level is retained based on a mesh grid sensitivity, as the grid allows for an accurate numerical solution at acceptable time costs.

4. Results and analysis

To evaluate the flow hydrodynamics in a three-dimensional multiphase turbulent flow, the identification of coherent structures is crucial, which allows a better understanding of the real space dynamics of turbulent flows. One of the most commonly used methods for categorizing three-dimensional flow structures is the Q-criterion proposed by Hunt et al. (1988). It is defined in terms of the instantaneous velocity gradient tensor as:

$$\boldsymbol{Q} = \mathbf{1/2((\nabla \cdot u)^2 - \nabla u : \nabla u^T) = 1/2\left((\nabla \cdot u)^2 + \| \Omega \|_2^2 - \| S \|_2^2\right)} \tag{7}$$

Here, S is the strain rate tensor and Ω is the rotation rate tensor. The turbulence associated with the flow through the inlet and across the stirred reactor is visualized at two different instants $t = 100$ s and $t = 190$ s in Figs. 2 and 3. As can be noted for the three SGS-models and as the flow is introduced from the nozzles into the tank, it is clear to have higher velocities near the inlet pipes. In addition, there is a velocity difference between the nozzle stream and the rotating flow in the tank, which creates local shear zones. As a consequence, inlet nozzles become the inherent turbulence generators. The intensity of the generated turbulence is less pronounced in the classical Smagorinsky model, while it seems to be more pronounced in the WALE and WMLES models.

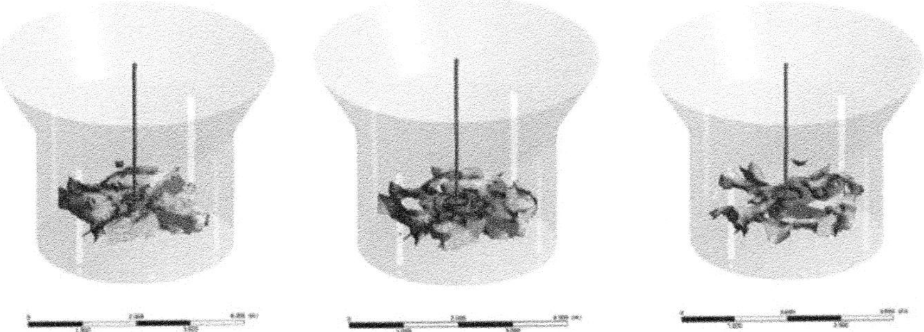

Figure 2: Turbulence structures for (left to right) Smagorinsky–Lilly model, WALE model, and WMLES model, respectively at t= 100 s

Furthermore, for the latter models, the instantaneous flow seems to be characterized by nearly concentric streamlines in a circular fluid motion, which is due to tangential inflow. The resulting vortex column around the central outlet structure is wrapped by the external turbulent filaments, leading to intensive momentum transport.

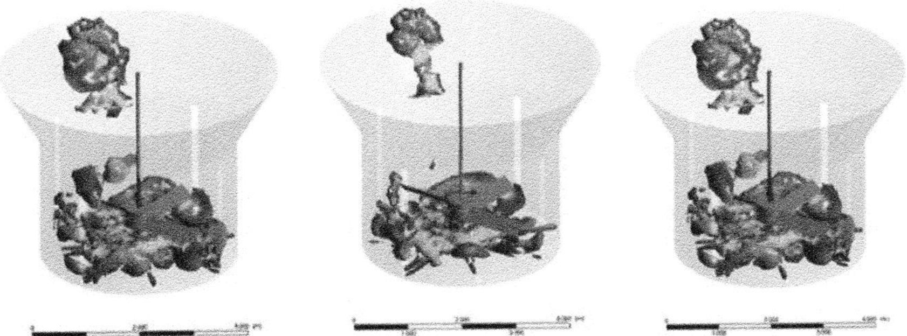

Figure 3: Turbulence structures for (left to right) Smagorinsky–Lilly model, WALE model and WMLES model, respectively at t = 190 s

Figure 5 presents the profiles of instantaneous magnitude velocities in two locations of the domain (denoted by P1 and P2 in Fig. 4).

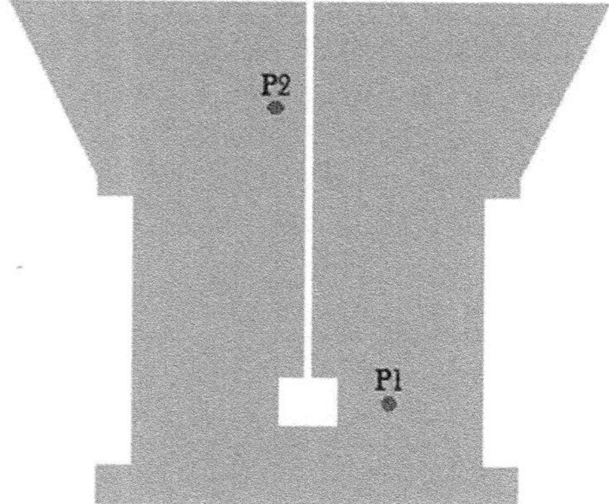

Figure 4: Axial projection plan in the middle of the domain

As a considerable throughput of liquid is injected near to position P2, a high velocity peak is observed in a short time. Then, as the mixing gets enhanced, the velocity decreases gradually until t = 150 s and a considerable flow amount of Ammonia gas achieves the top of the domain. This explains the emergence of new turbulent eddies in the P2 location.

LES models were able to capture the generated eddies coming from the liquid injection and from the mixing zone. WALE model gives higher values of the velocity magnitude than WMLES and Smagorinsky-Lilly models.

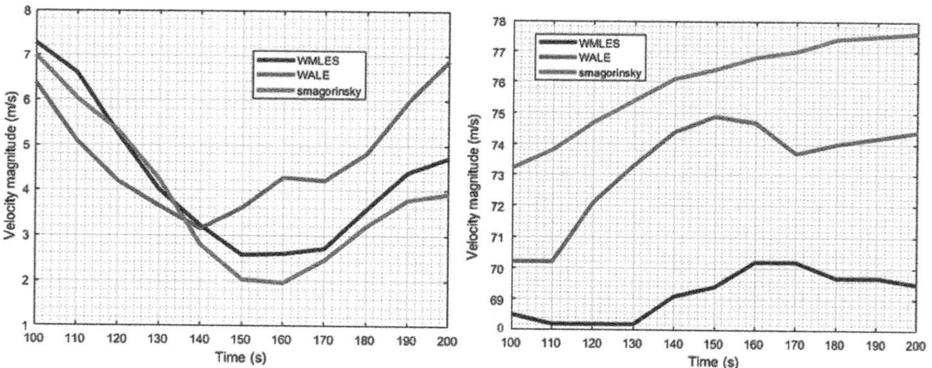

Figure 5: (a) & (b) Temporal evolution of velocity magnitude profiles in probes P1 and P2, respectively

However, based on the P2 probe position near to the shaft wall, the WALE is more appreciated to be used for near to the wall turbulence eddies prediction. As shown in Fig. 5b, the predicted velocity magnitude profiles fastly increase in the rotation zone (P2 probe). Strong eddies are developed by mixing, and interesting redistribution of the reactants in ensured by the circulation loop, which amplifies the magnitude of velocity. The occurrence of the gas flow from the nozzles with a considerable injection velocity considerably enhances the eddies and amplifies the circulation loops. The Smagorinsky-

Lilly model seems to capture the flow hydrodynamics with a good precision compared to WMLES and WALE models.

5. Conclusion

In this paper, the large eddy simulation method based on an Eulerian–Eulerian CFD model is used to simulate hydrodynamics of the gas–liquid reacting flow in the preneutralizer stirred reactor. The liquid and gas phase flow fields in the stirred tank simulated by the LES approach are more asymmetric in the case of WMLES and WALE models. Overall, each of the three models has its own specific ability for the flow prediction. However, the choice of an LES Turbulence model should be done carefully, for a good physical flow prediction.

References

G. R. Campbell, Y.-K. Leong, C. C. Berndt, J. L. Liow, 2006. Ammonium phosphate slurry rheology and particle properties—the influence of Fe (III) and Al (III) impurities, solid concentration and degree of neutralization. Chemical Engineering Science 61 (17), 5856–5866.

J. Derksen, 2001. Assessment of large eddy simulations for agitated flows. Chemical Engineering Research and Design 79 (8), 824–830.

S. Elmisaoui, S. Elmisaoui, S. Benjelloun, L. Khamar, M. Khamar, 2022. Cfd investigation of industrial gasliquid preneutralizer based on a bioreactor benchmark for spargers optimization. Chemical Engineering Transactions 93, 73–78.

S. Elmisaoui, L. Khamar, S. Benjelloun, M. Khamar, J. Ghidaglia, 2020. Modeling and study of hydrodynamic flow within the preneutralizer reactor using cfd approach. In: 30th European Symposium on Computer Aided Chemical Engineering. Elsevier, pp. 103 –108.

J. Hunt, S. Leibovich, K. Richards, 1988. Turbulent shear flows over low hills. Quarterly Journal of the Royal Meteorological Society 114 (484), 1435–1470.

Antonis Kokossis, Michael C. Georgiadis, Efstratios N. Pistikopoulos (Eds.)
PROCEEDINGS OF THE 33rd European Symposium on Computer Aided Process Engineering (ESCAPE33), June 18-21, 2023, Athens, Greece
 http://dx.doi.org/10.1016/B978-0-443-15274-0.50030-5

Mechanistic model for inkjet printing and applications in perovskite solar cells engineering

Naveen Bhati[a,*], Mohammad Khaja Nazeeruddin[b], and François Maréchal[a]

[a]*Industrial Process and Energy Systems Engineering, Ecole Polytechnique Fedérale de Lausanne, Sion, 1951, Switzerland*
[b]*Group for Molecular Engineering of Functional Materials, Institute of Chemical Sciences and Engineering, Ecole Polytechnique Fedérale de Lausanne, Sion, 1951, Switzerland*
**naveen.bhati@epfl.ch*

Abstract

Perovskite solar cells (PSCs) are the most exciting third-generation PV technology, which has shown an unprecedented increase in power conversion efficiencies from < 4% to > 25% in just over a decade. However, most of the research results indicating efficiencies of more than 20% are based on spin coating techniques with small cells with an active area < 20mm^2. With efficiency already been on par with established silicon solar cells, there is a need to upscale the technology to production scale and parallelly work on other facets of the technology like stability and toxicity. Among such techniques for mass production of PSCs is well-established inkjet printing, which has been explored to achieve PSC printing and can be used with various substrates and solvents. However, there is not much effort in modelling the inkjet printing process and understanding the impacts of different decision variables related to the process. Here, this article presents a mechanistic model to achieve perovskite printing (or entire stack of layers) using this technique and analyzing the influence of solvent selection and process parameters on film characteristic.

Keywords: Perovskite solar cells, inkjet printing. fabrication techniques, modelling.

1. Introduction

Inkjet printing is an attractive technology not only for research level but also for scaling the perovskite solar cells fabrication because of low cost, high production throughput and high material efficiency. Further, the non-contact nature of printing helps in depositing solutions on variety of materials. The technique has been discussed in literature with focus on droplet formation mechanism, travel to substrate and droplet interaction with substrate (Derby, 2010; Kang et al., 2020; Soltman and Subramanian, 2008). However, most of the studies in simulation regime were based on computational fluid dynamics which is time-consuming and resource intensive (Tofan et al., 2021; van der Bos et al., 2014) with less focus on process dynamics of the process. With the ever- increasing applications of inkjet printing in printable energy technologies like organic solar cells, polymer solar cells, fuel cells, energy storage devices, etc., it is important to understand the effect of different variables involved in the process on the final film quality.

Perovskite solar cells have shown unparalleled improvement in their efficiency over the last decade and have reached the levels of the established Si cells. Solution-based deposition methods have shown better results compared to their vacuum-based counterparts not only in efficiency but also provides cost effective, easily controllable, and scalable solutions. Among these solution-based deposition methods, inkjet printing has gained significant interest in the community where drop-by-drop printing leads to complete formation of film with control over droplet spacing and morphology via solvent engineering (Yang et al., 2022). The power conversion efficiency (PCE) of inkjet-printed devices has risen from 11.6% (Wei et al., 2014) to 21.6% (Abzieher et al., 2019) since its first application. Moreover, many researchers have also shown the complete fabrication of PSCs except the transparent conducting oxide layer using inkjet printing technique which shows the potential of complete fabrication of perovskite solar cells based on inkjet printing (Gao et al., 2021). However, the dynamics of inkjet printing is more complex as compared to other solution-based techniques like spin-coating (0-D) or slot-die coating (1-D) processes. Therefore, it is crucial to understand the influence of different parameters on the inkjet printing process to obtain homogeneous liquid films and to control the morphology of the final dry film (Huckaba et al. 2019, Mathies et al., 2016).

2. Model Description

The entire process of inkjet printing can be divided into four stages:

2.1. Droplet generation

Inkjet printing technique can be used only with certain solvents or solutions depending on the solution properties and is governed by coating window which can be represented in terms on non-dimensional numbers Eq. 1-3 (Derby, 2010).

$$0.1 \leq \mathrm{Oh}\,(= 1/\mathrm{Z}) \leq 1\,;$$ *Oh-Ohnesorge number*............Eq. 1

$$We > 4;$$ *We-Weber number*...............Eq. 2

$$We^{1/2}Re^{1/4} \leq 50;$$ *Re-Reynolds number*.............Eq. 3

2.2. Droplet travel to substrate

Once the drop is generated and leaves the nozzle, the set of differential equations describing the system in second stage include Eq. 4-7 where C_D is based on Kunii and Levenspiel(1991), velocity, v is based on Sloth(2007) and k_m is mass transfer coefficient.

$$\frac{dv}{dt} = \left(1 - \frac{\rho_a}{\rho_d}\right) g - \frac{C_D \rho_a v^2 \pi r_d^2}{2m_d} + \frac{F_t}{m_d};$$ *C_D-Drag coefficient, m_d:drop mass*...............Eq. 4

$$\frac{dm_{solvent}}{dt} = \left(-k_m 4\pi r_d^2 \frac{M_{solvent} P_{amb}}{RT_{avg}}\right)\left(y_{v,surface} - y_{v,amb}\right);$$ *y_v-mole fraction in vapor*.Eq. 5

$$\frac{dr_d}{dt} = \frac{1}{4\pi r_d^2 \rho_{solvent}} \frac{dm_{solvent}}{dt};$$ *r_d-droplet radius, t-time, ρ-density* ....Eq. 6

$$\frac{dT_d}{dt} = \frac{h_c[4\pi r_d^2 (T_{amb} - T_d)] + \Delta H_v \frac{dm_{solvent}}{dt}}{m_{solute} c_{p,solute} + m_{solvent} c_{p,solvent}};$$ *T-temperature, h_c-heat transfer coefficient*.....Eq. 7

The thermophoretic force component, F_t is a dominant force especially close to the surface when the temperature of substrate is very high (Filipovic et al., 2013).

2.3. Droplet impingement and positioning

Based on Stringer and Derby (2010), once the droplet reaches the surface, the equilibrium diameter of the droplet ($D_{eq,d}$) can be expressed in terms of diameter of droplet at impingement and the equilibrium contact angle (θ_{eqb}). For stable line formation without any defects like waviness or bulging, criteria for maximum and minimum droplet spacing is used. Further, width of the bead is deduced based on Stringer and Derby (2010). Afterwards, the line formed is assumed to be a flat rectangle with height depending on the width and total volume of the bead from initial state of flat cylinder for single droplet.

2.4. Film formation

The system of differential equations for tracking the solution parameters during line/film stages can be expressed as given in Eq. 8-10 where ΔH_v is enthalpy of vaporization.

$$\frac{dm_{solvent}}{dt} = \left[-k_m (A_{eff}) \frac{M_{solvent}P_{amb}}{RT_{avg}}\right] (y_{v,surface} - y_{v,amb});$$ *A_{eff}-mass transfer area*...Eq. 8

$$\frac{dH_{bead}}{dt} = \frac{1}{A_{eff'}\rho_{solvent}} \frac{dm_{solvent}}{dt};$$ *H_{bead}-bead height, $A_{eff'}$-heat transfer area from substrate.*Eq. 9

$$\frac{dT_{bead}}{dt} = \frac{h_c[A_{eff}(T_{amb}-T_{bead})]+\Delta H_v \frac{dm_{solvent}}{dt} + \frac{A_{eff'}(T_{subs}-T_{bead})(\kappa_{sol})}{H_{bead}/2}}{m_{solute}c_{p,solute}+m_{solvent}c_{p,solvent}};$$ *κ-thermal cond..* Eq. 10

The total time of the deposition process, t_{dp} can be calculated based on Eq. 11-12.

$$t_{dp} = \frac{cell_{x-dim}\, cell_{y-dim}\, dpi_{target-y}}{v_{ph}\, dpi_{native}\, width_{ph}};$$ *$cell_{x\text{-}dim}$, $cell_{y\text{-}dim}$ -x-and y-dimension of cell*.......Eq. 11

$$v_{ph} = \frac{25.4 f_{DOD}}{1000\, dpi_{target-x}};$$ *v_{ph}- printhead velocity, f_{DOD}-drop ejection frequency* ....Eq. 12

During film formation stage, line width (W_{line}) during multiple repetition (N_{rep-y}) is based on the lateral spacing (based on $1/dpi_{target-y}$), equilibrium diameter of droplet and initial line width ($W_{ini-line}$) (Eq. 13). When merging of line happens, the width ($W_{rem-bead}$) and length ($L_{rem-bead}$) of the new bead is expressed based on Eq. 14-15.

$$W_{line} = \frac{W_{ini-line}}{2} + \max\left(\frac{W_{ini-line}}{2}, (N_{rep-y} - 1) * \frac{0.0254}{dpi_{target-y}} + \frac{D_{eq,d}}{2}\right)$$ ....Eq. 13

$$L_{rem-bead} = N_d * D_{eq,d} - (N_d - 1) * \frac{0.0254}{dpi_{target-x}};$$ *N_d-number of drops deposited*....Eq. 14

$$W_{rem-bead} = \frac{0.0254}{dpi_{native}} - W_{line};$$ *dpi_{native}- nozzles per inch*............Eq. 15

3. Results

3.1. Validation

In this section, model is validated based on existing experimental data before extending it to elucidate film formation dynamics. Figure 1 a) shows the droplet velocity as a function of distance from nozzle which lies in the range of data collected by Kang et al. (2020). Further, Figure 1 b) and c) shows the evolution of droplet volume based on present model and comparison with experimental results obtained by Lim et al. (2009) which shows very good agreement in all range of initial droplet volume and substrate temperature with water (W) and ethylene glycol (EG) as solvents. Since, the analytical expression for equilibrium droplet diameter and first bead/line width were directly used, they are not validated here. With these validation results, the model can be used for further understanding of the process dynamics involved in the inkjet printing process including line formation and film formation and the impact of different decision variables.

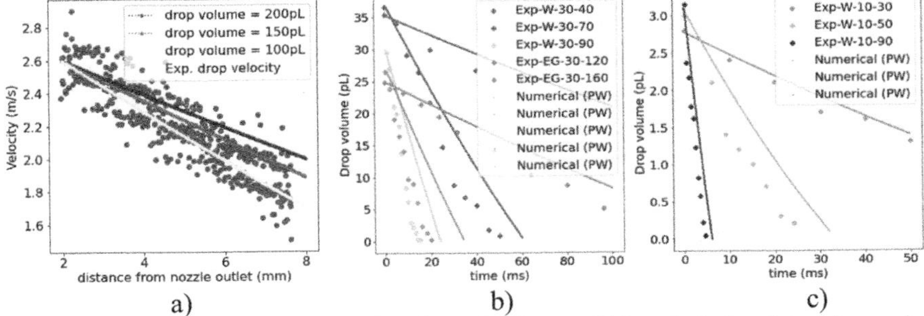

Figure 1: Validation results for a) droplet velocity, and b) and c) droplet volume during evaporation on substrate (PW-present work)

3.2. Screening decision variables space

In the first part, the screening of parameter space based on coating window available for the process can be investigated. The result shows that out of 84000 cases only 12600 fall within the coating window. Further, with constraints on the size of droplets with respect to nozzle and boiling point of solvents only 6480 cases remain feasible. Finally, with constraints on minimum spacing and maximum spacing between droplets only 4568 cases remain feasible for the inkjet printing process. Out of the 24 solvents commonly used in perovskite layer deposition, only 6 solvents have required properties with the given parameters selected for the analysis which include hexanol (1-HEX-01, 23), 2-2-ethoxyethoxy-ethanol (2-2-E-01, 21), 2-butoxyethanol (2-BUT-01, 17), alpha terpineol (ALPHA-01, 9), n-butanol (4) and ethylene glycol (ETHYL-02, 22).

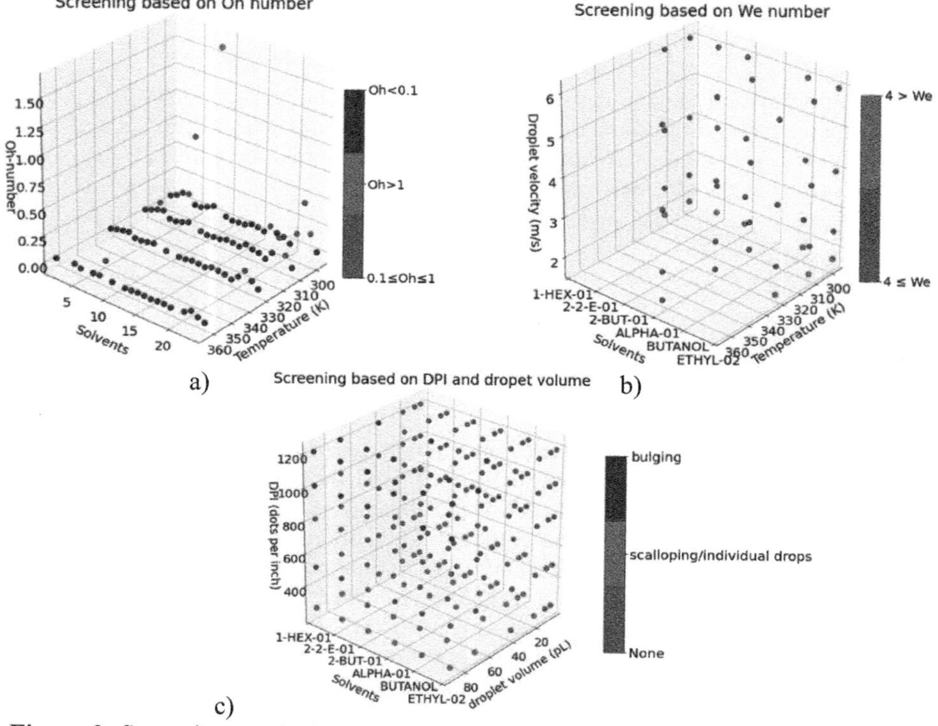

Figure 2: Screening results based on a) Oh-number b) We number and c) line defects

Ethanol (2) and IPA (3) are on the edges of the applicability window. Further DMSO (14), DMF (13), gamma-butyrolactone (15), o/m-Dichlorobenzene (24, 10) and NMP (19) are also lying close to the edges of the application window at low temperature. Other solvents used include water (1), chlorobenzene (5), methanol (6), 2-methoxyethanol (7), propanol (8), acetone (11), toluene (12), Methyl-phenyl-ether (16), tetrahydronaphthalene (18) and acetonitrile (20). Ambient conditions are 25°C and 40% RH.

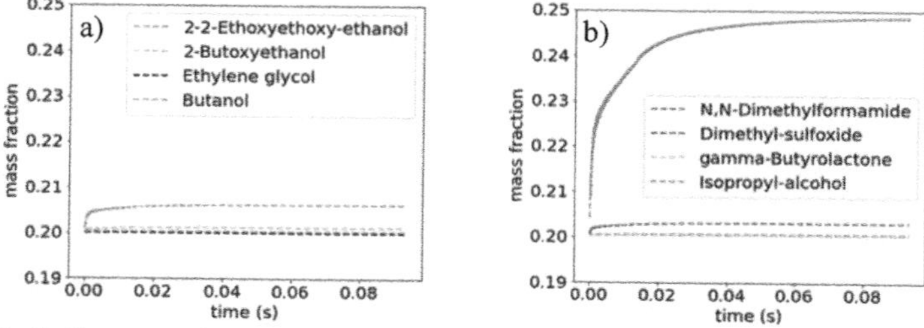

Figure 3: Influence of solvent properties a) within and b) outside coating window

3.3. Effect of solvent (T_{subs}=55°C, DPI=847, LPI =635, θ_{eqb} = 1°, V_d =10pL, f_{DOD} =5000)
Figure 3 shows the influence of different solvents on the wet film characteristics. IPA shows maximum change (>20%) in mass fraction over printing time (<100 ms). However, the solvents lying within the coating window don't show more than 5% change in the mass fraction over time with maximum change shown by Butanol. Therefore, mixing solvents with different properties can be used to influence the final film characteristics.

3.4. Effect of substrate temperature (T_{subs})
Figure 4 shows the effect of substrate temperature on film properties with Dimethyl-sulfoxide (DMSO) as solvent (for all remaining cases). The effect of temperature becomes more dominant at high temperature resulting in >10% difference in film height with progressive increase in mass fraction of solute. Here, the maximum temperature cannot be more than boiling point of solvent to avoid pin-hole formation in the deposited film.

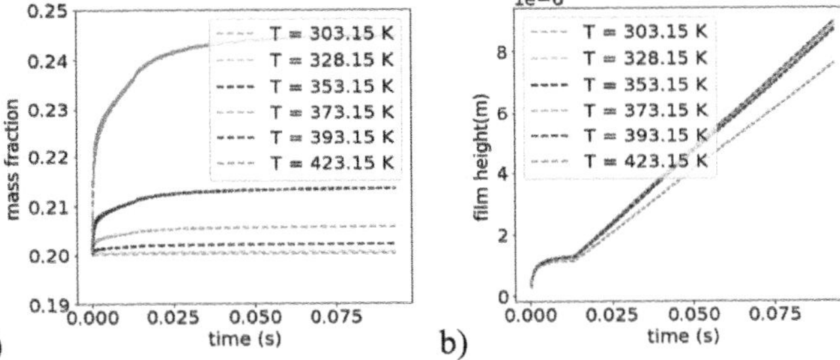

Figure 4: Influence of substrate temperature on a) mass fraction and b) film height

3.5. Effect of Target DPI ($dpi_{target-x}$) (LPI =51)
Figure 5 shows the effect of DPI (dots per inch, relates to spacing in direction of printing) for DMSO at substrate temperature of 55° C. Depending on the frequency the total time for printing changes with dpi. With increase in DPI, the total time increases, however, the change in mass fraction is very less. Further, with increase in DPI, film height increase as more droplets are deposited on the same area of substrate. With increase in dpi, the morphology of wet film changes from individual lines (from different nozzles) to merged film as closer droplets leads to broader lines which become greater than nozzle spacing.

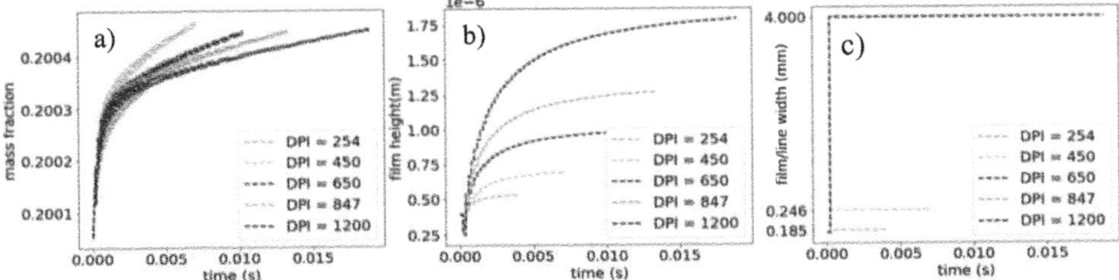

Figure 5: Influence of target DPI on a) mass fraction b) film height and c) film width

3.6. Effect of Target LPI (DPI =254)
In this subsection, influence of droplet spacing in lateral direction (lines per inch, lpi or $dpi_{target-y}$) is shown. With LPI increase, total time of process increases resulting in more solution deposited on the same area thus leading to thicker wet films. Further, change in mass fraction decreases and time for film formation increase with increase in LPI(Fig. 6).

4. Conclusions

Here a mechanistic model for inkjet printing is presented which can be used to understand the impact of different decision variables involved in the process of inkjet printing.

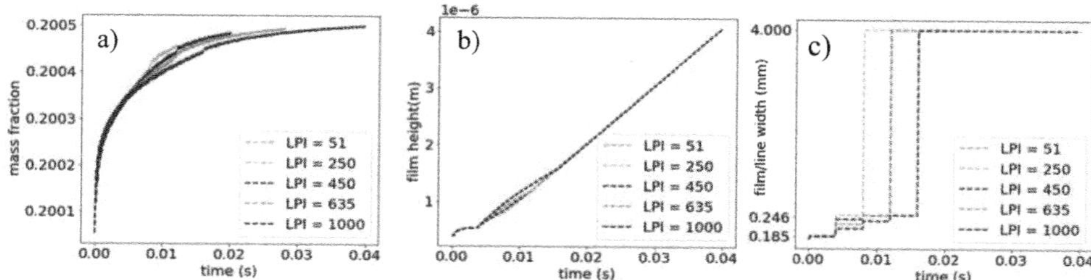

Figure 6: Influence of target LPI on a) mass fraction b) film height and c) film width

Screening solvents based on existing literature should be done with more lenient conditions. Most of the solvents show similar behavior during the printing process except a few which have high vapor pressure. Substrate temperature can have significant impact of final mass fraction of film. DPI and LPI have direct influence on the film thickness, processing time and film morphology with less impact on change in mass fraction.

5. Acknowledgements

This work is supported by the European Research Council (ERC) under the European Union's Horizon 2020 research and innovation programme under the Marie Skłodowska-Curie grant agreement no. 945363.

References

Li, X., Bi, D., Yi, C., Décoppet, J.D., Luo, J., Zakeeruddin, S.M., Hagfeldt, A. and Grätzel, M., 2016. Science, 353(6294), pp.58-62

Derby, B., 2010. Annual Review of Materials Research, 40, pp.395-414

Kang, S.H., Kim, S., Lim, J.W., Sohn, D.K. and Ko, H.S., 2020. Journal of Mechanical Science and Technology, 34(8), pp.3311-3315.

Soltman, D. and Subramanian, V., 2008. Langmuir, 24(5), pp.2224-2231

Stringer, J. and Derby, B., 2010. Langmuir, 26(12), pp.10365-10372

Tofan, T., Kruggel-Emden, H., Turla, V. and Jasevičius, R., 2021. Applied Sciences, 11(2), p.527

van der Bos, A., van der Meulen, M.J., Driessen, T., van den Berg, M., Reinten, H., Wijshoff, H., Versluis, M. and Lohse, D., 2014. Physical review applied, 1(1), p.014004

Yang, H., Wang, J., Yu, X., Feng, Y., Chen, X., Long, F., Ku, Z., Huang, F., Cheng, Y. and Peng, Y., 2022. Chemical Physics Letters, 807, p.140084

Wei, Z., Chen, H., Yan, K. and Yang, S., 2014. Angewandte. Chemie International. Edition, 53(48), pp.13239-13243

Abzieher, T., Moghadamzadeh, S., Schackmar, F., Eggers, H., Sutterlüti, F., Farooq, A., Kojda, D., Habicht, K., Schmager, R., Mertens, A. and Azmi, R., 2019.Advanced Energy Materials, 9(12), p.1802995.

Gao, B. and Meng, J., 2021. Solar Energy, 230, pp.598-604

Huckaba, A.J., Lee, Y., Xia, R., Paek, S., Bassetto, V.C., Oveisi, E., Lesch, A., Kinge, S., Dyson, P.J., Girault, H. and Nazeeruddin, M.K., 2019. Energy Technology, 7(2), pp.317-324

Mathies, F., Abzieher, T., Hochstuhl, A., Glaser, K., Colsmann, A., Paetzold, U.W., Hernandez-Sosa, G., Lemmer, U. and Quintilla, A., 2016. J. of Materials Chemistry A, 4(48), pp.19207-19213

Kunii, D. and Levenspiel, O., 1991. Fluidization engineering. Butterworth-Heinemann

Sloth, J., 2007. Formation of enzyme containing particles by spray drying (Doctoral dissertation, PhD Thesis. Department of Chemical Engineering, Technical University of Denmark)

Filipovic, L., Selberherr, S., Mutinati, G.C., Brunet, E., Steinhauer, S., Köck, A. and Schrank, F., 2013, July. In Proceedings of the world congress on engineering (Vol. 2, pp. 987-992).

Lim, T., Han, S., Chung, J., Chung, J.T., Ko, S. and Grigoropoulos, C.P., 2009. International Journal of Heat and Mass Transfer, 52(1-2), pp.431-441

Antonis Kokossis, Michael C. Georgiadis, Efstratios N. Pistikopoulos (Eds.)
PROCEEDINGS OF THE 33rd European Symposium on Computer Aided Process Engineering (ESCAPE33), June 18-21, 2023, Athens, Greece
 http://dx.doi.org/10.1016/B978-0-443-15274-0.50031-7

Modeling and Optimization of Low-carbon Oriented Urban Energy Systems Spatial Planning Considering Energy Networks Topology

Ruiyu Zhang[a], Pei Liu[a,*], Zheng Li[a]

[a]*State Key Lab of Power Systems, Department of Energy and Power Engineering, Tsinghua University, Beijing 100084, China*

liu_pei@tsinghua.edu.cn

Abstract

Spatial dimension planning of an energy system is critical at an urban scale due to the strong spatial heterogeneity of load levels and categories at different spots of a city and the high complexity of pipeline networks extension and connection topology. External requirements of low-carbon transition of a city may make its urban energy systems planning further complicated. In this paper we propose a novel urban energy systems spatial planning modeling and optimization approach considering energy systems form and topology. Coupling with integrated energy systems modeling and multiple energy pipeline networks topology modeling, the processes of energy generation, conversion, storage, and transmission are depicted synthetically. A case area is employed to illustrate feasibility of proposed method and the impacts of both low-carbon transition and user side participation in energy production on urban energy systems optimal layout. Results show that main energy supply technology shifts and corresponding networks topology changes significantly under the requirement of decarbonization and energy supply mode revolution. Decarbonization and participation of user side in power generation reduce the average network capacity by 19% and 26% respectively, indicating a more interconnected energy network with low capacity to balance energy between areas is a preferred form for a low-carbon city.

Keywords: urban energy systems; systems modeling and optimization; network topology; low carbon transition.

1. Introduction

An urban energy system consists of a variety of energy generating units and interconnected pipeline networks of multiple energy categories, including electricity, heat, or cold, thus has great potential to integrate multiple types of energy. However, the high complexity of such a system also brings great challenges to its planning. Moreover, urban areas are the main places of various kinds of social activities, in which might lead to massive carbon emissions and various environmental problems. It is expected more than 70% of the population will live in cities in 2050 (C40 Cites, 2017), and a considerable part of emissions may come from energy production and utilization. Therefore, it is of great importance to carry out low-carbon oriented urban energy systems planning, especially under the new trend of decarbonization and renewable energy utilization (Stoeglehner et al, 2016).

Currently, a series of urban energy related researches exist. One mainstream of concern is urban emissions measurement and analysis based on emissions inventory (Kennedy et al, 2010) and index decomposition method such as LMDI (Gu et al, 2019), IPAT and improved STRIPAT (Wang et al, 2017), to quantify the impact of various social and economic indicators on carbon emissions. Another group of scholars apply commercial software, for example LEAP (Huang et al, 2019), MARKAL (Isi et al, 2021) and Energy PLAN (Thellufsen et al, 2020), to implement simulation and designing for a low-carbon energy system. However, most of these studies are similar to decarbonization researches at a national scale (Sun et al, 2019), which generally apply lumped parameter method and regard objective into a whole, mainly discuss the impacts of relevant indicators change on carbon emissions. Spatial dimension, the most unique and vital aspect at a city scale, is neglected and heterogeneities and correlations within a city are thus not considered enough. Conversely, existing studies on urban space mainly focus on land use planning (Penazzi et al, 2019) and various location problems (Bélanger et al, 2019), where concerns of low-carbon oriented revolution of urban energy systems are absent.
As a result, the purpose of this study is to investigate the impacts of decarbonization as well as energy supply revolution represented by user side participation in energy supply on the optimal urban energy systems form and layout, aiming at facilitating spatial planning of low-carbon oriented urban energy systems.

2. Methodology

2.1. Overview of model structure

The aim of this study is to explore the impacts of decarbonization and low-carbon revolution on urban energy systems form and layout. Therefore, an urban energy systems spatial planning model is proposed, as illustrated in Figure 1.

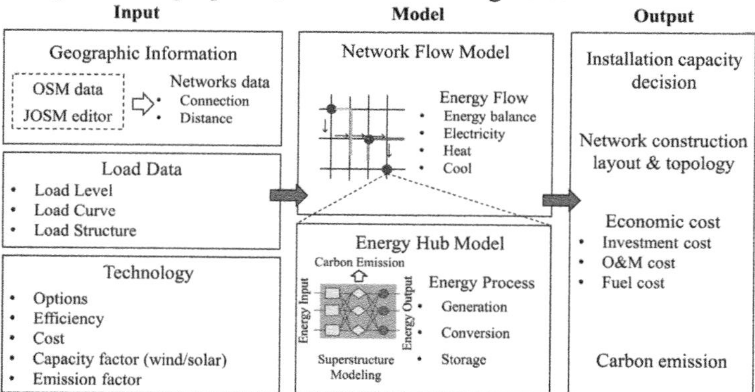

Figure 1 Model framework

Geographic information is derived from OpenStreetMap (OSM, 2022), a widely-used open-source map platform, and edited by Java OpenStreetMap (JOSM, 2022). Load data are used for depicting various energy demand and techno-economics parameters are applied for energy facilities modeling.
The core of the proposed model consists of two sub-models, namely the Network Flow Model and the Energy Hub Model. The purpose of the Network Flow Model is to characterize energy transmission in the energy network and energy balance at each node. The Energy Hub Model aims to establish a mathematical model for real energy facilities installed at energy hub nodes based on superstructure modeling method.

The planning goals are (a) which energy hub nodes should be selected, (b) which technologies and how much capacity to install in hubs, and (c) how to lay the energy pipe networks properly to satisfied all load nodes demand with minimum cost. By solving this problem, optimal installation decision, energy networks arrangement, detailed cost and carbon emissions can be obtained.

2.2. Mathematical formulation

2.2.1. Objective function

The optimization objective is to minimize the overall energy systems costs, including investment costs of technologies $cost_{tech}$ and networks $cost_{net}$, operation and maintenance costs $cost_{om}$ and fuel costs $cost_{fuel}$. The proportion coefficient of $cost_{om}$ refers to literature (Xu et al., 2020).

$$cost_{total} = cost_{tech} + cost_{net} + cost_{om} + cost_{fuel} \tag{1}$$

$$cost_{tech} = \sum_{tech,node} CAPEX_{tech} \cdot ic_{node} \cdot \frac{r(1+r)^T}{(1+r)^{T-1}} \tag{2}$$

$$cost_{net} = \sum_{tdn,road_{id}} DISTANCE_{road_{id}} \cdot (C_0 + C_{1_{tdn}}) \cdot icn_{tdn,road_{id}} \cdot \frac{r(1+r)^T}{(1+r)^{T-1}} \tag{3}$$

$$cost_{fuel} = \sum_{fuel,node_{hub}} FP_{fuel} \cdot \sum eimp_{node_{hub,s,dh,f}} \tag{4}$$

$$cost_{om} = coefficient \cdot cost_{tech} \tag{5}$$

2.2.2. Network flow modeling

Expression (6) and (7) describe the upper limit of flow and energy balance of the energy flow in network.

$$-icn_{tdn,road_{id}} \leq flow_{road_{id},s,dh,e,tdn} \leq icn_{tdn,road_{id}} \tag{6}$$

$$\sum flow_{in_{node}} - \sum flow_{out_{node}} = \begin{cases} LOAD_{node}, & if\ node \in\ load\ nodes \\ 0 & if\ node \in\ road\ nodes \\ -output_{node_{hub},s,dh,e} & if\ node \in\ hub\ nodes \end{cases} \tag{7}$$

2.2.3. Energy hub modeling

Nine types of electricity generation technologies are considered, including Super-critical Pulverized Coal (SPC), Ultra-Supercritical Pulverized Coal (UPC), Combined-cycle Gas Turbine (CCGT), Biomass power plant (BE) and options of the above technologies with CCS. Solar Photovoltaic (PV) is selected as the option of renewable technologies. Besides, six types of heat and cool generation/conversion technologies, namely Biomass Power Plant Combined Heat and Power (BE_CHP), Combined-cycle Gas Turbine Combined Heat and Power (CCGT_CHP), Gas Boiler (GB) and Air-source Heat Pump (ASHP), Air-conditioner (AC), Absorption chiller (ABS_CHILLER), and three kinds of energy storage technologies including Battery (BATTERY), Sensible Thermal Energy Storage (STES) and Ice Storage (ICES) are involved. Key expressions of energy generation, conversion and storage processes are presented in (8) to (11).

$$eimp_{node_{hub,s,dh,e}} = \sum_{gnt} ei_{gnt,node_{hub},gnt,s,dh,e} \tag{8}$$

$$\sum_e ei_{gnt,node_{hub},gnt,s,dh,e} \times eff_{gnt,ee} = eo_{gnt_{node_{hub},gnt,s,dh,e}} \tag{9}$$

$$est_{node_{hub},stt,s,dh+1,e} = est_{node_{hub},stt,s,dh,e} \times (1 - LOSS_{STT}) - \frac{eo_{stt_{node_{hub},stt,s,dh,e}}}{eff_{stt,ee}} + ei_{stt_{node_{hub},stt,s,dh,e}} \tag{10}$$

$$output_{node_{hub},s,dh,e} = \sum_{gnt} eo_{gnt_{node_{hub},gnt,s,dh,e}} + \sum_{cvt} eo_{cvt_{node_{hub},cvt,s,dh,e}} + \sum_{stt} eo_{stt_{node_{hub},stt,s,dh,e}} - \sum_{cvt} ei_{cvt_{node_{hub},cvt,s,dh,e}} - \sum_{stt} ei_{gnt_{node_{hub},stt,s,dh,e}} \quad (11)$$

2.2.4. Carbon emissions constrain

Emissions constrain is shown in (12), namely net emissions should be limited by quota. Specially, when quota is set as zero, it means carbon neutralization scenario.

$$\sum(eimp_{node_{hub},s,dh,f} - ei_{ccs,node_{hub},gnt,s,dh,f} \cdot 0.9) \cdot CEI_f \cdot DAY_s \leq QUOTA \quad (12)$$

Eventually, the model proposed above is built in General Algebraic Modelling System (GAMS, 2022) platform and solved via applying built-in CPLEX solver.

3. Case study

3.1. Area information and scenarios setting

A case area commonly used in several literatures (Chen et al., 2018; Xu et al., 2020) is applied in this study. As shown in Figure 2, case area is consisted of 205 nodes, including 20 load nodes colored in orange triangle, representing points where energy demands exist, and 8 candidate energy hub nodes in red square meaning where energy facilities can be constructed, and 345 roads consist of existing roads in solid line and candidate in dotted line. All energy pipes are limited to build along roads. Basic load data and road distance are derived from literature (Xu et al., 2020). In this study the load data are further split into hourly scale via load profiles from a widely-used energy system planning software HOMER (NREL, 2022) to reflect the characteristics and changes of different categories of loads including residents, industry, and commerce, as shown in Figure 3.

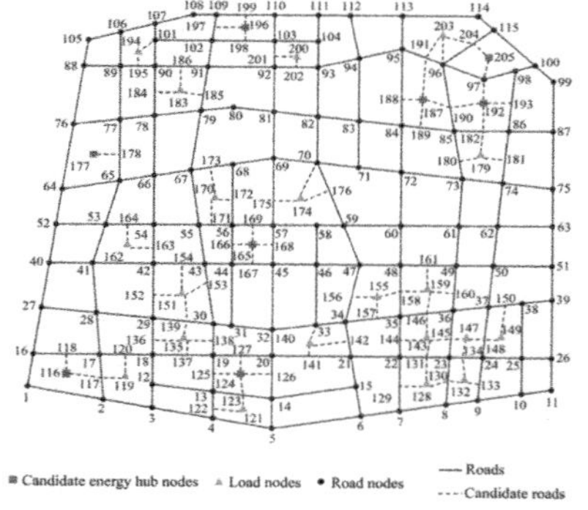

Figure 2 Networks of case area

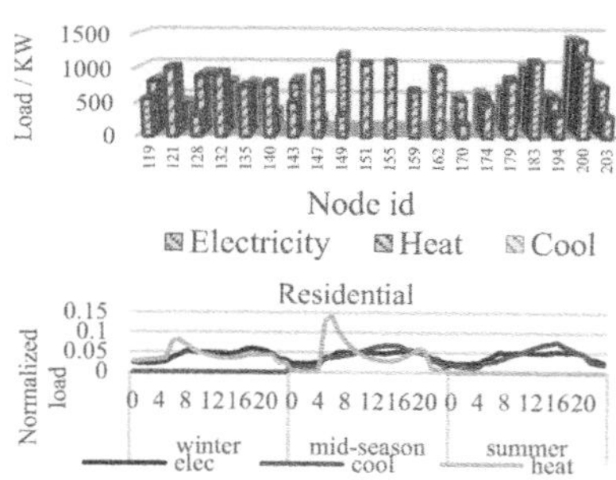

Figure 3 Load level and hourly-profile

To investigate the impacts of decarbonization on urban energy systems planning, three scenarios are set according to emissions constraint and energy supply mode, as shown in Table 1. **NC** is a reference scenario with no emissions constrain and follows the current mainstream idea of centralized energy supply, which means energy supply equipment except AC can only be installed in the energy hub. **DC-1** is a low-carbon scenario with the requirement of carbon neutralization but stays in traditional centralized energy supply mode, to illustrate the impacts of carbon neutralization. **DC-2** further indicates that distributed PVs are approved to be constructed at user side, namely load nodes and users play the role of both energy consumers and producers, which is an expected form of energy systems revolution in the process of decarbonization.

Table 1 Scenario setting and explanations

Scenario	Explanation
NC	No carbon emissions constrain; Centralized energy supply
DC-1	Decarbonization with carbon neutralization; Centralized energy supply
DC-2	Decarbonization with carbon neutralization; User side distributed PV available

3.2. Impacts of decarbonization on urban energy systems planning

3.2.1. Major energy technologies shifting

The choices of energy hubs in three scenarios are consistent, almost all candidate hub nodes are selected except node 205. However, the technology installing decisions are different as shown in figure 4 and figure 5. In DC-1, the capacity of PVs grows while UPC and ASHP descends markedly, and more types of technologies such as BE_CHP and ABS_CHILLER are used. The installed technologies at user side in these two scenarios are limited in AC only since scenarios setting. In DC-2 scenario, however, it can be found a considerable number of PVs are installed at the user side, and both batteries and thermal storage equipment are installed.

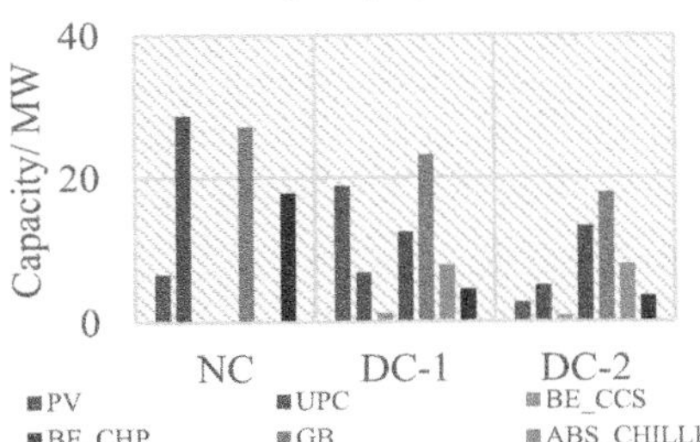

Figure 4 Installed capacity of energy hub

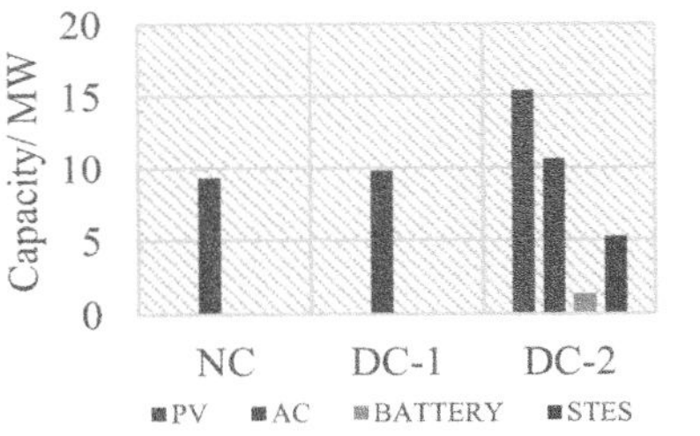

Figure 5 Installed capacity of user side

3.2.2. Energy networks topology

Figure 6 displays the electricity network and heat network topology between scenarios. Red, yellow, and blue nodes represent hub nodes, load nodes and road nodes respectively. Linkages between nodes mean where energy networks are constructed. It should be noticed that these topology diagrams only indicate the topological relationships, namely the connections between nodes, instead of real spatial direction and distance. Nodes unselected are not shown in diagrams.

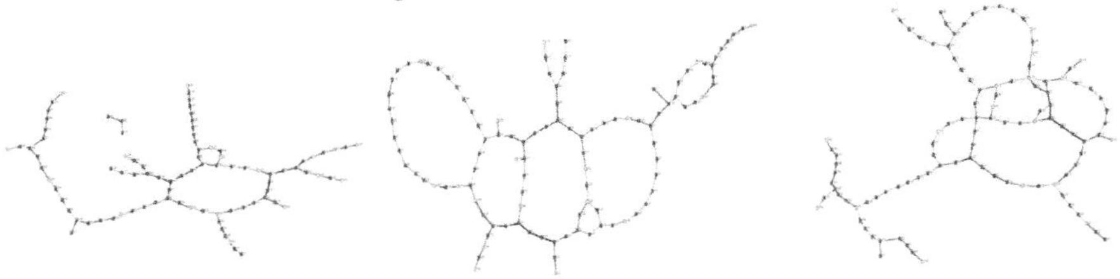

(6a) NC (6b) DC-1 (6c) DC-2

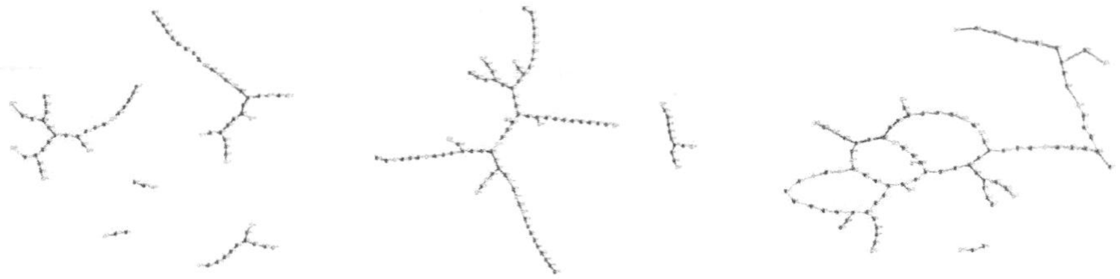

(6d) NC (6e) DC-1 (6f) DC-2

Figure 6 Electricity network (6a-6c) and heat network (6d-6f) topology

In detail, electricity network in NC scenario consists of two independent subnets. By comparison, the network exists as a whole and is more complex with six rings in DC-1. This is because abundant PVs are constructed under the constrains of carbon neutralization, which implies the mismatch between load and energy supply will be more severe due to the inherent volatility of renewable energy. Therefore, a more interconnected supply network is needed to balance the energy between nodes. And in DC-2 scenario, with the participation of user side distributed PVs, the number of latent energy supply nodes and the requirement for load-supply matching increase, explaining further complications on the network topology. Table 2 summarizes specific results of networks. It can be found that the results of total capacity in NC and DC-1 are almost the same, while the number of pipe lines increase 23%. Comparing DC-1 and DC-2, the number of lines is almost identical while total capacity decreases by 24%. A similar trend can be observed in heat network. However, the structure of heat network is more concise and its average capacity is approximately twice that of the electricity network.

Table 2 Energy networks capacity

Unit: MW	EN capacity	Number of lines	Average capacity	HN capacity	Number of lines	Average capacity
NC	128.3139	90	1.42571	235.1022	75	3.134695
DC-1	128.4098	111	1.156845	235.6052	89	2.64725
DC-2	97.06535	113	0.858985	188.0379	95	1.979346

4. Conclusions

Under the targets of decarbonization and energy systems revolution, the decision making and spatial layout of urban energy systems will change significantly. Results indicate that a dispatching and balancing oriented network with more complex topology and lower average line capacity is needed for the requirements of decarbonization and user side participation in energy supply, instead of a transmission-oriented net that carries energy from centralized plants to load nodes.

References

V. Bélanger et al, 2019, Recent optimization models and trends in location, relocation, and dispatching of emergency medical vehicles, European Journal of Operational Research, 272, 1, 1-23.

C40 Cites, 2017, Focused acceleration: A strategic approch for climate action in cities, www.mckinsey.com <accessed in 2022.11.13>.

J. Chen et al, 2018, Research on "Stations-Pipelines" Layout and Optimization of Regional Energy Internet, Proceedings of the CSEE, 38(03), 675-684. (In Chinese).

GAMS, www.gams.com <accessed in 2022.11.11>.

S. Gu et al, 2019, Coupled LMDI and system dynamics model for estimating urban CO2 emission mitigation potential in Shanghai, China, Journal of Cleaner Production, 240, 118034.

Y. Huang et al, 2019, Exploring potential pathways towards urban greenhouse gas peaks: A case study of Guangzhou, China, Applied Energy, 251, 113369.

M. Isik et al, 2021, Transportation emissions scenarios for New York City under different carbon intensities of electricity and electric vehicle adoption rates. Nature Energy 6, 92–104.

JOSM, josm.openstreetmap.de <accessed in 2022.11.11>.

C. Kennedy et al, 2010, Methodology for inventorying greenhouse gas emissions from global cities, Energy Policy, 38, Issue 9, 4828-4837.

NREL, www.nrel/gov/HOMER <accessed in 2022.11.11>.

OSM, www.openstreetmap.org <accessed in 2022.11.11>.

S.Penazzi et al, 2019, Planning low carbon urban-rural ecosystems: An integrated transport land-use model, Journal of Cleaner Production, 235, 96-111.

G. Stoeglehner et al, 2016, Chapter 3: Spatial Archetypes in the Energy Turn, Integrated Spatial and Energy Planning Supporting Climate Protection and the Energy Turn with Means of Spatial Planning, 35-51.

Z. Sun et al, 2019, China's carbon emission peak pre-2030: Exploring multi-scenario optimal low-carbon behaviors for China's regions, Journal of Cleaner Production, 231, 963-979.

J.Z. Thellufsen et al, 2020, Smart energy cities in a 100% renewable energy context, Renewable and Sustainable Energy Reviews, 129, 109922.

S. Wang et al, 2017, Examining the impacts of socioeconomic factors, urban form, and transportation networks on CO2 emissions in China's megacities, Applied Energy, 185, Part 1, 189-200.

C. Xu et al, 2020, Planning of Energy Station and Pipeline Considering Topological Characteristics of Regional Integrated Energy System , Automation of Electric Power Systems, 44(03), 74-82. (In Chinese).

Antonis Kokossis, Michael C. Georgiadis, Efstratios N. Pistikopoulos (Eds.)
PROCEEDINGS OF THE 33rd European Symposium on Computer Aided Process Engineering (ESCAPE33), June 18-21, 2023, Athens, Greece
 http://dx.doi.org/10.1016/B978-0-443-15274-0.50032-9

Multi-objective optimisation for microbrewery retrofitting

Ruosi Zhang [a,b], Melis Duyar[a], Lesley Foulkes[c], James Berrow[c], Michael Short[a*]

[a]*University of Surrey, Stag Hill, Guildford GU2 7XH, UK*
[b]*National Innovation Center for Digital Fishery, Beijing 100083, China*
[c]*Langham Brewery, The Granary, Lodsworth, West Sussex GU28 9BU, UK*
m.short@surrey.ac.uk

Abstract

In a highly competitive retail market, many microbreweries have attempted to maximise profits and decrease energy consumption through retrofitting their operations with renewable energy. This paper develops an optimisation model to minimise investment and operation costs of a microbrewery meeting the dynamic energy demands, via an integrated photovoltaic (PV) system with energy storage and different boiler choices to lower carbon emissions. A microbrewery in UK has been used for the case study to demonstrate the approach on real data, with challenges in implementation and real-world constraints and considerations discussed. A set of rigorous multi-objective optimisation and sensitivity analyses are performed to analyse the resulting system. For the particular brewery, a modern electric boiler combined with photovoltaic system is an economic and sustainable choice, due to the cooling and other electric requirements in the brewery, leading to a 33 percent reduction in operational costs with a payback time of 2.6 years.

Keywords: microbrewery operation, optimisation model, PV system, sustainable.

1. Introduction

Beer is a popular alcoholic drink throughout the world. Global beer consumption reached approximately 177.50 million kiloliters in 2020, equivalent to 537.9 billion 33 cl bottles (Kirin Holdings, 2022). The UK is ranked 8th with 60.2 liters beer consumption per-capita. In the past two decades, environmental concerns and increasing energy cost have impacted companies and beer consumers. The new challenges for the brewing industry related to energy efficiency, water consumption, emission management and waste generation, have attracted great attention from breweries. Beer production is an energy intensive process, and the energy demand is approximately 262 MJ/hL (59% thermal, 41% electrical) (Fadare et al., 2010). Compared with large breweries, microbreweries consume up to 1.5-2.0 times more for thermal and electric energy (Kubule et al., 2016). In 2020, the number of active microbreweries in the UK was 1,852, with a steady increasing trend (Statista, 2022). Approximately 8% of the production cost is used in thermal processes like boiling, cooling, and fermentation (Kubule et al., 2016). Local microbreweries are trying to find a way to reduce energy consumption. Breweries have considered sustainability and lowering carbon footprints in their processes, seeking the strategies to minimise emissions and impact on the environment. Shifting load to a lower tariffs period is potential saving in other industries, but not suitable for microbrewery because the load demand is non-deferrable. Renewable energy sources and energy storage system can offer a solution for reducing the operation costs. Muster-Slawitsch et al.(2011) have analysed the use of different renewable sources, and identified solar energy as the best option to support the brewery industry.

The studies of solar sources for the brewery are focused on large size of breweries (Eiholzer et al., 2017). Pino et al.(2019) used two case studies in Spain to model the integration of a hybrid PV system into thermal and electricity generation system. The simulation results showed that solar sources have potential benefits in microbreweries with payback period of 10.7 years. Another energy-related case study is based on a microbrewery in South Africa(Kusakana, 2020). A grid-connected tracking PV system and a battery were integrated into the microbrewery, and the payback period is 13.8 years based on simulation results. Very few published studies focused on the onsite small-scale PV panels electricity generation, and they have not looked at the application in the UK.
Given the popularity of microbreweries and needs to reduce energy costs, this paper develops an optimisation model to minimise investment and operation costs of a microbrewery meeting the dynamic energy demands, via an integrated PV system with energy storage and different boiler choices to lower carbon emissions. A microbrewery in UK has been used for the case study to demonstrate the approach on real data, with challenges in implementation and real-world constraints and considerations discussed. The thermal and electricity load profile of the brewing process has been collected in summertime and, based on the collected data, the dynamic energy demand profiles for brewing days has been characterised. After an initial analysis, the number of PV panels, installation and type of the battery, and type of boiler are used as decision variables in model. A set of rigorous multi-objective optimisation and sensitivity analyses are performed to analyse the resulting system. The proposed optimisation model and hybrid system can be used in industrial, residential, or other demand sectors to reduce their operational energy costs through a detailed multi-objective analysis and sensitivity study.

2. Case study description

The case study for the model is a microbrewery called Langham Brewery located at 50°59'N,0°41'W, which regularly produces six types of craft beers. Langham is a typical, small-scale British microbrewery. Electricity is provided from Scottish Power for the cooling process, and thermal processes use a steam boiler fueled by Furnace Flame. The brewery has a capacity of producing one batch of 1,650 L of beer four times per week. Therefore, the yearly demand is assumed at an average of 208 brewing days. Additionally, the cold storage and the fermentation chamber run as the baseload throughout the year.
The load demand has been recorded in a summer day in July 2022. The first brewing day is the most energy intensive one, with the different processes shown in Figure 1 where the brewery's load demand is the highest. During the night, from 18h30 to 6h30 next day, the baseload is assumed at 4.4 kW including boiler (turned on, but not producing steam and on standby), cold storage and fermentation chamber. Starting at 6h30 to 7h15, the 2.9kW boiler and 0.65 kW pump begin to work and prepare for the mashing process. The temperature of hot water in the hot liquor tank (HLT) is 77°C and 700 L of hot water is discharged into an insulated mash tank (mixer). Malted barley and specialty grains are conveyed into the mash tank. The mashing stage starts from 7h15 to 8h30, boiler and pump are kept on during this stage reheating water in HLT for the next process. After mashing, the temperature of mixture is 66.3°C.
At 8h30, the sparging process begins. In this process, 1,950 L hot water from the HLT is added to the mash tank over a period of about 100 min. Simultaneously, using a 0.65 kW pump, the dissolved sugars in the water (wort) extracted from the grain husks is slowly drained from the bottom and pumped into a boiler where the hops are added to the mixture (wort). The mixture is heated for an hour. At the end of the sparging process, the temperature of the mixture is 90°C. From 11h00 to 12h00, the mixture is heated from

90°C to 100°C, preparing for boiling. The boiling process starts from 12h00 to 13h00. After 15 min rest, the mixture is cooled by cold water (5°C) via heat exchanger to reach 20°C, desirable for the required fermentation. The whole preparation process ends at 18h30 after pumping the mixture into fermentation tanks and finishing the cleaning process. The beer wort is moved to the fermentation which constitutes the 4.4 kW base load together with the cold storage.

Based on our data collection, the daily operation cost including electricity and fuel costs on a brewing day is approx. £158. This value excludes maintenance costs. The daily CO_2 emission is 0.35 tonnes per brewing day(ScottishPower, 2022).

3. Methodology

After analysing Figure 1, it can be noticed that most of the peak power demand is occurring during the day, matching the solar irradiance profile. The implementation of a PV system can be of great advantage in maximising the electrical energy that can be generated from the panels while minimising the cost and amount of energy acquired from the grid. A battery storage system can be added to the supply system to store some energy during excess generation times and can also be used in the event the PV system cannot supply the demand. This will also increase the availability of energy supply as well as provide an opportunity of applying demand response strategies through optimal power dispatch. This section will describe the proposed system's operation modeling; the selected size and the basis of the economic analysis to be performed.

3.1. Optimal energy management model of the grid-connected PV system

The overall objective of this model is to minimize the total cost in the whole lifetime:

$$f = min\,(C_{Operation} + C_{Investment}) = (\textstyle\sum_{S=1,2,3,4} C_S^{grid} - I_S) * Lifetime + \sum_{Unit} AC_{Unit} \qquad (1)$$

where AC_{Unit} is the capital cost of the equipment including PV panels, battery, and boilers; C_S^{grid} is the seasonal cost of buying electricity from the grid; I_S is the seasonal income from selling electricity to the grid. There are two choices of the boiler, one is a gas boiler, and the other is an electric boiler. The satisfaction of electricity demand $P_{t,c}^{load}$ is described as below:

$$\sum_c P_{t,c}^{load} = P_t^{grid} + P_t^{PV} + P_t^{BAT} \qquad (2)$$

where P_t^{grid} is the power got from the grid to satisfy the loads. If it is positive, electricity is bought from grid to satisfy the load; if it is negative, electricity is sold back to main distribution. P_t^{PV} is the amount of power generated by the PV that is used to satisfy the demand, and P_t^{BAT} is the amount of power used by the batteries. If it is positive, the battery is discharging; if it is negative, the battery charges. Note that indices $t \in T, c \in C$ refer to the timestep and boiler type. Similarly, the heat demand is satisfied by the heat generated (note that heat storage or heating networks are not included in the model).

The total power generated by the PVs, is described below:

$$P_t^{PV} \leq N^{Panel} * P_{PV} \qquad (3)$$

where N^{Panel} is the number of panels installed (an integer variable) and needs to be decided, P_{PV} is the power generated each panel.

The model also ensures that electricity is not bought from or sold to the grid at the same time by the addition of the big-M constraints, which are similar to the constraints on the battery's charging or discharging status. The type of the boiler installed is limited by a binary variable K_c, constrained as follows:

$$\sum_{c} K_c = 1 \tag{4}$$

For proper operation, each power source must not be operated above its rated or maximum limit according to the manufacturer's specifications. The state variable also has boundaries where the minimum depends on the type of battery used.

4. Optimisation results

All the constraints in the optimisation model are linear with integer variables. The optimisation problem are solved using cbc (Forrest & Lougee-Heimer, 2005) solver from Julia JuMP (Dunning et al., 2017) optimisation toolbox. Simulations are performed to assess the effectiveness of the developed grid-connected system optimal operation model to minimise the daily energy cost of power purchased from the grid. The simulations reported in the sections below explain the system's behavior for a brewing day in four seasons with different boundaries on investment, which means disposable investment is gradually decreasing. The maximum number of PV panels is 1000. The upper boundary of investment is £500,000.

4.1. Different payback periods

We did not consider the maintenance costs for equipment, and there is no limit on the investment cost. In Figure 2, the relationships among daily operation cost, capacity of PV panels and payback period is shown. Under sufficient investment, the model chooses to install an electric boiler under all the payback periods. If the payback period is lower than 5 years, the brewery should only install an electric boiler, with no battery or PV panels. When the payback period is more than 5 years, with the installation of PV panels, operation costs decrease. To invest in an electric boiler, the operation cost will drop from approximately £158 per day to £102.6 per day, with an investment assumed to be £27,255, yielding a payback period of 2.6 years. This will also yield a CO_2 emissions reduction from the boiler system of 62 % and an overall business emissions reduction of 21 %.

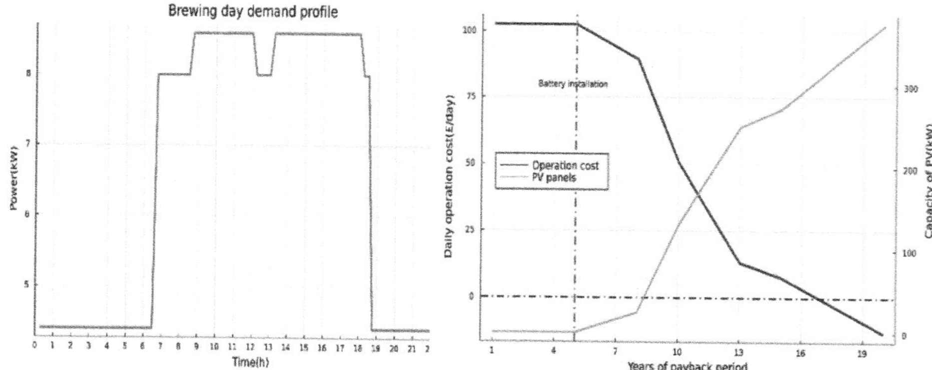

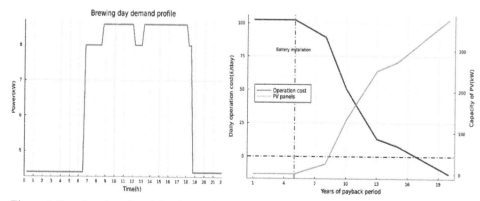

Figure 1 Brewing day electricity demand profile.

Figure 2 The relationships among daily operation cost, payback period, and capacity of PV panels.

The installation of PV panels has a great Impact on the daily operation cost of brewing days, especially in spring and summer, because the solar irradiance value is relatively high in these two seasons. The daily costs will decrease with longer payback periods especially 8-13 years if we have enough investment budget to install more PV panels. When the payback period is higher than 17 years, the operation costs are negative.

4.2. Different gas price and investment cost

To analyse whether gas boilers are ever considered, the payback period was set to 10 years gas prices from 4 to 40 p/kWh were simulated. When the price of natural gas is lower than 18 p/kWh, the model chooses to install a gas boiler not an electric one. Compared with the gas boiler, the CO_2 emissions can be reduced more than 50% through electrification. If the Smart Export Guarantee can increase the export price 20% (from 5p to 6p), carbon emissions can be reduced by 28%, though the overall cost is only reduced by 2%.

We also analyse the simulation results under different investment cost constraints, and the results are shown in Figure 3. Even under different investment conditions, the model always chooses an electric boiler due to high gas prices. Daily operation cost drops as number of solar panels increases, and a significant positive linear correlation between operation cost and installation of PV panels are shown in Figure 3 When the value of investment is lower than £42,000, the optimisation model chooses not to install a battery, the turning point is presented by a vertical dotted line with 18 kW PV capacity. On the right side of the line, the model chooses to install batteries. As the investment cost decreases, the operation cost gradually increases, with the increase in CO_2 emissions displayed by the orange line.

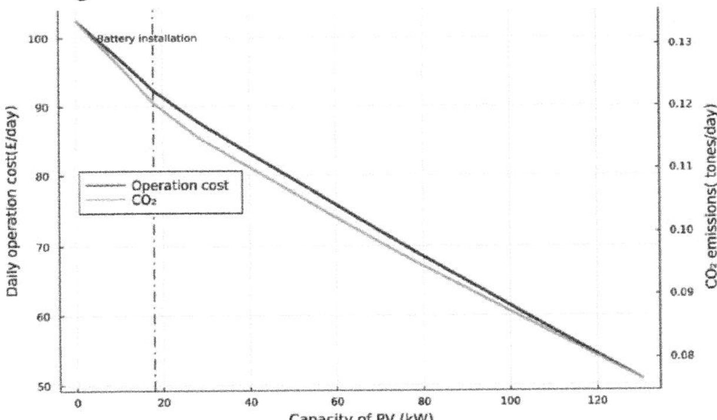

Figure 3 The relationships among daily operation cost, CO_2 emissions, and capacity of PV panels.

5. Conclusion

To be more competitive and sustainable, microbreweries need to reduce their energy cost without concession on the quality of final product. The aim of this work is to analyse the techno-economic benefits of using renewable sources by maximising profits and decreasing energy consumption through retrofitting their operations for a microbrewery. This paper develops an optimisation model to meet the dynamic energy demand based on data collection from a real microbrewery in UK, by minimising investment and operating costs and integrating PV systems (with energy storage and different boiler options to reduce carbon emissions). Along with the energy demand profiles, the solar resources, system components, and energy costs have been used as inputs to develop the model. Under the operational conditions and constraints, the model analyses the economic, technical, and sustainable performance of the system, together with company- and region-specific constraints related to budgets and incentives. The simulation results showed that the model can effectively optimise the capacity of PV system, installation of the battery, and type of boiler.

Compared to the scenario in which the grid is the only option for energy supply, the proposed system can significantly reduce energy costs. A potential energy cost saving of 33.3%, is possible when the optimal energy management model is applied to the proposed system, under the applicable operating condition. The length of the equipment life cycle has a great impact on the optimal investment strategy of the brewery, with a 7-year payback period important should the microbrewery wish to move towards electricification. Due to the installation of PV panels, there is a positive correlation between daily operation costs and carbon emissions. The carbon emissions of electric boilers are significantly lower than those of gas boilers, and the brewery prefers to install an electric boiler, unless the price of natural gas is lower than 18p/kWh. More PV panels installed brings lower daily operation cost and lower carbon emissions, but higher investment costs, and therefore the brewery needs to find a balance according to their own needs.

The satisfactory results obtained in this study demonstrate the potential techno-economic benefits of solar PV integration in the small-scale microbreweries. The proposed optimisation model and hybrid system can be used in industrial, residential, or other demand sectors to reduce their operational energy costs through a detailed multi-objective analysis and sensitivity study.

References

Dunning, I., Huchette, J., & Lubin, M. (2017). JuMP: A Modeling Language for Mathematical Optimization. *SIAM Review*, *59*(2), 295–320. https://doi.org/10.1137/15M1020575

Eiholzer, T., Olsen, D., Hoffmann, S., Sturm, B., & Wellig, B. (2017). Integration of a solar thermal system in a medium-sized brewery using pinch analysis: Methodology and case study. *Applied Thermal Engineering*, *113*, 1558–1568. https://doi.org/10.1016/j.applthermaleng.2016.09.124

Fadare, D., Nkpubre, D., Oni, A., Falana, A., Waheed, M., & Bamiro, O. (2010). Energy and exergy analyses of malt drink production in Nigeria. *Energy*, *35*(12), 5336–5346.

Forrest, J., & Lougee-Heimer, R. (2005). CBC User Guide. In *Emerging Theory, Methods, and Applications* (pp. 257–277). INFORMS. https://doi.org/10.1287/educ.1053.0020

Kirin Holdings. (2022). *Global Beer Consumption by Country in 2020 | 2022 | Kirin Holdings*. Kirin Holdings Company, Limited. https://www.kirinholdings.com/en/newsroom/release/2022/0127_04.html

Kubule, A., Zogla, L., Ikaunieks, J., & Rosa, M. (2016). Highlights on energy efficiency improvements: A case of a small brewery. *Journal of Cleaner Production*, *138*, 275–286. https://doi.org/10.1016/j.jclepro.2016.02.131

Kusakana, K. (2020). Optimal energy management of a grid-connected dual-tracking photovoltaic system with battery storage: Case of a microbrewery under demand response. *Energy*, *212*, 118782. https://doi.org/10.1016/j.energy.2020.118782

Muster-Slawitsch, B., Weiss, W., Schnitzer, H., & Brunner, C. (2011). The green brewery concept – Energy efficiency and the use of renewable energy sources in breweries. *Applied Thermal Engineering*, *31*(13), 2123–2134. https://doi.org/10.1016/j.applthermaleng.2011.03.033

Pino, A., Lucena, F. J. P., & Macho, J. G. (2019). Economic analysis for solar energy integration in a microbrewery. *2019 International Conference on Smart Energy Systems and Technologies (SEST)*, 1–6.

ScottishPower. (2022). *Where does our energy come from? | Gas and Electricity Company | ScottishPower*. https://www.scottishpower.co.uk/about-us/performance/fuel-mix

Statista. (2022). *Topic: Craft beer in the UK*. Statista. https://www.statista.com/topics/6456/craft-beer-in-the-uk/

Antonis Kokossis, Michael C. Georgiadis, Efstratios N. Pistikopoulos (Eds.)
PROCEEDINGS OF THE 33rd European Symposium on Computer Aided Process Engineering (ESCAPE33), June 18-21, 2023, Athens, Greece
 http://dx.doi.org/10.1016/B978-0-443-15274-0.50033-0

Thermodynamic equilibrium modelling for the optimal performance of a wood biomass downdraft gasifier for hydrogen production

Anamika Kushwah,[a] Tomas Ramirez Reina, [a,b] Michael Short,[a*]

[a]*University of Surrey, Guildford, GU27XH, UK*
[b]*Institute of Materials Science of Seville, Seville, 41092, Spain*

Abstract

The flexibility of gasification, which incorporates both characteristics of the kind of biomass and options for producing heat or fuel, is what stimulates interest in biomass gasification research and application possibilities. One of the challenges with gasification is the accurate prediction of gasification products from the wide array of feedstocks and operational conditions, making the design and optimisation challenging. There are several studies on equilibrium models that have been published, but most of the models examined the impact of moisture and heating value. Examining the other parameter effects within the same model can enhance the accuracy of a gasification model. This work aims to create stoichiometric equilibrium models that enable the analysis of parameter effects on a specific feedstock's gasification products for use in optimisation and system modelling. A mathematical model for fixed bed gasifiers that use downdraft gasification for wood as a feedstock is presented. When the algebraic model equations are solved, the conversion behaviour of wood biomass throughout a gasification process is predicted in terms of the composition of the end products. The model has been verified with published experimental data from the literature demonstrating good agreement. The created model may be regarded as a helpful tool to simulate the impact of numerous different biomass feedstocks and operating conditions on gas characteristics and the simple algebraic formulation allows for the application of the model for optimisation purposes.

Keywords: Biomass gasification, modelling, equilibrium models

1. Introduction

Global warming is regarded as one of the top global challenges for contemporary civilisation, notwithstanding COVID-19 crisis global challenges. More waste is produced because of growing urbanisation and economic growth. The primary problem is reducing environmental pollution and worldwide warming brought on by the extensive use of fossil fuels including coal, petroleum, and natural gas. These fossil fuels serve as the main source of energy generation in many nations, which results in continuous greenhouse gas emissions. Because of its widespread availability and practical processing, biomass is predicted to grow significantly among the many renewable energy sources in the near future (Yamany et al. 2016). There is ongoing research on methods for enhancing the production of energy from biomass, and several organisations are working on its development (Janajreh et al., 2021). One of the most adaptable industrial processes is biomass gasification, which converts biomass fuels into consumable gaseous fuels that may be used for heating, power production, and hydrogen fuel cells in a range of residential and commercial applications (Sidek et al., 2020). Due to its versatility, gasification - a thermochemical reaction that transforms solid organic resources into gaseous fuels at temperatures ranging from 700 to 900 °C is highly popular (Okolie et al.,

2021). It yields valuable products like CO, trace amounts of CH_4, H_2, and undesirable gases like N_2, CO_2, and other hydrocarbons. Because experimental conditions can affect the behaviour of reactions like water gas, steam reforming reactions, Boudouard and water gas shift reactions, and these all have an effect on syngas formation, stability, and thermal efficiency, restricting its future uses, it is crucial to use caution when simulating experimental conditions (Baruah & Baruah, 2014b).

Multiple studies have revealed that modelling and theoretical predictions based on equilibrium modelling have a great deal of similarities (Kushwah et al, 2022; Chaurasia, 2018). For instance, Pradhan et al. (2019) used varied equilibrium reaction rates to test five different biomass gasification models. When compared to other models, they discovered that the model that contained basic methanation and the water gas shift reaction generated syngas with the most accurate composition. The correlation output predictions resembled those of the equilibrium model, with the exception of CH_4. On the other hand, reliable correlations will require the integration of more empirical observations and the effect of other parameters.

Experiments generally consume significantly more time, money, and effort than modeling. Modeling has been shown to be a relatively quick and cost-effective option in situations when doing so is not always practicable or economically sensible (Galvanin et al., 2009). Thermodynamic equilibrium models forecast the highest possible yield of desired products without requiring any understanding of the conversion processes. Like this, downdraft gasifiers are widely employed in various energy platforms due to their straightforward construction, inexpensive maintenance requirements, and ability to produce gas with a low tar content for use in small-scale electrical generation. This paper proposes a thermodynamic stochiometric equilibrium model for a downdraft gasifier is used for the gasification of wood at 800 °C with moisture content from 0-40 % in order to predict the composition of the fuel gas produced and its associated heating values. The gasifier system is modelled using an optimisation modelling package which includes mass conservation equations, energy conservation equations and the equilibrium equations. The developed model, based on thermodynamic calculations, has been validated with experimental published data of other authors and provides opportunity to evaluate downdraft gasification processes in a fixed bed reactor, as well as effects of variations in biomass properties and operating conditions.

2. Methodology

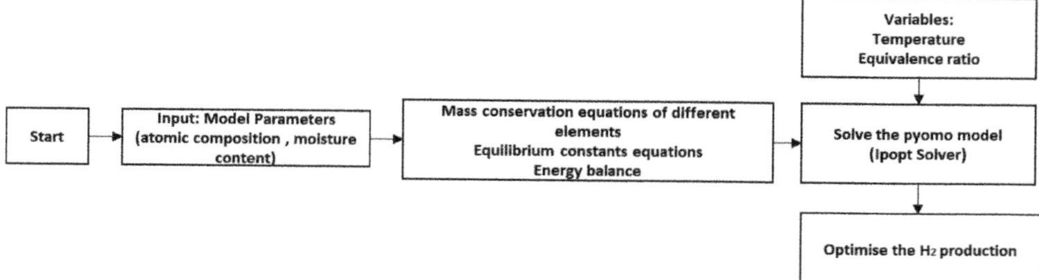

Figure 2.1: Structure of the simulation equilibrium model

In figure 2.1, the structure of the equilibrium model has been presented. In this, a mathematical model based on a single feedstock is created and developed in the algebraic modelling language, Pyomo (Hart et al. 2017) in order to determine the molar values of products. It includes the mass balance of carbon, hydrogen, oxygen, nitrogen in the woody biomass, relative mass of the moisture and the molar water content. In next step, the composition of the producer gas is estimated, minimising the Gibbs free energy for calculation of the equilibrium constants. A Pyomo model has been implemented that relates equations between inlet biomass and equilibrium constants with temperature to solve the model. The equivalency ratio and temperature are used as optimization variables when the model is optimised using the IPOPT nonlinear programming solver to get the maximum hydrogen production rate.

2.1 Biomass composition

To develop the model, the chemical formula of feedstock is defined as $CH_xO_yN_z$. The following is an equation for the global gasification reaction:

$$CH_xO_y + wH_2O + m\,(O_2 + 3.76N_2) \Rightarrow n_1\,H_2 + n_2\,CO + n_3\,CO_2 + n_4\,H_2O + n_5\,CH_4 + 3.76\,N_2 \quad (1)$$

Where x and y are the numbers of atoms of hydrogen and oxygen per atom of carbon in the feedstock, w is the amount of moisture per kmol of feedstock and m is the amount of oxygen per kmol of feedstock. On the right-hand side n_i are the numbers of moles of the species i that are unknown.

2.2 Chemical Equilibrium

Thermodynamic equilibrium is explained by minimisation of Gibbs free energy or by using an equilibrium constant. The two main equations (methane formation and water gas shift reaction) are considered in this case shown below:

Boudouard reaction: $C + CO_2 \rightarrow 2CO$ (2)

Water- gas reaction: $C + H_2O \rightarrow H_2 + CO$ (3)

Methane Formation: $C + 2H_2 \rightarrow CH_4$ (4)

Water gas shift reaction: $CO + H_2O \rightarrow CO_2 + H_2$ (5)

Equation (5) is obtained by subtracting equation (2) from (3). In this model, thermodynamic equilibrium was assumed for all chemical reactions takes place in gasification zone. All the gases were assumed to be ideal and all reaction form at 1 atm. Therefore, the equilibrium constants which are functions of temperature for the water-gas shift reaction and methane formation reactions are:

$$K1 = \frac{(nCH_4)}{(nH_2^2)} \text{ (Equilibrium constant for methane reaction)} \quad (6)$$

$$K2 = \frac{(nCO_2)(nH_2)}{(nCO)(nH_2O)} \text{ (equilibrium constant for water gas shift reaction)} \quad (7)$$

The value of K1 and K2 are calculated from Gibbs free energy

$$K = \exp(-\Delta G^{\circ}_T/RT) \quad (8)$$

Where R is the universal constant, 8.314 KJ/ (kmol. K)

$-\Delta G^{\circ}_T$ represents the Gibbs function of formation at given temperature which can be expressed by empirical equation (Green, 1984):

$$\frac{d(\Delta G^{\circ}/RT)}{dT} = -\frac{\Delta H}{RT^2} \quad (9)$$

Where ΔH is the heat of formation of biomass in KJ/kmol

Therefore: $$\frac{d\ln K}{dT} = -\frac{\Delta H}{RT^2} \quad (10)$$

Where, equation 9 shows the effect of temperature on the equilibrium constant; ΔH represents the enthalpy difference between any given state and at reference state. It can be calculated by

$$\Delta H = \int_{298}^{T} Cp(T)dt \quad (11)$$

Where $Cp(T)$ is specific heat at constant pressure in kJ/kmol K and is a function of temperature.

$$Cp(T) = a + bT + cT^2 + dT^3 \text{ where are a, b, c, d are specific gas coefficients} \quad (12)$$

Equations (8)-(12) is used to find the equilibrium constant for any reaction temperature T. Zainal et al. (2001), suggested the relationship for finding the equilibrium constant values for reaction temperature and calculating the lower heating value (LHV) of product gas that are:

$$K1 = \frac{7082.848}{T} - 6.567 * \ln(T) + 3.733 * 10^{-3} - 0.36066 * 10^{-6} + 0.3505 * \frac{10^{-5}}{T^2} + 32.54 \quad (13)$$

$$K2 = \frac{5870.53}{T} + 1.86 * \ln(T) + 2.7 * 10^{-4} - \frac{58200}{T^2} + 18.007 \quad (14)$$

$$LHV = n_1\, LHV(H_2) + n_2\, LHV(CO) + n_5 LHV(CH_4) \text{ MJ/Nm}^3 \quad (15)$$

However, the reference model is limited to moisture content and heating value but using an optimisation-based algebraic formulation can be used to study other parameter effects like equivalence ratio and gasification agent against temperature. For the desired product, this can also find the optimum parameters for gasification reaction. For different ranges of temperature, the values of equilibrium constant, numbers of moles and heating value have been calculated in a model, which is discussed in the next section.

3. Results and Discussion

The model developed in this study was used to optimise the production of hydrogen by varying temperature and equivalence ratio at 20% moisture content. Also, it was tested by comparing the calculation results of model with data from another researcher. To study the effect of moisture content of the wood biowaste, the amount of air was fixed at 0.3 of the stoichiometric requirement. Figure 3.1 shows the effect of moisture content in woody biomass on the product gas when the gasifier is operating at 800°C. It can be observed from the figure that the composition of methane produced is almost constant at a low percentage. The percentage of H_2 in the produced gas increases continuously with the moisture content from 0% to 40 %. A similar trend is also observed for the CO_2 however, the influence of the addition of more moisture will result in decrease in its concentration that might occur due to char, tar and soot formation, and water-gas shift reaction. Also, the concentration of CO reduces as moisture content increases. The results obtained from the model are in close agreement with experimental data from Zainal et al. (2001) which performed for wood biomass under similar operating conditions. Figure 3.2 shows the calorific value for wood biomass against moisture content. It is seen that the calorific values decrease with increases in moisture content. This is due to the greater reduction in carbon monoxide content compared to the increase in hydrogen content as moisture content increases.

Figure 3.3 explains the variation of calorific value against gasification temperature for the raw material wood with an initial moisture content of 20%. It can be observed that the predicted calorific values reduce with increase in temperature. Figure 3.4 shows the comparison between model and experimental results for wood at 18.5 % moisture content and gasification temperature of 800°C. It can be observed from graph that the predicted value percentage of hydrogen and carbon monoxide content is predicted as 16.87% & 22.92% against an experimental value of 12.23% & 21.00%. The hydrogen percentage predicted is more than the experimental value, while the carbon monoxide percentage predicted is lower than the experimental result. Also, higher equivalence ratio (ER) result in higher operating temperatures, which in turn increases the degree of oxidation of available carbon.

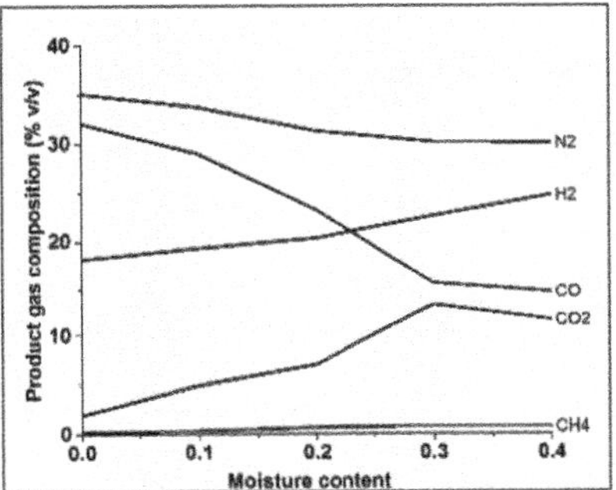

Figure 3.1: Effect of moisture content in wood biomass on product gas at 800°C

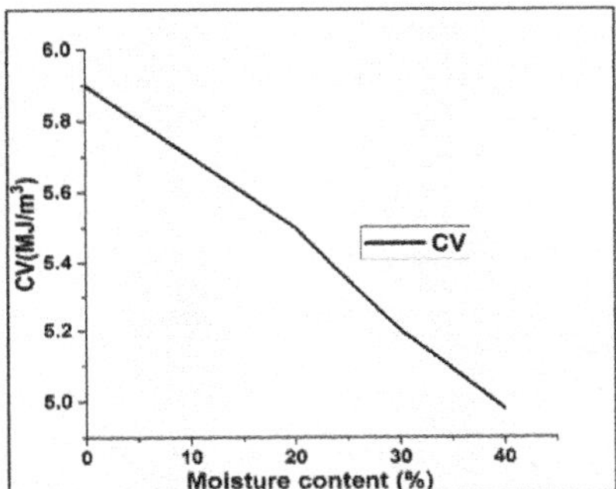

Figure 3.2: Effect of calorific value against moisture content at 800°C

The model is optimised using nonlinear programming solver IPOPT to find the maximum hydrogen production rate with the equivalence ratio and temperature as optimisation variables. Higher ER results in more CO_2 production because of oxidation of carbon; hence, the optimum ER, calculated by model is 0.1957 at 853.23°C and 20 % moisture, which yields 21.214% H_2. So, by combining the two above-mentioned gasification variables, temperature, and equivalence ratio, it is possible to optimise the operating conditions to magnify the H_2 production.

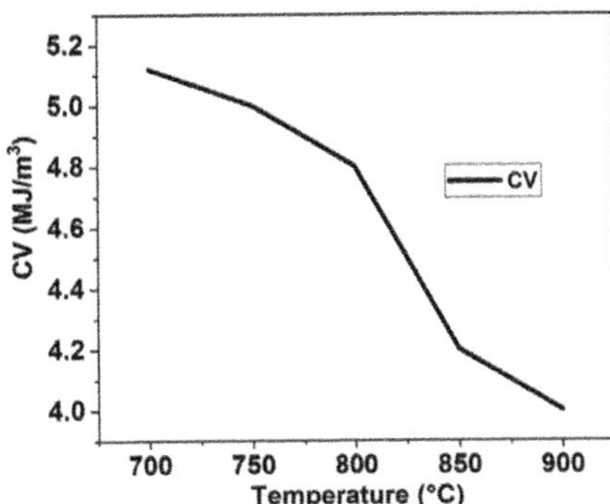

Figure 3.3: Calorific value against

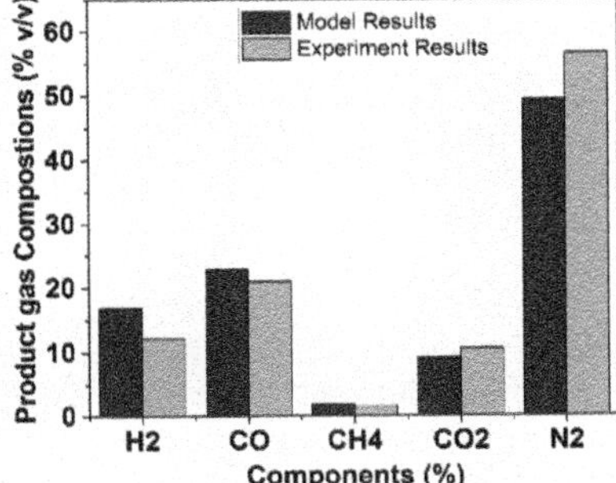

Figure 3.4: Comparison between model and experimental results at 800°C

4. Conclusion:

Stoichiometric thermodynamic equilibrium models for the biomass gasification process were developed in this work, with the entire conversion of carbon considered to estimate syngas composition.

For a wide range of temperatures, the equilibrium constants of the water gas shift process and the methanation reaction have been computed. The models are formulated in an algebraic modelling language for optimisation of the gasification process. This model can predict the composition of syngas from biomass sources and can also predict biomass gasification performance under a variety of operating parameters, such as temperature, moisture content and equivalence ratio. The optimum ER, calculated by model is 0.1957 at 853.23°C and 20 % moisture, which yields 21.214% H_2. The study leads to the conclusions that the amount of hydrogen in the product gas nearly rises linearly with the amount of moisture for biomass. However, carbon monoxide content in the product gas decreases with moisture content almost in a linear variation. The proportion of methane in the product gas is relatively low and increases linearly with moisture content. The calorific value of the producer gas decreases as the moisture content increase. In short, with a known ultimate analysis of any biomass and its calorific value, the model can predict the composition and the calorific value of the producer gas under different operating conditions.

References

Baruah, D., & Baruah, D. C. (2014). Modeling of biomass gasification: A review. Renewable and Sustainable Energy Reviews, 39, 806-815.

Chaurasia, A. (2018). Modeling of downdraft gasification process: Studies on particle geometries in thermally thick regime. Energy, 142, 991–1009.

Green, D. W. (1984). Perry's Chemical Engineers' Handbook, Eighth Edition by Don W. Green, McGraw-Hill Education. McGraw-Hill Education.

Galvanin, F., Barolo, M., & Bezzo, F. (2009). Online Model-Based Redesign of Experiments for Parameter Estimation in Dynamic Systems. Industrial &Amp; Engineering Chemistry Research, 48(9), 4415–4427.

Hart, W.E., Laird, C.D., Watson, J.P., Woodruff, D.L., Hackebeil, G.A., Nicholson, B.L. and Siirola, J.D., 2017. Pyomo-optimization modeling in python, 67, 277.

Janajreh, I., Adeyemi, I., Raza, S. S., & Ghenai, C. (2021). A review of recent developments and future prospects in gasification systems and their modeling. Renewable and Sustainable Energy Reviews, 138, 110505.

Kushwah, A., Reina, T. R., & Short, M. (2022). Modeling approaches for biomass gasifiers: A comprehensive overview. Science of The Total Environment, 155243.

Dogru, M. (2000) Fixed Bed Gasification of Biomass, Ph.D. thesis, University of Newcastle, UK.

Pradhan, P., Arora, A., & Mahajani, S. M. (2019), A semi-empirical approach towards predicting producer gas composition in biomass gasification. Bioresource Technology, 272, 535-544.

Sidek, F. N., Abdul Samad, N. A. F., & Saleh, S. (2020), Review on effects of gasifying agents, temperature, and equivalence ratio in biomass gasification process. IOP Conference Series: Materials Science and Engineering, 863(1).

Yamany, W., Fawzy, M., Tharwat, A., & Hassanien, A. E. (2015, December). Moth-flame optimization for training multi-layer perceptrons. In 2015 11th International computer engineering Conference (ICENCO), 267-272.

Zainal, Z. A., Rifau, A., Quadir, G. A., & Seetharamu, K. N. (2002). Experimental investigation of a downdraft biomass gasifier. Biomass and Bioenergy, 23(4), 283–289

Antonis Kokossis, Michael C. Georgiadis, Efstratios N. Pistikopoulos (Eds.)
PROCEEDINGS OF THE 33rd European Symposium on Computer Aided Process Engineering (ESCAPE33), June 18-21, 2023, Athens, Greece
 http://dx.doi.org/10.1016/B978-0-443-15274-0.50034-2

A model predictive control approach for recirculating aquaculture systems integrated with sustainable hybrid energy systems

Ruosi Zhang [a,b], Tao Chen[a], Michael Short[a*]

[a]*University of Surrey, Stag Hill, Guildford GU2 7XH, UK*
[b]*National Innovation Center for Digital Fishery, Beijing 100083, China*
m.short@surrey.ac.uk

Abstract

Aquaculture is a growing industry that provides high-quality protein, however, the growth of aquaculture production has a range of environmental concerns. Recirculating aquaculture systems (RAS) are a popular solution to accomplish intensifying fish production, and integrating renewable technologies is an attractive option that may provide low-carbon energy and promote sustainability. This paper presents an optimisation model for the control and operation of a grid-connected distributed energy system (DES) integrated with an RAS. A linear optimal power flow (OPF) combined with model predictive control (MPC) strategy is developed which simulates the thermal and electricity balances present throughout the renewable energy integrated electricity network and RAS, to make real-time optimal operation scheduling plans on different time horizons. The optimal results show that the MPC strategy improves the control performance of system operation and improves the process economics. With renewable energy sources (RES), operational costs can be reduced by up to 27%. Compared with a conventional scheduling plan, the rolling horizon approach can provide 3% energy cost saving, while maintaining fish well-being and system safety.
Keywords: recirculating aquaculture system, distributed energy systems, renewable energy, optimal power flow, model predictive control.

1. Introduction

Aquaculture is a main source of protein and income in many countries globally (FAO, 2020). Sustainable aquaculture is directly related to targets of the Sustainable Development Goals. In this context, the implementation of RAS are gaining more attention globally (Chen et al., 2021). RAS has the potential to provide a sustainable fish-rearing system since it is easily expandable, potentially environmentally friendly, and improves production by utilizing a small space. RAS is competitive since it is able to recycle 90-99% water for reuse. However, RAS has received little attention in terms of understanding the synergies between energy consumption and fish production. The application of RAS is difficult to commercialise. The reasons are that the high initial investments are hard to recover (Schneider et al., 2006) and the high energy consumption for operation (Ayer & Tyedmers, 2009). RAS integrating renewable technologies is an attractive option providing low-carbon energy and promoting sustainability.

The use of solar energy in RAS decreases the energy cost on fossil fuel, especially for fish farms with higher water temperatures. The PV system combined with a wind turbine (WT) can enhance the reliability of the system and smooth the power generation curve caused by the intermediate characteristics of RESs (Ruggles & Caldeira, 2022). Jamroen

(2022) developed design models to determine the appropriate size of PV and battery storage for fish farms. These models have common disadvantages. The first is that uncertainty of renewable energy is not considered, and another is the timescale considered. In real-time, the scheduling of the power system could be in minutes, and in these studies the scale of days is used, wherein it is possible that the system may not respond to changes quickly. Silvente et al. (2015) presented a linear operational model considering the prediction uncertainties from RESs, which is solved by an MPC rolling horizon strategy to correct the deviations from the initial conditions. MPC rolling horizon strategy is suitable for dealing with system control problems that are easy to model and have high stability. MPC includes four parts, a prediction model, rolling optimisation, feedback mechanism and reference trajectory. The feedback mechanism corrects the prediction model, and the error is compared with the reference trajectory. The final control decision applied to the system is calculated by the rolling optimisation.

This paper proposes a linear MPC approach for sustainable hybrid energy systems for RAS. The model simulates the thermal and electricity balances present throughout the renewable energy integrated electricity network and RAS, to make real-time optimal operation scheduling plans on different time horizons. First, a multi-time scale optimal scheduling problem is divided based on three scales: day-ahead long-term scale scheduling in 1 h, intra-day predictive control in 15 min and real-time adjustment in 5 min. Then, we establish an optimisation model based on the best economic operation of the system and the minimum penalty cost of start-up and shutdown. The inputs include the relevant climate parameters for renewable energy generation, the energy storage schedules, operation of facilities including climate and oxygen control equipment, as well as the fish biomass production quantity.

2. Model formulation

The components of the system are divided into load demand of RAS and power resources including generator, WT, PV and national grid as backup power in on-grid operation mode. The main load demand of the system is the thermal process to keep water temperature as required in RAS, and other demands such as lights, water pumps, water treatment system and aeration are treated as constant load demand in this model. The overall objective of this model is to minimise the daily operation cost:

$$OPERATION = COST_{gen} + COST_{grid} \tag{1}$$

$$COST_{grid} = \sum_{t} (P_t^{grid,\, buy} * C^{grid} - P_t^{grid,sold} * C^{tariff}) * \Delta t \tag{2}$$

$$COST_{gen} = \sum_{t} ((P_t^{gen} * F_1 + Y_{gen} * F_0) * C^{fuel} + S_{gen} \cdot u^t) * \Delta t \tag{3}$$

The cost of buying electricity from the grid is $COST_{grid}$, with C^{grid} and C^{tariff} the tariffs of electricity grid. $P_t^{grid,\, buy}$ and $P_t^{grid,sold}$ are the power of buying or selling electricity to the grid respectively, Δt is the time interval. The cost of generator's operation is $COST_{gen}$, P_t^{gen} is the output power of generator, Y_{gen} is the capacity of the generator, F_1 is generator fuel curve slope, F_0 is generator fuel curve intercept coefficient, C^{fuel} is the price of generator fuel, S_{gen} is the cost of start-up generator, and $\mathrm{u^t}$ is the start-up variable of generator.

$$\sum_{t} P_t^{gen} + W_t + P_{ess,t} + P_t^{grid} + PV_t - P_t^{HP} - P_t^{eb} = P_t^{load} \tag{4}$$

W_t, $P_{ess,t}$, P_t^{grid}, PV_t, P_t^{HP}, P_t^{eb}, P_t^{load} are the output power of WT, battery, main distribution, PV, heat pump, electric boiler, and load demand of RAS respectively. The output power of wind turbine and PV should be lower than the predicted output value according to the weather. Unit commitment:

$$\begin{aligned} x^{t-1} - x^t + u^t \geq 0 \\ x^t - x^{t-1} + v^t \geq 0 \end{aligned} \tag{5}$$

where x, u, v are binary variables. x is the unit commitment of generators (1 if the generator is committed at time t, and 0 otherwise). u^t is the start-up variable of generator (1 if generator starts at time t, and 0 otherwise). v^t is the start-up variable of generator (1 if the generator starts at time t and 0 otherwise).

The model also ensures that electricity is not bought from or sold to the grid at the same time through big-M constraints, which are similar to the constraints on the battery's charging or discharging status. For proper operation, each power source must not be operated above its rated or maximum limit according to the manufacturer's specifications. The state variable also has boundaries where the minimum depends on the type of battery used.

The water heat balance is used to calculate the energy flows of the water volume each hour. In other words, the water heating or cooling demand to maintain its temperature. The formulated simplified heat balance is:

$$\dot{Q}_{conv}(t) + \dot{Q}_{exchange}(t) = \dot{Q}_{demand}(t) \tag{6}$$

The convective heat transfer, $\dot{Q}_{conv}$ represents the heat exchange on the surface of the tank between water and air, which is determined by (Holman, 2009):

$$\dot{Q}_{conv} = h_{conv,w} \cdot A_{surface} \cdot (T_{air}(t) - T_w(t-1)) \tag{7}$$

where $h_{conv,w} (kW/(m^2 °C)$ is the convective heat transfer coefficient decided by indoor air velocity, T_w is the water temperature, T_{air} is the air temperature, and A_{surface} (m^2) is the surface area of the tanks. The water exchange that occurs in the system implies a heat gain or loss of $\dot{Q}_{exchange}$. The rate of water exchange in the system is assumed to be constant, with the same amount of water entering and leaving the system. The heat gain through water exchange can be calculated by:

$$\dot{Q}_{exchange}(t) = \dot{m}_{exhange} \cdot c_{p,w} \cdot \left(T_{w,in} - T_w(t-1)\right) \tag{8}$$

where $\dot{m}_{exhange}$is the mass flow of the water exchange, $c_{p,w}$ is the specific heat of the water and $T_{w,in}$ is the temperature of water entering the system.

The heating system will be active until a certain water temperature has been reached. When the temperature set point has been reached the heating system will be turned off, the water temperature will then drop to a lower limit where the heating system is activated again.

The core idea of MPC is rolling optimization and feedback correction, which solve the optimization problems with uncertainties, and can also conveniently include a variety of constraints. The rolling optimization process of MPC is shown in Figure 1. If the system operates at time t, based on the prediction information of the system disturbance variables in a future period, obtain the optimal outputs in the prediction domain with the objective of minimizing the operating cost. However, the optimization result is used only in one control domain, and the optimization process is repeated at time $t + 1$. The proposed scheduling is divided into three timescales: day-ahead, intraday, and real-time. The main decisions of the day-ahead dispatching are the unit start-up and shutdown plan in the next 24 hours, the charging and discharging state of energy storage, and the power purchase and sale to the main distribution network. The time interval in this model is 1 h. The intraday scheduling mainly corrects the output of the controllable units, the charging and discharging power of the energy storage and the purchasing and selling power of the main distribution grid in the intraday dispatching according to the latest renewable energy power output and thermal load forecast data. The time interval of this model is 15min and the dispatching period is 2h. The time interval of real-time dispatching is 5min, and the dispatching period is 15min. It is mainly based on the real-time forecast data of wind power, PV output and thermal load to finetune the intraday scheduling plan. In the intraday and real-time dispatching time, the unit start-up and shutdown plan, energy storage charging and discharging status and power purchase and sale status of the main distribution grid in the day-ahead scheduling results are kept unchanged. The level-by-level coordination of the three-time scales can better deal with uncertainties and make scheduling results more consistent.

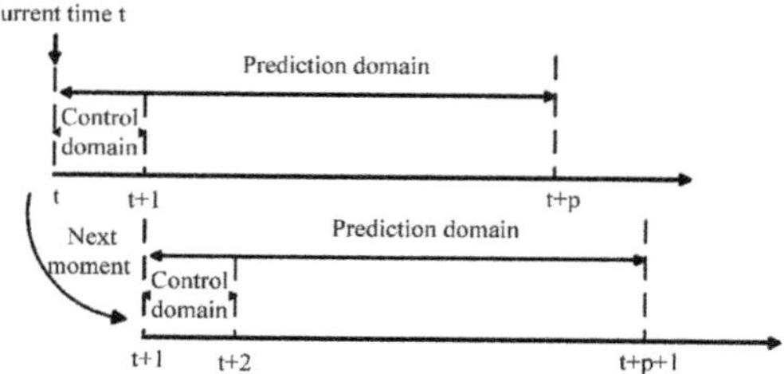

Figure 1 The rolling optimization process of MPC.

When optimising intraday and real-time scheduling, there is often a large error between the renewable energy output obtained by the short-term forecasting method and the daily forecast data, so the daily scheduling plan needs to be properly modified to ensure the effectiveness of the scheduling strategy. The objective function of intraday optimal scheduling model is:

$$minOperation_{intra} = min \sum_{t=1}^{N_T} \left[C_{G,t}^{intra} + C_{ESS,t}^{intra} + C_{M,t}^{intra} \right] \tag{9}$$

where N_T is intraday scheduling period, $N_T = 2h$; $C_{G,t}^{\text{intra}}$ is the generator operation cost at time t; $C_{ESS,t}^{\text{intra}}$ is the storage operation cost, $C_{\text{M},t}^{\text{intra}}$ is the operation cost of intraday purchasing or selling electricity from main distribution network.

During the intraday scheduling, the status of the generator is consistent with that of the day before, so the intraday scheduling cost of the controllable generator unit only includes the operation cost. The charging and discharging power of energy storage should follow the day-ahead scheduling results as much as possible. Therefore, intraday scheduling cost of energy storage is expressed as:

$$\begin{aligned} C_{ESS,t}^{\text{intra}} = \Bigg\{ K_{ESS,i} \left[\frac{1}{\eta_{E,\text{ dis}}} P_{E,t}^{\text{sum,dis}} + \eta_{E,\text{ ch}} P_{E,t}^{\text{sum,ch}} \right] \\ + \lambda_{ESS}^{\text{intra}} \left| SOC_t - SOC_t^{\text{ahead}} \right| \Bigg\} \Delta t \end{aligned} \tag{10}$$

where $\lambda_{ESS}^{\text{intra}}$ is the weight coefficient of energy storage adjustment during the intraday plan. Equation (10) has absolute value and auxiliary variables $S_{E,t}^{\text{au}}$ are introduced and linearized:

$$\begin{aligned} & C_{ESS,t}^{\text{intra}} = \left\{ K_{ESS,i} \left[\frac{1}{\eta_{E,\text{dis}}} P_{E,t}^{\text{sum,dis}} + \eta_{E,\text{ch}} P_{E,t}^{sum,ch} \right] + \lambda_{ESS}^{\text{intra}} S_{E,t}^{\text{au}} \right\} \Delta t \\ & \text{s.t.} \quad S_{E,t}^{\text{au}} \geq SOC_t - SOC_t^{\text{ahead}} \\ & S_{E,t}^{\text{au}} \geq -\left[SOC_t - SOC_t^{\text{ahead}} \right] \end{aligned} \tag{11}$$

Similarly, the cost of intraday power purchase and sale of the main distribution network is the same as that of energy storage and also follows the day-ahead scheduling results as much as possible. During intraday scheduling, to track the day-ahead charging and discharging power of storage energy, the following constraint should be satisfied:

$$SOC_{4t}^{Intra} = SOC_t^{ahead} \quad t = 1,2,\cdots,24 \tag{12}$$

where E_{4t}^{Intra} , E_t^{ahead} are the state of charge at the end of each hour for intraday and day-ahead scheduling respectively. Real-time optimal scheduling is similar to the intraday model, tracking the intraday dispatching plan as much as possible.
The process is summarized in Algorithm 1.

Algorithm 1 Model predictive control
1. Predict values for day ahead including renewable energy output, load demand. Input other parameters.
2. Solve the day-ahead model, with objective function (1), constraints are (2)-(8). Decide the binary variables of generator and battery.
3. Predict values for short-term scheduling, keep binary decision variables from day-ahead model, use rolling optimization, and objective function (9), solve intra-day model. Get the scheduling plan for intra-day.
4. Predict values for real-time, keep binary decision variables from day-ahead model, use rolling optimization, solve real-time model. Based on intra-day plan, adjust the output of each unit.

3. Case study

The test case considers a fictional RAS located in Lochinver, Scotland, where fish farming is a vital industry, as they are the largest producer of farmed salmon in the European Union. The system uses a heat pump (HP) and electric boiler to keep the water temperature in the feed tanks of RAS ranging 16°C to 17°C. The capacities of PVs, batteries, generator, and WT installed are decided by the solution of a design optimisation model, shown in Table 1. The results presented in Table 1 are used as inputs in this model, and a typical autumn day has been chosen specifically to test the model as renewable energy power generation is highest in this season.

Table 1 Model results summary.

PVs (kW)	Battery(kW)	Wind Turbines(kW)	Generator(kW)
13	7.7	19	35

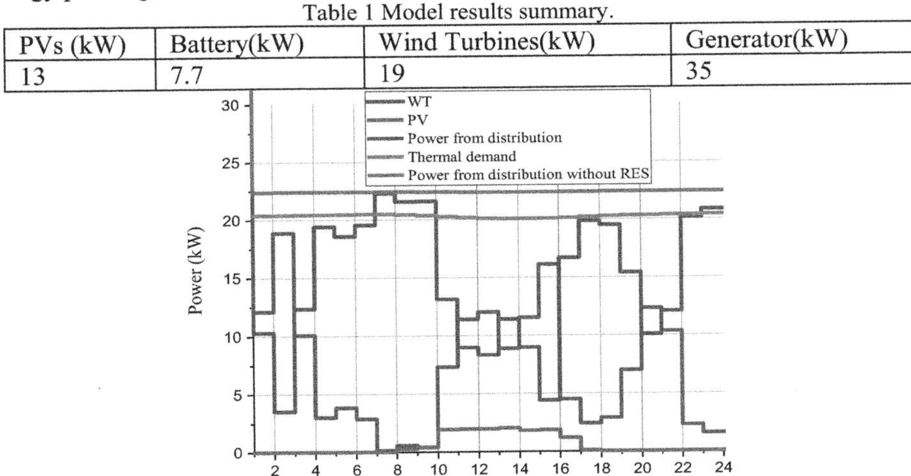

Figure 2 The scheduling plan of each unit in the system for day-ahead.

First, the day-ahead scheduling plan is shown in Figure 2. Due to the influence of geographical environment and weather, the output of the WT is far greater than that of PV. When the output of renewable energy is insufficient, the demand is met by purchasing power from the main distribution instead of starting up the generator, and the SOC of the battery is basically unchanged. If the parameters are changed, the electricity price is increased and the fuel price is reduced, the generator power generation will make up for the shortage of renewable energy, and the SOC of the battery will also change. The thermal demand is maintained at a relatively stable value, because of the high specific heat capacity. Compared with the original system without RESs, the daily operation cost can be reduced by about 27%, from £144.5 to £105.

Considering the disturbance of renewable energy output, rolling optimisation of MPC is introduced into the model. In intraday and real-time dispatching, to keep the energy balance of the network, the system prefers to purchase/sell the electricity from the main network rather than batteries, which is due to the lower electricity price compared with the use price of energy storage. Faced with renewable energy disturbance, the scheduling change of the MPC rolling algorithms is more stable and has less variation, which enhances the system's ability to cope with uncertainties. By using the rolling-horizon MPC model, the daily running expenses can be reduced by 3%.

4. Conclusion

RAS as a popular fish production system requires high energy demand due to thermal processes, water circulation, and aeration. This paper presents an optimisation model for the control and operation of a grid-connected distributed energy system integrated with an RAS, that is developed to optimise the performance of the system in economic, sustainable, and technical aspects in real time. An MPC strategy for a sustainable hybrid energy system for the RAS is proposed in this work. To analyse the performance of the model, the optimal results of the proposed system were compared with the traditional system in economic and environmental aspects. The optimal results show that the MPC strategy improves the control performance of system operation and improves the economy of system operation. Compared with a conventional RAS, the integration of renewable energy can save up to 27% operation cost in a typical day. With the MPC rolling-horizon algorithm, the system's ability to cope with uncertainties has been improved.

References

Ayer, N. W., & Tyedmers, P. H. (2009). Assessing alternative aquaculture technologies: Life cycle assessment of salmonid culture systems in Canada. *Journal of Cleaner Production*, *17*(3), 362–373.

Chen, Z., Ye, Z., Ji, M., Zhou, F., Ding, X., Zhu, S., & Zhao, J. (2021). Effects of flow velocity on growth and physiology of juvenile largemouth bass (Micropterus salmoides) in recirculating aquaculture systems. *Aquaculture Research*, *52*(7), 3093–3100.

FAO. (2020). *GLOBEFISH Highlights 2019 Statistics: A quarterly update on world seafood markets*. FAO. https://www.fao.org/documents/card/en/c/ca7459en/

Holman, J. (2009). *Heat Transfer* (10th edition). McGraw-Hill Education.

Jamroen, C. (2022). Optimal techno-economic sizing of a standalone floating photovoltaic/battery energy storage system to power an aquaculture aeration and monitoring system. *Sustainable Energy Technologies and Assessments*, *50*, 101862.

Ruggles, T. H., & Caldeira, K. (2022). Wind and solar generation may reduce the inter-annual variability of peak residual load in certain electricity systems. *Applied Energy*, *305*, 117773.

Schneider, O., Blancheton, J., Varadi, L., Eding, E., & Verreth, J. (2006). Cost price and production strategies in European recirculation systems. *Conference: Aqua 2006: Linking Tradition & Technology Highest Quality for the Consumer, Firenze (Florence), Italy, 9-13 May 2006*, 855.

Silvente, J., Kopanos, G. M., Pistikopoulos, E. N., & Espuña, A. (2015). A rolling horizon optimization framework for the simultaneous energy supply and demand planning in microgrids. *Applied Energy*, *155*, 485–501.

Antonis Kokossis, Michael C. Georgiadis, Efstratios N. Pistikopoulos (Eds.)
PROCEEDINGS OF THE 33rd European Symposium on Computer Aided Process Engineering (ESCAPE33), June 18-21, 2023, Athens, Greece
 http://dx.doi.org/10.1016/B978-0-443-15274-0.50035-4

Multi-stage optimization for marketing industrial flexibility

Martin Fischer[a], Karl-Wilhelm Schenzel[a], René Hofmann[a]

[a] *Institute of Energy Systems and Thermodynamics, TU Wien, Getreidemarkt 9/302, 1060 Vienna, Austria*

martin.fischer@tuwien.ac.at

Abstract

This work presents a method based on Mixed Integer Linear Programming (MILP) to consider wholesale energy markets in the operational optimization of industrial energy systems. A two-stage optimization framework is presented to consider price differences between the day-ahead and the intraday spot markets to optimize the electricity procurement in both markets. A representative use case of an industrial process is used to optimize the energy supply from the power generators and storage units and minimize energy costs as well as to benefit from participating in the wholesale market. The results show potential to save energy costs and make additional profits on the intraday market.

Keywords: Flexibility, Optimization, MILP, Intraday Market

1. Introduction

A rising share of renewables and an increasing number of volatile power generators and consumers are causing imbalances and congestions in the power grid. With decarbonization and the demise of conventional power plants, our energy system needs new flexibility measures and market participants. In most countries, the industrial sector is one of the largest electricity consumers and can therefore play a significant role in supplying flexibility to balance the power grid and provide security of supply. Previous work (Schenzel, et al., 2022) presented a method to identify existing flexibilities based on a MILP approach (Halmschlager et al., 2020), which is integrated into this work to consider the participation in wholesale markets, specifically focusing on the day-ahead and intraday markets.

2. Leveraging Energy Markets

Conventional optimization methods focus on high efficiency and low operational costs, but neglect opportunities by not considering energy markets. Most industrial customers procure their energy through energy suppliers, with fixed or floating electricity tariffs that reduce both the actual effort needed to operate on the wholesale market and the exposure to price volatilities (Schnorr 2019). At the same time, there are opportunities for industrial customers to make profits by marketing their flexibility - either by directly acting on the wholesale market or partnering with aggregators to have access to the wholesale markets. The advancement in energy storage technologies and the staggering increase in price spreads of wholesale markets support the case for industrial plant operators to act as a Flexibility Service Provider (FSP).

2.1. Flexibility

Villar, Bessa, and Matos (2018) define flexibility as the possibility of modifying generation and/or consumption patterns in reaction to an external signal (price or activation signals). Generally, industrial energy systems have three ways of providing flexibility. Firstly, they can adjust their production schedule. This is highly dependent on the processes involved and often needs more lead time and computational effort. Secondly, energy carriers can be substituted, meaning that in order to meet a certain electricity demand, it can either be consumed from the power grid or produced by generators such as gas turbines. Thirdly, energy storages can be used to shift energy demands in the process. While conventional power plants usually store electric energy by converting it to other forms of energy and backwards, with significant losses each conversion, industries can directly utilize the converted energy - like heat - to cover the demand of their production processes. Also, thermal heat storages are cheap in comparison to batteries and can be used to make industrial processes more efficient by reducing waste heat. Once identified, the electric industrial flexibility can be offered as bids on the intraday market at an electricity exchange like the European Power Exchange (EPEX) or as an ancillary service at the balancing markets of the Transmission System Operator (TSO).

2.2. Balancing Markets

These markets are used for the operating reserve to restore the frequency of the electricity grid in case of imbalances of the system. There are three different products available, Frequency Containment Reserve (FCR), automatic Frequency Restoration Reserve (aFRR), or manual Frequency Restoration Reserve (mFRR). Each product has their own requirements, pricing, and activation times. The renumeration model for aFRR and mFRR consists of two parts: one part is for keeping the capacity available (capacity price), and one is for the actual activation (energy price). Depending on the product, the activation must follow very shortly after the activation signal. Since these activations are very hard to predict they can interfere with production planning and safety.

2.3. Intraday Markets

To compensate for changes in energy production, often caused by insufficient weather forecasts on the day-ahead, energy companies can buy or sell energy to level their position. This can be done during the day and up to several minutes prior to delivery. With the growing number of volatile renewables, these markets have gained much importance in the last decade with more energy being traded at these markets each year. FSPs can provide their flexibility to the market, to increase liquidity, reduce the amount of balancing energy and to make additional profits.

3. Modelling Approach

Mathematical optimization represents a powerful method to maximize profits for power plant operators, often referred to as Unit Commitment (UC) problem. UC problems - a subclass of optimization problems - have been widely used in the energy market for many years. Industrial energy systems often face similar problems, when deciding when and how to utilize their power, heat, or steam generators to cover their demand in the most efficient way, especially since decentralized energy systems tend to get more diverse. In contrast to conventional UC problems of power plant operators, industrial plant operators are also constrained by their own operational targets, e.g., minimum production output, minimum up-times for batch processing, working hours, etc. For our purposes a fixed set of demand time series is sufficient, meaning that there is a predetermined production plan, the demand for different energy carriers such as steam on different pressure levels, gas and electricity can be derived. Scheduling is often impractical, since it means a bigger intervention in the operation and is therefore not considered. With the increase of digitization and the development of energy control systems, a focus on optimizing the energy management system is more applicable and reduces interference to operation to a minimum. In the following we present a method to include the intraday market in the optimization problem.

3.1. Optimization Method

The problem is formulated as a UC problem on the basis of MILP formulations, which are implemented in Matlab® and solved with a state-of-the-art solver (e.g. GUROBI). The model uses a combination of linear, continuous, semi-continuous and binary decision variables. They represent all operation trajectories of the energy supply system of the industrial plant, such as the power generated by the gas turbine, the charging and discharging of storages etc. Constraints are used to define the optimization problem, to depict operational constraints such as the satisfaction of energy demands, operational limits, power outputs or ramp-up/down constraints. The objective is to minimize operational costs, by optimizing the trajectories of the individual generating units of the industrial plant. This is done by utilizing the storages to meet the demands in the most efficient way and considering price curves of power spot markets for a cost-efficient operation plan. To better represent the market mechanisms of spot markets, the method is split in two optimization runs or "stages".

3.2. Two-stage Optimization

Figure 1 shows the two-stage optimization method to consider price signals between the day-ahead and intraday spot markets. Based on historical data, we developed a regression model to predict day-ahead market prices, based on weather forecasts and the previous day's market coupling results. In Stage 1 (D-1), the predictions for the day-ahead prices are used to calculate an optimal operation plan using MILP. Based on the results, we derive the electricity demand procured directly via placing price inelastic bids on the day-ahead market. The industrial plant acts as a price taker, meaning that the orders are executed independent the market price and all bids are accepted. In Stage 2 (D-0), the results from the day-ahead auction and the actual weighted average intraday market prices are used as input to solve a second optimization problem. We assume that price signals from the continuous intraday market are available at a single point in time and the weighted average result from each quarter hour can be used. Considering these price signals the plant operator can buy or sell additional electricity at a 15-minute resolution to make additional savings and reduce energy costs.

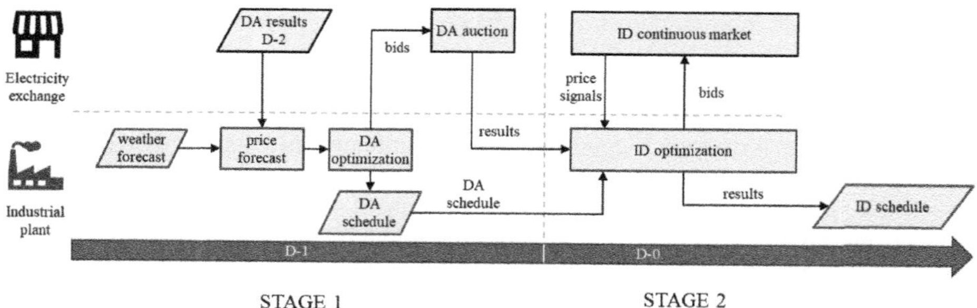

Figure 1: The two-stage optimization framework for day-ahead (DA) and intraday (ID) optimization.

4. Use Case

To demonstrate the proposed two-stage method, a simplified industrial energy system is modeled. The model consists of two demand curves, representing the power demand and steam demand, which result from a production plan. The power demand and the steam demand are fluctuating during the day but are set to a mean value of 1 MW. Both demands can be covered to a certain degree by a combined heat and power generator (CHP) with a maximal output of 0.5 MW_{el} and 1 MW_{therm}. A gas boiler can produce additional steam to cover the full steam demand. Additionally, a thermal and an electrical energy storage increase the efficiency and provide flexibility to the system.

4.1. Data

We used historical price data from the European Power Exchange (EPEX) spot market and the European Network of Transmission System Operators for Electricity (ENTSOE). We used hourly day-ahead market coupling results and weighted average results from the continuous intraday market for the Austrian bidding zone. Data from the year 2021 was used to train the linear regression model while data from the first half of 2022 was used to validate the regression model and to predict the daily day-ahead prices for the optimization framework.

4.2. Optimization Results

Figure 2 shows the optimization results of stage 2 for a single day in February 2022. The upper part of the plot shows the optimal trajectories to meet the electric power and steam demand of the production processes in the most cost-efficient way. Since the CHP has a limited electric power output, the industrial plant is dependent on power supplied by the public grid. The battery is charged and discharged to shift power demands to times of lower energy prices. Part B of Figure 2 shows the bids that were placed on the day-ahead and intraday markets to procure the power supply according to the optimal trajectories.

Figure 3 shows the potential energy cost savings for each day in correlation with the daily variance of the price spreads between the day-ahead and the intraday market. The savings are calculated by the difference of the cost function between stage 1 and stage 2. The mean value of the relative cost savings is 1.96 %. Generally, a higher price spread between these markets has shown a higher potential of saving energy costs.

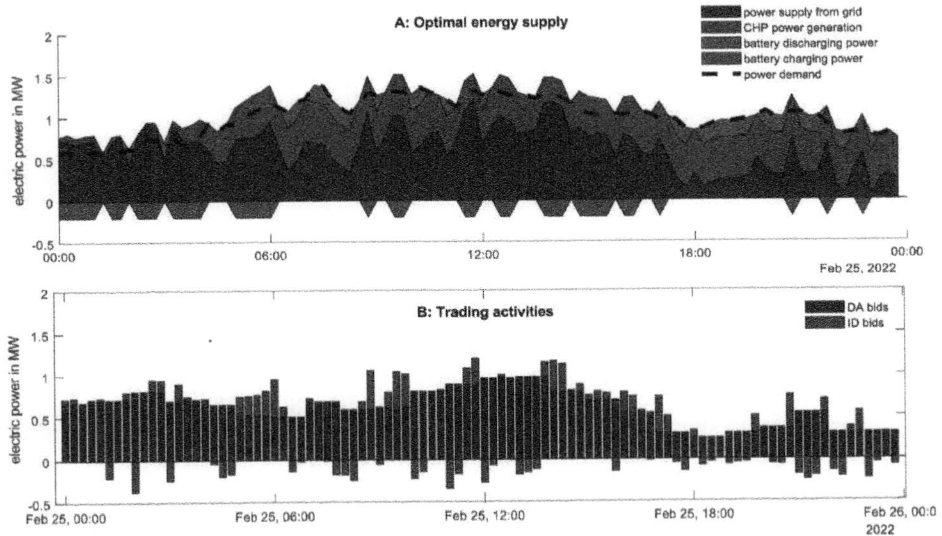

Figure 2: Optimization results for a single day. Part A shows the optimal operation trajectories of the energy supply system. Part B shows the corresponding bids on the day-ahead (DA) and intraday (ID) market.

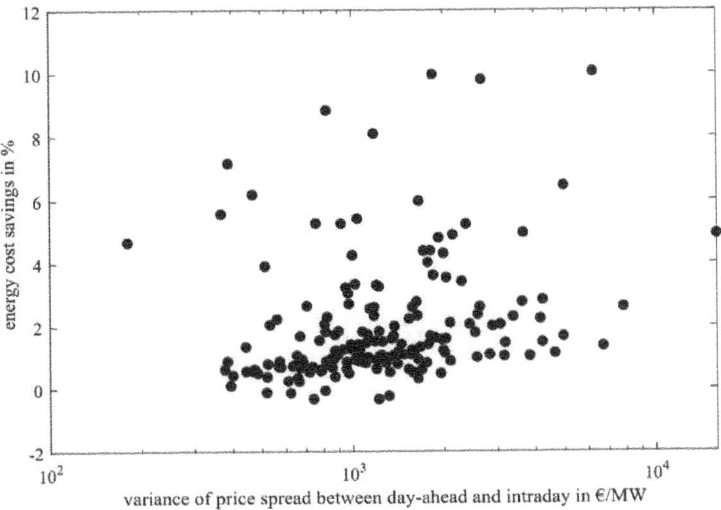

Figure 3: Relative energy cost savings per price spread

4.3. Discussions

The intraday bids in Figure 2 B are the corresponding reaction to price differences between the day-ahead and intraday markets. These can be interpreted as price signals from the current market situation and as reactions to deviations of the predicted day-ahead price curves that are used for optimization in stage 1.

By using the average result of the intraday market, we assume perfect foresight of the intraday market prices in the second stage. These prices are hard to predict though and are not traded at a single point in time, but continuously during the day. Thus, it may be possible to additionally exploit the volatility in this market by using the flexibilities of industrial energy systems.

5. Conclusion

The proposed two-stage optimization method enables a deeper integration of industrial energy systems in the short-term energy market. Hereby, industrial energy systems can react to fluctuations in the system in the form of price signals in the intraday market. Using real price data from the first half of 2022, our method also shows significant potential for reducing energy costs. Assuming direct access to wholesale market, the intraday market offers a supplementary opportunity for marketing flexibility, as the entry barriers are lower than in other markets such as balancing or redispatch markets. Further research will show the possible exploitation of volatility in the intraday market. In the prospect of reduced entry barriers for balancing markets and new opportunities for industrial energy systems to act as FSPs, mathematical optimization methods based on MILP may play an important role when participating in flexibility markets.

Acknowledgement

This work was supported by the project "Industry4Redispatch", which is part of the energy model region NEFI - New Energy for Industry and is funded by the Austrian Climate and Energy Fund [FFG, No.887780].

References

Halmschlager, Verena, und René Hofmann. 2020. „ Flexibility identification of an industrial production ". In *Computer Aided Chemical Engineering*, 48:1375–80. Elsevier. https://doi.org/10.1016/B978-0-12-823377-1.50230-5.

Schenzel, Karl, Fischer, Martin, Zlabinger, Erwin und René Hofmann. 2022. „Holistic Approach for the Optimization of Industrial Hybrid Energy Hubs with MILP". In *Proceedings of the 2nd NEFI Conference 2022*, 21:30. NEFI.

Schnorr, Stephan. 2019. *Energiebeschaffung in Industrieunternehmen: erfolgreiches Agieren am Energiemarkt.* Essentials. Wiesbaden [Heidelberg]: Springer Gabler. https://doi.org/10.1007/.

Villar, José, Ricardo Bessa, und Manuel Matos. 2018. „Flexibility Products and Markets: Literature Review". *Electric Power Systems Research* 154 (Januar): 329–40. https://doi.org/10.1016/j.epsr.2017.09.005.

Antonis Kokossis, Michael C. Georgiadis, Efstratios N. Pistikopoulos (Eds.)
PROCEEDINGS OF THE 33rd European Symposium on Computer Aided Process Engineering (ESCAPE33), June 18-21, 2023, Athens, Greece
 http://dx.doi.org/10.1016/B978-0-443-15274-0.50036-6

Intelligent Process Flowsheet Synthesis and Design using Extended SFILES Representation

Vipul Mann,[a] Rafiqul Gani,[b,c] Venkat Venkatasubramanian,[a]

[a]*Columbia University, New York, NY 10027, USA*
[b]*PSE for SPEED Company, Charlottenlund, DK-2920, Denmark*
[c]*Sustainable Energy and Environment Thrust, The Hong Kong University of Science and Technology (Guangzhou), Guangzhou, China*

Abstract

A central problem in chemical engineering (and several related areas) is evaluating the correct sequence of unit operations, their design aspects, and the continuous optimization of their operations to efficiently convert input materials to final products. The numerous decisions to be made at each problem-solving stage renders this problem combinatorically complex. In this work, we propose a hybrid, artificial intelligence-based multi-level framework to perform fast, efficient, and reliable flowsheet design and optimization. We build upon the previously proposed SFILES-based text representation of flowsheets to incorporate additional contextual details in the extended SFILES framework using hypergraph representations. We discuss our eSFILES framework using the well-known hyrdodealkylation (HDA) process.

Keywords: process design; flowsheet prediction; artificial intelligence; computer-aided flowsheet synthesis

1. Introduction

Process flowsheet synthesis and design is a challenging task that involves identifying the correct set of unit operations and their optimal sequence that enables the conversion of input materials to desired output products, while considering energy consumption, environmental impact, safety, operability, and many more. The goal of synthesis is to identify the tasks, the operations that will perform these tasks, and their sequence to determine the flowsheet. The goal of design is to add operational and equipment details to the flowsheet so that the process can be verified, optimized, built, and operated. The size of the synthesis problem varies with the number of tasks and the number of alternative operations for each task. The complexity of the design problem is related to matching the tasks and their associated operations with design parameters so that the designed process matches the desired process specifications. The design parameters are further optimized to obtain a sustainable process design.

Model-based methods are usually used for the synthesis and design stages. Although, in principle, the synthesis, design and optimization steps could be performed simultaneously, the current practice is to solve them separately due to the size and complexity of the resulting mathematical problem. The challenge here is how does one incorporate issues, such as, economics, environmental impacts, operability, safety, and sustainability in the early stages of process synthesis and/or design (Tula et al., 2017).

Computer-aided approaches using AI-inspired methods combined with fundamental concepts offer several advantages over purely data-driven methods, which may lead to infeasible flowsheets, or purely model-based methods, which may run into difficulties in the numerical solution step due to the complexity of the models used.

To facilitate the development of such hybrid AI-based approaches combining process knowledge with computational algorithms, an appropriate flowsheet representation is needed that is concise, complete, and accurate. The SFILES representations developed by (D'Anterroches, 2005), (Bommareddy et al., 2011), and (Tula et al., 2015) was the first step in this direction, and has been shown to have various applications, such as, flowsheet autocompletion (Vogel et al., 2022), piping and instrumentation diagram generation (Hirtreiter et al., 2022), and flowsheet pattern mining (Zhang et al., 2019). The SFILES strings represent correctly and consistently a wide range of process flowsheets, involving typical operations found in chemical and biochemical processes. Note that, like the SMILES strings for a molecule, a parser is needed to convert SFILES strings to the actual process flowsheet diagrams. As originally developed, a process flowsheet is first represented by a set of process groups, which are similar to the functional groups that represent a molecule. The process group representation is then converted to a SFILES representation. Unlike the molecular representation with SMILES, several additional details need to be considered. For example, information about the number of chemicals and their effect on the system behavior; the direction of flow-paths for reactants, products, inerts, solvents, etc. need to be tracked; and, start and end of the process need to be clearly marked. These additional issues give opportunities for a symbolic AI-based intelligent system (Venkatasubramanian & Mann, 2022) need to be incorporated to the current process synthesis-design methods such that the application of SFILES to represent process flowsheets as well as its use in computer-aided process synthesis and design can be extended.

The objective of this paper is to present a hybrid multi-level AI framework for fast, efficient, and reliable flowsheet synthesis and design taking into account concepts and theory on chemical process development, together with the knowledge and data for hundreds of process flowsheets that are known already. We present an extended SFILES representation (eSFILES) based on SFILES of process flowsheets ((D'Anterroches, 2005); (Bommareddy et al., 2011); (Tula et al., 2015)) and the annotated hypergraph representation, developed originally to study networks in organic chemistry (Mann & Venkatasubramanian, 2023). We highlight selected developments where we illustrate the hierarchical eSFILES representation framework for any chemical process flowsheet, also suitable for process synthesis, design, and innovation (Tula et al., 2015). Namely, we propose a three-level, extended SFILES framework where the lowest level contains flowsheet connectivity information which could be used to generate text-based flowsheet representations like SFILES but with additional information; the middle level contains additional details on process groups and streams providing information necessary information for process synthesis (superstructure-based optimization or process group based enumeration and test); and the top level contains process operational data in terms of design parameters for process simulation and innovation through optimization and intensification. At the core of our approach lies a hypergraph representation of flowsheet (Mann & Venkatasubramanian, 2023) that represents process groups as hyperedges and streams as nodes with annotations indicating contextual information.

2. Current state of flowsheet representation: SFILES

A text-based representation for chemical flowsheets reported by (Tula et al., 2015) is called simplified flowsheet input line entry scheme (or SFILES) analogous to the SMILES representation for molecules. Such a representation offers a concise way of representing flowsheets and could also facilitate not only storage of flowsheets, but also, enumeration, analysis, and deployment for interactive text-based algorithms process synthesis, design, and simulation of a given flowsheet. (Tula et al., 2015) developed separate rule-based algorithms for process synthesis and design, whose solutions were not added to SFILES strings. The underlying idea behind SFILES is to identify analogous process components similar to a SMILES string as follows –

- represent processes as *process groups* akin to functional groups in molecules.
- represent process flow streams akin to chemical bonds in molecules.
- represent various connections, recycle streams, and branching of streams akin to branching in molecules.
- unlike SMILES, SFILES need special unidirectional input nodes and output nodes that indicate the start and end of a process flowsheet, through special characters ‘i’ and ‘o’, respectively.

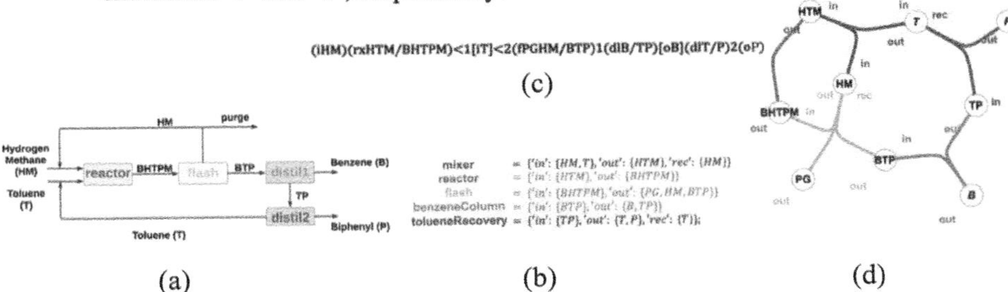

Figure 1: (a) HDA process flowsheet, (b) the various process groups, (c) the corresponding SFILES string, (d) HDA process represented by process groups. In the above subplots, the abbreviations refer to – mixture of hydrogen, methane (HM); mixture of benzene, hydrogen, toluene, biphenyl, methane (BHTPM); mixture of benzene, toluene, biphenyl (BTP); benzene (B); biphenyl (P); toluene (T); purge gas (purge). The hypergraph was generated using Wolfram Mathematica and algorithmic implementation would be done in Python.

We use the well-known hydrodealkylation (HDA) process to illustrate the basic concepts and the new developments. First the process flowsheet is represented by a set of process groups retrieved from the database (the process group representation is shown in Fig 1b) and the parser is used to generate the SFILES string (shown in Fig 1c). The reader is referred to (Tula et al., 2015) for detailed examples on SFILES generation.

3. Extended SFILES (eSFILES)

Hypergraph: The hypergraph representation of the HDA process is shown in Figure 1(d). While this new SFILES representation offer numerous benefits, there is scope for further improvement and incorporation of additional details that would aid in intelligent process flowsheet synthesis and design. The new information that, when incorporated, would generate additional, more complete information in the flowsheet representation such as:

- information on incoming/outgoing streams for each process group explicitly, especially when multiple streams enter or leave a process group.

- the minimum data needed to completely define a material and/or energy stream.
- information on driving force for the process groups indicating the ease/difficulty of conversion/separation.
- ability to systematically generate superstructure of flowsheet and enumerate valid alternatives aiding in the design problem.

Towards this end, we propose the extended SFILES (or eSFILES) representation that provides a framework for incorporation of this information in a hierarchical manner and further details on this are presented in the subsequent sections.

Hierarchical framework: A hierarchy is needed to efficiently organize process knowledge needed for different applications and to guide users to select, retrieve, and/or collect the data needed for a specific application. Namely, the eSFILES hierarchy has three levels of information -- first at the lower level, information needed to only represent the flowsheet (connectivity); next, at the middle-upper level, introduce additional details such as driving force and mixture composition that helps to establish the material and energy flows; and finally, the top level that contains information on composition and material/energy balance that would be need to solve a flowsheet among other applications. The eSFILES hierarchy is shown in Figure 2.

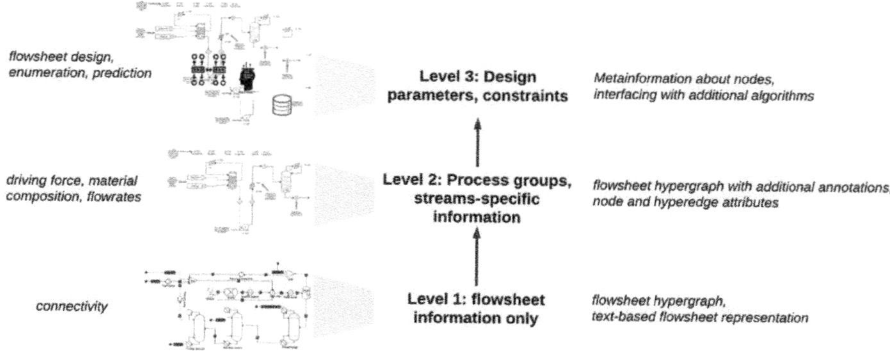

Figure 2: eSFILES framework with three levels of hierarchy

Implementation of hierarchical framework: The concept of hypergraphs is introduced to incorporate multi-level hierarchy of information with eSFILES. A hypergraph is a generalization of a graph where each edge is not limited to connecting only two vertices but could connect any number of vertices. Mathematically, a hypergraph consists of a set of hyperedges and vertices where each hyperedge consists of a non-empty subset of vertices. In addition, each vertex has a hyperedge-specific annotation indicating a predefined contextual information. The flexibility of hyperedges to connect more than one node and ability to have hyperedge-specific node annotations offers several advantages for flowsheet representation. Further details on the annotated hypergraph framework and its various applications in reaction engineering are presented in (Mann & Venkatasubramanian, 2023).

For instance, for the HDA process, the equivalent flowsheet hypergraph representation is shown in Figure 1(d) where each process group is represented as a colored hyperedge, process streams are represented as vertices, and the vertices have hyperedge-specific 'in'/'out' annotations indicating the incoming/outgoing nature of the process streams for a corresponding process groups. Moreover, recycle streams are indicated with an additional annotation ‘rec’ wherever applicable for the process streams. Figure 1(b) shows each process group as a hyperedge which essentially is a set of named vertices where the names (or annotations) indicate the streams' nature.

The three levels of information in the eSFILES framework is shown for the `toluene recovery' process group in Figure 3. In Figure 3(a), only the hyperedge representing the process group along with participating streams as vertices and role-specifying annotations are shown. In Figure 3(b), additional information for the entire process group (or hyperedge) such as driving force and stream (or vertex)-specific information such as mass fraction and composition is shown. Finally, information on design aspects and necessary information required for additional analyses such as mixture analysis, reaction analysis, and pure component properties are presented in Figure 3(c). Though the three levels in eSFILES are only shown for a process group brevity, the same could be done for an entire process using the same approach.

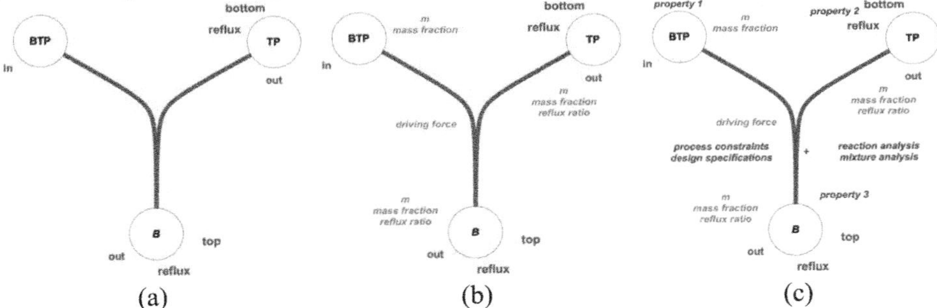

Figure 3: (a) Distillation column hyperedge with additional annotations. The corresponding text-based representation for this process groups is BTP(in)>dl>B(out,top,reflux)TP(out,bottom,reflux), (b) the driving force, material compositions, reflux rate, and conversion indicated as hyperedge attributes (for process groups) and hyperedge-specific node attributes (for streams) (c) the design parameters, properties, and constraints are represented as additional meta information.

Application of the eSFILES in sustainable process design: (Tula et al., 2017) proposed a 3-stages sustainable process design method consisting of 12 hierarchical steps. Stage-1 consists of steps 1-4 where the process synthesis problem is solved. Stage-2 consists of steps 5-9 where the base case process design problem is solved. Stage-3 consists of steps 10-12 where the base case design is further improved to find innovative and more sustainable alternatives. The original multi-stage method of (Tula et al., 2017) now retrieves the necessary data from the stored hypergraph. Also, different alternatives for stage-3 can easily be generated through different alternatives of the hypergraph, allowing the user to generate and analyze multiple alternatives as well as store the information for future use without having to repeat any of the previously solved steps.

4. Conclusions

The concept of hypergraph has been used to extend the original SFILES concept to multiple levels to store as well as retrieve information related to process synthesis, design, and innovation. In this way, an existing interactive multi-stage method has been made more intelligent and efficient through the adoption of an available AI method for sustainable process design. Therefore, the eSFILES-based framework for process synthesis and design incorporates a combination of artificial intelligence-based methods and well-known chemical engineering knowledge incorporated through an intelligent system facilitating fast, correct, and consistent decision-making related to process synthesis and design.

Current and future work involves the development of an extended manuscript providing more details of the eSFILES, its adoption in the multi-stage sustainable process design method, and its application in several case studies. In particular, we plan to provide rigorous examples to demonstrate the usability of eSFILES for new process flowsheets and also use them to address real challenges such as provide flowsheet alternatives, perform process intensification, and flowsheet property prediction based on the text-based flowsheet representation using a natural language framework similar to molecular property prediction (Mann et al., 2022) and chemical product design (Mann et al., 2023).

References

Bommareddy, S., Eden, M. R., & Gani, R. (2011). *Computer Aided Flowsheet Design using Group Contribution Methods* (pp. 321–325). https://doi.org/10.1016/B978-0-444-53711-9.50065-1

D'Anterroches, L. (2005). *Process flowsheet generation & design through a group contribution approach.* [CAPEC], Department of Chemical Engineering, Technical University of Denmark.

Hirtreiter, E., Balhorn, L. S., & Schweidtmann, A. M. (2022). *Towards automatic generation of Piping and Instrumentation Diagrams (P&IDs) with Artificial Intelligence.* http://arxiv.org/abs/2211.05583

Mann, V., Brito, K., Gani, R., & Venkatasubramanian, V. (2022). Hybrid, Interpretable Machine Learning for Thermodynamic Property Estimation using Grammar2vec for Molecular Representation. *Fluid Phase Equilibria, 561,* 113531. https://doi.org/10.1016/j.fluid.2022.113531

Mann, V., Gani, R., & Venkatasubramanian, V. (2023). Group contribution-based property modeling for chemical product design: A perspective in the AI era. *Fluid Phase Equilibria, 568,* 113734. https://doi.org/10.1016/J.FLUID.2023.113734

Mann, V., & Venkatasubramanian, V. (2023). AI-driven hypergraph network of organic chemistry: network statistics and applications in reaction classification. *Reaction Chemistry & Engineering.* https://doi.org/10.1039/D2RE00309K

Tula, A. K., Babi, D. K., Bottlaender, J., Eden, M. R., & Gani, R. (2017). A computer-aided software-tool for sustainable process synthesis-intensification. *Computers & Chemical Engineering, 105,* 74–95. https://doi.org/10.1016/j.compchemeng.2017.01.001

Tula, A. K., Eden, M. R., & Gani, R. (2015). Process synthesis, design and analysis using a process-group contribution method. *Computers & Chemical Engineering, 81,* 245–259. https://doi.org/10.1016/j.compchemeng.2015.04.019

Venkatasubramanian, V., & Mann, V. (2022). Artificial intelligence in reaction prediction and chemical synthesis. *Current Opinion in Chemical Engineering, 36,* 100749. https://doi.org/10.1016/j.coche.2021.100749

Vogel, G., Balhorn, L. S., & Schweidtmann, A. M. (2022). *Learning from flowsheets: A generative transformer model for autocompletion of flowsheets.* http://arxiv.org/abs/2208.00859

Zhang, T., Sahinidis, N. v., & Siirola, J. J. (2019). Pattern recognition in chemical process flowsheets. *AIChE Journal, 65*(2), 592–603. https://doi.org/10.1002/aic.16443

Antonis Kokossis, Michael C. Georgiadis, Efstratios N. Pistikopoulos (Eds.)
PROCEEDINGS OF THE 33rd European Symposium on Computer Aided Process Engineering (ESCAPE33), June 18-21, 2023, Athens, Greece
 http://dx.doi.org/10.1016/B978-0-443-15274-0.50037-8

Physics-informed Neural Network based Modeling of an Industrial Wastewater Treatment Unit

Tuse Asrav[a,b], Ece Serenat Koksal[a,b], Elif Ecem Esenboga[c], Ahmet Cosgun[c], Gizem Kusoglu[c] , Duygu Aydin[c] , Erdal Aydin[a,b,*]

[a]*Department of Chemical and Biological Engineering, Koç University, Istanbul 34450, Turkey*

[b]*Koç University TUPRAS Energy Center (KUTEM), Koç University, Istanbul, 34450, Turkey*
[c]*Turkish Petroleum Refineries Corporation, Körfez, Kocaeli 41790, Turkey*
eaydin@ku.edu.tr

Abstract

Wastewater treatment units consist of biological treatment with activated sludge and are subject to many disturbances such as influent flowrate, pollutant load and weather conditions bringing about many challenges for the modeling of such plants. Data-driven models may respond to these challenges at the cost of issues such as overfitting or poor fitting due to the lack of high-quality data. To benefit from the available physics-based knowledge and to eliminate the drawbacks of suboptimal and poor training, physics informed neural networks might be quite promising. In this work, artificial, recurrent and physics-informed neural network models are utilized for the wastewater plant in Tüpraş İzmit Refinery. For recurrent models with selected features based on correlation technique, test mean squared error is up to 82% smaller compared to the standard artificial neural network models. Physics-informed trained neural network models with selected features improved the test performance by decreasing mean squared error up to 87% with acceptable decreases in training performance which addresses its strength compared to fully data-driven models.

Keywords: wastewater treatment; physics-informed neural networks; recurrent neural networks; process optimization; wastewater control

1. Introduction

Wastewater treatment is the removal of contaminants from the wastewater before being discharged into the sea to minimize the damage to the environment. Pre-treatment of wastewater is a physical treatment that involves grid sieves, sand traps, equalization basins, and corrugated plate interceptor. Coagulants, flocculants, sodium carbonate, and hydrogen peroxide addition to the wastewater corresponds to the chemical treatment where disinfection, sanitization and purification is achieved. Then, the stream is further treated by dissolved air flotation (DAF) which reduces the amount of total suspended solids and oil and grease. Biological treatment involves the removal of biodegradable organic matter by making use of the bacteria which is called the activated sludge process occurring in aeration tanks. Here, bacteria use oxygen to break down the biodegradable organic matter. Activated sludge process is followed by a sedimentation tank also called clarifier. Suspended particles settle out of wastewater as it flows slowly through the tank therefore further purification is achieved. Part of the exiting flow from the clarifier with high mixed liquor suspended solids (MLSS) content is recycled to the aeration tanks (Benyahia et al., 2006.)

Raising concerns around the world about environment have started to shift the regulations for several process systems areas towards tighter windows. Wastewater treatment is among these areas where optimal operation will become inevitable. Since wastewater treatment is a complex, non-linear system including physical, chemical, and biological processes as expressed above and subject to many disturbances, mechanistic modeling of a wastewater treatment unit is challenging (Guo et al., 2015). Alternatively, data-driven models, in particular artificial neural networks (ANN), are widely used at the cost of outcomes of black-box models such as overfitting, underfitting and lack of high-quality data. Similarly, recurrent neural networks (RNN) are able to handle time-dependent responses. They are suited better than the ANNs for dynamic systems. On the other hand, RNNs are also subject to issues such as vanishing gradient problem. Long Short-Term Memory (LSTM) and Gated Recurrent Unit (GRU) introduce more complex recurrent units with gates to retain long-term dependencies by controlling which information is passed through. The presence of dynamic data in wastewater treatment units makes it appropriate to construct RNN, LSTM and GRU models (Quaghebeur et al., 2022).

In this work, ANN, simple RNN, LSTM, GRU and physics informed ANN models are constructed to predict dissolved oxygen concentration (mg/L) at the aeration tanks, sludge volume index (mL/g) and waste activated sludge (m^3/day) in Tüpraş İzmit Refinery. Prediction performance is further increased by removing some of the features based on a correlation technique. Section 2 involves the methodology. In Section 3, results are shown, and Section 4 concludes the results.

2. Methodology

2.1. Artificial and Recurrent Neural Networks

A feed forward, fully connected ANN can be expressed as follows:

$$\hat{y} = f_1(w_1 f_2(w_2 u + b_2) + b_1) \tag{1}$$

where f_1 and f_2 are activation functions, w_1 and w_2 are weights, b_1 and b_2 are bias terms at output and hidden layers, respectively. Moreover, u and $\hat{y}$ represents input and output terms, respectively. RNNs extend the feed forward ANNs to handle the sequential data by repeatedly transferring information to the cell states throughout the time series. The prediction at the current time step depends on the input at the current time step and the memory of the cell state. LSTM and GRU architectures can track longer dependencies better than the traditional RNNs due to their complex gated mechanisms controlling the flow of information.

2.2. Physics-Informed Neural Networks

The training of the neural networks can be expressed as an optimization problem where the objective function is the loss function given in Eq. 2. Typically, the purpose of training the neural networks is to find the appropriate values of weights and biases that minimize the loss function which can be achieved using the backward propagation algorithm. In order to improve the performance via physics-informed knowledge, neural networks can be trained using a bi-objective loss function which includes user-defined physics term. A bi-objective loss function can be proposed as follows:

$$L = \frac{1}{N}\sum_{i}^{N}(y_i - \hat{y}_i)^2 + W_p P \qquad (2)$$

where W_p is the scalar weight and P is the physics term.

3. Results

3.1. Comparison of Neural Networks for Prediction of Dissolved Oxygen Concentration at Aeration Tanks

Total suspended solids (TSS), chemical oxygen demand (COD), pH, oil and grease, ammonium (NH_4-N), phenol, sulfide, TSS at return activated sludge which are measured after dissolved air flotation (DAF) are taken as inputs, and dissolved oxygen concentration in two parallel aeration tanks (Aeration tank A and B) is taken as output. The data is normalized between -1 and 1. The first 90% of the data is used as training data and the remaining is used as test data. An ANN, a simple RNN, an LSTM, and a GRU networks are developed using 3 hidden layers with 50 neurons with the rectified linear unit (ReLU) activation function. The mean squared error loss function with the Adam optimization algorithm in Python is used to train the models. Results based on mean squared error (MSE) and mean absolute percentage error (MAPE) given in Table 1 show that RNNs predicted the test data better than the feed-forward artificial neural network. Among the RNNs, LSTM can be selected as the network delivering the best prediction over test data since simple RNN and GRU converges to more oscillatory predictions as shown in Figure 1, and it may be more useful to predict the direction. In addition, a correlation technique based on F-value for regression tasks in Sklearn library is used to select features for dissolved oxygen prediction in the aeration tanks based on the LSTM model. (Pedregosa et al., 2011). By this technique, COD, pH, NH_4-N, phenol, sulfide, and TSS at return activated sludge are chosen as inputs and TSS at DAF exit and oil are removed from features since the correlation between these features is low. The results are shown in Table 1 and Figure 1.

Table 1. Performance of the Data-driven Neural Network Models for Prediction of Dissolved Oxygen Concentration at Aeration Tanks.

	ANN	Simple RNN	GRU	LSTM	LSTM after feature selection
Train MSE	0.0194	0.0024	0.0028	0.0146	0.0089
Test MSE	0.1266	0.0374	0.0416	0.0327	0.0229
Train MAPE	1.7978	0.3580	0.3616	1.4553	1.224
Test MAPE	0.7006	0.3502	0.4316	0.4150	0.3293

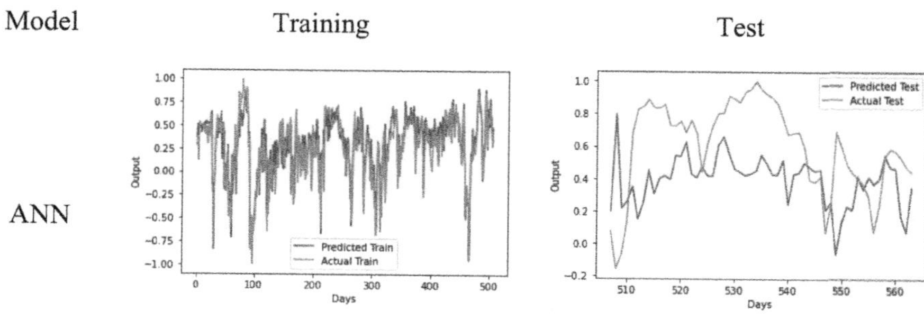

Figure 1.a. Figure 1.b.

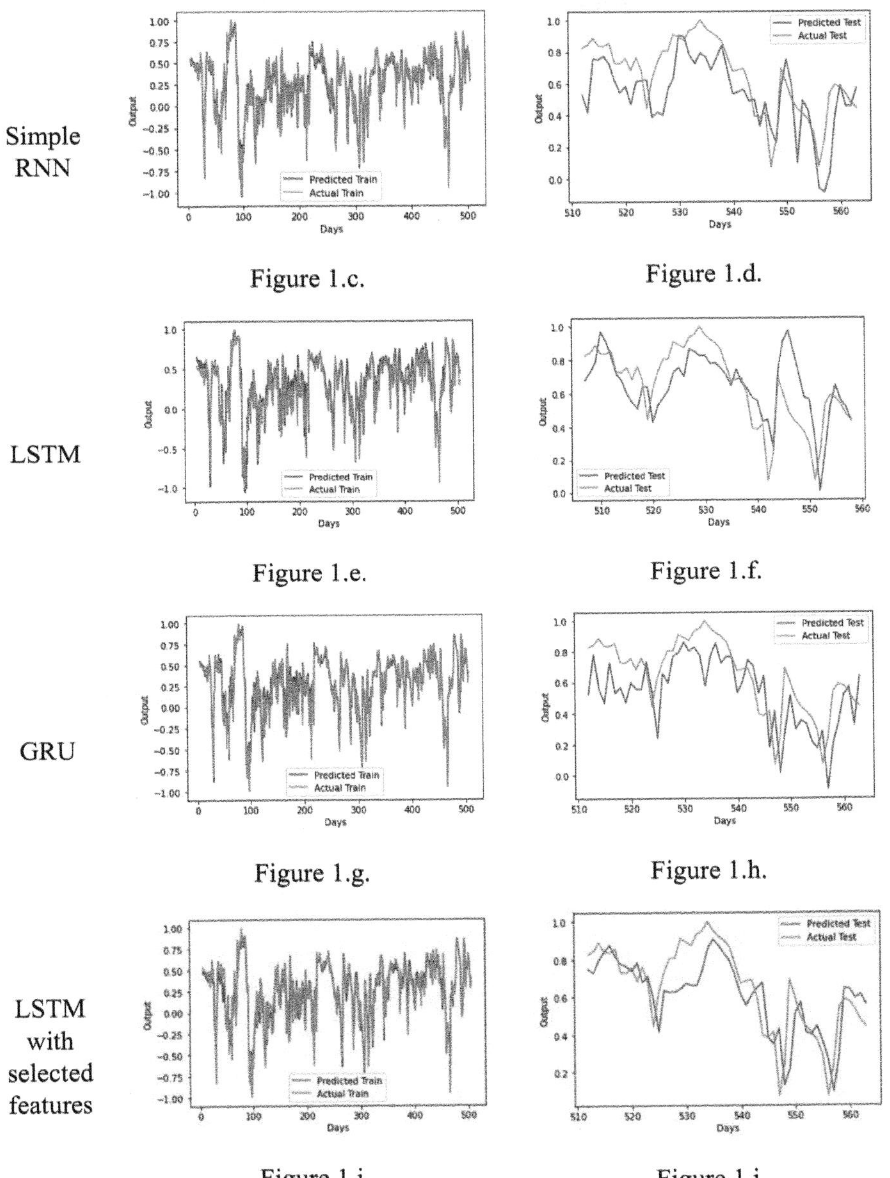

Figure 1.c. Figure 1.d.

Figure 1.e. Figure 1.f.

Figure 1.g. Figure 1.h.

Figure 1.i. Figure 1.j.

Figure 1. Training and test trends for the prediction of dissolved oxygen concentration at the aeration tanks.

3.2. *The effect of Physics-Informed Loss* Function

Feed-forward artificial neural networks trained with physics-informed (PI+) and physics-uninformed (PI-) loss functions are developed and the results are compared to observe the effect of bi-objective training.

3.2.1. Sludge Volume Index (SVI) Prediction

Dissolved oxygen concentration, pH, temperature, TSS, results of imhoff cone experiments where settleable suspended solid is measured (mL/L), waste activated sludge, and food to microorganism ratio (F/M) in the aeration tank A are taken as inputs, and sludge volume index (SVI) in aeration tank A is taken as output. First 60% of the data is selected as training. A single hidden layer with 5 neurons with the ReLU activation function is used. Physics-term related to the SVI is constructed by a rule-of-thumb used in the operation in the refinery. In addition, based on the results of the feature selection, a physics-informed ANN is developed taking dissolved oxygen concentration, pH, TSS, and settleable suspended solids in the aeration tank A as inputs since the correlation is only observed between these parameters and SVI. Results are shown in Table 2.

Table 2. Performance of the ANN Models for SVI Prediction.

	PI-	PI+	PI+ with feature selection
Train MSE	0.0053	0.0067	0.0072
Test MSE	0.0650	0.0181	0.0088
Train MAPE	0.1187	0.1661	0.0969
Test MAPE	0.1520	0.1144	0.0623

Even though the bi-objective physics-informed model delivers slightly higher training errors, the test performance of the model is improved when the physics term is introduced to the loss function. Moreover, the performance of the physics-informed trained ANN is further increased with selected features. Overall, the MSE is reduced 86.5%.

3.2.2. Prediction of Waste Activated Sludge

Oily water sewer feed, DAF exit properties (pH, TSS, COD, oil and grease, NH_4-N, phenol, sulfide), carbon-nitrogen ratio (C/N), COD load, aeration tanks' conditions (pH, dissolved oxygen concentration, temperature, TSS, F/M, settleable suspended solids, SVI), and TSS at return activated sludge are taken as inputs and waste activated sludge from both in aeration tank A and B are taken as outputs. First 80% of the data is selected as training data. There are 4 hidden layers with 50 neurons with the hyperbolic tangent (tanh) activation function. For the physics-term, as in the case for SVI, a rule of thumb related to the waste activated sludge is used. Furthermore, DAF exit properties are excluded from the inputs as a result of feature selection for physics-informed ANN model since they do not have a significant impact to predict the outputs. Training and test performances of the ANN models are evaluated and reported in Table 3. MSE values reduced 40.2% and 47.7% for tank A and B, respectively.

Table 3. Performance of the ANN Models for Prediction of Waste Activated Sludge.

	Waste Activated Sludge from tank A			Waste Activated Sludge from tank B		
	PI-	PI+	PI+ with selected features	PI-	PI+	PI+ with selected features
Train MSE	0.0005	0.0020	0.0004	0.0025	0.0022	0.0036
Test MSE	0.0169	0.0132	0.0101	0.0044	0.0029	0.0023
Train MAPE	0.0400	0.0882	0.0209	0.0338	0.0361	0.0171
Test MAPE	0.0307	0.0444	0.0308	0.0399	0.0259	0.0234

4. Conclusion

In this work, ANN, simple RNN, LSTM and GRU models are constructed to predict dissolved oxygen concentration at the aeration tanks in Tüpraş İzmit Refinery wastewater treatment unit. RNN-based models improved the prediction capability due to the dynamic nature of the data. LSTM is chosen as the model to be constructed since it predicts the direction better than the other models. Then, the prediction capability is further increased by selecting features based on correlation technique. As a result, the test MSE of LSTM with selected features is approximately 82 % better compared to standard ANN with a closer trend to the actual values. Accordingly, physics-informed knowledge is integrated into the training instances of the neural networks to improve test performance and reduce the possibility of suboptimal training for the prediction of SVI and waste activated sludge. Test MSE of physics-informed feature selected ANN models for the prediction of SVI, waste activated sludge in aeration tanks A and B are 86.5%, 40.2% and 47.7% smaller, respectively compared to the standard ANN. Finally, this study shows that RNNs may result in more accurate models especially when the process nature is dynamic. In addition, physics-informed knowledge can also be included to the standard ANN models to increase the prediction accuracy.

References

Benyahia, F., Abdulkarim, M., & Embaby, A. (n.d.). *Refinery wastewater treatment: a true technological challenge.*

Guo, H., Jeong, K., Lim, J., Jo, J., Kim, Y. M., Park, J. pyo, Kim, J. H., & Cho, K. H. (2015). Prediction of effluent concentration in a wastewater treatment plant using machine learning models. *Journal of Environmental Sciences (China)*, *32*, 90–101. https://doi.org/10.1016/j.jes.2015.01.007

Pedregosa, F., Varoquaux, Ga"el, Gramfort, A., Michel, V., Thirion, B., Grisel, O., … others. (2011). Scikit-learn: Machine learning in Python. *Journal of Machine Learning Research*, *12*(Oct), 2825–2830.

Quaghebeur, W., Torfs, E., de Baets, B., & Nopens, I. (2022). Hybrid differential equations: Integrating mechanistic and data-driven techniques for modelling of water systems. *Water Research*, *213*. https://doi.org/10.1016/j.watres.2022.118166

Antonis Kokossis, Michael C. Georgiadis, Efstratios N. Pistikopoulos (Eds.)
PROCEEDINGS OF THE 33rd European Symposium on Computer Aided Process Engineering (ESCAPE33), June 18-21, 2023, Athens, Greece
 http://dx.doi.org/10.1016/B978-0-443-15274-0.50038-X

Approaches to Reduce Optimization Time for Stochastic Optimization of Complex Chemical Processes

Fanyi Duanmu, Dian Ning Chia, Eva Sorensen*

Department of Chemical Engineering, University College London (UCL), Torrington Place, London WC1E 7JE, United Kingdom

e.sorensen@ucl.ac.uk

Abstract

Optimization is a crucial step to obtain an energy efficient and economically viable design in the chemical industry. The optimization of a chemical process is, however, often challenging due to the inherent complexity of chemical process designs. In this work, different combinations of speed up approaches – parallel computing (ParC), timeout function (TO), escape repeated simulations (ERS), dynamic bound (DB), and machine learning assisted optimization (ML) – are considered and their performances are compared. ParC is always activated as it has previously been found very effective to reduce optimization time. The results show that including DB as part of the speed up approach can ensure successful optimization for complex chemical processes. The TO and ML approaches can help reduce the optimization time significantly, but the ERS approach has no clear effect on the optimization performance. The approach with all speed up approaches activated together yields the best design in the shortest time.

Keywords: Stochastic optimization, genetic algorithm, parallel computing, machine learning, distillation

1. Introduction

Optimization is an important step in designing a chemical process, as an optimal design should not only have low capital and operating costs (CAPEX and OPEX) but also be energy efficient and sustainable. However, due to the complex nature of chemical process designs, their optimization is challenging, partly because of the difficulty in converging into a truly optimal design, but also due to the long optimization time. Different stochastic optimization methods (e.g., Genetic Algorithm (GA) and Particle Swarm Optimization (PSO)) and their corresponding settings (e.g., for GA, various types of parent selection, crossover, and mutation methods) are available that suit different designs and can help to improve convergence. However, the long optimization time is difficult to address as it is often caused by the high chance of failure of the initialization and re-initialization of the design (i.e., infeasible designs), which makes the solution time long. The optimization time can be significantly reduced if these infeasible designs could be avoided with the use of, for instance, dynamic bounds (Chia et al., 2021), removed by machine learning assisted optimization (Ibrahim et al., 2017), or timed-out using a timeout function. Moreover, escaping (i.e., skipping) repeated simulations is also a direct and efficient way to reduce optimization time. Finally, parallel computing (e.g., master-slave structure) is also widely used to speed up optimization.

2. Methodology

It has been reported that stochastic optimization, e.g., Genetic Algorithm (GA), is more robust but also more time-demanding compared with gradient-based deterministic optimization (Chia et al., 2021; Cavazzuti, 2013), and the stochastic optimization time of a chemical process may take up to a few hours. Therefore, in order to tackle long optimization times, different speed up approaches will be introduced in this work.

2.1. Parallel Computing (ParC)

The parallel computing approach (ParC) is one of the most successful and commonly applied methods. There are a few structures available to perform ParC, such as the master-slave model, island model, and cellular model, which are classified by different ways of communication within the model. Details of each type of parallel model can be found in Gong et al. (2015). In this work, the master-slave model is used. For this model, in each iteration, the individuals (e.g., chromosomes in GA) are assigned to available slaves (i.e., cores/processors) for evaluating the fitness and constraints. Once the results are obtained, the slaves return the results to the master to handle constraints and to generate the individuals for the next iteration (e,g., parent selection, crossover, and mutation in GA). Another important concept of ParC is the ways of assigning the tasks by either static or dynamic assignment (Mathew et al., 2014). In the case of static assignment, the tasks in a parallel job are assigned to a core/processor beforehand, while in the case of dynamic assignment, a task is assigned to whichever core/processor that is "free" at the moment. For the optimization of chemical processes, the dynamic assignment is preferred as the simulation time of each individual may vary considerably.

2.2. Timeout Function (TO)

From the authors' experience, some infeasible simulations (i.e., designs that fail to converge) may take a very long time before they fail to converge due to the multiple re-initialization performed to "confirm" that there is no numerical solution to that simulation. This is illustrated in Figure 1, which is one of the optimization results taken from the case study (10,000 simulations with 4,737 feasible simulations). Almost all the feasible simulations converged within 9 seconds but there is a proportion of infeasible simulations which take more than 9 s and may take up to about 38 s before they can be confirmed as "infeasible". A timeout function can therefore be used to rule out the simulations that take an unreasonably long time but have a high chance of being infeasible.

2.3. Escape Repeated Simulation (ERS)

Perhaps one of the most direct ways to reduce the optimization time is to escape any repeated simulations (i.e., skip the simulations that have the exact same input values). Repeated simulations can be present due to, for example, the elitism strategy adopted in GA. ERS can be achieved by saving and comparing the inputs to the simulation. In this work, a "short-term memory" strategy is adapted, where the inputs of the current iteration is saved and retained to the next iteration where the saved inputs and the "new" inputs generated are compared to identify any repeated simulations. The "old" inputs from the previous iteration are then discarded and replaced with the "new" inputs generated in this iteration. This strategy can reduce the need for memory allocation and avoid the slow down effect that will otherwise occur if a "long-term memory" strategy is used (i.e., saving and retaining all inputs that have occurred in all iterations so far).

2.4. Dynamic Bound (DB)

Dynamic bound (DB) is a function that changes (internally in the code) the upper and/or lower bound of a variable depending on the value of another variable. The benefit of DB may not be immediate in optimizing a purely mathematical problem, but the benefit is

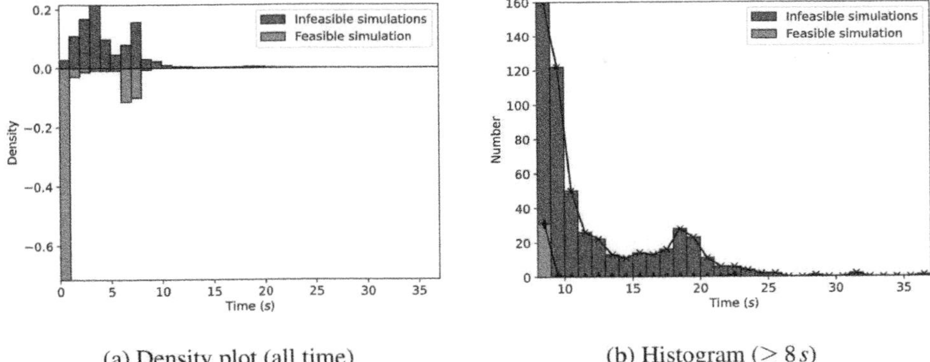

(a) Density plot (all time) (b) Histogram ($\geq 8\,s$)

Figure 1. Distribution of infeasible and feasible simulations depicted as:
(a) density plot for the whole range of simulations times and
(b) histogram as a function of simulation time from 8 s onwards.

tremendous when it comes to optimizing a chemical process. For example, it is obvious that the feed stage (N_f) of a distillation column should not be larger than the total number of stages (N_t). However, if the bounds of N_t have a significant overlap with the bounds of N_f (e.g., N_t: 10-40 and N_f: 5-35) due to lack of information about the process, there will be a very high chance that a design generated from the stochastic optimization has a higher value of N_f than N_t, which clearly leads to an infeasible simulation. The chance becomes even more significant if more feed locations and/or more distillation columns are involved in the optimization task. Therefore, if the rule (e.g., N_f should not be larger than N_t) is violated after generating new individuals (e.g., crossover in GA), then the DB function can be used to re-assign the feed location as a random integer within its lower bound and the total number of stages in the current design, thus reducing infeasibility.

2.5. Machine Learning Assisted Optimization (ML)

Due to the nature of the simulations of chemical processes, there are at least two sets of labels of a simulation, which are feasible/infeasible and on-/off-spec. The feasible/infeasible label describes if a simulation converges successfully or not, while the on-/off-spec label describes if a simulation stays within the constraint(s) or violates the constraint(s). Machine Learning (ML; e.g., Support Vector Machine (SVM)) is well suited for classification, which can be used to predict the status of a design before evaluating it. From the authors' experience, using ML to predict the on-/off-spec is more challenging as the training data requires a good balance between the number of on-spec and off-spec simulations, and for a complex chemical process, the proportion of on-spec simulation may be relatively low and even worse in the early iterations. Therefore, in this work, ML is applied to predict only the feasible/infeasible labels. ML will only collect the inputs and results of the simulations until the iteration reaches a user-defined iteration. After that, ML will use the information collected to train the model. Then, for the subsequent iterations, no training is required and only prediction with the trained model is carried out. For designs that are predicted to be infeasible, no evaluations will be carried out, thus the time required to process the "infeasible" designs can be saved.

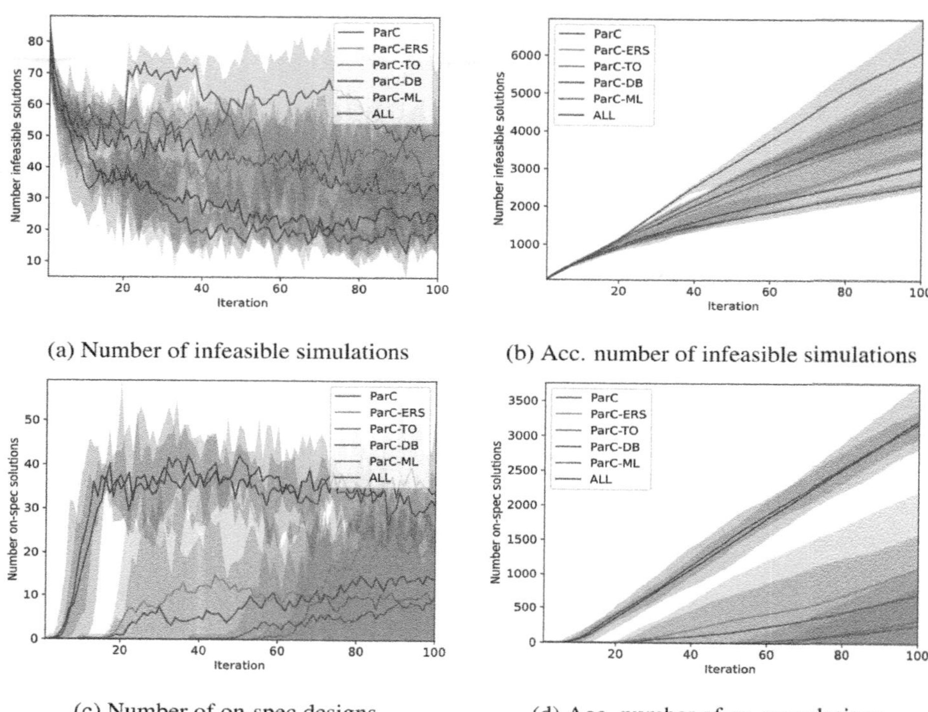

(a) Number of infeasible simulations (b) Acc. number of infeasible simulations

(c) Number of on-spec designs (d) Acc. number of on-spec designs

Figure 2. Performances of various combinations of speed up approaches on (a,b) reducing the number of infeasible simulations and (c,d) increasing the number of on-spec designs.

3. Case Studies

The case studies in this work were performed on a dual Intel Xeon Gold 6226R CPU with 16 Cores 2.90GHz (total 64 logic processors) for a highly integrated and complex hybrid dividing wall column (H-DWC) design to separate a mixture of ethyl acetate and ethanol, and the H-DWC details can be found in Chia et al. (2022). The simulation of a H-DWC involves two distillation columns (prefractionator and main column in the Petlyuk structure) and a membrane network, and the optimization variables are the design and operating variables of the distillation columns and membrane network. To compare the performance of different approaches, six optimization tasks are carried out: optimization with parallel computing (ParC); ParC with timeout function (ParC-TO); ParC with escaped repeated simulation (ParC-ERS); ParC with dynamic bound (ParC-DB); ParC with machine learning (ParC-ML); and all approaches activated together (ALL). ParC was always activated with 40 processors with the dynamic task assignment structure, as it had been found in previous work that ParC is very efficient in reducing optimization time (Chia et al., 2021). For TO, the timeout value was set as 10 s, decided through statistics shown in Figure 1. For ERS, the "short-term memory" strategy was adopted. For DB, a few rules related to column stages were defined, such as the feed locations should not be larger than its total number of stages (stage number is counted starting from the condenser), and the number of stages on both sides of the wall in the dividing wall column should be the same (although can of course be different). For ML, a support vector machine (SVM) library developed by Chang and Lin (2011) was used with a seventh-degree polynomial, C = 5, and $\varepsilon = 0.1$. The settings for SVM were decided through trial-and-error (not shown). The first 20 iterations were used for training the SVM. In this work, the H-DWC model was developed in gPROMS Process (Process Systems Enterprise, 2021). The Genetic Algorithm (GA) was used as external optimization (black-box optimization) and was coded in C# by the authors. GA was constructed with 100 iterations (generations) and 100 individuals (chromosomes), tournament with four players for the parent selection, flat crossover, and discrete random mutation.

Each optimization task was repeated five times and the results are plotted in Figure 2, where the solid lines are the average values. The bands bounded by the minimum and maximum values are plotted together. Figures 2a and 2b show the number of infeasible simulations and the number of accumulated infeasible simulations in each iteration, respectively. It can be seen that ParC-ML has the most infeasible simulations, and this is because the training of SVM is biased due to the fact that most of the simulations in the first 20 iterations are infeasible. ParC, ParC-ERS, and ParC-TO show similar trends, especially ParC and ParC-ERS. The finding is valid as for ERS, this function does not affect the operation of the GA, and for TO, a slightly larger deviation from ParC is also expected because there is still a chance that feasible simulations may be timed out and be considered as infeasible simulations, which may slightly reduce the performance of GA (considering the number of feasible and on-spec simulations). Considering ParC-DB and ALL, these have similar trends and the lowest number of infeasible simulations in each iteration. Compared with ParC, the total number of infeasible simulations (shown in Table 1) of ParC-DB and ALL drops from 4313 to 2596 (40 % reduction) and 3053 (29 % reduction), respectively, which indicates that DB is the key function to reduce the number of infeasible simulations. Considering the number of on-spec simulations shown in Figures 2c and 2d, it is not surprising that ParC-DB and ALL have higher number of on-spec simulations as many unreasonable (thus off-spec) designs are avoided by DB, so a higher chance of on-spec designs would be expected. Moreover, the use of DB balances the number of feasible and infeasible simulations in the first 20 iterations, thus the bias of ML in ALL is reduced compared to in ParC-ML.

To reflect the speed of the optimizations, three indicators are tabulated in Table 1, which are the convergence percentage (the number of successfully converged optimizations, i.e., the final optimal on-spec design divided by the total number of optimizations performed); quality of the fitness (e.g., the total annualized cost in the case studies); and optimization time. Starting with the convergence percentage, Table 1 shows that only ParC-DB and ALL could achieve 100 % convergence, which indicates that to obtain the same number of valid optimizations (i.e., converged optimizations), other approaches require more repeated optimizations leading to a longer total optimization time. Both ParC-DB and ALL yield designs with the lowest TACs (2.54 M \$ y^{-1}), indicating better optimization efficiency. Note that all optimizations stop after 100 iterations, which is the stopping criterion. Moving to the most direct indicator, optimization time, starting with ParC, the averaged optimization time is 1886 s. The time for each individual is recorded and the summation of all individual times is considered as the total time taken without parallel. The optimization time without parallel computing is 30,963 s, which indicates the optimization with parallel computing only is about 15 times faster. The table shows that ERS does not save CPU time as the elitism percentage is low (10 %), and repeated designs rarely appeared due to the applied Flat crossover method. Moreover, ParC-DB has minor improvement due to the lower infeasible rate. For the other three approaches, ParC-TO,

Table 1. Key performance indicators of the different speed up approaches. The values reported are the mean values obtained from five optimizations unless stated otherwise.

	ParC	ParC-ERS	ParC-TO	ParC-DB	ParC-ML	ALL
No. infeasible solutions	4314	4134	4902	2596	6086	3053
No. on-spec solutions	704	1090	321	3075	259	3225
Optimization time (s)	1885	1939	1397	1756	1452	1033
Convg. percentage (%)	60	80	60	100	40	100
TAC* ($M\,\$\,y^{-1}$)	2.65	2.63	2.80	2.54	2.67	2.54

* Considering only converged optimizations

ParC-ML, and ALL, they all show good time savings with about 26 %, 23 %, and 45 % reduction, respectively.

4. Conclusion

In conclusion, this work has compared five different approaches to speed up stochastic optimization. By considering the overall performance, the dynamic bound function is clearly the most effective function to ensure a successful optimization for a complex chemical process. Using a timeout function or machine learning has no clear improvement on the quality of the optimization results, but these functions show remarkable reduction in the optimization time. The escaping repeated simulation function shows no clear effect on the optimization speed. By applying all approaches together, the optimization achieves the best design, the same as for parallel computing with dynamic bound, but with the least time. In future work, a more comprehensive comparison will be performed with other stochastic optimization methods such as Particle Swarm Optimization and Simulated Annealing.

References

M. Cavazzuti, 2013. Optimization Methods. Vol. 53. Springer Berlin Heidelberg, Berlin, Heidelberg.

C.-C. Chang, C.-J. Lin, 2011. LIBSVM: A Library for Support Vector Machines. ACM Transactions on Intelligent Systems and Technology 2 (3), 1–27.

D. N. Chia, F. Duanmu, E. Sorensen, 2021. Optimal Design of Distillation Columns Using a Combined Optimisation Approach. In: M. Turkay, R. Gani (Eds.), 31st European Symposium on Computer Aided Process Engineering. Elsevier B.V., pp. 153–158.

D. N. Chia, F. Duanmu, E. Sorensen, 2022. Optimal Design of Hybrid Dividing Wall Columns for Azeotropic Separations. In: The 12th International Conference Distillation & Absorption, Toulouse, September 2022..

Y. J. Gong, W. N. Chen, Z. H. Zhan, J. Zhang, Y. Li, Q. Zhang, J. J. Li, 2015. Distributed evolutionary algorithms and their models: A survey of the state-of-the-art. Applied Soft Computing Journal 34 (2013), 286–300.

D. Ibrahim, M. Jobson, J. Li, G. Guillén-Gosálbez, 2017. Surrogate Models combined with a Support Vector Machine for the Optimized Design of a Crude Oil Distillation Unit using Genetic Algorithms. In: Computer Aided Chemical Engineering. Vol. 40. pp. 481–486.

T. Mathew, K. C. Sekaran, J. Jose, 2014. Study and analysis of various task scheduling algorithms in the cloud computing environment. In: 2014 International Conference on Advances in Computing, Communications and Informatics (ICACCI). IEEE, pp. 658–664.

Process Systems Enterprise, 2021. gPROMS Process version 2.2.

Antonis Kokossis, Michael C. Georgiadis, Efstratios N. Pistikopoulos (Eds.)
PROCEEDINGS OF THE 33rd European Symposium on Computer Aided Process Engineering (ESCAPE33), June 18-21, 2023, Athens, Greece
 http://dx.doi.org/10.1016/B978-0-443-15274-0.50039-1

Design of a CO_2 capture plant: A sustainable approach using deep eutectic solvents

Adrián Martínez-Lomovskoi,[a] Ana-Gabriela Romero-García,[a b] Eduardo Sánchez-Ramírez,[a] Juan-Gabriel Segovia-Hernández, [a] Cataldo De Blasio [b]

[a]*Departamento de Ingeniería Química, Universidad de Guanajuato, Noria Alta s/n, Guanajuato, Gto., 36050, México.*
[b]*Åbo Akademi University, Faculty of Science and Engineering, Laboratory of Energy Technology, Rantakatu 2, Vaasa, 65100, Finland.*

Abstract

In this work, a new design for post-combustion capture (PCC) of CO_2 is introduced, using green deep eutectic solvent Choline chloride/Urea (1:2). A multi-objective optimization using meta-heuristic algorithm differential evolution and tabu list (MODE-TL) considering environmental and economic **objectives** is preformed to obtain a sustainable process design. The proposed process was analyzed for treating flue gases from combustion of the most common fuels used in power plants. The process performs with a lower cost and environmental impact when treating flue gases from the combustion of coal. Through a simulation in Aspen Plus, it is shown that using ChCl/Urea (1:2) it's possible to obtain a process capable of recovering more than 95% of the CO_2 contained in the combustion gases with a purity of 95% molar. A sustainable capture processes according to the objectives of green chemical engineering and circular economy is developed.
Keywords: CO_2 capture, MEA, deep eutectic solvent

1. Introduction

The United Nation's 2030 Agenda for Sustainable Development has called for the taking urgent action to combat climate change and its impacts, mainly through Goals 7, 9, 12 and 13. Citing examples of the efforts companies can make to contribute by decarbonizing their operations and supply chains. Seeking to reduce the carbon footprint of their products, services and processes, and setting ambitious emissions reductions targets in line with climate science. As well as scaling up investment in the development of innovative low-carbon products and services. Post-combustion carbon capture (PCC) process is one of the most promising technologies due to its ability to treat high CO_2 concentration streams from emission points from fossil fuels or biomass in industrial facilities. Moreover it can be adapted as a complementary process in existing thermal power plants. Monoethanolamine (MEA) is conventionally used as solvent because of its high reactivity with CO_2 and relatively low cost. This technology has severe disadvantages including a high rate of corrosion of process equipment and degradation of amines due to the presence of SO_2, NO_2 and O_2 in the combustion gases (Luis, 2015). In addition, MEA is considered toxic, so its implementation entails a high environmental impact.

In recent years, research has focus in a new category of green solvents called Deep Eutectic Solvents (DES) with selective absorption capacity towards CO_2. DESs have the characteristics of low cost, low to no toxicity, and good biodegradation. In recent years the properties of several DES have been reported in literature. Among them, the eutectic combination of choline chloride and urea in a 1:2 (ChCl/Urea (1:2)) molar proportion has shown the highest absorption rate for CO_2 (García et al., 2015). This property, along with its low vapor pressure and high thermal stability, has made it a candidate for replacing MEA in CO_2 removal processes.

DESs industrial applications have been developed in recent years. Ma et al. modeled, simulated and evaluated the upgrading of biogas by removing CO_2 with aqueous ChCl/Urea (1:2), demonstrating that this solvent was suitable for this process. Using choline chloride/urea (1:2) as the absorbing agent, Luo et al. (2021) designed and simulated an industrial-scale PCC process, to treat flue gases from coal combustion. Evaluation of life cycle environmental sustainability showed that the CO_2 capture with DES has significantly lower values compared to MEA. highlighting the advantage of lower environmental impact. However, the application of DES in an optimization framework considering sustainable targets, is not yet developed. In other words, no study has been conducted to analyze the impact of post-combustion absorption process design variables and their effect on the sustainability of the process, using ChCl/Urea (1:2) as solvent. Furthermore, the capacity of the DES to treat flue gasses with different CO_2 concentrations needs to be studied, as traditional power plants require a wide operational range in terms of the fuel used for energy generation, and the associated CO_2 concentration.

In this work, a novel PCC process using green solvent ChCl/Urea (1:2) is designed and optimized for the first time under a sustainability scheme, using Aspen Plus. The impact of the design variables on environmental and economic metrics is analyzed. Furthermore, the plant's capacity for treating flue gases with different CO_2 concentrations is studied through four case studies, considering the use of Natural Gas (NG), Associated Gas (AG), Biogas (BG) and Coal.

2. Study Case

The study case presented in this work evaluates the scenario of a power plant coupled to a carbon capture plant using DES as solvent (Fig. 1). The power plant consists in a traditional steam-generating boiler applied in a combined-cycle configuration. A fuel feed flow of 1000 kmol/h is considered, for natural gas, biogas, and associated gas. For coal, a flue gas flow of the same order of magnitude as those obtained for the combustion of the gasses is considered. The combustion process simulation is based in the simplified model proposed by Luyben (2013). The Peng-Robinson (PR) thermodynamic model was used to calculate physical properties, and an Aspen RGibbs chemical-equilibrium reactor model is used for the combustor. The PCC process consists of an absorber/desorber system with two intermediate flash tanks. The RADFRAC module is used for the absorber and desorber columns, considering an equilibrium-based physical absorption. ChCl/Urea was modeled in Aspen Plus using a group contribution method, and critical and temperature dependent properties, as well as Henry constants for the gaseous components in the solvent were taken from experimental data. The NRTL thermodynamic model was used to calculate the activity coefficient in the liquid phase, for ChCl/Urea + water. PR thermodynamic model was used to calculate fugacity coefficient for the gaseous components, with parameters taken from Aspen databank.

The multi-objective optimization of the process modeled in Aspen Plus represents a non-linear and multivariable problem, where the objective function used as optimization criterion is generally non-convex with several local optimums. Stochastic global optimization methods, such as Genetic Algorithms, have been proven to be robust and can handle both MESH equations and phase equilibrium calculations with complete models. A multi-objective meta-heuristic optimization algorithm differential evolution and tabu-list (MODE-TL) is used to obtain an optimal design of the PCC process. This algorithm allows the comparison of multiple solutions in terms of the objective functions, described after the economic and environmental metrics considered. For the economic indices, total annual costs (TAC) were calculated with the module costing technique outlined by Turton *et al.* (2003). Return on Investment (ROI) was calculated according to Jiménez-Gutiérrez. Eco-Indicator 99, as a measure of environmental impact, was calculated according to the methodology developed by Goedkoop *et al.* (2001). Total Solvent Recovery Energy (TSRE) and Green House Gas Emissions (GHGE) environmental indices were also considered, according to the definitions proposed by Jiménez-Gonzláez *et al.* (2012). A general mathematical expression for the objective function and its respective decision variables involved in the optimization procedure are shown in Equation (1).

$$\min[TAC, -ROI, EI99, TSRE, GHGE] = f(F_{air}, x_{H_2O}, F_{DES}, T_{DES}, N_{abs}, N_{PS1}, N_{PS2}, P_{abs}, P_{F1}, \Psi_{F1}, P_{F2}, \Psi_{F2}, RR, D, N_{des}, N_{feed}, P_{des}) \quad (1)$$

$$s.t. \; y_{i,f} \geq x_{i,f}, and \; w_{i,f} \geq u_{i,f}$$

Where the solutions are restricted to satisfy the recovery of 95% of the CO_2 produced during the combustion and also to achieve a purity of 95% mol of CO_2. Where $y_{i,f}$represents the CO_2 recovered in the desorber column and $x_{i,f}$ represents the 95% of the CO_2 produced during the combustion. As well, $w_{i,f}$ represents the purity achieved at the desorber column and $u_{i,f}$ represents the purity expected of at least 95% mol of CO_2.

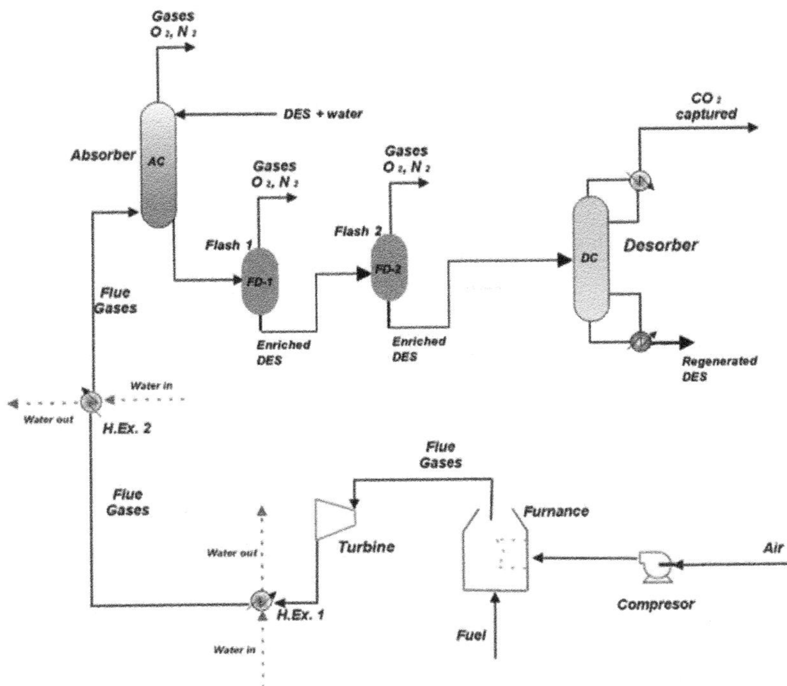

Figure 1. Process diagram of the combustion and PCC proposed scheme.

The decision variables and operation range considered for the optimization are described in Table 1. The optimization method was implemented using a hybrid platform that interconnects the Aspen Plus simulation with Excel through Visual Basic. The variable design vectors generated by the algorithm are fed to the process model, where the rigorous simulation is implemented. The outputs of the process model (flows, thermal loads, etc.) are fed back to Excel, where MODE-TL evaluates the objective functions and proposes new values of input vectors, according to the evolutionary nature of the algorithm, the process begins to iterate. The parameters used for the MODE-TL method were: 120 individuals, 800 maximum number of generations, taboo list of 50% of individuals, taboo radius of $1x10^{-4}$, and 0.9 and 0.3 crossover probability and mutation factor, based on previously reported works in literature for optimization of multicomponent separation schemes using the RADFRAC module.

Table 1. Decision variables and search intervals

Variable Name	Variable Type	Symbol	Range or Value
Feed air molar flow rate, kmol h^{-1}	Continuous	F_{air}	10025 – 35000
H_2O:ChCl/Urea molar proportion	Continuous	x_{H_2O}	1.6 – 5.5
DES molar flow rate, kmol h^{-1}	Continuous	F_{DES}	5000 – 120000
Absorber stages	Discrete	N_{abs}	20 – 100
1st packing section ending stage	Discrete	N_{PS1}	2 - $(N_{abs} - 1)$
2nd packing section starting stage	Discrete	N_{PS2}	$(N_{abs} + 1)$ - N_{abs}
Temperature of solvent feed, °C	Continuous	T_{DES}	40 - 60
Absorber top pressure, bar	Continuous	P_{abs}	1.1 – 14.0
Flash 1 pressure, bar	Continuous	P_{f1}	1.1 – 6.0
Flash 1 vapor fraction	Continuous	Ψ_{F1}	0 – 1
Flash 2 pressure, bar	Continuous	P_{f2}	1.01 – 6.0
Flash 2 vapor fraction	Continuous	Ψ_{F2}	0 – 1
Desorber reflux ratio	Continuous	RR	0.1 - 2.5
Distillate rate, kmol h^{-1}	Continuous	D	1378 – 1500
Desorber stages	Discrete	N_{des}	4 – 30
Desorber feed stage	Discrete	N_{feed}	2 - N_{des}
Desorber top pressure, bar	Continuous	P_{des}	1.1 - 14.0

3. Results

This section provides the multi-objective optimization results for the PCC process. For all designs, CO_2 recovery constraint is set at 95%, and 95% mol purity for the CO_2 at the distillate of the DC.

3.1. Pareto fronts

Two dimensional pareto diagrams obtained for the coal case study are presented in Fig. 2. The form of the pareto fronts for EI99 and TAC exhibits a trend of competing objective functions. This behavior indicates that the selection of a design with the lowest EI99 causes the TAC to increase, hence the solutions that offer the best trade-offs between the two objectives are those located in the curve zone of the Pareto chart. The pareto front for TSRE and TAC indicates that a design selection with the lowest TSRE causes TAC to diminish, therefore, designs that minimize both objectives are found in the lower left corner. The solvent recovery energy has a direct impact in TAC, given that the high use of vapor in the PCC process represents 70-80% of all annual costs. The pareto front for EI99 and GHGE exhibits a competing objective function trend, similar to EI99 vs TAC, where the best trade-off between objectives can be found in the curve area of the chart.

In terms of the impact of some design variables on the indices considered, a trend between Water:ChCl/Urea proportion and TAC was found. A higher water to ChCl/Urea proportion diminishes CO_2 absorption capacity, increasing the need for solvent flow. This causes process equipment topology to increase, as well as increasing TSRE requirements. As a consequence, GHGE also increases, and from the pareto chart trends in Fig. 2 EI99 decreases. Higher ChCl/Urea concentration will increase environmental impact and pumping costs considerably due to a higher viscosity of a poorly diluted ChCl/Urea. As well, the reflux ratio is directly related to the energy requirements. A higher reflux ratio causes energy requirements for the capture to increase.

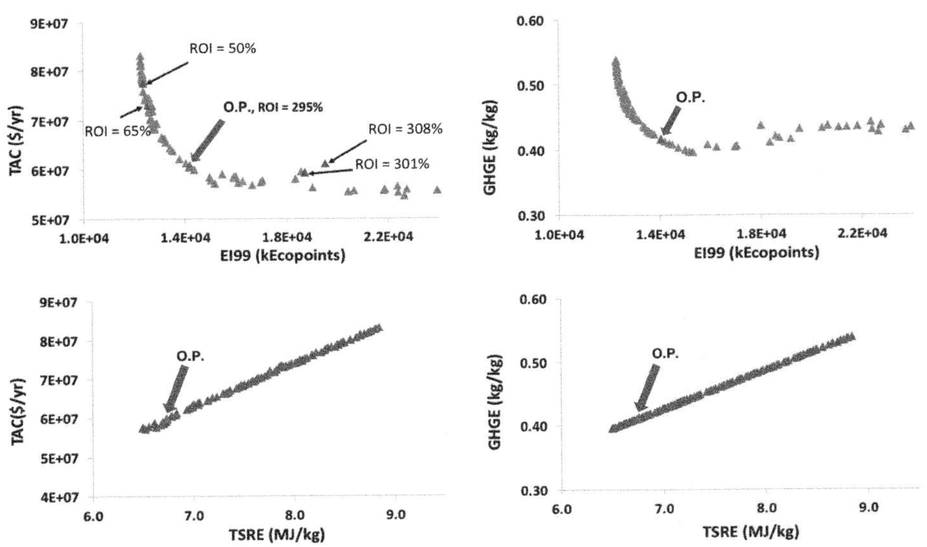

Figure 2. Pareto fronts for EI99 vs TAC, EI99 vs GHG, TSRE vs TAC and TSRE vs GHGE.

3.2. Optimal design performance indices

The optimal point (O.P.) was selected from the pareto fronts, considering a compromise between minimizing the objective functions. In the pareto front for EI99 and TAC, the highest and lowest ROI values found for the optimal designs were highlighted, along with the ROI value for the O.P. To showcase that a good performance in terms of this objective was also taken into consideration when searching for the optimal design.

The optimization results reveal that the Coal process provides the lowest energy usage per unit of CO_2 captured, with a TSRE and GHGE which is 12.3%, 27.7% and

31.7% lower than AG, NG and BG respectively. The energy savings are also reflected in the value of TAC for Coal (8.2%, 15.5% and 16.2% lower per tonne of CO_2 captured, compared with AG, NG and BG respectively). In terms of environmental impact, the Coal process presents a value that is 20.1%, 15.7% and 30.2% lower per tonne of CO_2 captured, as compared with AG, NG and BG respectively. The coal PCC process exhibits the best performance considering all environmental and economic indices simultaneously, although the highest ROI is obtained when AG is used as fuel.
The proposed scheme (Fig. 1) for solvent regeneration using two flash drums and a desorber column, allows the separation of CO_2 from the enriched solvent to occur at pressures between 1 and 5 bar, avoiding a costly vacuum pressure operation.

4. Conclusions

The simulation of the PCC process shows that the CO_2 capture of different flue gases is possible by use of green solvent ChCl/Urea (1:2).
Multi-objective optimization considered simultaneously economic and environmental objectives and resulted in optimal solutions according to sustainable design for the PCC process. Through the MODETL optimization method, design and operating conditions were found as to avoid vacuum pressure operation at the solvent regeneration steps while using a considerably lower absorption pressure (13.9 bar) compared to the PCC process proposed by Luo et al (45 bar).
The optimization results reveal that the Coal process provides the lowest energy use compared, a lower TAC and diminished environmental impact per tonne of CO2 captured, compared with other PCC processes. The highest ROI was obtained when using AG as fuel.

References

UN General Assembly, Transforming our world: the 2030 Agenda for Sustainable Development, 2015, A/RES/70/1, available at: https://www.refworld.org/docid/57b6e3e44.html [accessed 22 November 2022].

H. Eldardiry, E. Habib, Carbon capture and sequestration in power generation: review of impacts and opportunities for water sustainability. Energ Sustain Soc 8, 6 (2018).

P. Luis, Use of monoethanolamine (MEA) for CO_2 capture in a global scenario: Consequences and alternatives, Desalination, Volume 380, 2016, Pages 93-99.

G. García, S. Aparicio, R. Ullah, and M. Atilhan, Deep Eutectic Solvents: Physicochemical Properties and Gas Separation Applications, Energy Fuels 2015, 29, 4, 2616–2644

Ma, C.; Xie, Y.; Ji, X.; Liu, C.; Lu, X. Modeling, simulation and evaluation of biogas upgrading using aqueous choline chloride/urea. Appl. Energy 2018, 229, 1269−1283.

F. Luo, X. Liu, S. Chen, Y. Song, X. Yi, C. Xue, L. Sun, and J. Li., Comprehensive Evaluation of a Deep Eutectic Solvent Based CO_2 Capture Process through Experiment and Simulation, ACS Sustainable Chemistry & Engineering 2021 9 (30), 10250-10265.

R. Turton, R. C. Bailie, W. B. Whiting, and J. A. Shaeiwitz, Analysis, Synthesis, and Design of Chemical Processes, Second Edition, PrenticeHall, 2003.

M. Goedkoop, R. Spriensma, 2001, The Eco-Indicator 99: A Damage Oriented Method for Life Cycle Impact Assessment.

C. Jiménez-Gonzalez, D. J. C. Constable & C. S. Constable. Chem. Soc. Rev., 2012,41, 1485-1498

A.J. Gutiérrez, Diseño de procesos en ingeniería química, Reverte, 2021

W. L Luyben, Compressor heuristics for conceptual process design. Ind. Eng. Chem. Res. 2011, 50, 13984−13989.

J. R. Couper, W. R. Penney, J. R. Fair, S. M. Walas, 2005. Chemical Process Equipment – Selection and Design. Gulf Professional Publishing.

Antonis Kokossis, Michael C. Georgiadis, Efstratios N. Pistikopoulos (Eds.)
PROCEEDINGS OF THE 33rd European Symposium on Computer Aided Process Engineering (ESCAPE33), June 18-21, 2023, Athens, Greece
 http://dx.doi.org/10.1016/B978-0-443-15274-0.50040-8

Optimal Retrofitting of Conventional Oil Refinery into Sustainable Bio-refinery under Uncertainty

Lifeng Zhang,[a] Ana Inés Torres,[b] Bingzhen Chen,[a] Zhihong Yuan,[a*] Ignacio E. Grossmann[b*]

[a]Department of Chemical Engineering, Tsinghua University, Beijing, 100084, China
[b]Chemical Engineering Department, Carnegie Mellon UniVersity, Pittsburgh, Pennsylvania 15213, USA
Corresponding authors: zhihongyuan@mail.tsinghua.edu.cn (Z. Yuan), grossmann@cmu.edu (I.E. Grossmann)

Abstract

This paper focuses on a novel multistage retrofitting problem from conventional fossil-based refinery into biomass-based refinery with the aim of sustainable development of fuels. Given a typical refinery and potential biomass- based technologies, the problem has an objective to integrate the latter into the former by making use of existing units in the new process(es). A mixed-integer linear programming model that considers a ten-year long retrofit planning is formulated. Furthermore, the deterministic problem is extended to a multi-stage stochastic programming problem under both endogenous yield uncertainty and exogenous demand uncertainty and solved via a Lagrangean decomposition algorithm. The results provide flexible design alternatives by determining the units that should be added or modified for the selected biomass-based technologies. Regarding uncertainty, different schemes with strategic and operations decisions are determined. The results show advantages in considering retrofitting problem with detailed operation constraints for each year.

Keywords: Refinery; Biomass; Retrofit; Uncertainty; Lagrangean decomposition.

1. Introduction

Significant growth in renewable energy, such as solar and biomass, has taken place over the last 30 years. As the core part of energy supply, the petroleum refinery industry is increasingly gaining importance for producing more sustainable energy instead of fossil fuels. Among these novel energy forms, biomass-based energy is a promising alternative for the refinery industry since the biomass can be used to produce hydrocarbons that are drop-in replacements for the ones obtained from crude oils. This provides opportunities to repurpose existing crude oil refineries into biomass refineries. Such a plan has been under consideration in both industry and academia for a long time and has recently seen renewed interest (see for example Exxon Mobil (Sanicola, 2021), and in academia (Ericson et al., 2019)). The evaluation of such design is often executed with Life Cycle Assessment (LCA) (Garcia-Nunez et al., 2016) under predefined flowsheet and preselected biomass-based technologies. However, such framework usually ignores the flexibility to adopt flowsheets according to expected development in market and industry. This paper proposes a mixed linear programming (MILP) modeling framework to solve the long-term retrofit problem, while considering detailed production constraints in each year. Uncertainty is also addressed by extending the proposed model with stochastic programming.

2. Problem Statement

2.1. Crude-based Refinery Demonstration

The flowsheet of a prototypical crude-based refinery is shown in Fig. 1. The ranges of hydrocarbons are labelled to represent the corresponding components, typically considered in each stream/processed by each equipment. For example, the crude oil, is first distillated in crude distillation unit (CDU) into liquified petroleum gas (LPG, composed of hydrocarbons C_3 to C_5), naphtha (NAP composed of hydrocarbons C_5 to C_{10}), etc. These streams are then further hydrotreated, cracked and reformed in secondary processing units, such as continuous reformer unit (CRU), fluid catalytic cracking (FCC) and others. The final products, LPG, gasoline (GASO), jet fuel (JET), diesel (DIESEL) and fuel oil (OIL) are obtained by blending intermediates to satisfy quality specifications.

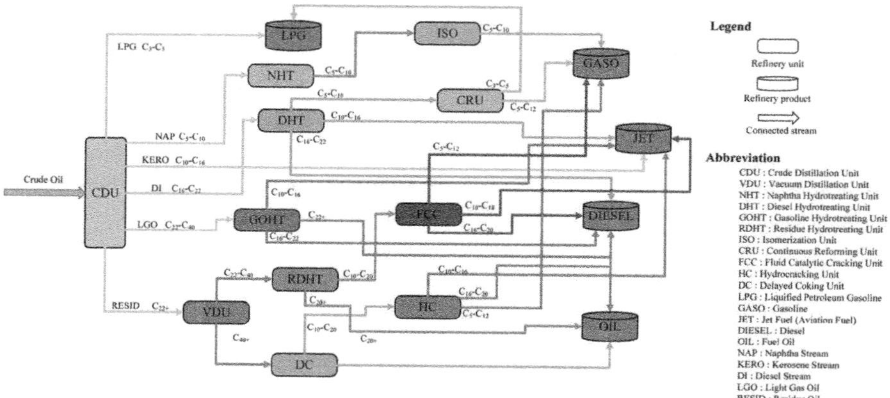

Fig. 1 Flowsheet of a given crude-based refinery. (Presented in Appendix)

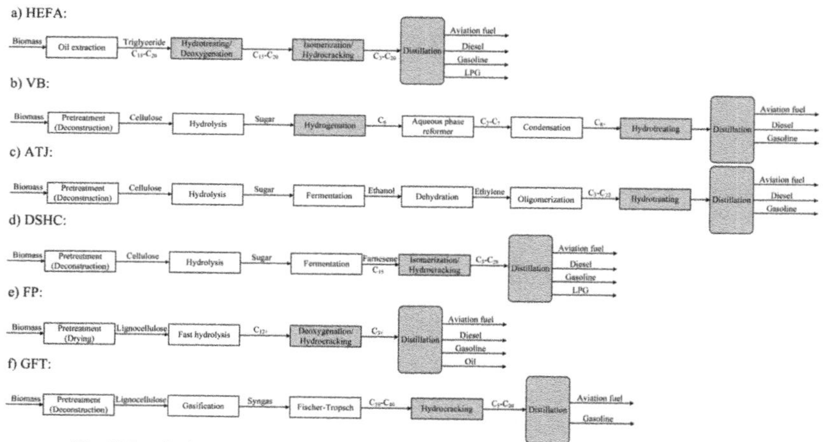

Fig. 2 Detailed process steps for potential biotechnologies. (Presented in Appendix)

2.2. Biomass-based Refinery Demonstration

To produce sustainable energy based on biomass, six biomass-based technologies are considered, namely, Hydroprocessed esters and fatty acids (HEFA), Virent's BioForming (VB), alcohol to jet (ATJ), direct sugar to hydrocarbon (DSHC), fast pyrolysis (FP), and gasification and Fischer-Tropsch (GFT) (Tanzil et al., 2021a, 2021b). Details of each technology path can be seen in Fig. 2 with illustration for every step and stream.

For example, the VB technology, which addresses the pine or stover as the feedstocks, first transforms the materials into cellulose. The cellulose is converted into sugar via hydrolysis. Then the sugar is hydrotreated into hydrocarbons and followed by aqueous phase reforming, condensation and hydrotreating. Finally, the biomass-based hydrocarbons are distilled to produce aviation fuel, gasoline and diesel. In the VB technology, the hydrotreating steps can be executed in a newly built unit or in an already existing refinery hydrotreating unit. We compare components of the streams entering each unit to find possible matches. In addition, the distillation step can also be done in the CDU instead of in new-installed units.

3. Mathematical Formulation

3.1. Deterministic Model

The goal is to minimize the total cost over ten years by subtracting sales income from the summation of material operation reconnection and new unit installation costs. For each year, the operation constraints on selection of materials and production are also considered. The problem is formulated as a MILP as follows and details for the mathematical formulation can be seen in Appendix.

$$\begin{aligned} & min\ z = a^T x + b^T y \\ & Ax + By \leq d \qquad\qquad \text{MILP} \\ & x \geq 0, y \in \{0,1\}^m \end{aligned}$$

3.2. Multistage Stochastic Programming Model

The problem is extended to a multistage stochastic programming (MSSP) under endogenous yield uncertainty and exogenous demand uncertainty (Apap and Grossmann, 2017). The scenario tree is formulated as Fig. 3, where each stage is formulated as an inner two-stage stochastic programing problem with endogenous ξ uncertainty and exogenous θ uncertainty (Lara et al., 2019). The decisions are divided into strategic decisions and operational decisions. The strategic decisions, including the selection of biomass-based technology, capacity expansion and unit substitution, are assumed only related to endogenous uncertainty. The operational decisions, such as type of crude oil, amount of feedstock and sale strategy for products in each year, are assumed related to demand uncertainty and are independent in each year.

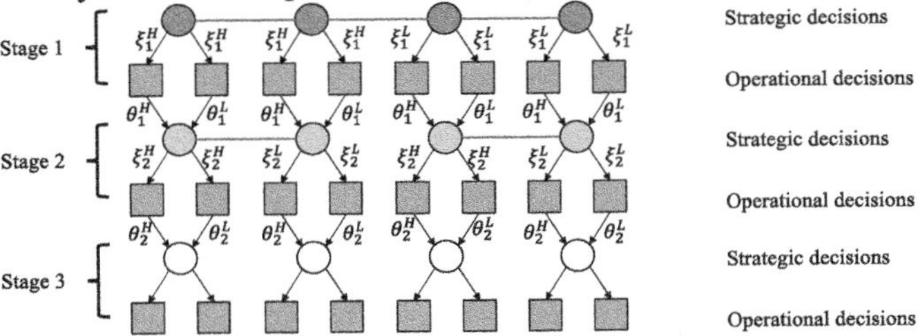

Fig. 3 Scenario tree illustration. (Presented in Appendix)

The mathematical formulation of the MSSP is given as follows. The set $i \in I$ denotes the outer scenarios corresponding to endogenous yield uncertainty and set $j \in J$ denotes the inner scenarios related to exogenous demand uncertainty appearing in each year. The problem is solved by reducing scenario pairs first, and then using a Lagrangean decomposition algorithm by dividing it into subproblems based on outer scenario pairs. Details on the modeling and algorithm can be found in Gupta and Grossmann (2014, 2011).

$$
\begin{aligned}
\min z = & \sum_{i \in I} p_i \left(\sum_{t \in T} c_{t,i} y_{t,i} + \sum_{j \in J} p_j d_{t,i,j} x_{t,i,j} \right) \\
s.t. \quad & \sum_{\tau \le t} a_{\tau,i} y_{\tau,i} \le b_{t,i} \quad \forall t \in T, i \in I \\
& e_{t,i} y_{t,i} + f_{t,i,j} x_{t,i,j} \le g_{t,i,j} \\
& y_{t,i} = y_{t,i'} \quad \forall t \in \{T_1\}, i, i' \in I \\
& Z_{t,i,i'} \Leftrightarrow F(y_{t1,i}, y_{t2,i} \cdots y_{t-1,i}) \quad \forall t \in T \backslash \{T_1\}, i, i' \in I \\
& \begin{bmatrix} Z_{t,i,i'} \\ y_{t,i} = y_{t,i'} \end{bmatrix} \vee \left[\neg Z_{t,i,i'} \right] \quad \forall t \in T \backslash \{T_1\}, i, i' \in I \\
& y_{t,i} \in \{0,1\} \ \forall t \in T, i \in I \\
& x_{t,i,j} \in R \quad \forall t \in T, i \in I, j \in J
\end{aligned} \qquad \text{MSSP}
$$

4. Results and Discussion

4.1. Computational Performance

Both the deterministic problem and the MSSP problem are solved in Gurobi 9.5.2 via GAMS 41.1.0 on Windows 11 with Intel(R) Core(TM) i7-9700 CPU @ 3.00GHz and 16GB memory. The model statistics and computational results are shown in Tables. 1 - 2 and Fig. 4.

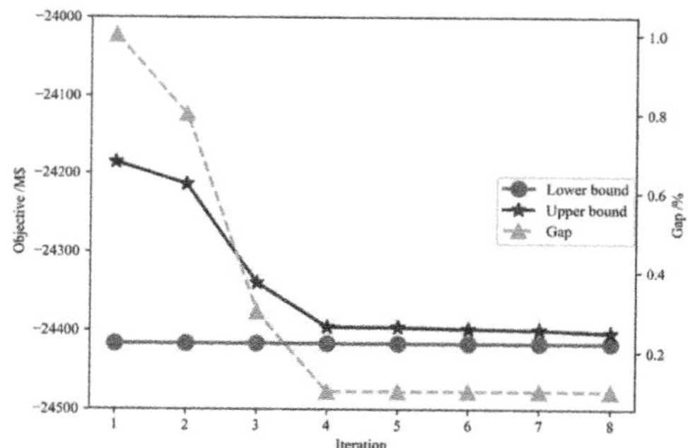

Fig. 4 Computational performance of Lagrangean decomposition algorithm. (Presented in Appendix)

Table.1 Model statistics.

Problem	# of continuous variable	# of binary variable	# of constraints
Deterministic model	14,197	7,146	13,954
MSSP	11,576,705	459,264	15,808,450

Table.2 Optimization results.

Problem	Solver	CPU time /s	Gap /%	Objective /M$
Deterministic model	CPLEX	138.5	0.01	-1020.85
	Gurobi	16.6	0.01	-1020.85
MSSP	CPLEX	7200.0	/	/
	Gurobi	7200.0	/	/
	Lagrangean decomposition	4511.3	0.1	-24402.6

The deterministic model is solved to optimality quite fast. For the MSSP model, 64 outer scenarios are included related to the combinations of endogenous yield uncertainty and in each year, 27 scenarios related to exogenous demand uncertainty are considered, leading to 270 inner scenarios in total. The scale of the MSSP increases exponentially as shown in Table. 1 and commercial solvers fail to give a feasible solution within 2 h of

CPU time; the Lagrangean decomposition algorithm, solves the problem to an optimality gap of 0.1% in ~ 75 min.

4.2. Results of Deterministic Model

From the deterministic model, HEFA, VB, FP, GFT are selected to produce the desired bioproducts. The retrofitted flowsheets in year 1, year 2, year 5 and year 9 are displayed in Fig. 5. For a better illustration, high qualitied figures are presented in Appendix.

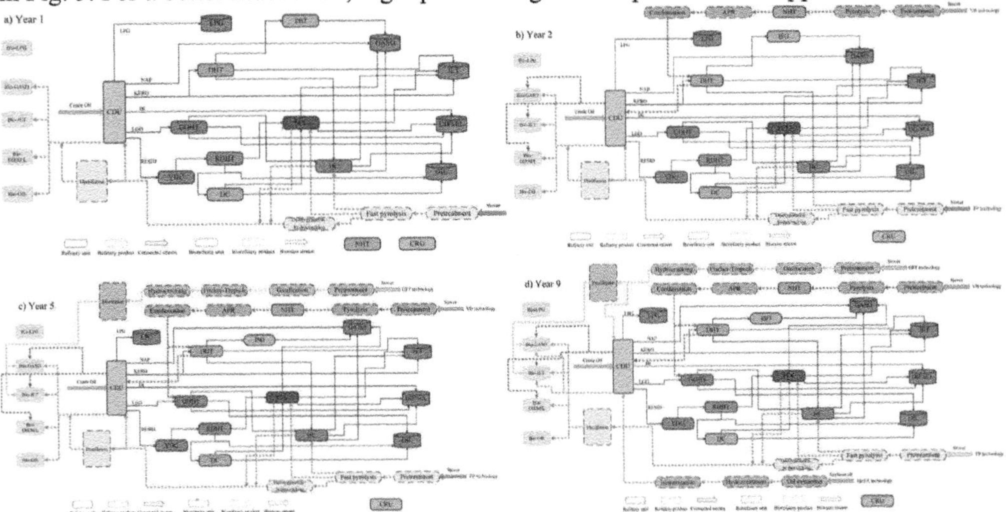

Fig. 5 Retrofitted flowsheet of the biorefinery over several years. (Presented in Appendix)

As can be seen VB is built in the first year to process corn stover. For each step in VB, a new operation unit is built. While in the hydrocracking step, the biomass-based hydrocarbon stream can be directly fed into the FCC and HC units for deep cracking. Although a distillation unit is built for VB, the CDU is still used to obtain bioproducts with the same time until year 7. The NHT and CRU units are removed from the refinery in the first year, and NHT is used in VB technology as the hydrotreating step as well as DHT unit in year 2. Later, the GOHT is also used to process the biomass-based hydrocarbons from VB as the final hydrotreating step in addition to oil-based stream. GFT is implemented in year 5 with all the units installed. In year 9, the HEFA technology is also installed and the products are obtained via the CDU. Note that the continuous reforming unit (CRU) is removed and never used during the time horizon.

Table.3 Statistics of result from MSSP

Bio-technology	Scenarios selecting bio-technology	Bio-technology	Scenarios selecting bio-technology
HEFA	28	DSHC	0
VB	64	FP	64
ATJ	0	GFT	36

4.3. Results of MSSP

With the MSSP model, the biomass-based technology selection statistics are presented in Table. 3. For a total of 64 scenarios, VB and FP are all selected as the main process to produce biomass-based hydrocarbons, while ATJ and DSHC are not. Comparing to VB and FP, the higher expense and lower yield of ATJ and DSHC prevent their applications. To compensate for the production demands, GFT and HEFA are the backup biomass-

based technologies to install in later years. The GFT is preferred since it is selected in 36 scenarios while HEFA is selected in 28 scenarios.
To evaluate the effect of MSSP, the value of stochastic solution (VSS) is also calculated by solving the problem stage by stage iteratively (Zhang et al., 2018). The definition of VSS and $\overline{VSS}$ are presented as follows.

$$VSS = z^{DP} - z^{MSSP} \quad (1)$$

$$\overline{VSS} = (z^{DP} - z^{MSSP}) / z^{DP} \quad (2)$$

z^{DP} is the optimal solution from solving problem for each stage iteratively. z^{MSSP} is the solution from MSSP model. The VSS equals to 522 M$ and the $\overline{VSS}$ is 2.15% which reveal the benefit of implementing stochastic programming.

5. Conclusion

This paper addressed the retrofit problem from a conventional refinery to a biorefinery by incorporating new biomass-based technologies into an existing oil-based refinery. The problem is extended into a MSSP model under both endogenous and exogenous uncertainties. Each stage is divided into strategic and operational steps. The Lagrangean decomposition algorithm is used to solve such problem. The results show that the FP and VB technologies are preferred, and a cost saving of 522 M$ can be obtained with stochastic programming with the evaluation of VSS.

Appendix

The Appendix can be seen at https://cloud.tsinghua.edu.cn/f/9d6288101611467d9601/

References

Apap, R.M., Grossmann, I.E., 2017. Models and computational strategies for multistage stochastic programming under endogenous and exogenous uncertainties. Comput. Chem. Eng. 103, 233–274.

Ericson, S.J., Engel-Cox, J., Arent, D.J., 2019. Approaches for Integrating Renewable Energy Technologies in Oil and Gas Operations (No. NREL/TP-6A50-72842). National Renewable Energy Lab. (NREL), Golden, CO (United States).

Garcia-Nunez, J.A., Rodriguez, D.T., Fontanilla, C.A., Ramirez, N.E., Silva Lora, E.E., Frear, C.S., Stockle, C., Amonette, J., Garcia-Perez, M., 2016. Evaluation of alternatives for the evolution of palm oil mills into biorefineries. Biomass Bioenergy 95, 310–329.

Gupta, V., Grossmann, I.E., 2014. A new decomposition algorithm for multistage stochastic programs with endogenous uncertainties. Comput. Chem. Eng. 62, 62–79.

Gupta, V., Grossmann, I.E., 2011. Solution strategies for multistage stochastic programming with endogenous uncertainties. Comput. Chem. Eng. 35, 2235–2247.

Lara, C.L., Siirola, J.D., Grossmann, I.E., 2019. Electric power infrastructure planning under uncertainty: stochastic dual dynamic integer programming (SDDiP) and parallelization scheme. Optim. Eng. 21, 1243–1281.

Sanicola, L., 2021. EXCLUSIVE Exxon, Chevron look to make renewable fuels without costly refinery upgrades -sources. Available at https://www.reuters.com/world/middle-east/exclusive-exxon-chevron-look-make-renewable-fuels-without-costly-refinery-2021-08-12/

Tanzil, A.H., Brandt, K., Wolcott, M., Zhang, X., Garcia-Perez, M., 2021a. Strategic assessment of sustainable aviation fuel production technologies: Yield improvement and cost reduction opportunities. Biomass Bioenergy 145.

Tanzil, A.H., Brandt, K., Zhang, X., Wolcott, M., Stockle, C., Garcia-Perez, M., 2021b. Production of Sustainable Aviation Fuels in Petroleum Refineries: Evaluation of New Bio-Refinery Concepts. Front. Energy Res. 9.

Zhang, Q., Bremen, A.M., Grossmann, I.E., Pinto, J.M., 2018. Long-Term Electricity Procurement for Large Industrial Consumers under Uncertainty. Ind. Eng. Chem. Res. 57, 3333–3347.

Antonis Kokossis, Michael C. Georgiadis, Efstratios N. Pistikopoulos (Eds.)
PROCEEDINGS OF THE 33rd European Symposium on Computer Aided Process Engineering (ESCAPE33), June 18-21, 2023, Athens, Greece
 http://dx.doi.org/10.1016/B978-0-443-15274-0.50041-X

A Combined D-optimal and Estimability Model-Based Design of Experiments of a Batch Cooling Crystallization Process

Xuming Yuan[a], Brahim Benyahia[a,*]

[a] *Department of Chemical Engineering, Loughborough University, Epinal Way, Loughborough, LE11 3TU, United Kingdom*
B.Benyahia@lboro.ac.uk

Abstract

In this work, a systematic methodology is proposed to help develop model-based design of experiments to build robust and reliable mathematical model of a batch crystallization process. The cooling crystallization of paracetamol in water and propanol is used as the case study. The mathematical model consists in the mass balance and a set of population balance equations, involving primary and secondary nucleation, growth, agglomeration, breakage and dissolution kinetics. Firstly, a structural identifiability approach is used to investigate whether the model parameters can be determined uniquely with an idealized input-output behavior of the process. The approach is also critical to determine the minimum set of required observable outputs and help discriminate model candidates. A novel Model-Based Design of Experiments (MBDoE) is then proposed based on the combination of the D-optimality criterion and the estimability analysis. The objective is to reduce the uncertainties in the model parameters by enhancing the data information content and help maximize the estimability potential of all model parameters while reducing correlation amongst them. Moreover, a new operating strategy based on temperature cycling is used in a sequential design of experiment to maximize data information content from one single experiment while reducing the experimental burden and inherent wastes.

Keywords: Structural Identifiability, Estimability, Sensitivity Analysis, Model-Based Design of Experiment (MBDoE), Temperature Cycling, Batch Crystallization.

1. Introduction

Crystallization is an important separation and purification technique that is widely applied in the pharmaceutical industry. The successful development of crystallization technologies is a painstaking and costly process. The availability of a reliable mathematical models is critical at all development stages to help explore more effectively the design space and deliver robust and cost-effective designs, operation, and control procedures (Benyahia et al., 2021; Liu and Benyahia, 2022). Most importantly, mathematical models can help achieve built-in quality insurance (e.g., Quality-by-Design and Quality-by-Control). However, the prediction capabilities and inherent uncertainties of the models strongly depend on the experimental data obtained at early development stages, their information content, and the parameter estimation procedure.

To obtain the parameter values, parameter estimation is performed, where the difference between the experimental data and the prediction is minimized. However, an excellent matching between the collected data and the prediction does not necessarily indicate that all the model parameter estimates are accurate and precise. For example, it can be found that wide ranges of parameter values may still result in similar model predictions (Balsa-Canto et al., 2010). Another common issue is the high uncertainty in the model parameter estimates. Such issues originate from the lack of structural identifiability or estimability (practical identifiability) of the model, and they may lead to the high uncertainty in model predictions which consequently affect the final deigns and product quality.

To address these issues, extensive studies focusing on estimability have been carried out. In 2003, Yao et al. proposed the use of a sequential orthogonalization algorithm to rank the parameters according to their estimability potential and least correlation. The method became very popular over the last two decades and found many successful applications in various research areas (Benyahia et al. 2011 and 2013). Different estimability methods were also used to capture more effectively the minimum subset of the most estimable parameters using more effective cut-off procedures (Fysikopoulos et al., 2019). Another method based on the mean-squared-error (MSE) and transformed parameters was used by Kim and Lee (2019) to help identify the cut-off value more effectively. Nonetheless, structural identifiability, as a necessary condition of estimability, was usually neglected in previous studies. Hence, Balsa-Canto et al., 2010 performed structural identifiability analysis prior to the estimability analysis to confirm that all model parameters have unique estimates theoretically. Recently, model-based design of experiment (MBDoE) is gaining more attention and it has been implemented in optimal experimental design to obtain high-quality parameter estimates. For instance, Maheshwari et al., 2013 applied to a double feedback gene switching model a multi-objective optimal experimental design that maximized the D-optimality criterion and minimized the correlation between the parameters simultaneously.

In this work, a novel approach that combines the estimability and MBDoE is proposed as a systematic approach to build more predictive models and design more effective and information rich experiments. The aim of this study is to demonstrate the benefits of combining D-optimal design with the estimability to reduce the uncertainty in crystallization models. This work systematically performs structural identifiability analysis, estimability analysis and D-optimal design in a seeded batch cooling crystallization process of paracetamol (PCM), which includes primary and secondary nucleation, growth, agglomeration, breakage and dissolution. Several new experimental strategies are proposed in the MBDoE which includes temperature cycling with holds.

2. Methodology

2.1 Process Model

A seeded batch cooling crystallization of paracetamol (PCM) is conducted in a 4:1 water/propanol solvent system, and the model is based on the quadrature method of moments (QMOM). Due to the temperature cycling, two mathematical models are used (hybrid model) as described below to capture the dissolution of crystals. When the concentration of PCM is higher than the concentration at saturation, primary nucleation, secondary nucleation, growth, agglomeration and breakage are considered, while dissolution is suppressed. The population balance of the kth moment is given by

$$\frac{d\mu_k}{dt} = 0^k(k_{b1}S^{b_1} + k_{b2}S^{b_2}\mu_2^{j_2}) + kk_gS^g\sum_{i=1}^{N_q}W_iL_i^{k-1} + \frac{1}{2}K_a\sum_{i=1}^{N_q}\sum_{j=1}^{N_q}W_iW_j(L_i^3 + L_j^3)^{1+\frac{k}{3}} - \frac{1}{2}K_a\sum_{i=1}^{N_q}\sum_{j=1}^{N_q}W_iW_jL_i^k(L_i^3 + L_j^3) + 2^{1-\frac{k}{3}}K_b\sum_{i=1}^{N_q}W_iL_i^{\gamma+k} - K_b\sum_{i=1}^{N_q}W_iL_i^{\gamma+k} \quad (1)$$

where $N_q = 2$ is the order of quadrature approximation, W and L are the weights and abscissa obtained from quadrature approximation, and S is the absolute supersaturation:

$$S = C - C^* \tag{2}$$

C^* is the concentration at saturation captured by a polynomial function of temperature T:

$$C^* = p_0 + p_1 T + p_2 T^2 \tag{3}$$

where p_0, p_1 and p_2 are the known polynomial coefficients.

Let $\bar{d}$, ρ and k_v be the mean crystal size, the density and the shape factor of PCM crystals. The evolution of the PCM concentration and mean crystal size along with time are mathematically expressed as

$$\frac{dC}{dt} = -3\rho k_v k_g S^g \sum_{i=1}^{N_q} W_i L_i^2 \tag{4}$$

$$\frac{d\bar{d}}{dt} = \frac{1}{\mu_0}\frac{d\mu_0}{dt} - \frac{\mu_1}{\mu_0^2}\frac{d\mu_0}{dt} \tag{5}$$

The cooling/heating rate $\frac{dT}{dt} = R$ is the decision variables in the optimal experimental design problem. The evolution of concentration at saturation is given by

$$\frac{dC^*}{dt} = (p_1 + 2p_2 T)\frac{dT}{dt} \tag{6}$$

In summary, the vectors of state $\boldsymbol{x}(t)$, observables $\boldsymbol{y}(t)$, control variables $\boldsymbol{u}(t)$ and unknown model parameters $\boldsymbol{\theta}$ are given by

$$\boldsymbol{x}(t) = [\mu_0(t), \mu_1(t), \mu_2(t), \mu_3(t), C(t), \bar{d}(t), C^*(t), T(t)]^T \tag{7}$$

$$\boldsymbol{y}(t) = [C(t), \bar{d}(t), \mu_0(t)]^T \tag{8}$$

$$\boldsymbol{u}(t) = \frac{dT(t)}{dt} = R(t) \tag{9}$$

$$\boldsymbol{\theta} = [k_{b1}, b_1, k_{b2}, b_2, j_2, k_g, g, K_a, K_b, \gamma, k_{ds}, ds]^T \tag{10}$$

When the PCM concentration is lower than its concentration at saturation, dissolution is activated; primary/secondary nucleation and growth are suppressed, while agglomeration and breakage still occur. The population balance of the kth moment becomes:

$$\frac{d\mu_k}{dt} = k k_{ds}(-S)^{ds} \sum_{i=1}^{N_q} W_i L_i^{k-1} + \frac{1}{2} K_a \sum_{i=1}^{N_q} \sum_{j=1}^{N_q} W_i W_j (L_i^3 + L_j^3)^{1+\frac{k}{3}} - \frac{1}{2} K_a \sum_{i=1}^{N_q} \sum_{j=1}^{N_q} W_i W_j L_i^k (L_i^3 + L_j^3) + 2^{1-\frac{k}{3}} K_b \sum_{i=1}^{N_q} W_i L_i^{\gamma+k} - K_b \sum_{i=1}^{N_q} W_i L_i^{\gamma+k} \tag{11}$$

In this scenario, the evolution of the PCM concentration becomes

$$\frac{dC}{dt} = -3\rho k_v k_{ds}(-S)^{ds} \sum_{i=1}^{N_q} W_i L_i^2 \tag{12}$$

while the evolution of mean crystal size and concentration at saturation can still be expressed using equations (5) and (6).

2.2 Formulation of the MBDoE Optimization Problem

Before performing MBDoE, a structural identifiability analysis was conducted, and it showed that the model is at least structurally locally identifiable. Afterwards, four operating strategies were proposed to achieve the intended systematic MBDoE namely: linear cooling, piecewise linear cooling with continuity, temperature cycling, and temperature cycling with holds (Figure 1 (A)). For the sake of brevity, only the formulation of the optimization problem using temperature cycling with holds is presented here.

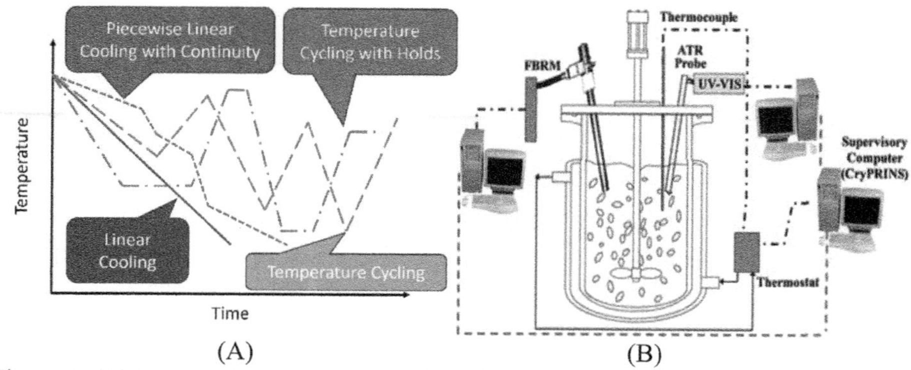

(A) (B)

Figure 1. (A) Proposed operating strategies using different temperature trajectories. (B) Experimental set up.

The mathematical formulation of the multiobjective MBDOE is given by

$$Obj1: \min_{R,\Delta t_R,t_s} \det(FIM^{-1})\,;\ Obj2: \max_{R,\Delta t_R,t_s} \min ||Z_j||$$

$$\begin{aligned}
&\text{s. t. } \dot{\boldsymbol{x}}(t) = f(t,\boldsymbol{x}(t;\boldsymbol{\theta});\boldsymbol{\theta}),\ \boldsymbol{x}(t_0) = \boldsymbol{x}_0(\boldsymbol{\theta})\\
&C1: -0.5 \le R_i \le 0 \quad (i = 1,3,5)\\
&C2: 0 \le R_i \le 1 \qquad (i = 2,4,6)\\
&C3: 0 \le \Delta t_{R,j} \qquad (j = 1,\dots,12)\\
&C4: \textstyle\sum_{j=1}^{12} \Delta t_{R,j} = 300\\
&C5: 21 \le 42 + \textstyle\sum_{i=1}^{6} R_i \Delta t_{R,2i-1} \le 42\\
&C6: 0 \le t_{s,1}\\
&C7: t_{s,10} \le 300\\
&C8: t_{s,k} - t_{s,k+1} \le 0\ (1 \le k \le 9)
\end{aligned} \tag{11}$$

where R denotes the cooling/heating rate, Δt_R represents the durations of different cooling/heating stages, and t_s refers to the sampling times.

The first criterion is the minimization of the D-optimality criterion, which aims at the minimization of the joint confidence regions of the model parameters; the second criterion maximizes the estimability score of the least estimable parameter (Max Min), so that its effect on the observables is enhanced, which may ensure higher quality parameter estimates.

3. Results & Discussion

Figure 2 shows a sample of results obtained with the multi-objective optimal experimental design. The Pareto Front in Figure 2 (A) shows the best MBDoE compromises. The star represents one selected Pareto solution that can be experimentally implemented. Figure 2 (B) shows the temperature trajectory of the selected Pareto optimal solution. The estimability analysis revealed that k_{b2} was the least estimable parameter, while γ was found to be the most estimable.

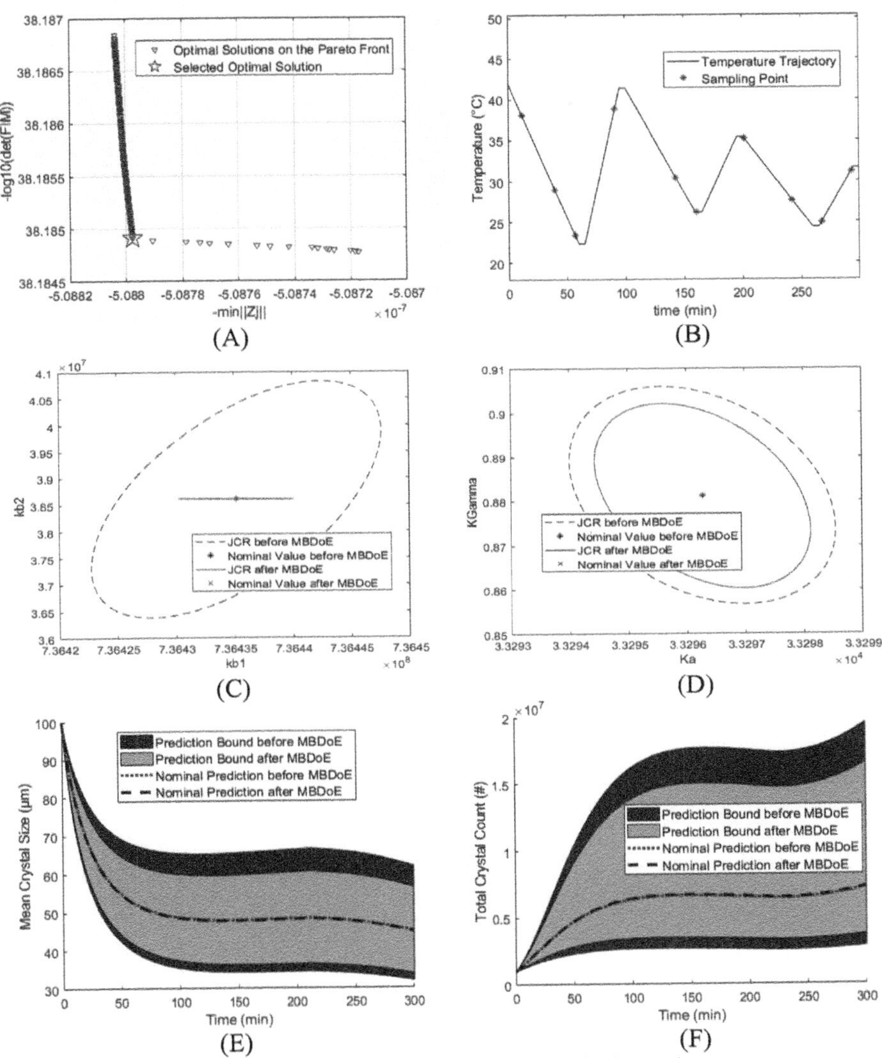

Figure 2. (A) Pareto Front of the optimal experimental design using temperature cycling and holds. (B) Optimal temperature operating trajectory. (C) 95% Joint confidence regions (JCR) of k_{b1} and k_{b2} before & after optimal experimental design. (D) 95% Joint confidence regions (JCR) OF K_a and γ before & after optimal experimental design. (E) Prediction bounds of mean crystal size before & after optimal experimental design. (F) Prediction bounds of total crystal count before & after optimal experimental design.

A basic single temperature cycle experiment was used as a reference case to show the benefits of the proposed MBDoE. Figure 2 (C) clearly shows that the joint confidence interval of k_{b2} and k_{b1} are significantly decreased and so is their correlation. The results shown in Figure 2 (D) demonstrate a decrease of the uncertainty in the most estimable model parameter γ. The decrease in parameter uncertainties resulted in more reliable predictions, as shown in Figure 2 (E) and (F), where the nominal predictions refer to the predictions using the nominal parameter estimates before and after applying the optimal experimental design. Several additional operating strategies were also investigated in this study as outlined in Figure 1 (A) (the results are not shown here for the sake of brevity). However, temperature cycling with holds resulted in the best parameter quality and the lowest uncertainties in model prediction. Thus, this operating strategy has the potential to be implemented in the development of MBDoE for similar crystallization systems.

4. Conclusions

A series of novel MBDoE strategies based on different operating trajectories were proposed and implemented to a batch crystallization process. Both D-optimal design of experiments and the estimability approaches were implemented to generate information-rich experiments and help reduce parameter uncertainties while increasing the estimability potential of the parameters. The combination of both strategies (multi-objective MBDoE) offered more optimal operating options and demonstrated the parameter estimates with reduced prediction uncertainties. The introduction of temperature cycling with holds was shown to be the best operating strategy in this study, which indicates that it has the potential to be a reliable alternative for the development and implementation of systematic MBDoE in crystallization.

Acknowledgements: This work was funded by the EPSRC (EP/V062077/1) Digital Medicines Manufacturing Research Centre.

References

Balsa-Canto, E., Alonso, A.A. and Banga, J.R. (2010). An iterative identification procedure for dynamic modeling of biochemical networks. *BMC Systems Biology*, 4(1).

Benyahia, B., Latifi, M.A., Fonteix, C. and Pla, F. (2011). Modeling of a batch emulsion copolymerization reactor in the presence of a chain transfer agent: Estimability analysis, parameters identification and experimental validation. *Computer Aided Chemical Engineering*, 29, pp.126–130.

Benyahia, B., Latifi, M.A., Fonteix, C. and Pla, F. (2013). Emulsion copolymerization of styrene and butyl acrylate in the presence of a chain transfer agent. Part 2: Parameters estimability and confidence regions. *Chemical Engineering Science*, 90, pp.110–118.

Benyahia, B., Anandan, P.D. and Rielly, C. (2021). Robust Model-Based Reinforcement Learning Control of a Batch Crystallization Process. In: *2021 9th International Conference on Systems and Control (ICSC)*, pp.89–94.

Fysikopoulos, D., Benyahia, B., Borsos, A., Nagy, Z.K. and Rielly, C.D. (2019). A framework for model reliability and estimability analysis of crystallization processes with multi-impurity multi-dimensional population balance models. *Computers & Chemical Engineering*, 122, pp.275–292.

Liu, J. and Benyahia, B. (2022). Optimal start-up strategies of a combined cooling and antisolvent multistage continuous crystallization process. *Computers & Chemical Engineering*, 159, p.107671.

Kim, B. and Lee, J.H. (2019). Parameter subset selection and biased estimation for a class of ill-conditioned estimation problems. *Journal of Process Control*, 81, pp.65–75.

Maheshwari, V., Kandpal, M. and Samavedham, L. (2013). Multi-Criteria Optimization Based Experimental Design for Parameter Estimation of a Double Feedback Gene Switching Model. *IFAC Proceedings Volumes*, 46(31), pp.333–337.

Yao, K.Z., Shaw, B.M., Kou, B., McAuley, K.B. and Bacon, D.W. (2003). Modelling Ethylene/Butene Copolymerization with Multi-site Catalysts: Parameter Estimability and Experimental Design. *Polymer Reaction Engineering*, 11(3), pp.563–588.

Antonis Kokossis, Michael C. Georgiadis, Efstratios N. Pistikopoulos (Eds.)
PROCEEDINGS OF THE 33rd European Symposium on Computer Aided Process Engineering (ESCAPE33), June 18-21, 2023, Athens, Greece
 http://dx.doi.org/10.1016/B978-0-443-15274-0.50042-1

A novel mathematical formulation for short-term scheduling of multipurpose batch plants in chemical manufacturing

Dan Li,[a] Taicheng Zheng,[a] Jie Li,[a,*]
[a]*Centre for process Integration, Department of Chemical Engineering, The University of Manchester, Manchester, M13 9PL, United Kingdom*

Abstract

In this work, a novel mixed-integer linear programming formulation is proposed based on the unit-specific event-based representation to address the short-term scheduling of multipurpose batch plants in network environment. Concepts of material transfers between processing units or storage tankers are introduced to conditionally sequence and align the related production and consumption tasks. In this case, material flows between the processing units can be explicitly tracked, resulting in decreases on appropriate number of event points and improvements on computational efficiency. Additionally, the processing units are allowed to temporarily store a batch after production over multiple event points. Computational studies shows that the developed continuous-time based formulation leads us to fewer numbers of event points in many cases and better objective results with a maximum improvement of 67%. More interestingly, all considered examples reach the optimality without any task spanning over multiple event points.

Keywords: Unit-specific event-based formulation, multipurpose batch process, material transfer, short-term scheduling

1. Introduction

Process scheduling is a key managerial tool for manufacturing, which seeks to allocate limited resources, determine task assignments and sequences to achieve production targets over time [Rakovitis et al. (2019)]. Multipurpose batch plants [Li et al. (2022)] widely exist in chemical industry for production of low-volume and high-value products. Unit-specific event-based formulation is an efficient continuous-time based approach to address the short-term scheduling of multipurpose batch plants. Numbers of event points have significant effects on the model size and computational performance. Attempts have been made to eliminate some unnecessary event points by relaxing some unconditional sequencing constraints enforced for the consumption and its related production tasks with the same state. Specifically, Seid and Majozi (2012) as well as Vooradi and Shaik (2013) relaxed the unconditional sequencing and alignments between the related production and consumption tasks if there is sufficient state inventory for consumption and storage space for production, respectively. However, the model of Seid and Majozi (2012) leads to real-time storage violation. Vooradi and Shaik (2013) introduced enormous variables, resulting in heavy computational burdens. More importantly, almost all existing models did not allow a batch of materials to be temporarily stored in processing units over multiple event points, leading to inefficient utilization of processing units.

In this work, a novel mixed-integer linear programming (MILP) model is proposed to address above mentioned limitations of the existing unit-specific event-based MILP formulations. Specifically, concepts on material transfer are introduced for the related production and consumption tasks at two adjacent event points. Material flows of states are monitored among units rather than specific tasks to decrease numbers of variables. And temporary holding of batches on units after production are modelled to yield better objective results for problems subject to finite intermediate storage (FIS).

2. Problem statement and material transfer

Multipurpose batch plants are formulated based on the state-task network [Kondili et al. (1993)]. Tasks $i \in \mathbf{I}$ are processed on units $j \in \mathbf{J}$. Units j performing task i are included in the set $\mathbf{J}_i$. States $s \in \mathbf{S}$ are consumed ($\rho_{si} < 0$) or produced ($\rho_{si} > 0$) by the consumption ($i \in \mathbf{I}_s^C$) or production ($i \in \mathbf{I}_s^P$) tasks with the proportion ρ_{si}. The formulation is proposed to minimize makespan to fulfill demands (D_s) of products ($s \in \mathbf{S}^P$) or maximize profit (z) over a given horizon (H). Material transfer between facilities is explicitly tracked to conditionally sequence the related production and consumption tasks of the same state. Four scenarios of material transfer (MT) are illustrated in Figure 1. In the first scenario (MT1), materials after production are transferred into its dedicated storage. In MT2, the batch of materials after production are first temporarily held on its processing unit and then flow into the storage. In MT3 and MT4, material after production at n are consumed by tasks at $(n+1)$. MT3 illustrates the indirect material transfer in which materials available at n would be first transferred to the storage and then to the downstream unit at $(n+1)$ as there are sufficient storage space. In MT4, materials are directly transferred to the downstream units due to the storage limitation, which is termed as the direct material transfer. Two sets of binary variables are introduced to formulate the indirect ($zI_{jj'n}$) and direct ($zD_{jj'n}$) material transfer between units $j \in \mathbf{J}_s^P$ and $j' \in \mathbf{J}_s^C$ performing the production and consumption tasks, respectively, of a state s. Specifically, $zI_{jj'n} = 1$, if a state produced on j at $(n-1)$ is indirectly transferred to j' at n. $zD_{jj'n} = 1$, if a state produced on j at $(n-1)$ is directly transferred to j' at n.

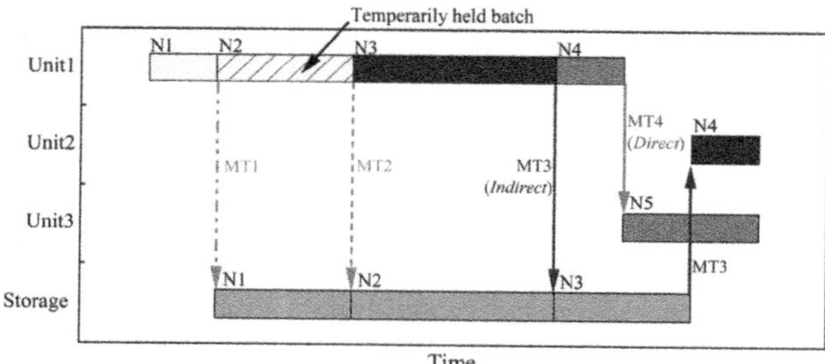

Figure 1. Four scenarios of material transfer between units and storage tankers

3. Mathematical model

The binary variable $w_{ijnn'}$ is one if task i is processed on j from event point n to n' ($n \leq n' \leq n + \Delta n$). Δn is the number of event points over which a task can span. Binary variables ys_{ijn} equal to one if one or partial batch of a task i is temporarily held on j at n. Allocation constraints demand at most one task to be processed or held on j at a time. Sets $\mathbf{I}_j$ and $\mathbf{I}^P$ include tasks that can be processed on j and held on units, respectively.

$$\sum_{i \in \mathbf{I}_j} \sum_{n-\Delta n \leq n' \leq n} \sum_{n \leq n'' \leq n'+\Delta n} w_{ijn'n''} + \sum_{i \in (\mathbf{I}_j \cap \mathbf{I}^P)} ys_{ijn} \leq 1 \qquad \forall j, n \qquad (1)$$

Batch sizes $b_{ijnn'}$ are bounded by maximum ($B_{ij}^{\max}$) and minimum ($B_{ij}^{\min}$) unit capacity.

$$B_{ij}^{\min} \cdot w_{ijnn'} \leq b_{ijnn'} \leq B_{ij}^{\max} \cdot w_{ijnn'} \qquad \forall j, i \in \mathbf{I}_j, n \leq n' \leq n + \Delta n \qquad (2)$$

Amounts of the batch of task i held on unit j at n are denoted by bs_{ijn}. Variables ST_{sn} express amounts of s stored at n, including units and its dedicated storage. Material balance of one state at n are formulated as Eqs. 3-4, where Eq. 3 is valid for the first event point. Variables $ST0_s$ and $bs0_{ij}$ express the initial inventory level of state s and initial amount for a batch of task i held in j, respectively. Amounts of states s stored at n should be always no less than its amounts temporarily held on the processing units (see Eq.5).

$$ST_{sn} = ST0_s + \sum_j \sum_{i\in(\mathbf{I}_j\cap\mathbf{I}_s^P)} \rho_{si}\cdot bs0_{ij} + \sum_j \sum_{i\in(\mathbf{I}_j\cap\mathbf{I}_s^C)} \sum_{n\le n'\le n+\Delta n} \rho_{si}\cdot b_{ijnn'} \quad \forall s, n=1 \tag{3}$$

$$ST_{sn} = ST_{s(n-1)} + \sum_j \sum_{i\in(\mathbf{I}_j\cap\mathbf{I}_s^P)} \sum_{n-1-\Delta n\le n'\le n-1} \rho_{si}\cdot b_{ijn'(n-1)} + \sum_j \sum_{i\in(\mathbf{I}_j\cap\mathbf{I}_s^C)} \sum_{n\le n'\le n+\Delta n} \rho_{si}\cdot b_{ijnn'} \quad \forall s, n>1 \tag{4}$$

$$\sum_j \sum_{i\in(\mathbf{I}_j\cap\mathbf{I}_s^P)} \rho_{si}\cdot bs_{ijn} \le ST_{sn} \quad \forall s\in\mathbf{S}^{FIS}, n \tag{5}$$

Start and finish time of a unit j at event point n are denoted by T_{jn}^{s} and T_{jn}^{f}, respectively. The unit j finishes the event point n after processing task i that is performed on j at n.

$$T_{jn}^{f} \ge T_{jn}^{s} + \sum_{i\in\mathbf{I}_j} \sum_{n\le n'\le n+\Delta n} (\alpha_{ij}\cdot w_{ijnn'} + \beta_{ij}\cdot b_{ijnn'}) \quad \forall j, n \tag{6}$$

Sequencing constraints require that on a unit j, an event point $(n+1)$ must start after the finish of it previous event point n (see Eq.7). Continuous variables T_{sjn} are the time for state s produced on unit j at n being available. Available time of the same state at the next event point $(n+1)$ should be larger than the available time at n, as Eq.8.

$$T_{j(n+1)}^{s} \ge T_{jn}^{f} \quad \forall j, n<N \tag{7}$$

$$T_{sjn} \le T_{sj(n+1)} \quad \forall s\in\mathbf{S}^{in}, j\in\mathbf{J}_s^P, n<N-1 \tag{8}$$

As formulated by Eqs. 9-10, the time when a state s is available at n should be after the finish of its production task on unit j and before the start of the next event point $(n+1)$.

$$T_{sjn} \ge T_{jn}^{f} - M\cdot\left[1 - \sum_{i\in(\mathbf{I}_j\cap\mathbf{I}_s^P)} \sum_{n-\Delta n\le n'\le n} w_{ijn'n}\right] \quad \forall s\in\mathbf{S}^{in}, j\in\mathbf{J}_s^P, n<N \tag{9}$$

$$T_{sjn} \le T_{j(n+1)}^{s} + M\cdot\left[1 - \sum_{i\in(\mathbf{I}_j\cap\mathbf{I}_s^P)} \left(\sum_{n-\Delta n\le n'\le n} w_{ijn'n} + ys_{ijn}\right)\right] \forall s\in\mathbf{S}^{in}, j\in\mathbf{J}_s^P, n<N \tag{10}$$

Indirect material transfer of a state takes place only with sufficient storage space for s. If it occurs from a unit j at n to j' at $(n+1)$, the unit j' at $(n+1)$ must start not earlier than the finish of j at n. Set $\mathbf{CJ}_j$ includes units j' consuming a state produced in unit j.

$$T_{jn}^{f} \le T_{j'(n+1)}^{s} + M\cdot[1 - zI_{jj'(n+1)}] \quad \forall j, j'\in\mathbf{CJ}_j, j\ne j', n<N \tag{11}$$

Continuous variables $bTi_{iji'j'n}$ express amounts of a state through indirect material transfer from j to j'. For any consumption task, its consumed amounts of s cannot exceed the total amounts stored in the storage and transferred from the production tasks.

$$\sum_{j'} \sum_{i'\in(\mathbf{I}_s^C\cap\mathbf{I}_{j'})} \left(-\rho_{si'}\cdot \sum_{n\le n'\le n+\Delta n} b_{i'j'nn'}\right) \le ST_{s(n-1)} + \sum_j \sum_{j'} \sum_{i\in(\mathbf{I}_s^P\cap\mathbf{I}_j)} \sum_{i'\in(\mathbf{I}_s^C\cap\mathbf{I}_{j'})} bTi_{iji'j'n} \quad \forall s\in\mathbf{S}^{in}, n>1 \tag{12}$$

$$\rho_{si}\cdot \sum_{n-1-\Delta n\le n'\le n-1} b_{ijn'(n-1)} \ge \sum_{j'} \sum_{i'\in(\mathbf{I}_{j'}\cap\mathbf{I}_s^C)} bTi_{iji'j'n} \quad \forall s\in\mathbf{S}^{in}, j, i\in(\mathbf{I}_j\cap\mathbf{I}_s^P), n>1 \tag{13}$$

$$-\rho_{si'}\cdot \sum_{n\le n'\le n+\Delta n} b_{i'j'nn'} \ge \sum_j \sum_{i\in(\mathbf{I}_j\cap\mathbf{I}_s^P)} bTi_{iji'j'n} \quad \forall s\in\mathbf{S}^{in}, j', i'\in(\mathbf{I}_{j'}\cap\mathbf{I}_s^C), n>1 \tag{14}$$

Amounts of s through indirect material transfer at n should not exceed the production amount finishing at $(n-1)$ and also the consumption amounts starting at n (see Eqs. 13-14). Without indirect material transfer, the corresponding variables on transfer amounts are zero, where $B_j^{max} = \max_{s,i\in(\mathbf{I}_j\cap\mathbf{I}_s^P)} [\rho_{si}\cdot B_{ij}^{max}]$ and $B_{j'}^{max} = \max_{s,i'\in(\mathbf{I}_{j'}\cap\mathbf{I}_s^C)} [-\rho_{si'}\cdot B_{i'j'}^{max}]$.

$$\sum_{i\in \mathbf{I}_j}\sum_{i'\in \mathbf{I}_{j'}} bTi_{iji'j'n} \le \min\left[B_j^{\max}, B_{j'}^{\max}\right]\cdot zI_{jj'n} \qquad \forall j,j'\in \mathbf{CJ}_j, j\neq j', n>1 \quad (15)$$

Constraint (16) enforces that a unit j' consuming a state at $(n+2)$ or higher event points should always start after the state is available on unit j at n to avoid storage violation.

$$T_{sjn} \le T_{j'(n+2)}^{s} + M\cdot\left[1-\sum_{i'\in\left(\mathbf{I}_{j'}\cap \mathbf{I}_s^C\right)}\ \sum_{n+2\le n'\le n+2+\Delta n} w_{i'j'(n+2)n'}\right]$$
$$\forall s\in \mathbf{S}^{in}, j\neq j', j\in \mathbf{J}_s^P, j'\in \mathbf{J}_s^C, n<N-1 \qquad (16)$$

When there is no sufficient storage space to store the production amount of a state s, this state must be transferred to the downstream unit directly or temporarily held on its processing unit, as indicated by Eq.17. Consequently, the unit j' should finish the processing at n before the end of unit j at n to avoid overlapping on j', if direct material transfer of one state takes place from j at n to j' at $(n+1)$, as constrained by Eq.18.

$$\sum_j \sum_{i\in\left(\mathbf{I}_j\cap \mathbf{I}_s^P\right)}\left[\rho_{si}\cdot \sum_{n-1-\Delta n\le n'\le n-1} b_{ijn'(n-1)}\right] + ST_{s(n-1)}$$
$$\le ST_s^{\max} + \sum_j \sum_{i\in\left(\mathbf{I}_j\cap \mathbf{I}_s^P\right)} \rho_{si}\cdot bs_{ijn}$$
$$+\sum_j\sum_{j'}\sum_{i\in\left(\mathbf{I}_j\cap \mathbf{I}_s^P\right)}\ \sum_{i'\in\left(\mathbf{I}_{j'}\cap \mathbf{I}_s^C\right)} bTd_{iji'j'n} \qquad \forall s\in \mathbf{S}^{FIS}, n>1 \qquad (17)$$

$$T_{j'n}^{f} \le T_{jn}^{f} + M\cdot\left[1 - zD_{jj'(n+1)}\right] \qquad \forall j,j'\in \mathbf{CJ}_j^F, j\neq j', n<N \qquad (18)$$

Similar to the indirect material transfer, amounts of a state s transferred at $(n+1)$ through direct material transfer should be bounded by the amount available (i.e. produced or temporarily held) at n and the consumption amount by tasks at $(n+1)$. And the amounts of state transferred should be zero without direct material transfer.

$$\rho_{si}\cdot\left[\sum_{n-1-\Delta n\le n'\le n-1} b_{ijn'(n-1)} + bs_{ij(n-1)}\right] \ge \sum_{j'}\sum_{i'\in\left(\mathbf{I}_{j'}\cap \mathbf{I}_s^C\right)} bTd_{iji'j'n} + \rho_{si}\cdot bs_{ijn}$$
$$\forall s\in \mathbf{S}^{FIS}, j, i\in\left(\mathbf{I}_j\cap \mathbf{I}_s^P\right), n>1 \qquad (19)$$

$$-\rho_{si'}\cdot \sum_{n\le n'\le n+\Delta n} b_{i'j'nn'} \ge \sum_j \sum_{i\in\left(\mathbf{I}_s^P\cap \mathbf{I}_j\right)} bTd_{iji'j'n} \qquad \forall s\in \mathbf{S}^{FIS}, j', i'\in\left(\mathbf{I}_{j'}\cap \mathbf{I}_s^C\right), n>1 \quad (20)$$

$$\sum_{i\in \mathbf{I}_j}\sum_{i'\in \mathbf{I}_{j'}} bTd_{iji'j'n} \le \min\left[B_j^{\max}, B_{j'}^{\max}\right]\cdot zD_{jj'n} \qquad \forall j,j'\in \mathbf{CJ}_j^F, j\neq j', n>1 \qquad (21)$$

Equation 22 ensures sufficient storage space for state s produced on j at n when $s\in \mathbf{S}^{FIS}$ is subject to FIS, by enforcing j' to finish at $(n-1)$ before the available time of s at n.

$$T_{sjn} \ge T_{j'(n-1)}^{f} - M\cdot\left[1-\sum_{i'\in\left(\mathbf{I}_{j'}\cap \mathbf{I}_s^C\right)}\ \sum_{n\le n'\le n+\Delta n} w_{i'j'nn'}\right]$$
$$\forall s\in \mathbf{S}^{FIS}, j\neq j', j\in \mathbf{J}_s^P, j'\in \mathbf{J}_s^C, 1<n<N \qquad (22)$$

$$ST_{sn} \le ST_s^{max} + \sum_{j\in \mathbf{J}_s^P}\left[\max_{i\in\left(\mathbf{I}_j\cap \mathbf{I}_s^P\right)}\left(\rho_{si}\cdot B_{ij}^{\max}\right)\right] \qquad \forall s\in \mathbf{S}^{FIS}, n \qquad (23)$$

$$bs_{ijn} \le B_{ij}^{\max}\cdot ys_{ijn} \qquad \forall j, i\in\left(\mathbf{I}_j\cap \mathbf{I}^P\right), n \qquad (24)$$

Amount of state $s\in \mathbf{S}^{FIS}$ stored on units and storage at n should satisfy Eq. 23. Amount of the batch temporarily held on j should be no larger than the unit capacity. Tightening constraints (refer Eqs. 34, 35 and 37 of Li et al. 2022) are formulated to handle binary variables on allocation and material transfer. Indirect material transfer tasks place if there is the direct material transfer between two units (Eq. 25). Bounds are added on variables as Eqs. 26-30. Both ys_{ijn} and bs_{ijn} are zero if states produced by i are subject to UIS.

$$zI_{jj'n} \geq zD_{jj'n} \quad \forall j,j' \in \mathbf{CJ}_j^F, j \neq j', n > 1 \quad (25)$$

$$T_{jn}^{s} \leq H, T_{jn}^{f} \leq H \quad \forall j,n \quad (26,27)$$

$$w_{ijnn'} = 0, b_{ijnn'} = 0 \quad \forall i,j,n,(n' < n) \cup (i \notin \mathbf{I}_j) \quad (28,29)$$

$$bTd_{iji'j'n} = bTi_{iji'j'n} = zI_{jj'n} = zD_{jj'n} = ys_{ijn} = bs_{ijn} = 0 \quad \forall i,i',j,j',n = 1, bs0_{ij} = 0 \quad (30)$$

$$ys_{ijn} = 0, bs_{ijn} = 0 \quad \forall i \notin \mathbf{I}^P, j, n \quad (31,32)$$

The objective is to maximize profit (see Eq. 33) or minimize makespan (see Eq. 34-35). Equation (36) is the tightening constraint for minimization of makespan.

$$z = \sum_{s \in \mathbf{S}^P} \left[p_s \cdot \sum_j \sum_{i \in (\mathbf{I}_j \cap \mathbf{I}_s^P)} \sum_n \sum_{n - \Delta n \leq n' \leq n} (\rho_{si} \cdot b_{ijn'n}) \right] \quad (33)$$

$$T_{jn}^{f} \leq MS \quad \forall j, n = N \quad (34)$$

$$ST0_s + \sum_j \sum_{i \in (\mathbf{I}_j \cap \mathbf{I}_s^P)} \sum_n \sum_{n \leq n' \leq n + \Delta n} (\rho_{si} \cdot b_{ijnn'}) \geq D_s \quad \forall s \in \mathbf{S}^P \quad (35)$$

$$\sum_{i \in \mathbf{I}_j} \sum_n \sum_{n \leq n' \leq n + \Delta n} (\alpha_{ij} \cdot w_{ijnn'} + \beta_{ij} \cdot b_{ijnn'}) \leq MS \quad \forall j \quad (36)$$

Development of the mathematical formulation is completed, which contains of equations (1-33) to maximize profit and (1-32) and (34-36) to minimize makespan.

Table 1. Computational results for examples 1-6 with FIS. (Exs1, 3 and 4 are addressed to minimize makespan. Exs2, 5 and 6 are addressed to maximize profit)

Ex	Model	Events	RMILP	MILP	CPU (s)	Binary	Continuous	Constraints
1	V&S	3	5.00	11.5	0.02	12	32	68
	L&F	3	5.00	11.5	0.02	6	29	43
	SLK2	4	5.00	11.5	0.02	12	83	99
	M	3	8.00	**8**	0.02	12	37	76
2	V&S	10	1795.48	989.03	2096	569	805	2670
	M	11	2208.35	**1201.39**[d]	3600	511	1801	3091
3	V&S	14	24.24	27.88[e]	3600	226	371	1031
	M	14	24.24	27.88	**1969**	213	450	1243
4	V&S	21[a]	47.38	47.68[f]	3600	1,080	1567	4955
	M	**21**	47.38	47.68	**82.0**	**648**	2449	3791
5	V&S	6[b]	400.00	400.00	0.16	132	197	596
	M	**6**	400.00	400.00	0.05	**79**	185	536
6	V&S	10[c]	400.00	400.00	1.77	316	421	1300
	M	**10**	400.00	400.00	0.16	**139**	317	956

a: $\Delta n = 2$, b: $\Delta n = 3$, c: $\Delta n = 7$. Relative gap, d: 5.74%, e: 0.08%, f: 0.64%.

4. Computational studies

Comparisons between the proposed model (M) and three existing continuous-time based models [V&S: Vooradi and Shaik (2013), L&F: Li and Floudas (2010), SLK2: Susarla et al. (2010)] are conducted. Considered examples 1 and 2 with FIS are the motivating example 2 and example 12 from Li et al. (2022). Examples 3-6 are examples 1a, 2b, 8 and 9 in their work. Models are solved using CPLEX 12.6.3/GAMS 24.6.1 on a desktop computer with AMD Ryzen 9 3900X 3.8 GHz, 48 GB RAM running Windows 10. Computational results of the examples 1-6 are listed in Table 1. Better objective results are obtained by M in examples 1 and 2 as the processing units can hold materials to share the storage burden under FIS. Computational time is reduced over 45% (1969 s vs. 3600 s) and 98% (82 s vs. 3600 s) by using M relative to model V&S for examples 3 and 4. That is attributed to the decreased number of binary variables. Concerning appropriate numbers of event points to reach optimality for examples 4-6, it proves that model M does not require any task to span over multiple event points (i.e. $\Delta n > 0$). Four instances (instances 9, 11, 14, and 15 in Vooradi and Shaik 2012) of the Kallrath example, are addressed using the developed model, as presented in Table 2. Results prove that the proposed model always generate lower makespan using less computational time. For instance, the model M obtains a 2.5% (39 h vs. 40 h) lower makespan using 45% (21890 s vs. 40000s) less computational time, compared with V&S.

Table 2. Computational results of the Kallrath examples with FIS to minimize makespan

In	Model	Events	RMILP	MILP	CPU (s)	Binary	Continuous	Constraints
1	V&S	10	16	32	10761	888	1386	5113
	M	10	16	32	**5377**	591	1267	4229
2	V&S	11	28	40[a]	40000	984	1530	5664
	M	11	28	**39**	21890	654	1401	4690
3	V&S	9	21.6	36	447	792	1242	4562
	M	9	21.6	36	348	528	1133	3768
4	V&S	15	40	56[b]	40000	1368	2106	7868
	M	15	40	**54**[c]	40000	906	1937	6534

Relative gap, [a]: 7.5%, [b]: 14.29%, [c]: 3.70%.

5. Conclusion

In this work, a novel unit-specific event-based MILP formulation is proposed to address the short-term scheduling of the multipurpose batch plants. Material transfer of states are explicated to conditionally sequence the related tasks on units. And batches of tasks after production are permitted to be held in the processing units. Computational results demonstrate that the proposed model leads us to better objective results and decreases the computational time up to 98%. More interestingly, the proposed model does not require any task to span over multiple event points to yield the global optima.

Acknowledgement

Dan Li and Taicheng Zheng appreciate financial support from China Scholarship Council - the University of Manchester Joint Scholarship (201908130170, 202106440020). Jie Li appreciates financial support from Engineering and Physical Sciences Research Council (EP/T03145X/1).

References

D. Li, N. Rakovitis, T. Zheng, Y. Pan, J. Li, G. Kopanos, 2022, Novel Multiple Time-grid Continuous-time Mathematical Formulation for Short-term Scheduling of Multipurpose Batch Plants. Ind. Eng. Chem. Res., 61, 16093-16111.

J. Li, C.A. Floudas, 2010, Optimal event point determination for short-term scheduling of multipurpose batch plants via unit-specific event-based continuous-time approaches, Ind. Eng. Chem. Res., 49, 7446-7469.

E. Kondili, C.C. Pantelides, R.W.H. Sargent, 1993, A general algorithm for short-term scheduling of batch operations-I MILP formulation. Comput. Chem. Eng., 17, 211-227.

N. Rakovitis, N. Zhang, J. Li, L.P. Zhang, 2019, A new approach for scheduling of multipurpose batch processes with unlimited intermediate storage policy, Front. Chem. Sci. Eng.,13, 784-802.

R. Seid, T. Majozi, 2012, A robust mathematical formulation for multipurpose batch plants. Chem. Eng. Sci., 68, 36-53.

N. Susarla, J. Li, I.A. Karimi, 2010. Novel approach to scheduling multipurpose batch plants using unit-slots. AIChE J., 56, 1859-1879.

R. Vooradi, M.A. Shaik, 2012, Improved three-index unit-specific event-based model for short-term scheduling of batch plants. Comput. Chem. Eng., 43, 148-172.

R. Vooradi, M.A. Shaik, 2013, Rigorous unit-specific event-based model for short term scheduling of batch plants using conditional sequencing and unit-wait times, Ind. Eng. Chem. Res., 52, 12950-12792.

Antonis Kokossis, Michael C. Georgiadis, Efstratios N. Pistikopoulos (Eds.)
PROCEEDINGS OF THE 33rd European Symposium on Computer Aided Process Engineering (ESCAPE33), June 18-21, 2023, Athens, Greece
 http://dx.doi.org/10.1016/B978-0-443-15274-0.50043-3

Heat integration and heat exchanger network design for oxyfuel cement plants

Leif E. Andersson,[a] Avinash Subramanian,[a] Mari Voldsund,[a] Rahul Anantharaman,[a] Kristina Fleiger,[b] Francisco Carrasco,[c] Mirko Weber[d]

[a]SINTEF Energy Research, Sem Sælands vei 11,7043 Trondheim, Norway
[b]VDZ Technology gGmbH, Toulouser Allee 71, 41476 Düsseldorf, Germany
[c]HeidelbergCement AG, Berliner Straße 6, 69120 Heidelberg, Germany
[d]Holcim Technology Ltd, Grafenauweg 10, 6300 Zug, Switzerland

leif.andersson@sintef.no

Abstract

The cement sector needs to reduce its CO_2 emissions. An oxyfuel CO_2 capture technology allows to considerably reduce the emission. However, heat recovery and energy efficiency measures are essential to make the technology economically feasible. An approach to design heat exchanger networks applied to a 1st generation oxyfuel cement plant is described in this article. The approach consists of two steps: preliminary targeting and heat exchanger network design. For the studied cement plant, the steam Rankine cycle was identified to be superior to organic Rankine cycles. In the ideal case about 10.5 MW of power can be recovered. However, in a cost-efficient simple heat exchanger network recovery of only about 8.7 MW is economically reasonable.

Keywords: Heat integration, Heat exchanger network design, Oxyfuel cement plants.

1. Introduction

The cement sector is responsible for about 7% global anthropogenic CO_2 emissions (IEA, 2018). Two-third of the CO_2 emission originate from calcination of limestone while one-third come from the combustion of fuel. It is, therefore, impossible to reach CO2 emission targets with fuel switching alone. CO_2 capture and storage is essential to become carbon neutral. An oxyfuel-based capture process is a promising candidate for capturing CO_2 from a cement plant (Voldsund et al., 2019). In the oxyfuel process the combustion is performed with oxygen mixed with recycled CO_2. The CO_2 enriched flue gas allows a relatively cost-efficient purification and separation. Nevertheless, CO_2 purification consumes additional power and process heat must be transferred to air streams to dry the raw material energy-efficiently. The temperature levels in an oxyfuel plant are higher than in conventional plants because of the increased oxygen concentration in the combustion, and flue gas has to be cooled before it is recirculated. Waste heat recovery and heat integration are important for an economic implementation of oxyfuel technology in cement plants.

An existing cement plants was investigated for retrofit of the 1st generation oxyfuel process. The process simulations are performed with VDZs in-house cement process model. Afterwards, the heat integration is performed using a two-step methodology.

This article focusses on a systematic approach to heat integration and applies it to an oxyfuel cement plant. In the following, a brief description of the model setup is given followed by the introduction of the two-step methodology for heat integration and its application the oxyfuel cement plant is described.

2. Methodology

2.1. Modelling approach

The retrofit of the 1st generation oxyfuel process to cement plant was performed with several models. The clinker burning process was accessed by a kiln process model. The model described the process from the kiln meal feed to the outlet of the clinker from the cooler. It is made up by individual linked models of preheater, calciner, bypass, kiln and cooler, where material and energy balances are calculated (Koring, 2013). The process model outputs are performance data, thermal energy demand, clinker quality and available excess heat. Input data about the process design and some plant specific data to make the process model representative was provided by the cement producer and equipment supplier.
A second model, the heat integration model, access the waste heat recovery and the heat integration of the CO_2 Processing Unit (CPU). Data was iteratively exchanged between the two models.

2.2. Heat integration

Stream data from the process engineering model by VDZ was used to create the heat integration model in Aspen HYSYS. In addition, a CPU model was created in Aspen HYSYS. These models were used to access the energy streams of the 1st generation oxyfuel cement plant. In the following the two-step methodology consisting of a preliminary targeting step and a Heat Exchanger Network (HEN) design step, is presented.

2.2.1. Preliminary targeting of the heat exchanger network design

The goal of the preliminary targeting is to identify the most promising heat to power cycles for the HEN design. A pinch analysis is performed which identifies the bottle neck of the plant regarding heat integration and allows to estimate an upper bound on the power production of the heat to power cycles. This step is also used to dimension the heat to power cycles. The stream data from this analysis is used in the HEN design phase.

2.2.2. Heat Exchanger Network design

In the second step after the preliminary targeting the HEN is designed which allows cost-efficient design of the heat recovery in the cement plant. The best HEN involves optimizing the trade-off between capital costs determined by the number of heat exchanger units and their areas and the operating costs determined by the amount of hot and cold utilities required by the process. The software tool termed SeqHENS (Sequential Framework for HEN Synthesis) developed at the Norwegian University of Science and Technology and SINTEF Energy was used to design the optimal HEN (Anantharaman, 2011). An overview of the four-step methodology in the toolbox is given next:

1) Given stream data on the relevant hot and cold streams given by the preliminary targeting the minimum amount of hot and cold utility required is determined. The Linear Programming (LP) transhipment model is used to solve the optimization (Papoulias et al., 1983).
2) The absolute minimum number of heat exchanger units is determined where the heuristic is used that the optimal number is close to the minimum number of

units. This problem is solved with the Mixed-Integer Linear programming formulation (Papoulias et al., 1983).

3) Given the numbers of units and using engineering judgement the correct matches between hot and cold streams is determined. This "Steam Match Generator" problem is formulated using the vertical MILP transportation model (Anantharaman, 2011). The result is a Heat Load Distribution which gives the amount of heat exchanged between hot and cold stream.
4) Finally, the optimal topology of the HEN is determined. This problem is solved with a nonconvex Nonlinear Program (Floudas et al., 1986). The objective is to minimize the total cost of the HEN, where engineering judgment is used to get a simple HEN.

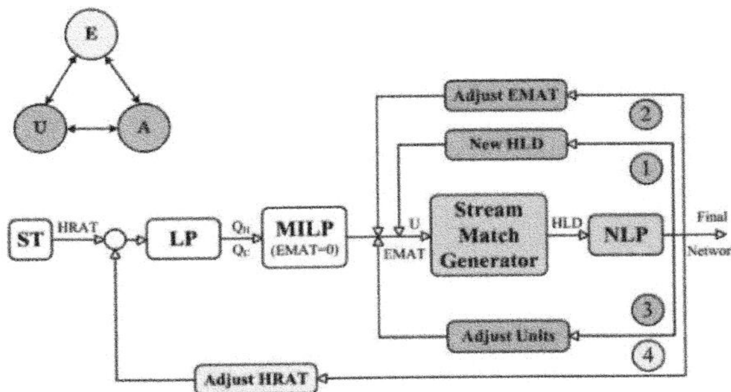

Figure 1. Overview of SeqHENS framework.

3. Case Study

The 1st generation oxyfuel cement plant consists of a preheater, pre-calciner, rotary kiln and cooler (Figure 2). The preheater exhaust is recirculated to the cooler, where a part of the recirculated stream is extracted and sent to the CPU. The Air Separation Unit (ASU) provides purified oxygen. An air stream is heated and used to dry the raw material.
The studied cement plant has an average annual clinker production of about 5600 t/d. The kiln line is a common dry process, and the raw material moisture is about 2-3%. The estimated drying demand for the raw material is about 300 $kJ/kg_{clinker}$. This heat must be supplied by the excess heat of the process before additional excess heat can be used in a heat to power cycle.
The hot streams in the oxyfuel cement plant are the preheater exhaust stream at a temperature of about 450°C, cooler exhaust stream at a temperature of about 180°C, and the bypass stream, which is pinched with parts of the recycling stream, after which the combined stream has a temperature of 400°C. Additional heat streams are available from the CPU at a temperature of about 120°C. The cold streams are the heat required for drying the raw material, heat required in the CPU and heat for the heat to power cycle.
It is necessary that the recirculated stream is cooled down to about 50°C to remove chlorine, sulfure and water. Moreover, bag filters are used to remove dust. In addition, the air stream existing the raw mill should have required heat left for evaporate water. Therefore, a minimum temperature of 120°C was chosen for this stream.

3.1. Preliminary targeting of 1st generation oxyfuel cement plant case study

The preliminary targeting phase was used to 1) find an optimal flue gas recirculation rate, 2) evaluate if heat from the Bypass should be used and 3) compare different heat to power cycles. The first and second step required iterations between the heat integration and

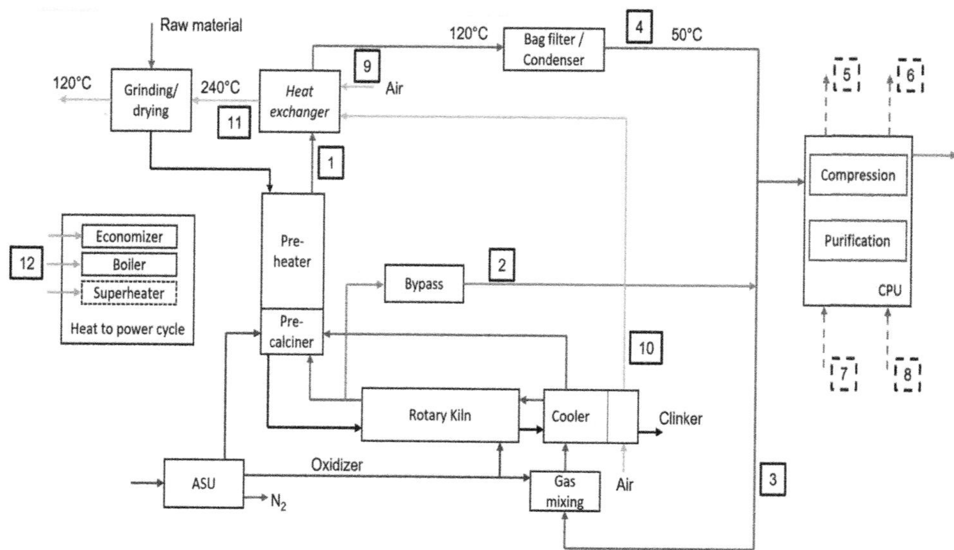

Figure 2. 1st generation oxyfuel cement plant. The numbers indicate hot and cold streams. Temperature constraints are displayed.

updating the process engineering model. It was concluded that no heat should be extracted from the Bypass since it increased the overall fuel use.

Organic Ranking cycles using Benzene or Butane and a steam Ranking cycle with different pressure levels were compared. The best performing Ranking cycle was the steam Rankine cycle at a pressure level of 15 bar, which was used for the HEN design (Table 1).

Table 1. Performance of different Rankine cycles.

Rankine cycle	Power production	Recirculation rate
Steam Rankine cycle (15 bar)	10.5 MW	0.49
ORC – Benzene	10.0 MW	0.49
ORC – Butane	8.4 MW	0.49

3.2. Heat exchanger network design of 1st generation oxyfuel cement plant case study

The preliminary targeting phase concluded that in the ideal case 10.4 MW power can be produced. However, this performance can only be reached with a complicated and expensive HEN. In the HEN design phase using SeqHENS stream splits and small heat exchangers which are not cost efficient were avoided. This resulted in a HEN which allows the required heat recovery and a power production in the steam Rankine cycle of about 8.7 MW (Figure 3). This is a 15% reduction compared to the ideal case.

In this study a retrofit is investigated, and the CPU will be located several hundred meters away from the kiln line. In the HEN obtained long pipelines are required between the CPU and the kiln line for only a small heat recovery. It was therefore decided to also investigate a case with the CPU excluded from the HEN design to create a simpler network. The design of the simpler network (Figure 4) increases the overall heat exchanger area but reduces complexity and piping, which is not included in the objective function of SeqHENS. The same amount of power as before (8.7 MW) can be produced.

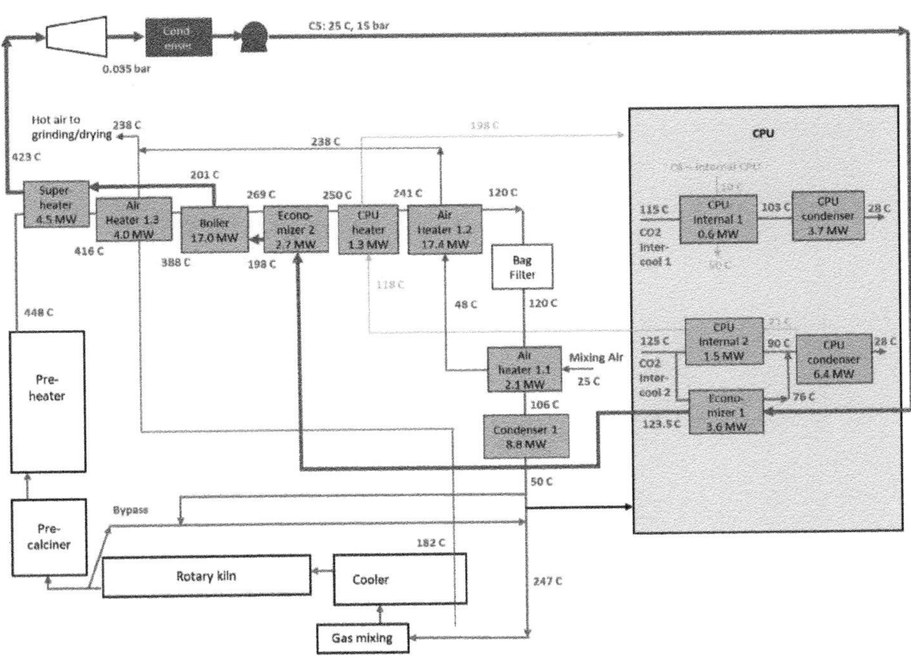

Figure 3. Heat exchanger network including heat integration of CPU.

In both HENs the Air Heater 1.2 is responsible for the majority of costs since it has a large duty, and it is a gas-gas heat exchanger with low heat transfer and large heat exchanger area.

4. Conclusion

This article shows how to apply heat integration and heat exchanger network design to a 1st generation oxyfuel cement plant. Heat integration is essential for economically implementation of the oxyfuel technologies to existing cement plants. required heat exchanger is a gas-gas heat exchanger which also is the largest cost driver. This heat exchanger is essential since air must be heated for drying the raw material. The direct use of the preheater exhaust as in conventional cement plants is impossible since it would emit CO_2. It is possible for this cement plant to recover about 8.7 MW of power which decreases the energy consumption of the plant. In future, a more detailed economic assessment of the steam Rankine cycle and required heat exchangers must be performed to evaluate the profitability.

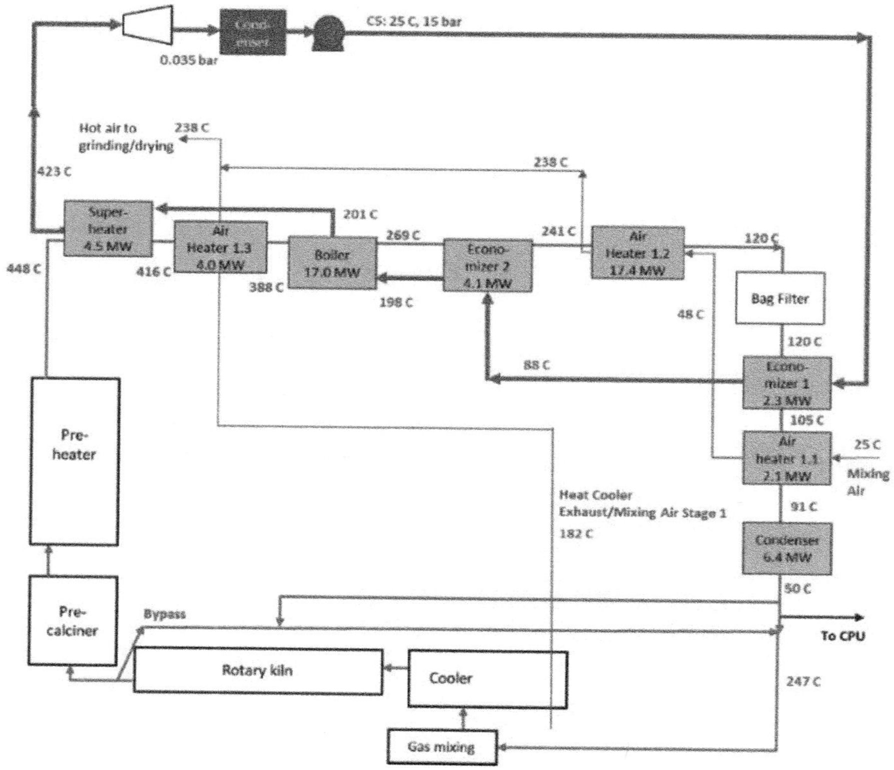

Figure 4. Heat exchanger network without heat integration of CPU.

Acknowledgements

This research is executed in scope of the AC²OCem project which is funded through the ACT program (Accelerating CCS Technologies, Horizon2020 Project No 299663). Financial contributions are made from; The Research Council of Norway, (RCN), Norway, Federal Ministry for Economic Affairs and Energy (BMWi), Germany, Swiss Federal Office of Energy (SFOE), Switzerland, General Secretariat for Research and Development (GSRT), Greece, French Environment & Energy Management Agency (ADEME), France, and TotalEnergies SE.

References

International Energy Agency (IEA), 2018, Technology Roadmap: Low-Carbon Transition in the Cement Industry, IEA: Paris, France.

M. Voldsund et al., 2019, Comparison of Technologies for CO2 Capture Cement Production – Part 1: Technical Evaluation, Energies, 12, 559.

K. Koring, 2013, Emissionsminderungspotential und technologische Auswirkungen der Oxyfuel- Technolgie im Zementklinkerbrennprozess, PhD-Thesis, Verlag Bau + Technik GmbH, Düsseldorf, Germany.

S. A. Papoulias et al., 1983, A structural optimization approach in process synthesis – II: Heat recovery networks, Computer & Chemical Engineering, 7(6), 707-721.

R. Anantharaman, 2011, Energy Efficiency in Process Plants with emphasis on Heat Exchanger Networks, PhD-Thesis, NTNU, Department of Energy and Process Engineering, Trondheim, Norway.

C. A. Floudas et al., 1986, Automatic synthesis of optimum heat exchanger network configurations, AIChE Journal, 32(2), 276-290.

Antonis Kokossis, Michael C. Georgiadis, Efstratios N. Pistikopoulos (Eds.)
PROCEEDINGS OF THE 33rd European Symposium on Computer Aided Process Engineering (ESCAPE33), June 18-21, 2023, Athens, Greece
 http://dx.doi.org/10.1016/B978-0-443-15274-0.50044-5

Technological Portfolio Optimisation for de-risking within the energy, water and food nexus

Maryam Haji, Mohammad Alherbawi, Sarah Namany, Tareq Al-Ansari

College of Science and Engineering, Hamad Bin Khalifa University, Qatar Foundation, Doha, Qatar
talansari@hbku.edu.qa

Abstract

Food insecurity is one of the major issues that have loomed as a result of rapid population growth. In recent studies addressing modern-day resource issues, the energy, water and food (EWF) nexus approach was introduced as a holistic method to aid decision-makers in pre-empting unintended, and possibly damaging consequences resulting from an imbalance made in any part of the nexus system. Existing studies have focused on centralised approaches to study the interlinkages amongst EWF systems by focusing mostly on two elements of the EWF nexus. However, the current and future transitions should address the entire EWF system, holistically and sustainably. Besides, the existing indicator frameworks do not expressly capture the key interactions between the EWF resources to address the main challenges and risk factors associated with the EWF systems. Consequently, the aim of this study is to tackle this knowledge gap by integrating the analytical hierarchy process (AHP) tool along with EWF risk-associated optimisation as means to assess the changes in EWF resource availability, and therefore identify the tradeoffs of the large-scale implementation of decentralised energy and water technologies. The proposed framework utilises a nexus approach to optimise the energy and water technological portfolios in an efficient way; thus, supporting the decision-making for resource management in high-risk environments, such as the State of Qatar. The methodology consists of designing a composite risk indicator using the AHP method to determine the weight of different risk criteria for energy and water technologies. The obtained values are then used as an input for multi-objective optimisation that aims at minimising the tradeoffs between two objectives (i.e. risk and GWP). Results of the AHP indicate that wind energy is 11.8% risker than solar energy, where the key issue is that the availability of these energy sources is far from constant but seasonal. In addition, treated wastewater was the most environmentally friendly, having the lowest impact on the ecosystem and resources, however, treated wastewater must be properly treated as it has the highest risk level (12.5%) and can lead to health issues. Finally, the multi-objective optimization model generates relatively 70 optimal solutions due to the mutual conflict between the two objectives.

Keywords: EWF nexus; food security; multi-objective optimisation; AHP.

1. Introduction

Energy, water and food systems, which are inherently interdependent, are essential for human health and survival, economic progress and resilience. These interrelated relationships were initially introduced at the Rio+20 Summit in 2012 as the "EWF" nexus. Developing a comprehensive decision-making framework necessitates the implementation of an effective EWF Nexus system that is adequate in tackling the multifaced issues within EWF systems to achieve resilient resource management. A common approach to deal with risks and uncertainties is building a framework based on composite indicators (Becker, 2017). In the EWF nexus, the establishment of composite indicators is a common method for a quantitative assessment. Despite the availability of numerous indicators within the EWF nexus, developing a resilience indicator using the AHP technique is extremely beneficial; specifically, since they are a valuable tool that deals with interdisciplinary boundaries (Albrecht, 2018). In this context, Haji et al., (2020) used the AHP to design a risk assessment indicator for greenhouse planning in arid regions. While Zhang et al., (2021) developed an integrated approach to assess the composite risk of WEF nexus systems using a generalized copula-based chance-constrained programming model and generate risk-based plans. The proposed approach has been applied to an agricultural WEF nexus system in northern China, where the shortage of water, energy and land affected agricultural outputs. A recent study by Singh et al., (2022) established an assessment framework to rank individual countries whilst still allowing decision-makers with the possibility to allocate resources sufficiently and incorporate other appropriate indicators in countries that struggle in regard to food security, wherein the set of food security indicators was assessed based on the future climate change scenario. In this study, an assessment tool based on the EWF Nexus approach will be developed to support addressing dynamic decision-making, especially in hazardous non-resilience environments, in order to reduce the effect of overall risk, thus improving the national adaptability of EWF systems. This will be achieved through implementing an integrated approach based on the AHP method coupled with a multi-objective optimization model. A composite risk indicator was established using the AHP method to determine the weight of different risk criteria for energy and water technologies. The considered technologies consist of solar, wind and bioenergy for energy generation, while groundwater, treated wastewater and desalinated water as water supplies. The obtained values are then used as an input for multi-objective optimisation that aims at quantifying each EWF nexus system in terms of risk and environmental impact. The multi-objective linear program is formulated to minimise the tradeoffs between two objectives (i.e. risk and GWP).

2. Dynamic decision-making framework for EWF Nexus

2.1. Development of Composite Risk Indicators using Analytic Hierarchy Process (AHP)

The AHP method was used to design a composite risk indicator that will determine the weight of four risk criteria (cost, availability, readiness and applicability) for each energy and water technology that affect the overall goal, as illustrated in Figure 1. The technologies considered consist of solar, wind and bioenergy for energy generation, while groundwater, desalinated water and treated wastewater as water supplies. The assessment process begins with two-level pairwise comparisons between criteria and alternatives by identifying which of the two is riskier and by how much. In which the preference is expressed on a semantic scale of 1 to 9. The information which is related to the importance of each factor in comparison to one and another is obtained from previous literature. For the 2nd level, there are 21 pairwise comparisons between seven alternatives considering each risk factor. However, for the 1st level, there are 6 pairwise comparisons between four risk factors. Then, the normalized relative weight of each factor and alternative is obtained. Finally, the normalized principal eigenvector of risk factor and alternative was determined by averaging across each row, as demonstrated in Table 1.

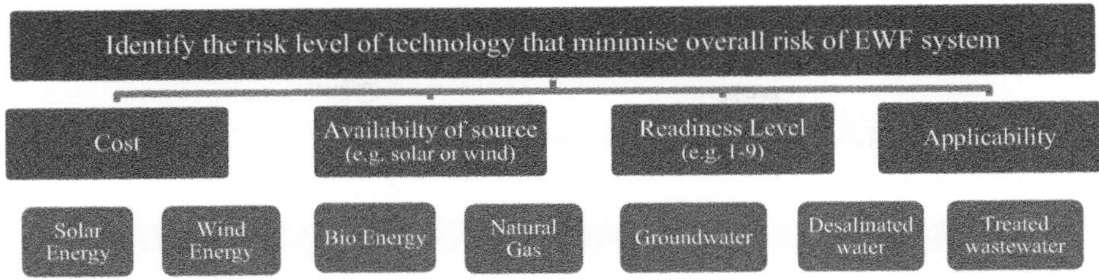

Figure 1: Hierarchy structure for determining the risk level of the energy and water technologies.

Table 1: Eigenvalues of four criteria and seven alternatives generated from the AHP method.

Risk Criteria	Eigen Vector of Criteria	Solar Energy (i=1)	Wind Energy (i=2)	Biomass Energy (i=3)	Natural Gas (i=4)	Groundwater (i=5)	Desalinated water (i=6)	Treated wastewater (i=7)
Cost	0.1067	0.1059	0.1237	0.0567	0.1269	0.1272	0.2133	0.2463
Availability	0.3576	0.1111	0.2390	0.1531	0.1261	0.1485	0.1111	0.1111
Readiness	0.3038	0.1250	0.2500	0.1250	0.1250	0.1250	0.1250	0.1250
Applicability	0.2320	0.1884	0.3259	0.1802	0.0540	0.0798	0.0798	0.0919

2.2. Multi-Objective Optimisation Model Development

The multi-objective optimisation model aims at minimising the risk and global warming potential (GWP) of different energy and water technologies required for food production. A summary of the mathematical model formulation is presented in Table 2. Two constraints were introduced to ensure that the sum of all contribution percentages of energy or water technologies must be 100%. The third constraint is to restrict groundwater to participate in the water mix when exceeding the annual renewable rate since groundwater consider as the scarcest water resource in the country studied. Moreover, the optimisation model assumed that the EWF nexus adopted does not involve the food sector explicitly in the technology selection, however, its contribution is translated through the water and energy required to produce food. The risk level of each technology generated from the AHP method will be used to run the minimisation of the risk objective function. However, as of GWP, the required data have been adapted from a study by Namany et al., (2019) which will be used to run the minimisation of the second objective function. A Solver was used to run and solve the optimization problem.

Table 2: Mathematical formulation for the optimisation model.

Objective Function:	
$\boldsymbol{Min\ Risk} = \sum_{i=1}^{4} R_i^e x_i^e + \sum_{i=5}^{7} R_i^w x_i^w$	Identifies the optimal energy and water technologies technology mix that minimises the total risk and global warming potential of the EWF system delivering 40% of Qatar's demand for food.
$\boldsymbol{Min\ Global\ Warming\ Potential} = \sum_{i=1}^{4} GWP_i^e x_i^e$	Identifies the optimal energy technologies mix that minimises GWP
Subject to:	
$\sum_{i=1}^{4} x_i^e \leq 100\%$	The sum of all contribution percentages of energy technologies must be 100%.

$\sum_{i=5}^{7} x_i^w \leq 100\%$	The sum of all contribution percentages of water technologies must be 100%.
$gx_6^w \leq A_{max}$	Restricts the participation of groundwater in the water mix, being the scarcest water resource in the country studied.
$R_i^e, R_i^w, GWP_i^e, GWP_i^w, x_i^e, x_i^w \geq 0$	It implies that all decision variables must be strictly positive.

Decision variables:
x_i^w: the percentage of contribution of each water source.
x_i^e: the percentage of contribution of each energy source

Parameters:
R_i^e: the risk level of using energy technology (i) where i=1,2,3 and 4
R_i^w: the risk level of using water technology (i) where i=5,6 and 7
GWP_i^e: the global warming potential of energy source. where i=1,2,3 and 4
GWP_i^w: the global warming potential of the water source. where i=5,6 and 7
g: the amount of groundwater utilised for food production with a 40% self-sufficiency level.
A_{max}: the annual renewable rate of groundwater in Qatar is assumed to be 58 million m^3 per year.

3. Result and Discussion

The result from the AHP method (shown in Table 3) indicates that wind energy (25.0%) is the most risker technology to be used for energy supply, where the key challenge is that the availability of this energy source is far from constant but seasonal. On the other hand, treated wastewater (12.5%) is slightly risker than groundwater (12.3%) mainly due to the fact that it should be properly treated to avoid any health issues. The risk values obtained from the AHP method and Namany et al., (2019) study which represent the GWP of each technology are summarised in Table 3 and were used as an input to run the multi-objective optimisation model that aims at minimising the tradeoffs between the two objectives (i.e., risk and GWP). The optimisation model has been conducted to determine the optimal energy and water supply share using different technologies within the existing EWF system. The result from the solver was practicing a tradeoff between the two mutually conflicting objectives and generated relatively 70 optimal solutions, as demonstrated in Figures 2 and 3. Thus, the decision maker can pick any of the optimal solutions based on additional criteria, such as which solution achieved the lowest risk/GWP. For instance, Table 4 illustrates that out of 70 solutions, Solution 1 achieved the lowest risk and solution 69 achieved the lowest GWP. However, by averaging all 70 solutions, then Solution 34 can achieve the balanced minimisation between risk and GWP.

Table 3: Risk and environmental data used for the optimisation model.

Technologies	Solar	Wind	Biomass	Natural Gas	Groundwater	desalinated water	Treated wastewater
Risk %	13.3%	25.0%	14.1%	10.9%	12.3%	11.9%	12.5%
GWP (kg of CO^2eq)	98,328	1,588	-	994,305,184	43,575,206	390,796,923	114,201,322

Table 4: The risk and GWP value of the three optimal solutions.

Solution #	Solar Energy	Wind Energy	Biomass Energy	Natural Gas	Groundwater	Desalinated water	Treated wastewater	Risk	GWP
1	33.3%	33.3%	33.3%	0.0%	4.74%	5.32%	89.84%	0.299	125,500,268
69	0.0%	0.00%	0.00%	99.99%	25.90%	26.08%	47.92%	0.232	1,162,161,486

34	22.9%	1.14%	26.65%	49.29%	8.87%	9.43%	81.68%	0.249	624,080,242

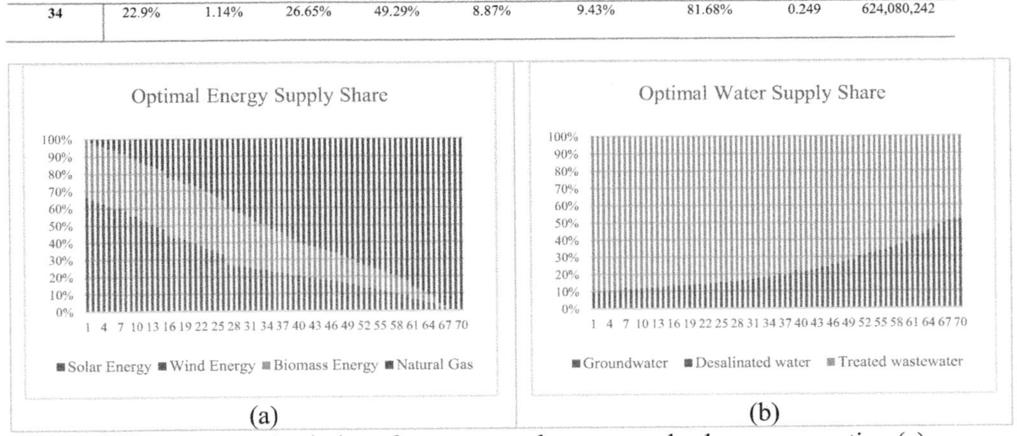

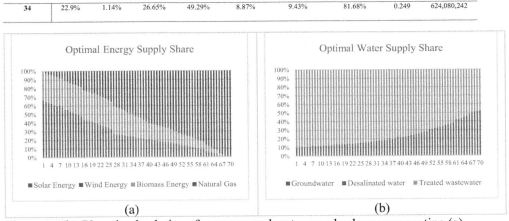

Figure 2: The 70 optimal solutions for energy and water supply share representing (a) energy technologies and (b) water technologies.

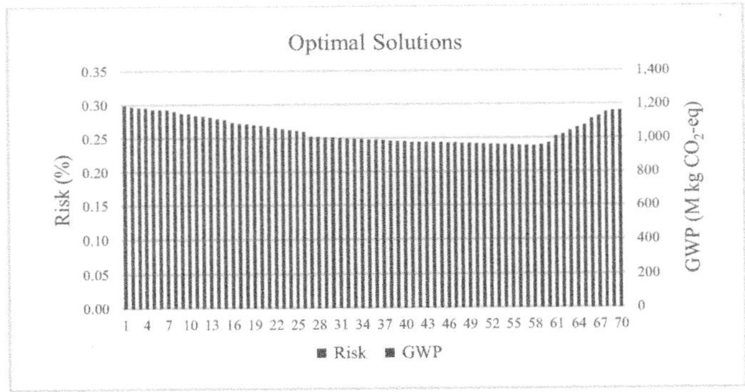

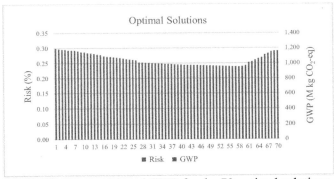

Figure 3: The tradeoff between risk and GWP for the 70 optimal solutions.

Moreover, due to the existence of mutually conflicting objectives, the problem solutions will not be a unique optimal but a set of non-dominated solutions referred to as the Pareto front due to the Pareto dominance concept (Pareto et al., 1896). Moreover, the solution will be Pareto-optimal if there is no other feasible solution that enhances one objective without weakening one another. In this study, the two objectives (risk % and GWP) of the non-dominated solutions set are demonstrated as Pareto Front in Figure 4 using excel.

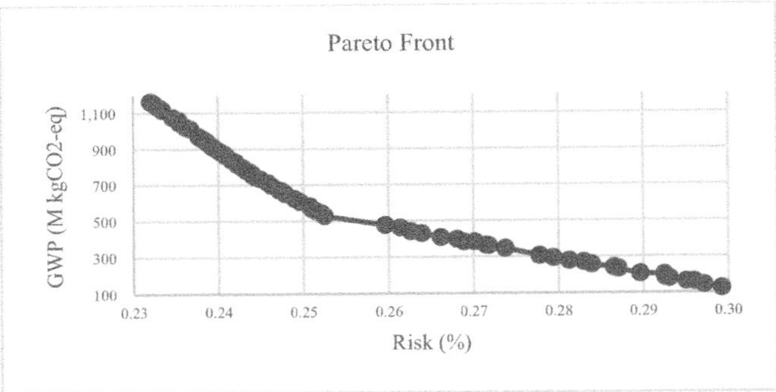

Figure 4: Schema of the Pareto front, for the minimization of two contradictory objectives risk and GWP.

4. Conclusion

This study introduced an EWF Nexus assessment framework to define the technological share needed in optimally supplying energy and water that will minimise the risk and GWP of the overall EWF system using the AHP method and solver. The AHP method has considered financial and operational aspects (e.g., cost, availability, readiness and applicability) of energy and water technologies. The model aimed at providing optimal energy and water supply share from various technologies that minimised overall EWF system risk and GWP. The multi-objective optimization model generates relatively 70 optimal solutions due to the mutual confliction between the two objectives. However, the results from the AHP indicate that wind energy is 11.8% more risker than solar energy, as the availability of these energy sources is seasonal. Furthermore, the treated wastewater was the most environmentally friendly, having the lowest impact on the ecosystem and resources, however, treated wastewater must be properly treated as it has the highest risk level (12.5%) and can lead to health issues. The assessment tool can support decision makers in understandings possible optimal ways of deploying various technologies to supply energy and water resources while minimizing overall risk and GWP of the EWF system.

Acknowledgments

This research was made possible by an Award (GSRA7-1-0407-20014) and supported by proposal number NPRP11S-0107-180216 from Qatar National Research Fund (a member of Qatar Foundation). The contents herein are solely the responsibility of the authors[s].

References

Albrecht, T. C. (2018). The Water-Energy-Food Nexus: a systematic review of methods for nexus assessment. *Environ. Res. Lett.*, *4*(13).

Becker, W. S. (2017). Weights and importance in composite indicators: closing the gap. *Ecol. Indicate.*, (80), 12–22.

Haji, M., Govindan, R., & Al-Ansari, T. (2020). Novel approaches for geospatial risk analytics in the energy–water–food nexus using an EWF nexus node. *Computers and Chemical Engineering*, *140*, 106936. https://doi.org/10.1016/j.compchemeng.2020.106936

Namany, S., Al-Ansari, T., & Govindan, R. (2019). Optimisation of the energy, water, and food nexus for food security scenarios. *Computers and Chemical Engineering*, *129*. https://doi.org/10.1016/j.compchemeng.2019.106513

Pareto, V., Politique, C. D., & Lausanne., F. R. *Volume I and II.* , (1896).

Singh, R. K., Joshi, P. K., Sinha, V. S. P., & Kumar, M. (2022). Indicator based assessment of food security in SAARC nations under the influence of climate change scenarios. *Future Foods*, *5*(January). https://doi.org/10.1016/j.fufo.2022.100122

Zhang, T., Tan, Q., Wang, S., Zhang, T., Hu, K., & Zhang, S. (2021). Assessment and management of composite risk in irrigated agriculture under water-food-energy nexus and uncertainty. *Agricultural Water Management*, *262*(August 2021), 107322. https://doi.org/10.1016/j.agwat.2021.107322

Antonis Kokossis, Michael C. Georgiadis, Efstratios N. Pistikopoulos (Eds.)
PROCEEDINGS OF THE 33rd European Symposium on Computer Aided Process Engineering (ESCAPE33), June 18-21, 2023, Athens, Greece
 http://dx.doi.org/10.1016/B978-0-443-15274-0.50045-7

Prediction of Melt Flow Rate (MFR) in polymer production, with auto-regressive Machine Learning algorithms, for multivariate time-series measured in irregular timestamps

Symon Doe[a], Christoforos Kassianides[a], Symeon Kassianides[a], Christos Christodoulou[b], Nikos Bakas[b]

[a]*Hyperion Systems Engineering Ltd., 36 Athalassis Street, Geri-Nicosia, 2201, Cyprus*
[b]*Computation-based Science and Technology Research Center, The Cyprus Institute, 20 Konstantinou Kavafi Street, Aglantzia-Nicosia, 2121, Cyprus*

Abstract

The purpose of this work is to use Machine Learning algorithms to implement soft sensors to predict the material quality (MFR – Melt Flow Rate) at polymer production plants. The goal is to identify off-spec material in real-time without needing specialist online analyzer equipment. The Machine Learning algorithms use historical time series data of production plant measurements. These historical data comprise a wide variety of associated features, potentially driving the evolution of the material over time. The features are the time series values of specific physical parameters, such as temperature, pressure, etc., measured on or before the polymerization reactors. We investigated a variety of mathematical formulations for predicting MFR and Machine Learning Algorithms accordingly. We present the dataset, as well as machine learning models results along with performance metrics and future steps.

Keywords: machine learning, multivariate time series, autoregressive model, feature selection, polymer production.

1. Introduction

The aim of this work is to predict the material quality (ASTM-MFR) at polymer production plants, based on historical time series data. These historical data comprise a variety of associated features, potentially driving the evolution of the material over time. Accordingly, the given dataset is comprised of approximately three thousand signals measured in different timestamps among them, as well as with the target variable (MFR). The research started with an investigation of the statistical properties of the studied variables, followed by feature selection, and predictive modelling with Machine Learning Algorithms. In any predictive modelling computation, the predicted values always differ from the given ones in the raw dataset (Bakas, 2019). Hence, and despite the extended effort made to create a robust model, the predictions deviate from the actual values of MFR, especially in certain timestamps, due to the high amount of statistical noise existing within the variables. The input variables, exhibit distributions with no known patterns or even chaotic ones, and we present their computational handling, and numerical results.

We had to deal with specific challenges, such as the sparseness of MFR results (low frequency, measured every 4 hours on average), while the corresponding features utilized as an input in the Machine Learning Models are measured more frequently, e.g. per minute or even per second. Furthermore, we had to smooth the signals' outliers and deal with their dynamic nature, as the correlations of the features with the target variable, though strong, vary significantly with time and disorientate the predictive modelling. Furthermore, we present how we approached the cleaning of the data along with heuristic features' selection algorithms (Bakas, et al., 2022), so as to obtain a structured dataset to feed the Machine Learning Models. The amount of noise in the dataset was significant; hence we utilized a variety of autoregressive models and time lags (Bakas, 2019), in order to attain the best possible prediction accuracy. We found a combination of XGBoost (Chen & Guestrin, 2016; Xu & Chen, 2014), with the gradients of the features ∇Fk at a time stamp k, to exhibit adequate accuracy. Furthermore, we investigate various ML models as new data are being obtained constantly. Particularly, we use Polynomial Regression (PR) (AlHamaydeh, et al., 2022), Random Forests (Breiman, 2001; Sadeghi, 2013), and Artificial Neural Networks (Bakas, et al., 2019).

To address the issue of noise and time-shift observations in the raw dataset, a hybrid approach was developed. This will use a simple model of the process equipment, which will precondition the data values by applying correct time offsets and filtering before sending it to the Machine Learning algorithms. This procedure removes the burden of processing noise and applying time offsets from the Machine Learning algorithms. As further development the model could identify different operating conditions, e.g., which catalyst is being used in the reactors, and/or change of raw material feeds to produce High/Low MFR product. This allows Machine Learning models to be trained for each condition with the expectation that each model will be more accurate within its defined range than a single combined model.

2. Specific Challenges

2.1. Sparseness of MFR results

We aim to predict MFR values which are measured every 4 hours on average (low frequency), while the corresponding features utilized as an input in the Machine Learning Models, are measured more frequently, e.g. per minute or even per second. We investigated a variety of models, initially we keep constant the previously known MFR value for the unknown regions, and ultimately, using only t timestamps with known values for all predictors and target.

2.2. Signal Outliers

The given dataset comprises a rich variety of potential features for the prediction of MFR. Twenty-one features exhibit high Pearson's Correlation with MFR (>0.9). However, these correlations change with time strongly, along with the corresponding outliers in the features. Signal outliers disorientate the predictions made by any of the utilized models. To confront this, we used robust-to-noise machine learning models and also tailor-made algorithms for feature selection, to obtain a final set of features which predicts MFR with the best possible accuracy.

3. Computational Approach

Exploratory Data Analysis was performed for the fundamental statistical properties, as well as inter-item correlation and clustering among the time-series. Predictive modelling has started using various time-lags and linear models to identify prediction accuracy with the given data. The predictive modelling of the database was implemented by utilizing four ML methods, and in particular: a. linear, b. higher-order regression, c. random forests, and d. gradient boosting. Machine learning methods exhibit diverse performance on the studied dataset, with respect to the error metrics each time utilized. The utilized error metrics were the Root Mean Squared Error, the Mean Absolute, the Mean Absolute Percentage Error, the Maximum Absolute Percentage Error, the slope of the Predicted vs Actual values, as well as the Pearson Correlation Coefficient. The studied data set contains a rich variety of input variables. However, many of them comprise extreme outliers. In order to identify an optimal subset of features, we use optimization algorithms (Bakas, et al., 2022; Plevris, et al., 2021).

3.1. Mathematical Formulation

Let Q be the sought Quality of MFR, which is intended to be predicted. We use symbol k as the index for the timestamps of Q. k, may iterate $\in \{1,2,\dots,\mathrm{m}\}$ values of all time stamps where the features are being measured, or iterate $\in \{1,2,\dots,n\}$ known values of Q, with $n \ll m$. Hence at each time stamp k we have a particular value Q_k, while the entire vector of the target variable is

$$Q := Q_k \tag{1}$$

Accordingly, we define $F_{k,j}$, with $j \in \{1,2,\dots,p\}$, for p features in total, the value of an input variable (feature) at time k, such that the vector of features at a time stamp k is

$$F_k := \{F_{k,1}, F_{k,2} \dots, F_{k,p}\}, \tag{2}$$

and hence the entire matrix of features is

$$F := F_{k,j} \tag{3}$$

Let $\mathcal{M}: R^p \to R$ be an continuous function of the features F_k returning the Quality values Q_k. We approximate $\mathcal{M}$with Machine Learning Algorithms, such that

$$Q_k = \mathcal{M}(F_k) + \epsilon_k, \tag{4}$$

and

$$\epsilon_k \sim \mathcal{U}(0, \sigma), \tag{5}$$

the regression errors, which should ideally follow a distribution $\mathcal{U}$ exhibiting a Gaussian shape with zero mean.

This is the baseline approach. We may enhance this model, by utilizing the moving average of each feature j, select a subset o of the p features, add the time-lagged features as new features, and also add the differences of $F_{k,j}$ along with the time passed until a difference occurs, as per the following formulation.

We define the differences of Q, at time k as

$$\Delta Q_k := Q_k - Q_{k-1}, \tag{6}$$

and the corresponding differences of$F_{k,j}$,

$$\nabla F_k := \{\Delta F_{k,1}, \Delta F_{k,2}, \dots, \Delta F_{k,p}\}. \tag{7}$$

Let $\mathcal{L}: R^{2p+1} \to R$ be an continuous function of the features F_k, such that

$$\Delta Q_k = \mathcal{L}(F_k, \nabla F_k, \Delta t_k) + \epsilon_k. \tag{8}$$

We now identify $\mathcal{L}$ with Machine Learning, taking into consideration the distribution of the errors ϵ_k as above-mentioned.

Afterwards, we may predictQ_k, by using

$$Q_k = Q_{k-1} + \Delta Q_k. \tag{9}$$

For the real-time, prediction stage, we iterate for each minute (or appropriate real-time interval), and sequentially add to the previous Q_{k-1}, the computed ΔQ_k, as predicted, with $\Delta t_k = 1$, and $k \in \{1,2,\dots,m\}$.

3.2. Latest data and Model

We retrain the models, after the data engineering as was automated by the HYPPOS program, enriched by new data, by using all raw features at MFR measurements timestamps as an input for the ML models, and target variable MFR. Particularly, this version is for ALL MFR values and both Catalysts. For Nonlinear Regression (NLR), XGBoost (XGB), and Random Forests (RF), we use exhaustive grid search with cross validation. The search parameters for tuning are presented in Table 1. For example, for XGBoost model, we run 36 combinations of Hyper-Parameters, times 100 intermediate rounds = 3600 different XGBoost models in total, and we select the best one. Furthermore, for each model, we run it 5 times in order to do cross-validation with a random permutation of the Train set. Hence the actual models are 18000.

Table 1. Hyperparameter Tunning Search Space.

Nonlinear Regression	
number of NLR folds	100
Percentage of NLR Cross Validation Samples per fold	0.8
Maximum Number of NLR Rounds	10
Maximum Number of NLR Features	∞
Polynomial Degree	2
XGBoost	
number of XGB folds	5
Percentage of XGB Cross Validation Samples per fold	0.8
Minimum Number of XGB Rounds	10
Maximum Number of XGB Rounds	1000
Eta	0.05, 0.2, 0.5
Percentage of Combinations to Check	1
Number of Best Models to keep	10
Depth	1, 7, 15
Number of Intermediate Rounds	100
colsample_bytree	0.5, 1
Subsample	0.5, 1
Random Forests	
number of RF folds	5
Percentage of RF Cross Validation Samples per fold	0.8
Percentage of RF Combinations to Check	1
Number of Best Models to keep	10
n_trees	10, 100, 1000
n_subfeatures	0.25, 0.5, 0.75
partial_sampling	0.25, 0.5, 0.75
max_depth	1, 10, 100
min_samples_leaf	1, 10, 100

4. Results

The best performing model for the new data after data engineering, was a combination of parameters of the Random Forests (Table 2). In Figure 1, we present the error analysis in the Test Set for Random Forests model. We may see that the errors are low on average, with some specific regions of high outliers (Figure 1).

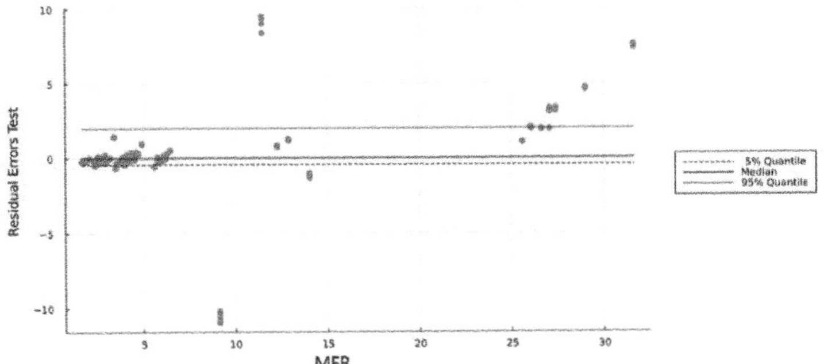

Figure 1: Prediction Errors in the Test Set, using the Random Forests model.

In order to explain statistically the drives of outliers in Figure 1, we performed an error analysis by following the next steps: (a) we split the Train Set into 1000 bins, equally distributed from the minimum value of MFR, up to its maximum value, (b) we counted the number of MFR values found per each particular bin, (c) for each bin, we computed the Mean Absolute Error (MAE), among the given MFR values, and the predicted ones, in the Test Set, (d) we plotted the results in a scatter plot (Figure 2), along with the 95% & 99% Quantiles of the MAE.

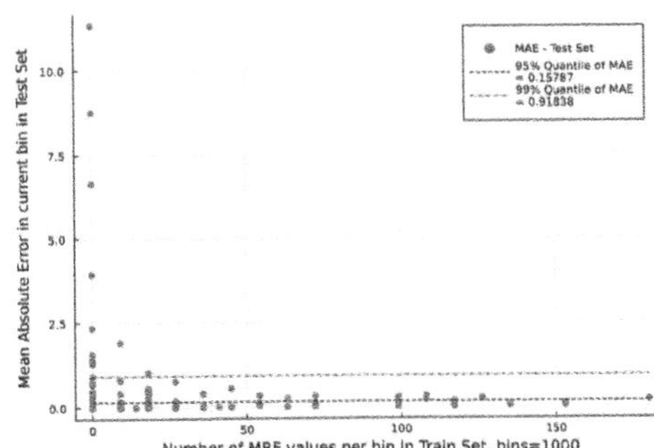

Figure 2: Number of MFR values per bin in the Train Set vs MAE in the Test Set.

Interestingly, we observe an inverse association among the number of MFR values per bin in the Train Set, and the corresponding MAE in the Test Set. Hence, we empirically deduce that the density of MFR values per MFR range corresponds to low errors and vice versa. Hence, e.g. for high values of MFR (>25) the high errors occur as a result of low number of MFR values in the training dataset. We foresee in the future steps of this work to gather more data and train new models, to resolve the data adequacy issue.

Table 2: Performance Metrics for all Models Used

Model	Dataset	Pearson	M.A.P.E.	M.A.E.	R.M.S.E
Linear	Train	0.988	0.165	0.791	1.176
Linear	Test	0.946	0.310	1.246	2.120
Polynomial	Train	0.992	0.127	0.650	0.998
Polynomial	Test	0.967	0.234	0.907	1.598
XGBoost	Train	1.000	0.000	0.001	0.002
XGBoost	Test	0.983	0.075	0.585	1.462
ANNBN	Train	0.999	0.054	0.260	0.350
ANNBN	Test	0.418	0.277	1.838	12.939
Random Forests	Train	1.000	0.005	0.032	0.080
Random Forests	Test	0.970	0.071	0.548	1.555

5. Conclusions

In this work, we developed machine learning models for the MFR time series prediction on real industrial conditions. The input for the models is a wide variety of industrial measurements, as obtained in the timestamps MFR was measured. We investigated various mathematical formulations, along with machine learning models. We found that hyperparameter tuning with exhaustive search, and tree-base models (XGBoost and Random Forests) yielded the best possible performance. The average errors are low for entirely out-of-sample data points. However, as outliers also exist in specific MFR timestamps, we foresee retraining the models with new data that will be obtained gradually in real-time industrial conditions. Despite the vast noise in data, irregular timestamps of MFR measurements and limited dataset, we developed a model exhibiting adequate accuracy on average.

Acknowledgements

This work received financial support by the EuroHPC-JU project EuroCC (G.A. 951732) of the European Commission, and the Research and Innovation Foundation of Cyprus programs for research, technological development and innovation "RESTART 2016–2020" (RIF PROPOSAL NUMBER: ENTERPRISES/0521/0175). Parts of the runs were performed on the Cyclone machine hosted at the HPC National Competence Center of Cyprus at the Cyprus Institute (project pro21b103).

References

M. AlHamaydeh, G. Markou, N. Bakas, M. Papadrakakis, 2022, AI-based shear capacity of FRP-reinforced concrete deep beams without stirrups, Engineering Structures, Volume 264.

N. Bakas, A. Langousis, M. Nicolaou, S. Chatzichristofis, 2019, A Gradient Free Neural Network Framework Based on Universal Approximation Theorem, s.l.:https://arxiv.org/abs/1909.13563.

N. Bakas, 2019, Numerical Solution for the Extrapolation Problem of Analytic Functions, Research - Science Partner Journal, Volume 2019.

N. Bakas, V. Plevris, A. Langousis, S. Chatzichristofis, 2022, Itso: A novel inverse transform sampling-based optimization algorithm for stochastic search, Stochastic Environmental Research and Risk Assessment, Volume 36, p. 67–76.

L. Breiman, 2001, Random forests, Machine Learning, Volume 45, p. 5–32.

T. Chen, C. Guestrin, 2016, XGBoost: A Scalable Tree Boosting System. s.l.:Proceedings of the 22nd ACM SIGKDD International Conference on Knowledge Discovery and Data Mining.

V. Plevris, N. Bakas, S. German, 2021, Pure Random Orthogonal Search (PROS): A Plain and Elegant Parameterless Algorithm for Global Optimization, Applied Sciences, Volume 11, p. 5053-5081.

B. Sadeghi, "DecisionTree.jl," 2013. [Online]. Available: https://github.com/JuliaAI/DecisionTree.jl.

B. Xu and T. Chen, "XGBoost.jl," 2014. [Online]. Available: https://github.com/dmlc/XGBoost.jl.

Antonis Kokossis, Michael C. Georgiadis, Efstratios N. Pistikopoulos (Eds.)
PROCEEDINGS OF THE 33rd European Symposium on Computer Aided Process Engineering (ESCAPE33), June 18-21, 2023, Athens, Greece
 http://dx.doi.org/10.1016/B978-0-443-15274-0.50046-9

Achieving Operational Excellence by Combining Material Tracking and On-line Polymer Analysis and Classification data in an all-in-one Integrated Decision Support System

Symeon Kassianides[a], Symon Doe[a], Sanja Micunovica[a], Artemis Theodoropoulou[a], Charis Komodromos[a], Sofia Louloudi[a]

[a]*Hyperion Systems Engineering Ltd., 36 Athalassis Street, Geri-Nicosia, 2201, Cyprus*

Abstract

The HYPPOS software has been developed to encapsulate experience gained over 15 years of building bespoke solutions for the polymer industry. HYPPOS not only is configurable and easily maintainable, but it also uniquely provides the ability to combine data analytics and Artificial Intelligence to support decision making.

We developed a mathematical material tracking system that discretizes continuous manufacturing processes into identifiable slices of material ("Quanta") and track them as they move through the manufacturing stages. To achieve this, HYPPOS is integrated with production and business management ICT systems. HYPPOS transforms data collected into information and applies different inferencing and Machine Learning algorithms, which calculate critical quality parameters, then tracking them in real-time. As the Quanta move past sensors, they collect KPI's and quality parameters making the history of the material visible.

Keywords: petrochemicals, sustainability, real-time monitoring, waste reduction, decision support

1. Introduction

Today polymer production and forming processes account for more than 80% of the carbon footprint of plastic manufacturing (Hydrocarbon Processing, 2018). That presents detrimental environmental challenges that need immediate attention and intervention. The International Energy Agency 's (IEA) "Future of Petrochemicals" report (IEA, 2018) recommends building a more sustainable and efficient petrochemicals industry. Early identification of potential process issues and their correct handling is critical to reducing material waste and energy consumption. This can be achieved through specialized software technology that enables automatic and real-time analysis. However, current software platforms do not offer the required specialization, flexibility, and information without extensive customization.

Identifying the gap in the currently available solutions, we developed HYPPOS - Hyperion Predictive Production Online Software. HYPPOS as presented at the 11th International Conference on Chemistry and Chemical Engineering (ICCCE, 2020) and at the 9th International Conference on Chemical and Process Engineering Materials (ICCPE, 2020) is an all in-one Artificial Intelligence driven decision support tool that addresses specific challenges of the polymers industry including quality traceability, real-time analysis, rapid and accurate reporting, process visualization and quality consistency. HYPPOS is positioned in Industry 4.0 Tactical Management Level, it collects data from systems at the Transactional Management Level, feeds and receives data from systems at the Strategical Management Level (e.g. Enterprise Resource Planning Systems such as SAP) and co-exists and interacts with other systems at the Tactical Management Level (e.g., Laboratory Information Management Systems).

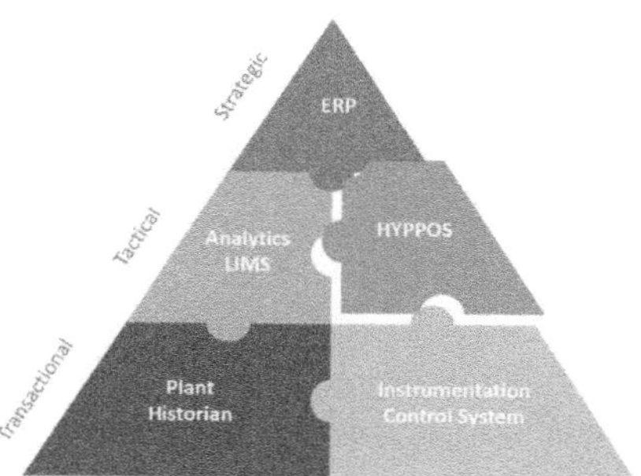

Figure 1: HYPPOS Positioning as per ISA 95 Industry Standard.

Polymer manufacturing requires quality monitoring at the different production stages from the polymerization reactors through to extrusion and batch assembly in blender silos. For this purpose, specialized On-Line Polymer Analysis equipment (OLPA) is available. Nevertheless, the majority of polymer plants are not equipped with such on-line analyzers primarily due to their high cost. Instead, process engineers and plant operators usually rely on off-line laboratory measurements of quality critical parameters. This does not allow real-time monitoring and early corrective actions, and as a result it leads to off-spec polymer product. Therefore, the aim of this work is to embed Machine Learning (ML) algorithms into HYPPOS software to create Soft Sensors to predict quality in real-time at early production stages eliminating the need for on-line physical sensors.

The Melt Flow Rate (MFR) is one of the most frequently used parameters in polymer manufacturing to control the material quality, and it is a measure of the viscosity of the polymer. Therefore, the MFR value range constitutes the most fundamental quality specification of the final polymer product since it determines the product grade.

2. Method

HYPPOS combines material flow models with AI/ML to create a digital twin of the production equipment and process. Its monitoring and inferential capabilities allow the online tracking of product quality as it moves through the production process and can forecast the final quality of polymer being produced enabling early corrective real-time actions. It uses a proprietary new method that allows the discretization of continuous processes and formulations of material into small 'Quanta' (depicted as *Qi* in Figure 2), which act as a discreet container of trackable information enabling the traceability of material and its properties throughout the entire production process. The integrated tracking system is based on a unique, in-house developed process flow model which moves the Quanta through the equipment on a FIFO basis. Using an analogy, the 'Quanta' are located on a conveyor belt that helps them move through the process equipment advancing at the same speed as the production stream.

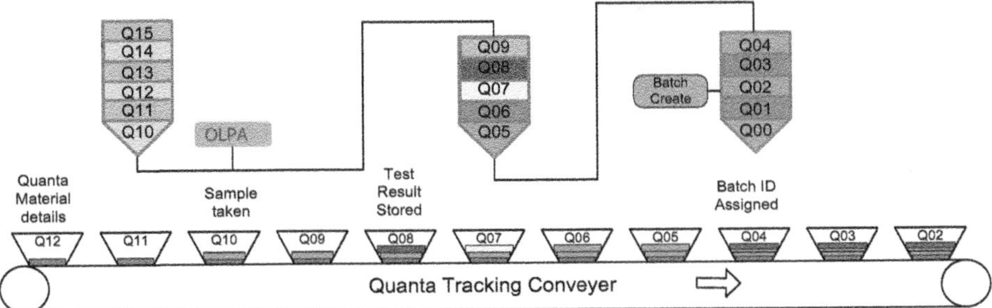

Figure 2: Analogy for Quanta Tracking System.

The model covers the main production stream from process feeds through to final product. The model is block based, with Equipment Blocks, Movements and Modifier blocks which merge/separate streams. The Movements transfer are driven by plant measurements, and transfer Quanta between Equipment using this information. The exact transfer rate calculation depends on the availability of plant measurements. Where instrumentation is missing, special proprietary algorithms calculate the movement rates based on measurements in surrounding equipment. The Equipment Block maintains details of the material holdup within the process equipment. Additionally analytical processes have been development to process the information collected by the model and produce detailed reports for the final product.

When real time OLPA instrumentation is available, the Quanta can collect and carry this information and can be used to display the product quality properties as it advances through the production process.

Efforts to predict the polymer quality through Soft Sensors that use ML models based on plant measurements have been reported before (Shi, et al., 2006; Jumari & Mohd-Yusof, 2016; Liu, et al., 2018; Abeykoon, 2018; Yan, et al., 2023;). The innovation of this work lies in embedding the ML models in the HYPPOS Quanta algorithm that is implemented at industrial scale and it allows the real-time tracking of material together with its quality characteristics as it moves through the production equipment.

We have identified the following challenges when implementing ML algorithms to a continuous process such as polymer production. The process measurements used as input in the ML model are recorded continuously in real-time, whereas the target variable (MFR) is measured on average every 4 hours. Since MFR measurements come off-line from the lab, they become available with a delay of around 2 hours. Therefore, they need to be time shifted to various points in the process.

Furthermore, the plants are using relatively infrequent Lab test results for the quality parameter measurements, and this makes it difficult to obtain useful quantities of data to train the ML system to predict these parameters.

While initial tests using information from the plant were promising, they were not working well in all situations, especially during process transients. To simplify the ML algorithms task, we developed a pre-conditioning algorithm which is based on a simple model description of the process stages. The model defines the measurement signals, and the time duration of each stage. This allows the pre-conditioner to correctly time shift the measurements and apply noise reduction techniques to the values.

3. Case Study Results and Discussion

The objective was to predict the MFR on the polymerization reactor of an industrial scale polypropylene plant. To train the ML system, 5 months of raw process data from the plant including Lab measurements were used. Features selection algorithm was implemented, and 47 process variables were found to correlate the best with MFR measurements including among others the hydrogen feed rate to the polymerization reactors, doner concentration, teal concentration, the catalyst type, the reactor pressure and reactor temperature, and input material flow rates.

The raw data was run through a specially developed tool that converted the raw data into a training data set. The training data set contained the target variable values (MFR lab measurements) and all the pre-conditioned values that affected the production at that instant.

Using the trained ML algorithms, predictions were made using pre-conditioned process test data. This resulted in a marked improvement in prediction accuracy and better handling of process transients.

Figure 3 below shows XGBoost prediction of MFR parameter without using pre-conditioned data. It clearly shows that the predictions anticipate changes in MFR, therefore the Orange prediction line generally changes before the Blue measured line.

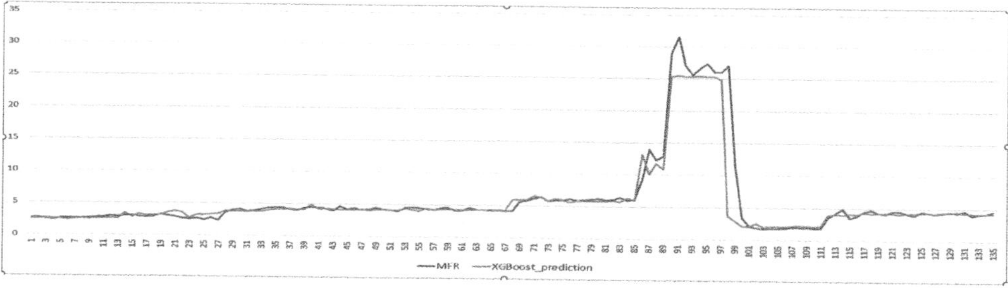

Figure 3: MFR Predictions without Preconditioning.

Figure 4 shows XGBoost prediction of MFR parameter with pre-conditioning. This shows great improvement in time shift and overall accuracy.

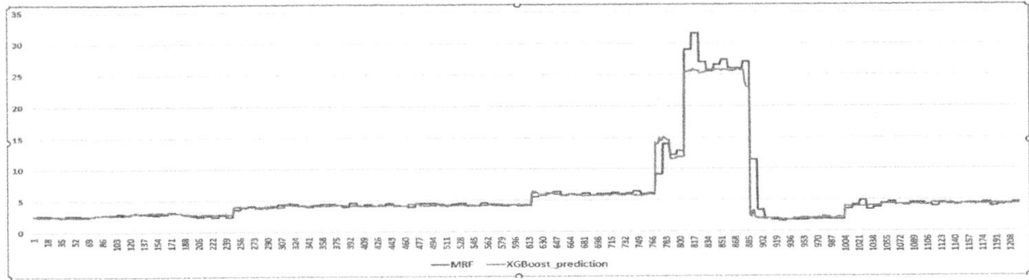

Figure 4: MFR Predictions with Preconditioning.

It is evident from the above graphs that there is a significant improvement in terms of tracking changes in operating conditions as now the Orange line is not time displaced from the Blue as in previous graph. There is also an improvement in prediction accuracy which is shown by the performance metrics calculated for each case as shown in the Table below.

Table 1. Performance Metrics for Methodologies Used.

	Pearson Correlation	Mean Absolute Percentage Error
No Preconditioning	0.92846	0.09523
With Preconditioning	0.98381	0.06697
Improvement	5.96%	-29.67%

The pre-conditioning technology has now been developed into a module within the HYPPOS software. This uses the process data values collected by HYPPOS in real-time, and at specified intervals it produces a table of pre-conditioned values. This table is passed to the ML prediction algorithm, which predicts the MFR, and the result is returned and passed to HYPPOS for use as part of the product quality management functionality.

It should be noted that these results were achieved with a relatively small quantity of training data, and that the installed system is expected to improve further on accuracy as it retrains itself based on continuous data collection.

4. Conclusions

Based on this, HYPPOS is expected to have a significant positive impact on production efficiency, profitability, and competitiveness while also reducing waste, with a consequential contribution in lowering the carbon footprint.

The implementation of ML pre-conditioning has been a significant step forward in improving accuracy of the quality parameters. And it demonstrates that it can produce useful results with relatively little training data, allowing rapid deployment on new plants. This technology gives HYPPOS users a real-time view of the quality of the product being produced, enabling rapid intervention and reduction in waste and off-spec production.

Acknowledgement

This work received financial support by the Horizon 2020 – 2nd Opportunity program of the European Commission, and the Research and Innovation Foundation of Cyprus programs for research, technological development, and innovation “RESTART 2016–2020” (RIF PROPOSAL NUMBER: OPPORTUNITY/0916/SME-II/0005).

References

https://www.hydrocarbonprocessing.com/news/2018/10/petrochemicals-set-to-be-the-largest-driver-of-world-oil-demand

https://www.iea.org/news/petrochemicals-set-to-be-the-largest-driver-of-world-oil-demand-latest-iea-analysis-finds

S. Louloudi, 2020, Software for Real-time Tracking, Analysis & Visualization of Polymer Data in Production Plants, Presentation at the 9th International Conference on Chemical and Process Engineering Materials (ICCPE 2020), 20-22 May, http://www.iccpe.org/ICIEM&ICCPE%20Conference%20Program-2020.pdf

S. Louloudi, 2020, Software for Real-time Tracking, Analysis & Visualization of Polymer Data in Production Plants, Presentation at the 11th International Conference on Chemistry and Chemical Engineering (ICCCE 2020), 08-10 July 2020.

Y. M. C. Abeykoon, 2018, Design and Applications of Soft Sensors in Polymer Processing: A Review, Sensors Journal, IEEE, [10.1109/JSEN.2018.2885609], https://doi.org/10.1109/jsen.2018.2885609

W. Yan, T. Dong, Y. Zhou, Z. Luo, 2023, Computational modeling toward full chain of polypropylene production: From molecular to industrial scale, Chemical Engineering Science, Volume 269.

N. Jumari, K. Mohd-Yusof, 2016, Comparison of Product Quality Estimation of Propylene Polymerization in Loop Reactors Using Artificial Neural Network Models, Jurnal Teknologi.

J. Shi, X. Liu, Y. Sun, 2006, Melt index prediction by neural networks based on independent component analysis and multi-scale analysis, ScienceDirect, Volume 70, pp. 280-287.

S. Liu, X. Gao,W. Qi, S. Zhang, 2018, Soft sensor modelling of propylene conversion based on a Takagi-Sugeno fuzzy neural network optimized with independent component analysis and mutual information, Transactions of the Institute of Measurement and Control.

Antonis Kokossis, Michael C. Georgiadis, Efstratios N. Pistikopoulos (Eds.)
PROCEEDINGS OF THE 33rd European Symposium on Computer Aided Process Engineering (ESCAPE33), June 18-21, 2023, Athens, Greece
 http://dx.doi.org/10.1016/B978-0-443-15274-0.50047-0

Markov Chain Monte Carlo simulation-based optimization for a production scale milk drying process

Adrián Ferrari[a], Soledad Gutiérrez[a], Gürkan Sin[b]

[a]*Chemical & Process Systems Engineering Group – Chemical Engineering Institute - Engineering School – Universidad de la República, Julio Herrera y Reissig 565, PC 11300, Montevideo, Uruguay*

[b]*Process and Systems Engineering Centre - Department of Chemical and Biochemical Engineering, Technical University of Denmark, Søltofts Plads, Building 229, DK-2800 Kgs. Lyngby, Denmark*

aferrari@fing.edu.uy

Abstract

Spray drying is widely used for dehydration of dairy products and among, the most energy-intensive unit operation in this field. An optimization problem for a production scale milk drying system was implemented and solved, considering plant capacity (maximization) and energy consumption (minimization) as the objective. Decision variables were inputs to the spray unit namely the inlet dry bulb air temperature, concentrate moisture, and dry solids flow rate. Product stickiness conditions and moisture content were the main constraints, which were modeled using mass and energy balances. A non-deterministic derivative free based optimization technique, namely Markov Chain Monte Carlo (MCMC) algorithm was chosen to solve the problem. The results showed that throughput maximization is achieved at the expense of a relative energy consumption penalization in the spray and revealed that the structure of the problem seems to be convex. This study shows a promising non-conventional use of MCMC algorithms in optimization studies.

Keywords: Milk drying process, process optimization, derivative free optimization, MCMC.

1. Introduction

The most widely used technique for dehydration of dairy products is spray drying after evaporation. These processes preserve food properties and allow storage of powders at an ambient temperature (Schuck et al., 2008). A drying plant normally presents three stages: 1) spray chamber; 2) internal fluid bed at the conical base of the spray chamber; and 3) external fluid bed to fine tune the outfeed product stream, as it is shown in Figure 1 (Schuck et al., 2005). As spray drying is the most energy-intensive unit operation in the dairy industry, process optimization becomes of critical importance to increase efficiency and reduce production costs (Petersen et al., 2015). As a continuation of previous works in a production scale system about modelling, uncertainty, identifiability, and sensitivity analysis from the point of view of identifying influential inputs/parameters and quantifying confidence intervals in model simulations (Ferrari et al., 2015; Ferrari et al., 2016; Ferrari et al., 2017), an optimization problem for the entire milk drying process is here implemented and solved.

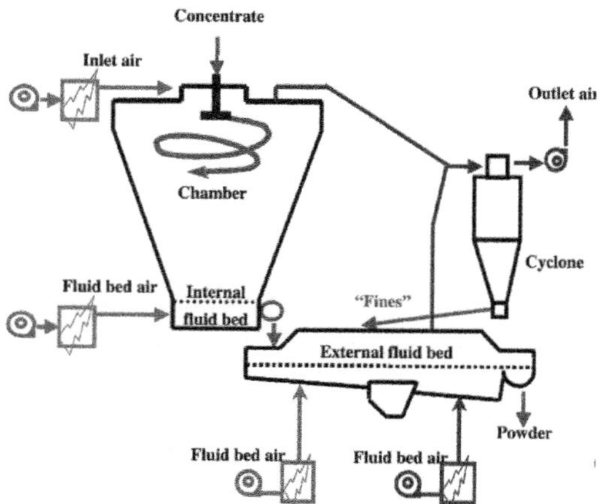

Figure 1. Physical system of three stage dryer scheme.

Maximization of plant capacity and minimization of energy consumption are considered in the objective function. Decision variables are all inputs at chamber inlet streams:

- $T_{A,1}$: Chamber inlet dry bulb air temperature [K], which is feasible to be manipulated at the feed to the spray unit.
- $M_{P,1}$: Concentrate moisture [$kg_{water}.kg_{total}^{-1}$], which is feasible to be manipulated by the output from the evaporation plant upstream of the drying processes.
- S: Dry solids flow rate [$kg_{dry\ solids}.h^{-1}$], which is feasible to manipulate via the feed to the evaporation plant.

Product stickiness conditions at internal fluid bed and moisture content in final powder are assumed as the main constraints in the problem. A global sensitivity analysis of the optimization results is also performed respect to normal and not controlled variations in the chamber inlet air humidity ($H_{A,1}$; range: 0.003 – 0.020 $kg_{water}.kg_{dry\ air}^{-1}$), finding the corresponding result for a different $H_{A,1}$ value at a time. A non-deterministic/derivative free based technique (Markov Chain Monte Carlo algorithm) is chosen to solve the problem to overcome issues when using a deterministic gradient-based method. The MCMC use is also interesting for trying to visualize non-convexity patterns in the model finding the posterior density functions of decision variables, representing a non-conventional use of MCMC. Details and results about the deterministic approach are not covered in this paper because of length restrictions. The work is structured as follows: first the conceptual optimization problem is described, then the optimization technique is introduced, afterwards the obtained results are analyzed and discussed.

2. Materials and methods

2.1. Optimization problem

To consider plant capacity maximization and energy consumption minimization, the objective function (OF) is structured as follows:

$$OF = \alpha.T_{A,1} - \beta.P_{out} - \gamma.W_{dried} \tag{1}$$

Being:

$$P_{out} = S.[1 + \frac{M_{P,4}}{1-M_{P,4}}] \quad (2)$$

$$W_{dried} = S.[\frac{M_{P,1}}{1-M_{P,1}} - \frac{M_{P,4}}{1-M_{P,4}}] \quad (3)$$

Where P_{out} is the total flow rate at system output [$kg.h^{-1}$]; W_{dried}, the total water dried in the system [$kg_{water}.h^{-1}$]; $M_{P,4}$, the final powder moisture [$kg_{water}.kg_{total}^{-1}$]; and $\alpha/\beta/\gamma$, dimensionless coefficients to have balanced terms in the objective function. It will be assumed $\alpha = 20$, $\beta = \gamma = 1$, all constant during the complete study. The main inequality constraints to be considered are related to the product stickiness [using glass transition temperature as indicator (Silalai and Roos, 2010)] at the internal fluid bed, and also with the residual moisture content in final powder, in accordance to Eq. (4) and (5).

$$T_{P,3} \leq T_{g,3} \quad (4)$$

$$M_{P,4} \leq M_{P,max} \quad (5)$$

Where $T_{P,3}/T_{g,3}$, are the product one and its glass transition temperature at the internal fluid bed [°C]; and $M_{P,max}$, is the maximum allowed powder moisture at the end of the system [$kg_{water}.kg_{total}^{-1}$]. It is assumed in this work a value of 0.030 $kg_{water}.kg_{total}^{-1}$ for this parameter (commercial constraint for whole milk powder). The decision variables for the optimization are $T_{A,1}$, $M_{P,1}$, and S, with the following bounds to screen an important range of operating conditions in the system.

$$453\ K \leq T_{A,1} \leq 503\ K \quad (6)$$

$$0.47\ kg_{water}.kg_{total}^{-1} \leq M_{P,1} \leq 0.53\ kg_{water}.kg_{total}^{-1} \quad (7)$$

$$5{,}000\ kg_{dry\ solids}.h^{-1} \leq S \leq 6{,}600\ kg_{dry\ solids}.h^{-1} \quad (8)$$

The optimization problem will minimize energy consumption and maximize throughput in the system, changing $T_{A,1}$, $M_{P,1}$, and S as decision variables, subject to mass and energy balances with also powder stickiness and residual moisture constraints. The detailed process and product stickiness models are described in Ferrari et al. (2015), Ferrari et al. (2016), and Ferrari et al. (2017). The complete structure of the optimization problem is summarized as follows:

$$\min_{T_{A,1};M_{P,1};S} \quad OF$$

$$s.t:$$

$$T_{P,3} \leq T_{g,3}$$

$$M_{P,4} \leq M_{P,max}$$

$$T_{A,1}; M_{P,1}; S \in R$$

2.2. Numerical methods

As a non-deterministic/derivative free technique, a Markov Chain based algorithm was used for solving the optimization problem and for detecting non-convex behaviours. This represents a non-conventional use of Markov Chain based algorithms, but it is interesting to analyze its capabilities and obtain first insights for finding posterior density functions of decision variables and trying to visualize non-convexity patterns. The Metropolis random walk algorithm (Gelman et al., 2004; Weinzierl, 2000) was also used as the method to accept/reject new samples from transition distributions. The algorithm proceeds as follows:

Step 1: randomly select a feasible initial vector of decision variables, θ^0. A uniform distribution is assumed for all of them.

Step 2: create a new trial vector, $\theta^* = \theta^{k-1} + \Delta\theta$, where $\Delta\theta$ is randomly sampled from a jumping distribution $q(\Delta\theta)$. This is assumed as multivariate normal centered at current iteration and calculated as follows:

$$\Delta\theta = u.\frac{2{,}4}{\sqrt{l(\theta)}} \tag{9}$$

Where u is a unit normal random number drawn for each decision variable; and $l(\theta)$, the length of θ vector.

Step 3: calculate the adapted Metropolis ratio (r):

$$r = \frac{OF_{Augmented}(\theta^k)}{OF_{Augmented}(\theta^{k-1})} \tag{10}$$

$$OF_{Augmented} = \begin{Bmatrix} OF & if \quad \theta \in Feasible\ Region \\ & OF_{Max} \quad Otherwise \end{Bmatrix} \tag{11}$$

$$OF_{Max} = \alpha.T_{A,1} \tag{12}$$

To compute the constraints of the optimization problem, an augmented objective function with a hard penalization term was defined in accordance to Eq. (11).

Step 4: accept the new vector by the following rule.

$$\theta^k = \begin{Bmatrix} \theta^* & if \quad a\ random\ probability < r \\ & \theta^{k-1} \quad Otherwise \end{Bmatrix} \tag{13}$$

To report the optimal values for decision variables, the following criterion is applied: if chains show that a bound constraint is active, such bound value is assumed as the optimal; otherwise, the mean for all chains data is reported as the optimal value. All mathematical models and numerical methods were implemented in Matlab R2015a using an Intel(R) Core(TM) i7-3770 CPU@3.4 GHz with 8 GB RAM memory.

3. Results and discussion

The results are obtained using between 10 and 20 Markov Chains with 5,000 samples per chain. Optimal $T_{A,1}$ and S values reach their upper bounds for all $H_{A,1}$ range (default values in the plant are 468 K and 5,760 $kg_{dry\ solids}.h^{-1}$ respectively), so throughput maximization is achieved at the expense of a relative energy consumption penalization in the spray. Respect to $M_{P,1}$ results (between 0.525 – 0.529 $kg_{water}.kg_{total}^{-1}$ covering the $H_{A,1}$ range), the values are slightly higher than the default one (0.52 $kg_{water}.kg_{total}^{-1}$) so this is aligned with water dried maximization in the system. Numerical issues were found during the deterministic approach, but the results here obtained confirmed main tendencies for decision variables. Figure 2 shows an example of the evolution for 20 chains, in which the convergence is achieved and requiring in most of the cases less than 5,000 samples. At the end of the chains upper bound is reached for $T_{A,1}$ and S. The computational burden was around 9 hours per optimization run and more representative results obtained from this example are presented in Table 1. Despite the acceptance ratio being low (0.021), most likely because of the constrained nature of the optimization problem, convergence was achieved for all decision variables (R-scale values close to 1), and the Monte Carlo estimation error presented low values for all cases (Gelman et al., 2004) as shown in Table 1.

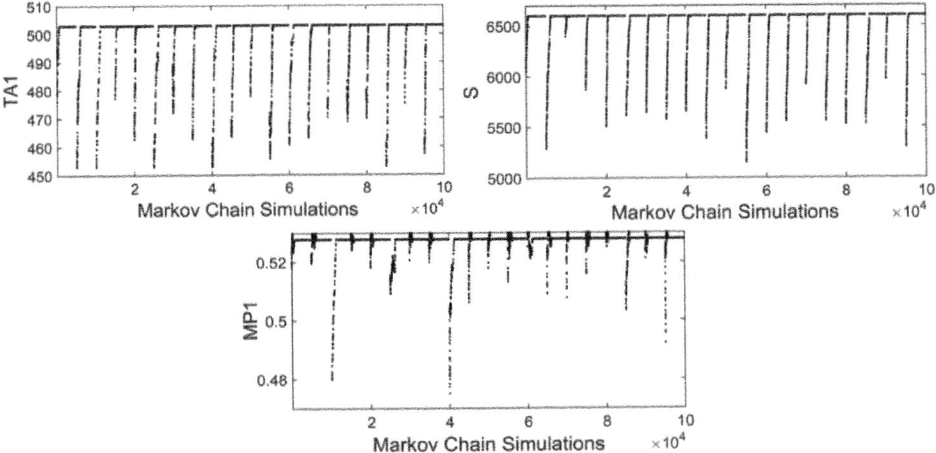

Figure 2. Simulation results for an example of 20 chains. $[T_{A,1}] = K$; $[M_{P,1}] = kg_{water}.kg_{total}^{-1}$; $[S] = kg_{dry\ solids}.h^{-1}$

Table 1. Summary of performance for the example shown in Figure 2

	$T_{A,1}$	$M_{P,1}$	S
R-scale	1.0331	1.0348	1.0240
Monte Carlo error	0.0007 % of mean	0.0006 % of mean	0.0004 % of mean

Figure 3 shows the densities obtained for decision variables in accordance with the example shown in Figure 2.

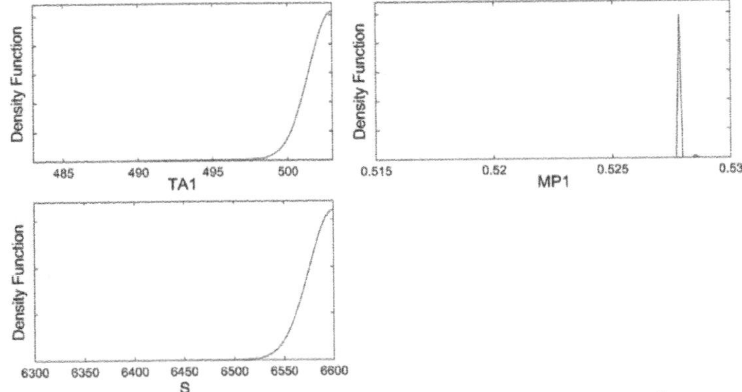

Figure 3. Density functions for decision variables. $[T_{A,1}] = K$; $[M_{P,1}] = kg_{water}.kg_{total}^{-1}$; $[S] = kg_{dry\ solids}.h^{-1}$

As can be interpreted, the nature of the problem seems to be convex since only one peak appeared in density functions. Hence, local optimum would also be global. In general MCMC explore the decision space using a random walk algorithm, which makes it robust against the initial conditions of decision variables. On the other hand, a prior distribution is required for them and the slow convergence can be improved using different methods such as differential evolution, parallel computing, etc. (Vrugt, 2016).

4. Conclusions

An optimization problem for a production scale milk spray and fluid bed drying was implemented and solved in Matlab. Results from a non-deterministic based technique (Markov Chain based random walk algorithm) revealed that dry bulb air temperature and dry solids flow rate could be controlled close to their upper bounds. These insights are aligned with the current situation in the production scale plant which this work is based on. It can also be observed that optimal concentrate moisture values are higher than the default one for the base problem, being this aligned with the maximization of total water dried in the system. The results involve an improvement in plant throughput but with a penalization in relative energy consumption in the spray dryer compared with the base operational point. The non-deterministic based results revealed that the structure of the problem seems to be convex. Despite it needs further theoretical analysis, the application case study shows promising results as a novel technique for non-deterministic/derivative free global optimization. This is especially suited for industrial applications, where the process models are not amenable for calculation of derivatives (Magnússon et al., 2020). Trying other non-deterministic tools such as Genetic Algorithms, Particle Swarm, etc., and doing a more comprehensive study on the multi-objective optimization problem, can be explored as future works. The developed model is expected to be a promising tool for decision-making at industrial scale.

References

P. Schuck; S. Mejean; A. Dolivet; and R. Jeantet, 2005, Thermohygrometric sensor: A tool for optimizing the spray drying process, Innovative Food Science & Emerging Technologies, 6(1), 45-50.

P. Schuck; A. Dolivet; S. Méjean; and R. Jeantet, 2008, Relative humidity of outlet air: the key parameter to optimize moisture content and water activity of dairy powders, Dairy Science & Technology, 88, 45–52.

L. N. Petersen; J. B. Jorgensen; and J. B. Rawlings, 2015, Economic optimization of spray dryer operation using nonlinear model predictive control with state estimation, IFAC-PapersOnLine, 48(8), 507-513.

A. Ferrari; S. Gutiérrez; and G. Sin, 2015, A comprehensive sensitivity and uncertainty analysis of a milk drying process, Computer Aided Chemical Engineering, 37, 2225-2230.

A. Ferrari; S. Gutiérrez; and G. Sin, 2016, Modeling a production scale milk drying process: parameter estimation, uncertainty and sensitivity analysis, Chemical Engineering Science, 152, 301-310.

A. Ferrari; S. Gutiérrez; and G. Sin, 2017, Powder stickiness in milk drying: uncertainty and sensitivity analysis for process understanding, Computer Aided Chemical Engineering, 40, 2743-2748.

A. F. Magnússon, R. Al, G. Sin, 2020, Development and application of simulation-based methods for engineering optimization under uncertainty, Computer Aided Chemical Engineering, 48, 451-456.

A. Gelman, J. B. Carlin, H. S. Stern, and D. B. Rubin, 2004. Bayesian data analysis. Texts in statistical science series.

S. Weinzierl, 2000. Introduction to Monte Carlo methods. arXiv preprint hep-ph/0006269.

J. A. Vrugt, 2016. Markov chain Monte Carlo simulation using the DREAM software package: Theory, concepts, and MATLAB implementation, Environmental Modelling & Software, 75, 273-316.

N. Silalai, and Y. H. Roos, 2010. Roles of water and solids composition in the control of glass transition and stickiness of milk powders, Journal of Food Science, 75(5), 285-296.

Antonis Kokossis, Michael C. Georgiadis, Efstratios N. Pistikopoulos (Eds.)
PROCEEDINGS OF THE 33rd European Symposium on Computer Aided Process Engineering (ESCAPE33), June 18-21, 2023, Athens, Greece
 http://dx.doi.org/10.1016/B978-0-443-15274-0.50048-2

Polymer Grade Transition Control via Reinforcement Learning Trained with a Physically Consistent Memory Sequence-to-Sequence Digital Twin

Zhen-Feng Jiang,[a] David Shan-Hill Wong,[a*] Jia-Lin Kang,[b*] Yuan Yao,[a*] Yao-Chen Chuang,[c]

[a]*Department of Chemical Engineering, National Tsing Hua University, Hsinchu 30043, Taiwan*

[b]*Department of Chemical and Materials Engineering, National Yunlin University of Science and Technology, Yunlin 64002, Taiwan*

[c]*Center for Energy and Environmental Research, National Tsing Hua University, Hsinchu 30043, Taiwan*

**dshwong@che.nthu.edu.tw, jlkang@yuntech.edu.tw, yyao@mx.nthu.edu.tw*

Abstract

In this work, a memory layer sequence-to-sequence digital twin (ML-StSDT) of a high-density polyethylene (HDPE) reactor simulated by ASPEN Dynamics™ was constructed using simulated grade transition and steady-state operating data. A reinforcement learning control (RLC) algorithm was developed by training with the ML-StSDT. The RLC was able to control both grade transition and steady-state operation of the simulated plant. The RLC performs better or equally well when compared with the direct application of ML-StSDT in nonlinear model predictive control (NLMPC) but substantially reduces the computation load. Our results demonstrate the feasibility of deep learning models serving as a digital twin for RLC training in nonlinear process control applications.

Keywords: Reinforcement learning; Sequence-to-Sequence with Memory Layer; Model Predictive Control, Grade transition.

1. Introduction

Model predictive control (MPC) has been used in the process industry to maximize the productivity of the process and reduce energy consumption. As the scope of the process being controlled increases, the following issues become more challenging. First, the number of state variables to be identified and decision variables to be optimized increases. Dynamic prediction horizons also become more and more complex and nonlinear. Hence a nonlinear model is required. Computation load becomes a challenging issue and control policies within the sampling period were increasingly difficult and suboptimal decisions may be made (Mayne et al. 2014) and The computation difficulty still limits control policy optimization over a less satisfactory horizon. The similarity between reinforcement learning (RL) and MPC has long been recognized, studied, and reviewed by Görges (2017). RL can drastically reduce online computing time. According to Nian et al. (2020), while RL methods such as deep deterministic policy gradient (DDPG) are suitable for online control, application of RL, application of RL to industrial control is mostly limited to a simulation environment. The Monte Carlo method is suitable for generating an infinite set of operations with a clear system terminal state, and the average return can be obtained after each trajectory calculation and does not require estimating values through previously estimated values to avoid the value function falling into local optima (Yoo et. al., 2021).

2. Memory Layer Sequence-to-Sequence Model

In our previous work, Jiang et al. (2022), we developed a sequence-to-sequence with memory layer (ML-StS) model (Figure 1) for a high-density polyethylene slurry reactor simulator (Figure 2) using ASPEN Dynamics^{TM}, based on the kinetic model by Khare et al. (2002). The sensors. manipulation and controlled variables (SV, MV, and CV) are shown in Table 1. The operator employed catalyst flow and hydrogen flow as MV to control the hydrogen-to-ethylene ratio (HER) and pressure. The ML-StS model was able to predict dynamic behavior with high accuracy as well as consistent directionality of MV/CV pairs u shown in Table 2.

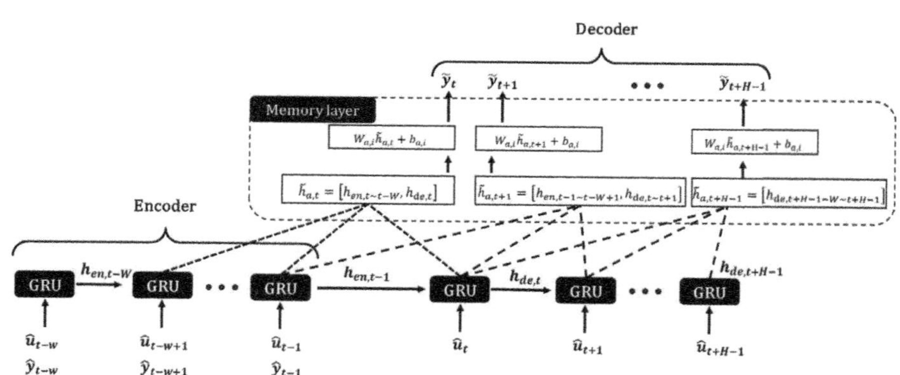

Figure 1. Structure of an StS model with the memory layer

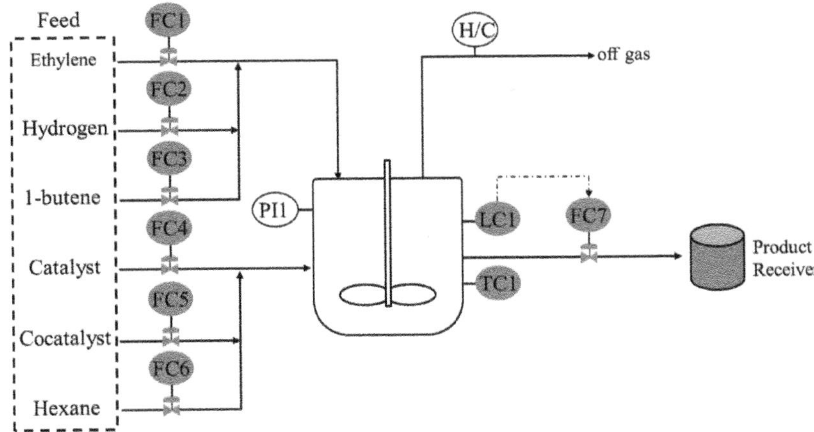

Figure 2. The schematic of the HDPE process.

Table 1: List of variables

Type	Tag
MV	The flow rate of ethylene
MV	The flow rate of hydrogen
MV	The flow rate of 1-butene
MV	The flow rate of catalyst
MV	The flow rate of cocatalyst
MV	The flow rate of hexane
MV	Temperature of reactor
CV	Pressure of reactor
CV	HER of purge gas

Table 2: Directionality of gain

	HER	Pressure
The flow rate of (	+	−
The flow rate of hydrogen	+	+

3. Nonlinear Model Predictive Control

The ML-StS model was then used in navigation as the predictive model used by nonlinear model predictive control (NLMPC) as shown in Figure 3. The differential evolution algorithm (DEA) was used to plan control actions in the future horizon and according to the future trajectory objective and operating constraints. While the NLMPC scheme is successful for both grade transition and steady-state operation, the computation load is too heavy to prevent frequent updating of manipulative variables.

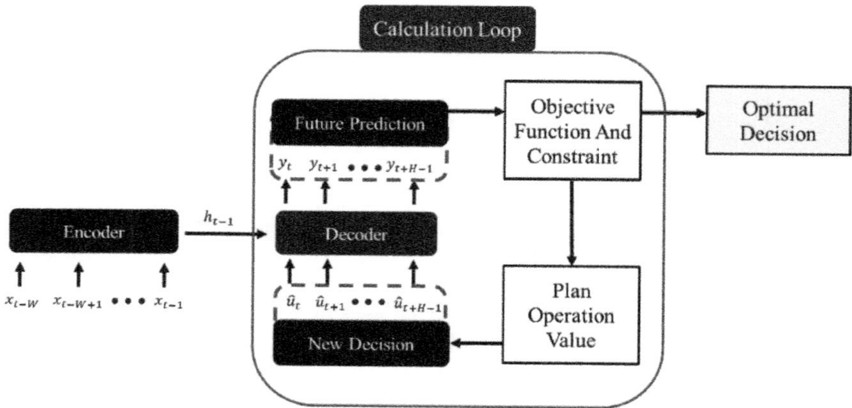

Figure 3. The schematic of the MPC calculation process.

4. Reinforcement Learning Control

Instead of direct application in NLMPC, the ML-StS model can also serve as a digital twin (DT) environment for training a reinforcement learning control (RLC). In this study, the "state" in RL was defined as the input to ML-StS encoder and future CV target setpoints trajectory. The "actor" was defined as the control action of the immediate future. The Monte Carlo Deep DeterministicPolicy Gradient (MC-DDPG) was used to train the state actor relation using the ML-StSDT as the "environment" (Figure 4). Eq. 1~8 are utilized to calculate the reward of action. Eq. 1 is the main target of the hydrogen-ethylene ratio. Eq. 2,3 make sure the current pressure (Pre.) is between the desired upper and lower limits (UL, LL). Eq. 4~7 limits the MV, catalyst flow (Cat.), and hydrogen flow (hyd.), within the allowable operating range. Eq. 8 allows RL models to achieve setpoint with minimal MV changes. After training using the ML-StSDT, RLC is then tested by the physical simulator.

$$R_1 = -\left|(\boldsymbol{HER_{setpoint,state}} - \boldsymbol{\tilde{y}_{HER,t}})\right| * \mathbf{10^5} \tag{1}$$

$$R_2 = -\left|(\boldsymbol{Pre._{UL}} - \boldsymbol{\tilde{y}_{Pre.,t}})\right| * \mathbf{10^7},\ \boldsymbol{if\ \tilde{y}_t > Pre._{UL}} \tag{2}$$

$$R_3 = -\left|(\boldsymbol{Pre._{LL}} - \boldsymbol{\tilde{y}_{Pre.,t}})\right| * \mathbf{10^7},\ \boldsymbol{if\ \tilde{y}_{Pre.,t} < Pre._{LL}} \tag{3}$$

$$R_4 = -\left|(\boldsymbol{Hyd._{UL}} - \boldsymbol{u_{hyd.,t}})\right| * \mathbf{10^6},\ \boldsymbol{if\ u_{hyd.,t} > Hyd._{UL}} \tag{4}$$

$$R_5 = -\left|(\boldsymbol{Hyd._{LL}} - \boldsymbol{u_{hyd.,t}})\right| * \mathbf{10^6},\ \boldsymbol{if\ u_{hyd.,t} < Hyd._{UL}} \tag{5}$$

$$R_6 = -\left|(\boldsymbol{Cat._{UL}} - \boldsymbol{u_{Cat.,t}})\right| * \mathbf{10^6},\ \boldsymbol{if\ u_{Cat.,t} > Cat._{UL}} \tag{6}$$

$$R_7 = -\left|(\boldsymbol{Cat._{LL}} - \boldsymbol{u_{Cat.,t}})\right| * \mathbf{10^6},\ \boldsymbol{if\ u_{Cat.,t} < Cat._{UL}} \tag{7}$$

$$R_8 = -\left|\boldsymbol{\Delta u_t}\right| * \mathbf{10^1} \tag{8}$$

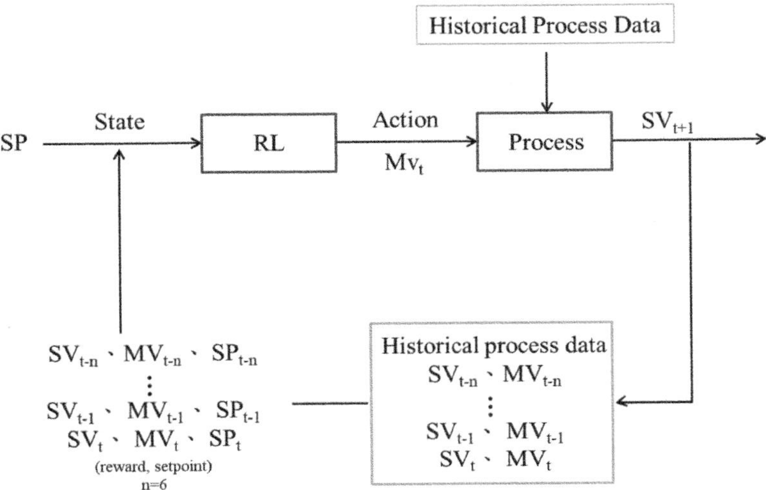

Figure 4. The schematic of the RL interacting with the HDPE process.

5. Results

Figure 5 and Figure 6 compared the results of two cases of grade transition directed by controlled LC and NLMPC. As shown in Figure 5 (a) the HER IAE of NLMPC is 34% lower in NLMPC but RLC produces a much smaller overshoot. In Figure 6 (a) the HER IAE of RLC and NLMPC are similar. However, NLNMC results in the oscillation of NLMPC being much more severe while RLC provides a smoother transition. Figure 5(b, d) and Figure 6 (b, d) showed that the action of NLMPC is much more aggressive while RLC provides a much smoother action due to higher frequency updating of control action. The changes in cumulative total and individual rewards of the RLC and NLMPC as shown in Figure 7. RLC also shows a much smoother trajectory. RLC computation time for each decision-making (18 milliseconds) is only about 1/1000 compared with the cost time of calculation of MPC (18 seconds). The 99.9% time reduction of RLC demonstrates the superior potential for the real complex process.

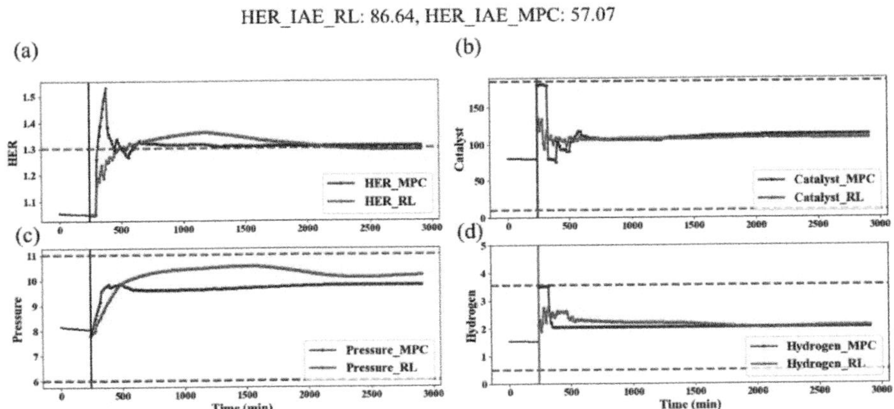

Figure 5.The result of MPC and RL of (a) the H2/C2H4 ratio (b) the flow rate of the catalyst (c) the pressure of the reactor (d) the flow rate of hydrogen.

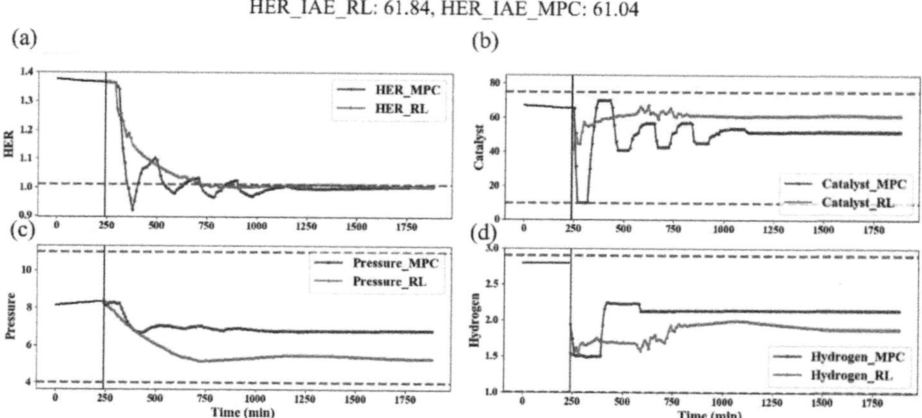

Figure 6. The result of MPC and RL of (a) the H2/C2H4 ratio (b) the flow rate of the catalyst (c) the pressure of the reactor (d) the flow rate of hydrogen.

6. Conclusion

In this work, we demonstrated that a memory layer sequence-to-sequence digital twin (ML-StSDT) of a high-density polyethylene (HDPE) reactor can serve as a model for NLMPC as well as a surrogate environment for training an RLC. The RLC performs better or equally well when compared with direct application in NLMPC but the online computation load is much reduced.

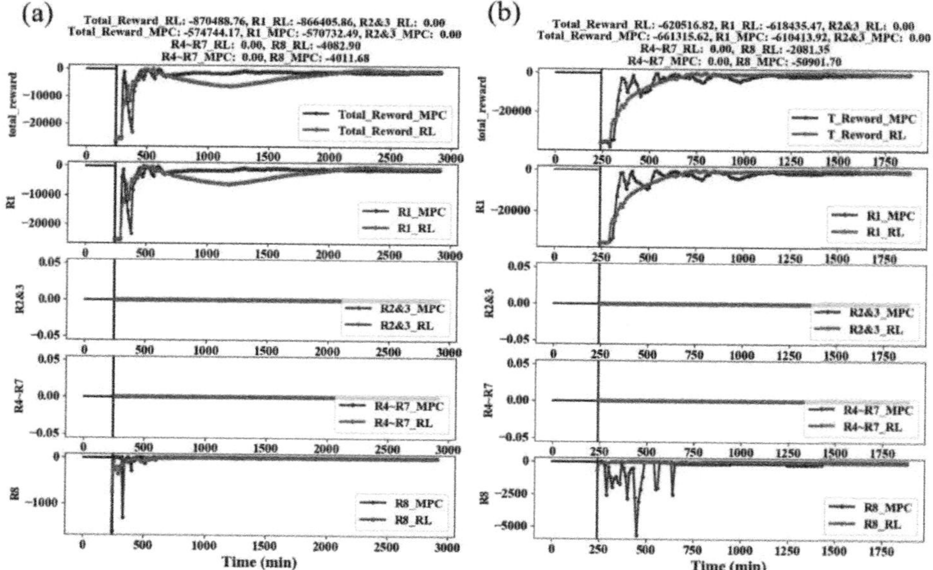

Figure 7. The result of total reward and R1~R8 reward of two different directions of grade transitions of MPC and RL.

Acknowledgment

This work was supported by the National Science and Technology Council of Taiwan (contract nos. NSTC 111-2221-E-007-005).

References

Görges, D. (2017). Relations between model predictive control and reinforcement learning. IFAC-PapersOnLine, 50(1), 4920-4928.

Jiang, Z. F., Wei, X. Z., Wong, D. S. H., Yao, Y., Kang, J. L., Chuang, Y. C., ... & Ou, J. D. Y. (2022). Model Predictive Control of Grade Transition with Attention Base Sequence-to-Sequence Model. In Computer Aided Chemical Engineering (Vol. 49, pp. 367-372). Elsevier.

Khare, N. P., Seavey, K. C., Liu, Y. A., Ramanathan, S., Lingard, S., & Chen, C. C. (2002). Steady-state and dynamic modeling of commercial slurry high-density polyethylene (HDPE) processes. Industrial & engineering chemistry research, 41(23), 5601-5618.

Mayne, D. Q. (2014). Model predictive control: Recent developments and future promise. Automatica, 50(12), 2967-2986.

Nian, R., Liu, J., & Huang, B. (2020). A review on reinforcement learning: Introduction and applications in industrial process control. Computers & Chemical Engineering, 139, 106886.

Yoo, H., Kim, B., Kim, J. W., & Lee, J. H. (2021). Reinforcement learning based optimal control of batch processes using Monte-Carlo deep deterministic policy gradient with phase segmentation. Computers & Chemical Engineering, 144, 107133.

Antonis Kokossis, Michael C. Georgiadis, Efstratios N. Pistikopoulos (Eds.)
PROCEEDINGS OF THE 33rd European Symposium on Computer Aided Process Engineering (ESCAPE33), June 18-21, 2023, Athens, Greece
 http://dx.doi.org/10.1016/B978-0-443-15274-0.50049-4

A Rolling Horizon Approach for Ethylene Production Scheduling with Daily Inventory Constraints

Haoran Li,[a,b] Tong Qiu,[a,b]

[a] *Department of Chemical Engineering, Tsinghua University, Beijing 100084, China*
[b] *Beijing Key Laboratory of Industrial Big Data System and Application, Tsinghua University, Beijing 100084, China*
qiutong@tsinghua.edu.cn

Abstract

The supply chain for ethylene production is becoming unstable, especially after Covid-19. This situation poses new challenges for ethylene plants to handle frequently changing feed supply conditions with more responsive scheduling strategies. This paper proposes a rolling horizon approach combined with moving average supply forecasting to assist in dynamic decision-making for ethylene production scheduling. When applied to an industrial case study, the proposed model shows superior calculation speed as well as the capability to find fairly good scheduling solutions under uncertain supply conditions.

Keywords: rolling horizon, scheduling, ethylene, inventory management

1. Introduction

Ethylene is one of the most important chemicals in the world and ethylene plants form one of the most profitable industrial chains (Li et al., 2022b). Owing to Covid-19, the global supply chain is experiencing ongoing turmoil, which increases the feedstock supply uncertainty in the ethylene industry. This brings pressing needs for ethylene plants to adopt a dynamic scheduling strategy to respond to unstable upstream supply conditions to guarantee optimal economic profits over time.

In recent years, ethylene production scheduling has been studied intensively. Ethylene cracking furnace systems process multiple feeds in multiple furnaces to produce ethylene and other olefin products. As the product yields vary with the furnace and the cracking duration as a result of coking inside the coil, the scheduling strategy, which determines the feedstock allocation between furnaces and the batch lengths in each furnace, has vital influences on the economic profits of the ethylene plants. And and Grossmann (And and Grossmann, 1998) first proposed a mixed-integer nonlinear programming (MINLP) model for the ethylene scheduling problem. After that, many modifications, for example, the non-simultaneous constraints (Liu et al., 2010) and integration with operation optimization (Li et al., 2022a) have been made to improve the model.

However, the dynamic response to the changing feedstock supply conditions has been overlooked in these previous scheduling studies. Facing this challenge, this paper developed a novel dynamic scheduling framework including (1) a Rolling horizon approach for the self-update of decision variables (2) Feed supply forecasting by moving average estimation (3) Mathematical formulation of the scheduling model with daily inventory constraints. This framework enables the model to receive new supply information from the environment and make quick responses by updating the scheduling variables. To demonstrate the efficacy of the developed new dynamic scheduling framework, a reference industrial case study taken from the literature is presented.

2. Problem statement

Fig. 1 shows the mass flow in a typical ethylene cracking furnace system (Li et al., 2022a). Feed i (1, 2, ..., NF) is cracked in furnace j (1, 2, ..., NF) and finally produces product l (1, 2, ..., NP) after quenching and separating.

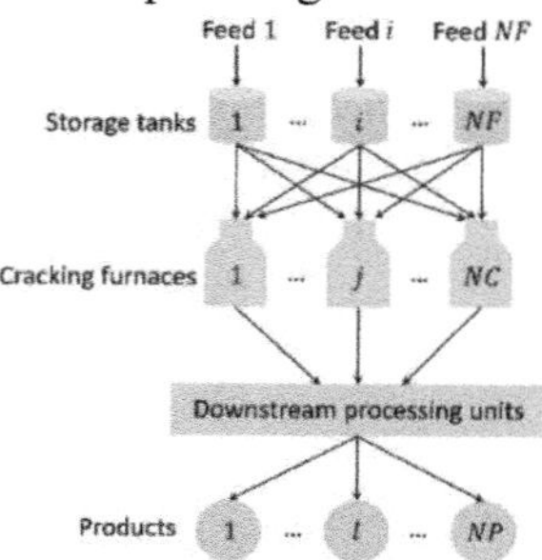

Figure 1. Mass flow of the ethylene cracking furnace system

However, because the feedstock supply conditions change frequently and decision-makers can't foresee the perfect information about the future feedstock supply conditions, updates of the following scheduling decision variables: (1) the inlet feed type in each furnace; (2) whether to perform decoking in a furnace given the visible future supply conditions, the inventory conditions, and the furnace conditions on the current day n are thus needed.

3. Methods

3.1. Rolling horizon approach

The rolling horizon approach is demonstrated in Fig. 2. The core idea is to optimize over a fixed horizon based on visible parameters and re-optimize the model repeatedly once the time step moves forward and new parameters are received. On the current day, the model will receive visible feed supply parameters from the market and optimize the schedule for the next H_1 days (lookahead horizon, LH). Before the optimization, the variables for the first H_2 days (fixed horizon, FH) will be fixed at variable solution levels from the previous schedule. In real cases, the time horizon of visible feed supply will be shorter than LH (referred to as imperfect information), which requires the model to estimate the future supply parameters based on the current information and generate the schedule. After the optimization, the schedule on the current day will be executed and the rest schedules in FH will be passed to the next optimization.

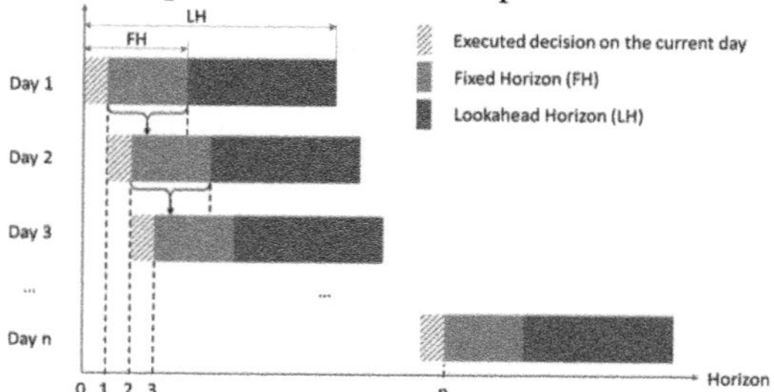

Figure 2. Illustration of rolling horizon approach

3.2. Supply forecasting by moving average estimation

Due to the existence of randomness, it's impossible to accurately foresee the long-term feed supply conditions needed for the scheduling optimization beforehand. We assume the feed supply conditions for the next H_2 days will be known (perfect information). One possible solution to this is to estimate the future supply based on the current visible information. Here, we proposed the moving average estimation to achieve this. As shown in Eq. (1), on the current day, the supply parameters will be observed as the true value for the first H_2 days. While for days between H_2 and H_1, the supply amount will be estimated as the average supply amount of the first H_2 days, as shown in Eq. (2).

$$supply_{i,k} = supply^{obs}{}_{i,k}, \quad \forall i, 1 \le k \le H_2 \tag{1}$$

$$supply_{i,k} = \frac{1}{H_2}\sum_{k=1}^{H_2} supply^{obs}{}_{i,k}, \quad \forall i, H_2 < k \le H_1 \tag{2}$$

3.3. Rolling horizon scheduling optimization model

3.3.1. Objective function

The objective of the rolling horizon scheduling model is to maximize the net profits over the next H_1 days. As shown in Eq. (3), the net profits are the product revenue minus the raw material costs, operation costs, and the decoking costs. Binary variable $y_{i,j,k}$ indicates whether feed i is processed in furnace j on day k. Binary variable $x_{j,k}$ denotes whether to perform decoking for furnace j on day k. $D_{i,j}$ is the flow rate of feed i in furnace j. P_l (the price of product l), Cr_i (raw material cost of feed i), $Cv_{i,j}$ (operation cost for processing feed i in furnace j), Cs_j (decoking costs for furnace j) are price-related parameters.

$$Max \quad J = \sum_i \sum_j \sum_k \sum_l P_l yd_{i,j,k,l} y_{i,j,k} D_{i,j} - (Cr_i + Cv_{i,j}) y_{i,j,k} D_{i,j} - Cs_j x_{j,k} \tag{3}$$

3.3.2. Fixed horizon constraints

$invent_{i,k}$ denotes the inventory level of feed i on day k. The integer variable $day_{i,j,k}$ is an indicator for the current batch lengths so far on day k. For each of the rolling horizon scheduling models, the decision variables in the fixed horizon will be fixed at the variable levels from the last scheduling model, as shown in Eq. (4). This guarantees decision consistency between adjacent days.

$$\begin{aligned} & invent_{i,k}, day_{j,k}, y_{i,j,k}, x_{j,k} \\ & = invent_{i,k}^{Fixed}, day_{j,k}^{Fixed}, y_{i,j,k}^{Fixed}, x_{j,k}^{Fixed}, \forall i, j, 1 \le k \le H_2 \end{aligned} \tag{4}$$

3.3.3. Feedstock allocation and decoking constraints

When allocating the feeds in different furnaces, only one type of feed is allowed to be cracked in a furnace at one time. If the furnace undergoes decoking, no feed will be allowed, as shown in Eq. (5). Besides, simultaneous decoking is restricted (Eq. (6)). Additionally, furnaces are not supposed to execute decoking on two consecutive days (Eq. (7)). Eq. (8) assures that if a furnace will not change the inlet feed type without performing a decoking operation.

$$\sum_i y_{i,j,k} \le 1 - x_{j,k}, \forall j, k \tag{5}$$

$$\sum_j x_{j,k} \le 1, \forall k \tag{6}$$

$$x_{j,k-1} + x_{j,k} \le 1, \forall j,k \tag{7}$$

$$(1 - x_{j,k-1} - x_{j,k})(y_{i,j,k} - y_{i,j,k-1}) = 0, \forall i,j,k \tag{8}$$

3.3.4. Batch length and yields constraints

If decoking happens on day k, the indicator $day_{i,j,k}$ will be 0. Otherwise, it will be $day_{i,j,k-1}$ plus 1 day (Eq. (9)). Owing to the limits of coking thickness, the batch lengths are not allowed to exceed the upper limit $tup_{i,j}$, as shown in Eq. (10). Besides, to avoid overly short batches, decoking is not allowed before the current batch length reaches the lower limit $tlo_{i,j}$ (Eq. (11)). The product yields $yd_{i,j,k,l}$ are determined by the exponential yielding model in Eq. (12). $a_{i,j,l}$, $b_{i,j,l}$, $c_{i,j,l}$ are yield-related parameters.

$$day_{i,j,k} = y_{i,j,k}(day_{i,j,k-1} + 1), \forall i,j,k \tag{9}$$

$$day_{i,j,k} \le tup_{i,j}, \forall i,j,k \tag{10}$$

$$day_{i,j,k} x_{j,k}(day_{i,j,k-1} - tlo_{i,j}) \ge 0, \forall i,j,k \tag{11}$$

$$yd_{i,j,k,l} = c_{i,j,l} + a_{i,j,l} e^{b_{i,j,l} day_{i,j,k}}, \forall i,j,k,l \tag{12}$$

3.3.5. Daily inventory constraints

The inventory level for each type of feed is determined by the inventory level of the previous day $invent_{i,k-1}$, the arrival supply on the current day $supply_{i,k}$ and the processed amount on the current day $\sum_j y_{i,j,k} D_{i,j}$, as shown in Eq. (13). Note that $invent_{i,k-1}$ should be non-negative.

$$invent_{i,k} = invent_{i,k-1} + supply_{i,k} - \sum_j y_{i,j,k} D_{i,j}, \forall i,k \tag{13}$$

4. Case study

4.1. Feedstock supply conditions Parameters

To evaluate the efficacy of the proposed model, a case study taken from the literature (Liu et al., 2010) with modification of feed supply conditions is used in this work. As shown in Fig. 3, the supply mode varies every 30 days. The other parameters remain the same as in the original literature.

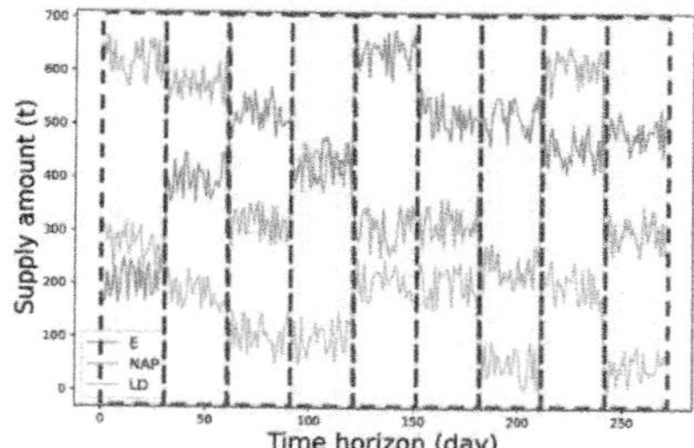

Figure 3. Feed supply conditions in 270 days.

4.2. Scheduling results

To demonstrate how our proposed rolling horizon model can respond to feed supply changes, Fig. 4 shows the supply parameters from day 114 to day 119. The solid line in each subplot represents the estimated parameters using moving average estimation, while the dotted line represents the future true supply parameters which are invisible on the current day.

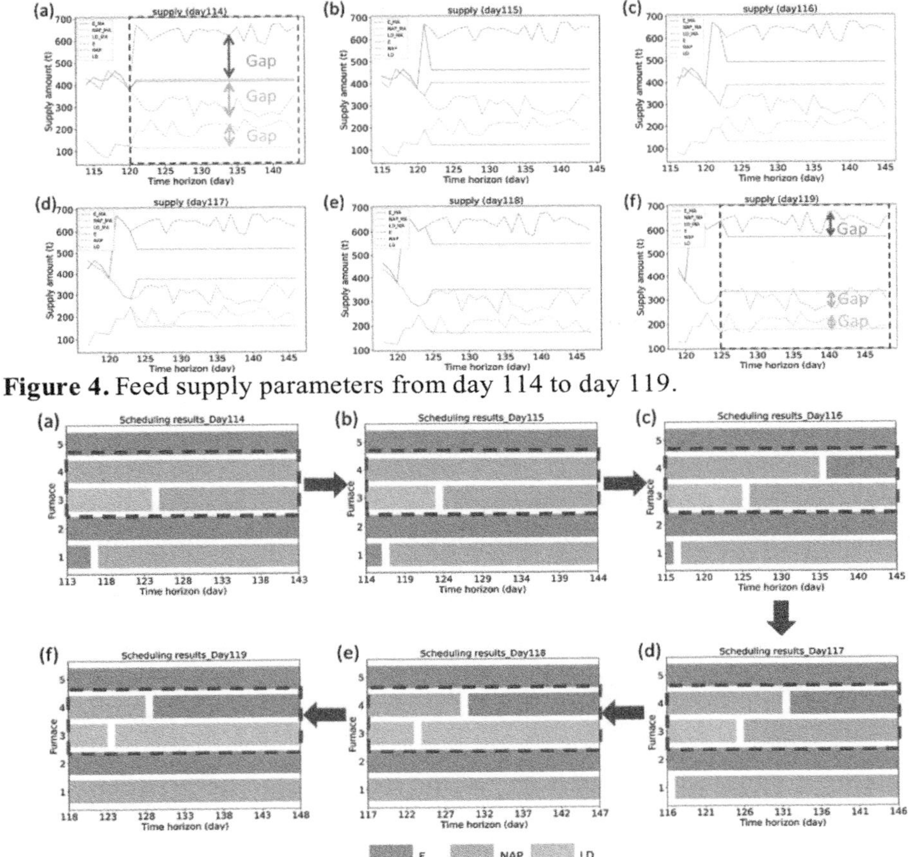

Figure 4. Feed supply parameters from day 114 to day 119.

Figure 5. Scheduling results from day 114 to day 119.

As can be observed in Fig. 4, the feed supply mode experiences a shift around day 120 with the ethane and light diesel supply increasing and the naphtha supply decreasing. On day 114, the estimated supply isn't aware of the change. As time step moving forward, the supply estimation for ethane and light diesel gradually upshifts because more observations are available now. Accordingly, the optimal schedule adjusts itself. As shown in Fig. 5(a) and (c), Furnace 4 planned to process naphtha from day 113 to 143 originally, but the schedule changed on day 116 by executing decoking on day 136 and processing ethane from day 137. Similarly, Furnace 3 planned to process light diesel followed by naphtha on day 114 but changes to process light diesel followed by light diesel on day 118 (Fig. 5(e)). The schedule changes in the two furnaces show how our model can track the supply conditions change and update the schedule accordingly. The executed optimal schedule for a long horizon of 270 days is shown in Fig. 6.

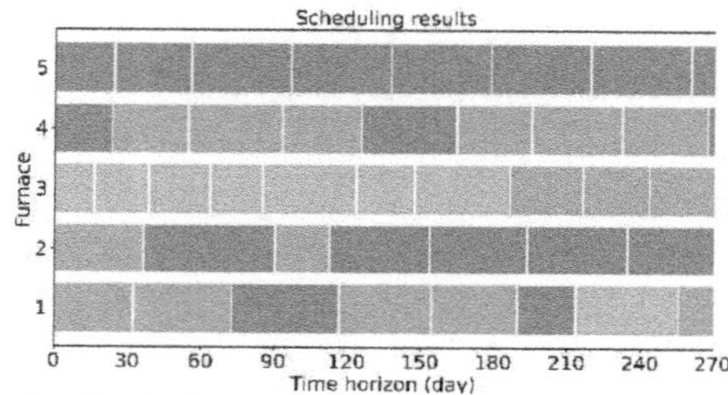

Figure 6. The executed scheduling results over 270 days.

4.3. Comparison of model size and solution with the original literature

The original model from the literature and our proposed model are developed in GAMS v38.2 on an Intel Core i7-9700CPU @ 3.00 GHz CPU with 16.0 GB RAM. Without daily inventory constraints, there's no guarantee that inventory constraints are satisfied at all times, as shown in Figure 7. Instead, the rolling model ensures feasible inventory levels.

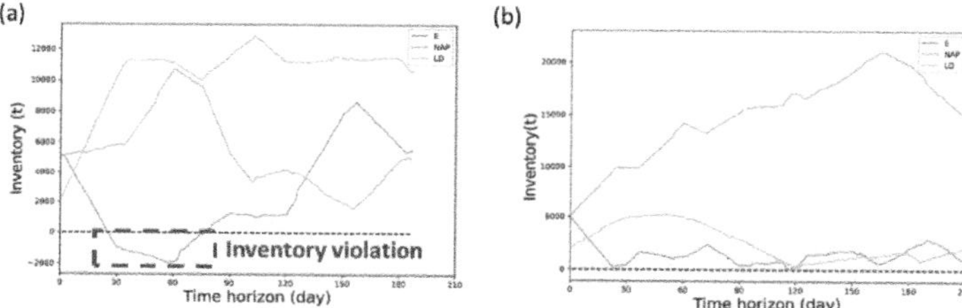

Figure 7. Inventory comparison of the original solution and rolling-horizon solution.

The model comparison results are listed in Table 1. The number of variables increases in the rolling model but the solution time is shortened. The objective value is quite close to the original solution, meaning good solution quality with supply uncertainties.

Table 1. Model size and solution comparison.

Variable	Original model	Rolling horizon model
Equations	8560	4442
Variables	1855	3392
Solution time	~ 32 min	~ 40 s
Objective value	151204 $/day	148495 $/day

5. Conclusion

This paper proposed a rolling horizon framework to deal with the ethylene production scheduling problem under feed supply uncertainty. When applied to a case study from the literature, our model shows superiority in reducing the model size as well as guaranteeing a good scheduling solution with consideration of daily inventory management.

References

And, V.J., Grossmann, I.E., 1998. Cyclic scheduling of continuous parallel-process units with decaying performance. AIChE J. 44, 1623–1636.

Li, H., Zhang, S., Qiu, T., 2022a. Two-Level Decoupled Ethylene Cracking Optimization of Batch Operation and Cyclic Scheduling. Ind. Eng. Chem. Res. acs.iecr.2c01682.

Li, H., Zhang, S., Zhou, J., Wang, Z., Zhao, X., Qiu, T., 2022b. Toward carbon-neutral ethylene production: assessment of the application potential of b IO-ETHYLENE production pathways in China. Biofuels Bioprod Bioref bbb.2406.

Liu, C., Zhang, J., Xu, Q., Li, K., 2010. Cyclic scheduling for best profitability of industrial cracking furnace system. Computers & Chemical Engineering 34, 544–554.

Antonis Kokossis, Michael C. Georgiadis, Efstratios N. Pistikopoulos (Eds.)
PROCEEDINGS OF THE 33rd European Symposium on Computer Aided Process Engineering (ESCAPE33), June 18-21, 2023, Athens, Greece
 http://dx.doi.org/10.1016/B978-0-443-15274-0.50050-0

Simulation and optimization of fixed bed and moving bed TSA processes for post-combustion CO_2 capture.

Shreenath Krishnamurthy[a*], Richard Blom[a], Giorgia Mondino[a], Schalk Cloete[b], John Morud[b], Abdelghafour Zaabout[b], Philip Llewelyn[c], Cecile Pereira[d], and Veronique Pugnet[c].

[a]*Process chemistry and functional materials, SINTEF Industry, Oslo, 0373, Norway*

[b]*Flow technology, SINTEF Industry, Trondheim, 7031, Norway*

[c]*TotalEnergies, Pau, 64000, France*

[d]*TotalEnergies, Palaiseau, 91120, France*

**Shreenath.Krishnamurthy@sintef.no*

Abstract

In this work three different TSA process configurations were chosen, and these processes were optimized using genetic algorithm. One of the processes was a moving bed TSA process while the other two process were fixed bed processes using a packed bed containing pellets and hollow fiber modules. With Lewatit as the sorbent chosen, the cycle configurations were optimized to identify minimum energy and maximum productivity values for CO_2 capture from a natural gas combustion cycle (NGCC) flue gas. The hollow fiber TSA process had the highest productivity of 4.25 mol CO_2/m^3 ads s, followed by the fixed bed and the moving bed TSA processes with values of 2.03 and 1.68 4.25 mol CO_2/m^3 ads s, respectively

Keywords: TSA process, optimization, CO_2 capture

1. Introduction

Adsorption processes using solid sorbents rely on the differences in the affinities of the different gases towards the solid and has been implemented on a large scale for several gas separation applications. Adsorption processes are broadly classified into two types based on the mode of regeneration of the sorbent: Pressure/vacuum swing adsorption (PSA/VSA) where the sorbent is regenerated by lowering the pressure of the column and temperature swing adsorption (TSA) process where the sorbent is regenerated by the supply of heat directly or indirectly. TSA processes are better suited to low concentrations of CO_2 than PSA/VSA processes such as a natural gas combustion (NGCC) cycle, where it is possible to achieve purity values >95% from a feed containing 3-5% CO_2. However, TSA processes applied to conventional fixed bed columns are usually slow due to the long heating and cooling times, and this is especially a drawback as it lowers the productivity of the process. Such a drawback could be mitigated through a combination of direct heating and the use of structured adsorbents like hollow fibres with better mass transfer and lower pressure drop, resulting in improved performance of the process in comparison with traditional packed bed systems [1,2]. At the core of an adsorption process is the adsorbent and the performance of an adsorbent depends on the specifics of the process. Therefore, each adsorbent must be mapped towards the best process configuration [3]. This information can only be obtained from detailed simulations and optimization of different adsorbent-process combination in a standardized manner to accurately reflect competitiveness. The present work employs a homogenized model to study the following processes: 1. A fixed TSA process with pellets 2. A fixed bed TSA process with hollow fibers 3. A moving bed TSA process with pellets. The adsorbent chosen for this study is a commercially sold benzyl amine polymer, Lewatit for a representative NGCC cycle with 5% CO_2 and 85% N_2. The three cycle configurations were optimized to identify minimum energy and maximum productivity values for purity-recovery targets of 95% and 90%, respectively.

2. Cycle design

For the fixed bed a 4-step TSA process was chosen. The four steps are as follows and the schematic of the process is shown in Figure 1.

1. Adsorption with feed: Feed enters the column at 313 K. Adsorption of the CO_2 takes place while nitrogen is collected at the product end. The outlet of the column is maintained at 1 atm. A part of the nitrogen is used to purge the column in the light reflux step.
2. Heavy reflux step with light reflux product (HRLRP): This step uses the product of the light reflux step. The duration of this step is the same as the light reflux (LR) step since the entire product of the latter is recycled back to the column. The CO_2 concentration in the column is enhance during this step. The column is also heated up slightly with the gases coming from the LR step.
3. Desorption with steam purge: The desorption of the CO_2 is aided by counter-currently purging the column with steam/water at a high temperature.
4. Light reflux step (LR): The remaining CO_2 in the column is then purged with the nitrogen product of the adsorption step form the counter-current direction. The product is fully recycled back in to the HRLRP step.

A similar concept was also considered for the moving bed temperature swing adsorption (MBTSA) process shown in Figure 1. The moving bed process is a simplified process with the adsorption and desorption sections. In the adsorption step, the solid flows counter-current to the gas. Once the adsorption step is complete, the HRLRP step is carried out with the solid staying constant. Then the solid is circulated to the desorption section. Here steam is fed to the column in a counter current direction and CO_2 is collected as a product. The solid also flows in the same direction as the steam. Once this step is completed the solid flow is stopped and the LR step is carried out with adsorption product. The product of the LR step is sent to the HRLRP step and the solid is now moved back into the adsorption step.

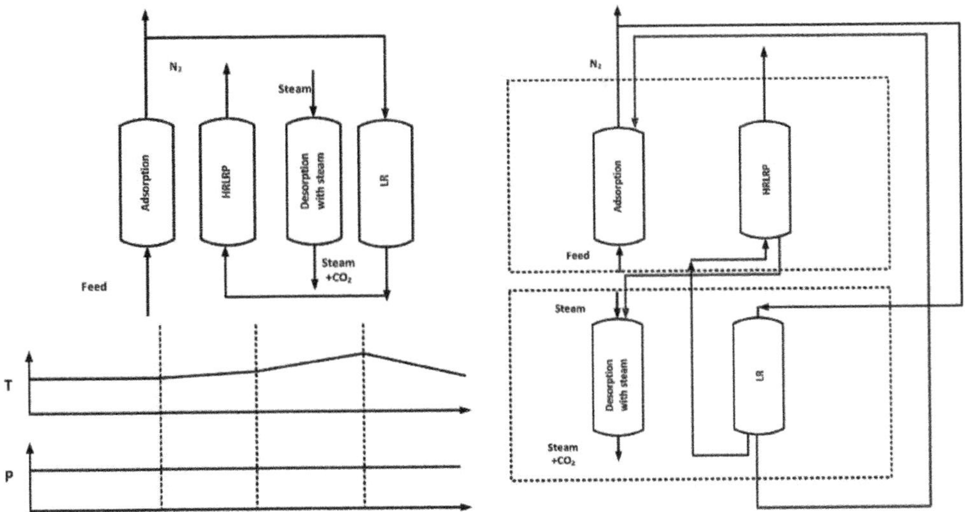

Figure 1: Schematic of the 4-step TSA process (left) and the 4-step MBTSA process (right). Solid blue lines represent the movement of solid, while black solid lines represent gas flow. The dotted lines in the figure on the right is for visual aid to separate the adsorption and desorption sections.

3. Model equations

The model equations used in this study are similar to the authors' earlier studies [4] on packed beds with the inclusion of additional terms to account for the solid movement. The equations are described below

$$\frac{\partial y_i}{\partial t}+\frac{y_i}{P}\frac{\partial P}{\partial t}-\frac{y_i}{T}\frac{\partial T}{\partial t}=\frac{T}{P}D_L\frac{\partial\left(\frac{P\partial y_i}{T\partial z}\right)}{\partial z}-\frac{T}{P}\frac{\partial\left(\frac{y_iP}{T}v\right)}{\partial z}-\frac{RT}{P}\frac{(1-\varepsilon)}{\varepsilon}\frac{\partial\bar{q}_i}{\partial t}-\frac{T}{P}V_S\frac{\partial\left(\frac{y_iP}{T}v\right)}{\partial z} \quad (1)$$

$$\frac{1}{P}\frac{\partial P}{\partial t}-\frac{1}{T}\frac{\partial T}{\partial t}=-\frac{T}{P}\frac{\partial(\frac{P}{T}v)}{\partial z}-\frac{(1-\varepsilon)}{\varepsilon}\frac{RT}{P}\sum_{i=1}^{n}\frac{\partial\bar{q}_i}{\partial t}-\frac{T}{P}V_S\frac{\partial\left(\frac{P}{T}v\right)}{\partial z} \quad (2)$$

$$\frac{\partial\bar{q}_i}{\partial t}=k_{LDF_i}\left({q_i}^*-\bar{q}_i\right)-\frac{T}{P}V_S\frac{\partial\bar{q}_i}{\partial z} \quad (3)$$

$$\left[\frac{(1-\varepsilon)}{\varepsilon}\left(\rho_sC_{ps}+C_{pa}\sum_{i=1}^{n}\bar{q}_i\right)\right]\left(\frac{\partial T}{\partial t}+V_S\frac{\partial T}{\partial z}\right)=\frac{K_z}{\varepsilon}\frac{\partial^2T}{\partial z^2}-\frac{C_{pg}v}{R}\frac{\partial T}{\partial z}-\frac{C_{pg}}{R}\frac{\partial P}{\partial t}+\frac{(1-\varepsilon)}{\varepsilon}\sum_{i=1}^{n}(-\Delta H_i-C_{pa}T)\frac{\partial\bar{q}_i}{\partial t}-\frac{2h_i}{\varepsilon r_i}(T-T_w) \quad (4)$$

$$\rho_wC_{pw}\frac{\partial T_w}{\partial t}=k_w\frac{\partial^2T_w}{\partial z^2}+\frac{2r_ih_i}{r_o^2-r_i^2}(T-T_w)-\frac{2r_oh_o}{r_o^2-r_i^2}(T_w-T_a) \quad (5)$$

$$\rho_fc_{p,f}\frac{\partial T_f}{\partial t}=-u_f\rho_fc_{p,eff}\frac{L_z}{L_x}\frac{\partial T_f}{\partial z}-\alpha_{f-HXw}h_f\left(T_f-T_s\right) \quad (6)$$

Equations 1-6 represent the component mass balance, total mass balance, adsorption rate, the energy balance, wall temperature and fluid temperature balances. The same set of equations were used for the fixed bed and the moving bed processes, with the solid velocity (V_S) being zero in the fixed bed case. Between the hollow fiber and the packed bed, the differences are in the equations used for the pressure drop and axial dispersion terms as shown in Table 1. For the MBTSA process the velocity of the solid was obtained from the desorption/adsorption sections and the duration of the adsorption and steam purge steps. For simplicity in comparison, the length of the column was kept constant in all the simulations. This is also true for the MBTSA process where the length of the adsorption and desorption sections were fixed to be 1m. Appropriate boundary conditions

were used for the different steps along with the initial conditions for each step to simulate a cyclic process. The TSA and MBTSA cycles were simulated until cyclic steady state (CSS) condition was achieved.
The model equations were converted to a dimensionless form and discretized using the finite volume method. The resultant system of equations was solved using the ode 15s solver in MATLAB. The performance of the process was defined by 4 performance indicators, namely CO_2 purity, CO_2 recovery, energy consumption and productivity. Detailed optimization of the different cycle configurations was carried out using genetic algorithm. In total 3600 simulations were carried out for the different cases

Table 1: Corrleations for axial dispersion and pressure drop in the simulations

Property	Hollow fiber	Packed bed/moving bed
Pressure drop	$\frac{-dP}{dZ} = 32\frac{\varepsilon v \mu}{d_c^2}$	$\frac{-dP}{dZ} = 180\left(\frac{(1-\varepsilon)}{\varepsilon}\right)^2 \frac{v\mu}{d_p^2}$
Axial dispersion	$D_L = \frac{20D_{mol}}{\varepsilon} + 0.5vd_p$	$D_L = D_{mol} + \frac{(d_{channel}v_0)^2}{192D_{mol}}$

4. Results and discussion

Figure 2 shows the pareto fronts for the fixed bed and the moving bed temperature swing adsorption processes. All the points on the pareto fronts satisfy 95% purity and 90% recovery constraints. The specific energy consumption is obtained from the energy consumed by the blower and compressor for the adsorption and HRLRP step and the sensible heat required to heat up the solid to the required temperature. It should be highlighted that the energy required to produce the steam is not accounted in the calculations. Based on the pareto plots, the hollow fiber TSA process was the best performing process in terms of productivity. Comparing the cycle times from Figure 3a reveal that the hollow fiber TSA process was able to operate at faster cycle times in comparison with the packed bed TSA process. This is due to the lower pressure drop of the hollow fiber process which helped to operate the cycle at higher flowrates. To minimize the CO_2 losses, the adsorption time is shorter and thereby the cycle time was reduced which improved the productivity. The energy consumption was higher for the hollow fibers due to the values of the regeneration temperatures chosen by the optimizer as seen from Figure 3b.

For the MBTSA process, the high energy and productivity values are a consequence of the high cycle time and higher steam temperature. In the MBTSA process, the feed velocity values were around 2.16-2.25 m/s. whereas for the packed bed it was between 1.8-2.3 m/s. The solid velocities for the MBTSA process were between 4-4.2 mm/s. The operating conditions corresponding to the minimum energy and maximum productivity values are shown in Table 2.

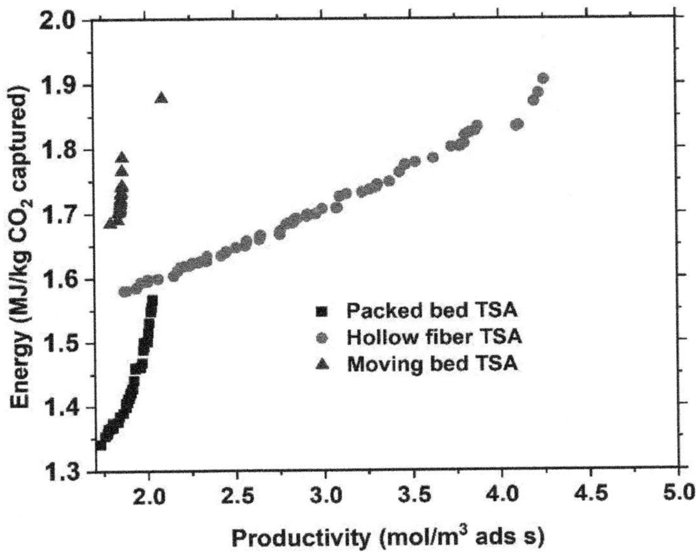

Figure 2: Specific energy vs Productivity plots for the different TSA processes

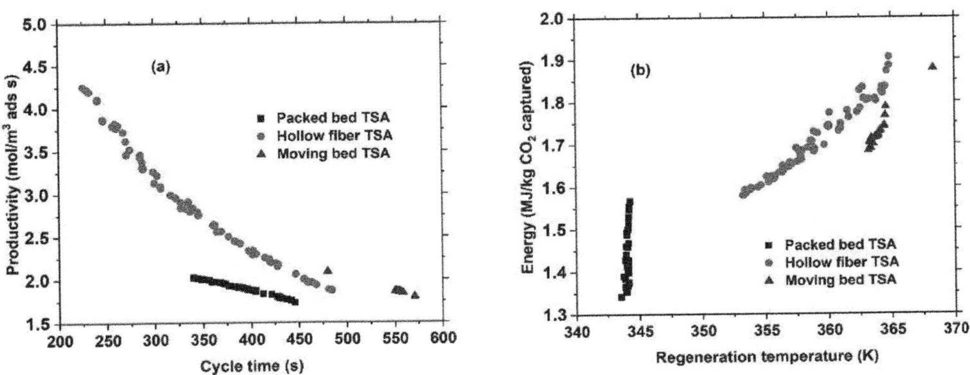

Figure 3: (a) cycle time vs productivity and (b) steam temperature vs Energy consumption

Process	t_{ads} (s)	t_{LR} (s)	$t_{steampurge}$ (s)	v_{feed} (m/s)	v_{steam} (m/s)	T_{steam} (K)
Minimum energy conditions						
Packed bed	242.10	4.93	193.55	1.79	0.99	343.53
Hollow Fiber	116.58	3.37	360.96	4.23	0.78	353.21
MBTSA	253.01	35.65	246.38	2.16	1.65	363.19
Maximum productivity conditions						
Packed bed	142.64	5.85	185.47	2.70	1.56	344.27
Hollow Fiber	111.60	1.51	111.47	4.66	2.13	364.85
MBTSA	240	20	200	2.25	1.56	368.32

5. Conclusions

Here we have used a homogenized model to simulate fixed bed and moving bed TSA processes. Detailed optimization of these processes revealed that the Lewatit sorbent was able to achieve 95% purity and 90% recovery values in all the three processes. The energy consumption was between 1.3-1.9 MJ/kg CO_2 captured for the different processes
It should be highlighted here that the MBTSA process is not the same cycle as that of Mondino et al., [5] and Morales Ospino et al., [6] where the solid moves between the adsorption, desorption and cooling sections. Further work is ongoing to improve the MBTSA process in order to simulate the existing cycles in literature. Additionally other optimization techniques like Bayesian optimization methods are also being explored by the authors

Acknowledgements: We acknowledge Gassnova SF and TotalEnergies E&P NORGE AS for financial support through the "Disruptive CO_2 capture" project (Grant nos. 620036 and 4100P20013). Discussions with Carlos Adolfo Grande are acknowledged.

References

1. Rezaei et al., 2014, Modeling of rapid temperature swing adsorption using hollow fiber sorbents, Chem Eng Sci, 113, 62-76
2. Keller et al., 2019, Electrical swing adsorption on functionalized hollow fibers, Chem Eng J, 371, 107-117
3. Rajagopalan et al., 2016 , Do adsorbent screening metrics predict process performance? A process optimisation based study for post-combustion capture of CO_2, Int J Greengh gas con, 46, 76-85
4. Krishnamurthy, S. 2022 Vacuum swing adsorption process for post-combustion carbon capture with 3D printed sorbents: Quantifying the improvement in productivity and specific energy over a packed bed system through process simulation and optimization, Chem Eng Sci 253 117585
5. Mondino et al., 2020, Production of MOF Adsorbent Spheres and Comparison of Their Performance with Zeolite 13X in a Moving-Bed TSA Process for Postcombustion CO_2 Capture, Ind Eng Chem Res, 59 7198-7211
6. Morales-Ospino et al., 2021, Parametric Analysis of a Moving Bed Temperature Swing Adsorption (MBTSA) Process for Postcombustion CO_2 Capture Ind Eng Chem Res, 60, 10736-10752

Antonis Kokossis, Michael C. Georgiadis, Efstratios N. Pistikopoulos (Eds.)
PROCEEDINGS OF THE 33rd European Symposium on Computer Aided Process Engineering (ESCAPE33), June 18-21, 2023, Athens, Greece
 http://dx.doi.org/10.1016/B978-0-443-15274-0.50051-2

AI-enhanced BoM Optimization with MILP for the EMS Industry

Hossein Mostafaei,[a,*] Eero Hiltunen,[a] Antti Liski[a]

[a]*AI and Machine Learning, Elisa IndustrIQ, Helsinki, Finland*
**hossein.mostafaei@elisa.com*

Abstract

This work explores how artificial intelligence and machine learning (ML) together with rigorous optimization models can help in optimizing electronic bill of materials. We propose a mixed-integer linear programming (MILP) model that considers different real-life restrictions required to complete a customer BoM. Furthermore, we implement a data-driven approach to find a set of alternates for a given component based on data mining and machine learning techniques. With these techniques, the BoM is completed with the most similar alternates that have the same form, fit, and functionality as the given component. The results from a real-life use case illustrate that the proposed AI-enhanced optimization model leads to superior results in terms of solution quality (lower price and shorter lead times) in comparison to the standalone MILP model, with the highlight being a 4.4% improvement in the total cost and a 24% reduction in the lead time.
Keywords: EMS industry, Electronic BoM, Optimization, Machine learning, MILP.

1. Introduction

The number of electronics used in day-to-day devices has increased in recent years which in conjunction with the global pandemic has caused disruptions in the supply chain of electronic components. According to industry experts, only one-third of components are available for order and the lead times have increased from days to several months. Obviously, this is problematic for companies whose business is dependent on sales of goods that include any electrical components. Therefore, the electronics manufacturing services (EMS) industry is looking for new ways to ensure order completion while maintaining the low costs of electronics manufacturing.

A common method for lowering costs in manufacturing industries is to use mathematical optimization. The quality, quantity, lead time, and price of goods from suppliers provide a profound impact on the overall performance of the supply chain (Pan, 2015) and can all be optimized using MILP models. The information is usually listed in the Bill of Materials (BoM) which is considered one of the critical documents for production planning (Tao and Chang, 2006). An optimized BoM can help companies to plan the purchases of raw materials at minimal prices, avoid excess amounts, manage inventory, stay on schedule, and control the budget.

Machine learning (ML) and data mining techniques have become increasingly useful in finding insights and value from a massive amount of data. ML models can learn patterns and rules automatically from the data that would be impossible for a human. For this reason, ML models are utilized in such applications where rules are not known or are too complex to understand and formulate. Data mining methods refer to techniques that utilize data to solve problems with algorithms (Leskovec et al., 2010). This definition covers methods ranging from clustering to different types of distance metrics.

Product matching is a well-known problem in e-commerce where vendors try to find duplicate items from their site and conduct price comparisons to other sites with different descriptions and part numbers of items. Proposed solutions incorporate both text-based (Shah et al., 2018) and image-based (Kiapour et al., 2015) models to find matching products. However, none of the proposed solutions address the problem of matching products in a domain where the same form, fit, and functionality of the matches are as crucial as in the electronics supply chain.

This paper proposes a MILP enhanced with an ML-based solution to optimize an electronic BoM. The optimization model strives to minimize the cost while fulfilling all restrictions, whereas the ML solution enables sufficient search space for components that are in stock or available with a short lead time. To experiment with the effect of this, we conduct a case study with real data from the EMS industry and compare our proposed AI-enhanced optimization to a standalone optimization model.

2. Problem Statement

A bill of materials (BoM) is a list of the raw materials, assemblies, components, parts, and quantities that are needed to manufacture a product. BoMs can also be enriched by some other information e.g., prices, lead times, minimum quantity that can be ordered, etc. In this work, we consider an enriched electronic BoM, as shown in Table 1. The BoM features 20 line items, with some having a few alternates. For example, line item I1 indicated in the light grey is an alternate (already determined by the customer) for $I1_{EMK107BJ105KA\text{-}T}$ (with the subscript indicating the manufacturing part number (MPN)), whereas the one in the dark gray is the one we are trying to find from a massive amount of data using ML techniques, which we will discuss later in Section 3.

Table 1: Example electronic BoM considered in this work.

LI[a]	QTY[b]	Supplier	MPN	MFr[c]	In-stock	Min	Mult	Price ($)	M-LT[d]
I1	9000	Digikey	EMK107BJ105KA-T	Taiyo Yuden	2400	500	1	0.018	250
I1	9000	Digikey	EMK107BJ105KA-T	Taiyo Yuden	2400	2000	200	0.015	250
I1	9000	Digikey	EMK107BJ105KA-T	Taiyo Yuden	2400	4000	4000	0.011	250
I1	9000	Digikey	GRM188R61C105KA12D	Murata	20000	4000	4000	0.008	300
I1	9000	Myarrow	GRT188R61C105KE13D	Murata	3000	4000	100	0.010	100
⋮	⋮	⋮	⋮	⋮	⋮	⋮	⋮	⋮	⋮
I20	5200	Avent	0603YC104KAT2A	Kyocera AVX	12000	2000	500	0.320	120

[a]Line item, [b]quantity required, [c]manufacturer, [d]manufacturing lead time. Note: Numbers are dummy and just for indications.

For the line item $I1_{EMK107BJ105KA\text{-}T}$, we have Min = (500, 2000, 4000), Mult = (1,100, 4000), and price = (0.018, 0.015, 0.011), meaning that any order for $I1_{EMK107BJ105KA\text{-}T}$ from Digikey should be a function of the Min and Mult of form Min + n × Mult where n is a positive integer number ($n \geq 0$). For example, if we want to order 2050 from $I1_{EMK107BJ105KA\text{-}T}$, we need to order 2000 with a unit price of \$0.015 and the remaining 50 with a unit price of \$0.018. Note that all 2050 cannot be ordered with the price of \$0.015, because there is no integer number n for which 2050 = 2000 + 100n. The requested quantity for I1 is 9000 and if we order all of them: i) from $I_{GRT188R61C105KE13D}$ supplied by Digikey, then we need to order 12000 (4000+4000×2), i.e., 3000 units more than the requested quantity, which is called *excess quantity*, ii) from $I_{GRT188R61C105KE13D}$ supplied by Myarrow, we need to wait at least 100 days (manufacturing lead time = 100) as the current in-stuck (3000) is not sufficient. In this work, we will develop a MILP to consider all combinations for order quantities while minimizing BoM cost.

3. Models and Architecture

The following section describes the models and architecture of our proposed system. First, we define the MILP model that is used to minimize the BoM while considering the restrictions. After that, we describe our alternative search solution that can find alternative components with the same form, fit, and functionality for the reference component.

3.1. MILP model

Here we develop a mixed integer linear programming (MILP) model to optimize a given BoM in terms of costs subject to some restrictions.

The binary variable $X_{i,m,s}$ is one if component i with MPN m is ordered from supplier s (hereafter as line item (i,m,s)). If line item (i,m,s) is not ordered, then its order quantity ($Q_{i,m,s}$) should be equal to zero, as imposed by Eq. (1).

$$Q_{i,m,s} \leq M.X_{i,m,s} \quad \forall(i,m,s) \in IMS \tag{1}$$

Note that the tuple (i,m,s) is based on BoM's columns (LI, MPN_i+MFr, Supplier), in which m is built by first indexing MPN column and then combining it with MFr column. For example, for the BoM in Table 1, we have IMS = {(I1, EMK107BJ105KA-T_1+Taiyo Yudan,Digikey),(I1,EMK107BJ105KA-T_2+TaiyoYudan,Digikey),(I1, EMK107BJ105KAT_3+TaiyoYudan,Digikey),(I1,GRM188R61C105KA12D_4+Murata, Digikey),…}.

If line-item (i,m,s) is ordered, the order quantity should be a function of minimum ($q_{i,m,s}^{\text{min}}$) and mult ($q_{i,m,s}^{\text{mult}}$) quantities. In Eq. (2) $Z_{i,m,s}$ is a continuous integer variable.

$$Q_{i,m,s} = q_{i,m,s}^{\text{min}}.X_{i,m,s} + q_{i,m,s}^{\text{mult}}.Z_{i,m,s} \quad \forall(i,m,s) \in IMS \tag{2}$$

As expressed in Eq. (3), the amount of line-item i ordered from all suppliers should be at least as large as the demanded quantity (dem_i) by the customer, otherwise, the backordered demand (or shortage) could take a positive value and result in delays in production. Note that Eq. (3) also considers the excess amount.

$$\sum_{(m,s)\in MS|(i,m,s)\in IMS} Q_{i,m,s} + ST_i \geq dem_i \quad \forall i \tag{3}$$

Eq. (4) controls the inventory, meaning that the total amount of component i_m that can be ordered must not exceed the in-stock at supplier s.

$$\sum_{i\in I} Q_{i,m,s} \leq in_{m,s} \quad \forall\ (m,s) \in MS \tag{4}$$

The objective function will minimize the component cost and backorder cost. Notice the shortages coefficient $bc = N \times max_{m,s}\{cc_{m,s}\}$, where $cc_{m,s}$ is the unit cost of a component, and N is an integer number ($N > \ max_{i,m,s} = |dem_i - q_{i,m,s}^{min}|$).

$$\min z = \sum_{(i,m,s)\in IMS} cc_{m,s} \times Q_{i,m,s} + \sum_{i\in I} bc \times ST_i \tag{5}$$

3.2. Alternatives search

Components that can be considered alternatives have the same form, fit, and functionality as the reference component. Such components are often called *Exact* or *Direct crosses.* The exact crosses are components that differ only in the manufacturer or packaging of the component, whereas direct crosses can contain alternate components that have higher specifications or compliances. Alternative components may have same or different MPN, description, and manufacturer compared to the known component.

Our alternatives search solution (Figure 1) utilizes rule-based search logic, machine learning and fundamental data mining techniques to find alternative components for reference component. Our system contains several machine learning models (e.g., random forest) and data processing steps which aim to improve the quality of the results.
First, the incoming MPN is cleaned from characters that are known to be irrelevant for MPNs (Kolhase 2019) and the MPN is normalized to same form as rest of the data. This means that only lower-case characters are allowed, and extra spaces are removed. Next, the system infers component *Description* from the database of known descriptions for the MPN. If the MPN is not found in the database, it matched is approximately to the closest MPN based on character level string distance metric (Gonzalo 2001). In our experiments, Levenshtein distance worked the best.

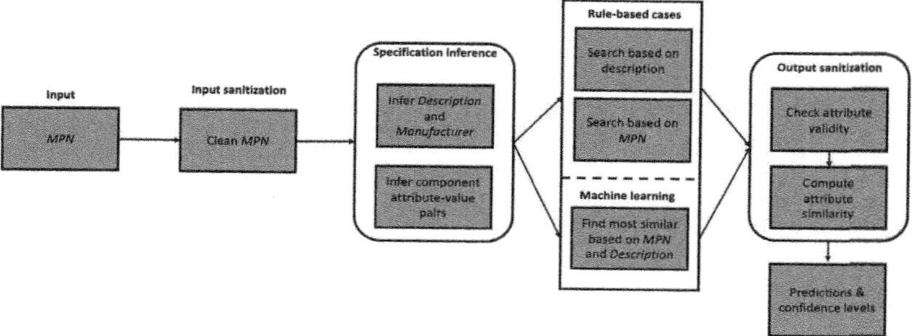

Fig. 1: Flowchart of our proposed alternatives search solution

After the component is inferred and both description and manufacturer for the MPN are found, the system extracts attribute-value pairs from the database, such as {resistance: 100-ohm, tolerance: 5%}. The inferred information is then utilized to find matching components based on *Rule-based cases* and two *Machine learning models*. The rule-based cases include: i) components that have same description but different MPN or manufacturer than the original component, and ii) components that have same MPN but different manufacturer.
Two machine learning models, one trained for description similarity and the other for MPN similarity, are utilized to find most similar components from the database of known MPNs and descriptions. The confidence of both models is averaged to obtain overall ML confidence for a candidate component.

$$Conf_{ML} = 0.5 \times (Conf_{Description} + Conf_{MPN}) \tag{5}$$

Rule-based and machine learning based matching components are then combined and output sanitization is carried out. First, components with conflicting attribute-value pairs (e.g., 1-khom vs. 10-kohm) are removed from the set of matching components and attribute similarity is computed for each matching component. The attribute similarity is utilized in the computation of the final confidence for a matching component:

$$Conf_{Final} = 0.5 \times (Conf_{ML} + Similarity_{Attributes}) \tag{6}$$

The output is then ordered according to the $Conf_{Final}$ which is used in the subsequent processing steps.

3.3. Solution procedure

Figure 2 depicts the solution procedure that consists of three steps: *preprocessing, alternatives addition,* and *optimization.* In the first step, the BoM is cleaned from items that are not viable in terms of client-specific restrictions, such as components from blacklisted suppliers or with incorrect packaging type. In addition, preprocessing step removes any items that have higher lead time that the client specified maximum lead time. This decreases the size of optimization space while maintaining the quality of the results.

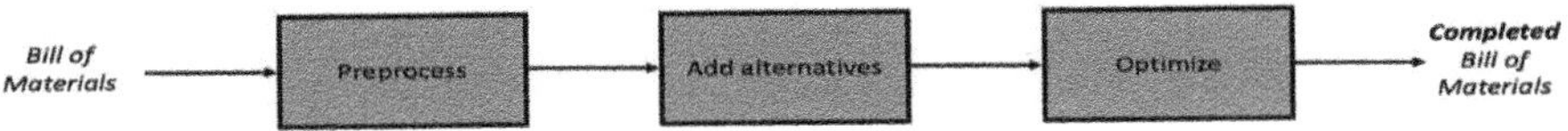

Fig 2: Illustration of three steps of our proposed solution procedure

Next, the BoM is complemented with new alternatives based on our alternatives search solution. Alternatives are queried for any component that have lower total inventory than total demand or no alternatives at all. Highest confidence alternative is picked from alternative search results that satisfies the requirements specified in preprocessing step. Finally, the MILP solver is used to complete the BoM while minimizing the cost and considering specified restrictions, such as inventory and lead time. The output of the solution is a completed BoM that may contain backorders.

4. Case Study

The optimization model was implemented with a Python-based software package *Pyomo 6.4.2* and solved using *Cbc 2.10.8* (Coin-or branch and cut) and *GLPK 4.32* (GNU Linear Programming Kit) open-source MILP solvers. We consider a real customer BoM with 181 components, some having a few alternates already determined by the customer. In total, the BoM consists of 5752 line items. We assume the client-specified maximum lead time is 1 year. Being solved to optimality using Cbc in 10s, the standalone optimization (with 0 lead time) involves 1702 constraints, and 10877 (7668) total variables (integers) and corresponds to an optimal cost of $183695.50 and a total backorder of 16671. Note that GLPK could not provide any solutions in 1 hour. After adding the alternate components using the techniques discussed in section 3, the BoM ended up with 7940 line items and the optimal solution (with 0 lead time) decreased by 44.4%. However, the solution CPU time was increased from 10s to 189s, due to enlarging the solution space of the MILP model. Figure 3 and Table 2 show the computational results for the case study (with different lead times) for the standalone MILP and its AI-enhanced counterpart. Summarizing the results, one can observe from Table 2 that the completed BoM with the standalone MILP would feature a lead time of at least 300 days, whereas the AI-enhanced MILP completes the BoM with a lead time of 240 days while still providing slight improvement in the BoM cost.

Table 2: Results obtained using standalone MILP optimization and its AI-enhanced version

Lead time	Optimization		AI + Optimization		The relative improvement in cost (%)
	Backorder	Cost ($)	Backorder	Cost ($)	
0	16671	183695	8384	102080	44.4
30	16671	183695	8384	102080	44.4
60	16671	183695	8384	102080	44.4
90	16671	183695	8384	101817	44.5
120	13082	148201	4932	67078	54.7
150	13082	148201	4932	67078	54.7
180	11082	147142	2932	65897	54.8

210	1250	50375	100	38272	24.0
240	250	40771	0	37256	8.6
270	250	40770	0	37256	8.6
300	0	38825	0	37151	4.3
330	0	38771	0	37096	4.3
360	0	38493	0	36818	4.4

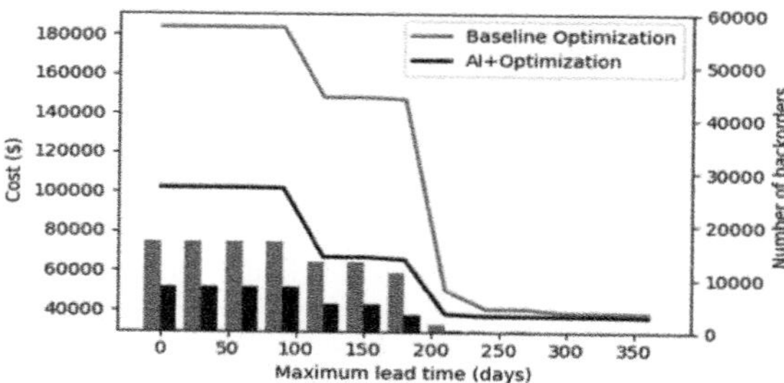

Fig 3: Comparison between standalone MILP and its AI-enhanced version

5. Conclusions

In this paper, we developed an AI-enhanced optimization model to complete an electronic bill of material with minimum cost and shorter lead time. The mathematical optimization is based on a mixed-integer linear programming (MILP) model that could consider different real-life restrictions required to complete a customer BoM. Furthermore, we implemented a data-driven approach to find a set of alternates for a given component with ML-powered alternative search solution. With this solution, the BoM is completed with the most similar alternates that have the same form, fit, and functionality as the given component. We considered a customer BoM and implemented the resulting optimization model in Pyomo. Compared to the stand-alone MILP model, the AI-enhanced MILP model provides better solutions in terms of solution quality and lead times but might deteriorate the computational time due to enlarging the solution space.

References

F. Pan, (2015), The optimization model of the vendor selection for the joint procurement from a total cost of ownership perspective, *Journal of Industrial Engineering and Management*,8,4, 1251-1269

H. Tao, C. Chen, (2006), Production Planning Based on Bom Optimization, *Journal of Electronic Science and Technology of China*, 4, 1, 89-92

J. Leskovec, A. Rajaraman, J.D. Ullman, (2010), Mining of Massive Datasets, http://mmds.org/

K. Kolhase, (2019), Part Number Building Tips, https://forum.digikey.com/t/part-number-building-tips/3460

K. Shah, S. Kopru, J-D. Ruvini, (2018), Neural Network based Extreme Classification and Similarity Models for Product Matching, *In Proceedings of the 2018 Conference of the North American Chapter of the Association for Computational Linguistics: Human Language Technologies*, Volume 3 (Industry Papers), 8–15

M. H. Kiapour, X. Han, S. Lazebnik, A. C. Berg and T. L. Berg, (2015), Where to Buy It: Matching Street Clothing Photos in Online Shops, *IEEE International Conference on Computer Vision (ICCV)*, 3343-3351

N. Gonzalo, (2001), A guided tour to approximate string matching, *ACM Computing Surveys*. 33 (1): 31–88

Antonis Kokossis, Michael C. Georgiadis, Efstratios N. Pistikopoulos (Eds.)
PROCEEDINGS OF THE 33rd European Symposium on Computer Aided Process Engineering (ESCAPE33), June 18-21, 2023, Athens, Greece
 http://dx.doi.org/10.1016/B978-0-443-15274-0.50052-4

Reduced cleaning model for highly viscous non-Newtonian fluids in pipelines

Vera Liebmann[a], Matti Heide[b], Hannes Köhler[b], Christian Golla[a], Jochen Fröhlich[a], Frank Rüdiger[a]

[a]*TU Dresden, Chair of Fluid Mechanics, George-Bähr-Str. 3c, 01069 Dresden, Germany*

[b]*TU Dresden, Chair of Processing Machines and Processing Technology, Bergstraße 120, 01069 Dresden, Germany*

vera.liebmann@tu-dresden.de

Abstract

To clean highly viscous fluids from pipework, flushing processes are used. The cleaning time for a pair of fluids, one being flushed and one used for flushing the other out, is typically estimated by using expensive experiments or costly fully resolved numerical simulations. In this work the authors propose a new approach to determine an accurate approximation of the cleaning time for a flushing process at substantially lower cost. This is done through a simplification of the cleaning process. The process is composed of two phases: core removal and layer removal. Sub-models for each phase are developed and then combined yielding the cleaning model. Subsequently, this model is validated using highly resolved CFD data of a cleaning process. The cleaning model is able to predict the cleaned radius over time with an error of less than 8% during core removal and less than 2% during layer removal compared to the CFD data. The computing time is negligible which makes the model highly suitable for practical application.

Keywords: Simulation, Cleaning, Highly Viscous Fluid, Flushing, Modelica.

1. Introduction

Cleaning is a vital process step during production to ensure that manufactured goods are safe for consumers and that they contain no contaminations from any other goods produced on the same line. During manufacturing of highly viscous products, such as oil, food and cosmetics, flushing processes are often employed to clean pipes of the product and at the same time replace it with a different product, thus ensuring quick product changeover, resulting in lower downtimes. The majority of such products are non-Newtonian fluids, i.e., they exhibit a viscosity dependent on the shear rate. A possible approach to classifying non-Newtonian fluids is to divide them into fluids with and without yield stress τ_0 and by shear-thinning and shear-thickening behavior. They can be modelled accordingly by using appropriate rheological models, e.g. Bingham, Herschel-Bulkley (Macosko, 1994) and Windhab model (Eischen and Windhab, 2002). For the majority of these models, analytical solutions of laminar single-phase flow exist in the literature, e.g. Bird et al. (1983) for Bingham, Kim and Lapushina (1968) and Grinchik and Kim (1972) for Herschel-Bulkley and Pitsillou et al. (2020) for Windhab.

Each flushing process is characterized by two fluids. The first is the flushed fluid, that is supposed to be displaced and the second the flushing fluid, which is used to displace the first. In the literature, flushing processes have been studied for various combinations of fluids. Hasnain et al. (2017) studied buoyant displacement of two Newtonian fluids with different densities and identical viscosity, Oladosu et al. (2020) and Swain et al. (2015)

studied the displacement of a non-Newtonian fluid by Newtonian fluid with different densities. Typically, highly complex experiments or simulations of flushing processes and their results are only transferable to systems, where the process parameters and fluids involved are similar. Since both experiments and simulations are time consuming, it is not practical to conduct and develop them for each individual flushing process. To obtain fast and yet accurate results, the current work proposes a new method, using only the fluids parameters, the geometry of the pipe and a mass flow rate as input to determine the flushing time necessary to obtain a cleaned radius r_c or a remaining mass of previous product.

The flushing process of two chocolates is selected as an example to showcase the applicability of the method. Before developing the method in more detail in section 3, the problem and its limitations are detailed in section 2. The obtained results are discussed in section 4 and finally, a conclusion is provided in section 5.

2. Problem description

In the present work flushed and flushing fluid are assumed to be incompressible, non-Newtonian and of similar density. The algorithm is developed and applied for the fully developed axisymmetric flow of two chocolates at $\vartheta = 40$ °C. They exhibit non-Newtonian behavior as described by the Windhab model (Eischen and Windhab, 2002), which relates shear stress τ and shear rate $\dot{\gamma}$ using a specific function $\tau(\dot{\gamma})$.

The pipe to be cleaned has a radius of $R = 13$ mm and a length of $L = 0.5$ m. White chocolate initially fills the entire pipe and is then flushed out by dark chocolate. Based on experience with detailed simulations, the cleaning process is divided into two phases. First, the core material is removed (Figure 1 a). This phase is characterized by a strong dependency of the cleaned radius r_c on the position x inside the pipe. The second phase occurs, when the cleaned radius r_c is nearly independent of x and the remaining layer of the previous fluid is slowly thinned out.

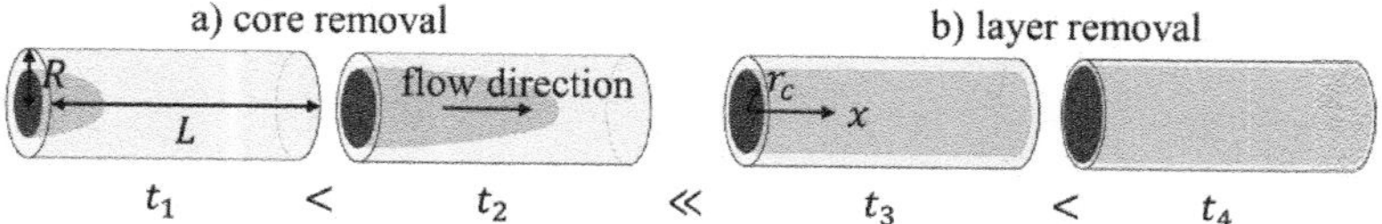

Figure 1 Schematic of the considered flushing process. a) first phase of cleaning – core removal; b) second phase of cleaning – layer removal

Following the distinction of the two phases, the authors first develop a model for core removal and then for layer removal, which is done in sections 3.1 and 3.2, respectively. A condition for the transition between the two phases is described in section 3.2.3.

3. Reduced Model

3.1. Core removal

The velocity profile $u(r)$ for an axisymmetric flow of a single fluid with Windhab viscosity can be derived analytically from the Navier-Stokes equations using the approach by Pitsillou et al. (2020). The Windhab model contains a parameter τ_0 to describe the yield stress of the fluid. For $\tau_0 > 0$ a plug flow region forms in the center of the pipe, i.e., for $r \leq r_0$, where $r_0 = 2\tau_0/f$ is called the plug flow radius, with $f = \Delta p/L$ being the pressure loss over pipe length. Starting from the wall, where a no-slip boundary condition is applied, $u(r = R) = 0$, the velocity $u(r)$ increases reaching a maximum velocity u_{max} at $r = r_0$, which remains constant until the center of the pipe at $r = 0$. Starting from $r =$

R, the velocity gradient $u'(r)$ decreases, from the maximum at the wall to zero within the plug flow radius $r \leq r_0$.
For $f \leq 2\tau_0/R$ no flow develops in the pipe, because the yield stress of the fluid is not overcome. With f increasing above that threshold the mass flow rate $\dot{m}$ and the maximum velocity u_{max} continuously increase, while the plug flow radius r_0 decreases.
The convective transport of a scalar α in one dimension can be described by

$$\partial_t \alpha + u_{tr} \partial_x \alpha = 0\,, \tag{1}$$

with the transport velocity u_{tr}. The method of characteristics may be used to solve this partial differential equation. For this paper only a characteristic of eq. (1), along which $\alpha = \text{const.}$ is used and given by

$$x = u_{tr}\, t\,. \tag{2}$$

Eq. (1) can be used to describe the core removal in a flushing process, by using α as a phase indicator. Then $\alpha(x,t) = 1$ indicates the presence of the first fluid at time t and location x, while $\alpha = 0$ indicates the absence of the first fluid and presence of the second fluid. Assuming both fluids have identical physical properties, the pipe flow is axisymmetric, laminar and no diffusion between the two fluids occurs, the transport velocity u_{tr} in eq. (2) can be replaced by the analytical solution $u(r)$ of the flushing fluid described above. The interface between the two fluids is identified here with $\alpha = 0.5$. The cleaned radius r_c for a given location x_0 along the flow axis x, can then be determined using with $x_0 = u(r_c)\, t$ (see eq. (2)). Defining a non-dimensional time using $\hat{t} = u_{max} t / x_0$ with the maximum velocity u_{max} and the spatial coordinate x_0 yields a universal curve for $r_c(\hat{t})/R$ as shown in Figure 2.

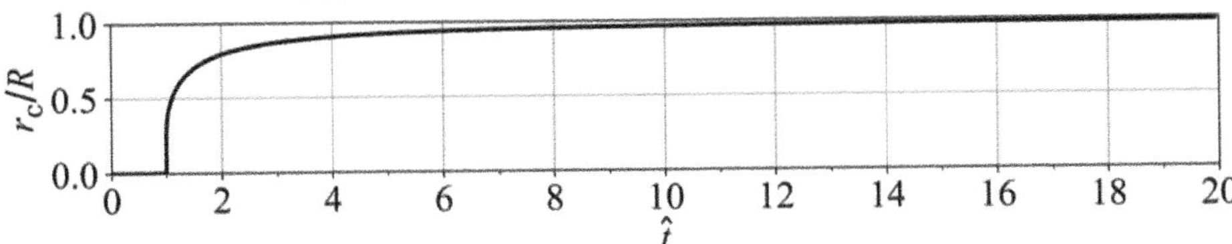

Figure 2 Cleaned radius r_c/R over dimensionless time $\hat{t} = u_{max} t / x_0$.

Cleaning starts after an offset time $\hat{t} = 1$. This is the time needed for the plug flow region of the flushing fluid traveling at u_{max} to reach the cross section at x_0. When the plug flow region passes, the cleaned radius instantly increases from $r_c = 0$ to $r_c = r_0$. From then on, the cleaned radius increases further. The rate of change, however, decreases since $u(r)$ decreases towards the wall.

3.2. Layer removal

3.2.1. Basic algorithm

Once the core has been removed, the remaining layer is transported away. A linear velocity profile is assumed in the remaining layer, i. e. for $r_c < r < R$

$$u_L(r) = u_c \left(1 - \frac{r_c - r}{r_c - R}\right), \tag{3}$$

where u_c is the velocity on the interface located at r_c. Integrating eq. (3) over the cross section yields the mass flow rate in the layer $\dot{m}_L$ as shown in eq. (4)

$$\dot{m}_L = 2\pi\rho \int_{r_c}^{R} u_L(r) r\, \mathrm{d}r = \frac{\pi}{3} \rho u_c (R - r_c)(R + 2r_c) \tag{4}$$

The mass of the fluid in the remaining layer m_L is calculated using eq. (5).

$$m_{\mathrm{L}} = \pi\rho(R^2 - r_c^2)L \quad (5)$$

Given these three equations, a simple algorithm to determine the cleaned radius r_c over time is derived. The problem is initialized with radius $r_c^0 = r_c(t = t_0)$. The determination of the starting value is discussed in section 3.2.3. Then, four steps are performed for each time step t^n. First, the velocity u_c^n is determined from the given cleaned radius at that time r_c^n, as detailed in section 3.2.2. Second, the mass flow rate $\dot{m}_{\mathrm{L}}^n$ is determined using eq. (4). Third, the mass of the remaining layer at the next time step is calculated from $m_{\mathrm{L}}^{n+1} = m_{\mathrm{L}}^n - \dot{m}_{\mathrm{L}}^n(t^{n+1} - t^n)$. Finally, the height of the remaining layer at the next time step, r_c^{n+1}, is determined by rewriting eq. (5). These steps are repeated until a desired radius is cleaned or a maximum time is reached.

3.2.2. Determining u_c

Figure 1 shows that, during layer removal the radius r_c is nearly independent of x, so that the flushing fluid is assumed to create a cylinder with radius r_c concentric to the pipe with radius R. Assuming a smooth interface between the two fluids, the flow of the flushing fluid can be treated like the flow inside a second pipe with a pipe radius equal to r_c and a velocity at the wall $u(r_c) = u_c$. The analytical solution by Pitsillou et al. (2020) is modified accordingly to yield the velocity profile $u_{\mathrm{fl}}(r)$ inside the flushing fluid. Finally, the value of u_c is determined from the condition that the shear stress at the interface is the same for the remaining layer of flushed and flushing fluid, i.e., $\tau_{\mathrm{L}}(r_c) = \tau_{\mathrm{fl}}(r_c)$. This implies a smooth interface between the phases. The shear stress at the interface is determined via the Windhab law $\tau(\dot{\gamma})$. For the flushing fluid the shear rate $\dot{\gamma}_{\mathrm{fl}}$ obtained by differentiating the velocity u_{fl} in the core region is used. A second expression for the shear stress at the interface is obtained from using the shear rate obtained from the linear velocity profile in the remaining layer (eq. (3)) to $\dot{\gamma}_{\mathrm{L}}$. Combining both expressions yields the unknown u_c.

3.2.3. Defining a starting value for r_c

The initial value of r_c to start the algorithm in section 3.2.1 must be sufficiently close to R and sufficiently far away from the plug flow radius r_0 to justify the approximation of a linear velocity profile inside the remaining layer. To find the starting value r_c^0 the analytical solution for the mass flow $\dot{m}_{\mathrm{L,an}}$ and the solution found by the linear approximation $\dot{m}_{\mathrm{L}}$ from (4) are compared. The former is found by integrating the analytical solution for the velocity profile $u(r)$ from Pitsillou et al. (2020) to

$$\dot{m}_{\mathrm{L,an}} = 2\pi\rho \int_{r_c}^{R} u(r)\, r\, \mathrm{d}r\,. \quad (6)$$

Figure 3 shows the relative error $\varepsilon = \dot{m}_{\mathrm{L}}/\dot{m}_{\mathrm{L,an}}$ for different total mass flow rates. Each mass flow rate corresponds to a different plug flow radius r_0, which is related to the cleaned radius r_c. The latter is varied as the choice affects the error ε. If $r_c/r_0 \approx 1$, the

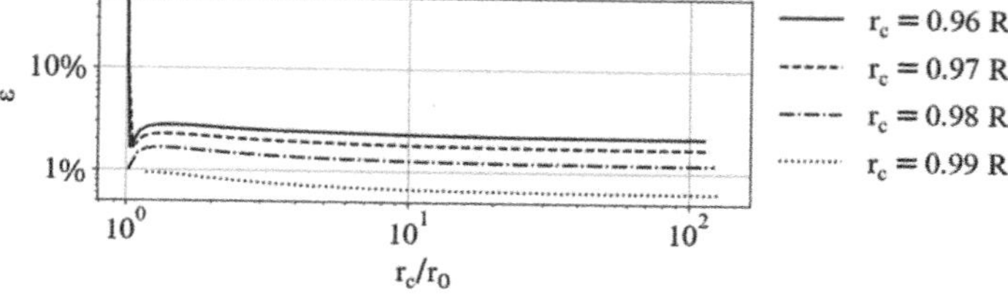

Figure 3 Influence of the choice of r_c on the error ε of the mass flow rate in the remaining layer that arises due the linear approximation of the velocity there. Parameters from Liebmann et al. (2022).

plug flow radius and the cleaned radius are approximately equal. In that case the error ε is high for all values of r_c. This is due to the fact, that at the plug flow radius r_0 the velocity $u(r)$ transitions from the high gradient at the wall to zero gradient within the plug flow region, which is not well approximated by a linear function. For values of $r_c/r_0 \gg 1$, the error is smaller, since the cleaned radius is further away from the plug flow radius, where the velocity gradient transitions to zero. The higher the ratio r_c/r_0 and the lower the value for r_c the more pronounced this effect becomes. For the second phase, the authors propose an initial cleaned radius of $r_c^0 = 0.97\ R$ which provides errors $\varepsilon < 2\%$ in this configuration. For more accurate results a higher value of r_c may be chosen.

3.3. Implementation in Modelica

The two-stage model described above was implemented in OpenModelica v1.19.2. The input parameters are the radius R, the length L of the pipe and the mass flow rate of the flushing fluid. For the first phase, the physical properties of the flushing fluid are used. For the second phase the flushed and flushing fluids properties are used. The values for the two chocolates are taken from Liebmann et al. (2022). The mass flow rate is used to determine the analytical solution of the Windhab velocity profile $u(r)$. The cleaned radius $r_c(t)$ during the core removal phase is determined according to section 3.1. Once the cleaned radius $r_c^0 = 0.97\ R$ described in section 3.2.3 is obtained during core removal, the algorithm in section 3.2.1 is used for layer removal.

4. Results

Figure 4 depicts the results obtained from a fully resolved OpenFOAM simulation, which took 20 hours to complete on 16 CPUs (see Liebmann et al., 2022) and the results obtained from the reduced model in OpenModelica. The latter required only 0.05 seconds on a single CPU to complete the calculation. Core removal starts when the plug flow region reaches the cross section at the considered position x. The time needed for that is shorter for smaller distances x. After that, r_c increases instantly by the plug flow radius and then continually grows. The cleaned radius in the OpenFOAM simulation is up to 8% higher during core removal. This may be due to additional effects at the interface not considered within the reduced model or because the physical properties of the two fluids are similar but not equal. At $r_c = 0.97\ R$ layer removal begins. During this phase, the reduced model shows very good agreement with the OpenFOAM simulations both in quality and quantity. The reduced model predicts r_c up to 2% smaller than the results from OpenFOAM. The model in OpenModelica runs about seven orders of magnitudes faster, offering a considerable speedup compared to the fully resolved simulation, with little error arising from the use of the reduced model, especially during layer removal.

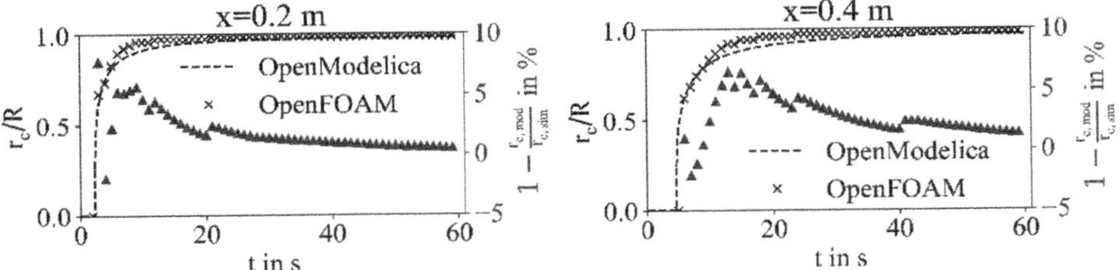

Figure 4 Comparison of results for the cleaned radius $r_c(t)$ obtained from a fully resolved simulation using OpenFOAM and the reduced model in OpenModelica introduced in this paper. The blue triangles with the right axis indicate the relative error between both results.

5. Conclusion and Outlook

The paper proposes an approach for a reduced model of a flushing process of two non-Newtonian fluids with yield stress in a straight pipe. The flushing process is divided into two phases. During the first phase (core removal) the presented results for the cleaned radius $r_c(t)$ are within 8% of the results obtained using a fully resolved numerical simulation. During the second phase (layer removal) this error is further reduced. The reduced model implemented in OpenModelica offers accurate results after less than a second. Compared to the current state of the art, i.e. conducting fully resolved simulations or experiments, which are time consuming, expensive and complex, the model can be applied in the rapid design of industrial flushing processes. It offers a method of designing new or adapting old processes to new process parameters while reducing the time and cost compared to current methods. Processes not adapted so far due to excessive cost of classical methods can now be optimized, thus saving resources, costs and time.

The approach presented for layer removal is only applicable if the velocity profile in the remaining layer is approximately linear. For cases where the layer is thicker, the authors were able to successfully employ an alternative approach to determine u_c from the analytical solution of the velocity within the remaining layer. However, this approach is specific to a particular combination of rheological models for the flushed and flushing fluid and, hence, not generally applicable. For these cases and for $r_c < r_0$ further development of the presented method is necessary. The approach can be extended for other combinations of fluids using the literature provided in section 1. This requires two adaptations of the present model: one for the velocity profile of the flushing fluid and one for the rheological behavior $\tau(\dot{\gamma})$ of the two fluids. The method has already been employed for a set of different fluid combinations. Further tests will be conducted to determine the limitations in the ratio of the two fluids physical properties.

References

R.B. Bird, G.C. Dai, B.J. Yarusso, 1983, The rheology and flow of viscoplastic materials. Reviews in Chemical Engineering, 1, 1, 1-70

J.-C. Eischen, E. J. Windhab, 2002, Viscosity of cocoa and chocolate products, Applied Rheology, 12, 1, 32-34

I.P. Grinchik, A.K. Kim, 1972, Axial flow of a nonlinear viscoplastic fluid through cylindrical pipes. Journal of Engineering Physics, 23, 2, 1039–1041

A. Hasnain, E. Segura, K. Alba, 2017, Buoyant displacement flow of immiscible fluids in inclined pipes, Journal of Fluid Mechanics, 824, 661-687

A.K. Kim, B.I. Lapushina, 1968, The motion of a liquid with variable rheological characteristics in a circular cylindrical tube, Journal of Engineering Physics, 14, 2, 159–161

V. Liebmann, M. Heide, K. Schoppmann, H. Köhler, J. Fröhlich, F. Rüdiger, 2022, Aspects of modelling the cleaning of a chocolate with yield stress in a pipe using CFD, Fouling and Cleaning in Food Processing 2022 Proceedings

C. W. Macosko, 1994, Rheology – Principles, Measurements and Applications, Wiley-VCH, 1

O. Oladosu, J. Bhakta, K. Alba, 2020, Density stable yield-stress discplacement flow of immiscible fluids in inclined pipes, Journal of Non-Newtonian Fluid Mechanics, 275, 4, 104203

R. Pitsillou, G.C. Georgiou, R.R. Huilgol, 2020, On the use of the Lambert function in solving non-Newtonian flow problems, Phys. Fluids, 32, 9, 093101

P.A.P. Swain, G. Karapetsas, O.K. Matar, K.C. Sahu, 2015, Numerical simulation of pressure-driven displacement of a viscoplastic material by a Newtonian fluid using the lattice Boltzmann method, European Journal of Mechanics B/Fluids, 49, A, 197-207

Antonis Kokossis, Michael C. Georgiadis, Efstratios N. Pistikopoulos (Eds.)
PROCEEDINGS OF THE 33rd European Symposium on Computer Aided Process Engineering (ESCAPE33), June 18-21, 2023, Athens, Greece
 http://dx.doi.org/10.1016/B978-0-443-15274-0.50053-6

Optimal Retrofit of Simple Distillation Sequences to Thermally Coupled Side-Stream Configurations

Anna S. Horsch,[a] Andres R. Acevedo,[a] Mirko Skiborowski[a]

[a]*Hamburg University of Technology - Institute of Process Systems Engineering, Am Schwarzenberg-Campus 4, 21073 Hamburg, Germany*
Mirko.Skiborowski@tuhh.de

Abstract

Retrofitting existing plant equipment to a more energy efficient setup is a promising and sustainable way to cut down on the energy demand and consequently the operating costs of a chemical processing plant. While this opportunity is attractive in terms of low capital expenditure and construction time, it is also highly challenging for the design engineers. Degrees of freedom in grassroot design transform into constraints in the retrofit project and finding an optimal design within the constraints of the existing equipment, requires a sophisticated model-based approach. In this work, such an approach is presented for the retrofit of distillation column sequences to thermally coupled side-stream configurations. Based on this approach, an easy evaluation of retrofit opportunities in terms of investment and operating costs is facilitated. The potential of this retrofit is illustrated in the current contribution by means of a case study, which compares the retrofit option to a fully thermally coupled dividing wall column grassroot design. While the dividing wall column generally allows for higher energy savings, the investment in the retrofit is significantly smaller, which yields a considerably quicker amortization.

Keywords: Distillation, Thermal Coupling, Retrofit, Optimization

1. Introduction

In light of rapidly increasing energy and resource prices as well as sociopolitical pressure due to the quickly progressing climate crisis, the arguably biggest goal of the chemical industry is to become more energy-efficient and sustainable. A focus in this effort lies on distillation, the most frequently applied fluid separation technique in the chemical industry, which is accountable for a major share of the energy consumption in chemical plants (Sholl and Lively 2016). A well-established approach to reduce the energy demand of distillation sequences is the thermal coupling of distillation columns in the form of side rectifiers (SR), side strippers (SS) and fully thermally coupled dividing wall columns (DWC). Out of these three configurations, the DWC is unmatched in terms of operating costs and investment, which sets the DWC as one of the frontrunners of distillation innovation with more than 300 industrial applications (Lukač et al. 2019).

However, constructing an energy and cost-efficient grassroot design is not the only promising option in the effort to re-engineer the chemical industry. Retrofitting existing plant equipment can facilitate a considerable decrease in energy consumption and related operating costs, combined with the opportunity to increase process capacity, with comparably little capital expenditure, effort and construction time. This becomes possible due to commonly applied design margins on plant equipment (Rangaiah 2016). Retrofitting and revamping projects account for the major share of design projects conducted in the chemical industry, offering potential for innovative retrofit options and optimal design methods (Rangaiah 2016).

One particularly interesting concept in that context is the so-called Liquid-Only-Transfer (LOT) configuration proposed by Agrawal (2000), which transforms the bidirectional vapor and liquid transfer in thermal couplings to an unidirectional liquid transfer. This can be achieved for SR and SS configurations by extending the side-columns with an additional column section and heat exchanger, which facilitates the elimination of the intercolumn vapor transfer. The resulting LOT configuration provides an excellent retrofitting option of an existing column sequence through integration of an additional side stream.

Although we are not aware of a dedicated industrial implementation of the LOT concept, it offers several advantages over the classical thermal coupling. Here, it is important to note that the LOT-configurations are thermodynamically equivalent to their thermally coupled counterparts (Madenoor Ramapriya et al. 2014) and as such hold the same potential energy savings, which can amount up to 30 % (Amminudin et al. 2001) for a ternary separation. In addition to these energy savings, the elimination of the vapor connection between the column allows for the hydrodynamic decoupling of both columns, enabling the application of thermal coupling to columns operated at different pressures. This presents the option for further heat integration in the thermally coupled columns through targeted pressure variation, providing potentially significant energy savings (Skiborowski 2020). Furthermore, replacing the generally uncontrolled vapor stream or vapor split with a controllable manipulated variable like the heat duty or reflux ratio, allows for a more precise control of the internal vapor streams and as such supports the control at the energy optimal operating point (Madenoor Ramapriya et al. 2014). Lastly, the structural similarity of LOT-sequences to uncoupled (in)direct sequences consisting of two regular distillation columns connected via a liquid stream, makes them perfect candidates for retrofitting simple distillation sequences with minimal effort.

In this contribution, a framework for the optimal retrofit of an uncoupled distillation sequence to thermally coupled LOT-side stripper and side rectifier configurations is presented. While these do generally not allow for the same energy savings as a fully thermally coupled DWC, they are oftentimes thermodynamically more efficient than the DWC (Agrawal and Fidkowski 1998). However, exploiting these benefits in an optimal retrofit design is a complex task, since structural constraints for the given equipment, including column height, vapor capacity, pressure drop and available heat exchanger area, have to be adhered in the design process. The current contribution proposes a superstructure based design approach, which builds on a modified mixed integer nonlinear programming (MINLP) problem originating from the grassroot design of column sequences as presented by Walterman and Skiborowski (2019). In an exemplary case study, the performance of this developed methodology is showcased, addressing the question if a retrofit design can compete with an optimally designed fully thermally coupled grassroot DWC.

2. Methods

In this chapter, the concept of thermally coupled Liquid-Only-Transfer distillation sequences is explained based on the example of the side rectifier configuration. Subsequently, the optimization procedure for the grassroot design and the retrofit is described.

2.1. Liquid-Only-Transfer (LOT) Sequences

As introduced by Agrawal (2000) and theoretically proven by Madenoor Ramapriya et al. (2014), a thermodynamically equivalent LOT configuration can be developed through elimination of the vapor transfer in thermal couplings in three steps: First, the stripping section of the second column (*S2*) including the reboiler is duplicated (*S2**) and placed at the bottom of the first column, where the recycle vapor stream (*V*) enters the first column. As the second step, the intercolumn vapor stream *V* is redirected to originate from the duplicated stripping section (*V**). Lastly, the duplicated section *S2** is connected to the stripping section of the first column (*S1*). This procedure is visualized in Figure 1. The development of the LOT-side stripper and the LOT-DWC is done in a similar manner.

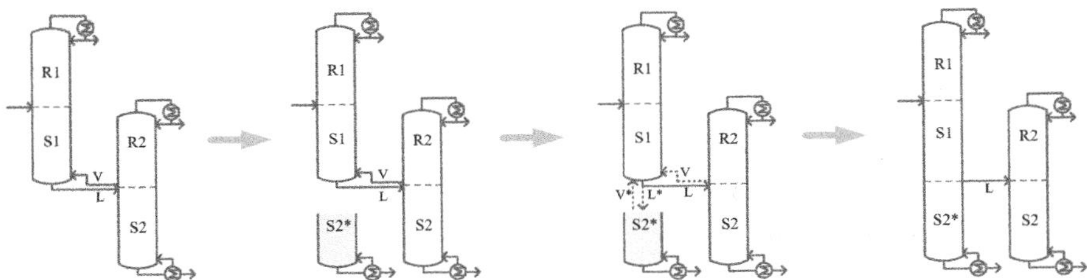

Figure 1: Development of thermodynamically equivalent LOT-side rectifier

2.2. Optimization-based design

Both the optimal grassroot design as well as the optimal retrofit design are obtained based on an equation-oriented superstructure-based optimization approach implemented in GAMS 31.2.0. The superstructure builds on an equilibrium-stage model of the distillation columns on the basis of the MESH-equations (**M**ass balance, **E**quilibrium conditions, **S**ummation constraint, **E**nthalpy balance). The thermodynamic properties are calculated by means of an external function programmed in C which communicates with GAMS via a dynamic link library, while the optimization problems are solved as a series of successively relaxed MINLP problems (Skiborowski et al. 2015). The grassroot design procedure is described in detail by Walterman and Skiborowski (2019). For retrofit design this approach is extended to account for pressure drop and separation efficiency and is subsequently adapted to calculate optimal retrofit designs for a predefined column sequence with given height and diameter.

2.2.1. Modelling of pressure drop and separation efficiency

In the original MESH-based grassroot design model described by Walterman and Skiborowski (2019), the pressure drop is assumed to be negligible and the height equivalent to one theoretical plate (HETP) is considered to be constant. The column height and diameter are determined in accordance with the required number of equilibrium stages and the vapor flowrates. While this is a reasonable simplification for conceptual grassroot designs, the pressure and separation efficiency need to be modelled more precisely for retrofit calculations, considering the utilization of existing equipment and the resulting changes in gas and liquid loads.

The pressure drop per height ($\Delta p/h$) as well as the separation efficiency, which is represented by the HETP, are calculated based on the vapor load. The vapor load is described by the F-factor $FF = \dot{V}/A_{column} \cdot \sqrt{\rho_V}$ which is calculated as the product of the superficial gas velocity, given as the ratio of vapor $\dot{V}$ per column section A_{column} and the square root of the vapor density ρ_V.

In this work, the properties of the Montz B1-500 packing are utilized. The correlation for the pressure drop is approximated with a third-degree polynomial $\Delta p/h = 0.83F^3 - 1.3F^2 + 1.6F - 0.3$ (R=0.999) and the $HETP$ is approximated with a second-degree polynomial $HETP = 0.01F^2 + 0.01F + 0.24$ (R=0.999). Both correlations are fitted to the vendor given and packing specific pressure drop and separation efficiency diagrams (Julius Montz GmbH 2022).

2.2.2. Optimal Retrofit

Feed properties, purity constraints and a maximum stage number of the superstructure are input information for both the grassroot design procedure as well as the retrofit design. Furthermore for the retrofit, structural constraints for the available column heights, column diameters and heat exchanger areas are considered. The model initialization and heat duty minimization is performed based on initial flash calculations and successive model refinement as described by Waltermann and Skiborowski (2019).

The grassroot design and the retrofit design optimization differ as described in the following. For the grassroot design, the total annualized costs (TAC)

$$TAC = C_{op} \cdot t_{OP} + C_{inv} \cdot \alpha \tag{1}$$

are minimized, considering an optimization of the equipment sizing and operating conditions, based on variable column heights, feed and side stream locations and heat duties. The TAC include the operating costs (C_{op}) considering heat duties and utility prices with an annual operation time (t_{op}) of 8000 h, and the investment costs (C_{inv}) which are based on a correlation from Biegler et al. (1997) with a depreciation factor $\alpha = f(z, t_a) = 0.37$, based on an interest rate (z) of 6 % and depreciation time (t_a) of 3 years. In the retrofit procedure only the operating costs C_{op} are minimized, if the existing equipment suffices. Therefore, the column structure is optimized on the basis of the superstructure model such that structural constraints on the available column height and diameter are obeyed and variable side stream and feed location result in maximum energy savings. In order to allow for more flexibility, a rearrangement of the columns for the LOT sequence is considered in the design process. If based on the resulting design the heat exchanger area of the available heat exchangers is insufficient, an additional, parallel heat exchanger is considered and depreciation costs are added to the objective. The heat exchanger costs are calculated based on the same correlation as in the grassroot case. The main steps of the grassroot design and the retrofit procedure are summarized in Figure 2.

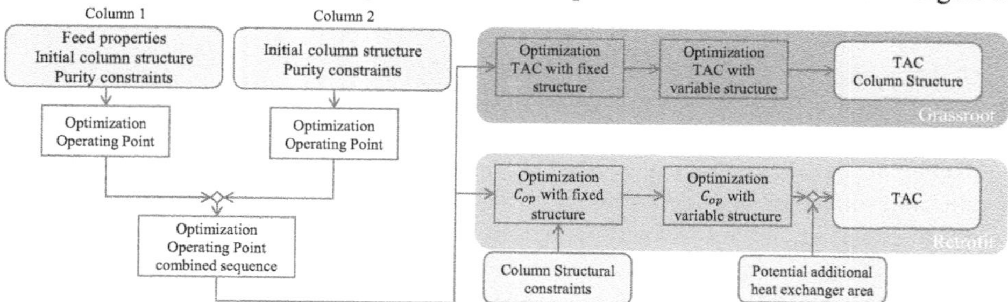

Figure 2: Modeling and solution approach for grassroot design and retrofit; "Optimization Operating Point" for individual columns and combined sequence according to Waltermann and Skiborowski (2019)

3. Case Study

As previous shortcut screening calculations highlighted the importance of feed compositions for the potential energy savings (Skiborowski, 2020), two exemplary feed compositions are evaluated for the separation of benzene, toluene and xylene, an equimolar feed F_1 and a second feed with low amount of mid boiling component F_2=(0.45/0.10/0.45) mol/mol. For both cases, a saturated liquid feed of 1 kmol/s at 1 atm shall be separated with product purities of at least 0.99 mol/mol. The liquid phase activities are modelled with the NRTL (Non-Random-Two-Liquid) model, while the vapor phase is assumed to behave ideal.

As a basis for the retrofit, an optimal grassroot design of a column sequence for the separation task is calculated. Since the direct sequence results in lower TAC than the indirect sequence, it is used as the reference case for both case studies. Design safety factors are applied to the height (+10 % stages), the cross-sectional area (80% of flooding velocity) and the heat exchanger area (+10 %) (Rangaiah 2016). The operating costs of this uncoupled sequence are the benchmark that will be compared to the grassroot DWC design and the retrofit design of the LOT-side rectifier. The costs for the direct sequence, as well as the retrofit and DWC for both study cases are summarized in Table 1.

Table 1: Summary of operating costs, investment, total annualized costs and possible amortization period for the initial direct sequence, the grassroot DWC and the retrofit LOT-SR.

		Operating Cost $C_{op} \cdot t_{op}$ [k€]	Investment C_{inv}[k€]	*TAC* [k€]	Amortization [a]
Case 1 F_1	DS	133	466	307	-
	Retrofit	124	18	131	2.2
	DWC	93	394	240	9.9
Case 2 F_2	DS	112	420	269	-
	Retrofit	90	30	101	1.3
	DWC	74	361	209	9.5

For the grassroot design DWC, both operating and investment cost and consequently the TAC are below that of the direct sequence for both cases. To guarantee comparability, the DWC design also contains safety factors of the same magnitude. The grassroot DWC is excellently suited for both cases with 30 % (F_1) and 34 % (F_2) of operating cost savings, which equals a saving potential S of 40 k€/year (F_1) and 38 k€/year (F_2). Based on the necessary investment, this yields an amortization period

$$A = C_{inv}/S \tag{2}$$

of 9.9 years and 9.5 years, respectively. Thus, it takes almost 10 years until the 30 %-34 % savings in operating costs amount to the necessary investment for the DWC.

For the retrofit LOT-side rectifier designs, investments of 18 k€ and 30 k€ are determined based on additional heat exchanger area that needs to be added for the retrofit. The retrofit reduces the operating costs by 6 % (F_1) and 20 % (F_2).

As to be expected, the energy saving potential of the retrofit remains lower than that of the fully thermally coupled DWC. However, since the necessary investment is much lower, the amortization period is noticeably shorter for the retrofit in both cases with only 2.2 years and 1.3 years, respectively. As a consequence, the TAC of the retrofit are considerably below the TAC of the DWC for both cases. The setup of the direct sequence, the retrofitted LOT-sequence and the DWC for the equimolar case F_1 is visualized in Figure 3.

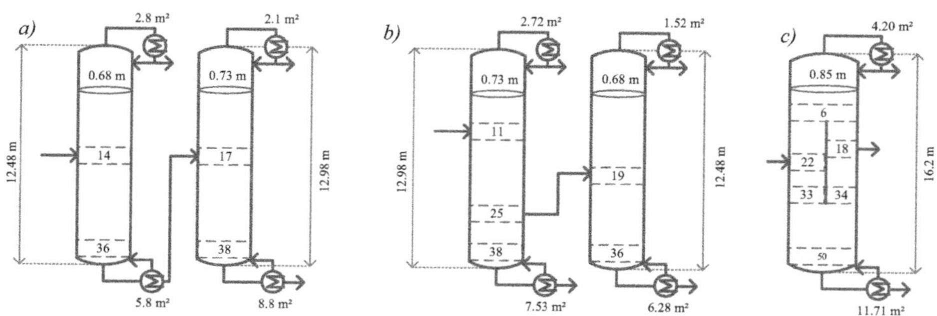

Figure 3: Results of (a) the grassroot design of the direct sequence, (b) the resulting retrofitted LOT-sequence and (c) the grassroot design of the DWC.

4. Conclusion

The current work presents an optimization-based methodology for the retrofit of distillation sequences to thermally coupled side stream configurations and illustrates the benefits of these configurations for the separation of two exemplary feed compositions of a benzene-, toluene- and xylene mixture. While the implementation of a new dividing wall column offers larger energy savings than the simple retrofit, the drastically lower investment of the side stream retrofit results in much lower pay out times. Considering the frontrunner status of the dividing wall column, this showcases the potential of such Liquid-Only-Transfer configurations when used for retrofit. For a final implementation of such a retrofit option further cost factors for reassembly and lost earnings for plant shutdown, which depend on the given plant, should be considered. However, the resulting retrofit configuration also enables further potential for process intensification by means of heat integration, which can even exceed the energy savings that are enabled by dividing wall columns, as has been illustrated by Skiborowski (2020).

5. References

Agrawal (2000): *AIChE J.* 46 (11), pp. 2198–2210.

Agrawal; Fidkowski (1998): *Ind. Eng. Chem. Res.* 37 (8), pp. 3444–3454.

Amminudin; Smith; Thong; Towler (2001): *Chem. Eng. Res. Des.* 79 (7), pp. 701–715.

Biegler; Grossmann; Westerberg, (1997): Upper Saddle River, NJ: Prentice-Hall (Prentice-Hall international series in the physical and chemical engineering sciences).

Julius Montz GmbH (20.10.2022): Separation Efficiency and Pressure Drop - Montz B1-500. Available online at https://www.montz.de/standard-blechpackung-typ-b1#&gid=1033229313&pid=2.

Lukač; Halvorsen; Olujić; Dejanović (2019): *Chem. Eng. Res. De.s* 79 147, pp. 367–377.

Madenoor Ramapriya; Tawarmalani; Agrawal, (2014): *AIChE J.* 60 (8), pp. 2949–2961.

Rangaiah (Ed.) (2016): Chichester, UK: John Wiley & Sons, Ltd.

Sholl; Lively (2016): *Nature* 532 (7600), pp. 435–437.

Skiborowski (2020): 30th European Symposium on Computer Aided Process Engineering, vol. 48: Elsevier (Computer Aided Chemical Engineering), pp. 991–996.

Skiborowski; Harwardt; Marquardt (2015): *Comput. Chem. Eng.* 72, pp. 34–51.

Waltermann; Skiborowski (2019): *Comput. Chem. Eng.* 129, p. 106520.

Antonis Kokossis, Michael C. Georgiadis, Efstratios N. Pistikopoulos (Eds.)
PROCEEDINGS OF THE 33rd European Symposium on Computer Aided Process Engineering (ESCAPE33), June 18-21, 2023, Athens, Greece
 http://dx.doi.org/10.1016/B978-0-443-15274-0.50054-8

Computer aided-design of energy efficiency tools on a microalgae biorefinery scheme

Araceli Guadalupe Romero-Izquierdo [a], Claudia Gutiérrez-Antonio [a], Fernando Israel Gómez-Castro [b], Salvador Hernández [b*], Juan Fernando García-Trejo [a]

[a] *Facultad de Ingeniería, Universidad Autónoma de Querétaro, Cerro de las Campanas S/N, Querétaro, Querétaro. 76010, México.*

[b] *Departamento de Ingeniería Química, Universidad de Guanajuato, Noria Alta S/N, Guanajuato, Guanajuato. 36050, México.* hernasa@ugto.mx

Abstract

The search of new alternatives of processing for the production of sustainable products, designed under energy efficiency tools, has gained the interest of scientific community. In this work is proposed the conversion of a microalgae (*Chlorella vulgaris*) through a biorefinery scheme to produce biojet fuel, other biofuels and value-added products. Process intensification and energy integration tools were applied as energy efficiency strategies to improve the system. The biorefinery is modelled in Aspen Plus. The application of both strategies increases the equipment cost up 7 %, but reduces in 70 % the thermal energy consumption and 45% the CO_2 emissions released.

Keywords: microalgae biorefinery, biojet fuel, process intensification, energy integration.

1. Introduction

Microalgae are micro-organisms able that covert sunlight and CO_2 to biomass, from which biofuels and value-added products can be produced; they do not require fertile soil, and can grow in wastewater (Romero-Izquierdo et al., 2022). In particular, the use of microalgae for the sustainable aviation fuel production has gained attention; this biofuel is the most resilient strategy to achieve the sustainable recovery of the sector (IEA, 2021). Various reports present the use of microalgae oil to produce biojet fuel (Gutiérrez-Antonio et al., 2017; O'Neil et al., 2019; Martinez-Villarreal et al., 2022), while a few ones present the conversion of whole-microalgae strains into a biorefinery scheme (de Carvalho et al., 2022; Siddiki et al., 2022; Elkelawy et al., 2022); however, the biojet fuel generated in these processes is not competitive with its fossil counterpart (NESTE, 2022). The microalgae strain cost and the high energy requirements for the processing are the main issues to reach its profitability (Romero-Izquierdo, 2020). Thus, the efforts must be oriented to the application of strategies to make more efficient the use of energy, such as process intensification and energy integration. Therefore, in this work the computer-aided design of intensified and integrated routes for the conversion of whole-microalgae (*Chlorella vulgaris*) to biojet fuel, other biofuels and high-value products in a biorefinery scheme is presented; the implementation of such strategies over the energetic, economic and environmental indicators of the production route is analyzed.

2. Modelling and simulation of microalgae biorefinery

Chlorella vulgaris is cultivated by 14 days in wastewater, estimating a microalgal biomass after the harvesting of 442,795.64 kg by production lot (Romero-Izquierdo, 2020). This biomass is composed by proteins (46 wt %), carbohydrates (18 wt %) and lipids (27 wt %) (Becker, 2007). The biorefinery (Figure 1) is divided in three zones: oil extraction (Z1), press cake treatment (Z2) and microalgal oil treatment (Z3); modelling each one using NRTL for Z1 and Z2, and Peng-Robinson and BK10 for Z3. Z1 considers the modelling of oil extraction by ultrasonication; Natarajan et al. (2014) reported a maximum cell disruption of 75 % is reached with 0.8 kWh/L. This process is modelled with *CFuge* module, from which oil and press cake streams are obtained. The press cake is processed in Z2: 10 wt % enters to pyrolysis process, 10 wt % to anaerobic digestion and 80 wt % to ethanol production. The pyrolysis is realized at 500 °C and air, obtaining bio-oil (mainly composed by levoglucosan, toluene and hexadecenoic acid) and biochar modelled by the empirical formula $C_{1.7}H_{8.2}N$ (Brown et al., 2012). This process was modelled by *RStoic* module. The anaerobic digestion is carried out at 55 °C with a water-biomass ratio of 1:1 (Li et al., 2014), and represented by the Buswell equation (Symons and Buswell, 1933). The biogas obtained is fed to isentropic turbine (module *Compr*) to produce power. In the case of ethanol production, the available biomass enters to a separate saccharification and fermentation process (SHF). The first step is the acid hydrolysis at 121 °C with 1 % of sulfuric acid; its outlet stream enters to the fermentation reactor (*RStoic* module) at 30 °C adding *Z. mobilis*. The non-converted biomass is removed (by *Sep* module) of the outlet stream from fermentation reactor to be fed to a distillation train with three columns. From the first column, 35 stages without condenser, the CO_2 is obtained by the top, extracting side stream rich in ethanol, which enters to second column, 13 stages and partial-vapor condenser, concentrating the ethanol at the top. The ethanol-rich stream enters to the third column, 15 stages and total condenser, breaking the ethanol-water azeotrope with glycerol. All columns are modelled with *RadFrac* module, reaching 99 % recovery for key components. The produced ethanol is inserted to the ATJ process on Z3, where are three consecutive reactors (modelled by *RStoic* modules): ethanol dehydration, oligomerization, and hydrogenation. The first one at 450 °C operates with a superheated steam stream with a 1:1 ethanol-steam ratio, reaching 99.4 % yield of ethylene; the second one at 120 °C and 35 bar reached 99 % yield to alkenes, and the last one operates at 110 °C with 99 % conversion to paraffins. The ethylene purification before the second reactive stage is realized using a heat exchanger (Heat-X module) and two flash tanks (Flash2 modules), reaching 99 % of ethylene purity. The outlet stream is conditioned until 30 °C and 1 bar before the separation, along with the hydrocarbons from hydroprocessing. The hydrotreating process uses 20 wt % of microalgae oil from Z1, and operates at 410 °C and 50 bar adding 1500 mL of hydrogen per oil mL (Verma et al., 2011), using a *RStoic* module. The outlet stream is conditioned until 27 °C and 1 bar using an isentropic turbine (*Compr* module); then, it is mixed with the ATJ hydrocarbons stream, to separate the hydrocarbons in a unique distillation train with two columns (conventional scheme). The first column, 52 stages, separates the naphtha fuel (C4-C7) by the top, whilst the bottom enters to second column, 106 stages, separating the biojet fuel (C8-C16) by the top and the green diesel (C17-C20) at the bottom. Both columns were designed with *DSTW* module, and later rigorously simulated with *RadFrac* module. The thermodynamic model used in both cases was BK10. The 80 wt % of the oil from Z1 is transesterified with methanol, at 62 °C, 1 atm and sodium methoxide as catalyst (Farooq et al., 2016), using *REquil* module and UNIFAC model. The biodiesel refining takes place on three stages; in the first one a

Decanter module is used to separate the FAME from glycerol, water and methanol. In the second stage, the FAME stream enters to a washing column (*RadFrac* module and UNIF-DMD model) with 2 stages, to remove the alcohol and glycerol traces. The washed stream enters to an evaporator (*Flash* module and UNIFAC model) to remove the water traces.

Two energy efficiency tools are implemented on the biorefinery scheme: process intensification (PI) on separation zone, and energy integration (EI) for the whole biorefinery. In this work, the PI through the Rong and Errico (2012) methodology is applied on the separation zone of hydrocarbons from ATJ and hydrotreatment processes. On the other hand, EI application is realized by the pinch point methodology (Kemp, 2006) on whole biorefinery, considering the scheme with the intensified distillation train.

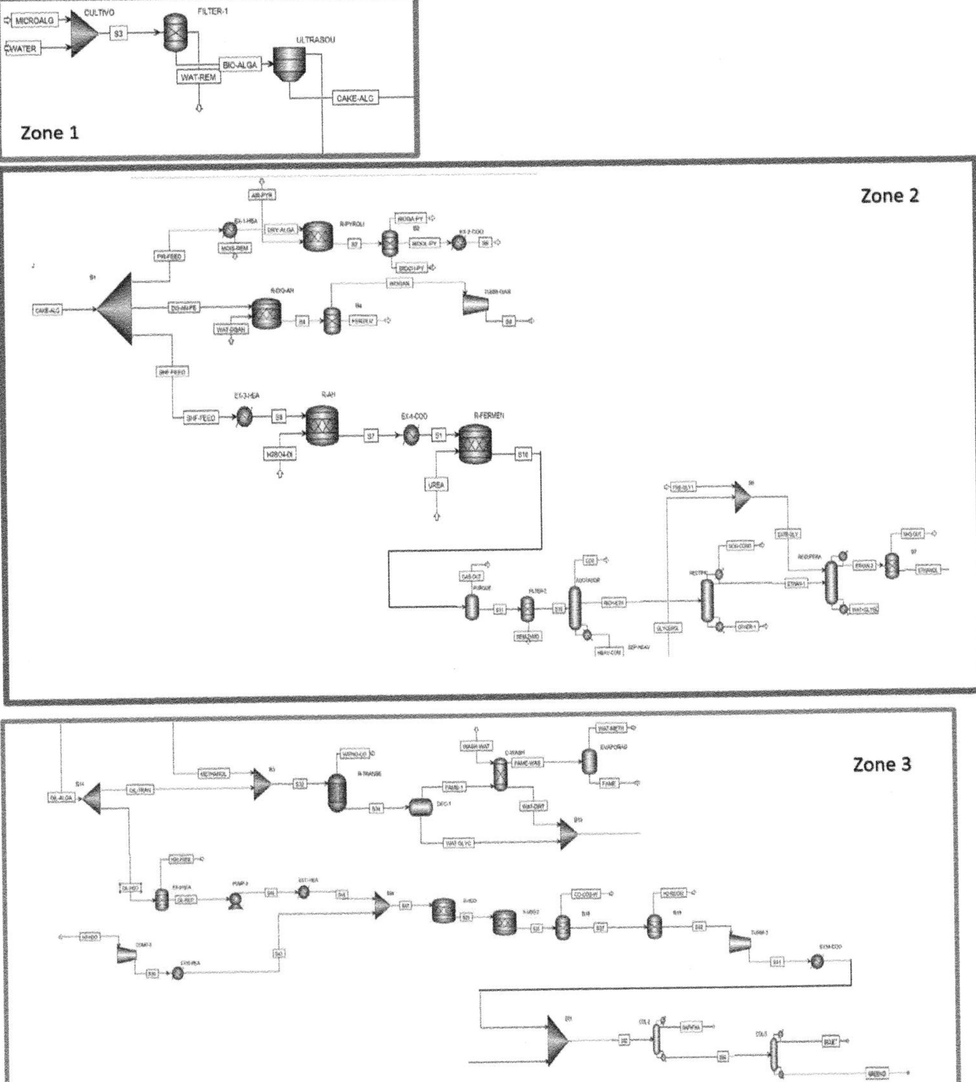

Figure 1. Biorefinery of microalgae by processing zones, non-intensified scheme.

3. Assessment of the biorefinery scheme

The assessment of the biorefinery is carried out with the total annual cost (TAC) and CO_2 emissions by processing. The TAC involves capital cost, calculated by Aspen Economics, adding 18% (A1) and 61% (A2) due to contingencies, installation fees and equipment maintenance (Turton et al., 2012). The operating cost involving utilities cost (heating and cooling), raw material cost (microalgae, hydrogen, methanol) and additional reagents cost (glycerol, urea, H_2SO_4, etc), excluding catalyst and filters. The CO_2 emissions considers the steam and electricity requirements, according to Gutiérrez-Antonio et al. (2016).

4. Analysis of results

In this section the main results obtained for the microalgae biorefinery are described. This biorefinery allows to obtain biojet fuel (16,538.18 kg/h), naphtha (10,055.994 kg/h), green diesel (277.025 kg/h) and biodiesel as fuels; also, glycerol (5,130.888 kg/h), acetic (113,329.49 kg/h), succinic (734.855 kg/h) and lactic (910.893 kg/h) acids, along with bio-oil (8,130.41 kg/h), bio-char (8,463.90 kg/h) of pyrolysis, power and water with nutrients. As can be seen the total biojet fuel produced is 22,844.07 kg/h (58 % from hydrotreatment and 42 % from ATJ); also, acetic acid, succinic and lactic acids, respectively, are produced as value-added products. Regarding to PI applied on the separation zone from ATJ and hydrotreating, the results of the conventional and intensified schemes are presented in Table 1.

Table 1. Results of conventional and intensified sequence.

Sequence	Q tot con (kW)	Q tot reb (kW)	No. Stages
Conventional	5,052.76	5,797.92	158
Intensified	5,007.29	5,722.55	157

The energy savings for the condenser and reboiler are 1.3 % and 0.9 %, regarding the conventional scheme. Also, the intensified scheme has two less stages than the conventional, thus the equipment cost could be reduced. With respect to energy integration, the biorefinery scheme was analyzed through two routes: the first one (R1) including the Z1 and Z2 processes, and the route 2 (R2) for Z3 processes. R1 has 5 and 4 cold and hot streams, respectively; whilst R2 6 and 5 cold and hot streams respectively. According to the pinch point methodology, and considering ΔT min of 10°C, it is possible to reach 23 % and 8.89 % of savings for heating and cooling, respectively, for R1. In the case of R2, the savings for heating and cooling are 75.43 % and 74.76 %, respectively. The exchanges between the process streams for R1 and R2 are presented in Figure 2. Both designs were constructed taking into account the minimum number of heat exchangers. The TAC estimation of the scheme with energy efficiency tools is 61,487.335 million USD/y. In the case of equipment cost, it is 7 % higher (30.479 million USD/y), regarding to original scheme due to the additional heat exchangers and the size increase of intensified column. Nevertheless, economic savings of 72.31 % (0.375 million USD/y) and 55 % (0.036 million USD/y) are obtained for heating and cooling costs. Additional factors involved in the estimation of TAC are the same for both schemes. Moreover, the original scheme releases 1,284.54 MTon CO_2/y, due to its steam requirements; whilst the scheme with energy efficiency tools releases 706.497 MTon CO_2/y, representing savings of 45 %.

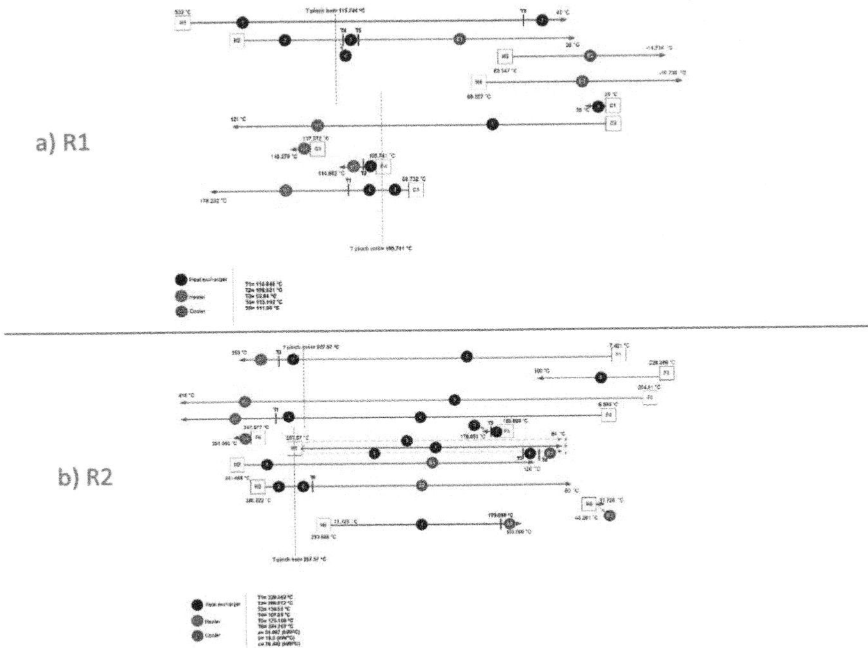

Figure 2. Heat exchanger networks: a) R1, b) R2.

5. Conclusions

The modeling and simulation of a microalgae biorefinery proposal with energy efficiency tools has been presented. The biorefinery produces biojet fuel, other biofuels and high value products. Based on results, the application of PI on the separation zone reduced 1.3 % of energy requirements at the reboilers. Likewise, the EI allowing reaching savings until 23 % and 75.43 % of heating utilities for R1 and R2, respectively. These energy savings enhanced the operational cost until 72.31 %, regarding to original scheme. However, microalgae cost has the higher contribution, which is the main opportunity area.

Acknowledgments

Financial support provided by CONACYT for the doctoral and post-doctoral studies of Araceli Guadalupe Romero-Izquierdo is acknowledged.

References

E. Becker, 2007. Micro-algae as a source of protein. Biotechnol. Adv. 25, 207–210.

R. Brown, R.C., Sidhu, S.S., Martinez, L., Wang, K., Homsy, S., 2012. Fast pyrolysis of microalgae remnants in a fluidized bed reactor for bio-oil and biochar production. Bioresour. Technol. 127, 494–9.

J.de Carvalho, D.T.M. Aulestia, M.A. de Carvalho, E.B. Sydney, A.I. Magalhães, C.R. Soccol, A. Ravishankar, R. Ambati, , 2022. 15 - Biorefinery approaches for integral use of microalgal biomass, in: Jacob-Lopes, E., Zepka, L.Q., Severo, I.A., Maroneze, M.M. (Eds.), 3rd Generation Biofuels, Woodhead Publishing Series in Energy. Woodhead Publishing, 321–344.

M. Elkelawy, H. Bastawissi, A. Radwan, M. Ismail, M. El-Sheekh, 2022. Chapter 15 - Biojet fuels production from algae: conversion technologies, characteristics, performance, and process simulation, in: El-Sheekh, M., Abomohra, A.E.-F. (Eds.), Handbook of Algal Biofuels. Elsevier, 331–361.

A. Farooq, U. Amin, Y. Abdullah, 2016. Transesterification of oil extracted from different species of algae for biodiesel production. African J. Environ. Sci. Technol. 7, 358–364.

C. Gutiérrez-Antonio, F. Gómez-Castro, J. de Lira-Flores, S. Hernández, 2017. A review on the production processes of renewable jet fuel. Renew. Sustain. Energy Rev. 79, 709–729.

C. Gutiérrez-Antonio, A. Romero-Izquierdo, F. Gómez-Castro, S. Hernández, 2016. Energy Integration of a Hydrotreatment Process for Sustainable Biojet Fuel Production. Ind. Eng. Chem. Res. 55, 8165–8175.

IEA.2021. IEA – COVID-19 Topics [WWW Document]. Int. Energy Agency. URL https://www.iea.org/topics/covid-19 (accessed 9.1.22).

I. Kemp, 2006. Pinch Analysis and Process Integration, Pinch Analysis and Process Integration. Elsevier Ltd.

H. Li, Z. Liu,Y. Zhang, B. Li, H. Lu, N. Duan, M. Liu, Z. Zhu, B. Si, 2014. Conversion efficiency and oil quality of low-lipid high-protein and high-lipid low-protein microalgae via hydrothermal liquefaction. Bioresour.

S. Martinez-Villarreal, A. Breitenstein, P. Nimmegeers, P. Perez Saura, B. Hai, J. Asomaning, A. Eslami, P. Billen, S. Van Passel, D. Bressler, D. Debecker, C. Remacle, A. Richel, 2022. Drop-in biofuels production from microalgae to hydrocarbons: Microalgal cultivation and harvesting, conversion pathways, economics and prospects for aviation. Biomass and Bioenergy 165, 106555.

R. Natarajan, W. Ang, X. Chen, M. Voigtmann, R. Lau, 2014. Lipid releasing characteristics of microalgae species through continuous ultrasonication. Bioresour. Technol. 158, 7–11.

NESTE, 2022. SAF for airlines and airports, [WWW Document]. Int. Energy Agency. URL https://www.neste.com/products/all-products/saf/key-benefits (accessed 9.1.22)

G. O'Neil, G. Knothe, C. Reddy, 2019. Chapter 15 - Jet biofuels from algae, in: Pandey, A., Chang, J.-S., Soccol, C.R., Lee, D.-J., Chisti, Y. (Eds.), Biofuels from Algae (Second Edition), Biomass, Biofuels, Biochemicals. Elsevier, pp. 359–395.

A. Romero-Izquierdo, 2020. Diseño, modelado y simulación de un esquema de biorefinería para el aprovechamiento integral de mezclas de materias primas renovables. Univ. Guanajuato PhD Thesis.

A. Romero-Izquierdo, C. Gutiérrez-Antonio, F. Gómez-Castro, S. Hernández, J. García-Trejo, 2022. 27 - Production of renewable aviation fuel from microalgae, in: Jacob-Lopes, E., Zepka, L.Q., Severo, I.A., Maroneze, M.M. (Eds.), 3rd Generation Biofuels, Woodhead Publishing Series in Energy. Woodhead Publishing. 639–664.

B. Rong, M. Errico, 2012. Synthesis of intensified simple column configurations for multicomponent distillations. Chem. Eng. Process. Process Intensif. 62, 1–17.

S. Siddiki, M. Mofijur, P. Kumar, S. Ahmed, A. Inayat, F. Kusumo, I. Badruddin, T. Khan, L. Nghiem, H. Ong, T. Mahlia, 2022. Microalgae biomass as a sustainable source for biofuel, biochemical and biobased value-added products: An integrated biorefinery concept. Fuel. 307, 121782.

G. Symons, A. Buswell, 1933. The Methane Fermentation of Carbohydrates1,2. J. Am. Chem. Soc. 55, 2028–2036.

R. Turton, R. Bailie, W. Whiting, J. Shaeiwitz, D. Bhattacharyya, 2012. Analysis, Synthesis, and Design of Chemical Processes, Fourth Edi. ed.

D. Verma, R. Kumar, B. Rana, A. Sinha, 2011. Aviation fuel production from lipids by a single-step route using hierarchical mesoporous zeolites. Energy Environ. Sci. 4, 1667–1671.

Antonis Kokossis, Michael C. Georgiadis, Efstratios N. Pistikopoulos (Eds.)
PROCEEDINGS OF THE 33rd European Symposium on Computer Aided Process Engineering (ESCAPE33), June 18-21, 2023, Athens, Greece
 http://dx.doi.org/10.1016/B978-0-443-15274-0.50055-X

Optimization of Simulated Moving Bed Chromatographic Processes using Surrogate Models

Rojiar Pishkari[a], Marcus Fechtner[a], Tobias Keßler[a], Achim Kienle[a,b]

[a]*Otto von Guericke University Magdeburg, Universitätsplatz 2, 39106 Magdeburg, Germany*

[b]*Max Planck Institute for Dynamics of Complex Technical Systems, Sandtorstraße 1, 39106 Magdeburg, Germany*

Rojiar.pishkari@ovgu.de

Abstract

In this paper, we investigate a surrogate-based optimization of Simulated Moving Bed (SMB) Chromatography with Langmuir adsorption isotherm using an iterative approach. Artificial neural networks are fitted in each iteration based on randomly distributed sampling points around the optimal solution of the previous iteration. Crucial (Hyper)parameters of this surrogate-based optimization are related to the sampling region, e.g. the size, the position, and the number of samples within. It is shown that for highly efficient chromatographic columns with a large number of theoretical stages, the surrogate-based optimization is much faster than the numerical optimization of the full-blown model.

Keywords: SMB, Surrogate model, Optimization, Neural network, Chromatography

1. Introduction

SMB is an advanced technology that realizes continuous chromatographic separations (Schmidt-Traub et al., 2020). In the last decade, biotechnology, fine chemistry, and pharmaceutical industries have benefited greatly from SMB applications. The main advantages of this process over conventional batch processes are increased productivity and reduced solvent consumption. Generally, chromatographic processes contribute significantly to the overall cost of the corresponding process chains. It is therefore essential to optimize their design and control through efficient process optimization.

Figure 1 illustrates a standard binary separation configuration. The process consists of four zones, each of which consist of at least one chromatographic column. They are interconnected and form a ring shape. The component with the higher affinity to the solid phase is obtained in the extract, whereas the component with the lower affinity to the solid phase is found in the raffinate. In order to achieve continuous separation, cyclic switching of the inlet- and outlet-ports in the direction of the fluid flow is used to simulate counter-current flow of the solid phase. SMBs are complex processes that involve periodic nonlinear dynamics governed by partial differential equations (PDEs). It is crucial to design and operate SMBs optimally, as this allows to maximize the economic potential of the process and helps with the successful implementation on an industrial scale.

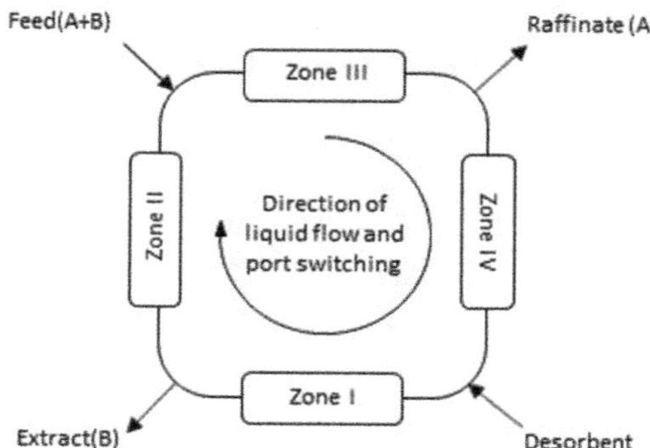

Figure 1. Schematic illustration of 4-zone SMB

The economic potential is maximized by finding optimal operating conditions regarding productivity and solvent consumption. Additionally, the process needs to fulfill certain purity requirements. For total separation, the well-known triangle theory developed by Storti et al. (1993) can be used in a first approximation to identify the optimal operating conditions. It is based on the assumptions of a true moving bed process, neglects axial dispersion and assumes thermodynamic equilibrium between the solid and the liquid phase. Productivity can be increased substantially if purity requirements can be relaxed, which depends on the specific type of application. For reduced purity requirements, Kaspereit et al. (2007) presented an extension of the triangle theory. However, the extension is involved, has limitations and is based on the same simplifying assumptions as mentioned above. By using genetic algorithms, Zhang et al. (2003) have optimized the productivity and purity of the SMB process based on a dynamic model, and compared it to the Varicol process, however computational time is rather high. In Li et al. (2014), a surrogate model was used to optimize SMB processes. They used two different types of surrogate models. The proper orthogonal decomposition (POD) method is employed to derive cost-efficient reduced-order models (ROMs) for the SMB process and the other one is a coarse model. To create a low-fidelity DAE model, they applied the Finite Element Method to the SMB model with coarser spatial discretization. In the present paper, a numerical optimization of a more detailed model is proposed instead. Due to the complexity of the used SMB model, rigorous numerical optimization is highly time-consuming. This paper adapts the methodology from Kessler et al. (2019) to develop a surrogate-based iterative approach by using a simple feed forward artificial neural network, requiring only minimal data, thereby increasing its numerical efficiency.

2. Description of the Optimization Problem

There are linear and nonlinear solvers in the optimizer toolbox in MATLAB (MathWorks, 2023) of which the 2021 version was used here. In this work the following types of optimization problems are considered:

$$\min \quad J(\boldsymbol{x}), \tag{1}$$

$$\boldsymbol{h}(\boldsymbol{x}) = 0,$$

$$\boldsymbol{g}(\boldsymbol{x}) \leq 0, \qquad x_i \in R,$$

$$lb_i \leq x_i \leq ub_i, \qquad \text{for i= 1,2,}$$

Where $J(\boldsymbol{x})$, is the objective function, $\boldsymbol{h}(\boldsymbol{x})$ are the equality constraints, e.g. material balances, $\boldsymbol{g}(\boldsymbol{x})$ are the inequality constraints, such as purity requirements, and $\boldsymbol{x}$ are the decision variables, which are the ratios between the liquid and solid phase flow rate. The well-known equilibrium dispersion model can be used to describe chromatographic columns.

$$\begin{cases} \varepsilon \dfrac{\partial c_i}{\partial t} + (1 - \varepsilon)\dfrac{\partial q_i}{\partial t} + \varepsilon v \dfrac{\partial c_i}{\partial z} = D_{ax}\varepsilon \dfrac{\partial^2 c_i}{\partial z^2} \\ q_i = \dfrac{H_i c_i}{1 + \sum_{i=1}^{n_c} b_i c_i} \end{cases} \tag{2}$$

In the first equation, which is the equilibrium model, v represents the interstitial velocity, while D_{ax} represents the apparent dispersion coefficient, which lumps together all effects that contribute to band broadening. Furthermore, c_i is the fluid phase concentration of component i, while q_i is the solid phase concentration and ε is the volume fraction of the fluid phase. In the second equation, which is the Langmuir isotherm, H and b are the adsorption constants of each component and n_c is the number of components. According to the adsorption isotherm, q_i depends on the fluid phase composition since the model assumes thermodynamic equilibrium between the two phases. There are linear and nonlinear solvers in the optimizer toolbox in MATLAB. In this case, we are using fmincon. The objective function to be maximized is the productivity for a given minimum product purity. Our decision variables, $\boldsymbol{x}$, are the m-values, which are the dimensionless ratios between the liquid and the solid phase flowrates in each column. Maximum productivity is obtained by maximizing the difference between m_3 and m_2, which is equivalent to find a feasible operating point farthest from the diagonal $m_2 = m_3$. At the beginning, the Henry constants are used as the starting point, i.e. the first m_3 and m_2. Since they are the optimal point of the linear adsorption isotherm for total regeneration, they can be used as initial guess for Langmuir adsorption isotherm. The domain for the optimization area is given by $m_2 \in [3, 7]$ and $m_3 \in [5.5, 9.5]$. In order to maintain total regeneration, m_1 and m_4 have a safety margin of 10% from Henry constants. A nonlinear inequality constraint of a minimum purity of 0.85 for both products is used for this maximization. In a cyclic steady state system, the purity of the raffinate and extract streams is defined as follows:

$$\begin{cases} pu_{raf} = \dfrac{\int_t^{t+tsw} c_2^{raf}\, dt}{\int_t^{t+tsw} c_1^{raf}\, dt + \int_t^{t+tsw} c_2^{raf}\, dt} \\ pu_{ex} = \dfrac{\int_t^{t+tsw} c_1^{ex}\, dt}{\int_t^{t+tsw} c_1^{ex}\, dt + \int_t^{t+tsw} c_2^{ex}\, dt} \end{cases} \tag{3}$$

Therein, tsw is the switching time. Variables pu_{raf} and pu_{ex} indicate the purity of raffinate and purity of extract, respectively, while c refers to the concentration of the components. The numerator indicates the quantity of target component collected during the switching period, and the denominator indicates the sum of two components collected during the switching period. Table 1 shows the parameters and operating conditions for SMB simulation. n_{col} is the number of columns, ϵ is the porosity and c-feed is the concentration of the feed. Optimizing is performed using Matlabs multi-start algorithm with 100 starting points and the sqp algorithm. The step tolerance option is set to 1e^{-8}. An extension of the optimization is possible for all m-values.

Table 1. Model Parameters of the reference model

SMB Parameters					
n_c	2	c-feed$_1$	0.5[gr. L^{-1}]	b_2	0.3
n_{col}	4	c-feed$_2$	0.4[gr. L^{-1}]	m_1	8.25
n_z	1000	H_1	5	m_2	x_1
tsw	99.5 s	H_2	7.5	m_3	x_2
ϵ	0.74	b_1	0.15	m_4	4.5

3. Surrogate Model

A surrogate model is developed to replace the detailed but expensive full-order SMB model for optimization purposes. In this work an artificial neural network is used as a surrogate model. Using the methodology introduced in Kessler et al. (2019), this paper proposes a surrogate-based iterative approach, requiring minimal data, which increases the numerical efficiency. The initial guess defined by the Henry constants and an area around it is defined. Some sample points are generated inside the area based on the Halton set algorithm from Kocis et al. (1997). Halton set creates points based on the Halton sequence in a quasi-random manner. To fill the space uniformly, the Halton sequence uses different prime bases in each dimension. The location of these points inside the sampling area is used as m-values (m_2 , m_3). A simulation is performed for each point inside the area and the output data, such as purity and productivity, are saved for every point (pair of m-values). Here 150 points are generated. The reference model is used to calculate the purity of the products, as well as the productivity for each point. The purity must be greater than 0.85 to be considered for the further calculation. Only the points that meet the purity criteria are used for the surrogate model optimization. During each iteration, the sample points are saved and reused if they are in the sampling area of the next iteration as well. Each iteration may have more or fewer samples than 150. The data obtained from this calculation are used to train the neural net. The neural net is feed forward and has one layer with 15 neurons. The surrogate model uses the same optimization algorithm as the full model, multi-start with 100 random starting points and step tolerance option set to 1e^{-8} to find an optimal point in each iteration. If the deviation of an optimal point at each iteration to the previously calculated optimum is bigger than the termination condition which is 0.019, the next calculation starts, otherwise the final optimal point is found. Using the current optimal point as an initial guess, the following formula defines the new sampling region (Kessler et al. (2019)):

$$lb/ub = x_{opt,k-1} \pm \frac{1}{2k} \tag{4}$$

The integer variable k corresponds to the iteration counter. The optimal point found by the previous iteration is the center of the current calculation area. The area is shrinking according to equation 4 and the density of samples inside it is increasing. In the present calculation procedure, the algorithm terminated after 4 iterations. The progress is shown in Figure 2. The first figure shows the optimal point as red star for the first iteration, while the red circle is illustrating the first initial guess. The black dots are the whole sample data generated by the Halton set, and the blue ones are the samples that are matching the purity requirement. A convex hull is created that includes all blue dots, and the optimizer searches for the solution inside this new area. It is clear that the shrinking effect is significant only in the first iteration indicating the importance of the first sampling region. The big difference from first to second region is due to the usage of equation 4. It can be seen from the second figure that the area decreases at a constant rate. Due to the reduced purity requirement that defines the convex hull, the optimal point lies on its boundary with maximal distance to the diagonal $m_2 = m_3$. In the last figure, the final optimal point found by surrogate-model (red star) almost coincides with the optimal point found by full-model optimization (yellow star). Further, the deviation from the previous optimum is less than the termination condition. The same initial guess point is used as the full-model optimization in the surrogate model.

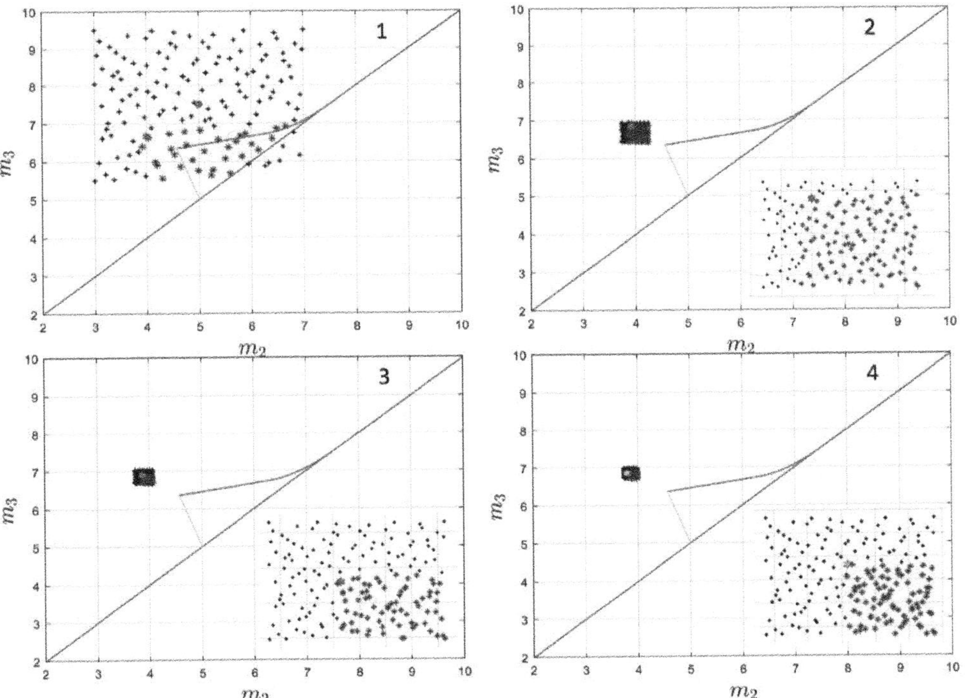

Figure 2. Optimal point of every iteration and how it converges to the final optimum.

4. Results and Conclusion

A surrogate based optimization approach for the SMB process is proposed in this work. In place of the time-consuming rigorous model, a surrogate model is used for optimization. Table 2 compares the final optimal points obtained by both models to show the effectiveness of this approach. Using the surrogate model, SMB productivity was optimized resulting in a reasonable computational effort. It was expected that the optimal point would be outside the triangle because of reduced purity. The concentration of the extract as well as the raffinate and the productivity are calculated by the reference model for the optimal points found by both models. It is shown that the purity requirements are met. Though the full-model's optimal point produces higher productivity than the surrogate-model, the difference is marginal. Despite almost identical results, the surrogate model's computational time is much faster than the full model for the presented optimization set-up, making it a better choice.

Table 2. CPU time comparison

	x_{opt}	cpu-time	pu_{ex}	pu_{raf}	productivity
Sur-model	3.8582, 6.8806	5423.75 s	0.8501	0.8504	3.0224
Full-model	3.8578, 6.8826	152453 s	0.8500	0.8500	3.0249

The proposed methodology could be particularly useful if deterministic global optimization is applied as in Kessler et al. (2019), which relies on a very large number of model evaluations but was beyond the scope of this paper. Future work will be concerned with minimizing the solvent consumption, which is achievable by minimizing $m_1 - m_4$. Furthermore, optimization of the SMB with Bi-Langmuir isotherm is also of interest.

References

M. Kaspereit, A. Seidel-Morgenstern, A. Kienle, (2007). Design of simulated moving bed process under reduced purity requirements. Journal of Chromatography A, 1162, 2-13.

T. Kessler, C. Kunde, K. McBride, N. Mertens, D. Michaels, K. Sundmacher, A. Kienle, (2019). Global Optimization of Distillation Columns using Explicit and Implicit Surrogate Models. Chemical Engineering Science, 197, 235-245.

L. Kocis, W. J. Whiten, (1997). Computational investigations of low-discrepancy sequences. ACM Transactions on Mathematical Software 23 (2), 266–294.

S. Li, L. Feng, P. Benner, & A. Seidel-Morgenstern, (2014). Using surrogate models for efficient optimization of simulated moving bed chromatography. Computers & Chemical Engineering, 67, 121-132.

MathWorks, (2023) MATLAB, https://uk.mathworks.com/products/matlab.html (Last accessed: 29.01.2023).

H. Schmidt-Traub, M. Schulte, A. Seidel-Morgenstern, (2020) Preparative Chromatography, 3rd edition.

G. Storti, M. Mazzotti, M. Morbidelli, S. Carrà, (1993). Robust design of binary countercurrent adsorption separation processes. AIChE J. 39(3), 471–492.

Z., Zhang, M., Mazzotti & M. Morbidelli, (2003). Multiobjective optimization of simulated moving bed and Varicol processes using a genetic algorithm. Journal of chromatography A, 989(1), 95-108.

Antonis Kokossis, Michael C. Georgiadis, Efstratios N. Pistikopoulos (Eds.)
PROCEEDINGS OF THE 33rd European Symposium on Computer Aided Process Engineering
(ESCAPE33), June 18-21, 2023, Athens, Greece
 http://dx.doi.org/10.1016/B978-0-443-15274-0.50056-1

Multi-Objective Model-based Design of Experiments of Pharmaceutical Tableting Process

Ilias Bouchkira[a], Brahim Benyahia,[a,*]

[a] *Department of Chemical Engineering, Loughborough University, Epinal Way, Loughborough, LE11 3TU, United Kingdom*
B.Benyahia@lboro.ac.uk

Abstract

The development of high-fidelity and predictive models is one of the cornerstones of process engineering. Often mathematical models involve several unknown parameters to be identified from experiments. Designing the minimum set of information rich experiments for precise estimate is a critical to reduce the examination costs and help improve model prediction capabilities. In this work, we present a novel Muti-Objective Model-Based Design of Experiments (MOMBDoE). The proposed approach is based on the simultaneous maximizing the D-optimal design of experiments criterion and the estimability potential of the model parameters (Estimability). Global sensitivity analysis is used to build the Fisher Information Matrix (FIM), which allows the maximization or minimization of MOMBDoE criteria. The Pareto optimal solutions which represent the best experimental compromises were ranked using a multicriteria decision aiding method to help identify the best alternatives for experimental validation. To validate the proposed MOMBDoE framework, a tablet lubrication process is used as a case study. Kushner and Moore's model is used to predict the tensile strength and hardness of the tablets as one of the main critical quality attributes in tablet manufacturing.

Keywords: Multi-objective Optimization, Estimability Analysis, Model-Based Design of Experiments, Tablet Lubrication Process.

1. Introduction

Pharmaceutical tablets are the most important dosage forms which commonly contain one or more active pharmaceutical ingredients and several excipients. The optimization of the process performance and product quality is painstaking as the aim is to produce tablets that meet a set of optimal critical quality attributes such as tensile strength, hardness, friability, dissolution, etc. The development of robust digital technologies in pharma requires systematic methodologies to build reliable and predictable mathematical models, soft sensors, and digital twins to help optimize the process/plant design and operation while delivering quality assurance. One of the most effective approaches to address these challenges is to implement a Model-Based Design of Experiments (MBDoE) to generate the minimum set of information-rich experiments.

Model-based design of experiments have been widely investigated for such purposes. Most of these methods are based on the Fisher Information Matrix which can be obtained from the sensitivity analysis. While the current model-based design of experiments aims at designing experiments with rich information, the estimability analysis is mainly used to assess whether the available experimental measurements contain sufficient information to reliably estimate all model parameters or only a subset of them. Combining these two approaches (i.e., MBDoE and estimability analysis), can consequently consolidate the advantages of each method and allow the design of a minimum set of experiments for precise estimates.

This work presents a novel multi-objective model-based design of experiments which combines a D-optimal design of experiments and global estimability analysis. The approach is formulated as a constrained multi-objective optimization problem that accounts for two objectives, the first is based on the D-optimal MBDoE criterion and the second is based on the estimability. The optimization problem is also subject to a set of operating constraints and bounds. The set of Pareto optimal solutions obtained from multi-objective optimization provide the best experimental compromises or tradeoffs. Here, the solutions are ranked using a multicriteria decision aiding method to help identify the best candidates for experimental implementation. To validate the proposed approach, a pharmaceutical tablet lubrication model is considered.

2. Model and methods

2.1. Lubrication model

In tablet press, tensile strength and hardness are very important Critical Quality Attributes (CQAs) which need to be tightly monitored and controlled. These CQA can be related the solid fraction of tablets to the powder lubrication extent through several expressions among which, are the ones developed by Kushner and Moore (2010) and Pitt et al. (1988):

$$\frac{ts}{ts_{sf=0.85,0}} = (1-\beta) + \beta\exp(-\gamma k) \tag{1}$$

$$H = \frac{ts\pi D^2}{10}\left(2.84\frac{\sigma}{D} - 0.126\frac{\sigma}{\sigma - 2\sigma_c} + 3.15\frac{\sigma - 2\sigma_c}{D} + 0.01\right) \tag{2}$$

Where $ts_{sf=0.85,0}$ is the initial tensile strength at 0.85 solid fraction, γ is the lubrication rate constant of the blend, and β is the total fraction of tensile strength that can be lost due to lubrication. D is the diameter of the tablet, σ is the total thickness of the tablet, and σ_c is the thickness of the convex cups.

To avoid dependence on the initial solid fraction, the following empirical equations were introduced (Cenci et al., 2022):

$$ts_{sf=0.85,0} = a_1\exp(b_1(1-sf)) \tag{3}$$

$$\beta = a_2(1-sf) + b_2 \tag{4}$$

The resulting model captures the impact of two main factors or inputs namely the solid fraction sf and lubricant extent k, and involves a vector of five unknown parameters to be identified from experiments ($\theta = [a_1(MPa);\ b_1(-);\ a_2(-); b_2(-), \gamma\ (dm^{-1})]$).

2.2. Multi-objective MBDoE

The proposed multi-objective model-based design of experiments relies on the Fisher Information Matrix (FIM), which is computed from the global sensitivity indices of the model unknown parameters for a given set of experimental recipes or operating conditions:

$$Z = \begin{pmatrix} S_{1,1}|m_1 & \cdots & S_{1,n_\theta}|m_1 \\ \vdots & \ddots & \vdots \\ S_{n_y,1}|m_x & \cdots & S_{n_y,n_\theta}|m_x \end{pmatrix}, \tag{5}$$

where m_i is the ith sampling point or experiment. n_θ is the number of model parameters and n_y is the number of observed/measured outputs. $S_{i,j}$ is the individual sensitivity coefficient of the ith output with respect to the jth parameter.

The FIM matrix is given by:

$$\boldsymbol{FIM} = Z.Z^T \tag{6}$$

$S_{i,j}$ is the individual sensitivity coefficient of the ith output with respect to the jth parameter which can be calculated using the algorithm developed by Saltelli (2012). The sensitivity matrix is then used to generate the FIM to optimize the D-optimal MBDoE criterion and to calculate the estimability magnitude of the least estimable parameter (Benyahia et al. 2011a; Fysikopoulos et al., 2019; Bouchkira et al., 2023). A regularization test is firstly performed to determine whether the FIM is invertible. If this condition is satisfied a multi-objective optimization is performed to determine the Pareto optimal solutions which can be ranked using a multicriteria decision-aiding method.

The mathematical formulation of the proposed MOMBDoE problem is given by:

$$\begin{aligned} &\max_u \boldsymbol{F} = [f_1, f_2]^T \\ &f_1 = det(FIM) \\ &f_2 = \min \sum_i \|r_{i,j}\| \\ &\text{St.} \quad 0 < Sf_{j-1} \leq Sf_j, \quad j \in [1;n] \\ &\qquad 0 < k_{r-1} \leq k_r, \qquad r \in [1;m] \\ &\qquad Sf_{j-1} - Sf_j \leq \omega_{sf} \\ &\qquad k_{r-1} - k_r \leq \omega_k \\ &\qquad Cond(FIM) \leq \delta \\ &\qquad \boldsymbol{u} = [Sf_1, \dots, Sf_n, k_1, \dots, k_m] \end{aligned} \tag{S.1}$$

where det and $Cond$ refer respectively to square diagonal of FIM and conditioning number of FIM. $r_{i,j}$ are orthogonalization residuals calculated as in Benyahia et al. 2013, ω_{sf} and ω_k represent the difference between two solid fractions and lubricant extents, used to avoid duplicated experiments, and δ is a cut-off criterion to guarantee that FIM is invertible, in this work $\delta = 10$ is considered as in Cenci et al., 2022.

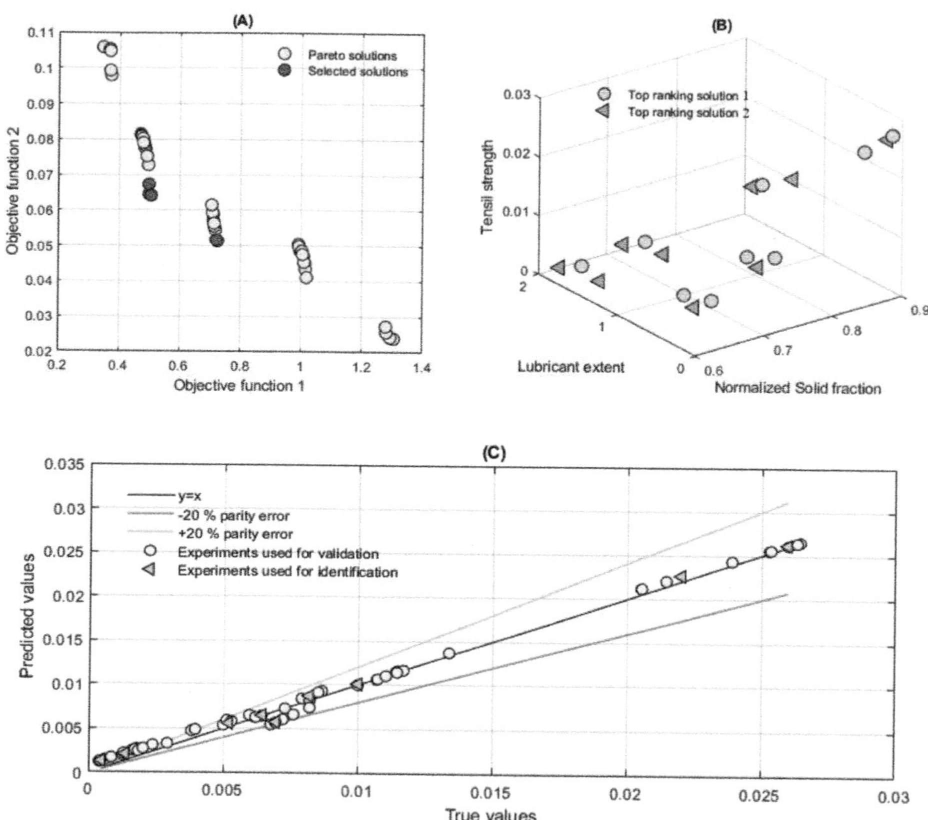

Figure 1: MOMBDoE results. (A) Pareto front and top-ranking solutions. (B) Experimental strategies /recipes for the 2 top ranking solutions from Pareto front. (C) Model validation showing normalized tensile strength and hardness data.

Table 1: Sample of four top ranking Pareto optimal solutions (ranked using MAUT).

k_1	k_2	k_3	Sf_1	Sf_2	Sf_3	Objective 1	Objective 2
0.132	0.368	1.863	0.615	0.704	0.848	0.499	0.064
0.179	1.376	1.883	0.618	0.710	0.868	0.724	0.051
0.287	1.058	1.753	0.726	0.807	0.886	1.304	0.023
0.249	1.373	1.818	0.640	0.737	0.897	1.011	0.045

3. Results and discussions

The global sensitivity analysis based on Sobol approach was first performed. A Quasi-Monte Carlo sampling of size 8000 is used to achieve convergence of the global sensitivity indices. Sensitivity indices are then used to build the Z sensitivity matrix and FIM. The latter is used in the proposed MOMBDoE. The resulting multi-objective optimization problem was solved using a Genetic Algorithms solver under MATLAB to generate a large set of Pareto optimal solutions which represent the best experimental compromises.

The optimal profiles i.e., X^{opt} from the multi-objective optimization approach are used to generate information rich data, to be used for the identification of the unknown model parameters and model validation. The approach used in this work consists of identifying model unknown parameters from two sets of experiments: (i) from true measurements generated using X^{opt}. (ii) and from noisy measurements, then to compare model performances in both scenarios to see if it keeps very good prediction capabilities, even from noisy measurements. It is worth noting that that noisy data used in this work is created by adding gaussian noise to true data from Cenci et al., 2022.

The optimal solutions (i.e., X^{opt}) are given in the Pareto front as shown in Figure 1A. Indeed, in the Pareto front, each point corresponds to an optimal profile. For demonstration purposes, four points labeled "1-4" in the Pareto front are given in Table 1. The Pareto solutions represent the best tradeoffs which are commonly considered as equally good. However, the experimental validation requires a small set of solutions to be implemented. Consequently, it is important to implement a decision-aiding method to select the best profile.

The Multi-Attribute-Utility-Theory (MAUT), which has been successfully used as a robust decision aiding method in many engineering problems (Benyahia et al. 2011b; Bouchkira et al., 2022) was used to rank the Pareto optimal solutions based on the utility functions that capture the distance from the best single objective and its criticality. The Pareto solutions were ranked accordingly from the best to the worst which allows the identification of the best experimental strategies /recipes to demonstrate the benefits of the proposed MOMBDoE approach.

The Pareto solutions are shown in Figure 1A with a sample of the top-ranking solutions highlighted in a different color. The experimental conditions/recipes associated with 2 selected top-ranking solutions (Table 1) are shown in Figure 1B. Each of the 2 identified optimal experimental campaigns require 9 experiments at various lubricant extents and solid fractions. The top-ranking experimental campaign, identified as the best experimental candidate using the proposed approach above, was implemented to generate true and noisy data, based on which model performances are compared in Figures 1C. It is shown that the predicted values from the novel MOMBDoE are in good agreement with the true values from the literature (Cenci et al., 2022). The data in blue triangles are used in the parameter identification, while the data in yellow circles represents measurements that were not used in the identification process. The parity errors between the two compared sets of outputs lie between -20% and 20% which reflects good model performances and highlight the reliability of the proposed approach.

4. Conclusion

In this work, a novel multi-objective model-based design of experiments is presented and implemented to a case study relevant to a pharmaceutical tableting process. The proposed MOMBDoE relies on global sensitivity analysis to combine traditional MBDoE with global estimability analysis. This new approach consequently allows the design of experiments which guarantee that all model unknown parameters are estimable and exhibit higher estimate precision (lower uncertainties). As the approach proposes several optimal solutions, decision-aiding is used to select a set of experimental candidates to be implemented. The results from the case study showed that the model developed based on the proposed MOMBDoE approach shows enhanced prediction capabilities and accuracy.

Acknowledgements: This work was funded by the EPSRC (EP/V062077/1) Digital Medicines Manufacturing Research Centre.

References

A. Saltelli, (2002), Making best use of model evaluations to compute sensitivity indices, Computer physics communications, 145(2), 280-297.

B. Benyahia, M. A. Latifi, C. Fonteix, F. Pla, (2011a). Modeling of a batch emulsion copolymerization reactor in the presence of a chain transfer agent: Estimability analysis, parameters identification and experimental validation. Computer Aided Chemical Engineering, 29, 126-130.

B. Benyahia, M. A. Latifi, C. Fonteix, F. Pla, (2013). Emulsion copolymerization of styrene and butyl acrylate in the presence of a chain transfer agent. Part 2: Parameters estimability and confidence regions. Chemical Engineering Science, 90, 110-118.

B. Benyahia, M. A. Latifi, C. Fonteix, F. Pla. (2011b). Multicriteria dynamic optimization of an emulsion copolymerization reactor. Computers & Chemical Engineering, 35(12), 2886-2895.

D. Fysikopoulos, B. Benyahia, A. Borsos, Z. K. Nagy, C. D. Rielly, (2019), A framework for model reliability and estimability analysis of crystallization processes with multi-impurity multi-dimensional population balance models. Computers & Chemical Engineering, 122, 275-292.

F. Cenci, G. Bano. C. Christodoulou, Y. Vueva, S. Zomer, M. Barolo, F. Bezzo, P. Facco, (2022). Streamlining tablet lubrication design via model-based design of experiments, International Journal of Pharmaceutics, 614, 121435.

I. Bouchkira, A. M. Latifi, L. Khamar, S. Benjelloun, (2021), Global sensitivity based estimability analysis for the parameter identification of Pitzer's thermodynamic model, Reliability Engineering & System Safety, 207, 107263.

I. Bouchkira, A. Latifi, L. Khamar, S. Benjelloun, 2022. Modeling and multi-objective optimization of the digestion tank of an industrial process for manufacturing phosphoric acid by wet process. Computers & Chemical Engineering, 156, 107536.

I. Bouchkira, S. Benjelloun, L. Khamar, A. Latifi, 2023. Thermodynamic modeling and parameter estimability analysis of a wet phosphoric acid process with impurities. Fluid Phase Equilibria, 564, 113594.

IV, J. Kushner, F. Moore, (2010), Scale-up model describing the impact of lubrication on tablet tensile strength, International journal of pharmaceutics, 399(1-2), 19-30.

K. G. Pitt, J. M. Newton, P. Stanley. Tensile fracture of doubly-convex cylindrical discs under diametral loading, Journal of materials science 23 (1988), 2723-2728.

Antonis Kokossis, Michael C. Georgiadis, Efstratios N. Pistikopoulos (Eds.)
PROCEEDINGS OF THE 33rd European Symposium on Computer Aided Process Engineering (ESCAPE33), June 18-21, 2023, Athens, Greece
 http://dx.doi.org/10.1016/B978-0-443-15274-0.50057-3

Water bodies restoration as an optimal control problem with integrated internal and external strategies. The inclusion of artificial floating islands

Amira Siniscalchi[a,b], Laura J. Fritz [a], Vanina Estrada[a,b], Maria Soledad Diaz[a,b]

[a]Planta Piloto de Ingeniería Química (PLAPIQUI CONICET-UNS), Camino La Carrindanga km. 7, Bahía Blanca, Argentina

[b]Departamento de Ingeniería Química, Universidad Nacional del Sur (UNS), Bahía Blanca, Argentina

Abstract

In this work, we propose an integrated water quality model of an artificial reservoir that is the drinking water source for two cities and undergoes recurrent algal blooms. The integrated model is formulated within a dynamic optimization framework as an optimal control problem to address its restoration through the combination of different restoration techniques. The model includes a global mass balance that takes into account inflows from tributaries and rain, and outflows for drinking water and evaporation. Composition gradients are considered along the water column, rendering a Partial Differential Algebraic Equations (PDAE) system. Dynamic mass balances are formulated for the main biogeochemical variables in both reservoir and wetland, as well as artificial floating islands (AFIs) platform. The PDAE is then transformed into a DAE. Numerical results allow analyzing the performance of the different restoration strategies and their joint application on a eutrophic reservoir, showing the efficiency of AFIs as complement of artificial wetlands.

Keywords: eutrophication control, optimal control problem, AFI, dynamic optimization

1. Introduction

Eutrophication is a severe environmental problem caused by anthropogenic nutrient enrichment (mainly nitrogen and phosphorus) that causes Cyanobacteria Harmful Algal Blooms (CyanoHABs), and biodiversity and economic losses (WHO, 2020). In 2007 a severe algal bloom in Taihu Lake (China) caused economic losses of US$ 6.5 billons (Tao and Xin, 2014). CyanoHABs affect public health, commercial and sports fishing, tourism and recreational activities around the world. The potential production of toxins by cyanobacteria (cyanotoxins) is nowadays one of the main environmental and public health concerns (Figueiredo *et al.* 2004). Many water quality restoration strategies have been applied to attempt to mitigate eutrophication and control cyanobacteria growth (Hamilton *et al.* 2016). Since the 90s, there has been a lot of debate about the relative importance of control phytoplankton blooms by nutrients decrease (bottom-up control) (Siniscalchi *et al.* 2020) or by zooplankton grazing (top-down control) (Beklioğlu, 1999; Estrada *et al.* 2011). Constructed Wetlands (CW) and Artificial Floating Islands (AFIs) has been both proposed as bottom-up strategies. CW are portions of land covered with macrophytes (plants that can live in flooded lands), built next to the banks of the tributaries to reduce nitrogen and phosphorus loading by runoff. On the other hand, in AFIs, the macrophytes are sown in buoyancy structures on the surface of the water body and the N and P (and micronutrients) are up-taken by roots directly from the water column.

In this work, we propose an integrated water quality mechanistic model of an artificial reservoir (Siniscalchi *et al.*, 2020) that is the drinking water source for more than 450,000 people and undergoes recurrent algal blooms. The integrated model includes the water quality model for Paso de las Piedras Reservoir (PPR), CWs model for the two reservoir´s tributaries, and the AFIs model. It is formulated within a dynamic optimization framework as an optimal control problem to address its restoration through the combination of CW and AFIs strategies. The optimization integrated model includes CWs and AFIs areas as design variables and the fractions of tributaries that are diverted through the wetlands as control variables.

2. Case Study

Paso de las Piedras Reservoir, is an eutrophic freshwater reservoir, that supplies drinking water to more than 450,000 inhabitants of two cities and for industrial purposes at a petrochemical complex nearby. It has two tributaries that run through an important agricultural area in the country, with a drainage basin area of 1620 km^2. The lake has a surface area of 36 km^2, while its mean depth is 8.2 m and its retention time is 4 years. The eutrophic characteristic of the reservoir is mainly caused by the external discharge of nutrients (mainly P and N). This fact causes recurrent algal blooms, which produce several problems in the potabilization process.

3. Mathematical Model description

A first principle-based ecological water quality model for Paso de las Piedras Reservoir (PPR model) was previously formulated, calibrated and validated (Estrada *et al.*, 2009 a,b, 2011; Di Maggio *et al.*, 2016). This model includes dynamic mass balances for main biogeochemical components in the water body: three taxonomical groups of phytoplankton (cyanobacteria, chlorophytes and diatoms), two groups of zooplankton (copepoda and cladocera), three nitrogen species (organic nitrogen, nitrate and ammonia), two phosphorus species (organic phosphorus and orthophosphate), dissolved and particulate organic carbon and dissolved oxygen. and calibrate the model with experimental data (Estrada *et al.*, 2009a), respectively. The model was formulated within an optimization framework to evaluate the application of top-down restoration strategies (Estrada *et al.*, 2011). Also, the model includes a global mass balance that takes into account inflows from tributaries and rain, and outflows for drinking water, and evaporation. Composition gradients are considered along the water column, rendering Partial Differential Algebraic Equations (PDAE) system. The PDAE is then transformed into an ordinary Differential Algebraic Equation system (DAE) by spatial discretization in horizontal layers. Siniscalchi *et al.* (2020) integrated the PPR model with a constructed wetland model (CWs model) which includes the possibility of installing CW in both tributaries of the lake. The model includes mass balances for: macrophyte biomass, three nitrogen species (organic nitrogen, nitrate and ammonia), two phosphorus species (organic phosphorus and orthophosphate), dissolved and particulate organic carbon and dissolved oxygen. For a detailed description of this model see Siniscalchi *et al.* (2020).

In this work we formulated the AFIs model integrated with PPR model and CWs model in order to evaluate different restoration strategies. AFIs model includes a mass balance for macrophytes (Eq. 1). The experimental data to set the parameters used for plant growth and harvest both in AFIs and in CWs correspond to the species *Senecio bonariensis*, a native plant with known remedial potential (Sinscalchi, 2013). The *S. bonariensis* biomass balance ($C_{Biomass}$ in mgC L^{-1})includes a growth term ($R_{Biomass,\,growth}$) (d^{-1}) (Eq. 2) that considers the maximum growth of the planted species multiplied by the main factors that affect photoautotrophic organisms: temperature ($f(T)$) (Eq. 3), solar radiation ($F(I)$) (Eq. 4), P and N concentrations ($f(T)$) (Ec 5) from the upper layer of the reservoir; a mortality term ($R_{Biomass,death}$) (d^{-1}) (Eq. 6) dependent on temperature, and a harvest term ($R_{Biomass,\,harvest}$) (d^{-1}) (Eq. 7) that is performed only when the plant reaches its maximum growth.

$$\frac{dc_{Biomass}}{dt} = r_{Biomass} = R_{Biomass,\,growth} - R_{Biomass,death} - R_{Biomass,\,harvest} \quad \text{Eq. 1}$$

$$R_{Biomass,\,growth} = k_{Biomass,\,growth}\, f(T) f(I) f(N) C_{Biomass} \quad \text{Eq. 2}$$

$$f(T) = \left(\frac{T_{emp}}{T_{opt}}\right) exp\left(1 - \left(\frac{T_{emp}}{T_{opt}}\right)\right) \quad \text{Eq. 3}$$

$$f(I) = \left(\frac{I_o}{I_{opt}}\right) exp\left(1 - \left(\frac{I_o}{I_{opt}}\right)\right) \quad \text{Eq. 4}$$

$$f(N) = \left(\frac{C_{upper\ layer,PO_4}}{K_P + C_{i,PO_4}}\right)\left(\frac{C_{upper\ layer,NO_3}}{K_N + C_{i,NO_3}}\right) \quad \text{Eq. 5}$$

$$R_{Biomass,death} = k_{Biomass,\,death}\, \theta_{Biomass,death}^{(T-20)} C_{Biomass} \quad \text{Eq. 6}$$

$$R_{Biomass,\,harvest} = \begin{cases} 0, & if\ R_{Biomass,growth} < k_{Biomass,growth\,max} \\ \beta * C_{Biomass}, & if\ R_{Biomass,\,growth} = k_{Biomass,growth\,max} \end{cases} \quad \text{Eq. 7}$$

Were $Temp$ and T_{opt} are the water temperature and the optimal growth temperature (°C), respectively; I_o and I_{opt} are irradiance and optimal irradiance for growth (wm^{-2}), respectively; K_P and K_N are the half saturation constants (mgL^{-1}) for P and N uptake; $k_{Biomass,\,growth}$ and $k_{Biomass,\,death}$ are the maximum growth rate and the mortality rate (d^{-1}); $\theta_{Biomass,death}$ is the temperature for mortality rate; and β is the proportion of harvested biomass. As the Paso de las Piedras Reservoir model is spatially discretized in two horizontal layers, and the AFIs implementation leads to a vertical difference in components concentrations (i.e., Areas covered and non-covered by AFIs), the model was again discretized including two vertical compartments in the upper layer.

4. Optimization Problem

The objective function (Eq 8, Table 1) follows the dual criteria (P and N) recommended by Environmental Protection Agency of U.S. (EPA, 2015) and is the minimization of the sum of integrals of the square differences between inorganic phosphorus (P-PO_4) and inorganic nitrogen (N-NO_3) concentrations and a desired value below eutrophication (0.02 for P and 0.3 mg L^{-1} for N) (Wetzel, 1983). As degrees of freedom of the control problems are the area of the constructed wetlands for El Divisorio Stream (CW_{ED}) and Sauce Grande River (CW_{SG}) (design variables) (Eqs. 11 and 12, Table 1), the area covered by AFIs (A_{AFI}) (design variable) (Eq. 18, Table 1) (Saviolo Osti *et al.*, 2020), and the fraction of the tributaries diverted to the wetlands ($F_{CW_{SG}}$and $F_{CW_{ED}}$) (control variables) (Eqs. 14 and 15 Table 1). Also, a path constraint is included (Eq- 13, Table 1) to satisfy that the sum of the areas (A_{CW_Total}) of both wetlands ($A_{CW_{SG}}$and $A_{CW_{ED}}$) does not exceed a quarter of the total area of the reservoir (9 km^2). To solve the formulated optimal control and design problems, a control vector parameterization approach is applied within an equation-oriented framework (gPROMS, PSEnterprise Ltd., 2019).

Table 1. Design and optimal control problems

$\min z \int_0^{tf} (C_P(t) - 0.02)^2 dt + \int_0^{tf} (C_N(t) - 0.3)^2 dt$			Eq. 8
Subject to			
DAE Water Quality Model (PPR model) + CWs model + AFIs model			
$C(0) = C^0$			Eq. 9
$C^L \leq C \leq C^U$			Eq. 10
CWs constraints		AFIs constraints	
$V_{CW_{SG}} = A_{CW_{SG}} h_{CW_{SG}}$	Eq. 11	$A_{AFI\_Total} = A_{AFI} - A_{no\_AFI}$	Eq. 16
$V_{CW_{ED}} = A_{CW_{ED}} h_{CW_{ED}}$	Eq. 12	$V_{AFI} = A_{AFI} * h_{upper\_layer}$	Eq. 17
$0 \leq A_{CW_{SG}} + A_{CW_{ED}} \leq 0.25\, A_{CW\_Total}$	Eq. 13	$0 \leq A_{AFI} \leq 0.5\, A_{Total}$	Eq. 18
$0 \leq F_{CW_{SG}} \leq 0.1$	Eq. 14		
$0 \leq F_{CW_{SG}} \leq 0.5$	Eq. 15		

5. Numerical Results

The resulting integrated (PPR+AFIs+CWs) model has 64 differential equations and more than 200 algebraic ones. The model is implemented in a personal computer with an Intel® Core™ i7-4700 K processor, operating at a CPU frequency of 3.6 GHz, and with 8 Gb RAM. The total CPU time is 4.469s. Numerical results provide optimal areas of AFIs and CWs, as well as nutrients, phytoplankton, zooplankton, fish, dissolved oxygen, *S. bonariensis* biomass in both CWs and AFIs profiles along a five-year time horizon. Numerical results (Figures 1 and 2, Table 2) allow assesses the performance of the different bottom-up strategies and their combination to restore the water quality in a eutrophic reservoir, showing the synergy between both combined strategies (CWs and AFIs). With the application of this approach the theoretical reduction of inorganic P and N is 61 % and 92 %, respectively. It is important to note that when only CWs strategy is applied, the reduction of 10 % in P (Table 2) is achieved with 3 and 6 km^2 of CW_{SG} and CW_{ED}, respectively, which seems inapplicable in practice. However, with an area covered by AFIs of 1.8 km^2 (5% of the total area of the reservoir, Saviolo Osti *et al.*, 2020) the necessity for CWs area is halved and it is only necessary to construct a wetland on El Divisorio Stream. With this combination P reduction is 56 % (Table 2). Figures 1 and 2 shows that the best strategy is the combination of restoration techniques (AFIs + CWs) approach for the substantial reduction in nutrients and, also, for phytoplankton and cyanobacteria concentration.

Table 2. Nutrient removal and decrease in phytoplankton and cyanobacteria biomass concentrations for each restoration strategy in percentage, as compared to no restoration

	Phosphorous	Nitrogen	Phytoplankton	Cyanobacteria
AFIs (%)	56	14	5	1
CW_{SG} and CW_{ED} (%)	10	48	3	1
AFIs + CW_{ED} (%)	61	92	50	20

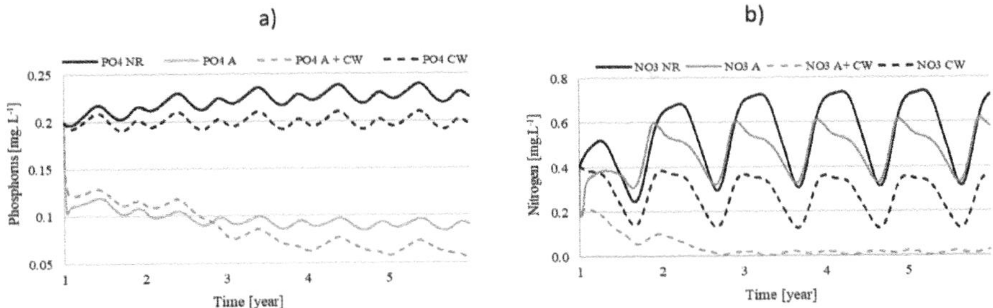

Figure 1: Concentration time profiles for **a)** inorganic phosphorus and **b)** inorganic nitrogen concentrations. NR: without restauration; A: AFIs; CW: Constructed wetlands.

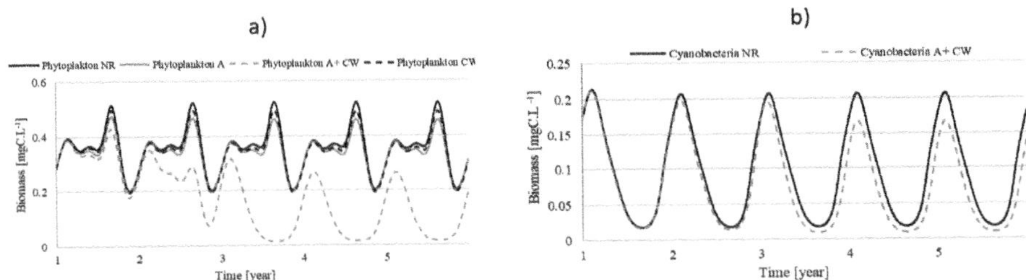

Figure 2: Concentration time profiles for **a)** total phytoplankton biomass and **b)** cyanobacteria biomass concentrations. NR: without restoration; A: AFIs; CW: Constructed wetlands.

6. Conclusions

In this work we study the water quality improvement after combined restoration strategies application formulated as optimal control problems. The integrated optimization model includes a water quality model of Paso de las Piedras Reservoir, and models for two bottom-up restoration strategies, constructed wetlands and floating artificial islands. Numerical results indicate that implementation of AFIs would be an effective green technology to restore contaminated ecosystems by reducing excess of nutrients and controlling cyanobacteria and total phytoplankton growth in combination with constructed wetlands. Finally, integrated models used with dynamic optimization approaches are powerful management tools for rational evaluation and planning of bottom-up strategies for water quality restoration.

References

M. Beklioğlu, 1999, A review on the control of eutrophication in deep and shallow lakes. Turkish Journal of Zoology, 23, 327-336.

J. Di Maggio, C. Fernández, E.R. Parodi, M.S. Díaz, V. Estrada, 2016, Modeling phytoplankton community in reservoirs. A comparison between taxonomic and functional groups-based models, Journal of Environmental Management, 165, 31-52.

V. Estrada, E.R. Parodi, M.S. Diaz, 2009a, Determination of biogeochemical parameters in eutrophication models with simultaneous dynamic optimization approaches, Computers and Chemical Engineering, 33, 1760-1769.

V. Estrada, E.R., Parodi, M.S., Diaz, 2009b, Addressing the control problem of algae growth in water reservoirs with advanced dynamic optimization approaches, Computers and Chemical Engineering, 33, 1598-1613.

V. Estrada, J. Di Maggio, M.S. Díaz, 2011, Water sustainability: a systems engineering approach to restoration of eutrophic lakes, Computers and Chemical Engineering, 35, 1598-1613.

EPA, 2015, Preventing eutrophication: scientific support for dual nutrient criteria. Environmental Protection Agency EPA- 820-S-15-001.

D.P. Hamilton. N. Salmaso, H.W. Paerl, 2016, Mitigating harmful cyanobacterial blooms: strategies for control of nitrogen and phosphorus loads, Aquatic Ecology, 50, 351-366

D.R. de Figueiredo, U.M. Azeiteiro, S.M. Esteves, F. Goncalves, M.J. Pereira, 2004, Microcystin-producing blooms—a serious global public health issue, Ecotoxicology and Environmental Safety, 59, 151–163.

PSEnterprise Ltd., 2019. gPROMS Advanced User Guide-release 2.3. Process Systems Enterprise Ltd., London.

J.A. Saviolo Osti, C. Ferreira do Carmo, M. A. Silva Cerqueira, M.T. Duarte Giamas, A.C. Peixoto, A. Vaz-dos-Santos, C.T. Mercante, 2020, Nitrogen and phosphorus removal from fish farming effluents using artificial floating islands colonized by Eichhornia crassipes, Aquaculture Reports, 17, 100324.

A. Siniscalchi, 2013, Fitorremediación de un arroyo eutrofizado mediante el cultivo de dos especies autóctonas: *Senecio bonariensis* (Compositae) y *Cladophora surera* (Chlorophyta). http://repositoriodigital.uns.edu.ar/handle/123456789/507.

A. Siniscalchi, J.A. Di Maggio, V. Estrada, M.S. Díaz, 2020, Integrated mathematical models for drinking water reservoirs and constructed wetlands as a tool for restoration planning. Journal of Hydrology, 586, 124867.

T. Tao, K.L. Xin, 2014, A sustainable plan China's drinking water, Nature, 511, 527-528.

R.G. Wetzel, 1983, Limnology, second ed. Saunders College Publishing, pp. 860.

WHO, 2020, Cyanobacterial toxins: microcystins. Background document for development of WHO Guidelines for drinking-water quality and Guidelines for safe recreational water environments. Geneva: World Health Organization; 2020. Licence: CC BY-NCSA 3.0 IGO. Switzerland.

Antonis Kokossis, Michael C. Georgiadis, Efstratios N. Pistikopoulos (Eds.)
PROCEEDINGS OF THE 33rd European Symposium on Computer Aided Process Engineering
(ESCAPE33), June 18-21, 2023, Athens, Greece
 http://dx.doi.org/10.1016/B978-0-443-15274-0.50058-5

Developing robust hybrid-models

Peter Jul-Rasmussen[a,b], Xiaodong Liang[a], Xiangping Zhang[b], Jakob Kjøbsted Huusom[a]

[a]*Department of Chemical and Biochemical Engineering, Technical University of Denmark, Søltofts Plads 228A, 2800 Kgs. Lyngby, Denmark.*
[b]*Institute of Process Engineering, Chinese Academy of Sciences, 1 North 2nd street, Zhongguancun, Haidian District, Beijing, PR China*

Abstract

With the increased availability of process data, data-driven methods are becoming more appealing to use in process modelling. Data-driven methods are especially interesting in systems where the existing first-principles models are not adequate in describing the full system dynamics. This work compares a serial semi-parametric hybrid-modelling approach, integrating Machine Learning and first-principles modelling, to a stochastic greybox modelling approach proposed by (Kristensen et al. 2004). Through a CSTR case study both modelling approaches shows accurate predicting performance and robustness towards noisy data. The stochastic greybox modelling is found to be more robust towards measurement frequency as the true model structure is provided when fitting the parameters. Through this work, it is found that good modelling performance can be achieved without providing an explicit model structure by utilizing process data.

Keywords: Process Simulation, Machine Learning, Hybrid-Modelling, Kalman Filter

1. Introduction

With Industry 4.0 the accessibility and quality of data from the production lines in the chemical and biochemical industries is improving due to digital advances such as better data infrastructure and more advanced sensors for on-line/at-line measurements. The increased availability of process data along with improved computational power and algorithmic developments is improving the possibility for introducing data-driven modelling methods in chemical engineering applications (Udugama et al. 2020). The use of data-driven methods enables the modelling of systems where the existing physical and chemical knowledge is limited and, as a result, the existing first-principles models are inaccurate. A problem arising when using data-driven methods is the process data not being truly "big-data" (Venkatasubramanian 2019), a usual requirement for the development of accurate Machine Learning (ML) models. Problems often also arise in predicting the states, interpreting the models, and extrapolating the models when using data-driven modelling. The introduction of semi-parametric hybrid-models, in which process knowledge is combined with data-driven methods, has shown promising performance in mitigating the issues in "purely" data-driven modelling. Semi-parametric hybrid-models shows improvements in state prediction, interpretability, and extrapolability compared to "purely" data-driven modelling when using the same data (von Stosch et al. 2014). Semi-parametric hybrid-models have previously been used in chemical reactor modelling (Bellos et al. 2005) and for modelling flocculation processes (Nazemzadeh et al. 2021). Experimental knowledge can also be included in modelling through stochastic greybox modelling in which maximum likelihood or maximum a posteriori parameter estimates are used for modelling using a system of stochastic differential equations (Kristensen et al. 2004). Stochastic greybox modelling has proven accurate in modelling several different processes including fed-batch cultivation (Rasmussen et al. 2006) and alternating activated sludge process (Halvgaard et al. 2015).

2. Serial semi-parametric hybrid-modelling

Lack of process knowledge is mitigated in semi-parametric hybrid-models by modelling unknown or uncertain phenomena in the system using ML. In this work, a serial semi-parametric hybrid-model is introduced using the workflow given in figure 1.

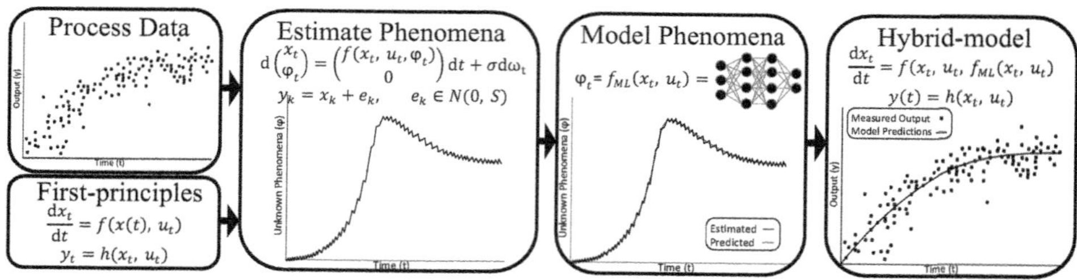

Figure 1 Workflow for serial semi-parametric hybrid-modelling using ML model trained on unknown phenomena, φ_t, estimated for training and validation data using a Continuous-Discrete Extended Kalman Filter.

Considering a process, which is described by a set of ODEs as given in equation (1).

$$dx_t = f(x_t, u_t, \varphi_t)dt \tag{1}$$

In which $t \in \mathbb{R}$ is the time, $x_t \in \mathbb{R}^n$ is the state variables, $u_t \in \mathbb{R}^m$ is the input variables, $f(\cdot) \in \mathbb{R}^n$ is a non-linear function based on first-principles, and $\varphi_t \in \mathbb{R}^p$ is some unknown phenomena in the system. The unknown phenomena, φ_t, is modelled by introducing a ML model, $f_{ML}(\cdot) \in \mathbb{R}^p$, such that $f_{ML}(x_t, u_t) = \varphi_t$ resulting in the system equations (2).

$$dx_t = f(x_t, u_t, f_{ML}(x_t, u_t))dt \tag{2}$$

A problem arise in training the ML model using supervised learning as the unknown phenomena, φ_t, often cannot be measured. This problem is mitigated by estimating φ_t for the training and validation data before training the ML model. φ_t can be estimated using a Continuous-Discrete Extended Kalman Filter (CDEKF) in which φ_t is included as an additional state (Cox 1964) with system equation (3) and measurement equation (4).

$$d\begin{pmatrix} x_t \\ \varphi_t \end{pmatrix} = \begin{pmatrix} f(x_t, u_t, \varphi_t) \\ 0 \end{pmatrix} dt + \sigma(u_t, \varphi_t)d\omega_t \tag{3}$$

$$y_k = h(x_k, u_k, \varphi_k) + e_k \tag{4}$$

In which $k = 0, \dots, N$ is the discrete time index, $y_k \in \mathbb{R}^l$ is the measured output variables, $\sigma(\cdot) \in \mathbb{R}^{n \times n}$ and $h(\cdot) \in \mathbb{R}^l$ are functions, ω_t is a n-dimensional standard Wiener process, and $e_k \in N(0, S)$ is an l-dimensional white noise process. A ML model can now be trained using the CDEKF estimates of φ_t before introducing the ML model to the system equations (2).

3. Case Study: Synthetic CSTR

To investigate the performance of different data-driven modelling methods, a CSTR case study (Figure 2) is considered using simulated data. A first-order exothermic conversion of A to B is taking place in the CSTR system with the material- and energy balance (5) being the governing equations.

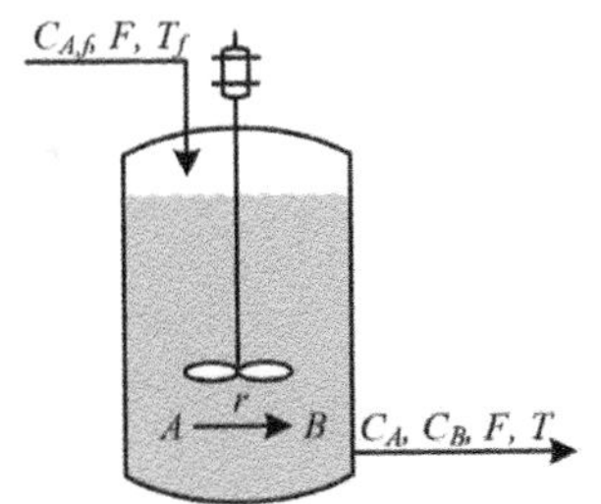

Figure 2 Systematic diagram of CSTR

$$\mathrm{d}\begin{pmatrix} C_A \\ C_B \\ T \end{pmatrix} = \begin{pmatrix} \frac{F}{V}\left(C_{A,f}(t) - C_A(t)\right) - r(t) \\ -\frac{F}{V}C_B(t) + r(t) \\ \frac{F}{V}\left(T_f(t) - T(t)\right) - \frac{\Delta H}{\rho C_p} r(t) \end{pmatrix} \mathrm{d}t \qquad (5)$$

In which C_A and C_B are the concentrations of A and B respectively, T is the temperature in the CSTR, F is the feed flowrate, V is the volume of the CSTR. $C_{A,f}$ is the concentration of A in the feed, T_f is the temperature in the feed. ΔH is the heat of reaction, ρ is the density of the mixture, and C_p is the heat capacity coefficient of the mixture. For the CSTR case, both stochastic greybox modelling and serial semi-parametric hybrid-modelling is applied, introducing data-driven methods for modelling $r(t)$.

3.1 Stochastic Greybox Modelling

Three different model structures are introduced for the reaction rate in the stochastic greybox modelling: 1) assuming 0th order reaction rate, $r(t) = r$, 2) assuming 1st order reaction, $r(t) = k \cdot C_A(t)$, and 3) using an Arrhenius expression, $r(t) = k_0 \exp\left(-\frac{E_a}{RT(t)}\right) C_A(t)$.

Table 1 Maximum likelihood estimates of parameters for stochastic greybox models.

Model	Parameter
0th Order	$r = 1.78 \cdot 10^{-1}$
1st Order	$k = 1.81 \cdot 10^{-1}$
Arrhenius	$k_0 = 3.50 \cdot 10^{7}$
	$E_a = 1.18 \cdot 10^{4}$

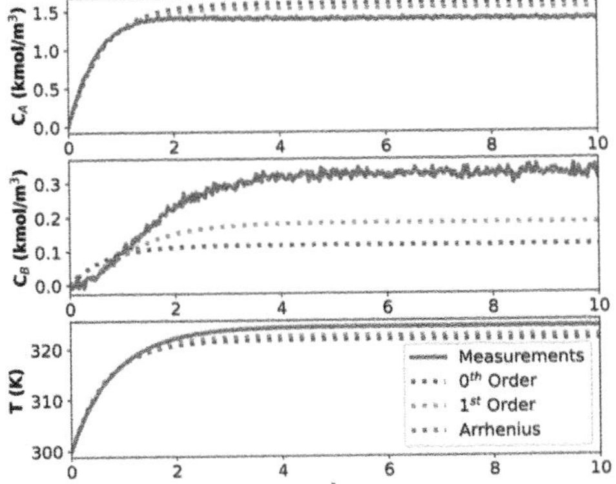

Figure 3 Stochastic greybox model predictions of output variables.

Plotting the model predictions along with the measurements (Figure 3) it is evident that the models using 0th- and 1st order reaction rates are performing significantly worse than the model using an Arrhenius expression.

3.2 Serial semi-parametric Hybrid-modelling

A semi-parametric hybrid-modelling structure is constructed by introducing a feedforward artificial neural network (f_{NN}) for modelling the reaction rate resulting in the system equations (6).

$$\mathrm{d}\begin{pmatrix} C_A \\ C_B \\ T \end{pmatrix} = \begin{pmatrix} \frac{F}{V}\left(C_{A,f}(t) - C_A(t)\right) - f_{NN}(C_A(t), T(t)) \\ -\frac{F}{V}C_B(t) + f_{NN}(C_A(t), T(t)) \\ \frac{F}{V}\left(T_f(t) - T(t)\right) - \frac{\Delta H}{\rho C_p} f_{NN}(C_A(t), T(t)) \end{pmatrix} dt \tag{6}$$

The reaction rate is first estimated using a CDEKF in which the reaction rate is included as an additional state resulting in the system- and measurement equations (7).

$$\mathrm{d}\begin{pmatrix} C_A \\ C_B \\ T \\ r \end{pmatrix} = \begin{pmatrix} \frac{F}{V}\left(C_{A,f}(t) - C_A(t)\right) - r(t) \\ -\frac{F}{V}C_B(t) + r(t) \\ \frac{F}{V}\left(T_f(t) - T(t)\right) - \frac{\Delta H}{\rho C_p} r(t) \\ 0 \end{pmatrix} dt, \qquad \begin{pmatrix} y_1 \\ y_2 \\ y_3 \end{pmatrix}_k = \begin{pmatrix} C_A \\ C_B \\ T \end{pmatrix}_k + e_k \tag{7}$$

The CDEKF is applied to the training data, for training a neural network (NN) with one hidden layer, and to the validation data, for selecting hyper-parameters using a grid search. The possible number of neurons, n, are specified in the grid search to be $n \in [2, 4, 8, 16, 32, 64, 128]$. The best prediction performance on the validation data set was achieved in a NN with 2 neurons in the hidden layer (Figure 4).

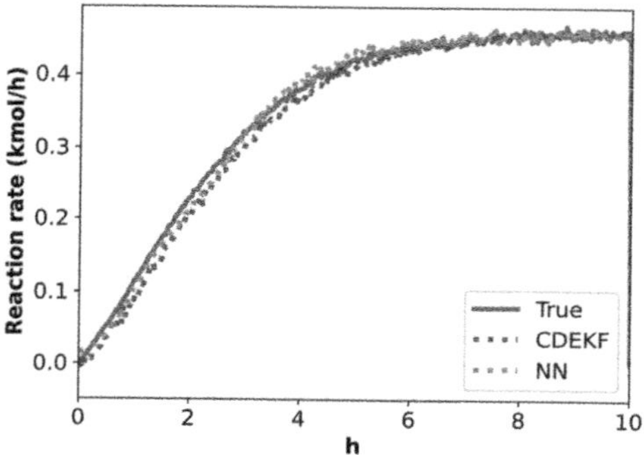

Figure 4 Predictions, estimates, and true reaction rate.

Predictions of the three output variables are obtained after introducing the trained NN model to the system equations (6). Plots of the output predictions (Figure 5) shows that the serial semi-parametric hybrid-model provides accurate predictions for all three output variables.

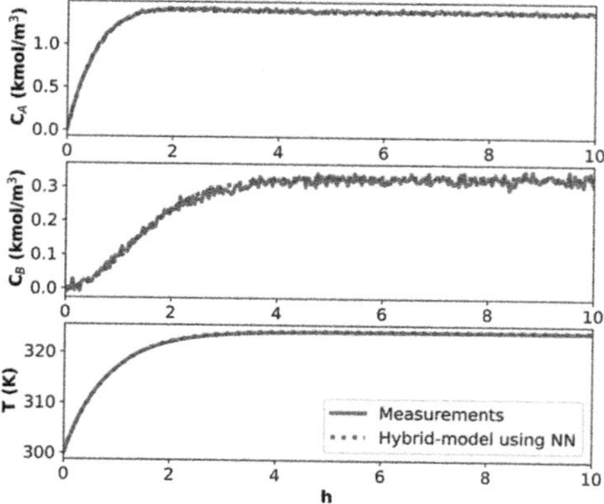

Figure 5 Serial semi-parametric hybrid-model predictions of output variables.

4. Robustness

When modelling real production processes the available data may be noisy and the measurement frequency may be low, especially when using off-line measurements. In the following an assessment of the robustness of the two different modelling approaches towards noise and measurement is provided.

4.1. Noise

White noise is added in the generated data sets as both output measurement noise and input noise on the three input parameters, $C_{A,f}$, T_f, and F, with the same standard deviation for all noise terms in the individual datasets. Three different levels of noise is investigated using standard deviation of 0.01, 0.02, and 0.05. Both stochastic greybox models using an Arrhenius expression as the reaction rate, and serial semi-parametric hybrid-models using a NN as reaction rate are trained using each of the noisy data sets.

Table 2 fitness of prediction with different noise levels for the three output variables using both stochastic greybox model with Arrhenius expression as reaction rate and serial semi-parametric hybrid-model.

Noise std.	Arrhenius C_A fitness MAE (R^2)	 C_B fitness MAE (R^2)	 T fitness MAE (R^2)	Serial semi-parametric hybrid-model C_A fitness MAE (R^2)	 C_B fitness MAE (R^2)	 T fitness MAE (R^2)
0.01	0.008 (0.99)	0.008 (0.87)	0.014 (0.99)	0.009 (0.99)	0.009 (0.85)	0.053 (0.99)
0.02	0.017 (0.97)	0.016 (0.69)	0.016 (0.99)	0.018 (0.97)	0.018 (0.66)	0.095 (0.99)
0.05	0.041 (0.88)	0.039 (0.38)	0.041 (0.99)	0.041 (0.88)	0.040 (0.36)	0.093 (0.99)

The MAE is approximately the same as the noise standard deviation for all output predictions at each of the noise levels with the exception of the T predictions when using a serial semi-parametric hybrid-model (Table 2). This indicates that the two different approaches are able to predict the trend in the data, while the increase in error is a result of the increase in the noise.

4.2. Measurement Frequency

The robustness towards measurement frequency of the two modelling approaches is investigated using data sets generated using different sampling time. The NN in the serial semi-parametric hybrid-model is sensitive to sparse data, posing a problem when training the model. This problem is mitigated in the data sets with low measurement frequency by interpolating between the measurements using cubic smoothing splines when preprocessing the training data, generating more data for training.

Table 3 fitness of prediction with different measurement frequency for the three output variables using both stochastic greybox model with Arrhenius expression as reaction rate and serial semi-parametric hybrid-model.

Freq. (hr^{-1})	Arrhenius C_A fitness MAE (R^2)	 C_B fitness MAE (R^2)	 T fitness MAE (R^2)	Serial semi-parametric hybrid-model C_A fitness MAE (R^2)	 C_B fitness MAE (R^2)	 T fitness MAE (R^2)
30	0.008 (0.99)	0.008 (0.87)	0.014 (0.99)	0.009 (0.99)	0.009 (0.85)	0.053 (0.99)
15	0.008 (0.99)	0.008 (0.87)	0.008 (0.99)	0.010 (0.99)	0.009 (0.84)	0.066 (0.99)
3	0.007 (0.99)	0.008 (0.89)	0.008 (0.99)	0.008 (0.99)	0.009 (0.86)	0.041 (0.99)
1	0.006 (0.99)	0.007 (0.92)	0.012 (0.99)	0.018 (0.99)	0.016 (0.86)	0.190 (0.99)

The stochastic greybox model is found to be robust towards measurement frequency, as the fitness does not decrease when lowering the measurement frequency (Table 3). The serial semi-parametric hybrid-model is found to be less robust towards measurement frequency, as the MAE is increasing when the measurement frequency is low. The higher robustness of the stochastic greybox model towards measurement frequency may be a result of the higher utilization of process knowledge through the introduction of the "true" structure for the reaction rate in the model. Effectively data is only used to fit two parameters in the stochastic greybox model, whereas the full structure of the reaction rate has to be learned by the NN in the serial semi-parametric hybrid-model.

5. Conclusion

Two different approaches for applying hybrid-models, stochastic greybox modelling and serial semi-parametric hybrid-modelling, were used in this paper. The two approaches were applied to a CSTR case study using simulated data. Both approaches were successful in developing models with accurate output predictions and robustness towards noisy data, with the stochastic greybox model being slightly more accurate. The stochastic greybox model was found to be more robust towards measurement frequency due to the "true" structure of the reaction rate being provided for the model. While the stochastic greybox modelling provides a better model, it has higher requirements for process knowledge, as an appropriate model structure has to be provided for good performance to be achieved. It is found that good prediction performance can be achieved without providing an explicit model structure by introducing a ML part for modelling unknown or uncertain phenomena to the first-principles model.

Acknowledgements

This study was financially supported by the Sino-Danish Center for Education and Research (SDC).

References

G. D. Bellos, L. E. Kallinikos, C. E. Gounaris, N. G. Papayannakos, 2005, Modelling of the performance of industrial HDS reactors using a hybrid neural network approach, Chemical Engineering and Processing: Process Intensification, 44 (5), 505-515

H. Cox, 1964, On the estimation of state variables and parameters for noisy dynamic systems, Ieee Transactions on Automatic Control, Volume AC-9 (1), 5-12

R. F. Halvgaard, L. Vezzaro, M. Grum; T. Munk-Nielsen, P. Tychsen, H. Madsen, 2015, Stochastic Greybox Modeling for Control of an Alternating Activated Sludge Process, 9th IWA Symposium on Systems Analysis and Integrated Assessment (Watermatex 2015)

N. R. Kristensen, H. Madsen, S. B. Jørgensen, 2004, A Method for Systematic Improvement of Stochastic Grey-Box Models, Computers and Chemical Engineering, 28 (8), 1431-1449

N. Nazemzadeh, A. A. Malanca, R. F. Nielsen, K. V. Gernaey, M. P. Andersson, S. S. Mansouri, 2021, Integration of first-principle models and machine learning in a modeling framework: An application to flocculation, Chemical Engineering Science, 245, 116864

J. K. Rasmussen, H. Madsen, S. B. Jørgensen, 2006, Grey-box stochastic modelling of industrial fed-batch cultivation, Computer Aided Chemical Engineering, 21 (C), 421-426

I. A. Udugama, C. L. Gargalo, Y. Yamashita, M. A. Taube, A. Palazoglu, B. R. Young, K. V. Gernaey, M. Kulahci, C. Bayer, 2020, The Role of Big Data in Industrial (Bio)chemical Process Operations, Industrial and Engineering Chemistry Research, 59 (34), 15283-15297

V. Venkatasubramanian, 2019, The promise of artificial intelligence in chemical engineering: Is it here, finally?, Aiche Journal, 65 (2), 466-478

M. von Stosch, et. al., 2014, Hybrid semi-parametric modeling in process systems engineering: Past, present and future, Computers and Chemical Engineering, 60, 86-101

Antonis Kokossis, Michael C. Georgiadis, Efstratios N. Pistikopoulos (Eds.)
PROCEEDINGS OF THE 33rd European Symposium on Computer Aided Process Engineering (ESCAPE33), June 18-21, 2023, Athens, Greece
 http://dx.doi.org/10.1016/B978-0-443-15274-0.50059-7

A gPC-based Global Sensitivity Analysis for Phosphate Slurry Flow in Pipelines

Marwane Elkarii,[a,b] Radouan Boukharfane,[b] Saad Benjelloun, [b] *Chakib Bouallou,[a] Nabil Elmocayd,[c,d]*

[a] *Mohammed VI Polytechnic University (UM6P), MSDA Group, Benguerir, Morocco*

[b] *MINES ParisTech, PSL Research University, Center of Energy Efficiency of Systems (CES), Paris, France*

[c] *Mohammed VI Polytechnic University (UM6P), IWRI & IAP, Benguerir, Morocco*
marwane.elkarii@mines-paristech.fr

Abstract

The present work focuses on a new application of the Generalized Polynomial Chaos (gPC) approach: the stochastic response of a three-dimensional simulation of slurry pipe flows in pipelines subject to parametric uncertainties. Initial and boundary conditions (e.g., the slurry flowrate used at the pipe entrance, initial solid concentration), material properties (e.g., particles size), model parameters (e.g., specularity coefficient when the kinetic theory of granular flow (KTGF) is coupled with the Eulerian-Eulerian model), and geometry-related factor (e.g., pipe inclination) are considered as random parameters. gPC surrogate model is built through a least angle regression (LAR) methodology in order to perform uncertainty quantification and global sensitivity analysis following a variance-based approach. The use of gPC is motivated based on its ability to estimate Sobol' indices efficiently. These variance-based sensitivity indices are effective to perform sensitivity analysis without any assumptions about the model's linearity or monotony. Retaining the gPC technique has the advantage of giving the global sensitivity Sobol' indices in a straightforward manner at a lower computing cost than the usual Monte Carlo (MC) method. The first order and total Sobol' indices of the pressure drop along the pipe are calculated and their inspection show that the variability of the pressure gradient is mainly due to the principal effects of the inlet velocity, followed by the inclination of the pipe and then the size of particles. Within the framework of uncertainty quantification, the gPC expansions will also be applied as a surrogate model, as its objective is to recreate the global behavior of the CFD model in a manner that is consistent with a polynomial decomposition.

Keywords: Phosphate slurry flow, Generalized polynomial chaos, CFD, Global sensitivity analysis

1. Introduction

With the advancement of computational methods, numerical simulation has become a popular method for predicting the behavior of physical systems. Upon its simulation with a numerical model, uncertainties are ubiquitous and arise from a variety of sources that may contain substantial randomness in their description. Hence, the results in model outcomes are necessarily subject to uncertainty. To improve prediction reliability, it is consequently necessary to quantify the related uncertainties. This is generally achieved through the non-linear propagation of the initial parameter's uncertainties in CFD simulations (Walters and Huyse, 2002). Global Sensitivity Analysis (GSA) can be carried out to attribute to each input its contribution to the resulting output uncertainty. In the previous two decades, several methodologies have been developed to assess the uncertainty propagation in CFD simulation. The reader is referred to Saltelli et al. (2006) who provides a thorough state-of-the-art of the methods. As sensitivity metrics, the Sobol' indices are extensively used in this context. Several approaches can be employed for calculating the Sobol indices (Sobol, 2001; Kucherenko et al., 2005). Nevertheless, the cornerstone of this method is the highly expensive cost of computations needed to estimate these indices (Sudret, 2008). To overcome the computational requirement associated with the stochastic simulations, new methodologies based on the use of surrogate models have been introduced. gPC expansions have been demonstrated to be effective for Uncertainty Quantification (UQ) and GSA (Blatman and Sudret, 2011), among other options.

To this purpose, the issue of slurry flows in pipes is tackled using GSA together with the gPC expansion approach, to rank the input parameters based on their influence on quantity of interest. As model inputs, five variables are investigated and sampled within a physical range. Initially, we address the physical parameters defining the transport of the intended slurry, including the solids particle size distribution, solid concentration in the flow, and the entrance velocity of the flow mixture. In slurry transport, the inclination of the pipe, a geometry-related factor, is also taken into consideration as a crucial component. Furthermore, the specularity coefficient, which is a wall boundary coefficient related to the kinetic theory for granular flows (Zhong et al., 2015), is considered as a modeling parameter. After identifying the model's uncertain inputs and modeling them within a probabilistic framework, the second stage involves propagating the uncertainty across the model. Finally, a sensitivity analysis is performed, in which the input parameters are ranked according to their effect on the prediction uncertainty.

2. Polynomial Chaos Expansion for Sensitivity Analysis

The gPC expansion aims at recreating the global behavior of a simulation in a manner that is consistent with a polynomial decomposition. The latter are multivariate orthogonal polynomials and serve as the basis functions that are selected in accordance with the joint probability distributions of the stochastic input variables following the so-called Askey scheme of polynomial (Xiu and Karniadakis, 2002). In the present study, if our variables (shown in Tab. 1) are stored in the vector $\mathbf{X} \in R^d$, and $\mathcal{Y}$ represents the model responses discussed earlier, then one may write, according to the gPC expansion:

$$\mathcal{Y} = \mathcal{M}(\mathbf{X}) = \sum_{\alpha \in N^d} \pi_\alpha \Psi_\alpha(\mathbf{X}), \tag{1}$$

Where $\mathbb{N}^d$ denotes a multi-index $\alpha = \{\alpha_1, \ldots, \alpha_d\}$, $\{\pi_\alpha, \alpha \in \mathbb{N}^d\}$ are the expansion coefficients that has to be computed. $\{\Psi_\alpha(\mathbf{X}), \alpha \in \mathbb{N}^d\}$ are multivariate polynomials that are orthonormal with regards to the joint (pdf) $f_\mathbf{X}$ of $\mathbf{X}$, i.e., $E[\Psi_\alpha(\mathbf{X})\Psi_\beta(\mathbf{X})] = 1)$ if $(\alpha = \beta)$ and 0 otherwise. The multivariate polynomials Ψ_α are constructed by taking the tensor product of their corresponding univariate polynomials, i.e.,

$$\Psi_\alpha(\mathbf{X}) = \prod_{i=1}^{d} \phi_{\alpha_i}^{(i)}(X_i), \tag{2}$$

where $\phi_{\alpha_i}^{(i)}$ is a polynomial in the i-th variable of degree α_i. gPC expansion in Eq.1 must be truncated after p terms for computational purposes. Typically, the polynomials Ψ_α with total degree up to p are maintained as $\Psi_\alpha(\mathbf{X}) = \sum_{\alpha \in \mathcal{A}^{d,p} \subset \mathbb{N}^d} \pi_\alpha \Psi_\alpha(\mathbf{X})$. The set $\mathcal{A}^{d,p}$ is defined here based on a total polynomial degree.

The next step is the computation of the polynomial chaos coefficients. The regression technique can be used to estimate the spectral coefficients π_α. This could be achieved by solving a least- square (LS) minimization problem in ℓ_2-norm (Berveiller et al., 2006). We first define the metric error ϵ_{PC} as the difference between the (exact) model's evaluation ($\mathcal{Y}$) and the gPC surrogate estimates for a finite training set of randomly sampled input variables $\mathbf{X}$. A set of N_{ls} realization of the input vector, $\mathbf{X} = X(1), \cdots, X(N_{ls})$, is then needed, called experimental design (ED). The input ED is achieved in the present work using the well-known method Latin Hypercube sam- pling (LHS) approach. The regression problem is solved using a Least-angle-regression (LAR) approach, which offers a sparse gPC representation. This allows us to overcome the curse of dimensionality. Here, we follow the procedure as suggested in Blatman and Sudret (2011). Evaluating its quality is of crucial interest, therefore, Leave-One-Out (LOO) error is employed as it gives an excellent measure of accuracy by enabling error estimates at a reasonable processing. The definition of the relative LOO error is as follows:

$$\epsilon_{\mathrm{LOO}} = \frac{\sum_{i=1}^{N} \left(\frac{\mathcal{M}(x^{(i)}) - \mathcal{M}^{\mathcal{PC}}(x^{(i)})}{1 - h_i} \right)^2}{\sum_{i=1}^{N} (\mathcal{M}(x^{(i)}) - \widehat{\mu_Y})^2} \tag{3}$$

where h_i is the ith diagonal term of matrix $\Psi(\Psi^T\Psi)^{-1}\Psi^T$, where $\Psi = \{\Psi_{ij} = \Psi_j(X^i)\}) and (\widehat{\mu_Y} = \frac{1}{N}\sum_{i=1}^{N} \mathcal{M}(x^{(i)})$. Once the PCE is built, the mean μ and the total variance D can be obtained using properties of the orthogonal polynomials, such that $\mu = y_0$ and $D = \sum_{\alpha \in \mathcal{A} \setminus 0} y_\alpha^2$. As mentioned above, the Sobol' indices of any order can be computed in a straightforward manner. The first order and total Sobol' indices are then given by $S_i = \sum_{\alpha \in \mathcal{A}_i} y_\alpha^2 / D$ and $\mathcal{A}_i = \{\alpha \in \mathcal{A}: \alpha_i > 0, \alpha_{j \neq i} = 0\}$, and $S_i^T = \sum_{\alpha \in \mathcal{A}_i^T} y_\alpha^2 / D$ with $\mathcal{A}_i^T = \{\alpha \in \mathcal{A}: \alpha_i > 0\}$, respectively.

3. CFD modeling procedure

The system under consideration is a three-dimensional pipeline of diameter $D = 0.103m$ and of length $L = 140D$, thus ensuring a fully developed flow. The transported slurry consists of a mixture of phosphate and water with densities of $\varrho_s = 2600$ and $\varrho_w = 1000$ kg/m^3, respectively. The current CFD modeling approach is based on the Euler-Euler two-fluid formulation, which treats both solid and fluid phases as interpenetrating continua (Gidaspow, 1994). We explore this two-phase slurry flow in pipes numerically using the open-source continuum mechanics suite Open∇FOAM®, which is a C++ toolkit for developing numerical solutions. The CFD simulations are run using the solver TwoPhaseEulerFoam, which has been adapted to accommodate mixtures of solids and liquids (Elkarii et al., 2020). The reader is directed to (Elkarii et al., 2022) for further information on the current CFD modeling technique as well as to review the numerical validation. In the present work, the head loss parameter will be estimated using the CFD simulations. This will serve as the quantity of interest for the GSA, since it constitutes a good proxy for the energy requirements of the pipe flow.

4. UQ methodology

To perform UQ and GSA for the problem of slurry flows in pipelines, one should identify the various sources of uncertainty, which will be considered as model inputs. As stated in the introduction, the slurry pipe flows behavior is often defined through the properties of the materials (El Moçayd and Seaid (2021)). This motivates the choice of $\mathbf{X} = \{\phi_s, u_m, d_p, \theta, \mathcal{SC}\}$ as random parameters. These random variables are presented according to their respective pdf's (cf. table1). In order to assess the quality of the gPC expansion, four design of experiments of sizes N = {200,500,900,1000} are considered. They are generated using the LHS method. Once the design of experiments are defined it remains to run the computational model for each sampling node in those sets to obtain the vector $\mathcal{Y}$ of observations of the quantity of interest (QOI).

Table 1: Phosphate slurry flow: parameters of the model

Parameters	Type of PDF	Orthogonal polynomials	Range of variability
Inlet velocity u_m [m/s]	Uniform	Legendre	[1.75, 5]
Solid concentration ϕ_s [-]	Uniform	Legendre	[20%, 40%]
Particles size d_p [μm]	Uniform	Legendre	[44, 250]
Pipe inclination θ°	Gaussian	Hermite	[0, 4]
Specularity coefficient SC[-]	Uniform	Legendre	[0, 1]

5. Results and discussion

To ensure results convergence, the LOO error (Eq. 3) is calculated for each design of experiment and portrayed in figure 1. The LOO error convergence rate indicates that 1000 samples are sufficient to build a high quality gPC expansion.

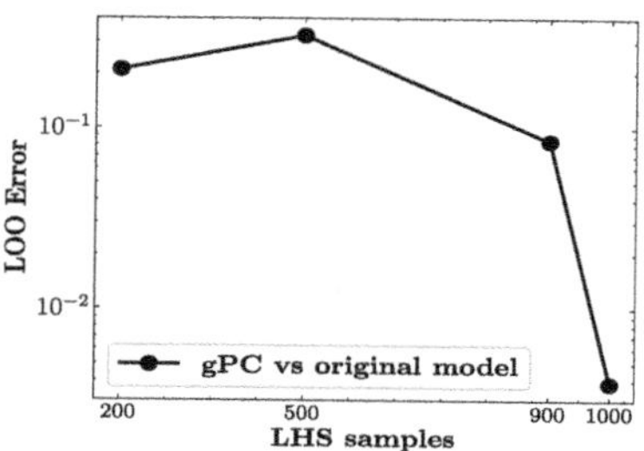

A further investigation of gPC accuracy is performed. First, the sensitivity of the gPC with respect to the polynomial degree is evaluated as depicted in Fig. 2a. A polynomial degree $p = 6$ is sufficient to carry on UQ and GSA using the surrogate model. Then, the ability of the surrogate model to mimic the physical behavior of the CFD model is presented in Fig. 2b. This constitutes a good validation for the gPC ability to serve as a low-cost meta-model.

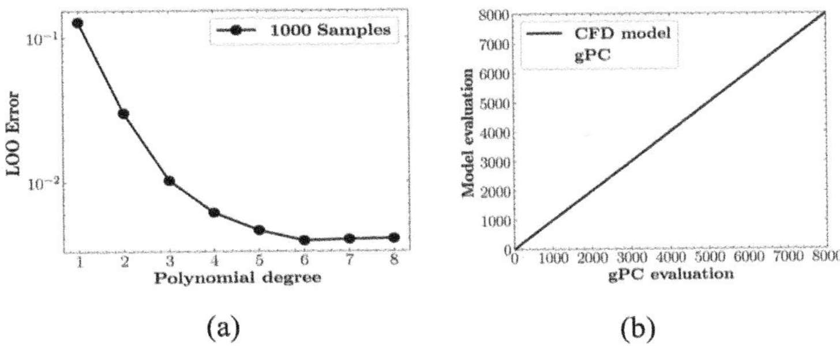

Figure 2: (a) Optimal gPC degree with respect to gPC degree and (b) Comparison between gPC and the exact model using 1000 samples.

Next, we examine the marginal effect of the uncertain parameters on the model response. It is observed that there is a good fit, which is highlighted by a minor LOO error (less than 0.001) (Fajraoui et al., 2017), and which corresponds to an optimal PCE of degree P = 6. We carry out this analysis on the total head loss, in order to better understand how energy loss in pipelines is impacted, since the total head loss is a measure of the energy reduction produced by the sum of elevation head, velocity head and frictions head of the fluid as it passes through a fluid system. The sensitivity of the head loss to the variability of the random parameters $\{\phi_s, u_m, d_p, \theta, SC\}$, can be assessed by means of the Sobol' indices. Figure 3 shows bar plots of the computed first and total Sobol' indices.

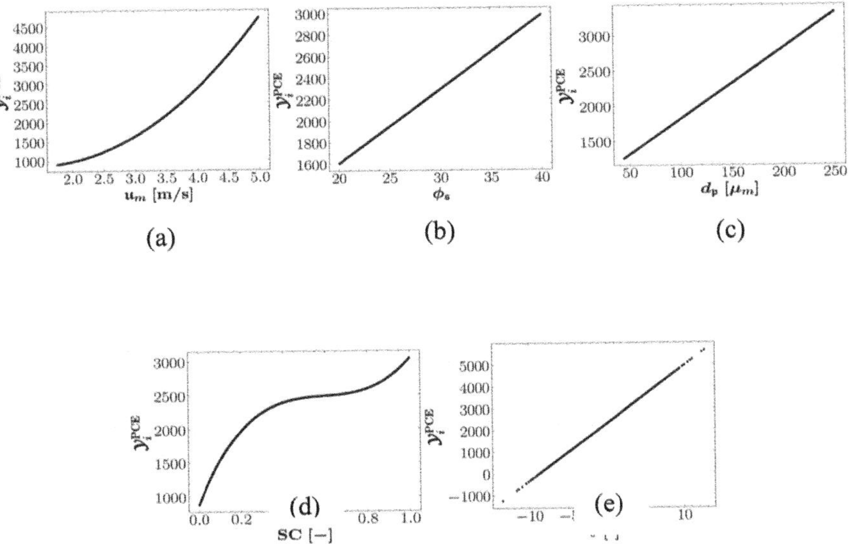

around $u_m \leq 1.75$. This occurs as a result of approaching the limit deposition velocity, at which point a stationary regime emerges, culminating in total particle sedimentation. As for θ (cf Fig. 3e), ΔH_{Total} increases with positive slopes and decreases on the other hand with the negative ones. At positive slope, the longitudinal resultant force acts as an extra resistivity force for the wall contact layer and tends to reduce its velocity compared to horizontal flow. However, as a suspension goes down a pipe, flow is increased due to elevation effects, and pressure levels rise downstream. This extra kinetic energy is utilized to compensate for frictional consequences. With respect to the solids concentration and the particle sizes, ΔH_{Total} grows as both of these variables rise, as noticed in Figs. 3b and 3c. Finally, when it comes to the specularity coefficient *SC*, it is remarked that the pressure drop increases with the latter (cf Fig. 3d). We recall here that *SC* quantifies the momentum and energy transfer caused by colliding particles with the pipe wall. As a result, a larger specularity coefficient implies that it is fair to expect a considerable loss of kinetic energy once particles collide with the pipe wall, resulting in an increase in pressure drop. Finally, we evaluate the sensitivity of the head loss uncertainty to the variability of the random parameters $\{\phi_s, \mathrm{u}_m, d_p, \theta, SC\}$. This will be assessed using the Sobol' indices resulting from the gPC expansion. Figure 4 shows bar plots of the computed first and total Sobol' indices. Inspection of the sensitivity indices showed that the variability of the total head loss ΔH_{Total} is mainly due to the principal effects of u_m followed by θ and then d_p. A small but not negligible influence of ϕ_s and SC, is also observed. There is a notable difference between first and total order sobol' indices, indicating interactions among the random parameters.

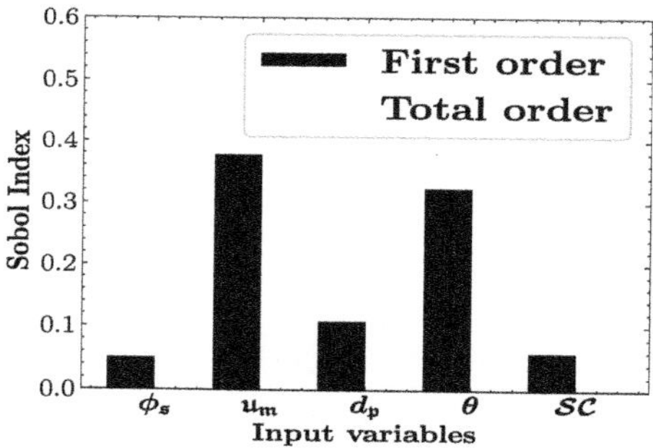

6. Conclusion

In this study, a sensitivity analysis of turbulent phosphate slurry flow in the pipeline has been performed. The effects of our input parameter on the pressure drop along the pipe have been investigated. The choice of the parameters is performed with the aim to improve the physical understanding of the considered flow configuration. Results showed that the variability of the pressure drop is largely influenced by the effect of $\mathbf{u_m}$, $\boldsymbol{\theta}$, an $\boldsymbol{d_p}$. It can be also deduced, that a good calibration of SC in the numerical validation step is crucial, to ensure reliable predictions of the pressure drop along the pipe.

References

M. Berveiller, B. Sudret, M. Lemaire, 2006. Stochastic finite element: a non intrusive approach by regression. European Journal of Computational Mechanics/Revue Europe´enne de Me´canique Nume´rique 15 (1-3), 81–92.

G. Blatman, B. Sudret, 2011. Adaptive sparse polynomial chaos expansion based on least angle regression. Journal of computational Physics 230 (6), 2345–2367.

N. El Moc ayd, M. Seaid, 2021. Data-driven polynomial chaos expansions for characterization of complex fluid rheology: Case study of phosphate slurry. Reliability Engineering & System Safety 216, 107923.

M. Elkarii, C. Bouallou, A. Ratnani, 2020. Towards modelling a diphasic flow using the CFD technique to achieve a digital twin of a phosphate slurry piping process. Chemical Engineering Transactions.

M. Elkarii, R. Boukharfane, S. Benjelloun, C. Bouallou, 2022. A cfd-based surrogate model for predicting slurry pipe flow pressure drops. Particulate Science and Technology, 1–11.

D. Gidaspow, 1994. Multiphase flow and fluidization: continuum and kinetic theory descriptions. Academic press.S. Kucherenko, et al., 2005. Global sensitivity indices for nonlinear mathematical models, review. Wilmott Mag 1, 56–61.

A. Saltelli, M. Ratto, S. Tarantola, F. Campolongo, et al., 2006. Sensitivity analysis practices: Strategies for model-based. Reliability Engineering & System Safety 91 (10-11), 1109–1125.I.

M. Sobol, 2001. Global sensitivity indices for nonlinear mathematical models and their Monte Carlo estimates. Mathematics and computers in simulation 55 (1-3), 271–280.

B. Sudret, 2008. Global sensitivity analysis using polynomial chaos expansions. Reliability engineering & system safety 93 (7), 964–979.

R. W. Walters, L. Huyse, 2002. Uncertainty analysis for fluid mechanics with applications.

D. Xiu, G. E. Karniadakis, 2002. The wiener–askey polynomial chaos for stochastic differential equations. SIAM journal on scientific computing 24 (2), 619–644.

H. Zhong, X. Lan, J. Gao, Y. Zheng, Z. Zhang, 2015. The difference between specularity coefficient of 1 and no-slip solid phase wall boundary conditions in cfd simulation of gas–solid fluidized beds. Powder Technology 286, 740–743.

Antonis Kokossis, Michael C. Georgiadis, Efstratios N. Pistikopoulos (Eds.)
PROCEEDINGS OF THE 33rd European Symposium on Computer Aided Process Engineering (ESCAPE33), June 18-21, 2023, Athens, Greece
 http://dx.doi.org/10.1016/B978-0-443-15274-0.50060-3

Hybrid dynamic model of monoclonal antibody production using CHO cells

Mariana Monteiro,[a] Cleo Kontoravdi[a]

[a]*Sargent Centre for Process Systems Engineering, Department of Chemical Engineeering, Imperial College London, Exhibition Road, London SW7 2AZ, United Kingdom*

Abstract

Mammalian cells are used to produce up to 80% of the commercially available therapeutic proteins, with Chinese Hamster Ovary (CHO) cells being the main production host. Antibody production occurs in a train of bioreactors typically operated in fed-batch, i.e., semi-continuous, mode. Real-time monitoring of antibody concentration is costly and involves sampling and offline analytics. Current reactor models lack extrapolation power while being highly reliant on extensive amounts of data for parameterization. This work aims to leverage knowledge of the cell's metabolism and incorporate it in the production reactor model, thus creating a hybrid formulation with increased predictive capability and generality. This novel formulation can make predictions without requiring kinetic parameter estimations. The applicability of the proposed formulation is demonstrated with a soft sensor for antibody production.

Keywords: Biomanufacturing, Hybrid Modeling, Fed-Batch, CHO cells, monoclonal antibodies

1. Introduction

The global market of biologics reached an annual value of US$265 billion in 2020, with an expected increase to US$856 billion by 2030 (Precendence Research, 2021). Mammalian cells are used to produce up to 80% of the commercially available therapeutic proteins, with Chinese Hamster Ovary (CHO) cells representing the major faction of that (Fan, *et al.*, 2015), because they produce safe and efficacious protein products. As a result of high demand, there is pressure to optimize the manufacturing process even further. Current upstream optimization techniques focus on medium composition, feeding strategies and process conditions (e.g., temperature and pH) (Bassem, *et al.*, 2021). Given the number of possible operating parameters to manipulate, it is both time-consuming and costly to perform experiments covering the entire design space. However, this burden can be lessened using a combination of predictive process models and targeted experimentation.

Current mathematical models describing the production reactor assume saturation kinetics and typically treat the cell as a black box. Although this formulation tends to fit cell culture data well, it offers little extrapolating power given that most parameters heavily rely on the process conditions for which the model was parameterized, such as temperature and pH. In this work, we propose a hybrid modeling approach comprising a stoichiometric model of cell metabolism and a dynamic reactor model, in which the rate of change of each metabolite concentration is calculated using the metabolic model instead of treating the cell as a black box.

Metabolic models depict the network of chemical reactions occurring inside the cell and describe the consumption and production of metabolites. CHO cell metabolic network models can range from a few tens of reactions (Nolan and Lee, 2011) to over 6000 reactions and 4000 metabolites (Hefzi, *et al.*, 2016). Even when using a relatively small-scale network, it is not possible to experimentally obtain reaction rates of all reactions. As such, given that the system is underdetermined, it is necessary to solve the system of equations as an optimization problem and constrain the solutions to feasible and biological relevant spaces. The typical way to solve these problems is to define a feasible objective function that describes the cell's objective; examples include the maximization of growth rate or ATP production, minimization of ATP consumption, or minimization of NADH production (Carvalho *et al.*, 2019; Höffner *et al.*, 2019).

In this work we take advantage of our knowledge of cell metabolism to define a metabolic network model and we incorporate it in a dynamic model of the fed-batch production reactor. Since the core reactions in a cell's metabolic network always hold true, our novel formulation offers increased generality. We apply this approach to describe an antibody-producing CHO cell culture and demonstrate its applicability as a soft-sensor of antibody concentration (one of the costliest sensors).

2. Methodology

2.1. Hybrid Model

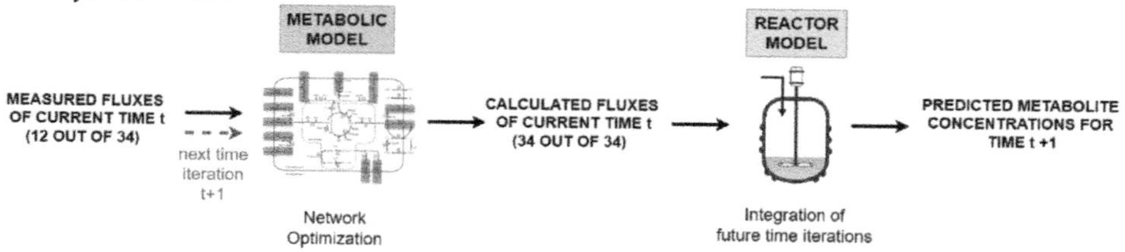

METABOLIC MODEL
REACTOR MODEL
MEASURED FLUXES OF CURRENT TIME t (12 OUT OF 34)
next time iteration t+1
CALCULATED FLUXES OF CURRENT TIME t (34 OUT OF 34)
PREDICTED METABOLITE CONCENTRATIONS FOR TIME t +1
Network Optimization
Integration of future time iterations

Figure 1 Proposed hybrid model

Figure 1 depicts the proposed hybrid modeling algorithm. The dynamic simulation loops over the following steps:

Step 1. We calculate the measured fluxes by subtracting the current (time t) and past (time t-1) concentrations of each metabolite. As shown in figure 1, we have available a subset of the total number of fluxes required (12 out of the 34 fluxes), none of which are intracellular.

Step 2. We perform the metabolic network optimization using the measured fluxes (graphically represented as the pink boxes in the metabolic model). As a result, we calculate the fluxes of the entire network for the current time t.

Step 3. We perform an integration of future time steps, in the reactor model, to calculate the predicted metabolite concentrations for time $t+1$, using the 34 calculated fluxes.

Step 4. We repeat each step iteratively, for each daily measurement available.

Python library Scipy (Virtanen, *et al.*, 2020) was used for both optimization and simulation of the proposed hybrid model.

2.2. Metabolic model

The implemented optimization routine is shown below (equation 1), and it is formulated as a Metabolic Flux Analysis (MFA) problem, i.e., a minimization of the squared difference between measured and calculated fluxes:

$$\min_{v_k^{calc}} \sum \left(\frac{v_k^{calc} - v_k^{meas}}{v_k^{meas}}\right)^2 \; \forall\, k \in measured\ fluxes \tag{1}$$

$$S.v = 0$$
$$0 < v_{irrev}$$

where $v_k^{calc}$are the calculated fluxes, $v_k^{meas}$are the measured fluxes, S is the stoichiometric matrix, v are the fluxes of the entire networks, $v_{irrev}$are the irreversible fluxes and k represents the number of the flux. The measured fluxes, $v_k^{meas}$are the inputs for the optimization and the calculated fluxes, v_k^{calc}, are the optimization decision variables. The stoichiometric matrix is a mxn matrix, where m is the total number of metabolites and n is the total number of reactions of the network. As such, there is a row per metabolite and a column per reaction. The stoichiometric matrix used contains 24 metabolites and 34 reactions. The network and reactions were taken from previous published work (Nolan and Lee, 2011).

This optimization routine is subject to two constraints. The first one enforces a *quasi*-steady state for each metabolite, meaning that it is assumed that the total rate of production equals the total rate of consumption for each metabolite, during each short time iteration. The second constraint provides information of which fluxes are irreversible and bounds them to biologically meaningful values.

2.3. Fed-batch model

The fed-batch model is implemented as a system of differential and algebraic equations, describing the mass balances for each metabolite, of the following form:

$$\frac{dMetabolite}{dt} = ([Metabolite]_{t=0} - [Metabolite]_{t=t}) * \frac{F(t)}{V} - flux_{Metabolite}(t) * X_v(t) \tag{2}$$

where Xv is the cell volumetric density, $F(t)$ is the inlet flow to the fed-batch and V is the working volume. Given that there is no outlet flow of the reactor, the working volume is solely a function of the inlet:

$$\frac{dV}{dt} = F_{in}(t) \tag{3}$$

2.4. Soft Sensor Validation

The applicability of the hybrid model formulation to soft sensing is demonstrated using published data from a fed-batch experiment (Kyriakopoulos and Kontoravdi, 2014). The cell line used is CHO GS46 and the expressed protein is a chimeric IgG4 monoclonal antibody. The cells were grown in a working volume of 250 mL for 14 days and fed on alternate days. Available daily measurements include viable cell density, viability and extracellular glucose, lactate, ammonia, and amino acid concentrations. The soft sensor capability is demonstrated by rerunning the implemented algorithm with no measurement data of antibody concentration except for the initial time point.

3. Results

In this paper, we describe the formulation of a hybrid model comprising a metabolic network and a reactor model, which can use early measurement data to predict key metabolite concentrations throughout the culture. The model is then used as a soft sensor for antibody concentration. Figure 2 compares the hybrid model's predictions for key extracellular metabolite concentrations to experimental data. In addition, the graphs show the soft sensor results, for which no antibody measurements were used to solve the metabolic model except for the initial time point.

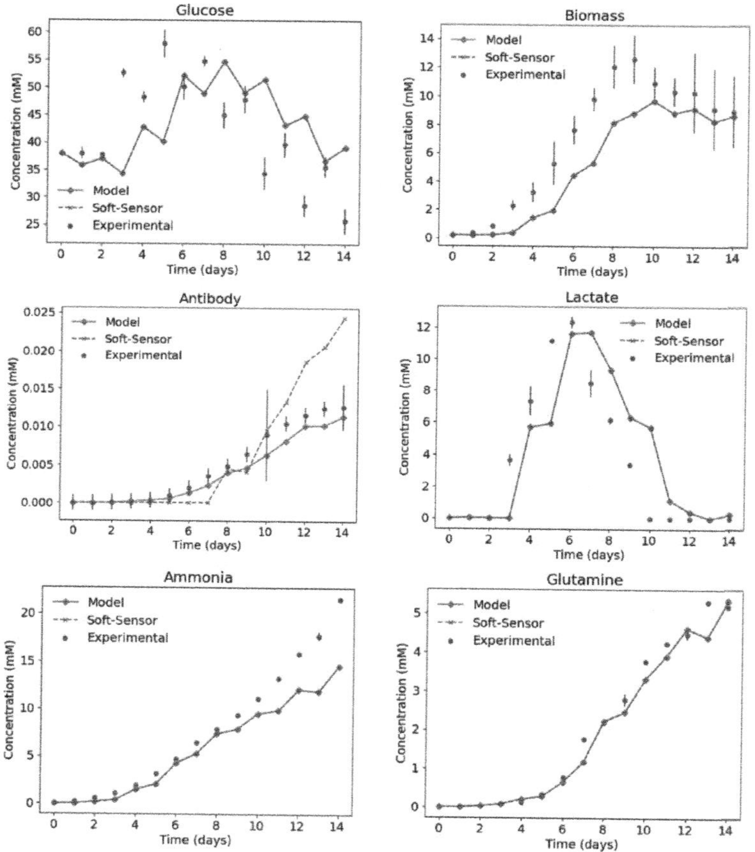

Figure 2 Model Validation Results: Measured vs Predicted Concentrations (in mM) of glucose biomass, antibody, lactate, ammonia, and glutamine, with (Hybrid) and without (Soft-Sensor) antibody, as a function of fed-batch time (in days)

Overall, the proposed hybrid model shows good agreement for the exponential growth phase (initial stages of the batch). Furthermore, the model can capture the lactate shift from consumption to production. The model shows deviations in glucose consumption in the later stages of the culture, which have a cascading effect on the other metabolites. Given that the model is predicting a lower glucose consumption, the cells do not produce as much energy, so they cannot grow as much or produce as much antibody product. The main reason for this underprediction of glucose lies in the *quasi*-steady state assumption, which is one of the constraints of metabolic network optimization. This assumption is valid for the early stages of the culture, when there is a high-nutrient uptake, resulting in fast cell growth. However, it proves insufficient to explain cellular behavior in the later stages, when cells stop growing at a high rate and significant metabolic changes occur, which are dependent on process conditions. As such, the assumption of *quasi*-steady state for the duration of one day (in the metabolic model) leads to higher deviations in the model predictions (Coulet, *et al.*, 2022). Additionally, metabolic network models are known to underpredict glucose consumption as there are energy requirements not captured in the stoichiometry of the reactions of the model, resulting in an underprediction of energy demands. As an improvement, we hope to remove the glucose steady-state constraint in the later stages of the batch and replace it with a minimum glucose uptake rate, backed by experimental data of multiple batches.

The soft sensor is able to estimate monoclonal antibody concentration in the first half of the batch, overpredicting it in the later stages of the batch. The overprediction is linked to an underestimation of the energy requirements of the cell. Given the simplicity of the metabolic model, the entirety of glucose consumed is directed mostly to cell growth and antibody production. There are, however, other metabolic functions that also take up energy, that are not accounted for in the model used herein. This shortcoming could be overcome by using a more comprehensive metabolic network.

The soft-sensor formulation shows the highest deviation from experimental values. When we compare the deviation of the soft sensor from the data and that from the hybrid model, the difference is not as pronounced, reflecting how much closer the hybrid model predictions are to the data. The soft sensor does not show any significant changes when compared to the original hybrid model formulation, for the remaining amino acids. This seems to indicate that removing the measured flux of antibody does not increase the deviation error of the predictions for the remaining amino acids. A possible reason for the low impact of antibody is its low flux value when compared to the remaining amino acids in the network optimization.

The novelty of the proposed model is the estimation of antibody from metabolite concentrations readily available online without the requirement of kinetic parameters estimation. We envisage this formulation as a technological tool towards Industry 4.0, as it paves the way for advanced process control without the need for online monitoring of hard-to-measure data. Our formulation can be integrated into a real-time optimization framework and act as feedback for a control system. Defining control objectives such as maximizing yield or minimizing the batch duration can be possible with the proposed framework, as the soft-sensor would provide the current state of the product.

4. Concluding Remarks

In this paper, we propose a hybrid formulation which integrates a metabolic network in the reactor model that avoids parametrization by using knowledge of cell metabolism to predict key metabolite concentrations. Initial results show that our hybrid formulation offers good predictions for the early stages of the batch, where cells are showing a high nutrient consumption and growing at a fast pace. We demonstrate the applicability of the proposed formulation as a soft sensor for antibody concentration.

Acknowledgements

The authors gratefully acknowledge funding from United Kingdom's Biotechnology and Biological Sciences Research Council (BBSRC) and GlaxoSmithKline (GSK).

References

M. Carvalho, A. Nikdel, J. Riesberg, D. Lyons, H. Budman, 2019, Identification of a dynamic metabolic flux model for a mammalian cell culture, 12th IFAC Symposium on Dynamics and Control of Process Systems

M. Coulet, O. Kepp, G. Kroemer, S. Basmaciogullari, 2022, Metabolic Profiling of CHO Cells during the Production of Biotherapeutics, Cells, 11. 10.3390/cells11121929.

Y. Fan, I. Jimenez Del Val, C. Müller, J. Wagtberg Sen, S. K. Rasmussen, C. Kontoravdi, D. Weilguny, M.R. Andersen, 2015, Amino acid and glucose metabolism in fed-batch CHO cell culture affects antibody production and glycosylation. Biotechnol. Bioeng., 112: 521-535.

H. Hefzi, K.S. Ang, M. Hanscho, *et al.*, 2016, A consensus genome-scale reconstruction of chinese hamster ovary cell metabolism, Cell Systems, 3(5):434–443

K. Höffner, S. M. Harwood, P. I Barton, 2019, A reliable simulator for dynamic flux balance analysis, Biotechnology and bioengineering, 110(3):792–802, 2013.

Y. M. Huang, W. Hu, E. Rustandi, K. Chang, H. Yusuf-Makagiansar, T. Ryll, 2010, Maximizing productivity of CHO cell-based fed-batch culture using chemically defined media conditions and typical manufacturing equipment, Biotechnology progress, 26(5), 1400–1410

S. Kyriakopoulos, C. Kontoravdi, 2014, A framework for the systematic design of fed-batch strategies in mammalian cell culture, Biotechnology and bioengineering, 111(12), 2466–2476

R. P. Nolan, K. Lee, 2011, Dynamic model of CHO cell metabolism, Metabolic engineering, 13(1), 108–124

Precendence Research. Biopharmaceuticals Market - Global Industry Analysis, Size, Share, Growth, Trends, Regional Outlook, and Forecast 2021 – 2030, 2021, https://www.precedenceresearch.com/biopharmaceutical-market accessed on 30/11/2022

P. Virtanen, R. Gommers, T. E. Oliphant, *et al.*, 2020, SciPy 1.0: Fundamental Algorithms for Scientific Computing in Python. Nature Methods, 17(3), 261-272.

B. B. Yahia, L. Malphettes, E. Heinzle, 2021, Predictive macroscopic modeling of cell growth, metabolism and monoclonal antibody production: Case study of a CHO fed-batch production, Metabolic Engineering, 66, 204-216

Antonis Kokossis, Michael C. Georgiadis, Efstratios N. Pistikopoulos (Eds.)
PROCEEDINGS OF THE 33rd European Symposium on Computer Aided Process Engineering (ESCAPE33), June 18-21, 2023, Athens, Greece
 http://dx.doi.org/10.1016/B978-0-443-15274-0.50061-5

Combined Bayesian optimization and global sensitivity analysis for the optimization of simulation-based pharmaceutical processes

Niki Triantafyllou[a], Nilay Shah[a], Maria M. Papathanasiou[a], Cleo Kontoravdi[a,*]

[a] *The Sargent Centre for Process Systems Engineering, Imperial College London, London, United Kingdom, SW72AZ*

**cleo.kontoravdi98@imperial.ac.uk*

Abstract

We propose an efficient framework that employs Bayesian optimization and global sensitivity analysis for the optimization of detailed pharmaceutical flowsheets. Global sensitivity analysis based on quasi-random sampling is utilized to reduce the dimensionality of the problem by identifying critical process and economic parameters that contribute significantly to the variability of Key Performance Indicators (KPIs) such as batch size and OpEx. Then, Bayesian optimization is performed in the previously identified critical input space based on gaussian process surrogate models and a number of different acquisition functions to find the optimal critical operating conditions that minimize the aforementioned KPIs. We apply this framework to the manufacture of plasmid DNA (pDNA), which is a critical raw material for advanced therapeutics, leading to a surge in demand for pDNA for clinical or commercial use. Optimized manufacturing recipes identified with the proposed framework are projected to achieve an up to 170% increase in the batch size and a 34.7% decrease in the OpEx per batch.

Keywords: flowsheet optimization, Bayesian optimization, global sensitivity analysis

1. Introduction

Plasmid DNA (pDNA) is a carrier of genetic information, which makes it a critical raw material in the manufacturing and delivery of Advanced Therapy Medicinal Products (ATMPs). Some of the main pharmaceuticals that use pDNA include DNA and mRNA vaccines and cell and gene therapies (Gosse *et al.*, 2022). The fast-growing interest in ATMPs has led to a growing demand for the production of pDNA for clinical and commercial use. Scaling up the pDNA manufacturing process to meet current and future demands is one of the major challenges in the cell and gene therapy market (Ohlson, 2020). Simultaneously, pDNA production costs and manufacturing timeframes need to be minimized in an attempt to decrease the cost of goods (COGs) and maximize the production of these innovative products. The pDNA manufacturing process is described by long batch times, high capital, operating and labour expenditures. Nevertheless, pDNA manufacturing is considered a well-known process and limited process development and optimization are observed compared to other similar products such as viral vectors and monoclonal antibodies.

Manufacturers need to develop strategies for the establishment of scalable and cost-effective manufacturing processes that will accelerate the wider adoption of ATMPs. Process systems engineering tools can be employed for the development of (bio)pharmaceutical processes, which mostly rely on batch processes that may not be operated at optimal conditions (Boukouvala & Ierapetritou, 2013). Techno-economic analysis using flowsheet modelling is an efficient way to verify the feasibility of a process and compare different technologies. In addition, flowsheet optimization can significantly improve process efficiency by identifying the optimal operation conditions that minimise/maximise key performance indicators (Wang *et al.*, 2017), while data-driven techniques can be employed for simulation-based flowsheet optimization (van de Berg *et al.*, 2022).

In this study, we present a framework for (bio)pharmaceutical process flowsheet optimization that utilizes Bayesian Optimization (BO) and variance-based Global Sensitivity Analysis (GSA). GSA is used as means for uncertainty quantification in the flowsheet. This information is utilized to rank variables and decrease the dimensionality of the black-box model that is then optimized with BO. BO is a sample-efficient optimization method that utilizes a probabilistic model to represent the uncertainty in the objective function. The proposed framework is used for the identification of optimal operating conditions in the production of pDNA.

2. Materials & Methods

The proposed framework combines techno-economic assessment (TEA), global sensitivity analysis and Bayesian optimization. The underlying algorithm is displayed below:

Algorithm 1 Combined TEA, GSA, and BO.

Start: Design process and gather process and economic data.

Input: Converged flowsheet, an initial set of N uncertain parameters $\boldsymbol{P} = \{p_1, p_2, \dots, p_N\}$ and their acceptable ranges $[\boldsymbol{P}_{min}, \boldsymbol{P}_{max}]$.

while $threshold \leq S_{T,min}$ **do**

1. *Screening*: Select subset $n \subseteq N$ of uncertain parameters $\boldsymbol{P} = \{p_1, p_2, \dots, p_n\}$.
2. *Sampling*: Draw 4,096 input samples P based on Sobol sequence and run the converged flowsheet to obtain the target outputs $\boldsymbol{y} = \{y_1, y_2, \dots, y_n\}$.
3. *GSA*: Compute the total sensitivity indices S_T based on the RS-HDMR metamodel.

end while

BO initialization: train $\mathcal{GP}^0$ with an initial data set $D_0 = (\boldsymbol{P}_0, y_0)$

for $i = NumberOfIterations$ **do**

1. Find $\boldsymbol{P}_i$ by optimizing the acquisition function over the $\mathcal{GP}^i$, where $\boldsymbol{P}_i = argmin_P\, u(\boldsymbol{P}|D_{1:i-1})$.
2. Run the flowsheet to sample the objective function $y_i = f(\boldsymbol{P}_i) + \varepsilon_i$.
3. Augment the data set $D_{1:i} = \{D_{1:i-1}, (\boldsymbol{P}_i, y_i)\}$ and update the $\mathcal{GP}^i$.

end for

2.1. Techno-economic Modeling & Global Sensitivity Analysis

A fed-batch production process for pDNA is modelled using SuperPro Designer (Intelligen, Inc) (Ferreira & Petrides, 2021). SuperPro Designer uses built-in process and economic models based on sets of algebraic and differential equations to compute equipment sizes, operation scheduling, CapEx and OpEx. The pharmaceutical pDNA production flowsheet includes cell cultivation, pDNA recovery, purification, formulation and fill & finish as shown in Fig. 1. Information on process and economic parameters and their proposed accepted ranges are based on literature, cGMP manufacturers, suppliers and expert feedback for the design of the specific process at hand.

Global sensitivity analysis is then performed to quantify manufacturing uncertainty based on the contribution of uncertain input parameters to the output variance of specific KPIs, namely batch size, formulation concentration, throughput, batch time, cycle time, CapEx, OpEx, and production cost. GSA was performed using a SobolGSA (Kucherenko, 2013)-Matlab-Component Object Model (COM)-SuperPro interface with 4,096 quasi-randomly generated samples of the uncertain input variables based on Sobol quasi-random sequence within the pre-defined ranges shown in Table 1. Critical parameters have, for at least one of the KPIs, total sensitivity indices exceeding a user-specified threshold ($S_{T,min} = 1\%$). A higher $S_{T,min}$ would lead to a lower-dimensional design space but for the purpose of this work, the capabilities of the algorithm are tested for a high-dimensional space. Non-influential parameters are fixed after each global sensitivity run until the algorithm converges and all the critical parameters are identified. The use of GSA enables the ranking of uncertain parameters and therefore dimensionality reduction of the design space by only considering a subset of the most influential inputs.

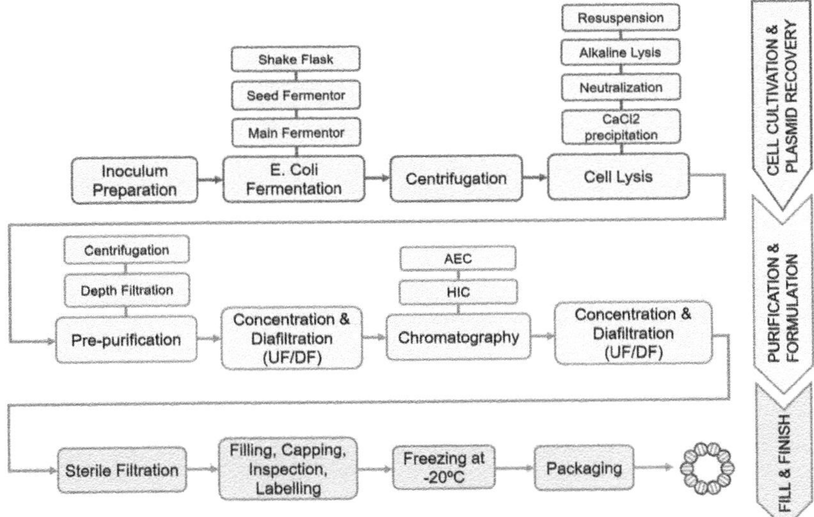

Figure 1. Process flowsheet for the production of plasmid DNA.

2.2. Bayesian Optimization (BO)

Once the critical design space has been identified, BO is used to derive the optimal manufacturing recipes. The pDNA flowsheet is treated as a black-box function ($f(\boldsymbol{P})$) that can be sampled at the critical inputs ($\boldsymbol{P} \in \mathbb{R}^n$) to obtain the optimal solution ($\boldsymbol{P}^*$). Gaussian processes (GPs) with zero mean and the Matérn kernel function are used as the surrogate model. GPs are chosen because they can reliably quantify uncertainty and point to unexplored regions of the search space and due to their ability to incorporate prior knowledge into the model (Rasmussen & Williams, 2006). Initially, the GP surrogate is trained with 5 samples based on Sobol sequence and then BO iteratively updates the GP based on adaptive sampling. Adaptive sampling is achieved through the optimization of an acquisition function that can balance exploration versus exploitation. In this work, we test four of the most well-known acquisition functions– probability of improvement (PI), expected improvement (EI), upper (or lower) confidence bound (UCB or LCB), and max-value entropy search (MES). UCB and LCB are tested for two different confidence values that control the level of exploration. Before the GP is trained, data features are normalized, and data outputs are standardized. The BO part of the algorithm is built using the Python packages PyTorch and BoTorch (Balandat *et al.*, 2020).

3. Results & Discussion

The screening for critical parameters considers an initial set of more than 25 parameters including task lengths, conversions, buffer volumes, working volumes, rejection and concentration coefficients, failure rates, etc. The parameters that had an insignificant effect on the output metrics were fixed to their nominal values without adding extra uncertainty to the model. The GSA part of the algorithm was repeated until all remaining parameters were critical. The final subset of critical parameters that are responsible for at least 1% of the output variability for at least one output and their corresponding acceptable ranges are displayed in Table 1.

Table 1. Critical parameters in pharmaceutical pDNA production, their acceptable ranges, nominal values, and optimal values for the maximization of the batch size and the minimization of OpEx.

Input	Process step	Range	Nominal	$\text{Optimal}_{\text{batch-size}}$	$\text{Optimal}_{\text{OPEX}}$	Unit
Fermentation duration (seed)	Seed Fermentor	12–30	30	23.25	12	*h*
Fermentation duration (main)	Main Fermentor	18–40	40	20.75	18	*h*
Fed-batch medium volume		1.5–4.5	3	2.63	1.5	*L*
Starting working volume		8.1–24.2	16.1	14.1	8.1	*L*
Resuspension buffer volume	Resuspension	22.78–68.36	45.57	28.48	22.79	*L*
Lysis buffer volume	Lysis	22.78–68.36	45.57	28.48	22.79	*L*
Neutralization buffer volume	Neutralization	22.78–68.36	45.57	28.48	22.79	*L*
Diafiltration volumes 1	1st UF/DF	10–50	50	42	10	–
Flush volume per filter area 1		1.5–4.5	3	4.12	1.5	L/m^2
Diafiltration volumes 2	2nd UF/DF	10–50	40	43	10	–
Flush volume per filter area 2		0.67–2	1.33	1.17	0.67	L/m^2

The BO algorithm is performed for all critical parameters in a 9-dimensional space. However, the critical parameters are ranked and contribute differently to each output metric. Fig. 2 illustrates the variance as defined by the GSA algorithm by summarizing the first (S_i) and second (S_{ij}) order sensitivity indices for two of the output metrics. Fig. 3 illustrates the total variance of the two output metrics when all critical inputs are simultaneously varied within their acceptable ranges.

The BO algorithm leads to (a) an increase of 170% in the batch size and (b) a decrease of 34.7% in the OpEx per batch from the nominal points while operating within the acceptable ranges for all critical inputs (Fig. 3). Optimal manufacturing recipes are shown in Table 1. The nominal point of the OpEx lies far up from the median because the base case flowsheet operates on the upper bounds of the seed and main fermentor task durations.

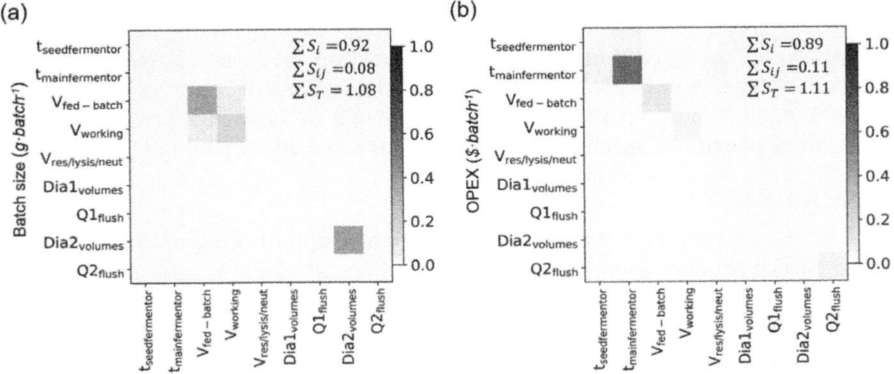

Figure 2. GSA of the nine critical input parameters on the batch size and OpEx for the production of pharmaceutical pDNA. First-order effects (S_i) are displayed in the diagonal and second-order effects (S_{ij}) are presented in the upper and lower triangular.

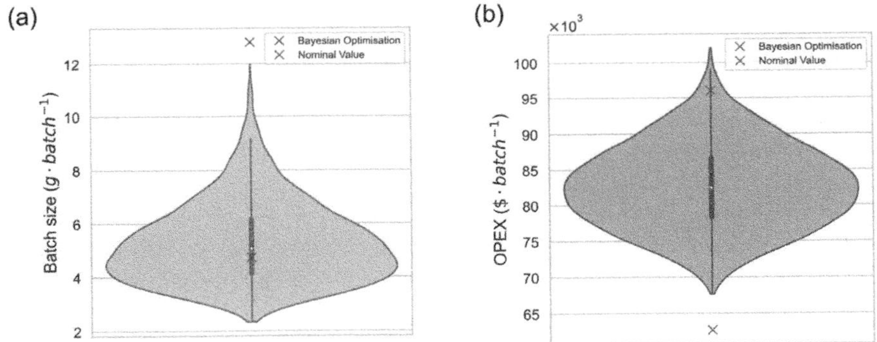

Figure 3. Violin plots for two of the key output metrics. The black bar in the center is the interquartile range, the white dot inside the black bar is the median value, and the black lines stretched from the bar are the lower/upper adjacent values. Bayesian optimization is used to (a) maximize the batch size and (b) minimize the OpEx.

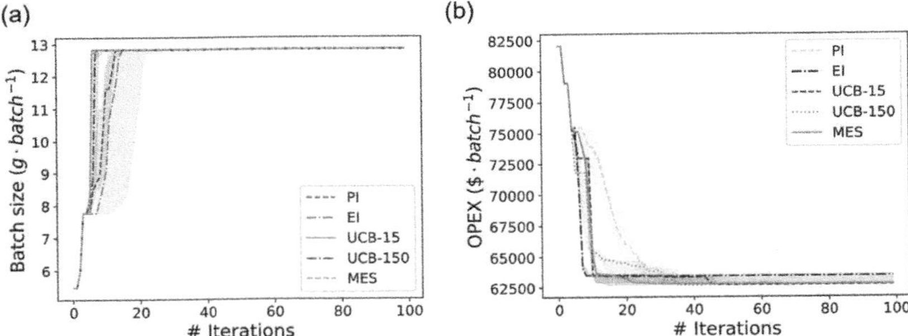

Figure 4. Bayesian optimization progress for different acquisition functions for (a) batch size maximization and (b) OpEx minimization. Plot presents the median and confidence intervals (1st and 3rd quartiles) from 10 random seeds.

Fig. 4 shows the progress of the BO algorithm for the different acquisition functions for 100 iterations. Most acquisition functions converge to the optimum after a few iterations. In the batch size maximization problem, the best-performing acquisition function is UCB, for both confidence values, resulting in convergence to the optimum in less than 10 iterations. On the contrary, EI performs worse compared to the others, converging after 17 iterations. In the OpEx minimization, MES performs the best, reaching the optimum after 25 iterations, with UCB-150 following at 20 iterations. It can be observed that the algorithm requires more iterations to find the optimum for OpEx because this metric has a wider range of critical variables contributing to the overall variability (Fig. 2).

4. Conclusions

In this work, we present a framework for black-box optimization of simulation-based pharmaceutical processes that relies on the principles of global sensitivity analysis and Bayesian optimization. The case study shows that the proposed algorithm performs reliably well and leads to optimum solutions after a few iterations. We use four different acquisition functions that use different strategies for exploration and exploitation and the algorithm converges in all cases. The right acquisition function for each output metric is key for optimum algorithm performance, with cases of acquisition functions performing differently for different objective functions being observed.

Acknowledgements

This research is funded by the Department of Health and Social Care using UK Aid funding and is managed by the Engineering and Physical Sciences Research Council (EPSRC; grant EP/R013764/1). The views expressed in this publication are those of the author(s) and not necessarily those of the Department of Health and Social Care.

References

M. Gosse, C. Jones, D. Jesus, S. D'Costa, 2022, Regulatory & supply chain implications for plasmids as critical starting materials in the manufacture of viral vector gene therapy products. *Cell and Gene Therapy Insights*, 8(2), 279–286.

J. Ohlson, 2020, Plasmid manufacture is the bottleneck of the genetic medicine revolution, *Drug Discovery Today*, *25*(11), 1891–1893.

F. Boukouvala, M. Ierapetritou, 2013, Surrogate-Based Optimization of Expensive Flowsheet Modeling for Continuous Pharmaceutical Manufacturing, *Journal of Pharmaceutical Innovation*, 8, 131–145.

Z. Wang, M. Sebastian Escotet-Espinoza, M. Ierapetritou, 2017, Process analysis and optimization of continuous pharmaceutical manufacturing using flowsheet models, *Computers and Chemical Engineering*, 107, 77–91.

D. van de Berg, T. Savage, P. Petsagkourakis, D. Zhang, N. Shah, E. A. del Rio-Chanona, 2022, Data-driven optimization for process systems engineering applications, *Chemical Engineering Science*, 248, 117135.

R. Ferreira, D. Petrides, 2021, Plasmid DNA (pDNA) Large Scale Manufacturing – Process Modeling and Techno-Economic Assessment (TEA) using SuperPro Designer, doi: 10.13140/RG.2.2.12780.28800.

S. Kucherenko, 2013, SobolHDMR: A general-purpose modeling software. *Methods in Molecular Biology*, 1073, 191–224.

C. E. Rasmussen, C. K. I. Williams, 2006, Gaussian Processes for Machine Learning, *The MIT Press*.

M. Balandat, B. Karrer, D. R. Jiang, S. Daulton, B. Letham, A. Gordon Wilson, E. Bakshy, 2020, BoTorch: A Framework for Efficient Monte-Carlo Bayesian Optimization, *Advances in Neural Information Processing Systems*, 33.

Antonis Kokossis, Michael C. Georgiadis, Efstratios N. Pistikopoulos (Eds.)
PROCEEDINGS OF THE 33rd European Symposium on Computer Aided Process Engineering (ESCAPE33), June 18-21, 2023, Athens, Greece
 http://dx.doi.org/10.1016/B978-0-443-15274-0.50062-7

Competitive adsorption of copper, nickel, and chromium ions onto EDTA modified SBA-15

Bawornpong Pornchuti,[a] Yuttana Phoochahan,[a] Prarana Padma,[a] Suchada Ruengrit,[a] Pravit Singtothong,[b]

[a]*Department of Process and Industrial Engineering, School of Engineering and Industrial Technology, Mahanakorn University of Technology, Bangkok 10530, Thailand*

[b]*Department of Chemistry, Faculty of Science, Mahanakorn University of Technology, Bangkok 10530, Thailand*

Abstract

Mathematical modeling is an important tool in chemical engineering. Not only to understand processes, it is also used in many fields such as design and scale-up, process control, optimization, experimental design, trouble shooting, etc. One application in adsorption process is to represent adsorption isotherm. In contrast to the real processes, most of researches studied single-component isotherm. Wastewater from electroplating industry always comprises of several heavy metal ions which are harmful to human and ecosystem. In this work, EDTA modified SBA-15 was used to treat copper, nickel, and chromium ions in multicomponent system. It was found that adsorption of heavy metals preferred higher pH. Moreover, adsorption of nickel ions was very low. The multicomponent isotherm models were applied. The RMSEs (Residual root mean square errors) of Freundlich, Langmuir, extended Langmuir, and modified competitive Langmuir models were 2.02, 3.37, 9.12, and 10.72, respectively.

Keywords: adsorption, heavy metals, multicomponent isotherm, EDTA, SBA-15.

1. Introduction

The COVID-19 pandemic extremely affected this world. Million people died. Many cities were locked down. Human activities dropped dramatically. This pandemic still had positive impacts such as bluer skies, fewer crimes, lower accidents, better hygiene, stronger relationships, etc. Considering environmental remediation, it was evident that pollution was mainly caused by human. Many pollutants were contaminated in air, water, and soil.

Heavy metals are hazardous pollutants. They are always contaminated in wastewater from electroplating processes. One effective way treating heavy metal ions is adsorption. Adsorbents are the key of adsorption processes. Novel adsorbents are developed and investigated.

Surface modification is used to improve adsorption performance. Silica is favored in adsorbent production because it is easy to adjust its surface and has large surface area. Silica materials can have different pore structures. However, our previous study found that adsorption capacities of modified silica do not depend on pore structure (Pornchuti et al., 2021). As a consequence, SBA-15 was selected in this work since it is convenient to synthesize. In addition, EDTA was grafted onto SBA-15 surface to catch copper, nickel, and chromium ions in multicomponent system.

Generally, real wastewater contains various heavy metal ions. There are interactions among these ions (Girish, 2017). Consequently, the study of multi-ion adsorption is essential. In case of single ion adsorption, Freundlich and Langmuir models are always chosen for the representation of adsorption isotherm. These two models are simple but in case of multi-ions system they do not include the effect of other ions. For multicomponent system, extended Langmuir and modified competitive Langmuir models were developed to cover their interaction. As a consequence, Freundlich, Langmuir, extended Langmuir, and modified competitive Langmuir models were selected in this study. In summary, this work was set up to study the modeling of competitive adsorption of copper, nickel, and chromium ions onto EDTA modified SBA-15.

2. Materials and Methods

2.1. Materials

TEOS (Tetraethyl orthosilicate), Pluronic P123, APTES (Aminopropyltriethoxysilane), EDTA disodium salt, ethanol, toluene, HCl, acetic acid, and ammonium hydroxide were purchased from Sigma Aldrich or Merck.

2.2. Synthesis of SBA-15

SBA-15 was synthesized by following the method of Naik et al. (2011). In brief, 6.64 g of Pluronic P123 and 13.5 mL of HCl solution were agitated with 202 mL of deionized water. Then 13.86 g of TEOS was dropped into the solution. The mixture was kept for 24 h at 90°C. After vacuum filtration, the solid was washed and dried. SBA-15 was obtained by calcination at 550°C for 3 h.

2.3. Integration of amino functional groups

Amino functional group was grafted onto SBA-15 surface by applying the method of Parida and Rath (2009). At first, 2 g of SBA-15 was suspended in 60 mL of toluene. Next, 1.2 mL of APTES was added dropwise. The mixture was refluxed at 110°C for 8 h. The solid was dried after washing with ethanol and deionized water.

2.4. Incorporation of EDTA

The method of Repo et al. (2009) was adapted to the incorporation of EDTA. Briefly, 2 g of amine functionalized SBA-15, 5.4 g of EDTA disodium salt, and 42.0 mL of ethanol-acetic acid solution were mixed together. The solid was filtered and washed with ammonium hydroxide solution, ethanol-acetic acid solution, deionized water, and ethanol, respectively. The EDTA modified SBA-15 was received after dried at 105°C.

2.5. Multicomponent Adsorption

Normally, multi-ion adsorption experiments were conducted as follows. 100 mg of EDTA modified SBA-15 was shaken in 50 mL of multi-ion solution at 105 rpm for 48 h. The solid was separated by vacuum filtration. The filtrate was analyzed by atomic absorption spectroscopy (AAS). Eq. (1) was applied to calculate adsorption capacity.

$$q_e = \frac{(C_i - C_e)V}{m} \tag{1}$$

q_e is equilibrium adsorption capacity; C_i is metal concentration at start; C_e is equilibrium metal concentration; V is the volume of metal solution; m is the adsorbent mass. To compare the suitability between models, the RMSE was determined from Eq. (2).

$$\mathrm{RMSE}=\sqrt{\frac{1}{n}\sum_{i=1}^{n}\left(q_{i,exp}-q_{i,model}\right)^2} \quad (2)$$

$q_{i,exp}$ is the experimental adsorption capacity of component i and $q_{i,model}$ is the adsorption capacity of component i calculated from the model.

3. Results and Discussion

3.1. Effect of pH

The amount of metal ions adsorbed on EDTA modified SBA-15 at various pH was shown in Figure 1. Effect of pH on competitive adsorption was similar to those of single metal adsorption. Metal adsorption was hindered by hydrogen ion at low pH. As a result, the amount of adsorbed metals increased with an increase of pH.

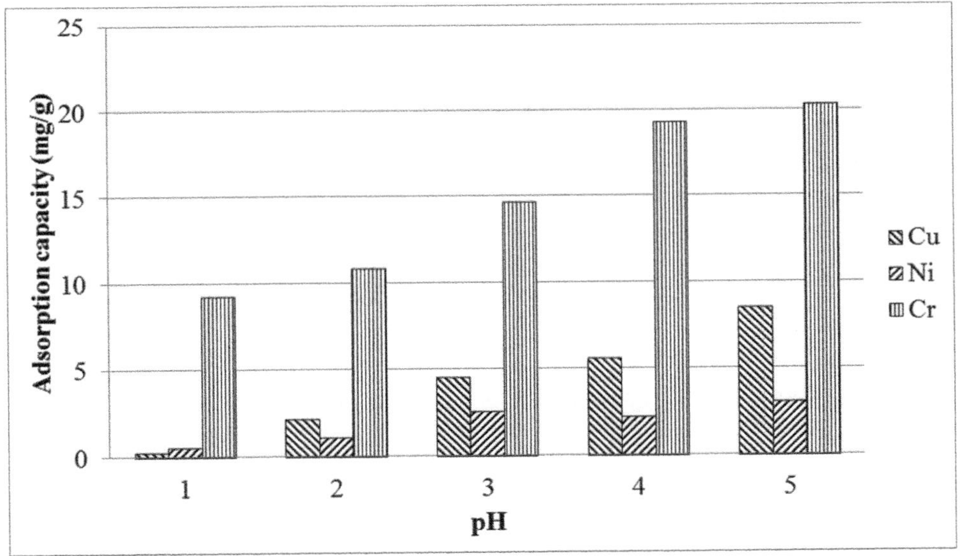

Figure 1. pH Effect.

3.2. Mathematical Modeling

Adsorption isotherm was illustrated as Figure 2. It can be seen that adsorption of nickel ion was very low comparing to copper and chromium ions. It was suppressed by other ions. Other adsorbents were also found this antagonistic effect such as adsorption by amine functionalized SBA-15 (Pornchuti et al., 2022), olive stones, date stones, peat, sphagnum moss, laterite, and bentonite (Girish 2018). Since the amount of nickel adsorbed was low, the mathematical models were applied only with copper and chromium ions.

Although Freundlich and Langmuir models are designed for single component adsorption, their utilization in multicomponent adsorption can be found such as the investigation of Qi and Pichler (2017). Freundlich model can be represented as Eq. (3).

$$q_{e,i}=K_{F,i}C_{e,i}^{1/n_i} \quad (3)$$

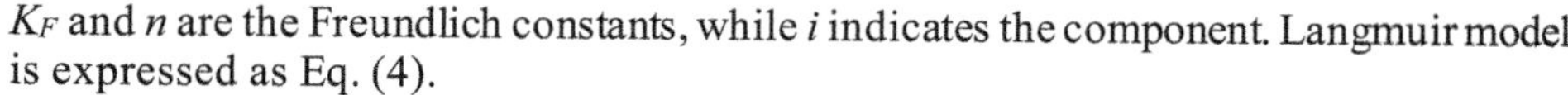

K_F and n are the Freundlich constants, while i indicates the component. Langmuir model is expressed as Eq. (4).

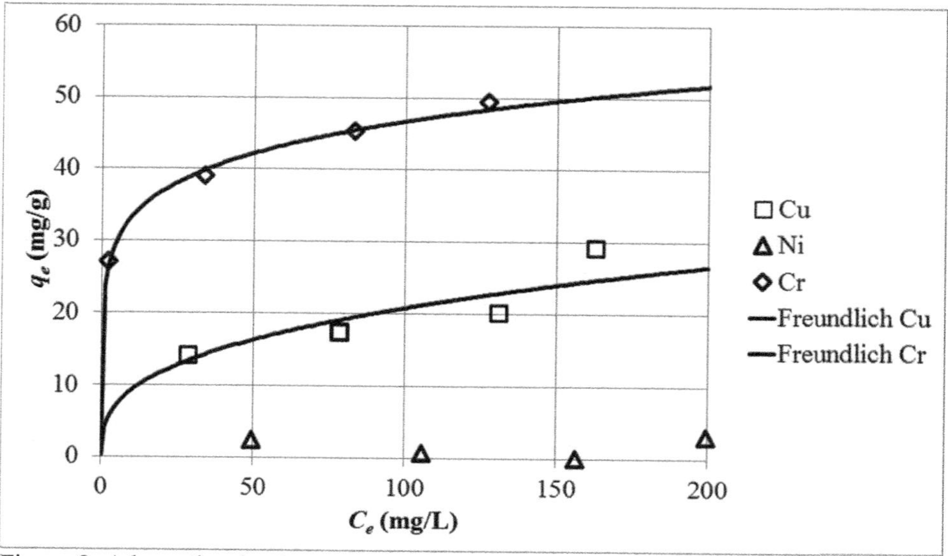

Figure 2. Adsorption isotherm.

$$q_{e,i} = \frac{q_{m,i} b_{L,i} C_{e,i}}{1 + b_{L,i} C_{e,i}} \quad (4)$$

q_m is the maximum loading and b_L is the Langmuir constant. The linearized form of both models were applied to compute their parameters and listed in Table 1.

Table 1. Parameters of isotherm models.

Parameters of isotherm models	Cu	Cr
Freundlich isotherm		
$K_{F,i}$	4.1505	23.7739
n_i	2.8531	6.8213
Langmuir isotherm		
$q_{m,i}$	26.5252	45.2489
$b_{L,i}$	0.0382	0.6500
Extended Langmuir isotherm		
$q_{m,i}$	23.6173	206.3216
$b_{L,i}$	13.1572	5.9948
Modified competitive Langmuir isotherm		
$q_{m,i}$	97.0874	163.9344
$b_{L,i}$	0.0258	0.0667
$\eta_{L,i}$	5.8333	8.9594

Extended Langmuir model is shown as Eq. (5).

$$q_{e,i} = \frac{q_{m,i} b_{L,i} C_{e,i}}{1+\sum_{j=1}^{N}(b_{L,j} C_{e,j})} \tag{5}$$

Total number of ions is represented as N, while j indicates the component. The parameters of this model were computed by minimization of the error in non-linear regression analysis and shown in Table 1. Modified competitive Langmuir model can be written as Eq. (6).

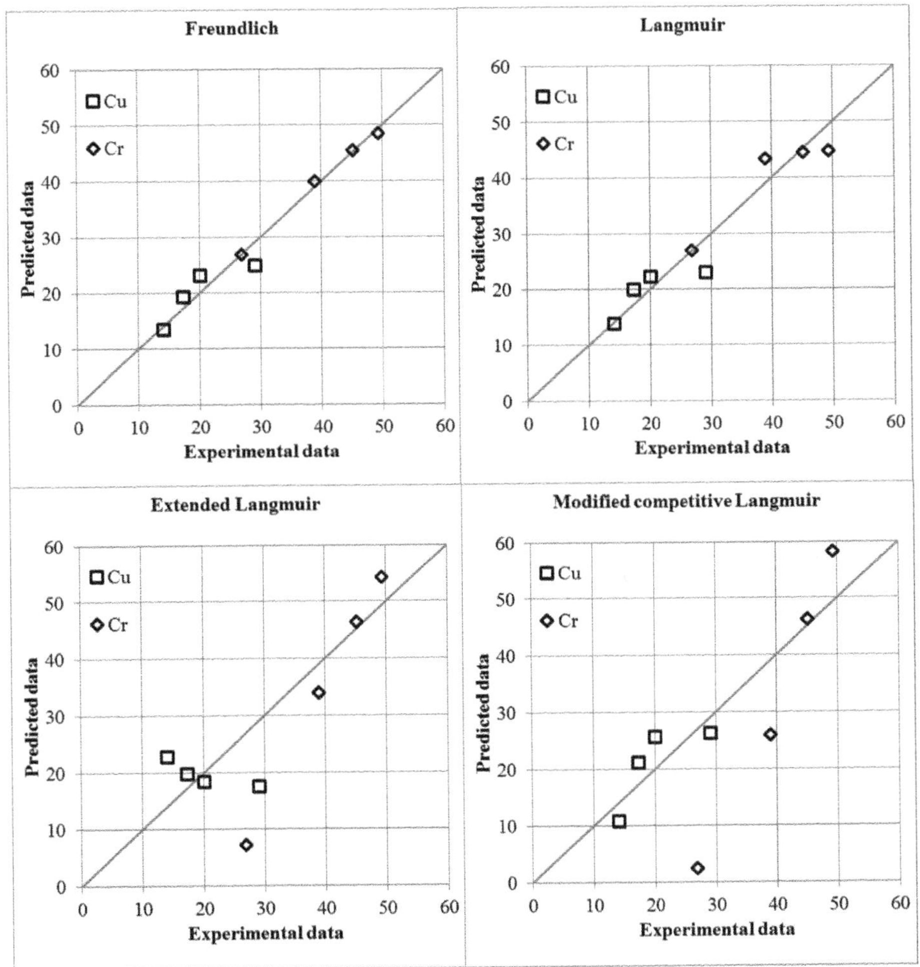

Figure 3. Comparison between predicted adsorption capacities and experimental data.

$$q_{e,i} = \frac{q_{m,i} b_{L,i} \left(C_{e,i}/\eta_{L,i}\right)}{1+\sum_{j=1}^{N}\left(b_{L,j}\left(C_{e,j}/\eta_{L,j}\right)\right)} \tag{6}$$

q_m and b_L in this model is Langmuir parameters of single-ion adsorption. They were obtained from our previous study (Pornchuti et al., 2021). η_L is interaction factor. They were computed from experimental data of multi-ion adsorption by minimization of the error in non-linear regression analysis. All parameters of this model were listed in Table 1.

Each isotherm models were utilized to predict adsorption capacities and compare with the experimental data as demonstrated in Figure 3. The RMSEs of Freundlich, Langmuir, extended Langmuir, modified competitive Langmuir models were 2.02, 3.37, 9.12, and 10.72, respectively. It obviously indicated that Freundlich and Langmuir were more appropriate in the representation of experimental data in this case. Many researches illustrated isotherm in 2D like Figure 2 which lead to erroneous interpretation. The concentration of each ion was related to the others. It should be realized that both models cannot explain the effect of other ions. For extended Langmuir and modified competitive Langmuir models, they took the concentration of other ion in the model but gave high RMSEs. This might be caused by the complexity of the models and more observational data were required to improve their RMSEs.

4. Conclusions

Simultaneous adsorption of copper, nickel, and chromium ions onto EDTA modified SBA-15 was conducted. It was figured out that adsorption of these ions favored high pH. Moreover, nickel ion was hindered by other ions. Finally, fitting of Freundlich model was better than those of Langmuir, extended Langmuir, and modified competitive Langmuir models.

References

C. R. Girish, 2017, Various Isotherm Models for Multicomponent Adsorption: A Review, Journal of Civil Engineering and Technology, 8, 80-86

C. R. Girish, 2018, Multicomponent Adsorption and the Interaction between the Adsorbent and the Adsorbate: A Review, International Journal of Mechanical Engineering and Technology, 9, 177-188

B. Naik, V. Desai, M. Kowshik, V. S. Prasad, G. F. Fernado, N. N. Ghosh, 2011, Synthesis of Ag/AgCl-Mesoporous Silica Nanocomposites using a Simple Aqueous Solution-Based Chemical Method and a Study of Their Antibacterial Activity on E. coli, Particuology, 9, 243-247

K. M. Parida, D. Rath, 2009, Amine Functionalized MCM-41: An Active and Reusable Catalyst for Knovenagel Condensation Reaction, Journal of Molecular Catalysis A: Chemical, 310, 93-100

B. Pornchuti, B. Pongpattananurak, D. Sutthiard, P. Singtothong, 2021, Adsorption of Copper, Nickel, Chromium Ions Using EDTA Modified Silica Adsorbents, The 30th TIChE Conference (TIChE2021), Nakhon Ratchasima, Thailand, 6-7 May 2021, 744-749

B. Pornchuti, Y. Phoochahan, P. Padma, S. Ruengrit, P. Singtothong, 2022, Competitive Adsorption of Copper, Nickel, and Chromium Ions onto Amine Functionalized SBA-15, Computer Aided Chemical Engineering, 49, 2065-2070

P. Qi, T. Pichler, 2017, Competitive Adsorption of As(III), As(V), Sb(III), and Sb(V) onto Ferrihydrite in Multi-Component Systems: Implications for Mobility and Distribution, Journal of Hazardous Materials, 330, 142-148

E. Repo, T. Agustiono Kurniawan, J. K. Warchol, M. E. T. Sillanpää, 2009, Removal of Co(II) and Ni(II) Ions from Contaminated Water using Silica Gel Functionalized with EDTA and/or DTPA as Chelating Agents, Journal of Hazardous Materials, 171, 1071-1080

Antonis Kokossis, Michael C. Georgiadis, Efstratios N. Pistikopoulos (Eds.)
PROCEEDINGS OF THE 33rd European Symposium on Computer Aided Process Engineering (ESCAPE33), June 18-21, 2023, Athens, Greece
 http://dx.doi.org/10.1016/B978-0-443-15274-0.50063-9

Auto-extractive reactive distillation for the production of fatty esters

Hernán D. Muñoz,[a,b] Christian Hoffmann,[b] Markus Illner,[b] Álvaro Orjuela,[a] Jens-Uwe Repke[b]

[a]*Department of chemical and Enviromental Engineering, Universidad Nacional de Colombia, Bogotá D.C. 111321, Colombia*
[b]*Technische Universität Berlin, Sekr. KWT 9, Process Dynamics and Operations Group, Str. des 17. Juni 135, D-10623 Berlin, Germany*
hdmunozg@unal.edu.co

Abstract

This work proposes a systematic methodology based on graphic methods for the conceptual design of simultaneous reactive distillation-extractive distillation processes in one single apparatus (named extractive-reactive distillation). The configuration is useful to overcome azeotropes that reactive distillation alone cannot. A key assumption of the methodology is the use of an entrainer that is part of the components consumed in the reaction (i.e. an auto-entrainer). In the methodology, the analysis of the statics is used to generate a reactive distillation configuration, while an iso-volatility and bifurcation analysis is used to generate an extractive distillation configuration. For the combined system, we propose to extend the analysis of the statics to include the auto-entrainer effect in reactive distillation. The methodology is applied to the production of isopropyl myristate from acid oleochemical streams. A new column configuration that has the potential to reduce additional azeotropic separation units is proposed.

Keywords: Reactive distillation; extractive distillation; isopropyl myristate

1. Introduction

Reactive distillation (RD) is a well-known method for process intensification. In RD, reactants are fed to a reactive section and the products are separated via distillation in the same apparatus at the same time. In some special cases, however, an RD process may also show an auto-extractive behavior (Coker, 2010), i.e., one of the reactants behaves like an internal mass separation agent (entrainer), so that at least one of the azeotropes in the system is overcome. In such cases, the entrainer is the component with the highest boiling point in the system and does not form azeotropes with the other components. The double functionality reactant-entrainer generates a synergetic effect, which is not always obvious. This effect can be exploited in the synthesis of intensified process configurations that have so far not been reported in the literature. In spite of this potential benefit, available methodologies only consider the reactive phenomena (Almeida-Rivera et al., 2004). This contribution presents a methodology that explicitly takes the auto-extractive behavior into account. The methodology was applied exemplarily on the esterification of fatty acids, e.g., myristic acid (MA), with isopropanol (IPA) to produce isopropyl myristate (I-MA). Besides its auto-extractive behavior, the system is of interest because it can be used for the valorization of acid oleochemical streams, which are an industrial waste rich in free fatty acids. In these systems, the alcohol (i.e., IPA) forms a minimum-boiling azeotrope with water (W) produced by the reaction **(Bock et al, 1997)**. In addition, the system contains a liquid phase split that is considered during the analysis.

Table 1. Methodology outline.

Step	Purpose	Input	Output
Topology (quaternary system)	Determination of the thermodynamic characteristics of the system.	Component boiling points. Thermodynamic model.	Graphical representation of phase equilibria. Residue curves. Characteristic distillation maps.
Analysis of the statics (quaternary system)	Determination of the initial single-feed structure of the RD..	Stoichiometric coefficients. Chemical equilibrium. Residue curves.	Limiting steady states. Trial trajectory. Localization of the reactive zone.
Isovolatility (ternary subsystem)	Determination of the extractive distillation configuration (direct or indirect).	Thermodynamic model.	Locus of the constant relative volatility in the composition space.
Bifurcation analysis (ternary subsystem)	Determination of feasible entrainer flows in extractive distillation	Pinch equations. Initial guess entrainer/feed ratio.	Pinch branches. Pitchfork bifurcation.
Coupled synthesis (quaternary-ternary system)	Determination of the single-feed reactive distillation- extractive distillation configuration	Feasible reactive distillation trial trajectories. Feasible extractive distillation mass balance lines.	Reactive distillation feed composition that matches bottoms composition of extractive distillation in the same composition space. Feasible entrainer to feed ratio

2. Methodology

An outline of the proposed methodology is presented in **Table 1**, in which the motivation, the input, and the output of each step is given. The individual steps are described in more detail in the following sections.

2.1. Thermodynamic Topological analysis

Most of the graphic methods in distillation begin with a thermodynamic characterization of the systems based on its fixed points and boundaries. The case study in this contribution is a reactive quaternary system: IPA (21,15°C), W (100,00°C), I-MA (314,85°C) and MA (326,20°C). There are four ternary subsystems, from which the subsystem IPA-W-MA presents a residue curve map type 1-0-0 according to the nomenclature reported by Matsuyama et al. (1977). This is the characteristic map of extractive distillation with heavy entrainer and without additional azeotrope formation (Foucher et al., 1991). There are six binary subsystems, one of them with a homogeneous binary azeotrope (IPA-W). The map does not include first-order separatrices (distillation boundaries), but it has a second-order separatrix (geometric) that subdivides the composition space in two subregions with a different saddle node each one. There is a liquid-liquid equilibrium in three of the four ternary subsystems and in the quaternary region near the water node.

2.2. Analysis of the statics (AS)

The method is based on the geometric theory of distillation and was recently reviewed by Cardona et al (2020). The AS determines the limiting steady states (LSS) of a reactive distillation system, which are defined as the steady states that maximize the yield of a target product. The AS uses a reactive distillation process model that assumes a) kinetically reversible chemical reaction in the liquid phase, b) an infinitely high separation efficiency (∞/∞), c) a sufficiently high volume of the reaction zones in the column, and d) catalyst placed anywhere in the column. The AS requires little initial information, such as the reaction stoichiometry and parameters for the physical and chemical equilibria. The procedure that is followed in the analysis of the statics consists of the following steps. For more details, the reader is please referred to (Cardona et al, 2020) or the original paper by Pisarenko et al. (2001) in which the terms are explained more extensively.

1. *Complete set of steady states* These steady states are lines representing mass balances of distillation columns where for a defined top or bottoms, the conjugated product falls in the boundary of the composition space. Each steady state is determined for a given feed composition and conversion. For a direct arrangement in the production of isopropyl myristate, a representative sample of the steady states is shown in **Figure 1**a.
2. *Limiting steady states based on complete set of steady states*. From the lever rule it is known that the steady states have a corresponding relation of product flows P/W (top flow/bottom flow). This relation depends on conversion and feed composition. By computing P/W as function of the conversion, it is possible to find the LSS. In the case study, the LSS were grouped in the a, b, c, d types shown in **Figure 1**b.
3. *Segregation of LLS based on feasibility criterion of distillation to ensure the occurrence of all chemical species for the separation*. For the case study, only the LSS type "b" were found to be feasible because only this line contains all four components. The other three lie on the ternary sub-planes.
4. *Determination of the reactive distillation process design able to develop the feasible LSS*. For the case study, this is exemplified in **Figure 1**c for the feasible LSS corresponding to a stoichiometric feed composition X_F. This feed follows a reaction along a stoichiometric line until the feasible steady state is achieved, which is characterized by the conversion, ξ. For this steady state, X^* represents the overall composition of the column products in mixture that are joined by a mass balance between the top product X_p and the bottom product X_w of the column. A "trial distillation trajectory" (Pisarenko et al., 2001) joins the top and bottoms product of the mass balance. This trial trajectory is a simplified representation of a residue curve starting at the top composition, and therefore it is similar to the trajectory followed by a column with infinite reflux and number of separation stages ∞/∞.
5. *Flowsheet synthesis*. The synthesis of the RD flowsheet begins with the definition of the reactive and non-reactive sections of the column. The location of the trial trajectory with respect to the equilibrium defines those profile sections, in which the reaction should be promoted with a catalyst addition. In the case study, the entire trial trajectory lays in a region of the composition space where the presence of catalyst promotes the forward reaction. Consequently, the whole column is a catalytic section, see **Figure 2**.

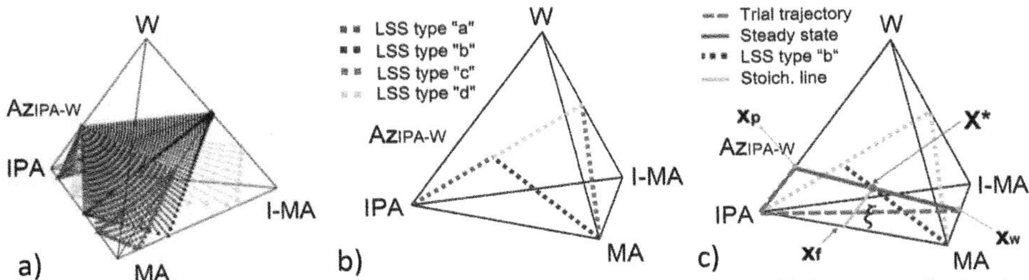

Figure 1: Analysis of the statics for the esterification of myristic acid with isopropanol, a) set of steady states, b) set of limiting steady states, c) trial trajectory for the limiting steady stated type b that corresponds to an equimolar feed.

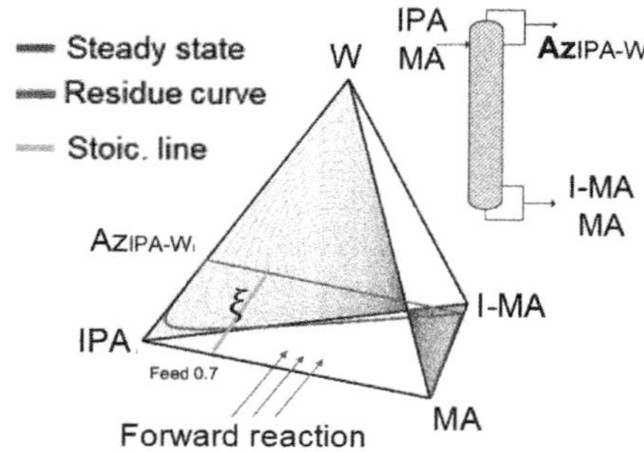

Figure 2: Smoothing of the trial trajectory to a residue curve, configuration.

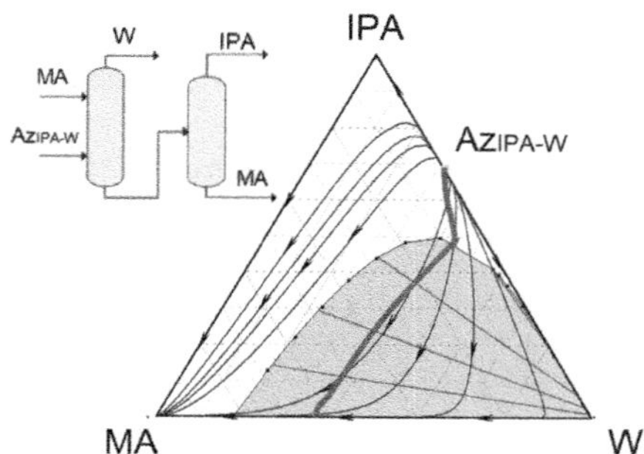

Figure 3: Residue curve map (blue lines) for the ternary system IPA-W-MA. Univolatility line (red).

2.3. Iso-volatility curve

An iso-volatility curve is the locus of the ternary composition space in which the relative volatility of two defined components is constant. The iso-volatility curve of constant value one is a uni-volatility line that starts in the azeotrope composition and ends in one of the opposite edges of the ternary composition space. In the ternary system IPA-W-MA, the univolatility line falls into the edge W-MA, see **Figure 3**. This indicates that the distillate of the extractive distillation column is almost pure water (indirect configuration). Consequently, the bottoms product is a mixture IPA-MA that requires an additional fractionation column for separation. The two-column arrangement is shown in the upper left side of the **Figure 3**. For a given MA/azeotrope feed ratio, it is possible to define a global mass balance and specify all the streams. However, the MA/azeotrope feed ratio is sometimes difficult to obtain and requires the more detailed analysis introduced in the next section.

2.4. Bifurcation analysis

A method for the definition of the minimum entrainer flow (minimum MA/azeotrope feed ratio) that is required for separation via extractive distillation was previously proposed by Knapp et al. (1994). The method is based on the bifurcation analysis of the pinch branches of the middle section in an extractive distillation column. The bifurcation analysis observes the behavior of the middle section pinch branches with respect to small additions of entrainer. Thus, the minimum amount of entrainer is determined with the help of the phenomenon called "pitchfork", which precedes a strong change in the topology of the pinch branches (please refer to Figure 14 presented by Knapp et al. (1994) for a visualization of this behavior). For the ternary system IPA-W-MA, different additions of entrainer generated the pinch branches changes shown in **Figure 4**. For an entrainer-to-feed ratio Fr=0.5, **Figure 4** shows that the end points of the pinch branches have achieved a reflux ratio of zero, and even with more additions of entrainer it is not possible to find the pitchfork phenomena. This indicates that for the case study, extractive distillation alone is not feasible because it is not possible to find a continuous profile joining the striping-middle section to the rectifying section of the column in absence of the pitchfork phenomena (**see Figure 5,b).** However, if the rectifying section is replaced by a top decanter as is shown in **Figure 5**a, the LLE helps to join the top product water in a heterogeneous extractive distillation column profile (**Figure 5**,b). As the bottom product is mainly a mixture of IPA-MA, it is possible to separate this mixture and send the IPA back to the RDC and the MA to EDC (**Figure 5**a).

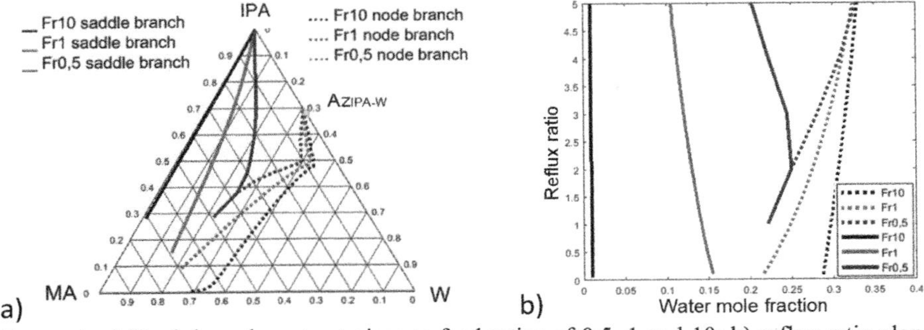

Figure 4: a) Pinch branches at entrainer to feed ratios of 0.5, 1 and 10; b) reflux ratio along the pinch branches

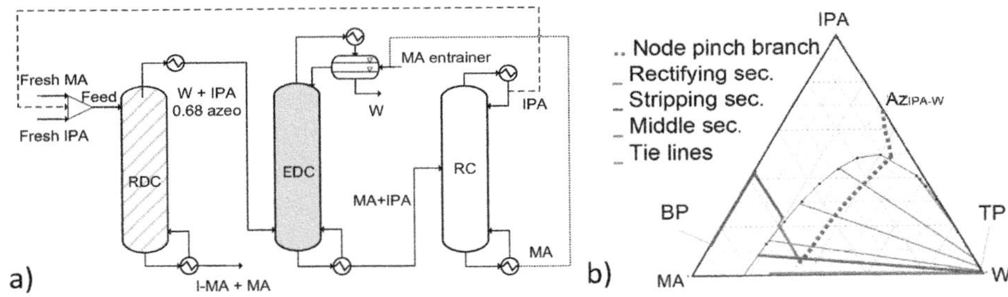

Figure 5: a) Heterogeneous extractive distillation column (EDC) with recovery column (RC) to overcome the azeotrope formed in the reactive distillation column (RDC) in **Figure 2**. b) Composition profiles of the EDC (BP=Bottom product; TP=top product).

2.5. Coupled system

Both the reactive distillation and the extractive distillation can be represented in the same quaternary composition space. Assuming the top product of the extractive column is almost pure water, it is possible to plot a mass balance line that starts at the water vertex and ends in the binary edge IPA-MA, see **Figure 6**a. As the bottom composition of the EDC is similar to the feed composition of the RDC, they can be recirculated, and with this, the recovery column is not necessary anymore, see **Figure 6**b. The bottoms binary composition depends on the amount of entrainer that has been added in the extractive distillation column, but this amount can be varied to achieve different stoichiometric lines in the reactive column.

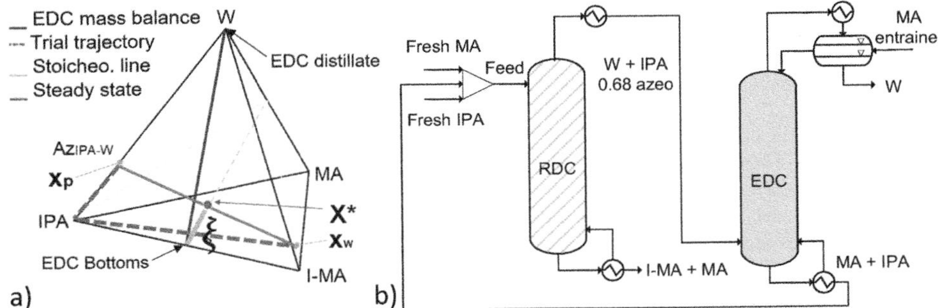

Figure 6: a) Representation of the overall extractive distillation mass balance and a feasible reactive distillation steady state, b) corresponding flowsheet.

3. Results and discussion

Based on the proposed methodology, the coupled arrangement showed in **Figure 6** reduces one of the columns in the **Figure 5** (the RC). On the other hand (see **Figure 6**), as the bottoms product of the EDC is similar to the RDC feed, and the top product of the RDC is similar to the EDC feed, the two totally integrated configurations shown in **Figure** 7a and b, are proposed. The main difference between both configurations is the IPA feed location in the reactive section (RS). When the feed stage is located above the reactive section (**Figure** 7a), the section is a single-feed reactive section and follows the behavior of a reactive residue curve. Such a behavior limits the quality of the products because the I-MA falls in a saddle node, and therefore it cannot be obtained as pure component. This drawback is avoided by feeding the alcohol at the bottom of the reactive section (**Figure** 7b). In this case a double-feed reactive section is generated (the section is defined by the location of the catalyst) and, in analogy to extractive distillation, the area of possible products is expanded.

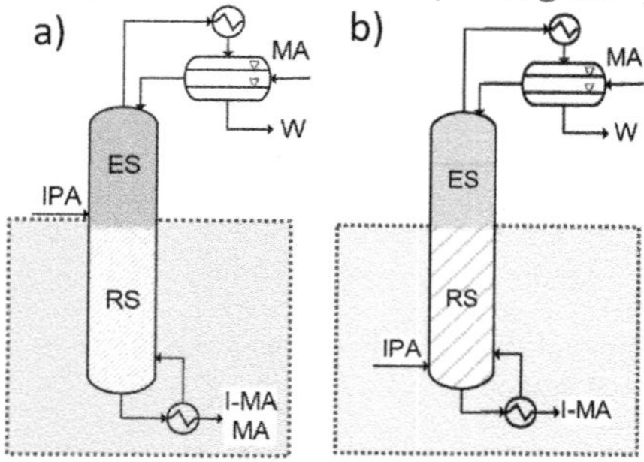

Figure 7: a) Extractive distillation with reactive stripping, b) reactive distillation with extractive rectification.

4. Conclusions

A method for designing an auto-extractive reactive distillation column has been presented. This method contributes to synthesis of configurations able to overcome azeotropes that reactive distillation alone cannot. The method was applied for the esterification of myristic acid with isopropanol. It was demonstrated that this system presents an auto-extractive phenomenon. Given the heterogeneous nature of such phenomenon, a decanter was considered by the methodology. An extension of the analysis of the statics was presented to include the auto-extractive phenomenon. The analysis was used to generate a process flowsheet that avoids the use of a secondary column in the recovery of the auto-entrainer. The extended analysis of the statics was also used to propose a configuration with a two-feed, heterogeneous, auto-extractive reactive distillation that ensures pure product streams. To our understanding, this configuration has not been proposed in previous studies for the production of IMA. A reduction in capital and energy cost is expected with the new configuration.

References

Almeida-Rivera, C.P., Swinkels, P.L.J., Grievink, J., 2004. Designing reactive distillation processes: Present and future. Comput Chem Eng 28, 1997–2020.

Bock, H., Wozny, G., Gutsche, B., 1997. Design and control of a reaction distillation column including the recovery system. Chemical Engineering and Processing 36, 101–109.

Cardona Alzate, C., Ortiz Sanchez, M., Pisarenko, Y.A., 2020. Reactive Separation for Process Intensification and Sustainability. CRC Press, New York.

Foucher, E.R., Doherty, M.F., Malone, M.F., 1991. Automatic Screening of Entrainers for Homogeneous Azeotropic Distillation. Ind Eng Chem Res 30, 760–772.

Knapp, J.P., Doherty, M.F., 1994. Minimum entrainer flows for extractive distillation : a bifurcation theoretic approach. AIChE Journal 40, 243–268.

Matsuyama, H., Nishimura, H., 1977. Topological and themodynamic classification of ternary vapor-liquid equilibria. J. Chem. Eng. Japan 10, 181–187.

Pisarenko, Y.A., Serafimov, L.A., Cardona, C.A., Efremov, D.L., Shuwalov, A.S., 2001. Reactive distillation design: analysis of the process statics. Reviews in Chemical Engineering 17, 253.

Antonis Kokossis, Michael C. Georgiadis, Efstratios N. Pistikopoulos (Eds.)
PROCEEDINGS OF THE 33rd European Symposium on Computer Aided Process Engineering (ESCAPE33), June 18-21, 2023, Athens, Greece
 http://dx.doi.org/10.1016/B978-0-443-15274-0.50064-0

Long-term Capacity Management of Pharmaceutical Manufacturing Networks

Simon B. Lindahl,[a,b] Deenesh K. Babi,[b] Gürkan Sin[a]

[a] *Process and Systems Engineering Center (PROSYS), Department of Chemical and Biochemical Engineering, Technical University of Denmark, Building 228 A, 2800 Kgs. Lyngby, Denmark*

[b] *PS API Manufacturing Development BIO, Novo Nordisk A/S, Novo Allé 1, 2880 Bagsværd, Denmark*

Abstract

This work presents a stochastic capacity and production planning model for network environments with multiple stages and shared manufacturing lines. Network-wide capacity decisions are modelled under uncertainties in demand and objective function coefficients with uncorrelated capacity expansions. A framework is presented for evaluating the model under uncertainty and it is applied to a case study from pharmaceutical manufacturing. First, the relevant uncertainties are defined through a sensitivity analysis and their effects on the optimal capacity decisions are explored. A sensitivity analysis of the input distributions is used to systematically study the required locations and total expansion amounts subject to different means, standard deviations, trends, and distribution types for the relevant uncertain demands. It is shown that capacity requirements are underestimated when the uncertainties are not considered, and that uncertainty sets with larger demand volatilities lead to larger capacity expansions.

Keywords: Capacity expansion, production planning, stochastic programming, sensitivity analysis

1. Introduction

Capacity planning (CP) is essential to ensure the long-term ability to deliver medicines and increase revenues from exclusive sales. Most pharmaceuticals are made in multi-stage processes where intermediates are produced for storage and each manufacturing line can produce multiple products. As such, production planning (PP) details should be considered in the CP problem to describe the allocation of time on shared manufacturing lines and the inventories of all materials (Marques et al. (2020)). Capacity decisions are resource intensive and irreversible, so the decisions should be made while considering relevant parameter uncertainties. An early work used 2-stage stochastic programming to define which new facilities to construct in a single-stage environment under clinical trial uncertainty (Levis & Papageorgiou (2004)). C. M. Marques et al. (2018) used Monte Carlo Simulations to find candidate process designs for a simplified flowsheet which were passed on to a second model that decided the optimal design per period for each product. Uncertainties in clinical trials and product demands were considered. Vieira et al (2020) applied a 2-stage stochastic programming model for design and scheduling of multiproduct plants under demand uncertainty to achieve a trade-off between profits and financial risk. Chou et al. (2014) used a dynamic programming model to determine capacity requirements subject to uncertain demands with constant trends and different volatilities. They conclude that higher demand volatilities increase the capacity requirements for problems with demand satisfaction. Sensitivity and uncertainty analysis can be used to identify how individual parameter uncertainties affect the model output, which is the sensitivity analysis problem, or what uncertainty in the model outputs can be expected subject to an uncertain input space, which is the uncertainty analysis problem (Saltelli et al. (2019)). Different methods have been developed to consider uncertainties directly in optimization models (Gabrel et al. (2014)) but even if a representation of uncertainty is considered that representation might itself be uncertain. This work therefore studies the effects of different uncertainty sets on the here-and-now capacity decisions to provide a recommendation on where and how much to expand capacity.

1.1. Problem Statement

The engineering context for the problem is defined as follows: a company wants to determine the required capacity expansions in its manufacturing network to provide a tradeoff between inventory risk and expansion cost. The studied system is a general network with multiple stages and multiple lines on each stage. Each line can produce a set of materials with given initial capacities. The goal is to determine the location and amount of capacity expansion as well as material inventory profiles subject to uncertain demands and capacity expansion costs. The company policy requires that the inventories be maintained above the safety level and as close to the strategic level as possible. Capacity expansions for different materials are assumed independent with costs defined through economies of scale.

2. Methodology

2.1. Model

The developed model follows the state-task-network (STN) structure (Kondili et. al (1993)) that has been expanded to include variable capacities and uncertainty through its formulation as a stochastic program. The main variables are inventory for material m at time t in scenario s, $I_{mts} \in \mathbb{R}^{\geq 0}$, safety inventory violation, $B_{mts} \in \mathbb{R}^{\geq 0}$, inventory distance above/below the strategic inventory, $I^{+}_{mts}/I^{-}_{mts} \in \mathbb{R}^{\geq 0}$, manufacturing amount of task i on line j, $P_{ijts} \in \mathbb{R}^{\geq 0}$, fractional time allocation $W_{ijts} \in \mathbb{R}^{\geq 0}$ and the choice of capacity level n, $Y_{ijn} \in \{0,1\}$. The allocation variable from the original STN has been relaxed and a variable capacity has been added to choose the required capacity level κ_{ijn} through the binary variable Y_{ijn} for all task-line combinations under demand and cost coefficient uncertainties. Capacity decisions are modelled as here-and-now variables that are chosen once for each task-line combination across all scenarios and time periods. This approach assumes independent capacity upgrades or in other words no synergies between capacity expansions. The PP details are wait-and-see variables under the assumption that the production scheduling on each line can be modified based on realized demand.

$$\textbf{minimize}\ C^{I} + C^{r} \tag{1}$$

$$\begin{aligned} I_{mts} = I_{m,t-1,s} &+ \sum_{i\in\mathbf{I}_m^+}\sum_{j\in\mathbf{J}_i} \rho^{+}_{im} P_{i,j,t-\tau_i,s} \\ &- \sum_{i\in\mathbf{I}_m^-}\sum_{j\in\mathbf{J}_i} \rho^{-}_{is} P_{i,j,t-\tau_i,s} + B_{mts} - B_{mt-1,s} - \delta_{mts}, \forall\, mts \end{aligned} \tag{2}$$

$$\sigma^{sfty}_{mt} \leq I_{mts}, \qquad \forall\, mts \tag{3}$$

$$1 - \frac{I_{mts}}{\sigma^{strat}_{mt}} \leq I^{-}_{mts}, \qquad \forall\, m \in \mathbf{M}_p, ts \tag{4}$$

$$\frac{I_{mts}}{\sigma_{mt}^{strat}} - 1 \leq I_{mts}^{+}, \qquad \forall\, m \in \mathbf{M}_p, ts \tag{5}$$

$$\sum_{i \in \mathbf{I}_j} \sum_{n \in \mathbf{P}_{ij}} \sum_{t'=t}^{t-\tau_i+1} W_{ijt'ns} \leq 1, \quad \forall\, jts \tag{6}$$

$$P_{ijts} \leq \sum_{n \in \mathbf{P}_{ij}} \kappa_{ijn} W_{ijtns}, \quad \forall j, i \in \mathbf{I}_j, ts \tag{7}$$

$$\sum_{n \in P_{ij}} Y_{ijn} \leq 1, \qquad \forall j, i \in \mathbf{I}_j \tag{8}$$

$$\sum_{t \in \mathbf{T}} W_{ijtns} \leq Y_{ijn} |\mathbf{T}|, \qquad \forall j, i \in \mathbf{I}_j, n \in \mathbf{P}_{ij}, s \tag{9}$$

$$\sum_{m \notin \mathbf{M}_p} \sum_{t \in \mathbf{T}} B_{mts} = 0, \forall s \tag{10}$$

$$C^I = \sum_{s \in \mathbf{S}} \alpha_s \sum_{m \in \mathbf{M}_p} \sum_{t \in \mathbf{T}} I_{sts}^{+} \upsilon_m^{I+} + I_{sts}^{-} \upsilon_m^{I-} + B_{mts} \upsilon_m^{I,bl} \tag{11}$$

$$C^r = \sum_{j \in \mathbf{J}} \sum_{i \in \mathbf{I}_j} \sum_{n \in \mathbf{P}_{ij}} Y_{ijn} \upsilon_{ij}^{cb} \kappa_{ijn}^{\ \ \nu_{ij}^{ce}} \tag{12}$$

The objective function is defined in Eq. (1) as the minimization of costs for inventory and capacity expansion. Eq. (2) sets the inventory balance with slack variables to allow inventories to go below the safety level in Eq. (3). Eqs. (4-5) define the distance below/above the strategic safety level. Eq. (6) limits the maximum allocation to each line, Eq. (7) constraints the maximum production based on the chosen capacity level, Eq. (8) ensures that only one capacity level is chosen, and Eq. (9) forces all capacity allocations for capacity levels not chosen to 0. Eq. (10) ensures that safety violations are only possible for products. Eq. (11) defines the inventory cost for not being at setpoint with three terms for being either above/below the strategic inventory or below the safety inventory. Individual parameters are used to define their relative importance where it generally applies that $\upsilon_m^{bl} > \upsilon_m^{I-} > \upsilon_m^{+}$. Finally, Eq. (12) defines the capacity expansion cost through a sum of individual capacity expansions in which an economy of scale exponent ν_{ij}^{ce} is used to ensure that few larger expansions are preferred over many small expansions.

2.2. Uncertainty Treatment

The model uses an uncertainty set composed of discrete scenarios for demands, base capacity costs and economy of scale exponents as input. Its solution provides here-and-now CP decisions which are shared across all scenarios and wait-and-see PP decisions for each individual scenario. The demands are time series, and the cost factors depend on the required changes to achieve a given capacity expansion. The available information on the cost factors depends on the amount of work that has been carried out to investigate expansion possibilities while for demands, both the levels, trends and volatilities are uncertain. In this work a 4-step solution framework is used: (1) baseline deterministic model solution with an uncertainty set of one scenario with expected values for all parameters, (2) sensitivity analysis of the deterministic model with different single-scenario uncertainty sets to determine parameters for further analysis, (3) baseline stochastic programming solution for an uncertainty set with scenarios sampled from expected parameter distributions, (4) sensitivity analysis of the stochastic program with different uncertainty sets sampled from different parameter distributions to provide recommendations in the presence of incomplete distribution information. Scenarios are created by combining samples from each parameter distribution. For demands, a sample is drawn for each time point and Figure 1 shows 100 example paths created by independent and identically distributed samples from a normal distribution with mean 1 and standard deviation 0.2. The paths with highest, lowest, and median mean demands are highlighted and the right plot shows the distribution of mean demands, which approaches normal distribution as sampling number increases.

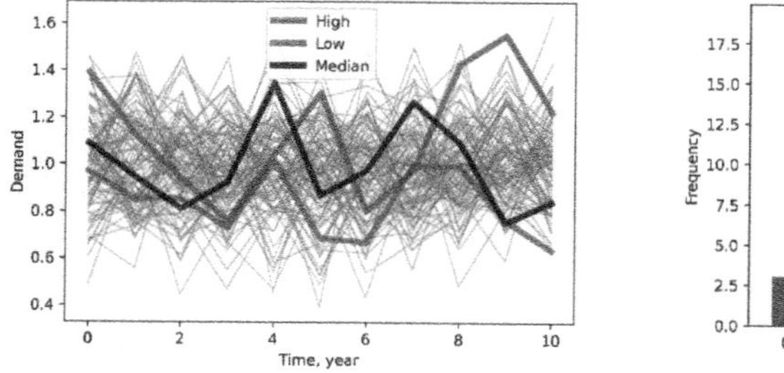

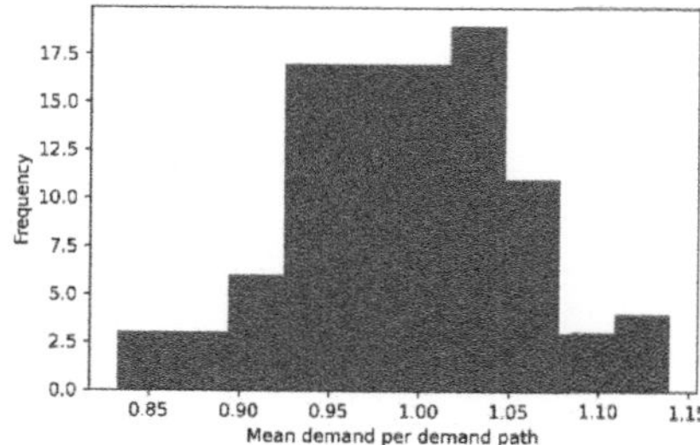

Figure 1: Demand paths (left) and mean demand distribution (right)

3. Case Study

The proposed model is applied to a case study abstracted from a real pharmaceutical manufacturing network consisting of three stages with three lines on the first stage, five lines on the second stage and two lines on the third stage. Six products are produced, and the total number of materials is 14. The time horizon is 10 years divided in a discrete yearly time grid. Capacity expansions of 0%, 10% or 20% were allowed.

3.1. Baseline Deterministic Solution

The baseline deterministic model solution was created by solving the stochastic program with the nominal scenario. In this case, no capacity expansions are required.

3.2. Sensitivity Analysis of Deterministic Model

The sensitivities to product demands were tested by sequentially solving the deterministic model with demands increased by a factor 10% or 20% at all time points for one product at a time. No demand decreases were tested as no capacity expansions were seen in the baseline solution and lower demands require less capacity. The sensitivities to the capacity base cost υ_{ij}^{cb} and the capacity exponent v_{ij}^{ce} were tested by varying them around their nominal point for each demand increase of each product. Table 1 shows the total capacity expansion for each combination of increased demand and cost parameters. The results show relevant demand uncertainties for products B, C and F. Both capacity cost uncertainties are relevant for demand increases of 20%.

Table 1: Total capacity expansion requirement for increased demand of each product and different cost parameters (change from nominal value given in parenthesis)

Demand Factor	Capacity Base cost	Capacity Exponent	Product demand varied					
			A	B	C	D	F	G
	υ_{ij}^{cb}	v_{ij}^{ce}	Total capacity expansion, %					
+10%	150	0.8	0	10	10	0	0	0
+10%	125 (-25)	0.8	0	10	10	0	0	0
+10%	175 (+25)	0.8	0	10	10	0	0	0
+10%	150	0.7 (-0.1)	0	10	10	0	0	0
+10%	150	0.9 (+0.1)	0	10	10	0	0	0
+20%	150	0.8	0	10	20	0	0	0
+20%	125 (-25)	0.8	0	10	20	0	10	0
+20%	175 (+25)	0.8	0	10	20	0	0	0

+20%	150	0.7 (-0.1)	0	40	20	0	10	0
+20%	175 (+25)	0.9 (+0.1)	0	10	20	0	0	0

3.3. Baseline Stochastic Programming Solution

Without specific distribution information, the costs were assumed uniformly distributed with the bounds from Table 1 and the demands for products B, C and F at each time point were created by multiplying the nominal demand δ_{mt}^{nom} with a sample from a uniform distribution with a lower/upper bound of 0.8/1.2. An uncertainty set with 100 equiprobable scenarios was created by random sampling, and the stochastic program was solved which resulted in no capacity expansions. Also, the deterministic model was solved for each scenario, giving in an average total expansion of only 2.4% distributed over two different product-line combinations which indicates that capacity expansions might be achieved through optimizations of the existing process and equipment.

3.4. Sensitivity Analysis of Uncertain Parameter Distributions

A sensitivity analysis was carried out on the demand distributions to identify locations in scope for capacity expansions and to test the robustness of the recommendations. A computational study with 54 uncertainty sets was defined with variations along the demand dimensions of distribution type D (uniform, normal), base level μ (1.0, 1.1, 1.2), standard deviation σ (baseline, +25%, +50%) and trend ψ (1.0, 1.01, 1.02). The demand in each period was created as $\delta_{mts} = \delta_{mt}^{nom} \cdot \psi^t \cdot D(\mu, \sigma)$ and the stochastic program was solved for each uncertainty set while the deterministic model was solved for each scenario in each set. Figure 2 shows the total capacity expansion for each uncertainty set based on the overall mean and standard deviation of all the uncertain demands combined. Higher mean demands require larger expansions and the middle region for mean demands is affected by larger demand volatilities. The uniform and normal distributions produce similar results although larger expansions are needed in three cases for uniformly distributed demands. The deterministic model results are comparable to those of the stochastic model, but for some of the uncertainty sets it underestimates the capacity expansion due to its larger flexibility which contrasts with its higher capacity increase in section 3.3. Now, larger capacity expansions are preferred by the stochastic program to benefit the high demand scenarios while lower demand scenarios will experience overcapacity. The effect of demand trends at a given base demand can be seen by comparing the three points on the upward diagonal starting from any point with mean demand in the set (1.0, 1.1, 1.2). As such, demand trends become important if the base demand level is also increased.

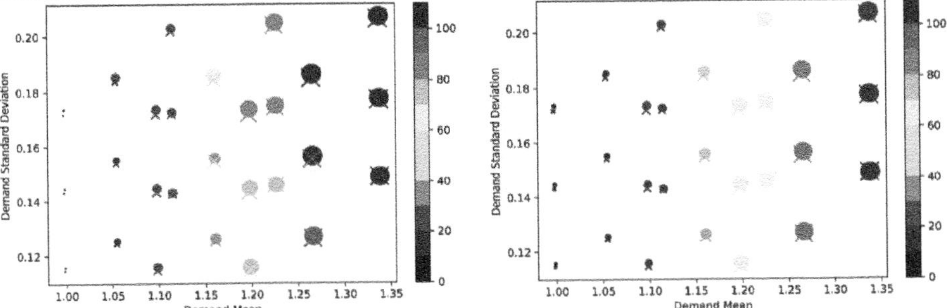

Figure 2: Total capacity expansion for uniform (o) and normal (x) demands at different means and standard deviations. Stochastic model (left) and deterministic model (right).

The model determines the locations that should be expanded, and we suggest to use this information to guide investigations into optimization or retrofit projects. Table 2 shows the frequency of expansion choices based on the 54 uncertainty set solutions of the stochastic program. Table 2 shows that five lines may require capacity expansions if demand increases are realized, and it provides the product on that line that is relevant to expand. Naturally, an expansion for one product on a line might affect the other products as well but this interaction can be included in subsequent solutions of the provided model.

Table 2: Frequency of expansion size choice across parameter combinations

Line	A1	A2	B4	C1	C2
Product	B	C	F	B	C
0%	0.57	0.44	0.67	0.19	0.13
10%	0.0	0.26	0.17	0.37	0.37
20%	0.43	0.3	0.17	0.44	0.50

4. Conclusion and Perspectives

A stochastic programming model was presented to determine the location and size of capacity expansions under uncertainty. A framework was used to find and explore the input uncertainties that influence the capacity decisions. The relevant parameters were determined through sensitivity analysis of a deterministic model. Then, a sensitivity analysis of the stochastic program with uncertainty sets from different distributions was used to study the decisions subject to incomplete distribution information. The framework was applied to a case study which shows that the capacity expansion is underestimated by the deterministic model and that higher demand volatilities increase the expansion need at moderate demand increases. Generally, the same expansion was needed regardless of the distribution type but in some cases, the requirements were larger for uniformly versus normally distributed demands. The proposed framework can be used to identify where to expand capacity to meet long-term demand given limited information on the input uncertainties. As such, the results can be used to identify the need for retrofit and optimization projects as well as determine which product demands that should be monitored more closely since their increase will require capacity expansions.

References

Y. C. Chou, W. C. Sung, G. Lin & J. Jahn, (2014). “A comparative study on the performance of timing and sizing models of capacity expansion under volatile demand growth and finite equipment lifetime”, Computers & Industrial Engineering, 76, 98-108.

V. Gabrel, C. Murat, A. Thiele, (2014), “Recent advances in robust optimization: An overview”, Europena Journal of Operational Research, 235, 471-483.

A. A. Levis & L. G. Papageorgiou, (2004), “A hierarchical solution approach for multi-site capacity planning under uncertainty in the pharmaceutical industry”, Computers & Chemical Engineering, 28(5), 707-725

C. M. Marques, S. Moniz and J. P. de Sousa, (2018). "Strategic decision-making in the pharmaceutical industry: A unified decision-making framework." Computers & Chemical Engineering, **119**, 171-189

C. M. Marques, S. Moniz, J. P. de Sousa, A. P. Barbosa-Povoa & G. Reklaitis, (2020), “Decision-support challenges in the chemical-pharmaceutical industry: Findings and future research directions”, Computers & Chemical Engineering, 134, 106672.

A. Saltelli, K. Aleksankina, W. Becker, P. Fennell, F. Ferretti, N. Holst, S. Li & Q. Wu, (2019), “Why so many published sensitivity analyses are false: A systematic review of sensitivity analysis practices”, Environmental Modelling & Software, 114, 29-39.

M. Vieira, H. Paulo, T. Pinto-Varela and A. P. Barbosa-Póvoa, (2020), “Assessment of financial risk in the design and scheduling of multipurpose plants under demand uncertainty”, International Journal of Production Research, 59 (20), 6125-6145

Antonis Kokossis, Michael C. Georgiadis, Efstratios N. Pistikopoulos (Eds.)
PROCEEDINGS OF THE 33rd European Symposium on Computer Aided Process Engineering (ESCAPE33), June 18-21, 2023, Athens, Greece
 http://dx.doi.org/10.1016/B978-0-443-15274-0.50065-2

Modeling of a trickle bed reactor: Study of arabinose oxidation on gold catalyst extrudates

Mouad Hachhach[a,*], Vincenzo Russo[a,b], Dmitry Yu. Murzin[a], Tapio Salmi[a]

[a] *Laboratory of Industrial Chemistry and Reaction Engineering (TKR), Johan Gadolin Process Chemistry Centre (PCC), Åbo Akademi University, Åbo-Turku FI-20500, Finland*

[b] *Università di Napoli Federico II, Chemical Sciences, Complesso Universitario Monte S. Angelo, Napoli IT-80126, Italy*

**Corresponding Author:mouad.hachhach@abo.fi*

Abstract

A comprehensive trickle bed reactor model for simulation purposes was developed in this work As an industrially relevant case study, the model was applied on selective sugar oxidation to sugar acid on gold catalyst extrudates. The obtained results revealed that the maximum conversion for extrudates is only 25% for a reactor length of 0.5 m and diameter of 1 cm, which is due to slow mass transfer inside the catalyst layer and external mass transfer limitations. Variation of the residence time and the reactor length revealed that a reactor length of 2m minimum is required to achieve complete conversion of the sugar (arabinose) to the corresponding sugar acid (arabinonic acid).

Keywords: Trickle Bed Reactor, Oxidation, Extudates, gProms, Dynamic model.

1. Introduction

Trickle bed reactors (TBR) are continuous packed beds where liquid moves in downwards through a catalyst packing whereas gas can move either in a downward or upward direction (Lange et al., 2004). These reactors are solid-gas-liquid contacting devices in which the gas-liquid stream is in continuous contact with the catalyst particles (Hamdan and Abdullah, 2022). Trickle beds are the classical workhorses of chemical industry: in fact these reactors have been used extensively in a wide range of chemical industries from petro-chemical and refinery processes such as hydrodesulfurization and hydrogenation, to the oxidation of biomass and organics in wastewaters, volatile organic compound abatement in air pollution control, enzymatic reactions and bio-methanation (Aleksandrov et al., 2022; Dudukovic et al., 1999; Jønson et al., 2022), in addition to the oxidation of sugar (glucose) (Diallo et al., 2017; Tsukamoto et al., 1982). Trickle beds offer several advantages: simplicity in operation, flexibility for a variety of materials, high catalyst loading per unit volume, and low capital and operating costs as well as lower energy consumption and higher gas conversion rate and higher productivities compared to tank and bubble column reactors (Klasson et al., 1992; Li et al., 2013; Ranade et al., 2011). In contrast to slurry reactors, trickle beds enable safe continuous operation, whereas slurry reactors are in practise limited to batch and semibatch operation.

In this work, the goal was to develop a general and comprehensive trickle bed reactor model, and use it to study the catalytic oxidation of arabinose on gold based catalyst extrudates. The model can be used as framework for the simulation of various reaction systems as well as a tool to assist process design, scale-up and life cycle assessment for arbitrary reaction systems carried out in trickle beds.

2. Materials and methods

The trickle bed reactor model was developed to be as general as possible. The concentration profiles inside the reactor tube and within the catalyst particles were computed by solving simultaneously the mass and energy balances of the components including rate equations (kinetic laws) as well as semi-empirical correlations for pressure drop and fluid dynamics. It was assumed that all the reactions take place inside the catalysts particles only, therefore the rate term is present only in the mass balance of the solid phase, and the active metal of the catalyst is assumed to be distributed evenly inside the catalyst particles. The kinetics of sugar oxidation from batch experiments (Herrero Manzano et al., 2021) was used as an input for the model and realistic values of physical properties were selected. The most essential parameters are summarized in Table 1.

Table 1. List of parameters that characterize the simulation

Parameters	Value	Unit	Parameters	Value	Unit	Parameters	Value	Unit
ρ_{cat}: Catalyst density	400	Kg/m^3	Reactor radius	5	cm	Reactor length	0.5	m
ρ_L: Density of liquid phase	950	Kg/m^3	CO_2S :saturation concentration of O_2	0.103	mol/m^3	Effective diffusivity (D_{eff}) of O_2	2.403 10^{-9}	m^2/s
m(O_2): Molar mass	32	g/mol	Pe_G: Péclet number of gas phase	100		k: rate constant	53.99	m^6/(s mol.kgcat)
σ_{H2O}: Water surface tension	72.8 10^{-3}	N/m	Pe_L: Péclet number of liquid phase	10		Effective diffusivity (D_{eff}) of Arabinose	2.474 10^{-9}	m^2/s
η_{H2O}: Water viscosity	0.2 10^{-3}	Pa s	Effective diffusivity (D_{eff}) of arabinoic acid	9.851 10^{-9}	m^2/s	Arabinose diffusivity in water (D_L)	4 10^{-8}	m^2/s
a*:limiting activity factor	0.47282							

2.1. Mass balances

The mass balance of the gas-phase components displayed in equation (1) affirm that the accumulation term is the sum of gas-liquid mass transfer with a negative sign, axial and radial dispersion alongside the convection terms (Russo et al., 2015).

$$\frac{\partial Ci,g(t,z,r)}{\partial t} = \frac{-k_{i,L}a}{\varepsilon_G}\left(C_{i,L}^{*} - C_{i,L}\right) - \frac{\frac{\partial\left(u_G(z,r)Ci,g(t,z,r)\right)}{\partial z}}{\varepsilon_G} + D_{z,g}(z,r)\frac{\partial^2 C_{i,g}(t,z,r)}{\partial z^2} + D_{r,g}(z,r)\left(\frac{\partial^2 Ci,g(t,z,r)}{\partial r^2} + \frac{1}{r}\frac{\partial Ci,g(t,z,r)}{\partial r}\right) \quad (1)$$

The liquid-phase mass balance equation is shown in equation (2): the accumulation is the sum of gas-liquid mass transfer term with a positive sign in opposition to the gas phase, in addition to the convection and axial and radial dispersion terms, and finally the mass transfer flux from/to the solid catalyst phase as represented in the last term.

$$\frac{\partial Ci,L(t,z,r)}{\partial t} = \frac{k_{i,L}a}{\varepsilon_L}\left(C_{i,L}^{*} - C_{i,L}\right) - \frac{uL}{\varepsilon_L}\frac{\partial(Ci,L(t,z,r))}{\partial z} + D_{z,L}\frac{\partial^2 C_{i,L}(t,z,r)}{\partial z^2} + D_{r,L}\left(\frac{\partial^2 Ci,L(t,z,r)}{\partial r^2} + \frac{1}{r}\frac{\partial Ci,L(t,z,r)}{\partial r}\right) - \frac{s\,D_{eff,i}}{\varepsilon_L\,Rp}\frac{\partial Ci,s(t,z,r,r_p)}{\partial r_p}\Big|_{r_p = R_p} \quad (2)$$

For the solid catalyst phase, the concentration field depends on the effective diffusivity, catalyst shape factor and the holdups of the liquid and solid phases. Moreover, the reaction term is included comprising the catalyst density, ratio of holdups, stoichiometric coefficients and reaction rate, as shown in equation (3) below.

$$\frac{\partial Ci,s(t,z,r,r_p)}{\partial t} = \frac{D_{eff,i}}{\varepsilon_p}\left(\frac{\partial^2 Ci,s(t,z,r,r_p)}{\partial r_p^2} + \frac{s}{r_p}\frac{\partial Ci,s(t,z,r,r_p)}{\partial r_p}\right) + \rho_{cat}\sum\left(\vartheta_{i,j}r_j(t,z,r,r_p)\frac{\varepsilon_L}{\varepsilon_p}\right) \quad (3)$$

2.2. Physical properties and reactor parameters

The liquid viscosity, which is required to calculate the Reynolds number, was included in the model, the correlation adopted in equation (4) was based on arabinose hydrogenation experiments (Sifontes Herrera et al., 2016) and taken into account the dependence of the viscosity on both temperature and arabinose concentration (x_a).

$$\eta_{fluid} = 10^{-3}\exp\left(\frac{1.54x_a}{T_{in,L}-273} + 3.81.10^{-4}x_a^2 - 1.10.10^{-2}x_a - 2.85 + \frac{194}{T_{in,L}-273} - \frac{3890}{(T_{in,L}-273)^2}\right) \quad (4)$$

The gas-liquid mass transfer coefficient is taken into account in the mass balance as shown previously in equations (1) and (2). The equation 6 adopted from Ranade et al. (2011) was used in this study.

$$k_La(z,r) = \gamma 0.45\left(\frac{DL}{dK^2}\right)\left(X_G(z,r)^{0.65}ReL^{1.04}WeL^{0.26}ScL^{0.65}\right)\left(\frac{asp\ dK}{1-\varepsilon_p}\right)^{0.325} \quad (5)$$

The axial dispersion coefficient was calculated by using the Péclet number as shown in equations (7) and (8), while the radial dispersion coefficient was taken equal to one third of the axial one as shown in the equations (9) and (10) as has been reported previously

(Russo et al., 2015). According to our previous work by (Durante et al., 2014; Kilpiö, 2013) the order of magnitude of the liquid phase Péclet number was 2-10, while for the gas phase the Péclet number can go up to 100. For numerical simulations, the Péclet number for liquid-phase was taken 10 and for the gas phase 100.

$$Dz = uL.\frac{L}{Pe} \quad (6) \qquad DzG(z,r) = uG(z,r).\frac{L}{PeG} \quad (7)$$

$$Dr = \frac{Dz}{3} \quad (8) \qquad DrG = \frac{DzG(z,r)}{3} \quad (9)$$

The shape factor is evaluated by equation (10) proposed by Salmi et al., (2019), where V_p and A_p are respectively the volume and surface area of the catalyst particle.

$$s = 1 + R_p \frac{A_p}{V_p} \quad (10)$$

2.3. Kinetics of the case study

The case study is the selective oxidation of arabinose to arabinoic acid on gold catalyst (Herrero Manzano et al., 2021; Kusema et al., 2012, 2010). The kinetics has been described by classical Langmuir-Hinshelwood rate expression as shown in the equations (12) and (13). The kinetic expression and parameters values were adapted from the work of Herrero Manzano et al. (2021).

$$Arabinose + \frac{1}{2}O_2 \rightarrow Arabinoic\ Acid \quad (11) \qquad \frac{da}{dt} = -kC_{00}{}^{\alpha-1}(a - a*)^{\alpha} \quad (12)$$

$$r(t,z,r,r_p) = a^2 \frac{K(t,z,r,rp)C_{O2,s}(t,z,r,r_p)C_{Ara,s}(t,z,r,r_p)}{1+K_{ara}C_{Ara,s}(t,z,r,r_p)+K_{O2}C_{O2,s}(t,z,r,r_p)} \quad (13)$$

The model equations were implemented in the software gPROMS, which was chosen due to its high convergence speed and its user-friendliness: In fact, in this case study the convergence time was in order of 5-10 minutes. The partial differential equations (PDEs) describing the mass balances (1)-(3) were converted to ordinary differential equations (ODEs) by discretization of the radial and axial coordinates with finite differences. A stiff differential equation algorithm was used for solving the ODEs, the initial value problem.

3. Results and discussion

The obtained results are summarized in Figure 1. The simulations reveal that the maximum conversion for the catalyst extrudates at the reactor output is only 25% for a reactor length of 0.5 m and diameter of 1 cm, which can be due to slow mass transfer inside the catalyst particles but also the existence of external mass transfer limitations, which can be surmounted by using smaller particle sizes, slurry reactor technology or structured catalysts, such as monoliths or solid foams. The results show that the radial position is not affecting the sugar conversion as illustrated by Figure 1, where the conversion is practically the same for various radial positions across the reactor tube, in contrast to the axial position, where the sugar conversion increases as the length of the reactor increases.

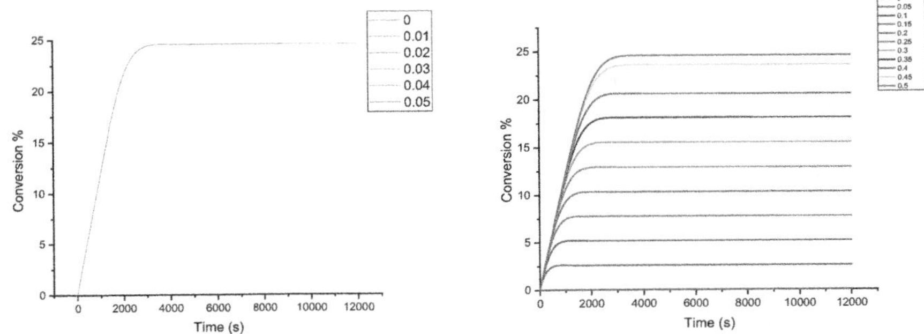

Figure 1: Conversion rate at various axial (left) and radial (right) positions

In order to increase the conversion, the residence time and the reactor length were varied and the conversion at the reactor outlet was computed as displayed in Figure 2, which confirms that a reactor length of 2m minimum is required to achieve complete conversion of the sugar (arabinose) to the corresponding sugar acid (arabinonic acid).

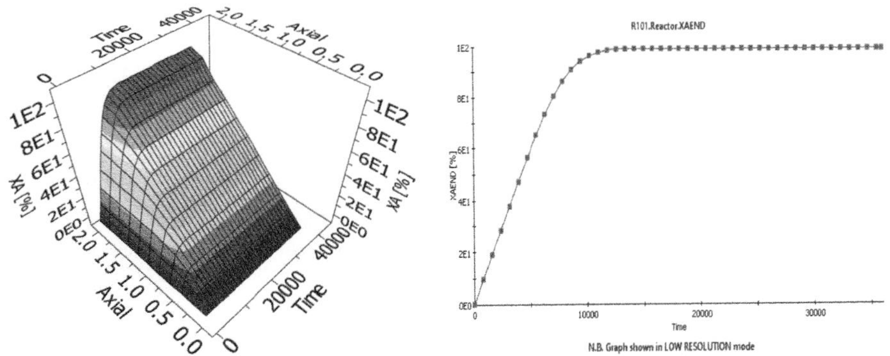

Figure 2: Effect of Simultaneous increase of reaction time and length of the reactor on the Conversion Rate (left). Conversion rate at the outlet of the scaled-up reactor (2m) (right)

The obtained results demonstrate quite interesting modelling stability that might pave the way to the investigation of the scale-up of various reaction systems, although more accurate simulations would require further kinetic and flow experiments using trickle bed reactors.

Acknowledgment

This work is partially financed by the Academy of Finland, the Academy Professor grant 319002 (T. Salmi). The financial support obtained from PCC, Finnish Cultural Foundation (SKR) , Walter Ahlstrom Foundation, and from The Society of Swedish Literature in Finland (SLS) is gratefully acknowledged (M. Hachhach).

References

Aleksandrov, P. V, Reshetnikov, S.I., Bukhtiyarova, G.A., Noskov, A.S., 2022. Deep hydrodesulfurization of gas oils with high sulfur content: Experiment and kinetic modeling. Chem. Eng. J. 446, 137059. https://doi.org/https://doi.org/10.1016/j.cej.2022.137059

Diallo, M.M., Derrien, E., Marion, P., Perret, N., Pinel, C., Besson, M., 2017. Base-free aerobic oxidation of glucose to glucaric acid over Pt-Au/ZrO2 catalysts in batch and trickle-bed reactors.

Dudukovic, M.P., Larachi, F., Mills, P.L., 1999. Multiphase reactors - revisited. Chem. Eng. Sci. 54, 1975–1995. https://doi.org/10.1016/S0009-2509(98)00367-4

Durante, D., Kilpiö, T., Suominen, P., Herrera, V.S., Wärnå, J., Canu, P., Salmi, T., 2014. Modeling and simulation of a small-scale trickle bed reactor for sugar hydrogenation. Comput. Chem. Eng. 66, 22–35. https://doi.org/10.1016/j.compchemeng.2014.02.025

Hamdan, H.Y., Abdullah, G.H., 2022. A novel trickle bed electrochemical reactor design for efficient hydrogen peroxide production. Chem. Eng. Process. - Process Intensif. 181, 109123. https://doi.org/https://doi.org/10.1016/j.cep.2022.109123

Herrero Manzano, M., Eranen, K., Freites Aguilera, A., Warnå, J., Franz, S., Peurla, M., Serna, J.G., Murzin, D., Salmi, T., 2021. Interaction of intrinsic kinetics, catalyst durability and internal mass transfer in the oxidation of sugar mixtures on gold nanoparticle extrudates. Ind. Eng. Chem. Res. 60, 6483–6500. https://doi.org/10.1021/acs.iecr.0c05305

Jønson, B.D., Tsapekos, P., Tahir Ashraf, M., Jeppesen, M., Ejbye Schmidt, J., Bastidas-Oyanedel, J.-R., 2022. Pilot-scale study of biomethanation in biological trickle bed reactors converting impure CO2 from a Full-scale biogas plant. Bioresour. Technol. 365, 128160. https://doi.org/https://doi.org/10.1016/j.biortech.2022.128160

Kilpiö, T., 2013. Mathematical modeling of laboratory scale three-phase fixed bed reactors. Doctoral thesis, Åbo Akademi University.

Klasson, K.T., Ackerson, M.D., Clausen, E.C., Gaddy, J.L., 1992. Bioconversion of synthesis gas into liquid or gaseous fuels. Enzyme Microb. Technol. 14, 602–608. https://doi.org/10.1016/0141-0229(92)90033-K

Kusema, B.T., Campo, B.C., Mäki-Arvela, P., Salmi, T., Murzin, D.Y., 2010. Selective catalytic oxidation of arabinose - A comparison of gold and palladium catalysts. Appl. Catal. A Gen. 386, 101–108. https://doi.org/10.1016/j.apcata.2010.07.037

Kusema, B.T., Mikkola, J.-P.P., Murzin, D.Y., 2012. Kinetics of L-arabinose oxidation over supported gold catalysts with in situ catalyst electrical potential measurements. Catal. Sci. Technol 2, 423–431. https://doi.org/10.1039/c1cy00365h

Lange, R., Schubert, M., Dietrich, W., Grünewald, M., 2004. Unsteady-state operation of trickle-bed reactors. Chem. Eng. Sci. 59, 5355–5361. https://doi.org/10.1016/J.CES.2004.09.007

Li, D., Li, Z., Li, W., Liu, Q., Feng, Z., Fan, Z., 2013. Hydrotreating of low temperature coal tar to produce clean liquid fuels. J. Anal. Appl. Pyrolysis 100, 245–252. https://doi.org/10.1016/J.JAAP.2013.01.007

Ranade, V., Chaudhari, R., Gunjal, P.R., 2011. Trickle Bed Reactors, Trickle Bed Reactors. Elsevier. https://doi.org/10.1016/C2009-0-16553-4

Russo, V., Kilpiö, T., Di Serio, M., Tesser, R., Santacesaria, E., Murzin, D.Y., Salmi, T., 2015. Dynamic non-isothermal trickle bed reactor with both internal diffusion and heat conduction: Sugar hydrogenation as a case study. Chem. Eng. Res. Des. 102, 171–185. https://doi.org/10.1016/j.cherd.2015.06.011

Salmi, T., Mikkola, J.-P., Wärnå, J., 2019. Chemical Reaction Engineering and Reactor Technology, Chemical Reaction Engineering and Reactor Technology. Chapman and Hall/CRC. https://doi.org/10.1201/9781315200118

Sifontes Herrera, V.A., Rivero Mendoza, D.E., Leino, A.R., Mikkola, J.P., Zolotukhin, A., Eränen, K., Salmi, T., 2016. Sugar hydrogenation in continuous reactors: From catalyst particles towards structured catalysts. Chem. Eng. Process. - Process Intensif. 109, 1–10. https://doi.org/10.1016/J.CEP.2016.07.007

Tsukamoto, T., Morita, S., Okada, J., 1982. Oxidation of Glucose on Immobilized Glucose Oxidase in a Trickle-Bed Reactor : Effect of Liquid-Solid Contacting Efficiency on the Global Rate of Reaction. Chem. Pharm. Bull. (Tokyo). 30, 1539–1549. https://doi.org/10.1248/cpb.30.1539

Antonis Kokossis, Michael C. Georgiadis, Efstratios N. Pistikopoulos (Eds.)
PROCEEDINGS OF THE 33rd European Symposium on Computer Aided Process Engineering (ESCAPE33), June 18-21, 2023, Athens, Greece
 http://dx.doi.org/10.1016/B978-0-443-15274-0.50066-4

Numerical simulation of flow in porous media using the SPH method and Ergun force

Carlos E. Alvarado-Rodriguez,[a,b] Lamberto Diaz-Damacillo,[c] Eric Plaza,[d,e]

[a]*Departamento de Ingeniería Química, DCNE, Universidad de Guanajuato, Noria Alta S/N, Guanajuato, 3605, México.*
[b]*Consejo Nacional de Ciencia y Tecnología, Avenida Insurgentes Sur 1582, Crédito Constructor, Ciudad de México, 03940, México.*
[c]*Departamento de Ciencias Básicas, Universidad Autónoma Metropolitana Azcapotzalco (UAM-A), Av. San Pablo 180, Ciudad de México 02200, Mexico.*
[d]*ESPOL Polytechnic University, Escuela Superior Politécnica del Litoral, ESPOL, Departamento de Física, Laboratorio de Fluidos, km. 30.5, Vía Perimetral Guayaquil, 09-01-5863, Ecuador.*
[e]*Laboratorio de Física de Fluidos y Plasmas, Instituto Venezolano de Investigaciones Científicas (IVIC), Altos de Pipe, 21827, Venezuela.*

Abstract

This work presents the implementation of the Ergun force term on the DualSPHysics code, to simulate flow in porous media using a representative element volume (REV). In the study of microscopic seepage and fluid-solid interactions, a typical approach is a pore-scale simulation due to its precision; however, it involves a high computation cost, making it more suitable to adopt the REV scale. Here are presented nine simulations comparing the velocity, at the outlet of the system, obtained by a pore-scale geometry and the Ergun force term and REV simulations using SPH. The results have a good fit between the pore-scale simulations and the Ergun force term simulations, where the lowest and highest errors are 2.92×10^{-3} and 1.43×10^{-2}, respectively. These results present novel and alternative simulations in the study of the flow in porous media using the SPH method where the pore-scale geometry is usually used.

Keywords: Porous media, SPH, Ergun force, CFD.

1. Introduction

Smoothed particle hydrodynamics (SPH) is a fully Lagrangian particle method introduced by Lucy (1977). Since then, due to its ease of implementation and flexibility, the method has spread into numerous branches of computational physics and engineering. SPH has a Lagrangian nature making enabling simulations of complex systems that would be very difficult to carry out with the use of more conventional grid-based schemes. In SPH, the discretization is carried out by computing an interpolating kernel function that uses the properties of the surrounding particles over the studied one to compute its physical properties. Due to the characteristics of discretization in SPH, the dynamic of the fluid volume element is controlled by the local variations of pressure and density, making it possible to simulate the direct inclusion of boundaries, treatment of the free surface, and recently the treatment of multiphase flows. SPH can be used in several branches of science, including the flow in porous media (Tartakovsky et al., 2007; Kunz et al., 2015; Klapp et al., 2022), which presents a great interest to the industry due to its connection with crude oil extraction, the dynamics of flow in aquifers, agricultural irrigation processes, and into materials that can be used as filters. In general, the dynamic of fluid is controlled by the principles of conservation of momentum, energy, and mass. These differential equations do not allow analytical solutions for this kind of problem, which opens the way for the introduction of numerical simulations as a tool for prediction. In the case of flow in porous media, Darcy's Law has been widely used, when the Reynolds number is small. This equation, obtained through experimental measurements, approximates the behavior of the fluid in the porous media, using empirical relationships that associate the permeability of the medium and geometric variables. The purpose of this work is to extend the empirical relation of Darcy and provide a more accurate representation of flow in porous media. To do this, the Ergun term was added to the movement equation, which includes a friction expression and allows for the quantification of the pressure drop and energy required for pumping the flow, (Ergun, 1952). A novel SPH model was developed to simulate flow in porous media, making use of the Ergun force term in the momentum equation. The model avoids the need to depict the internal shapes of the porous media in the simulation and instead uses parameters that describe the media's specifications. This enables the simulation of large length scales and provides a better understanding of the internal flow in porous media. The work is structured as follows: Section 2 presents the complete SPH numerical model. Section 3 validates the velocity profile in a Poiseuille flow. Section 4 compares numerical cases to demonstrate the effectiveness of using the Ergun force term in the momentum equation. Finally, the conclusions of the work are presented in Section 5.

2. The SPH Numerical Model

Smoothed Particle Hydrodynamics (SPH) is a meshless Lagrangian method that recently has been applied to simulate a wide range of applications within the field of Computational Fluid Dynamics.

For viscous incompressible flows, the governing equation is the Navier–Stokes expressed in SPH form as

$$\frac{d\mathbf{v}_a}{dt} = -\sum_b m_b\left(\frac{p_a + p_b}{\rho_a\rho_b} + \Gamma\right)\nabla_a W_{ab} + \mathbf{g} \tag{1}$$

where ρ is the density, p pressure, $\mathbf{v}$ is the velocity vector, W is the kernel function, $\mathbf{g}$ is the gravity acceleration and Γ is the viscosity dissipative term. In a standard SPH formulation, where the pressure is given as a function of the density, local variations of the pressure gradient may induce local density fluctuations in the flow. Therefore, flow is modelled by an artificial fluid that is slightly compressible. The particle mass remains constant and only is associated with density fluctuations. Such density fluctuations are calculated by solving the continuity equation expressed in Eq. (2). The dynamical pressure p, is calculated using the relation expressed in Eq. (3).

$$\frac{d\rho_a}{dt} = \sum_{b=1}^{N} m_b(\mathbf{v}_a - \mathbf{v}_b)\cdot\nabla_a W_{ab}, \tag{2}$$

$$p = p_0\left[\left(\frac{\rho}{\rho_0}\right)^{\gamma} - 1\right] \tag{3}$$

where γ=7, $p_0 = c_0^2\rho_0/\gamma$, ρ_0 is a reference density, and c_0 is the sound speed at the reference density.

The viscous stress tensor Γ is described using the laminar viscosity model and the Sub Particle Scale (SPS) model, proposed by Darlympe & Rogers, (2006) and reported in the DualSPHysics code as

$$\Gamma_{ab} = \sum_b m_b \left(\frac{4v_0 \boldsymbol{r}_{ab} \cdot \nabla_a W_{ab}}{(\rho_a + \rho_b)(r^2{}_{ab} + \eta^2)} \right) \mathbf{v}_{ab} + \sum_b m_b \left(\frac{\vec{\tau}^{j}_{ab}}{\rho_b{}^2} + \frac{\vec{\tau}^{i}_{ab}}{\rho_a{}^2} \right) \nabla_a W_{ab}, \tag{4}$$

where v_0 is la kinematic viscosity and $\vec{\tau}^{j}_{ab}$ is the shear stress.
In the field of fluid dynamics in porous media, numerical simulations are a crucial tool for predicting fluid flow behavior. When conducting these simulations, two scales are typically considered: the pore scale and the Representative Elementary Volume (REV) scale. Pore-scale simulations provide more accurate results and have an advantage in the study of microscopic seepage and fluid-solid interactions. However, they are computationally expensive and not feasible for large systems, such as those of an oil reservoir. In these cases, REV-scale simulations are the more practical option, despite their limitations in terms of accuracy.
Since the geometry of the porous media is not used in the REV-scale simulations, the porous media are modeled by adding the resistance term in the momentum equation (Eq. 5), and the interaction of the solid matrix with the fluid is out of consideration, which reduces the computation time.

$$\frac{d\mathbf{v}_a}{dt} = -\sum_b m_b \left(\frac{p_a + p_b}{\rho_a \rho_b} + \Gamma \right) \nabla_a W_{ab} + \mathbf{g} + \frac{F_a}{m_a}. \tag{5}$$

The Ergun formula is widely used to calculate the additional resistance term due to its simple form, which is

$$\boldsymbol{F} = -\frac{\epsilon v}{K}\mathbf{v} - \frac{\epsilon F_e}{\sqrt{K}}|\mathbf{v}|\mathbf{v}, \tag{6}$$

where $\mathbf{v}$ is the velocity vector, v is the viscosity of the fluid; K is the permeability of the porous medium, ϵ is the porosity and F_e is a structure-function, the parameters shown in Eq.(6) are reported in detail by Lai et al., (2020).
In this work, Eq. (6) is applied in the last term of Eq. (5) by particle, for the cases of a REV scale. The implementation was performed using version 5 of the DualSPHysics code (Dominguez et al., 2022).

3. Validation test.

The SPH model presented in section 2 is validated with the simulation of the Poiseuille flow which has an analytical solution. The SPH numerical results are compared with the analytical solution of the system. For this case was considered the geometry shown in Figure 1, where the distance between plates is 1 mm and the length is 5 mm. The properties of the fluid are according to the liquid water, ρ = 100kg/m^3, ν = 1x10^{-6} Pa*s, and the particles have an acceleration of 1x10^{-4} m/s^2 in the right direction. This case was simulated with different resolutions changing the initial distance between particles dp = 0.1, 0.05, 0.025, and 0.01 mm. The velocity profiles obtained from the numerical results using the model reported in section 2 and their comparison with the analytical solution are shown in Figure 1. The errors (RMSE) for each simulation are reported in Table 1 where is clear that the error decrease when the resolution is increased.

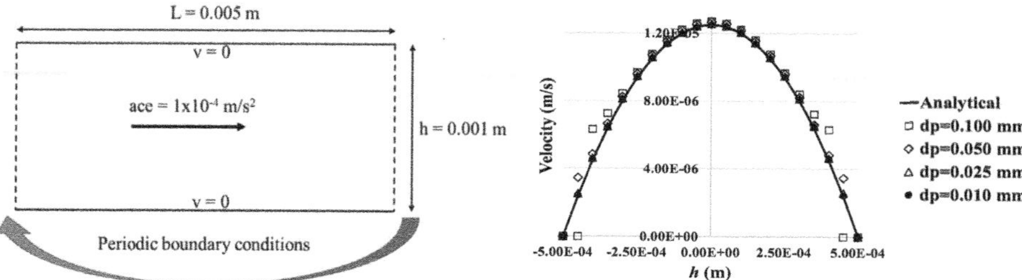

Figure 1. The left figure represents the dimensions and boundary conditions for the Poiseuille flow simulation, and the right one the velocity profiles for the different initial distances between particles (dp).

Table 1. Root mean square error between the numerical results and the analytical solution in the validation test cases.

Dp (mm)	0.1	0.05	0.025	0.01
RMSE	1.01403×10^{-6}	4.04818×10^{-7}	1.10592×10^{-7}	5.28235×10^{-8}

4. Comparative cases.

The pore-scale simulations were conducted in 2D with porosities of 0.55, 0.44, and 0.3, represented by three different arrays of circles named square, diagonal, and random (as shown in Figure 2). A total of 9 study cases were carried out. The porous media were inserted in the center of a rectangular channel with dimensions of 10 cm x 5 cm. The inlet boundary was located at the left side of the channel and was set to a constant velocity profile (as shown in Figure 2). The outlet boundary was located at the right side of the channel, where the velocity value of the particles was computed (as shown in Figure 2). The top and bottom boundaries were no-slip boundary conditions, using the dynamics particle method described in (Crespo et al., 2007). The fluid was characterized with the same parameters used in the validation case, with an initial particle distance of dp = 0.001 mm for all cases reported in this section. In the simulations using the Ergun Force, the geometry of the porous media was changed to a rectangular section (as shown in Figure 2) and the Ergun force (Eq. 6) was imposed in the momentum equation (Eq. 4), while maintaining the same parameters and channel dimensions as the pore-scale simulations.

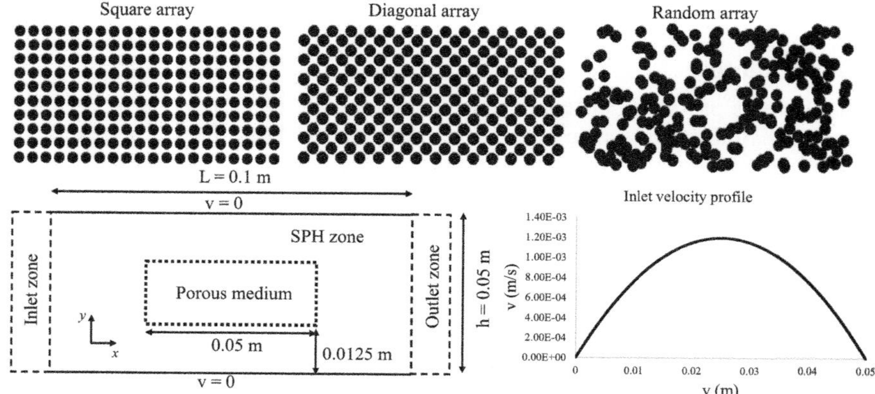

Figure 2. The three different left arrays, shown an example of the PM used for each porosity value. Right frame: at top is shown the geometry used for all cases, and the bottom is the inlet velocity profile.

The velocity profiles at the outlet obtained in the pore-scale and REV scale were compared to prove the accuracy of the Ergun force to represent the flow in different porous media. The velocity fields obtained from both scale simulations are shown in Figure 3, only for the case with $\epsilon = 0.55$ and the random array, both results are very similar in qualitative analysis. The quantitative fit was obtained using the velocity profiles at the outlet shown in Figure 4 for all cases. Here, the root-mean-square error (RMSE) is used as the error of the models. RMSE is a measure of accuracy, typically used to compare forecasting errors of different models for a particular dataset and not between datasets, as it is scale dependent. The results presented in Figure 4 have very similar profiles, between the simulations carried using the Ergun force term and those using the porous geometry for different porosities and geometries. The error between the profiles shown in Figure 4 are reported in Table 3. The errors reported in Table 3 are two order of magnitude less than the maximum velocity shown in Figure 4 for all cases.

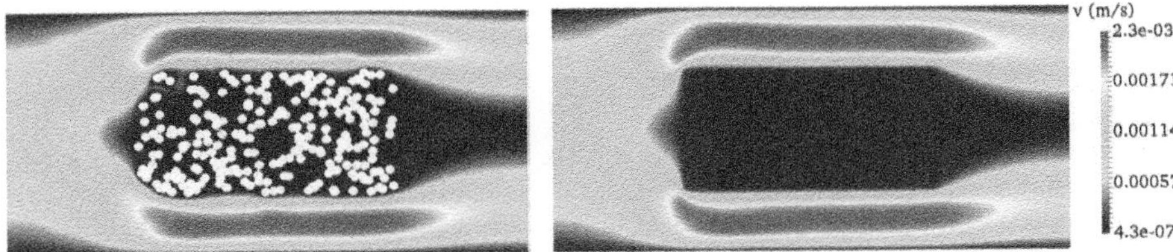

Figure 3. Comparison of the velocity field between the pore-scale and REV simulations: left frame pore-scale, right frame Ergun force.

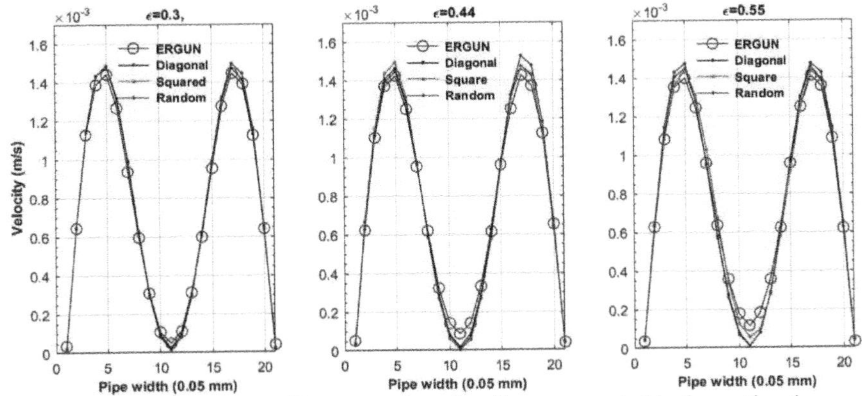

Figure 4. Velocity profiles comparison for all cases reported in the section 4.

Table 2. Root mean square error between the numerical results and the Ergun force term simulations.

	$\epsilon = 0.33$	$\epsilon = 0.44$	$\epsilon = 0.55$
RMSE (Squared)	2.84x10^{-5}	3.82x10^{-5}	4.36x10^{-5}
RMSE (Diagonal)	3.20x10^{-5}	5.23x10^{-5}	5.77x10^{-5}
RMSE(Random)	3.19x10^{-5}	5.12x10^{-5}	5.91x10^{-5}

5. Conclusions.

In this work was implemented the Ergun force term in the momentum equation using the Smoothed Particle Hydrodynamics (SPH) method. The implementation was validated through a Poiseuille flow simulation and a comparison of the numerical results with the analytical solution. Additionally, a convergence test was conducted and the error between the numerical and analytical data decreased as the resolution increased, reaching a minimum error of 5.28235 x 10^{-8} in the simulation with a particle distance of 0.01 mm and 501501 total particles. The results demonstrate the successful implementation of the Ergun force using the DualSPHysics code. Comparing the pore-scale and REV simulations, the velocity profiles at the outlet were similar, with a maximum error of 5.91 x 10^{-5} in the case of a porosity of 0.55 and the random array and a minimum error of 2.84 x 10^{-5} in the case of a porosity of 0.3 and the square array. The square geometry resulted in better emulation of the Representative Elementary Volume (REV) area, with a lower error in all cases, due to the similarity of the squared area to the area where the Ergun force is applied. The error increased with the porosity, but further simulations with a wider range of porosities are necessary to confirm the proper application of the Ergun force in porous media simulations.

References

L. Lucy, 1977, A numerical approach to testing the fission hypothesis, Journal Astronomical, 82:1013 1924.

A. M. Tartakovsky, P. Meakin, T. D. Scheibe, B. D. Wood, 2007, A smoothed particle hydrodynamics model for reactive transport and mineral precipitation in porous and fractured porous media. Water Resour. Res. 43, W05437 (18 pages).

J. Klapp, L. D. G. Sigalotti, C. E. Alvarado-Rodríguez, 2022, Approximately consistent SPH simulations of the anisotropic dispersion of a contaminant plume. Comp. Part. Mech. 9, 987–1002.

P. Kunz, I. M. Zarikos, N. K. Karadimitriou, M. Huber, U. Nieken, S. M. Hassannizadeh, 2015, Study of multiphase flow in porous media: comparison of SPH simulations with Micro-model experiments, Transport in Porous media, Springer.

S. Ergun, 1952, Fluid flow through packed columns . Chemical Engineering Progress, 48, 89–92.

A. M. Tartakovsky, A. Panchenko, 2016, Pairwise Force Smoothed Particle Hydrodynamics model for multiphase flow: Surface tension and contact line dynamics, Journal of Computational Physics, 305, 1119–1146.

R. A. Dalrymple, B. Rogers, 2006, Numerical modeling of water waves with the SPH method. Coastal Engineering, 53:141 147.

A. J. C. Crespo, M. Gómez-Gesteira, R. A. Dalrymple, 2007a, Boundary conditions generated by dynamic particles in SPH methods, Computers, Materials & Continua ,5,173–184.

J. M. Domínguez, G. Fourtakas, C. Altomare, R. B. Canelas, A. Tafuni, O. García-Feal, I. Martínez-Estévez, A. Mokos, R. Vacondio, A. J. C. Crespo,B. D. Rogers, P. K. Stansby, M. Gómez-Gesteira, 2022, DualSPHysics: from fluid dynamics to multiphysics problems. Computational Particle Mechanics, 9(5), 867-895.

T. Lai, X. Liu, S. Xue, J. Xu, M. He, Y. Zhang, 2020, Extension of Ergun quation for the calculation of the flow resistance in porous media with higher porosity and open-celled structure, Applied Thermal Engineering, 173, 115262.

Antonis Kokossis, Michael C. Georgiadis, Efstratios N. Pistikopoulos (Eds.)
PROCEEDINGS OF THE 33rd European Symposium on Computer Aided Process Engineering (ESCAPE33), June 18-21, 2023, Athens, Greece
 http://dx.doi.org/10.1016/B978-0-443-15274-0.50067-6

Impact of structural change in electricity and hydrogen end use on low-carbon transition of an energy system: A case study of China

Nuobei Zhang,[a] Pei Liu,[*] Zheng Li

[a]*State Key Lab of Power Systems, Department of Energy and Power Engineering, Tsinghua-BP Clean Energy Center, Tsinghua University, Beijing 100084, China*
liu_pei@tsinghua.edu.cn

Abstract

Driven by global targets to reduce greenhouse gas emissions, energy systems are expected to undergo fundamental changes, and the proportion of electricity and hydrogen may greatly increase. Scenarios analysis based on optimal planning is beneficial to exploring how this growth affects the structure of an energy system, whilst at the same time reducing costs, better fulfilling the carbon emission target and ensuring energy security. This study sets two groups of different scenarios based on share of electricity and hydrogen in end use respectively, and uses China Regional Energy Supply System Optimization Model (CRESOM) as an analysis tool. China is chosen as a case study due to the large scale, complex structure, fast growth rate, strict yet changing emissions constraints of its energy system. Results show that with an increase in electrification rate, cumulative carbon dioxide emissions over a certain planning horizon could reduce, at a priced of increased unit reduction cost. By increasing the share of electricity and hydrogen in the end use from 71% to 80% in 2060, the unit carbon reduction cost rises by 17%. Increasing share of hydrogen could move the carbon emissions peak time by approximately five years.

Keywords: Scenarios analysis; Low-carbon transition; China; Electricity; Hydrogen

1. Introduction

Low carbon development has become a worldwide consensus, but the task is arduous. Fossil energy is still the main part of the global energy consumption. According to BP's scenario forecast (BP, 2020), the proportion of fossil energy will drop from 84% (2018) to 21.7% (2050) to achieve near zero emissions. The global energy system is expected to undergo a profound low-carbon transition inevitably. In the next 30 years, with the promotion of decarbonization of the energy system, the end use of electricity and hydrogen could grow significantly (BP, 2020). Therefore, it is necessary to conduct research to get guidance such as determining the transition objectives and reducing the total cost in the transition process.

At present, there are many studies on transition of the energy system. Dai et al. (2016) built two scenarios from the perspective of renewable development, and assessed the impact of large-scale renewable development on economy and environment by 2050. Guo et al. (2017) established the LoMLoG model, gave the most likely scenario in the future, proved the importance of the goal of clean energy for reducing carbon emissions in the power sector and promoting the use of renewable energy. Guo et al. (2021) conducted scenario analysis on building energy consumption, taking China as a case, and pointed out that the carbon peak time is expected to be 2020-2035. Duan et al. (2021) compared the results of various models, pointed out that China will reduce carbon emissions by 90% with the target of 1.5 ℃ under the scenario of policies, explained the importance of negative emission technology in the future. Zhang et al. (2022) set different scenarios, analyzed the contribution of emission reduction measures in different periods, and analyzed the uncertainty of key parameters.

However, previous studies paid little attention to the impact of the increase in the end-using electricity proportion and end-using hydrogen proportion in the energy system. Ignoring these two factors may well leads to unclear description of energy substitution process, so it is necessary to consider in the setting of scenarios.

Based on the existing optimization model, this study analyzes the impact of the increase in the end use of electricity and hydrogen in the transition of energy structure by setting different scenarios.

The structure of this paper is as follows. The methodology will be introduced in Section 2. The case study and the results will be introduced in Section 3, and Section 4 is the conclusions.

2. Methodology

Scenario analysis based on optimization model is chosen as a method. Through scenario analysis, we can get some valuable transition guidance for countries with fossil energy as the main energy.

2.1. The introduction of CRESOM

The model used in this study is China Regional Energy Supply System Optimization Model (CRESOM) developed by Tsinghua-BP Clean Energy Research and Education Center, which is mainly used for the optimal planning of energy supply system under the given low-carbon transformation strategy.

The main inputs of the model include the historical data used to describe the current energy supply and demand and infrastructure construction, prospective data used to describe future economic growth, energy intensity and energy technology cost, scenario data used to describe the emission reduction policies and carbon policies.

The model first calculates the energy demand by category of each terminal energy department in each region from 2016 to 2060 through the terminal energy demand forecasting model, then the final planning scheme is obtained by minimizing the total cost through the power-thermal system optimization model.

The output data includes the forecast of the future and the planning scheme at different stages. The structure of the model is shown in Figure 1.

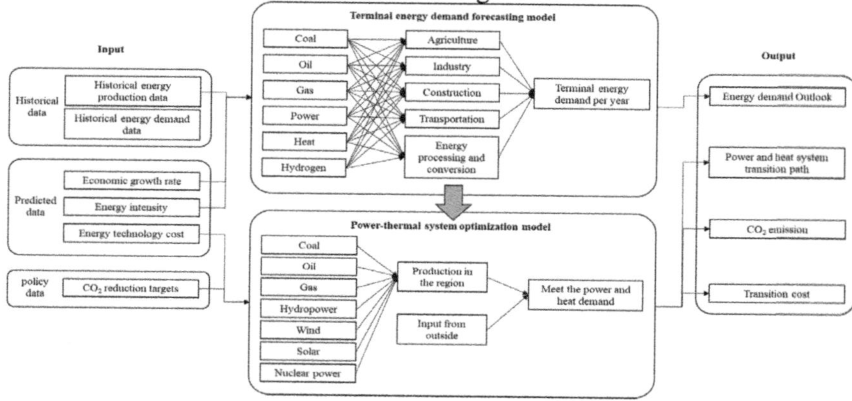

Figure 1. The structure of CRESOM

2.2. Scenario setting

This study focuses on the impact of electricity and hydrogen in the process of energy transition, and constructs two groups of scenarios. One is the high proportion electricity scenario, the other one is high proportion hydrogen scenario. The relevant settings of the scenarios are shown in the tables below.

Table 1. The setting of high proportion electricity scenario

Scenario	EH65	EH68	EH71	EH74	EH77	EH80
End-using electricity proportion in 2060 (%)	55	57.5	60.1	62.6	65.2	67.7
End-using hydrogen proportion in 2060 (%)	10	10.5	10.9	11.4	11.8	12.3
Total proportion (%)	65	68	71	74	77	80

Table 2. The setting of high proportion hydrogen scenario

Scenario	H12	H15	H18	H21	H24	H27	H30
End-using electricity proportion in 2060 (%)	65	62	59	56	53	50	47
End-using hydrogen proportion in 2060 (%)	12	15	18	21	24	27	30
Total proportion (%)				77			

3. Case Study

As the largest developing country in the world, China's energy structure needs to be transformed pressingly, thus its energy transition research is essential. This study takes China as a case to study the impact of the increase of end-using electricity and hydrogen proportion on the energy system through multi scenario calculations, aiming to provide guidance for its transition and provide experience for other countries with fossil energy as the main energy source.

3.1. High proportion electricity scenario results

3.1.1. Total emissions and system costs

As shown in Figure 2A, the carbon emissions show a decline after reaching the peak in 2025, and the promotion of electrification makes the carbon emissions decline faster. The $costyear$ is the sum of the costs of all subsystems each year, as shown in Eq. (1). The $costyear$ increases by 26.4% in 2060 (EH65 to EH80), mainly due to the large-scale energy substitution in 2040-2060 (Figure 2B).

$$costyear = costyear_c + costyear_o + costyear_g + costyear_h + costyear_e \quad (1)$$

The $cost_{CUTCRC}$ (Cumulative Unit Transition Carbon Reduction Cost) is shown in Eq. (2). The molecular part is the total cost of all scenarios compared with the scenario without energy substitution policy. The denominator part is the carbon dioxide reduction of all scenarios compared to the scenario without energy substitution policy. It reflects the cost performance of the system's carbon reduction.

$$cost_{CUTCRC} = \frac{\sum_{2020}^{2060} costyear_{EH} - \sum_{2020}^{2060} costyear_{BAU}}{\sum_{2020}^{2060} CO_{2BAU} - \sum_{2020}^{2060} CO_{2EH}} \quad (2)$$

As the end-using electricity proportion increases, the $cost_{CUTCRC}$ continues to rise, especially when the total proportion is greater than 74%, where the proportion of end-using electricity in EH80 is 6% higher than that in EH74, but the $cost_{CUTCRC}$ is 17% higher. Therefore, excessive promotion of electrification leads to a decline in the emission reduction cost performance.

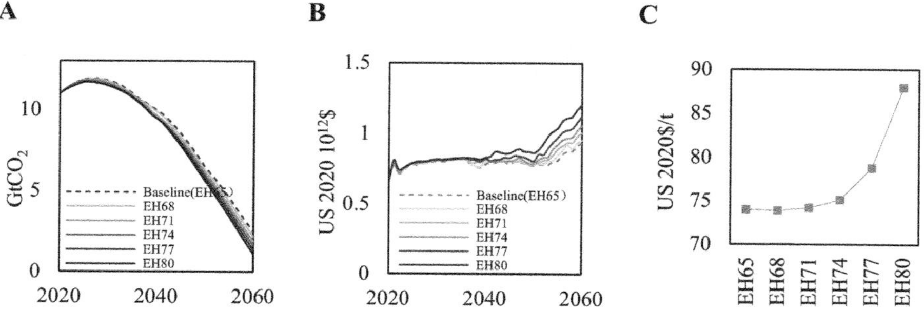

Figure 2 Changes in carbon emissions and system costs under the high proportion electricity scenario. (A) Total carbon emission. The darker the color, the higher the end-using electricity proportion. The following figures are the same. (B) $costyear$. (C) $cost_{CUTCRC}$.

3.1.2. Power supply structure

Non fossil power shows a significant increase, as shown in Figure 3A. The EH80 (2060) and the EH65 (2060) have increased by 9.3 times and 8.5 times respectively compared with 2020. For fossil power, the capacity of fossil power remains stable from 2020 to 2040, and declines after 2040 (Figure 3B). However, with the increase of end-using electricity proportion, the installed capacity of fossil power declines slowly. In 2060, fossil power installations are dominated by coal power with CCS and gas power with CCS, which mainly play a peak shaving role. This phenomenon indicates that the rising non fossil power installations lead to the growth of power fluctuations. From the perspective of cost, the fossil power installation with CCS brings more construction costs and operating costs than the non fossil power installation, thus it brings an increase in the system cost of unit power generation (Figure 3C), and this cost of EH80 is 20.3% higher than that of EH65.

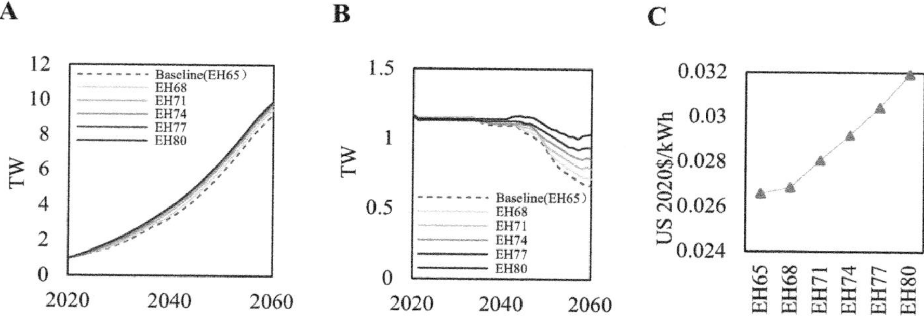

Figure 3: Changes in power supply structure under the high proportion electricity scenario. (A) Non fossil power installations capacity (B) Fossil power installations capacity (C) System cost of unit power generation

3.1.3. High proportion hydrogen scenario results

3.1.4. Total emissions and system costs

Under the setting of high proportion hydrogen scenario, the sum of end-using electricity and hydrogen proportion among different scenarios is the same, so there is little difference in carbon emission trajectory and annual system cost close to 2060. The difference in results is mainly reflected in the advance of peak year due to the increase of hydrogen (Figure 4A) and the change of total accumulated cost ($cost_{TAC}$). The expression of $cost_{TAC}$ is as follows:

$$cost_{TAC} = \sum_{2020}^{2060} (costyear_c + costyear_o + costyear_g + costyear_h + costyear_e) \quad (3)$$

The $cost_{TAC}$ of H15 decreased by 0.5% compared with H12, while the $cost_{TAC}$ of H30 increased by 2.7% compared with H12(Figure 4B). From the perspective of cost performance of emission reduction. The $cost_{CUTCRC}$ increases first and then decreases with the increase of hydrogen, it shows that both lower and higher end-using hydrogen proportions have good emission reduction cost performance.

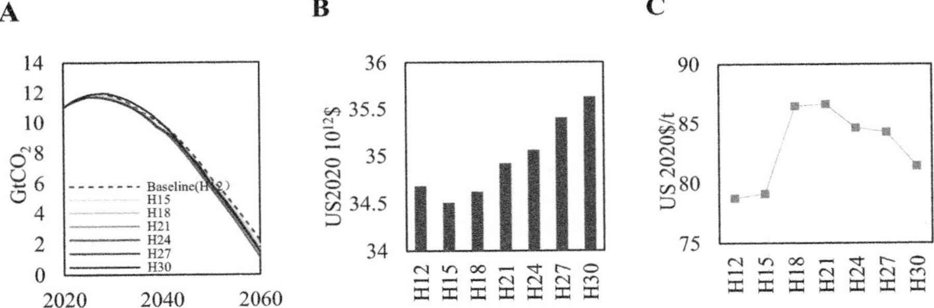

Figure 4: Changes in carbon emissions and system costs under the high proportion hydrogen scenario. (A) Carbon emissions. (B) $cost_{TAC}$ (C) $cost_{CUTCRC}$

3.1.5. Power supply structure

The installed fossil capacity shows a decline after a steady change in time (Figure 5A). With the increase of hydrogen, the decline time of the fossil capacity is advanced. The decline time of H12 is 2046, while the decline time of H30 is 2025. At the same time, the decline speed of the scenario with a higher hydrogen proportion is relatively slow. A higher hydrogen proportion means a higher demand for wind power and photovoltaic installation (Figure 5B). The installed capacity in H12 (2060) and H30 (2060) has increased by 15.76 times and 20.15 times respectively compared with that in 2020.

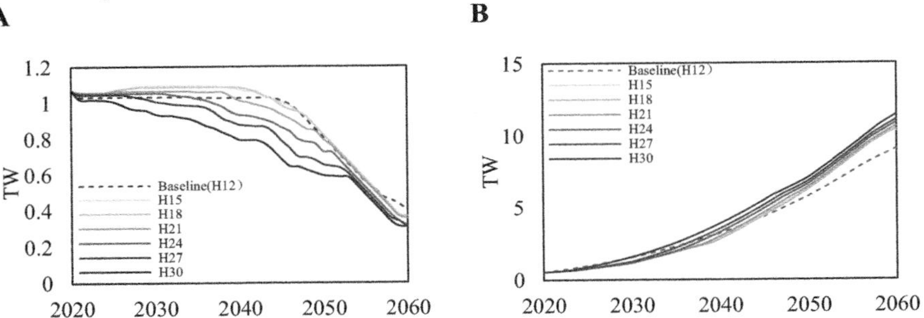

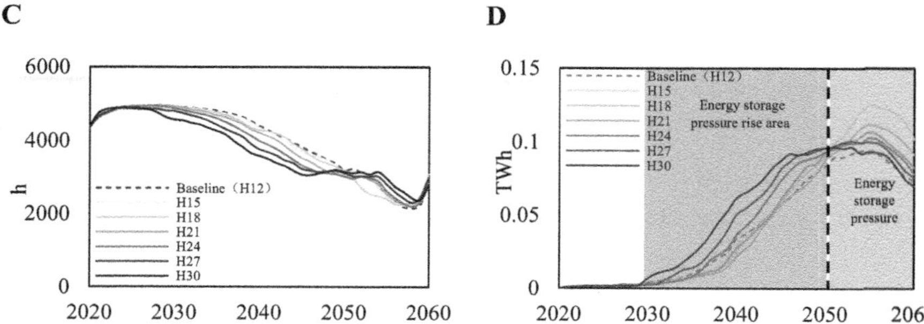

Figure 5: Changes in power supply structure under the high proportion hydrogen scenario. (A) Fossil capacity. (B) Non fossil capacity. (C) Operating hours of coal power. (D) Energy storge capacity

The operating hours of coal power decrease in advance with the increase of hydrogen (Figure 5C), which means that hydrogen promotes the transition of coal power from the main power generation energy to the energy with peak shaving function. This advance also shows that the development of hydrogen in 2030-2050 brings a large number of wind power and photovoltaic installation that need a large amount of energy storage for peak shaving (Figure 5D). From 2050 to 2060, it can be seen that electric energy storage decreases with the increase of hydrogen in the last ten years. Therefore, the increase of hydrogen causes the increase of energy storage pressure from 2030 to 2050, but the decrease of electricity reduces the energy storage pressure from 2050 to 2060.

4. Conclusions

Based on CRESOM, this study reveals the impact of electricity and hydrogen on the low-carbon transition of energy supply system through multi scenario calculations.

Under the high proportion electricity scenario, the promotion of electrification can promote the reduction of carbon emissions, whilst excessive promotion of the end-using electricity proportion brings about an increase in the $cost_{CUTCRC}$, this also means the reduction of the cost performance of emission reduction the system. The higher non fossil power installations capacity, the higher the demand for peak shaving. Therefore, more fossil power with CCS is retained, but system cost of unit power generation rises.

Under the high proportion hydrogen scenario, the increase of hydrogen brings forward the peak year of carbon. For $cost_{CUTCRC}$, both lower and higher end-using hydrogen proportions have good emission reduction performance. The increase of wind power and photovoltaic installation brought about by the increase of hydrogen demand has led to more peak shaving demand. Fossil energy has changed to peak shaving energy in advance, and the demand for energy storage has increased in advance.

References

BP, 2020, BP Energy Outlook 2020 edition.

H. Dai, X. Xie, Y. Xie, et al., 2016, Green growth: The economic impacts of large-scale renewable energy development in China[J]. Applied Energy, 162: 435-449.

Z. Guo, L. Ma, P. Liu, et al., 2017, A long-term multi-region load-dispatch model based on grid strutures for the optimal planning of China's power sector. Computers & Chemical En-gineering, 102:52-63.

S. Guo, D. Yan, S. Hu, et al., 2021, Modeling building energy consumption in China under different future scenarios[J]. Energy, 214: 119063.

H. Duan, S. Zhou, K. Jiang, et al., 2021, Assessing China's efforts to pursue the 1.5°C warming limit. Science, 372: 378–385.
X. Zhang, X. Huang, D.Zhang, et al., 2022, Research on the Pathway and Policies for China's Energy and Economy Transformation towardCarbon Neutrality. Management World, 1:35-51.

Antonis Kokossis, Michael C. Georgiadis, Efstratios N. Pistikopoulos (Eds.)
PROCEEDINGS OF THE 33rd European Symposium on Computer Aided Process Engineering (ESCAPE33), June 18-21, 2023, Athens, Greece
 http://dx.doi.org/10.1016/B978-0-443-15274-0.50068-8

Optimal planning for thermal power plants with CCS under source–sink matching constraints: A case study of China

Binglin Du,a Pei Liu,* Zheng Li

[a]*State Key Lab of Power Systems, Department of Energy and Power Engineering, Tsinghua-BP Clean Energy Center, Tsinghua University, Beijing 100084, China*
liu_pei@tsinghua.edu.cn

Abstract

The low-carbon transition pathway of the power sector, especially thermal power plants is significant for reducing greenhouse gas emissions worldwide. Carbon capture and storage (CCS) technology is identified as a critical decarbonization technology, especially in a thermal power-dominated power system. Previous studies have focused on the potential of CCS in low-carbon transformation of large-scale coal-fired power plants, but rarely considered the fact that many thermal power plants locate far from suitable CO2 storage places, which may lead to deviations in the CCS potential. To quantify the application potential of CCS technology in power sector at a regional level, a long-term power generation expansion planning model is proposed in this paper, featuring detailed technical and economic characteristics of various source–sink matching options, retrofitting options and carbon emission constraints, which can get the optimal planning for thermal power plants with CCS. The power supply system of China is selected as a case study, due to its large capacity and relatively young average life of thermal power plants. Results show that with consideration of source–sink matching, technically and economically viable capacity of power plants that can be retrofitted with CCS can be reduced by 50 percent with a more concentrated distribution in North China.

Keywords: Carbon capture and storage; Coal-fired power plants; Source–sink matching; China

1. Introduction

The decarbonization of power sector is decisive to tackle climate change challenge. Carbon emissions from electricity and heat generation sector account for 40 % of global energy-related carbon emissions, and the share of electricity in energy consumption will keep rising in the future (IEA, 2019). Besides, the power sector is projected to attain carbon neutral or even negative carbon emissions earliest among all sectors, due to its higher concentration of carbon emissions and more advanced carbon emission reduction technologies (Zhang et al.,2013).

Carbon capture and storage (CCS) technology is identified as a critical decarbonization technology, especially in a thermal power-dominated power system, for the reason that CCS is the only technology choice to achieve net-zero emission of fossil energy. If the large fleet of existing thermal power plants, especially coal-fired power plants are not retrofitted with carbon capture and storage (CCS) technologies, they will retire earlier than designed lifetime, which will bring about high stranding cost (Zhang et al., 2021). Besides, when penetration of renewable energy becomes high, more flexibility is needed to ensure the reliability of the power supply (Chen et al., 2021). CCS technology is an important technical choice for coal-fired and gas power stations to participate in balancing the fluctuation of renewable energy in the carbon-neutral power system (Xiao et al.,2022).

Many researchers have studied the potential of CCS from the perspective of power sector development and carbon emission constraints. In the Intergovernmental Panel on Climate Change (IPCC) Special Report on Global Warming of 1.5°C, it is proposed that CCS technology plays an important role in the three scenarios with a 1.5°C target (IPCC, 2018). For the power industry, CCS can contribute 14% of the emission reduction by 2050 (IEA, 2017). However, these studies pay little attention to the fact that many thermal power plants locate far from suitable CO2 storage places, which may lead to rather high costs for transportation and storage of CO2 (Wei et al. 2021). The lack of CCUS source-sink matching constraints will lead to deviations in the CCS potential.

Relevant studies rarely focus on CCUS source-sink matching compared with other topics, such as cost and investment evaluation (Fan et al.,2020). Wang et al. (2020) proposed a source–sink matching model for optimal CCUS deployment in China's existing coal power plants to achieve the 2 °C target, and found 175 GW coal-fired power plants need to be retrofitted with CCS under the 2 °C constraint. Fan et al. (2020) evaluated the deployment potential of CCUS in existing coal-fired power plants by A high precision CCUS source-sink matching model. However, these studies on CCUS source-sink matching lack consideration of the whole power system development, such as the power supply mix and the scale of power transmission.

To address the research gap, a long-term power generation expansion planning model with source-sink matching options is proposed, which can give the optimal planning of thermal power plants with CCS with the overall consideration of power sector development and source-sink matching constraints. Seven scenarios are set to quantitatively analyze the impact of source-sink matching constraints on coal power development.

The structure of this paper is as follows. Section 2 introduces the methodology, section 3 introduces the case study and its results, and section 4 is the conclusion.

2. Methodology

2.1. Model structure and assumptions

The long-term power generation expansion planning model with source-sink matching options proposed here aims to optimize long-term planning both for power systems and CCS. The model structure is shown in Fig.1. The input part includes the existing installed capacity, costs, and efficiency of different power generation technologies, predicted future power demand, policy target, and resource endowment of renewable energy and fossil fuels. The constraints can be divided into four parts: (1)the balance of power supply and demand; (2)the potential of each energy source; (3)the technical factors of power plants; (4)the carbon target. The model objective is to minimize the total system cost in the planning horizon. The output can provide the power supply mix, power, and CO_2 transmission, the annual investment. Among them, the spatial distribution of sources, sink storage potential and the cost of CO_2 transport cost constitute the regional source-sink matching options, which will get the distribution of CCS technology and the transmission of CO_2.

The model includes sixteen different power generation technologies: Sub and super-critical coal power plants (SPC), Ultra-supercritical coal power plants (UPC), Sub and super-critical coal power plants with CCS(SPCC), Ultra-supercritical coal power plants with CCS(UPCC), Sub and super-critical coal-biomass co-firing power plants with CCS(SPCCOCCS), Ultra-supercritical coal-biomass co-firing power plants with CCS(UPCCOCCS), Natural gas combined cycle(NGCC), Nuclear(NU), Hydropower(HD), Wind onshore(WD), Wind offshore(WDOFF), Centralized solar photovoltaic(PV), Distributed solar photovoltaic(PVDIS), Biomass(BE), Biomass with CCS(BECCS). Among them, coal-fired power plants and natural gas combined cycle power plants can be retrofitted with CCS or decommissioned earlier than the anticipated lifetime, and all power plants with CCS will be constrained by source-sink matching. The model is a multi-regional model that includes 17 regions based on resources and grid structure. It also takes into account short-term planned cross-region transmission lines. Each year in the planning horizon is split into 24 time slices in the temporal module, with four typical days for each season.

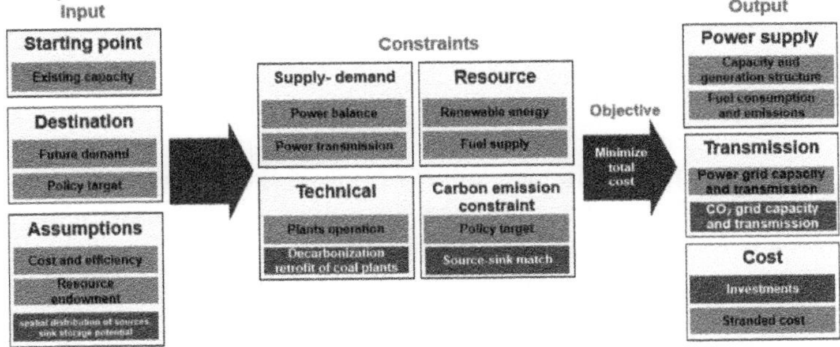

Fig.1 Model structure

2.2. *Mathematical formulation*

The six subscripts (t, r, b, g, f, s) are set respectively to cover different dimensions (year, region, basin, technology, fuel type, and time slice) for the parameters and variables.

2.2.1. *Objective function*

The model objective is minimizing the accumulated total system cost (*atoc)* in the planning horizon (from 2020 to 2060), including capital cost $tocac_{r,t}$, capital cost for retrofit $torfcac_{r,t}$, operation, and maintenance cost $toomc_{r,t}$, fuel cost $tofc_{r,t}$, power transmission cost $totrc_{r,t}$, and CO2 transmission cost $toctrc_{r,t}$, as shown in Eq. (1) – (7).

$$atoc = \sum_r \sum_{t=2020}^{2060} \left(\cdot \frac{tocac_{r,t}+torfcac_{r,t}+toomc_{r,t}+tofc_{r,t}+totrc_{r,t}+toctrc_{r,t}}{(1+I)^{t-2020}} \right) \tag{1}$$

$$tocac_{r,t} = \sum_g \sum_{t'=t-ELT_g+1}^{t} \left(CAC_{r,t',g} \cdot nbca_{r,t',g} \cdot \frac{I \cdot (1+I)^{-1}}{1-(1+I)^{-ELT_g}} \right) \tag{2}$$

$$torfcac_{r,t} = \sum_g \sum_{t'=t-ELT_g+1}^{t} \sum_{t''=t'+1}^{t} \left(RFCAC_{r,t',g} \cdot rf_{r,t',g} \cdot \frac{I \cdot (1+I)^{-1}}{1-(1+I)^{-(t'+ELT_g-t'')}} \right) \tag{3}$$

$$toomc_{r,t} = \sum_g OMC_{t,g} \cdot inca_{r,t,g} \tag{4}$$

$$tofc_{r,t} = \sum_f FEULP_{f,r,t} \cdot \sum_{g,s} feuld_{f,r,t,g,s} \tag{5}$$

$$totrc_{r,t} = \sum_s \sum_{r' \neq r} idpt_{r,r',t,s} \cdot PTRC_{r,r'} \tag{6}$$

$$toctrc_{r,t} = \sum_s \sum_b cot_{r,b,t,s} \cdot CTRDIS_{r,b} \cdot CTRC_{r,b} \tag{7}$$

2.2.2. *Operational constraints*

Reginal power demand is needed to satisfy by regional power supply, which equal regional power generation $pge_{r,t,g,s}$ (installed capacity $inca_{r,t,g}$ multiply hours of operation $H_{r,t,s,g}$) coupled with net power import $npi_{r,t,s}$ (power import $pit_{t,r',r}$ minus power export $pit_{t,r'',r}$) and the net electricity provided by energy storage (electricity discharged $dchar_{r,t,s}$ minus electricity charged $char_{r,t,s}$), as expressed in Eq. (8)-(10). Ideal power transmission from r to r' equals the negative value of that from r' to r and is limited by transmission capacity.

$$PD_{r,t,s} \le \sum_g egs_{r,t,g,s} + npi_{r,t,s} + dchar_{r,t,s} - char_{r,t,s} \tag{8}$$

$$egs_{r,t,g,s} \le inca_{r,t,g} * H_{r,t,s,g} \tag{9}$$

$$npi_{r,t,s} = \sum_{r'',r' \ne r}[pit_{t,r',r} - (1 - TRLOSS_{r,r'}) * pit_{t,r'',r}] \tag{10}$$

2.2.3. Construction constraints

The installed capacity of technologies equals the sum of newly built capacity ($nbca_{r,t,g}$) each year during the past years (i.e. lifetime of each power plants) minus the early retired capacity ($rfca^{t'}_{r,t,g}$) and retrofitted capacity ($erca^{t'}_{r,t,g}$) plus the capacity retrofitted by other power plants ($erfca^{t'}_{r,t,g'}$), as shown in Eq. (13).

$$inca_{r,t,g} = \sum_{t'=t-ELT_g+1}^{t} nbca_{r,t',g} - \sum_{t'=t-ELT_g+1}^{t} \sum_{t''=t'+1}^{t} (rfca^{t''}_{r,t',g} + erca^{t''}_{r,t',g}) + \sum_{t'=t-ELT_{g'}+1}^{t} \sum_{t''=t'+1}^{t} \left(erfca^{t''}_{r,t',g'}\right) \tag{11}$$

Other inequality constraints such as the upper bound of regional total installed capacity of renewable energy, annual newly-built capacity are also considered.

2.2.4. Carbon emission and source-sink matching constraints

The annual net CO2 emission ($tocoe_t$) equals CO2 emission from fuel energy combustion minus the amount of CO2 capture, which equals the sum of CO2 stored in each basin. The total emission in the planning horizon cannot exceed the 1.5°C carbon target ($COET$). Each basin's injection capacity is limited, including storage potential (CSP_b) and annual injection rate ($ACSP_b$).

$$tocoe_t = \sum_f tofueld_{f,t} \cdot COEF_f - \sum_r \sum_s \sum_b cot_{r,b,t,s} \tag{12}$$

$$\sum_{t=2020}^{2060} tocoe_t \le COET \tag{13}$$

$$\sum_t \sum_r \sum_s cot_{r,b,t,s} \le CSP_b \tag{14}$$

$$\sum_r \sum_s cot_{r,b,t,s} \le ACSP_b \tag{15}$$

3. Case Study

The power supply system of China is selected as a case study, due to its large capacity and relatively young average life of thermal power plants. At the same time, the source-sink matching of China's CCS technology is not ideal, which makes it difficult to reduce fossil energy emissions and achieve carbon neutrality target. This study will apply the above method to explore the impact of consideration and non-consideration of CCS source-sink matching and the number of storage sites and transportation costs on the trajectory of thermal power unit development during the planning period of 2020–2060.

The economic and technical parameters of each power plant are incorporated from previous work (Chen et al., 2021). The transmission cost of carbon dioxide is set based on the report about China's CCUS development (Zhang et al., 2021). And the injection capacity of each basin and the distance from emission source to sink refer to the research of current China CCUS source–sink matching situation (Chen et al.,2021).

7 scenarios (S1–S7) are set in this study. S1 don't take the CCS source-sink matching restrictions and relative cost into account. S2 consider the CCS source-sink matching constraints with only 9 basins (Tarim, Junggar, Ordos, Songliao, Erlian, Tuopan-Hami, Sanjiang, Qaidam, Hailaer), which are far from populated areas for security reasons. S3-S7 consider the CCS source-sink matching with 16 basins, which also includes 7 basins in Central and South China (Sichuan, Qinshui, Bohai Bay, Suibei, Hehuai, Jianghan, Nanxiang). In S2-S7, the cost of transporting CO2 will decrease by 50%, 90%, 70%, 50%, 30%, and 10% in 2060 compared to 2020.For modelling and optimization, the

General Algebraic Modelling System (GAMS, GAMS Development Corporation, Washington, DC, USA) Linear Programming Solver was employed.

3.1. Installed capacity of thermal power plants with CCS in 2060

As shown in Figure 1 the constraint of source-sink matching will have a great impact on the total installed capacity of thermal power plants with CCS. When there are only 9 optional basins and the cost of CO2 transportation drops by 50 %, the total installed capacity falls from 592 GW to 145 GW, with a decrease of 76.1 %, and the overall installed capacity was 236GW when there were 16 chosen basins, with a decrease of 60.1%. Among these thermal power plants, the most affected is the UPCC, with a decreased by 86.3 % and 77.8 % in S2, S5 relative to S1. Analysis of S3-S7 shows that for every 20 % reduction in carbon dioxide reduction rate, the installed capacity of thermal power CCS units also decreases by 29-36 GW.

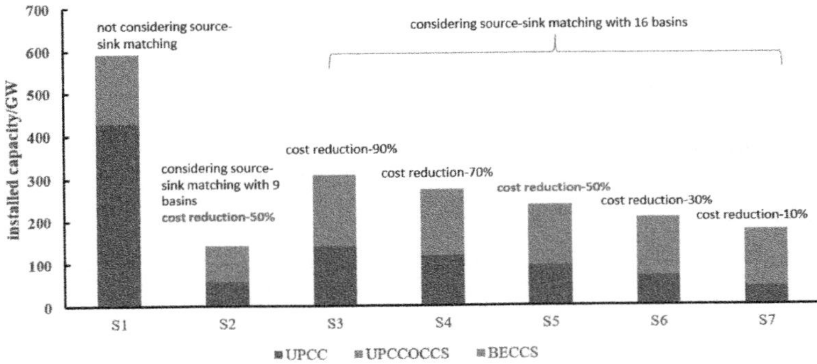

Figure 1 Installed capacity of thermal power plants with CCS in 2060

3.2. Spatial distribution of thermal power plants with CCS in 2060

With the consideration of CO2 transmission cost, the installed capacity of coal plants with CCS in Inner Mongolia and East and BECCS in Chuan-Yu, Central and Henan will decline significantly. There will be more thermal power plants with CCS in places near to storage basins, such as Xinjiang, Northeast and Shandong.

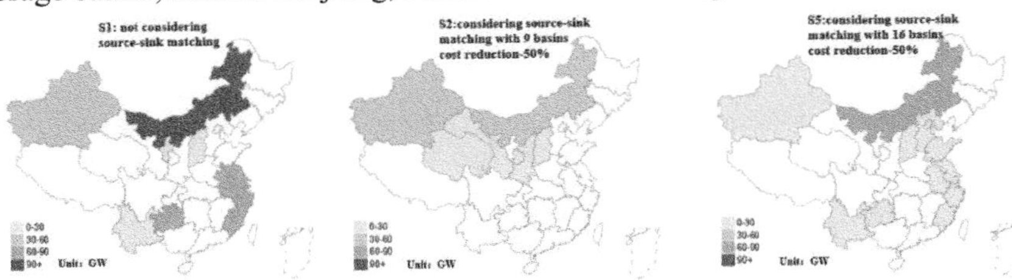

Figure 2 Spatial distribution of coal plants with CCS (UPCC and UPCCOCCS) in 2060

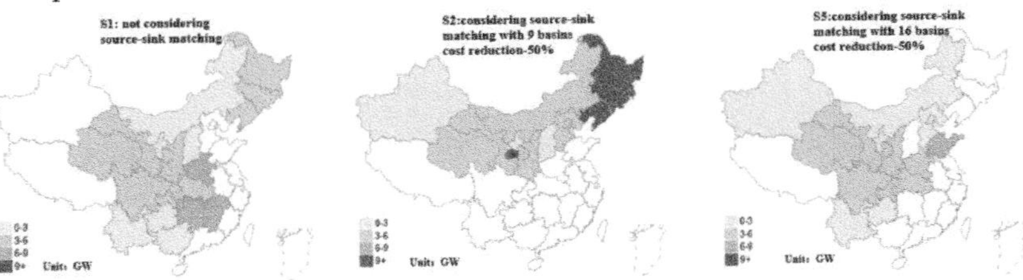

Figure 3 Spatial distribution of BECCS in 2060

3.3. CO2 storage amount in each basin

As shown in Figure 4, CO2 will be buried in Ordos, Junggar, Songliao, Erlian, and Turpan basins in the S2 scenario. The two basins with the most CO2 storage are Ordos Basin and Junggar Basin, with 4.3 Gt and 2.9 Gt, respectively. CO2 will also be stored in Sichuan, Qinshui, Bohai Bay, Subei, and Nanxiang basins in the S5 scenario. Among them, the Ordos Basin, Sichuan Basin, and Bohai Bay Basin have the largest sequestration amount, with 4.3Gt, 3.7Gt, and 3.5Gt respectively.

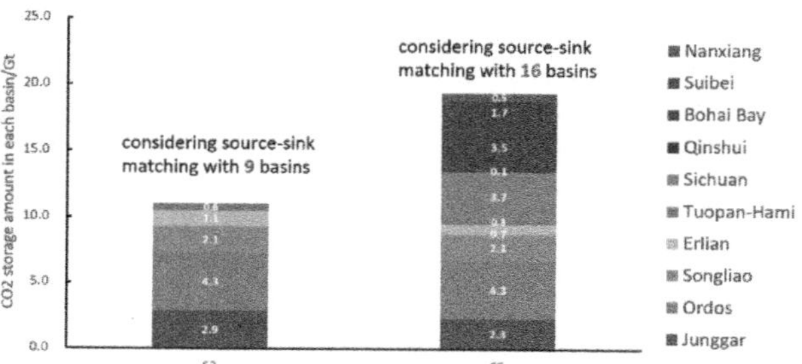

Figure 4 CO2 storage amount in each basin

4. Conclusions

This study proposes a low-carbon transition model of power system with source-sink matching module, which can be used to quantitatively analyze the development path of thermal power plants with CCS with minimum transition costs based on both the power system development and CCS source-sink matching constraints. The results of scenario analysis show that the total installed capacity of thermal power plants with CCS will drop significantly when source-sink matching is taken into account. The spatial distribution of CCS will change greatly, and the distance to the storage sinks will also become a key factor in the layout of CCS. The Ordos, Junggar and Bohai Bay Basin will become important CO2 storage locations in the future.

References

IEA, 2019, World Energy Outlook.

D. Zhang, P. Liu, L. Ma, Z. Li, 2013, A multi-period optimization model for planning of China's power sector with consideration of carbon dioxide mitigation-The importance of continuous and stable carbon mitigation policy, Energy Policy, 58, 319-328.

X. Zhang, K. Li, Q. Ma, J. Fan, 2021, Orientation and prospect of CCUS development under carbon neutrality target, China Population,Resources and Environment, 31, 29-33.

S. Chen, Z. Li, W. Li, 2021, Integrating high share of renewable energy into power system using customer-sited energy storage, Renewable & Sustainable Energy Reviews, 143.

K. Xiao, B. Yu, L. Cheng, F. Li, D. Fang, 2022, The effects of CCUS combined with renewable energy penetration under the carbon peak by an SD-CGE model: Evidence from China, Applied Energy, 321.

IPCC, 2018, Summary for Policymakers in: Global Warming of 1.5°C.

IEA, 2017, Energy Technology Perspectives.

Y.-M. Wei, J.-N. Kang, L.-C. Liu, Q. Li, P.-T. Wang, J.-J. Hou, Q.-M. Liang, H. Liao, S.-F. Huang, B. Yu, 2021, A proposed global layout of carbon capture and storage in line with a 2 degrees C climate target, Nature Climate Change, 11.

J.-L. Fan, M. Xu, S. Wei, S. Shen, Y. Diao, X. Zhang, 2021, Carbon reduction potential of China's coal-fired power plants based on a CCUS source-sink matching model, Resources Conservation and Recycling, 168.

P.-T. Wang, Y.-M. Wei, B. Yang, J.-Q. Li, J.-N. Kang, L.-C. Liu, B.-Y. Yu, Y.-B. Hou, X. Zhang, 2020, Carbon capture and storage in China's power sector: Optimal planning under the 2 degrees C constraint, Applied Energy, 263.

C. Wen-Hui, L. Xi, 2022, The optimal layout of CCUS clusters in China's coal-fired power plants towards carbon neutrality, Climate Change Research, 18, 261-271.

Antonis Kokossis, Michael C. Georgiadis, Efstratios N. Pistikopoulos (Eds.)
PROCEEDINGS OF THE 33rd European Symposium on Computer Aided Process Engineering (ESCAPE33), June 18-21, 2023, Athens, Greece
 http://dx.doi.org/10.1016/B978-0-443-15274-0.50069-X

On discrete time chemical production scheduling MILP models containing record keeping variables

Amin Samadi,[a] Nathan Adelgren,[b] Christos T. Maravelias,[a,b]

[a]*Department of Chemical and Biological Engineering, Princeton University, Princeton, NJ 08540, USA*

[b]*Andlinger Center for Energy and the Environment, Princeton University, Princeton, NJ 08540, USA*

Abstract

In this paper, we present strategies for reformulating discrete time-based mixed-integer programming models for chemical production scheduling. We introduce several new integer variables, which we refer to as record keeping variables, that mixed-integer linear programming solvers are able to exploit in order to reduce total solution time. We consider these record keeping variables in the context of both batch process networks and continuous process networks. Results of several computational tests are provided that demonstrate the utility of incorporating record keeping variables in chemical production scheduling models.

Keywords: batch process, continuous process, scheduling, discrete time, reformulation.

1. Introduction

We consider a chemical processing facility that converts raw materials into higher value products. In this context, production scheduling involves assigning tasks to processing units and determining the times at which the processing of these tasks should begin, possibly subject to a set of restrictions. As these task assignments and timing decisions ultimately govern a significant portion of a facility's operations, scheduling is a crucial decision-making step with applications in a broad range of systems, from batch production of low-volume products such as pharmaceuticals (Papavasileiou et al., 2007) to crude oil blending (Castro and Grossmann, 2014). Determining an appropriate schedule requires knowledge of the processing duration of each task, any limitations on the units to which a given task can be assigned, material conversion coefficients, and many other complex, application-specific details. As such, much research in the field of chemical production scheduling has been devoted to developing strategies for modeling various process characteristics such as limits on utility and/or resource availability (Méndez et al., 2001), changeovers (Wolsey, 1997), and material transfer restrictions (Giménez et al., 2009). In considering this wide array of process characteristics, researchers aim to develop highly general models that can be readily employed in a wide range of industrial applications. However, discrete-time models grow at least linearly with the number of time periods (Velez et al., 2017). As such, a major research direction in production scheduling has been the development of computationally efficient procedures for solving existing scheduling models. Research efforts to reduce the computational costs of solving chemical production scheduling models have included the development of reformulations (Velez and Maravelias, 2013a), decomposition-based algorithms (Harjunkoski and Grossmann, 2002), parallel computing tools

(Subrahmanyam et al., 1996), and tightening methods using valid inequalities (Velez, 2013b). In this work we present a reformulation strategy for discrete time-based mixed-integer linear programming (MILP) models for production scheduling. We divide our discussion into two categories: (i) reformulations for MILP models related to batch processes, and (ii) reformulations for MILP models related to developments for continuous processes.

2. Preliminaries

2.1. Batch and Continuous Process Modeling Differences

There are significant modeling differences between batch and continuous processes, primarily due to the fact that continuous processes produce (output) and consume (input) materials continuously and simultaneously, whereas batch processes are assumed to consume all required materials at the start of a task and produce all outputs at the end. The amount of material produced/consumed in continuous production depends on both the rate and duration of task execution, where the duration is not strictly fixed. This implies a different number of degrees of freedom (DOF). Batch processes have one DOF: the batch size. Continuous processes, however, have two DOF: the processing rate and duration of a task. Because of this, MILP models for continuous processes are often carefully constructed extensions of MILP models for batch processes. Hence, we first discuss model reformulations for batch process and later discuss them for continuous processes.

2.2. Notation

We employ the following convention for notation: (i) sets are indicated using bold, upper-case, Roman letters, (ii) indices are indicated using lower-case Roman letters, (iii) parameters are indicated using lower-case Greek letters, and (iv) variables are indicated using non-bold, upper-case, Roman letters. Throughout the work we utilize $i \in \mathbf{I}$ to represent tasks, $j \in \mathbf{J}$ to represent processing units, $k \in \mathbf{K}$ to represent materials, and $n \in \mathbf{N}$ to represent discrete time points. We note that, while the typical use of $n \in \mathbf{N}$ is to represent a discrete time *point* that is $n\delta$ time units beyond the start of the scheduling horizon, it can also be used to represent the time *period* $[(n-1)\delta,\ n\delta)$. For a given $i \in \mathbf{I}$, we use $\mathbf{J}_i \subseteq \mathbf{J}$ to represent the set of units capable of processing task j. Additionally, for a given $k \in \mathbf{K}$, we use $\mathbf{I}_k^+$, $\mathbf{I}_k^- \subseteq \mathbf{I}$ to represent the sets of tasks producing and consuming material k, respectively. The MILP models we employ also rely on the parameters: η - the scheduling horizon; $\delta = \frac{\eta}{|\mathbf{N}|}$ - the discretization time step; $\tau_{i,j}$ - the time required to process task $i \in \mathbf{I}$ in unit $j \in \mathbf{J}_i$; $\beta_j^{\mathrm{MAX}}/\beta_j^{\mathrm{MIN}}$ - the maximum/minimum capacity of unit $j \in \mathbf{J}$; χ_k^{MAX} - the maximum amount of material $k \in \mathbf{K}$ that can be stored; $\rho_{i,k}$ - the conversion coefficient of material $k \in \mathbf{K}$ produced or consumed by task $i \in \mathbf{I}$; and $\xi_{k,n}$ - the net shipment of material $k \in \mathbf{K}$ at time $n \in \mathbf{N}$. We also use the variables: $X_{i,j,n} \in \{0,1\}$, which equals 1 if task $i \in \mathbf{I}$ begins in unit $j \in \mathbf{J}_i$ at time point $n \in \mathbf{N}$; $B_{i,j,n} \in \mathbb{R}_+$, the batch size of task $i \in \mathbf{I}$ processed in unit $j \in \mathbf{J}_i$ at time point $n \in \mathbf{N}$; and $S_{k,n} \in \mathbb{R}_+$, the inventory level of material k during time period t. The MILP models we employ can utilize various objectives, but all include at least the following sets of constraints:

$$\sum_{i \in \mathbf{I}_j} \sum_{t=0}^{\lceil \tau_{i,j}/\delta \rceil - 1} X_{i,j,n-t} \leq 1 \qquad \forall\, j \in \mathbf{J}, n \in \mathbf{N}$$

$$\beta_{i,j}^{MIN} X_{i,j,n} \leq B_{i,j,n} \leq \beta_{i,j}^{MAX} X_{i,j,n} \qquad \forall\, i \in \mathbf{I}, j \in \mathbf{J}_i, n \in \mathbf{N}$$

$$S_{k,n+1} = S_{k,n} + \sum_{i\in \mathbf{I}_k^+}\sum_{j\in \mathbf{J}_i} \rho_{i,k} B_{i,j,n} + \sum_{i\in \mathbf{I}_k^-}\sum_{j\in \mathbf{J}_i} \rho_{i,k} B_{i,j,n} + \xi_{k,n} \le \chi_k \qquad \forall\, k \in \mathbf{K}, n \in \mathbf{N}$$

3. Batch Processes

Beginning with the MILP model employing the three sets of constraints given above, Velez and Maravelias (2013a) propose a reformulation that involves adding an integer variable $N_{i,j}$ that represents the number of times task i is carried out in unit j. Specifically, for each $i \in \mathbf{I}$ and $j \in \mathbf{J}_i$ the authors add the constraint $\sum_{n\in\mathbf{N}} X_{i,j,n} = N_{i,j}$ and bound $N_{i,j}$ as $0 \le N_{i,j} \le \left\lfloor \frac{\eta}{\tau_{i,j}} \right\rfloor$.

We propose the use of several additional integer variables and associated constraints. Consider the variables $N_i \in \mathbb{Z}$ - the number of times task i is performed; $N_j \in \mathbb{Z}$ - the number of times unit j performs a task; $N_n \in \mathbb{Z}$ - the number of tasks performed at time n; and $N \in \mathbb{Z}$ - the total number of tasks performed. These can be incorporated into the MILP model using the following constraints and bounds:

$$N_i = \sum_{j\in\mathbf{J}_i}\sum_{n\in\mathbf{N}} X_{i,j,n}; \qquad 0 \le N_i \le \sum_{j\in\mathbf{J}_i}\left\lfloor \frac{\eta}{\tau_{i,j}} \right\rfloor \qquad \forall i \in \mathbf{I},$$

$$N_j = \sum_{i:j\in\mathbf{J}_i}\sum_{n\in\mathbf{N}} X_{i,j,n}; \qquad 0 \le N_j \le \left\lfloor \frac{\eta}{\min\limits_{i:j\in\mathbf{J}_i}\{\tau_{i,j}\}} \right\rfloor \qquad \forall j \in \mathbf{J},$$

$$N_n = \sum_{i\in\mathbf{I}}\sum_{j\in\mathbf{J}_i} X_{i,j,n}; \qquad 0 \le N_n \le |\mathbf{J}| \qquad \forall n \in \mathbf{N},$$

$$N = \sum_{i\in\mathbf{I}}\sum_{j\in\mathbf{J}_i}\sum_{n\in\mathbf{N}} X_{i,j,n}; \qquad 0 \le N \le \min\left\{\sum_{i\in\mathbf{I}}\sum_{j\in\mathbf{J}_i}\left\lfloor \frac{\eta}{\tau_{i,j}} \right\rfloor, \sum_{j\in\mathbf{J}}\left\lfloor \frac{\eta}{\min\limits_{i:j\in\mathbf{J}_i}\{\tau_{i,j}\}} \right\rfloor\right\}.$$

As each of the variables $N_{i,j}$, N_i, N_j, N_n, and N serves to *keep record* of a quantity of interest, we refer to each as a *record keeping variable* (RKV). We now summarize the results of a study in which we compare the performance of branch-and-bound (BB) when a carefully chosen subset of these RKVs are included within our model. We note that this, as well as the remainder of the tests described in this work, were conducted using GAMS v36.1 for modeling and CPLEX 20.1 as the MILP solver. The tests described here are performed using a combination of randomly generated instances and instances obtained from minlp.org and from Velez and Maravelias (2013a). In total, we utilize 115 instances and consider cost minimization and profit maximization objectives. For brevity, and because the results for both objective types displayed relatively similar patterns, we only present results for cost minimization here. We report results for all instances that at least one of the considered formulations was able to solve in under 5 hours. The results are summarized in the performance profile displayed in Figure 1.

In Figure 1, we use the following notation: O - no RKVs are added; B - $N_{i,j}$ is added; I - N_i is added; BIA - $N_{i,j}$, N_i, and N are added; BIJA - $N_{i,j}$, N_i, N_j, and N are added; and BIJNA - $N_{i,j}$, N_i, N_j, N_n, and N are added. Here O is selected as it represents the original formulation with no RKVs, B is selected as it represents the most promising reformulation considered in (Velez and Maravelias, 2013a), and the rest of the considered reformulations are selected for their relatively strong performance. Of the 115 instances considered, there were 11 that no considered formulation was able to solve in 5 hours, 8 that were solved in under 5 hours by B and BIJA but not by O (B - 4667.8s avg; BIJA - 170.4s avg), and 22 that were solved in under 5 hours by BIJA but not be either O or B (BIJA - 399.4s avg). Moreover, of the remaining 74 instances, there were 44 that at least one of the formulations O, B, or BIJA took more than 5 seconds to solve, and on average BIJA solved these in 1.16% of the time used by O and 2.63% of the time used by B.

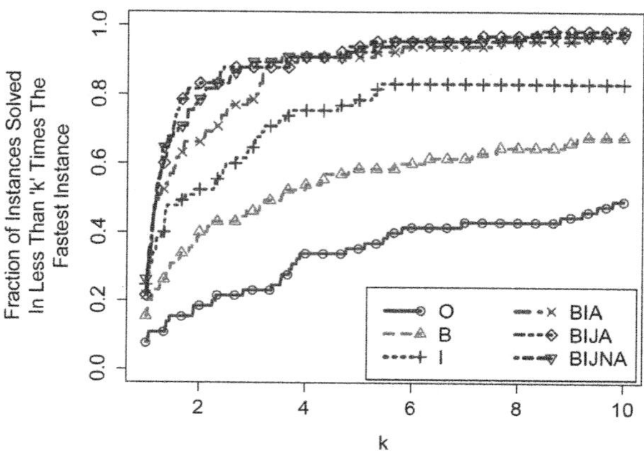

Figure 1: Performance Profile for Batch Process Formulations

4. Continuous Processes

In the field of chemical production scheduling, less focus has been given to continuous processes relative to batch processes. The simultaneous production and consumption of materials in conjunction with the flexibility to choose the duration of task execution adds a layer of processing complexity to continuous process models. Strategies to more efficiently model continuous processes, such as introducing RKVs, would prove to be valuable, especially when considering transient operations such as startups, shutdowns, and direct transition tasks that notably escalate model complexity; transient operations can result in the addition of up to four more binary variables per task as well as the ancillary constraints needed for their modeling. A general optimization framework that accurately represents system dynamics, specifically transient operations, is utilized; however, the reader is directed to the literature for the extensive list of modeling constraints (Wu and Maravelias, 2021).

From a modeling perspective, a continuous task is represented as a series of identical subtasks, each having a processing time of one discrete time step. Any set of consecutive executions of a single subtask are then referred to as a *run* of the associated continuous task. The minimum/maximum processing rates of a task $i \in \mathbf{I}$ executed in a unit $j \in \mathbf{J}_i$ during each time period are comparable to batch sizes in the batch process model, so they are represented similarly ($\beta_{i,j}^{\mathrm{MIN}}/\beta_{i,j}^{\mathrm{MAX}}$). In terms of run duration, parameters to enforce run length restrictions ($\tau_{i,j}^{\mathrm{MIN}}/\tau_{i,j}^{\mathrm{MAX}}$) are also applied. In addition to $X_{i,j,n}$, two new binary variables containing information about the start and end times of a run are employed; $Y_{i,j,n}^{\mathrm{S}}$and $Y_{i,j,n}^{\mathrm{E}}$ indicate that a run of task i in unit j starts, or ends, at time n, respectively. The values of these variables are enforced as:

$$Y_{i,j,n}^{\mathrm{S}} = X_{i,j,n} - X_{i,j,n-1} + Y_{i,j,n}^{\mathrm{E}} \qquad \forall\, i \in \mathbf{I}^{\mathrm{P}}, j \in \mathbf{J}_i, n \in \mathbf{N}$$

The above equation ensures that $Y^{\mathrm{S}}_{i,j,n} = 1$ whenever a run of a production task $i \in \mathbf{I}^{\mathrm{P}} \subseteq \mathbf{I}$ begins. We note that production tasks are simply tasks that are not transition tasks, i.e., their role is to produce material, as opposed to transitioning to/from other tasks.

The same RKVs introduced in Section 3, i.e., those defined in terms of the $X_{i,j,n}$ binary variables, can be written for continuous process models. However, with the availability of the $Y^{\mathrm{S}}_{i,j,n}$ binary variable, an additional class of RKVs can be as follows:

$$N^{\mathrm{Y}}_{i,j} = \sum_{n \in \mathbf{N}} Y^{\mathrm{S}}_{i,j,n}; \qquad 0 \le N^{\mathrm{Y}}_{i,j} \le \left\lfloor \frac{\eta}{\tau^{\mathrm{MIN}}_{i,j} + \alpha^{\mathrm{MIN}}_{i,j}} \right\rfloor \qquad \forall i \in \mathbf{I}^{\mathrm{P}}, j \in \mathbf{J}_i,$$

$$N^{\mathrm{Y}}_{i} = \sum_{j \in \mathbf{J}_i} \sum_{n \in \mathbf{N}} Y^{\mathrm{S}}_{i,j,n}; \qquad 0 \le N^{\mathrm{Y}}_{i} \le \sum_{j \in \mathbf{J}_i} \left\lfloor \frac{\eta}{\tau^{\mathrm{MIN}}_{i,j} + \alpha^{\mathrm{MIN}}_{i,j}} \right\rfloor \qquad \forall i \in \mathbf{I}^{\mathrm{P}},$$

$$N^{\mathrm{Y}}_{j} = \sum_{i: j \in \mathbf{J}_i} \sum_{n \in \mathbf{N}} Y^{\mathrm{S}}_{i,j,n}; \qquad 0 \le N^{\mathrm{Y}}_{j} \le \left\lfloor \frac{\eta}{\min\limits_{i: j \in \mathbf{J}_i}\{\tau^{\mathrm{MIN}}_{i,j}\}} \right\rfloor \qquad \forall j \in \mathbf{J},$$

$$N^{\mathrm{Y}}_{n} = \sum_{i \in \mathbf{I}} \sum_{j \in \mathbf{J}_i} Y^{\mathrm{S}}_{i,j,n}; \qquad 0 \le N^{\mathrm{Y}}_{n} \le |\mathbf{J}| \qquad \forall n \in \mathbf{N},$$

$$N^{\mathrm{Y}} = \sum_{i \in \mathbf{I}} \sum_{j \in \mathbf{J}_i} \sum_{n \in \mathbf{N}} Y^{\mathrm{S}}_{i,j,n}; \qquad 0 \le N^{\mathrm{Y}} \le \min\left\{ \sum_{i \in \mathbf{I}} \sum_{j \in \mathbf{J}_i} \left\lfloor \frac{\eta}{\tau^{\mathrm{MIN}}_{i,j} + \alpha^{\mathrm{MIN}}_{i,j}} \right\rfloor, \sum_{j \in \mathbf{J}} \left\lfloor \frac{\eta}{\min\limits_{i: j \in \mathbf{J}_i}\{\tau^{\mathrm{MIN}}_{i,j}\}} \right\rfloor \right\}.$$

In the above, $\alpha^{\mathrm{MIN}}_{i,j}$ represents the minimum transition time needed to transition to and from a production task i. If a production task has no transitions associated with it, $\alpha^{\mathrm{MIN}}_{i,j} = 0$. To test the utility of continuous models that incorporate RKVs, we include various combinations of the above RKVs that track the number of runs of a task ($Y^{\mathrm{S}}_{i,j,n}$) and the RKVs that track the task execution ($X_{i,j,n}$) within our cost-minimization MILP model. A total of 96 instances were solved to optimality by all formulations within a 24-hour time limit, and a performance chart of the results is given in Figure 2.

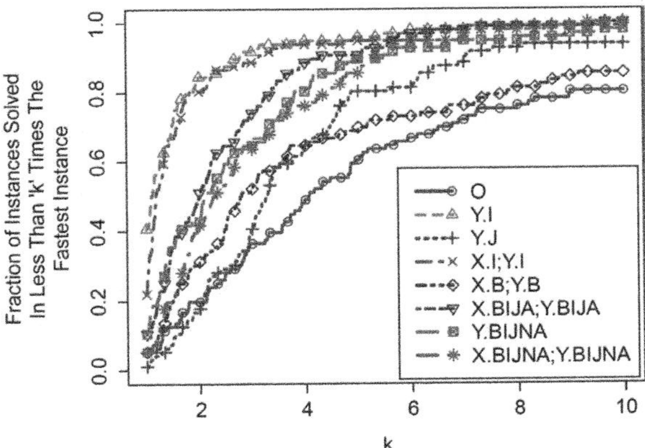

Figure 2: Performance Profile for Continuous Process Formulations

In Figure 2 we use similar notation to that of Figure 1, but we prefix the run-related variables with "Y." and the task execution-related variables with "X.". We note that we considered several combinations of RKVs beyond those for which results are displayed. However, for brevity, only formulations that showed strong performance relative to the original formulation with no RKVs are illustrated in the figure. From our results, it appears that the run-related RKVs perform better than the task execution-related RKVs. Additionally, we observe from the curves labeled "X.I;Y.I" and "Y.I" in Figure 2 that N_i^{Y} was the most impactful of our considered RKVs. We also point out that formulation Y.I solved about 40% of the instances the fastest. One possible explanation for these results is that by branching on N_i^{Y}, the solver is able to bound the number of runs of each task i, thus efficiently eliminating suboptimal or infeasible schedules and closing the optimality gap more quickly.

5. Conclusion

We have presented results that demonstrate the large computational improvements that can be achieved by simply incorporating RKVs into chemical production scheduling MILP models. Moreover, we have provided evidence that such improvements can be obtained in the context of both batch processes and continuous processes. However, this reformulation technique is not limited to only chemical production scheduling models but can also be extended to various other types of MILP models that utilize a large quantity of integer variables.

References

P. M. Castro and I. E. Grossmann, 2014, Global optimal scheduling of crude oil blending operations with RTN continuous-time and multiparametric disaggregation, Industrial and Engineering Chemistry Research, 53, 39, 15127–15145.

D. M. Giménez, G. P. Henning, and C. T. Maravelias, 2009, A novel network-based continuous-time representation for process scheduling: Part I. Main concepts and mathematical formulation. Computers and Chemical Engineering, 33, 9, 1511–1528.

I. Harjunkoski and I. E. Grossmann, 2002, Decomposition techniques for multistage scheduling problems using mixed-integer and constraint programming methods. Computers and Chemical Engineering, 26, 1533–1552.

C. A. Méndez, G. P. Henning, and J. Cerdá, 2001, An MILP continuous-time approach to short-term scheduling of resource-constrained multistage flowshop batch facilities, Computers and Chemical Engineering, 25, 701–711.

V. Papavasileiou, A. Koulouris, C. Siletti, and D. Petrides, 2007, Optimize manufacturing of pharmaceutical products with process simulation and production scheduling tools, Chemical Engineering Research and Design, 85, 7A, 1086–1097.

S. Subrahmanyam, G. K. Kudva, M. H. Bassett, and J. F. Pekny, 1996, Application of Plant Distributed Design and Computing to Batch Scheduling, AIChE Journal, 42, 6, 1648–1661.

S. Velez and C. T. Maravelias, 2013, Reformulations and branching methods for mixed-integer programming chemical production scheduling models, Industrial and Engineering Chemistry Research 52, 10, 3832–3841.

S. Velez, A. Sundaramoorthy, and C. T. Maravelias, 2013, Valid Inequalities Based on Demand Propagation for Chemical Production Scheduling MIP Models, AIChE Journal, 59, 3, 872–887.

S. Velez, Y. Dong, and C.T. Maravelias, 2017, Changeover formulations for discrete-time mixed-integer programming scheduling models, Eur J Oper Res, 260, 3, 949–963.

L. A. Wolsey, 1997, MIP modelling of changeovers in production planning and scheduling problems, European Journal of Operational Research 99, 154–165.

Y. Wu and C. T. Maravelias, 2021, A general framework and optimization models for the scheduling of continuous chemical processes, AIChE Journal, 67, 10, 1–15.

Antonis Kokossis, Michael C. Georgiadis, Efstratios N. Pistikopoulos (Eds.)
PROCEEDINGS OF THE 33rd European Symposium on Computer Aided Process Engineering (ESCAPE33), June 18-21, 2023, Athens, Greece
 http://dx.doi.org/10.1016/B978-0-443-15274-0.50070-6

ESTAN – A toolbox for global sensitivity based estimability analysis

Ilias Bouchkira,[a,b] Abderrazak M. Latifi,[b,*] Brahim Benyahia,[a,*]

[a] *Department of Chemical Engineering, Loughborough University, Epinal Way, Loughborough, LE11 3TU, United Kingdom.*

[b] *Laboratoire Réactions et Génie des Procédés, CNRS-ENSIC, Université de Lorraine, 1 rue Grandville, BP20451, Nancy Cedex, 54001, France.*

B.Benyahia@lboro.ac.uk abderrazak.latifi@univ-lorraine.fr

Abstract

Recently in process engineering field, there is an increasing demand for high fidelity, large and multi-scale mathematical models. In most cases, these models involve several unknown parameters whose identifiability from experimental measurements is often not guaranteed. It is therefore necessary to carry out an estimability analysis to determine which parameters can be reliably estimated. This task is however laborious and is still neglected in most studies. Most importantly, its wide adoption is hampered by the lack of standardized tools or methodologies. To address these issues, a new estimability toolbox, ESTAN, was developed to make the estimability analysis accessible to a broader community of specialist and non-specialist users. ESTAN can handle different types of mathematical models including dynamic and non-dynamic models. It uses a Quasi-Monte Carlo method to sample the unknown model parameters within their range of variation. Then, depending on whether the studied model is computationally cheap or expensive, global sensitivity indices are calculated using either the Sobol method or the Fourier Amplitude Sensitivity Test. The sensitivities are exploited within an orthogonalization algorithm to rank the parameters from the most to the least estimable followed by the identification of the subset of the most estimable parameters based on a preset estimability threshold. Finally, more reliable parameter estimates are obtained for the subset of the most estimable parameters. To validate the toolbox and demonstrate its capabilities, ability analysis of three models is performed using the developed toolbox. They are given by a non-dynamic, a dynamic, and a computationally expensive model. The results for the case studies are found to be very promising, showing how the presented toolbox simplifies the investigation of the estimability analysis, and significantly improves the model's precision.

Keywords: Sensitivity analysis, Estimability analysis, Orthogonalization, Toolbox.

1. Introduction

For many years, mathematical models have been employed extensively in various branches of science and engineering. They are frequently utilized in the design of processes and plants, risk analysis, process optimization, and control. These models typically consist of thermodynamic, kinetic, and hydrodynamic equations as well as momentum, mass, and energy balances. Recently, with the advent of industry 4.0 and inherent digital transformation, there is an increasing need for more precise, high fidelity, large, and multi-scale models. However, these models always involve several parameters to be identified from available experimental measurements. The presence of a large number of model parameters to be identified combined with poor data information content and data redundancy, model structure, correlation between parameters, non-linearity with respect to model states and parameters etc., result in poor prediction capabilities, and poor model interpolation and extrapolation. To address some of these critical issues, it is necessary to implement an estimability analysis approach to assess whether the model parameters can be identified from the available or planned experimental data. The current local/global estimability approaches involve complex and laborious calculations, among which, Monte/Quasi-Monte Carlo sampling, estimation of local and/or global sensitivity indices, assessing sensitivity indices convergence, orthogonalization etc. These techniques are difficult to develop, implement, or verify, and quite often, do not necessarily belong to the modeler's skills. Consequently, the estimability part is commonly neglected by simply assuming that the available data contain the necessary information to identify all the unknown model parameters, which is very often not true.

Several estimability approaches have been investigated in the literature (Bouchkira et al.,2021, Fysikopoulos et al., 2017). These techniques are not unified, and currently, no single toolbox is available. ESTAN toolbox addresses these gaps and suggests a user-friendly interface to help address estimability in the cases of dynamic and non-dynamic models (i.e., ODEs, DAEs, Algebraic expressions). The objective is also to propose a versatile and standard tool for all engineering and scientific subjects and make the estimability analysis accessible to non-specialist users (i.e., modelers with only basic background in sensitivity and estimability analysis and/or MATLAB). The toolbox is proposed as an open-source package to allow experienced modelers to easily use it, customize and even integrate it to other digital tools.

2. ESTAN structure

Figure 1 shows the framework of the parameters estimability analysis toolbox ESTAN. It is based on computing the global sensitivity indices of the model inputs (with respect to the outputs) and then process the sensitivity matrix Z using an orthogonalization algorithm to determine which parameters can be reliably estimated, and in which order.

The estimability matrix Z below represents the global sensitivity indices of the model inputs with respect to the outputs at a given set of operating conditions, for steady models, and sensitivity indices at given sampling points for the dynamic models.

$$Z = \begin{pmatrix} S_{1,1}|_{m_{1,k}} & \cdots & S_{1,d}|_{m_{1,k}} \\ \vdots & \ddots & \vdots \\ S_{n,1}|_{m_{n,k}} & \cdots & S_{n,d}|_{m_{n,k}} \end{pmatrix}, k \in [1; M] \quad (1)$$

where $m_{i,k}$ is the ith sampling point of the kth experiment (dynamic models) or the set of operating conditions (steady state models) and M is number of observed/measured outputs, $ntheta$ is the number of model parameters and n is either the number of sampling points (dynamic) or number of repeated experiments (steady state or algebraic models). $S_{i,j}$ is the individual sensitivity coefficient of the ith output with respect to the jth parameter.

Among several sensitivity analysis approaches available in the literature, Saltelli's approach (Saltelli, 2008) and Fast approach (Cukier et al., 1973) have been adopted in the proposed toolbox, as they are very reliable, precise, and widely accepted. In both approaches, the variance-based sensitivity indices are given by partial and total output variances:

$$S_{i,k} = \frac{V_i}{Var(Y)}|_k \,, \qquad k \in [1; M] \tag{2}$$

ESTAN toolbox requires several user inputs for sensitivity analysis performance, among which, the number of parameters and measured outputs, the variability bounds of the unknown parameters, the initial conditions etc. Before performing sensitivity analysis, well informed variability bounds are required which can be obtained from the literature for existing models. Moreover, depending on whether the model to be evaluated is computationally cheap or expensive, the toolbox computes the $S_{i,k}$ either based on the approach developed by Saltelli (2008), using Quasi-Monte Carlo sampling and statistical estimates to approximate the average and the variance of the model outputs, or the Fast method developed by (Cukier et al., 1973), which is based on Fourier series to evaluate the partial and total variances in $S_{i,k}$ (Test 1). Once the sensitivity indices $S_{i,k}$ are computed, they are used to build the Z estimability matrix, to be processed by the orthogonalization algorithm as in (Fysikopoulos et al., 2019). The later allows to determine the most estimable parameters as well as their ranking according to their estimability from the available experimental measurements (Test 2).

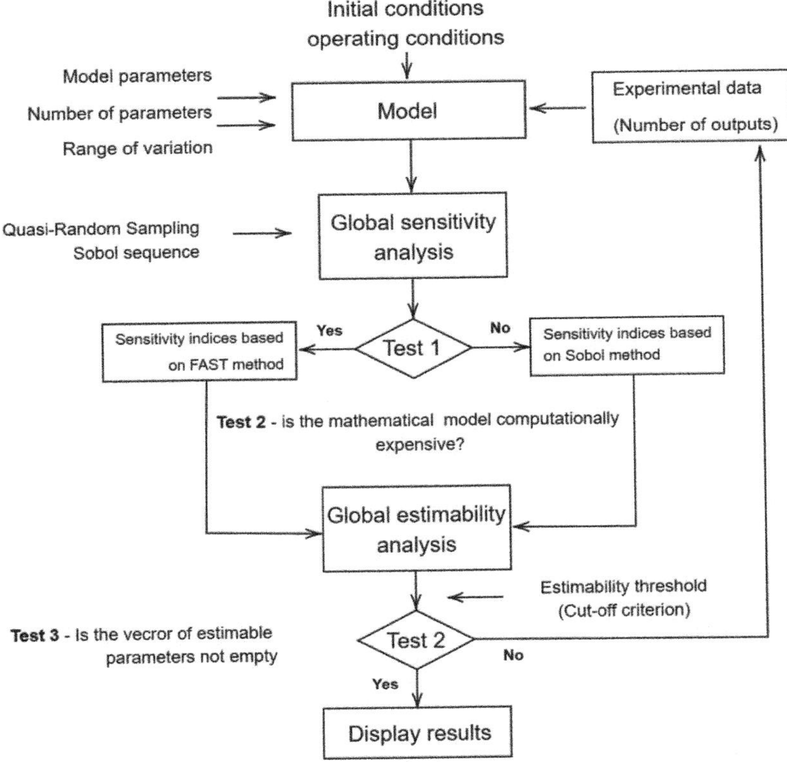

Figure 1: Simplified Framework of the parameters estimability analysis using the toolbox.

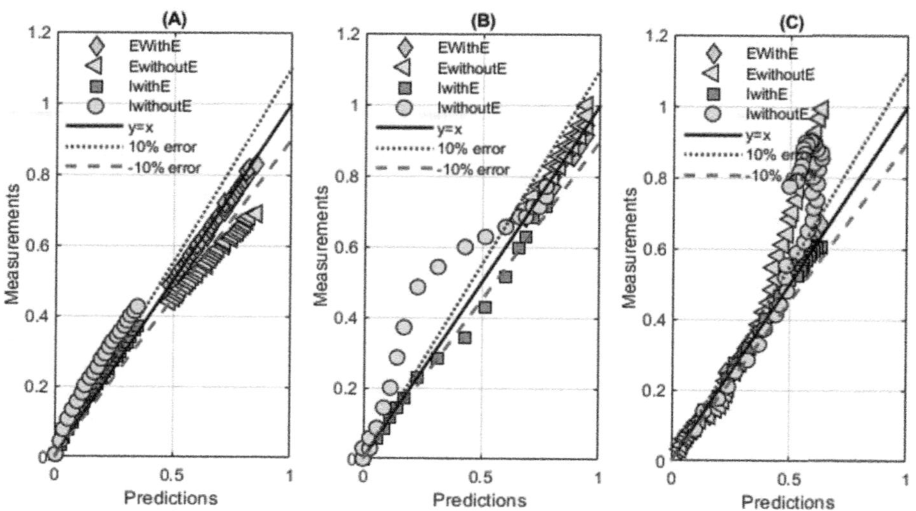

Figure 2: Results for model extrapolation and Interpolation. (A): Case of the non-dynamic model. (B): Case of the dynamic model. (C): Case of an expensive model. E/Iwith/withoutE: Extrapolation/Interpolation with/without estimability

Table 1: Summary of the outcomes of the GSA and estimability analysis.

	Interpolation	Extrapolation
	Non-dynamic model	
Estimable parameters	3 out of 6	
Fixed values (3)	(Bedel et al., 2017)	
CPU time (Sobol)	173.02 seconds	
MPE with estimability (%)	4.114	4.541
MPE without estimability (%)	33.10	13.45
	Dynamic model	
Estimable parameters	3 out of 4	
Fixed values (1)	(Benyahia et al., 2017)	
CPU time (Sobol)	187.21 seconds	
MPE with estimability	0.226	0.441
MPE without estimability	0.238	39.51
	Expensive model	
Estimable parameters	4 out of 10	
Fixed values (6)	(Bouchkira et al., 2022;2023)	
CPU time (FAST)	359 seconds	
MPE with estimability	2.331	2.121
MPE without estimability	17.91	31.74

3. Case studies

3.1. Models

To validate the developed toolbox, three case studies are considered. They include (1) an adsorption isothermal model from (Bedel et al., 2017) involving 6 unknown parameters. (2) A dynamic cooling batch crystallization model from (Benyahia et al., 2021) involving 4 unknown parameters. (3) A thermodynamic model from (Bouchkira et al., 2023) involving 10 unknown parameters.

3.2. Experimental measurements

A database of experimental measurements was developed mainly from our previous works. For the estimability of the parameters of the adsorption isothermal model, measurements of carbon dioxide adsorption on a commercial adsorbent are taken from (Bedel et al., 2017). For the estimability of the batch cooling crystallization model, experimental measurements concerning particle size distribution of crystallized paracetamol are taken from (Anandan al., 2021). Finally, for the thermodynamic model, equilibrium measurements are taken from (Bouchkira et al., 2022;2023).

3.3. Computational method

The present toolbox was developed based on App Designer MathWorks and involves several MATLAB built-in functions which are available only in MATLAB v.2020 onwards. A reasonable computational power is required to cope with GSA and inherent simulation (e.g., Intel or AMD x86-64 processor and > 4 GB RAM). All calculations in this work were performed using an ASUS workstation with 12th Gen Intel(R) Core (TM) i9-12900K 3.20 GHz 128GB Ram processor.

4. Results and discussion

The estimability analysis results for the three case studies based on ESTAN toolbox are summarized in Table 1 and Figure 2. The sensitivity analysis for the first two cases was performed based on Saltelli (2008) in view of their non-expensive computing nature. To this end, a Quasi-Monte Carlo sampling of size 5000 was used. Whilst, for the last case, where the model is moderately expensive, FAST method based on Cukier et al. (1973) was investigated. The CPU times to achieve convergence of global sensitivity indices were estimated to 173 seconds for the non-dynamic model, 187 seconds for the dynamic model, and 359 for the algebraic expression-based model. The computed global sensitivity indices were then used to fill the Z estimability matrices for all tree cases, and these matrices were used to determine the number of estimable parameters in each case study, as well as their estimability-based ranking. Based on the collected data, it has been found that 3 parameters out of 6 are estimable in the first case, 3 parameters out of 4 are estimable in the second case and 4 parameters out of 10 are estimable in the last case. For each case, two scenarios are carried out for comparison purposes. In the scenarios "with estimability", only estimable parameters are identified from the available measurements and non-estimable parameters are fixed from previous works. While in the scenario "without estimability", all model parameters are identified neglecting the estimability analysis results. Moreover, for both scenarios, in each case study, the resulting models are used for model interpolation and extrapolation.

As expected, the results show how carrying out estimability analysis can significantly improve the model precision. This can be shown quantitatively from the computed Mean Percentage Errors (MPE) between models' predictions and experiments. It has been found that taking into account estimability analysis results reduced the MPE from 33.10% to 4.11% and from 13.45% to 4.54% respectively for interpolation and extrapolation in the case of the non-dynamic model, from 0.24% to 0.22% and from 39.51% to 0.44% respectively for interpolation and extrapolation in the case of the dynamic model, from 17.91 to 2.33% and from 31.74% to 2.12% respectively for interpolation and extrapolation in the case of the algebraic expression based model. These results can finally be presented in a normalized parity diagrams as in figure 2, where it can be seen that the models keep good prediction capabilities when taking estimability analysis results into account, while they fail to keep good interpolation and extrapolation performances when neglecting estimability.

5. Conclusion

ESTAN toolbox was developed to standardize the estimability analysis, and to help improve model prediction capabilities and precision. As demonstrated in the three case studies, ESTAN can be used for both cheap and expensive computationally models, through its ability to switch between Sobol and FAST methods for the calculation of sensitivity indices, to help enhance model prediction capabilities. Moreover, it is developed to avoid all laborious tasks involved in estimability analysis. It is worth mentioning that the current ESTAN version uses an arbitrary estimability criterion to determine the set of estimable parameters, and it is possible to implement a systematic approach to determine the optimal value of this cutoff estimability criterion as in (Bouchkira et al. 2021). This is a very interesting perspective that we will address in our future works. The estimability open-source toolbox will be freely available from the authors for non-commercial research and educational purposes.

References

A. Saltelli, 2002, Making best use of model evaluations to compute sensitivity indices. Computer physics communications, 145(2), 280–297.

B. Benyahia, M. A. Latifi, C. Fonteix, F. Pla, (2011). Modeling of a batch emulsion copolymerization reactor in the presence of a chain transfer agent: Estimability analysis, parameters identification and experimental validation. In Computer Aided Chemical Engineering (Vol. 29, pp. 126-130). Elsevier.

B. Benyahia, P. D. Anandan, & C. Rielly, (2021). Robust Model-Based Reinforcement Learning Control of a Batch Crystallization Process. In 2021 9th International Conference on Systems and Control (ICSC) (pp. 89-94). IEEE.

D. Fysikopoulos, B. Benyahia, A. Borsos, Z. K. Nagy, C. D. Rielly, (2019). A framework for model reliability and estimability analysis of crystallization processes with multi-impurity multi-dimensional population balance models, Computers & Chemical Engineering, 122, 275-292.

I. Bouchkira, A. Latifi, L. Khamar, S. Benjelloun, 2021, Global sensitivity based estimability analysis for the parameter identification of Pitzer's thermodynamic model. Reliability Engineering & System Safety, 207, 107263.

I. Bouchkira, A. Latifi, L. Khamar, S. Benjelloun, 2022. Modeling and multi-objective optimization of the digestion tank of an industrial process for manufacturing phosphoric acid by wet process. Computers & Chemical Engineering, 156, 107536.

I. Bouchkira, S. Benjelloun, L. Khamar, A. Latifi, 2023. Thermodynamic modeling and parameter estimability analysis of a wet phosphoric acid process with impurities. Fluid Phase Equilibria, 564, 113594.

R. Cukier, C. Fortuin, K. Shuler., A. Petschek, J. Schaibly, 1973. Study of the sensitivity of coupled reaction systems to uncertainties in rate coefficients. I Theory. The Journal of chemical physics, 59(8), 3873–3878.

S. Bedel, C. Vallieres, A. Latifi, 2017, Parameters estimability analysis and identification for adsorption equilibrium models of carbon dioxide. Adsorption, 23(2), 373–380.

Antonis Kokossis, Michael C. Georgiadis, Efstratios N. Pistikopoulos (Eds.)
PROCEEDINGS OF THE 33rd European Symposium on Computer Aided Process Engineering (ESCAPE33), June 18-21, 2023, Athens, Greece
 http://dx.doi.org/10.1016/B978-0-443-15274-0.50071-8

Multi-technology separation network synthesis

Garry S.P. Taifan,[a] Christos T. Maravelias,[a,b]

[a]*Department of Chemical and Biological Engineering, Princeton University, 50-70 Olden St, Princeton and 08540, USA*
[b]*Andlinger Center for Energy and Environment, Princeton University, 86 Olden St, Princeton and 08540, USA*
maravelias@princeton.edu

Abstract

Considering multiple technologies when synthesizing separation networks can lead to a more efficient separation. We present a superstructure-based approach to synthesize separation networks while considering multiple technologies. The approach uses key separation properties to generate potential separation sequences for each technology and employs richly connected superstructure to represent numerous potential configurations. Using a generalized problem statement, the proposed model can handle variable inlet and outlet streams thereby allowing the coupling with reactor network synthesis models.
Keywords: process synthesis, separation, global optimization, superstructure.

1. Introduction

In many chemical processes, a sequence of separation tasks is typically performed to recover unconverted raw materials and to purify final products. Synthesizing an appropriate separation network can result in significant savings for the overall process. Most of the available studies on separation network synthesis revolve around distillation with other technologies being an afterthought. Consequently, the benefits of having multiple technologies in the separation network are not fully exploited. Nevertheless, attention toward non-thermal separations has recently gained traction due to their potential for energy-efficient processes and positive environmental impact (Sholl and Lively, 2016). The incorporation of different technologies in separation network synthesis expands the search space, and thus allows potentially better networks to be found.

The separation network synthesis problem has been addressed using various approaches. Heuristic approaches rely on rules based on engineering knowledge and judgment, whereas evolutionary approaches generate a network by sequentially improving several initial flowsheets based on heuristics. Alternatively, most potentially useful units and their relevant interconnections can be embedded in a superstructure (Sargent and Gaminibandara, 1976). Superstructure-based approaches consider multitudes of promising network configurations in an optimization problem. Agrawal (2003) introduced a systematic method to generate all possible basic distillation configurations, and Heckl et al. (2006) considered different technologies explicitly. However, the former approach used a problem statement where complete information on the feed stream is required. Ryu et al. (2020) discussed a generalized problem statement which allows the seamless integration of the separation network and reactor network synthesis.

Accordingly, we propose a superstructure-based approach for separation network synthesis with multiple technologies where the feed streams may be variable, enabling coupling with reactor network synthesis. The approach employs a superstructure that encompasses numerous promising configurations comprising multiple separation technologies. For each technology, we construct a separation matrix (Shah and Agrawal, 2010) based on the key properties that enable separation. We identify all possible separations and connections using these matrices and generate the corresponding superstructure. The synthesis problem is formulated as a mixed-integer nonlinear programming (MINLP) model. We demonstrate the applicability of the proposed approach through a multicomponent mixture separation case study.

2. Framework

2.1. Problem statement

For all technologies, we consider simple sharp split separators with no distributed components. This separation is performed by exploiting the difference in the magnitude of some key properties (e.g., volatility for distillation). Consequently, a component ranking system based on a specific key property is used to generate possible separation splits. In zeotropic distillation network synthesis, for example, all possible sequences of the separations can be enumerated if the ranking of components' relative volatilities is given. Analogously, we can generate sequences of separations for other technologies. The difference in the rankings between technologies provides an opportunity for more efficient separation. If two components are adjacent in technology A's ranking but not in technology B's ranking, separating them using technology B may be easier. It is worth noting that the actual separation may favor the former technology depending on the magnitudes of the key properties. If components have the same ranking, then it would be impractical to separate these components using that technology.

Traditionally, reactor and separation systems are synthesized sequentially; therefore, the conventional separation network synthesis problem requires full information on the inlet stream. When these two systems are considered simultaneously, the flowrates of the separation network's inlet stream vary depending on the decisions made in the reactor system. In this case, the inlet stream is termed variable inlet stream. The generalized separation network synthesis problem can be stated as follows. We are given variable inlet and outlet streams (with some specifications), the ranking of postulated components for all technologies, and cost functions for the separation tasks. The objective is to find the least cost separation network.

2.2. Superstructure generation

We use the concepts of nodes and arcs to represent the superstructure (Kong and Maravelias, 2020). We introduce set $\boldsymbol{N}$ to denote all nodes and set $\boldsymbol{I} = \{A, B, C, \dots\}$ to denote the components in the superstructure. Set $\boldsymbol{N}$ can be split into three subsets: (1) source nodes $\boldsymbol{N}^{SO}$ (inlets to the network); (2) sink nodes $\boldsymbol{N}^{SI}$ (outlets of the network); and mixture nodes $\boldsymbol{N}^{M}$ (nodes where separation or direct transfer to the sink nodes can take place). Each mixture node n is uniquely associated with a set of present components $i \in \boldsymbol{I}_n^M$. Throughout the manuscript, mixture nodes are labeled by the present components in alphabetical order (e.g., $\boldsymbol{I}_{ABC}^M = \{A, B, C\}$).

To enumerate potential sequences from an initial mixture, we utilize the matrix method of Shah and Agrawal (2010), originally developed to find a distillation sequence to separate a $|\mathbf{I}|$-component mixture into $|\mathbf{I}|$ pure product streams. The construction of the matrix for technology $t \in \mathbf{T}$ requires the ranking of the components, $\rho_{t,i}$, and an initial mixture. For an element (or equivalently mixture) located in row j and column k, the candidate top products are in the same row (i.e., $j' = j, k' > k$) while the candidate bottom products are in the same diagonal (i.e., $j' = j + k' - k, k > k$). To illustrate, in Fig 1 (left panel), we construct matrices for initial mixture ABCD for three technologies with different rankings. The solid/dotted arrows represent top/bottom arcs connecting a mixture to a candidate top/bottom product, and the dashed arrows connect equivalent mixtures in different matrices. One possible separation sequence for mixture ABCD is split AC/BD in technology T2 followed by split A/C in technology T1 and split B/D in technology T3. Note that since component B (or A) is missing from the top (or bottom) product of split AC/BD, mixture AC (or BD) is equivalent to mixture ABC (or ABD) in the matrix for technology T1 (or T2). The equivalent sequence in the aggregated matrix is shown in Fig 1 (right panel). Here, both mixtures AC and BD are assigned unique labels.

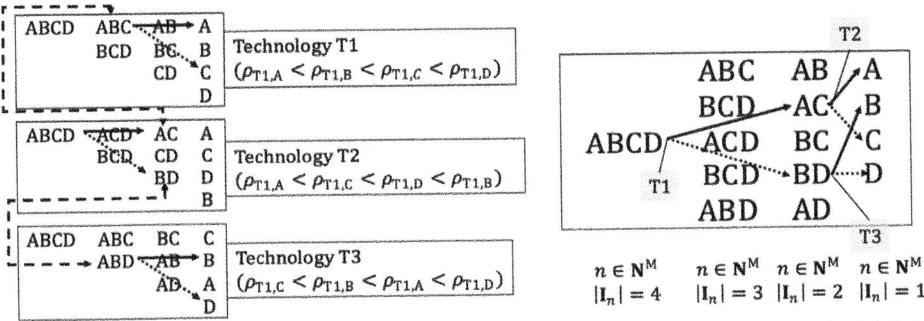

Figure 1: Different technology separation matrices (left) and aggregated matrix (right).

Similar to the aggregated matrix, in the final superstructure, mixtures from different technology matrices are combined into unique nodes, $n \in \mathbf{N}^M$, (see Fig 2); and the top and bottom arcs are included in set $\mathbf{A}_t^{\mathrm{TECH,T}}/\mathbf{A}_t^{\mathrm{TECH,B}}$. The overall top/bottom arcs are contained in set $\mathbf{A}^{\mathrm{T}}/\mathbf{A}^{\mathrm{B}}$, the union of $\mathbf{A}_t^{\mathrm{TECH,T}}/\mathbf{A}_t^{\mathrm{TECH,B}}$ over $t \in \mathbf{T}$.

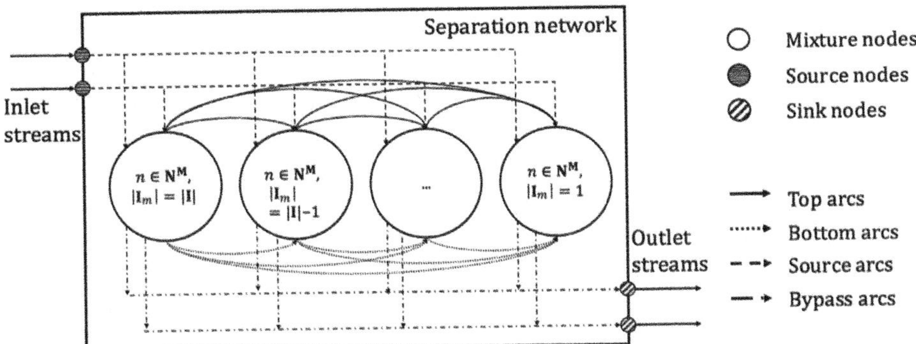

Figure 2: Superstructure representation of the separation network.

2.2.1. Mixture nodes

In each mixture node, streams from (1) sources, (2) top products, and (3) bottom products are mixed (see Fig. 3, left panel). A fraction of this mixed stream can be sent to the outlets directly, thereby bypassing unnecessary separations, while the remaining stream is sent to a separation block where technology t is performed leading to the outgoing top and bottom products. A separation block in this approach represents a set of operations that enables a sharp split separation. In solvent extraction, for instance, a separation block consists of an extraction column and the solvent recovery system (see Fig. 3, right panel). Similarly, in reverse osmosis, we utilize multi-stage membrane units to achieve a sharp split separation. We also introduce subset $\mathbf{N}^{\mathrm{M,P}} \subset \mathbf{N}^{\mathrm{M}}$ to denote mixture nodes containing only pure component (i.e., $|\mathbf{I}_n^{\mathrm{M}}| = 1$). In these nodes, no separation will be performed; hence, the mixed stream is directly sent to the outlets.

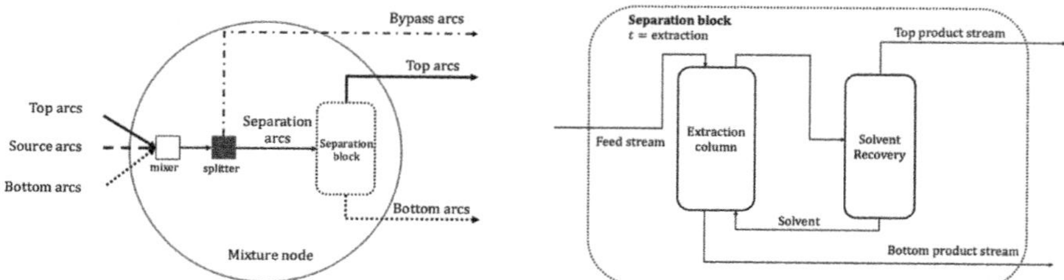

Figure 3: Mixture node (left) and separation block for $t =$ extraction (right). Separation block and its streams are relevant only for $n \in \mathbf{N}^{\mathrm{M}} \setminus \mathbf{N}^{\mathrm{M,P}}$.

2.2.2. Connectivity

Using the previously defined sets, we employ $X_n^{\mathrm{M}} \in \{0,1\}$ to indicate if mixture node n is active and $Y_{\hat{n},n}^{0} \in \{0,1\}$ to denote the activation of the source arc connecting source node $\hat{n} \in \mathbf{N}^{\mathrm{SO}}$ to mixture node n. From each source node, there can only be one active arc entering an active mixture node. We employ $Z_i \in \{0,1\}$ to denote the existence of component i in the separation network, and using this variable, we deactivate several mixture nodes when a component is not present.

To represent connections between mixture nodes, we employ $Y_{n,\dot{n}}^{\mathrm{T/B}} \in \{0,1\}$ to denote the active top/bottom arc connecting mixture node n to its top/bottom product mixture node $\dot{n}$. The active pairs of top and bottom arcs represent the performed separation splits in the network. We employ $Y_{n,\tilde{n}}^{\mathrm{BP}} \in \{0,1\}$ to denote active arcs from the splitter which bypasses separation and directly goes to sink node $\tilde{n} \in \mathbf{N}^{\mathrm{SI}}$. For node $n \in \mathbf{N}^{\mathrm{M}} \setminus \mathbf{N}^{\mathrm{M,P}}$, the remaining mixed stream can undergo separation which produces top and bottom products. The selection of separation technology in this node is done via variable $Y_{n,t}^{\mathrm{C}} \in \{0,1\}$. On the other hand, all outgoing arcs from mixture node n will go to only sink nodes $\tilde{n} \in \mathbf{N}^{\mathrm{SI}}$.

2.2.3. Active streams

Arcs in the superstructure essentially represent streams in the network; hence, we associate each arc (binary variable) with component molar flowrates (continuous variables). The component molar flowrate of the inlet stream from source node $\hat{n}$, $F_{i,\hat{n}}^{0} \in \mathbb{R}^+$, is disaggregated into $F_{i,\hat{n},n}^{0,\mathrm{D}} \in \mathbb{R}^+$ for each mixture node n. The value of $F_{i,\hat{n},n}^{0,\mathrm{D}}$ is positive only when its corresponding arc is active (i.e., $Y_{\hat{n},n}^{0} = 1$). We use the inlet streams' flowrates to detect the existence of a component in the network. Moreover, we disaggregate the top and bottom product streams, $F_{i,n}^{\mathrm{T}} \in \mathbb{R}^+$ and $F_{i,n}^{\mathrm{B}} \in \mathbb{R}^+$, into $F_{i,n,\dot{n}}^{\mathrm{T,D}} \in \mathbb{R}^+$ and $F_{i,n,\dot{n}}^{\mathrm{B,D}} \in \mathbb{R}^+$, respectively, to represent the material transfer from one mixture node to another. Similarly, only flow variables associated with active $Y_{n,\dot{n}}^{\mathrm{T/B}}$ have positive values. Finally, we use split fraction $K_{n,\tilde{n}}^{\mathrm{BP}} \in [0,1]$ to calculate the amount of the mixed stream which bypasses separation and goes directly to the sink node $\tilde{n}$, $F_{i,n,\tilde{n}}^{\mathrm{BP}} \in \mathbb{R}^+$. Both $F_{i,n,\tilde{n}}^{\mathrm{BP}}$ and $K_{n,\tilde{n}}^{\mathrm{BP}}$ can be deactivated by variable $Y_{n,\tilde{n}}^{\mathrm{BP}}$ via big-M constraints.

2.3. Separation block model

Since only sharp separations are considered, we characterize a separation by the selection of a pair of key components in a mixture. The candidate pairs can vary depending on the chosen separation technology. We adopt the terms "light" and "heavy" keys and generalize them by defining the light/heavy key as the component in the separation block feed stream with the highest/lowest ranking $\rho_{t,i}$ that is not present in the bottom/top product stream. In each mixture node, we find candidate light and heavy keys according to a specific technology. We introduce $Y_{i,n}^{\mathrm{LK/HK}} \in \{0,1\}$ to indicate if component i is the light/heavy key in the separation performed in mixture node n. Using these binary variables, we write constraints to describe the relationship between the light and heavy keys and the top and bottom arcs. Since we consider only adjacent sharp splits, components with a ranking lower/higher than or equal to the ranking of the light/heavy key will not be present in the bottom/top products.

3. Examples

Our approach can be applied to separations where alternative technologies are considered. One such example is xylene isomers separation for which distillation is energy-intensive and thus non-thermal technologies (e.g., reverse osmosis and extraction) are investigated. All MINLP models are solved using BARON 38.3.0 to global optimality (optcr = 10^{-6}).

3.1. Conventional separation network synthesis

A mixture of 1 kmol/s containing four components (15% A, 30% B, 35% C, 20% D) is to be separated into pure components (Fig 3, left panel). We consider three technologies: distillation, reverse osmosis, and solvent extraction. Components A and C have similar polarities and molecular sizes, which makes separation split between these two components impractical. The objective is to minimize the annualized separation cost:

$$\Phi_1 = \sum_{n,t} \gamma_{n,t}^{\mathrm{FC}} Y_{n,t}^{\mathrm{C}} + \sum_{\hat{\imath},i,t,n} \gamma_{i,t,n,\hat{\imath}}^{\mathrm{VC}} F_{i,n,t,\hat{\imath}}^{\mathrm{F,LK}} \tag{1}$$

where $F_{i,n,t,\hat{\imath}}^{\mathrm{F,LK}}$ is the molar flowrate of the stream undergoing separation using technology t with $\hat{\imath}$ as the light key, and $\gamma_{n,t}^{\mathrm{FC}}/\gamma_{i,t,n,\hat{\imath}}^{\mathrm{VC}}$ is the parameter for fixed/variable cost. The cost parameters can be obtained by doing offline calculations.

The resulting model has 2,567 equations and 1,115 variables (236 binaries) and is solved in 4 s. In the optimal solution (Fig 3, right panel), all three technologies are used. The reverse osmosis block separates B and sends the bottom product (ACD) to the distillation block. The distillation then separates A taking advantage that component B is missing, resulting in an easier separation. Finally, mixture CD is separated using solvent extraction.

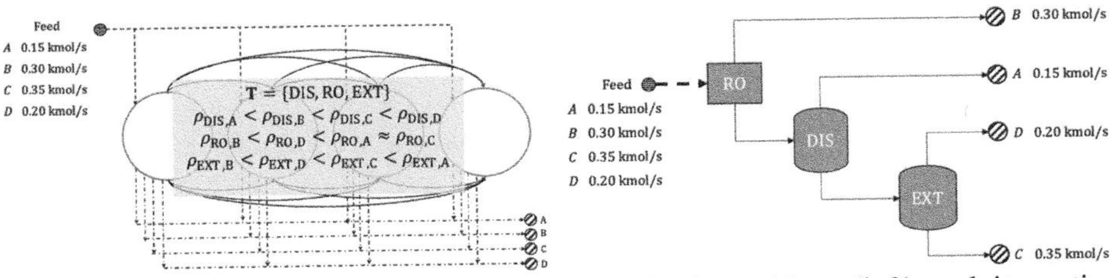

Figure 4: Conventional separation network synthesis problem (left) and its optimal solution (right). DIS: distillation, RO: reverse osmosis, EXT: solvent extraction.

3.2. Integrated reactor-separation network synthesis

Raw materials *B* and *C* are purchased and fed into the reactor network to undergo one of three reactions: 2C + B → 3A (80% conversion); C + 2B → 3D (90% conversion); C + B → A + D (80% conversion) (Fig 4, left panel). The reactor effluent is then fed into the separation network to recover valuable product A or D. For each product, the reactor can only produce up to 10 kmol/s. Depending on the chosen reaction, the reactor effluent may contain different combinations of components at different concentrations. Due to the partial conversions, there will be unconverted reactants that can be recycled. The technologies along with the associated rankings are the same as in the previous example. Our goal is to identify the raw material purchases, reactor selection, and separation network which result in the maximum profit:

$$\Phi_2 = \sum_{i,\hat{n}} \gamma_{\hat{n}}^{\mathrm{SELL}} F_{i,\hat{n}}^{\mathrm{P}} - \left(\sum_i \gamma_i^{\mathrm{RM}} R_i^{\mathrm{RM}} + \sum_{rx} \gamma_{rx}^{\mathrm{RX}} R_{rx}^{\mathrm{FEED}} + \Phi_1\right) \quad (2)$$

where R_i^{RM} denotes the purchased raw material, R_{rx}^{FEED} denotes the flowrate of the feed undergoing reaction rx, and $\gamma_{\hat{n}}^{\mathrm{SELL}}/\gamma_i^{\mathrm{RM}}/\gamma_{rx}^{\mathrm{RX}}$ are the product price/raw material price/reaction cost. The influence of feed compositions on the separation cost is considered via the use of cost parameter $\gamma_{i,t,n,\hat{\imath}}^{\mathrm{VC}}$ which depends on the component molar flowrate in the separation block feed stream.

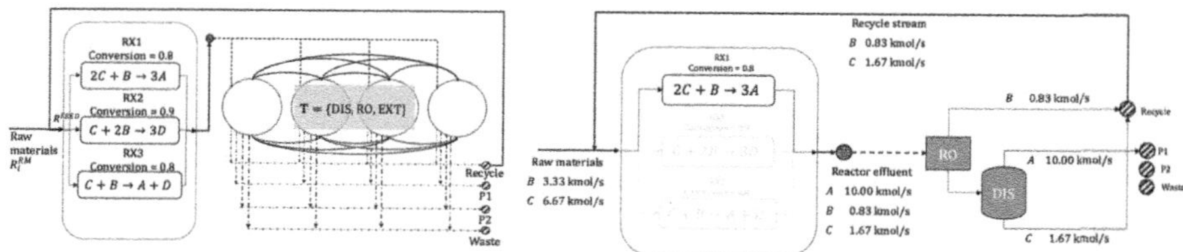

Figure 5: Integrated reactor-separation network synthesis problem (left) and optimal solution (right).

The resulting MINLP model has 2,623 equations and 1,134 variables (239 binaries) and is solved within 6 s. In the solution (Fig 5, right panel), reaction 1 is chosen, and the effluent (mixture ABC) is first separated by reverse osmosis followed by distillation. Unconverted reactants are recovered, mixed, and recycled. Finally, raw materials purchases are such that, given the recycle stream, the reactor feed stream is sufficient to produce 10 kmol/s of A.

4. Conclusion

We proposed a superstructure-based approach to synthesize separation networks with multiple technologies. We utilize rankings based on key properties to generate possible separations using each technology. To account for interactions between different separation technologies, rich interconnections between mixture nodes are considered (e.g., bypass streams). The adopted generalized problem statement allows variable feed to the network, enabling the simultaneous synthesis of separation and reactor networks.

References

R. Agrawal, 2003, Synthesis of multicomponent distillation column configurations, AIChE journal, 49, 2, 379-401.

I. Heckl, Z. Kovács, F. Friedler, L. T. Fan, J. Liu, 2007, Algorithmic synthesis of an optimal separation network comprising separators of different classes, Chemical Engineering and Processing: Process Intensification, 46, 7, 656-665.

L.Kong, C. T. Maravelias, 2020, Expanding the scope of distillation network synthesis using superstructure-based methods, Computers & Chemical Engineering, 133, 106650.

J. Ryu, L. Kong, A. E. P. de Lima, C. T. Maravelias, 2020, A generalized superstructure-based framework for process synthesis, Computers & Chemical Engineering, 133, 106653.

R. W. H. Sargent, K. Gaminibandara, 1976, Optimum design of plate distillation columns, in Optimization in action, New York: Academic Press, 267-314.

V. H. Shah, R. Agrawal, 2010, A matrix method for multicomponent distillation sequences, AIChE journal, 56, 7, 1759-1775.

D. S. Sholl, R. P. Lively, 2016, Seven chemical separations to change the world, Nature, 532, 7600, 435-437.

Antonis Kokossis, Michael C. Georgiadis, Efstratios N. Pistikopoulos (Eds.)
PROCEEDINGS OF THE 33rd European Symposium on Computer Aided Process Engineering (ESCAPE33), June 18-21, 2023, Athens, Greece
 http://dx.doi.org/10.1016/B978-0-443-15274-0.50072-X

A Reinforcement Learning Development for The Exact Guillotine with Flexibility on Cutting Stock Problem

Jie-Ying Su,[a] Chia-Hsiang Liu,[c] Cian-Shan Syu,[c] Jia-Lin Kang,[b*] Shi-Shang Jang,[a*]

[a]*Department of Chemical Engineering, National Tsinghua University, Hsinchu, 300, Taiwan ROC*

[b]*Department of Chemical and Materials Engineering, National Yunlin University of Science and Technology, Yunlin 64002, Taiwan ROC*

[c]*Department of Industrial Engineering and Engineering Management, National Tsinghua University, Hsinchu, 300, Taiwan ROC*

[*] *jlkang@yuntech.edu.tw; ssjang@mx.nthu.edu.tw*

Abstract

The two-dimensional cutting stock problem (CSP) is critical in several industries. Reinforcement learning (RL) is a novel method to obtain a quality solution of two-dimensional CSP in a short computation time. In this research, we applied a model-free off-policy RL algorithm to an industrial example of exact guillotine two-dimensional CSP, and compared the results with mixed-integer programming (MIP), which is a common traditional mathematical method for optimization. The results showed that RL had a much lower computation time than MIP with a solution closed to optima, and the ability to make a trade-off between waste, inventory level, and back order.

Keywords: Reinforcement learning, machine learning, cutting stock problem

1. Introduction

The CSP is a problem of minimizing stock material waste while cutting stock material into pieces of required sizes. The two-dimensional CSP has been applied to a variety of industrial applications, such as paper, metal, and glass manufacturing. It is critical for industries to minimize costs, including waste costs, inventory costs, and back-order costs. The two-dimensional CSP is divided into guillotine cutting and non-guillotine cutting. Guillotine cutting can be classified into an exact guillotine and a non-exact guillotine (Andrade et al. (2016)), which increases the complication and difficulty of the two-dimensional CSP.

The computation time of traditional mathematical optimization methods such as MIP grows rapidly as the amount of data increases and the problem becomes more complex. The exact guillotine that caused some items cannot be scheduled, leading to back-order increases when there are no matching items due to the inflexible limitations of traditional MIPs.For addressing the scheduling problem, RL is a promising method that is expected to provide solutions close to optimal within a very short period after training. (Hubbs. et al. (2020)) However, only one previous research is committed to solving CSP by RL, but for one-dimensional. (Pitombeira-Neto el al. (2022))

The purpose of this study is to provide a RL model named Monte Carlo-Deep Deterministic Policy Gradient (MC-DDPG) to solve a two-dimensional exact guillotine CSP in the paper-cutting factory, with the flexibility to switch to the non-exact guillotine mode for no matching items to reduce back-order and accelerate the scheduling. In the industrial case, there are several bins with different length, width, and weight per area (called base weight) to cut into various items which corresponding to the orders for daily scheduling. Each order may contain one or more items with varying specifications, and total weight of the items finally produced must meet the order's demand. Furthermore, the width requirement of the items is fixed, but the length may be a range, thus the length is also a variable that must be determined. There are several constraints that must be satisfied due to physical and procedural limitations.

This paper is divided into four sections. Section 2 describes the details of our proposed approach. Section 3 and 4 discuss the results and conclusions.

2. Method

2.1. Reinforcement learning (RL)

RL involves an agent to interact with the environment. The agent determines an *action* according to a given *state*, and the environment returns next state and *reward* (Fig 1). Sets of state, action, reward, and next state at time step t can be denoted as the following form: (s_t, a_t, r_t, s_{t+1}). The (s_t, a_t, r_t, s_{t+1}) sequence of an episode with several time steps is called a *trajectory*. The goal of the agent is to maximize the reward, thus RL develops a *policy* for choosing action depending on the state.

RL can be divided into on-policy and off-policy. The main difference between on-policy and off-policy is that whether the policy which is improved (target policy) is same as the policy that is used to select actions (behavior policy). Same for on-policy and not for off-policy. Thus, current data and historical data can both be used for off-policy training.

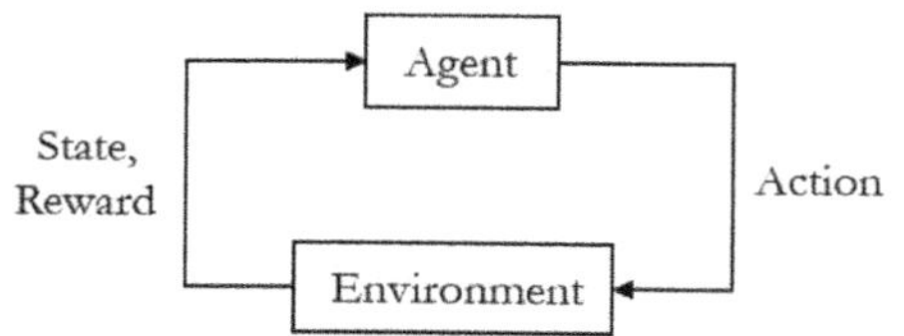

Fig 1. Reinforcement learning system diagram.

2.2. Monte Carlo- Deep Deterministic Policy Gradient (MC-DDPG)

DDPG is an off-policy method in RL, composed of two neural networks: actor and critic. Actor determines the actions, and the critic estimates the expected return value which indicates how good are the actions of corresponding states. Actor is trained by maximizing the value output by critic, and the critic is trained by minimizing the difference with the expected return value. (Casas (2017))

Monte Carlo (MC) is a common method for critic updating that estimate the expected return value as:

$$G_t=\sum_{k=t}^{T} r_k \tag{1}$$

However, we assume that the current action has a negligible effect on the reward of future actions in our problem, thus we proposed an expected return value function which is explained in the later chapter instead. We store the transition (s_t, a_t, r_t, s_{t+1}) after the end of an episode instead of step by step due to the calculation of G_t.

2.3. Purposed method

The proposed framework of MC-DDPG is composed of two parallel parts: replay buffer management (solid line in Fig 2) and critic and actor training (dashed line in Fig 2). After an episode of interaction between the actor and the environment, the expected return values G_t were calculated, and the trajectory was stored into the replay buffer. It should be noted that whether an episode ends or not is according to the parameter D_{target}, which was adjusted when replay buffer stored more than a set amount of data for the training to start, ensuring that there was enough data with action that met the constraints to improve the policy. Once the training began, a batch size of data was sampled from replay memory at each time step and used to update both actor and critic.

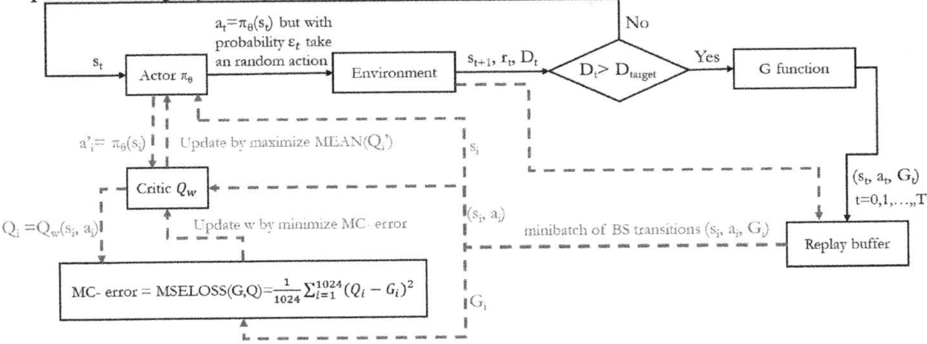

Fig 2. Proposed process diagram.

2.3.1. Environment (Sur el al. (2022))

At the beginning of each episode, we sample M bins and N orders from the database which is provided by a real-world factory, where M= DiscUnif(1,100) and N= DiscUnif(1,50).

Fig 3 and Table 2 showed detailed information in state and action provided by actor. Noted that the size of state is fixed to 190 and the size of action is fixed to 6 ×18=108. There have 4 types of action:

Table 1. Design parameters and variables in process diagram.

parameter/ variable	value/statement
D_{target}	2, while replay buffer size < 50000 1, otherwise
ε_t	1, while replay buffer size < 50000 0.1, while $\varepsilon_{t-1} \times 0.999 \leq 0.1$ $\varepsilon_{t-1} \times 0.999$, otherwise
BS	batch size = 1024
π_θ	actor network following parameter matrix θ
Q_w	critic network following parameter matrix w
T	total time step of the episode

I. Slit: Select one of the 15 items to schedule on the current bin in the width direction (called as slit), and also determine the length l of the item. To determine the length, we define a variable called ratio, then the length of item would be:

$$l = ratio \times (h_{max} - h_{min}) + h_{min} \quad (2)$$

We may select a ratio of 0, 0.25, 0.5, 0.75, or 1. In addition, we have another option of setting l equal to the length of the previous slit item without using the ratio.

II. Cut: Finish the scheduling in width direction, and cut the length at the longest length of the slit items after previous cut.

III. Finish: Finish the scheduling of the current bin.

IV. Abandon: Abandon the current bin without scheduling.

The constraints of the system were categorized as physical constraints, process constraints, and environmental constraints. Physical constraints contained the total width of slit items should be no larger than the width of the bin, the total length of cuts should be no larger than the length of the bin, and the item length should be between maximum and minimum length of the item. Process constraints contained the number of slit items should be no larger than 6 and the total produced weight of the order should be no larger than the demand weight. Environmental constraints contained there should be no action if the current bin is to be abandoned, there should be a cut action just before finish action, there should be some slit items before cut action, there should be a previous slit item if l is assigned the length of the previous slit item, the item should exist if it is selected to slit.

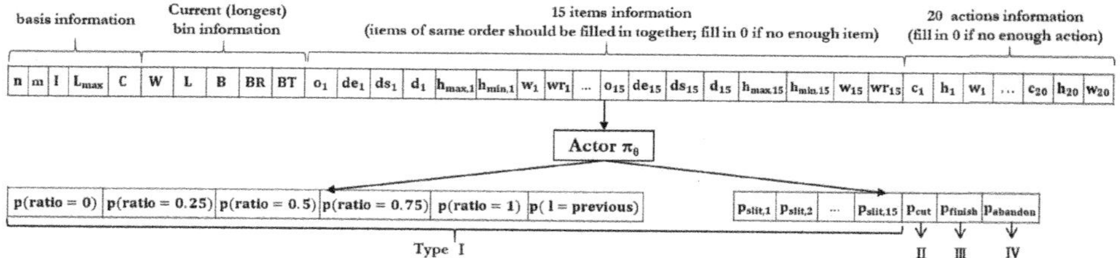

Fig 3. State and action of our proposed method.

Table 2. Definitions of variables in state.

variable	statement
n/m	number of items/bins in state
I	number of the slit items after previous cut
L_{max}	the maximum length in the slit items after previous cut
C	number of cuts in the current bin
W/ L	residual width/ length of the current bin after I slits/ C cuts
B/ BT	base/ total weight of the current bin
BR	residual weight of the current bin after C cuts
o_i/ de_i	number/ allowable demand error of corresponding order of i-th item
ds_i/ d_i	total produced/ demand weight of corresponding order of i-th item
$h_{max/min,i}$	maximum/minimum length of i-th item
w_i/	width of i-th item
wr_i	total produced weight of i-th item / d_i
c_k	number of cuts at k-th action (slit or cut)
h_k / w_k	$\begin{cases} \text{length/width of items chose by k-th action}, & \text{if k-th action=slit} \\ -1, & \text{if k-th action=cut} \end{cases}$

A cut reward would be observed if a cut action is executed, cut reward can be formulated as:

$$r_{cut} = \frac{\text{utilization area of current cut}}{\text{total area of current cut}} \quad (3)$$

A bin reward would be observed when taking a finish action, bin reward can be formulated as:

$$r_{bin} = \frac{\text{utilization area of current bin}}{\text{total area of current bin}} \quad (4)$$

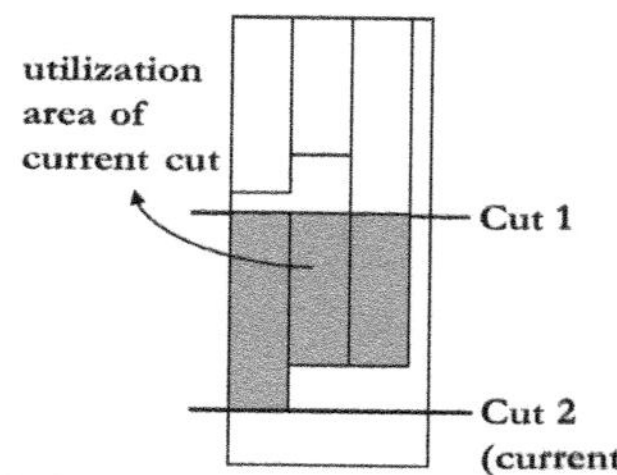

Fig 4. Schematic diagram of the scheduled bin.

A reward of -1 would be observed if the action violates the constraints, and 0 otherwise.

There is a variable D_t to determine whether the episode ends. The value of D_t is 2 if all of the bins or items are done, 1 if the action violates the constraints, and 0 otherwise.

The expected return value of the environment is estimated as:

$$G_{j,k} = \bar{r}_{cut,j} \times r_{bin,j}, \bar{r}_{cut,j} = \frac{\sum_z^{Z_{T,j}} L_{z,j} \times r_{cut,z,j}}{\sum_z^{Z_{T,j}} L_{z,j}} \quad (5)$$

where $G_{j,k}$ is the expected return value for k-th action in bin j; $\bar{r}_{cut,j}$ is the mean utilization rate of all of the cuts in bin j based on length; $Z_{T,j}$ is the total number of cuts in bin j; $L_{z,j}$ is the length of z-th cut in bin j; $r_{cut,z,j}$ is utilization rate of z-th cut in bin j.

3. Results and discussion

3.1. Training curve.

The higher the mean bin used ratio the better ability for agent to continuously provide actions that satisfied the constraints. After training, the mean bin used ratio achieved about 0.4 (Fig 5-(a)), and the mean utilization rate achieved about 0.7 which is the mean value in the factory (Fig 5-(b)).

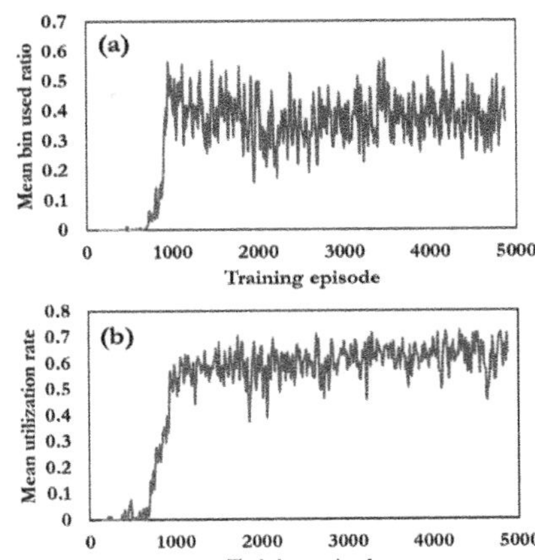

Fig 5. Training curve of (a) mean bin used ratio and (b) mean utilization rate (average of the last 20 episodes) for each training episode.

3.2. Testing

4 weeks data was used to evaluate the performance of our proposed RL agent. To efficiently reduce the inventory level and the back-order, we force the agent to select the action with the largest probability and satisfying the constraint, so that the agent would never terminate by taking an action that violates the constraints. We compare the result of inventory level, back-order, and utilization rate with MIP. There is an addition constraint in MIP that the utilization rate of each bin should be larger than 0.7, to ensure the mean utilization rate achieve 0.7. To compare our proposed method with MIP fairly, we also add the constraint to our agent and called as "RL_0.7".

Although we forced the agent to take an action, avoiding violate the constraints, there are still some reasons lead to a bin used ratio less than 1, such as all of the orders are satisfied and some bins are abandoned (type IV action).Surprisingly, our RL agent solved most bins with a mean utility rate closed to 0.7 (Fig 6).

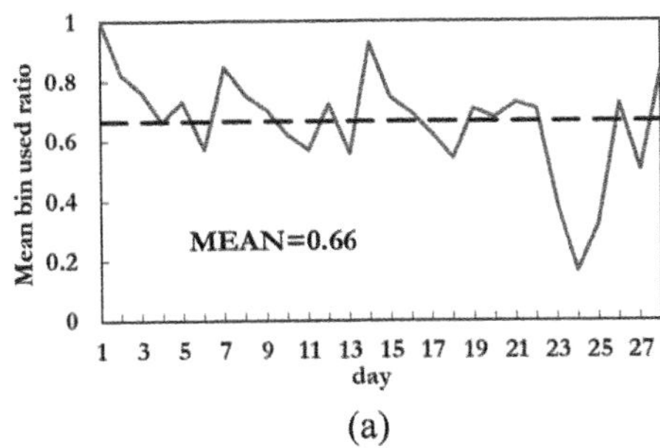

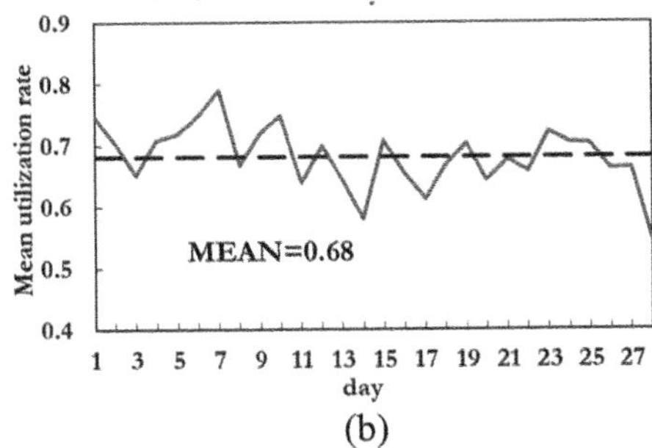

Fig 6. RL testing result of (a) Mean bin used ratio and (b) mean utilization rate for each day.

Table 3. Computation time of testing.

Model	Computation time
MIP	23min 50s
RL	2min 36.4s
RL_0.7	9min 11s

According to Fig 7-(a) and Table 3, MIP had the highest mean utilization rate, but also the highest computation time. RL_0.7 showed no utilization rate lower than 0.8 with quite lower computation time than MIP which is important in real-world factory for rescheduling. Although RL had the lowest mean utilization rate, the value was also close to 0.7 which is the mean value in the factory and with the lowest computation time. The inventory level of MIP gradually increased every day. RL and RL_0.7 both showed a lower inventory level compared to

MIP and ability to decrease inventory from a high level. (Fig 7-(b)).Fig 7-(c) showed that RL produced the most orders result in the least back-order. RL_0.7 had less back-order compared to MIP.

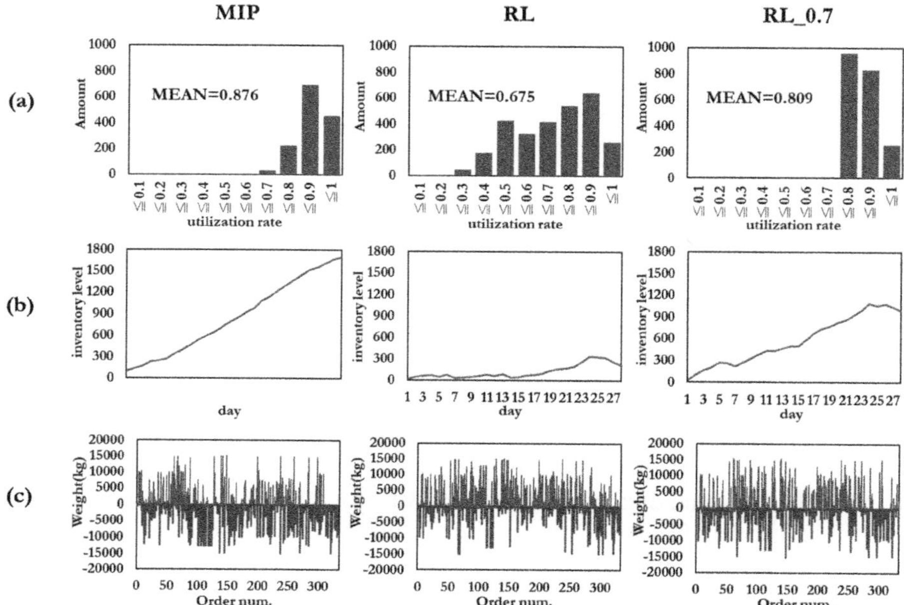

Fig 7. (a) Utilization rate statistics histogram (b) Inventory level for each day (c) Satisfied order demand weight (blue) and back-order weight (red) for each order of MIP, RL and RL_0.7.

4. Conclusion

The proposed RL showed a much better result of inventory level and back-order compared to MIP. Although the mean utilization rate of MIP is the best, the computation time of MIP is also the longest. Furthermore, RL may easily and flexibly make a trade-off between waste, inventory level, and back order, but MIP must add more constraints to take these issues into account, which may make the problem more difficult to solve. The surprising result indicated that RL ability to deal with complex practical industrial problems.

References

Andrade, R., Birgin, E. G., & Morabito, R. (2016). Two-stage two-dimensional guillotine cutting stock problems with usable leftover. International Transactions in Operational Research, 23(1-2), 121-145.

Hubbs, C. D., Li, C., Sahinidis, N. V., Grossmann, I. E., & Wassick, J. M. (2020). A deep reinforcement learning approach for chemical production scheduling. Computers & Chemical Engineering, 141, 106982.

Pitombeira-Neto, A. R., & Murta, A. H. (2022). A reinforcement learning approach to the stochastic cutting stock problem. EURO Journal on Computational Optimization, 10, 100027.

Casas, N. (2017). Deep deterministic policy gradient for urban traffic light control. arXiv preprint arXiv:1703.09035.

Sur, G., Ryu, S. Y., Kim, J., & Lim, H. (2022). A Deep Reinforcement Learning-Based Scheme for Solving Multiple Knapsack Problems. Applied Sciences, 12(6), 3068.

Antonis Kokossis, Michael C. Georgiadis, Efstratios N. Pistikopoulos (Eds.)
PROCEEDINGS OF THE 33rd European Symposium on Computer Aided Process Engineering (ESCAPE33), June 18-21, 2023, Athens, Greece
 http://dx.doi.org/10.1016/B978-0-443-15274-0.50073-1

Real-time optimization of a chemical plant with continuous flow reactors via reinforcement learning

Min Wu[a], Furkan Elmaz[b], Ulderico Di Caprio[a], Dries De Clercq[c], Siegfried Mercelis[b], Peter Hellinckx[b], Leen Braeken[a], Florence Vermeire[d] and M. Enis Leblebici[a,*]

[a]*Center for Industrial Process Technology, KU Leuven, Agoralaan Building B, 3590 Diepenbeek, Belgium*
[b]*Faculty of Applied Engineering, University of Antwerp, Groenenborgerlaan 171, 2020 Antwerp, Belgium*
[c]*Ajinomoto Bio Pharma Services, Cooppallaan 91, 9230 Wetteren, Belgium*
[d]*Chemical Reactor Engineering and Safety, KU Leuven, Celestijnenlaan 200f, 3001 Leuven, Belgium*
[*]*muminenis.leblebici@kuleuven.be*

Abstract

Reinforcement learning (RL) has many new applications in recent years, and its results often exceed human performance, especially in environments where the action space is discrete. However, it is challenging to use RL in the chemical industry, where variables are often continuous and various constraints are complex. This study applies RL with continuous actions to maximize the productivity of a continuous process. The RL agent provides optimal setpoints of flow rates and temperatures while the concentrations of raw materials are changing. Two environments with one and six actions were established after the sensitivity analysis. In the one-action environment, the agents SAC, PPO and A2C showed similar performances, but A2C needed fewer timesteps for training. SAC outperforms PPO and A2C in the environment with six actions. This paper shows the successful RL applications in a continuous process and the high applicability of SAC in both low-dimension and high-dimension environments.

Keywords: reinforcement learning, real-time optimization, continuous flow reactors, continuous action space

1. Introduction

Reinforcement learning (RL) is a machine learning technique that learns an optimal decision policy based on defined rewards and observations of an environment. Because of its flexibility, it gained popularity in many applications in recent years. The RL results often exceed human performance, especially in the environments (e.g. game) where the action spaces are discrete and the reward is straightforward. However, chemical processes are more difficult to use RL as the common settings should be approached in a continuous way (e.g. flow rates, temperatures). Usually, more trials are done and more time is required compared to discrete spaces at the training stage. Nevertheless, the calculation of the RL agent at the application stage is much faster than classic methods (e.g. linear or nonlinear programming), as no models are used. In the open literature, a RL agent needs 87.7% less calculation time than the nonlinear programming when optimizing a continuous stirred tank reactor (CSTR) (Powell et al., 2020). Moreover, various constraints in chemical plants are challenging to define in the reward function. Because the RL agent encourages the actions which can lead to higher reward values, the reward function should be designed wisely to meet different constraints. Père et al. (2022) have studied the impact of rewards for the microgrid control. However, the open literature lacks the studies about real-time optimization of continuous flow reactors (CFR) via reinforcement learning and the corresponding designs of reward functions.

This study used RL to optimize the productivity of a chemical plant with six continuous action spaces. The applicability of three RL algorithms, namely Soft actor-critic (SAC) (Haarnoja et al., 2018), proximal policy optimization (PPO) (Schulman et al., 2017), and advantage actor-critic (A2C) (Mnih et al., 2016), in the low- and high-dimension environments was investigated. The following section explains the details of the chemical plant and how RL works.

2. Methodology

The plant consists of four identical CFRs in series (see Figure 1). The first three are equipped with jackets, and the last one is air-cooled. The real-time optimization provides six setpoints, which are raw materials A and B flow rates ($F_{A,SP}, F_{B,SP}$), feed temperature ($T_{f,SP}$) and three jacket temperatures ($T_{J1,SP}, T_{J2,SP}, T_{J3,SP}$). The optimization aims to maximize the throughput of product C. A simulation model of this plant for steady states was developed via swarm intelligence (Wu et al., 2022), which was used as the environment in the RL training phase.

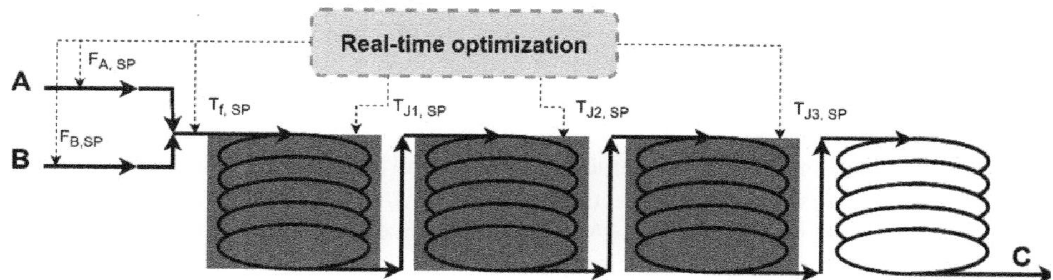

Figure 1: A schematic drawing of the chemical plant. This plant consists of four identical flow reactors. In this system, the feeding raw materials are A and B, and the product is C. The real-time optimization provides six setpoints: A and B flowrates ($F_{A,SP}, F_{B,SP}$), feed temperature ($T_{f,SP}$) and three jacket temperatures ($T_{J1,SP}, T_{J2,SP}, T_{J3,SP}$).

RL consists of two main parts: Agent and Environment. The agent can be structured via different RL algorithms. In this paper, SAC, PPO and A2C are utilized and compared. The simulation environment is only required during the training phase. When the deployment of the trained RL agent is done, the simulation model is not needed any longer (i.e. morel-free algorithms). Also, the agent calculation time is usually much less than classic model-based methods because of the high computational complexity of accurate models involved in model-based approaches (Alizadeh et al., 2020). The interaction loop between the agent and the environment in RL is shown in Figure 2.

At each timestep, the agent provides actions after getting observations from the environment. After actions are executed, the environment gives a reward and transits to a new state with new observations. The observations in this paper are set to the changing concentrations of raw materials A and B. The reward value is the normalized outlet throughput of product C, which is the multiplication of the C concentration and the total flow rate.

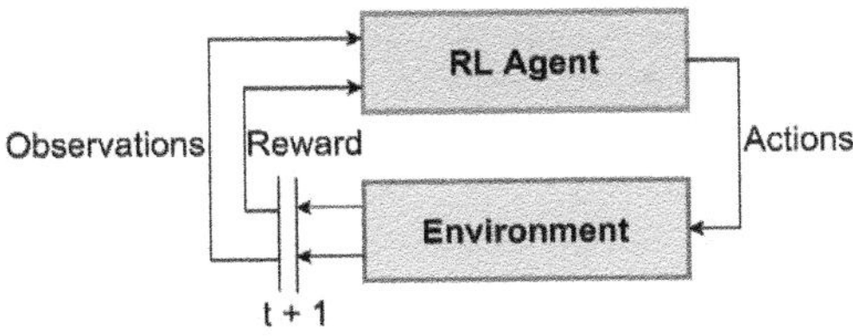

Figure 2: Interactions between the environment and the agent in reinforcement learning. At each time step (t), after getting observations from the environment, the agent provides actions. After the actions are executed, the environment gives a reward and transits to a new state.

Because of the environment complexity, the RL training can require more than 100,000 environment evaluation steps. Therefore, to reduce the system complexity, a sensitivity study is conducted to check the importance of six actions to the productivity (Li et al., 2010). The RL agent is first trained for the most important action. Then with similar structures, the agent for six actions is developed. The calculation time and the agent performance are also analyzed to obtain the suitable methodology to develop the RL agent. The environments are encoded in "OpenAI Gym v0.21.0" (Brockman et al., 2016) and the python library "Stable-Baseline3 v1.6.2" is utilized (Raffin et al., 2021).

3. Results

A steady-state model for the studied was used to establish the training environment. The agent learning is based on the environment, and the utilized model accuracy is crucial. The model in the paper has high accuracy and fast calculation as demonstrated in the previous work. To reduce the model system complexity, a sensitivity study is conducted to check the importance of six actions.

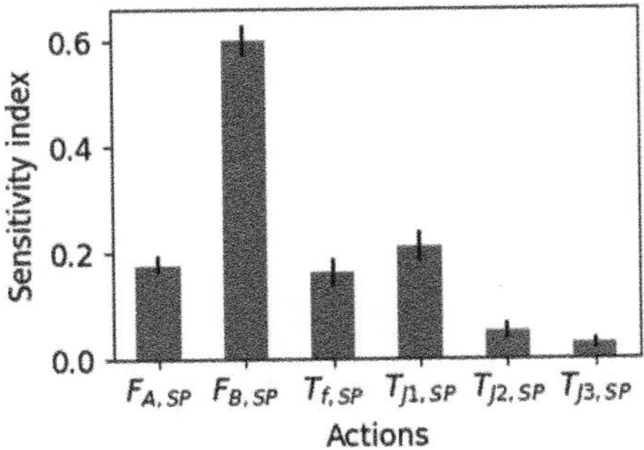

Figure 3: Sensitivity indices of actions to the productivity in the environment. The black vertical lines at the top of the bars show the confidence intervals, and the small ranges of the confidence interval confirm the robustness of the utilized simulation model.

3.1. Sensitivity analysis of available actions

The aim of the global sensitivity analysis is to rank the importance of six available actions with respect to the productivity. The actions with higher sensitivity indices are more important than the others, which means these actions should be considered first if the computation power or the time is limited. Latin Hypercube Sampling generated 1,000 analysis samples within 30% deviations from the baseline setpoints. The baseline setpoints were provided by plant operators and they are 97 kg/h ($F_{A,SP}$), 80 kg/h ($F_{B,SP}$),

20 °C ($T_{f,SP}$), 32.5 °C ($T_{J1,SP}$), 32.5 °C ($T_{J2,SP}$), and 32.5 °C ($T_{J3,SP}$). Moreover, the feeding raw material concentrations ($C_{f,A}$, $C_{f,B}$) are kept constant at 91 wt% and 40 wt%.

The calculated global sensitivity indices of actions are shown in Figure 3. The analysis returned $F_{B,SP}$ to be the most important action. Physically, this is because B is the raw material in both main and side reactions. $T_{J1,SP}$ gives a relatively high sensitivity index resulting from that the fact that most reactions are fast and can complete in the first reactor. Therefore, $F_{B,SP}$ is selected as the most important action in this paper to test three algorithms – SAC, PPO, A2C firstly. After the initial development for the one-action environment, the RL agent for six actions in the high-dimension environment is developed as well.

3.2. Performance analysis of agents

The training progresses of SAC, PPO, A2C with one and six actions are illustrated in Figure 4a. The reward value is the normalized productivity, and in all cases the RL training stops when the total timesteps reach 100,000. For the one-action system, all three algorithms reached the maximum reward at the end of the training, and A2C already approached it with less than 20,000 steps. However, A2C got the lowest ending reward in the six-action environment where SAC outperforms the other two algorithms.

Figure 4b gives the required training time to reach the maximum reward value in different approaches. It can be seen that the agents trained in the one-action environment reached the maximum reward faster than the ones trained in the high-dimension environment with six actions, which meets the initial hypotheses. For A2C, the six-action agent requires 25.1% more time than the one-action agent, and that number is 154.5% in SAC. PPO did not even reach the maximum reward value at the end of the training. It shows the high applicability of SAC in both low- and high-dimension environments. A2C is more suitable for the low-dimension environments compared to SAC and PPO, which requires a low amount of training time.

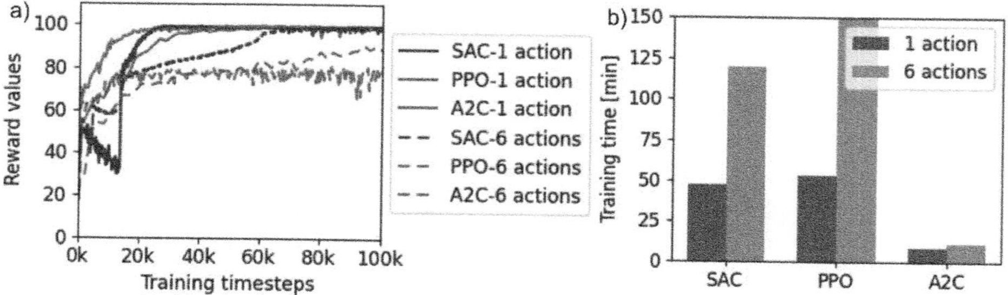

Figure 4: a) The training progress of SAC, PPO, A2C with one and six actions. The reward value is the normalized C productivity. The maximum number of training steps is the same, which is 100,000. b) The required training time of SAC, PPO, A2C with one and six actions to reach the maximum reward values. The used processor is Intel® Xeon® Gold 5220 CPU @2.20GHz.

After completing the training, the developed agent was applied, where the concentrations of the raw materials A and B ($C_{f,A}$, $C_{f,B}$) changed randomly over 50 hours, as shown in Figure 5a. The agents proposed one or six actions to reach a high reward value, and the corresponding reward values are given in Figure 5b and Figure 5c. In the one-action system, SAC, PPO and A2C offered similar performances, which follows the findings in their training progress. Only SAC gave a slightly low reward at the beginning. In the six-action system, SAC outperformed PPO and A2C. In other words, SAC returned a better optimization than PPO and A2C in the high-dimension environment. Figure 6 shows the behaviors of the six actions from SAC agent. As the concentrations of raw materials changed, the SAC agent gave six setpoints in order to obtain the highest productivity at the new conditions.

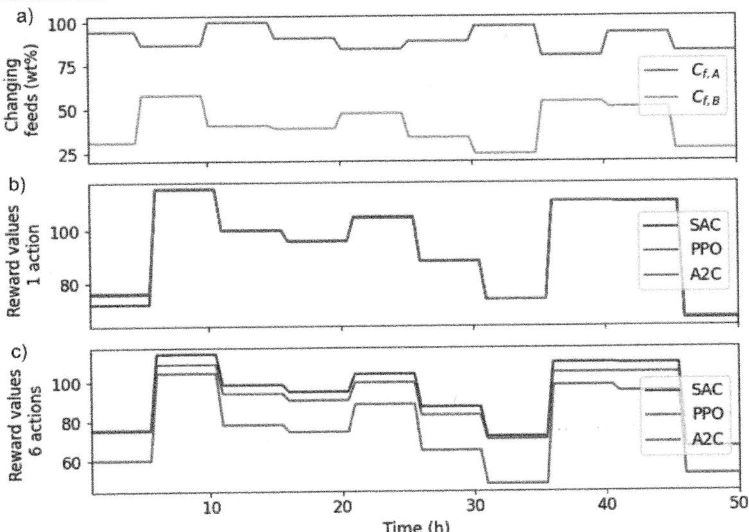

Figure 5: The reward values of RL agents for one and six actions over 50 hours. The upper plot shows how the concentrations (in weight mass percentage wt%) of the raw materials A and B ($C_{f,A}$, $C_{f,B}$) change with time. The RL agents were developed via three algorithms (SAC, PPO, A2C) for one and six actions, respectively.

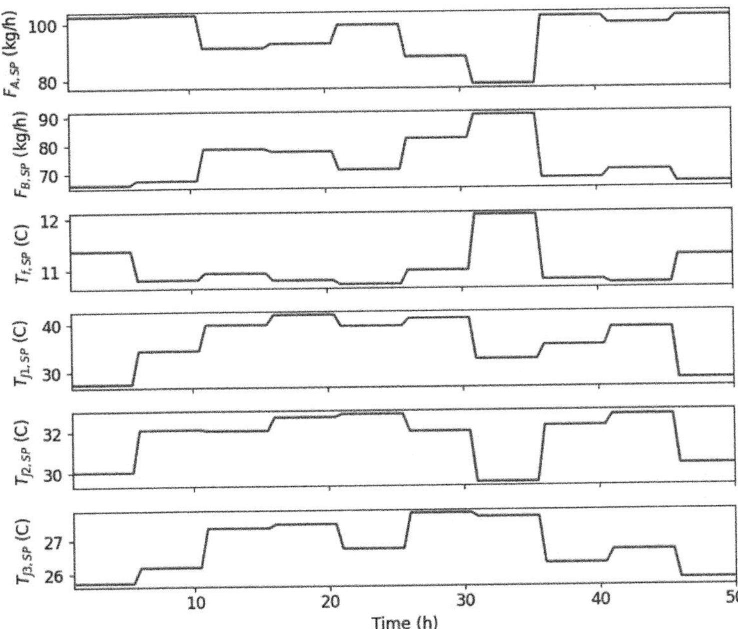

Figure 6: The provided six actions ($F_{A,SP}, F_{B,SP}, T_{f,SP}, T_{J1,SP}, T_{J2,SP}, T_{J3,SP}$) from SAC agent, which obtains higher reward values than PPO and A2C.

4. Conclusions

This study successfully applied reinforcement learning (RL) in a chemical plant with continuous flow reactors for the real-time optimization. In the plant, the concentrations of the raw materials changed over time. In order to obtain the maximum productivity, six setpoints of flow rates and temperatures were available to be optimized. The developed RL agents provided optimal setpoints based on their learned strategies. The agent learning or training process relies on the interactions between the environment and the agent. Two environments with one and six actions and three RL agent algorithms (soft actor-critic (SAC), proximal policy optimization (PPO), and advantage actor-critic (A2C)) were investigated. The one-action environment was established after the sensitivity analysis. In the one-action environment, SAC, PPO and A2C showed similar performances, but A2C needed less training time. This indicates that A2C is more suitable in the low-dimension environment due to its high efficiency. SAC outperformed PPO and A2C in the environment with six actions, which illustrates the high applicability of SAC in both low-dimension and high-dimension environments. This paper showed the performance of RL agents in the chemical industry. The future work will compare the employed RL algorithms with classic model-based methods. Also, the applied reward functions in both environments for steady states are simply the normalized productivity in this paper. The future work will design new reward functions considering complex constraints, and will study the transit behaviors of the plant from dynamic models.

Acknowledgements

The authors acknowledge the funding from VLAIO, DAP²CHEM: Real-time data-assisted process development and production in chemical applications (HBC.2020.2455).

References

Alizadeh, R., Allen, J.K., Mistree, F., 2020. Managing computational complexity using surrogate models: a critical review. Res. Eng. Des. 31, 275–298.

Brockman, G., Cheung, V., Pettersson, L., Schneider, J., Schulman, J., Tang, J., Zaremba, W., 2016. OpenAI Gym 1–4.

Haarnoja, T., Zhou, A., Abbeel, P., Levine, S., 2018. Soft actor-critic: Off-policy maximum entropy deep reinforcement learning with a stochastic actor. arXiv.

Li, G., Rabitz, H., Yelvington, P.E., Oluwole, O.O., Bacon, F., Kolb, C.E., Schoendorf, J., 2010. Global Sensitivity Analysis for Systems with Independent and / or Correlated Inputs. J. Phys. Chem. A 6022–6032.

Mnih, V., Badia, A.P., Mirza, L., Graves, A., Harley, T., Lillicrap, T.P., Silver, D., Kavukcuoglu, K., 2016. Asynchronous methods for deep reinforcement learning. 33rd Int. Conf. Mach. Learn. ICML 2016 4, 2850–2869.

Père, V., Baillon, F., Milhe, M., Dirion, J.L., 2022. The Impact of Reward Shaping in Reinforcement Learning for Agent-based Microgrid Control. Comput. Aided Chem. Eng. 51, 1459–1464.

Powell, B.K.M., Machalek, D., Quah, T., 2020. Real-time optimization using reinforcement learning. Comput. Chem. Eng. 143, 107077.

Raffin, A., Hill, A., Gleave, A., Kanervisto, A., Ernestus, M., Dormann, N., 2021. Stable-baselines3: Reliable reinforcement learning implementations. J. Mach. Learn. Res. 22, 1–8.

Schulman, J., Wolski, F., Dhariwal, P., Radford, A., Klimov, O., 2017. Proximal Policy Optimization Algorithms 1–12.

Wu, M., Di Caprio, U., Elmaz, F., Metten, B., De Clercq, D., Van Der Ha, O., Mercelis, S., Hellinckx, P., Braeken, L., Leblebici, M.E., 2022. A comparative study of swarm intelligence and artificial neural networks applications in modeling complex reaction processes, Comput. Aided Chem. Eng. 51, 175-180

Antonis Kokossis, Michael C. Georgiadis, Efstratios N. Pistikopoulos (Eds.)
PROCEEDINGS OF THE 33rd European Symposium on Computer Aided Process Engineering (ESCAPE33), June 18-21, 2023, Athens, Greece
 http://dx.doi.org/10.1016/B978-0-443-15274-0.50074-3

From problem specification to converged flowsheet simulations: Integrated workflow in an industrial context

Mirko Skiborowski[a], Norbert Asprion[c], Sergej Blagov[c], Michael Bortz[b], Dennis Manuel Heim[b], Kai Fabian Kruber[a], Thulasi Rolland[a], Tobias Seidel[b]

[a]*Hamburg University of Technology, Institute of Process Systems Engineering, Am Schwarzenberg-Campus 4, 21073 Hamburg, Germany*

[b]*Fraunhofer Institut für Techno- und Wirtschaftsmathematik, Optimierung, Fraunhofer Platz 1, 67663 Kaiserslautern, Germany*

[c]*BASF SE, Process Modeling & Cheminformatics, Carl-Bosch-Str. 38, 67063 Ludwigshafen am Rhein, Germany*

mirko.skiborowski@tuhh.de

Abstract

Although distillation is considered the most mature and computationally easiest to analyze and optimize fluid separation process, support for flowsheet synthesis for the separation of multicomponent azeotropic mixtures is still limited. Commercial simulation software like Aspen Plus mainly offers a graphical analysis of residue curve maps for ternary (sub)systems or an iterative evaluation based on tedious simulation studies. The current work presents an integrative algorithmic approach that pursues an automatic synthesis of distillation processes for azeotropic mixtures, building on a topological analysis of the distillation regions and evaluation of residue curves, pinch lines, and rectification bodies. The derived flowsheet structures and the estimated minimum reflux ratios are the basis for optimization-based initialization of rigorous column models and full-scale simulations of the integrated flowsheets in the industrial in-house simulation tool Chemasim. Thus, the method provides an algorithmic fast-track from a thermodynamic mixture model to a rigorous simulation of automatically derived flowsheets.

Keywords: process synthesis, simulation, distillation, conceptual design, azeotropic mixtures.

1. Introduction

The conceptual design of separation processes has been subject to continuous research for more than 40 years, with various developments, especially for the design of distillation-based processes. While the separation of zeotropic mixtures is well understood and all basic and complex column sequences can be generated (Madenoor Ramapriya et al., 2018) and evaluated efficiently by systematic and optimization-based methods (Jiang et al., 2019), the synthesis of distillation processes for the separation of azeotropic mixtures appears to present a more challenging problem, which has not been satisfactorily solved jet. The most established tools for the generation of process concepts rely on graphical analysis by means of residue curves or distillation line maps, which represent the vapor-liquid equilibrium characteristics, including the location of azeotropes and the illustration of distillation boundaries. While available in many flowsheet simulators, such as Aspen Plus, such tools are limited in applicability to ternary and quaternary mixtures. Few algorithmic methods have been proposed to automate this process since the early 2000s. Superstructure optimization methods either consider linear approximations of distillation boundaries and the treatment of azeotropes as pseudo-components (Yang et al., 2011) or, e.g., a limitation to binary azeotropes, which can be separated by extractive distillation (Tsai and Ward, 2020). A more general concept builds on an analysis of split feasibility considering topological distillation regions, as proposed by Rooks et al. (1998), which was the basis for the synthesis methods proposed by Tao et al. (2003) and the work of Thong et al. (2004), who illustrated the potential application to multicomponent mixtures with several azeotropes and distillation regions. Yet, only the method of Waslykiewicz and Castillo (2001), which was developed as part of DISTIL, was shortly available in Aspen Plus after the acquisition of Hyprotech by Aspen Tech. Yet, to our knowledge, there is no documented case of an industrial application of these methods. One potential reason for the lack of adoption is the lack of a consistent connection between the process synthesis methods and a rigorous process simulation, which is the main tool used for process development in industrial practice and which, given sufficiently accurate thermodynamic models, can be seen as a validation of the feasibility of the derived process concept.

In order to bridge this gap, the current study presents an integrated workflow for the automated synthesis of distillation processes for the separation of homogeneous azeotropic mixtures that allows for a direct transfer and validation of the results by means of a converged rigorous steady-state simulation of the derived flowsheets. Flowsheet alternatives are generated using a recently developed automatic process synthesis (APS) tool (Sasi and Skiborowski, 2020), building on a topological analysis, residue curve and pinch line computations, and the rectification body method (Bausa et al., 1998), providing an estimate of the minimum energy demand of the individual separations. The resulting flowsheet configurations, mass balance information, and energy estimates are further processed in a FlowsheetExplorer, which determines initial composition profiles and actual reflux ratios for individual columns with finite stage numbers that are finally transferred to the BASF in-house flowsheet simulator Chemasim, where the full process flowsheet is generated and optimized providing a converged rigorous flowsheet simulation, which satisfies the product specifications. The method is demonstrated for the separation of a well-known quaternary mixture of acetone-chloroform-benzene and toluene, which exhibits a strongly curved distillation boundary. The results illustrate the potential benefits of an algorithmic flowsheet synthesis and the possible speed-up of the conceptual design process.

2. Method

The proposed integrated workflow that allows for a direct application in an industrial context is schematically illustrated in Figure 1. It consists of two main steps for an automated process flowsheet synthesis (APS Tool, Fig. 1 left) and an optimization-based column and flowsheet initialization (FlowsheetExplorer, Fig. 1 right), which are pursued in a sequence of steps further outlined in the following subsections. The combination allows for the derivation of converged flowsheet simulations on the basis of rigorous equilibrium tray models of the individual distillation columns, which satisfy the desired product specifications, starting from a definition of the separation problem and the thermodynamic property models.

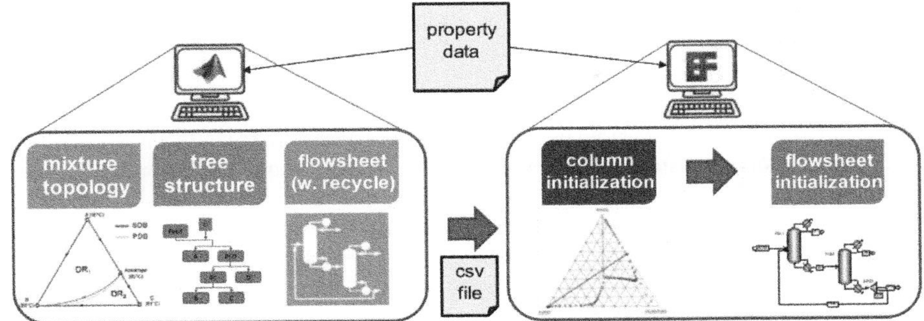

Figure 1: Illustration of the individual steps of the topology-based process synthesis (APS Tool) and optimization-based initialization and flowsheet simulation (FlowsheetExplorer).

2.1. Automated process synthesis

The APS tool aims at the generation of feasible process configurations based on the specification of a thermodynamic model for the phase equilibria of the considered mixture and the specification of the feed and desired products. In the first step, the topology of the mixture is analyzed, i.e., the computation of azeotropes and the characterization of all singular points regarding their stability, as well as the identification of the topological distillation regions (DR), based on the concept of the adjacency and reachability matrices (Rooks et al., 1998). The knowledge of the topological distillation regions is the foundation on which process variants are derived in the subsequent step, building on an algorithmic split feasibility check, which combines repeated computations of residue curves (total reflux distillation) and pinch lines (reversible distillation) to check if the products are located in the same DR for at least one mode of limiting operation. Further application of the rectification body method (RBM) (Bausa et al., 1998) evaluates the minimum reflux ratio of individual splits, given an intersection of the linear approximation of the potential manifolds of the individual column section profiles. Based on the derived tree structure, potential recycle streams and locations are identified following the ideas of Tao et al. (2003), closing the respective recycle loops in an iterative approach with additional RBM evaluations. The final flowsheets are thereby characterized by the respective mass balances for the individual columns and the estimates of the required heat duties, which not only allow for an initial ranking of alternative flowsheet configurations, but also provide essential information for the initialization of more rigorous flowsheet simulations based on consistent equilibrium tray models. This information is further transferred to the FlowsheetExplorer for the subsequent initialization and flowsheet optimization in a specific CSV file format. Refer to the article of Sasi and Skiborowski (2020) for further details on the process synthesis method, including potential pressure modifications. The MATLAB tool can be evaluated via the open GitLab project (APS tool), including respective examples.

2.2. Optimization-based column and flowsheet initialization

The second part of the integrated workflow aims at the initialization and subsequent optimization of rigorous flowsheet models in BASF's in-house flowsheet simulation software Chemasim. The results of the variant generation provide the necessary connectivity and mass balance information to build the individual flowsheet models for the derived candidates in the FlowsheetExplorer.

Since the minimum heat requirements derived from the RBM build on the computation of pinch points alone, they do not provide continuous composition and temperature profiles for the individual columns. To derive appropriate stage numbers for each column

section and to provide reasonable estimates for the composition and temperature profiles, individual optimization problems are first solved for all columns. These optimizations aim to find a composition profile such that the compositions of the bottom and top streams of the column are close to the compositions found by the APS tool. The optimizations are constructed as described by Seidel et al. (2020). The profiles are partially calculated using tray-to-tray calculations starting from the top and bottom streams and calculated towards the feed stage. The equations of the feed stage are included as constraints in the optimization problem. This means that after the optimization has converged, all equations for the rigorous simulation of the column are satisfied. For the optimization of each column, a constant molar overflow is assumed, and the optimizer can vary the number of stages and the reflux ratio for each individual column.
The goal of the individual column optimizations is to find good initial profiles quickly and robustly. However, in the transfer, the compositions calculated by the APS tool and this initialization procedure will not match perfectly, such that the feed streams of columns in the flowsheet will potentially change, and the profiles calculated by a rigorous flowsheet simulation may deviate due to the connection of the different columns. In order to further improve the flowsheet initialization, a second optimization can be started in the FlowsheetExplorer to find an initial profile that matches the concentrations, but this time with respect to the connected closed-loop flowsheet. Since the first optimization already found good initial profiles, this second optimization can be done within Chemasim. The FlowsheetExplorer automatically sets up and executes this second optimization, which not only verifies the feasibility of the derived process configurations, but also allows for further derivation of more informative cost estimates for the different process variants.

3. Case Study

As a representative case study, the separation of the four-component mixture of acetone, chloroform, benzene, and toluene (ACBT) with an equimolar composition and a flow rate of 10 mol/s, as investigated in the article of Liu et al. (2004), is considered. While the algorithmic approach of Liu et al. (2004) was able to derive a variety of flowsheets with at least four distillation columns, Kraemer et al. (2009) proposed three additional process configurations with only three distillation columns by means of graphical analysis of the pitchfork distillation boundary (PDB), which were shown to be favorable in terms of the minimum energy demand estimates of the RBM. The APS tool automatically derives 12 process configurations as feasible flowsheets, including the three configurations with only three distillation columns proposed by Kraemer et al. (2009). This becomes feasible due to the consideration of both limiting operation modes by computation of residue curves and pinch lines. The computations take less than five minutes of computational time on a standard desktop computer. Figure 2 summarizes the respective minimum heat duties estimated by means of the RBM in the APS tool (right) and illustrates the four most promising process variants (left), which include all three of the previously mentioned three-column configurations. All remaining process variants consist of either four or five column flowsheets, with an increase in the accumulated heat requirement of up to 135%.

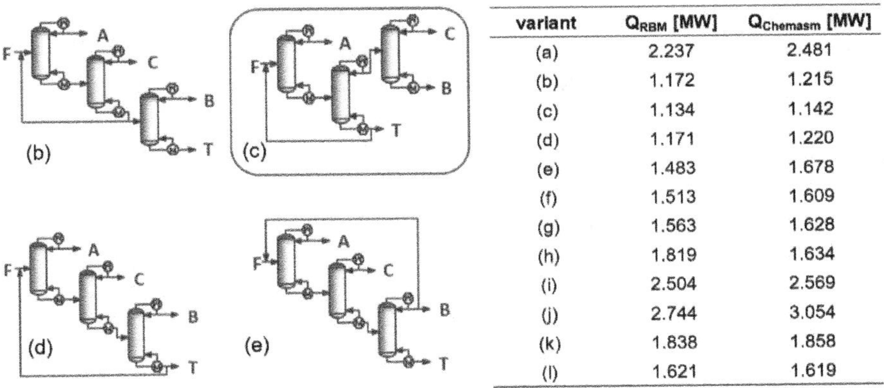

variant	Q_{RBM} [MW]	$Q_{Chemasm}$ [MW]
(a)	2.237	2.481
(b)	1.172	1.215
(c)	1.134	1.142
(d)	1.171	1.220
(e)	1.483	1.678
(f)	1.513	1.609
(g)	1.563	1.628
(h)	1.819	1.634
(i)	2.504	2.569
(j)	2.744	3.054
(k)	1.838	1.858
(l)	1.621	1.619

Figure 2: Illustration of the four most promising process flowsheets for the ACBT separation (left) and overview of the cumulative heat requirement of all 12 derived process variants (right).

Based on the results of the APS tool, adequate tray numbers and column profiles are first determined for each individual column and then for the entire flowsheets by means of the FlowsheetExplorer, which takes around 16 minutes of computational time. Converged closed-loop flowsheet simulations were obtained for all 12 process variants. Figure 3 exemplarily illustrates the composition profiles of the individual components in the respective columns 1-3 for process configuration (c), which provides the lowest energy requirements. The resulting flowsheets and the obtained starting profiles were passed over to Chemasim to pursue a heat minimization with rigorous equilibrium tray models and product purities of 99 mol% set as constraints. The optimization of all 12 flowsheets takes around 52 minutes. The excellent agreement of the resulting accumulated heat duties of the rigorous flowsheet optimization and the initial shortcut evaluation are illustrated by the side-by-side comparison in Figure 2 (right).

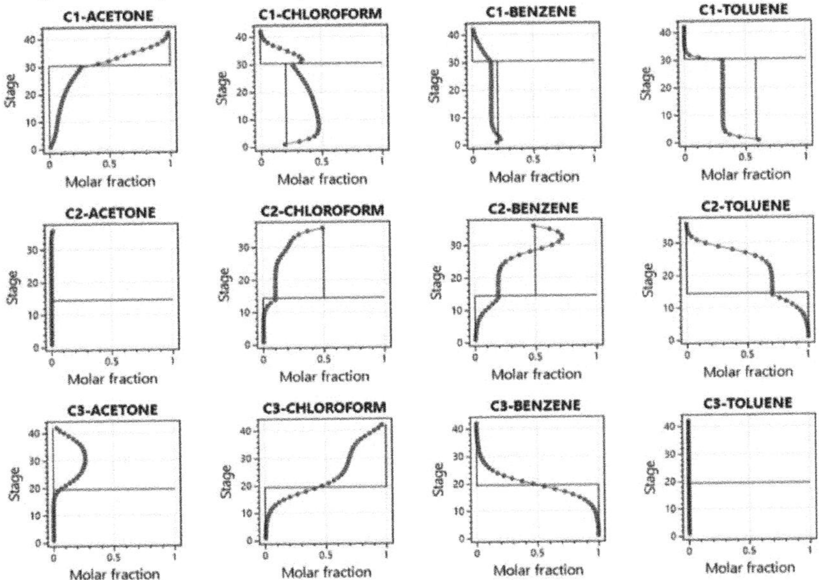

Figure 3: Illustration of the resulting composition profiles of the three columns in the flowsheet variant (c) after initialization and optimization in Chemasim.

Especially for the three best variants, a mean deviation of only 3% is reached. Overall, even larger deviations, which can be explained by the linear approximations of the RBM and variations in the recycle compositions, are limited (< 15%) and do not significantly affect the ranking of the variants. Furthermore, it can be noted that the results of the rigorous flowsheet evaluation of the best flowsheet variant fit very well with the optimization results reported by Kraemer et al. (2009).

4. Conclusion and outlook

The current article illustrates how a set of multi-purpose tools for the conceptual design of distillation-based separation processes can be integrated for a fully automated synthesis and rigorous simulation of distillation processes to separate homogeneous azeotropic mixtures. The basis for the integrative approach is a specification of the separation problem and a consistently applied thermodynamic model, which is the fundamental requirement for any meaningful flowsheet simulation. Since the applied topological synthesis tool of Sasi and Skiborowski (2020), which also allows for the consideration of pressure variations, does not build on constant relative volatilities, a graphical interpretation or a geometric approximation of the distillation boundaries, it is practically not limited with respect to the number of components or complexity of the mixture. The derived flowsheet structures and the estimated minimum reflux ratios build the basis for the subsequent optimization-based initialization of rigorous equilibrium tray models and the final simulation of the integrated closed-loop flowsheet in Chemasim.

Future investigation will not only evaluate the application to much larger multicomponent mixtures, but also extend the design space to consider liquid-liquid separations, as in heteroazeotropic distillation (Sasi and Skiborowski, 2020) and liquid-liquid extraction (Sasi and Skiborowski, 2022), as well as extractive distillation. Furthermore, the possibility for process intensification and energy integration may also be explored in an automated form by further extending the synthesis method with adapted shortcut-screening methods (Skiborowski, 2018). Also, to speed up the closure of multiple recycle streams simultaneously, we are currently exploring the option to adaptively determine points on the distillation boundary and formulate a linear optimization problem.

References

APS tool GitLab project, 2023, https://collaborating.tuhh.de/v-4/psi-public/aps-tool.

J. Bausa, R.v. Watzdorf, W. Marquardt, 1998, AIChE J., 44(10), 2181-2198

Z. Jiang, T.J. Mathew, H. Zhang, J. Huff, U. Nallasivam, M. Tawarmalani, R. Agrawal, 2019, Comput. Chem. Eng., 126, 249–262.

K. Kraemer, S. Kossack, W. Marquardt, 2009, Ind. Eng. Chem. Res., 48, 6749–6764.

G. Liu, M. Jobson, R. Smith, O. Wahnschafft, 2004, Ind. Eng. Chem. Res., 43, 3908-3923.

G. Madenoor Ramapriya, M. Tawarmalani, R. Agrawal, 2018, AIChE J., 64, 649–659.

R.E. Rooks, V. Julka, M.F. Doherty, M.F. Malone, 1998, AIChE J., 44, 1382–1391.

T. Sasi, M. Skiborowski, 2020, Ind. Eng. Chem. Res., 59(47), 20816-20835

T. Sasi, M. Skiborowski, 2020, Comp. Aid. Chem. Eng., 48, 1009-1014

T. Sasi, M. Skiborowski, 2022, Comp. Aid. Chem. Eng., 51, 751-756

M. Skiborowski, 2018, Chem. Eng. Trans., 69, 199-204.

L. Tao, M.F. Malone, M. F. Doherty, 2003, Ind. Eng. Chem. Res., 42, 1783–1794.

D. Thong, G. Liu, M. Jobson, R. Smith, 2004, Chem. Eng. Proc., 43, 239–250.

T. Tsai, J.D Ward, 2020, Ind. Eng. Chem. Res., 59, 17964–17976

S.K. Wasylkiewicz, F.J.L. Castillo, 2001, Comput. Aided Chem. Eng., 9, 591–596.

X. Yang, H.G. Dong, I.E. Grossmann, 2011, AIChE J., 58, 1487–1502.

T. Seidel, A. Hoffmann, M. Bortz, A. Scherrer, J. Burger, N. Asprion, K.-H. Küfer, H. Hasse, 2020, Chem. Eng. Sci. X, 7, 100063

Antonis Kokossis, Michael C. Georgiadis, Efstratios N. Pistikopoulos (Eds.)
PROCEEDINGS OF THE 33rd European Symposium on Computer Aided Process Engineering (ESCAPE33), June 18-21, 2023, Athens, Greece
 http://dx.doi.org/10.1016/B978-0-443-15274-0.50075-5

A Hybrid Modelling Framework for Dynamic Modelling of Bioprocesses

Haiting Wang[a], Cleo Kontoravdi[a], Ehecatl Antonio del Rio Chanona[a]

[a] *Sargent Centre for Process Systems Engineering, Department of Chemical Engineering, Imperial College London, London, United Kingdom SW72AZ*

Abstract

One of the main hurdles in the computer-aided design and optimization of industrial bioprocesses is the limited capability of models to accurately represent biosystems. On one hand, mechanistic models have drawbacks in terms of expressing all potential biomechanisms. On the other hand, data-driven models have limited extrapolation ability. Thus, hybrid models may represent the best of both worlds. However, it is challenging to build a hybrid model that can accurately balance its data-driven and first principles components, hence allowing for extrapolation ability and accuracy.

In this work, we propose a framework for the development of hybrid models that includes a mechanistic backbone for its construction and judiciously adds the data-driven components. Statistical methods e.g., Bayesian Information Criterion (BIC), Akaike Information Criterion (AIC) and Hannan Quinn Criterion (HQC), are employed to choose the statistically best hybrid model structure. The proposed framework is tested on a simplified microalgae cultivation case study and shows good prediction capabilities under different noise levels, especially by applying BIC for hybrid model selection.

Keywords: Model selection, Hybrid modelling, Bioprocess modelling, Information criteria.

1. Introduction

Industrial bioprocesses have experienced fast development in many fields. However, most bioprocesses suffer from low production yields, and experimental optimization of these systems is both resource-intensive and time-consuming. Therefore, building mathematical models which can accurately describe the dynamic behavior of biosystems is crucial for process design, scale-up and optimization. Constructing a mechanistic (first principles) model based on physical and biological knowledge has been widely investigated with very good success, but also room for improvement (Tsopanoglou and Jiménez del Val, 2021). Because investigating and quantifying cellular mechanisms is very challenging, especially considering the intracellular metabolic reaction networks of living cells and how they are influenced by variations in operating conditions. To address these challenges, data-driven models have attracted a lot of attention given that they do not require physical knowledge. However, the biggest drawback of data-driven modelling is the requirement for high quantity and quality of measured data for model training, and their inability to extrapolate, which is critical for optimization and control.

To take advantage of both mechanistic and data-driven models while avoiding their drawbacks, hybrid models are considered good potential solutions. They have mechanistic backbones to promote extrapolation performance. At the same time, complicated biomechanisms are modelled with properly selected data-driven components that avoid heavy mechanistic expressions. Therefore, when training a hybrid model with the same dataset, it has better extrapolation and interpretation ability compared to the data-driven model, and it is easier to construct than a mechanistic model.

However, to build a good performing hybrid model, it is important to balance the weight of the data-driven sub-model and the mechanistic backbone. To achieve this, empirical model selection methods such as the L1 and L2 regularizations have been widely applied (Tibshirani, 1996). However, these methods are non-statistical, and the model selection results are not consistent. Alternatively, in this work, statistically-derived information criteria are applied for model selection, specifically, the Akaike Information Criterion (AIC) (Stoica and Selen, 2004), Bayesian Information Criterion (BIC) (Stoica and Selen, 2004), and Hannan Quinn Criterion (HQC) (Hannan and Quinn, 1979).

AIC is derived from information theory, which defines the best model as the one that can minimize the information divergence between the observed data and simulation results. The equation of AIC can be represented as below:

$$J = 2n_{\theta} + 2NLL(\boldsymbol{\theta}) \tag{1}$$

Where $NLL(\boldsymbol{\theta})$ is the negative log-likelihood (NLL) function based on parameter $\boldsymbol{\theta} \in \mathbb{R}^{n_\theta}$, which represents the goodness of fit between the model prediction and the measured data, and the total number of parameters in the model candidate, n_θ, is penalized.

On the other hand, BIC is derived from the Bayes' rule, which assumes that a 'true' model exists among all candidate models, and the best model is the one which has the maximum posterior probability value to be the best model (Stoica and Selen, 2004):

$$J = 2n_{\theta}ln(n_d) + 2NLL(\boldsymbol{\theta}) \tag{2}$$

It can be seen from Eq.2 that BIC penalises model complexity by considering not only n_θ, but also n_d which is the number of data points in the training set. Like the BIC, the HQC is derived to select the 'true' model instead of the one that minimizes the information loss from data, and the expression can be shown below:

$$J = 2n_{\theta}ln(ln(n_d)) + 2NLL(\boldsymbol{\theta}) \tag{3}$$

In terms of penalty terms, the HQC also considers the impact of the training dataset size, while it penalises the model complexity less strongly than BIC. By using these information criteria for model selection, the best model is chosen by minimizing the value of these information criteria.

Because these information criteria are derived from different statistical perspectives, they show different model selection behaviors. It can be found from the literature that BIC tends to consistently choose the 'true' model and penalizes model candidates with higher model complexity more than the 'true' model, while AIC prefers to select the model with better fitting performance even if it is more complex than the 'true' model (Vrieze, 2012). In this work, we propose a hybrid modelling framework with the application of statistical information criteria to efficiently construct hybrid models with good extrapolation ability for the dynamic modelling of bioprocesses.

2. Hybrid modelling framework

The proposed hybrid modelling framework comprises 4 steps: mechanistic backbone construction, data-driven sub-modelling, hybrid model integration and model validation as shown in Figure 1. To find the best hybrid model that balances the fitting and extrapolation performance, AIC, BIC and HQC are used for statistical model selection and embedded into the framework. During this step, model structure selection and parameter estimation are conducted. In the end, the parameter uncertainty analysis and robustness test are carried out on the constructed hybrid model.

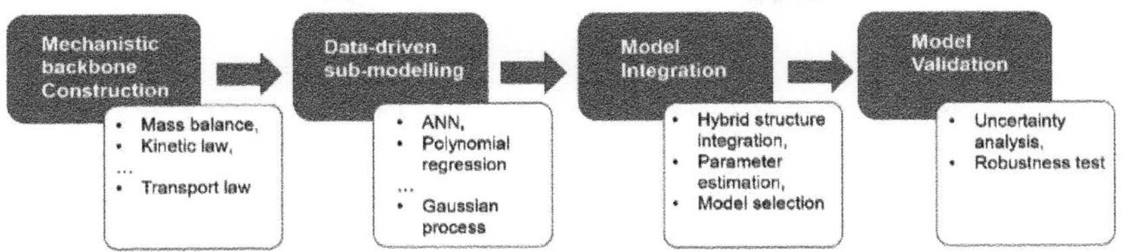

Figure 1 Illustration of the hybrid modelling construction framework.

3. Case study –Microalgae Cultivation

3.1. Microalgae cultivation system

A simplified microalgae cultivation process under fed-batch operating conditions was used as a case study. This model is a simplification from the model which describes the biomass growth and nitrate consumption in a bioreactor (del Rio-Chanona et al., 2015). A mechanistic model is introduced to describe the dynamic change of biomass concentration C_X and nitrate concentration C_N with a constant feeding rate $F_{in}N_{feed}$:

$$\frac{dC_X}{dt} = \mu_M C_X \frac{C_N}{C_N + K_N} - \mu_d {C_X}^2 \quad (4)$$

$$\frac{dC_N}{dt} = -Y_{N/X}\mu_M C_X \frac{C_N}{C_N + K_N} + F_{in}N_{feed} \quad (5)$$

Where μ_M is the specific biomass growth rate, μ_d is the biomass death rate, $Y_{N/X}$ is the specific nitrate uptake ratio, and K_N is the half-velocity coefficient.

3.2. In-silico data generation

The mechanistic model above is used for in-silico data generation for model training. In this work, three computational experiments are generated with different initial concentrations for C_X and C_N. These computational experiments simulate the dynamic change of C_X and C_N in 384 hours of operational time, and the generated data for C_X and C_N is collected in 10-hour intervals. Different levels of Gaussian noise are applied to the training datasets to mimic real-life measurements. Noise levels of C_X and C_N are certain percentages of their mean values respectively. At the same time, three validation datasets are generated with different initial concentrations, but the same data collection intervals compared to training datasets.

3.3. Hybrid model construction

Following the first step in the framework, the mechanistic backbone is designed first. By applying simple mass balance on C_X and C_N, the mechanistic backbone is shown below:

$$\frac{dC_X}{dt} = \mu_m C_X - \mu_d {C_X}^2 \quad (6)$$

$$\frac{dC_N}{dt} = -Y_{N/X}\mu_m C_X + F_{in}N_{feed} \quad (7)$$

Among all variables in the mechanistic backbone, it is challenging to accurately describe the biomass growth rate μ_m dependence on metabolite concentrations and bioreactor operating conditions. Thus, to avoid the complex mechanistic expression, we choose a data-driven method to represent it in terms of the variable it depends on.

In this work, the data-driven sub-model aims to determine the relationship between μ_m, C_X and C_N. Third-order polynomial regression is chosen for this purpose as shown below:

$$\mu_m = p_0 + y_0 p_1 C_{X\_norm} + y_1 p_2 C_{N\_norm} + y_3 p_4 {C_{X\_norm}}^2 + \cdots, + y_8 p_9 {C_{N\_norm}}^3 \quad (8)$$

$$C_{X\_norm} = \frac{C_X}{C_{X_{max}}} \quad (9)$$

$$C_{N\_norm} = \frac{C_N}{C_{N_{max}}} \quad (10)$$

Where p_i are coefficients of each polynomial term and y_i are binary variables. Binary variables are used to describe whether their corresponding polynomial term is selected or not when carrying out statistic model selection. Moreover, $C_{X_{max}}$ and $C_{N_{max}}$ are the maximum measured value of C_X and C_N respectively and they are introduced to normalize the dependencies. Other data-driven methods can also be applied instead of polynomial regression, especially for highly nonlinear systems, such as Artificial Neural Network (ANN), Partial Least Square regression (PLS) and Gaussian Process (GP).

By integrating the mechanistic backbone with the polynomial expression of μ_m we can get the basic structure of the hybrid model. Then, to find the best structure of the hybrid model, statistical model selections are applied, which is the core of the model integration step. During this step, parameter estimation and model selection can be carried out by solving a MINLP problem to obtain the best hybrid model. Finally, the best hybrid model is validated using 3 sets of validation data as described in section 3.2.

4. Results and Discussion

First, a simple task is carried out to demonstrate the concept of statistical selection of the best-performing hybrid model. In this case, we assume the ideal behavior of each state variable is described by the mechanistic model that we used for data generation, thus the best hybrid model is the one that gives the best agreement with the mechanistic model even if it is trained by noisy data. The agreement is judged based on the mean square error from validation datasets. Then, to further test the proposed hybrid model construction framework, we assess its performance when it is trained with datasets with different noise levels, from 5% of the mean of C_X and C_N to 20%.

4.1. Proof of concept

To empirically study whether the information criteria can effectively select the best hybrid model structure that can avoid overfitting and achieve the best extrapolation capability, we generate 9 hybrid model structures by adding polynomial terms in the sequence in Eq.8. The existence of polynomial terms is represented by setting their corresponding y_i to 1. Then, each hybrid model is trained with a dataset with 5% noise. All three information criteria and NLL values are estimated for each structure. Moreover, each model is validated using validation datasets and the results are shown in Figure 2:

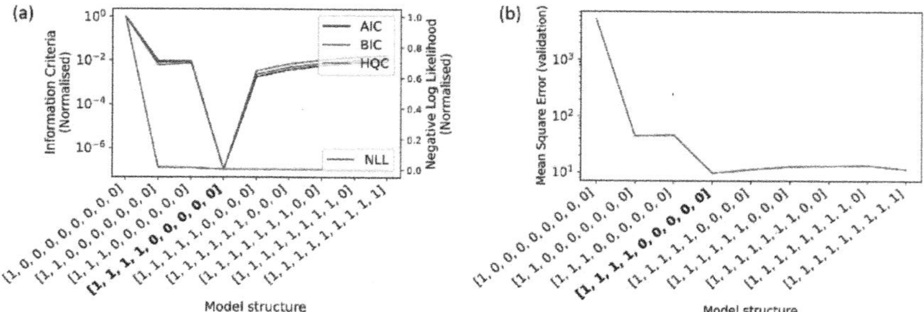

Figure 2. Training and validation results of 9 hybrid model structures. (a) Estimation of AIC, BIC, HQC and NLL for each hybrid model structure. (b) Validation results of each hybrid model structure.

As shown in Figure 2, the increase of polynomial terms in sequence of the hybrid model decreases the corresponding NLL values, which indicates an improvement in the fitting. However, the improvement becomes much less significant after a certain stage, indicating the occurrence of overfitting. By calculating the value for AIC, BIC and HQC, we can see from Figure 2(a) that all three information criteria start to penalize the over-complexed structure when the model has more than the first 4 polynomial terms. The validation results show a strong agreement on the best hybrid model structure selected by information criteria among these 9 candidates, highlighting the effectiveness of statistical model selection methods.

4.2. Robustness test under different noise levels

As mentioned in Section 1, AIC, BIC and HQC have different statistical properties. Thus, we need to evaluate their robustness, especially under different noise levels since low-quality data is one of the most common problems for bioprocess modelling. Therefore, 4 training datasets are generated with the noise level of 5%, 10%, 15% and 20% respectively. Also, under each noise level, 10 datasets are generated and used for hybrid model construction to quantify result uncertainties.

The total number of polynomial terms (n_p) in Eq.8 can vary from 1 to 8. Therefore, we can first find 7 potential best hybrid models with certain n_p value (models with n_p = 8 are significantly overfitted and not presented here). Then the corresponding information criteria's values and the validation results can be estimated for each model with a certain standard deviation as shown in Figure 3.

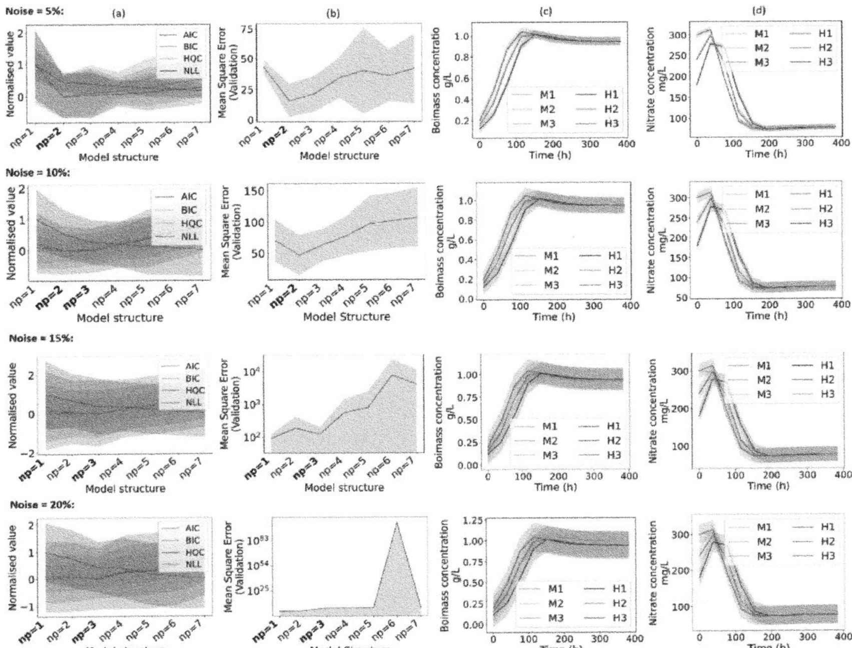

Figure 3. Training and validation results of all potential hybrid model structures under different noise levels. (a) Estimation of AIC, BIC, HQC and NLL for each hybrid model structure. (b) Validation of each hybrid model structure. (c) and (d) Validation results of biomass and nitrate concentration prediction given by the best hybrid model respectively. (M1, 2 & 3 represent validation datasets from the mechanistic model, while H1, 2 & 3 represents validation results given by the hybrid model)

It can be seen from the validation results in Figure 3(b) that the best hybrid model structure varies with the increase of noise level because the model is easier to overfit when data is highly noisy. The best hybrid model is the one with $n_p = 2$ under 5% and 10% noise while it becomes the one with $n_p = 1$ when noise increases to 15% and 20%.

Statistical model selection results are shown in Figure 3(a) at each noise level. It can be seen that all information criteria choose the same model structure under 5% noise. However, they show different model selection results under different noise levels. When noise increases to 10%, BIC and HQC still successfully select the best hybrid model structure, while AIC chooses a model structure with worse performance ($n_p = 3$). Under a 15% noise level, both HQC and AIC select the wrong model structure ($n_p = 3$) while BIC still selects the correct one. By increasing the noise level to 20% only BIC and HQC can find the best model structure.

In summary, our results suggest that BIC is the most robust information criterion which shows stable performance on choosing the best model structure, while HQC and AIC are not as reliable since they put more weight on getting the better fitting performance and penalise model structure complexity weaker. Additionally, it can be seen from Figure 3 (c) and (d) that by using information criteria for model selection, even if the hybrid model is trained under strong noise, the model can maintain good extrapolation ability and shows good agreement with the purely mechanistic model.

5. Conclusion

In this work, a hybrid model construction framework was proposed and tested on a biomass cultivation case study. The results show that by applying this framework, and especially by using statistical information criteria, the best hybrid model structure can be efficiently selected. Among all three information criteria, BIC shows the most consistent performance against the increase of noise level from the training dataset. In the future, we will apply this framework to more complex case studies such as the photo-production process, and the polynomial regression will be replaced with a more complex data-driven model such as ANN. Finally, an uncertainty analysis of the constructed hybrid model will be carried out.

References

DEL RIO-CHANONA, E. A., DECHATIWONGSE, P., ZHANG, D., MAITLAND, G. C., HELLGARDT, K., ARELLANO-GARCIA, H. & VASSILIADIS, V. S. 2015. Optimal Operation Strategy for Biohydrogen Production. *Industrial & Engineering Chemistry Research,* 54, 6334-6343.

HANNAN, E. J. & QUINN, B. G. 1979. The determination of the order of an autoregression. *Journal of the Royal Statistical Society: Series B (Methodological),* 41, 190-195.

STOICA, P. & SELEN, Y. 2004. Model-order selection: a review of information criterion rules. *IEEE Signal Processing Magazine,* 21, 36-47.

TIBSHIRANI, R. 1996. Regression Shrinkage and Selection Via the Lasso. *Journal of the Royal Statistical Society: Series B (Methodological),* 58, 267-288.

TSOPANOGLOU, A. & JIMÉNEZ DEL VAL, I. 2021. Moving towards an era of hybrid modelling: advantages and challenges of coupling mechanistic and data-driven models for upstream pharmaceutical bioprocesses. *Current Opinion in Chemical Engineering,* 32, 100691.

VRIEZE, S. I. 2012. Model selection and psychological theory: a discussion of the differences between the Akaike information criterion (AIC) and the Bayesian information criterion (BIC). *Psychological methods,* 17, 228.

Antonis Kokossis, Michael C. Georgiadis, Efstratios N. Pistikopoulos (Eds.)
PROCEEDINGS OF THE 33rd European Symposium on Computer Aided Process Engineering (ESCAPE33), June 18-21, 2023, Athens, Greece
 http://dx.doi.org/10.1016/B978-0-443-15274-0.50076-7

Simultaneous selection and optimization of reaction pathways and downstream separation network

Shuang Xu[a], Selen Cremaschi[a], Mario R. Eden[a] and Anjan K. Tula[b*]

[a]*Department of Chemical Engineering, Auburn University, Auburn, AL 36849, USA*
[b]*College of Control Science & Engineering, Zhejiang University, Hangzhou, China*
[*]*anjantula@zju.edu.cn*

Abstract

Synthesis of separation networks has been researched for many years because of their potential in improving profit and reducing carbon emissions. However, most of the presented work concentrated on identifying novel solutions for the downstream section such as improving thermal efficiency by using complex heat exchanger networks, applying advanced separation techniques, and optimizing the design/operating parameters, etc., without considering the effect of upstream reaction pathways. This work aims to develop a mixed integer non-linear programming (MINLP) optimization model, which integrates reaction pathway selection and downstream separation optimization. To ensure both process profit and sustainability, the objective is to maximize the process profit per carbon emission. Decision variables such as binary variables, which represent the selection of reaction pathway and separation task, and continuous variables, which influence the separation mass/energy balance, are considered. By giving parameters like reaction conversion rate, component relative volatility, component enthalpy, raw material/product price, etc., this optimization model can simultaneously identify the optimal reaction pathways and their separation network. The optimization model was applied to isobutylene and methanol utilization case studies. The results indicate that the model could identify the best product and the associated best performing process.

Keywords: Reaction pathway identification, separation synthesis, optimization.

1. Introduction

The downstream separation network is defined as a process that comprises different unit operations, such as separators, heat exchangers and mixers, etc., to handle multicomponent separation and produce the desired products. In general, several configurations based on different separation techniques and separation sequences can be generated, and the selected options significantly influence the overall process cost and sustainability. Therefore, the optimal configuration has to be identified to ensure overall process profitability and sustainability. One established approach is the superstructure-based model, where all feasible separation configurations are generated, and the optimal choice is identified through mathematical optimization. Despite significant research on this approach, most work has been limited to downstream separation optimization, such as adding intensified equipment into the separation configurations (Caballero and Grossmann 2001, 2013), systematically identifying novel intensified unit operations (Demirel et al., 2017), simultaneous optimization of distillation networks and heat exchanger networks (Yeomans and Grossmann, 2000), but the effect of upstream conditions/decisions was not considered explicitly. Ryu et al. (2020) pointed out that the separation network synthesis efforts should also consider the upstream reaction because the choice of the reaction influences the decisions on the downstream separation.

Integrating different reaction pathways and separation process configurations to determine the optimal products/feedstocks and process configuration has been widely researched in the biochemistry area. Kong and Shah (2016) proposed a mixed integer linear programming (MILP) model to determine the optimal reaction and separation process for utilizing bio-based materials. Zondervan et al. (2011) reported a superstructure-based optimization model to determine the optimal reaction and separation steps for biorefinery systems under different optimization objectives. Kelloway and Daoutidis (2014) developed an optimization model to identify the optimal feedstocks, products, and process configurations by maximizing the net present value and carbon efficiency for the biorefinery systems. However, with the progress of sustainable design, it is also necessary to construct a generalized optimization model for the conventional process industry to simultaneously select/optimize reaction pathways and downstream separation networks.

2. Methodology

The main goal is to construct an optimization model to identify the best process for producing a specific product or utilizing raw materials, which includes two parts: 1) selection of reaction pathways that convert raw materials into products. 2) identification of separation configurations that separate the reaction outlet into the final product. An example superstructure representation of this problem is illustrated in Figure 1. Given a low-value raw material, various reaction pathways can be applied to convert the raw material into different high-value products, and the outlet mixture can be separated by passing through the separation networks. The objective is to identify the most valuable product and the optimal separation configuration that ensure both process profit and sustainability.

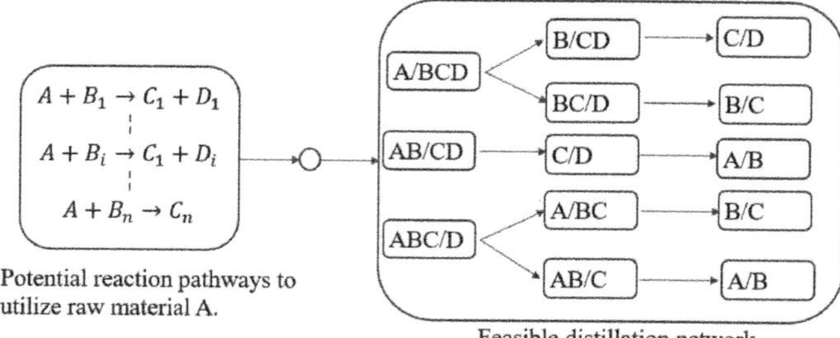

Figure 1. Superstructure representation of the optimization model.

2.1. Optimization model formulation

The objective is to identify the optimal reaction pathway and separation network that has the maximum process profit per carbon emission (Eq. 1). The whole process profit involves product sales and process operating costs. Also, to ensure the process sustainability, carbon emissions is considered. Reaction pathways selection and distillation mass balance constraints are given in Eqs. (2) and (3). To calculate the distillation operating cost, a short-cut distillation model (Ryu et al., 2020) that is based on distillation column vapor flowrate and enthalpy balance is given in Eqs. (4) – (7). Note here, because the number of inlet components varies according to the selection of different reaction pathways, a pseudo-component that has zero inlet flowrate is defined in this model. To find feasible solutions of the short-cut model when the inlet component has zero flowrate, a disjunctive constraint is formulated in Eq. (4). The carbon emission constraint is given in Eq. (8). Here, natural gas is set as the primary fuel for generating the utility. By converting the heat duty into the amount of standard natural gas, the carbon emissions can be calculated based on the standard natural gas conversion factor (50.15 Kg CO_2/GJ) (Shi et al., 2022). In this model, the purity of the product is specified at 99.5%, which is given in Eq. (9).

$$max: \quad \frac{\sum_k P_k F_k - \sum_j Q_j}{\sum_j C_j} \tag{1}$$

$$s.t. \quad \sum_i y_i = 1 \ , \ y_i \in \{0,1\} \tag{2}$$

$$yd_j F_{j,k} = F_{j,k,0} + F_{j,k,1} \ , \ yd_j \in \{0,1\} \ \forall j \in J, k \in K \tag{3}$$

$$\begin{bmatrix} F_{j,k} F_{j,k+1} = 0 \\ \phi_l = 0, l = k \end{bmatrix} \vee \begin{bmatrix} F_{j,k} F_{j,k+1} \neq 0 \\ \sum_k \frac{\alpha_k F_{j,k}}{\alpha_k - \phi_l} = 0 \\ \alpha_k < \phi_l < \alpha_{k-1} \end{bmatrix}, \ \forall k \in K, j \in J, l \in L \tag{4}$$

$$V_{j,0} \geq \sum_k \frac{\alpha_k F_{j,k,0}}{\alpha_k - \phi_l}, \ \forall j \in J, l \in L \tag{5}$$

$$V_{j,1} \geq \sum_k \frac{-\alpha_k F_{j,k,1}}{\alpha_k - \phi_l}, \ \forall j \in J, l \in L \tag{6}$$

$$Q_j = U_{j,0} V_{j,0} H_{j,0} + U_{j,1} V_{j,1} H_{j,1}, \ \forall j \in J \tag{7}$$

$$C_j = \eta V_{j,1} H_{j,1}, \ \forall j \in J \tag{8}$$

$$g(F_k) \geq 0.995, \ \forall k \in K \tag{9}$$

Here, the sets $k \in K, j \in J, i \in I$, and $l \in L$ represent components, separation tasks, reaction pathways, and active roots. P is the product price. F is flowrate. Q is distillation heat duty. U is the utility price (Kalakul et al., 2014). C is the amount of carbon emission. y is a binary decision variable for reaction pathway selection. yd is a binary decision variable for separation task selection. α is relative volatility. ϕ is active root. V is the vapor flowrate inside the stripping (1) and rectifying (0) section. H is latent heat. η is carbon emission coefficient.

3. Case study

The developed optimization model was applied to two case studies, which aim to find the best product and process configurations for utilizing isobutylene and methanol. In these two case studies, two different optimization objectives (process profit and process profit per carbon emission) are considered. All the optimization problems are solved by using BARON. Note here, for all the case studies, the separation pressure is adjusted to guarantee that cooling water is used for condensation. For temperature-sensitive products such as pivalic acid, the separation pressure is close to vacuum.

3.1. Isobutylene utilization

Isobutylene was widely applied for methyl t-butyl ether (MTBE) production, which is an important gasoline additive. However, with the development of the electric car, the market demand for isobutylene may drop in the future, so three alternative isobutylene utilization reaction pathways that produce pivalic acid (PA), prenol, and tert-butylamine, are considered in this case study. The parameters like reaction conversion rate, component relative volatility, price data (www.echemi.com, www.sunsirs.com/uk), and latent heat, which are the input for the optimization model, are listed in Table 1. In this case study, the isobutylene inlet flowrate is 150 kmol/h and the co-reactant inlet flowrate is calculated based on reaction stoichiometry.

Table 1. Parameters for the isobutylene utilization case study.

	Components	Conversion	Rel. Vol.	Price, \$/kmol	Latent heat, kJ/mol
R1	A: isobutylene	0.8	5548.36	51.79	25.31
	B: water	-	25.5	0	44.02
	C: PA	-	1	510.65	58.91
	D: *pseudo*	-	1	0	0
R2	A: formaldehyde	-	16.86	5.76	18.63
	B: isobutylene	0.8	7.82	51.79	16.99
	C: prenol	-	1	301.46	36.11
	D: *pseudo*	-	1	0	0
R3	A: ammonia	-	5.94	13.42	19.86
	B: isobutylene	0.52	2.61	51.79	16.99
	C: tert-butylamine	-	1	234.048	20.97
	D: *pseudo*	-	1	0	0

The optimal solution, including the reaction pathway, process configuration, detailed stream flowrate, carbon emissions, etc., are shown in Figure 2 and Table 2. The results indicate that using isobutylene for PA production has both the highest overall process profit and the highest process profit per amount of carbon emission (291.6 \$/(h, kg CO_2)). The corresponding optimal separation configuration is based on the sequence of component boiling point.

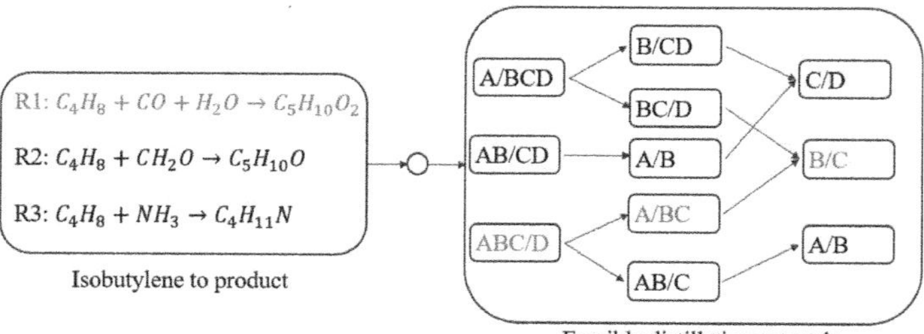

Figure 2. Optimal reaction pathway and configuration for isobutylene utilization.

Table 2. Optimization results of using isobutylene for PA production.

		Separation task					
		ABC/D		A/BC		B/C	
		Distillate	Bottom	Distillate	Bottom	Distillate	Bottom
Component flowrate, kmol/h	A	30	0	29.85	0.15	0.15	0
	B	30	0	0.15	29.85	29.25	0.6
	C	120	0	0	120	0	120
	D	0	0	0	0	0	0
Carbon emissions, kg CO_2/h		0		84.23		104.35	
Operating cost, \$/h		0		31.44		33.62	

3.2. Methanol utilization

In this case study, four reaction pathways are considered. R1 uses methanol for dimethyl ether (DME) production. R2 converts methanol into gasoline with liquefied petroleum (LPG) gas as a by-product. R3 directly converts methanol into acetic acid (AA). And R4 utilizes methanol and propylene carbonate (PC) to produce dimethyl carbonate (DMC) with by-product propylene glycol (PG). The model input parameters are listed in Table 3. In this case study, the methanol inlet flowrate is 150 kmol/h and the co-reactant inlet flowrate is calculated based on reaction stoichiometry.

Table 3. Parameters for the methanol utilization case study.

	Components	Conversion rate	Relative volatility	Component price, $/kmol	Latent heat, kJ/mol
R1	A: DME	-	81.28	25.7	17.91
	B: methanol	0.82	3.38	11.7	30.24
	C: water	-	1	0	36.9
	D: *pseudo*	-	1	0	0
R2	A: LPG	-	10	39.1	16.58
	B: methanol	0.98	3.27	11.7	29.56
	C: gasoline	-	1	139	35.45
	D: *pseudo*	-	1	0	0
R3	A: methanol	0.95	5.91	11.7	35.27
	B: AA	-	1	26.4	23.38
	C: *pseudo*	-	1	0	0
	D: *pseudo*	-	1	0	0
R4	A: methanol	0.78	175.95	11.7	35.27
	B: DMC	-	37.94	81.5	33.67
	C: PG	-	7.1	103.9	52.14
	D: PC	-	1	15.8	49.14

The solution for this case study is shown in Figure 3 and Table 4. The results show that using methanol for gasoline production has the highest process profit per amount of carbon emission (78.2 \$/(h, kg CO_2)). The process configuration, which separates gasoline first, has the best performance. However, without considering the process carbon emission, using methanol for DMC production has the highest overall process profit.

Table 4. Optimization results of using methanol for gasoline production.

		Separation task					
		AB/CD		A/B		C/D	
		Distillate	Bottom	Distillate	Bottom	Distillate	Bottom
Component flowrate, kmol/h	A	13.08	0	13.07	0.01	0	0
	B	2.7	0.3	0.07	2.64	0.3	0
	C	0	58.8	0	0	58.8	0
	D	0	0	0	0	0	0
Carbon emissions, kg CO_2/h		58.83		29.96		0	
Operating cost, $/h		11.75		5.99		0	

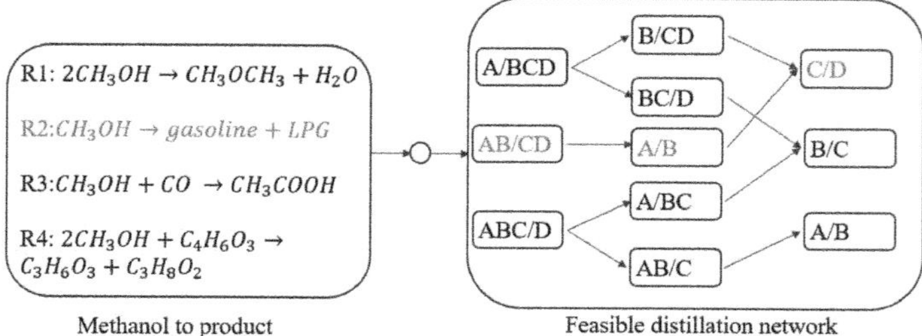

Figure 3. Optimal reaction pathway and configuration for methanol utilization.

Conclusions

This paper developed a generalized MINLP model for simultaneous reaction pathway selection and separation network optimization. Given a product/raw material and the potential reaction pathways, this model can find the most valuable product or promising raw materials and the optimal separation configuration that ensures both process profit the sustainability.

The model was applied to isobutylene and methanol utilization case studies. The results show that the best process for isobutylene utilization, which guarantees both process profit and process sustainability, is to convert isobutylene into PA and separate the mixture based on the sequence of boiling point. For the methanol utilization case study, the best product is gasoline, and the configuration, which separates gasoline first, gives us the most promising process performance. Future work will focus on solving multistep reaction pathway selection and separation network optimization problems.

References

Caballero, J. A., & Grossmann, I. E. (2001). Generalized disjunctive programming model for the optimal synthesis of thermally linked distillation columns. Industrial & Engineering Chemistry Research, 40(10), 2260-2274.

Caballero, J. A., & Grossmann, I. E. (2013). Synthesis of complex thermally coupled distillation systems including divided wall columns. AIChE Journal, 59(4), 1139-1159.

Demirel, S. E., Li, J., & Hasan, M. F. (2017). Systematic process intensification using building blocks. Computers & Chemical Engineering, 105, 2-38.

Yeomans, H., & Grossmann, I. E. (2000). Disjunctive programming models for the optimal design of distillation columns and separation sequences. Industrial & Engineering Chemistry Research, 39(6), 1637-1648.

Ryu, J., Kong, L., de Lima, A. E. P., & Maravelias, C. T. (2020). A generalized superstructure-based framework for process synthesis. Computers & Chemical Engineering, 133, 106653.

Kong, Q., & Shah, N. (2016). An optimisation-based framework for the conceptual design of reaction-separation processes. Chemical Engineering Research and Design, 113, 206-222.

Shi, T., Liu, Y., Yu, H., Yang, A., Sun, S., Shen, W., & Ren, J. (2022). Improved design of heat-pump extractive distillation based on the process optimization and multi-criteria sustainability analysis. Computers & Chemical Engineering, 156, 107552.

Kalakul, S., Malakul, P., Siemanond, K., & Gani, R. (2014). Integration of life cycle assessment software with tools for economic and sustainability analyses and process simulation for sustainable process design. Journal of cleaner production, 71, 98-109.

Antonis Kokossis, Michael C. Georgiadis, Efstratios N. Pistikopoulos (Eds.)
PROCEEDINGS OF THE 33rd European Symposium on Computer Aided Process Engineering (ESCAPE33), June 18-21, 2023, Athens, Greece
 http://dx.doi.org/10.1016/B978-0-443-15274-0.50077-9

Workflow for adaptation, analysis and application of mechanistic models for experimental planning of protein refolding processes

Jan Niklas Pauk,[a,b] Chika Linda Igwe,[a,b] Christoph Herwig,[b]

[a]*Competence Center Chase GmbH, Altenberger Straße 69, 4040 Linz, Austria*
[b]*Institute of Chemical, Environmental and Bioscience Engineering, TU Wien, Gumpendorfer Straße 1a, 1060 Vienna, Austria*

Abstract

Mechanistic models are an important tool as they allow storage of process knowledge and enable model-based techniques to plan, monitor, and control biological processes. Fundamental to these operations are models that can accurately describe their in vivo counterpart. Therefore, good modelling practices have been described including different analysis steps to ensure and maintain identifiability of model parameters during model calibration and the model life cycle (Daume et al., 2020).

Here, a workflow to analyze, adapt and apply a mechanistic model for experimental planning is suggested and its application demonstrated with a set of fed-batch refolding experiments conducted in the controlled environment of a stirred tank reactor. The model is based on the folding kinetics of a recombinant protein solubilized from aggregated proteins accumulated inside bacterial inclusion bodies. Along the adaptation cycle the model is analyzed for unique identifiability of model parameters both from simulations and noisy measurements.

Keywords: Mechanistic Model, Identifiability, Fisher Information, Protein Refolding, MBDOE

1. Introduction

In light of the current quality by design (QbD) paradigm and the process analytical technology (PAT) framework, introduced by the FDA, the use of process models for process monitoring and control are drawn into focus (FDA, 2004). Such models can either be black box data-driven models based on large data sets. Alternatively, they can be more deterministic models such as mechanistic models based on first principles or mass-, energy and momentum balances, or hybrid combinations thereof.

The focus of this work are purely mechanistic models, as they offer some important advantages: i) They require significantly smaller data sets compared to data-driven models. This is relevant, as measurements of biological processes are often expensive, involve long evaluation times such as chromatographic methods and sometimes, as is the case with cell specific reaction rates, there are no direct measurements available. ii) They serve as a good storage medium for process knowledge since they contain easily interpretable information due to correlation of data and physical parameters and physical equations. iii) Furthermore, this knowledge can be used for process monitoring (Kager et al., 2018; Sinner et al., 2019), process control (Kager et al., 2020; Ulonska et al., 2018) and optimal experimental design.

Model-based design of experiments (MBDOE) aims to generate optimal and valid models from as few experiments as possible. Thus, techniques exist to either discriminate different model variants statistically against each other to find the model that best suits the process (Brik Ternbach et al., 2005; Franceschini and Macchietto, 2008) or to design an experiment with maximal information content - usually done by maximizing D- or E-optimality of the fisher-information-matrix - to uniquely identify model parameters and achieve a valid model with good fit (Daume et al., 2019; Franceschini and Macchietto, 2008).

MBDOE generally starts with the use of a basic model structure to define different possible model variants, however, despite the advantages mentioned above, mechanistic models often require expert knowledge for the setup of the model. This contribution focuses on the generation of different model or kinetic substructures for mechanistic models. Therefore, a workflow is developed on how to adapt and improve a mechanistic model in accordance with defined good modelling practices including analysis of model sensitivity and model parameter identifiability. This workflow is applicable during the whole life cycle of the model, including the initial process development, adaptation to new products or process modes, integration of new measurement systems leading to improved parametrization potential and reoccurring model validations in the production of biopharmaceuticals. As an example, a fed-batch protein refolding process was used, optimization potential of the model was analyzed, and an experiment was designed to specifically analyze dependencies on the reaction kinetics. Furthermore, the superiority of this optimized process compared to the classical fed-batch process was shown.

This publication is organized as follows: First, the experimental and computational methods are explained. Then the model adaptation workflow is introduced, and its application discussed using the protein refolding example. Finally, some thoughts about the further application potential are given.

2. Methods

2.1. Protein Refolding Model

The basic non-linear process model of the protein refolding process consists of four protein states: solubilized protein S, folding intermediates I, native protein N and aggregated protein A. It is assumed, that solubilized protein instantly starts the folding process upon entering the refolding buffer and thus state S is removed from the model. Furthermore, extension of the model by the changing denaturant concentration D and reaction volume V due to the feeding during the fed-batch dilution is necessary. The reaction kinetics are described by the following rates depending on the concentration of the denaturant:

$$k_n = a_n \cdot (1 + D)^{b_n} \qquad a_n\colon [0, \infty), \quad b_n\colon (-\infty, 0] \qquad (1)$$

$$k_a = a_a \cdot (1 + D)^{b_a} \qquad a_a\colon [0, \infty), \quad b_a\colon (-\infty, 0] \qquad (2)$$

Here, k_n denotes the reaction rate of native protein formation and k_a the reaction rate of aggregate formation. For a more thorough review of the model see (Pauk et al., 2021).

2.2. Model Analysis & Optimization

Model Fitting: Fitting of the basic model on experimental data is performed using the genetic algorithm from the global optimization toolbox in MATLAB (2020a, MathWorks, Natick, USA) by minimizing the residual sum of squares between simulation and measured data, weighted by the measurement variances:

$$WRSS(\theta) = \sum_{p=1}^{m} \sum_{k=1}^{d} \frac{\left(y_{p,k} - \hat{y}_{p,k}\right)^2}{\sigma_{p,k}^2} \qquad (3)$$

Here, $y_{p,k}$ denotes the measured concentration, $\hat{y}_{p,k}$ denotes the simulated concentration, $\sigma^2_{p,k}$ denotes the variance of state p at time point k. m is the number of states and d is the number of sampling times.

Structural & practical identifiability: Structural and practical identifiability are important attributes of a model describing if fitting of model parameters yields a unique solution from the model structure itself or the noisy measurements, respectively. Local structural identifiability analysis is performed by calculating the discrete-time scaled sensitivity matrix S with entries,

$$s_{ij} = \theta_j \cdot \frac{\partial y_i}{\partial \theta_j} \tag{4}$$

and checking for full matrix rank (Daume et al., 2020). Here, s_{ij} denotes the local sensitivity of model parameter j on the model output y at measurement i. Practical identifiability focuses on the assessment of measurement noise on the unique identifiability of all model parameters. Furthermore, correlations between model parameters resulting in multiple possible solutions where parameter value changes counteract each other need to be analyzed. One possibility to calculate parameter co-linearity is to calculate the fisher information matrix (FIM) from the sensitivity matrix and measurement variance. From the Cramer-Rao theorem follows, that the minimum parameter covariance matrix is the inverse of the FIM and following normalization yields the parameter correlation matrix (Daume et al., 2020). Here, values close to one or negative one on the off-diagonal entries imply high correlation between these parameters and therefore problems with the practical identifiability. (Brun et al., 2001) suggested two parameters which can help finding suitable parameter subsets to facilitate practical parameter identifiability. The parameter importance index can be used to rank model parameters based on their influence on model outputs due to changes in the parameter. The co-linearity index quantifies the linear dependency of model parameters of a given subset of the model parameters in regard of their practical identifiability.

Experimental optimization & model fitting: Optimization of the experiment was done using the genetic algorithm NSGA2. It is embedded in the pyOptSparse framework, can handle non-convex and non-smooth global optimization problems using a tournament selection-based strategy (Wu et al., 2020).

2.3. Protein Refolding

Protein refolding was conducted via dilution in fed-batch mode. Solubilized lactate dehydrogenase (7 g L-1 protein, 4 M guanidine hydrochloride (GuHCl), 150 mM phosphate buffer, pH 6.0) originating from bacterial fermentation with recombinant Escherichia coli BL21 (DE3) was fed with constant feed-rate of 6.48 mL h-1 for 9 hours into a Labfors 5 stirred tank reactor (Infors AG, Bottmingen, Switzerland) containing 0.8 L refolding buffer (150 mM phosphate buffer, 1 mM EDTA, 20 μm NADH, pH 6.0). Refolding was carried out under controlled conditions at a temperature of 10 °C, constant stirring with a Rushton impeller at 200 rpm, headspace aeration with pressurized air at a flow-rate of 1 vvm. The pH was controlled to 6.0 by addition of 0.2 M HCl and 0.5 M NaOH .

Offline samples for the analysis of folding intermediates and native protein in the soluble fraction were measured via SEC-HPLC (Dionex UltiMate 3000, Thermo Fisher, Waltham, USA) with a BEH 200A SEC 1.7 μm, 4.6 x 300 mm, 3.5 μm column (Waters Corporation, Milford, USA). 2 μL of sample were injected and eluted at a flow-rate of

0.5 mL min-1 via isocratic elution (80 mM phosphate buffer, 250 mM KCl, pH 6.8) at a runtime of 18 minutes. The column oven was controlled at 25 °C and absorbance was monitored at a wavelength of 214 nm and 280 nm. The samples were further analyzed for the concentration of soluble proteins via RP-HPLC on a Polyphenyl BioResolve-RP-mAb 2.7 µm 3.0 x 100 mm column (Waters Corporation, Milford, USA) according to Kopp et al. (2020). The concentration of aggregated proteins was determined by BCA-assay as the difference of unprocessed and centrifugated sample.

3. Results & Discussion

3.1. Workflow

The workflow illustrated in Figure 1 shows a cyclic model-based optimization approach to find suitable model structures during the model development process. An initial model either from literature or preliminary experiments (1) is used to analyze for optimization potential of the process model (2). The model-based optimization is performed (3) and the experiment conducted (4). During the evaluation process the model is analyzed regarding sensitivity, identifiability and suitable parameter subsets before the new model structure is fitted on the data set (5). The fit can now be compared to the original model using the WRSS, NRMSE of single states or a specifically targeted behavior (6). Improved fits result in adapted process models, otherwise the process can be analyzed for new optimization potential.

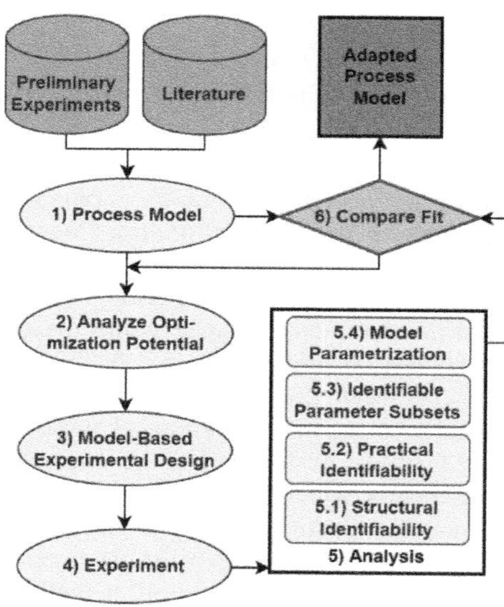

Figure 1: Experimental design cycle.

3.2. Adaptation of a Protein Refolding Model

From a preliminary refolding experiment (data not shown), it was observed that the refolding rate started to stagnate after a total protein concentration of 0.6 g L-1 and a denaturant concentration of 0.2 mol L-1 was reached while aggregation started to increase. From equation 1 follows, that indeed the rate decreases with increasing denaturant concentration. However, a clear dependency with the total protein concentration is not described. Therefore, the rates are extended by a term penalizing increasing total protein concentration,

$$k_n = a_n \cdot (1 + D)^{b_n} \cdot (1 + P)^{c_n} \qquad c_n: (-\infty, 0] \qquad (5)$$

$$k_a = a_a \cdot (1 + D)^{b_a} \cdot (1 + P)^{c_a} \qquad c_a: (-\infty, 0] \qquad (6)$$

with P the total protein concentration in the refolding vessel. Furthermore, an experiment is designed to analyze the dependency of rate and protein concentration by changing one of the parameters while keeping the other constant. The process consists of three phases: (I) an initial fed-batch phase, (II) a phase of constant protein and denaturant concentration, and (III) a phase with constant denaturant but decreasing protein concentration (see Figure 2, left). These phases are realized by using three different reservoirs and feed pumps with reservoir one containing the solubilized protein (7 g L^{-1}, 4 M GuHCl), reservoir two containing pure refolding buffer, and reservoir three containing refolding

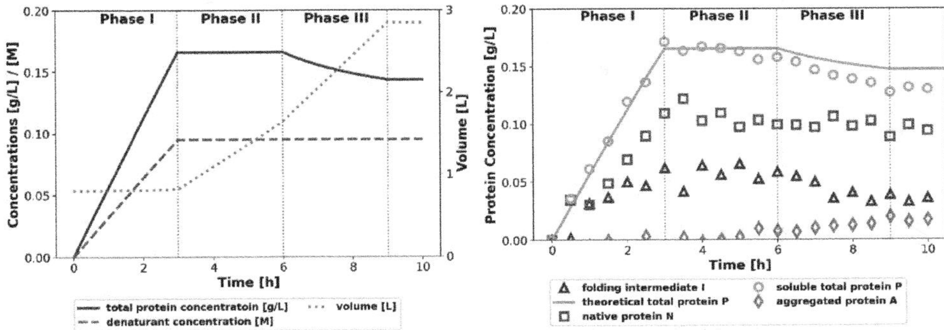

Figure 2: Feeding strategy (left) and data of the experiment optimized through model-based optimization (right). Three phases (vertical dashed lines): initial fed-batch phase, phase with constant protein and denaturant concentrations, phase with constant denaturant and declining protein concentration.

buffer spiked with 0.031 M GuHCl. The first feed rate and the concentrations in reservoir one was set according to section 2.3, the other two feed rates and the concentration of the denaturant (GuHCl) in reservoir three were optimized by the NSGA2 algorithm (F_{R2} = 0.2667 L h^{-1}; F_{R3} = 0.3962 L h^{-1}).

Figure 2 (right) shows measurements of the protein states and the total soluble protein concentration as symbols and the theoretical total protein concentration as solid line. Although the total soluble protein concentration decreases during the later stages of the process due to aggregation, the protein concentration follows the optimized trajectory well and the process results in a final yield of 64 % compared to 28 % in the preliminary experiment.

Analysis of the model by calculating the rank of the sensitivity matrix and the FIM suggested that all model parameters are structurally and practically identifiable. However, the correlation matrix and calculation of the co-linearity index showed moderate correlation of the model parameters a_n to b_n and a_a to b_a in case of the initial rates as well as high correlation of b_n to c_n and b_a to c_a for the adapted rates. Thus, a subset analysis or an optimization of the information density by optimizing sampling times would be necessary to accurately estimate the model parameters (Daume et al., 2019).

Table 1: NRMSEs comparing the fit of the model with initial (I) and adapted rates (II).

States	I	N	A	D	V
NRMSE (I)	0.1437	0.0878	0.1410	0.0027	0.0167
NRMSE (II)	0.1414	0.0775	0.1443	0.0027	0.0165
Improvement	1.61 %	11.73 %	-2.33 %	0.73 %	1.09 %

Nonetheless, the fit of the model with old and adapted rates is compared using the NRMSE of each state (see Table 1). A slight overall improvement is visible and a large improvement specifically in state N, the native protein. N is the most important state regarding a protein refolding process, therefore, the conducted optimized experiment results in a sufficiently different adapted process model structure for further analysis in the model discrimination process. The designed experiment was able to identify the potential of the adapted rates by decoupling protein and denaturant concentrations during the different phases of the experiment.

4. Conclusion

This paper suggested a workflow to generate new model or kinetic structures as prerequisite for a model discrimination step. The potential of model-based experimental design was shown. An experiment specifically targeted to identify dependencies of the reaction kinetics on protein and denaturant concentration was designed, resulting in a new set of reaction kinetics to better describe the refolding process, specifically the concentration of native protein. The example also showed difficulties with co-linear model parameters. Furthermore, due to the low protein concentration resulting from the high dilutions in phase II and III, the yield of native protein was increased by almost two-fold compared to a fed-batch experiment with only one feed.

5. Acknowledgements

The authors acknowledge financial support through the COMET Centre CHASE, funded within the COMET Competence Centers for Excellent Technologies program by the BMK, the BMDW and the Federal Provinces of Upper Austria and Vienna. The COMET program is managed by the Austrian Research Promotion Agency (FFG). The authors also acknowledge insights and support provided by the project partner Bilfinger Life Science.

References

Brik Ternbach, M., Bollman, C., Wandrey, C., Takors, R., 2005. Application of model discriminating experimental design for modeling and development of a fermentative fed-batch L-valine production process. Biotechnol. Bioeng. 91, 356–368.

Brun, R., Reichert, P., Künsch, H.R., 2001. Practical identifiability analysis of large environmental simulation models. Water Resour. Res. 37, 1015–1030.

Daume, S., Kager, J., Herwig, C., 2019. Time Resolved Sensitivity & Identifiability Analysis for Directed Parametrization of Highly Dynamic Models, in: Computer Aided Chemical Engineering. Elsevier, pp. 1111–1116.

Daume, S., Kofler, S., Kager, J., Kroll, P., Herwig, C., 2020. Generic Workflow for the Setup of Mechanistic Process Models, in: Pörtner, R. (Ed.), Animal Cell Biotechnology: Methods and Protocols. Springer US, New York, NY, pp. 189–211.

FDA, 2004. Guidance for Industry PAT - A Framework for Innovative Pharmaceutical Development, manufacturing, and Quality Assurance.

Franceschini, G., Macchietto, S., 2008. Model-based design of experiments for parameter precision: State of the art. Chem. Eng. Sci. 63, 4846–4872.

Kager, J., Herwig, C., Stelzer, I.V., 2018. State estimation for a penicillin fed-batch process combining particle filtering methods with online and time delayed offline measurements. Chem. Eng. Sci. 177, 234–244.

Kager, J., Tuveri, A., Ulonska, S., Kroll, P., Herwig, C., 2020. Experimental verification and comparison of model predictive, PID and model inversion control in a Penicillium chrysogenum fed-batch process. Process Biochem. 90, 1–11.

Pauk, J.N., Raju Palanisamy, J., Kager, J., Koczka, K., Berghammer, G., Herwig, C., Veiter, L., 2021. Advances in monitoring and control of refolding kinetics combining PAT and modeling. Appl. Microbiol. Biotechnol. 105, 2243–2260.

Sinner, P., Kager, J., Daume, S., Herwig, C., 2019. Model-based Analysis and Optimisation of a Continuous Corynebacterium glutamicum Bioprocess Utilizing Lignocellulosic Waste. IFAC-Pap. 52, 181–186.

Ulonska, S., Waldschitz, D., Kager, J., Herwig, C., 2018. Model predictive control in comparison to elemental balance control in an E. coli fed-batch. Chem Eng Sci 191, 459–467.

Wu, N., Kenway, G., Mader, C.A., Jasa, J., Martins, J.R., 2020. pyOptSparse: A Python framework for large-scale constrained nonlinear optimization of sparse systems. J. Open Source Softw. 5, 2564.

Antonis Kokossis, Michael C. Georgiadis, Efstratios N. Pistikopoulos (Eds.)
PROCEEDINGS OF THE 33rd European Symposium on Computer Aided Process Engineering (ESCAPE33), June 18-21, 2023, Athens, Greece
 http://dx.doi.org/10.1016/B978-0-443-15274-0.50078-0

Mechanistic modeling of product formation in recombinant *Escherichia coli* cultivations

Don Fabian Müller[a], Philipp Pably[a], Daniel Wibbing[c], Julian Kager[d] and Christoph Herwig*[a,b]

[a]*Institute of Chemical, Environmental and Bioscience Engineering, TU Wien, Gumpendorfer Straße 1a, 1060 Vienna, Austria*
[b]*Competence Center CHASE GmbH, 4040 Linz, Austria*
[c]*Festo SE & Co. KG, 73770 Denkendorf, Germany*
[d]*Department of Chemical and Biochemical Engineering, Technical University of Denmark, Building 228A, 2800 Kgs. Lyngby, Denmark*
christoph.herwig@tuwien.ac.at

Abstract

In this work a nonlinear mechanistic model was developed predicting the product formation of recombinant lactate dehydrogenase (LDH) using fed-batch cultivations of Escherichia coli. The derived product formation kinetics consider the inclusion body (IB) fraction as well as the soluble protein fraction. After the structural identifiability of the model parameters was analyzed, five fed-batch cultivations were conducted in total in order to identify the model parameters and to validate the model subsequently. The accuracy of the final model was overall good with NRMSE values of around 0.1 to 0.2 for both product fractions. The derived model can be seen as an enabler for advanced monitoring and control strategies like model predictive control and other advances technologies defined in the process analytical technology (PAT) initiative.

Keywords: *Escherichia coli*, bioprocess control, product formation, mechanistic model, process analytical technology

1. Introduction

In contrast to the chemical process industry, industrial bioprocess control strategies are still mainly based on linear closed-loop control schemes like PID control or simple predefined open-loop trajectories (Narayanan et al. 2020). So far in bioprocess development, feed rate optimization is often realized by design of experiments (DoE) approaches. Instead of optimizing the feed rate directly, modern approaches aim to optimize the biomass specific substrate uptake rate (q_S) as a physiological parameter resulting in a more generic approach independent of the bioreactor scale (Wechselberger et al. 2012). However, these solutions are assuming a direct linear relationship between q_S and the specific product formation rate (q_P), which might not be valid throughout the course of the process. Furthermore, batch-to-batch variability is an inherent nature of bioprocesses and therefore open-loop control strategies come to their limits as they cannot react on process disturbances. Advantages of using more advanced model-based control algorithms were discussed previously (Ulonska et al. 2018; Wechselberger, Sagmeister, and Herwig 2013; Kager et al. 2022). Especially due to the complex multivariate and nonlinear relationships in bioprocesses, model predictive control (MPC), which is taking the interrelations between process variables into account, offers higher flexibility than decoupling the control task into several single setpoint tracking control loops (Mears et al. 2017). Kager et al. (2022) developed a two-degree of freedom control strategy using nonlinear feedback linearization in order to directly control the specific product formation rate q_P in cultivations of recombinant *Escherichia coli* and stated increased biomass specific titers with this model-based controller compared to optimal constant feed-rate setpoints. The prerequisite of such advanced control strategies is a process model representing the dynamics of the system with sufficient accuracy. In this work we developed a simple mechanistic model for the product formation of a recombinant Escherichia coli strain producing LDH. In five fed-batch strain characterization experiments we analyzed the soluble product content as well as the aggregated inclusion body (IB) fraction of the product and included both as separated states in our model, which to our knowledge has not been done before.

2. Methods

2.1. Cultivation of *Escherichia coli*

Table 1: Conducted experiments

Name	$q_{S,mean}$
Process 1	0.25 $gg^{-1}h^{-1}$
Process 2	0.14 $gg^{-1}h^{-1}$
Process 3	0.35 $gg^{-1}h^{-1}$
Process 4	0.23 $gg^{-1}h^{-1}$
Process 5	0.14 $gg^{-1}h^{-1}$

Five cultivations were conducted in total. Four out of five experiments (1, 2, 4, 5) were used as a training dataset for parameter estimation and one experiment (3) was used as validation dataset. Recombinant Escherichia coli BL21(DE3) was cultivated in fed-batch mode on DeLisa (DeLisa et al. 1999) minimal medium. The target product was the enzyme L-lactate dehydrogenase (LDH). The bioreactor system Labfors 5 (Infors, Bottmingen, Switzerland) with a total volume of 3.3 L was used for cultivation. 1 L medium was supplemented with 100 mM Amplicillin and 15 gL^{-1} glucose and inoculated with 100 mL of preculture (1 L shake flask, 37 °C, 230 rpm, 12 h) The controlled value for pH was 6.75 (by addition of $NH4^+$ and H_3PO_4), for temperature 37 °C before induction and 31.5 °C after induction, and for pO_2 30% via stirrer speed (400 rpm to 1200 rpm). After the batch phase was finished a fed-batch phase was initiated by controlling the specific substrate uptake rate q_S to 0.25 $gg^{-1}h^{-1}$. After 24 h of process time the product formation was induced with 1 mM isopropyl β-d-1-thiogalactopyranoside (IPTG). Then, different values for q_S were applied. Table 1 shows the mean values of q_S after induction of the product formation.

2.2. Analytical methods

Online measurements of pH, dissolved oxygen tension (pO_2), liquid temperature (ϑ_L), substrate feed rate (F_R), airflow rate (F_{AIR}), oxygen flow rate (F_{O2}) and offgas analysis (BlueInOne, BlueSens, Herten, Germany) were recorded and used for real-time calculations. For offline reference analytics samples of the culture broth were taken every 2 h during the bioprocess. For biomass quantification the dry cell weight (DCW) was analyzed. After centrifugation of 1 mL sample (20000 g, 10 min), the pellet was dried at 105 °C and gravimetrically quantified in 1.5 mL tubes. The optical density (OD) was determined at 600 nm and the substrate content was obtained through an automated photometric assay using a hexokinase kit (BioHT, Roche, Germany). The live-dead analysis was carried out via a Flow Cytometry (CyFlow® Cube 6, Sysmex Partec GmbH, Germany) with the staining protocol described in (Langemann et al. 2016). The recorded forward-scatter and fluorescence signals were correlated with the amount of dead cells. In order to measure the intracellularly produced LDH the cell pellets were resuspended and disrupted by a combination of chemical lysis and a sonication rod. After centrifugation the supernatant was analyzed for its soluble content of the target enzyme and the pellet containing the IBs was further processed. The latter was washed and then solubilized using a solution of Guanidinium chloride, denaturing and dissolving the aggregated protein chains. By diluting the resulting protein solution the enzyme refolded into its active form. The enzymatic activity of the soluble and the refolded LDH was determined using an adaptation of the enzyme assay procedure described in (Ponsoda et al. 1991), utilizing the catalytic nature of LDH for the reduction of pyruvate to lactate. The change in adsorption was measured at 340 nm with the automatic Spark® multimode microplate reader (Tecan Trading AG, Switzerland). Additionally, several samples were analyzed with a RP-HPLC (Thermo Fisher Scientific Inc., USA) using the method of (Kopp et al. 2020) together with a BSA-standard curve. The activities of the same samples were correlated to their protein content and a conversion factor of 51400U g−1 was determined. This was further used to translate all activity measurements to protein concentrations.

2.3. Bioprocess model

The model structure which was used is an unstructured model of six states, namely liquid volume (V), living biomass (X_v), dead biomass (X_d), substrate (S), total product (P) and inclusion body product (P_{IB}).

$$\frac{dV}{dt} = F_{in} - F_{out} \qquad \frac{dX_v}{dt} = Y_{X/S} \cdot q_S \cdot X_v - k_d \cdot X_v - F_{out} \cdot \frac{X_v}{V}$$

$$\frac{dX_d}{dt} = k_d \cdot X_v - F_{out} \cdot \frac{X_d}{V} \qquad \frac{dS}{dt} = F_R \cdot c_{SR} - q_S \cdot X_v - F_{out} \cdot \frac{S}{V} \tag{1}$$

$$\frac{dP}{dt} = q_P \cdot X_v - k_d \cdot P - F_{out} \cdot \frac{P}{V} \qquad \frac{dP_{IB}}{dt} = q_{PIB} \cdot X_v - k_d \cdot P_{IB} - F_{out} \cdot \frac{P_{IB}}{V}$$

Three kinetic equations were used to describe the cell specific rates. Monod kinetics were used to describe specific substrate uptake q_S, a combination of Monod and Haldane kinetics were used to describe total product formation q_P and Moser kinetics were used to describe IB formation q_{PIB}. q_P is modeled dependent on q_S and on the cumulated metabolized substrate S_{met}, which is obtained as the integral of $q_S \cdot X_v$ since the induction.

$$q_S = q_{Smax} \cdot \frac{c_S}{k_S + c_S} \qquad q_{PIB} = q_{P_{IB},max} \cdot \frac{q_P^N}{q_P^N + K_{p,q_{PIB}}^N} \tag{2}$$

$$q_P = q_{P,max} \cdot \frac{q_S}{q_S + K_{s,qs}} \cdot \frac{S_{met}/X}{\frac{(S_{met}/X)^k}{K_{I,qp}} + S_{met}/X + K_{s,qp}}$$

2.4. Model analysis and parameterization

Before parameter estimation a structural identifiability analysis was performed. Therefore, the sensitivity matrix

$$S = \frac{\partial(x_1,\dots,x_m)}{\partial(\theta_1,\dots,\theta_n)}, \qquad S_{i,p} = \frac{\partial x_i}{\partial \theta_p} \tag{3}$$

of the model was derived and the rank was calculated. If S has full rank, the model is structurally identifiable, which means that all model parameters θ can be uniquely identified from the measured states x.

2.5. Real-time implementation

Data analysis and model development was executed with MATLAB® and Simulink® (Version R2022a, MathWorks, Natick, Massachusetts, USA). Real-time process control and communication to the equipment was realized through the Lucullus Process

Information Management System PIMS (Securecell, Urdorf, Switzerland). Lucullus was connected via OPC to a computer running MATLAB®, where the data preprocessing, the feed control and the state estimation model were executed for the different experiments. The numerical derivative of the feed weight signal was computed and smoothed with a Savitzky-Golay-Filter approach (Savitzky and Golay 1964) to obtain the feed rate F_R used as model input.

3. Results & Discussion

The exploration of the collected data yielded a correlation between the productivity q_P, the specific substrate uptake q_S and the metabolized substrate S_{met} as an inhibitor, which acts as an indicator for cell age. This hypothesis is supported by the data as seen in fig. 1 for the biomass specific product titer $P_{specific}$.

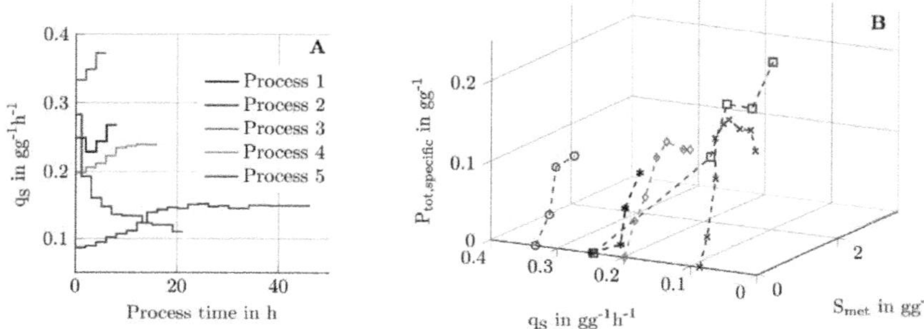

Figure 1: Cell-specific total product titer in gg^{-1} dependent on substrate uptake rate q_S and metabolized substrate Smet for all six processes.

These states were further used to find a kinetic equation modeling the cell productivity after induction. Several kinetic formulations, combining Monod and other substrate uptake equations with inhibition terms were tested for their suitability and performance. Additionally, the structural identifiability of the resulting models were analyzed to proof that the selected structure can be uniquely parameterized when no measurement error is assumed. The same was done for the subsequent inclusion body formation, expressed through q_{PIB}. Here, a linear connection to the overall productivity q_P was observed in the data and was therefore chosen as the variable for the kinetic equations. The structural analysis and parameterization with the data sets yielded two optimal kinetic equations by comparing their scores for the $R^2_{adjusted}$ and the NRMSE values. The specific productivity q_P of choice was a combination of a Monod term with a Haldane inhibition term triggered by S_{met}. For the product localization the best specific inclusion body formation q_{PIB} among the tested kinetics was one of Moser type (eq. 3). The performance of the final model against the given data sets together with the respective scores can be seen in fig. 2. The identifiability analysis yielded that the chosen kinetics with 5 parameters for the total productivity and 3 parameters for the IB formation yield a structural identifiable model. A practical analysis of the identifiability also revealed a sensitivity towards the process noise, which can create problems for the parameterization and lead to unstable solutions for the parameter vector or leave some positions as unidentifiable.

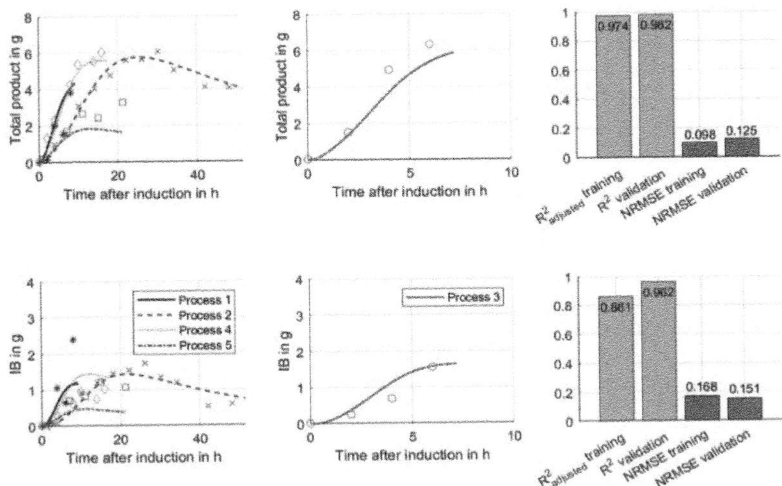

Figure 2: Results of the parameterization with the final model kinetics for the total product and the IBs. Model parameters: $q_{P,max} = 0.0396$ $gg^{-1}h^{-1}$, $K_{s,qs} = 0.0727$ $gg^{-1}h^{-1}$, $k = 5.56$, $K_{s,qp} = 0.4669$ gg^{-1}, $K_{I,qp} = 3.6960$, $q_{PIB,max} = 0.0195$ $gg^{-1}h^{-1}$, $K_{p,qPIB} = 0.0282$ $gg^{-1}h^{-1}$, $N = 1.9627$

The simulation shows that the final model structure and choice of variables allows a good tracking of the productivity, both in the strong surge right after the induction, as well as the decline over the progress of the process. The NRMSE values for the parameterization datasets as well as for the validation datasets range between 0.1 and 0.2 indicating an overall satisfactory model calibration. The model validity spans in a high range of feed rate inputs (q_S from 0.14 $gg^{-1}h^{-1}$ to 0.35 $gg^{-1}h^{-1}$) and the model can probably be easily transferred to other feeding strategies. Overall, the impact of the chosen q_S-setpoints on the product and IB formation is well represented enabling the usage of this model in future prediction and control algorithms.

4. Conclusion

We developed a mechanistic model for the product formation of *Escherichia coli.* The model is based on macro-kinetic principles describing product formation dependent on the specific rate of substrate uptake q_S and the cumulative metabolized substrate S_{met}, which were considered as indicators for the metabolic load of the cell. In contrast to some other unstructured models described in literature (Neubauer, Lin, and Mathiszik 2003; Kim et al. 2022; Kager et al. 2022) here we describe the formation of soluble product as well as the aggregation into inclusion bodies which opens up the possibility to separately control the two states of the product and possibly steer the production in a specific direction. Model predictive control (MPC) can serve as a framework to implement this strategy and control directly on the product formation and its status. Further, transferability of the identified mechanism to other products and organism needs to be proven to see the developed model as a platform.

5. Acknowledgements

The authors acknowledge the TU Wien Bibliothek for financial support through its Open Access Funding Program. The authors acknowledge financial support through the COMET Centre CHASE, funded within the COMET – Competence Centers for Excellent Technologies programme by the BMK, the BMDW and the Federal Provinces of Upper

Austria and Vienna. The COMET programme is managed by the Austrian Research Promotion Agency (FFG).

References

DeLisa, M. P., J. Li, G. Rao, W. A. Weigand, and W. E. Bentley. 1999. "Monitoring GFP-Operon Fusion Protein Expression during High Cell Density Cultivation of Escherichia Coli Using an on-Line Optical Sensor." *Biotechnology and Bioengineering* 65 (1): 54–64.

Kager, Julian, Johanna Bartlechner, Christoph Herwig, and Stefan Jakubek. 2022. "Direct Control of Recombinant Protein Production Rates in E. Coli Fed-Batch Processes by Nonlinear Feedback Linearization." *Chemical Engineering Research and Design* 182: 290–304.

Kim, Jong Woo, Niels Krausch, Judit Aizpuru, Tilman Barz, Sergio Lucia, Peter Neubauer, and Mariano Nicolas Cruz Bournazou. 2022. "Model Predictive Control and Moving Horizon Estimation for Adaptive Optimal Bolus Feeding in High-Throughput Cultivation of E. Coli." arXiv. http://arxiv.org/abs/2203.07211.

Kopp, Julian, Florian B. Zauner, Andreas Pell, Johanna Hausjell, Diana Humer, Julian Ebner, Christoph Herwig, Oliver Spadiut, Christoph Slouka, and Reinhard Pell. 2020. "Development of a Generic Reversed-Phase Liquid Chromatography Method for Protein Quantification Using Analytical Quality-by-Design Principles." *Journal of Pharmaceutical and Biomedical Analysis* 188 (September): 113412.

Langemann, Timo, Ulrike Beate Mayr, Andrea Meitz, Werner Lubitz, and Christoph Herwig. 2016. "Multi-Parameter Flow Cytometry as a Process Analytical Technology (PAT) Approach for the Assessment of Bacterial Ghost Production." *Applied Microbiology and Biotechnology* 100 (1): 409–18. https://doi.org/10.1007/s00253-015-7089-9.

Mears, Lisa, Stuart M. Stocks, Mads O. Albaek, Gürkan Sin, and Krist V Gernaey. 2017. "Mechanistic Fermentation Models for Process Design, Monitoring, and Control." *Trends in Biotechnology* 35 (10): 11.

Narayanan, Harini, Martin F. Luna, Moritz Stosch, Mariano Nicolas Cruz Bournazou, Gianmarco Polotti, Massimo Morbidelli, Alessandro Butté, and Michael Sokolov. 2020. "Bioprocessing in the Digital Age: The Role of Process Models." *Biotechnology Journal* 15 (1): 1900172.

Neubauer, P., H. Y. Lin, and B. Mathiszik. 2003. "Metabolic Load of Recombinant Protein Production: Inhibition of Cellular Capacities for Glucose Uptake and Respiration after Induction of a Heterologous Gene InEscherichia Coli." *Biotechnology and Bioengineering* 83 (1): 53–64.

Ponsoda, Xavier, Ramiro Jover, José Vicente Castell, and María José Gómez-Lechón. 1991. "Measurement of Intracellular LDH Activity in 96-Well Cultures: A Rapid and Automated Assay for Cytotoxicity Studies." *Journal of Tissue Culture Methods* 13 (1): 21–24.

Savitzky, Abraham., and M. J. E. Golay. 1964. "Smoothing and Differentiation of Data by Simplified Least Squares Procedures." *Analytical Chemistry* 36 (8): 1627–39.

Ulonska, Sophia, Daniel Waldschitz, Julian Kager, and Christoph Herwig. 2018. "Model Predictive Control in Comparison to Elemental Balance Control in an E. Coli Fed-Batch." *Chemical Engineering Science* 191 (December): 459–67. https://doi.org/10.1016/j.ces.2018.06.074.

Wechselberger, Patrick, Patrick Sagmeister, Helge Engelking, Torsten Schmidt, Jana Wenger, and Christoph Herwig. 2012. "Efficient Feeding Profile Optimization for Recombinant Protein Production Using Physiological Information." *Bioprocess and Biosystems Engineering* 35 (9): 1637–49. https://doi.org/10.1007/s00449-012-0754-9.

Wechselberger, Patrick, Patrick Sagmeister, and Christoph Herwig. 2013. "Model-based Analysis on the Extractability of Information from Data in Dynamic Fed-batch Experiments." *Biotechnology Progress* 29 (1): 285–96. https://doi.org/10.1002/btpr.1649.

Antonis Kokossis, Michael C. Georgiadis, Efstratios N. Pistikopoulos (Eds.)
PROCEEDINGS OF THE 33rd European Symposium on Computer Aided Process Engineering (ESCAPE33), June 18-21, 2023, Athens, Greece
 http://dx.doi.org/10.1016/B978-0-443-15274-0.50079-2

Physics-informed neural networks for optimization of polymer reactor design

Yubin Ryu[a,b], Sunkyu Shin[c], Jay J. Liu[d], Wonbo Lee[c], Jonggeol Na[a,b,*]

[a] *Department of Chemical Engineering and Materials Science, Ewha Womans University, Ewhayeodae-gil 52, Seodaemun-gu, Seoul 03760, Republic of Korea*
[b] *Graduate Program in System Health Science and Engineering, Ewha Womans University, Ewhayeodae-gil 52, Seodaemun-gu, Seoul 03760, Republic of Korea*
[c] *School of Chemical and Biological Engineering, Seoul National University, Gwanak-ro 1, Gwanak-gu, Seoul 08826, Republic of Korea*
[d] *Department of Chemical Engineering, Pukyong National University, Yongso-ro 45, Nam-gu, Busan 48513, Republic of Korea*
jgna@ewha.ac.kr

Abstract

Computational fluid dynamics (CFD) is utilized for chemical reactor design/analysis and enables accurate predictions by solving numerical methods. However, real-time prediction is currently infeasible due to its high computational cost. As a result, physics-informed neural networks (PINNs), which integrate data-based machine learning and physical principles, have recently emerged as a surrogate model of CFD (Raissi, M., et al. (2019)). But previous PINNs can only predict properties in terms of location or time. Here, we offer novel PINNs that can also handle operating conditions as input variables to evaluate the impact of various design variables for reactor design optimization (Raissi, M., *et al.* (2020)). The proposed model is applicated to an autoclave reactor for free radical polymerization of ethylene and demonstrated to be capable of interpolating/extrapolating solutions accurately under various settings.

Keywords: Computational fluid dynamics; deep learning; optimization; reactor design; artificial intelligence

1. Introduction

Computational fluid dynamics (CFD) solves fundamental nonlinear differential equations using numerical methods to analyze flow characteristics. Especially in chemical process engineering, CFD optimizes operating conditions and considerably reduces costs. However, for the physics residual of the governing equations to converge, a huge computational cost is required. As a result, a data-based model called physics-informed neural networks (PINNs) emerged as a surrogate model of CFD.

To integrate the data-driven model and physics-based model, PINNs solve partial differential equations (PDEs) by transforming the issue of finding the solution into a loss function optimization problem. The physical residual terms are added to the loss function and act like a penalty that limits the space of the solution (Cuomo, S., *et al.* (2022)). Therefore, PINNs can consider governing principles, rather than inferring solutions based only on training data. In addition, PINNs are based on a mesh-free approach and can be applied to complex geometries satisfying physical details with enhanced generalization, training speed, and accuracy (Karniadakis, G.E., *et al.* (2021)). Initial PINNs models have been developed to solve mathematical models or inverse problems, solving nonlinear PDEs such as Schrödinger, Burgers, and Allen-Kahn equations (Raissi, M., et al. (2017)). Moreover, PINNs have also been applied to fluid dynamics to solve high-dimensional problems dominated by the Navier-Stokes equation. However, since previous PINNs models could only predict properties in terms of space and time, we intend to develop a novel PINNs model that can even handle operating conditions as input variables.

This study uses PINNs to analyze the flow and reaction in an autoclave reactor where ethylene is polymerized. Unlike previous models, the proposed model was constructed by adding two operating conditions to search for optimal conditions. The potential of PINNs as a surrogate model of CFD was demonstrated by comparing the accuracy of prediction according to location and two driving conditions (inlet temperature and concentration of initiator).

2. Methodology

2.1. Polymerization reactor modeling via CFD

Figure 1 depicts the geometry of the 3D autoclave reactor, with a pitched blade impeller rotating at a speed of 100 rpm. In this reactor, the monomer flows in from the inlet at the top, the initiator flows in from the side, and generated polymer flows out through the outlet at the bottom. The reactor model is divided into two domains: the inner domain and the outer domain. The governing equations depend on the domain. In the case of the inner domain, the governing equation is determined by the moving reference frame (MRF), since it is a region directly affected by the impeller. In comparison, the outer domain is the part excluding the inner domain, so the governing equation is determined by the stationary frame. The governing equations used in the reactor model are shown in Table 1. Details of variables used in partial differential equations are listed in the appendix.

To generate data, polymerization reactor modeling was performed using ANSYS-Fluent v21.2. Each component's velocity, pressure, temperature, and concentration were extracted according to location and operating conditions. A total of 357 combinations of operating conditions were used as training data, 17 for inlet temperature and 21 for inlet concentration.

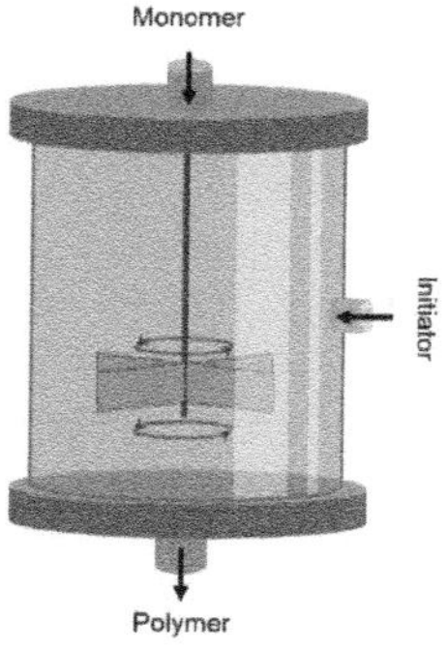

Figure 1. Schematic of autoclave reactor

Table 1. Governing Equations of Inner Domain

Continuity	$u_x + v_y + w_z$
Momentum	$\rho\left(u_t + u_x u + u_y v + u_z w - \omega(u_x y - u_y x - v)\right) + p_x - \mu(u_{xx} + u_{yy} + u_{zz})$
	$\rho\left(v_t + v_x u + v_y v + v_z w - \omega(v_x y - v_y x + v)\right) + p_y - \mu(v_{xx} + v_{yy} + v_{zz})$
	$\rho\left(w_t + w_x u + w_y v + w_z w\right) + p_z - \mu(w_{xx} + w_{yy} + w_{zz})$
Species Transport	$\left(uI_x + u_x I + vI_y + v_y I + wI_z + w_z I\right) - k_i e^{-\frac{E_a}{T}}$

Table 2. Governing Equations of Outer Domain

Continuity	$u_x + v_y + w_z$
Momentum	$\rho\left(u_t + u_x u + u_y v + u_z w\right) + p_x - \mu(u_{xx} + u_{yy} + u_{zz})$
	$\rho\left(v_t + v_x u + v_y v + v_z w\right) + p_y - \mu(v_{xx} + v_{yy} + v_{zz})$
	$\rho\left(w_t + w_x u + w_y v + w_z w\right) + p_z - \mu(w_{xx} + w_{yy} + w_{zz})$
Species Transport	$\left(uI_x + u_x I + vI_y + v_y I + wI_z + w_z I\right) - k_i e^{-\frac{E_a}{T}}$

2.2. Physics-informed neural networks for reaction systems

2.2.1. Model Description

The architecture of the PINNs model is shown in Figure 2. Input variables are coordinates (x, y, z) and inlet temperature, inlet concentration of the initiator. Output variables are velocity, temperature, pressure, and concentration of initiator and monomer. The loss function of PINNs is summed up of empirical loss and physics loss. Empirical loss refers to the mean square error (MSE) between prediction and target data. On the other hand, the physics loss is the residual of the governing equations obtained by automatic differentiation. This loss term forces PINNs to obey the laws of physics. To consider the scale between the governing equations (continuity, momentum, and species transport), physics loss is calculated by assigning a weight corresponding to each loss. PINNs are trained by updating the neural network parameters to minimize total loss.

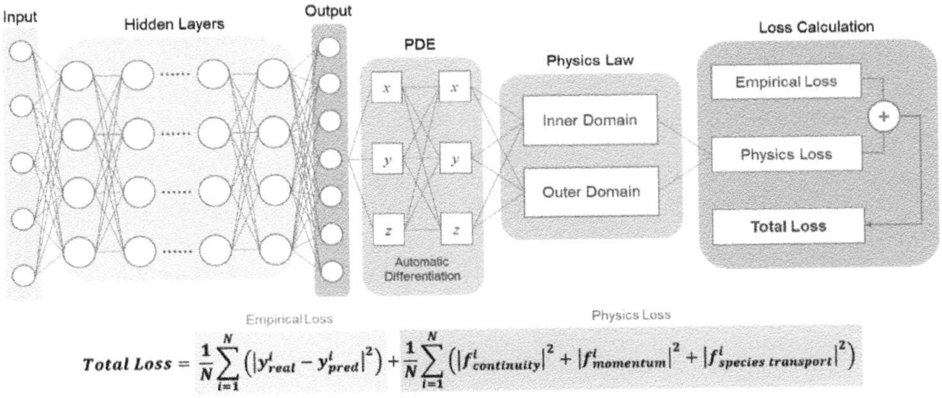

Figure 2. Schematic of PINNs: The architecture of PINNs consists of 13 layers including 100 neurons for each layer. Empirical loss measures how well the neural network reflects the training data. Physics loss ensures that the solution is consistent with the known physics.

2.2.2. Learning Strategy

For training data, a total of 200000 data with 4000 collocation points were used. We split the CFD simulation data into 3 sets: a train set used to train the model, a validation set used to limit overfitting, and a test set used to test the performance of our model. Learning rate reduction and early stopping by monitoring validation loss were also applied to our model to prevent overfitting. Additionally, we normalized our dataset to consider the scale between input and output variables and improve model stability. To reduce learning time, the proposed model is trained using distributed data-parallel (DDP) utilizing Pytorch v1.7.1 and GPU of NVIDIA GeForce RTX 3090.

In the process of obtaining the derivative for PDE calculation, the computational cost of PINNs increases exponentially. Therefore, we implemented space adaptive sampling to increase the efficiency of training. For every 1000 epochs, we added the top 200 data with the greatest error to the training dataset. Space adaptive sampling was compared with random sampling, which continuously adds 200 random data. The experiment in figure 3 verified that space adaptive sampling had a 30% accuracy improvement compared to random sampling.

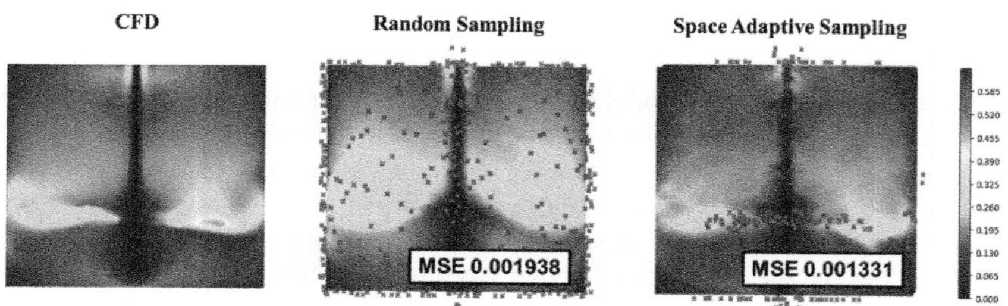

Figure 3. Comparison of CFD, random sampling, and space adaptive sampling model for velocity: Red points are the top 200 points with the largest empirical loss and are added to the training data.

3. Results

3.1. Performance of physics-informed neural networks

To check the validity of the proposed model, the target data and prediction of PINNs were compared by contour plot of velocity and monomer concentration. Figure 4 shows that PINNs prediction has high accuracy for both velocity and monomer concentration, and MSE between the target and predicted value was within 0.01%. In addition, to demonstrate that PINNs can reflect changes in driving conditions, we compared the conversion value between PINNs and CFD simulation data (Figure 5). The predicted conversion and target value were comparable, and the conversion graph form based on operating conditions was equivalent to CFD. The higher the inlet concentration of the initiator and the lower the inlet temperature of the initiator, the higher the conversion. To summarize, PINNs can accurately predict fluid flow and reaction.

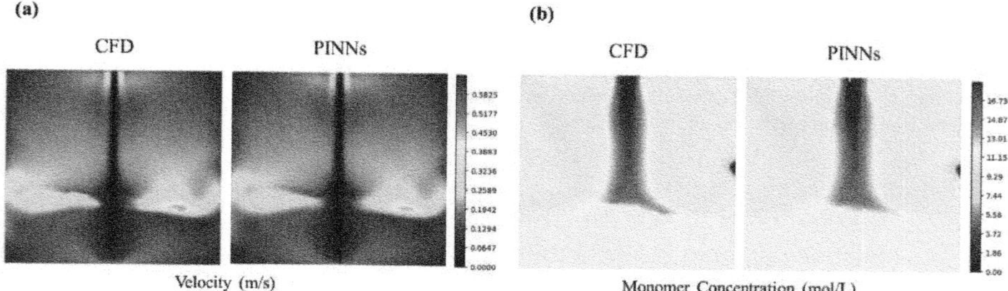

Figure 4. Comparison of CFD and PINNs for (a) velocity and (b) monomer concentration

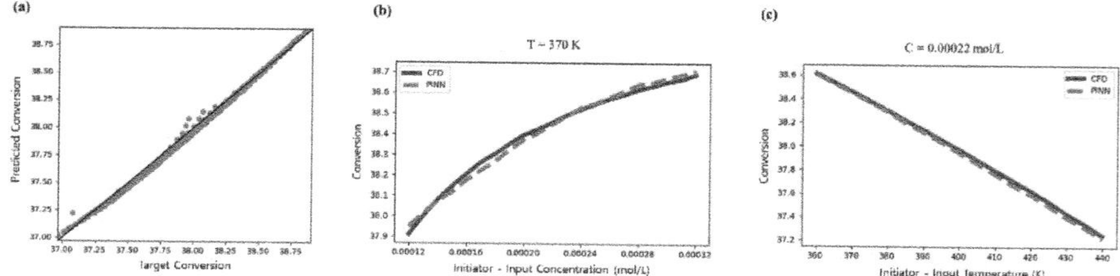

Figure 5. Comparison of CFD and PINNs for conversion: (a) Parity plot of the predictions for conversion with the proposed PINNs over the corresponding CFD results, (b) Conversion corresponding to input concentration of the initiator, (c) Conversion corresponding to input temperature of the initiator.

3.2. Operating condition interpolation & extrapolation

As the results show above, the suggested model is possible to search for optimal operating conditions, when applied to chemical reactors. However, there was no substantial difference between PINNs and neural networks (NNs). Therefore, to obtain the distinction of PINNs through additional experiments, the physical residual, interpolation, and extrapolation performance of PINNs were compared with NNs.

Figure 6 (a) compares the physical residuals of PINNs and NNs. The fact that PINNs have smaller physical residuals in governing equations indicates that it is trained to obey physical rules more than NNs. The bar plot in figure 6 (b) compares the interpolation and extrapolation performances of PINNs and NNs. Interpolation was performed inside the training data range, and extrapolation was derived using 264 simulated data outside of the training data range.

The accuracy of conversion prediction was tested, and PINNs outperformed NNs in both interpolation and extrapolation. The performance difference was greater in extrapolation than in interpolation. In the case of interpolation, the area to be predicted existed inside the training data range, hence the training data had a significant influence on interpolation results. Extrapolation, on the other hand, amplified the effect of the physics law, resulting in a considerable difference between PINNs and NNs.

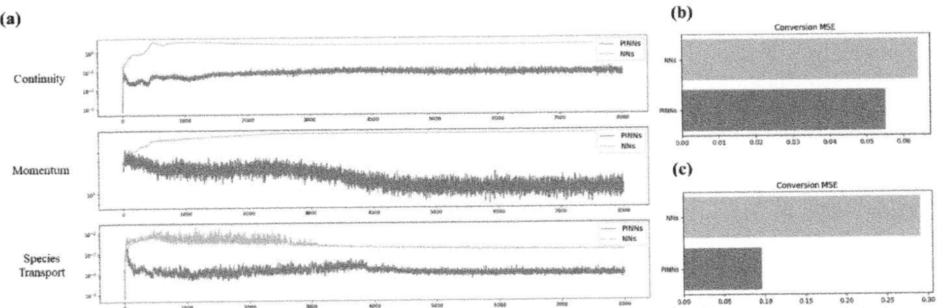

Figure 6. Comparison of PINNs and NNs: (a) Physics loss of continuity, momentum, species transport equation, (b) Conversion MSE based on interpolation results, (c) Conversion MSE based on extrapolation results

Conclusion

In this study, we proposed a novel PINNs model that can predict changes based on driving conditions. According to experimental results, the suggested method can explore optimal operating conditions by resolving nonlinear problems and interpolating solutions under various settings. Overall, we emphasize the significance of PINNs when it comes to data-driven system optimization and control. Our future work will be focused on geometric optimization to complete the PINNs-based reactor design framework. Following that, the PINNs-based framework is expected to be a key component of digital twin or multi-scale simulations.

Notation

x	x coordinate
y	y coordinate
z	z coordinate
u	x-velocity
v	y-velocity
w	z-velocity
T	Temperature
P	Pressure
I	Initiator Concentration
M	Monomer Concentration
E_a	Activation Energy
ρ	Density
k_i	Reaction constant
ω	Viscosity

References

Raissi, M., Perdikaris, P., & Karniadakis, G. E., 2019, Physics-informed neural networks: A deep learning framework for solving forward and inverse problems involving nonlinear partial differential equations, *Journal of Computational Physics*, 378, 686-707.

Raissi, M., Yazdani, A., & Karniadakis, G. E., 2020, Hidden fluid mechanics: Learning velocity and pressure fields from flow visualizations, *Science*, 367(6481), 1026-1030.

Cuomo, S., Di Cola, V. S., Giampaolo, F., Rozza, G., Raissi, M., & Piccialli, F., 2022, Scientific Machine Learning through Physics-Informed Neural Networks: Where we are and What's next, *arXiv preprint arXiv:2201.05624.*

Karniadakis, G. E., Kevrekidis, I. G., Lu, L., Perdikaris, P., Wang, S., & Yang, L., 2021, Physics-informed machine learning, *Nature Reviews Physics*, 3, 422-440.

Raissi, M., Perdikaris, P., & Karniadakis, G. E., 2017, Physics informed deep learning (part i): Data-driven solutions of nonlinear partial differential equations, *arXiv preprint arXiv:1711.10561.*

Antonis Kokossis, Michael C. Georgiadis, Efstratios N. Pistikopoulos (Eds.)
PROCEEDINGS OF THE 33rd European Symposium on Computer Aided Process Engineering (ESCAPE33), June 18-21, 2023, Athens, Greece
 http://dx.doi.org/10.1016/B978-0-443-15274-0.50080-9

Extending the SI Decomposition to Continuous-Time Two-Stage Scheduling Problems

Daniel Montes[a*], José Luis Pitarch[b], César de Prada,[a,c]

[a]*Department of Systems Engineering and Automatic Control, Universidad de Valladolid, C/ Prado de la Mergelina 3-5, Valladolid (47011), Spain.*
[b]*Instituto Univ. de Atutomática e Informática Industrial (ai2), Universitat Politècnica de Valencia, Camino de Vera s/n, Valencia (46022), Spain.*
[c]*Institute of Sustainable Processes, Universidad de Valladolid, C/ Dr. Mergelina s/n, Valladolid (47011), Spain.*
*danielalberto.montes.lopez@uva.es

Abstract

Scheduling often involves making decisions in presence of uncertainty, which governs the pricing of raw materials, energy, resource availability, demands, etc. A common approach to incorporate uncertainty in the decision-making process is using two-stage stochastic formulations. Unfortunately, the mathematical complexity of the resulting problems grows exponentially with the number of uncertainty scenarios, which is further complicated by the presence of binary variables The authors have recently proposed a method using the so-called Similarity Index for discrete-time two-stage scheduling problems that enable scenario-based decomposition. This paper extends this method for scheduling problems formulated on a continuous-time basis. The fundamental idea is to use the Similarity Index to meet non-anticipation in the binary variables and Progressive Hedging on the continuous ones. The proposal is tested on a literature case study that consists of a multiproduct plant with a single processing unit. The combined SI-PH decomposition managed to solve the problem much faster than its monolithic counterpart.

Keywords: Similarity Index, Progressive Hedging, Optimization under uncertainty.

1. Introduction

The authors recently presented a decomposition method based on the idea of the Similarity Index (SI) to efficiently solve two-stage stochastic scheduling problems (TSSP) formulated using a discrete-time basis (Montes et al., 2022a, 2022b). TSSP handles uncertainty by discretizing the underlying probability distribution of uncertain parameters in a set of scenarios with associated probabilities. Then, the decision variables of the problem are grouped into two sets: first-stage and second-stage ones. In the first stage, the actual value that uncertain parameters will have is still unknown, but some decisions must be made regardless. Hence, the schedule within this first stage cannot be tailored to each scenario, as the uncertainty is not revealed yet. This is known as non-anticipation, i.e., decisions in the first stage must be unique. In the second stage, the values of the uncertain parameters are realized and assumed known, so the decisions can be adjusted to each realization accordingly.

Two-stage formulations are challenging to solve, as the problem size grows exponentially with the number of scenarios considered. This fact is further complicated by the presence of many binary variables, usual in scheduling formulations. One approach to tackle this issue is to solve each scenario as an independent subproblem of reduced size, but the non-anticipation criterion for the first stage must be satisfied. The Progressive Hedging Algorithm (PHA) was originally proposed for convex problems involving only continuous variables and has convergence and optimality guarantees (Rockafellar & Wets, 1991). However, although the assumptions and guarantees vanish in presence of discrete variables, this method has been used successfully as a heuristics for mixed-integer problems (Bashiri et al., 2021; Peng et al., 2019). In the PHA, the non-anticipation constraints are relaxed, which enables scenario decomposition. Non-anticipation is then progressively enforced by penalizing the deviation from the average schedule in the first stage, i.e., the decision-variables average values computed from the subproblems solutions in a previous iteration.

Likewise, the Similarity-Index decomposition also enables scenario decomposition by approximating, or estimating, the degree of non-anticipation in the discrete decisions of the first stage. The scenario subproblems are solved independently and the similarity among their solutions is then measured using the SI. A weighing parameter is used to progressively increase the importance of maximizing the similarity. Eventually, all scenario solutions are equal in the first stage so non-anticipation is met. However, the SI decomposition was originally devised for discrete-time scheduling formulations. In this work, the authors aim to combine the SI decomposition with Progressive Hedging to decompose TSSP based on continuous-time formulations. The SI handles the binary variables of the problem, while Progressive Hedging handles the continuous ones. The proposal is tested successfully on a literature case study.

The paper is organized as follows: Section 2 presents a summary of the SI concept and its extension to continuous-time scheduling formulations. Section 3 describes the proposed SI-PH decomposition algorithm. The case study is presented in Section 4. A comparison between the proposed approach and the standard monolithic approach is summarized in Section 5. Conclusions and open issues are outlined in Section 6.

2. Similarity Index in Continuous-Time Formulations

In discrete-time basis, the Similarity Index is relies on the idea of fuzzifying the binary decision variables along the surrounding time periods. The generated areas from the fuzzification for each scenario are compared and their overlap is used as a measure of similarity. Consequently, even if the decisions are not made in the same period among scenarios, one could quantify their similarity by temporal proximity. As the non-anticipation constraints (NAC) require the scenario schedules to be equal in the first stage, the SI can be used to remove the NACs from the formulation, enabling scenario decomposition. The SI is can be pushed up through adding it to the objective function.

Two main difficulties arise for extending the Similarity-Index Decomposition to continuous-time TSSP: Time synchronization of decisions among scenarios; and weighing the slot duration as part of the SI computation. To get around these limitations, we propose: 1) not to consider the slot duration in the fuzzification process, and 2) to fuzzify in a slots basis instead of time periods. Then, the SI formula becomes:

$$SI := \sum_{l \in \mathcal{L}_1} \frac{\min_{e \in \mathcal{E}} \left\| y_{le} + 0.5 y_{(l+1)e} + 0.5 y_{(l-1)e} \right\|_1}{2|\mathcal{L}| - 1} \tag{1}$$

Where the time slots l belong to an ordered set $\mathcal{L}$. The subset $\mathcal{L}_1 \subset \mathcal{L}$ implies that the Similarity Index is only computed for the decisions made in the first stage of the problem. y_{le} are the decision variables indexed by the slots and the scenarios $e \in \mathcal{E}$. In essence, the Similarity Index is computed as the intersection of the generated areas from fuzzifying the decisions divided by the maximum intersection possible. Formula (1) implies that the similarity is only weighed in the two surrounding slots. As in the time-basis case (Montes et al., 2022b), this formula can be generalized for different fuzzification lengths.

Note that (1) does not cover the slot duration and other continuous variables that might be present in the problem (such as inventory levels, production levels, etc.). We propose to use Progressive Hedging (PH) to handle these variables, which results in a hybrid SI-PH method. Incorporating both the SI and the PH in a general TSSP, each subproblem formulation is given by:

$$\begin{aligned}
&\min_{x,y} J_e - \lambda\, SI_e + \sum_{l\in\mathcal{L}_1}\left(\omega_{le}x_{le} + \frac{\rho}{2}\left(\frac{x_{le}-\bar{x}_l}{\bar{x}_l}\right)^2\right)\\
&\text{s.t.:}\\
&\text{Specific process constraints}\\
&SI_e = \frac{\sum_{l\in\mathcal{L}_1}|s_{le}|_1}{2|\mathcal{L}|-1}\\
&s_{le} \le y_{le} + 0.5y_{(l+1)e} + 0.5y_{(l-1)e} \quad \forall l \in \mathcal{L}_1\\
&s_{le} \le \bar{y}_l + 0.5\bar{y}_{l+1} + 0.5\bar{y}_{l-1} \quad \forall l \in \mathcal{L}_1\\
&x \in \mathbb{R}, y \in \{0,1\}
\end{aligned} \tag{2}$$

Note that the exact/global SI cannot be computed inside the subproblems to be solved independently, and that the element-wise operator **min**{·} in (**1**) is non-linear, so it cannot be used in the usual MILP scheduling formulations. Thus, the SI is estimated locally in each subproblem (**2**) by replacing the intersection of the fuzzified variables with a set of slack variables that are upper-bounded. The slack variables $\boldsymbol{s_{le}}$ are maximized in the objective function so this bound is tight. The $\boldsymbol{SI_e}$ is then an approximatin of the local SI by scenario, computed as the similarity of the scenario schedule with a reference schedule defined by fixed values $\bar{\boldsymbol{y}}_l$. For details about this estimation refer to (Montes et al., 2022b). $\boldsymbol{\lambda}$ and $\boldsymbol{\omega_{le}}$ are the weighing parameters for the SI and the PH. They are updated in each iteration $\boldsymbol{k}$ from the scenario solutions, as follows, where $\bar{\boldsymbol{x}}_l$ is the expected value of x, using probabilities $\boldsymbol{p_e}$, in the first stage:

$$\lambda^{(k+1)} = \lambda^{(k)} - \alpha(SI-1);\ \omega_{le}^{(k+1)} = \omega_{le}^{(k)} + \rho\left(\frac{x_{le}-\bar{x}_l}{\bar{x}_l}\right) \forall l \in \mathcal{L}_1, e \in \mathcal{E}$$
$$\bar{x}_l = \sum_{e\in\mathcal{E}} p_e x_{le} \ \forall l \in \mathcal{L}_1 \tag{3}$$

The parameters $\boldsymbol{\alpha}$ and $\boldsymbol{\rho}$ are the step-size for $\boldsymbol{\lambda}$ and $\boldsymbol{\omega}$. They greatly influence the algorithm behavior, both in terms of optimality and convergence, as discussed later.

3. SI-PH Decomposition

Non-anticipation can be enforced progressively using the SI and PH ideas. Hence, the original problem can be decomposed into scenario subproblems (2) by dropping the non-anticipation constraints. Each subproblem is solved independently and in parallel. The subproblems solutions are later used to update the required parameters for the SI and PH with (3). The procedure is repeated until either the convergence criteria are met, or the maximum number of iterations is reached. Algorithm 1 below shows the pseudo-code for the SI-PH decomposition.

Algorithm 1

Require $\alpha, \rho, k_{\max}, \Delta, \text{tol}$

$k \leftarrow 0, \lambda^{(0)} \leftarrow 0, \omega_{le}^{(0)} \leftarrow 0, \overline{y}_l \leftarrow 0, \overline{x}_l \leftarrow 0$

repeat

  for $e \in \mathcal{E}$ **do**

    $y_{le}^*, x_{le}^* \leftarrow \underset{y_{le}, x_{le}, s_{le}}{\arg\min} J_e - \lambda^{(k)} SI_e + \sum_{l \in \mathcal{L}_1} \left(\omega_{le}^{(k)} x_{le} + \frac{\rho}{2} \left\| x_{le} - \overline{x}_l^{(k)} \right\|_2^2 \right)$

    $SI_e^* \leftarrow SI_e(s_{le}^*)$

  end for

  $SI \leftarrow SI(y_{le}^*)$ using (1)

  $\overline{y}_l \leftarrow \underset{y_{le}^*}{\arg\min}\, SI_e^*$

  $\lambda^{(k+1)} \leftarrow \lambda^{(k)} - \alpha\left(SI^{(k)} - 1\right)$

  $\overline{x}_l^{(k+1)} \leftarrow \sum_{e \in \mathcal{E}} \mathrm{p}_e x_{le}^{(k)} \quad \forall l \in \mathcal{L}_1$

  $\omega_{le}^{(k+1)} \leftarrow \omega_{le}^{(k)} + \rho\left(x_{le}^{(k)} - \overline{x}_l^{(k+1)}\right) \quad \forall l \in \mathcal{L}_1, e \in \mathcal{E}$

  $k \leftarrow k + 1$

until $\left(SI = 1 \wedge \left\| x_{le}^{(k)} - \overline{x}_l^{(k+1)} \right\|_2 \leq \text{tol}\right) \vee k = \mathrm{k}_{max}$

return y_{le}^*, x_{le}^*

The convergence criteria need to check that non-anticipation is met. For binary variables, that is easy to verify, as the SI needs to reach the value of 1. For continuous variables, non-anticipation can be confirmed if the 2-norm of the difference between the scenario solutions and the reference value is beyond a small enough tolerance.

4. Case Study

The SI-PH decomposition was tested on a case study originally presented by Dogan & Grossmann (2006). The model uses a continuous-time representation for simultaneously integrating both the scheduling and planning of a continuous multiproduct plant with a single processing unit. The planning horizon is divided into fixed-duration time periods $t \in \mathcal{T}$ (weeks), which are further subdivided into variable-duration time slots $l \in \mathcal{L}$. Products $\mathrm{i} \in \mathcal{I}$ are committed to each slot using binary variables W_{ilt}. Only the modifications to the original formulations are listed in this work for brevity and not the original equations.

The original formulation of Dogan & Grossmann (2006) is here extended to consider the production rates r_i as uncertain. Consequently, an additional index $e \in \mathcal{E}$, corresponding to the set of uncertainty realization scenarios, is added to all variables and equations of their model. As the time horizon is divided into both weeks and slots, additional variables W_{ije}^{aux}, with an additional auxiliary slot index j, are introduced to compute the Similarity Index as in (1). This variable is equal to the product assignment W_{ilte} but indexed by j. The index j enumerates the slots from 1 to $N \cdot |\mathcal{T}_1|$. Where N is the number of slots per time period and $\mathcal{T}_1$ is the subset of time periods belonging to the first stage. For instance, given time periods made up of three slots, the first slot of the second time period corresponds $j = 4$.

$$W_{ije}^{aux} = W_{ilte} \qquad \forall j = l + |\mathcal{L}| \cdot (t-1), l, t, e \tag{4}$$

The original formulation of the problem did not explicitly consider the slot duration ϕ_{lte}. As the non-anticipation criterion of the two-stage formulation requires the slot duration variables to be equal among the scenarios in the first stage, some additional constraints are added to compute them from the starting time of the slots Ts_{lte}:

$$\phi_{lte} = Ts_{l+1,te} - Ts_{lte} \qquad \forall l \neq l_N, t \in \mathcal{T}_1, e \tag{5}$$

$$\phi_{l_N te} = \mathrm{HT_t} - Ts_{l_N te} \qquad \forall t \in \mathcal{T}_1, e \tag{6}$$

Where $\mathrm{HT_t}$ is the elapsed time (in hours) from the beginning of the time horizon, at the end of each time period.

In addition to the economic function of the problem z^P, the objective function shall include the terms corresponding to the SI and the PH. The SI is computed for the product assignment variables W_{ije}^{aux} while the PH is applied to the slot duration ϕ_{lte}, and the inventory levels INV_{ite}. The objective function to minimize in each subproblem is:

$$\begin{aligned} J_e := z_e^P - \lambda SI_e + \sum_{i,t \in \mathcal{T}_1, e} \omega_{ite} \frac{INV_{ite}}{\overline{\mathrm{INV}}_{it}} + \sum_{l,t \in \mathcal{T}_1, e} \mu_{lte} \frac{\phi_{lte}}{\overline{\phi}_{lt}} \\ + \frac{\rho_1}{2} \sum_{i,t \in \mathcal{T}_1, e} \left(\frac{INV_{ite} - \overline{\mathrm{INV}}_{it}}{\overline{\mathrm{INV}}_{it}} \right)^2 + \frac{\rho_2}{2} \sum_{l,t \in \mathcal{T}_1, e} \left(\frac{\phi_{lte} - \overline{\phi}_{lt}}{\overline{\phi}_{lt}} \right)^2 \bar{x}_l \end{aligned} \tag{7}$$

After solving each subproblem, the multipliers λ, ω_{ite} and μ_{lte}, and the expected values $\overline{\mathrm{INV}}_{it}$ and $\overline{\phi}_{lt}$ need to be updated, as follows:

$$\lambda^{(k+1)} = \lambda^{(k)} - \alpha(SI - 1)\,; \qquad \overline{\mathrm{INV}}_{it} = \sum_{e \in \mathcal{E}} p_e INV_{ite} \ \forall i, t \in \mathcal{T}_1;$$

$$\omega_{ite}^{(k+1)} = \omega_{ite}^{(k)} + \rho_1 \frac{INV_{ite} - \overline{INV}_{it}}{\overline{\mathrm{INV}}_{it}} \ \forall i, t \in \mathcal{T}_1, e; \quad \overline{\phi}_{lt} = \sum_{e \in \mathcal{E}} \pi_e \phi_{lte} \ \forall l, t \in \mathcal{T}_1;$$

$$\mu_{lte}^{(k+1)} = \mu_{lte}^{(k)} + \rho_2 \frac{\phi_{lte} - \overline{\phi}_{lt}}{\overline{\phi}_{lt}} \quad \forall l, t \in \mathcal{T}_1, e \tag{8}$$

Values α, ρ_1, and ρ_2 are the tuning parameters of the algorithm.

5. Preliminary results

The combined SI-PH decomposition algorithm was tested on a case study instance of five products (A, B, C, D, E), eight weeks, and five slots. A total of eight uncertainty realization scenarios were considered. Problem 2a parameters from (Dogan & Grossmann, 2006) were used as a basis for the instance. The resulting problem had 17601 variables, 2240 of which were binary, and 17827 equations. GAMS 40.2.0 was used to code the model, and Gurobi 10.0 to solve it. All calculations were performed on a 2-CPU Xeon Gold 6130 computer. The optimality gap for Gurobi was set to 0.5% in all cases.

The monolithic instance was assigned 32 threads. The solver failed to provide an optimal solution after 10 hours of computation. The reported objective value of the best feasible solution found till that moment was $z^P = 57643$, reporting a 6.06% optimality gap.

Each of the subproblems in the decomposed instance was assigned 8 threads. Note that although the machine only has 32 cores, some scenarios are solved much faster than others, which frees cores for other threads. Note that, if too few threads are assigned per subproblem, the overall CPU capacity may be underutilized.

The tuning parameters greatly affected the quality of the solution. The values $\rho_1 = 95$, $\rho_2 = 120$, and $\alpha = 190$ provided the highest quality solution. The decomposition approach arrived at an objective value of $z^P = 54682$ in 1552 seconds. Compared to the objective value of the monolithic approach, this solution is around 5% worse. However, just setting $\rho_2 = 115$ reaches a much suboptimal $z^P = 48717$, and with $\rho_2 = 125$ the algorithm did not converge after 2000 iterations.

6. Summary

This paper presented a hybrid decomposition method for TSSPs formulated on a continuous-time basis. The non-anticipation constraints are replaced by a combination of the Similarity Index and Progressive Hedging. This allows decomposing the original problem into smaller subproblems that are easier to solve. Their solutions are combined to iteratively build a high-quality feasible solution to the original problem.

Unfortunately, tuning the algorithm is not smooth due to the PH part, and small parameter variations lead to very different objective values. Some reported solutions are even far from optimality. Future work will focus on rethinking the PH decomposition strategy.

Acknowledgments

These results are funded by the Spanish MCIN/AEI as part of the a-CIDiT (PID2021-123654OB-C31, PID2021-123654OB-C32) and LOCPU (PID2020-116585GB-I00) research projects. The first author has received financial support from the 2020 call for pre-doctoral contracts of the University of Valladolid and Banco Santander.

References

Bashiri, M., Nikzad, E., Eberhard, A., Hearne, J., & Oliveira, F. (2021). A two stage stochastic programming for asset protection routing and a solution algorithm based on the Progressive Hedging algorithm. *Omega (United Kingdom)*, *104*.

Dogan, M. E., & Grossmann, I. E. (2006). A decomposition method for the simultaneous planning and scheduling of single-stage continuous multiproduct plants. *Industrial and Engineering Chemistry Research*, *45*(1), 299–315.

Montes, D., Pitarch, J. L., & de Prada, C. (2022a). Decomposition of Two-stage Stochastic Scheduling Problems via Similarity Index. *Computer Aided Chemical Engineering*, *51*, 985–990.

Montes, D., Pitarch, J. L., & de Prada, C. (2022b). The Similarity Index to Decompose Two-Stage Stochastic Scheduling Problems. *IFAC-PapersOnLine*, *55*(7), 821–826.

Peng, Z., Zhang, Y., Feng, Y., Rong, G., & Su, H. (2019). A Progressive Hedging-Based Solution Approach for Integrated Planning and Scheduling Problems under Demand Uncertainty. *Industrial and Engineering Chemistry Research*, *58*(32), 14880–14896.

Rockafellar, R. T., & Wets, R. J. B. (1991). Scenarios and Policy Aggregation in Optimization Under Uncertainty. *Mathematics of Operations Research*, *16*(1), 119–147.

Antonis Kokossis, Michael C. Georgiadis, Efstratios N. Pistikopoulos (Eds.)
PROCEEDINGS OF THE 33rd European Symposium on Computer Aided Process Engineering (ESCAPE33), June 18-21, 2023, Athens, Greece
 http://dx.doi.org/10.1016/B978-0-443-15274-0.50081-0

Modelling of water vapor adsorption/desorption on hopcalite in an industrial medical air dryer

Cristian Cardenas,[a,b,*] & David Farrusseng[a]

[a]*Catalysis and Environment Research Institute Lyon, 2 Avenue Albert Einstein, 69626 Villeurbanne, France.*

[b]*MIL'S Society, 15 rue de Genève, 69746 Genas, France.*

**cristian.cardenas@ircelyon.univ-lyon1.fr*

Abstract

A dynamic multiscale model was developed to simulate the adsorption and desorption process in a fixed bed for air drying. This model is described by equations of mass balance, thermodynamics, and adsorption/desorption kinetics. The temperature-dependent Toth-Aranovich-Donohue equation was applied to predict the water vapor amount adsorbed on hopcalite. Experimental adsorption and desorption breakthrough curves at different water vapor concentrations and gas flow rates are measured and used to identify the different kinetic resistances, particularly the axial dispersion, external film mass transfer, pore diffusion, and internal mass transport. The model was implemented and solved within COMSOL Multiphysics®. The identified model is validated using different water vapor adsorption and desorption breakthrough curves from those used for parameter identification. The model predictions and the measurements showed a good agreement, quantified using performance indices and confirmed by a Kolmogorov-Smirnov statistic test.

Keywords: Water vapor adsorption, Hopcalite, Breakthrough curves, Modeling and simulation, Multiscale model.

1. Introduction

The air treatment system is an essential part of medical air installations: it ensures the drying and purification of the compressed air produced, to provide healthcare facilities with medical air that complies with the European Pharmacopeia n°1238-medical air in NF EN ISO 7396-1. A medical compressed air dryer is produced by pressure swing adsorption (PSA) using water-selective adsorbents. The dew point of the air treatment system reaches a minimum of -45°C which allows for high air quality. The production of dry medical air is ensured by air compressors and a column operating in PSA mode with a counter-current regeneration mode. The common industrial column is a fixed bed with various moisture adsorbents such as alumina, hopcalite and activated carbon in series in the form of granules. The state of the art is based on the choice of the nature of the adsorbents, their arrangement in the column and their respective volume (Cardenas et al, 2022). Qualitatively, the adsorbents placed at the bottom of the column are the least hydrophilic and adsorb the most moisture, while the adsorbents placed at the top of the column are the most hydrophilic and allow moisture adsorption at low concentrations. In the literature some models to describe the dynamic adsorption of water vapor on alumina are found, contrary to hopcalite where there is limited. Therefore, this study proposes to develop a fickian multiscale model to predict the adsorption and desorption breakthrough fronts on hopcalite as a function of gas flow rate, concentration, and temperature. This model is described by mass balance, thermodynamic, and adsorption/desorption kinetics equations. Several parameters are needed to implement the model, thus the hopcalite is characterized, and water vapor adsorption isotherm measurements are carried out. Experimental adsorption and desorption breakthrough curves at different water vapor concentrations and gas flow rates are measured and used to estimate the axial dispersion, external film mass transfer, mixture diffusion and internal mass transport coefficients involved in the model equations, which are implemented, and solved within COMSOL Multiphysics®. The identified model is validated using different water vapor adsorption and desorption breakthrough curves from those used for parameter identification. The model predictions and the measurements showed a good agreement, quantified using performance indices and confirmed by a Kolmogorov-Smirnov statistic test. The validated model can be used as a predictive tool for the design and optimization of an industrial air dryer.

2. Adsorption isotherms

In the model development of the water vapor adsorption/desorption process on hopcalite, one of the main issues is to select the adsorption equilibrium isotherm that best describes the process. Experimental adsorption isotherms of water vapor on hopcalite in a temperature range of 288-313 K and pressures up to the saturated concentration were measured using a manometric BelsorpMax 2 instrument.

Toth-Aranovich–Donohue (Toth-AD) equation

The temperature-dependent Toth-Aranovich-Donohue equation was applied to approximate the equilibrium relationship of water vapor on hopcalite. The model equation is as follows:

$$q_e = \frac{q_m \cdot b \cdot c}{\left[1+(b \cdot c)^{t_o}\right]^{\frac{1}{t_o}}} \left[\frac{1}{(1-b_2 \cdot c)^{n_2}}\right] \tag{1}$$

where q_e (mol.kg^{-1}) is the equilibrium amount adsorbed, c (mol.m^{-3}) is the water vapor concentration, q_m (mol.kg^{-1}) is the maximum amount adsorbed, b and b_2 (m^3.mol^{-1}) are the equilibrium constants at low and high concentrations, respectively, t_o is the Toth's heterogeneity factor and n_2 is an adjustable dimensionless factor. The temperature-dependent parameters q_m, b, b_0 and t_0 are detailed in the study by Cardenas et al. (2022a).

3. Dynamic adsorption/desorption modeling

A comprehensive one-dimensional fickian multiscale model is developed to simulate the dynamic process of adsorption and desorption of water vapor. The model is based on mass balance, thermodynamic, and kinetic equations.

3.1. Model assumptions

The assumptions of the model are as follows: (i) the gas mixture obeys the ideal gas law, (ii) water vapor is the only adsorbed molecule, (iii) the pressure drop along the bed is negligible, (iv) the resistance to mass transfer through the boundary layer surrounding the solid particles is characterized by an external film mass transfer coefficient, (v) the mass transfer within the particles is due to Fick's diffusion and characterized by an effective diffusion coefficient in the macropores, (vi) the mass transport of water vapor in the mesopores is due to mesopore diffusion, (vii) the Toth-Aranovich-Donohue equation describes the equilibrium isotherm of the gas phase with the adsorbent, (viii) the adsorbent particles are assumed to be spherical and homogeneous in size and density, (ix) the temperature and physical properties of the adsorbent are assumed to be constant, (x) only the axial dimension is considered (one-dimensional model), and (xi) the adsorption and desorption process is reversible.

3.2. Fickian multiscale model

The multiscale model considers the bed scale and the particle scale. The particle is divided into two phases: a gas and a solid phase.

(i) In the bed column, the equation of the water vapor mass balance in the gas phase is expressed as follows:

$$\frac{\partial c}{\partial t} - D_{ax}\frac{\partial^2 c}{\partial z^2} + \frac{\partial (vc)}{\partial z} = -\frac{1-\varepsilon_b}{\varepsilon_b}\frac{3}{r_{pe}}k_f\left(c - c_{pe|r_p=r_{pe}}\right)\left(\frac{1}{1+B_i}\right) \tag{2}$$

(ii) In the particle, assuming the absence of lateral and azimuthal variations of the concentration, the mass balance in the radial direction can be written as:

$$\varepsilon_{pe}\frac{\partial c_{pe}}{\partial t} + \left(1-\varepsilon_{pe}\right)\frac{\partial q}{\partial t} = \frac{D_e}{r_p^2}\frac{\partial}{\partial r_p}\left(r_p^2\frac{\partial c_{pe}}{\partial r_p}\right) \tag{3}$$

(iii) In the particle, the water vapor adsorbed on the solid surface is approximated by the linear driving force LDF model and is written as:

$$\frac{\partial q}{\partial t} \approx \frac{\partial \overline{q}}{\partial t} = k_1\left(q_e^* - \overline{q}\right) \tag{4}$$

where c (mol.m^{-3}) is the water vapor concentration in the gas phase, $v = u/\varepsilon_b$ (m.s^{-1}) is the interstitial gas velocity, D_{ax} (m^2.s^{-1}) is the axial dispersion coefficient, ε_b (m$_{gas}^3$.m$_{col}^{-3}$) is the bed porosity, r_{pe} (m) is the mean radius of the adsorbent particle, k_f (m.s^{-1}) is the external film mass transfer coefficient, $c_{p|rp=rpe}$ (mol.m^{-3}) is the water vapor concentration at the surface of the particle, $B_i = r_{pe}k_f/(5D_e)$ is the material Biot number, $D_e = \varepsilon_{pe}^{4/3}D_M$ (m^2.s^{-1}) is the effective diffusion coefficient, which depends on the porosity, and the mixture diffusion D_M (m^2.s^{-1}) in the particle, $\overline{q}$ (mol.m^{-3}) is the average water vapor amount adsorbed on the solid surface, $q_e^*=q_e\,\rho_s$ (mol.m^{-3}) is the equilibrium water vapor amount adsorbed, ρ_s (kg.m^{-3}) is the solid density, and k_1 (s^{-1}) is the internal linear driving force factor. This equation is completed by the following initial and boundary conditions:

- For $0 \le z \le L$, at $t = 0$: $c = 0$, $c_{pe} = 0$, $\overline{q} = 0$
- For t and $r_p > 0$, at $z = 0$: $-D_{ax}\frac{\partial c}{\partial z} = v\left(c_0 - c\right)$, $\overline{q} = 0$, and at $z = L$: $\frac{\partial c}{\partial z} = 0$
- For t and $z > 0$, at $r_p = 0$: $\frac{\partial c_{pe}}{\partial r_p} = 0$, and $r_p = r_{pe}$: $-D_e\left(\frac{\partial c_{pe}}{\partial r_p}\right)_{r_p} = k_f\left(c - c_{pe|r_p=r_{pe}}\right)$

Because of this reversibility, we can propose to treat the desorption breakthrough curve as if it were an adsorption breakthrough curve. Only the initial conditions and the boundary conditions are different. Therefore, the same model as described above is used along with the following modified initial and boundary conditions:

- For $0 \le z \le L$, at $t = 0$: $c = c_0$, c_{pe}, $c_{pe} = c_{pe|r_p=r_{pe}}$, $\overline{q} = q_e$
- For t and $r_p > 0$, at $z = 0$: $\frac{\partial c}{\partial z} = 0$

3.3. Implementation and simulation of the model

The external mass transfer coefficient is estimated using the Wakao and Funazkri (1978) correlation:

$$k_f = \frac{D_{AB}Sh}{2r_{pe}} \text{ with } Sh = c_{m1} + c_{m2}Re_p^{0.6}Sc^{1/3} \text{ and } c_{m1} = 2 \tag{5}$$

For the axial dispersion coefficient, the correlation of Rastegar and Gu (2017) was used:

$$D_{ax} = \frac{2r_{pe}v}{Pe'} \text{ with } \frac{1}{Pe'} = \frac{0.7D_{AB}}{2r_{pe}v} + \frac{\varepsilon_b}{c_{m3} + c_{m4}\left(Re_p\right)^{0.59}} \quad (6)$$

where Sh, R_{ep}, Sc, and Pe' are the dimensionless Sherwood, Reynolds, Schmidt and Peclet numbers respectively, D_{AB} ($m^2.s^{-1}$) is the molecular diffusion coefficient. It is important to point out that D_{ax} and k_f increase with increasing gas flow rate. To keep this property, c_{m2}, c_{m3} and c_{m4} are also considered as unknown parameters to be identified from the measurements (Esposito et al, 2022). A total number of five estimable parameters is therefore identified, i.e., k_1, D_M, c_{m2}, c_{m3} and c_{m4}. Comsol Multiphysics® version 6 is used to implement and solve the set of partial differential equations of the model. The estimability analysis algorithm using the orthogonalization algorithm described by Yao et al. (2003) is applied to the multiscale model and the value of the estimability threshold was set equal to 0.04 as suggested by Bouchkira et al., (2021). The following estimability order: $k_1 > D_M > c_{m2} > c_{m3} > c_{m4}$. It should be noted that all parameters are estimable based on available measurements.

3.4. Model prediction performance

The prediction performance of the model is quantitatively assessed by means of criteria which compares model predictions with the experimental measurements. In this study, it is evaluated according to two different criteria: The root mean-square error function (*RMSE*) and the Pearson correlation coefficient (r).

4. Results and discussion

4.1. Parameters identification

The experimental data used consist of breakthrough curve measurements carried out at a room temperature of 20 °C, at gas flow rates of 90, 120, and 150 $mL.min^{-1}$ with a water vapor concentration of 0.67 $mol.m^{-3}$, and at a gas flow rate of 135 $mL.min^{-1}$ with four water vapor concentrations of 0.19, 0.38, 0.48, 0.72, and 0.77 $mol.m^{-3}$ i.e., eighth adsorption and desorption experimental measurements to identified parameters. The optimal parameters identified and estimated during adsorption are used in the desorption simulation, except for the parameter k_1, which is estimated simultaneously from available desorption experimental measurements. Figures *1* compare the experimental measurements to the model predictions obtained using the optimized parameters in the adsorption and desorption process. The water vapor/hopcalite isotherm model used is the Toth-Aranovich-Donohue.

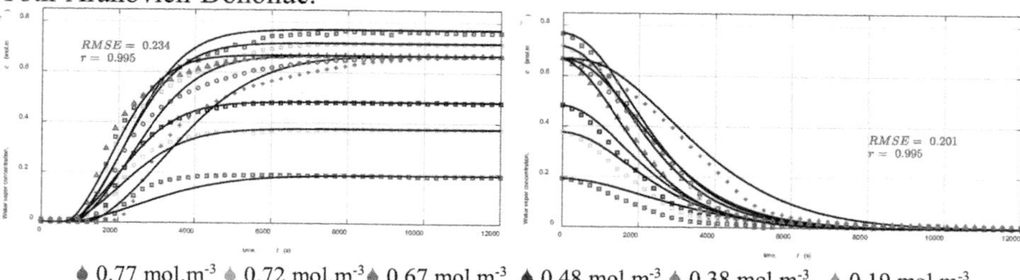

Figure 1: Comparison of water vapor adsorption (left) and desorption (right) breakthrough curves predicted with the measurements using the multiscale model.

The figure show that the predictions of the multiscale model are consistent with the experimental measurements because the Pearson correlation coefficient is close to 1

(r=0.99) and the root mean square error ($RMSE$<0.23) is low. The optimized parameter values as well as their 95% confidence intervals (CI), and the dimensionless material Biot number (Bi) of is reported in the Table *1*.

Table 1. Optimized parameter values for water vapor adsorption on hopcalite

B_i=4.63					*Adsorption*	*Desorption*
Parameters	c_{m2}	c_{m3}	$c_{m4}\times10^{-5}$	D_M, (m^2.s^{-1})	$k_{1,ads}$ (s^{-1})	$k_{1,des}$ (s^{-1})
Value	5.93	0.005	1.01	4.3×10^{-5}	0.065	0.0018
CI (95%)	±1.26	±0.002	±0.09	$\pm8.5\times10^{-6}$	±0.005	±0.0006

The fast saturation of the adsorbent bed at higher flow rates is since as the flow rate increases, the thickness of the film surrounding the particles decreases, reducing the resistivity value of the film for mass transfer. The value of the material Biot number (see Table *1*) is greater than 3. Therefore, the adsorption mechanism is controlled by internal pore diffusion. Moreover, the breakthrough time, defined as the time required to reach 10% of the water vapor concentration in the bed column, decreases with increasing water vapor concentration at the inlet. It can also be observed that an increase in water vapor concentration at the inlet results in bed saturation being reached earlier due to relatively faster transport. Since the confidence interval values are lower than the parameter values reported in Table *1*, the parameter identification can be considered accurate given the excellent fit indices. The coefficients for external film mass transfer and axial dispersion could be estimated with a different correlation than the one used, for example, by Delgado (2006) and Valdés-Solís et al., (2004) to improve the agreement between measurements and predictions. Despite this discrepancy, the values of the estimated coefficients at different gas flow rates reported in Table *1* are consistent with the magnitudes reported in the literature (Cardenas et al. 2021).

4.2. Model validation

The identified model is then validated with additional measurements of water vapor adsorption and desorption breakthrough curves obtained on hopcalite at different temperatures, i.e., 15, 30, 40, and 50 °C, corresponding to the water vapor concentrations of 0.48, 0.47, 0.46, 0.45 mol.m^{-3} and a gas flow rate of 150 mL.min^{-1}. The results are shown in Figure *2*. It is important to note that these operating conditions (concentration, gas flow rate, and temperature) are different from those used in parameter identification. The Kolmogorov-Smirnov statistical test is used and is based on the calculation of the divergence (D_n) between two distributions. For more details, see the study by Cardenas et al. (2022b).

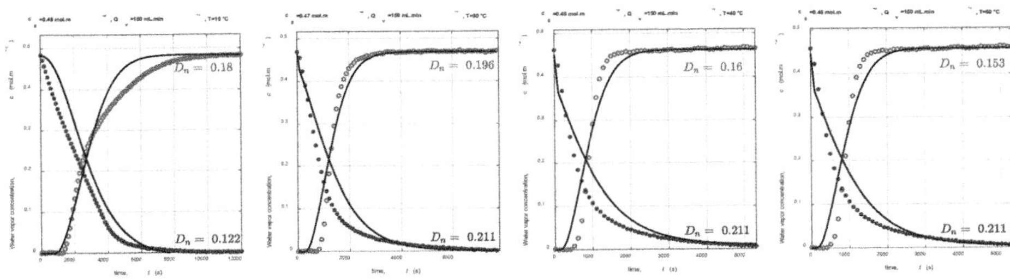

○ Adsorption ● Desorption – Model

Figure 2: Comparison between predicted and measured breakthrough curves on hopcalite at four temperatures of 15, 30, 40, and 50 °C corresponding to water vapor concentrations of 0.48, 0.47, 0.46, 0.45, mol.m^{-3} and a gas flow rate of 150 mL.min^{-1}.

The calculated divergence values that are below the critical value (D_n^*=0.2) allow us to accept the null hypothesis (H_0), i.e., we can claim with 95% probability that the predicted and measured breakthrough curves have the same distribution. As shown in Figure *2*, the predictions of the water vapor adsorption model guarantee this null hypothesis in almost all cases, so we can fully validate the multiscale model prediction of water vapor adsorption on hopcalite. On the other hand, the predictions of the water vapor desorption model, despite being very close to the critical value, cannot be guaranteed to be predicted.

5. Conclusions

The accuracy of the one-dimensional dynamic adsorption and desorption model using the temperature-dependent Toth-Aranovich-Donohue equation developed, identified, and validated for the simulation of a packed bed column in which water vapor is physically adsorbed on hopcalite was demonstrated. The fickian multiscale model allowed the finding that the adsorption process is controlled by the diffusion of water vapor in the macropores by a value of the material Biot number higher than 3. The statistical validation and the performance of the model performance indices ensure the quality of the model predictions, which is not always the case in modeling studies. This model coupled with an energy balance to improve the design and optimization of commercial adsorption columns using multilayer beds.

References

I. Bouchkira, A. M. Latifi, L. Khamar, S. Benjelloun, 2021. Global sensitivity based estimability analysis for the parameter identification of pitzer's thermodynamic model. Reliability Engineering & System Safety, 207, 107263.

C. Cardenas, A. M. Latifi, C. Vallières, S. Marsteau, L. Sigot, 2021. Analysis of an industrial adsorption process based on ammonia chemisorption: Modeling and simulation. Computers & Chemical Engineering. 154, 107474.

C. Cardenas, D. Farrusseng, C. Daniel, R. Aubry, 2022. Modelling of Water Vapor Adsorption and Desorption in an Industrial Medical Air Dryer. Available at SSRN 4259051.

C. Cardenas, D. Farrusseng, C. Daniel, R. Aubry, 2022a. Modeling of equilibrium water vapor adsorption isotherms on activated carbon, alumina and hopcalite. Fluid Phase Equilibria, 561, 113520.

C. Cardenas, A. M. Latifi, C. Vallières, S. Marsteau, 2022b. Analysis of an industrial adsorption process based on ammonia chemisorption: model validation. Computer Aided Chemical Engineering. 51, 457-462.

J. Delgado, 2006. A critical review of dispersion in packed beds. Heat Mass Transf. 42, 279-310.

F. Esposito, C. Cardenas, A. M. Latifi, S. Marsteau, 2022. CFD modeling and simulation of an ammonia adsorption process. Computer Aided Chemical Engineering. 51, 463-468.

S. O. Rastegar, T. Gu, 2017. Empirical correlations for axial dispersion coefficient and peclet number in fixed-bed columns. Journal of Chromatography. 1490, 133-137.

T. Valdés-Solís, M. Linders, F. Kapteijn, F, G. Marbán, & A. Fuertes, 2004. Adsorption and breakthrough performance of carbon-coated ceramic monoliths at low concentration of n-butane. Chemical Engineering. 59, 2791-2800.

N. Wakao, T. Funazkri, 1978. Effect of fluid dispersion coefficients on particle-to-fluid mass transfer coefficients in packed beds. Chemical Engineering Science. 33, 10, 1375-1384.

K. Z. Yao, B. M. Shaw, B. Kou, K. B. McAuley, D. W. Bacon, 2003. Modeling ethylene/butene copolymerizationwith multi-site catalysts: Parameter estimability and experimental design. Polymer Reaction Engineering. 11, 3, 563-588.

Antonis Kokossis, Michael C. Georgiadis, Efstratios N. Pistikopoulos (Eds.)
PROCEEDINGS OF THE 33rd European Symposium on Computer Aided Process Engineering (ESCAPE33), June 18-21, 2023, Athens, Greece
 http://dx.doi.org/10.1016/B978-0-443-15274-0.50082-2

Multi-fidelity Bayesian Optimisation of Reactor Simulations using Deep Gaussian Processes

Tom Savage,[a] Nausheen Basha,[a] Omar K. Matar,[a] Antonio del Rio Chanona,[a]
[a]*Imperial College London, London SW7 2BX, United Kingdom.*

Abstract

Coiled tube reactors under pulsed-flow conditions have been shown to provide promising mixing characteristics. In order to validate performance in an industrial setting, and investigate the underlying physics of successful mixing, coiled tube reactors must be optimised. In this work, we apply a novel framework to locate optimal solutions to this nonlinear, computationally expensive, and derivative-free problem. Our optimisation framework takes advantage of deep Gaussian processes to learn a multi-fidelity surrogate model. We apply this model within a novel Bayesian optimisation algorithm, using faster, less accurate, and potentially biased lower-fidelity simulations to enable faster reactor optimisation. We subsequently investigate the physical insights into the swirling flows of these optimal configurations, directly informing the design of future coiled-tube reactors under pulsed-flow conditions. We demonstrate our design framework to be extensible to a broad variety of expensive simulation-based optimisation problems, supporting the design of the next-generation of highly parameterised chemical reactors.

Keywords: Optimisation, Computational Fluid Dynamics, Multi-fidelity Modelling

1. Introduction

1.1. Background

Achieving plug flow performance in chemical reactors is critical for processes involving the manufacture of chemicals, pharmaceuticals, biodiesel production, and waste treatment to minimise energy consumption and costs, and maximise product quality. Plug-flow conditions are dependent upon the reactor geometry, with product distributions of low variance being desired. Here, we study coiled-tube reactors, which have been shown to demonstrate promising plug flow performance in previous computational and experimental studies (Mansour, Liu et al. 2017, McDonough, Ahmed et al. 2019, McDonough, Murta et al. 2019, de Brito Lira, Riella et al. 2022).

Chemical reactors including coiled-tube reactors have been extensively explored through computational fluid dynamics (CFD) simulations, where numerical solutions are solved iteratively for a system of partial differential equations (PDEs) with large degrees of freedom resulting in large computational cost. In addition to being expensive, gradient information is practically unavailable. Derivative-free optimisation has found significant application in domains where mathematical expressions or gradients are unavailable. Examples include the optimisation of proprietary chemical process software (Caballero and Grossmann 2008, Savage, Basha et al. 2022), chemical reaction optimisation (Felton, Rittig et al. 2021), and topology optimisation of two-dimensional chemical reactor channels (Cai, Guo et al. 2021). With the advent of new technologies in reactor design, reactor geometries are becoming more highly parameterised, resulting in higher-dimensional, more complex derivative-free optimisation problems.

In certain situations, computational expense can be traded off with accuracy via one or multiple fidelity parameters. Examples include training epochs in the context of hyper-parameter optimisation (Wu, Toscano-Palmerin et al. 2019, Schmucker, Donini et al. 2020), mesh fidelities in the context of finite element analysis (Huang, Allen et al. 2006), or combining real-time measurements and predictions in industrial processes (Petsagkourakis, Chachuat et al. 2021). Including fidelity control within a Bayesian optimisation framework enables optimisation with fewer computational resources whilst gaining a 'high fidelity' solution (Lindauer, Eggensperger et al. 2019). Equation 1 describes this multi-fidelity problem

$$x^* = \arg\max_{x \in X} f(x, z^*). \tag{1}$$

where potentially M different fidelities, $z \in \mathbb{R}^M$ become controllable parameters, and z^* indicates the highest-fidelity evaluation that is the one of interest.

Contribution: In this work, we present the real-world problem of optimising the geometry of a coiled-tube reactor to maximise its plug-flow performance. We apply a state-of-the-art deep GP (DGP)-based, multi-fidelity Bayesian optimisation algorithm to identify novel reactor configurations using an amalgam of different fidelity simulations, modelled using a DGP. Figure 1 demonstrates how our approach takes advantage of lower-fidelity simulations. Our approach contains no additional hyper-parameters when compared to standard Bayesian optimisation. Having identified an optimal geometry, we investigate the physical insights to inform future design of pulsed flow coiled tube reactors.

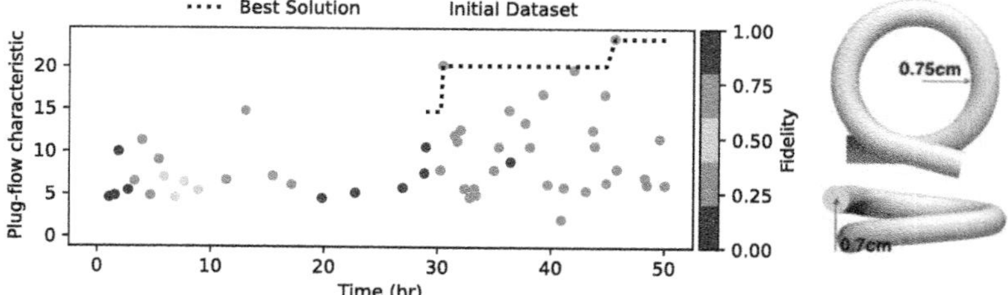

Figure 1: Left: The progression of multi-fidelity Bayesian optimisation using DGPs as a multi-fidelity model. Multiple different fidelity levels are selected for evaluation throughout optimisation, reducing optimisation time. Right: the best coil geometry which has a relatively large coil radius and low pitch.

2. Method

By applying DGPs within a Bayesian optimisation framework, we enable an end-to-end model of all fidelities. A more accurate model of higher fidelities should enable the optimisation procedure to make more evaluations at lower, less expensive fidelities.

2.1. Parameterisation

A coiled tube reactor is parameterised by a coil radius, and a coil pitch. Inversions within coiled tube reactors have been shown to provide effective mixing properties (Singh and

Nigam 2016, Rossi, Gargiulo et al. 2017, McDonough, Murta et al. 2019). Therefore, in addition to coil radius and coil pitch, we include an additional inversion parameter δ to

investigate geometries which involve a change of coil direction. δ takes values between 0 and 1, and specifies where along the coil the inversion takes place. Figure 2 demonstrates the effect of δ on helical tube geometry.

(a): The effect of δ for a helical tube with fixed length, viewed from above.

(b): The effect of δ for a helical tube with fixed length, viewed from side.
Figure 2: From left to right: δ is evaluated at 0, 0.15 0.3, 0.6, and 0.75 with coil radius and pitch remaining constant.

Two fidelity aspects, axial and radial, can be varied when meshing is performed given a tube geometry. (Savage, Basha et al. 2022) demonstrates how these two fidelities affect the final mesh. Five discrete fidelities were simulated and compared with experimental data, in order to test assumptions regarding the accuracy of a simulation. Figure 3 validates the tracer concentration profile of simulations at five fidelities against two sets of experimental data. The objective of the optimisation is to maximise the number of equivalent tanks-in-series (McDonough, Murta et al. 2019), where high values of N are an indicator for good radial mixing and poor axial mixing. Figure 3 also demonstrates how increasing fidelity (and therefore cell count) results in a closer approximation to the experimental value of N, derived from each concentration profile. The accuracy of prediction when increasing cell-count in experiment 2 drops slightly.

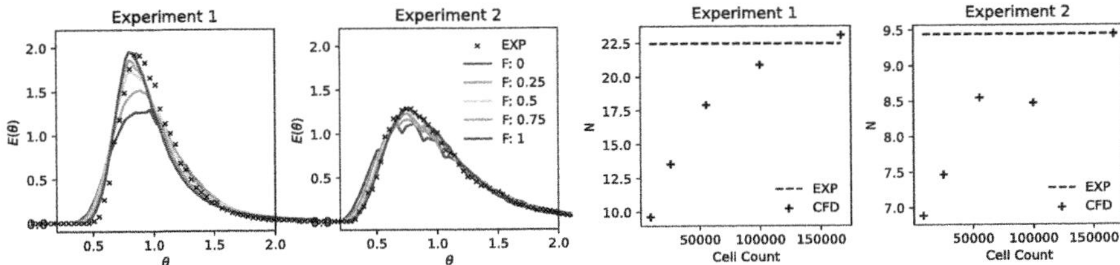

Left: The concentration profile of a tracer injection at five fidelity levels between 0 and 1 against experimentally obtained data. $E(\theta)$ represents dimensionless concentration as a function of dimensionless time θ

Right: The value of N corresponding to the concentration profile from each fidelity, which has been converted to cell count. N represents plug-flow characteristic which is the quantity optimised for.

Figure 3: Validation of five discrete mesh fidelities corresponding to different cell counts, across two sets of experimental data under different conditions.

2.2. Model Specification

To model each fidelity level we apply Deep Gaussian processes as demonstrated by (Cutajar, Pullin et al. 2019). DGPs provide a natural extension to sequentially trained multi-fidelity models, where a generating function f at a given discrete fidelity z is modeled as a linear or nonlinear function of lower fidelities plus a mismatch term. For example, in the nonlinear case, $f_z(x)$ is given by

$$f_z(x) = \rho_{z-1}(f_{z-1}(x), x) + \phi_{z-1}. \tag{2.1}$$

Multi-fidelity DGPs combine ρ_{z-1} and ϕ_{z-1} into a single term g_{z-1} which is modelled using a GP and trained end-to-end, resulting in a composition of GPs. For example, for Z different fidelity levels, and observations x at each fidelity, the highest fidelity is modeled as

$$f_z(x_z, \ldots, x_1) = g_z(\ldots g_{z-1}(g_1(x_1), x_{z-1}), x_z). \tag{2.2}$$

In this work, we assume five discrete fidelity levels. Thus the resulting DGP has five layers ($Z = 5$), corresponding to simulations at fidelities equal to 0, 0.25, 0.5, 0.75, and 1. MF-DGPs are implemented in Python using Emukit (Paleyes, Pullin et al. 2019).

2.3. Discrete Multi-fidelity Bayesian Optimisation using Deep Gaussian Processes

Whilst continuous approximations have been applied for multi-fidelity Bayesian optimisation (Kandasamy, Dasarathy et al. 2017), we present an algorithm that takes advantage of the discrete fidelity predictions of a deep Gaussian process. Our approach is similar in spirit to that of (He, Tuo et al. 2017) and the experimental design work of (Thodoroff, Kaiser et al.). We perform a standard Bayesian optimisation step on the layer corresponding to the highest fidelity. Highlighting our interest only in the high-fidelity model. We then subsequently evaluated the variance of this solution with respect to the variance of other fidelities, providing insight into potential information gain. These variances are weighted by simulation cost enabling us to trade off information-gain with computational expense. Algorithm 1 highlights this approach.

Algorithm 1 Deep GP-based Multi-fidelity Bayesian Optimisation

Require: $f_1(x) \ldots f_{z^\bullet}(x)$, $\mathcal{X}$, n
 for z in $1, \ldots, z^\bullet$ **do**
  Generate n samples, $\mathbf{x}_z$, and evaluate $f_t(\mathbf{x})$ resulting in $\mathbf{y}_z$.
  $\tau_z \leftarrow$ average simulation time
 end for
 while Budget not exhausted **do**
  Train DGP using $\mathbf{x}_1, \ldots, \mathbf{x}_{z^\bullet}$ and $\mathbf{y}_1, \ldots, \mathbf{y}_{z^\bullet}$
  Solve UCB for highest-fidelity: $x^* \leftarrow \arg\max_x\{\mu_{z^\bullet}(x) + \beta^{1/2}\sigma_{z^\bullet}(x) | x \in \mathcal{X}\}$
  Choose fidelity based on variance of DGP and simulation cost: $z^* \leftarrow \text{argmax}_z\{\gamma_z \beta^{1/2}\sigma_z(x^*)\}$ where $\gamma_z = \max(\tau)/\tau_z$
  Evaluate $f_{z^*}(x^*)$, add x^* to $\mathbf{x}_{z^*}$ and $f_{z^*}(x^*)$ to $\mathbf{y}_{z^*}$
 end while

We note that in choosing γ_z based on the points sampled to construct the initial DGP, the algorithm contains no additional hyper-parameters than standard UCB Bayesian optimisation. In addition, γ_z may be updated after a simulation has been performed, more accurately reflecting the simulation time at that fidelity.

Experimental Results

Figure 1 demonstrates the MF-DGP based optimisation, which shows the selection of multiple simulation fidelities throughout optimisation. We note that initialising the DGPs for multi-fidelity modelling requires simulations at each fidelity. This presents a disadvantage in discrete over both single-fidelity and continuous fidelity settings, as the number of fidelities grows so does the initial data required. The number of initial data points and at what respective fidelities is the subject of future work.

2.4. Recommendations

Overall, we find that it may not be beneficial in situations where a large number of discrete fidelities are available, to apply all fidelities within an optimisation framework. A large number of fidelities results in a large number of hyper-parameters, more difficult to train multi-fidelity models such as DGPs with more layers resulting in longer inference times, and potentially slower exploration. If a discrete multi-fidelity Bayesian optimisation method is to be used, we make the recommendation to apply 2 or 3 discrete fidelities in systems with no prior knowledge, despite more being available. Alternatively, if more fidelities are required, modelling multi-fidelity data in a continuous space and applying an appropriate algorithm such as the one presented in (Kandasamy, Dasarathy et al. 2017) may be necessary. Future work will apply multi-fidelity Bayesian optimisation approaches in continuous domains, as well as apply similar methods on more industrial case studies for engineering discovery. Further experiments will be performed to correlate dimensionality of the problem with the number of discrete fidelities.

4. Conclusions

The optimisation of coiled tube reactor geometry is critical to maximise plug-flow behaviour and investigate industrial viability. The optimisation problem is formulated as an expensive black-box problem. In this paper, we propose multi-fidelity Bayesian optimisation using Deep Gaussian processes to find good reactor configurations, taking advantage of less accurate but faster simulations. We demonstrate experimental validation of five discrete fidelities and present a modified multi-fidelity Bayesian optimisation algorithm which relies on fewer hyper-parameters than existing approaches. A multi-fidelity DGP provides correct quantification and propagation of uncertainty, which we use to select not only the next experimental design, but also the fidelity of the evaluation within the algorithm. Our approach can be extended to other problems involving parameterised CFD simulations. This work demonstrates an industrially relevant use case of multi-fidelity deep Gaussian processes for the optimisation of expensive black-box functions. We hope it provides insight and inspiration for people in the chemical engineering community to apply multi-fidelity methods across a variety of potential problems.

Acknowledgements

The authors acknowledge funding by the Engineering \& Physical Sciences Research Council, United Kingdom through the PREMIERE (EP/T000414/1), and thank Ilya Orson Sandoval for providing insights on this work. The authors would also like to thank Dr Jonathan McDonough, Newcastle University. Tom Savage acknowledges the support of the Imperial College President's scholarship.

References

Caballero, J. A. and I. E. Grossmann (2008). "An algorithm for the use of surrogate models in modular flowsheet optimization." AIChE journal **54**(10): 2633-2650.

Cai, H., K. Guo, H. Liu, C. Liu and A. Feng (2021). "Derivative-free level-set-based multi-objective topology optimization of flow channel designs using lattice Boltzmann method." Chemical Engineering Science **231**: 116323.

Cutajar, K., M. Pullin, A. Damianou, N. Lawrence and J. González (2019). Deep Gaussian Processes for Multi-fidelity Modeling.

de Brito Lira, J. O., H. G. Riella, N. Padoin and C. Soares (2022). "Fluid dynamics and mass transfer in curved reactors: A CFD study on Dean flow effects." Journal of Environmental Chemical Engineering **10**(5): 108304.

Felton, K. C., J. G. Rittig and A. A. Lapkin (2021). "Summit: benchmarking machine learning methods for reaction optimisation." Chemistry-Methods **1**(2): 116-122.

He, X., R. Tuo and C. J. Wu (2017). "Optimization of multi-fidelity computer experiments via the EQIE criterion." Technometrics **59**(1): 58-68.

Huang, D., T. T. Allen, W. I. Notz and R. A. Miller (2006). "Sequential kriging optimization using multiple-fidelity evaluations." Structural and Multidisciplinary Optimization **32**(5): 369-382.

Kandasamy, K., G. Dasarathy, J. Schneider and B. Póczos (2017). Multi-fidelity bayesian optimisation with continuous approximations. International Conference on Machine Learning, PMLR.

Lindauer, M., K. Eggensperger, M. Feurer, A. Biedenkapp, J. Marben, P. Müller and F. Hutter (2019). BOAH: A Tool Suite for Multi-Fidelity Bayesian Optimization; Analysis of Hyperparameters.

Mansour, M., Z. Liu, G. Janiga, K. D. P. Nigam, K. Sundmacher, D. Thvenin and K. Z\"hringer (2017). "Numerical study of liquid-liquid mixing in helical pipes." Chemical Engineering Science **172**: 250-261.

McDonough, J. R., S. M. R. Ahmed, A. N. Phan and A. P. Harvey (2019). "The development of helical vortex pairs in oscillatory flows } A numerical and experimental study." Chemical Engineering and Processing - Process Intensification **143**: 107588.

McDonough, J. R., S. Murta, R. Law and A. P. Harvey (2019). "Oscillatory fluid motion unlocks plug flow operation in helical tube reactors at lower Reynolds numbers (Re $$ 10)." Chemical Engineering Journal **358**: 643-657.

Paleyes, A., M. Pullin, M. Mahsereci, C. McCollum, N. Lawrence and J. González (2019). Emulation of physical processes with Emukit. Second Workshop on Machine Learning and the Physical Sciences, NeurIPS.

Petsagkourakis, P., B. Chachuat and E. A. del Rio-Chanona (2021). Safe Real-Time Optimization using Multi-Fidelity Gaussian Processes.

Rossi, D., L. Gargiulo, G. Valitov, A. Gavriilidis and L. Mazzei (2017). "Experimental characterization of axial dispersion in coiled flow inverters." Chemical Engineering Research and Design **120**: 159-170.

Savage, T., N. Basha, O. M. Ehecatl and A. D.-R. Chanona (2022). "Deep Gaussian Process-based Multi-fidelity Bayesian Optimization for Simulated Chemical Reactors." arXiv preprint arXiv:2210.17213.

Schmucker, R., M. Donini, V. Perrone and C. Archambeau (2020). Multi-objective multi-fidelity hyperparameter optimization with application to fairness. NeurIPS 2020 Workshop on Meta-learning.

Singh, J. and K. Nigam (2016). "Pilot plant study for effective heat transfer area of coiled flow inverter." Chemical Engineering and Processing: Process Intensification **102**: 219-228.

Thodoroff, P., M. Kaiser, R. Williams, R. Arthern, S. Hosking, N. Lawrence, J. Byrne and I. Kazlauskaite "Multi-fidelity experimental design for ice-sheet simulation."

Wu, J., S. Toscano-Palmerin, P. I. Frazier and A. G. Wilson (2019). Practical Multi-fidelity Bayesian Optimization for Hyperparameter Tuning.

Antonis Kokossis, Michael C. Georgiadis, Efstratios N. Pistikopoulos (Eds.)
PROCEEDINGS OF THE 33rd European Symposium on Computer Aided Process Engineering (ESCAPE33), June 18-21, 2023, Athens, Greece
 http://dx.doi.org/10.1016/B978-0-443-15274-0.50083-4

Design of a sustainable biodiesel production process by a multi-objective optimization

Barcia-Quimi A.F.[a,*], Risco-Bravo A.[a] , Alcivar-Espinoza K.[a] , Tinoco-Caicedo D.L.[a,b]

[a] *Facultad de Ciencias Naturales y Matemáticas (FCNM), Escuela Superior Politécnica del Litoral Ecuador, 090903 Guayaquil, Ecuador.*

[b] *Centro de Energías Renovables y Alternativas (CERA), Escuela Superior Politécnica del Litoral Ecuador, 090903 Guayaquil, Ecuador.*

**abarcia@espol.edu.ec*

Abstract

A multi-objective optimization is proposed for determining the optimal operating conditions for biodiesel production. The objective functions considered were the production cost, the carbon footprint, and the yield of palm oil transesterification. The kinetic model proposed by Narváez et al. (2015) was validated with state-of-the-art experimental data. The optimization was conducted using Python 3.8.8 and openLCA 1.11.0 to determine environmental parameters. Four-decision variables were optimized by using 3 methodologies: Pareto Set, R-Method, and a Techno-economic analysis. The optimal solution resulted in a maximum yield of 80.29% with a minimum production cost of 0.44$/kg, and a minimum carbon footprint of 0.144 kg CO_2-eq/kg.

Keywords: Multi-objective Optimization, Kinetic model, Carbon footprint, Biodiesel

1. Introduction

Nowadays, the predominant energy source is fossil fuels. However, research on alternative fuels has begun due to increased energy demand and limited fossil resources. One major alternative is biodiesel production which is now implemented in many countries. Nevertheless, the drawback of the industrialization of this process is the high volume of reactants used such as methanol and conventional catalysts, which leads to a high production cost and a high carbon footprint. For instance, an Ecuadorian company produces 144 thousand tons per year (Roca Meza, 2014), and the production costs exceed $1000/ton. This causes the price of biodiesel to be 5.56 $/gal (Clickgasoil, 2021) higher than diesel 4.33 $/gal (EP PETROECUADOR, 2022). Several experimental studies have focused on finding the optimal conditions to maximize performance. Castillo Gónzalez et al. (2020) shows that, for the formation of alkyl esters, a molar ratio of alcohol:oil 6:1 result in a high yield of biodiesel. Narváez et al. (2007) performed experimental tryouts on the kinetics of palm oil transesterification, obtaining high-yield results with elevated temperatures and molar ratios. Suzihaque et al. (2022) state that the catalyst's concentration is an important variable to consider since low concentrations imply a smaller catalytic surface area to conduct the transesterification reaction, while high concentrations could lead to saponification. Either way, yield is affected negatively. Research on biodiesel optimization (Pugazhendhi et al., 2020) is often focused on maximizing yield. However, an optimization also including production cost and carbon footprint would be a suitable approach toward sustainable production. Therefore, the scope of this study is to perform a multi-objective optimization that minimizes the production cost and the carbon footprint and maximizes the yield of the transesterification reaction of palm oil.

2. Methodology

2.1. System description

Figure 1 shows the process flow for the biodiesel batch production proposed. A NaOH-methanol mixture (Stream 1) and palm oil (Stream 2) enter the batch-type reactor (R-101) for the transesterification reaction. R-101 uses steam supplied (Stream 6) by the boiler (E-101). E-101 works with liquefied natural gas (Stream 5) as fuel and air intake (Stream 4) at ambient conditions for combustion.

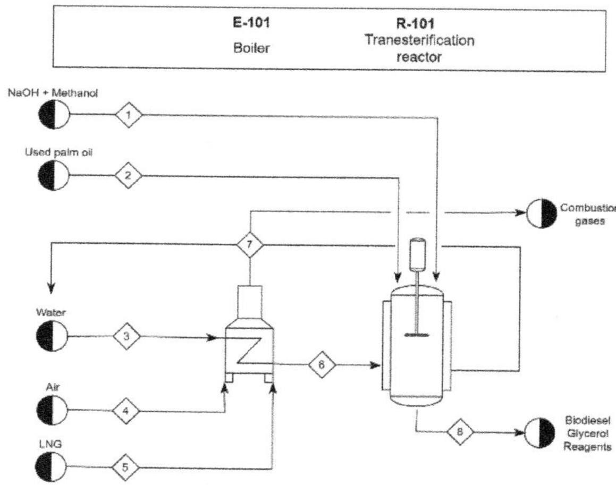

Figure 1. Process flow diagram to produce biodiesel from waste oil

2.2. Kinetic model and Validation

The kinetic model for the palm oil's transesterification reaction was implemented in Python 3.8.8. The model was based on a previous study from Narváez et al. (2015) which assumes that palm oil's transesterification reaction involves three reversible chain reactions that follow an elementary rate law. Equations 1 to 6 present the kinetic model,

$$\frac{d[TG]}{dt} = -k_1\, e^{\frac{E_1}{RT}} \left(\frac{C_{cat}}{k_{1\,cat} + C_{cat}}\right)[TG][M] + k_{-1}\, e^{\frac{E_{-1}}{RT}} \left(\frac{C_{cat}}{k_{-1\,cat} + C_{cat}}\right)[DG][EM] \tag{1}$$

$$\begin{aligned}\frac{d[DG]}{dt} = {} & k_1\, e^{\frac{E_1}{RT}} \left(\frac{C_{cat}}{k_{1\,cat} + C_{cat}}\right)[TG][M] - k_{-1}\, e^{\frac{E_{-1}}{RT}} \left(\frac{C_{cat}}{k_{-1\,cat} + C_{cat}}\right)[DG][EM] \\ & - k_2\, e^{\frac{E_2}{RT}} \left(\frac{C_{cat}}{k_{2\,cat} + C_{cat}}\right)[DG][M] + k_{-2}\, e^{\frac{E_{-2}}{RT}} \left(\frac{C_{cat}}{k_{-2\,cat} + C_{cat}}\right)[MG][EM]\end{aligned} \tag{2}$$

$$\frac{d[MG]}{dt} = k_2\, e^{\frac{E_2}{RT}} \left(\frac{C_{cat}}{k_{2\,cat} + C_{cat}}\right)[DG][M] - k_{-2}\, e^{\frac{E_{-2}}{RT}} \left(\frac{C_{cat}}{k_{-2\,cat} + C_{cat}}\right)[MG][EM] - k_3\, e^{\frac{E_3}{RT}} \left(\frac{C_{cat}}{k_{3\,cat} + C_{cat}}\right)[MG][M] + k_{-3}\, e^{\frac{E_{-3}}{RT}} \left(\frac{C_{cat}}{k_{-3\,cat} + C_{cat}}\right)[G][EM] \quad (3)$$

$$\frac{d[EM]}{dt} = k_1\, e^{\frac{E_1}{RT}} \left(\frac{C_{cat}}{k_{1\,cat} + C_{cat}}\right)[TG][M] - k_{-1}\, e^{\frac{E_{-1}}{RT}} \left(\frac{C_{cat}}{k_{-1\,cat} + C_{cat}}\right)[DG][EM] + k_2\, e^{\frac{E_2}{RT}} \left(\frac{C_{cat}}{k_{2\,cat} + C_{cat}}\right)[DG][M] - k_{-2}\, e^{\frac{E_{-2}}{RT}} \left(\frac{C_{cat}}{k_{-2\,cat} + C_{cat}}\right)[MG][EM] + k_3\, e^{\frac{E_3}{RT}} \left(\frac{C_{cat}}{k_{3\,cat} + C_{cat}}\right)[MG][M] - k_{-3}\, e^{\frac{E_{-3}}{RT}} \left(\frac{C_{cat}}{k_{-3\,cat} + C_{cat}}\right)[G][EM] \quad (4)$$

$$\frac{d[G]}{dt} = k_3\, e^{\frac{E_3}{RT}} \left(\frac{C_{cat}}{k_{3\,cat} + C_{cat}}\right)[MG][M] - k_{-3}\, e^{\frac{E_{-3}}{RT}} \left(\frac{C_{cat}}{k_{-3\,cat} + C_{cat}}\right)[G][EM] \quad (5)$$

$$\frac{d[M]}{dt} = -\frac{d[EM]}{dt} \quad (6)$$

where *TG* represents triglycerides, *DG* diglycerides, *MG* monoglycerides, *G* glycerol, *EM* methyl esters, and *M* methanol. This model considers the influence of the reaction temperature and catalyst concentration on the kinetics of the reaction.

To validate the kinetic model, the degree of fit of the experimental kinetic data of Narváez et al. (2007) and Leevijit et al. (2004) were compared with the curve predicted by the model. Then, experimental data from Narváez et al. (2007), Cheng et al. (2004), Silva et al. (2021) and Wongjaikham et al. (2021) were used to compare their reported yield (Y%) or conversion (C%) results with those predicted by the kinetic model according to Equations 7 and 8 respectively. The relative errors between the experimental and model-predicted results were determined.

$$C\,\% = \frac{[TG]_0 - [TG]_f}{[TG]_0}\, x\, 100\% \quad (7) \qquad Y\,\% = \frac{[EM]_f}{3\,[TG]_0}\, x\, 100\% \quad (8)$$

2.3. Multi-objective optimization

For the multi-objective optimization, an algorithm was developed and implemented in Python 3.8.8. The independent variables considered were *reaction temperature (T), reaction time (t), NaOH catalyst concentration (C_{cat}),* and *methanol: oil molar ratio (RM).* Therefore, the vector with e decision variables is presented in Equation 9:

$$x = (T\,, t\,, C_{cat}\,, RM) \quad (9)$$

The optimization algorithm started by evaluating each of the independent variables within the objective functions included in *Table 1*. The openLCA 1.11.0 program along with the *ef_secondarydata_202202* database was used to propose the carbon footprint objective function.

Three different strategies were used to discriminate the results of the solution set. First, the Pareto set was determined using the Python package *paretoset 1.2.0.* This package is based on the skyline operator methodology that identifies the non-dominated solutions of the set, i.e., those that are the best in all or at least one of the objectives. Then, the R-Method proposed by Rao & Lakshmi (2021) was used. This is a ranking algorithm to select the *best* solution based on the level of importance of each of the objectives. In this case, yield and production costs were set as objectives with the same degree of interest. Finally, the ten solutions with the highest score from the R-Method analysis were selected to perform a techno-economic analysis. The methodology proposed by Turton et al.

(2018) was used to determine the cash flow, Internal Rate of Return (IRR), Net Present Value (NPV), and Payback Period for each solution. The solution with the highest IRR, NPV, and shortest payback period was selected as the optimal solution. This solution would not only maximize yield, minimize production costs, and carbon footprint but also would have the best economic potential.

Table 1. Objective functions

Yield* **(%)**	**Production Cost**** **(\$/kg)**	**Carbon footprint**** **(kg CO_2-eq/kg)**
$\max\limits_{x} \frac{[EM]_f}{3 * [TG]_0} x\ 100\%$	$\min\limits_{x} \frac{C_M + C_{Cat} + C_E + C_C}{M_{bio}}$	$\min\limits_{x} \frac{CF_M + CF_{Cat} + CF_E + CF_F}{M_{bio}}$

Restrictions

$45 \leq \boldsymbol{T}\ (°C) \leq 60$	$30 \leq \boldsymbol{t}\ (\boldsymbol{min}) \leq 120$	$3{:}1 \leq \boldsymbol{RM} \leq 10{:}1$	$0.5 \leq \boldsymbol{C_{cat}}\ \left(\%\frac{w}{w}\right) \leq 1.5$

$*[EM]_f$ the final concentration of methyl ester, $[TG]_0$ *the* initial concentration of triglycerides.

$**C$ and CF represent the cost and carbon footprint of methanol (M), catalyst (Cat), electricity (E), and fuel (F). M_{bio} is the mass of biodiesel produced.

3. Results

The kinetic model for the palm oil's transesterification reaction was validated. As shown in *Figure 2,* the theoretical kinetic model in comparison with the data from Narváez et al. (2007) (a) and Leevijit et al. (2004) (b) have a suitable approximation with R^2 values of 0.983 and 0.924 respectively.

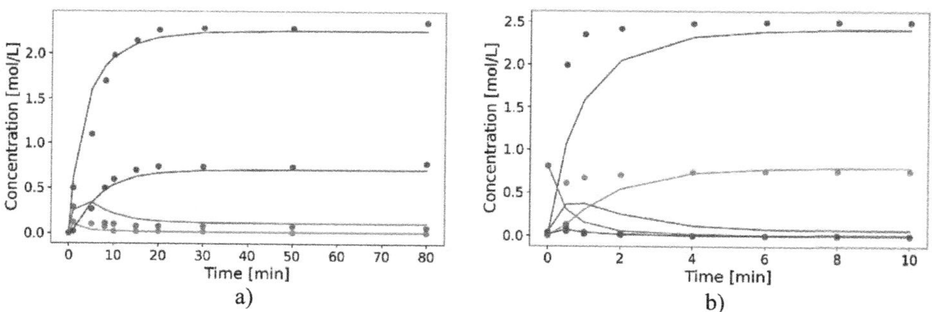

Figure 2. Comparison between experimental data (•) and the kinetic model (—) for palm oil transesterification at RM 6:1 and a) T = 50 °C and C_{cat} = 0.2% and b) T = 60 °C and C_{cat} = 1.0%

Table 2 presents the experimental and model-predicted results for conversion or yield. The relative errors are below 5%. Based on this, it can be concluded that the model proposed by Narváez et al. (2015) is highly reliable to describe the kinetics of palm oil transesterification. Also, that is possible to extrapolate the model for the selected operating ranges and obtain acceptable results.

The solution set consisted in 12544 results for *Yield* (%), *Production Cost* (\$/kg) and *Carbon Footprint* (kg CO_2-eq/kg). As yield increases, production costs and carbon footprint also increase as shown in *Figure 3*. The Pareto set contained 286 solutions that are on the lower boundary of the solution set showing that these are the non-dominated solutions according to the criteria of this optimization.

Table 2. Kinetic model validation results

Reference	T (°C)	t (min)	C_{cat} (% w/w)	RM	Parameter	Experimental	Model	% Error
Cheng et al. (2004)	60.5	7	0.5	10:1	C %	99.00	99.77	0.78
Narváez et al. (2007)	60.0	80	1.0	6:1	Y %	97.60	96.94	0.68
Silva et al. (2021)	45.0	60	1.0	4:1	Y %	88.50	90.08	1.75
Wongjaikham et al. (2021)	62.5	40	1.5	3:1	Y %	77.00	80.41	4.39

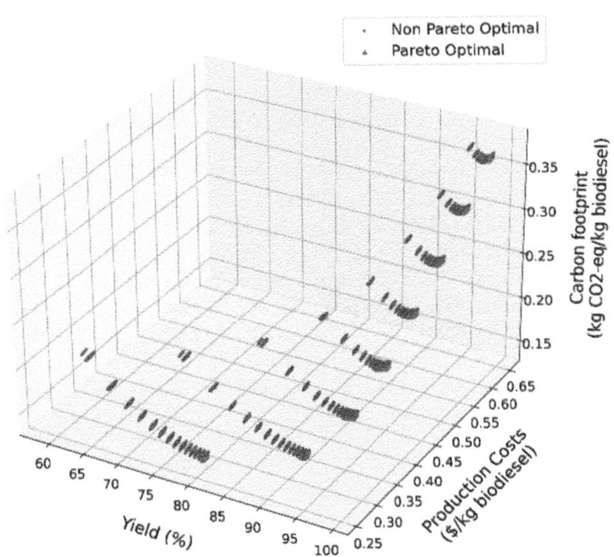

Figure 3. Yield, production cost and carbon footprint results of the solution set

The operating conditions that maximize yield while minimizing production costs and carbon footprint are T = 45°C, t = 30 min, RM = 3:1 and C_{cat} = 1.5%. Under these conditions, a biodiesel production yield of 80.29%, a production cost of 0.26 $/kg, and a carbon footprint of 0.144 kg CO_2-eq/kg is achieved. This production cost only includes raw materials and reactor utilities as proposed in its objective function. However, in the techno-economic analysis, a final cost of 0.44 $/kg was estimated considering other costs such as operating labor and maintenance.

Similar proposals for biodiesel production via vegetable oil transesterification obtained a biodiesel cost between 0.58 $/kg (El-Gharbawy, 2017) and 1.35 $/kg (Mohammadshirazi et al., 2014). The product cost can be reduced up to 67.37% when the process operates at the obtained optimal conditions. Therefore, this solution turns out to be the most favorable not only in the studied range but also against other proposals. In addition, the optimal solution makes the proposal attractive and viable, with an IRR of 128.94% and a NPV of $1.42 million. The invested capital would be recovered in 8 months and 12 days.

4. Conclusions

The multi-objective optimization algorithm determined the optimal temperature, molar ratio, catalyst concentration and reaction time. This solution not only maximizes yield while minimizing production costs and carbon footprint but also has the best economic potential. In addition, this solution reduces production costs by up to 67.37% compared to other similar proposals. The proposed framework could be a potential tool for developing countries to target the biofuel market with sustainable production. In future work, a different optimization algorithm could be used in order to compare and further evaluate the obtained results.

References

Castillo Gónzalez, J. P., Álvarez Gutiérrez, P. E., Adam Medina, M., López Zapata, B. Y., Ramírez Guerrero, G., & Vela Valdés, L. (2020). Effects on Biodiesel Production Caused by Feed Oil Changes in a Continuous Stirred-Tank Reactor. *Applied Sciences*, *10*(3), 992.

Cheng, S. F., Choo, Y. M., Ma, A. N., & Chuan, C. H. (2004). Kinetics study on transesterification of palm oil. *Journal of Oil Palm Research*, *16*(2), 19–29.

Clickgasoil. (2021). *Evolución del precio de Biodiesel*. https://www.clickgasoil.com/c/evolucion-del-precio-biodiesel.

El-Gharbawy, A. (2017). Cost Analysis for Biodiesel Production from Waste Cooking Oil Plant in Egypt. *INTERNATIONAL JOURNAL of SMART GRID*, *1*(1), 16–25.

EP PETROECUADOR. (2022). *Precios de venta en terminal para las comercializadoras calificadas y autorizadas a nivel nacional*. https://www.eppetroecuador.ec/wp-content/uploads/downloads/2022/05/estructura-de-precios-mayo-2022.pdf

Leevijit, T., Wisutmethangoon, W., Prateepchaikul, G., Tongurai, C., & Allen, M. (2004). A Second Order Kinetics of Palm Oil Transesterification. *The Joint International Conference on "Sustainable Energy and Environment (SEE),"* *3*, 277–281.

Mohammadshirazi, A., Akram, A., Rafiee, S., & Bagheri Kalhor, E. (2014). Energy and cost analyses of biodiesel production from waste cooking oil. *Renewable and Sustainable Energy Reviews*, *33*, 44–49.

Narváez, P. C., Noriega, M. A., & Cadavid, J. G. (2015). Kinetics of palm oil ethanolysis. *Energy*, *83*, 337–342.

Narváez, P. C., Rincón, S. M., & Sánchez, F. J. (2007). Kinetics of Palm Oil Methanolysis. *Journal of the American Oil Chemists' Society*, *84*(10), 971–977.

Pugazhendhi, A., Alagumalai, A., Mathimani, T., & Atabani, A. E. (2020). Optimization, kinetic and thermodynamic studies on sustainable biodiesel production from waste cooking oil: An Indian perspective. *Fuel*, *273*, 117725.

Rao, R. V., & Lakshmi, R. J. (2021). Ranking of Pareto-optimal solutions and selecting the best solution in multi- and many-objective optimization problems using R-method. *Soft Computing Letters*, *3*, 100015.

Roca Meza, L. G. (2014). *Valoración Financiera de la empresa La Fabril S.A.* Escuela Superior Politécnica del Litoral.

Silva, F. C. da, Guardiola, J. F. H., Teixeira, L. P., Maria, A. C. L., Souza, L. A. de, & Belém, A. L. (2021). Optimization of palm oil biodiesel production using response surface methodology. *Revista Brasileira de Ciências Ambientais (Online)*, *56*(2), 274–285.

Suzihaque, M. U. H., Alwi, H., Kalthum Ibrahim, U., Abdullah, S., & Haron, N. (2022). Biodiesel production from waste cooking oil: A brief review. *Materials Today: Proceedings*.

Turton, R., Shaeiwitz, J. A., Bhattacharyya, D., & Whiting, W. B. (2018). *Analysis, Synthesis, and Design of Chemical Processes* (5th ed.). Prentice Hall.

Wongjaikham, W., Wongsawaeng, D., Ratnitsai, V., Kamjam, M., Ngaosuwan, K., Kiatkittipong, W., Hosemann, P., & Assabumrungrat, S. (2021). Low-cost alternative biodiesel production apparatus based on household food blender for continuous biodiesel production for small communities. *Scientific Reports*, *11*(1), 13827.

Antonis Kokossis, Michael C. Georgiadis, Efstratios N. Pistikopoulos (Eds.)
PROCEEDINGS OF THE 33rd European Symposium on Computer Aided Process Engineering (ESCAPE33), June 18-21, 2023, Athens, Greece
 http://dx.doi.org/10.1016/B978-0-443-15274-0.50084-6

Modeling of an Industrial Delayed Coker Unit

Anne Firstauthor,[a] Tim B. Secondauthor,[b] James Q. Thirdauthora,[b]

[a]First affiliation, Address, City and Postcode, Country
[b]Second affiliation, Address, City and Postcode, Country

Gizem Kuşoğlu,[a,b] Yaman Arkun,[b]

[a] Turkish Petroelum Refinery, Körfez, Kocaeli, 41790, Turkey
[b] Chemical and Biological Engineering, Koç University, Rumelifeneri, İstanbul, 34450, Turkey
gkusoglu16@ku.edu.tr, gizem.kusoglu@tupras.com.tr

Abstract

Delayed Coker Unit (DCU) converts the vacuum residual feedstock to lighter and more valuable products such as motor fuels and eliminates the low-order and environment-damaging streams. Thus, optimal operation of this unit provides great economic return. In this direction, we have modeled an industrial DCU which exists in the TUPRAS Refinery. The steady-state model consists of the furnace and coke drums and implemented using MATLAB. Physical properties are determined by Aspen HYSYS. The obtained model was used for predicting the coke level in the coke drums where the reaction takes place and the distribution of the products. Both furnace and coke drum models were verified by comparing obtained results with actual plant data.

Keywords: Delayed coking, furnace and coke drum models.

1. Introduction

Delayed Coker Unit (DCU) is a critical refinery unit that enables the conversion of heavy materials such as vacuum residue into valuable products. This conversion occurs in coke drums where thermal cracking reactions take place. In the literature, there are a limited number of studies about modeling of the unit. Among these studies, Lima et al. (2016) and Borges et al. (2015) have modeled coke drums as plug flow reactor (PFR) by preparing in-house codes or by using simulation programs such as Aspen HYSYS/PLUS. On the other hand, Bozzano and Dente (2005) and Zhou et al. (2007) focused on the estimation of product yields with kinetic models as well. With this study, separate models of the coke drums and the furnace were created and the content of the coke drum outlet stream was estimated close to the reality.

2. Process Description

The main purpose of Delayed Coker Unit is thermal cracking the Vacuum Residue (VR) feed into valuable products such as Coker Fuel Gas, LPG, Stabilized Naphtha, Light Coker Gas Oil (LCGO), Heavy Coker Gas Oil (HCGO), and Coke. A simplified diagram of the delayed coking section is shown in Fig. 1. It consists of a fractionator tower, four coke drums, and coke heaters. Fresh vacuum residue is preheated and enters the bottom of the main fractionator where it is combined with HCGO recycle. The combined fresh feed and recycle is pumped through the heaters in order to increase the temperature to a point at which coke formation occurs in the coke drums. As can be seen in Fig. 1, coking occurs only at one of the coke drum pairs, while the other drum is being cleaned and prepared for the next coking cycle. The mixture leaving the heaters passes through the coke drums where cracking reactions take place and solid coke accumulates. Product vapors obtained by cracking are returned to the bottom of the main fractionator above the fresh feed entry tray.

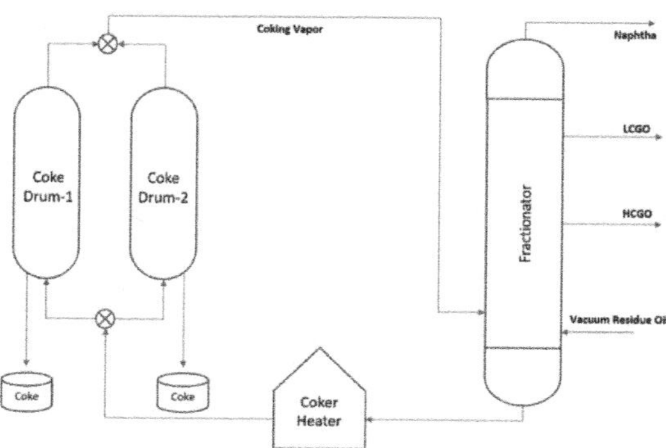

Figure 1. General representation of the Delayed Coker Unit (DCU)

3. Delayed Coker Furnace Model

The main role of the furnace is to increase the temperature of the VR feed before sending it to coke drums. VR enters into the furnace as a subcooled liquid and is transformed into a vapor-liquid mixture containing fresh vacuum residue and some products by some cracking reactions in the furnace tubes. While the two-phase stream flows through the furnace tubes, significant pressure drop can develop. In this study, the rigorous furnace model is generated based on the assumptions that there is two-phase flow with endothermic reaction under pressure drop effect, and there is no coke formation. For modeling of the coker furnace, the model structure proposed by Skogestad et al. (1986) has been used with Vapor-Liquid Equilibrium (VLE) for phase split calculations. It is basically based on the mass, energy and momentum conservation equations.

The mass balance is obtained for plug flow reactor (PFR) based on the kinetic model of 5 discrete lumps proposed by Singh et al. (2005). For each discrete lump, kinetic parameters are estimated based on feed and product properties by using the correlations proposed by Castellanos et al. (1991). The reactions only take place in liquid phase. Therefore, the reaction rate expressions have been derived by assuming first-order power-law approach for the PFR as following.

$$\frac{dF_i}{dz} = -r_i A_{liq} \tag{1}$$

where i = VR, HCGO, LCGO, Naphtha, Gas lumps; A_{liq}, liquid fraction in the tube (m^2); r_i, reaction rate of ith lump (kmol/sm^3); F_i, molar flowrate (kmol/s).

From the perspective of energy, it is assumed that temperature inside the tubes is only changing with convective heat transfer for getting rid of complex equations. Effects of the conduction and radiation are neglected. Resulting enthalpy balance is given in Eq. (2).

$$H(z) = H(0) + \int_0^z Q(z).\pi.D.dz \tag{2}$$

where Q is the heat duty (kW/m²); $H(0)$ enthalpy of the feed at the inlet of the furnace (kW). Pressure drop values along the tubes are being calculated based on three different terms such as frictional, acceleration and gravity gradients. Frictional pressure gradient is calculated with Martinelli correlation with the assumption that two phases can have different properties (Thome and Cioncolini, 2015). Gradient is basically defined as functions of liquid-only or gas-only pressure multiplier and liquid-only or gas-only pressure gradients. Two-phase flow frictional pressure gradient is calculated as in Eq. (3).

$$\left(\frac{dP}{dz}\right)_{fr} = \left(\frac{dP}{dz}\right)_{lo} \emptyset_{lo}^2 = \left(\frac{dP}{dz}\right)_{go} \emptyset_{go}^2 \tag{3}$$

where $\left(\frac{dP}{dz}\right)_{lo}$,single phase frictional pressure gradients for the liquid-only; $\left(\frac{dP}{dz}\right)_{go}$, single phase frictional pressure gradients for the gas-only; $\emptyset_{lo}$, two-phase liquid-only multiplier; $\emptyset_{go}$, two-phase gas-only multiplier.

Acceleration pressure drop gradient is explained on the basis of momentum density ρ_m(kg/m³) gradient and mass flux G(kg/m²s). Momentum density term is a function of liquid density ρ_L(kg/m³), gas density ρ_G(kg/m³), quality(x), void fraction (ε). Acceleration pressure drop gradient is computed as follows:

$$\left(\frac{dP}{dz}\right)_{acc} = G^2 \frac{d}{dz}\left(\frac{1}{\rho_m}\right) = G^2 \frac{d}{dz}\left[\frac{(1-x)^2}{\rho_L(1-\varepsilon)} + \frac{x^2}{\rho_G \varepsilon}\right] \tag{4}$$

where void fraction is calculated based on Lockhart and Martinelli correlation. The last term left for gradient pressure calculation is gravitational pressure drop which is expressed in terms of density and void fraction, such that (Thome and Cioncolini, 2015):

$$\frac{\Delta P_g}{\Delta z} = [\rho_L(1-\varepsilon) + \rho_G \varepsilon] g sin\theta \tag{5}$$

where $\frac{\Delta P_g}{\Delta z}$, gravitational pressure drop; θ, the angle of channel inclination from horizontal; g, acceleration gravity (9,81 m/s²).

4. Coker Drum Model

The VR feed enters into the coker drums as two phase composed of vapor and liquid and it exposes to the cracking reactions after a while. Cracking reactions result in vapor products and solid coke which means that there are three phases as coke, residue and distillate in the reactor. From this point of view, it can be modeled similar to the trickle bed reactor. Liquid vacuum residue phase is the primary phase and the gas distillate phase is secondary phase. The system can be modeled in the Eulerian-Eulerian approach. The coke phase is assumed as homogeneous porous medium and accumulates with time.

4.1.1. Material Balance

Vacuum residue is decomposed into gas, distillate products (HCGO, LCGO, Naphtha) and coke. Coke amount changes with time but does not have any positional dependency. Therefore, vacuum residue is assumed to crack only to gas and distillate for steady-state model and coke level is associated with void fraction. Reaction kinetic expression and parameters are calculated based on the correlation proposed by Esfahani et al. (2019). General balance and component-based mass balances are given in Eq. (6a-c).

$$\frac{d(\gamma\alpha_R\rho_R v_{z,R})}{dz} = \gamma\alpha_R\rho_R\Gamma_R \tag{6-a}$$

$$\frac{d(\gamma\alpha_D\rho_D v_{z,D})}{dz} = \gamma\alpha_R\rho_R\Gamma_D \tag{6-b}$$

$$\frac{d(\gamma\alpha_R\rho_R v_{z,R} Y_{i,R})}{dz} = \gamma\alpha_R \dot{m}_{i,R} \tag{6-c}$$

where i represents Gas, Distillate, Vacuum Residue lumps; γ, coke void fraction; α_R, residue volume fraction; α_D, distillate volume fraction; $v_{z,R}$, velocity of the residue phase (m/s); $v_{z,D}$, velocity of distillate phase(m/s);Γ, reaction velocity(1/s); $\dot{m}_{i,R}$, mass transfer of component i from another phase (kg/m^{3}s); $Y_{i,R}$, weight fraction of ith lump.

4.1.2. Momentum Balance

Even though the assumption of no mass transfer between phases, the phase interactions encountered in the trickle beds are involved. The momentum balance of each phase is provided in Eq. (7a-b).

$$\frac{d(\gamma\alpha_R\rho_R v_{z,R} v_{z,R})}{dz} = \gamma F_G - \gamma\alpha_R\frac{dP}{dz} + \gamma\alpha_R\rho_R g \tag{7-a}$$

$$\frac{d(\gamma\alpha_D\rho_D v_{z,D} v_{z,D})}{dz} = \gamma F_L - \gamma\alpha_D\frac{dP}{dz} + \gamma\alpha_D\rho_D g \tag{7-b}$$

where F_G, F_L, phase interaction terms (kg/m^{2}s^2); g, acceleration gravity (9,81 m/s^2); $\frac{dP}{dz}$, pressure drop (kg/m^{2}s^2); γ, coke void fraction. Phase interactions are calculated by the equations provided in the Alopaeus et al. (2006)'s study.

4.1.3. Energy Balance

Assuming that the coke, vacuum residue and product vapors are in close contact, the temperatures of the three phases are assumed to be the same throughout the reactor. Therefore, the energy balance is given as following.

$$\frac{d(\gamma\alpha_R\rho_R v_{z,R} C_{P,R} T + \gamma\alpha_D\rho_D v_{z,D} C_{P,D} T)}{dz} = \gamma\alpha_R\rho_R\Gamma_R\Delta\dot{H}_R + U\frac{A}{V_r}(T_C - T) \tag{8}$$

where $C_{P,R}$, $C_{P,D}$ are heat capacity of residue and distillate phases, respectively (J/kgK); $\Delta\dot{\boldsymbol{H}}_R$, heat of reaction (J/kg); $\boldsymbol{A}$, surface area(m2); $\boldsymbol{V_r}$, volume of the reactor(m3); U, overall heat transfer coefficient(W/Km^2); $\boldsymbol{T_C}$, temperature of surrounding(K); T, temperature of the fluid(K).

5. Results

The models are run for vacuum residue inlet stream of three different days that differed from each other in terms of density (API) and concarbon (CCR). API and CCR values are given in Table 1. The CCR content of the VR affects the yield of the products and so the accuracy of the model predictions was evaluated against the variation of the content.

Table 1. Properties of vacuum residue stream for three different days

Date	API	CCR
1st Day	4,35	23,55
2nd Day	6,50	22,00
3rd Day	4,80	18,38

Models were coded over MATLAB for only one dimension (z-direction). The balance equations of furnace model are solved sequentially until values of variables converge to a point. The coke model is computed by ode15s solver. All required properties (density, phase splits, heat capacity, enthalpy) of each lump are calculated by using Aspen HYSYS. Since the vacuum residue (VR) was first fed to the furnace, the furnace model was run first and then the coke drum model was triggered. The furnace model was run with the VR stream of the 1st day and the results in Fig. 2 were obtained. After that, the results of the heater model are used as the input of the coke drum model. The obtained coke drum model results for $\gamma = 0{,}7$ are plotted as in Fig. 3.

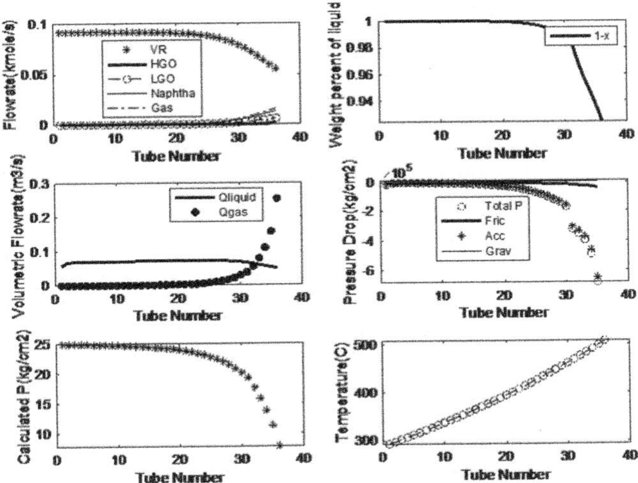

Figure 2. Results of the coker heater model for 1st Day

According to Fig. 2 and Fig. 3, it is seen that the estimated values of outlet pressures are quite close to the real outlet values. Real outlet values are represented as Actual T and Actual P in the following graph. The relative error values with respect to temperature, pressure and distillation weight fraction in Table 2 also support this. Despite varying CCR contents of VR, the maximum relative error is 15,77% with steady-state models.

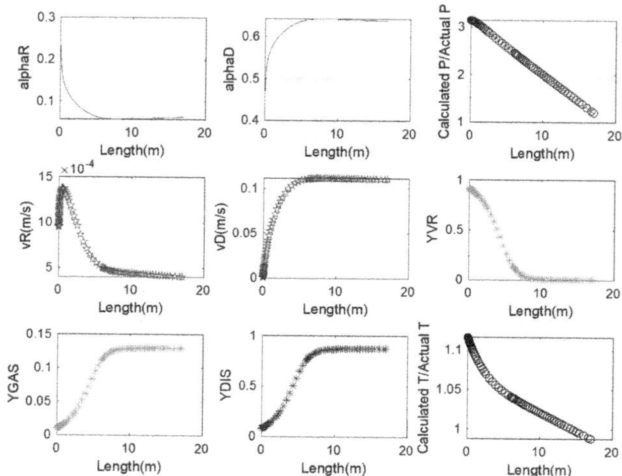

Figure 3. Results of the coke drum model for 1st Day (for $\gamma = 0.7$)

Table 2. Relative Error with respect to Coke Drum Outlet Conditions

Date	Abs Relative Error (%) Based on Temperature	Abs Relative Error (%) Based on Pressure	Abs Relative Error (%) Based on Y_{DIS}
1st Day	0,88	12	6,33
2nd Day	0,91	11	5,79
3rd Day	0,92	15,77	2,79

To obtain these results, correlations proposed in the literature for reaction kinetic parameters were used. On the other hand, only diameter of the coke particle was adjusted to capture the real results in the coke drum model. As shown in Fig. 4, the coke diameter increases with the coalescence of the coke particles while coke level in the drum increases.

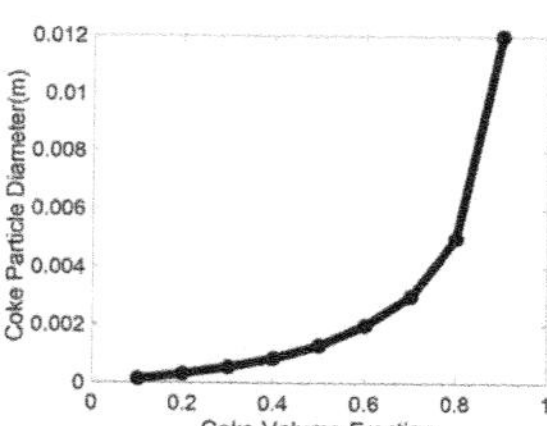

Figure 4. Coke diameter

6. Conclusion

Models of both coker heater and coke drums were developed to predict the performance of DCU and verified with the real data sets from Tupras Refinery. The coker heater model is designed as two-phase plug-flow reactor based on 5 discrete lumps (VR, HCGO, LCGO, Naphtha, Gas). The coke drums are assumed as the three-phase trickle-bed reactor with 3 discrete lumps (VR, Distillate, Gas). It has been observed that models running sequentially give realistic results regardless of both density and CCR content of the VR.

References

N.M.N. Lima, P. Sbaite, C. B. Batistella, R. M. Filho, M. R. W. Maciel, F. Manenti, M. Embiruçu, 2016, Simulation, Optimization and Control of Heavy Oil Upgrading Processes: Application to a Delayed Coking Unit, Computer Aided Chemical Engineering, Elsevier, 38, 1671-1676.

C. N. Borges, M. A. Mendes, R. M.B. Alves, 2015, Mathematical Modeling of an Industrial Delayed Coking Unit, Computer Aided Chemical Engineering, Elsevier, 37, 515-520.

G. Bozzano, M. Dente, 2005, A mechanistic approach to delayed coking modelling, Computer Aided Chemical Engineering, Elsevier, 20, 2005, 529-534.

X.L. Zhou, S.Z. Chen, C. L. Li, 2007, A Predictive Kinetic Model for Delayed Coking, Petroleum Science and Technology, 25, 1539-1548.

S. Skogestad, T. Gundersen, O. Johnsen, 1986, Compositional simulation of a refinery coker furnace-An industrial example of two-pase with chemical reaction, Modeling, Identification and Control, 7, 25-44.

J. Singh, M.M. Kumara, A.K. Saxena, S. Kumar, 2005, Reaction pathways and product yields in mild thermal cracking of vacuum residues: A multi-lump kinetic model, Chemical Engineering Journal, 108,239-248.

J. Castellanos, J.L. Cano, R. Del Rosal, V.M. Briones, R.L. Mancilla, 1991, Kinetic model predicts visbreaker yields, Oil & Gas Journal, 89, 76-82.

J. R. Thome and A. Cioncolini, 2015, Encyclopedia of two-phase heat transfer and flow, Set 1: Fundamentals and Methods, Volume 3: Flow Boiling in Macro and Microchannels, World Scientific Publishing, Chapter 6: Two-Phase Pressure Drop.

F. T. Esfahani, M.R. Ehsani, J. Ivakpour, 2019, A Kinetic Model for Delayed Coking Process of Iranian Vacuum Residues, Petroleum Science and Technology, 37, 2049-2057.

V. Alopaeous, K. Hynynen, J. Aittamaa, M. Manninen, 2006, Modeling of Gas-Liquid Packed-Bed Reactors with Momentum Equations and Local Interaction Closures, Industrial Engineering Chemistry Research, 2006, 45, 8189-8198.

Antonis Kokossis, Michael C. Georgiadis, Efstratios N. Pistikopoulos (Eds.)
PROCEEDINGS OF THE 33rd European Symposium on Computer Aided Process Engineering (ESCAPE33), June 18-21, 2023, Athens, Greece
 http://dx.doi.org/10.1016/B978-0-443-15274-0.50085-8

A sustainable energy supply system in smart cities: an integrated natural gas and electricity network operation

Ali Ahmadian,[a] Ali Almansoori,[b] Ali Elkamel,[c]

[a]*Department of Electrical Engineering, University of Bonab, Bonab, Iran.*
[b]*Department of Chemical Engineering, Khalifa University, Abu Dhabi 127788, United Arab Emirates*
[c]*Department of Chemical Engineering, University of Waterloo, 200 University Avenue West, Waterloo, ON, N2L 3G1, Canada*

Abstract

In this paper, a stochastic approach is presented for optimal planning of an integrated natural gas and electricity network. The planned network can play an important role in the smart cities in order to reduce the emission pollution and operation cost. In addition, various resource and elements of a sustainable green energy system including renewable energy units and energy storage are modeled. A comprehensive investigation is carried out based on different scenarios and the extracted simulation results are discussed. In order to guarantee the global optimum results, the modeling equations of the studied problem have been linearized and the problem became to a mixed-integer linear programming (MILP). The linearized problem optimized using a mathematical based optimization algorithm.

Keywords: Sustainable green energy, integrated energy networks, smart cities.

1. Introduction

Today's, the sustainable green energy systems play an important role in the smart cities. Reducing the emission pollution and the operation cost of energy systems is one of the main purpose of the sustainable cities' planner. Several ideas have been proposed in the literature in order to optimal management of energy systems. Optimal energy management in a micro-grid [1], optimal management of electric vehicles charging demand in terms of smart charging [2], optimal energy storage planning and operation in the energy systems [3], etc. are the most famous approaches that are presented in the literature. Although the optimal energy management in the electric network could reduce the operation cost, it can be more beneficial if the electric grid and the natural gas are integrated together [4]. Hence, the integrated energy networks are introduced to optimal operation and planning of both natural gas and electricity networks, simultaneously. In order to study the integrated network, the energy systems' elements including the power plants, gas fired units, gas storage, energy storage, renewable energy resources, etc. should be modeled mathematically. In addition, the power flow in the electricity network and the gas flow in the natural gas network should be modeled and their constraints should be considered.

In this paper, an approach is presented for optimal planning of integrated electricity and gas networks. In addition, the stochastic parameters are modeled based on the Copula method which can model the stochastic parameters more efficiently than other stochastic methods such as Monte Carlo. Furthermore, to decrease the computational burden of the scenario-based method the Kantorovich method is utilized. A comprehensive study by modeling different resources such as renewable energy resources, and different storages are presented. Based on this method, the system operator can get the most out of these resources. The optimization problem turns into a mixed-integer linear programming (MILP) approach by using appropriate methods which can guarantee the global optimal solution.

Nomenclature	
Indices	
c	CHP units index
e	Emission type index
n	Index of electric and gas storages
g	Gas supplier index
u	Unit Index
gs	Gas storages index
p, q	Gas network nodes indices
s	Scenarios index
t	Time index
Parameters	
a_u, b_u, c_u	Unit's cost coefficients
EC_e	Externality cost of each emission type ($/lb)
$EF_{u,e}$	Emission factor of each unit (lb/MWh)
$ES_n^{CH,MAX}$, $ES_n^{DIS,MAX}$	Maximum rate of charging and discharging in storages
$H_c^A, H_c^B, H_c^C, H_c^D$	Coefficients related to the amount of produced heat by CHP units (MW)
$HD_{s,t}$	Total heat demand (MW)
$P_c^A, P_c^B, P_c^C, P_c^D$	Coefficients related to the amount of produced power by CHP units (MW)
P_u^{MIN}, P_u^{MAX}	Minimum and maximum generated active power by units (MW)
GP_p^{MIN}, GP_p^{MAX}	Minimum and maximum pressure limit of nodes (Psig)
Q_u^{MIN}, Q_u^{MAX}	Minimum and maximum generated reactive power by units (MVar)
SU_u, SD_u	Startup and shutdown cost of each unit ($)
$\eta_n^{CH}, \eta_n^{DIS}$	The efficiency of charging and discharging in storages
λ_g^G	Gas price ($/kcf)
π_s	Probability of each scenario

Variables	
$GF_{p,q,s,t}$	Gas flow in pipelines (kcf)
$G^{G}_{g,s,t}$	Produced gas by gas suppliers (kcf)
$P_{u,s,t}, Q_{u,s,t}$	Generated active (MW) and reactive (MVar) power by each unit
$ES^{CH}_{n,s,t}, ES^{DIS}_{n,s,t}$	Amount of storages charge and discharge
$GP_{p,s,t}$	Pressure in nodes of the gas network (Psig)
$SOC_{n,s,t}$	Stored power (MWh) or gas (kcf) in storages
$U_{u,s,t}$	Binary variable for on/off of units
$U^{CH}_{n,s,t}, U^{DIS}_{n,s,t}$	Binary variables to control charging and discharging states

2. Problem Formulation

An integrated electricity and gas network as shown in Fig. 1 is considered in this paper.

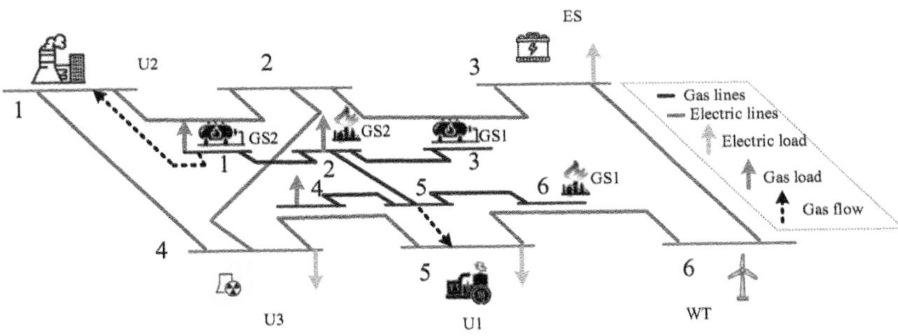

Fig. 1: overall structure of the proposed method.

This study aims to minimize the total cost of both the electricity and gas networks together as expressed in equation (1). This cost consists of three terms: the first term is the emission cost of fossil-fueled based units, the second term explains the operation cost of units to generate electricity, and the third term is related to the cost of natural gas.

$$\min \sum_s \pi_s \sum_t \Big(\underbrace{\sum_e \sum_u EC_e \, EF_{u,e} P_{us,t}}_{Emission\ Cost} + \underbrace{\sum_u \left(a_u P_{u,s,t}{}^2 + b_u P_{u,s,t} + c_u + SU_u U_{u,s,t}(1 - U_{u,s,t-1}) + SD_u U_{u,s,t-1}(1 - U_{u,s,t}) \right)}_{Operation\ Cost\ of\ Generation\ units} + \underbrace{\sum_g \lambda^G_g G^G_{g,s,t}}_{Gas\ Cost} \Big) \quad (1)$$

Constraints related to the generation units are described in equations (2) and (3); these constraints limit the total active and reactive power generated by units. The minimum up and down time limits and ramp rate of units are modeled based on reference [5].

$$P^{MIN}_u U_{u,s,t} \le P_{u,s,t} \le P^{MAX}_u U_{u,s,t} \quad (2)$$

$$Q^{MIN}_u U_{u,s,t} \le Q_{u,s,t} \le Q^{MAX}_u U_{u,s,t} \quad (3)$$

For the CHP units which are responsible for satisfying heat demand besides the above constraints, constraints are investigated to control the amount of generated heat by these units [6]. Moreover, in order to model the electric and gas storages equations are used. Linear power flow equations are used to check the electric network constraints based on reference as presented in [7]. To meet the limitation of the gas network, the presented equations in [8] are utilized.

2.1. Equation linearization

In the equations of the system, there are some equations that make the problem a mixed-integer nonlinear programming (MINLP). Solving MINLP problems has a high computational cost. In addition, it is possible to stick to local optimal points. To overcome these issues, the appropriate linearization methods are utilized. For linearization, the multiplication of binary variables, and continuous problems the methods presented in [9] are used. For the linearization of quadratic operation cost of generation units, the piecewise linear method is utilized [10]. Finally, the Weymouth equation is linearized.

3. Numerical results and discussion

3.1. Input data

To show the effectiveness of the proposed method the electric-gas network which is shown in Fig. 1 is implemented. AP^{DR} is considered as 10%. η_n^{CH} and η_n^{DIS} are 0.8. Emission data are considered based on [7]. Generation units constraints are presented in references [4] that show the parameters of the gas network.

3.2. Uncertainty Modeling

In this paper, the Copula method is investigated for modeling uncertainties. The advantage of copula against other generative methods such as Monte Carlo is its ability to consider the relationship between different features.

3.3. Simulation Results

The results after the implementation of the proposed method are shown in figures 2-5 and Table 1. Fig. 2 shows the generated power by three units. The first unit which is responsible for heat demand too generated its maximum power during the day. However, unit 3 is not operated at all hours because the emission cost is implemented for units in this study; in this regard, unit 3 is not economical. Unit 2 produces its maximum power in the middle of the day when we have the maximum power and gas demands.

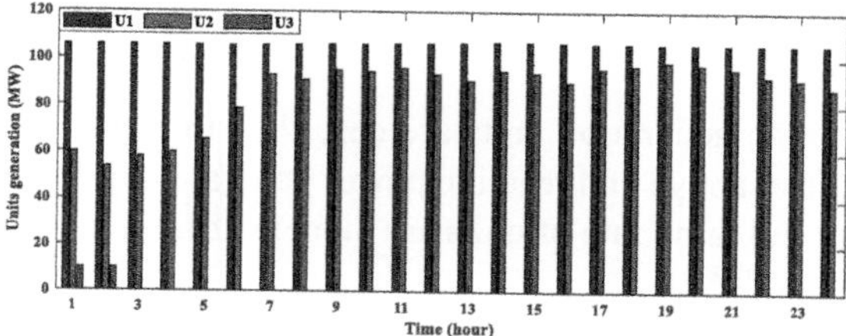

Fig. 2: Generated power by different units.

Fig. 3 demonstrates the total gas demand in the network which is the sum of gas demand and consumed gas by units. This gas demand is supplied by two gas suppliers as shown in Fig. 1, the amount of generated gas by two units are presented in Fig. 4.

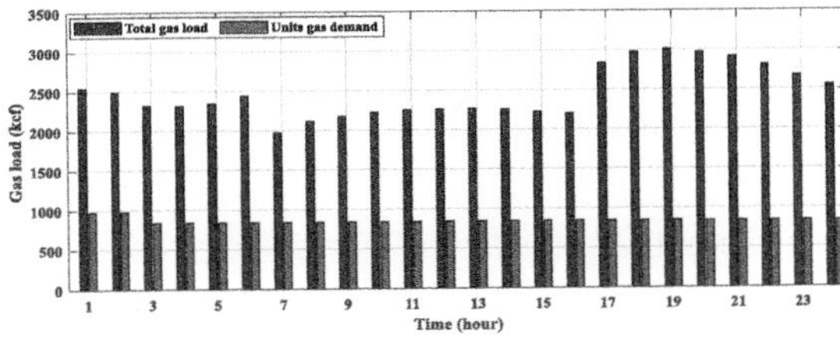

Fig. 3: Total gas demand.

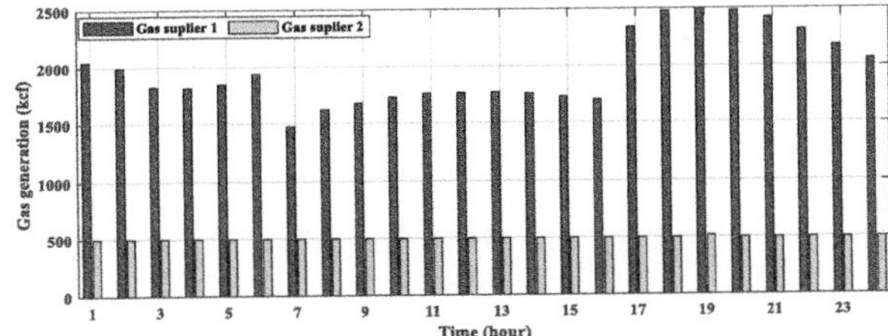

Fig. 4: The amount of generated gas by gas suppliers.

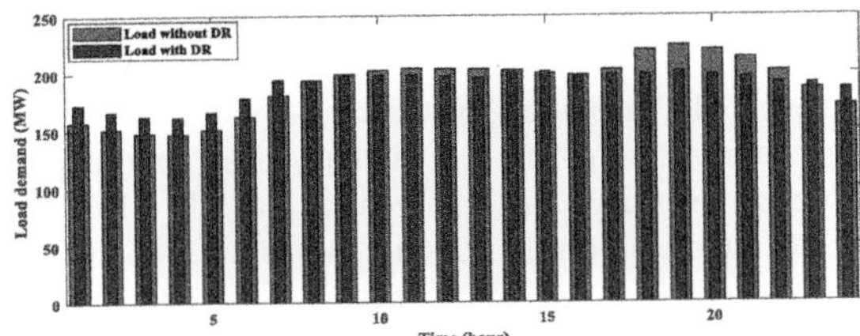

Fig. 5: Electric demand before and after implementation of DRP.

To better clarify the results the detail of the results is presented in Table 1 for each source.

Table 1: Different costs of integrated electricity and gas networks.

Parameters	Value of Cost ($)
Unit 1	20483
Unit 2	21519
Unit 3	286
Gas cost	154106
Emission cost	4821
Total cost	201215

4. Conclusion

In this paper, an integrated natural gas and electricity network was investigated. The elements and resources of the both networks were modeled mathematically and the power flow in the electricity network and the gas flow in the natural gas network were modeled. A mixed integer linear programming (MILP) approach is presented to optimal planning and operation of the studied integrated network. The simulation results verified the effectiveness and robustness of the proposed methodology in reducing of the emission pollution and operation cost.

References

[1] Thirunavukkarasu, Gokul Sidarth, Mehdi Seyedmahmoudian, Elmira Jamei, Ben Horan, Saad Mekhilef, and Alex Stojcevski. "Role of optimization techniques in microgrid energy management systems—A review." Energy Strategy Reviews 43 (2022): 100899.

[2] Sadeghian, Omid, Arman Oshnoei, Behnam Mohammadi-Ivatloo, Vahid Vahidinasab, and Amjad Anvari-Moghaddam. "A comprehensive review on electric vehicles smart charging: Solutions, strategies, technologies, and challenges." Journal of Energy Storage 54 (2022): 105241.

[3] Sedghi, Mahdi, Ali Ahmadian, and Masoud Aliakbar-Golkar. "Optimal storage planning in active distribution network considering uncertainty of wind power distributed generation." IEEE Transactions on Power Systems 31, no. 1 (2015): 304-316.

[4] AlHajri I, Ahmadian A, Elkamel A. Techno-economic-environmental assessment of an integrated electricity and gas network in the presence of electric and hydrogen vehicles: A mixed-integer linear programming approach. J Clean Prod 2021:128578.

[5] Sadeghi S, Jahangir H, Vatandoust B, Golkar MA, Ahmadian A, Elkamel A. Optimal bidding strategy of a virtual power plant in day-ahead energy and frequency regulation markets: A deep learning-based approach. Int J Electr Power Energy Syst 2021;127:106646. https://doi.org/10.1016/j.ijepes.2020.106646.

[6] Gougheri SS, Dehghani M, Nikoofard A, Jahangir H, Golkar MA. Economic assessment of multi-operator virtual power plants in electricity market: A game theory-based approach. Sustain Energy Technol Assessments 2022;53:102733.

[7] Gougheri SS, Jahangir H, Golkar MA, Ahmadian A, Aliakbar Golkar M. Optimal participation of a virtual power plant in electricity market considering renewable energy: A deep learning-based approach. Sustain Energy, Grids Networks 2021;26:100448. https://doi.org/10.1016/J.SEGAN.2021.100448.

[8] Jiang Y, Xu J, Sun Y, Wei C, Wang J, Liao S, et al. Coordinated operation of gas-electricity integrated distribution system with multi-CCHP and distributed renewable energy sources. Appl Energy 2018;211:237–48.

[9] Gougheri SS, Jahangir H, Golkar MA, Moshari A. Unit Commitment with Price Demand Response based on Game Theory Approach. 2019 Int. Power Syst. Conf., IEEE; 2019, p. 234–40.

[10] Luo S, Yang L, Zhang X, Chen W, Wang K, Xu Y. A fully linear-constrained optimal electricity-gas flow in an integrated energy system. 2018 2nd IEEE Conf. Energy Internet Energy Syst. Integr., IEEE; 2018, p. 1–6.

Antonis Kokossis, Michael C. Georgiadis, Efstratios N. Pistikopoulos (Eds.)
PROCEEDINGS OF THE 33rd European Symposium on Computer Aided Process Engineering (ESCAPE33), June 18-21, 2023, Athens, Greece
 http://dx.doi.org/10.1016/B978-0-443-15274-0.50086-X

A hybrid forecasting framework for the energy sector

Stefanos G. Baratsas,[a,b] Funda Iseri,[a,b] Efstration N. Pistikopoulos[a,b]

[a]*Artie McFerrin Department of Chemical Engineering, Texas A&M University, College Station, TX 77843, United States*

[b]*Texas A&M Energy Institute, Texas A&M University, College Station, TX 77843, United States*

Abstract

Energy prices are nonlinear, noisy, non-stationary time-series, which are sensitive and volatile to supply and demand imbalances, changes in monetary and fiscal policies, new environmental regulations, major political and social events, as well as technological breakthroughs. Despite the fact that numerous methods have been proposed in the literature to forecast the prices of the several types and sources of energy, there is no single forecasting method that systematically performs better over the other methods for any given data set. To this respect, Energy Price Index (EPIC) extended forecasting capabilities are presented here. Statistical and machine learning forecasting methods of different nature have been incorporated into the framework, enabling the forecasting of energy prices up to 14 months in the future. The historical prices of each energy product are studied independently, so as to identify seasonality, patterns, trends and outliers. The results reveal the superiority of the deep neural networks, especially those with feedback loops, over statistical forecasting methods.

Keywords: forecasting framework, energy prices, machine learning, hyperparameters tuning.

1. Introduction

A time-series is a sequence of observations made over a period of time at equal time intervals. An inherent characteristic of time-series is that adjacent observations are often dependent. Determining the probability law that governs a time-series is a crucial task towards unraveling the underlying dynamics, and eventually forecasting future events.

Traditional methods such as autoregressive integrated moving average (ARIMA) (Box et al., 2015) and exponential smoothing (Gardner et al., 1985; Winters, 1960), focus mainly on parametric models which are approximations of the real stochastic dynamics that govern a data set. The increase in data availability and computing power over the last decades led to a new class of methods that are capable of identifying nonlinear patterns, learning by trial-and-error complex data structures, and improving their performance over time, without requiring the underlying knowledge or an analytical model (Montgomery et al., 2015). As such, various machine learning (ML) methodologies have been proposed for a variety of applications (Lim et al., 2021). However, the prevalence of the more complicated ML methodologies in fitting and learning the temporal dynamics of the available data, does not necessarily lead to more accurate forecasts (Makridakis et. al.,

2018). To this respect, hybrid models that combine both statistical and ML methodologies have illustrated superior performance against their individual components in a range of applications (Smyl, 2020).

In this work, we extend the forecasting capabilities of the Energy Price Index (EPIC) (Baratsas et al., 2021, 2022) with the incorporation of 35 statistical and machine learning forecasting methods of different natures, enabling the forecasting of the energy prices up to 14 months in the future. Since EPIC consists of 56 different energy products with 33 unique time-series data, the extended forecasting framework is applied to each one of these univariate time-series individually and the forecasting model that demonstrates the most accurate forecasting results over a testing period is then utilized for the future forecasts.

2. Hybrid Forecasting Framework for the Energy Sector

The large number of unique time-series that demonstrate unique patterns, trends, and cycles and need to be predicted for different forecasting horizons, introduce major challenges in the forecasting process. Figure 1 highlights these unique characteristics on a subset of the time-series.

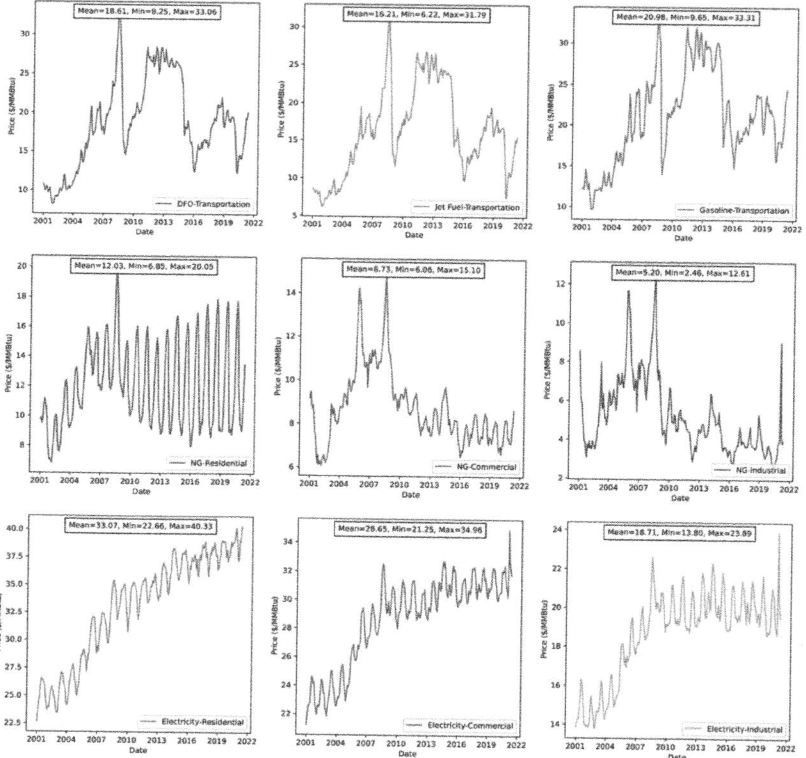

Figure 1: 20 years of monthly prices of the 9 energy products with the highest energy demand in 2017-2021

The 33 time-series are monthly indexed with data sets of 161 to 425 months. The forecasting framework incorporates 35 statistical and ML forecasting methods into 7 distinct groups, which are summarized in Table 1. The first forecasting group *"Benchmark Forecasting"* consists of five rather simple but quite effective methods (Hyndman, 2018). They are used to provide a benchmark on forecasting performance against the rest of the forecasting groups which are more advanced and computationally expensive. The second forecasting group *"Forecasting with Decomposition"* utilizes STL decomposition (Gardner et al., 1985) and then the forecasts of the STL objects are obtained by applying a non-seasonal forecasting method to the seasonally adjusted data and re-seasonalizing using the last year of the seasonal component. The third forecasting group *"Exponential Smoothing"* (Winters, 1960; Bergmeir et. al., 2016) consists of nine forecasting methods, where forecasts are weighted averages of past observations with the weights though decaying exponentially for the older observations. A complementary approach to exponential smoothing is the fourth forecasting group, namely "ARIMA", which combines autoregressive and moving average models, while allowing differencing of the data series. Seasonal and non-seasonal ARIMA models ($ARIMA_{(p,d,q)(P,D,Q)m}$), are used. Due to the large number of hyperparameters to be determined, grid search is conducted in the last two methods (fc4d, fc4e) and the configuration with the smallest AICc and AIC value (Cleveland et. al, 1990) respectively is selected. The fifth forecasting group consists of advanced statistical forecasting methods, including dynamic regression and TBATS models (Cleveland et. al, 1990; Crone et. al., 2010), along with bootstrapped time series that use the Box-Cox and Loess-based decomposition bootstrap (De Livera et. al, 2011).

Table 1: Hybrid forecasting framework for energy prices

Group	Forecasting Method		Description of Forecasts
Benchmark Forecasting	fc1a	Average	based on the historical mean
	fc1b	Random Walk (RW) – Naïve	naïve w/ RW model
	fc1c	Seasonal Naïve	ARIMA(0,0,0)(0,1,0) model
	fc1d	RW w/ Drift	RW w/ drift
	fc1e	RW w/ Drift Transformed & Bias Adjusted	fc1d adj Box-Cox transformation
Forecasting with Decomposition	fc2a	Automated STL	automated STL decomposition
	fc2b	ARIMA Automated STL	ARIMA seasonally adjusted data
	fc2c	Naive Automated STL	naïve seasonally adjusted data
	fc2d	RW w/ Drift Automated STL	RW w/ drift seasonally adjusted data
	fc2e	ETS Automated STL	STL decomposition, w/ ETS seasonally adjusted back Box-Cox transformed data
Exponential Smoothing	fc3a	Simple Exponential Smoothing	
	fc3b	Holt's Damped trend	
	fc3c	Holt-Winters' (HW) Additive	
	fc3d	HW Multiplicative	
	fc3e	HW Damped Additive	
	fc3f	HW Damped Multiplicative	
	fc3g	Error, Trend, Seasonal (ETS)	exponential smoothing state space

	fc3h	ETS Transformed Bias Adjusted	fc3g w/ adj Box-Cox transform
	fc3i	Bagged ETS	ETS to all bootstrapped series
ARIMA	fc4a	Manual ARIMA	manually selected terms
	fc4b	Auto. Arima Non-Seasonal	automated non-seasonal ARIMA
	fc4c	Auto. Arima Seasonal	automated seasonal ARIMA
	fc4d	Grid Search Arima	grid searching of hyperparameters
	fc4e	Grid Search Arima	grid searching of hyperparameters
Advanced Forecasting Methods	fc5a	Dynamic Regression model	w/ Fourier terms
	fc5b	TBATS	
	fc5c	Bootstrapping ARIMA	Box-Cox & Loess-based decomposition bootstrap
Neural Networks (NN)	fc6a	Feed-forward NN	single hidden layer (hl), lagged inputs
	fc6b	MLP_1	MLP w/ 5 hl & 20 training reps
	fc6c	MLP_2	MLP w/ 5 hl & 50 training reps
	fc6d	MLP_3	fc6c w/ 12 autoregressive lags
Grid Search NN	fc7	MLP	MLP model w/ 2 hl, followed by corresponding dropout layers, and a final output layer
	fc8	RNN	RNN model w/ 2 hl layers and a final output layer
	fc9	LSTM	LSTM model w/ a single LSTM hl, followed by a dropout layer, a dense fully connected layer followed by a dropout layer, and a final output layer
	fc10	CNN-LSTM-DNN	Hybrid CNN-LSTM model w/ 2 convolutional layers for 1D inputs, followed by a pooling & a flattened layer, 2 LSTM hl, and a final output layer

The last two forecasting groups consist of ML methods. In the sixth forecasting group "NN", the hyperparameters of the models are pre-selected, while in the seventh forecasting group "Grid Search NN", a grid search is applied for tuning the hyperparameters. More details about the grid searches are given in Table 2. In the last forecasting group, each configuration is evaluated 3 times. The average of these values for each accuracy measure is considered as the final one.

Table 2: Hyperparameters tuning through grid search

#	Details	Hyperparameters
fc4d	Grid Search Arima	p=[1...5], d=1, q=[1...3], P=[1...5], D=[0...1], Q=[1...3], m=18
fc4e	Grid Search Arima	
fc7	Dense class of Keras API v2.4.3 for Python 3.9 with TensorFlow v2.5.0	inp=[h, 2h], nod=[32, 64], Dropout rate = [0.1], Learning rate = [1e-4], epoch=[250], batch=[8, 16], Differences = [0, 1, h], Standardization = [T, F]
fc8	RNN and Dense class of Keras API v2.4.3 for Python 3.9 with TensorFlow v2.5.0	
fc9	LSTM and Dense class of Keras API v2.4.3	

	for Python 3.9 with TensorFlow v2.5.0	
fc10	Dense class of Keras API v2.4.3 for Python 3.9 with TensorFlow v2.5.0	nodes=[64, 128], Learning rate = [1e-4, 1e-6], epochs=[500, 1000], batches=[64], seq=[3] Steps: # of timesteps within each subsequence, steps=[12], filters=[128, 256], ker=[3, 6]

3. Data Preprocessing and Evaluation Metrics

Each time-series is preprocessed to achieve stationarity in its mean and variance, and is decomposed so as to extract time-series patterns such as trend, seasonality and cycles. This is done using Seasonal and Trend decomposition using Loess (STL), Box-Cox transformation, first order, and seasonal differencing so as to remove trend and seasonality. The residuals between the historical data and the corresponding fitted values, should be uncorrelated and have zero mean. The data sets are split into 80%-20% sets, with the first 80% of the data used for training/validating the models, and the last 20% used for testing their forecasting accuracy. Three accuracy measures are used in this paper: Root Mean Squared Error (RMSE), symmetric Mean Absolute Percentage Error (sMAPE) and Mean Absolute Error (MAE), and are defined as follows:

$$RMSE = \sqrt{\frac{1}{h}\sum_{t=1}^{h}(y_t - \hat{y}_t)^2} \qquad sMAPE = \frac{2}{h}\sum_{t=1}^{h}\frac{|y_t - \hat{y}_t|}{|y_t| + |\hat{y}_t|} * 100\%$$

$$MAE = \frac{1}{h}\sum_{t=1}^{h}|y_t - \hat{y}_t|$$

where h is the forecasting horizon, y_t are the historical data, and $\hat{y}_t$ are the forecasts produced by the model at point i.

4. Results

The forecasting framework is applied to each one of the 33 time-series. In each case, the forecasting model with the highest forecasting accuracy based on the three accuracy measures is selected for the future forecasts. If different models result from the accuracy measures, then the following process is followed: i) if the same model results from two accuracy measures, this one is selected, ii) if a different model results from each accuracy measure, then the model that has the highest overall ranking among all three accuracy measures is selected e.g. fc3a: 1st RMSE, 2nd sMAPE, 3rd MAE. In case there is still no clear winner, the model with the lowest RMSE is selected. Table 3 summarizes the forecasting results for the 33 time-series.

Table 3: Best forecasting models and accuracy measures

Energy Product	Selected Method	Average RMSE ($/MMBtu)	Average sMAPE (%)	Average MAE ($/MMBtu)
Distillate Fuel Oil (DFO) [R]	fc7	1.436	4.760	1.002
Kerosene [R, C, I]	fc9	3.403	8.982	1.962
Hydrocarbon Gas Liquids (HGL) [R]	fc3	1.430	5.003	1.111
DFO [C, I]	fc7	1.025	4.262	0.775
HGL [C, I, T]	fc7	1.014	4.241	0.767
Gasoline [C, I]	fc8	1.050	4.130	0.824
Petroleum Coke [C, I]	fc9	0.232	9.792	0.194
Residual Fuel Oil (RFO) [C, I, T]	fc8	0.708	5.856	0.545

Asphalt and Road Oil [I]	fc9	1.069	7.492	0.834
Lubricants [I, T]	fc9	1.044	7.443	0.829
Other Petroleum Products [I]	fc7	1.085	6.601	0.817
Aviation Gasoline [T]	fc9	1.668	4.982	1.229
DFO [T]	fc8	0.836	3.857	0.663
Jet Fuel [T]	fc9	1.071	5.765	0.771
Motor Gasoline [T]	fc9	1.019	3.950	0.813
Geothermal Energy [R, C, I]	fc9	0.527	0.510	0.125
Solar [R]	fc9	0.484	0.280	0.157
Biomass Energy [R, C, I]	fc9	0.176	0.299	0.082
Hydroelectric Power [C, I]	fc7	0.468	1.601	0.206
Solar [C, I]	fc7	1.035	0.997	0.349
Wind Energy [C, I]	fc8	0.088	0.365	0.043
Biomass (Fuel-Ethanol), [C, I, T]	fc9	1.094	3.719	0.856
Biomass (Bio-Diesel) [T]	fc9	1.444	5.064	1.165
Natural Gas, [R]	fc8	0.596	4.183	0.467
Natural Gas, [C]	fc8	0.263	2.685	0.204
Natural Gas, [I]	fc8	0.891	8.846	0.393
Natural Gas, [T]	fc9	0.611	4.223	0.401
Electricity [R]	fc10	0.551	1.125	0.430
Electricity [C]	fc4a	0.651	1.115	0.354
Electricity [I]	fc8	0.810	1.915	0.389
Electricity [T]	fc2d	0.664	1.855	0.532
Coal, [C]	fc9	0.059	0.953	0.042
Coal, [I]	fc9	0.037	0.704	0.027

R: Residential, C: Commercial, I: Industrial, T: Transportation

The above results clearly demonstrate the superiority of the last forecasting group "Grid Search NN" over the other forecasting groups, since 31 out of the 33 energy products are modeled more accurately using one of its four forecasting methods. In particular, LSTM is the most accurate method since it is used for 15 energy products, followed by RNN, MLP, and CNN-LSTM-DNN which are used for 9, 6, and 1 energy products respectively. Just two models are modeled using forecasting with decomposition and ARIMA groups. It is worth mentioning, that even without considering the last forecasting group, the NNs of the sixth forecasting group would only be used for 5 out of 33 energy products. Thus, NNs require special attention and proper tuning for forecasting time-series.

5. Conclusions

The historical prices of each energy product are analyzed independently considering seasonality, patterns, trends, cycles and outliers. Each time-series is pre-processed to establish stationarity in the mean and variance before being split into training and testing datasets. Then, the forecasting framework that consists of 35 statistical and ML forecasting methods is applied to each time-series individually, and the forecasting model that produces the best accurate results throughout a testing period is used for predictions up to 14 months in the future. The results highlight the critical importance of accurately tuning a NN's hyperparameters and demonstrate that deep NNs, especially those with feedback loops, perform better than statistical forecasting techniques.

6. Acknowledgments

Financial support from Texas A&M Energy Institute is also gratefully acknowledged.

References

Baratsas, S. G., Niziolek, A. M., Onel, O., Matthews, L. R., Floudas, C. A., Hallermann, D. R., et al., 2021. A framework to predict the price of energy for the end-users with applications to monetary and energy policies. Nature Communications 12, 1–12.

Baratsas, S. G., Niziolek, A. M., Onel, O., Matthews, L. R., Floudas, C. A., Hallermann, D. R., et al., 2022. A novel quantitative forecasting framework in energy with applications in designing energy-intelligent tax policies. Applied Energy 305, 117790.

Bergmeir, C., Hyndman, R. J. and Benítez, J. M., 2016. Bagging exponential smoothing methods using STL decomposition and box–cox transformation, International journal of forecasting, vol. 32, no. 2, pp. 303–312.

Box, G. E., Jenkins, G. M., Reinsel, G. C., & Ljung, G. M., 2015. Time series analysis: forecasting and control. John Wiley & Sons.

Cleveland, R. B., Cleveland, W. S., McRae, J. E. and Terpenning, I., 1990. Stl: A seasonal-trend decomposition, J. Off. Stat, vol. 6, no. 1, pp. 3–73.

Crone, S. F., and Kourentzes, N., 2010. Feature selection for time series prediction–a combined filter and wrapper approach for NN, Neurocomputing, vol. 73, no. 10-12, pp. 1923–1936.

De Livera, A. M., Hyndman, R. J., and Snyder, R. D., 2011. Forecasting time series with complex seasonal patterns using exponential smoothing, Journal of the American statistical association, vol. 106, no. 496, pp. 1513–1527.

Gardner Jr, Everette S., 1985. Exponential smoothing: The state of the art. J. Forecasting 4, 1–28

Géron, A., 2019. Hands-on machine learning with Scikit-Learn, Keras, and TensorFlow: Concepts, tools, and techniques to build intelligent systems. O'Reilly Media.

Goodfellow, I., Bengio, Y., and Courville, A., 2016. Deep learning. MIT press.

Hyndman, R. J., and Athanasopoulos, G., 2018. Forecasting: principles and practice. OTexts.

Lim, B., & Zohren, S., 2021. Time-series forecasting with deep learning: a survey. Philosophical Transactions of the Royal Society A 379, 20200209.

Makridakis, S., Spiliotis, E., & Assimakopoulos, V., 2018. Statistical and machine learning forecasting methods: Concerns and ways forward. PloS one 13, e0194889.

Montgomery, D. C., Jennings, C. L., & Kulahci, M., 2015. Introduction to time series analysis and forecasting. John Wiley & Sons.

Smyl, S., 2020. A hybrid method of exponential smoothing and recurrent neural networks for time series forecasting. International Journal of Forecasting 36, 75–85.

Winters, P. R., 1960. Forecasting sales by exponentially weighted moving averages. Management Science 6, 324–342

Antonis Kokossis, Michael C. Georgiadis, Efstratios N. Pistikopoulos (Eds.)
PROCEEDINGS OF THE 33rd European Symposium on Computer Aided Process Engineering (ESCAPE33), June 18-21, 2023, Athens, Greece
 http://dx.doi.org/10.1016/B978-0-443-15274-0.50087-1

A Chance-Constrained Programming Approach to Optimal Planning of Low-carbon Transition of a Natural Gas Supply System: A Case Study of China

Jiaqi Zhang [a], Pei Liu*, Zheng Li

[a] *Tsinghua BP Clean Energy Research and Education Centre, Department of Energy and Power Engineering, Tsinghua University, Beijing 100084, China*
liu_pei@tsinghua.edu.cn

Abstract

Low-carbon transition of energy systems is an inevitable trend to address climate change challenges, featuring more natural gas consumption to replace coal and oil. Planning of a natural gas supply system with multiple supply sources, end-consumers, and large infrastructures are challenging tasks. Uncertainty and spatial and seasonal mismatch of natural gas supply and demand makes the natural gas supply infrastructure planning problem more complex. In this study, a chance-constrained programming approach to optimal planning of low-carbon transition of a natural gas supply system is presented. This approach considers the uncertainties of end-consumer demand, pipe aging and import fluctuation to ensure natural gas supply reliability. China is taken as a case study. Results show that with consideration of uncertainty, an additional 40.75 % of LNG ports capacity and an additional 25.59 % of gas storage capacity need to be installed per year on average, with a 38.10 % increase in the transition cost. Finally, sensitivity analysis results are provided, showing that a rapid increase of transition costs occurs when confidence level exceeds 99.9 %. The results in this study provide references for decision makers to plan the low-carbon transition of natural gas supply systems as well as weighing transition costs against natural gas supply stability.

Keywords: chance-constrained programming, natural gas supply system, optimization, decarbonization

1. Introduction

Decarbonisation is an inevitable trend to address climate change challenges. In recent years, global natural gas consumption has grown rapidly to replace coal and oil, and the share of natural gas in primary energy consumption has reached a high level of 24% (BP, 2022). Natural gas supply highly depends on specialized and high-cost infrastructure, including pipeline networks, liquid natural gas (LNG) ports, and storage facilities. Spatial mismatch between natural gas resources and natural gas demand requires a large amount of infrastructure. China surpassed Japan as the world's largest LNG importer and accounted for close to 60% of global LNG demand growth in 2021 (BP, 2022).

For the low-carbon transition in developing regions, more natural gas is needed at the current stage. Infrastructure costs are a major part of the cost of the gas supply system, and if infrastructure costs are too high, the low-carbon transition will slow, disrupt or even reverse. In addition, natural gas probably be replaced by zero-carbon energy sources in the future. Therefore, developing an appropriate planning framework to obtain the optimal planning by minimizing the overall cost would facilitate the transition and avoiding investment waste.

Modelling and optimization of a natural gas supply system has been studied in many researches. Li et al. (2019) propose a monthly-scale, multi-period and multi-regional modelling and optimization framework for planning and operation of a natural gas supply system at a transient stage. Huang et al. (2022) proposes a method based on Bayesian networks (BNs) to optimize the reliability of gas supply in natural gas pipeline networks. Dieckhoner et al. (2013) and Chaudry et al. (2014) developed a pipeline network model and obtained an expansion plan to address increasing demand of natural gas based on existing infrastructure. An interconnected natural gas and electrical transmission model was developed for the natural gas system in Illinois, considering spatiotemporal changes in energy demand (Chiang, 2016). Zhang et al. (2016) established a multi-regional natural gas distribution model of China, and optimized gas flow and infrastructure deployment by minimizing the total supply cost.

However, previous studies focus less on the impact of uncertainty in natural gas supply system optimization planning. In recent years, international turmoil led to huge fluctuations in natural gas import. There also have uncertainties about pipeline aging and natural gas demand in a natural gas supply system, which need to be considered in planning.

In this study, a chance-constrained programming approach to optimal planning of a natural gas supply system is established. It can consider the natural gas infrastructure condition of a developing region, as well as using chance-constrained programming (Charnes and Cooper, 1959) to consider the uncertainty of import volatility, aging pipelines and natural gas demand. Under the optimization goal of minimum transition cost, the optimal planning of low-carbon transition of a natural gas supply system is given. In addition, this study also quantitatively analyses the cost increase caused by considering the degree of uncertainty, so as to provide reference for policy makers.

The structure of this paper is as follows. Section 2 introduces the methodology, section 3 introduces the case study and its results, and section 4 is the conclusion.

2. Methodology

Regional natural gas supply system optimization model is used to quantitatively optimize the natural gas supply system transition path with the consideration of uncertainty. Input parameters include the information of existing natural gas infrastructure, monthly natural gas demand, upper limit of infrastructure construction ability, and infrastructure investment price. Decision variables of model are used to describe the capacity size and construction location of liquefied natural gas (LNG) ports, natural gas storage system and natural gas pipeline as well as how they change from year to year. The structure of the model is shown in Figure 1.

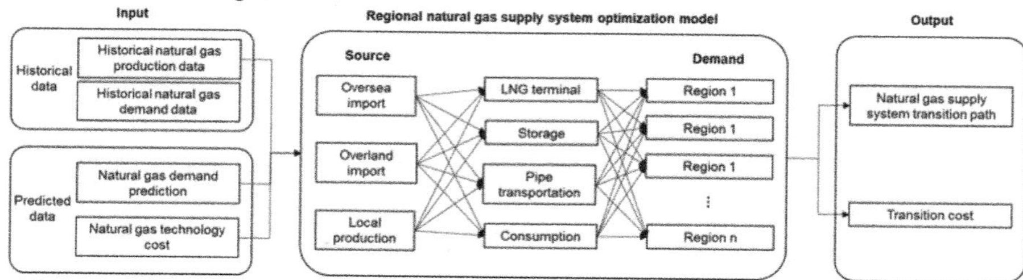

Figure 1. The structure of regional natural gas supply system optimization model

2.1. Mathematical formulation

In Regional natural gas supply system optimization model, the core is to meet the supply of natural gas. In each region and in each month, total supply equals total demand. Total supply comprises of domestic production, net import, net transmission from other regions and net storage reduction, as shown in Eq. (1).

$$qua_{dem,r,t,m} = qua_{pro,r,t,m} + qua_{sto,r,t,m} + qua_{imp,r,t,m} + qua_{tra,r,t,m} \tag{1}$$

Domestic production and net import make up total natural gas supply. Natural gas import is constrained by import capacity as shown in Eq. (2). Natural gas can only be transmitted between two adjacent regions, and the transmission quantity cannot exceed the upper bound constrained by infrastructure, as shown in Eq. (3). A binary variable $y_{r,rr}$ is introduced to represent geographic constraints, where $y_{r,rr}$ equals to zero representing two regions are not adjacent. Natural gas storage is constrained with capacity and period, where storage quantity cannot exceed its maximum capacity as shown in Eq. (4), and ending stock of this period equals the beginning of the next period, as shown in Eq. (5).

$$qua_{imp,r,t,m} \leq capa_{imp,r,t}/12 \tag{2}$$

$$qua_{tra,r,rr,t,m} \leq capa_{tra,r,rr,t} \times y_{r,rr}/12 \tag{3}$$

$$qua_{sto0,r,t,m} \leq capa_{sto,r,t} \tag{4}$$

$$qua_{sto0,r,t,m} = qua_{sto1,r,t,m-1} \tag{5}$$

Along with the increase of the natural gas demand, more pipelines, LNG ports and storage facilities would be constructed. Expansions of natural gas storage facilities are continuous design variables, whilst expansions of pipelines and ports are discontinuous variables. The new capacity is an integer multiple of the standard capacity, as shown in Eqs. (6~7). Annual capacities of them are calculated by Eq. (8).

$$inc_{imp,r,t} = int_{imp,r,t} \times std_{imp,r,t} \leq 3 \times 41.4 \tag{6}$$

$$inc_{tra,r,t} = int_{tra,r,t} \times std_{tra,r,t} \leq 5 \times 300 \tag{7}$$

$$capa_{i,r,t} = capa_{i,r,t-1} + inc_{i,r,t} - dec_{i.r,t} \tag{8}$$

A chance-constrained programming method is established to consider the uncertainty of import fluctuation, pipeline aging and natural gas demand. At the time of natural supply, the actual values of import, pipeline capacity and natural gas demand can be seen as the sum of the predicted value and random fluctuation, as shown in Eq. (9).

$$\begin{cases} qua_{imp,t,m} = qua_{imp0,t,m} + \delta_{imp,t,m} \\ capa_{tra,t,m} = capa_{tra0,t,m} + \delta_{tra,t,m} \\ qua_{dem,t,m} = qua_{dem0,t,m} + \delta_{dem,t,m} \end{cases} \tag{9}$$

Considering chance-constraint, Eq. (1) can be expressed as a probabilistic constraint. In Eq. (10), α is the confidence level, which shows the extent of considering uncertainty.

$$\Pr\{qua_{pro,r,t,m} + qua_{imp,r,t,m} + qua_{tra,r,t,m} + qua_{sto,r,t,m} - qua_{dem,r,t,m} \geq 0\} \geq \alpha \tag{10}$$

This study assumes these variables with uncertainty follow a normal distribution, Eq. (10) can be converted into Eq. (11).

$$qua_{pro,r,t,m} + +qua_{sto,r,t,m} \geq \mu qua_{imp,r,t,m} + \mu qua_{tra,r,t,m} - \mu qua_{dem,r,t,m} + z_\alpha \times \sqrt{\sigma^2_{imp,r,t,m} + \sigma^2_{tra,r,t,m} + \sigma^2_{dem,r,t,m}} \tag{11}$$

The objective function of regional natural gas supply system optimization model describes the total cost of transition path, including the cost of infrastructure investment as well as import, gasification, transportation and storage of natural gas, as shown in Eq. (12).

$$cost_t = cost_{inv,t} + cost_{imp,t} + cost_{ify,t} + cost_{tra,t} + cost_{sto,t} \quad (12)$$

The mixed integer nonlinear programming optimization problem is implemented in General Algebraic Modelling System (GAMS).

3. Case Study

China is the largest developing country in the world. It aims to peak carbon emissions by 2030, and strive to achieve carbon neutral goal by 2060. In this study, China is taken as a case. The above approach is used to optimize the transition path of natural gas supply system of China. At the same time, the impact of consideration and non-consideration of uncertainty on planning is compared. Finally, Sensitivity of confidence level is analyzed. This model takes 2020 as base year and the transition path is optimized until 2060.

3.1. Supply structure of natural gas

Figure 2 shows the changes in natural gas supply structure of China. For China, its natural gas supply will mainly rely on imports until 2050, which proportion will exceed 40% by 2050. Natural gas is mainly supplied from Central Asia import and oversea import with other import methods also account for a proportion. Multi-source imports improve energy security.

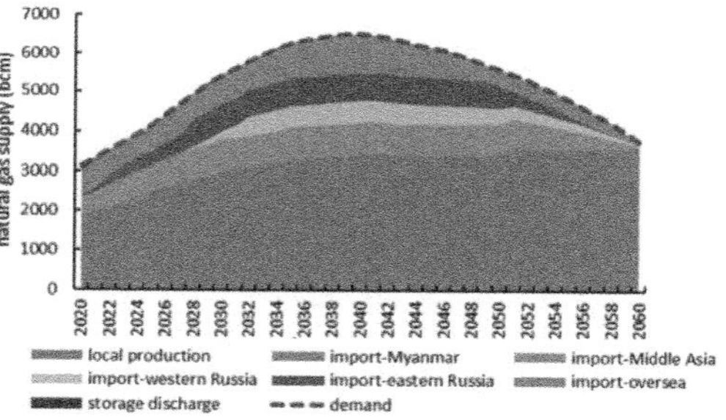

Figure 2 Natural gas supply structure of China

3.2. Comparing results between considering and non-considering uncertainty

The confidence level used in the scenario of considering uncertainty is 99%. In both scenarios of considering and not considering uncertainty, the capacity of LNG ports will peak in 2040. However, after considering uncertainty, an additional 40.75% of LNG ports capacity need to be installed per year on average. Figure 3 shows the distribution of LNG ports in 2040 and 2060 with consideration of uncertainty. LNG ports will mainly be built in Shandong, Zhejiang and Guangdong. The first two will also remain the main provinces importing natural gas in 2060.

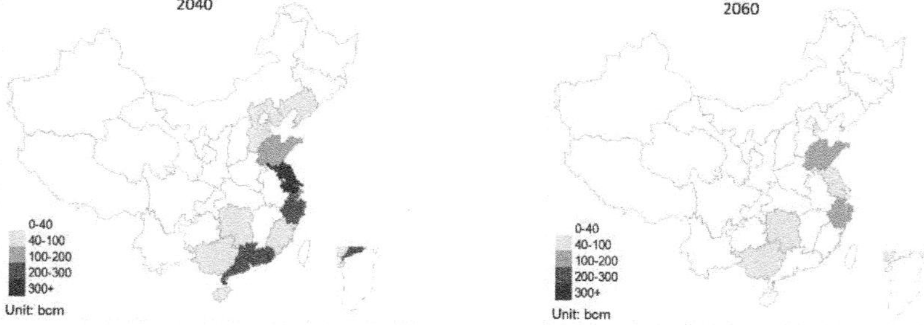

Figure 3 Distribution of LNG ports of China considering uncertainty

For nature gas storage facilities, its capacity will peak in 2040 in both scenarios. There is an additional 25.59% of storage facilities need to be installed per year on average after considering uncertainty, the distributions of which in 2040 and 2060 are shown in Figure 4. Natural gas storage is mainly built in Xinjiang, Heilongjiang and Shaanxi.

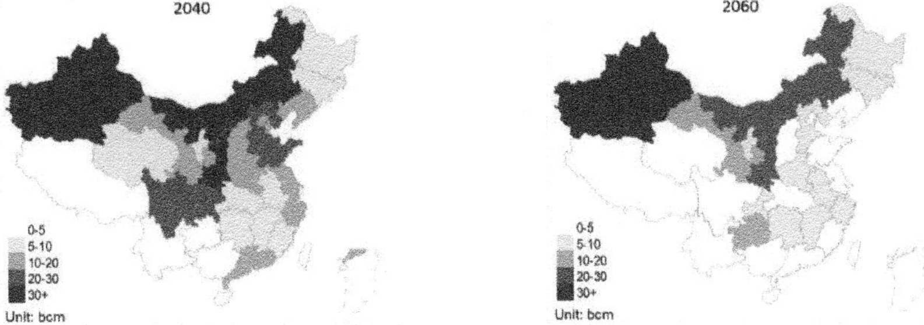

Figure 4 Distribution of storage facilities of China considering uncertainty

For natural gas networks, considering uncertainty hardly changes the total capacity of the network, but makes the capacity of each pipeline more uniform. Figure 5 shows the distribution of natural gas pipeline networks in 2040 under both scenarios. The orange lines represent the main pipelines with an annual transmission capacity of more than 100bcm, which is less than 100bcm in the grey lines. It shows that the distribution of the main pipe network is wider after considering the uncertainty.

Figure 5 Distribution of pipeline of China in 2040

3.3. Sensitivity Analysis

When considering uncertainty, different confidence levels may lead to different energy system optimization results. The results of sensitivity analysis are shown in Figure 6. when the confidence level is 90 % ~ 99 %, considering the uncertainty will not bring great changes to the natural gas supply system structure and transition cost. When the confidence level is higher than 99.9 %, the impact of capacity of LNG ports and the cost of transition begin to show an exponential rise. The results can provide a reference for policy makers to build what size of natural gas infrastructure in response to the uncertainty while taking the cost of transition into account.

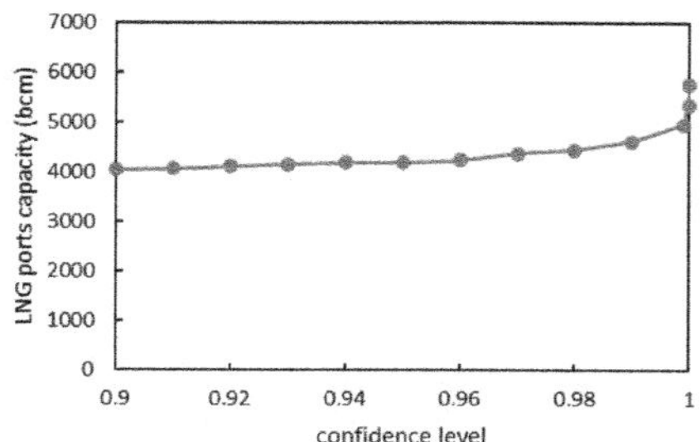

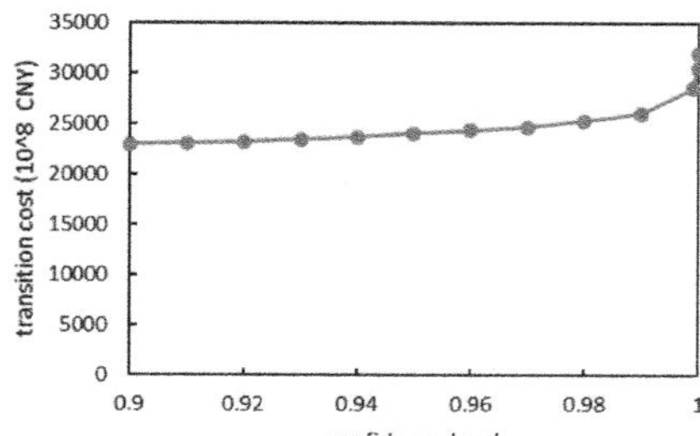

Figure 6 Sensitivity analysis of confidence levels

4. Conclusions

This study proposes a regional natural gas supply system optimization model, which can be used to quantitatively analyze the optimal transition path for a natural gas supply system with minimum transition costs. A chance-constrained programming approach is adopted to consider the impact of uncertainty of import fluctuation, pipeline aging and natural gas demand. Finally, sensitivity analysis of confidence level is carried out to show the impacts of it on the capacity planning of natural gas infrastructure as well as the transition costs.

The results show that increasing capacity of LNG ports and natural gas facilities can enhance the ability to cope with uncertainty. With consideration of uncertainty, under confidence level as 99 %, an additional 40.75 % of LNG ports capacity and an additional 25.59 % of gas storage capacity need to be installed per year on average. In addition, Sensitivity analysis of confidence level shows that rapid growth in transition cost occurs when the confidence level exceeds 99.9 %, which provides a reference for decision makers to make energy system planning while weighing transition costs against natural gas supply stability.

For future research, more uncertainties can be taken into account, such as price uncertainty. In addition, the coupling research of natural gas system and hydrogen system is meaningful. Natural gas is a transition period energy, and hydrogen will be promising in the future. Researching on the coupling of them that natural gas infrastructure can be converted into hydrogen infrastructure can reduce investment costs.

References

BP, 2022, Statistical Review of World Energy.

T. Li, P. Liu, and Z. Li, 2019, Modelling and optimization of a natural gas supply system at a transient stage: a case study of China. BMC Energy. 1: 5.

Huang W., Li Y., Yu W., Yu H., Shan X., Wang H. and Gong J., 2021, An evaluation index system of the user satisfaction for the natural gas pipeline network. Journal of Pipeline Science and Engineering. 1(4): 452-458.

Dieckhoener C, Lochner S, Lindenberger D, 2013, European natural gas infrastructure: the impact of market developments on gas flows and physical market integration. Appl Energy.102: 994–1003.

Chaudry M, Jenkins N, Qadrdan M, Wu JZ, 2014, Combined gas and electricity network expansion planning. Appl Energy. 113:1171–87.

Chiang N-Y, Zavala VM, 2016, Large-scale optimal control of interconnected natural gas and electrical transmission systems. Appl Energy. 168:226–35.

Zhang Q, Li Z, Wang G, Li H., 2016, Study on the impacts of natural gas supply cost on gas flow and infrastructure deployment in China. Appl Energy, 162, 1385–98.

Charnes A., Cooper W. W., 1959, Chance-Constrained Programming, Management Science, 6(1), 73-79.

Antonis Kokossis, Michael C. Georgiadis, Efstratios N. Pistikopoulos (Eds.)
PROCEEDINGS OF THE 33rd European Symposium on Computer Aided Process Engineering (ESCAPE33), June 18-21, 2023, Athens, Greece
 http://dx.doi.org/10.1016/B978-0-443-15274-0.50088-3

Efficient 1D modelling of hot melt extrusion process for pharmaceutical applications

Michaela Vasilaki,[a] Umair Zafar,[b] Ioannis .S. Fragkopoulos,[c] Ioannis K. Kookos,[a]

[a]*Department of Chemical Engineering, University of Patras, Rio, 26504, Patras Greece*
[b]*Oral Formulation Research, Novo Nordisk A/S, Måløv, Denmark*
[c]*Oral Drug Product Process Development, Novo Nordisk A/S, Måløv, Denmark*

Abstract

Hot melt extrusion (HME) involves the melting of materials and mixing and pushing of various solids under elevated temperature and pressure. It is a mature technology that has been extensively used for a broad spectrum of applications in the polymers, food, and pharmaceutical industries. HME can be a promising downstream processing method due to its high efficiency and economic value, however mathematical modelling of HME processes has always been challenging due to the limited understanding and the demanding experimental techniques. In this work, an 1D model is developed for the efficient simulation of the steady state and dynamic operation of HME processes. It is shown that the steady state operation can be described as a classical complementarity problem. The solution of the resulting complementarity problem can be performed efficiently in available software. The determination of the consistent initial conditions facilitates the solution of the system of differential and algebraic equations (DAEs).

Keywords: Hot melt extrusion, Pharmaceutical industry, Simulation, Complementarity problem.

1. Introduction

HME is a mature technology, that has been used extensively in the polymers industry, and involves the melting of materials and/or mixing and pushing of various solids under elevated temperature and pressure. There has been an active interest in extending the application of HME technology in the polymers, food, and pharmaceutical industry (Crowley et al., 2007). Mathematical modelling of HME processes has always been challenging due to the limited understanding of the material behavior, its complex geometry, the complex interactions between the material processed and the equipment, the existence of multiple zones (solids zone and melt zone) and the demanding experimental techniques necessary to validate theoretical models.

There has been extensive research in developing efficient mathematical models which can be used for process synthesis. Most of the existing modelling approaches have been reviewed recently (Grimard et al., 2016; Kohlgruber, 2020). The objective of this study is to propose a novel approach to the solution of 1D models for HME processes. The proposed methodology is based on transforming the algebraic model that describes the steady state operation using material, energy, and momentum balances as well as logical conditions, into an equivalent complementarity problem (a problem that has been extensively studied in the mathematical programming literature). The proposed novel approach is proven to be computationally efficient and is extended into the dynamic model of the HME process.

2. 1-D model of the HME process

2.1. Steady state model

Figure 1 shows the general pressure-throughput relationship for extruders that is discussed extensively by Kohlgrüber (2020). The following model is proposed for creeping flows of Newtonian fluids:

$$\dot{V}^* = \dot{V}^*_{max} - \frac{\dot{V}^*_{max}}{\Delta\dot{p}^*_{max}}\Delta p^* \tag{2.1}$$

where the dimensionless pressure drop $\Delta p^*$and flowrate $\dot{V}^*$ are defined as follows:

$$\Delta p^* = \frac{D}{\eta\mathcal{N}}\cdot\frac{\Delta p}{\Delta z}, \qquad \dot{V}^* = \frac{\dot{V}}{\mathcal{N}\cdot D^3} \tag{2.2}$$

where D is the diameter of the barrel, η is the viscosity, $\mathcal{N}$ is the rotational frequency, Δp is the pressure drop observed in an element with axial length Δz and $\dot{V}$ is the volumetric flowrate. After substituting Equations (2.2) into Equation (2.1) we obtain:

$$\dot{m} = m_{fb} + m_p = \dot{V}^*_{max}\cdot\rho(\mathcal{N}\cdot D^3) - \frac{\dot{V}^*_{max}}{\Delta\dot{p}^*_{max}}D^4\frac{\rho}{\eta}\cdot\frac{\Delta p}{\Delta x} \tag{2.3}$$

where ρ is the density, m_{fb} denotes the mass flowrate for forward (m_f) or backward (m_b) moving elements and m_p the pressure induced mass flowrate. Equation (2.3) holds true for completely filled elements. For partially filled elements with filing ratio f:

$$\dot{m} = m_{fb} + m_p \tag{2.4}$$

$$m_{fb} = \dot{V}^*_{max}\cdot f\cdot\rho(\mathcal{N}\cdot D^3) \tag{2.5}$$

$$m_p = -\frac{\dot{V}^*_{max}}{\Delta\dot{p}^*_{max}}D^4\frac{\rho}{\eta}\cdot\frac{dp}{dz} \tag{2.6}$$

In order to develop a mathematical description of the twin-extruder, the extruder is divided into i={1, 2,..., n} discrete elements and material balances are applied in each element. The discrete representation of the solution space is shown in Figure 2. A material balance can be written for each discrete element as follows:

$$m_f(i-1) + m_b(i+1) + m_p(i-1) = m_f(i) + m_b(i) + m_p(i) \tag{2.7}$$

$$m_{fb}(i) = \dot{V}^*_{max}(i)\cdot\big(1-\varepsilon(i)\big)\cdot\rho(i)\cdot(\mathcal{N}\cdot D^3) \tag{2.8}$$

$$m_p(i) = -\frac{\dot{V}^*_{max}(i)}{\Delta\dot{p}^*_{max}(i)}D^4\frac{\rho(i)}{\eta(i)}\cdot\left(\frac{p(i+1)-p(i)}{z(i+1)-z(i)}\right) \tag{2.9}$$

where we have introduced the void fraction $\varepsilon(i) = 1 - f(i)$. The pressure-induced mass flowrate implied by Equation (2.9) involves a conditional statement: pressure buildup or pressure generation can only be developed in completely filled elements (f=1 or ε=0 $\Rightarrow$ $p(i)$>0). For partially filled elements the pressure is equal to zero (f<1 or ε>0 $\Rightarrow$ $p(i)$=0, apparently the pressure is defined relative to the atmospheric pressure).

Equation (2.7), using Equations (2.8) and (2.9), can be written at steady state in the following form where the unknowns $(\varepsilon(i), p(i))$ appear linearly:

$$\mathbf{A}\big(\dot{V}^*_{max}(i), \Delta\dot{p}^*_{max}(i), D, \mathcal{N}; \rho, \eta\big)\cdot\boldsymbol{x} = \boldsymbol{b}\big(\dot{V}^*_{max}(i), D, \mathcal{N}; \rho\big) \tag{2.10}$$

$$x = \begin{bmatrix} \varepsilon \\ p \end{bmatrix} = [\varepsilon(1) \quad \varepsilon(2) \quad \dots \quad \varepsilon(n) \quad p(1) \quad p(2) \quad \dots \quad p(n)]^T \tag{2.11}$$

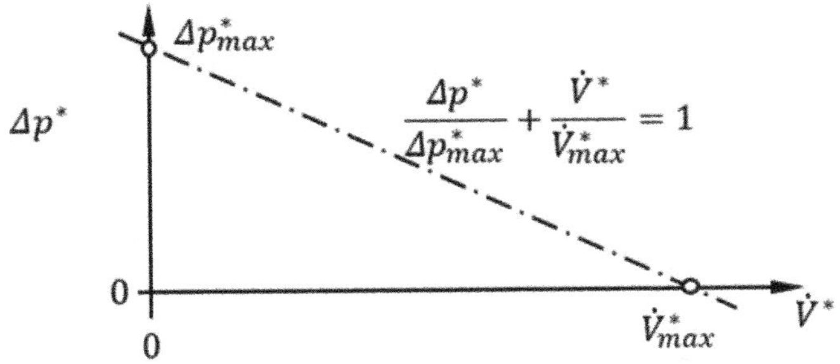

Figure 1. Pressure versus volumetric flowrate for a typical extruder element.

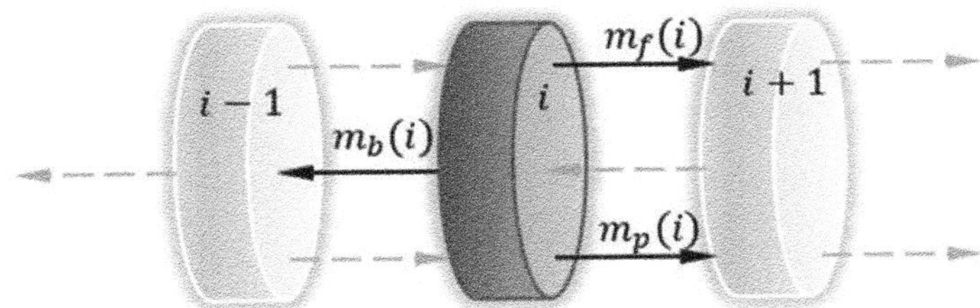

Figure 2. Building block of the 1D model for a twin-screw extruder.

For constant thermophysical properties (ρ, η) and given flow characteristic and operating parameters $(\dot{V}^*_{max}(i), \Delta p^*_{max}(i), D, \mathcal{N})$, Equation (2.10) is a linear underdetermined system of n equations in $2n$ unknowns (see also Choulak et al., 2004 and Eitzlmayr et al., 2014). The system of linear equations can be solved by adding n specifications to the mathematical description. This has been done in previous studies by first establishing the state of each element (partially filled/completely filled) and then calculating the resulting pressure profile. Numerical problems have been reported in previous studies which have been attributed to inefficiencies of the commercial software used. A systematic methodology is presented in what follows that does not suffer from methodological or numerical issues and allows solution of the problem using arbitrarily fine discretization of the solution space.

The important observation towards developing an efficient methodology for the steady state simulation of extruders is that, apart from the system of n underdetermined Equations (2.10), there is for each element i an additional logical condition that needs to be satisfied. This condition can be stated as: either the element is partially filled and the pressure is zero or the element is completely filled, and the void fraction is zero. This can be expressed mathematically as follows:

$$\varepsilon(i) \cdot p(i) = 0, i = 1,2, \dots, n \tag{2.12}$$

this is a set of n additional nonlinear equations. When Equation (2.12) is written together with Equation (2.10), then the system of n equations in n unknowns is obtained:

$$\mathbf{A}\cdot\begin{bmatrix}\boldsymbol{\varepsilon}\\ \boldsymbol{p}\end{bmatrix}=\boldsymbol{b},\qquad \boldsymbol{\varepsilon}^T\cdot\boldsymbol{p}=0,\qquad \mathbf{0}\leq\boldsymbol{\varepsilon}\leq\mathbf{1}_n,\qquad \mathbf{0}\leq\boldsymbol{p}\leq\mathbf{1}_n P_{max} \tag{2.13}$$

Formulation (2.13) is well known in the mathematical programming literature as the linear complementarity problem (LCP), for which efficient solution methodologies are now available (Kookos, 2022, Cottle, 1992). One of the alternative methodologies can be based on the definition of binary variables $\delta(i)$:

$$\delta(i)=\begin{cases}1, & if\ \varepsilon(i)=0\\ 0, & if\ p(i)=0\end{cases},\qquad i=1,2,\dots,n \tag{2.14}$$

Using the definition (2.14) the LCP problem can be postulated as a mixed-integer, linear programming problem (MILP):

$$\min_{\boldsymbol{\varepsilon},\boldsymbol{p},\boldsymbol{\delta}} \mathbf{1}_n^T\boldsymbol{\delta}$$

$$s.t. \tag{2.15}$$

$$\mathbf{A}\cdot\begin{bmatrix}\boldsymbol{\varepsilon}\\ \boldsymbol{p}\end{bmatrix}=\boldsymbol{b},\qquad \mathbf{0}\leq\boldsymbol{\varepsilon}\leq\mathbf{1}_n-\boldsymbol{\delta},\qquad \mathbf{0}\leq\boldsymbol{p}\leq\boldsymbol{\delta}\cdot P_{max},\qquad \boldsymbol{\delta}\in\{0,1\}^n$$

2.2. Dynamic model of an extruder

The dynamic model of an extruder consists of the dynamic material balances:

$$\frac{dM(i)}{dt}=m_f(i-1)+m_b(i+1)+m_p(i-1)-\left(m_f(i)+m_b(i)+m_p(i)\right) \tag{2.16}$$

$$\frac{dm(i)}{dt}=\left(m_f(i-1)+m_p(i-1)\right)\cdot\omega(i-1)-\left(m_f(i)+m_b(i)+m_p(i)\right)\cdot\omega(i)+m_b(i+1)\cdot\omega(i+1) \tag{2.17}$$

where $M(i)=\rho(i)\,Acr\,\delta z\,f(i)$ is the total mass holdup in element i, $m(i)=M(i)\omega(i)$ is the mass holdup of component in element i, $\omega(i)$ is the mass fraction of the component of interest, A_{cr} is the cross-sectional area available for materials flow and δz the axial length of the discrete element. The dynamic model also includes the algebraic equations (2.8) and (2.9). There are as many differential equations as differential variables but there are n more algebraic variables than algebraic equations. To avoid using the complementarity constraint in the Differential-Algebraic Equations (DAE) model, that will give rise to a high index problem, we are assuming that the pressure dynamics is significantly faster than the flow dynamics. If the solution of the DAE model is available at time t, then the filling ratio is known, and the pressure profile can be established. With the pressure profile known the DAE model can be solved and obtain the dynamics of the material flow through the extruder. In this way the DAE model is, in effect, simplified to an ordinary differential equation (ODE) model.

3. Case study and discussion

We consider the twin-screw extruder studied experimentally by Puaux et al., (2000) (screw configuration B) and theoretically by Eitzlmayr et al., (2014). The experimental extruder is a twin-screw, corotating extruder with a centerline distance of C_L=0.021 m, an external screw diameter of D=0.025 m and a screw core diameter of D_C=0.016 m. The extruder consists of seven screw type elements arranged in series with lengths (from hopper to die) 50/50/100/50/150/200/300 mm and pitches (T_S) 16/25/33/25/16/25/33 mm. All elements are right-handed elements (forward pushing and/or pressure generating elements) apart from the fourth one which is left-handed (backward pushing or pressure consuming element). Several combinations of operating conditions with mass flowrates in the range $\dot{m}_0 \in [3.8 \text{ kg/h}, 10.8 \text{ kg/h}]$ and screw speeds of $\mathcal{N} \in [200 \text{ rpm}, 500 \text{ rpm}]$ are considered in the experimental equipment and the results of residence time distribution determination experiments are presented in Puaux et al., (2000). The parameters of the pressure/volumetric flowrate characteristics ($\dot{V}^*_{max}, \Delta\dot{p}^*_{max}$) of the various screw elements need to be estimated. For the maximum dimensionless volumetric flowrate Kohlgrüber (2020) proposes that $\dot{V}^*_{max} = T_s/2D$. Parameter $\Delta\dot{p}^*_{max}$, which needs to be determined experimentally, was assumed to be equal to 2000, a value that appears to be reasonable for the screw diameter and pitch of the elements in the experimental extruder (according to experimental curves reported by Kohlgrüber (2020)).

The LCP formulation (2.15) is solved first for the case of $\mathcal{N} = 400\ rpm$ and $\dot{m}_0 =$ 6.5 kg/h using CPLEX/GAMS with an increasing number of discretization elements. The overall length of the extruder is 900 mm and 90, 900, 9000 and 90000 discrete elements are selected corresponding to discrete element lengths of 10 mm, 1 mm, 0.1 mm, and 0.01 mm, correspondingly. In Figure 3.a the profile for the filling ratio for 90 and 90000 discrete elements, are presented and compared (the results for 900 and 9000 discrete elements coincide with that of the 90000 elements). The necessary computational time was 0.07 s for the 90 elements and 20.1 s for the 90000 elements (Intel(R) Core(TM) i5-7200U CPU @ 2.50GHz, 6GB memory computer). The computational time is found to scale with n^2 for the intermediate discretizations. As the formulation as an LCP problem was proven to be extremely efficient, we were able to solve repeatedly the formulation by varying parameters such as the feed mass flowrate or the speed of rotation. In Figure 3.b the results obtained by varying systematically the mass flowrate of the feed for a constant speed of rotation is shown. Similar plots can be obtained by varying the speed of rotation for a constant mass flowrate of the feed stream.

To demonstrate the ability of the proposed formulation to predict the dynamic response of the experimental extruder, the dynamic model was solved for a unit pulse experiment in MATLAB using the ode15s DAE solver (the pressure dynamics is assumed faster than the flow dynamics). The volume of each element of the discrete formulation was determined based on the geometry, by calculating the cross-sectional area based on the work by Booy (1980). The experimental results are compared with the simulation results in Figure 4. The agreement is more than satisfactory and proves that the proposed formulation is sufficient for simulating the dynamics of complex twin-screw extruders.

4. Summary and Conclusions

Preliminary results on a novel approach for solving the steady state and dynamic model for complex twin-screw extruders has been presented and validated with experimental results. The proposed formulation that is based on the linear complementarity problem has been proven to be a powerful approach for solving the steady state determination problem in twin-screw extruders. The model, at its present form, has been developed for constant thermophysical properties that are independent from pressure and temperature (density and viscosity). The model needs to be extended to incorporate the temperature dynamics that involves the energy balance for the material, barrel, and screw. It also involves the use of constitutive equations for the density and viscosity to link material flow dynamics with energy dynamics. The preliminary results are promising enough to guarantee that further investigation will improve the predictive capabilities of the model and will allow us to obtain a better and more detailed understanding of the complex and multiscale dynamics of twin-screw extruders. Finally, discrete element simulations can be used to determine the characteristic parameters of the pressure/throughput curve so as to reduce our reliance on experimental results for developing high fidelity models.

Acknowledgement

M. Vasilaki & I.K. Kookos were supported for this work by the project "INVALOR: Research Infrastructure for Waste Valorization and Sustainable Management" (MIS 5002495) which is implemented under the Action "Reinforcement of the Research and Innovation Infrastructure", funded by the Operational Program "Competitiveness, Entrepreneurship and Innovation" (NSRF2014–20) and co-financed by Greece and the EU (EU Regional Development Fund).

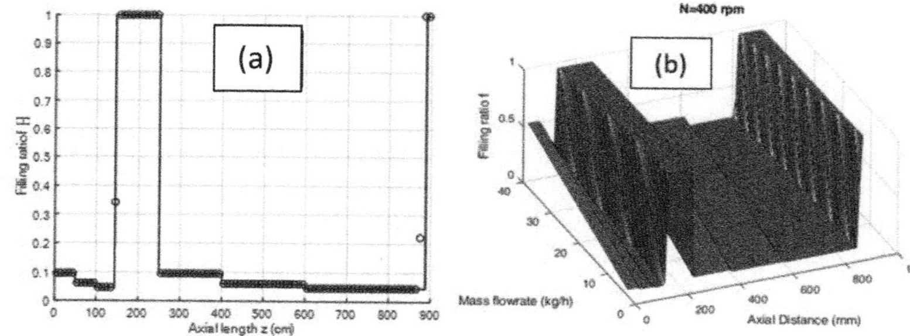

Figure 3. Filling ratio for the experimental extruder for (a) different number of discrete elements (continuous line 90000 elements, o 90 elements) and (b) several values of the mass flowrate and fixed frequency of rotation ($\mathcal{N}$ = 400 rpm).

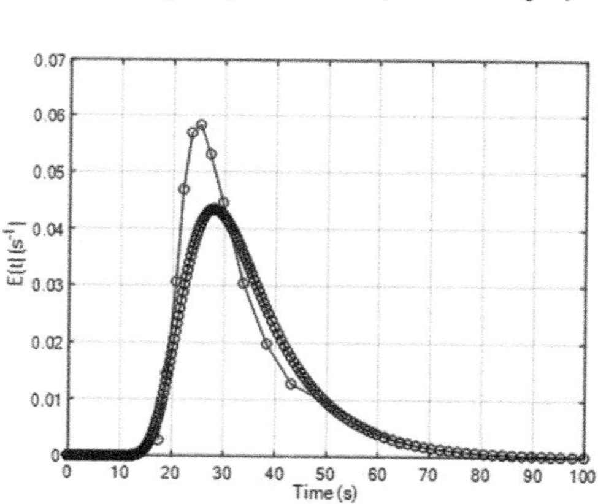

Figure 4. Comparison of the experimental results and simulation results for the RTD experiment ($\mathcal{N} = 400\ rpm$, $\dot{m}_0 = 6.5$ kg/h).

References

M.L. Booy, M.L., 1980, Isothermal Flow of Viscous Liquids in Corotating Twin Screw Devices, Polymer Engineering and Science, Vol 20(18), pp. 1220-1228.

S. Choulak, Couenne, F., Gorrec, Y.L., Jallut, C., Cassagnau, P., Michel, A., 2004, Generic Dynamic Model for Simulation and Control of Reactive Extrusion, Industrial Engineering Chemistry Research, 43, pp. 7373-7382.

R.W. Cottle, Pang, J.S., Stone, R.E., 2009, The linear Complementarity Problem, Siam Philadelphia.

M.M., Crowley, Zhang, F., Repka, M.A., Thumma, S., Upadhye, S.B., Kumar Battu, S., McGinity, J.W. and Martin, C., 2007, Pharmaceutical Applications of Hot-Melt Extrusion: Part I., Drug Development and Industrial Pharmacy, 33(9), pp.909–926.

A. Eitzlmayr, Koscher, G., Reynolds, G., Huang, Z., Booth, J., Shering, P., Kninast J., 2014, Mechanistic modelling of modular co-rotating extruders, International Journal of Pharmaceutics, 474, pp. 157-176.

J. Grimard, Dewasme, L., Vande Wouwer, A., 2016. A Review of Dynamic Models of Hot-Melt Extrusion. Processes, 4(2), p.19.

K. Kohlgruber, 2020, Co-Rotating Twin-Screw Extruders: Fundamentals, Carl Hanser, Munich.

I.K. Kookos, 2022, Practical Chemical Process Optimization, Springer-Nature.

J.P. Puaux, Bozga, G., Aisner, A., 2000, Residence Time Distribution in a Co-rotating Twin-Screw Extruder, Chemical Engineering Science, 55, pp. 1641-1651.

Antonis Kokossis, Michael C. Georgiadis, Efstratios N. Pistikopoulos (Eds.)
PROCEEDINGS OF THE 33rd European Symposium on Computer Aided Process Engineering (ESCAPE33), June 18-21, 2023, Athens, Greece
 http://dx.doi.org/10.1016/B978-0-443-15274-0.50089-5

Towards Machine Learning of Power-2-Methanol Processes

Carl Julius Martensen,[a] Christoph Plate,[a] Tobias Keßler,[b] Christian Kunde,[c] Lothar Kaps,[c] Achim Kienle,[b,c] Andreas Seidel-Morgenstern,[c,d] Sebastian Sager[a]

[a] *Institute of Mathematical Optimization, Otto-von-Guericke University, Universitätsplatz 2, 39106 Magdeburg, Germany*
[b] *Institute for Automation Engineering, Otto-von-Guericke University, Universitätsplatz 2, 39106 Magdeburg, Germany*
[c] *Max Planck Institute for Dynamics of Complex Technical Systems, Sandtorstraße 1, 39106 Magdeburg, Germany*
[d] *Institute of Process Engineering, Otto-von-Guericke University, Universitätsplatz 2, 39106 Magdeburg, Germany*

Abstract

Many dynamic models in process engineering rely on uncertain or even largely unknown mechanism. These parts are often modeled using best-practice knowledge or heuristics, which may result in 1) overly complex models, 2) models which do not reproduce the expected predictive outcome or 3) ill posed optimization problems. A promising approach is the use of machine learning surrogate models inside a partially known mechanistic model. In this study the impact of measurement noise and sample rate on the predictive performance of such a hybrid model is investigated. As an example, process methanol synthesis is used. Additionally, it is shown that the surrogate model can learn even unobservable states from the data.

Keywords: Scientific Machine Learning, Methanol Synthesis, Automated Modeling, System Identification

1. Introduction

Methanol is an important basic chemical. In the frame of the upcoming energy revolution, power-2-methanol processes become an interesting option for methanol production. These processes rely on green hydrogen, which is produced from regenerative electrical energy by water electrolysis and are therefore usually operated under dynamic conditions due to inevitable fluctuations in the regenerative energy supply. One approach is to generate methanol through the hydrogenation of CO and CO_2 (Olah et al., 2009). Although methanol synthesis is a commonly employed process, its kinetic modeling is still a field of ongoing research. The influence of the dynamic catalyst state on the reaction rates is an important part of the model, which is, however, hard to model from first principles. A promising approach to overcome this challenge is to replace this sub-model with an artificial neural network (ANN) and combine it with the remaining first principle model to obtain a hybrid model. It is not clear a priori how much data is required to successfully train the resulting model. This study investigates the influence of data quality on such a data-driven model. Artificial datasets of reactor output trajectories with different measurement frequencies and amounts of added noise are generated from a rigorous model (Seidel et. al., 2018, 2021). A hybrid model is then derived by replacing the sub-model that describes the influence of the catalyst state on the reaction rates with an ANN. Hyperparameter optimization of different surrogate architectures, objectives, and penalty functions is conducted, and the best-performing model structure is identified. The hybrid model is trained on the different datasets and the results are evaluated by comparing the state trajectories with those of the rigorous model. Additionally, the trained ANNs are compared with the original sub-model to evaluate the ability to reconstruct the original sub-model behaviour from the different synthetic datasets. The contributions of this analysis are twofold: First, it leads to a better understanding of the necessary data quality for training hybrid models of heterogeneously catalysed methanol synthesis. Second, it generates insights into the parsimony of available rigorous models depending on the operating conditions.

2. Materials and Methods

The overall dynamics of the continuous stirred tank reactor (CSTR) for methanol synthesis can be described as a system of differential algebraic equations (DAE) of the form (Seidel et. al., 2018, 2021)

$$\begin{aligned} f(\dot{x}, x, p, t, u, c) &= 0 \\ h(x, p, t, u, c) &= y \end{aligned} \tag{1}$$

where $x \in \mathbb{R}^{n_x}$ denotes the states of the system and $\dot{x} \in \mathbb{R}^{n_x}$ its time derivative subject to the set of DAE given by f. The states of the system consist of the mole fractions of the individual species $x_i \in (0,1)$ with the additional constraint $\sum_i x_i = 1$, $i \in$ (CH_3OH, CO_2, CO, H_2, H_2O, N_2) and the percentage of reduced centres of the catalyst $x_\phi \in (0,1)$ (Seidel et. al., 2018). $p \in \mathbb{R}^{n_p}$ is a vector of parameters, $t \in \mathbb{R}$ the time, $u \in \mathbb{R}^{n_u}$ the external inputs to the system, representing an incoming feed of species and a specific volumetric flow rate. The function h maps the states onto the observed variables $y \in \mathbb{R}^{n_y}$. The constant operating conditions $c = (T, P)$ collect the information about temperature and pressure $T \in \mathbb{R}$ and $P \in \mathbb{R}^+$, respectively. Three reaction pathways are used to describe the synthesis of methanol from CO, CO_2, and H_2 over CuO/ ZnO/ Al_2O_3 (unreduced state):

$$\begin{aligned} \mathrm{CO} + 2\,\mathrm{H_2O} &\leftrightarrows \mathrm{CH_3OH} \\ \mathrm{CO_2} + 3\mathrm{H_2O} &\leftrightarrows \mathrm{CH_3OH} + \mathrm{H_2O} \\ \mathrm{CO_2} + 3\mathrm{H_2} &\leftrightarrows \mathrm{CO} + \mathrm{H_2O} \end{aligned} \tag{2}$$

where the first two reactions describe the methanol synthesis via hydrogenation and the last reaction is the reverse water-gas shift reaction (RWGS). The dynamical system following from Eq. (2) is still not fully understood. Especially the influence of the catalyst on the reaction rate is not clear. Following (Seidel et. al., 2018, 2021) a scaled reaction rate is used to take this effect into account:

$$\bar{r} = \left[\gamma_{CO} r_{CO} \ \gamma_{CO_2} r_{CO_2} \ \gamma_{RWGS} r_{RWGS}\right]^T \quad (2)$$

where $\bar{r}_i \in \mathbb{R}$ represents the effective reaction rate of the reaction $i \in$ as the product of the influence of the catalyst $\gamma_i \in \mathbb{R}^+$ and the reaction rates based upon a Langmuir–Hinshelwood mechanism $r_i \in \mathbb{R}$.

The influence of the catalyst state on the effective reaction rates is of special interest, given that it is a challenging reaction over a distributed parameter space. The current implementation assumes a dependence on the catalyst state and is modeled based on practical heuristics rather known first principles

$$\gamma = \left[1 - x_\phi \ \ x_\phi^2 \ \ \frac{x_\phi}{1 - x_\phi}\right]^T \quad (4)$$

Within our study an existing model of an isochoric, isobaric, isothermal CSTR for methanol synthesis is used, which is based on the dynamics of a steady-state experiment. Its goal is to investigate the influence of data quality and sample distance on the approximative performance of an embedded ANN. While the idea of using ANNs itself is not new, see e.g., (Ljung et. al., 2020), data-driven inference of dynamical systems has gained much attention during the past decade, e.g., extracting (physical) signals (Cranmer et. al., 2020), operator learning (Kovachki et. al., 2021), and approximating solutions to PDE, ODE and DAE systems (Raissi et. al., 2021).

The rigorous model (Seidel et. al., 2018, 2021) has been implemented using Julia (Benzanson et. al., 2015). To assess the use of ANN as a tool to extract and learn unknown parts of our process model, the approach described in (Rackauckas et. al., 2021) is followed. Here, an ANN is embedded inside the known model, effectively extending the previously introduced model with a universal approximator $\hat{\gamma} : \mathbb{R}^{n_x} \times \mathbb{R}^{n_\theta} \mapsto \mathbb{R}^3$ containing trainable parameter $\theta \in \mathbb{R}^{n_\theta}$. The hybrid model takes the form

$$\begin{aligned} \hat{f}(\dot{x}, x, p, t, u, c, \hat{\gamma}(x, \theta)) &= 0 \\ h(x, p, t, u, c) &= \hat{y} \end{aligned} \quad (3)$$

where $\hat{f}$ denotes the known parts of the model in composition with the ANN $\hat{\gamma}$ replacing γ within the effective reaction rates $\bar{r}$.

3. Numerical Experiments

The fully known ground-truth model was used to generate a dataset $\mathcal{D} = (d_1, \ldots, d_{30})$ of 30 trajectories $d_i = (X, t_X)$. Here, $X \in \mathbb{R}^{n_x \times n_m}$ denotes the collected state trajectory over m measurements in a matrix form and $t_X \in \mathbb{R}^{n_m}$ the corresponding time points. The initial conditions of the different species mole fractions were sampled randomly from a uniform distribution and normalized. Additionally, the condition that $x_{H_2} \in [0.35, 0.8]$ has been imposed by rejecting invalid samples. The initial conditions of the catalyst state x_ϕ are sampled uniformly within the interval [0.2, 0.85]. The operating conditions are sampled from intervals [460.15, 490.15] K for the temperature and [55.0, 65.0] bar for the total pressure. Over all datasets, the incoming species and volumetric flow rate is kept constant. Using the forward solution of the underlying DAE system, dense solutions were created which can be subsampled. Each trajectory was simulated within the time interval $t \in [0, 3000]$ s and a measurement sample rate $t_M \in \mathbb{R}$ of 1 s. Synthetic noise has been added to each observable state[1] $y = \left[x_{CH_3OH}\, x_{CO_2}\, x_{CO}\, x_{H_2}\, x_{H_2O}\, x_{N_2}\right]^T$ with a multiplicative error $\tilde{y}_i = (1 + \epsilon_i) y_i$. The error is sampled from a truncated normal distribution $\epsilon_i \sim \mathcal{N}(0.0, \sigma_M)$, such that $\epsilon_i \in [-0.1, 0.1]$. To investigate the impact of the noise, the variance $\sigma_M \in \mathbb{R}^+$ is varied. To optimize the parameters θ of the ANN under varying noise and sampling rates, a maximum likelihood approach is used

$$\min_{\theta,\sigma,\sigma_\theta} -\sum_{d \in \tilde{\mathcal{D}}} \ln \bar{\mathcal{L}}(d, \theta, \sigma, \sigma_\theta), \tilde{\mathcal{D}} \subseteq \mathcal{D}, s.t.\ \sigma_i \geq 0, i = 1, \ldots, n_y \tag{4}$$

Here $\bar{\mathcal{L}} : \mathbb{R}^{n_y \times n_m} \times \mathbb{R}^{n_\theta} \times \mathbb{R}^{n_y} \times \mathbb{R}^+ \mapsto \mathbb{R}^+$ denotes the likelihood function for the overall problem for a subset $\tilde{\mathcal{D}}$ of the training data

$$\begin{aligned} \ln \bar{\mathcal{L}}(d, \theta, \sigma, \sigma_\theta) = & \overbrace{\sum_{(y_i, t_i) \in d} \ln \mathcal{L}_\Sigma(y_i, \hat{y}_i, \sigma)}^{Data\ fitting} + \overbrace{\sum_{\theta_i \in \Theta} \ln \mathcal{L}_\theta(\theta_i, \sigma_\theta)}^{Parameter\ penalty} \\ & + \underbrace{\sum_{\sigma_i \in \sigma} \ln \mathcal{L}_\sigma(\sigma_i)}_{Prior\ of\ the\ scales} \end{aligned} \tag{7}$$

where the error distributions likelihood is given as $\mathcal{L}_\Sigma : \mathbb{R}^{n_y} \times \mathbb{R}^{n_y} \times \mathbb{R}^{n_y} \mapsto \mathbb{R}^+$. $\mathcal{L}_\Theta : \mathbb{R}^{n_\theta} \times \mathbb{R}^+ \mapsto [0,1]$ represents a prior on the parameters of the ANN. The parameter $\sigma_\theta \in \mathbb{R}^+$ represents the scaling of the prior distribution of the tuneable parameters θ,

---

[1] Strictly speaking x_{H_2O} is not observable. Given that $\sum_i y_i = 1$, we can compute it if all other mole fractions are known.

which may be interpreted as a parameter penalty to avoid overfitting[2]. As it is common, σ_θ is introduced as a hyperparameter rather than a true optimization variable. Finally, $\mathcal{L}_\sigma : \mathbb{R}^+ \mapsto [0,1]$ denotes a prior on the scale parameter of the error distribution $\sigma \in \mathbb{R}^{n_y}$, assuming $\sigma_i \sim \text{Lognormal}(0,1)$.

4. Results

By common practice, hyperparameter optimization was performed to find a suitable architecture for the ANN, the right distribution for the prediction error, and for the prior distribution of the parameters θ and their scale σ_θ. Furthermore, it is assumed that all initial conditions are known, given that only the fitting capabilities of the ANN are investigated. The activation of the output layer has been fixed to be softplus, enforcing positive definite scaling $\hat{\gamma}$. Using a Monte Carlo approach, different candidate configuration are sampled and trained using Eq. (6) with ADAM (Kingma and Ba, 2014). As an evaluation metric the Bayesian Information Criterion (BIC) has been choosen. Hyperparameter optimization on the generated dataset was performed using a sample distance $t_M = 30.0$ s and noise variance $\sigma_M = 5e^{-2}$ with a train-test-split of 0.8. In addition, the training dataset has been batched to reduce training time while keeping a relatively good performance. In total, 300 samples were evaluated. The best architecture consists of the input layer with n_X neurons followed by a single hidden layer with 2 neurons, both using a selu activation. The error and prior distribution have been chosen to be Laplacian distributed. Figure 1 shows the result of the hyperparameter optimization for trajectories from both train and test set. Since the ground truth Eq. (4) is known, we can directly compare the learned model. A statistical evaluation of the approximation error for the data can be seen in Table 1, a plot of the approximation of the missing equations can be seen in Figure 2 (A). The ANN learns the scaling for the hydrogenation pathways over CO_2 close to ideal but fails generalize on CO and RWGS. This is in good agreement with available a priori knowledge that for the conditions employed here, RWGS is close to chemical equilibrium and CO_2 is the dominating path for methanol synthesis.

---

[2] E.g., for a normal distributed variable x with mean μ and variance σ we can write the loglikelihood (up to a constant) as $-\frac{1}{2\sigma^2}(x-\mu)^2 = -\lambda(x-\mu)^2$ which can be thought of as a L_2 penalty with a penalty parameter $\lambda = \frac{1}{2\sigma^2}$.

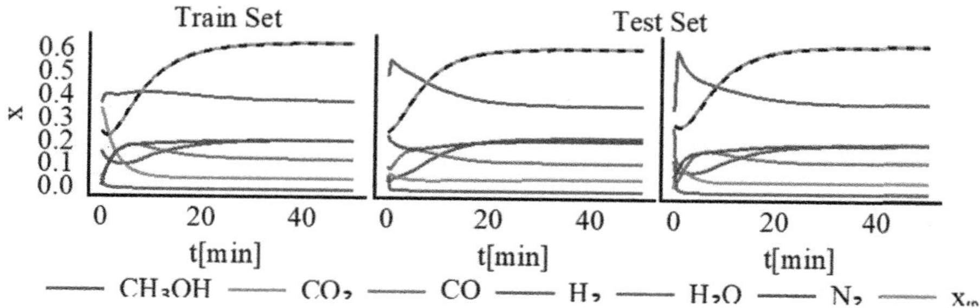

Figure 1 Example trajectories from train and test set. The trained ANNs prediction (full line) from the noisy data (scattered) match the ground truth trajectory of the unobserved state (dashed black).

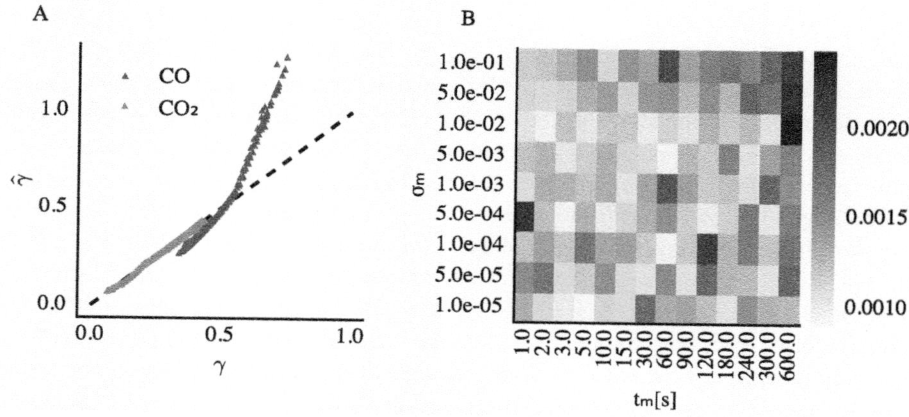

Figure 2 (A) Parity of the ground truth and approximation of the scaling of the hydrogenation pathways evaluated on the test set. RWGS is not shown. (B) Minimum L_1 norm error of the unknown state over varying noise levels and sampling distances evaluated on the test set using the minimum sample distance.

Table 1 : Summary of the L_1 error between the ground truth and approximation of the scaling γ

	Train set				Test set			
	Min	Max	Mean	Std.	Min	Max	Mean	Std.
CO - Hydrogenation	1.3e-4	7.2e-1	7.9e-2	7.0e-2	6.8e-4	5.3e-1	8.0e-2	6.8e-2
CO_2- Hydrogenation	8.5e-7	1.5e-1	5.0e-3	6.3e-3	5.1e-5	2.5e-1	5.5e-3	3.8e-3
RWGS	3.0e-4	2.0e0	7.9e-2	8.0e-2	4.4e-3	1.8e-1	7.4e-2	4.1e-2

For the optimal architecture, the sample distance and noise variance are varied to study the impact of the data quality on the predictive performance. For each pair (σ_M, t_M) the optimal structure is retrained five times and evaluated on the ground truth data with no noise and fully observable measurements. Figure 2 (B) shows the results of the noise variance study. Overall, the ANN performs well with an overall median error 0.0017 on the test set. Only high noise and high measurement rates perform bad.

5. Conclusion

It has been shown that an embedded ANN is able to learn missing components of a physics-based model efficiently using a loglikelihood driven approach. The influence of measurement noise and rate on the predictive performance of the ANN has been

investigated and the resulting model compared to the known ground truth, showing good performance over a wide range of measurement rates and noise levels for the state trajectories. Even if some components cannot be approximated sufficiently, a suitable output is achieved. Several promising future directions open for investigation, e.g., using (optimal) design of experiments, active learning approaches, and learning with additional constraints and unknown parameters like initial conditions. In addition, further testing of our approach on real data collected via an experimental setup will be conducted.

Acknowledgments

The work was funded by the German Research Foundation DFG within the priority program 2331 'Machine Learning in Chemical Engineering' under grants KI 417/9-1, SA 2016/3-1, SE 586/25-1. Carl Julius Martensen was supported by the European Regional Development Fund (ERDF) within the Research Center of Dynamic Systems (CDS).

References

Bezanson, J., Edelman, A., Karpinski, S., Shah, V.B., 2015. Julia: A Fresh Approach to Numerical Computing arXiv:1411.1607.

Cranmer, M., Sanchez Gonzalez, A., Battaglia, P., Xu, R., Cranmer, K., Spergel, D., Ho, S., 2020. Discovering symbolic models from deep learning with inductive biases. Adv Neur In 33, 17429–17442.

Kingma, D.P., Ba, J., 2014. Adam: A method for stochastic optimization. doi:10.48550/ARXIV.1412.6980.

Kovachki, N., Li, Z., Liu, B., Azizzadenesheli, K., Bhattacharya, K., Stuart, A., Anandkumar, A., 2021. Neural Operator: Learning Maps Between Function Spaces arXiv:2108.08481.

Ljung, L., Andersson, C., Tiels, K., Schön, T.B., 2020. Deep learning and system identification. IFAC-PapersOnLine 53, 1175–1181. doi:10.1016/j.ifacol.2020.12.1329.

Olah, G.A., Goeppert, A., Prakash, G.K.S., 2009. Beyond Oil and Gas: The Methanol Economy. Wiley. doi:10.1002/9783527627806.

Rackauckas, C., Ma, Y., Martensen, J., Warner, C., Zubov, K., Supekar, R., Skinner, D., Ramadhan, A., Edelman, A., 2021. Universal Differential Equations for Scientific Machine Learning arXiv:2001.04385.

Raissi, M., Perdikaris, P., Karniadakis, G.E., 2019. Physics-informed neural networks: A deep learning framework for solving forward and inverse problems involving nonlinear partial differential equations. J Comput Phys 378, 686–707. doi:10.1016/j.jcp.2018.10.045.

Seidel, C., Jörke, A., Vollbrecht, B., Seidel-Morgenstern, A., Kienle, A., 2018. Kinetic modeling of methanol synthesis from renewable resources. Chemical Engineering Science 175, 130–138. doi:10.1 016/j.ces.2017.09.043.

Seidel, C., Nikolić, D., Felischak, M., Petkovska, M., Seidel-Morgenstern, A., Kienle, A., 2021. Opti- mization of methanol synthesis under forced periodic operation doi:10.3390/PR9050872.

Antonis Kokossis, Michael C. Georgiadis, Efstratios N. Pistikopoulos (Eds.)
PROCEEDINGS OF THE 33rd European Symposium on Computer Aided Process Engineering (ESCAPE33), June 18-21, 2023, Athens, Greece
 http://dx.doi.org/10.1016/B978-0-443-15274-0.50090-1

Water resource portfolio design for optimum water productivity in agriculture

Fatima-Zahra Lahlou,[a] Rajesh Govindan,[a] Bilal Hazrat,[a] Hamish R. Mackey,[b] Tareq Al-Ansari[a*]

[a]*Division of Sustainable Development, College of Science and Engineering, Hamad Bin Khalifa University, Qatar foundation, Doha, Qatar*
[b]*Department of Civil and Natural Resources Engineering, University of Canterbury, Private Bag 4800, Christchurch 8140, New Zealand*

**Corresponding Author: talansari@hbku.edu.qa*

Abstract

As the demand for food production increases, it is crucial to reassess current water management practices to improve and optimize water use. In this study, the potential of treated sewage effluent in replacing part of the conventional water resources in the agricultural sector is assessed using a mixed integer linear programming optimization model. The proposed model is applied to a case scenario in the state of Qatar, where water from different sources is allocated to four different agricultural subsectors to minimize the carbon footprint while limiting the water stress to a safe global average level which will improve the water productivity of the agricultural sector. Results demonstrate that the proposed model emits twice as much carbon footprint as compared to the base scenario; however, the water stress level is reduced from 323% to 25%, and the water productivity of the agricultural sector increases from \$0.0027 m^{-3} to \$0.036 m^{-3}.

Keywords: Treated Wastewater; Agriculture; Optimization; Water Productivity; Carbon Footprint.

1. Introduction

The global water demand has increased by 600% in the last century and it is now estimated at 4,600 km^3 (Ritchie and Roser, 2018). With the ever-increasing population and living standards, the demand is expected to increase further by at least 20% in the next 30 years (Wada et al., 2016). While there are enough freshwater resources in the world to meet the global demand, the resources are not equally distributed which leads to water scarcity in some parts of the world. Heading the list of water-scarce regions is the middle east, where countries have been relying on different solutions to reduce water stress including desalination and wastewater reuse. The state of Qatar happens to be the most water-scarce country in the world. It has an annual renewable freshwater resources rate as low as 21 $m^3.capita^{-1}.year^{-1}$; nevertheless, it has one of the highest water consumption rates globally with over 182 $m^3.capita^{-1}.year^{-1}$ (Planning and Statistics Authority, 2019). Whether it is groundwater (GW) abstraction, desalination or wastewater reuse, Qatar must take a holistic approach to find the most suitable solutions. Given its status as one of the most water-scarce countries globally, Qatar must continue to prioritize policies and strategies aimed at reducing its reliance on finite freshwater.

The two main direct consequences of the high-water usage in Qatar are the over-abstraction of GW resources and the heavy reliance on energy-intensive desalination to meet the population's needs. The overuse of GW has resulted in an increase in well salinity to different levels across the country, especially near the shore, such that Reverse Osmosis (RO) is required to treat GW before its safe usage. The state of Qatar also produces a large amount of treated sewage effluent (TSE) of which only a small percentage is reused. That said, TSE represents an opportunity to alleviate the water scarcity in Qatar. Currently, around 50% of the produced TSE is reused and is directed for green space irrigation and fodder production (Planning and Statistics Authority, 2019). If public perception about TSE can be overcome, the reuse of TSE represents a promising solution for Qatar as it can reduce the energy consumed in desalination, reduce water stress and increase water productivity (Lahlou et al., 2022, 2021). That said, around 95 million m^3 of TSE is available and can be injected into the agricultural sector which heavily relies on GW resources. Introducing TSE in the rest of the agricultural subsectors would help reduce water stress and increase the level of water productivity.

This study suggests a water planning framework that aims to optimally allocate the available water resources in Qatar which consist of GW resources, desalinated water (DW) and TSE, to four agricultural sub-sectors which currently rely 100% on wells to meet their needs. The objectives are to do so at a reduced carbon footprint and water stress, for an enhanced water productivity. The developed model is generic and can be applied to other regions.

2. Methodology

2.1. Case Study

The agricultural subsectors considered in this study are dairy, red meat, egg and poultry production. The annual GDP from these subsectors is 1,86 million Qatari Ryal (Q.R.) (Planning and Statistics Authority, 2020). The water consumption of these sectors is estimated using the annual production information provided by the local authorities and the global average water required to produce the commodities (Mekonnen and Hoekstra, 2010). The water resources considered are GW, DW, and TSE. Groundwater resources have an annual safe yield of 57 million $m^3.year^{-1}$ and have salinity levels ranging between 1000 $mg.L^{-1}$ up to 12000 $mg.L^{-1}$ based on the well distance from the shore (Baalousha and Ouda, 2017). It is assumed that the energy required for GW pumping is 0.84 $kWh.m^{-3}$ and for RO to get rid of the excess Total Dissolved Solids (TDS) level is between 2.6-5.5 $kWh.m^{-3}$, and that the reject water rate ranges between 30-50% (Chen et al., 2019; Enikeeva and Khadra, 2020; Philippe, 2021). The DW is produced by 13 different plants across the country. The three technologies used are Multi-Stage Flash (MSF), Multi-Effect Distillation (MED) and RO. The least energy-intensive technology is RO as it requires as little as 5$kWh.m^{-3}$ as opposed to MSF and MED which require 21 $kWh.m^{-3}$ (Darwish et al., 2016). The wastewater treatment plants considered in this study are the six plants that receive and treat at least 1% of the total wastewater produced in the country. All six plants use tertiary treatment in addition to Nitrogen & Phosphorus removal (Planning and Statistics Authority, 2019). The supply of DW is carried out using pipeline distribution. As for TSE, two types of distribution are considered, pipeline and truck tankers. Pipeline water distribution requires 0.0055 $Kwh.m^{-3}.km^{-1}$ (Tow et al., 2021). As for truck tankers distribution, it required an average of 0.4 $kWh.km^{-1}$ for every truck tanker with a capacity of 38 m^3 (Lahlou et al., 2020).

2.2. Problem Formulation

The objective of the model is to reduce the carbon footprint associated with the supply and treatment of the three water resources considered in this study and it is described in eq (1) while limiting the water stress level of the GW resources to a safe average of 25% which would enhance the water productivity of the agricultural sector. The water stress level and water productivity are calculated using eq (2) and (3).

$$CF_t = (e^p \sum_{i=1}^{i=4}\sum_{j=1}^{j=6} W_i . x_{p\,i,j}^{TSE} . d_{p\,i,j}^{TSE} + \sum_{i=1}^{i=4} e^{TSE} . W_i \sum_{j=1}^{j=6} x_{p\,i,j}^{TSE} + x_{r\,i,j}^{TSE} + e^p \sum_{i=1}^{i=4}\sum_{j=1}^{j=13} W_i . x_{p\,i,j}^{DW} . d_{p\,i,j}^{DW} + \sum_{i=1}^{i=4} W_i \sum_{j=1}^{j=13} e_j^{DW} . x_{p\,i,j}^{DW}). CF_u + e^{GWp} \sum_{i=1}^{i=15} W_i . x_i^{GW} . (1 + \frac{GW_{RO-i}^{\%} . R_i^{\%}}{1-R_i^{\%}}) + \sum_{i=1}^{i=15} W_i . e_i^{GWt} . x_i^{GW}). CF_u + (\sum_{i=1}^{i=4}\sum_{j=1}^{j=6} ceil\left(\frac{W_i . x_{r\,i,j}^{TSE}}{V_c}\right) . e^r . d_{r\,i,j}^{TSE}) . \frac{CF_d}{E_d} \quad (1)$$

$$WS = \frac{WWD_t}{WA_t - WWI} \quad (2)$$

$$WP = \frac{GDP}{WWD_t} \quad (3)$$

$$WWD_t = \sum_{i=1}^{i=15} W_i . x_i^{GW} . (1 + \frac{GW_{RO-i}^{\%} . R_i^{\%}}{1-R_i^{\%}}) \quad (4)$$

Such that:

- CF_t is the total carbon footprint which needs to be minimized [ton-$CO_{2\text{-eq}}$.month^{-1}],
- $x_{p-i,j}^{DW}$ is the percentage of the water requirement of sinks i that is sourced from the desalination plant j [%],
- $x_{p-i,j}^{TSE}$ represents the percentage of the water requirement of sink i that is sourced from WWTP j via pipeline [%],
- $x_{r-i,j}^{TSE}$ represents the percentage of the water requirement of sink i that is sourced from WWTP j via road [%],
- $d_{p-i,j}^{DW}$ pipeline distance between the desalination plant j and the sink I [km],
- $d_{p-i,j}^{TSE}$ & $d_{r-i,j}^{TSE}$distances between sink i and WWTP j via pipeline and via road respectively [km],
- $x_i^{GW}$is the percentage of the water requirement of sink i that is extracted from GW [%],
- CF_u is the carbon footprint of 1 kWh depending on the country energy mix [g-$CO_{2\text{-eq}}$.kWh^{-1}],
- E_d is the Energy in 1 litre of diesel [kWh.L^{-1}],
- CF_d is the Carbon footprint associated with burning 1 litre of diesel [g-$CO_{2\text{-eq}}$.L^{-1}],
- e^p is the energy required to transport 1 m3 of water through 1 km via pipe [kWh.m^{-3}.km^{-1}],
- e^p is the energy required for truck-tanker to travel 1 km [kWh.km^{-1}],
- e^{TSE} is the energy required to treat 1 m3 of TSE [kWh.m^{-3}],
- e^{DW} is the energy required to produce 1 m3 of DW [kWh.m^{-3}],
- e^{GWp} is the energy required to pump 1 m3 of GW [kWh.m^{-3}],
- $R_i^{\%}$ is the % of rejected water as brine from the RO unit for sink i [%],
- e_i^{GWt} is the energy required to produce 1 m3 of freshwater from GW using RO in sink i [kWh.m^{-3}],
- $GW_{pure-i}^{\%}$ is the percentage of untreated GW required with salinity level TDS_{in-i} [%],
- $GW_{RO-i}^{\%}$ is the percentage of treated GW required with salinity level TDS_f [%],
- $R_i^{\%}$ is the percentage of GW that is rejected as brine after reverse osmosis [%],
- GDP is the total gross domestic product [QAR],
- WWD_t is the total water withdrawals by all economic sectors [million m^3],
- WA_t is the total renewable water availability [million m^3],
- WWI is the water withdrawal intensity [million m^3],

The model is constrained by the capacities of the desalination and wastewater treatment plants, the water requirements of the subsectors, in addition to logical constraints as described by eq (5-11).

$$\sum_{i=1}^{i=15} W_i.(x_{p\,i,j}^{TSE} + x_{r\,i,j}^{TSE}) \leq C_j^{TSE} \quad ,\forall j \qquad (5)$$

$$\sum_{i=1}^{i=15} W_i.x_{p\,i,j}^{DW} \leq C_j^{DW} \quad ,\forall j \qquad (6)$$

$$x_i^{GW} + \sum_{j=1}^{j=6} x_{p\,i,j}^{TSE} + x_{r\,i,j}^{TSE} + \sum_{j=1}^{j=13} x_{p\,i,j}^{DW} = 1\,,\forall i \qquad (7)$$

$$0 \leq x_i^{GW} \leq 1 \quad ,\forall i \qquad (8)$$

$$0 \leq x_{p\,i,j}^{TSE} \leq 1 \quad ,\forall i \qquad (9)$$

$$0 \leq x_{p\,i,j}^{TSE} \leq 1 \quad ,\forall i \qquad (10)$$

$$0 \leq x_{p\,i,j}^{DW} \leq 1 \quad ,\forall i \qquad (11)$$

3. Results and Discussion

The total GW usage of the base scenario is estimated at over 184 million $m^3.year^{-1}$. Assuming that the considered agricultural subsectors are the only users of GW resources, the excessive usage puts the water stress level at over 300%, while the global water stress, which is considered to be safe, is around 18.6% (United Nations, 2022). The carbon footprint of the base scenario is found to be 121.3 thousand ton-$CO_{2\text{-eq}}$. As for the water productivity it is estimated to be as little as 0.01 Q.R.m^{-3} which is equivalent to \$0.0027 m^{-3}.

For limited water stress that is set below or equal to 25%, and enhanced water productivity, the minimum carbon footprint that can be achieved is 253.2 thousand ton-$CO_{2\text{-eq}}$ which is more than twice that of the base scenario. This is due to the limited GW and TSE resources and the need to meet the rest of the demand via energy-intensive desalination. In fact, more than 70% of the calculated carbon footprint comes from the sea water desalination process (Figure 1). Although the carbon footprint of the base scenario is lower than what can be achieved using the proposed scenario, the water stress level is reduced from 323% to 25%, and the water productivity level increases from 0.010 Q.R.m^{-3} to as much as 0.13 Q.R.m^{-3}, equivalent to \$0.036 m^{-3}. That said, TSE has the potential to alleviate the water stress as well as increase water productivity in the state of Qatar. The water allocation required to achieve these results is described in Figure 2.

Given that all wastewater produced in the State of Qatar has to be treated before being reused or disposed of, it can be argued that the carbon footprint associated with the process of treatment should not be borne by the agricultural sector. This would result in a 19% reduction in the carbon footprint. In addition to that, sludge produced from WWTPs can be used to offset part or all of the carbon footprint associated with the practice, if converted to power.

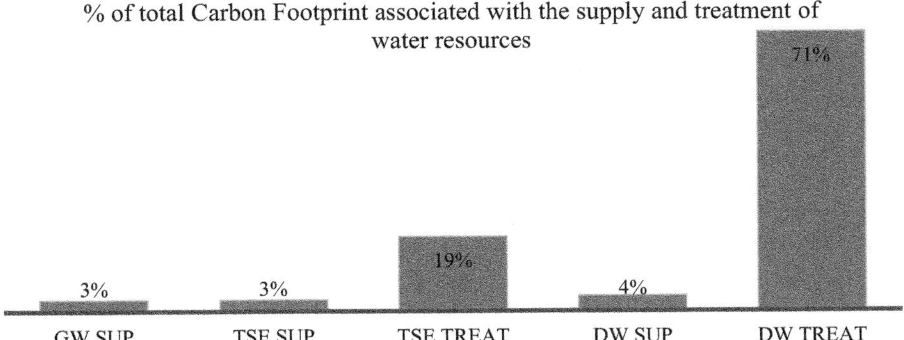

Figure 1: Percentage of the carbon footprint associated with the supply (SUP) and treatment (TREAT) of groundwater (GW) resources, TSE, and desalinated water (DW).

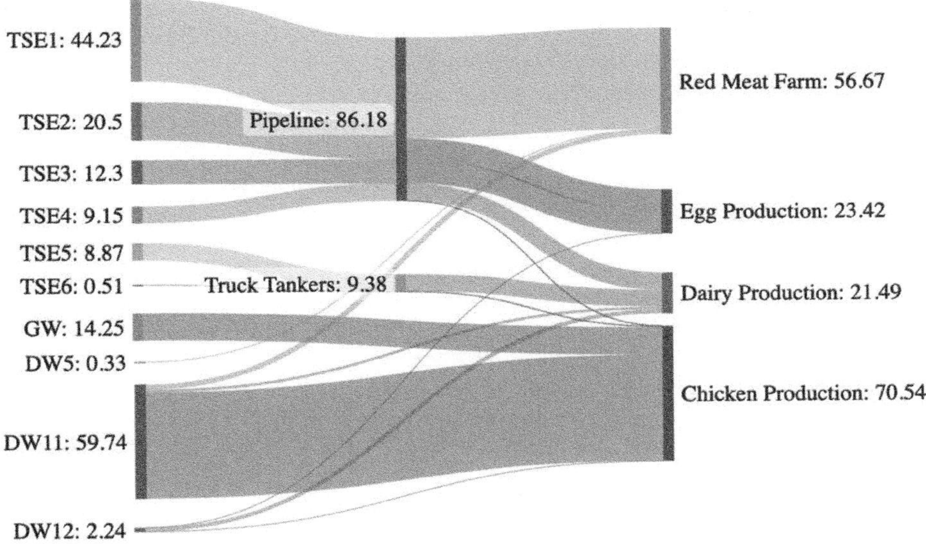

Figure 2: Water resources allocation for the four agricultural subsectors in million $m^{-3}.year^{-1}$.

4. Conclusion

The state of Qatar is one of the most water-scarce countries in the world but also one of the highest users of water. For this reason, there is a need for sustainable water budget planning in the country. One of the possible solutions that can alleviate the water stress in the State of Qatar is the reuse of TSE. This work investigates the optimum water planning framework for GW, DW and TSE water use in four agricultural subsectors for a reduced carbon footprint, limited water stress and improved water productivity. Results suggest that limiting the use of GW resources will increase the carbon footprint associated with the supply and treatment of the agricultural sector's water requirements, as the needs will need to be met through desalinating seawater. However, this scenario is still considered an improved one since the water stress is reduced from 323% to 25% and the resulting water productivity level is 13 times higher than the base scenario. For future work, it would be interesting to include the other subsectors of the agricultural sector such as outdoor and indoor farming. In addition to that, it would be interesting to look at how the use of sludge, another important byproduct of WWTPs, can further improve the environmental aspects of the agricultural sector.

5. Acknowledgement

This publication was made possible by grant reference MME01-0922-190049 from the Ministry of Municipality and Environment in the State of Qatar. The findings herein reflect the work and are solely the responsibility of the authors.

References

Baalousha, H.M., Ouda, O.K.M., 2017. Domestic water demand challenges in Qatar. Arab. J. Geosci. 10, 537. https://doi.org/10.1007/s12517-017-3330-4

Chen, X., Thorp, K.R., Ouyang, Z., Hou, Y., Zhou, B., Li, Y., 2019. Energy consumption due to groundwater pumping for irrigation in the North China Plain. Sci. Total Environ. 669, 1033–1042. https://doi.org/10.1016/j.scitotenv.2019.03.179

Darwish, M.A., Abdulrahim, H.K., Hassan, A.S., 2016. Realistic power and desalted water production costs in Qatar. Desalin. Water Treat. 57, 4296–4302. https://doi.org/10.1080/19443994.2014.992977

Enikeeva, K., Khadra, C., 2020. Water saving greenhouse, trial and demonstration center. Doha.

Lahlou, F.-Z., Mackey, H.R., Al-Ansari, T., 2022. Role of wastewater in achieving carbon and water neutral agricultural production. J. Clean. Prod. 339, 130706. https://doi.org/10.1016/j.jclepro.2022.130706

Lahlou, F.-Z., Mackey, H.R., McKay, G., Onwusogh, U., Al-Ansari, T., 2020. Water planning framework for alfalfa fields using treated wastewater fertigation in Qatar: An energy-water-food nexus approach. Comput. Chem. Eng. 141, 106999. https://doi.org/10.1016/j.compchemeng.2020.106999

Lahlou, F.Z., Mackey, H.R., McKay, G., Al-Ansari, T., 2021. Reuse of treated industrial wastewater and bio-solids from oil and gas industries: Exploring new factors of public acceptance. Water Resour. Ind. 26, 100159. https://doi.org/10.1016/j.wri.2021.100159

Mekonnen, M.M., Hoekstra, A.Y., 2010. The green, blue and grey water footprint of farm animals and animal products.

Philippe, B., 2021. Desalination of seawater and brackish water. Encycl. l'énergie.

Planning and Statistics Authority, 2020. Agricultural Statistics. Doha.

Planning and Statistics Authority, 2019. Water Statistics in the State of Qatar. Doha.

Ritchie, H., Roser, M., 2018. Water use and stress.

Tow, E.W., Hartman, A.L., Jaworowski, A., Zucker, I., Kum, S., AzadiAghdam, M., Blatchley, E.R., Achilli, A., Gu, H., Urper, G.M., Warsinger, D.M., 2021. Modeling the energy consumption of potable water reuse schemes. Water Res. X 13, 100126. https://doi.org/10.1016/j.wroa.2021.100126

United Nations, 2022. Ensure availability and sustainable management of water and sanitation for all.

Wada, Y., Flörke, M., Hanasaki, N., Eisner, S., Fischer, G., Tramberend, S., Satoh, Y., van Vliet, M.T.H., Yillia, P., Ringler, C., Burek, P., Wiberg, D., 2016. Modeling global water use for the 21st century: the Water Futures and Solutions (WFaS) initiative and its approaches. Geosci. Model Dev. 9, 175–222. https://doi.org/10.5194/gmd-9-175-2016

Antonis Kokossis, Michael C. Georgiadis, Efstratios N. Pistikopoulos (Eds.)
PROCEEDINGS OF THE 33rd European Symposium on Computer Aided Process Engineering (ESCAPE33), June 18-21, 2023, Athens, Greece
 http://dx.doi.org/10.1016/B978-0-443-15274-0.50091-3

S-GNN: State-Dependent Graph Neural Networks for Functional Molecular Properties

Adem R.N. Aouichaoui, Alessandro Cogliati, Jens Abildskov, Gürkan Sin*

Process and Systems Engineering Centre (PROSYS), Department of Chemical and Biochemical Engineering, Technical University of Denmark, Søltofts Plads, Building 228A, 2800 Kgs. Lyngby, Denmark

*gsi@kt.dtu.dk

Abstract

Property models are an integral part of many chemical engineering applications and have been the subject of a lot of interest, especially with recent advancements in deep learning such as graph neural networks. Despite being of major importance, little effort has been dedicated to functional properties where the property dependency goes beyond the molecular structural information and depends on the state variables such as temperature and pressure. In this work, we demonstrate a flexible framework to extend graph neural networks to account for such use cases. A total of 13 different temperature-dependent properties were modeled covering enthalpy of vaporization, various heat capacities, densities, and thermal conductivities as well as surface tension and vapor pressure. While many were successfully modeled with some reaching an average absolute relative error below 6%, some still require further attention such as surface tension and thermal conductivity to achieve good accuracy.

Keywords: Graph neural networks, Thermophysical properties, Molecules, QSPR.

1. Introduction

In-silico evaluation of the various thermal, physical, safety, and transport properties of molecules is an integral element in many process simulation and product design software. These properties are usually the product or derived from experimental measurements and as such predictive models are necessary to perform in-silico evaluation of the properties. Quantitative structure-property relationships (QSPRs) are mathematical models that can provide such a service by leveraging the structural information of the molecule in inferring its property (Aouichaoui et al., 2023, 2022). Molecular properties of interest in the chemical engineering field can generally be divided into primary, secondary, and functional properties (Kontogeorgis and Gani, 2004). The primary properties are single-value properties that can be determined solely from the structural information, while secondary properties cannot be explicitly calculated uniquely from the structural information but rather as a function of other properties. An example of primary properties is the critical point properties as well as the normal boiling point, while the secondary properties could be the acentric factor or the solubility parameters. Both types of properties have long been the subject of various QSPR modeling through e.g. the use of group-contribution methods (Hukkerikar et al., 2012) as well as more advanced deep learning models (Aouichaoui et al., 2023, 2022). While much focus has been on taking advantage of the recent development in the field of deep learning (DL) to develop increasingly better QSPR models for primary properties and secondary properties, little attention has been given to functional properties. Functional properties on the structural information as well as a set of intensive variables such as temperature, pressure, or composition. Such properties include saturated liquid density, vapor pressure, heat capacities, and thermal conductivity. Predictive models that can infer these properties are an integral part of any process simulation software as many unit operations are operated at varying temperatures and pressures. These properties are usually predicted through correlations that include compound-specific constants and as a function of other properties of the compounds. An example of the correlation of saturated vapor pressure (P^{sat}) using the Antoine equation as seen in Eq.(1), where T is the temperature and A, B, and C are compound-specific constants determined through regression. A further example is a correlation for the liquid density (ρ^{sat}) using the modified Racket equation as seen in Eq.(2), where R is the universal gas constant, T_c and P_c are the ciritcal temperature and pressure respectively, T_r is the reduced temperature and Z_{RA} is a compound specific constant determined through regression.

$$log_{10}P^{sat} = A - \frac{B}{T + C} \tag{1}$$

$$\frac{1}{\rho^{sat}} = \frac{R\,T_c}{P_c} Z_{RA}^{(1+(1-T_r)^{2/7})} \tag{2}$$

The two examples show that the property is dependent on other inputs that are either determined experimentally or through other correlations (e.g. the use of other property models to evaluate the critical properties). One major drawback of this is the fact that uncertainties become untraceable due to many input parameters. A further issue is when the property evaluation requires other QSPR models. As such, we advocate the leveraging end-to-end learning framework from the field of deep learning and extending it to build property models for functional properties that solely rely on the molecular structure and the intensive state variable such as the temperature. The developed framework is denoted state-dependent graph neural networks (S-GNN) and is elaborated on in the following.

2. S-GNN

Graph neural networks (GNNs) provide a unified approach that combines the feature extraction process and the regression process (Aouichaoui et al., 2022). GNNs takes as input a graph representation of the molecule where nodes correspond to atoms and edges correspond to the bonds. The node and edge are associated with feature vectors with information about the atoms and bond e.g. atom type, number of hydrogen attached, and bond type. Using a series of operations (also called convolutional layers) involving matrix multiplication, the graph representation is updated and converted into a numerical vector that is used as the molecular descriptor. This process is denoted message-passing and governs the feature extraction process and distinguishes GNN models from each other.

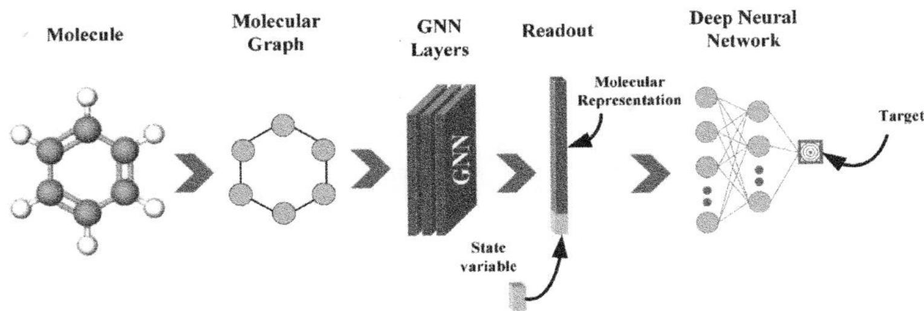

Figure 1: Schematic of the workflow of S-GNN

The S-GNN extends on the classical GNN models by considering the state variables as global variables that are concatenated to the learned molecular representation resulting from the GNN feature extraction step as shown in Figure 1. In this work, we demonstrate the concept using a GNN model known as *attentiveFP*, which in recent studies has demonstrated state-of-the-art performance across many properties (Aouichaoui et al., 2022; Xiong et al., 2020). The model combines the attention mechanism and recurrent neural networks on two levels: node-level like any other GNN model and as part of the readout by operating on a graph where the molecule is considered a super node, while all atoms are linked to it. The node operations can be seen in Eq.(3) to Eq.(8). The initialization step aims to dimension the atom features (x_v^{atom}) of node v to the desired hidden dimension through a learnable weight matrix W and a non-linear activation function. The edge features x_v^{atom} of neighboring atoms are combined with the node features to produce the latent representation of the neighboring atoms. The alignment aims to combine the node representation with that of its neighbors, which is used as input to a softmax to produce a weight coefficient. The coefficient is used as a weight for the importance/influence the hidden representation of the neighboring node should have when updating the hidden representation of a given node as seen in the context.

Initialization

$$h_v^0 = ReLU(W \cdot x_v^{atom}) \quad for\ t = 0 \tag{3}$$

$$h_w^0 = ReLU\left(W \cdot [x_w^{atom}, x_{v,w}^{bond}]\right) \quad for\ \mathrm{t} = 0 \tag{4}$$

Alignment

$$\epsilon_{v,w}^{t+1} = leaky_{ReLU}(W \cdot [h_v^t, h_w^t]) \ \ \mathrm{for\ t+1 \leq T} \tag{5}$$

Weighting

$$\alpha_{v,w}^{t+1} = softmax\left(\epsilon_{v,w}^{t+1}\right) \ \ \mathrm{for\ t+1 \leq T} \tag{6}$$

Context:

$$C_v^{t+1} = elu\left(\sum_{w \in \mathrm{Neighbor}(v)} \alpha_{v,w} \cdot W \cdot h_w^t\right) \tag{7}$$

Update

$$\mathrm{h}_v^{t+1} = GRU(h_v^t, C_v^{t+1}) \ \ \mathrm{for\ t+1 \leq T} \tag{8}$$

Similar reasoning is applied for the graph level embedding seen in Eq.(9) to Eq.(13), where L is the number of graph embedding layers and s denotes the supernode.

Initialization $$h_s^0 = \sum_{v\ in\ G} h_v^T \text{ for } l = 0 \tag{9}$$

Alignment $$\epsilon_{s,v}^{l+1} = leaky_{ReLU}(W \cdot [h_s^l, h_v^T]) \text{ for } \mathrm{l} + 1 \leq \mathrm{L} \tag{10}$$

Weighting $$\alpha_{s,v}^{l+1} = softmax(\epsilon_{s,v}^{l+1}) \text{ for } \mathrm{l} + 1 \leq \mathrm{L} \tag{11}$$

Context $$C_s^{l+1} = elu\left(\sum_{v \in \mathrm{Neighbor}(s)} \alpha_{s,v} \cdot W \cdot h_v^t \text{ for } \mathrm{l} + 1 \leq \mathrm{L}\right) \tag{12}$$

Update $$h_s^l = GRU(h_s^l, C_s^{l+1}) \text{ for } l + 1 \leq \mathrm{L} \tag{13}$$

3. Case study: modeling of 13 temperature-dependent properties

3.1. Data

A total of 13 temperature-dependent properties were investigated: ideal gas heat capacity (ICP), liquid heat capacity (LCP), solid heat capacity (SCP), liquid and solid density (LDN, SDN), enthalpy of vaporization (HVP), surface tension (ST), liquid-solid and vapor thermal conductivity (LTC, STC, VTC), vapor pressure (VP) and liquid-vapor viscosity (LVS, VVS). All data were retrieved from the AIChE DIPPR 801 Database and only considered experimental data, smoothed or derived (Rowley et al., 2019). Chemical diversity analysis showing the different species (hydrocarbons, oxygenated, nitrogenated, chlorinated, fluorinated, brominated, iodinated, sulfonated, phosphorated, silicon-containing, or multi-functional) present can be seen in Table 1 where the number of total distinct species (N_{spec}) and the total amount of observations (N_{obs}) are also reported.

Table 1: Chemical diversity analysis of the dataset

	CH	O	N	Cl	F	Br	I	P	S	Si	Mul	N_{spec}	N_{obs}
ICP	30	23	4	3	9	1	1	0	7	0	10	88	1,529
LCP	219	285	59	26	16	13	6	0	44	0	73	741	13,831
SCP	167	163	33	17	12	5	0	0	38	0	52	487	16,567
LDN	412	626	107	63	43	17	7	0	70	0	194	1,539	25,485
SDN	57	129	18	11	2	4	3	0	5	0	50	279	718
HVP	397	601	105	61	36	18	7	0	66	0	183	1,474	24,279
ST	104	337	61	42	25	14	7	0	42	0	104	736	6,814
LTC	89	157	26	28	16	10	5	0	2	0	42	375	5,104
STC	2	9	1	0	0	0	0	0	0	0	4	16	132
VTC	47	34	9	6	18	2	1	0	1	0	17	135	1,550
VP	411	629	105	65	39	18	7	0	71	0	195	1,540	50,232
LVS	200	392	71	45	23	15	7	0	35	0	117	905	12,192
VVS	49	26	4	7	15	2	1	0	3	0	13	120	1,734

3.2. Model training

For the model training, the data were split randomly with 70% allocated for training and the remaining evenly split between validation and testing. No data scaling was performed. The model was trained over a maximum of 500 epochs using the ADAM optimizer, with mean absolute error as a loss function and the early stopping policy as a contingency against overfitting. The training-specific hyperparameters and their bounds consisted of: the batch size ([16, 1024]), the initial learning rate ([10^{-6}, 10^{-1}]), the weight decay ([0, 0.25]), and the dropout ([0, 0.25]). The model-specific hyperparameters covered: the number of node embedding layers ([1,4]), the number of graph embedding layers ([1,4]), the length of the graph feature ([50,512]), and the number of DNN layers ([1,3]). The hyperparameter optimization was conducted using Bayesian optimization with a Gaussian process (GP) model with Matern32 kernel as covariance function, with 100 random exploration steps and 100 BO steps while the acquisition strategy chosen was as the upper confidence bound (UCB). The BO was performed using the "*Bayes_Opt*" toolbox in python (Jiménez and Ginebra, 2017). The node and edge features used in this work are similar to those in previous studies (Aouichaoui et al., 2022; Xiong et al., 2020). The model can be accessed via: https://github.com/gsi-lab/s-afp. Since the data are proprietary, they are not made available, but the workings of the model are illustrated using another freely available dataset.

3.3. Results

The performance of the models in terms of the R^2 and mean absolute percentage error and the mean absolute error and vis á vis the test set as well as the complete dataset can be seen in Table 2, an example of the parity plot for the liquid density and the saturated vapor pressure can be seen in Figure 2.

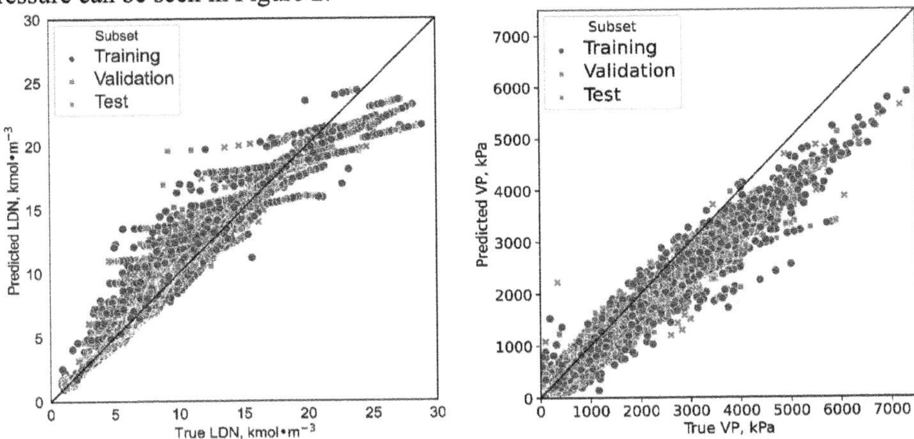

Figure 2: Parity plots for liquid density (LDN-left) and vapor pressure (VP-right)

Table 2: Model performance considering the test set and the overall data

Property	Unit	R^2		MAPE		MAE	
		test	all	test	all	test	all
ICP	[$J \cdot kmol^{-1} \cdot k^{-1}$]	0.99	0.99	1.2	1.1	1,958	1,678
LCP	[$J \cdot kmol^{-1} \cdot k^{-1}$]	0.99	0.99	3.6	3.3	7,810	7,260
SCP	[$J \cdot kmol^{-1} \cdot k^{-1}$]	0.98	0.98	110.4	67.9	10,360	10,370
LDN	[$kmol \cdot m^{-3}$]	0.95	0.96	5.9	5.9	0.50	0.50
SDN	[$kmol \cdot m^{-3}$]	0.98	0.98	4.3	4.7	0.49	0.46

HVP	[J·kmol^{-1}]	0.96	0.96	5.6	8.6	2.3e6	2.2e6
ST	[N·m^{-1}]	0.63	0.63	16.3*	15.8*	5e-3	5e-3
LTC	[W·m^{-1}·k^{-1}]	0.96	0.97	3.7	3.4	0.005	0.004
STC	[W·m^{-1}·k^{-1}]	0.12	0.44	30.1	50.4	0.06	0.10
VTC	[W·m^{-1}·k^{-1}]	0.98	0.88	12.1	12.4	3e-3	3e-3
VP	[kPa]	0.95	0.95	35.6	35.9*	58.6	60.31
LVS	[Pa·s]	N.A	0.35	204e2*	203e2*	1.2	2.7
VVS	[Pa·s]	0.90	0.91	9.9	5.7	1.1e-6	9.4e-7
*: Median absolute percentage error was used instead							

Many properties achieve good accuracy with 7 datasets reaching below 10% MAPE. In some cases, the use of the MAPE is not suitable as some datapoint are very small and close to zero and as such even small errors lead to a very large relative error which is misleading instead we suggest using the median absolute percentage error as it is robust to such outliers in the sample data. Especially the results obtained for the LDN, suggest the model results are similar to the state-of-the-art reported in previous studies using the modified Racket equation (Spencer and Adler, 1978). However, the absence of overall performance metrics makes it difficult to perform a comparison. However, noteworthy is also the fact that some datasets present a challenge to achieve a good accuracy performance, especially considering STC, LVS, and to some extent ST. One possible explanation for the STC dataset is the fact that the chemical diversity is very low with only 16 distinct molecules, which makes it difficult to learn any suitable representation. For the LVS, we observed some experimental data that were 7 times larger in order of magnitude, which might have been data that were tagged as derived. This highlights the requirement of addressing the data quality used during the model training as these are detrimental in allowing the model to extract knowledge it can correlate to the target property. For the ST, we observed a wide range of data with a notable bias (an intercept on the y-axis) when constructing the parity plot. This could suggest that the learning of the model has not achieved convergence enough to be able to reduce the variance observed. Another piece of evidence is the observation on the magnitude of the data points was between 0 and 0.06, which would require an even smaller learning rate than the one used during BO. One possible solution would be to perform data scaling as this will limit the spread of the data in addition to performing outlier detection to remove odd data present in the dataset.

4. Conclusions

In this work, we provide an extension to graph neural networks that can generalize to any graph neural network and allow it to evaluate properties that depend on system state variables such as temperature and pressure in addition to the structural information of the molecule. Out of 13 properties evaluated, the majority of the properties were modeled with high accuracy (R^2>0.9), while surface tension and thermal conductivity properties need further focus as they prove challenging to predictively model.

References

Aouichaoui, A.R.N., Fan, F., Mansouri, S.S., Abildskov, J., Sin, G., 2023. Combining Group-Contribution Concept and Graph Neural Networks Toward Interpretable Molecular Property Models. J Chem Inf Model.

Aouichaoui, A.R.N., Mansouri, S.S., Abildskov, J., Sin, G., 2022. Uncertainty estimation in deep learning-based property models: Graph neural networks applied to the critical properties. AIChE Journal 68.

Hukkerikar, A.S., Sarup, B., ten Kate, A., Abildskov, J., Sin, G., Gani, R., 2012. Group-contribution + (GC +) based estimation of properties of pure components: Improved property estimation and uncertainty analysis. Fluid Phase Equilib 321, 25–43.

Jiménez, J., Ginebra, J., 2017. pyGPGO: Bayesian Optimization for Python. The Journal of Open Source Software 2, 431–433.

Kontogeorgis, G.M., Gani, R., 2004. Computer Aided Property Estimation for Process and Product Design: Computers Aided Chemical Engineering. Elsevier.

Rowley, R.I., Wilding, W.V., Oscarson, J.L., Giles, N.F., 2019. DIPPR Data Compilation of Pure Chemical Properties.

Spencer, C.F., Adler, S.B., 1978. A critical review of equations for predicting saturated liquid density. J Chem Eng Data 23, 82–89.

Xiong, Z., Wang, D., Liu, X., Zhong, F., Wan, X., Li, X., Li, Z., Luo, X., Chen, K., Jiang, H., Zheng, M., 2020. Pushing the boundaries of molecular representation for drug discovery with the graph attention mechanism. J Med Chem 63, 8749–8760.

Antonis Kokossis, Michael C. Georgiadis, Efstratios N. Pistikopoulos (Eds.)
PROCEEDINGS OF THE 33rd European Symposium on Computer Aided Process Engineering (ESCAPE33), June 18-21, 2023, Athens, Greece
 http://dx.doi.org/10.1016/B978-0-443-15274-0.50092-5

Switching Multi-Objective Dynamic Optimization (MODO) for the Production of Value-Added Products

Juan C. Acosta-Pavas,[a] Carlos. E Robles-Rodríguez,[a] Jérôme Morchain,[a] David Camilo Corrales,[b] Claire Dumas,[a] Arnaud Cockx,[a] César A. Aceves-Lara,[a]

[a]*TBI, Université de Toulouse, CNRS, INRAE, INSA, Toulouse, France*
[b]*ToulouseWhite Biotechnology (TWB), INRAE, 135 Avenue de Rangueil, 31077 Toulouse, France*

Abstract

A Multi-Objective Dynamic Optimization (MODO) applied to the biomethanation process is developed. The MODO strategy was designed to find the trade-off between the maxima yield and productivity of methane and acetate modifying the inlet liquid and gas flow rates. First, a multi-objective optimization was applied to find the Pareto optimal set between productivity and yield of methane and acetate independently. Then, Pareto optimal points (POPs) were selected to develop five cases in dynamic optimization, which used a Model Predictive Control (MPC) approach. Cases 1-2 corresponded to the model simulation and the use of the POPs directly in simulation. Cases 3-4 addressed the POPs to maximize the Euclidean length for methane and acetate. Case 5 was performed to demonstrate a switch between these objectives. The ability of the MODO strategy to perform the switch was demonstrated. Additionally, the dynamic optimization allowed the reduction of the inlet gas flow rate to $1.5 m^3/d$.

Keywords: Biomethanation, Multi-objective Optimization, Pareto Optimal Set, Dynamic Optimization, Process Control

1. Introduction

In Anaerobic Digestion (AD) organic matter is transformed by the synergistic work of different microorganisms into CH_4, and CO_2 through four steps: hydrolysis, acidogenesis, acetogenesis, and methanogenesis. The produced biogas contains 50-75% of CH_4, and 25–50 % of CO_2 (Iglesias et al., 2021). Biomethanation uses microorganisms to convert the CO_2 in the biogas from AD and syngas (a mix of H_2, CO, and CO_2) into CH_4. This process, however, could also be used to produce value-added products such as acetate (Chaikitkaew et al., 2021). Acetate serves as a chemical platform in the textile, polymer, pharmaceutical, and food industries (Martín-Espejo et al., 2022). Nevertheless, controlling this type of process is an arduous task due to the multiple reactions and microorganisms involved. As a result, obtaining desired performances of yields or productivities at an industrial scale remains difficult, particularly when you need to optimize several of them simultaneously. The use of dynamic models plays a crucial role in the design of control strategies. For instance, Model Predictive Control (MPC) (Morales-Rodelo et al., 2020) is implemented to maintain or optimize several variables. MPC refers to control actions that optimize a criterion in a future behavior of the system, which is determined by the dynamic model (Camacho and Bordons, 2007). Multi-

Objective Optimization (MOO) implies optimizing problems where there is more than one objective to be optimized simultaneously. These objectives are usually conflictive (Vertovec et al., 2021).
This work aims at proposing a Multi-objective Dynamic Optimization (MODO) to maximize yield and productivity of the biomethanation process regarding two of the potential products: CH_4 and acetate. The proposed dynamic optimization approach used a MPC with two control variables corresponding to the inlets of liquid and gas flow rates. MPC uses the Pareto optimal sets, where each solution is considered as a Pareto Optimal Point (POP). The robustness of the MODO strategy is analyzed by performing a switch between the optimized values for CH_4 and acetate.

2. Multi-Objective Dynamic Optimization as Control Strategy

A MODO strategy was proposed in a previous work (Acosta-Pavas et al., 2022a) to determine the optimal values of the objectives. This strategy entails the following steps:
Step 1-Model definition. Proposition of the dynamic model to represent the biological process.
Step 2-Definition of the objective functions $Y_1^*, \dots, Y_m^*$ to be maximized/minimized in the MOO. A MOO can be formulated as a minimization problem, Eq (1).

$$\min_{x,u,p,t} \{Y_1^*(x,u,p), \dots, Y_m^*(x,u,p)\} \tag{1}$$

$$Subject\ to \begin{cases} dy/dt = \xi(x,u,p,t) & \\ f_i(x,u,p,t) \leq 0 & i = 1,2,\dots,n \\ g_i(x,u,p,t) = 0 & i = 1,2,\dots,k \\ u^L \leq u \leq u^U & \end{cases} \tag{2}$$

where $Y_1^*, \dots, Y_m^*$ are the m objective functions, x the state variables, u the control variables and p the control variables. The dynamic model is represented by dy/dt; f_i and g_i indicate inequality and equality constraints. u^L and u^U corresponds to the lower and upper bounds of the control variables.
Step 3-Selection of the POP. Determination of the Pareto optimal set $Y_1^*, \dots, Y_m^*$ and selection of the POP to be used as the reference trajectory in the dynamic optimization.
Step 4-Definition of the dynamic problem with a single weighted objective. Formulation of an objective function considering the previously identified POP in terms of a MPC problem. The dynamic optimization determines the input variables that minimize the following objective function,

$$\min_u \left(\sum_{j=t}^{t+H_p} \left(\frac{|Y^* - y(t+j|t)|}{Y^*} \right)^2 + \sum_{j=t}^{t+H_c} W_u\, \Delta u(t+j|t)^2 \right) \tag{3}$$

where u is the vector of the control variables; H_p and H_c are the prediction and control horizons; $y(t+j|t)$ is the output prediction calculated at time instant $t+j$ using the information available at time instant t. Y^* is a reference trajectory that enables to reach the set point and is determined by the MOO. The term $\Delta u(t+j|t)$ is the control move at time instant $t+j$ calculated using information available at time instant t. The problem in Eq. (3) is also subject to Eq. (2).

3. Multi-objective Dynamic Optimization in Biomethanation Process

The main goal was to optimize yields and productivities of two value-added products, CH_4 and acetate, in order to demonstrate that a control strategy could attained larger values than the ones obtained from the model simulation. Two manipulated variables were proposed for the optimization: the inlet gas $\left(q_{gas}^{in}\right)$ and liquid (q_{liq}^{in}) flow rates.

Step 1. The model used for simulation was proposed in a previous study by Acosta-Pavas et al., (2022b). This model was built upon experimental data from (Sun et al., 2021) and it considered the conversion of H_2 and CO to enhance CH_4 production. The experimental data from Sun et al. (2021) was carried out in a bubble column reactor with a working volume of 3.75×10^{-2} m^3 and a hydraulic retention time (HRT) of 20 days. The reactor operated at 37°C for 207 days. The organic loading rate (OLR) corresponded to 0.53 $kg_{COD}/m^3\,d$ with q_{liq}^{in} of $1.9\times10^{-3}\ m^3/d$. Gas addition was carried out in five stages (I - V) in which q_{gas}^{in} and the gas loading rate (GLR) were varied in time. The model (Acosta-Pavas et al., 2022b) can be rewritten as,

$$\frac{dS_{gas,i}}{dt} = \frac{q_{gas}^{in}}{V_{gas}} S_{gas,i}^{in} + N_i\left(\frac{V_{liq}}{V_{gas}}\right) - \frac{q_{gas}}{V_{gas}} S_{gas,i} \tag{4}$$

$$\frac{dS_{liq,j}}{dt} = \frac{q_{liq}^{in}}{V_{liq}}(S_{liq,j}^{in} - S_{liq,j}) + \sum_k Y_k f_{j,k}\mu_k - N_i \tag{5}$$

$$\frac{dX_k}{dt} = \frac{q_{liq}^{in}}{V_{liq}}(X_k^{in} - X_k) + Y_k\mu_k - \mu_{k,dec} \tag{6}$$

Sub-indices j = 1-8 denote glucose, butyrate, propionate, acetate, H_2, CH_4, CO, and CO_2 in liquid phase; sub-indices $i = H_2$, CH_4, CO, and CO_2 in the gas phase; while sub-indices k = biomass that degraded glucose, butyrate, propionate, acetate, H_2, and CO in liquid phase. q_{gas} is the outlet gas flow rate, V_{liq} and V_{gas} are the liquid and gas volume, respectively. $S_{liq,j}^{in}$, $S_{gas,i}^{in}$, and X_k^{in} are the inlet concentrations of components j in the liquid phase, the inlet concentration of components i in the gas phase, and the inlet concentration of biomass k in the liquid phase. Y_k is the yield of biomass k, $f_{j,k}$ are the stoichiometric coefficients, μ_k and $K_{k,dec}$ are the growth rate and the decay constant for biomass k. Finally,N_i is the mass transfer rate of component i.

Step 2. Two objectives were considered to optimize: the yields Y_{CH4}, Y_{ac} and productivities P_{CH4}, P_{ac}, for CH_4 and acetate, respectively. These are defined as,

$$Y_{CH4} = \frac{q_{gas,CH4}}{kgCOD_{added}}\ ;\ P_{CH4} = \frac{q_{gas,CH4}}{V_{liq}} \tag{7}$$

$$Y_{ac} = \frac{S_{liq,ac}}{kgCOD_{added}};\ P_{ac} = \frac{S_{liq,ac}}{V_{liq}} \tag{8}$$

The MOO for the maximization of Y_h and P_{hi} was proposed as,

$$\max_{\{q_{gas}^{in},\, q_{liq}^{in}\}} (P_h, Y_h) \tag{9}$$

$$Subject\ to \begin{cases} Equations\ 4-6 \\ Y_{CH4} \leq Y_{CH4}^{max} = 0.39\ m_{CH4}^3/kgCOD_{added} \\ 1\times10^{-3} \leq q_{gas}^{in} \leq 10\ m^3/d \\ 1\times10^{-3} \leq q_{liq}^{in} \leq 0.1\ m^3/d \end{cases} \tag{10}$$

where $h = CH_4, acetate$, Y_{CH4}^{max} corresponds to $0.39\ m_{CH4}^3/kgCOD_{added}$ which represents the theoretical yield of CH_4 that could be obtained by kg of COD at 37°C. This

corresponds to $0.35 m^3_{CH4}/kgCOD_{added}$ at standard temperature and pressure conditions (only for P_{CH4}, and Y_{CH4} maximization).
In this study, the simulations were run using an Intel® Core i7 8665U 2.11 GHz, 16 GB RAM computer. The *paretosearch* function from MATLAB® was used to obtain the Pareto optimal set for each stage.
Step 3. The Pareto fronts were computed for each stage. Concerning CH_4, 60 Pareto points were calculated for all stages. Regarding acetate, 60 Pareto points were calculated for stages I, II, and V, while 29 and 35 Pareto points were obtained for stages III and IV, respectively.

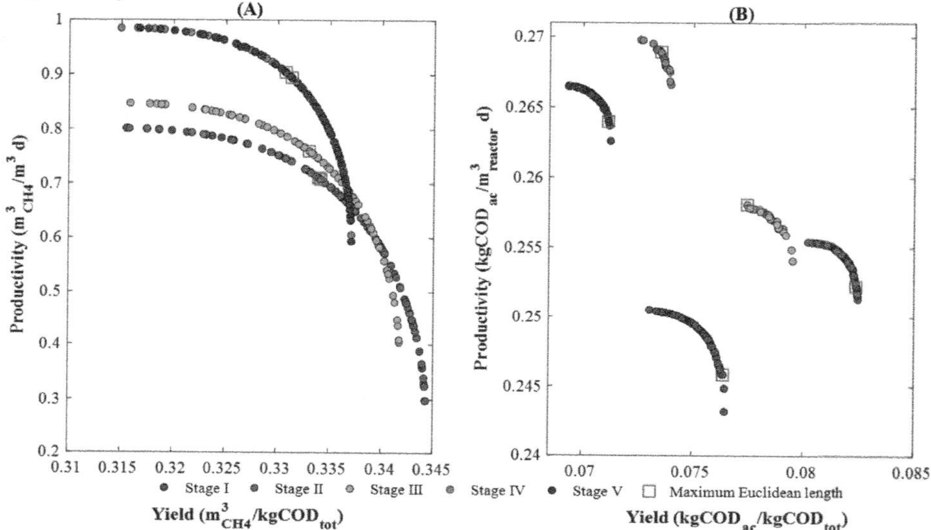

Figure 1. Pareto optimal sets for CH_4 (A), and acetate (B) at each stage.

Figure 1 shows the Pareto optimal sets for CH_4 and acetate at each stage. A progressive increase in P_{CH4} was observed, while Y_{CH4} decreased slightly.
Y_{ac} and P_{ac} increased from stage I to II, then Y_{ac} decreased for the subsequent stages. However, P_{ac} continued to increase until stage IV to encounter a decrease in stage V.
At each stage, the POP were selected to maximize the Euclidean length (d_{max}) for CH_4 and for acetate (red squares in Fig. 1). For both cases, d_{max} was calculated as the distance from the origin, using a normalization as in Eq. (11).

$$d_{max} = max\left(\sqrt{\left(\frac{Y^*_{CH4} - min(Y^*_{CH4})}{max(Y^*_{CH4}) - min(Y^*_{CH4})}\right)^2 + \left(\frac{P^*_{CH4} - min(P^*_{CH4})}{max(P^*_{CH4}) - min(P^*_{CH4})}\right)^2}\right) \quad (11)$$

Step 4. The dynamic optimization for Y_h and P_h was proposed as follows ($h = CH_4$, acetate),

$$\min_{\{q^{in}_{gas}, q^{in}_{liq}\}} \left(\sum_{j=t}^{t+H_p} \left(\frac{|Y^*_h - Y_h(t)|}{Y^*_h}\right)^2 + \left(\frac{|P^*_h - P_h(t)|}{P^*_h}\right)^2 + \sum_{j=t}^{t+H_c} W_{u,1}\, \Delta q^{in}_{gas}(t)^2 + W_{u,2}\, \Delta q^{in}_{liq}(t)^2 \right) \quad (12)$$

Eq. (12) is subject to the constraints in Eq (10). Y^*_i and P^*_i denote the POP values for yield and productivity computed by the MOO, $\Delta q^{in}_{gas}(t)^2$ and $\Delta q^{in}_{liq}(t)^2$ are the differences between the inlet gas and liquid flow rates, respectively, before and after each step of the dynamic optimization. $W_{u,1}$ and $W_{u,2}$ are the parameters that weight the importance of the control effort term in the optimization. In all cases, the initial values for both manipulated variables were $1 \times 10^{-3}\ m^3/d$. H_p and H_c were equal and equivalent to the time length of each stage. Optimization was performed with the *patternsearch* algorithm in MATLAB®.

Step 5. Five cases were studied to assess the dynamic optimization. Case 1 regarded model simulation (data without control). The other cases correspond to: the direct use of the POP in simulation (case 2); the inclusion of POP into a MODO for the maximization for d_{max} CH_4 (case 3) and d_{max} acetate (case 4), and a switch strategy aiming at maximizing CH_4 in stages I-III and then maximizing acetate in stages IV-V (case 5). The weights $W_{u,1}$ and $W_{u,2}$ (for cases 2-5) were manually adjusted to values of 1×10^{-6}.

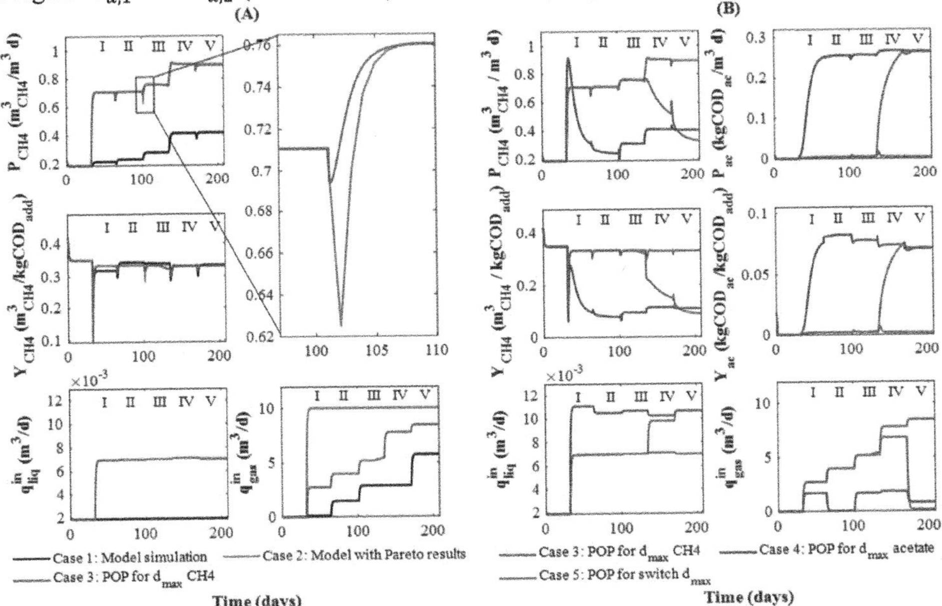

Figure 2. Yields, productivities, injected and inlet gas flow rates in the MODO. Cases 1-5 (A); Case 6 (B).

Figure 2-A presents the results of cases 1 to 3. It is observed that Y_{CH4} was maximized in the experimental part. On the other side, P_{CH4} can be enhanced through the utilization of MODO.

The advantages to use the dynamic optimization were observed when cases 2 and 3 are compared. Both of them achieved similar results for P_{CH4}. However, the transition between stages was smoother and faster in case 3, due to the dynamic part of the MODO (zoom in Figure 2-A). For instance, between stages II and III the time to reach 95% of the steady state decreased from 104 days in case 2 to 102 days in case 3, which could translate into less substrate consumption and thus a higher profit. Regarding Y_{CH4}, it was similar in all cases, while q^{in}_{liq} varied from 7.1×10^{-3} to 6.9×10^{-3} m^3/d at stage V. For q^{in}_{gas}, a value of 10 m^3/d was obtained for all stages in case 2. Nonetheless, this value was reduced in case 3 where q^{in}_{gas} ranged between 2.73 and 8.44 m^3/d.

Figure 2-B displays the cases 3 to 5. In stage V of case 3, a value of 0.89 $m^3_{CH4}/m^3\ d$ was obtained for P_{CH4}, while 0.33 $m^3_{CH4}/kgCOD_{added}$ was obtained for Y_{CH4}. In stage V of case 4, values of 0.27 $kgCOD_{ac}/m^3\ d$ and 0.07 $kgCOD_{ac}/kgCOD_{added}$ were achieved for P_{ac} and Y_{ac}, respectively.

In case 5, a switch was applied at the beginning of stage IV, where the reference values for P_{ac} and Y_{ac} were achieved at the end of the stage. Then, in stage V, values of 0.27 $kgCOD_{ac}/m^3\,d$ and 0.07 $kgCOD_{ac}/kgCOD_{added}$ were obtained for P_{ac} and Y_{ac}, respectively, while the control variables were calculated to maintain the values from cases 3 and 4. This case shows that the MODO strategy is robust when switching multi-objectives and attaints the optima founded by the MOO. When comparing the performance of the MODO for CH4 with the model simulation, (case 3 vs case 1), it is observed that P_{CH4} increased between 3.26 to 2.13 times from stage I to V concerning case 1, while Y_{CH4} was maintained similar.

4. Conclusion

A MODO strategy was successfully applied over a biomethanation process based on the dynamic model ADM1_ME. The feasibility of using Paretos to find the trade-off between objective functions, such as yields and productivities of CH_4 and acetate was demonstrated. The application of a dynamic optimization allowed an improvement in the response by reducing approximately 2 days the time in which the steady state was reached in the stage changes. Additionally, a reduction in gas flow rates of 1.5 m^3/d was achieved with the dynamic optimization. The robustness of the MODO strategy was demonstrated with a successful switch between the products of interest. These results showed that both control variables (q_{liq}^{in}, and q_{gas}^{in}) have a key role at each case and stage. Although these results show the feasibility of the proposed strategy, it is important to note that these are simulation results, and the adaptation of the microorganisms to the propose switching conditions might be different. However, this work showed the possibility to optimize in a smoother way the production of these two products and the change of objectives.

References

J.C. Acosta-Pavas, C.E. Robles-Rodríguez, C.A. Méndez Suarez, J. Morchain, C. Dumas, A. Cockx, C.A. Aceves-Lara, 2022a. Dynamic Multi-Objective Optimization Applied to Biomethanation Process. Chemical Engineering Transactions 96, 319–324.

J. C. Acosta-Pavas, J. Morchain, C. Dumas, V. Ngu, A. Cockx,, C.A. Aceves-Lara, 2022b. Towards Anaerobic Digestion (ADM No. 1) Model's Extensions and Reductions with In-situ Gas Injection for Biomethane Production. IFAC-PapersOnLine 55, 635–640.

E.F. Camacho, C. Bordons, 2007. Model Predictive control, Advanced Textbooks in Control and Signal Processing. Springer London, London.

S. Chaikitkaew, J. Seengenyoung, C. Mamimin, N.-K. Birkeland, A. Reungsang, S. O-Thong, 2021. Simultaneous biogas upgrading and acetic acid production by homoacetogens consortium enriched from peatland soil. Bioresource Technology Reports 15, 100701.

R. Iglesias, R. Muñoz, M. Polanco, I. Díaz, A. Susmozas, A.D. Moreno, M. Guirado, N. Carreras, M. Ballesteros, 2021. Biogas from Anaerobic Digestion as an Energy Vector: Current Upgrading Development. Energies 14, 2742.

J.L. Martín-Espejo, J. Gandara-Loe, J.A. Odriozola, T.R. Reina, L. Pastor-Pérez, 2022. Sustainable routes for acetic acid production: Traditional processes vs a low-carbon, biogas-based strategy. Science of The Total Environment 840, 156663.

K. Morales-Rodelo, M. Francisco, H. Alvarez, P. Vega, S. Revollar, 2020. Collaborative control applied to bsm1 for wastewater treatment plants. Processes 8, 1–22.

H. Sun, Z. Yang, Q. Zhao, M. Kurbonova, R. Zhang, G. Liu, W. Wang, 2021. Modification and extension of anaerobic digestion model No.1 (ADM1) for syngas biomethanation simulation: From lab-scale to pilot-scale. Chemical Engineering Journal 403, 126177.

N. Vertovec, S. Ober-Blöbaum, K. Margellos, 2021. Multi-objective minimum time optimal control for low-thrust trajectory design, in: 2021 European Control Conference (ECC), Delft, Netherlands,1975-1980.

Antonis Kokossis, Michael C. Georgiadis, Efstratios N. Pistikopoulos (Eds.)
PROCEEDINGS OF THE 33rd European Symposium on Computer Aided Process Engineering (ESCAPE33), June 18-21, 2023, Athens, Greece
 http://dx.doi.org/10.1016/B978-0-443-15274-0.50093-7

Dynamic modeling of a waste incinerator furnace in view of supervision

Lionel Sergent,[a,b] François Lesage,[a] Abderrazak M. Latifi,[a] Jean-Pierre Corriou,[a] Alexandre Grizeau[b]

[a]*Laboratoire Réactions et Génie des Procédés, CNRS-ENSIC, Université de Lorraine, 1 rue de Grandville, 54000 Nancy, France*
[b]*SUEZ RV France, 4 rue du Clausenhof, 67590 Schweighouse-sur-Moder, France*
lionel.sergent@suez.com

Abstract

Roughly 30% of municipal waste is incinerated in the European Union. Because of the heterogeneity of the waste and lack of local measurements, the industry relies on traditional control strategies. However, advanced modeling strategies have been used to gain insights on the design of such facilities. In this paper, a first-principles model of the waste combustion based on the simplification of past literature works is developed. Simulations are carried out using the walking column approach, reproducing the major trends observed in other works and on site. The dynamic responses resulting from the model will serve as the centrepiece of a global model to be used for supervision assistance.

Keywords: Waste Incineration, First-principles Model, CFD Simulation, Supervision.

1. Introduction

The first municipal incinerator appeared in London in 1870 as a response to fast urbanization and sanitation issues (Bertolini, 2000). To minimize pollution and cost, and to maximize the recovery of energy, the process was improved over time.

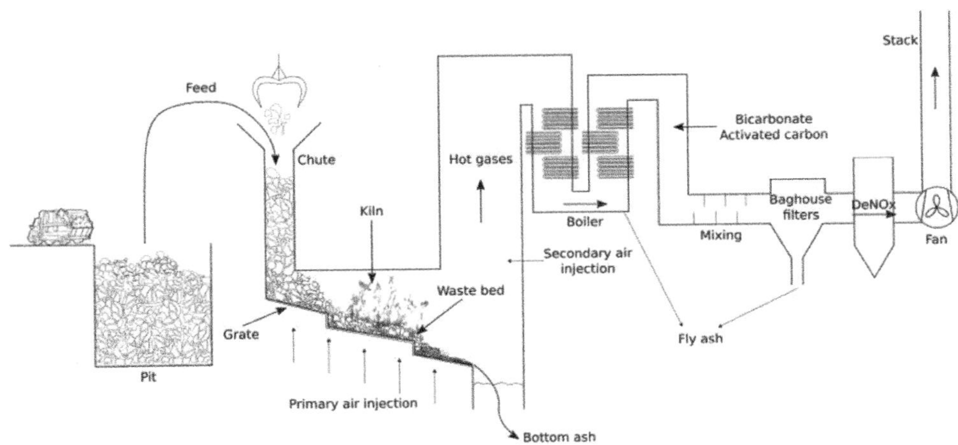

Figure 1: Schematic view of a typical incineration process.

Nowadays, the process can be roughly separated in three main stages: the feeding stage, hot stage, and flue-gas treatment stage, each one containing several units. A typical incineration process is given in Figure 1. For municipal waste, the most common type of kiln is the reciprocating grate, as it provides good mixing and a steady movement of the solids while being robust.

Regarding chemical engineering, the difficulties inherent to the incineration process are twofold. First, the waste heterogeneity and lack of control of the feeding imply strong uncertainties right after the process inlet. Second, there are few sensors in or near the kiln due to harsh mechanical and thermal conditions. There are many more sensors towards the process outlet. Therefore, it is relevant to develop a model that considers the process from inlet to outlet. Nevertheless, as the furnace is both the heart of the process and the host of the most complicated phenomena, it deserves specific attention. This paper presents a model of the behavior of a burning bed of waste to be integrated in a complete model of a real plant. Although greatly inspired by the literature, it differs by providing significant simplifications to a family of models only dedicated to the kiln.

2. Literature survey

2.1. Experiments

Because in industrial facilities only global fluid flows and temperatures can be measured, it is useful to rapidly present the 3 main experiments conducted in the literature as they give information on the local behavior.

1. The thermogravimetric analysis gives a classification of the fractions of waste known as the immediate analysis (humidity, volatile matter, char and ash). The ultimate analysis may be obtained to get a breakdown by elements. Thermal analysis also provides data for pyrolysis kinetics. Despite a high number of experiments, it must be noted that the resulting kinetics laws are quite unreliable (White et al., 2011).
2. The so-called pipe-bowl experiments (Menard, 2003) are similar to cylindrical wood stoves. They feature an array of thermocouples and gas analyzers that allow to follow locally the combustion of medium size samples. This experiment unravels the propagation of reaction fronts through the combustible.
3. Cold grate experiments study how solid particles travel and mix along an industrial grate. In the absence of combustion, it is much easier to track objects by means of visual hints or RFID tags (Džiugys et al., 2006).

2.2. Modeling approaches

Few works present models geared towards the control of incinerators. Van Kessel (2003) uses a 0D CSTR model and Magnanelli et al. (2020) present a 1D horizontal model for that purpose. However, these models cannot explain the behavior observed during the 3 mentioned experiments. Therefore, other models were considered, generally developed to gain insight on the design and pollutants behavior. Lagrangian approaches offer detailed simulations at the cost of tedious mechanical descriptions and high computation time (Brosch et al., 2014). Eulerian approaches are more common, generally solved in 2D (Menard, 2003). Neglecting horizontal gradients leads to a method introduced by Goh et al. (2001) as the "walking column" approach, which is mostly retained for the present work in order to alleviate computation time.

3. Proposed model

3.1. Model features

The waste present on the grate is subdivided in *N* columns which are modeled as a porous medium. It evolves dynamically while moving towards the bottom-ash pit at a constant speed. The porous medium is represented by a reduced set of 13 vertically distributed state variables: 4 solid mass fractions (humidity, volatile matter, residual carbon, inert matter), 7 gas mass fractions $(N_2, O_2, H_2O, CH_4, CO, CO_2, H_2)$, the solid and gas temperatures. It can be noted that in the literature, generally more state variables are necessary.

The evolution of state variables is determined by partial differential equations derived from the general transport equation, expressed as

$$\partial_t(\rho\phi) + \partial_z(\rho v\phi) = \partial_z(\Gamma\partial_z\phi) + S \tag{1}$$

where ρ is the apparent density, ϕ a conservative variable corresponding to a state variable, Γ the diffusion coefficient, v the vertical velocity, and S the source term.

The bottom boundary condition is imposed by the primary air injection, which varies along the grate, while the top boundary condition is given by the pressure generated by the induced draft fan.

The gas phase properties are estimated by correlations (Poling et al., 2001). The properties of the solid phase are generally constant or linear with temperature, and the values are the standard estimates used in the industry.

Equation (1) written for the total mass of gas is rearranged into a pressure equation by assuming Darcy's and ideal gas laws. Its dynamics are fast and neglecting the accumulation term had no influence on the resulting pressure field. Removing the accumulation term allows to analytically solve the flow field, offering a significant simplification compared to the more general momentum equation widely used. It also makes the numerical solution more stable.

3.2. Main phenomena involved

The source terms are constituted by a combination of various rates describing the main phenomena involved in the process:

- The drying rate is calculated through a classical Sherwood correlation. The driving force is the difference between the partial and saturation pressures of water.
- The pyrolysis rate is assumed to be a first order reaction. The pre-exponential factor and energy activation have been fitted with thermogravimetric experiments.
- The residual carbon burning rate is modeled by a first-order reaction and transport of oxygen through the particle boundary layer.
- The rates of light gases (CH_4, CO, H_2) combustion are calculated with kinetic laws, which are all monomial in the gas concentrations, with mixing and channelling limitations.
- The radiative exchange rate with the flame and the kiln structure is modeled by Bouguer's law, while the radiative exchange between particles is treated using the radiative diffusion approximation.
- The convective heat exchange rate between the solid and gas phases is calculated with a classical Nusselt correlation.

The proportions of gases produced by pyrolysis are precalculated by a routine which ensures that the measured lower heating value, immediate analysis and ultimate analysis of the waste are all respected. The atomic and energy balances used motivate the choice of exactly 7 gas mass fractions, as a smaller number would not allow the respect of the mentioned constraints.

3.3. Numerical solution

Simulations were carried out using MATLAB environment. Discretization is based on the finite volume method and the FVTool library (Eftekhari and Schüller, 2022). The advection term is discretized with the upwind scheme. The used grid is denser near the boundaries. The time integration is done with an adaptive implicit Euler scheme. A dynamic simulation with fixed input parameters to reach a steady state typically takes a few minutes on a personal computer, although the calculation time is highly sensitive to several parameters.

Because the vertical velocity of the solid is neglected, the volume reduction effect does not appear in transport equation (1). Instead, the volume of each cell is modified at each time step. This calculation reflects the fact that drying and devolatilization mostly leave internal pores in the particles, while char burning consumes the remaining structure and rapidly decreases the size of particles, resulting in a collapse of the solid bed.

4. Results and discussion

4.1. 1D simulation of a single column

Figure 2 presents time profiles of the gas outputs of a single column using a one-dimensional model. To make this simulation comparable with most pipe-bowl experiments found in literature, the air supply is kept constant at under-stoichiometric conditions, while the height of the waste bed is 30 cm.

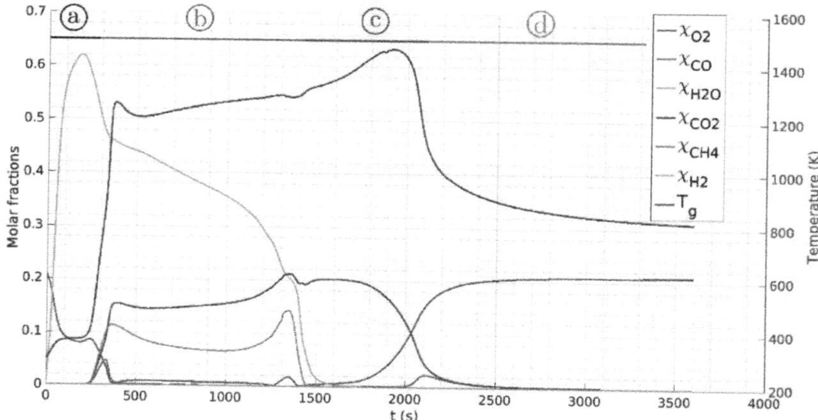

Figure 2: Evolution of the gas output of a 1D column model with constant air supply.

In phase a (Figure 2), the radiative flux heats up and dries the top of the bed, characterized by a short temperature plateau around 373 K. In phase b, the pyrolysis starts. Light gases are emitted and quickly burned, resulting in a stiff temperature increase and a near-depletion of oxygen. Some of the heat generated is used to dry the lower layers that will in turn pyrolyze and burn. When this reaction front reaches the bottom, the remaining carbon is burned out in phase c which results in a temperature peak. In phase d, the combustion is completed, and a thermal equilibrium is reached between radiative heating and convective cooling by the primary air injection.

4.2. 2D simulation of the entire bed

For a waste bed traveling on a grate, the state variables can be represented in maps such as the one given in Figure 3. The phases that were observed against time on Figure 2 can now be roughly observed with respect to position, despite large differences in primary air and bed height. After half a meter of pure drying, the pyrolysis starts early in zone 1, but not enough air is present to burn the volatile gases yet. Because the conditions are over-stoichiometric in zone 2, pyrolysis, light gas combustion and carbon combustion happen mostly simultaneously.

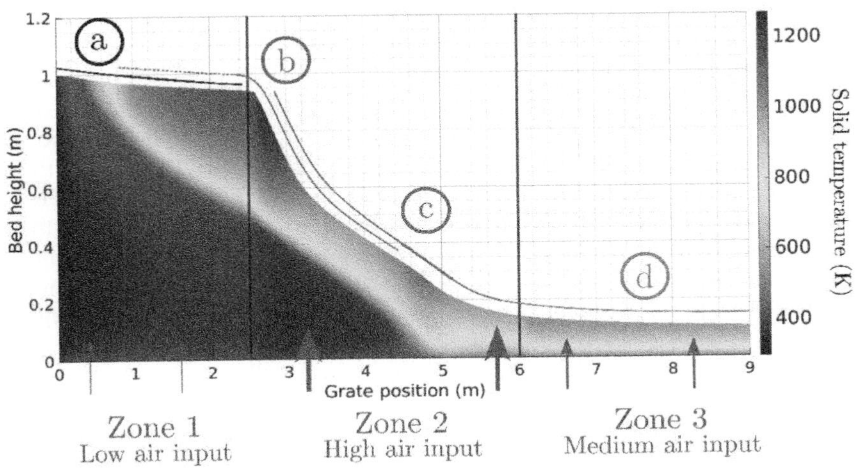

Figure 3: Solid temperature (K) distribution of a simulated waste bed. The grate and primary air injection parameters are taken from the incinerator of Schweighouse-sur-Moder (France) operated by the SUEZ company. The computation grid is 100 by 100.

4.3. Waste bed response to a step change

Figure 4 displays the waste bed response to a step decrease of primary air injection.

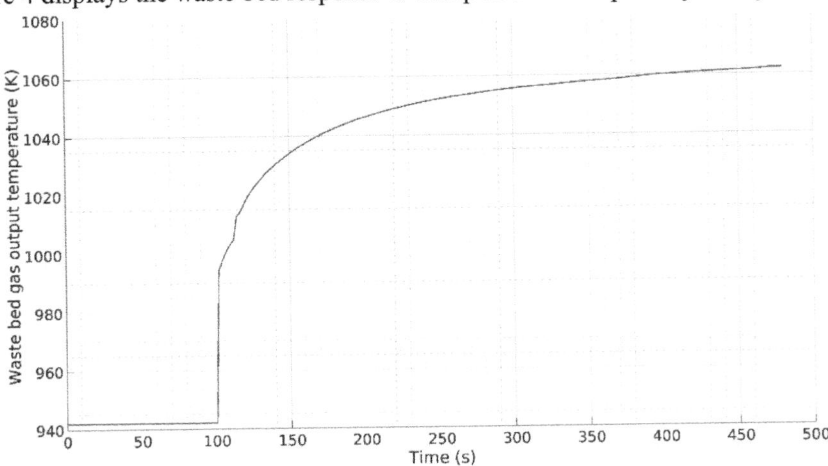

Figure 4: Response of the kiln to a primary air step decrease of 30% at time t=100s.

The waste bed gas output temperature is computed by integrating the gas enthalpy and mass flows at the bed surface before any secondary combustion can happen. Note that because the walking columns algorithm produces nearly cyclic computation artefacts, a moving average has been applied to obtain a smoother curve (Figure 4). Since the conditions are over-stoichiometric, a decrease in primary air results in an immediate increase in temperature as less dilution occurs. Then, the temperature continues to increase closely to a 1st order response. This is mainly due to the solid thermal inertia. If the conditions were under-stoichiometric, a decrease in primary air would result in a temperature decrease due to incomplete combustion.

5. Conclusion and perspectives

A simplified first-principles model of the behavior of a burning bed of waste in an incinerator furnace has been developed and used to carry out simulations. Major trends observed on site and in the literature were successfully reproduced. For the first time, this type of model is used to study the dynamic responses of a burning waste bed. However, the results appear to be very sensitive to the model parameters, including the solid representation (particles, porosity), kinetic rates and heat transfer. Improvements could be made on the numerical resolution, such as an adaptive grid to follow the reaction front, the use of Total Variation Diminishing advection schemes and higher order integrators.

The next step consists in modeling the freeboard which will allow partial validation of the kiln combustion model with plant data, thanks to the secondary combustion chamber temperature and O_2 sensors. Then, the boiler, with its thermal inertia, will be modeled and offer steam production responses. Once the flue gas treatment stage is modeled, comparison with online gas analyzers will be possible. The final process model will be used as a basis for the prototype of a supervision tool whose exact form will depend on the characteristics of the global dynamic model, including its ability to treat uncertainty.

References

G. Bertolini, 2000, Aperçu historique (jusqu'en 1950) du développement de l'incinération des ordures ménagères dans le monde, Déchets, sciences et techniques, 20, 3–8.

B. Brosch, V. Scherer, S. Wirtz, 2014, Simulation of municipal solid waste incineration in grate firing systems with a particle based novel Discrete Element Method, VGB Powertech, 75–83.

A. Džiugys, B. Peters, H. Hunsinger, L. Krebs, 2006, Evaluation of the Residence Time of a Moving Fuel Bed on a Forward Acting Grate, Granular Matter, 8, 125–135.

A.A. Eftekhari, K. Schüller, 2022, FVTool: Finite volume toolbox for Matlab.

Y.R. Goh, Y.B. Yang, R. Zakaria, R.G. Siddall, V. Nasserzadeh, J. Swithenbank, 2001, Development of an Incinerator Bed Model for Municipal Solid Waste Incineration, Combustion Science and Technology, 162, 1, 37–58.

E. Magnanelli, O.L. Tranås, P. Carlsson, J. Mosby, M. Becidan, 2020, Dynamic modeling of municipal solid waste incineration, Energy, 209, 118426.

Y. Menard, 2003, Modélisation de l'incinération sur grille d'ordures ménagères et approche thermodynamique du comportement des métaux lourds, PhD Thesis, Institut National Polytechnique de Lorraine, Nancy, France.

B.E. Poling, J.M. Prausnitz, J.P. O'Connell, 2001, The properties of gases and liquids, 5th ed, McGraw-Hill, New York.

R. van Kessel, 2003, Stochastic disturbances and dynamics of thermal processes, PhD Thesis, Technische Universiteit Eindhoven, Netherlands.

J.E. White, W.J. Catallo, B.L. Legendre, 2011, Biomass pyrolysis kinetics: A comparative critical review with relevant agricultural residue case studies, Journal of Analytical and Applied Pyrolysis, 91, 1, 1–33.

Antonis Kokossis, Michael C. Georgiadis, Efstratios N. Pistikopoulos (Eds.)
PROCEEDINGS OF THE 33rd European Symposium on Computer Aided Process Engineering (ESCAPE33), June 18-21, 2023, Athens, Greece
 http://dx.doi.org/10.1016/B978-0-443-15274-0.50094-9

Reinforcement learning combined with model predictive control to optimally operate a flash separation unit

Dean Brandner[a*], Torben Talis[b*], Erik Esche[b], Jens-Uwe Repke[b], Sergio Lucia[a]

[a]*Technische Universität Dortmund, Laboratory of Process Automation Systems, Emil-Figge-Straße 70, Building G2, Dortmund 44227, Germany*

[b]*Technische Universität Berlin, Process Dynamics and Operations Group, Str. des 17. Juni 135, Berlin 10623, Germany*

**dean.brandner@tu-dortmund.de, t.talis@tu-dortmund.de*

Abstract

Model predictive control (MPC) and reinforcement learning (RL) are two powerful optimal control methods. However, the performance of MPC depends mainly on the accuracy of the underlying model and the prediction horizon. Classic RL needs an excessive amount of data and cannot consider constraints explicitly. This work combines both approaches and uses Q-learning to improve the closed-loop performance of a parameterized MPC structure with a surrogate model and a short prediction horizon. The parameterized MPC structure provides a suitable starting point for RL training, which keeps the required data in a reasonable amount. Moreover, constraints are considered explicitly. The solution can be obtained in real-time due to the surrogate model and the short prediction horizon. The method is applied for control of a flash separation unit and compared to a MPC structure that uses a rigorous model and a large prediction horizon.

Keywords: Reinforcement Learning, Model Predictive Control, Flash Separation Unit

1. Introduction

Model predictive control (MPC) is a powerful control concept as it can deal with process constraints and nonlinear multivariable systems. However, the control performance mainly depends on the accuracy of the system model as well as its prediction horizon. Using a rigorous first principles model and a long prediction horizon may lead to limited real-time capability due to large computational effort. One possibility to reduce the computational effort is to use surrogate models that are less complex than the original first principles ones. In addition, the prediction horizon can be kept short to reduce the number of decision variables of the optimal control problem. However, both strategies will lead to worse closed-loop performance.

Due to the digitalization of the process industry, an increasing amount of data becomes available. Data can improve process operation in different manners, for example with data-driven surrogate models or via reinforcement learning (RL). However, traditional RL approaches require very large amounts of data, and they cannot deal easily with process constraints.

In this work, we follow the success of recent approaches and utilize a parameterized MPC structure within an RL scheme to improve the closed loop performance of an MPC controller with an imperfect surrogate model and to consider process constraints explicitly. We first show how the parameterized MPC controller can be used to approximate the action-value function of the optimal control problem and illustrate afterwards the results of the application on a flash unit.

2. Background on surrogate modelling for dynamic systems

Constrained dynamic systems are usually described by differential algebraic equation (DAE) systems. They consist of both ordinary differential equations and algebraic equations and can be written in semi-explicit form with state variables $x = (y, z)$, the control variables u, the parameters p, and time t:

$$\dot{y} = f(y, z, u, p, t) \tag{1}$$

$$0 = g(y, z, u, p, t). \tag{2}$$

High index problems frequently arise in chemical engineering, e.g., by assuming negligible vapor holdup. These systems cannot be integrated by explicit methods and are usually rather large DAE systems, which are slow and difficult to solve. This is limiting the applicability of optimization-based operation and control, such as MPC. Therefore, data-driven surrogate models are increasingly introduced.

There are many different approaches to model dynamic systems using data, e.g., Gaussian processes (Kocijan, 2016), different recurrent neural network architectures (Esche et al., 2022), and many more. In case a system is Markovian, a feedforward neural network is sufficient and can predict the state of the system at the next timestep, based on the current state and controls (Himmelblau, 2008).

3. Q-Learning for parameterized MPC structures

Reinforcement learning (RL) is a tool to solve Markov decision process (MDP) problems via the interaction of an agent (controller) and an environment (plant) (Sutton and Barto, 2018). The agent computes an action $\boldsymbol{a}_k$ based on the current state of the environment $\boldsymbol{s}_k$. The action is applied to the system, which replies with the next state $\boldsymbol{s}'_k$ and a reward r_k (stage cost). The combination gives an element of a Markov chain $\mathcal{M} = \{(\boldsymbol{s}_k, \boldsymbol{a}_k, r_k, \boldsymbol{s}'_k)\}_{k=0}^{N}$. The state of the environment $\boldsymbol{s}_k$ does not have to be the system state $\boldsymbol{x}_k$ only, for example the state of the environment can also be made up of past control actions $\boldsymbol{u}_{k-1}$. The agent's action is computed according to its policy $\pi(\boldsymbol{s})$. The policy can be learned implicitly from the optimization of the action-value function.

Q-learning focusses on finding the optimal action-value function $Q^*(\boldsymbol{s}, \boldsymbol{a})$ of a MDP. The action-value function is a measure of how good an action $\boldsymbol{a}$ in a given state $\boldsymbol{s}$ is, when following a certain policy π

$$Q(\boldsymbol{s}, \boldsymbol{a}) = r(\boldsymbol{s}, \boldsymbol{a}) + \gamma \cdot Q^{\pi}(\boldsymbol{s}', \pi(\boldsymbol{s}')) \tag{3}$$

where $\gamma \in [0; 1]$ is a so-called discount factor. The optimal control policy can then be derived by optimization over the optimal action-value function. For control, the optimal policy π^* minimizes the expected cost. In the solution process for continuous control tasks, the action-value function is approximated using a function approximator $Q_{\boldsymbol{\theta}}$.

A common choice to approximate that function is to choose deep neural networks (Mnih et al., 2015). However, they may suffer from lack of structure, therefore requiring an excessive amount of data. One way to reduce the amount of data is to use a more structured function approximator like a parameterized MPC structure (Gros and Zanon, 2020). MPC takes an underlying system model f_d and predicts future states with it to derive an optimal control input. A parameterized MPC formulation can be written as:

$$Q_{\theta}(s,a) = \min_{\substack{x_{[0:N]} \\ u_{[0:N-1]} \\ \sigma_{[0:N]}}} \quad \gamma^N \left(V_{f,\theta}(x_N) + w_f^{\top}\sigma_N\right) + \sum_{i=0}^{N-1} \gamma^k \left(\ell_{\theta}(x_k, u_k) + w^{\top}\sigma_k\right) \tag{4}$$
$$\text{s.t.} \quad x_{k+1} = f_{d,\theta}(x_k, u_k), \;\; x_0 = s, \;\; u_0 = a.$$
$$g(u_k) \leq 0, \;\; \sigma_k \geq 0$$
$$h_{\theta}(x_k, u_k) \leq \sigma_k, \;\; h_{f,\theta}(x_N) \leq \sigma_N.$$

The parameterization is performed using the parameter vector $\theta \in \mathbb{R}^{n_\theta}$, which can parameterize the objective function, the system model, and the constraints. For classic MPC, the discount factor is chosen as $\gamma = 1$. The parameterized MPC structure consists of the objective function, which is compound of the stage cost ℓ_θ, the terminal cost $V_{f,\theta}$, and penalties for the slack variables σ, but also input and state constraints. The state constraints are considered as soft constraints.

Classic Q-learning updates may cause instabilities (Sutton and Barto, 2018). Mnih et al. introduce robustness via a replay buffer and target function $Q_{\hat{\theta}}$ (Mnih et al., 2015). The replay buffer stores data from the past. In each update step, past samples are taken to perform a batchwise update. The target function is updated slower than the actual function. The target parameters are updated as $\hat{\theta} \leftarrow (1-\tau)\hat{\theta} + \tau\theta$ with $0 \leq \tau \leq 1$.

The Q-learning update requires the gradient of the Q-function. For the parametric MPC, the gradient of Q-function equals the gradient of the Lagrangian $\mathcal{L}$ of the underlying optimization problem evaluated at the optimal points (Büskens and Maurer, 2001):

$$\nabla_\theta Q_\theta(s,a) = \nabla_\theta \mathcal{L}\left(x^*_{[0:N]}, u^*_{[0:N-1]}, \lambda^*, \nu^* \middle| s, a\right). \tag{5}$$

The main idea of our work is to adapt the MPC structure (4) via the parameters θ, using Q-learning to counteract the imperfect surrogate model $f_{d,\theta}$ and short prediction horizon.

4. MPC parameter optimization with RL on a flash simulation

4.1. Dynamic pressure-driven flash model

In order to build a realistic environment, a first principles model of a dynamic flash unit is created, see Figure 1. It describes the binary mixture of ethanol and water. The vapor liquid equilibrium is described assuming an ideal gas, but non-ideal liquid phase using Wilson's activity coefficient. Pressure and level are controlled via distillate and bottom stream respectively.

A surrogate model of the first principles model will be used within the parameterized MPC. For this, the rigorous model is sampled to generate data, which is used for training of a dynamic surrogate model. The control variables and their bounds for the model are the supplied heat flux $\dot{Q}$ (0-30 kW), feed flow F^F (1-2 mol s^{-1}), and feed composition x^F (0.01-0.99). The feed temperature is constant at 365 K.

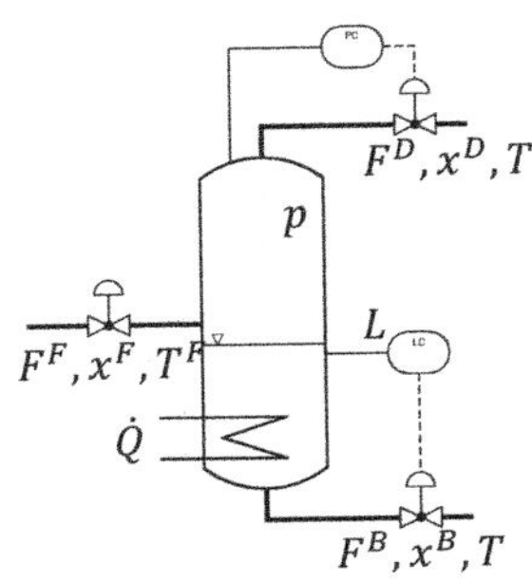

Figure 1. Flowsheet of the flash unit with 4 inputs: F^F, x^F, T^F, $\dot{Q}$ and 7 outputs p, T, L, F^D, y^D, F^B, x^B.

To create training data, the former two are excited with an amplitude modulated pseudo-random binary sequence (APRBS) - which can be interpreted as a sequence of step signals with different holding times with a frequency of 1 s^{-1}. The feed composition is held constant for 50 s and then increased by 0.01. A second trajectory is created to later test the surrogate model, which uses a lower frequency for the APRBS while allowing the same number of steps.
The 50'000 samples of the first trajectory are shuffled and split into (50 %, 75 %, or 95 %) training data, while the rest is used for validation. These points are used to train three similar feed forward neural network with 10 nodes in one hidden layer for 10 epochs with the Adam optimizer (Kingma and Ba, 2014). The hyperbolic tangent was used for the activation function. The MSE on the validation data is $5.7 \cdot 10^{-5}$, $5.0 \cdot 10^{-5}$ and $1.6 \cdot 10^{-5}$ for the three different training scenarios. The second trajectory is then used for testing and gets predicted with an MSE of $3.1 \cdot 10^{-4}$, $5.8 \cdot 10^{-5}$, and $1.5 \cdot 10^{-5}$, respectively. In the following section the first model (model 1) is used to show that the presented approach works well, even for surrogate models with a rather large plant model mismatch.

4.2. Design of the RL-environment based on the rigorous dynamic flash model

The RL environment consists of the rigorous first principles model. The current state of the environment $\boldsymbol{s}_k$ is composite of all system states $\boldsymbol{x}_k = (p, T, L, F^D, y^D, F^B, x^B)_k^\mathsf{T}$ at time k and the previously applied control action $\boldsymbol{u}_{k-1} = (F^F, Q)_{k-1}^\mathsf{T}$ at time $k-1$. x^F and T^F are kept constant. Hence, it is built as $\boldsymbol{s}_k = (\boldsymbol{x}_k^\mathsf{T}, \boldsymbol{u}_{k-1}^\mathsf{T})^\mathsf{T}$. The control action applied to the environment is $\boldsymbol{a}_k = \boldsymbol{u}_k$. The stage cost $\ell(\boldsymbol{s}_k, \boldsymbol{a}_k)$ of the environment is a composite of three subterms:

$$\ell(\boldsymbol{s}_k, \boldsymbol{a}_k) = L(\boldsymbol{x}_k, \boldsymbol{u}_k) + r(\boldsymbol{u}_k, \boldsymbol{u}_{k-1}) + \epsilon(\boldsymbol{x}_k, \boldsymbol{u}_k). \quad (6)$$

$L(\boldsymbol{x}_k, \boldsymbol{u}_k)$ is a quadratic cost term, which penalizes deviations of the system states to the setpoints. $r(\boldsymbol{u}_k, \boldsymbol{u}_{k-1})$ penalizes changes in the control inputs, while $\epsilon(\boldsymbol{x}_k, \boldsymbol{u}_k)$ penalizes constraint violations. The terms are defined as follows with the values from Table 1

$$L(\boldsymbol{x}_k, \boldsymbol{u}_k) = 10 \cdot (y_k^D - y_{SP}^D)^2 + 1 \cdot (F^D - F_{SP}^D)^2 \quad (7)$$

$$r(\boldsymbol{u}_k, \boldsymbol{u}_{k-1}) = \Delta \boldsymbol{u}_k^\mathsf{T} \begin{pmatrix} 1 & 0 \\ 0 & 30 \end{pmatrix} \Delta \boldsymbol{u}_k \quad (8)$$

$$\epsilon(\boldsymbol{x}_k, \boldsymbol{u}_k) = \boldsymbol{w}_x^\mathsf{T} \boldsymbol{\sigma}_{x,k} + \boldsymbol{w}_u^\mathsf{T} \boldsymbol{\sigma}_{u,k}. \quad (9)$$

Table 1: Setpoints (SP), lower bounds (lb), upper bounds (ub) and constraint penalties (w_i) for the states and control actions within the environment.

	p	T	F^D	y^D	F^B	x^B	$\dot{Q}$	F^F
	bar	°C	mol s^{-1}	–	mol s^{-1}	–	kW	mol s^{-1}
SP	–	–	0.2	0.45	–	–	–	–
lb	0	20	0	0.4	0	0	0	1
ub	1	90	0.5	1	2	1	30	2
w_i	20	20	20	20	20	20	250	250

4.3. Parameterized MPC structure for optimal control of the flash separation unit

The structure from *(4)* is used to learn the optimal control policy. Model 1 from section 4.1 is used to predict future states. The model is kept unchanged. The parameterization occurs solely in the terminal cost $V_{f,\theta}$, which is defined as

$$V_{f,\theta}(x_N) = 10^2 \cdot \theta_{11} \cdot (F_N^D - F_{SP}^D)^2 + 10^4 \cdot \theta_{22} \cdot (y_N^D - y_{SP}^D)^2 \qquad (10)$$

with initial values of $\theta_{11} = \theta_{22} = 0$. The MPC acts subject to the constraints given in Table 1. The considered stage cost is the same as given in equation (6). The prediction horizon is chosen to be $N = 10$. A Q-learning update is then conducted at each time instance using a minibatch of $m = 16$ samples taken from the replay buffer of size 4'096 with a learning rate of $\alpha_\theta = 10^{-4}$. The target MPC is directly updated using $\tau = 1$.

4.4. Control performance of parameterized MPC

Figure 2 shows the control performance of a benchmark MPC (rigorous model, $N = 20$, solid line) compared to the parameterized MPC (section 4.3, dashed line). The figure shows the controlled states in the left half of each subplot and the control actions on the right half. The dotted lines demonstrate the respective setpoints. The left plot compares the control performance of the benchmark MPC and the parameterized MPC before any training took place. This is similar to applying just MPC with the surrogate model. It can be seen that y^D is not tracked properly by the parameterized MPC. Also, the control input trajectories differ from the benchmark especially for F^F. However, it provides a very suitable starting point for RL training in contrast to conventional RL with neural networks, for which initialization is critical for performance but very difficult to choose properly. The right plot shows the performance of the parameterized MPC controller after 100 episodes of training. One episode is a closed loop trajectory of 5 s, each consisting of 50 samples. An update is performed after each time step. The setpoints are now both tracked better. Furthermore, also the control trajectories are closer to those obtained with the benchmark controller. The benchmark achieved a closed-loop cost after 5 s of 0.21367, while the parameterized MPC reduces from initially 0.22063 (untrained) to 0.21749 (100 episodes).

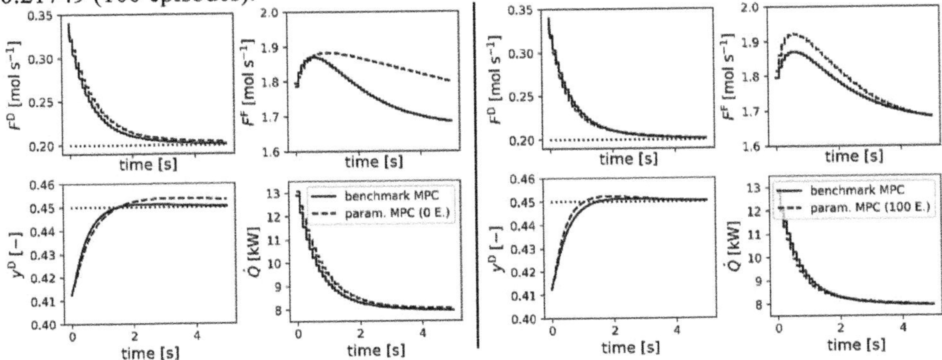

Figure 2. Left: Application of the MPC controller without RL improvements. Right: Application of the RL improved MPC controller. The dotted lines specify the setpoints. 0 E. and 100 E. specify the number of trained episodes.

Both models are also compared with respect to their computation time. The computations were conducted on an intel core i9-10900K CPU, using Python 3.9.13 and the toolbox do-mpc with CasADi and ipopt. The computation of a single control input for the benchmark MPC took 0.414 ± 0.081 s on average, which is approximately four times the sampling interval of 0.1 s. For the parameterized MPC with the surrogate model, the computation only took 0.015 ± 0.003 s, which is approximately 0.15 times the sampling interval and clearly illustrates the computational advantages of the proposed approach.

5. Conclusion

We showed that MPC and RL can be combined to improve controllers in chemical engineering applications. Just by using a simple parameterization of the MPC structure within the terminal cost and a small amount of data, the performance of the controller could be improved to resemble the MPC with a full detailed model, while reducing the computation time by over an order or magnitude.

Future work will include the study of more detailed parametrizations, comparisons with other RL techniques and the extension to more complex control tasks.

Acknowledgements

This work was funded by the Deutsche Forschungsgemeinschaft (DFG, German Research Foundation) – 466380688 – within the Priority Program "SPP 2331: Machine Learning in Chemical Engineering".

References

Büskens, C. and Maurer, H. (2001) 'Sensitivity Analysis and Real-Time Optimization of Parametric Nonlinear Programming Problems', in Grötschel, M., Krumke, S. O. and Rambau, J. (eds) Online Optimization of Large Scale Systems, Berlin, Heidelberg, Springer Berlin Heidelberg, pp. 3–16.

Esche, E., Weigert, J., Brand Rihm, G., Göbel, J. and Repke, J.-U. (2022) 'Architectures for neural networks as surrogates for dynamic systems in chemical engineering', Chemical Engineering Research and Design, vol. 177, pp. 184–199.

Gros, S. and Zanon, M. (2020) 'Data-Driven Economic NMPC Using Reinforcement Learning', IEEE Transactions on Automatic Control, vol. 65, no. 2, pp. 636–648.

Himmelblau, D. M. (2008) 'Accounts of Experiences in the Application of Artificial Neural Networks in Chemical Engineering', Industrial & Engineering Chemistry Research, vol. 47, no. 16, pp. 5782–5796.

Kingma, D. P. and Ba, J. (2014) Adam: A Method for Stochastic Optimization.

Kocijan, J. (2016) Modelling and Control of Dynamic Systems Using Gaussian Process Models, Cham, Springer International Publishing.

Mnih, V., Kavukcuoglu, K., Silver, D., Rusu, A. A., Veness, J., Bellemare, M. G., Graves, A., Riedmiller, M., Fidjeland, A. K., Ostrovski, G., Petersen, S., Beattie, C., Sadik, A., Antonoglou, I., King, H., Kumaran, D., Wierstra, D., Legg, S. and Hassabis, D. (2015) 'Human-level control through deep reinforcement learning', Nature, vol. 518, no. 7540, pp. 529–533.

Sutton, R. S. and Barto, A. (2018) Reinforcement learning: An introduction, Cambridge, Massachusetts, London, England, The MIT Press.

Antonis Kokossis, Michael C. Georgiadis, Efstratios N. Pistikopoulos (Eds.)
PROCEEDINGS OF THE 33rd European Symposium on Computer Aided Process Engineering (ESCAPE33), June 18-21, 2023, Athens, Greece
 http://dx.doi.org/10.1016/B978-0-443-15274-0.50095-0

Architecture and Design of a Modern Commercial Process Simulator

Ian Boys,[a] Jochen Steimel[b]
[a]*AVEVA, 26561 Rancho Parkway South, Lake Forest, CA 92630, USA*
[b]*AVEVA, Mainzer Landstraße 178-190, 60327 Frankfurt, Germany*
Ian.Boys@aveva.com

Abstract

Over the past 10 years AVEVA (formerly SIMSCI, Invensys) has been developing a next generation process simulator, with a clean design from the ground up. In this article we will share the experience developing and marketing this software and reflect on how early design ideas have evolved and matured into a commercially available software application. The software is based on a novel concept for setting up, parametrizing, initializing, and solving equation-based simulation models targeting the next generation of process engineers. We will discuss some of the successes and challenges in realizing this next-generation concept. We will furthermore discuss customer reaction and feedback, and how the software has been received in the marketplace.

Keywords: Simulation, Software, Numerical Methods, Modeling

1. Introduction

In recent years, two contributions were published, one from industry Broecker (2021), and one from academia Pistikopoulus (2021), which both highlight the need for process simulation software to consider advances in computing technology as well as modern algorithms and user experience expectations. Both publications highlight the need for open interfaces, enabling cross-domain applications, which combine real-world data with first principles and data-driven models. Broecker points out that future innovation will arise at the intersection and combination of tools, as complex problems demand cross-domain expertise and no tool can fill all needs. This vision of course requires open tools that allow interaction and integration into larger, more complex workflows. There seems to be a consensus that established tools have so far been lacking in this regard.

In their review article, Mitsos et. al (2018) spelled out the following conflict: "[in the petrochemical industry] many of the developed chemical processes are prototypical applications […]. Often, there is no good business case to justify the implementation of a special solution to a very specific problem class in commercial simulation software". Further in the article, they lament the limited interaction between software vendors and the industry and the slow adoption of novel academic techniques into commercial software. As a possible solution, they conjecture that third parties could fill the gap of capabilities lacking in commercial tools, but also admit that the limited accessibility of those tools often prohibits the implementation of new process systems engineering methods into an integrated offering acceptable for industrial use. The authors want to address this gap, by discussing a tool that is open for extension and research by academia, yet fit for every-day use by industrial practitioners.

We observe: the world of process engineering is changing fast, and next-generation tools must have an outstanding degree of flexibility and customizability to serve the needs of a future generation of process engineers, to empower them to tackle the various challenges induced by the energy transition and the need to switch to sustainable feedstocks. With this contribution we want to discuss the salient features of one commercial simulation software that was developed from the ground up with extensibility and future use-cases in mind, made possible by strong customer interaction and short development cycles.

2. Properties of a Next-Generation Simulator

2.1. Overall architecture

To address the needs discussed above AVEVA started work on developing a next-generation process simulator about 10 years ago. The main goal of this project was to create a platform (called AVEVA Process Simulation, further referred to as the software) that would be powerful enough to support all kinds of process engineering, simulation, and design workloads, while also being in line with modern standards of user experience, scalability, and responsiveness. The software is designed to drive the digital transformation in process engineering through lifecycle simulation and a holistic approach, as demonstrated in de Beer (2021).

A central pillar is our modern client/server architecture. In the software, client interactivity is separated from simulation analysis, model instantiation and solving. This design liberates the software from the requirement that both parts are being run on the same machine, enabling scalability and cloud computing.

For the end-user the simulation platform is flexible and can be adapted to changing use cases. This flexibility is mainly provided by an open-equation custom modeling environment powered by an equation-oriented solver. Instead of implementing process equations in subroutines, as sequential-modular process simulators might do, in the software, (nearly) every equation, variable, model and library can be inspected and modified by the end-user. The only exceptions to this are external equation sets, which allow external code to be connected to the platform (see sec. 2.4.). The availability of all equations and variables as first-class objects enables the dynamic restructuring and re-interpretation of the equation system depending on the problem statement. For example, a seamless transition from steady-state to dynamic simulation is supported by the platform, as only the time derivatives need to be handled differently between the solver and the dynamic integrator. Thus, it is possible to go back-and-forth between steady-state and dynamic mode for the same model, without the need to manually adjust equations or specifications.

When developing new modeling capabilities, the development team is using the same tools available to the end-user. Everything in our libraries can be reconstructed by the user as there are no hidden authoring tools, making the software an open and transparent modeling tool. A screenshot illustrating the model editor, for a model describing the heat transfer through a pipe wall, making use of steady-state and dynamic equations side-by-side is given in fig. 1. An example for this extensibility can be found in Bishop (2020), where the author developed a model for a membrane reactor using our software and performed operability studies.

In line with the platform design, thermodynamic calculations and databases are kept separate from the client/server. AVEVA Process Simulation offers a wide set of industry standard thermodynamic routines for equation of states or activity coefficient models, as well as thermodynamic codes for specialized systems (e.g., electrolytes, cryogenic hydrogen, or supercritical CO2).

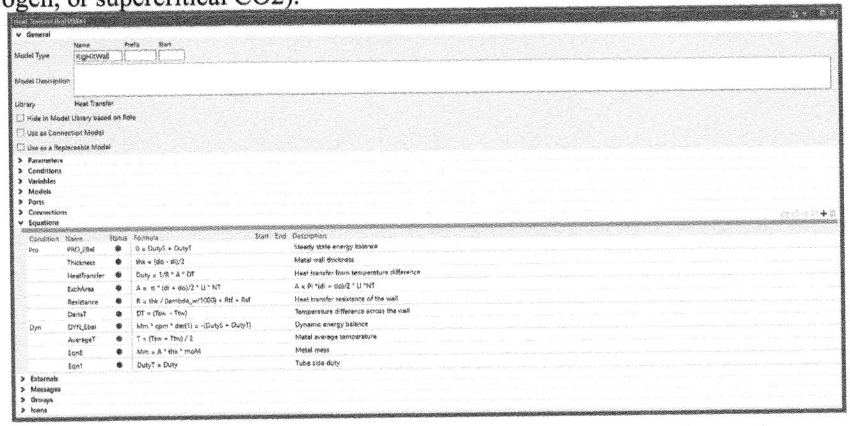

Figure 1: Model equation editor showing steady-state and dynamic equations

2.2. Numerics

The numerical engine powering the platform is a general-purpose equation solving system consists of an instantiation algorithm that translates a model into a flat equation system, a block decomposition algorithm that reduces a large equation system into smaller equation systems that can be solved independently, and a solver factory that applies tailored solvers to each of the equation systems. The block decomposition system is based on the graph algorithms of Tarjan (1972) and Dulmage (1958). Each well-specified block is analyzed, and if the convergence criteria are not met, a solver suitable for the block is selected to converge the residuals. If the block decomposition determined that any blocks are independent from each other, as often happens with phase equilibrium equation sets, they can be solved in parallel to reduce execution time.

2.3. Interactivity

Over the last years, expectations on interactivity in computer programs have shifted dramatically. Apps and smartphones have reached widespread adoption, which led to an increased demand in application ergonomics. In the past, simulators often followed a four-step procedure: define flowsheet topology, provide specifications, trigger solve and inspect output. Many times, users needed to backtrack when it became apparent that a certain topology would not allow the desired set of specifications, but that fact only became apparent during the inspection of the log files. Today users expect instant response and an exploration-driven user interface where making mistakes is cheap and easy to remedy and information is not hidden deep in nested dialogs. To address this shift in expectations for industrial software, AVEVA Process Simulation supports three paradigms that were quite uncommon in simulation tools so far: solve-as-you-go, undo/redo and a simulation repository.

Empowered by the state-of-the-art algorithms, the simulation building core of the software can identify parts of a flowsheet that have changed after a user interaction and need an update. This results in smaller equation systems to be rebuilt and solved and a very fast response and presentation of results. Likewise, the undo/redo system can undo changes to parameter and variable values and even impactful operations like optimization runs. In addition to reverting an action, the system uses implicit snapshots of the "state before" to forego the need for an additional convergence pass after undo, resulting in an even faster response. The simulation repository is an abstraction on the file system. Once a simulation is imported from the file system into the repository, it is managed by the software. Any change to the flowsheet is persisted instantaneously ("auto-save"). Combined with undo/redo, this leads to a seamless modeling experience, without the fear of losing data (e.g., caused by a crash of the software).

Together, these three features empower a special way of approaching equation-oriented modeling and simulation that is fundamental to our software: incremental initialization. It is well-known that finding good initial guesses for an equation-oriented simulation is a non-trivial task and crucial in simulators relying on a general equation solver. In the software, the flowsheet is built up one unit operation at a time. By following a few basic rules (fully specify each unit before moving to the next, always have a square simulation with the same number of unknown variables and equations, undo as soon as the solver cannot solve a problem) it is very easy, also for inexperienced users, to bootstrap large-scale process flowsheet models, even with recycle loops. At the time when a recycle needs to be closed, the returning stream already is at a converged open-loop solution, so convergence is often quick and robust. In addition, a snapshot of the current simulation can be taken, which allows the user to re-initialize to a good solution.

2.4. Extensibility

As mentioned before the software is designed as an open platform, ready to be extended to fit the user's need in process simulation. To support those use cases, the platform offers two main entry points for extension beyond the already flexible open-equation modeling framework.

As mentioned briefly in section 2.1 the software offers an external equation interface which is part of a software development kit (SDK). With the SDK users can provide custom calculations that cannot be expressed in algebraic form (e.g., thermodynamic algorithms, machine-learning models). Users need to provide callback methods that return equation residuals and partial derivatives to fit into our equation-oriented solver framework, but how they calculate those values is fully under their control. In this way it becomes possible to extend modeling to non-smooth equations, conditional logic, or algorithmic structures. To exemplify the power of the SDK, the default thermodynamic methods provided by the software are implemented using the external equation toolkit, keeping in line with our philosophy that any platform user should be able to use the same tools that our developers are using.

The second extension point is the scripting interface. As a side effect of the client/server architecture, any action that can be performed in the GUI client is accessible as a function call for a scripting client. The scripting interface is bundled with the product as a Python module and a .NET library. Any clients, such as Excel/VBA, Python scripts or full native desktop applications can automate our software to implement complex computing workloads such as structural optimization, training of surrogate models and uncertainty analysis.

3. Experiences in Developing a Next-Generation Simulator

As part of the go-to-market strategy, the software benefitted greatly from a customer-centric approach. Instead of developing against an internally defined roadmap, customers were engaged early on and asked to provide feedback on the priority of development activities. Instead of our treating us as a regular software vendor, formal partnerships allowed both parties to synchronize migration and development activities, so that important features were made available when they were needed. We have mapped our target audience against archetypal personas, representing a wide range of users, including students, junior/senior design engineers, thermodynamicists, modelling experts and control engineers. When new features are proposed, we make sure to consult our list of personas to make sure the functionality fits with their needs.

The agile software development process underlying the development activities made it possible to release incremental improvements early and rapidly. This in turn allowed customers to give feedback quickly so that features could be fine-tuned while still in development resulting in a shorter turnover time for new features.

The users of the software are very engaged and are given ample opportunities to be heard, from user meetings over advisory boards to direct meetings with the development team in voice-of-customer sessions. The level of trust resulting from this close contact, allowed developers to get first-hand feedback from real users, a crucial requirement for our agile process. Some of that feedback was expected, some was surprising. In the following, we want to share the aspects that were most important to us.

Not surprisingly, adapting to the equation-oriented paradigm was a challenge to users accustomed to sequential-modular simulators. The benefits coming from it, e.g., near instant results and being able to select any variable to be specified or calculated were highly regarded, but especially finding the root cause of an unconverged simulation proved to be very hard for many users. Interestingly, we found out that younger users, who were not as used to the sequential-modular paradigm, found it easy and intuitive to learn our software, whereas experienced engineers struggled more often.

To address this feedback, we greatly expanded the information about the solution that was reported to the user, in the form of log files and small badges shown over flowsheet icons. Whenever possible this information was translated from the mathematical domain into the process engineering domain. For example, instead of reporting a singular solution, the software reports that for a certain variable no unique value can be determined and that a different specification set should be tried. In another example, instead of just reporting degrees of under- or overspecification, the system visually marks potential variable candidates that are part of the under- or over-specified equation sets. Our goal is to provide a list of small, actionable suggestions that could lead to a solution of the problem instead of a general “solve failed” message.

A well-liked feature is the customizability of the flowsheet canvas. The software supports variable references to be added next to unit operations, which allow variables to be displayed and changed without the need to open a new dialog. Similarly, equations can be added to the flowsheet, so that custom design specifications can be shown next to the process equipment. This feature enables customers to document their flowsheets and simulations within the software itself and reduces the need for external explanations, improving reuse and collaboration.

4. Conclusion

In this article, we have shared the fundamental architecture and design philosophy for our next-generation process simulation tool. Our conclusion is that the next generation concepts envisaged at the outset have proven viable in practice and are able to sustain an industrially relevant and capable process simulator. We base this conclusion upon customer feedback, both from industry and academia (research and teaching). Interactivity is a huge factor in attaining a better understanding of a process and is appreciated by all customers.

AVEVA Process Simulation is reaching wider adoption in the chemical industry and a feature maturity that will allow the software to expand into modern usage scenarios, such as design optimization or uncertainty quantification. We therefore think that the next generation of process engineers can look forward to new ways of working with process simulation, where the workflow is driven more by the goals of the engineer and less by the algorithms of the software.

References

S. Broecker et. al., 2021, Process Simulation – Fit for the future?, Position Paper of the Process Net Working Committee, DECHEMA e.V.

E.N. Pistikopoulus et. al., 2021, Process system engineering – The generation next?, Computers & Chemical Engineering, 147, 107252

A. Mitsos, 2018, Challenges in process optimization for new feedstocks and energy sources, Computers and Chemical Engineering, 113, 209-221

J. de Beer et. al., 2021, The Role of Process Engieering in the Digital Transformation, Computers & Chemical Engineering, 154, 107423

B. Bishop et. al., 2020, Modeling, Simulation, and Operability Analysis of a Nonisothermal, Countercurrent, Polymer Membrane Reactor, Processes, 8, 78

R. Tarjan, 1972, Depth-First Search and Linear Graph Algorithms, SIAM Journal on Computing, 1, 14-160

A.L. Dulmage et. al., 1958, Coverings of Bipartite Graphs, Canadian Journal of Mathematics, 10, 517-534

Antonis Kokossis, Michael C. Georgiadis, Efstratios N. Pistikopoulos (Eds.)
PROCEEDINGS OF THE 33rd European Symposium on Computer Aided Process Engineering (ESCAPE33), June 18-21, 2023, Athens, Greece
 http://dx.doi.org/10.1016/B978-0-443-15274-0.50096-2

High and Robust Fault Detection via Polynomial Approximated Isomap Embeddings

Burak Alakent[a]

[a]: *Department of Chemical Engineering, Bogazici University, Istanbul, 34342, Turkiye*
burak.alakent@boun.edu.tr

Abstract

In Multivariate Statistical Process Monitoring (MSPM), a large number of measured variables is monitored online usually in a latent variable space. To this end, various linear/ nonlinear, Gaussian/non-Gaussian, and static/dynamic methods have been proposed. While implementations of deep learning methods in MSPM are frequently seen nowadays, less emphasis is given to unsupervised nonlinear manifold learning (ML) methods, such as Isomap. It is important to note that ML takes the geometry of the data into consideration while reducing dimensions, hence may have an advantage, particularly, over Kernelized methods. However, requirement of approximations for out-of-sample (test) points renders ML methods less practical. To remedy this issue, we have recently proposed independent component analysis of polynomial approximation to Isomap embeddings coupled with principal component analysis (ICA_{pIso}-PCA) method, and showed that fault detection and isolation performances are drastically improved compared to traditional methods. In the current study, we include lagged process measurements in ICA_{pIso}-PCA, i.e. dynamic ICA_{pIso}-PCA ($dICA_{pIso}$-PCA), and show that fault detection rate is further increased, and more robust with respect to number of selected components compared to various methods on Tennessee Eastman plant.

Keywords: Fault detection, dynamic PCA, ICA, Manifold Learning, Tennessee Eastman.

1. Introduction

Occupational and process safety issues are nowadays more of concern than ever, hence fault detection, isolation, identification and prognosis have recently gained utmost importance. To achieve this aim, Multivariate Statistical Process Monitoring (MSPM) is one of the most frequently employed methods mainly due to i) difficulties in obtaining accurate mechanistic mathematical models for chemical industries, and ii) developments in statistical and machine learning methods making it possible to extract information from high dimensional, nonlinear and collinear process variables (Ji and Sun; 2022). Though Principal Components Analysis (PCA) and Projection to Latent Structures (PLS) are historically the benchmark methods for MSPM (Kourti and MacGregor; 1995), both methods are known to suffer from various limitations, such as extracting only static relations, and relatively poor performance under nonlinear process dynamics and non-Gaussian variable distributions. Various combinations of Kernel methods (Lee et al., 2004) and Independent Components Analysis (ICA) (Kano et al., 2003; Lee and Qin, 2007) are commonly employed to overcome these limitations. Furthermore, applications of deep Artificial Neural Network (ANN) models are shown to increase the fault detection performance in MSPM applications (Qian et al., 2022).

With the new millennium, there has been a surge of research activity in manifold learning, such as Isomap (Tenenbaum, de Silva, and Langford, 2000), and locality preserving projections (LPP) (He and P. Niyogi, 2003), and various applications have been implemented for MSPM purposes. While manifold learning methods, unlike Kernel and ANN methods, offer the capability of determining the geometry of a low-dimensional manifold, on which the process data is deemed to lie, nonlinear manifold learning methods lack explicit mapping function to transform the future query points into embedding space. A number of solutions have been proposed to tackle this issue. One solution method is to approximate the nonlinear embedding using a linear transformation in the optimization objective function (Luo, 2014; Miao. Song, and Shen, 2015). Another method is to approximate the nonlinear transformation using various learners, such as ANN (van der Maaten, 2009). Recently, we have proposed ICA of polynomial approximation to Isomap embeddings coupled with PCA (ICA_{pIso}-PCA) method, in which embedding approximations are performed using a high order (regularized) polynomial, rendering both high fault detection rates, and contribution plots with highly reduced smearing effect (Alakent, 2022). In the current study, we aim to investigate whether fault detection rate can further be improved including lagged process measurements; the resulting method is named dynamic ICA_{pIso}-PCA ($dICA_{pIso}$-PCA). Applications on Tennessee Eastman simulation plant show that fault detection rates of $dICA_{pIso}$-PCA are generally higher compared to those of traditional and state-of-the-art MSPM methods, and ICA_{pIso}-PCA. The most significant advantage of the suggested $dICA_{pIso}$-PCA method is shown to be its high fault detection accuracy over a wide range of number of components selected, making it easier to construct a reliable monitoring model without using a very large dataset.

2. Dimensional Reduction Tools used in SPM

2.1. PCA

The aim in PCA is to determine linear projections of the original coordinates with the highest variance. Given the mean-centered (and scaled) data matrix $\mathbf{X} \in \mathbb{R}^{N \times D}$, where N and D correspond to number of samples and process variables, or features, respectively, eigenvalue decomposition of the covariance matrix $\mathbf{C}$ yields the loading matrix $\mathbf{P} \in \mathbb{R}^{p \times p}$, on which $\mathbf{X}$ is projected to obtain $\mathbf{T} \in \mathbb{R}^{N \times D}$ scores matrix:

$$\mathbf{C} = \mathbf{X}^T\mathbf{X}/(N-1) = \mathbf{P}\boldsymbol{\Lambda}\mathbf{P}^T \quad (1)$$

$$\mathbf{T} = \mathbf{X}\boldsymbol{P} \quad (2)$$

Usually, the first d ($< D$) principal components (PCs) with the highest eigenvalues (variances) are used to construct a lower dimensional subspace, in which essential process variability is deemed to be captured:

$$\mathbf{T}_d = [\mathbf{t}_1\, \mathbf{t}_2 \ldots \mathbf{t}_d] = \boldsymbol{\Psi}\mathbf{P}_d = \boldsymbol{\Psi}[\mathbf{p}_1\, \mathbf{p}_2 \cdots \mathbf{p}_d] \quad (3)$$

2.2. ICA

The generative model of ICA is $\boldsymbol{x} = \mathbf{A}\boldsymbol{s}$, in which $\boldsymbol{s} \in \mathbb{R}^m$, consisting of m independent components each with unit variance, and $\boldsymbol{x} \in \mathbb{R}^D$ is a zero-mean random variable vector, and $\mathbf{A} \in \mathbb{R}^{D \times m}$ is the mixing matrix. A demixing matrix $\mathbf{W} \in \mathbb{R}^{m \times D}$ is to be determined:

$$\mathbf{Y} = \hat{\mathbf{S}} = \mathbf{W}\mathbf{X} \quad (4)$$

The problem stated above may be solved via Fast-ICA procedure, in which each vector in **W** and **Y** is determined iteratively (Hyvärinen, Karhunen and Oja, 2001). Similar to PCA, a reduced d (< D) number of ICs ($\mathbf{W}_d$) may be used. However, ranking ICs is not straightforward as that in PCA, since variance of each IC is constrained to be equal to unity. In the current study, ranking ICs is based on the negentropy of **y** values.

2.3. Isomap

In Isomap, sample points are assumed to lie on a smooth manifold, and approximate geosedic distances, i.e. distances measured via shortest tracks on the manifold, are used to represent pairwise sample distances. The K-Isomap method involves connecting K nearest neighbors (K=5 used currently) to each sample point, and storing the squares of the geodesic distances, determined via summing the distances along the shortest paths, in a $\mathbf{S} \in \mathbb{R}^{N\times N}$ matrix. Then, eigenvalue decomposition is employed on **S** mean-centered via $\mathbf{H} = \mathbf{I} - \mathbf{1}\mathbf{1}^T/_N$, in which **I** is the identity matrix, and $\mathbf{1} = [1\ 1 \ldots 1]^T \in \mathbb{R}^N$:

$$-\frac{1}{2}\mathbf{HSH} = \mathbf{V\Lambda V}^T \tag{5}$$

Similar to PCA, the eigenvalues are ranked in decreasing order, and the largest d eigenvectors of V are used to construct the Isomap coordinates:

$$\mathbf{Y} = \mathbf{\Lambda}_d^{1/2}\mathbf{V}_d^T \tag{6}$$

Unlike $\mathbf{P}_d$ and $\mathbf{W}_d$ in PCA and ICA, respectively, Isomap does not yield an explicit transformation matrix that may be used for embedding novel samples; hence, approximations are required to determine the Isomap coordinates of future points.

2.4. *Dynamic PCA (dPCA)*

PCA is based on the assumption of time-independent samples, that is easily violated in industrial processes. In order to include the autocorrelation in process measurements in constructing the PC space, dPCA was proposed (Ku, Storer and Georgakis, 1995). The dPCA involves augmenting the features matrix $\mathbf{\Psi}$ using the current (sample n) and τ time lagged (samples n-1, n-2, … n-τ) measurements as follows:

$$\mathbf{\Psi} = [\mathbf{X}(n)\ \ \mathbf{X}(n-1) \ldots \mathbf{X}(n-\tau)] \in \mathbb{R}^{(N-\tau)\times D(\tau+1)} \tag{7}$$

The rest of dPCA involves applying eigenvalue decomposition on the covariance matrix of $\mathbf{\Psi}$ and employ traditional PCA, as in Eqns. (1-3). The dPCA method was shown to improve fault detection performance since deterioration of the time-dependence of the variables could also be monitored. Including time lagged measurements in other dimension reduction tools, such as dynamic ICA (dICA), have been widely used.

3. Principal Fault Detection Indices in SPM

Based on the traditional application of PCA, T^2 and SPE are the most commonly used surrogate variables (indices), which measure the proximity of the test samples to origin of normal operation in the reduced subspace, and the magnitude of the residual component of the test sample, unaccounted by the reduced space, respectively.

$$T^2 = \sum_{i=1}^{d}\frac{t_i^2}{\lambda_i} = \sum_{i=1}^{d}\frac{(\mathbf{x}^T\mathbf{p}_i)^2}{\lambda_i} \tag{8}$$

$$SPE = \left\|\boldsymbol{x} - \boldsymbol{P}_d^T\boldsymbol{t}\right\|_2^2 \tag{9}$$

In Eqn. 8, the coordinates of the query sample in the original (mean-centered) and projected spaces are shown with $\boldsymbol{x}$ and $\boldsymbol{t} = [t_1\ t_2 \ldots t_d]^T$, respectively, while $\boldsymbol{\lambda}_i$ denotes the variance of scores of normal training data. In Eqn. 9, the squared norm of the residuals of the query point is determined. Phase I of MSPC consists of projecting historical normal operation data to a lower dimensional space, followed by determining the control limits

(CLs) of surrogate variables, i.e. T_{lim}^2 and SPE_{lim}^2, at a specific Type-I error rate. Phase II consist of checking whether the surrogate indices of novel measurements (T^2 and SPE^2) fall within CLs.

4. The Proposed Method: dICA$_{pIso}$-PCA

We recently proposed ICA$_{pIso}$-PCA, in which Isomap embeddings are approximated using ridge regression of fourth to sixth ordered polynomials, followed by ICA and PCA (Alakent, 2022), and, in the current study, we aim to improve the fault detection capability of ICA$_{pIso}$-PCA via addition of lagged variables. The dICA$_{pIso}$-PCA starts with employing PCA and Isomap separately on $\mathbf{\Psi}$, which is the features matrix consisting of time lagged measurements (see Eqn. 7), and all the PC scores and the first I_p (I_p=8 used currently) embedding coordinates are stored. Then, the following regressor matrix $\mathbf{\Psi}^*$ is formed:

$$\mathbf{\Psi}^* = [\mathbf{\Psi} \quad (\mathbf{\Psi})^{o2} \dots (\mathbf{\Psi})^{or}] \in \mathbb{R}^{(N-\tau)\times Dr(\tau+1)} \tag{10}$$

Here, $(.)^{oa}$ is used to raise the power of each element in the matrix to $a = 1,2,\dots r$, and r is taken to be equal to six. Taking $\mathbf{F} \in \mathbb{R}^{N\times \mathrm{I}_p}$ as the matrix of Isomap embeddings, ridge regression ($\alpha = 10^{-4}$) is employed to determine the approximated Isomap embeddings:

$$\left[\hat{\boldsymbol{f}}_1\, \hat{\boldsymbol{f}}_2 \dots \hat{\boldsymbol{f}}_{\mathrm{I}_p}\right] = \mathbf{\Psi}^*\left(\mathbf{\Psi}^{*T}\mathbf{\Psi}^* + \alpha\mathbf{I}\right)^{-1}\mathbf{\Psi}^{*T}\mathbf{F} \tag{11}$$

Highest linearly correlated PC scores with Isomap approximations are replaced by each of $\left[\hat{\boldsymbol{f}}_1\, \hat{\boldsymbol{f}}_2 \dots \hat{\boldsymbol{f}}_{\mathrm{I}_p}\right]$, i.e. PC scores, most similar to high variance approximated Isomap embeddings, are replaced by these approximations, yielding a "nonlinear scores" matrix of reduced dimensions of the training data, $\boldsymbol{T}_{nl} \in \mathbb{R}^{(N-\tau)\times D'}$, $D' \geq D(\tau+1)$. Then, ICA and PCA are sequentially employed on $\boldsymbol{T}_{nl}$, i.e. application of ICA yields the following.

$$\hat{\mathbf{S}} = [\hat{\mathbf{s}}_1\, \hat{\mathbf{s}}_2 \dots \hat{\mathbf{s}}_N] = \mathbf{W}\boldsymbol{T}_{nl}^T \tag{12}$$

Here, $\mathbf{W}$ is the demixing matrix in Eqn. (4), and used to transform the nonlinear (correlated) scores matrix to independent scores $\hat{\mathbf{S}}$. The IC scores are computed via $I^2 = \hat{\mathbf{S}}^T\hat{\mathbf{S}}$, and 99% CL for I^2 indices (I_{lim}^2) is computed using kernel density estimator. The I^2 value of a new query point is compared with I_{lim}^2, and out-of-control alarm is given if $I^2 > I_{lim}^2$. The same procedure is repeated with T^2 and SPE indices, which are obtained via employing PCA on the residuals of the nonlinear scores, unexplained by the ICA. For a more detailed explanation of the procedure, please refer to Alakent (2022).

5. Results and Discussion

We employed dICA$_{pIso}$-PCA on the simulated Tennessee Eastman plant (TEP) (Downs and Vogel, 1993). The main products in TEP are G and H with the purged byproduct F from a feed consisting of reactants A, C, D, E and inert component B. In our study, data from 33 process variables (excluding concentration measurements) sampled at 3 min intervals were used. Normal operation data is comprised of 960 samples, while test data consists of one normal operation, and 21 different faults of 960 samples (Chiang, Russell, and Braatz, 2000). The process diagram, variables and fault descriptions of TEP, omitted here for space considerations, can be found in the sources given in the References. We compared Missed Detection Rate (MDR) values, i.e. percent of faulty samples deemed to be normal, of dICApIso-PCA with the best results obtained from dynamic and/or Kernel methods, such as KICA, dPCA, dICA, dICA$_{PCA}$ and dKPCA. CLs of the monitoring indices in all methods were adjusted to yield a false detection rate of 0.01 (rate of false signals for normal query points) in order to render an unbiased comparison of MDR values (Mutlu and Alakent, 2019). We used two time lags in all dynamic dimension reduction methods. To save space, we here show only the minimum MDR obtained from all monitoring indices in Table 1. We included a state-of-the-art method, namely Global Local Preserving Projection (GLPP), which utilizes the NPP to capture both the local and global structure of data (Miao, Song and Shen, 2015).

In Table 1, dICA$_{\text{pIso}}$-PCA gives the smallest average MDR, significantly smaller than ICA$_{\text{pIso}}$-PCA, indicating that including time-lagged measurements improves fault detection rate of our method. Looking in more detail, dICA$_{\text{pIso}}$-PCA yields the top and second best performances in 14 and the remaining 7 faults, respectively. On the other hand, dGKPP has the best and second best performance in 9 and 6 of the faults, respectively, while exhibiting a moderate performance in the remaining 6 faults, showing a poorer fault detection capability compared to that of dICA$_{\text{pIso}}$-PCA.

Table 1. MDRs of various MSPC models against 21 faults in TEP.

Fault no.	KICA[i]	GKPP[ii]	ICA$_{\text{pIso}}$-PCA[iii]	dPCA[iv]	dKICA[v]	dKPCA[vi]	dGKPP[vii]	dICA$_{\text{pIso}}$-PCA[viii]
1	**0**[ix]	0.25	**0**	0.500	0.125	0.125	0.125	0.125
2	1.5	1.5	1.5	1.75	1.75	1.625	1.5	**1.125**
3	96.9	97	97.1	98.4	96.1	96.3	95.5	**95.1**
4	**0**	0.625	**0**	0.249	**0**	**0**	**0**	**0**
5	**0**	**0**	**0**	72.1	**0**	0.125	**0**	**0**
6	**0**	**0**	**0**	0.249	**0**	**0**	**0**	**0**
7	**0**	**0**	**0**	0.249	**0**	**0**	**0**	**0**
8	2	2.13	2.13	2.63	1.88	1.63	1.63	**1.5**
9	96.6	97.5	96	98.1	97	96	96.5	**95.9**
10	32.6	11.1	11	63.6	7.875	13.75	**5.75**	6.38
11	19.8	28.4	17.5	12.5	11.9	6.75	19.6	**4.88**
12	0.25	0.125	0.125	0.998	**0**	0.25	0.125	**0**
13	4.75	5.38	4.75	4.99	4.25	4.38	**3.63**	3.88
14	**0**	**0**	**0**	0.249	**0**	**0**	0.125	**0**
15	95.6	94.1	94	96.6	83.4	92.1	**60.9**	65
16	38.6	8	8.63	70.1	6.25	8.375	**3.13**	4.00
17	3.75	7.50	2.75	3.74	2.38	2.13	2.63	**2.00**
18	10	10.1	9.50	10.1	9.375	**9.25**	9.50	9.38
19	46.6	13.9	5.13	28.8	0.75	1.88	0.375	**0.125**
20	33	9.13	8.75	46.9	11.1	28.8	**8**	8.38
21	43.8	47.8	43.6	49.1	54.4	51.3	49.9	**43.5**
Avg[ix]	*25.0*	*20.7*	*19.2*	*31.5*	*18.5*	*19.7*	*17.1*	***16.3***

[i] 40 components, [ii] 19 components with K=10 and $\sigma = 2$, [iii] 1 IC and 4 PCs, [iv] 21 components, [v] 10 ICs, [vi] 65 PCs, $\sigma = 300$·, [vii] 29 components with K=10 and $\sigma = 10$,[viii] 6 ICs and 4 PCs.

[ix] Bold values correspond to smallest MDR values obtained for each fault type.

[x] Average MDR values for each model shown in the last row.

Additionally, Fig. 1 shows that average MDR of dICA$_{\text{pIso}}$-PCA does not significantly depend on the number of selected components, unlike dKPCA and dGKPP, which, in practical applications, would require separate datasets comprising faulty operating conditions to adjust the number of components. Hence, selecting the "right" number of components, that is still an unsolved issue in MSPM applications, seems to be less of an issue for dICA$_{\text{pIso}}$-PCA.

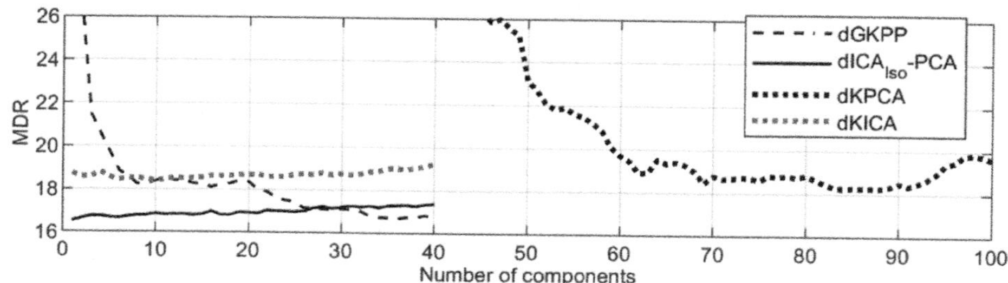

Figure 1. Change of average MDR with respect to number of PCs or ICs.

References

C. Ji and W. Sun (2022), A Review on Data-Driven Process Monitoring Methods: Characterization and Mining of Industrial Data, *Processes*, 10, 335.

T. Kourti and J.F. MacGregor (1995), Process analysis, monitoring and diagnosis, using multivariate projection methods, *Chemom. Intell. Lab. Sys.*, 28, 3-21.

J.-M. Lee, C.K. Yoo, S.W. Choi, P.A. Vanrolleghem, and I.-B. Lee (2004), Nonlinear process monitoring using kernel principal component analysis, Chem. Eng. Sci., 59, 223-234.

M. Kano, S. Tanaka, S. Hasebe, I. Hashimoto, and H. Ohno (2003), Monitoring independent components for fault detection, *AIChE J.*, 49, 969-976.

J.-M. Lee and S.J. Qin (2007), Fault Detection of Non-Linear Processes Using Kernel Independent Component Analysis, *Can. J. Chem. Eng.*, 85, 526-536.

J. Qian, Z. Song, Y. Yao, Z. Zhu, X. Zhang (2022), A review on autoencoder based representation learning for fault detection and diagnosis in industrial processes, *Chemom. Intell. Lab. Syst.*, 231, 104711.

B. Tenenbaum, V. de Silva, and J.C. Langford (2000), A global geometric framework for nonlinear dimensionality reduction, *Science*, 290, 2319-23.

X.F. He and P. Niyogi (2003), Locality preserving projections. *In Proceedings of the Conference on Advances in Neural Information Processing Systems*, Dec 8–13, Vancouver, Canada; MIT Press: Cambridge, MA.

L. Luo (2014), Process Monitoring with Global-Local Preserving Projections, *Ind. Eng. Chem. Res.*. 53, 7696–7705.

E.C. Mutlu and B. Alakent (2019), Revisiting reweighted robust standard deviation estimators for univariate Shewhart S-charts, *Qual. Reliab. Eng.*, 35, 995-1009.

A. Miao, Z. Ge, Z. Song and F. Shen (2015), Nonlocal structure constrained neighborhood preserving embedding model and its application for fault detection. *Chemom. Intell. Lab. Syst.*, 142, 184-196.

L. van der Maaten (2009), Learning a parametric embedding by preserving local structure, *AISTATS*, volume 5 of JMLR Proceedings, 384-391.

B. Alakent (2022), Reducing Smearing Effect in Contribution Plots and Improving Fault Detection via Polynomial Approximated Isomap Embeddings, *Can. J. Chem. Eng.*, DOI: 10.1002/cjce.24689.

A. Hyvärinen, J. Karhunen and E. Oja (2001), *Independent Component Analysis*, John Wiley & Sons, Inc..

W. Ku, R.H. Storer and C. Georgakis (1995), Disturbance detection and isolation by dynamic principal component analysis, *Chemom. Intell. Lab. Syst.*, 30, 179-196.

J.J. Downs and E.F. Vogel (1993), A plant-wide industrial process control problem, *Comput. Chem. Eng.*, 17, 245-255.

L.H. Chiang, E.L. Russell, and R.D. Braatz (2000), Fault diagnosis in chemical processes using Fisher discriminant analysis, discriminant partial least squares, and principal component analysis, *Chemom. Intell. Lab. Syst.*, 50, 243–252.

Antonis Kokossis, Michael C. Georgiadis, Efstratios N. Pistikopoulos (Eds.)
PROCEEDINGS OF THE 33rd European Symposium on Computer Aided Process Engineering (ESCAPE33), June 18-21, 2023, Athens, Greece
 http://dx.doi.org/10.1016/B978-0-443-15274-0.50097-4

A comparative study of distributed feedback-optimizing control strategies

Vegard Aas,[a] Risvan Dirza,[a] Dinesh Krishnamoorthy,[a] Sigurd Skogestad,[a*]

[a] *Department of Chemical Engineering, Norwegian University of Science and Technology, NO-7491 Trondheim, Norway*
**sigurd.skogestad@ntnu.no*

Abstract

In industry, the processes often consist of several subsystems with a common constraint, for example, a shared resource. This paper considers the problem of steady-state real-time optimization (RTO) for a subsea gas-lifted oil production network with multiple wells and constrained access to shared gas-lift supply. Such problems can be solved by a centralized numerical optimization, which can be computationally expensive. To avoid the use of numerical optimization, one can utilize either online primal or dual decomposition instead, where the problem is converted into a feedback-based problem. The main benefit of primal method is that it distributes local setpoints which complies with primal feasibility, however, the dual method is more general. Both the primal and dual methods allow for a distributed implementations. The dual method is more general in terms for allowing for many constraints, but as shown for the simulations with uncertainty and measurement noise, the primal method may give better dynamic constraint satisfaction.

Keywords: Distributed optimization, Feedback control, Production optimization.

1. Introduction

Industrial process often consists of several subsystems with a common constraint, for example, a shared resource. Problems like this can be decomposed and solved using distributed optimization. However, this can be computationally expensive as it solves several rounds of numerical optimization problems online at each sample time. This can be addressed by indirectly moving the optimization problem into the control layer (Morari et al. (1980)). Such problems are known as feedback-optimizing control, which can be implemented using simple tools such as Proportional-Integral-Derivative (PID) controllers.

In our previous work (Dirza et al., 2022(b)), we have experimentally validated a recently developed method of feedback-optimizing control called distributed feedback-based RTO. This method is developed based on dual decomposition and optimally handles steady-state changes in active constraints. However, the constraints are controlled in a slower time scale by updating the dual variables. This leads to the need for significant "back-off" strategy, which could lead to profit loss in the long run. To eliminate or reduce the "back-off," Dirza et al. (2022(a)) introduces an alternative distributed feedback-optimizing control based on online primal decomposition using feedback and constraint controller(s) which distribute local setpoints without violating the common constraint to avoid or minimize use of a "back-off" strategy.

In this paper, we provide a comparative analysis of the two distributed feedback-optimizing control approaches. The model we consider represents our lab-scale experimental rig that consider uncertainty and measurement noise. The rig emulates a subsea oil production network with gas-lift rate as the manipulated variables. The experimental results will be provided as the future work.

2. Feedback-optimizing control

Consider the optimization problem for the entire system built by a network of N subsystems. We assume each subsystem is optimized locally and that we always have active constraint. Thus, the steady-state optimization problem can be expressed as follows,

$$\min_{u_i, \forall i \in N} \quad J_N = \sum_{i \in N} J_{N,i} \tag{1a}$$

$$s.t. \qquad \boldsymbol{f}_i(\boldsymbol{x}_i, \boldsymbol{u}_i, \boldsymbol{d}_i) = 0 \quad i \in [1, N] \tag{1b}$$

$$\sum_{i \in N} g_i(\boldsymbol{x}_i, \boldsymbol{u}_i, \boldsymbol{d}_i) - g^{max} = 0 \tag{1c}$$

where $\boldsymbol{x}_i, \boldsymbol{u}_i$ *and* $\boldsymbol{d}_i$ denote the vector of states, inputs and disturbances respectively. Constraint (1b) is related to the entire system model, and (1c) is an equality constraint.

2.1. Distributed Feedback-Optimizing Control using Online Primal Decomposition

To solve problem (1) using simple feedback controller, it is possible to construct a method based on online primal decomposition (Dirza et. al (2022(a))). The main motivation for this method is that we want to achieve optimal steady-state operation in a distributed manner, with minimal dynamic constraint violation and without demand of solving numerical optimization problems online. By introducing an initial value of local constraint g_i^{sp}, (1c) can be written as

$$g_i(\boldsymbol{x}_i, \boldsymbol{u}_i, \boldsymbol{d}_i) - g_i^{sp} = 0 \quad i \in [1, N] \tag{1d}$$

$$\sum_{i \in N} g_i^{sp} = g^{max} \tag{1e}$$

instead. As long as (1e) is satisfied, primal feasibility is guaranteed. Each subproblem estimates local Lagrange multipliers, which is used in central constraint controllers. These controllers update the setpoints iteratively, where the goal is to provide setpoints that satisfy the primal feasibility. In this paper we assume that the constraint is always active, therefore we have two types of approaches of updating the local setpoints. To ensure that (1e) is satisfied, one of the local setpoints is updated as follows,

$$g_N^{sp,k+1} = g^{max} - \left(g_1^{sp,k+1} + \cdots + g_{N-1}^{sp,k+1}\right) \tag{2}$$

This is called the compensator subsystem, and the objective is to ensure primal feasibility. For the remaining subsystems, $i = \{1, \ldots, N-1\}$, the local setpoints is updated by

$$g_i^{sp,k+1} = g_i^{sp,k} + K_{I,i}(-\lambda_1^k + \lambda_N^k) \tag{3}$$

We may use integrating controllers with integral gain $K_{I,i} = \frac{1}{K_i(\tau_{c,i})}$, where K_i is the step response gain and $\tau_{c,i}$ is the closed-loop time constant. However, a proportional integral (PI) controller can also be used. The local Lagrange multipliers, λ_i, can be estimated

$$\lambda_i = -\nabla_{u_i}\hat{J}_{N,i}\left(\nabla_{u_i}\hat{g}_i(x_i, u_i, d_i)\right)^{-1} \tag{4}$$

Where $\nabla_{u_i}\hat{J}_{N,i}$ and $\nabla_{u_i}\hat{g}_i$ are the estimated gradient of local cost and local setpoints respectively. In figure 2.1 the online primal decomposition framework is illustrated. The central constraint controllers, which contains both normal and compensator subsystems, provide new set points for the local constraints. As the constraints always are active, these set points are considered as inputs to the subsystems. When there is a presence of disturbance, we can use the plants current information to estimate the current state and parameters by implementing a local dynamic estimator, for instance Extended Kalman Filter (EKF). By applying the inputs, estimated states and parameters, we can estimate cost gradient as well as constraint gradient to compute the local Lagrange multipliers. The multipliers are then implemented in the central constraint controller to calculate the new set points.

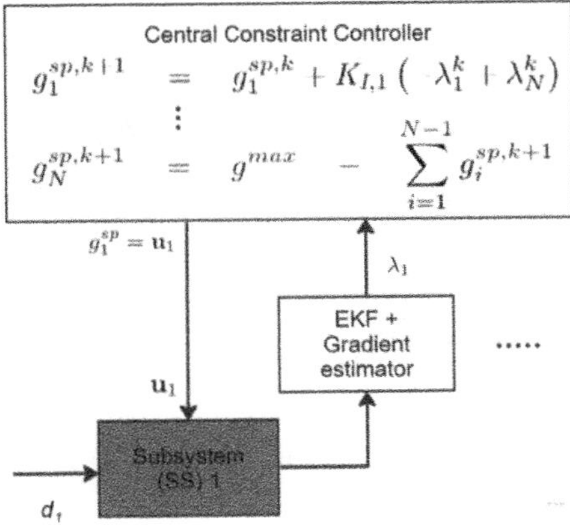

Figure 2.1 The online primal decomposition control structure.

3. Simulation descriptions

3.1. Simulation model setup

To emulate the subsea gas-lifted oil production system, we use a model created in MATLAB R2021a based on a lab-scale experimental rig from a paper by Matias et al. (2022) and has been tested and used in earlier papers related to the lab-rig. It is implemented with noise similar to what is expected from the lab-rig. The reservoir section of the model is implemented by valve openings $\boldsymbol{p} = [\boldsymbol{p_1}, \boldsymbol{p_2}, \boldsymbol{p_3}]$ which determine the liquid flow. These valves represent the disturbances from the reservoir. In the MATLAB model the gas-lift flow controllers are given $\mathbf{u}_i$ as setpoints. They are implemented as a 5 second delay plus input noise to simulate controller action instead of coding the actual controllers. The setpoint controllers in the central constraint controller are integral controllers tuned using SIMC tuning rules (Skogestad (2003)).

3.2. Optimization problem setup

The objective of the optimization problem in this model setup is to maximize the liquid flow rate, which equals the sum of the liquid production of the three wells, with a limited amount of gas-lift injection, which is input shared constraint. Considering problem (1). the economic objective can be expressed as below,

$$J(\boldsymbol{u},\boldsymbol{p}) = \sum_{i=1}^{3} f_i(u_i,p_i) = -20Q_{l,1}(u_1,p_1) - 25Q_{l,2}(u_2,p_2) - 30Q_{l,3}(u_3,p_3) \quad (5)$$

where $Q_{l,1}$, $Q_{l,2}$ and $Q_{l,3}$ are the produced liquid flow rates of well 1, 2 and 3 respectively. We assume different values of the hydrocarbon flows as shown in eq. 6, to illustrate how different values affect the behavior of the subsystems. The input vector is defined as $\boldsymbol{u} = [Q^{\mathbf{sp}}_{gl,1} \quad Q^{\mathbf{sp}}_{gl,2} \quad Q^{sp}_{gl,3}]^{\boldsymbol{T}}$ where $Q^{sp}_{gl,1}$, $Q^{sp}_{gl,2}$ and $Q^{sp}_{gl,3}$ are the gas-lift set points from the central controller of well 1, 2 and 3 respectively. In addition, the reservoir valve $\mathbf{p}$ is time varying.

3.3. Comparative method

We consider our previous work, distributed feedback-based optimization with dual decomposition, to compare with primal decomposition. The implemented dual decomposition method is based on Dirza et al. (2022(b)) and Krishnamoorthy et al. (2021). In figure 3.1 the implemented dual decomposition framework is shown.

4. Results and Discussion

Figure 4.1 shows the disturbance in this simulation, which corresponds to the reservoir valve openings **p** in the lab-rig model. The first disturbance occurs when p_1 gradually decreases from t = 6 to t = 12.5 min. During this interval, we expect the gas-lift injection in well 1 to decrease, and redistribution of the gas to the other wells. The second disturbance occurs when also p_3 gradually decreases from t = 14 to t = 18 min. As before, we expect that the gas-lift injection rate in well 3 will go down. At the same time, we expect the other wells will gain a higher gas-lift injection. The third and fourth disturbance occur from t = 21 to t = 24.5 and t = 28 to t = 34.5 respectively. During this time p_3 and p_1 are gradually increased back up to the initial values. This is because we want to see how the controllers behave for both decrease and increase in the disturbance.

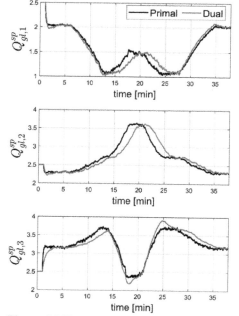

Figure 3.1 Block diagram of dual decomposition control structure for one well. The area within the dashed gray box is duplicated N times. For a more detailed version the reader is referred to Dirza et al. (2022b).

In this paper subsystem 3 has been used as the compensator for the primal decomposition. This subsystem was chosen because it has the highest gain magnitude, which results in most profit for cases with active shared constraint.

In figures 4.2 – 4.7 we compare the simulation results from primal decomposition and dual decomposition. The calculated input setpoints are shown in figure 4.2, the actual gas-lift flow rate will deviate slightly from the setpoints due to the implemented measurement noise in the model. Another cause is how the gas flow rate controllers are implemented in the model, which is described in section 3.1.

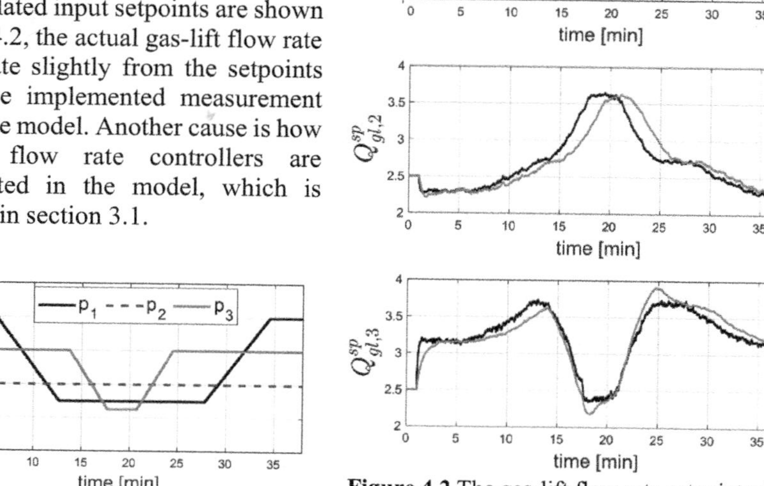

Figure 4.1 Disturbance profile during the simulation.

Figure 4.2 The gas-lift flow rate setpoint $\boldsymbol{u}^{sp} = \boldsymbol{Q}_{gl}^{sp}$ of every well due to disturbance from the simulation.

In figure 4.2 and 4.3 we see that the dual control is responding slower to the disturbances than the primal control for correcting the gas-lift setpoints. This is a consequence of the time-scale separation between the central controller and the gradient controllers needed in the dual structure (Dirza et al (2022b)).

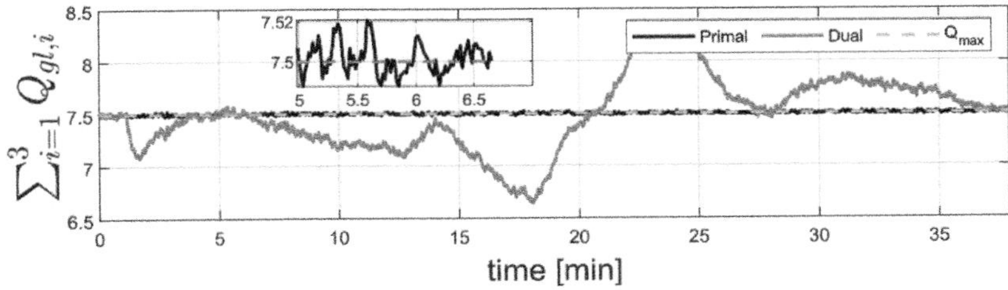

Figure 4.3 The constraint satisfaction of both primal and dual. There is a magnifying plot in time window 5 to 6.5 min showing only constraint satisfaction for primal.

Figure 4.3 shows the constraint satisfaction. As expected, the Primal decomposition performs much better than the dual decomposition here. This is because the compensator system, subsystem 3, "absorbs" the deviations from the constraint. The absorbing ability comes from how the compensator set point is calculated, see eq. 2, and is the reason for the primal decomposition's capability to maintain the active constraint. In terms of what this means for the operation, with the use of primal decomposition we can run the system without any significant back-off and remain feasible at all times. On the other and, for the dual decomposition we must implement back-off, especially from t = 21 when the disturbances start to increase again.

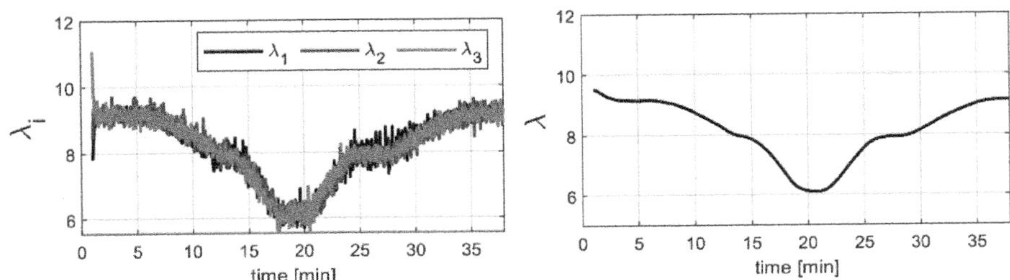

Figure 4.4 Local Lagrange multipliers for primal decomposition.

Figure 4.5 Lagrange multiplier for dual decomposition

Figure 4.4 depicts that the local Lagrange multipliers for Primal Decomposition converge to the same value. Figure 4.5 show the Lagrange multiplier for the dual decomposition. We can observe that the multiplier converges slower and is smoother for this method. This is because the central controller for dual decomposition operates on a slower timescale. When the Lagrange multiplier converges around t = 3.5 we can see that the active constraint is controlled almost as good as for the Primal case. It is also worth to notice that the Lagrange multipliers converges to the same value for both Primal and Dual.

To analyze the optimization performance of primal and dual decomposition, we compare the profit obtained by the two methods with a naive approach, where it is considered fixed inputs, $\boldsymbol{u}_i = {Q_{gl}^{max}}/{3}$. The naive approach represents the case where no information is available, therefor, the best approach is to divide the available gas equally among the

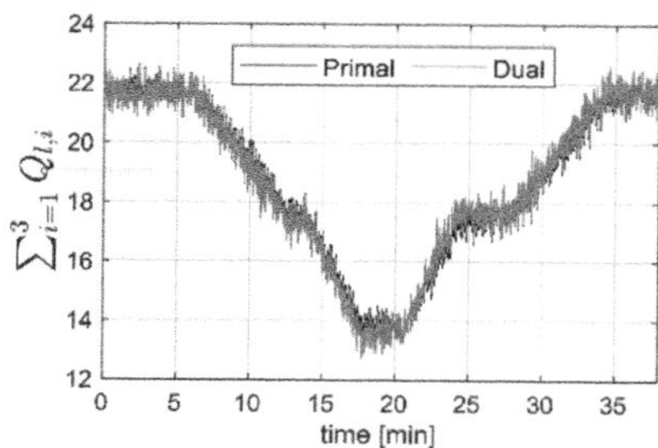

Figure 4.6 Total liquid flow of the system

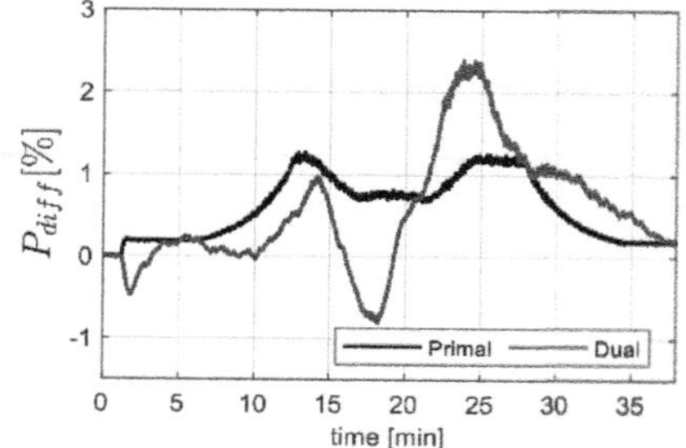

Figure 4.7 Instantaneous profit compared to naive approach.

wells. This additional approach is used as benchmark to show how primal and dual decomposition compares to not do any optimization at all. To display the performance, we plot the difference in percentage between the instantaneous profit of primal/dual and the naive approach. The difference is calculated as $J_{diff} = \frac{J - J_{naive}}{J_{naive}} \cdot 100$, where J is the profit of the method of interest and $\boldsymbol{J_{naive}}$ is the profit of the naive approach. Figure 4.7 shows that primal is more profitable than dual until t = 21 min, after this dual appears to be favorable. However, in figure 4.3 we see that after t = 21 dual does not achieve primal feasibility and therefore the profit here is not viable.

5. Conclusion

In this work, we have done simulation with uncertainty and measurement noise to compare feedback-based real-time optimization with online primal and dual decomposition. Based on the results we can conclude that primal decomposition is able to effectively ensure primal feasibility for separable systems with an active shared input constraint. It performs better than dual decomposition in this aspect, mainly because of the timescale difference of the central constraint controllers. This significantly reduces the need for any significant back off, which results in more profitable operations. While primal ensures constant feasibility, it is less general than dual. As the continuation of this work, we consider obtaining the experimental result, implementing dual with override and develop a primal structure able to switch between active constraints.

References

Dirza, R., Rizwan, M., Skogestad, S., & Krishnamoorthy, D. (2022a). Real-time Optimal Resource Allocation using Online Primal Decomposition. *IFAC-PapersOnLine*, *55*(21), 31-36.

Matias, J., Oliveira, J. P., Le Roux, G. A., & Jäschke, J. (2022). Steady-state real-time optimization using transient measurements on an experimental rig. *Journal of Process Control*, *115*, 181-196.

Dirza, R., Matias, J., Skogestad, S., & Krishnamoorthy, D. (2022b). Experimental validation of distributed feedback-based real-time optimization in a gas-lifted oil well rig. *Control Engineering Practice*, *126*, 105253.

Krishnamoorthy, D. (2021). A distributed feedback-based online process optimization framework for optimal resource sharing. *Journal of Process Control*, *97*, 72-83.

M. Morari, Y. Arkun, G. Stephanopoulos, 1980. Studies in the synthesis of control structures for chemical processes: Part i: Formulation of the problem. process decomposition and the classification of the control tasks analysis of the optimizing control structures. AIChE Journal 26, 220 – 232.

Skogestad, S. (2003). Simple analytic rules for model reduction and PID controller tuning. Journal of Process Control, 13(4), 291–309.

Antonis Kokossis, Michael C. Georgiadis, Efstratios N. Pistikopoulos (Eds.)
PROCEEDINGS OF THE 33rd European Symposium on Computer Aided Process Engineering (ESCAPE33), June 18-21, 2023, Athens, Greece
 http://dx.doi.org/10.1016/B978-0-443-15274-0.50098-6

Superstructure configuration and optimization of the biofuel production perspectives in Austria

Safdar Abbas,[a] Ada Josefina Robinson[a], Stavros Papadokonstantakis[a]

[a]*Institute of Chemical, Environmental and Bioscience Engineering*
TU Wien,1060 Wien, Austria
Safdar.abbas@tuwien.ac.at

Abstract

This paper presents a superstructure model of mid to long-term biofuel production systems in Austria and determines the optimal configurations that simultaneously satisfy the profit maximization and environmental impact minimizations. A Linear Programming (LP) mathematical model formulation is developed based on data collected from different studies on the biofuel production process. The superstructure-based model considers the availability of Austrian second-generation feedstock, conversion technologies, and final products and by-products. The economic objective function considers the total net profit. The environmental objective function considers the overall environmental impact measured through the Prospective life cycle assessment. The multi-objective model is solved in GAMS software using the ε-constraint method. The results demonstrate the solution for minimum Environmental Impact is zero and profit also zero which means nothing is produced then the model solved for maximum profit which is 284 M€/Year and an overall Environmental Impact is 0.67 Mt of CO_2-Eq.

Keywords: Biomass, Biofuels, Prospective Life Cycle Assessment (PLCA), Superstructure, Optimization.

1. Introduction

Renewable energy resources are emerging as an attractive option for ensuring future energy security around the world. Biomass has been demonstrated to have the potential for fulfilling energy and chemical needs as a good alternative to petroleum because of its rising prices and uncertain availability. There has been a lot of concern about the sustainability of several first-generation biofuels, which has led to discussions about the possibility of producing biofuels from non-food crops, known as second-generation biofuels (Sharma et al., 2013). Second-generation biofuels offer many potential benefits, such as converting waste oils into fuel, reclaiming abandoned land and reducing food crop competition. Also, second-generation feedstock made from locally available sources can be very beneficial to rural and emerging regions of a country's economy. A high yielding feedstock ensures a reasonable price for commercial scale biofuel production. This is a very important aspect of feedstock selection. The benefits of biomass over other renewable energy sources are that it can be stored for long periods of time until needed and that it is an excellent alternative to liquid fuels in today's transportation system. Fuels obtained can be blended to replace classic petroleum-based fuels, or used as (without blending) in existing vehicles (Kowalski et al., 2022).

Biofuels are rapidly gaining interest in society and policies for secure, sustainable, and circular economy concepts, and thus designing and optimizing the entire biorefinery value chains from various biomass feedstock types to biofuel production and end-use has become a significant research challenge(Yue et al., 2014). Superstructure-based modelling and optimization is a systematic approach when considering the commercial feasibility and perspectives of biofuel production, diverse technological options, supply chain, market scenarios, and environmental objectives. Superstructure based approach also helps to setting the boundaries for the scope of the analysis at a country level (e.g., for Austria in this study) or for a union of countries (e.g., for EU, if extending the scope) to lead the results of higher practical importance for decision makers and industrial stakeholders that can influence policies and investments, respectively. This approach is typically influence by spatial aspects with respect to feedstock supply chains and market conditions. To this end, currently the first-generation feedstock is used to produce biofuel in Austria but their overall carbon footprint, land-use change, food availability, price, and biodiversity have all been criticized(The European Parliament and the Council of the European Union, 2015).

2. Problem Statement

The problem addressed in this work concerns the superstructure-based model of mid to long-term second-generation biofuel production system in Austria while optimizing two objectives: the profit and the environmental impact. Figure 1shows the superstructure-based model accounting the feedstock available in Austria according to RED II Annex IX, conversion pathways, final products, and by-products. The conversion technologies have different technology module because these are designed using a different catalyst (Morales et al., 2021). The environmental objective function considers the overall environmental impact in terms of a global warming potential including feedstock extraction, processing and final disposal of products. The economic aspects are considered in terms of total net profit. The goal of this study is to determined the optimal configuration of second-generation biofuel production system in Austria that simultaneously maximize the profit and minimize the environmental impact.

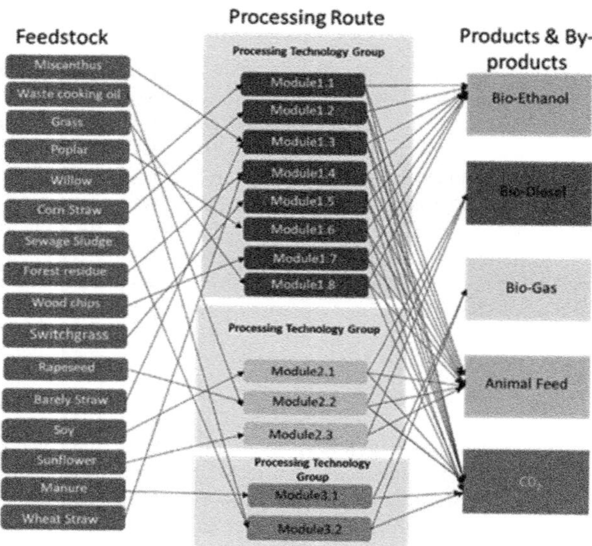

Figure 1: Superstructure for the second-generation biofuel production system

3. Model Formulation

The mathematical formulation is developed based on the work performed by Santibañez-Aguilar et al., 2011 to determine the optimal configuration as presented in Table1.The mathematical model is coded in the GAMS software and solved using the solver CPLEX. The ε-constraint method (Soroudi, 2017) is used to determine the set of optimal solutions that satisfy both objectives and that are used to build the Pareto front. In order to solve the multi-objective problem by ε-constraint method, first the maximum and minimum values of the profit are calculated. These border values are determined by maximizing and minimizing profit separately as the objective function. Then, the profit is assumed as a constraint, and the environmental impact is calculated as the sole objective function. A Pareto optimal front is then derived for both objective functions.

3.1. Economic Analysis

The economic analysis considered in terms of profit is expressed in Table 1. The C_p^{value} is the selling price of products and C_b^{value} is the selling price by-products in €/Ton (IEA, 2022). The annual operating cost involve mainly the feedstock cost in Austria (Ruiz et al., 2019) and utilities consumed during the processing (including energy, catalyst, solvents, etc.). The total capital investment is calculated by process blocks build up method. This method is adopted from Tsagkari et al., 2020 and it consists of a modular approach to estimate the processing cost of different blocks inside the battery limit. The cost of each process block is calculated by the power law equation (Tsagkari et al., 2020). The sum of the process block costs is the total capital investment. The equation, which follows the power law functions, shows a decrease in unit production costs with an increase in plant capacity. Therefore, nonlinear cost functions are linearized over different intervals between given processing limits to obtain a linear function.

Table1: Mathematical formulation of superstructure configuration and optimization of the biofuel production.

Sets, Positive Variables, Decisions Variables, Parameters, Objective Function, and Constraints

Sets
q: Feedstock
p: Products
b: by products
r: Processing routes

Positive Variables
u_{pqr}: Flowrate of products
F_{pqr}: Flowrate of feedstock

Decision Variable
Profit: Net annual gain
EI: Environmental Impact

Parameters
Cost(p): Product Cost
Cost(q): Feedstock Cost
Cost(b): By-product Cost
Processing Cost
Max(q): Maximum available resources
Max(p): Maximum demand of products
EI(q): Environmental Impact of feedstock
EI(p): Environmental Impact of products
EI(r): Environmental Impact of processing routes
ω_{pqr}: Conversion factor to product
γ_{pqrb}: Conversion factor to by product

Objective Function

Objective Function = [**max profit; min EI**]
Where,
Profit = (Annual Sales – Annual Operating cost – Total Capital investment)
Annual Sales = $(\sum_p \sum_q \sum_r u_{pqr} C_p^{value} + \sum_p \sum_q \sum_r \sum_b \gamma_{pqrb} C_b^{value})$
Total Capital Investment = sum of the total capital investment cost of all process block
Total Capital cost of one process block = $I_{rk}^{ref} \left[\frac{Output\ capacity\ of\ a\ new\ plant}{Output\ capacity\ of\ a\ reference\ plant}\right]^{n_{r,k}}$
EI= (LCA of feedstock + LCA of processing route + LCA of final product)

Constraints

$\sum_p \sum_r f_{pqr} \leq f_q^{max} \qquad q \in Q$

A first set of inequality constraints are the maximum availability of the feedstock. it is restricted to use more than the existing quantity of feedstock for its processing to the corresponding final products.

$\sum_q \sum_r u_{pqr} \leq u_p^{max} \qquad p \in P$

A second constraints is the product demand.This constraint helps to prevent the higher production rate than demand and also avoid to waste of sources.

$u_{pqr} \leq u_{pqr}^{max} \qquad q \in Q, r \in R, p \in P$

As a final constraint, the maximum processing limit for feedstocks also considered. This is only for upper limit which are the maximum amount of feedstock for each processing route.

3.2. Environmental Analysis

The Environmental analysis includes the overall environmental impact measured through the prospective Life Cycle Assessment (PLCA) methodology. Prospective Life Cycle Assessment is useful tool to determine the environmental impact of current and emerging technologies in the future. The PLCA requires an inventory database that reflects the anticipated changes in technology and the environment at a particular point in time, based on a specific socio-techno-economic path. In this study PLCA contains an Ecoinvent 3.8 database along with Integrated assessment model (IAM)(Sacchi et al., 2022). IAM describes transformation pathways of the interlinked energy-economy-land-climate system. IAM is based on Shared Socioeconomic Pathways (SSPs) which create a framework for the study of climate-related scenario outcomes. The background provides for quantification of SSP2 as a reference or 'marker' implementation (Fricko et al., 2017).

The SSP2 narrative describes a middle-of-the-road development in the mitigation and adaptation challenges space. There has been an extension of the historical experience in many dimensions of the SSP2 marker implementation, especially in terms of improving the baseline carbon and energy intensity. SSP2 can serve as a starting point to further explore integrated solutions for achieving multiple societal objectives in light of the climate adaptation and mitigation challenges that society could face over the 21st century (Fricko et al., 2017). An open source software Activity browser is used to conduct the overall environmental impact in this study. The PLCA is carried out using an attributional approach because the background scenario does not change. The goal of this LCA is to identify the environmental improvement of different technologies. The results of this analysis are used to build a superstructure optimization model. The results of Environmental impact analysis are presented in terms of GWP (kg CO_2 Eq) and are shown in Table 2 and 3, respectively.

Table 1: Environmental Impact Analysis for Each Feedstock using PLCA

Feedstock	GWP (Kg CO_2 Eq)	Feedstock	GWP (Kg CO_2 Eq)
Miscanthus	0.08	Switchgrass	0.08
Willow	0.051	Rapeseedoil	0.76
Poplar	0.072	Waste cooking oil	0.083
Wheatstraw	0.038	Soy	0.3
Cornstraw	0.086	Sunflower	0.55
Barleystraw	0.030	Manure	0
Forest residue	0.034	Sewage sludge	0.0099
Wood chips	0.039	Grass	0.1

Table 2: Environmental Impact Analysis for Each Processing Route PLCA

Feedstock	Processing Routes	GWP (Kg CO_2Eq)	Feedstock	Processing Routes	GWP (Kg CO_2 Eq)
Miscanthus	Fermentation and saccharification	0.74	Switchgrass	Fermentation and saccharification	0.74
Willow	Fermentation and saccharification	0.32	Rapeseedoil	Transesterification	2.00
Poplar	Fermentation and saccharification	0.39	Waste cooking oil	Transesterification	0.38
Wheatstraw	Fermentation and saccharification	0.37	Soy	Transesterification	0.42
Cornstraw	Fermentation and saccharification	0.42	Sunflower	Transesterification	0.40
Barleystraw	Fermentation and saccharification	0.37	Manure	Biodigestion	0.67
Forest residue	Fermentation and saccharification	0.23	Sewage sludge	Biodigestion	0.40
Wood chips	Fermentation and saccharification	0.57	Grass	Biodigestion	0.16

4. Results & Discussions

The proposed methodology is tested through a case study to establish a biorefinery system in Austria. This study involves the 16 types of biomass, mainly agriculture and forest residue available in Austria (Ruiz et al., 2019) used as a feedstock to obtained bioethanol, biodiesel and biogas as a product and cattle feed and carbon dioxide as by product. The conversion efficiency of processing route that converts the feedstock into products and byproducts were obtained by previous results (Morales et al., 2021). Additional required data such as selling price of products byproducts and demand of the product were taken from governmental institutions (IEA,2022; Biofuels,2022.).The solution obtained by solving this multi-objective problem is presented in Figure 2. First, the solution for the minimum profit is obtained when nothing is produced, so the profit is equal to zero which is shown at Point A in Figure 2. The solutions for point B and C in Pareto curve are shown in Figure 3 and 4 respectively. Regarding the production of bioethanol at point B 76% demand is fulfil while the biodiesel and biogas is only 20% and 2% of the total demand respectively. The maximum profit at point B is 173.3 M€/Year and EI is 0.39 Mt of CO_2-Eq. Figure 4 shows that 100% bioethanol demand is fulfil with only 60% feedstock of woodchips available in Austria. The Production of biodiesel and biogas remains the same as point B. The overall Profit at point C is 284 M€/Year and EI is 0.67 Mt of CO_2-Eq.

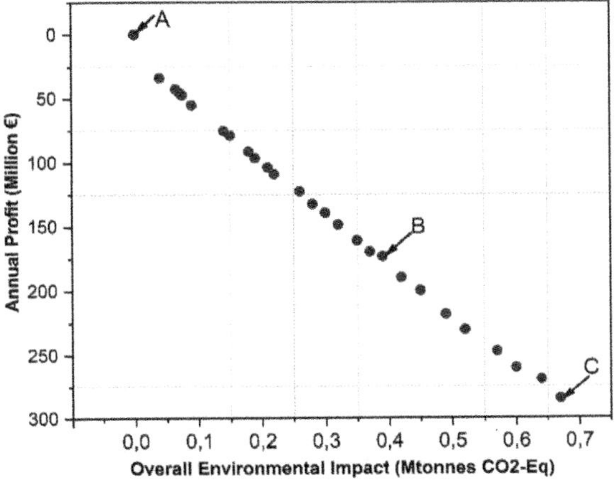

Figure 2: Pareto front of optimal solutions

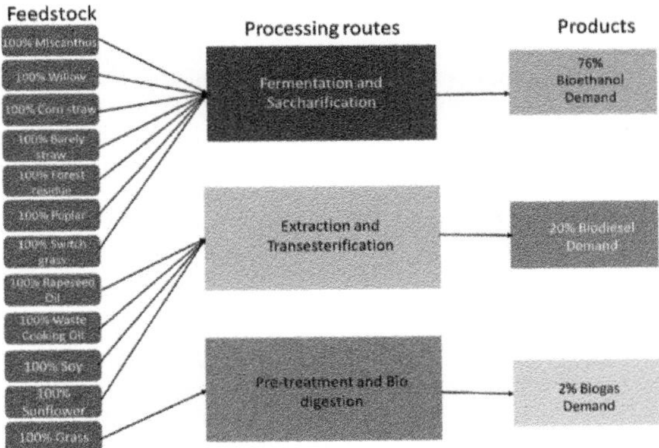

Figure 3: Solution for point B of Pareto front

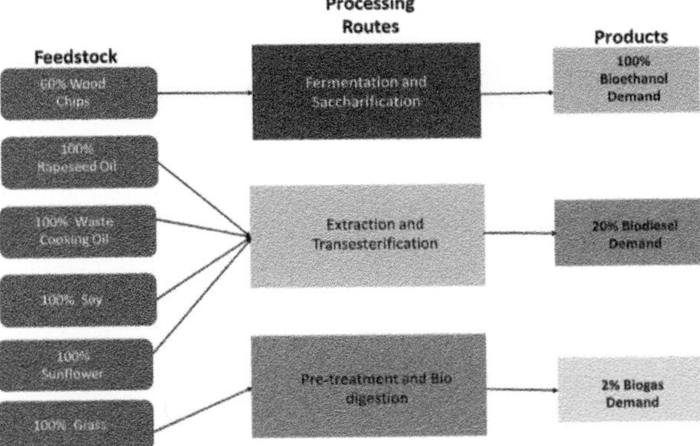

Figure 4: Solution for point C of Pareto front

5. Conclusions

This paper presents a systematic approach to the design of superstructure for optimal configuration of biofuel considering simultaneously maximum profit and minimum environmental impact. An Austrian second-generation feedstock mainly agriculture residue and forest residue considered. In this study we considered different constraints the maximum feedstock available, maximum demand of biofuel in Austria prospective. The multi-objective optimization is solved using the ε-constraints method to establish the tradeoff among the economic and environmental objectives. Results shows that model can help to solve the problem of biofuel production from agriculture residue or forest residue.

6. References

2022, I. (2022) https://www.iea.org/countries/austria

Biofuels, I. (2022) https://www.iea.org/fuels-and-technologies/bioenergy

Fricko, O., Havlik, P., Rogelj, J., Klimont, Z., Gusti, M., Johnson, N., Kolp, P., Strubegger, M., Valin, H., Amann, M., Ermolieva, T., Forsell, N., Herrero, M., Heyes, C., Kindermann, G., Krey, V., McCollum, D. L., Obersteiner, M., Pachauri, S., … Riahi, K. (2017). The marker quantification of the Shared Socioeconomic Pathway 2: A middle-of-the-road scenario for the 21st century. *Global Environmental Change*, *42*, 251–267. https://doi.org/10.1016/j.gloenvcha.2016.06.004

Kowalski, Z., Kulczycka, J., Verhé, R., Desender, L., de Clercq, G., Makara, A., Generowicz, N., & Harazin, P. (2022). Second-generation biofuel production from the organic fraction of municipal solid waste. *Frontiers in Energy Research*, *10*(August), 1–15. https://doi.org/10.3389/fenrg.2022.919415

Morales, M., Arvesen, A., & Cherubini, F. (2021). Integrated process simulation for bioethanol production: Effects of varying lignocellulosic feedstocks on technical performance. *Bioresource Technology*, *328*(December 2020), 124833. https://doi.org/10.1016/j.biortech.2021.124833

Ruiz, P., Nijs, W., Tarvydas, D., Sgobbi, A., Zucker, A., Pilli, R., Jonsson, R., Camia, A., Thiel, C., Hoyer-Klick, C., Dalla Longa, F., Kober, T., Badger, J., Volker, P., Elbersen, B. S., Brosowski, A., & Thrän, D. (2019). ENSPRESO - an open, EU-28 wide, transparent and coherent database of wind, solar and biomass energy potentials. *Energy Strategy Reviews*, *26*(June 2019), 100379. https://doi.org/10.1016/j.esr.2019.100379

Sacchi, R., Terlouw, T., Siala, K., Dirnaichner, A., Bauer, C., Cox, B., Mutel, C., Daioglou, V., & Luderer, G. (2022). PRospective EnvironMental Impact asSEment (premise): A streamlined approach to producing databases for prospective life cycle assessment using integrated assessment models. *Renewable and Sustainable Energy Reviews*, *160*(April 2021), 112311. https://doi.org/10.1016/j.rser.2022.112311

Sharma, B., Ingalls, R. G., Jones, C. L., & Khanchi, A. (2013). Biomass supply chain design and analysis: Basis, overview, modeling, challenges, and future. In *Renewable and Sustainable Energy Reviews* (Vol. 24, pp. 608–627). Elsevier Ltd. https://doi.org/10.1016/j.rser.2013.03.049

The European Parliament and the Council of the European Union. (2015). Amending Directive 98/70/EC relating to the quality of petrol and diesel fuels and amending Directive 2009/28/EC on the promotion of the use of energy from renewable sources. *Official Journal of The European Union*, *2014*(September), 20–30.

Tsagkari, M., Kokossis, A., & Dubois, J. L. (2020). A method for quick capital cost estimation of biorefineries beyond the state of the art. *Biofuels, Bioproducts and Biorefining*, *14*(5), 1061–1088. https://doi.org/10.1002/bbb.2114

Yue, D., You, F., & Snyder, S. W. (2014). Biomass-to-bioenergy and biofuel supply chain optimization: Overview, key issues and challenges. *Computers and Chemical Engineering*, *66*, 36–56. https://doi.org/10.1016/j.compchemeng.2013.11.016

Antonis Kokossis, Michael C. Georgiadis, Efstratios N. Pistikopoulos (Eds.)
PROCEEDINGS OF THE 33rd European Symposium on Computer Aided Process Engineering (ESCAPE33), June 18-21, 2023, Athens, Greece
 http://dx.doi.org/10.1016/B978-0-443-15274-0.50099-8

Optimal operation of an industrial low pressure carburizing furnace

Fatima Matamoros[a,b], Pierre-Alexandre Glaude[a], Roda Bounaceur[a], Hubert Monnier[b], Abderrazak Latifi[a,*]

[a] *Université de Lorraine, CNRS, LRGP, 54000, Nancy, France*

[b] *Institut National de Recherche et de Sécurité, 54500, Vandœuvre-lès-Nancy, France*

abderrazak.latifi@univ-lorraine.fr

Abstract

This study presents the dynamic optimization of an industrial low-pressure gas carburizing (LPC) furnace. The objective is to determine the optimal operating conditions that minimize the production of harmful substances, i.e., polycyclic aromatic hydrocarbons (PAH) and soot, while meeting the industrial and mechanical requirements of the treated steel parts. A dynamic optimization problem is proposed and solved by means of a control vector parametrization approach and a gradient-based method using the software MATLAB. Optimized operating conditions are proposed and they are experimentally tested on an industrial LPC furnace. The results confirm that the proper treatment of steels parts is ensured and show that the PAH generated are reduced.

Keywords: Low-pressure gas carburizing, polycyclic aromatic hydrocarbons, dynamic hybrid optimization, control vector parameterization.

1. Introduction

Low-pressure gas carburizing (LPC) is a process that aims to improve the mechanical properties of steels by increasing the carbon content on their superficial layers. Since its introduction in the 20th century, LPC has become quite widespread in the heat treatment industry as it is an environmentally friendly process that reduces carburizing time and causes no intergranular oxidation of steel (Zajusz et al., 2014). This is achieved by carrying out the process at high temperatures and low pressures while supplying a cyclic feed of carbon-bearing gases like acetylene. Like other carburizing processes however, LPC has proven to be a significant source of occupational exposure to toxic substances like soot and polycyclic aromatic hydrocarbons (PAH) (Champmartin et al., 2017). The challenge is therefore to determine the best sequence of acetylene feed that minimizes the production of toxic compounds without compromising the carbon content constraints on the steel parts. The present paper addresses this challenge by formulating a hybrid dynamic optimization problem involving two modes (acetylene feed on and off), where each mode is characterized by model equations, a transition condition and transition functions. A control vector parametrization (CVP) approach is used to transform the problem into a nonlinear programming (NLP) problem that is solved by means of a gradient-based method. The results of the optimization and the experimental studies carried out to verify the validity of the findings are presented.

2. Materials and methods

The experimental installation used for the carburizing of gear box pinions (shown in

Figure 1) is presented in Figure 2. It consists of an industrial furnace made up of two independent chambers, i.e., a quenching chamber and a carburizing chamber, an acetylene and nitrogen feeding system, pressure and temperature control and regulation systems and a gas collection system consisting of a sampling rod and a cold trap connected to a pump. Nitrogen is used to inert the gas collection system to keep oxygen out of the setup.

Figure 1: gear box pinion (23MnCrMo5 steel)

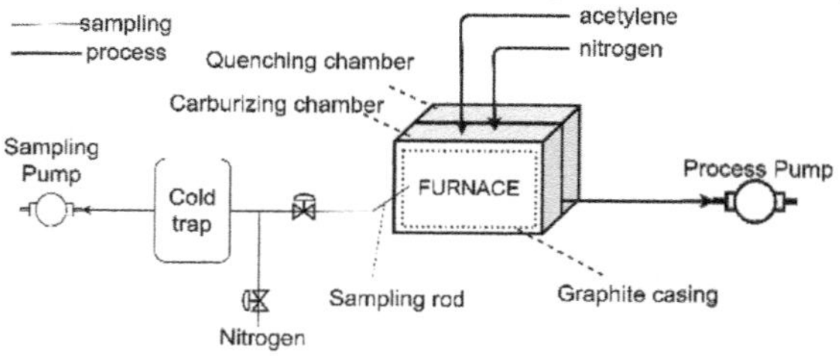

sampling
process
Quenching chamber
acetylene
nitrogen
Carburizing chamber
Sampling Pump
Cold trap
FURNACE
Process Pump
Sampling rod
Graphite casing
Nitrogen

Figure 2 : experimental setup

The experiments consist in programming the operating conditions (the "recipe") used for carburizing, i.e., temperature, pressure, gases inlet flowrates and feeding durations, then in loading the steel parts into the furnace. The gas collection system is used to collect the gases generated all throughout the process. At the end of carburizing, the metal parts undergo a 15-minute nitrogen quench under 15 bars of pressure. The contents of the cold trap are analyzed offline by gas chromatography coupled to a mass spectrometer (GC-MS). The results of the Vickers hardness tests conducted on the carburized steel parts to determine their micro hardness profiles are presented in this paper. Two recipes are used for the carburizing of the steel parts: an industrial recipe and a recipe obtained by solution of the optimization problem presented in the following section.

3. Formulation of the optimization problem

3.1. Process model

The process model considers the formation of gaseous and solid species due to the thermal decomposition of acetylene, as well as the interactions between the fluid and solid phases. The pyrolysis of acetylene is modeled by means of a 140 species and 444 reactions kinetic mechanism obtained by reduction of the detailed mechanism proposed by Bensabath (2017). It is completed by the modeling of soot by means of the method of moments (Frenklach, 2002) and the modeling of pyrocarbon by assuming pseudo-homogenous deposition kinetics (Ziegler, 2004). The external transfer of acetylene to the surface of the steel and the diffusion of carbon in the steel are modeled by means of Fick's second law. The interaction between the fluid and the steel is described by a Langmuir-Hinshelwood Hougen Watson model. Temperature and pressure are assumed to be constant throughout the reaction zone. The gas mixture is assumed to obey the ideal gas law and the reactor is assumed to be a perfectly mixed stirred tank reactor (PSTR). Under these assumptions, the mass balances on gas species and the differential equations for the moments of soot particles are given by equations (1) and (2) respectively:

$$\rho_g V \frac{dY_k}{dt} = \dot{m}_{in}\left(Y_{k,in} - Y_k\right) + \dot{m}_{\text{prod},k} - Y_k(\dot{m}_{\text{surf}} - \dot{m}_{\text{soot}}) \quad (1)$$

$$\frac{dM_r}{dt} = R_r + G_r + W_r - \frac{M_r Q_{out}}{V} \quad (2)$$

where ρ_g is the mass density of the gas mixture, V is the volume of the reactor, Y_{kin} and Y_k are respectively the inlet mass fraction and the mass fraction of gaseous species k in the reactor, $\dot{m}_{in}$ is the inlet mass flowrate, $\dot{m}_{\text{prod},k}$ is the mass production rate of species k by all reactions, $\dot{m}_{\text{surf}}$ is the total mass production rate of gaseous species by the reaction at the surface of the steel, $\dot{m}_{\text{soot}}$ is the mass production rate of soot, M_r is the moment r, R_r, G_r and W_r are soot source terms due to the phenomena of nucleation, coagulation and surface growth respectively, and Q_{out} is the outlet volumetric flowrate. The mass balances in the fluid/solid boundary layer and in the steel lead to set of equations (3) and (4) respectively:

$$\begin{gathered}\frac{\partial[\mathrm{C_2H_2}]}{\partial t}=D_{C_2H_2}\frac{\partial^2[\mathrm{C_2H_2}]}{\partial z^2}\\ [C_2H_2](z,t_0)=0\\ [C_2H_2](z=0,t)=\left[C_2H_{2(PSTR)}\right]\\ -D_{C_2H_2}\left.\frac{\partial[C_2H_2]}{\partial z}\right|_{z=z_L}=\frac{V_p}{A_p}r_{\text{surf}}\end{gathered} \tag{3}$$

$$\begin{gathered}\frac{\partial\omega_C}{\partial t}=D_C\frac{\partial^2\omega_C}{\partial y^2}\\ \omega_C(y,t_0)=\omega_{C,0}\\ -D_C\left.\frac{\partial\omega_C}{\partial y}\right|_{y=0}=r_{surf}\frac{V_p}{A_p}\frac{100W_C}{\rho_p}\\ D_C\left.\frac{\partial\omega_C}{\partial y}\right|_{y=y_L}=0\end{gathered} \tag{4}$$

where $[\mathrm{C_2H_2}]$ is the concentration of acetylene in the boundary layer, $[\mathrm{C_2H_2}_{PSTR}]$ is the concentration of acetylene in the PSTR, ω_C is carbon weight percentage in the steel, $D_{C_2H_2}$ and D_C are the diffusion coefficients of acetylene in the gas phase and of carbon in the steel respectively, z and y are acetylene and carbon diffusion directions, W_C is carbon molar weight, V_p, A_p and ρ_p are the volume, surface and density of the steel respectively, and r_{surf} is the volumetric rate of the reaction at the surface of the steel given by:

$$r_{\text{surf}}=k_{\text{app}}K_1\frac{\left[C_2H_{2(surf)}\right]}{\left(1+K_1\left[C_2H_{2(surf)}\right]+K_2\left[C_{(fe)}\right]\right)^2} \tag{5}$$

where $\left[C_2H_{2(surf)}\right]$ is the concentration of acetylene in the gas phase at the surface of the steel and $\left[C_{(fe)}\right]$ is the concentration of carbon at the surface of the steel. Parameters K_1 and K_2 are Langmuir equilibrium constants of the adsorption of acetylene and carbon respectively, and k_{app} is an apparent kinetic constant. A hybrid model is proposed to take into account the phases where acetylene feeding is on (boost stages, designed by mode 1) and off (diffusion stages, designed by mode 2). The model equations on state variables x are given by:

$$\dot{x}=f^{(1)}(x) \tag{6}$$

$$\dot{x}=f^{(2)}(x) \tag{7}$$

where f are the relations described above. The transition conditions L and transition times t_{boost} and t_{diff} are given by:

$$L^{(2k)}=t_{\text{boost}}^{(k+1)}-t=0,\qquad \text{where } t_{\text{boost}}^{(k)}=\sum_{i=1}^{k-1}(a_i+b_i),\quad k=1,\dots N_C-1, \tag{8}$$

$$L^{(2k-1)} = t_{\text{diff}}^{(k)} - t = 0, \;\; \text{where } t_{\text{diff}}^{(k)} = a_k + \sum_{i=1}^{k-1} (a_i + b_i), \qquad k = 1, \dots N_C \tag{9}$$

where a_k and b_k are the lengths of boost and diffusion stage i respectively and N_c is the number of cycles, where each cycle corresponds to one boost stage followed by a diffusion stage. The transition functions are given by:

$$x_{\text{boost}}^{(k+1)}\left(t_0^{(k+1)}\right) - x_{\text{diff}}^{(k)}\left(t_f^{(k)}\right) = 0, \qquad k = 1, \dots N_C - 1 \tag{10}$$

$$x_{diff}^{(k)}\left(t_0^{(k)}\right) - x_{\text{boost}}^{(k)}\left(t_f^{(k)}\right) = 0, \qquad k = 1, \dots N_C \tag{11}$$

3.2. Performance index

The performance index corresponds to the total mass of PAH and soot generated:

$$J = \int_{t_0}^{t_f} (\dot{m}_{HAP} + \dot{m}_{\text{suies}})dt = \int_{t_0}^{t_f} \dot{x}_{obj}dt = x_{obj}(t_f), \quad \text{where } x_{obj}(t_0) = 0$$

3.3. Decision variables and constraints

The decision variables considered are the operating conditions, i.e., the number of cycles N_c, the lengths of the boost and diffusion stages $a_1 \dots a_{N_C}$ and $b_1 \dots b_{N_C}$, and the inlet flowrate of acetylene Q. The lengths of boost and diffusion stages vary in the [1-60] min range and acetylene flowrate in the [800-3500] NL/h range.

3.4. Constraints on the state variables

Constraints on the state variables related to the industrial requirements on the steel parts are considered and are given by the following relations:

$$\omega_c^{surf}\left(t_f\right) = 0.6\ \% \tag{12}$$

$$\omega_c^{0.4\ mm}\left(t_f\right) = 0.34\ \% \tag{13}$$

$$\omega_c^{surf}(t) \leq 1.2\ \%, \; t_0 \leq t \leq t_f \tag{14}$$

where ω_c^{surf} and $\omega_c^{0,4mm}$ are the carbon weight percentage at the surface of the steel and a depth of 0.4 mm respectively, t_f is the duration of the process and the value of 1.2 % corresponds to the saturation of carbon in the steel at the current operating conditions.

3.5. Optimization algorithm

To solve the optimization problem, it is necessary to deal with some difficulties that arise due to the nature of the decision variables. Firstly, the number of cycles N_C is an integer variable, making the optimization problem a mixed integer dynamic optimization problem. To overcome the ensuing difficulty in trying to solve such a problem, a decomposition of the original problem is proposed. Various sub-problems are thus established and for each one, the number of cycles is fixed and is thus not considered as a decision variable. Several sub-problems are then solved, each with a different number of cycles. The optimal number of cycles is then determined by selecting the sub-problem that leads to the minimum of the optimization criterion. Another difficulty that arises is the fact that the inlet flowrate of acetylene Q is a time-dependent variable of infinite dimension. A CVP approach is adopted to transform the problem into a finite dimension problem. The time interval is therefore divided into $2N_C$ segments and Q is approximated by a piecewise constant function.

$$\begin{aligned} Q(t) &= Q_k, t \in \left[t_{\text{boost}}^{(k)}, t_{\text{diff}}^{(k)}\right], \qquad k = 1,2, \dots N_c \\ &= 0 \text{ otherwise} \end{aligned} \tag{15}$$

The resulting NLP problem is solved by means of *fmincon* optimizer in the MATLAB environment. The hybrid model is integrated by means of *odes15s* solver.

4. Results and discussion

4.1. Optimization results

Figure 3 shows the results of the optimization : Figure 3(a) shows the performance index (PI), the quantity of acetylene consumed and the carbon integrated into the steel parts; Figure 3(b) shows the total durations of boost and diffusion stages. Six different sub-problems were considered and N_c varies between 5 and 10. For each N_c, the results are normalized by the results obtained with the industrial recipe R0.

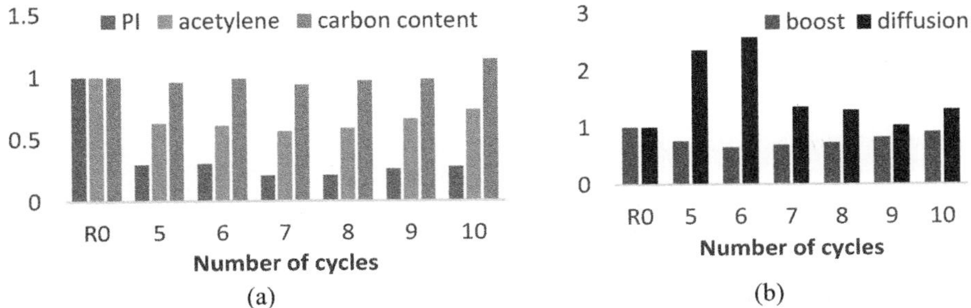

Figure 3 : Results of the optimization of the different sub-problems and predictions with the industrial recipe R0

The results show that the optimization of the different sub-problems leads to a considerable reduction of the optimization criterion when compared to the industrial recipe (Figure 3 (a)). This can be directly related to the reduction of the quantity of acetylene that is fed to the reactor, as the toxic substances are generated from its thermal decomposition. However, even if less acetylene is used, the carbon added to the steel parts is of the same order of magnitude for all sub-problems and for R0. This suggests that acetylene is in excess with R0 and thus decomposes to produce PAH and soot. Figure 3(b) shows that all sub-problems lead to an increase in the total duration of the diffusion stages and a decrease in the total duration of boost stages when compared to R0. Recipes with shorter boost times and longer diffusion times thus lead to a decrease of toxic compounds.

4.2. Experimental results

Among the different sub-problems, the operating conditions of sub-problem $N_c = 7$ were chosen as the optimal conditions. Compared to the industrial recipe, the total process duration is 33% longer but leads to a 44% decrease in acetylene consumption. The results of the hardness tests conducted on a pinion carburized with the industrial recipe and one carburized with the optimized recipe are shown in Figure 4(a) and Figure 4(b) respectively. The tests were performed according to different directions along the teeth of the pinions to study the homogeneity of carburizing in space. The results show that both recipes lead to homogeneous carburizing and to very similar hardness profiles. The results obtained for PAH production with both recipes are shown in Figure 4(c). They show a clear decrease in the production of PAH with the optimized recipe as compared to the industrial one. The experimental results obtained thus show that it is possible to significantly reduce toxic compounds without compromising the industrial quality of steel parts.

5. Conclusions

In this study, a first-principles modeling and optimization methodology was proposed to optimize an industrial low-pressure gas carburizing furnace. Experiments were performed on an industrial scale to quantify PAH formation and to evaluate the quality of the steel parts carburized with an optimized recipe. The results show that it is possible to ensure industrial quality while reducing toxic compounds by carefully selecting the acetylene feed profile. In future works, although the objective will remain the minimization of the process toxicity, the methodology adopted here will be used to consider other aspects to ensure the feasibility of the approach from an industrial viewpoint. Multi-objective optimization will thus be applied to deal with different case studies, e.g., problems that consider the total process duration or even the deposition of pyrocarbon on the furnace walls. Indeed, the former entails higher energy demands, and the latter leads to an increased number of maintenance operations, which not only expose the operators to toxic compounds, but also represent higher operating expenses.

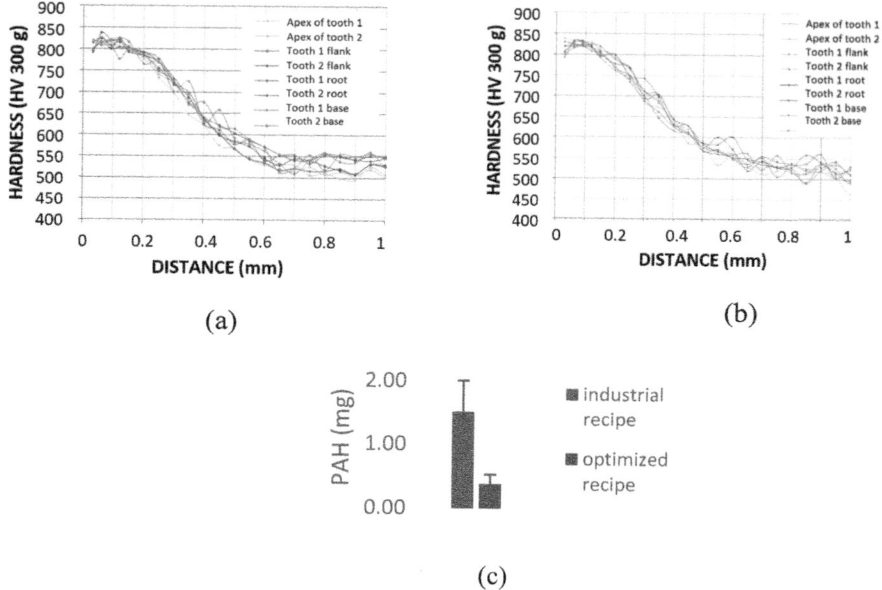

Figure 4: Experimental results (a) hardness profile of a pinion carburized with the industrial recipe (b) hardness profile of a pinion carburized with the optimized recipe and (c) PAH generated with the industrial and optimized recipes

References

T. Bensabath, 2017, Approche préventive pour une réduction des Hydrocarbures Aromatiques Polycycliques (HAP) dans les fours à pyrolyse, PhD thesis, Université de Loraine, France.

C. Champmartin, F. Jeandel, H. Monnier, 2017, Maintenance of Low-Pressure Carburising Furnaces: A Source of PAH Exposure. Annals of Work Exposures and Health, 61(3), 321–332, https://doi.org/10.1093/annweh/wxw024.

M. Frenklach, 2002. Method of moments with interpolative closure. Chemical Engineering Science, 57(12), 2229–2239. https://doi.org/10.1016/S0009-2509(02)00113-6.

M. Zajusz, K. Tkacz-Śmiech, M. Danielewski, 2014, Modeling of vacuum pulse carburizing of steel. Surface and Coatings Technology, https://doi.org/10.1016/j.surfcoat.2014.08.023.

Ziegler, 2004. Modélisation cinétique des dépôts de pyrocarbone obtenus par pyrolyse d'hydrocarbures (Issue 2004INPL045N) [Université de Lorraine]. https://hal.univ-lorraine.fr/tel-01750078.

Antonis Kokossis, Michael C. Georgiadis, Efstratios N. Pistikopoulos (Eds.)
PROCEEDINGS OF THE 33rd European Symposium on Computer Aided Process Engineering (ESCAPE33), June 18-21, 2023, Athens, Greece
 http://dx.doi.org/10.1016/B978-0-443-15274-0.50100-1

Development of a holistic Python package for optimal selection of experimental design criteria in kinetic model discrimination

Maerthe Theresa Tillmann[a,b], Federico Galvanin[a*]

[a] *Department of Chemical Engineering, University College London, Torrington Place, London WC1E 7JE, UK*

[b] *Faculty of Mechanical Engineering, RWTH Aachen University, Eilfschornsteinstraße 18, 52062 Aachen, Germany*

* *f.galvanin@ucl.ac.uk*

Abstract

Starting with a candidate set of kinetic models for a reaction, model-based design of experiment (MBDoE) techniques can be used to determine experimental conditions for fast model identification of reaction kinetics using the minimum number of experimental runs to specify both model structure and corresponding parameters. However, practically, determining optimal settings, including criteria and selection methods for model discrimination for efficient model identification under consideration of parametric uncertainty in the whole identification procedure is still an open and challenging task.

In this work, a holistic Python package is presented which comprises MBDoE for model discrimination and, subsequently, MBDoE for parameter precision. The new package is tested on in-silico experiments for the identification of a Baker's Yeast model to evaluate and compare: *i*) the total number of experiments required for kinetic model identification considering different experimental design criteria; *ii*) the rate of correct model selections using different model selection methods.

Keywords: model-based design of experiments, model discrimination, experimental design criteria

1. Introduction

In chemical engineering, kinetic models are used to mathematically describe and predict the time-dependent behavior of reaction systems whereby mechanistic models, in contrast to data-driven models, can provide easy interpretability and reliable extrapolations. Starting with several candidate model formulations and assuming insufficient information to statistically determine the best formulation among the proposed ones, model-based design of experiment (MBDoE) can be a promising strategy to select further experimental conditions to facilitate model identification (Asprey and Macchietto, 2000). In sequential MBDoE strategies, the optimal experimental design problem is formulated as an iterative optimization problem through the maximization of a design criterion related to model identification based on the current state of knowledge from collected data. The MBDoE strategy reflects the motivation of pursuing model identification with the minimum number of experiments to avoid experimental costs, resources and time.

The classical procedure for model identification using MBDoE according to Asprey and Macchietto (2000) comprises three stages: *i*) a theoretical preliminary identifiability analysis of the set of proposed model candidates, *ii*) MBDoE for model discrimination

and *iii*) parameter precision for the selected model determined in *ii*). Several MBDoE design criteria for model discrimination, used in stage *ii*) as design objective function, have been proposed in literature in terms of statistical approaches of discrimination and complexity of the formulation (Michalik et al., 2010, Olofsson et al., 2019). Multiple MBDoE design criteria also exist for improving parameter precision, but the practical differences have been discussed in literature (Franceschini and Macchietto, 2008). So far, comparative studies of different design criteria for model discrimination (Burke et al., 1997, Olofsson et al., 2019) are limited to the stage of model discrimination only, so that the impact of parametric uncertainty on the whole identification procedure, and thus the effect on the number of experiments required in stage *iii*), is neglected. Practically, selection of settings for model identification is necessary, including experimental design criteria and probabilistic model selection settings. However, it is generally complex to determine which settings for the stage of model discrimination would provide the best prerequisites for model identification with the minimum number of experiments.
In order to address the question of optimal settings for model discrimination, a Python package is presented in this paper where different MBDoE criteria for model discrimination and for parameter precision are implemented as well as different model selection strategies from literature (cf. Olofsson et al., 2019). Applying this tool to the case study of yeast fermentation demonstrates a new holistic approach for comparing MBDoE design criteria and model selection settings to achieve fast model identification with the minimum number of experimental runs.

2. Model Identification using Model-Based Design of Experiments

The proposed framework of comparative study using the holistic model identification function (*HoliMI*) implemented in the Python package HoliMI is illustrated in Figure 1. For each combination of settings, the model identification function (*HoliMI*) is called repeatedly considering different scenarios for preliminary experiments.

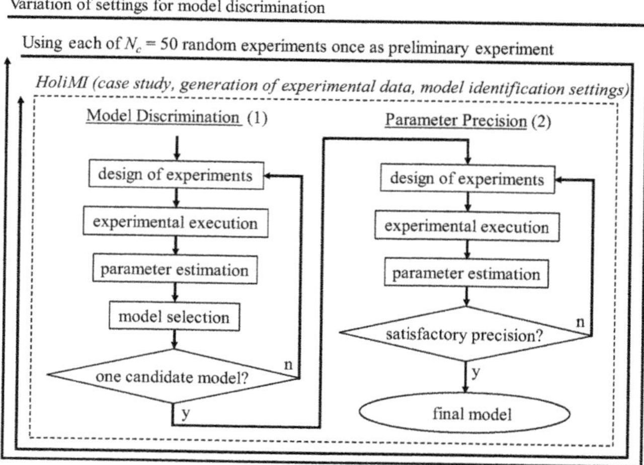

Figure 1 *Proposed framework of comparative study and model identification function (HoliMI) implemented in the Python package including variants of settings for model discrimination.*

The sequence of operations implemented as main function (*HoliMI*) in the Python package follows the MBDoE procedure for holistic model identification presented in Asprey and Macchietto (2000). Function inputs are: *i*) specifications of the case study, *ii*) configurations defining in-silico experiments and *iii*) settings of the implemented sequence of operations for model identification. In the comparative study, settings for *iii*) related to model discrimination (criteria, model selection settings) are investigated.

2.1. Bioprocess Case Study

The fermentation of Baker's Yeast is considered as in-silico system for the comparative study (Chen and Asprey, 2003). The best approximating model structure, which describe the biomass (X) and substrate (S) concentration in a fed-batch reactor, i.e. $\hat{\boldsymbol{y}} = [\hat{S}, \hat{X}]$, should be selected among four proposed candidates. An experimental budget of $N_{\text{exp max}} = 10$ experiments limits the model identification task. Each experiment is defined by a design vector $\boldsymbol{\varphi} = [D, S_f, X_0]$ including dilution rate $D\,[\text{h}^{-1}]$, substrate concentration in the feed $S_f\,[\text{g/L}]$ and initial concentration of biomass $X_0\,[\text{g/L}]$ while samples are taken at fixed time points $\boldsymbol{t_{sp}} := [10\,\text{h}, 20\,\text{h}, 30\,\text{h}, 40\,\text{h}]$. Candidate model formulations, initial conditions of the system, initial parameter guesses and experimental design space are chosen in accordance with Chen and Asprey (2003).

2.2. In-Silico Experiments

For the generation of in-silico experiments, Monod kinetics including a term for biomass mortality are assumed as true model of the system. In addition to the mathematical solution of the model, a gaussian-distributed measurement error affects the true model predictions, i.e. the in-silico generated experimental data is defined by $\boldsymbol{y} = N(\hat{\boldsymbol{y}}(\boldsymbol{\varphi}), \boldsymbol{\Sigma})$, where $\boldsymbol{\Sigma} := \begin{pmatrix} 0.12 & 0 \\ 0 & 0.12 \end{pmatrix}$ is the time-constant variance-covariance matrix used.

2.3. Sequence of Operations for Model Identification

The holistic model identification function (*HoliMI*), see Figure 1, comprises the iterative procedure for model discrimination (step 1) and the subsequent iterative procedure for parameter precision of the selected model (step 2) which approximates the experimental data best among the proposed candidates. Using MBDoE, experiments are designed such that the design criterion Ψ for model discrimination is maximized in each iteration and the specific formulation of the design criterion for model discrimination is defined by the settings as function input. Based on the set of experimental data, the parameters of each model are then estimated so that the weighted sum of squared residuals χ^2as measure for the lack of fit is minimized (Chen and Asprey, 2003). The lack of fit χ^2 is then used to assign the models a probability p of being the best approximating model structure. These relative probabilities are used by Schwaab et al. (2006) to reject model candidates with low probabilities. If one of the model probabilities reaches 99.9 %, model discrimination is terminated and parameter precision for the selected model is pursued until the confidence interval of all parameter estimates fall below 0.05. D-optimality is used as design criterion for the subsequent MBDoE for parameter precision.

3. Comparative Study

Besides following a holistic approach for comparing the number of experiments required for correct model identification including a precise parameter estimation, the comparative study is characterized by taking different parametric initializations for the first MBDoE iteration into account. Results are averaged over multiple runs considering accurate and inaccurate parameter estimates scenarios based on preliminary system knowledge.

3.1. Characterization of Preliminary Knowledge on Model Parameters

In order to provide the most general comparison of settings for model identification, numerous scenarios to describe the preliminary knowledge on model parameters are considered. A set of 50 randomly selected experimental conditions and corresponding measurements from in-silico experiments is generated. For each combination of settings, the model identification function (*HoliMI*) is called $N_c = 50$ times to consider a different preliminary experiment from the generated set in each realization so that different scenarios of parametric uncertainty in the subsequent MBDoE steps are considered.

3.1.1. Settings Used in the Comparative Study

Investigated settings in the comparative study are *i*) the MBDoE design criterion used for model discrimination as objective function of optimization problem; *ii*) the probability calculation method for model selection; *iii*) the probability-based model rejection method. The design criterion of Hunter and Reiner (1965), Eq. (1), is a summation of differences between model predictions over all pairs of M candidate models. In contrast to this basic formulation, the criterion of Box and Hill (1967), Eq. (2), takes model probabilities p and the variance of model predictions $\hat{\boldsymbol{V}}$ into account. A statistically different approach has been proposed by Buzzi-Ferraris and Forzatti (1990), Eq. (3). Michalik et al. (2010) have derived the Akaike's weights design criterion, Eq. (4), from the Akaike Information Criterion of the candidate models with N_θ as number of estimated model parameters. Recently, Olofsson et al. (2019) presented the Jensen-Rény divergence as design criterion, Eq. (5), for N_y measurement variables. Furthermore, the comparison of different MBDoE design criteria is set in relation to a random design of experimental conditions.

$$\Psi_{\mathrm{HR}} = \sum_{i=1}^{M-1}\sum_{j=i+1}^{M}\left(\hat{\boldsymbol{y}}_i - \hat{\boldsymbol{y}}_j\right)^{\mathrm{T}}\left(\hat{\boldsymbol{y}}_i - \hat{\boldsymbol{y}}_j\right) \tag{1}$$

$$\Psi_{\mathrm{BH}} = \sum_{i=1}^{M-1}\sum_{j=i+1}^{M} p_i p_j \left(\mathrm{tr}\left\{\frac{\hat{\boldsymbol{V}}_i}{\hat{\boldsymbol{V}}_j} + \frac{\hat{\boldsymbol{V}}_j}{\hat{\boldsymbol{V}}_i} - 2\boldsymbol{I} + \left(\hat{\boldsymbol{y}}_i - \hat{\boldsymbol{y}}_j\right)^{\mathrm{T}}\left(\frac{1}{\hat{\boldsymbol{V}}_i} + \frac{1}{\hat{\boldsymbol{V}}_j}\right)\left(\hat{\boldsymbol{y}}_i - \hat{\boldsymbol{y}}_j\right)\right\}\right) \tag{2}$$

$$\Psi_{\mathrm{BF}} = \sum_{i=1}^{M-1}\sum_{j=i+1}^{M}\frac{\left(\hat{\boldsymbol{y}}_i-\hat{\boldsymbol{y}}_j\right)^{\mathrm{T}}\left(\hat{\boldsymbol{y}}_i-\hat{\boldsymbol{y}}_j\right)}{\hat{\boldsymbol{V}}_i+\hat{\boldsymbol{V}}_j}\ \mathrm{tr}\left\{\frac{2\boldsymbol{\Sigma}}{\hat{\boldsymbol{V}}_i+\hat{\boldsymbol{V}}_j}\right\} \tag{3}$$

$$\Psi_{\mathrm{AWDC}} = \sum_{i=1}^{M}\frac{1}{\sum_{j=i}^{M}\exp\left\{-\frac{1}{2}\left(\hat{\boldsymbol{y}}_i-\hat{\boldsymbol{y}}_j\right)^{\mathrm{T}}\hat{\boldsymbol{V}}_i^{-1}\left(\hat{\boldsymbol{y}}_i-\hat{\boldsymbol{y}}_j\right)+N_{\theta_i}-N_{\theta_j}\right\}} \tag{4}$$

$$\Psi_{\mathrm{JRD}} = \mathrm{H}_2\left(\sum_{i=1}^{M} p_i g_i\right) - \sum_{i=1}^{M} p_i \mathrm{H}_2(g_i) \text{ with}$$

$$\mathrm{H}_2\left(\sum_{i=1}^{M} p_i g_i\right) = \frac{N_y}{2}\log\{2\pi\} - \log\left\{\sum_{i=1}^{M}\frac{p_i^2}{2^{\frac{N_y}{2}}\left|\hat{\boldsymbol{V}}_i\right|^{\frac{1}{2}}} + 2\sum_{j=1}^{i-1} p_i p_j \exp\left\{-\frac{1}{2}\phi_{ij}\right\}\right\},$$

$$\phi_{ij} = \frac{\hat{\boldsymbol{y}}_i^{\mathrm{T}}\hat{\boldsymbol{y}}_i}{\hat{\boldsymbol{V}}_i} + \frac{\hat{\boldsymbol{y}}_j^{\mathrm{T}}\hat{\boldsymbol{y}}_j}{\hat{\boldsymbol{V}}_j} - \frac{\left(\frac{\hat{\boldsymbol{y}}_i}{\hat{\boldsymbol{V}}_i}+\frac{\hat{\boldsymbol{y}}_j}{\hat{\boldsymbol{V}}_j}\right)^{\mathrm{T}}\left(\frac{\hat{\boldsymbol{y}}_i}{\hat{\boldsymbol{V}}_i}+\frac{\hat{\boldsymbol{y}}_j}{\hat{\boldsymbol{V}}_j}\right)}{\frac{1}{\hat{\boldsymbol{V}}_i}+\frac{1}{\hat{\boldsymbol{V}}_j}} + \log\left|\hat{\boldsymbol{V}}_i\right| + \log\left|\hat{\boldsymbol{V}}_j\right| + \log\left|\frac{1}{\hat{\boldsymbol{V}}_i}+\frac{1}{\hat{\boldsymbol{V}}_j}\right| \text{ and} \tag{5}$$

$$\mathrm{H}_2(g_i) = \frac{N_y}{2}\log\{4\pi\} + \frac{1}{2}\log\left|\hat{\boldsymbol{V}}_i\right|$$

Four different calculation methods of probability weights w with $p_i = w_i/\sum_{k=1}^{M} w_k$ are compared: M1 as inverse of lack of fit; M2 resulting from comparing the lack of fit with the χ^2 distribution including the degrees of freedom of the model; M3 being based on the Akaike Information Criterion and M4 being similarly based on the likelihood function $L_i = \prod_{q=1}^{N_{exp}}\prod_{r=1}^{N_{sp}}\prod_{s=1}^{N_y}\left(1/\,2\pi\boldsymbol{\Sigma}_{s,s}\right)^{1/2}\exp\left\{-\left(y_{q,r,s} - \hat{y}_{i\,q,r,s}\right)^2/\,2\boldsymbol{\Sigma}_{s,s}\right\}$ (Michalik et al., 2010) as M3 but without penalty term to include the degrees of freedom of the model.

$$\text{M1: } w_i = \frac{1}{\chi_i^2} \tag{6}$$

$$\text{M2: } w_i = 1 - F_{DoF_i}\left(\chi_i^2\right) \tag{7}$$

$$\text{M3: } w_i = \frac{1}{\sum_{k=1}^{M}\exp\{AIC_i - AIC_k\}} \quad \text{with } AIC_i = -2\log\{L_i\} + 2DoF_i \tag{8}$$

$$\text{M4: } w_i = \frac{1}{\sum_{k=1}^{M}\exp\{\pi_i - \pi_k\}} \quad \text{with } \pi_i = -2\log\{L_i\} \tag{9}$$

The probability calculation method and preferences for model rejection define the settings used for model selection: either probability-based model rejection is deactivated ("off") or candidate models with $p \leq 3\,\%$ are rejected as indicated by Schwaab et al. (2006).

3.2. Evaluation Metrics

The model identification function (*HoliMI*) is not deterministic since the measurement error generated for the in-silico experiments is randomly distributed. Possible return values are: *i*) correctly selected model with precise parameter estimation; *ii*) correctly selected model with interruption (i.e. maximum number of experiments reached) of parameter precision; *iii*) interruption of model discrimination and *iv*) false model selection. Thus, the different setting variants are compared by means of effectiveness $EFN \propto \frac{number\ of\ correct\ model\ selections}{number\ of\ model\ identifications}$ and efficiency $EFC \propto \frac{1}{total\ number\ of\ experiments}$ of cases with identification of correct model structure and precise parameters, return *i*).

3.3. Results and Discussion

Figure 2 shows the distribution of returns from the model identification function (*HoliMI*) averaged over all realizations with respective design criterion (*2.a*) and model selection settings (*2.b*). The average number of experiments required for correctly selected models and precise parameter estimation with respective settings is also presented (*2.c*, *2.d*). The required number of experiments is split in number of experiments required for model discrimination and those required for parameter precision.

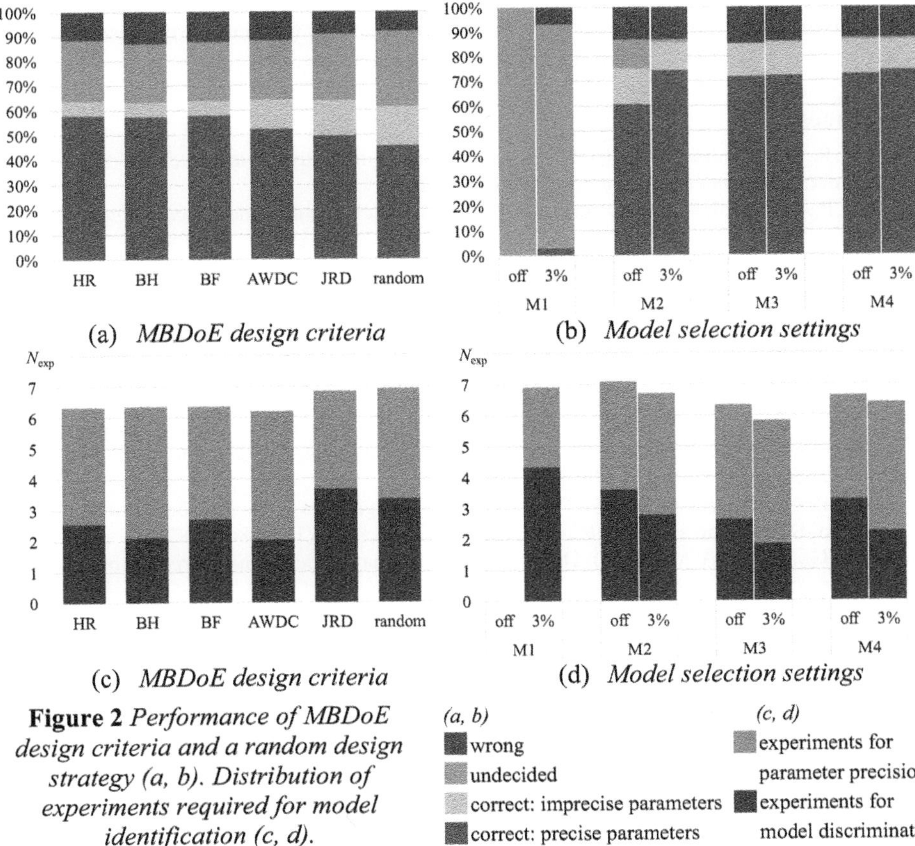

Figure 2 *Performance of MBDoE design criteria and a random design strategy (a, b). Distribution of experiments required for model identification (c, d).*

In general, MBDoE in contrast to a random design is the more efficient strategy to design experiments for the identification of both model structure and parameters. However, MBDoE effectiveness is affected by parametric uncertainty as can be seen by the small improvement compared to a random design (*2.a*, *2.c*) when averaging over multiple parametric scenarios. In contrast, the probability calculation method M1 is significantly less effective than the other methods (*2.b*), with M3 performing most efficiently (*2.d*). Furthermore, rejecting models at a probability of 3 % resulted in faster model identification than in probability-free model rejection methods and in lower rates of false model selections due to the risk of rejecting the true model.

4. Conclusion

The proposed Python package HoliMI allows a holistic comparison of settings for model identification using MBDoE techniques for both model discrimination and improvement of parameter precision. Results on the application of the package to a benchmark case study related to the identification of kinetics in a bioreactor system showed that the selection of model selection settings can affect the model identification performance substantially. Depending on the number of candidate model, a threshold for probability-based model rejections should be considered to enhance model identification with minimum number of experiments. Furthermore, it is recommended to use probability weights based on the probability likelihood function (M3 or M4) for efficient correct model selections. General guidelines on optimal MBDoE design criteria for model discrimination can be derived from the application to further study cases and the implementation of the proposed package in autonomous reactor platforms.

References

Asprey, S.P., Macchietto, S. (2000). Statistical tools for optimal dynamic model building. *Computers & Chemical Engineering*, 24, 1261–1267

Box, G.E.P., Hill, W.J. (1967). Discrimination among mechanistic models. *Technometrics*, 9 (1), 57–71

Buzzi-Ferraris, G., Forzatti, P. (1990). An improved version of a sequential design criterion for discriminating among rival multiresponse models. *Chemical Engineering Science*, 45 (2), 477–481

Burke, A.L., Duever, T.A., Penlidis, A. (1997). Discriminating between the terminal and penultimate models using designed experiments: An overview. *Industrial & Engineering Chemistry Research*, 36 (4), 1016–1035

Chen, B.H., Asprey, S.P. (2003). On the design of optimally informative dynamic experiments for model discrimination in multiresponse nonlinear situations. *Industrial & Engineering Chemistry Research*, 42 (7), 1379–1390

Franceschini, G., Macchietto, S. (2008). Model-based design of experiments for parameter precision: State of the art. *Chemical Engineering Science*, 63 (19), 4846–4872

Hunter, W.G., Reiner, A.M. (1965). Designs for discriminating between two rival models. *Technometrics*, 7 (3), 307

Michalik, C., Stuckert, M., Marquardt, W. (2010). Optimal experimental design for discriminating numerous model candidates: The awdc criterion. *Industrial & Engineering Chemistry Research*, 49 (2), 913–919

Olofsson, S., Hebing, L., Niedenführ, S., Deisenroth, M.P., Misener, R. (2019). Gpdoemd: A python package for design of experiments for model discrimination. *Computers & Chemical Engineering*, 125, 54–70

Schwaab, M., Silva, F.M., Queipo, C.A., Barreto, A.G., Nele, M., Pinto, J.C. (2006). A new approach for sequential experimental design for model discrimination. *Chemical Engineering Science*, 61 (17), 5791–5806

Antonis Kokossis, Michael C. Georgiadis, Efstratios N. Pistikopoulos (Eds.)
PROCEEDINGS OF THE 33rd European Symposium on Computer Aided Process Engineering (ESCAPE33), June 18-21, 2023, Athens, Greece
 http://dx.doi.org/10.1016/B978-0-443-15274-0.50101-3

A Model-based Computational Approach to Optimal Design of Nose-to-Brain Drug Delivery Systems

Athina Vasileiadou,[a] Costas Kiparissides,[a,b]

[a]*Department of Chemical Engineering, Aristotle University of Thessaloniki, 54124, Thessaloniki, Greece*
[b]*Chemical Process & Energy Resources Institute, 6th km Harilaou-Thermi rd., P.O. Box 60361, 57001, Thessaloniki, Greece*

Abstract

In the present study, a dynamic polymer degradation-drug diffusion model is developed to calculate the drug release rate from a population of size-distributed drug-loaded biodegradable polymer particles and analyze the effects of particle size distribution (PSD), drug loading distribution (DLD), and polymer molecular properties on degradation kinetics and drug release rate. The derived model, comprising a system of partial differential and algebraic equations, is numerically solved using the method of lines and validated against published experimental data on the drug release rate from drug-loaded PLGA microspheres. By controlling the PSD, DLD, molecular and morphological properties of the polymeric carriers, a desirable drug release profile can be realized. Toward this objective, a model-based dynamic optimization approach is developed to determine the optimal PSD and DLD of the biodegradable polymer carriers to achieve a drug release rate of zero-order over an administration period of three weeks.

Keywords: Controlled drug delivery; polymer degradation kinetics; drug diffusion; model-based optimal design of DDS.

1. Introduction

Neurological diseases encompass a broad spectrum of brain pathological conditions, i.e. Parkinson's and Alzheimer's diseases, multiple sclerosis, meningitis, etc., which can cause alterations to neural functions and progressive loss of neural tissue. Despite the recent advances in both drug delivery research and improved understanding of the pathogenesis of neurological diseases, effective treatment options are still missing due to the complexity of the central nervous system (CNS) and the alleged multifactorial pathogenic mechanisms. Thus, currently available therapeutic formulations have been generally proven ineffective because they are unable to reverse or/and completely restore normal neural functions. Moreover, the delivery of therapeutics to the brain by conventional systemic approaches is hindered by several barriers, including the Blood brain barrier (BBB) and the Blood-Cerebrospinal Fluid Barrier (BCSFB). In all cases, the delivery of active pharmaceutical ingredients (APIs) to the brain in sufficient quantities and in a highly controlled and sustainable way (e.g., over a time period of three weeks) is required in order to achieve a desired therapeutic effect.

In a recent EU Horizon collaborative research program, www.n2b-patch.eu, a novel formulation was developed for nose-to-brain controlled delivery of biopharmaceutics for the chronic treatment of CNS disorders. The novel drug formulation consisted of biodegradable PLGA particles loaded with a cognition enhancing drug, embedded into a biodegradable hydrogel matrix. The hydrogel patch was deposited, via a nasal endoscopic applicator as a thin liquid-gelling film, onto the olfactory region (Kiparissides et al., 2019). The research goal was to obtain a desirable drug release profile over a long period of time (e.g., 3-4 weeks) by controlling the PSD, drug loading, molecular and morphological properties of the embedded PLGA drug carriers into a hydrogel patch. By controlling the molecular and morphological properties of the biodegradable hydrogel and embedded PLGA carriers, a desirable drug release profile from the embedded particles into the hydrogel matrix to the mucus layer and then to the olfactory bulb can be established.

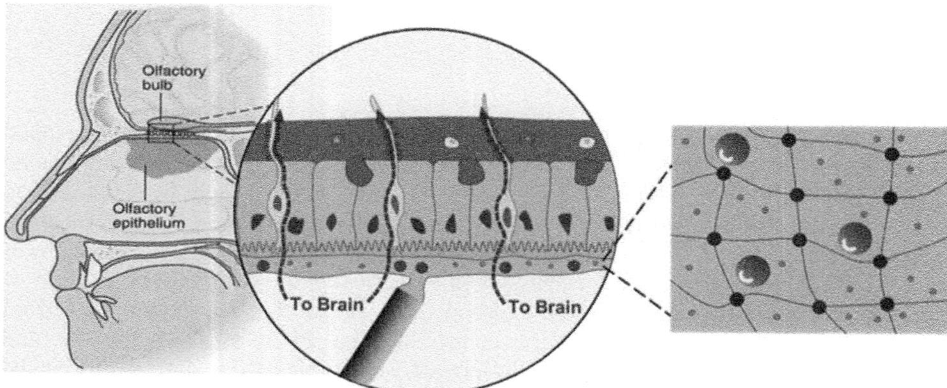

Figure 1. The basic concept of the proposed nose-to brain drug delivery system. Illustration of local hydrogel formation at the olfactory cleft via the aid of an endoscopic applicator. Drug-loaded nanoparticles are embedded into the deposited hydrogel patch. Blue arrows show the potential intracellular and extracellular pathways through which the API is transferred to the brain (www.n2b-patch.eu).

In the present contribution, a process systems computational approach, comprising models at different time and length scales, is developed to elucidate the fundamental physical, transport and biological processes in relation to the nose-to-brain delivery of selected APIs from drug-loaded biodegradable PLGA carriers, embedded into a hydrogel film deposited at the olfactory mucosa, to the epithelium and then to the olfactory bulb (Kiparissides et al. 2020). A polymer degradation-drug diffusion model is derived to calculate the drug release rate from a population of size-distributed drug-loaded particles and analyze the effects of particle size distribution (PSD), drug loading distribution (DLD), polymer molecular and morphological properties on polymer degradation kinetics on drug release profile. The model takes into account the spatial-temporal variation of the drug and water diffusion coefficients due to polymer degradation that results in a decrease of the molecular weight of the polymer. The unknown model parameters are estimated by fitting model predictions to the experimental data of Berchane et al. (2007). Finally, a multi-parametric optimization problem is formulated to calculate the optimal PSD and DLD of drug-loaded polymer carriers to obtain a desired drug release profile over a drug administration period of several weeks.

2. Model-based optimal design of drug-loaded polymeric carriers

Frequently, a zero-order release profile is required for sustainable, controlled delivery of a specific drug dose to a patient in order to achieve a maximum therapeutic outcome. As shown in Figure 2a, the experimentally prepared drug delivery formulations of Berchane et al. (2007) exhibit a tri-phasic drug release profile (see blue, red and green lines). The corresponding drug release rates of the three formulations are depicted in Figure 2b. As can be seen the respective drug release rate exhibits an initial burst followed by an almost zero drug release rate and at approximately 20-30 days, the release rate shows a second drug release burst. This means that the prepared formulations cannot maintain a sustainable drug release rate of zero-order over a long period and thus cannot yield an optimal therapeutic effect. In the present study, a model-based dynamic optimization approach was developed to determine the optimal particle size and drug-loading distributions to obtain a sustainable drug release rate profile of zero-order over a specified period of time.

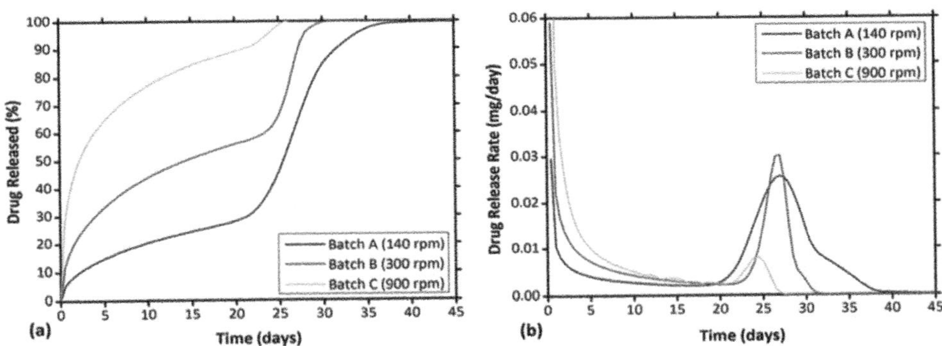

Figure 1. (a) Experimental profiles of cumulative drug amount released from three size-distributed populations of piroxicam-loaded PLGA particles. **(b)** Respective drug release rates for the three experimental formulations (Berchane et al. 2007)).

Selection of the optimal particle size distribution

In this study, it was assumed that the optimal drug-loaded PLGA particles followed a log-normal volume probability distribution, $N_V(d)$, given by

$$N_V(d) = \frac{1}{d\sigma\sqrt{2\pi}} \exp\left(-\frac{1}{2}\left(\frac{\ln d - \mu}{\sigma}\right)^2\right); \qquad 0 < d < \infty \tag{1}$$

Where μ is the mean value and σ is the standard deviation of the log-normal distribution. Thus, the values of μ and σ were selected as control variables in the optimization problem.

Selection of the optimal drug loading distribution

Regarding the drug loading distribution (i.e., drug mass per unit of polymer mass), it was assumed to follow a size-dependent logarithmic function with respect to particle size. Moreover, the initial particle concentration at t=0 was assumed to be independent of particle radius.

$$f_{DL}(d) = -\alpha \cdot \ln(d) + \beta \tag{2}$$

Thus, the two drug loading parameters α and β were treated as two additional control variables in the optimization problem.

Drug release from a size distributed population of drug-loaded particles

To calculate the drug release rate from a monodisperse population of biodegradable drug-loaded PLGA microspheres, two partial differential equations were formulated and numerically solved to calculate the spatial-temporal variation of drug and water concentrations in the PLGA microspheres. Accordingly, the following partial differential equation and associated initial and boundary conditions were derived to describe the drug diffusion and release from the PLGA microspheres.

$$\frac{\partial C_d(r,t)}{\partial t} = \frac{1}{r^2}\frac{\partial}{\partial r}\left(r^2 D_{d,eff}(t,r)\frac{\partial C_d(r,t)}{\partial r}\right) = D_{d,eff}\left(\frac{\partial^2 C_d}{\partial r^2} + \frac{2}{r}\frac{\partial C_d}{\partial r}\right) + \frac{\partial C_d}{\partial r}\frac{\partial D_{d,eff}}{\partial r}, \tag{3}$$

$$\left.\frac{\partial C_d(r,t)}{\partial r}\right|_{r=0} = 0 \tag{4}$$

$$D_{d,eff}(t,r)\left.\frac{\partial C_d(r,t)}{\partial r}\right|_{r=R} = -k_{c,d}\left(C_d(R,t) - C_{out}(t)\right) \tag{5}$$

$$C_d(r,0) = C_{in} \tag{6}$$

$$V_{out}\frac{dC_{out}(t)}{dt} = N_p k_{c,d} A_p \left(C_d(R,t) - C_{out}(t)\right) \tag{7}$$

The symbols $C_d(r,t)$ and $C_{out}(t)$ denote the drug concentration in the particles and release medium, respectively. C_{in} is the initial drug concentration in the particles at $t=0$, independent of particle radius. $D_{d,eff}(r,\ t)$ and $k_{c,d}$ are the effective intra-particle drug diffusion coefficient and drug mass transfer coefficient from the particle's external surface (A_p) to the continuous release medium, respectively. R is the particle radius, V_{out} is the volume of the release medium, N_p, is the total number of drug-loaded particles. Note Eq. (7) is a dynamic mass balance on the drug concentration in the continuous release medium (sink).

Similarly, one can derive a partial differential equation describing the temporal-spatial variation of water in the PLGA particles

$$\begin{aligned}\frac{\partial C_w(r,t)}{\partial t} &= \frac{1}{r^2}\frac{\partial}{\partial r}\left(r^2 D_{w,eff}(t,r)\frac{\partial C_w(r,t)}{\partial r}\right) - k_{a,1}C_w(r,t)\mu_0(r,t) \\ &\quad + k_{a,2}C_{H^+}(r,t)\lambda_0(r,t) \\ &\quad - k_d C_{H^+}(r,t)C_w(r,t)\left(\mu_1(r,t) - \mu_0(r,t)\right) \\ &= D_{w,eff}\left(\frac{\partial^2 C_w}{\partial r^2} + \frac{2}{r}\frac{\partial C_w}{\partial r}\right) + \frac{\partial C_w}{\partial r}\frac{\partial D_{w,eff}}{\partial r} - k_{a,1}C_w\mu_0 \\ &\quad + k_{a,2}C_{H^+}\lambda_0 - k_d C_{H^+}C_w(\mu_1 - \mu_0)\end{aligned} \tag{8}$$

$$\left.\frac{\partial C_w(r,t)}{\partial r}\right|_{r=0} = 0 \tag{9}$$

$$D_{w,eff}(t,r)\left.\frac{\partial C_w(r,t)}{\partial r}\right|_{r=R} = -k_{c,w}\left(C_w(R,t) - C_{w,eq}\right) \tag{10}$$

$$C_w(r,0) = 0 \tag{11}$$

Where $C_w(r, t)$ and $C_{w,eq}$ are the water concentration in the particles and at equilibrium, respectively. $D_{w,eff}(r,t)$ is the effective intra-particle water diffusion coefficient and $k_{c,w}$ is the mass transfer coefficient of water from the release medium to the particle's external surface.

Note that in order to reduce the computational effort (i.e., the number of differential equations to be integrated), the particle size distribution was discretized into four particle classes, N_c. The final model comprised a system of 2549 non-linear ordinary differential equations (i.e., 9 variables for each discrete particle class times 71 or 70 discretization points in the radial coordinate plus the C_{out} variable) and 28 algebraic equations.

Definition of the objective function and numerical solution

Based on the above stated optimization objective (i.e., of zero-order drug release rate), the following dynamic optimization problem was formulated.

$$minJ_{obj,2} = \left\|\boldsymbol{R}_p - \boldsymbol{R}_d\right\|_2, \qquad w.r.t.(\sigma, \mu, \alpha, and\ \beta) \tag{12}$$

Where the vectors $\boldsymbol{R}_d = (R_{d,1}, R_{d,1}, \ldots, R_{d,max})$ and $\boldsymbol{R}_p = (R_{p,1}, R_{p,1}, \ldots, R_{p,max})$ denote the desired and model-based optimal values of the drug release rate, respectively. The maximum number of discrete drug release rate values was set equal to 50 over a drug release horizon of 25 days, which corresponds to one drug release rate value per ½ day.

It is important to note that the above optimization problem was solved together with the following equality constrain regarding the conservation of the total drug mass in the system.

$$\left(\sum_{j=1}^{N_c} m_{d,j}\,(t=0)\right) = m_{d,total} \tag{13}$$

where $m_{d,total}$ is the total drug dose (i.e., 0.4 mg) and $m_{d,j}$ is the drug mass in the "jth" particle class at t=0.

Thus, our optimization objective was to deliver a total therapeutic drug dose of 0.4 mg over a time period of 25 days by maintaining an almost zero-order drug release rate, 0.0128 mg/day. The above stated optimization problem was numerically solved in MATLAB®, employing the ode15s solver for the integration of the differential-algebraic equations and the "MultiStart" global optimizer (a MATLAB® built-in function) for the estimation of the four control variables. From the numerical solution of the optimization problem the parameters μ and σ of the log-normal PSD (Eq. (1)) and the parameters α and β of the drug loading function (Eq. (2)) were estimated so that a desired zero-order drug release rate profile could be realized. The optimal estimates of the four control variables are reported in Table 1 as well as the optimal dug loading per discrete class of particles.

Table 1. Estimated optimal values of the four control variables.

Control Variables				Drug Loading Per Particle Class			
μ_X (μm)	σ_X (μm)	α	β	Class 1	Class 2	Class 3	Class 4
80.15	16.60	0.0139	0.1475	9.30%	8.59%	8.12%	7.77%

In Figure 3a, the calculated optimal PSD and optimal drug loading distribution are depicted. As can be seen the calculated particle size distribution varies from 40 μm to 140 μm while the drug loading per particle class (i.e., % of drug mass per polymer mass of a class) decreases as the particle size increases. In Figure 3b, the optimal and desired percentage of cumulative drug released and respective drug release rate are plotted with respect to the drug release time. As can been seen the optimal formulation of the drug-loaded PLGA carriers (see optimal values of $\sigma, \mu, \alpha, and\ \beta$ in Table 1) closely follows the desired drug release rate of zero-order from the second to the twenty third day of the total release period. Note that the observed deviations in the calculated drug release rate from the desired zero-order profile are primarily due to the inherent initial burst in drug release rate while the observed deviation after the 23rd day is due to the depletion of drug concentration in the PLGA carriers. The results of the present investigation do show that the proposed model-based optimal design of drug-loaded polymer carriers can aid the development of new drug delivery systems exhibiting a desired drug release profile that can largely enhance the in vivo drug bioavailability and efficacy.

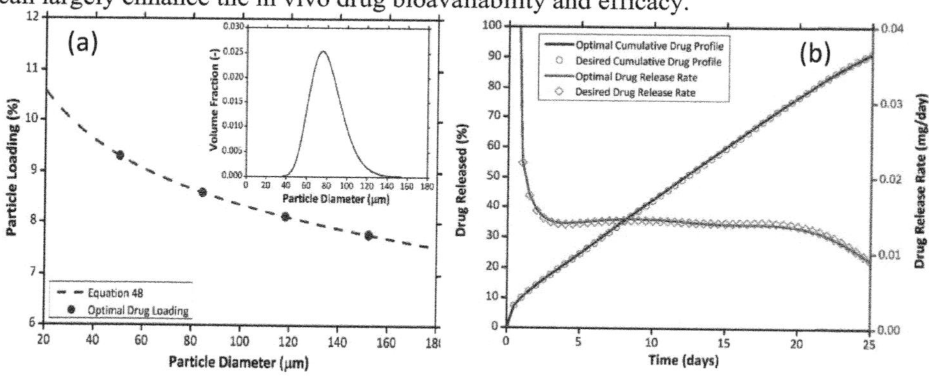

Figure 3. (a) Calculated optimal drug loading values per particle class. In the inset graph, the calculated optimal particle volume fraction distribution is depicted. **(b)** Total drug released and drug release rate from a size-distributed population of particles.

References

N.S. Berchane, K.H. Carson, A.C. Rice-Ficht, M.J. Andrews, 2007, Effect of Mean Diameter and Polydispersity of PLG Microspheres on Drug Release: Experiment and Theory, International Journal of Pharmaceutics, 337, 1-2, 118-126.

C. Kiparissides, A. Vasileiadou, F. Karageorgos, S. Serpetsi, 2020, A Computational Systems Approach to Rational Design of Nose-to-Brain Delivery of Biopharmaceutics, Industrial & Engineering Chemistry Research, 59, 6, 2548–2565.

T.D. Knab, S.R. Little, R.S. Parker, 2015, A Systems Approach to Modeling Drug Release from Polymer Microspheres to Accelerate in vitro to in vivo Translation, Journal of Control Release, 211, 74-84.

S.H.S Koshari, D.P. Chang, N.B. Wang, I.E. Zarraga, K. Rajagopal, A.M. Lenhoff, N.J. Wagner, 2019, Data-Driven Development of Predictive Models for Sustained Drug Release, Journal of Pharmaceutical Sciences, 108, 11, 3582-3591.

Antonis Kokossis, Michael C. Georgiadis, Efstratios N. Pistikopoulos (Eds.)
PROCEEDINGS OF THE 33rd European Symposium on Computer Aided Process Engineering (ESCAPE33), June 18-21, 2023, Athens, Greece
 http://dx.doi.org/10.1016/B978-0-443-15274-0.50102-5

Global estimability analysis and model-based design of experiments in surrogate modeling

Ilias Bouchkira,[a,b] Sanae Elmisaoui,[b,c] Saad Benjelloun,[c] Abderrazak M. Latifi,[b,*]

[a] *Department of Chemical Engineering, Loughborough University, Epinal Way, Loughborough, LE11 3TU, United Kingdom.*

[b] *Laboratoire Réactions et Génie des Procédés, CNRS-ENSIC, Université de Lorraine, 1 rue Grandville, BP20451, Nancy Cedex, 54001, France.*

[c] *Mohammed VI Polytechnic University, Hay Moulay Rachid, Ben Guerir 43150, Morocco.*

abderrazak.latifi@univ-lorraine.fr

Abstract

In this work, a novel approach for the development of surrogate models is presented. It aims to combine traditional model-based design of experiments and global estimability analysis. The approach is based on the Fisher Information Matrix to set two objective functions to be maximized. The first is given by D-optimal design of experiments criterion that maximizes the determinant of FIM, thus minimizing the volume of the confidence region and resulting in more precise parameter estimate. The second is given by the sum of Euclidian norms of FIM's columns to improve the estimability of all the unknown parameters. The resulting multi-objective optimization problem is solved to determine the Pareto front of the optimal solutions. The Multi-Attribute Utility Theory is used as a decision making aid method to select the best solution, needed for the development of the surrogate model. For demonstration purposes, a polynomial Response Surface Model (RSM) based surrogate is developed. It mimics a computationally expensive first-principles model that predicts the conversion rate of a phosphate ore digestion by phosphoric acid. The results are very promising, showing how the approach allows to select the most estimable model parameters from intelligently designed data points. High performance of the developed surrogate is demonstrated by comparing its predictions and computation time to those obtained by first-principles model.

Keywords: Surrogate modeling, Model-based design of experiments, Sensitivity and Estimability analysis, Multi-objective optimization, Decision making aid.

1. Introduction

The development of reliable first-principles models is of utmost importance in process engineering. Recently, there is an increasing demand for high fidelity, large and multi-scale mathematical models. However, in most cases, the process models usually involve sets of ODEs, PDEs etc., that take time to integrate making the developed models very slow/expensive to converge. To overcome these problems, high-fidelity and accurate surrogate models that allow to quickly compute the required model outputs from the inputs are needed. Indeed, surrogate models have been increasingly investigated recently (Rabhi et al., 2018; Elmisaoui et al., 2022). They are developed to mimic the behavior of a given simulation model or an outcome of interest that cannot be easily measured or computed, as closely as possible, while being computationally cheap to evaluate.
Surrogate modeling is usually based on a limited number of intelligently chosen data points (black-box modeling), and on choosing an appropriate mathematical expression of the surrogate, which involves in most cases many unknown parameters to be identified.

For the development of accurate surrogates, it is very important to design the most optimal data points that maximize the estimability of the surrogate model parameters in order to determine the most estimable parameters to be identified.
These two objectives are generally overlooked in most of the current published papers. The main purpose of the present work is to develop a novel approach that enables to (i) firstly, select intelligently the dataset of points to mimic by the surrogate model, and (ii) secondly to determine with accuracy, the unknown surrogate model parameters to be identified from the selected dataset of points.
This approach is based on a constrained multi-objective optimization problem that accounts for two objectives, one from Model-based design of experiments and one from parameter estimability, while the constraints are mainly given by the range of variation of the operating conditions. The solution of the formulated problem will be a set of optimal profiles, i.e., a Pareto front, from which, the best solution can be chosen using a decision making aid method. For demonstration purposes, the approach is applied to the case of polynomial Response Surface Model (RSM)-based surrogate, to mimic the behavior of a computationally expensive first-principals model that predicts the conversion rate of the digestion reaction of phosphate ore by phosphoric acid.

2. Models and methods

2.1. First-principals model of phosphate ore digestion

The case study in this work deals with a phosphate ore digestion model previously developed by Elmisaoui et al. (2021). It describes the dissolution behavior of a pure tri-calcium phosphate (TCP) in phosphoric acid to produce mono-calcium phosphate (MCP). The dissolution reaction is given by:

$$Ca_3(PO_4)_2 + H_3PO_4 \rightarrow 3\ Ca(H_2PO_4)_2 \quad (1)$$

The conversion rate of this reaction depends on several operating conditions such as, the reaction temperature, the initial phosphoric acid concentration, the tri-calcium phosphate particle size distribution and porosity, the hydrodynamic conditions, the solid/liquid ratio, the residence time etc. In this work, four sensitive parameters are considered. They include the initial TCP particle size (R_0), the initial phosphoric acid concentration (C_0), the tank stirring speed through a hydrodynamic parameter (α), and the residence time (τ) inside the reactor. Moreover, to predict the dissolution rate, the shrinking core model is used considering three main phases: (i) the liquid bulk, (ii) the liquid film that surrounds the solid particles, (iii) and the solid phase. Some assumptions are made for the development of the first-principals model and can be found in Elmisaoui et al. (2021). The resulting model is given by the following system of mass balance equations:

$$\frac{dX}{dt} = \frac{3M_{TCP}D_{MCP}}{x_{TCP}R_0\delta\rho_s}(1-X)^{\frac{2}{3}}(C_{MCP}|_{r=R} - C'_{MCP}) \quad (2)$$

$$\frac{\partial C_t}{\partial t} = \frac{1}{r^2}\frac{\partial}{\partial r}\left(D_i r^2 \frac{\partial C_i}{\partial r}\right) \quad i = H_3PO_4, Ca(H_2PO_4)_2 \quad (3)$$

$$r = R: -D_i \frac{\partial C_i}{\partial r}\bigg|_{r=R} = v_i k_r C_{H_3PO_4}\big|_{r=R} \quad (4)$$

$$r = R + \delta: C_i|_{r=R+\delta} = C'_i \quad (5)$$

$$\delta = R\left[1 + \alpha\left(\frac{R}{R_0}\right)^{8/9} D_{MCP}^{-1/3}\right]^{-1} \tag{6}$$

$$C'_{MCP} = \frac{3n^0_{TCP}}{V_L} \text{ and } C'_{H_3PO_4} = C^0_{H_3O_4} - \frac{4n^0_{TCP}}{V_L} \tag{7}$$

In the system of equations above, (2) describes the solid phase, (3) describes the liquid film surrounding the solid particles, (4 and 5) are the two main boundary conditions, (6) is used to predict the film thickness δ, and (7) describes the liquid bulk. m_s and R are respectively the mass and radius of the solid phosphate ore particles, x_{TCP} and M_{TCP} are the mass fraction of TCP and its molecular weight. $C_{MCP}|_{r=R}$ and C'_{MCP} are the concentrations of MCP at the solid surface and in the liquid bulk. k_r is a rate constant, D_i and v_i are the diffusion coefficient and stoichiometric coefficient of component i, respectively. n^0_{TCP} is the initial number of moles of TCP in the solid, V_L is the liquid volume in the tank, α is a hydrodynamic parameter calculated from the stirring speed of the tank. The solution of (2-6) allows to predict the conversion rate $X(\tau, C_0, R_0, \alpha)$ to be mimicked by the surrogate model.

2.2. Polynomial Response Surface Model-based surrogate

Surface Response model-based surrogates have been extensively used recently. They proved to be accurate, flexible, and available in many commercial data analysis software. The third degree polynomial RSM-based surrogate model used in this work writes as:

$$\tilde{Y} = \underbrace{y_0}_{Cst} \underbrace{+\sum_i a_i X_i +}_{Linear\ terms} \underbrace{\sum_i b_i X_i X_i + \sum_{i \neq j} d_{i,j} X_i X_j +}_{Binary\ interaction\ terms} \underbrace{\sum_i c_i X_i X_i X_i + \sum_{i \neq j \neq k} e_{i,j,k} X_i X_j X_k}_{Ternary\ interaction\ terms} \tag{7}$$

Considering four main sensitive operating parameters, i.e., $u = \{\tau, C_0, R_0, \alpha\}$, the surrogate model above involves 23 unknown parameters (table 1). The main challenges that the proposed approach addresses are *(i)* which data points (from the first-principals model) to mimic, and *(ii)* which unknown model parameters to estimate.

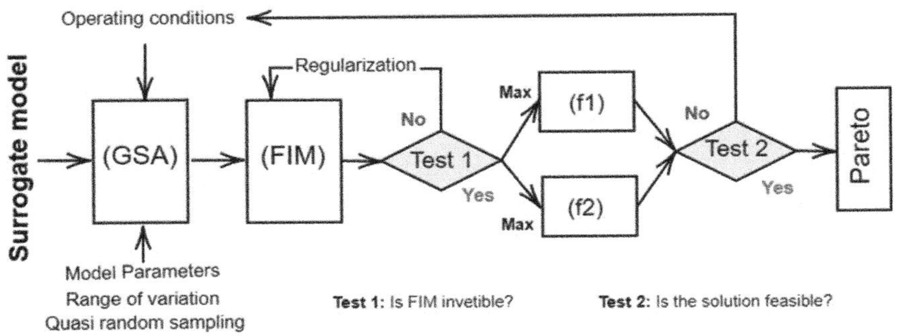

Figure 1: Framework of the multi-objective model-based design of experiments. (GSA): Global Sensitivity Analysis. (FIM): Fisher Information Matrix.

2.3. Multi-objective model-based design of experiments

Figure 1 shows the proposed approach for the development of the surrogate models. It is based on the Fisher Information Matrix (FIM), which is computed from the global sensitivity indices of the model's unknown parameters as:

$$Z = \begin{pmatrix} S_{1,1}|m_1 & \cdots & S_{1,d}|m_1 \\ \vdots & \ddots & \vdots \\ S_{n,1}|m_n & \cdots & S_{n,d}|m_n \end{pmatrix}, \qquad \boldsymbol{FIM} = Z.Z^T \tag{8}$$

where m_i is the ith sampling point, d is the number of model parameters and n is the number of measurements. $S_{i,j}$ are the individual sensitivity indices of the ith output with respect to the jth parameter. They are calculated by means of the method developed by Saltelli (2012). FIM is then used to compute the two objective functions f_1 and f_2. They represent the D-optimal model-based design of optimal experiments (MBDoE) criterion and the Euclidian norm of FIM's columns, respectively (Bouchkira et al., 2021; 2022; 2023). A regularization test is performed to ensure that the FIM is invertible (Test 1). Then, multi-objective optimization is performed to maximize the two objective functions and to determine the Pareto front of the optimal solutions (Test 2). The multi-objective optimization problem is formulated as:

$$\begin{aligned}
&\max_{\varphi} \boldsymbol{F} = [f_1, f_2]^T \\
&f_1 = det(FIM) \\
&f_2 = \sum_{i,j} \|S_{i,j}\| \\
&FIM = Z(X,\theta).Z(X,\theta)^T \\
&\varphi = [C_{0,1}, \dots, C_{0,n}, R_{0,1}, \dots, R_{0,m}, \alpha_1, \dots, \alpha_k, t_1, \dots, t_w] \\
&\text{s.t.} \quad 0 < C_{0,j-1} \leq C_{0,j}, j \in [1;n]; \ 0 < R_{0,r-1} \leq R_{0,r}, r \in [1;m]; \\
&\qquad 0 < \alpha_{s-1} \leq \alpha_s, s \in [1;k]; \ 0 < t_{p-1} \leq t_p, p \in [1;w] \\
&\qquad C_{0,j-1} - C_{0,j} \leq \omega_C; \quad R_{0,r-1} - R_{0,r} \leq \omega_R \\
&\qquad \alpha_{s-1} - \alpha_s \leq \omega_\alpha; \quad t_{p-1} - t_p \leq \omega_t \\
&\qquad Cond(FIM) \leq \theta
\end{aligned} \tag{S.1}$$

where ω_C, ω_R, ω_α and ω_t are used to avoid duplicated experiments. θ is a cut-off criterion to guarantee that FIM is revertible, $\theta = 10$ is considered as in (Cenci et al., 2022).

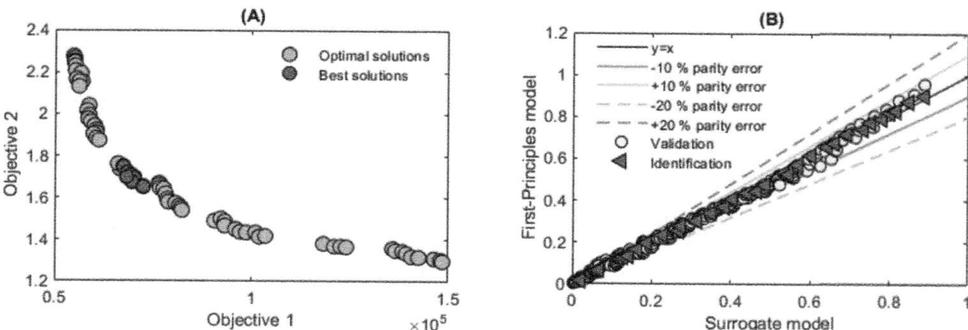

Figure 2: Optimization and identification results. (A): Pareto front. (B): Parity diagram.

Table 1: Surrogate modeling and estimability results. Parameters with (*) are not estimable

Parameter	Terms	Value	Precision (±)	Estimability magnitude	Rank
x_0	cst	−0.076	0.0361	0.156	3
a_1	C_0	−9.187	$7{,}72.10^{-3}$	0.051	11
a_2	R_0	$-5.88.10^5$	149.1	0.061	8
a_3*	α	-	-	0.015	13
a_4	t	$0.15.10^{-2}$	$7.19.10^{-4}$	0.487	2
b_1	C_0, C_0	$-2.02.10^{-2}$	$8.23.10^{-6}$	0.094	4
b_2	R_0, R_0	$7.22.10^5$	$2.03.10^4$	0.054	10
b_3*	α, α	-	-	0.003	20
b_4	t, t	$-1.79.10^{-1}$	$2.05.10^{-4}$	0.705	1
c_1*	C_0, C_0, C_0	-	-	0.002	21
c_2	R_0, R_0, R_0	$-2.20.10^{13}$	$9.03.10^{10}$	0.066	7
c_3*	α, α, α	-	-	0.012	14
c_4	t, t, t	$0.01.10^{-7}$	$1.78.10^{-9}$	0.081	5
d_1*	C_0, R_0	-	-	0.008	15
d_2*	C_0, α	-	-	0.008	16
d_3	C_0, t	$2.90.10^{-1}$	$1.88.10^{-4}$	0.045	12
d_4*	R_0, α	-	-	0.001	23
d_5	R_0, t	−0.509	0.0138	0.061	9
d_6*	α, t	-	-	0.004	19
e_1*	C_0, R_0, α	-	-	0.001	22
e_2	C_0, R_0, t	−0.111	$2.239.10^{-2}$	0.073	6
e_3*	C_0, α, t	-	-	0.004	18
e_4*	R_0, α, t	-	-	0.006	17

3. Results and discussion

Global sensitivity analysis is first performed based on the method of Saltelli (2002), and a Quasi-Monte Carlo sampling of size 10.000 was sufficient to achieve sensitivity indices convergence. The sensitivity indices were then used to build the FIM which is used in the multi-objective optimization problem. The latter is then implemented and solved under MATLAB, and a Genetic Algorithms solver was used to achieve global optimality.

Resolution of the multi-objective optimization problem allowed to generate the Pareto front (Figure 2A) of the optimal compromises to be mimicked by the surrogate model. It is worth mentioning that all the points on the Pareto front are optimal. However, for further development of the surrogate model, the best compromises of the Pareto front should be selected to be implemented. To this end, the Multi-attribute Utility Theory method was used as a decision making aid method as in our previous works (Bouchkira et al. 2022). It allowed to classify the solutions and to select the best one. Five best solutions are highlighted in red color in figure 2A. The best solution was selected to be implemented and suggests to mimic 32 outputs of the first-principles model predictions, based on the combination of two initial phosphoric acid concentrations i.e., $C_0 = 115$ and $C_0 = 453$; two initial particle radius, i.e., $R_0 = 6.9\ 10^{-4}$ and $R_0 = 5.1\ 10^{-4}$; two hydrodynamic parameters, i.e., $\alpha = 2.73$ and $\alpha = 3.18$; and four residence times, i.e., $\tau = 155$, $\tau = 215$, $\tau = 403$, and $\tau = 600$.

Table 1 summarizes the results obtained in the development of the polynomial RSM-based surrogate model using the best solution. It shows first the ranking of the surrogate unknown parameters according to their estimability from the selected best solution. Moreover, the results show that only 12 unknown parameters out from 23 are estimable, and this is deduced from the estimability magnitudes of the parameters that were computed within the optimization approach, and their comparison to an estimability cut-off criterion proposed in (Yao et al. 2011). The values of the estimable parameters are provided along with their confidence intervals.

It should be noted here that the proposed approach has the advantage of identifying only the estimable model parameters, which avoids the model overfitting and improves the model precision by ignoring uncertain parameters.
Figure 2B shows a comparison between the predictions of the developed surrogate model and those of the expensive first-principals model. The results show the accuracy and good performance of the developed surrogate model, and this can be shown in the parity errors that do not exceed -10% and +10% for all the data points used for the identification of the model parameters (red triangle), while the errors on other data points that were not used during the identification process (validation: yellow circles) lie between -20% and +20%. It is finally worth mentioning that the developed surrogate model proved to be more than three times faster than the first-principals model, and this was deduced by running both models several times on an ASUS workstation with 12th Gen Intel(R) Core (TM) i9-12900K 3.20 GHz128GB Ram processor.

4. Conclusion

A novel multi-objective optimization-based approach for the development of accurate surrogate models is presented. It has the advantage of designing intelligently the data points from the computationally expensive model to be mimicked by the surrogate model, while selecting only its estimable unknown parameters. To demonstrate the effectiveness of the proposed approach, it is used in this work for the development of a surrogate model to mimic the predictions of a computationally expensive first-principals model that predicts the conversion rate of a phosphate ore digestion using phosphoric acid. High performance of the developed surrogate model was finally demonstrated by comparing its predictions and computation time to those obtained by the first-principals model.

References

I. Bouchkira, A. Latifi, L. Khamar, S. Benjelloun, 2021, Global sensitivity based estimability analysis for the parameter identification of Pitzer's thermodynamic model. Reliability Engineering & System Safety, 207, 107263.

I. Bouchkira, S. Benjelloun, L. Khamar, A. Latifi, 2023. Thermodynamic modeling and parameter estimability analysis of a wet phosphoric acid process with impurities. Fluid Phase Equilibria, 564, 113594.

I. Bouchkira, A. Latifi, L. Khamar, S. Benjelloun, 2022. Modeling and multi-objective optimization of the digestion tank of an industrial process for manufacturing phosphoric acid by wet process. Computers & Chemical Engineering, 156, 107536.

F. Cenci, G. Bano. C. Christodoulou, Y. Vueva, S. Zomer, M. Barolo, F. Bezzo, P. Facco, (2022). Streamlining tablet lubrication design via model-based design of experiments, International Journal of Pharmaceutics, 614, 121435.

S. Elmisaoui, A. M. Latifi, L. Khamar, M. Salouhi, (2021), Analysis of the dissolution mechanism in the phosphoric acid manufacturing process: modelling and simulation. Computer Aided Chemical Engineering 50, 891–897.

S. Elmisaoui, S. Benjelloun, A. Chkifa, A.M. Latifi, (2022), A sparse polynomial surrogate model for the shrinking core model in phosphate ore digestion, Computer Aided Chemical Engineering , 51,1291-1296.

A. Rabhi, A. Chkifa, S. Benjelloun, A. Latifi, (2018), Surrogate-based modeling in flotation processes, Computer Aided Chemical Engineering , 43, 229-234.

A. Saltelli, (2002), Making best use of model evaluations to compute sensitivity indices. Computer physics communications, 145(2), 280–297.

K.Z. Yao, M.B. Shaw, B. Kou, K.B. McAuley, D.W. Bacon, (2003), Modeling ethylene/butene copolymerization with multi-site catalysts: parameter estimability and experimental design, Polymer Reaction Engineering, 11(3), 563-588.

Antonis Kokossis, Michael C. Georgiadis, Efstratios N. Pistikopoulos (Eds.)
PROCEEDINGS OF THE 33rd European Symposium on Computer Aided Process Engineering (ESCAPE33), June 18-21, 2023, Athens, Greece
 http://dx.doi.org/10.1016/B978-0-443-15274-0.50103-7

Tractable Data-driven Solutions to Hierarchical Planning-scheduling-control

Damien van de Berg[a], Roberto Xavier Jimenez Jimbo[a], Nilay Shah[a], Ehecatl Antonio del Rio-Chanona[a,*]

[a]*Sargent Centre for Process Systems Engineering, Roderic Hill Building South Kensington Campus. London, SW7 2AZ, United Kingdom*

Abstract

Using numerical optimization for the hierarchical integration of decision-making units is crucial to provide feasibility and optimality of all levels. However, realistically modelling hierarchical decision-making calls for multilevel formulations, which are numerically intractable and mathematically difficult. In this work, we show how to leverage two data-driven techniques – derivative-free optimization and optimality surrogates – to decrease the computational burden of multilevel problems. We reformulate a tri-level planning-scheduling-control problem into a single-level black-box problem wherein each evaluation calls a scheduling instance with embedded optimal control surrogates. We show that solving this integrated problem instead of the single-level instance leads to changes in the optimal production planning and scheduling sequence, and discuss trade-offs associated with both techniques.

Keywords: Derivative-free optimization; Optimization with embedded surrogates; Enterprise-wide optimization; Hierarchical decision-making

1. Introduction

Chemical enterprises can capture significant value by using mathematical optimization for the integration of hierarchical levels of operations from supply chain and process design down to control. Traditionally, in *sequential* decision-making, upper-level decisions are taken while disregarding lower-level considerations, and then fed as setpoints to the lower levels. In the *monolithic* approach, lower-level feasibility considerations are included into upper-level optimization problems accompanied by a significant drop in computational tractability. In the *hierarchical* approach, upper-level decision-makers consider lower-level optimality, accounting for how lower-level optimization problems affect upper-level objectives. This results in multi-level formulations, which are often numerically intractable and mathematically difficult (Chu and You, 2015). In this work, we show how we can leverage derivative-free optimization (DFO) (van de Berg et al., 2022) and surrogate modelling techniques to efficiently approach the tri-level optimum of an integrated planning-scheduling-control optimization problem.

To solve hierarchically integrated optimization problems, we reformulate multiple nested levels of decision-making into a single-level formulation. To this end, we can leverage data-driven techniques in a bottom-up or top-down approach. In the bottom-up approach, we first construct 'optimality surrogates' that map the optimal response variables of the lower level appearing in the upper levels as a function of the upper-level variables appearing in lower levels. Similar techniques are known under various names across different fields, such as multi-parametric model predictive control, approximate explicit optimal control laws, amortized learning in Machine Learning, or policy-based methods in Reinforcement Learning (Avraamidou and Pistikopoulos, 2019; Sacchio et al., 2021). The top-down approach relies on derivative-free optimization (DFO), which is closely related to black-box and simulation-based optimization. The reader is referred to van de Berg et al. (2022) for a review on DFO for process systems engineering applications and to Beykal et al. (2022) for an exposition of how DFO could be used to solve multilevel problems.

Figure 1 shows our approach to the planning-scheduling-control problem: The bottom-up technique constructs optimality surrogates for the lower-level control problem. This surrogate is created offline and included as a constraint in the middle scheduling problem. As such, we collapse the two lower levels of scheduling and control into a single level. We then use DFO to find the optimal upper-level planning variables: DFO iteratively proposes a set of planning-level variables which are fixed and fed as setpoints to the lower-level integrated scheduling-control problem. The lower-level is then solved via traditional optimization software (Pyomo) which returns the complicating quantities that inform the upper-level objective to be optimized via DFO. As such, we manage to find a tractable solution to an otherwise intractable trilevel integrated planning-scheduling-control problem. It is worth noting that there will always be an approximation error when using surrogates, therefore, these must be trained and cross-validated appropriately.

2. Methodology

2.1. Problem Statement

We consider canonical planning, scheduling, and planning problems as described in Chu and You (2015). Using the hierarchical approach to integrate these decision levels yields the following tri-level optimization problem:

$$
\begin{aligned}
\min_{\substack{\mathbf{x}_p,\mathbf{x}_{prod},\mathbf{x}_s^*,c_s^*,\\ \mathbf{x}_b^*,c_c^*,t^*,\mathbf{x}_c^*}} \quad & \mathrm{f_p}(\mathbf{x}_p,\mathbf{x}_{prod}) + c_s^* + c_c^* \\
\text{s.t.} \quad & \mathbf{h}_p(\mathbf{x}_p,\mathbf{x}_{prod}) = \mathbf{0}, \\
& \mathbf{g}_p(\mathbf{x}_p,\mathbf{x}_{prod}) \le \mathbf{0}, \\
& \begin{matrix}\mathbf{x}_s^*,c_s^*,c_c^*,\\ \mathbf{x}_b^*,t^*,\mathbf{x}_c^*\end{matrix} \in \operatorname*{arg\,min.}_{\substack{\mathbf{x}_s,\mathbf{x}_{batch},c_s,\\ \mathbf{x}_{ctrl}^*,t_f^*,c_{ctrl}^*}} \ \mathrm{f_s}(\mathbf{x}_s,\mathbf{x}_{batch},t_f^*;\mathbf{x}_{prod}) \\
& \qquad \text{s.t.} \quad \mathbf{h}_s(\mathbf{x}_s,\mathbf{x}_{batch},c_s,t_f^*;\mathbf{x}_{prod}) = \mathbf{0}, \\
& \qquad \qquad \mathbf{g}_s(\mathbf{x}_s,\mathbf{x}_{batch},t_f^*;\mathbf{x}_{prod}) \le \mathbf{0}, \\
& \qquad \qquad \mathbf{x}_{ctrl}^*,t_f^*,c_{ctrl}^* \in \operatorname*{arg\,min.}_{\mathbf{x}_c,t_f,c_{ctrl}} \ \mathrm{f_c}(\mathbf{x}_c,t_f;\mathbf{x}_{batch}) \\
& \qquad \qquad \qquad \text{s.t} \quad \mathbf{h}_c(\mathbf{x}_c,t_f,c_{ctrl};\mathbf{x}_{batch}) = \mathbf{0} \\
& \qquad \qquad \qquad \qquad \mathbf{g}_c(\mathbf{x}_c,t_f;\mathbf{x}_{batch}) \le \mathbf{0}
\end{aligned}
\tag{1}
$$

Eq. (1) denotes a tri-level optimization problem where $\mathrm{f_p}(\cdot),\mathrm{f_s}(\cdot),\mathrm{f_c}(\cdot)$ denote the objective functions, $\mathbf{h}_p(\cdot),\mathbf{h}_s(\cdot),\mathbf{h}_c(\cdot)$ denote the equality constraints, and $\mathbf{g}_p(\cdot),\mathbf{g}_s(\cdot),\mathbf{g}_c(\cdot)$ denote the inequality constraints of the planning, scheduling, and control problems respectively. $\mathbf{x}_p,\mathbf{x}_{prod}$ denote the planning-specific and the production targets. The highest-level objective consists not only of $\mathrm{f_p}(\cdot)$ but also the optimal scheduling and operating costs appearing as c_s^*, c_c^* in the planning, c_s, c_{ctrl}^* in the scheduling, and c_c^* in the control (no scheduling cost in the control problem) that are obtained as the result of the scheduling optimization and its nested control optimization in the constraints. The mid-level scheduling problem takes the planning variables as setpoints $\mathbf{x}_{\text{prod}}$ and optimizes the scheduling and batch variables $\mathbf{x}_s, \mathbf{x}_{\text{batch}}$ which appear as optimal response variables in the planning level as $\mathbf{x}_s^*, \mathbf{x}_b^*$. The scheduling problem is also influenced by the optimal batch processing time as defined in the lower-level control problem. The control problem takes as setpoint the batch variables from the scheduling $\mathbf{x}_{\text{batch}}$ and determines the optimal control variables, batch processing times, and operating cost $\mathbf{x}_c^*, t^*, c_c^*$ (planning), $\mathbf{x}_{\text{ctrl}}^*, t_f^*, c_{\text{ctrl}}^*$ (scheduling), and $\mathbf{x}_c, t_f, c_{ctrl}$ (control) for said batch. It is worth noting that our considered planning, scheduling, and control problems would, in isolation, represent linear, mixed-integer linear, and optimal control problems, respectively.

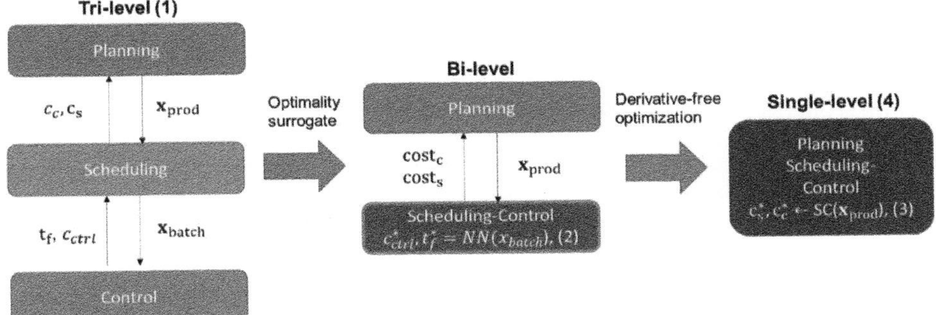

Figure 1. We reformulate tri-level planning-scheduling-control problem (1) by collapsing the control problem as an optimality surrogate into the scheduling problem ((2), and then using derivative-free optimization to solve the remaining bi-level problem as a single-level black-box optimization problem (4), where each evaluation calls the scheduling model with the embedded optimal control surrogate (3).

As illustrated in Figure 1, we will start by reformulating the bi-level scheduling and control problems into a single-level formulation using the bottom-up approach, before collapsing the integrated scheduling-control level into the planning level.

2.2. Collapsing scheduling and control levels using optimality surrogates

By inspection of Problem (1), we notice that the optimal operating cost and batch processing time c_{ctrl}^* and t_f^* are the only control-level variables appearing in upper levels. Consequently, we can formulate surrogates that map the optimal control response variables c_{ctrl}^* and t_f^* as a function of the batch production targets $\mathbf{x}_{\text{batch}}$ as the only setpoints feeding into the control level.

$$c_{\text{ctrl}}^*, t_f^* = \text{NN}(\mathbf{x}_{\text{batch}}) \tag{2}$$

While there are many potential surrogates we could use (Dias and Ierapetritou, 2020), we use neural networks for their general approximation qualities , their potential for MILP reformulation and their ease of incorporation into optimization pipelines (Ceccon et al., 2022). We uniformly sample batch sizes and solve the corresponding optimal control problem, logging the optimal response variables. Then, we train a small neural network of two hidden layers with five nodes each using Pytorch to predict the optimal response variables given the setpoints. We can then reformulate the neural network surrogate into mixed-integer linear constraints that can be incorporated as a Pyomo block into the planning Pyomo model using OMLT (Ceccon et al., 2022). Since we are using rectified linear unit (ReLU) activation functions, we can use the full-space formulation to ensure that the bi-level scheduling-control model can be reformulated into a single-level mixed-integer linear program (3), which is crucial for computational tractability. This allows us to use specialized solvers (Gurobi). Following the monolithic approach or using other formulations would require less efficient mixed-integer *nonlinear* solvers.

$$\begin{aligned} \mathbf{x}_s^*, c_s^*, c_c^*, \mathbf{x}_b^*, t^*, \mathbf{x}_c^* \in \mathrm{SC}(\mathbf{x}_{\mathrm{prod}}) = \underset{\mathbf{x}_s, \mathbf{x}_{\mathrm{batch}}, c_s, t_f^*, c_{\mathrm{ctrl}}^*}{\arg\min.} \quad & f_s(\mathbf{x}_s, \mathbf{x}_{\mathrm{batch}}, t_f^*; \mathbf{x}_{\mathrm{prod}}) \\ \text{s.t.} \quad & \mathbf{h}_s(\mathbf{x}_s, \mathbf{x}_{\mathrm{batch}}, c_s, t_f^*; \mathbf{x}_{\mathrm{prod}}) = \mathbf{0}, \\ & \mathbf{g}_s(\mathbf{x}_s, \mathbf{x}_{\mathrm{batch}}, t_f^*; \mathbf{x}_{\mathrm{prod}}) \le \mathbf{0}, \\ & c_{\mathrm{ctrl}}^*, t_f^* = \mathrm{NN}(\mathbf{x}_{\mathrm{batch}}) \end{aligned} \tag{3}$$

2.2.1. Single-level black-box optimization given optimal scheduling-control

After collapsing the bi-level scheduling-control problem into an integrated single-level formulation, we need to incorporate the integrated model with the higher-level planning, resulting in bi-level Formulation (4):

$$\begin{aligned} \min_{\substack{\mathbf{x}_p, \mathbf{x}_{\mathrm{prod}}, \mathbf{x}_s^*, c_s^*, \\ \mathbf{x}_b^*, c_c^*, t^*, \mathbf{x}_c^*}} \quad & f_p(\mathbf{x}_p, \mathbf{x}_{\mathrm{prod}}) + c_s^* + c_c^* \\ \text{s.t.} \quad & \mathbf{h}_p(\mathbf{x}_p, \mathbf{x}_{\mathrm{prod}}) = \mathbf{0}, \\ & \mathbf{g}_p(\mathbf{x}_p, \mathbf{x}_{\mathrm{prod}}) \le \mathbf{0}, \\ & \mathbf{x}_s^*, c_s^*, c_c^*, \mathbf{x}_b^*, t^*, \mathbf{x}_c^* \in \mathrm{SC}(\mathbf{x}_{\mathrm{prod}}) \end{aligned} \tag{4}$$

Inspired by Beykal et al. (2022), we intend to use derivative-free optimization (DFO) to find the planning variables that optimize the planning-level objective. DFO is closely related to black-box and simulation-based optimization. Its aim is to find input variables that optimize a system whose gradient expressions cannot be cheaply obtained. DFO solvers can be loosely categorized into direct methods, which handle function evaluations directly, and model-based methods, which rely on the intermediate construction and optimization of surrogates. As opposed to traditional numerical optimization where all variables are considered 'decision variables', in DFO, we only optimize the degrees of freedom, a subset of the variables that when fixed allow for the remaining quantities to be determined using the remaining equality constraints. To this end, we first decompose the planning-specific variables $\mathbf{x}_p$ into $\mathbf{x}_{\mathrm{store}}, \mathbf{x}_{\mathrm{transp}}, \mathbf{x}_{\mathrm{sales}}$, the storage, transportation, and sales variables respectively. We notice that when fixing $\mathbf{x}_{\mathrm{prod}}, \mathbf{x}_{\mathrm{transp}}, \mathbf{x}_{\mathrm{sales}}$, we can obtain the storage variables $\mathbf{x}_{\mathrm{store}}$ by leveraging $\mathbf{h}_p(\mathbf{x}_{\mathrm{store}}, \mathbf{x}_{\mathrm{transp}}, \mathbf{x}_{\mathrm{sales}}, \mathbf{x}_{\mathrm{prod}})=\mathbf{0}$. We can then obtain all quantities of the planning-level objective: $c_s^*, c_c^* \in \mathrm{SC}(\cdot)$ and $f_p(\cdot)$. In the DFO, we augment the planning-level objective with a penalty term on the inequalities $g_{\mathrm{viol}} = \max\left(0, \mathbf{g}_p(\cdot)\right)^2$.

$$\begin{aligned} \min_{\mathbf{x}_{\mathrm{prod}}, \mathbf{x}_{\mathrm{transp}}, \mathbf{x}_{\mathrm{sales}}} \quad & f_p(\mathbf{x}_{\mathrm{store}}, \mathbf{x}_{\mathrm{transp}}, \mathbf{x}_{\mathrm{sales}}, \mathbf{x}_{\mathrm{prod}}) + c_s^* + c_c^* + g_{\mathrm{viol}} \\ \text{where} \quad & \mathbf{x}_{\mathrm{store}} \leftarrow \mathbf{h}_p(\mathbf{x}_{\mathrm{store}}, \mathbf{x}_{\mathrm{transp}}, \mathbf{x}_{\mathrm{sales}}, \mathbf{x}_{\mathrm{prod}}) = \mathbf{0}, \\ & g_{\mathrm{viol}} \leftarrow \mathbf{g}_p(\mathbf{x}_{\mathrm{store}}, \mathbf{x}_{\mathrm{transp}}, \mathbf{x}_{\mathrm{sales}}, \mathbf{x}_{\mathrm{prod}}) \le \mathbf{0}, \\ & c_s^*, c_c^* \leftarrow \mathrm{SC}(\mathbf{x}_{\mathrm{prod}}) \end{aligned} \tag{4}$$

As such, we have reformulated the tri-level Problem (1) into a single-level black-box optimization problem (4), where in each evaluation of the DFO input we evaluate the remaining planning-level variables necessary to compute the planning-specific objective and inequality penalty term. We obtain the lower-level optimal costs by calling the integrated planning-scheduling optimization instance.

2.3. Notes on the derivative-free optimization

Since the black-box evaluations are expensive, we only have a restricted budget of 400 evaluations to optimize the 178-dimensional problem. This makes the DFO problem very challenging to be solved without *a priori* insights. As such, we hot-start the solution to the problem by extracting the relevant DFO variables from the solution of the following planning-only problem:

$$\begin{aligned} \min_{\mathbf{x}_p, \mathbf{x}_{prod}} \quad & f_p(\mathbf{x}_p, \mathbf{x}_{prod}) \\ \text{s.t.} \quad & \mathbf{h}_p(\mathbf{x}_p, \mathbf{x}_{prod}) = \mathbf{0}, \\ & \mathbf{g}_p(\mathbf{x}_p, \mathbf{x}_{prod}) \leq \mathbf{0}, \end{aligned} \tag{5}$$

Given the expensive nature of the evaluations, we only have enough budget to perform a 'local search' around the optimum obtained from (5) rather than perform extensive exploration of the solution space. Informed by van de Berg et al. (2022), we use the trust-region method Py-BOBYQA given its highly exploitative nature.

3. Results

The alternative to solving the tri-level optimization problem is solving the planning-only problem in (5) and sending its optimal planning target as a setpoint to the scheduling problem which is optimized and finally fed to control (sequential approach). We use the solution of the planning problem of the sequential approach as an initial guess for the DFO and decrease the tri-level objective from a cost of 4.762 to -0.132 million, turning the tri-level objective profitable. The tri-level solution also translates into tangible production and scheduling changes: Figures 2A and 2B demonstrate that going from the initial guess to the DFO solution finetunes the production targets of some products (in red case) given more accurate accounting of storages costs from the scheduling level. Figures 2C and 2D illustrate that the production assignment of products P1 and P2 changes (in red case) due to a more accurate estimate of the processing time from the control problem.

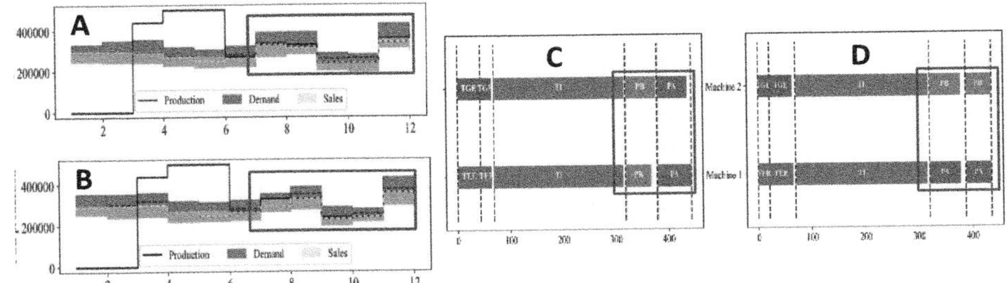

Figure 2. Figures A and B represent the production target of one product in kg over 12 weeks for the planning-only initial guess (A) and the tri-level solutison (B) respectively. Figures C and D show the scheduling job assignments for all products in a particular week for the planning-only initial guess (C) and the tri-level solution (D) respectively. Changes highlighted in red.

4. Discussion

Our proposed methodology lends itself well to the decision-making frameworks of the process industries. Given the expensive nature of monolithic or hierarchical decision-making, planning, scheduling, and control problems are usually solved sequentially. Our proposed integrated approach leverages an initial guess to the DFO problem using the single-level planning problem. As such, we can always fall back on the sequential approach if the DFO fails to make significant progress. This way, when applying exploitative DFO solvers, our proposed methodology can be interpreted as 'fine-tuning' the single-level planning optimum towards a tri-level objective in the available evaluation budget available to decision-makers. Our methodology begs the question if the two techniques (DFO and optimality surrogates) can be used interchangeably. There are clear trade-offs associated with both techniques. DFO for bi-level optimization has the advantage that it does not have to sacrifice solution accuracy through approximation errors. However, DFO performs best in lower dimensions. To use DFO for bi-level optimization requires all degrees of freedom in the upper-level optimization problem to be determined, which scales rapidly with the problem size, especially given our tight evaluation budget due to the expensive scheduling-control solution at each iteration. On the other hand, while optimality surrogates present the more tractable alternative, they inevitably introduce approximation error. As such, the choice of optimality surrogates should be followed by strategies that mitigate approximation error, such as cross-validation of the surrogate's performance. Future work will apply this methodology to different case studies, investigate the potential of the surrogate approach to the planning-scheduling integration and of decision trees to be used as optimality surrogates.

5. Conclusion

We leverage DFO and optimality surrogates to reformulate a tri-level optimization problem into a single-level black-box optimization problem that we solve using the derivative-free optimizer Py-BOBYQA where each evaluation calls a scheduling instance with an embedded optimal control surrogate. We suggest hot-starting the solution to our problem using the solution to the planning-only optimization problem. As such, we use the remaining computational budget to approach the tri-level optimum from the hierarchical approach, which most accurately reflects decision-making. Nonetheless, we can always fall back on the 'status quo', i.e. the solution of the planning-only optimization problem from the sequential method. Our approach leads to tangible changes in the optimal production planning and scheduling sequence from the 'status quo'.

References

S. Avraamidou, E. N. Pistikopoulos, 2019, A Bi-Level Formulation And Solution Method For The Integration Of Process Design And Scheduling. Proceedings of the 9th International Conference on Foundations of Computer-Aided Process Design, 47, 17-22

B. Beykal, S. Avraamidou, E. N. Pistikopoulos, 2022, Data-driven optimization of mixed-integer bi-level multi-follower integrated planning and scheduling problems under demand uncertainty, Computers & Chemical Engineering 156, 107551

F. Ceccon, & J. Jalving, J. Haddad, A. Thebelt, C. Tsay, C. Laird, R. Misener, 2022, OMLT: Optimization & Machine Learning Toolkit, arXiv preprint arXiv:2202.02414

Y. Chu, F. You, 2015, Model-based integration of control and operations: Overview, challenges, advances, and opportunities, Computers & Chemical Engineering, 83, 2–20

L.S. Dias, M.G. Ierapetritou, 2020, Integration of planning, scheduling and control problems using data-driven feasibility analysis and surrogate models. Computers & Chemical Engineering, 134, 106714

S. Sachio, M. Mowbray, M. Papathanasiou, E.A. del Rio-Chanona, P. Petsagkourakis, 2021, Integrating process design and control using reinforcement learning. Chemical Engineering Research and Design, 183, 160-169

D. van de Berg, T. Savage, P. Petsagkourakis, D. Zhang, N. Shah, E. A. del Rio-Chanona, 2022, Data-driven optimization for process systems engineering applications. Chemical Engineering Science, 248, 117135

Antonis Kokossis, Michael C. Georgiadis, Efstratios N. Pistikopoulos (Eds.)
PROCEEDINGS OF THE 33rd European Symposium on Computer Aided Process Engineering (ESCAPE33), June 18-21, 2023, Athens, Greece
 http://dx.doi.org/10.1016/B978-0-443-15274-0.50104-9

A graph classification approach to determine when to decompose optimization problems

Ilias Mitrai,[a] Prodromos Daoutidis[a*]

[a]*Department of Chemical Engineering and Materials Science, University of Minnesota, Minneapolis, 55455 MN, USA*

[*] *Corresponding author, e-mail: daout001@umn.edu*

Abstract

Decomposition-based optimization algorithms are widely used for the solution of optimization problems. However, the choice of whether a decomposition-based solution approach should be selected over a monolithic one is not apparent in general. In this work, we propose a graph classification approach for determining a-priori when to use a decomposition-based solution approach for the solution of convex mixed integer nonlinear optimization problems. We apply the proposed approach to benchmark optimization problems and analyze the predictive performance of the classifier.

Keywords: Graph Classification, Decomposition-based solution algorithms, convex MINLP

1. Introduction

Decomposition-based solution algorithms have been widely used to solve optimization problems by exploiting the underlying structure of the problem. Although these algorithms are usually employed for the solution of large-scale problems (Conejo, 2006) for some cases, such as convex Mixed Integer Nonlinear Programs (MINLPs), they are more efficient than monolithic solution approaches even for problems with few variables and constraints (Duran and Grossmann, 1986). However, the effectiveness of a decomposition-based solution approach can only be determined after an optimization problem is solved with both approaches. Although this is a reasonable task for a simple problem, it is computationally expensive for large-scale problems. Hence, a method to determine a-priory when to apply a decomposition-based solution algorithm is necessary. In this work, we will pose this as an algorithm selection problem (Rice, 1976) in which given an optimization problem $\mathcal{P}$ and a set of algorithms $\mathcal{A} = \{\alpha_1, \alpha_2, ..., \alpha_n\}$, one must find the algorithm α^* that solves problem $\mathcal{P}$ in the minimum computational time. Mathematically this problem can be stated as follows:

$$\alpha^* \in \arg\min_{\alpha \in \mathcal{A}} m(\mathcal{P}, \alpha), \tag{1}$$

where $m : \mathcal{P} \times \mathcal{A} \rightarrow R$ is the solution time of problem $\mathcal{P}$ using algorithm α. This is a black box optimization problem since the function m is not known explicitly. Machine learning techniques have been used to approximate either the solution time or the solution of the problem itself with a surrogate model based on some features of the problem (Kerschke et al. (2019), Xu et al. (2011), Bonami et al. (2022), Kruger et al. (2017)). These approaches considered the selection of monolithic solution algorithms for Mixed

Integer Linear (MILPs) and Mixed Integer Quadratic Programs (MIQPs) and determined if a MILP should be solved using the Dantzig-Wolfe (DW) reformulation or Branch and Cut. In all these cases, the regression or classification is performed based on problem features obtained from the problem formulation and the preprocessing step. However, the features aggregate information about the problem and do not explicitly consider the structural coupling between the variables, which is the basis for the application of decomposition-based solution algorithms (Conejo (2006), Allman et al. (2019), Mitrai and Daoutidis (2021), Mitrai et al. (2022)).

In this paper, we propose a graph classification approach to determine if a convex MINLP should be solved using a monolithic or a decomposition-based solution approach. Specifically, we use graph convolutional neural networks (Kipf and Welling, 2016) to capture the structural and functional interactions among the variables of the problem. Given this representation, a graph classifier is trained to determine if a convex MINLP should be solved using Branch and Bound (B&B) (Gupta and Ravindram, 1985), a monolithic solution approach, or the Outer Approximation (OA) algorithm (Duran and Grossmann, 1986) a commonly used decomposition-based solution approach. The rest of the paper is organized as follows: In Section 2 we present the combined structural and functional graph representation of an optimization problem. In Section 3 we present the graph classification-based algorithm selection solution approach and in Section 4 we analyze the classifier's performance on benchmark convex MINLPs.

2. Graph representation of an optimization problem

We assume that the following convex MINLP must be solved:

$$\begin{aligned} \mathcal{P} : \min_{x,y} \ & f(x,y) \\ s.t. \ & g(x,y) \leq 0 \\ & x \in \mathbb{Z}^{n_x}, y \in \mathbb{R}^{n_y} \end{aligned} \tag{2}$$

where $n_x + n_y = n$, and g, f are convex in the variables y and in the relaxed integer variables x. Based on our previous work (Mitrai et al. (2022), Allman et al. (2019)), we will represent problem $\mathcal{P}$ as a graph.

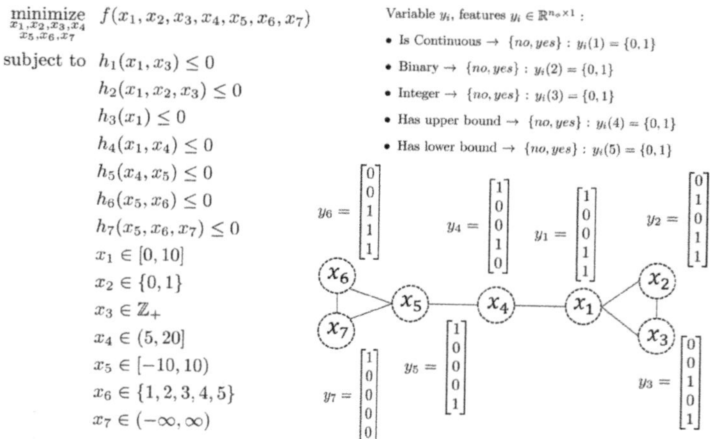

Figure 1: Graph and feature representation of an optimization problem

Three graphs can be generated to capture the structural coupling among the variables and constraints, the bipartite variable-constraint graph, the constraint graph, and the variable graph. For the rest of the paper, we will use the variable graph $\mathcal{G}_n(V_n, E_n)$ $(|V_n| = n)$, where every node represents a variable and the edges represent constraints that couple two variables. In this work, we extend the aforementioned representation to a combined structural and functional graph representation of an optimization problem. Specifically, for every node in a graph, we add a set of features that capture different characteristics of the node, such as the domain of the variable and the upper and lower bounds (see Fig. 1). These features form the feature matrix F. Given this representation, a graph is represented by the adjacency matrix A and the feature matrix F, i.e., $\mathcal{G}_n(A, F, E)$.

3. Algorithm selection via graph classification

3.1. Graph classification approach

Given an optimization problem $\mathcal{P}$ and the variable graph $\mathcal{G}_n(A, F, E)$ the goal is to develop a classifier $\mathcal{C}\colon F(\mathcal{P}) \times A(\mathcal{P}) \mapsto p \in \mathbb{R}^{N_a}$ to determine if problem $\mathcal{P}$ should be solved using Branch and Bound (B&B) or the Outer Approximation (OA) algorithm $(\mathcal{A} = \{B\&B, OA\}, |\mathcal{A}| = N_a = 2)$. The inputs in the classifier are the adjacency $A(\mathcal{P})$ and feature $F(\mathcal{P})$ matrix (which depend on the problem $\mathcal{P}$), and the output is a vector $p \in \mathbb{R}^{N_a}$, where p_i is equal to the probability that algorithm α_i solves the problem in the minimum computational time. Under this setting, the algorithm selection problem is transformed into a graph classification problem, where a given problem $\mathcal{P}$ is solved using the algorithm with the maximum probability $a^* = \arg\max p$ with $p = \mathcal{C}(F(\mathcal{P}), A(\mathcal{P}))$. The overall framework is presented in Fig. 2.

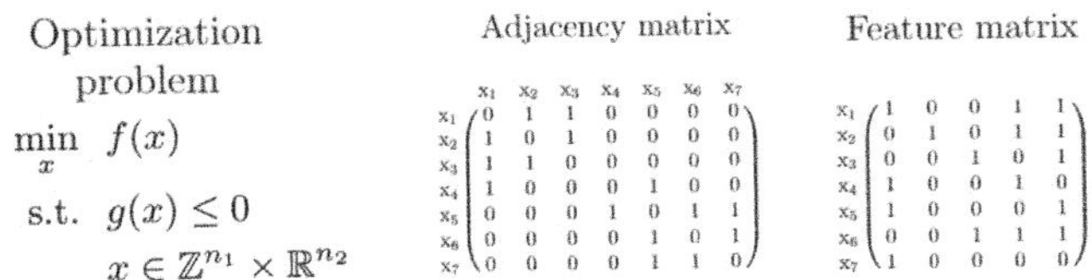

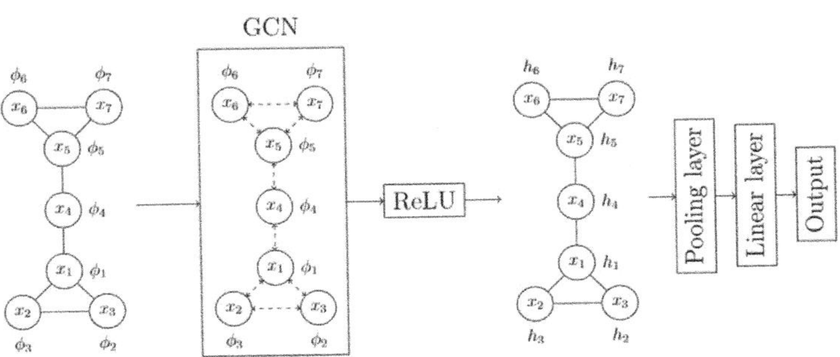

Figure 2: Learning when to decompose framework

3.2. Graph classifier architecture

The graph classification task is performed using graph convolutional neural (GCN) networks (Kipf and Welling, 2016). Given an optimization problem $\mathcal{P}_i$ (see Eq. 2) with N_v variables and N_m constraints, we can generate the variable graph $\mathcal{G}_v\,(V_v, E_v)$ $(|V_v| = N_v)$ and obtain the adjacency $A \in \mathbb{R}^{N_v \times N_v}$ and feature $F \in \mathbb{R}^{N_v \times N_\phi}$ matrices where N_ϕ

is the number of features per node. We define as ϕ_i the features of node i (ϕ_i is the ith row of matrix F). We will assume that L GCN layers are stacked sequentially and the features of the nodes are updated using a GCN layer as follows:

$$H^{(1)} = \sigma\left(\widehat{D}^{-\frac{1}{2}}\hat{A}\,\widehat{D}^{-\frac{1}{2}} F\, W^{(1)}\right)$$
$$H^{(l)} = \sigma\left(\widehat{D}^{-\frac{1}{2}}\hat{A}\,\widehat{D}^{-\frac{1}{2}} H^{(l-1)}\, W^{(l)}\right)\ \forall\, l \in \{2, L\},$$

where $l \in \{1, L\}$ denotes the layer number, σ is the activation function, $\hat{A} = A + I$ (I is the identity matrix of appropriate dimension), $\widehat{D}_{ii} = \sum_{j=0}^{n_v} \hat{A}_{ij}$ is the degree of node i and $\widehat{D} = diag(\{\widehat{D}_{ii}\}_{i=1}^{N_v})$. $W^{(1)} \in \mathbb{R}^{N_\phi \times N_h}$ are the weights for the first layer, $W^{(l)} \in \mathbb{R}^{N_h \times N_h}$ are the weights for layer l, and N_h is equal to the number of hidden features for each node. The output of the last layer L is passed through a pooling layer which creates a graph-level feature $r \in \mathbb{R}^{N_h \times 1}$ which is equal to

$$r_i = \frac{1}{N_v}\sum_{j=1}^{N_v} H_{ij}^{L} \quad \forall\, i = 1, \ldots, N_h.$$

Finally, the graph level feature is an input to a linear transformation layer, where the output is $y \in \mathbb{R}^{N_a \times 1}$ (y_i is the probability that algorithm i solves the problem in the minimum computational time) and is equal to $y = \Theta r + b$, where $\Theta \in \mathbb{R}^{N_A \times N_h}, b \in \mathbb{R}^{N_A \times 1}$ are parameters.

This classifier can classify problems with different number of variables and constraints since the pooling layer after the last convolution layer creates the vector r which has $N_h \times 1$ dimension. Therefore, for the final classification step, every problem is represented in the hidden features dimension space N_h.

3.3. Training of the graph classifier

We assume that a set of optimization problems $\{\mathcal{P}_i\}_{i=1}^{N_{data}}$ is available and for these problems we can obtain the variable graphs $\{\mathcal{G}_i\}_{i=1}^{N_{data}}$, the adjacency and feature matrix $\{A_i, F_i\}_{i=1}^{N_{data}}$, and labels $\{z_i\}_{i=1}^{N_{data}}$ where z_i is the algorithm that solves problem $\mathcal{P}_i$ in the minimum computational time. Given these data, the task is to learn the classifier's parameters that will maximize its accuracy by optimizing some loss function, such as the cross entropy.

4. Deciding when to decompose convex MINLPs

In this section, we apply the proposed approach to benchmark convex MINLPs from the MINLPLib (2022) to determine if a given convex MINLP should be solved using Branch and Bound (Gupta and Ravindram, 1985) or the Outer Approximation algorithm (Duran and Grossmann, 1986) as implemented in the BONMIN solver (Bonami et al., 2008).

4.1. Feature representation of the problem and data generation for classification

We will consider the following features for every node in the variable graph: variable domain (continuous, binary, integer) and bound (upper, lower). We use one-hot encoding for the representation of these features as follows:

$\phi(1) \in \{0,1\}$: 1 if the variable is continuous and 0 otherwise
$\phi(2) \in \{0,1\}$: 1 if the variable is binary and 0 otherwise

$\phi(3) \in \{0,1\}$: 1 if the variable is integer and 0 otherwise
$\phi(4) \in \{0,1\}$: 1 if the variable has an upper bound and 0 otherwise
$\phi(5) \in \{0,1\}$: 1 if the variable has a lower bound and 0 otherwise

Under this representation, every node has five features ($N_\phi = 5$).

4.2. Data generation for classification

The MINLPLib library contains 295 convex optimization problems which are solved with both algorithms (B&B and OA) using BONMIN 1.8.8 with a maximum computational time of 3000 seconds. From the 295 problems, 227 are solved with at least one solver (151 problems are solved faster with OA and 57 with B&B), and for each problem we obtain the adjacency and feature matrix (A_i, F_i). Overall, we obtain the dataset $\mathcal{D} = \{(A_i, F_i), y_i\}_{i=1}^{N_{data}}$, where 66% of the data points have label OA and 34% B&B.

4.3. Graph classification architecture and implementation

We split the dataset $\mathcal{D}$ at random into a training set and a testing set. The training set has 197 data points and the testing set has 30 data points picked at random (15 random data points have label OA and 15 have label B&B). The graph classifier has four convolution layers ($L = 4$), a global mean pooling layer, a linear layer (with dropout probability equal to 0.5), the number of hidden features is 12 ($N_h = 12$), and ReLU is used as the activation function. For the training, we use the Adam algorithm (Kingma and Ba, 2014) for 50 iterations with random initialization, the learning rate equal to 0.005, and batch size equal to 10. The loss function is the cross entropy and different weights are assigned to the two classes to account for the imbalance in the training dataset. The weights are computed as follows: $\omega_i = N_{data}\, N_{\alpha}^{-1}\left(\sum_{j=1}^{N_{data}} 1_i(y_j)\right)^{-1}$, where 1_i is the indicator function, i.e., $1_i(y_j) = 1$ if $y_j = i$ and 0 otherwise. The weight for class B&B is 1.6148 and for class OA is 0.7243. The GCN is implemented in PyTorch Geometric (Fey and Lenssen, 2019) and the training is done using PyTorch (Paszke et al., 2015).

4.4. Classification results

The performance metrics of the classifier on the testing data set are presented in Table 1.

Table 1: Performance metrics for the graph classifier on the testing data set

Accuracy	0.9	F1 score	0.9	Confusion matrix		
					Predicted label	
Label	Precision	Recall	F1 score	True label	OA	B&B
OA	0.83	1.00	0.91	OA	15	0
BB	1.00	0.80	0.89	B&B	3	12

From the results, we observe that the classifier has 90% accuracy and the F1 score is 0.9. A few data points with label B&B are misclassified as OA. This misclassification could be caused by the small size and the imbalance in the training set. Finally, we observe that the classifier can accurately predict all the problems that are solved faster with the OA algorithm.

5. Conclusions

The selection of the appropriate solution method for a general optimization problem can be a challenging and time-consuming step. In this work, we propose a graph classification approach to determine whether a convex MINLP should be solved using a decomposition-based or a monolithic-based solution approach by considering simultaneously the structural and functional coupling between the variables and the constraints of a problem. Application of this approach to benchmark problems shows that the graph classifier can achieve 90% accuracy. However, more data are necessary to reduce the imbalance in the dataset and improve the performance of the classifier.

6. Acknowledgment

This work was supported by the National Science Foundation (NSF-CBET, award number 1926303) and a Doctoral Dissertation Fellowship (DDF) from University of Minnesota

References

A. Allman, W. Tang, P. Daoutidis, DeCODe: a community-based algorithm for generating high-quality decompositions of optimization problems, Optimization and Engineering 20 (4) (2019) 1067–1084.

A. J. Conejo, E. Castillo, R. Minguez, R. Garcia-Bertrand, Decomposition techniques in mathematical programming: engineering and science applications, Springer Science & Business Media, 2006.

A. Paszke, et al., Pytorch: An imperative style, high-performance deep learning library, in: H. Wallach, H. Larochelle,A. Beygelzimer, F. d'Alch´e-Buc, E. Fox, R. Garnett (Eds.), Advances in Neural Information Processing Systems 32, Curran Associates, Inc.,2019, pp. 8024–8035.

D. P. Kingma, J. Ba, Adam: A method for stochastic optimization, arXiv, preprint arXiv:1412.6980 (2014).

I. Mitrai, W. Tang, P. Daoutidis, Stochastic blockmodeling for learning the structure of optimization problems, AIChE Journal 68 (6) (2022) e17415.

I. Mitrai, and P. Daoutidis, 2021. Efficient solution of enterprise-wide optimization problems using nested stochastic blockmodeling. Industrial & Engineering Chemistry Research, 60(40), pp.14476-14494.

J. R. Rice, The algorithm selection problem, in: Advances in computers,Vol. 15, Elsevier, 1976, pp. 65–118.

Kerschke, P., Hoos, H.H., Neumann, F. and Trautmann, H., 2019. Automated algorithm selection: Survey and perspectives. Evolutionary computation, 27(1), pp.3-45.

L. Xu, F. Hutter, H. H. Hoos, K. Leyton-Brown, Hydra-mip: Automated algorithm configuration and selection for mixed integer programming, in: RCRA workshop on experimental evaluation of algorithms for solving problems with combinatorial explosion at the international joint conference on artificial intelligence (IJCAI), 2011, pp. 16–30.

M. A. Duran, I. E. Grossmann, An outer-approximation algorithm for a class of mixed-integer nonlinear programs, Mathematical programming 36 (3) (1986) 307–339.

M. Fey, J. E. Lenssen, Fast graph representation learning with pytorch geometric, arXiv preprint arXiv:1903.02428 (2019).

M. Kruber, M. E. L¨ubbecke, A. Parmentier, Learning when to use a decomposition, in: International conference on AI and OR techniques in constraint programming for combinatorial optimization problems, Springer, 2017, pp. 202–210.

MINLPLib (2022) Mixed-integer nonlinear programming library. http://www.minlp lib.org/. Accessed 27 May 2022

O. K. Gupta, A. Ravindran, Branch and bound experiments in convex nonlinear integer programming, Management science 31 (12) (1985) 1533–1546.

P. Bonami, A. Lodi, G. Zarpellon, A classifier to decide on the linearization of mixed-integer quadratic problems in cplex, Operations Research (2022).

P. Bonami, L. T. Biegler, A. R. Conn, G. Cornu´ejols, I. E. Grossmann, C. D. Laird, J. Lee, A. Lodi, F. Margot, N. Sawaya, et al., An algorithmic framework for convex mixed integer nonlinear programs, Discrete optimization 5 (2) (2008) 186–204.

T. N. Kipf, M. Welling, Semi-supervised classification with graph convolutional networks, arXiv preprint arXiv:1609.02907 (2016).

Antonis Kokossis, Michael C. Georgiadis, Efstratios N. Pistikopoulos (Eds.)
PROCEEDINGS OF THE 33rd European Symposium on Computer Aided Process Engineering (ESCAPE33), June 18-21, 2023, Athens, Greece
 http://dx.doi.org/10.1016/B978-0-443-15274-0.50105-0

An Enhanced Group-Interaction Contribution Method for the Prediction of Freezing Point and Enthalpy of Fusion for Pure Organic Compounds

Khaled Mansour[a,(*)], Mourad Korichi[a]

[a]*Univ. Ouargla, Fac. des Sciences Appliquées, Lab. Dynamique, Interactionset Réactivité des Systèmes, BP 511 Ouargla, Algeria.*
khaledmansour47@yahoo.fr

Abstract

In this work, we developed new models using an enhanced group-interaction contribution method for the prediction of freezing point and enthalpy of fusion of pure organic compounds. These enhancements proved their efficiency by results presented on an earlier published paper, (Mansour and Korichi (2016)), which includes the extension of the training dataset by new experimental data, add 85 new structural groups and additional correction groups to distinct between isomers including Cis or Trans as well as Ortho, Meta or Para structures. 1124 and 501 experimental data points used for the new (T_{fp}) and (ΔH_f) models development and testing respectively.

Keywords: Thermophysical Properties Prediction, Group-Interaction Contribution, Freezing point, Enthalpy of Fusion, Pure Organic Compounds.

1. Introduction

Freezing point is of high importance property for pure organic compounds characterization; it is influenced by the presence of other substances (impurity) made it an important criterion for purity.

The enthalpy of fusion also known as the latent heat of fusion (ΔH_f) is an important physical property of the solid state, and can be useful in correcting thermo-chemical data to a standard state when combined with other thermodynamic properties, Wang et, *al.*, (2010).

The knowledge of these two properties for pure organic compounds is a great challenge for any new process and/or products design, especially when taking in consideration the enormous number of molecules, and not forgetting time and costs. An alternative solution available to solve the lake of experimental data and the request for these properties, prediction methods. Numerous based group-contribution methods and models developed for the prediction of freezing point and enthalpy of fusion in order to fill the gap between new available molecules and the properties values, Boethling et *al.*,(2000); Poling et *al.*, (2004).

Group contribution methods divide the molecules in to groups (fragments; atoms, bonds or group of atoms, etc) each group has a partial value contribute in the total value of the calculated property. These contributions calculated from known experimental data. Property of a compound obtained by a summing up the values of all contributions presented in the molecule via the developed mathematical model. Kolská, et *al.*, (2012).

Marrero and Pardillo (1999) proposed, the group-interaction contribution (GIC) which defined a fragment as the interaction between two simple groups e.g the interaction between $-CH_3$ and $>CH_2$ make a group, this approach was not applied for the development of models allowing the prediction of freezing point and enthalpy of fusion.

The appearance of new experimental data helps the development of new models, an enhanced group interaction contribution method published, Mansour and Korichi (2016) presenting new models using the (GIC) approach. In this paper, the enhanced approach applied for the development of new models for the prediction of freezing point and enthalpy of fusion at the freezing point.

2. Models development

2.1. Approach

The mean benefits of the new developed models are accuracy and isomers distinction (due to new available data, which conduct to new added groups). In this work, the (GIC) approach adopted for the prediction of these properties with some modifications in order to enhance the models prediction efficiency:

- Initially, after studying the utility of the groups, some of them were eliminated because they are useless, e.g. ($=CH_2$ and $=CH_2$), which appears only in ethene, this group has no use because it will not be used in the prediction of new compounds properties, and we have the properties of the mentioned compound.
- Secondly, modify some structural groups, such as ($-CH_3$ and -CN) and ($-CH_2$ and -CN), which they will be replaced by only one group (-C≡ and ≡N) to reduce the groups number in order to simplify the method.
- Then, add new groups according to the new available data in our database taking in consideration the approach philosophy in generating groups, for example (>CH- and -Br), (>CH- and -C≡) are new added groups.
- Also, add some groups based on the functional chemical groups e.g. –OH and –OR for the distinction between alcohols and ethers, O=C-OH and O=C-O-R for the distinction between acids carboxylic and esters...etc, to enhance the method isomers distinction ability.
- Finally, introduce the differentiation between the aromatic and double bonds, which was not the case for the method of Marrero and Pardillo (1999).

2.2. Groups selection

Selection of contributions is based mainly on the interaction between two groups and the type of this interaction (bonds type; simple, double, triple or aromatic), these groups are similar to the groups used by Marrero and Pardillo (1999) with new added groups according to the modifications mentioned above, Fig. (1) shows the group's selection methodology. The groups used for the generation of contributions illustrated in table (1).

Fig. 1: Groups selection methodology.

Table 1: Groups used for the generation of group-interaction ((a), aromatic; (s), single; (d), double; (r), rig. For the correction groups, we have proposed cis, trans, ortho, meta and para.)

Groups with single bond					
-CH_3	>CH_2	>CH–	>C<		
Groups with double bond					
=CH_2	=CH- (s)	-CH= (d)	=C< (s)	>C= (d)	=C=
Groups with triple bond					
≡CH	≡C–				
Groups within rig with single bond					
>CH_2(r)	>CH-(r)	>C<(r)			
Groups within rig with double bond					
=CH-(s) (r)	-CH= (d) (r)	>C= (d) (r)			
Groups within aromatic rig					
=CH– (a)	=C< (a)				
Oxygen group with single bond	**Amine group with single bond**				
–O–	>NH (s)				
Sulfur group with single bond	**Amine group within rig with double bond**				
–S– (s)	–N= (r)				
Acetone group within rig	**Acetone groups with single bond**				
>CO(r)	>CO				
Isocyanate group with double bond					
=N- (s)	–N= (d)				

2.3. Database

Special care was taking while the collection of the experimental data by choosing the right sources, the book of Yaws and Narasimhan (2008) was the main sources of our data base, beside, other sources were also used as verification for the database; table (2) resumes the available data points for each property, for every class.

Table 2: Properties and their experimental data points in the database

Class*	CH	O	N	S	H	multi	Total
T_{fp}	422	299	123	60	192	29	1124
ΔH_f	265	95	37	33	62	9	501

*CH, hydrocarbons; O, oxygenated; N, nitrogenated; S, sulfur containing; H, halogenated; multi, multifunctional compounds.

2.4. Proposed models

Two types of regression used for the obtaining of the models. The first is a nonlinear regression using 100% of the data points, which allows as getting the adjusted parameters a_1 and b_1. The second one is a linear regression, the data points was divided in two parts, 90% was used as training set for the obtaining of the contributions values and the adjusted parameters c_1 and c_2, and 10% reserved for the testing of models performances.

Several models with different types of functions passed through the processes of models development and testing as mentioned above, and the adopted models for the prediction of (T_{fp}) and (ΔH_f) are respectively equations (1) and (2):

$$\frac{T_{fp}}{(K)} = a_1 * \left(\sum_i N_i C_i + c_1\right)^{b_1} \quad (1)$$

Where a_1=10.944, b_1= 1.522 and c_1=3.6155

$$\frac{\Delta H_f}{(kJ/mol)} = T_{fp} * \left(\sum_i N_i C_i + c_2\right)^2 \quad (2)$$

Where c_2=0.14327

3. Results and discussion

The mean tools for the validation of our models are: MAPE; Mean Absolute Percentage Error, MAE; Mean Absolute Error and SD; Standard Deviation Qiang et *al.* (2009) which also will be used for the judgment of the proposed models performances, beside the two determination coefficients R^2 and adjusted R^2. The values of these parameters calculated for the total data points and for each class, as illustrated in table (3).

Table 3: Performances of the proposed models for the prediction of T_{fp} and ΔH_f

	T_{fp}				ΔH_f			
Class	Data points	MAPE (%)	MAE (K)	SD (K)	Data points	MAPE (%)	MAE (kJ/mol)	SD (kJ/mol)
CH	421	8.19	14.76	18.48	265	17.32	1.24	1.63
O	299	6.44	14.79	18.77	95	10.25	1.07	1.56
N	123	6.12	14.55	18.74	37	10.11	1.15	1.6
S	60	6.93	11.82	15.90	33	10.7	0.84	1.3
H	192	8.27	16.76	20.30	62	17.76	1.01	1.43
Multi	29	4.60	11.69	15.68	9	11.06	1.31	2.03
Testing set	112	8.35	16.78	20.36	50	16.52	1.55	1.98
Total	1124	7.35	14.85	18.72	501	14.95	1.15	1.58
	R^2 = 0.901		Adj. R^2 = 0.879		R^2 = 0.905		Adj. R^2 = 0.868	

The new proposed models present accurate results for all the classes according to the validation parameters presented in the table (3), the hydrocarbon classes for both models offer the most accurate results among all the classes.

The value of the two determination coefficients R^2 and adjusted R^2 for the two models are very high and close to one unit, which represent another indication for the accuracy and reliability of the suggested models.

Beside validation parameters, distributions between experimental values and estimated values for both properties presented in Fig. (2) as validation tool for the chosen models. As illustrated, we can observe the accuracy obtained using the proposed models, these figures testify to an excellent reliability presented by the new models.

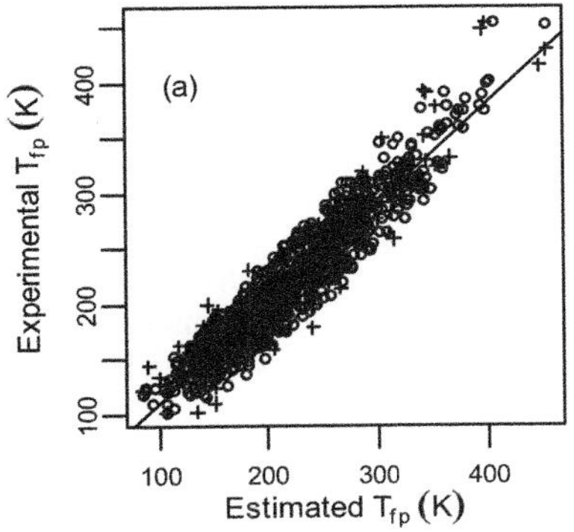

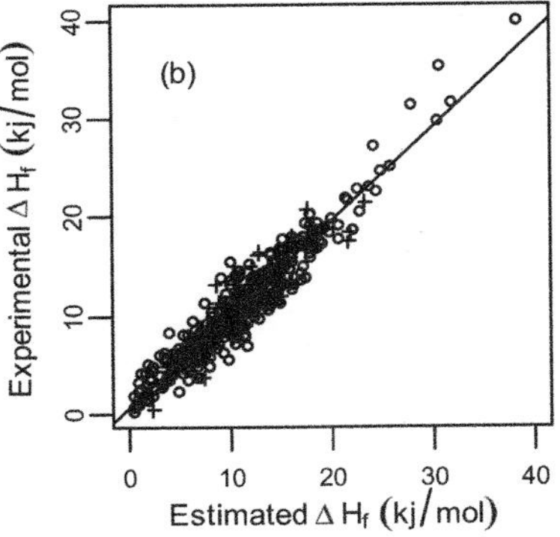

Fig. 2: Distributions of experimental and estimated values for (a) T_{fp}; (b) ΔH_f for training set (o) and testing set (+).

Table 4: Comparison between our model and earlier published models for the prediction of T_{fp}

Methods	Data Points	Groups number	MAPE (%)	MAE (K)	SD (K)
Constantinou and Gani (1994)	312	66 1st order	8.90	17.39	22.51
		41 2sd order	7.23	14.03	18.28
Marrero and Gani (2001)	1547	166 1st order	9.30	24.90	33.87
		117 2sd order	7.90	21.41	29.52
		62 3rd order	7.60	20.22	27.67
Wang, et al., (2008) CPG	730	93	6.67	14.46	NA
This work	**1124**	**203**	**7.35**	**14.85**	**18.71**

Table 5: Comparison between our model and earlier published models for the prediction of ΔH_f

Methods	Data points	Groups number	MAPE (%)	MAE (kJ/mol)	SD (kJ/mol)
Marrero and Gani (2001)	711	131 1st order	18.30	2.58	4.16
		68 2sd order	16.40	2.32	3.88
		40 3rd order	15.70	2.17	3.65
This work	**501**	**143**	**14.95**	**1.15**	**1.58**

Tables 4 and 5 presents the comparison between the proposed models and the models available in the literature, using the validation parameters MAPE, MAE and SD. Although, the difference between the accuracy of new models and old ones is not great, but the extension of the application range and isomers distinction gives more power to the new models, beside the simplicity of the method.

4. Conclusion

We presented new models allowing the estimation of (T_{fp}) and (ΔH_f); these models developed via an enhanced group-interaction contribution method. Comparing with other methods available in literature, the proposed models showed excellent results concerning reliability; accuracy and application range.

Although, the (GIC) method has no theoretical background on how the groups are selected, but it bases on the bound type as criteria for the groups selection, which enhance its performance concerning the distinction between structural isomers Cis, Trans and Ortho, Meta and Para, which offers more accuracy for the prediction of the mentioned properties.

5. References

B.E. Poling, J.M. Prausnitz and J.P. O'Connell, 2001, The Properties of Gases and Liquids, 5th Edition, Mc GRAW HILL, New York.

C.L. Yaws and P.K. Narasimhan, 2008, Critical properties and acentric factor-organic compounds, Thermophysical Properties of Chemicals and Hydrocarbons, Texas, William Andrew, pp 01-809.

K. Mansour and M. Korichi. 2016, An Enhanced Group-Interaction Contribution Method for the Prediction of Normal Boiling and Critical points, 26th European Symposium on Computer Aided Process Engineering 1st Edition, Vol, 38, pp1237-1242.

K.G. Joback and R.C. Ried, 1987, Estimation of Pure-Component Properties from Group-Contributions, Chemical Engineering Communications, Vol.57, pp 233-243.

Kolská, Z., Zábranský, M., & Randova, A. (2012). Thermodynamics - Fundamentals and Its Application in Science. IntechOpen. Chapter 6: Group Contribution Methods for Estimation of Selected Physico-Chemical Properties of Organic Compounds. pp. 135-162

L. Constantinou and R. Gani, 1994, New Group - Contribution Method for the Estimation of Properties of Pure Organic Compounds, AIChE Journal, Vol 10, pp 1697-1710.

M.J. Marrero and F.E. Pardillo. 1999. Estimation of Pure Compounds Properties Using Group-Interaction Contribution, AIChE Journal, Vol 45, pp 615-621.

R.S. Boethling, D. Mackay and with a foreword by W.J. Lyman, 2000, Handbook of Property Estimation Methods for Chemicals: Environmental and Health Sciences, CRC Press LLC. Florida.

W. Qiang, J. Qingzhu, Z. Zhenyan and D.Yu, 2010, Position Group Contribution Method for the Prediction of the Fusion Enthalpy of Organic Compounds, 2010 AIChE Annual Meeting, New York, USA.

W. Qiang, M. Peisheng and N. Shifeng, 2009, Position Group Contribution Method for Estimation of Melting Point of Organic Compounds, Chinese Journal of Chemical Engineering, Vol 17, N° 3, pp 468-472.

Antonis Kokossis, Michael C. Georgiadis, Efstratios N. Pistikopoulos (Eds.)
PROCEEDINGS OF THE 33rd European Symposium on Computer Aided Process Engineering (ESCAPE33), June 18-21, 2023, Athens, Greece
 http://dx.doi.org/10.1016/B978-0-443-15274-0.50106-2

Single and Multi-Objective Superstructure Optimization of an Integrated Continuous Multistage Reaction-Crystallization-Filtration Process with Recycles

Jiaxu Liu, Brahim Benyahia*

Department of Chemical Engineering, Loughborough University, Loughborough, LE11 3TU, United Kingdom

B.Benyahia@lboro.ac.uk

Abstract

A superstructure optimization methodology is developed to identify optimal design and operation strategies of a combined continuous multistage reaction and crystallization process. The production of aspirin is used as a case study. The main objective is to maximize yield and minimize wastes while meeting tight critical quality constraints. The synthetic pathway involves a two-step reaction, which runs continuously, followed by a multistage seeded continuous cooling antisolvent crystallization process, and finally a wash filtration stage. To reduce the environmental footprint of the process and enhance its circularity, the filtrate (mother liquor) is purified using a multistage purification process then made available as recycle streams to the reaction and crystallization steps. The superstructure also considers that part of the filtered crystals can be recycled to the crystallization stages. A purge is used to control the level of impurities present in the system. Based on the proposed approach, several optimization scenarios were developed with a large set of decision variables, including jacket temperatures, antisolvent flow rates, recycle flow rates, residence times, and recycled crystal mass flow rate. The objective functions include maximizing the mean crystal size, and yield and minimizing the coefficient of variation (CV). It was shown that a maximum mean crystal size of 628 μm, a maximum yield of 87.91% and a minimum CV of 0.3596 can be achieved using the single objective optimization options. In addition, a multi-objective mixed integer nonlinear optimization problem was proposed to identify the best design and operation compromises captured by the Pareto front. Finally, a multicriteria decision-aiding approach based on the multi-attribute utility theory was used to rank the optimal Pareto trade-offs and help identify the most suitable solution.

Keywords: Multistage Reaction and Crystallization, Superstructure Optimization, MINLP, Multi-Objective Optimization, Multicriteria Decision Aiding.

1. Introduction

Over the last decade, continuous processing and flow chemistry have gained increasing interest in pharma as the most cost-effective, flexible, and environmentally friendly operation option. Despite the increased adoption of continuous processing, the development of fast and systematic approaches to synthesize integrated continuous processes and the identification of robust operating strategies is still an open topic in the pharmaceutical industry. Despite the high costs of pharma grade raw materials, which includes high purity reagent, intermediates and solvents, the implementation of recycles to rationalize resources consumption and minimize wastes is still not a common practice in pharma, and only a few investigations have been published to date (Benyahia 2018).

However, the presence of impurities and the inherent stringent regulatory requirements make the design and implementation of recycling even more challenging in pharma. As such, the development of more rigorous and systematic design and operation strategies is critical to demonstrate the benefits of the recycles and reduce inherent risks.

Pharmaceutical upstream processing, also referred to as drug substance manufacturing, commonly involves several reaction and purification steps which often involve a crystallization step followed by a wash-filtration. Often, the filtrate (mother liquor), which in the current practice is considered as waste, contains significant amounts of residual API, reagents, and solvents. However, the filtrate may also contain substantial amounts of impurities. Consequently, the design of an effective and environmental process with recycles requires effective purification technologies of the filtrates, and most importantly, optimal and robust design and operating strategies.

To address the optimal design problems from a more general perspective while embedding circularity principles, several operating and design options were introduced to the system along with systematic recycling strategies, which include recycling both solid products (crystals) and purified filtrate (mother liquor). In addition, optimal seeding policies were also considered as a possible option in all crystallization stages. These generic design options provide a solid basis for a systematic and effective process synthesis of multistage reaction and crystallization processes based on a superstructure optimization framework. The objective is to maximize flexibility and enhance the targeted key performance indicators such as the mean crystal size, yield, and coefficient of variation (CV). The generic optimal design problem, which captures all possible design scenarios, will be introduced in the following part. The resulting optimization problem is a mixed integer nonlinear programming problem (MINLP) which will deliver the optimal design options and the most effective recycling strategies. In addition to the single objective MINLP problem, a multi-objective MINLP is developed to identify the best design and operation compromises. Furthermore, the resulting set of optimal Pareto solutions is analyzed and ranked based on a multicriteria decision-aiding method to help determine the best design and operation option.

2. Method

The synthesis and purification of aspirin is used as a case study. This involves two continuous reaction steps (Joiner, Billeter et al. 2014), a multistage continuous crystallization (Liu and Benyahia 2022), and a filtration stage. The mathematical models and experimental data were obtained from the literature and the proposed superstructure is shown in figure 1. The fresh feed and antisolvent are added to each stage separately at a fixed total mass flowrate, used as a constraint in the optimization problems. Crystal seeds and recycles (both liquid and crystals) can be introduced at any crystallization stage. Before recycling, the filtrate or mother liquor, obtained at the filtration stage, goes through a multistage membrane purification process to increase the concentration of API (Espinoza-Gomez, Saucedo-Castillo et al. 2018) but must be delivered at undersaturated level to avoid crystallization in the recycle stream. A purge stream is also used to control the level of impurities in the process.

Combining the reaction and crystallization allows us to investigate the impact of recycling unreacted reagents and even API and solvents on the overall process performance. Here the recycle is connected to the second reactor due to the fact that reaction 1 is a water-free reaction and the addition of water will significantly reduce reaction yield. The potential advantage is to reduce the antisolvent (water) in the recycled mother liquor and recycle the API in the mother liquor, which will increase the overall yield and decrease the SA waste (all added water will end up in the SA waste stream).

Based on these generic settings, a series of optimization scenarios were developed by manipulating a large set of decision variables, namely: the jacket temperatures, antisolvent flow rates, feed flow rates, seeding policies, liquid recycles, crystals recycles, and a set of binary (integer) variables which determine whether a stream is required or not (0/1). The objective of the superstructure optimization problem is to maximize the mean crystal size, yield or minimize the CV. These criteria are optimized individually using single objective optimization strategies, and then a combination of these criteria is addressed using a multi-objective optimization approach to deliver the best design and operation compromises. Finally, the Pareto optimal solution are classified using the Multi-Attribute Utility Theory (MAUT) (Benyahia et al. 2011).

2.1. Formulation of the optimization problems

Firstly, a MINLP is proposed based on the generic flow diagram depicted in figure 1. Each of the recycles, fresh feeds and seeding streams is allocated a real variable, to determine the required flow rate, and an integer variable, to determine whether the stream is needed. The mathematical formulation of the optimization problem is shown in equation 1. The objective function can be the mean crystal size, CV, Yield (for single objective optimization) or a combination of the above (multi-objective optimization).

In the mathematical formulation below (equation 1), $T_{J,i}$ is the temperature of jacket temperature (°C), $F_{AS,i}$ is the antisolvent flow rate (g/min), $F_{feed,i}$ is the feed flow rate (g/min), and $F_{seed,i}$ is the suspended seed flow rate (g/min). $\mu_{0,seed}$, d_{seed} are the number density of the suspended seed solution and crystal size (μm) respectively. $R_{l,i}$, $R_{p,i}$ are the recycles flow rate of mother liquor (g/min) and crystals (g/min). I_i is the binary (integer) variable associated with the ith stage.

The constraints C1 to C9 represent the upper bound and lower bounds of the decision variables. C10 is used to allow cooling only, and C11-C13 defined the final constraints of the jacket temperature and antisolvent ratio. The supersaturation should be maintained above 1 during the process as no dissolution is allowed.

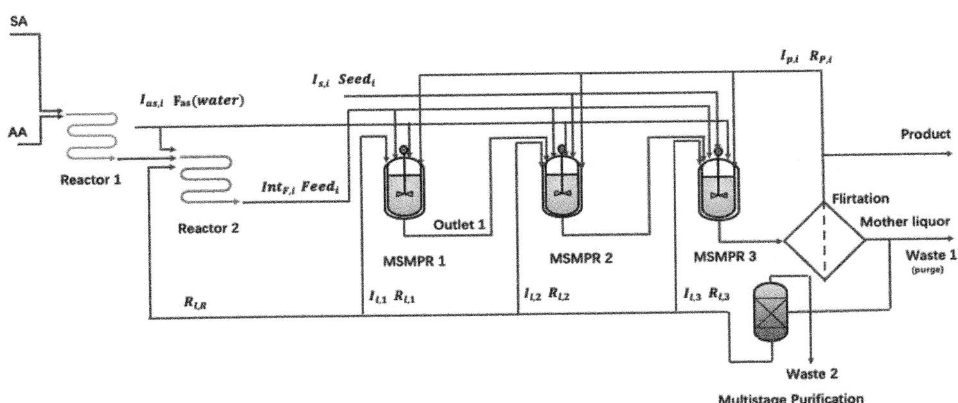

Figure 1 Generic superstructure of the multistage reaction and purification process with recycles.

$$\min_{T_{J,i},F_{AS,i},F_{feed,i},F_{seed,i},\mu_{0,seed},d_{seed},R_{l,i},R_{p,i},I_{S,i},I_{F,i},I_{as,i},I_{p,i},I_{l,i}} \{-d, -Yield, CV\}$$
$$\begin{aligned}
s.t. \quad & \dot{\boldsymbol{x}} = g(\boldsymbol{x},\boldsymbol{y},\boldsymbol{u},\boldsymbol{p},t) \qquad \boldsymbol{x}_{t=0} = \boldsymbol{x}_0 \\
& 0 = h(\boldsymbol{x},\boldsymbol{y},\boldsymbol{u},\boldsymbol{p},t) \\
& C1{:}\ 20 \le T_{J,i} \le 40 \\
& C2{:}\ 0 \le F_{AS,i} \le 20 \\
& C3{:}\ 0 \le F_{feed,i} \le 30 \\
& C4{:}\ 0 \le F_{seed,i} \le 20 \\
& C5{:}\ 0 \le \mu_{0,seed} \le 10^6 \\
& C6{:}\ 0 \le R_{l,j} \le 30 \\
& C7{:}\ 0 \le R_{p,i} \le 5 \\
& C8{:}\ 50 \le M_{l,max,k} \le 500 \\
& C9{:}\ 50 \le M_{l,max,3} \le 10000 \\
& C10{:}\ T_{i+1} \le T_i \\
& C11{:}\ \omega_{as,k} \le 60\% \\
& C12{:}\ \sum F_{feed,i} = 25 \\
& C13{:}\ \omega_{as,3} = 60\% \\
& C14{:}\ S_i \ge 1 \\
& C15{:}\ I_{S,i}, I_{F,i}, I_{as,i}, I_{p,i}, I_{l,i} = 0 \text{ or } 1 \\
& i = 1,2,3,\ k = 1,2
\end{aligned} \tag{1}$$

3. Results and discussion

3.1. Single objective optimization

The MINLP problem described above was solved using a genetic algorithm (GA) in MATLAB. The maximum mean crystal size of 628 μm, 87.91% yield, and 0.36 CV were obtained based on the three single objective optimization scenarios. Here, for the sake of brevity, only the first scenario which maximizes the mean crystal size, is discussed in detail. The corresponding optimal process design shows that the optimal solution, shown in Table 1, indicates that all fresh feed must be delivered to the first crystallization stage along with the entire liquid recycle stream (purified mother liquor) to promote growth. Compared to the feed, the purified mother liquor has a lower API concentration. Therefore, its introduction into the second and third stage may dilute the crystallization medium and significantly reduce supersaturation, which in turn leads to reduced crystal growth. The optimal design scenario, in the case maximization of the mean crystal size, recommends that larger residence times are required in the second and third crystallization stages to promote growth, while the first crystallization is used as a nucleator. Furthermore, the crystals recycle streams are implemented at all crystallization stages at their maximum allowed flow rates (upper bound) to prioritize the production of very large crystals. However, the implementation of the crystal recycles strongly depend on the targeted CV value and may be removed under different CV considerations.

3.2. Multi-objective optimization

A multi-objective optimization problem is developed to optimize three critical objectives to help identify the best design and operation compromises. Here, the mean crystal size and yield are maximized while the CV is minimized. The optimization problem is subject to the same set of constraints as those in the problem formulated above. The corresponding Pareto solutions are shown in the 2D bubble map, where the grey scale and

Table 1. Optimal solutions (real and integer variables) for scenario 1.

	T (C°)	F_{as} (g/min)		R_{ml} (g/min)		R_p (g/min)		F_{feed} (g/min)		M_{max}(g)	F_{seed}(g/min)	
MSMPR 1	$T_{j,1}$	$I_{as,1}$	$F_{as,1}$	$I_{l,1}$	$R_{ml,1}$	$I_{p,1}$	$R_{p,1}$	$I_{F,1}$	$F_{feed,1}$	$M_{max,1}$	$I_{S,1}$	$F_{seed,1}$
	33.87	1	9.81	1	0.98	1	4.94	1	20.83	354.56	1	1.44
MSMPR 2	$T_{j,2}$	$I_{as,2}$	$F_{as,2}$	$I_{l,2}$	$R_{ml,2}$	$I_{p,2}$	$R_{p,2}$	$I_{F,2}$	$F_{feed,2}$	$M_{max,2}$	$I_{S,2}$	$F_{seed,2}$
	33.45	1	5.90	0	0	1	5.00	0	0	500	0	0
MSMPR 3	$T_{j,3}$	$I_{as,3}$	$F_{as,3}$	$I_{l,3}$	$R_{ml,3}$	$I_{p,3}$	$R_{p,3}$	$I_{F,3}$	$F_{feed,3}$	$M_{max,3}$	$I_{S,3}$	$F_{seed,3}$
	27.00	1	2.81	1	0.01	1	4.97	0	0	1000	1	2.95
Optimal Seeding properties	μ_0					23909						
	size (μm)					24.40						
	Mean crystal size (μm)					628						

bubble size represents the mean crystal size (third objective function). To help the decision maker identify few alternative optimal designs and operation options for the experimental verification, the multi attribute utility theory (MAUT) is used to rank all Pareto optimal solutions. The 210 Pareto solutions were ranked according to the utility function scores. The size of the circle represents the mean crystal size. It can be found in the 2D Pareto solutions distribution shape (figure 2 (largest)) that the mean crystal size and CV performance go in opposite directions. The crystals lie at the right side of the shape, where CV is the worst. However, the yield doesn't vary significantly but overall stays within acceptable range.

3.3. Combined multistage reaction and crystallization process

Here the reactions stages are also optimized. By adding the ratio of SA and AA, represented by r, as one of the decision variables, the conversion of SA can be manipulated. With the optimal operation profile, a mean crystal size of 307 μm can be obtained based on the proposed superstructure optimization.

Compared to scenario 1 (maximize mean crystal size), the optimal solution requires higher seeding flowrates, but the required seed concentration/density is lower compared to the previous case. It is because the seed is suspended in a saturated solution, which has a higher solubility than the outlet flow from the reactor. It means that the dissolved API is a priority to generate more growth. Besides, the optimal solution also indicates that recycling the purified mother liquor is not advised to maximize the mean crystal size

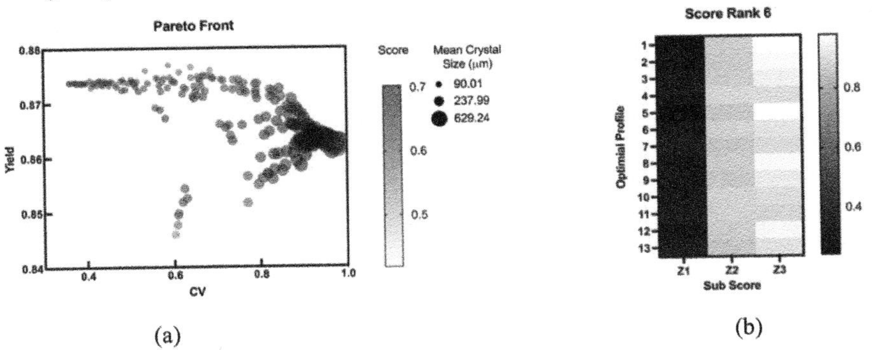

Figure 2 Ranking of the Pareto solutions based on MAUT (a) and sub-scores (b)

when the optimization is based on the combined reaction and crystallization. Because of the conflicting effect between a better reaction conversion and better crystallization yield, the recycle stream may dilute the crystallization system more significantly far beyond what was observed in scenario 1, when the concentration of API is higher than the nominal/initial concentration in the outlet flowrate (fresh feed) from the reactor.

4. Conclusion

A superstructure optimization approach was proposed to design an integrated reaction-crystallization-filtration process with recycles. All possible options for the recycle streams, crystal seeding, and fresh feed additions were considered and allocated binary and real decision variables. Other critical decision variables were also considered which include the design and operating options such as cooling temperatures and crystallizer capacity (residence time). Firstly, single objective optimization problems were solved separately to (1) maximize mean crystal size, (2) minimize the CV, (3) maximize yield. This was followed by a combination of three objectives to address a mixed integer multi-objective optimization problem. The resulting Pareto solutions were analyzed and ranked using MAUT. The exploration of the combination of the reaction and crystallization also indicate that recycling the purified mother liquor to the second and third stage in crystallization are not recommended. And with the recycles to the reactor, the mean crystal size was maximized to 307 μm with less water added in the second reactor. The proposed systematic approach shows that wastes can be significantly reduced, and resources consumption may be rationalized without compromising on product quality.

Acknowledgements

The authors acknowledge funding from the UK Engineering and Physical Sciences Research Council (EPSRC), for Made Smarter Innovation – Digital Medicines Manufacturing Research Centre (DM2), EP/V062077/1.

References

Benyahia, B. (2018). Applications of a plant-wide dynamic model of an integrated continuous pharmaceutical plant: Design of the recycle in the case of multiple impurities. Computer Aided Chemical Engineering, Elsevier B.V. **41:** 141-157.

Benyahia, B., M. A. Latifi, C. Fonteix and F. Pla (2011). "Multicriteria dynamic optimization of an emulsion copolymerization reactor." Computers & Chemical Engineering **35**(12): 2886-2895.

Espinoza-Gomez, H., E. Saucedo-Castillo, L. Z. Flores-López, E. Rogel-Hernandez, M. Martínez and F. T. Wakida (2018). "Ethanol:water blends separation using ultrafiltration membranes of poly(acrylamide-co-acrylic acid) partial sodium salt and polyacrylamide." The Canadian Journal of Chemical Engineering **96**(3): 763-769.

Joiner, D. E., J. Billeter, M. E. P. Mcnally, R. M. Hoffman and P. J. Gemperline (2014). "Comprehensive kinetic model for the dissolution, reaction, and crystallization processes involved in the synthesis of aspirin." Journal of Chemometrics **28**(5): 420-428.

Liu, J. and B. Benyahia (2022). "Optimal start-up strategies of a combined cooling and antisolvent multistage continuous crystallization process." Computers and Chemical Engineering **159**.

Mencarelli, L., Q. Chen, A. Pagot and I. E. Grossmann (2020). "A review on superstructure optimization approaches in process system engineering." Computers and Chemical Engineering **136**.

Antonis Kokossis, Michael C. Georgiadis, Efstratios N. Pistikopoulos (Eds.)
PROCEEDINGS OF THE 33rd European Symposium on Computer Aided Process Engineering (ESCAPE33), June 18-21, 2023, Athens, Greece
 http://dx.doi.org/10.1016/B978-0-443-15274-0.50107-4

Prediction of product distribution using machine learning techniques

Praveen Kumar Bommineni,[a] Manohar Kakunuri,[a] Sarat Babu Anne,[a]

[a]*Department of Chemical Engineering, National Institute of Technology Warangal, Warangal – 506004, India*

Abstract

The product distribution of multi-product processes involving series-parallel reactions depends on several variables. Hence, traditional experimental and computational approaches are laborious, time consuming and the results may not be satisfactory. Data driven approaches serve as an alternative to the traditional approaches in predicting and optimizing the product distribution. Machine learning (ML) provides reliable and accurate predictions. In this work, application of ML techniques to predict product distribution of one such complex process is demonstrated. Chlorination of methane has been chosen as case study for this purpose. Data required for the ML models is generated by simulating the process flow sheet in aspenONE. Of the several ML models used, k-nearest neighbors (kNN) method provides most accurate product distribution with high R-square value and low mean absolute error (MAE).

Keywords: Series-parallel reactions, product distribution, aspenONE, Machine Learning.

1. Introduction

Reactions and processes are typically desired to operate at optimum conditions of process variables such as temperature, pressure, flow rates, compositions etc. (Tabor 2018, Gupta 2021). Therefore, the chief goal is to find the optimal conditions so as to maximize performance in terms of yields, purity etc. However, it is not a trivial task to find optimum conditions even at lab scale as many variables/parameters effect the performance and it requires iterative costly procedures. Data-driven approaches serve as an alternative and reliable way to achieve the same (Shields 2021, Marcus 2021, Hein 2021, Joung 2021).
Machine learning is proving to be useful tool for applications in various fields including materials discovery from failed experiments, optimizing reaction conditions, reaction classification, synthesis of small molecules (Park 2022, Raccuglia 2016, Zunyun 2020, Schwaller 2021, Probst 2022, Segler 2018). In this article we demonstrate prediction of product distribution using ML techniques. First we present the process simulation followed by application of machine learning techniques to predict product flow rates and compositions.

2. Methane chlorination process simulation

2.1. Steady state process simulation

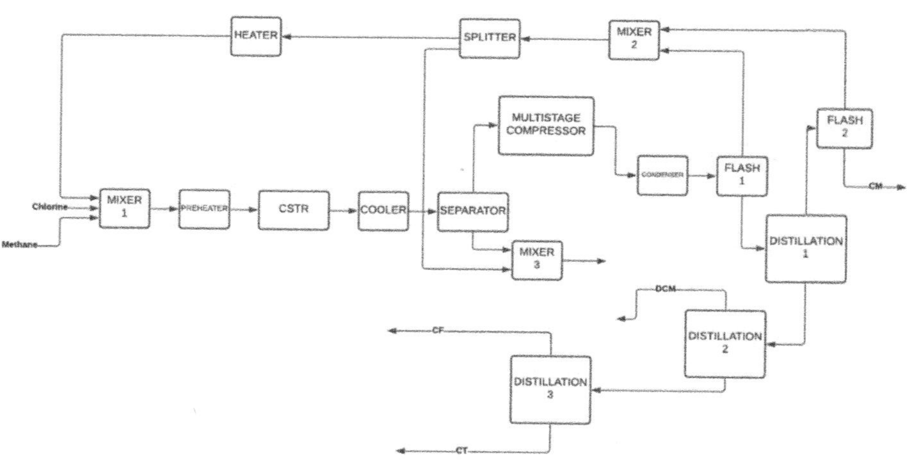

Figure 1. Methane chlorination process block diagram

Methane chloride can be produced by hydrochlorination of methanol (Thyagarajan 1966, Habata 1976) and chlorination of methane (Dantus 1996, Dantus 1999) processes. In this work, chlorination of methane which is one of the widely studied systems since 1930's (Robert 1931, Dantus 1996, Dantus 1999, Rozanov 2010) has been simulated. For industrial application, thermal chlorination is generally employed which is highly exothermic in nature. Methane undergoes following chlorination reactions to produce chloromethane (CH_3Cl), dichloromethane (CH_2Cl_2), trichloromethane ($CHCl_3$) and tetrachloromethane (CCl_4).

$$CH_4 + Cl_2 \rightarrow CH_3Cl + HCl \quad \Delta H = -103.5 \text{ kJ/mol} \quad \rightarrow (1)$$

$$CH_3Cl + Cl_2 \rightarrow CH_2Cl_2 + HCl \quad \Delta H = -102.5 \text{ kJ/mol} \quad \rightarrow (2)$$

$$CH_2Cl_2 + Cl_2 \rightarrow CHCl_3 + HCl \quad \Delta H = -99.2 \text{ kJ/mol} \quad \rightarrow (3)$$

$$CHCl_3 + Cl_2 \rightarrow CCl_4 + HCl \quad \Delta H = -94.8 \text{ kJ/mol} \quad \rightarrow (4)$$

Of these four reactions, the main reaction that takes place is methane and chlorine reacting to form methyl chloride. Based on the operating conditions, the product distribution varies by a wide margin. The data required for ML techniques is generated by simulating the process flow sheet in aspenONE (Dantus 1996, Dantus 1999). Methane chlorination process block diagram is shown in Figure 1.

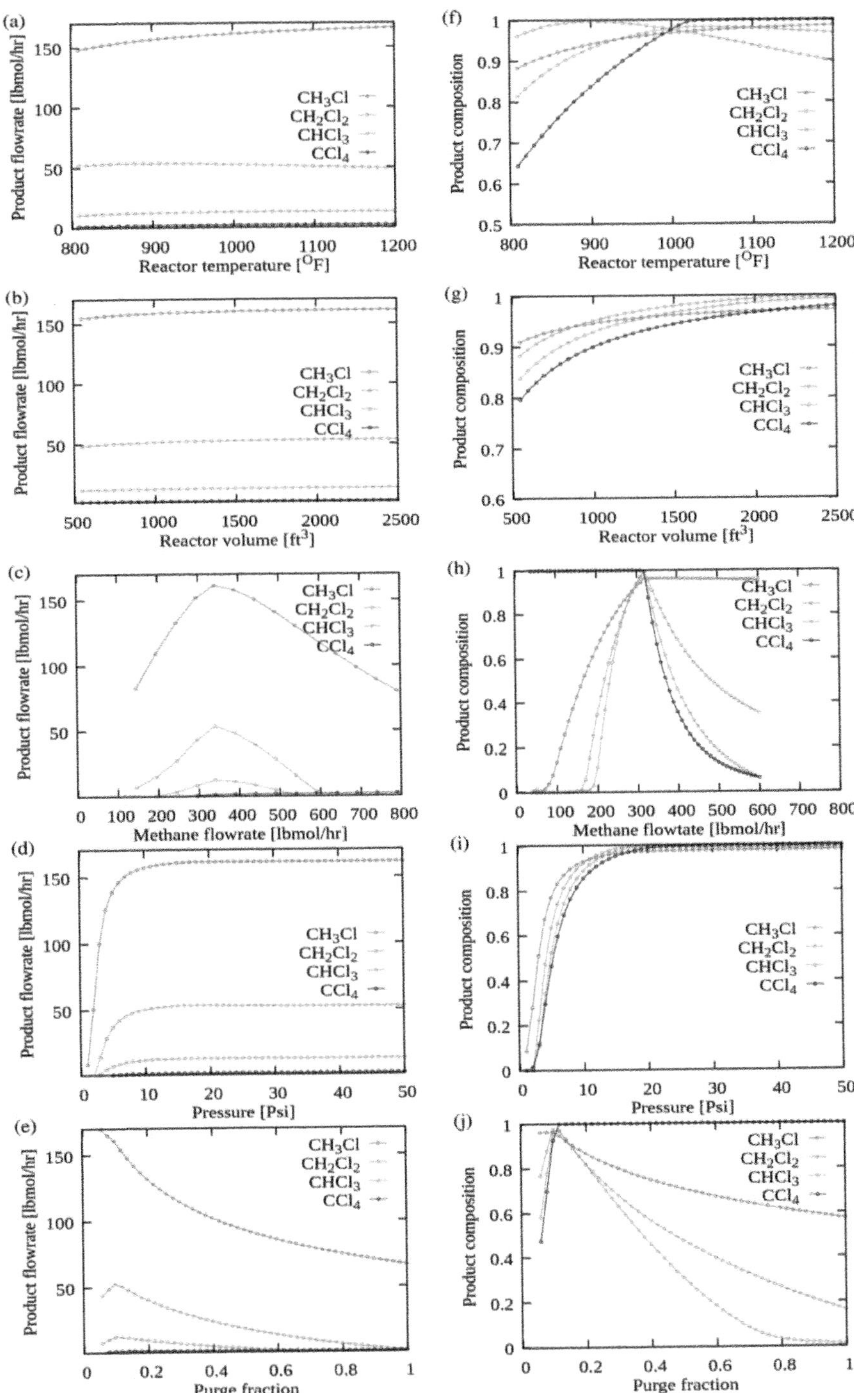

Figure 2. Effect of all five variables on (a-e) Product flow rates, (f-j) Product compositions.

2.2 Sensitivity analysis

Once the basic flow sheet is simulated, sensitivity analysis has been carried out in order to identify the parameters on which product flow rates and compositions are highly dependent. From the sensitivity studies, it is found out that the 5 parameters, i.e., reactor temperature, reactor volume, methane feed flow rate, reactor pressure, and purge fraction are the key parameters. Varying the above key parameters within the effective range for each of these variables, simulations are carried out to obtain product flow rates and compositions. The parameter ranges for which data generated is listed in Table 1. All the four product flow rates and compositions as a function of the five variables are shown in Figure 2(a-e) & (f-j) respectively.

Tabel 1. Effective parameter ranges (in fps units) in which sensitivity analysis is carried out.

Variable	Range
1. Reactor temperature	800 – 1200 °F
2. Reactor volume	500 – 2500 ft^3
3. Methane feed flow rate	30 – 600 lbmol/hr
4. Reactor pressure	0.1 – 25 psia
5. Purge fraction	0.04 – 0.8

3. Machine learning models

The data generated using sensitivity analysis is used to train five different ML models. These models include multi-output linear regression (MLR), support vector regression (SVR), decision trees, kNN and artificial neural networks (ANN). Data set from sensitivity analysis generated consisting of 2500 data points for each variable is divided in the ratio of 80:20 for training and testing ML models respectively. These models train by mapping the function relating input variables to the output variables in the given training data set. The trained model is then evaluated by predicting the output for the inputs given in the test set and comparing the predicted output to actual output. The predicted flow rates and compositions from five ML models are compared against actual values. However, for one ML model i.e., kNN, product flow rates obtained from simulations and values predicted by the model are shown in Figure 3 for four products. Two metrics i.e., R-square and mean absolute error (MAE) are used to measure the goodness of fit and accuracy of the models. These values are given in Table 2. The results indicate that the kNN model is found to be more accurate compared to the other models with high R-square value of 0.995 and low MAE of 23.38 for the process chosen to study. The next best model to kNN is decision trees. MLR has a low R-square and SVR has high MAE compared to other models. However, it is observed that increasing number of data points to train SVR and ANN results in low MAE.

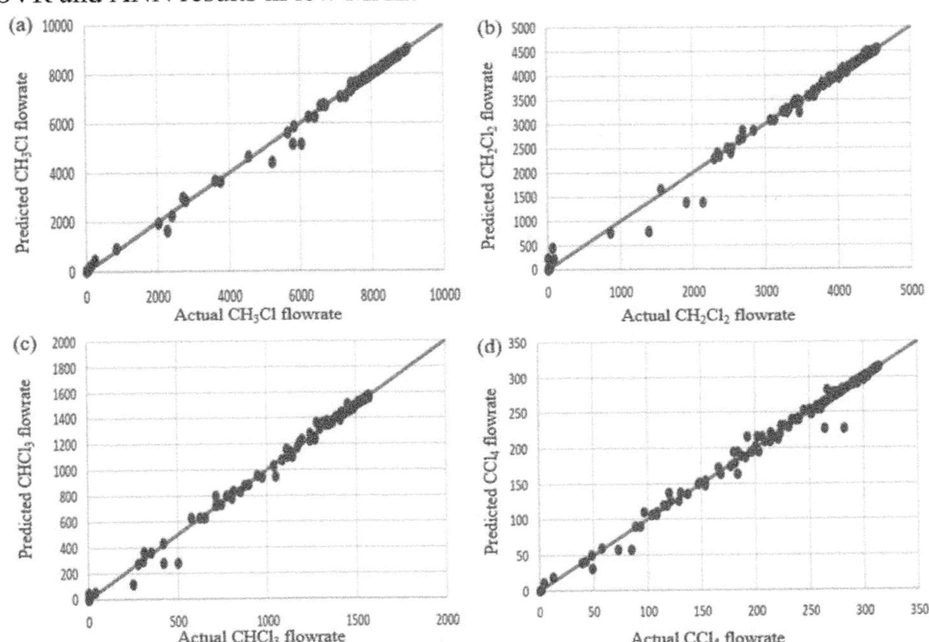

Figure 3. Parity charts of flowrates obtained from simulation and kNN model for all four products (a) CH_3Cl (b) CH_2Cl_2 (c) $CHCl_3$ and (d) CCl_4

Table2: Comparison of R-square and MAE for different ML models.

ML Model	R-Square	MAE
1. Multi output linear regression	0.57	396.58
2. SVR	-	2777.61
3. Decision trees	0.98	39.27
4. kNN	0.995	23.38
5. ANN	-	345.40

4. Conclusions

In conclusion, we have shown the application of ML techniques to predict the product distribution for multi-product series-parallel reactions. Data generated from simulations is used to train and test different models. R-square and MAE are used as model performance metrics. For the methyl chloride process considered in this study, kNN model performs better than other techniques considered. Better performance of the kNN model may be attributed to the large data set used to train the model. Our work demonstrates that ML models can be successfully used to predict the product distribution our multi product series-parallel reactions with good accuracy and without the involvement of cumbersome experiments or simulations.

References

1 Tabor, D.P. et al (2018), Accelerating the discovery of materials for clean energy in the era of smart automation. Nat Rev Mater 3, 5–20.

2 Jason E. Hein, (2021), Machine learning made easy for optimizing chemical reactions, Nature 590, 40-41.

3 Markus Meuwly, 2021, Machine Learning for Chemical Reactions. Chemical Reviews 121 (16) , 10218-10239

4 Udit Gupta, Dionisios G. Vlachos, 2021, Learning Chemistry of Complex Reaction Systems via a Python First-Principles Reaction Rule Stencil (pReSt) Generator. Journal of Chemical Information and Modeling 61 (7) , 3431-3441.

5 Joonyoung F. Joung, Minhi Han, Jinhyo Hwang, Minseok Jeong, Dong Hoon Choi, Sungnam Park., 2021, Deep Learning Optical Spectroscopy Based on Experimental Database: Potential Applications to Molecular Design. JACS Au , 1 (4) , 427-438.

6 Andrzej M. Żurański, Jesus I. Martinez Alvarado, Benjamin J. Shields, Abigail G. Doyle, 2021, Predicting Reaction Yields via Supervised Learning. Accounts of Chemical Research , 54 (8) , 1856-1865.

7 Hanyu Gao, Jean Pauphilet, Thomas J. Struble, Connor W. Coley, Klavs F. Jensen, 2021, Direct Optimization across Computer-Generated Reaction Networks Balances Materials Use and Feasibility of Synthesis Plans for Molecule Libraries. Journal of Chemical Information and Modeling 2021, 61 (1) , 493-504.

8 Shields B. J. et al., (2021), Bayesian reaction optimization as a tool for chemical synthesis, Nature 590, 89–96.

9 S. Park, H. Han, H. Kim, S. Choi,(2022) Machine learning applications for chemical reaactions, Chem. Asian J. , 17, e202200203.

10 Raccuglia, P., Elbert, K., Adler, P. et al. (2016) Machine-learning-assisted materials discovery using failed experiments. Nature 533, 73–76.

11 Zunyun et al., (2020), Optimizing chemical reaction conditions using deep learning: a case study for the Suzuki-Miyaura Cross coupling reaction, Org. Chem. Front. 7, 2269-2277.

12 Schwaller, P., Vaucher, A. C., Laino, T., & Reymond, J. L. (2021). Prediction of chemical reaction yields using deep learning. Machine learning: science and technology, 2(1), 015016.

13 Daniel Probst, Jean-Louis Reymond et al. (2022) Reaction classifi cation and yield prediction using the diff erential reaction fi ngerprint DRFP, Digital Discoverry, 1, 91.

14 Segler, M., Preuss, M. & Waller, M. (2018), Planning chemical syntheses with deep neural networks and symbolic AI. Nature 555, 604–610.

15 Thyagarajan, M. S., Kumar, R., and Kuloor, N. R. (1966), Hydro chlorination of Methanol to Methyl Chloride in Fixed Catalyst Beds, I & E C Process Design and Development 5(3): 209 213.

16 Habata, K., Tanaka, S., Araki, H., (1976), Shin-Etsu Chemical Co. Ltd. "Process for Preparing Methyl Chloride", USA, US 3,983,180A.

17 Mauricio M. Dantus and Karen A. High., (1996), An Economic Tool for the Retrofit of Chemical Processes Through Waste Minimization and Process Integration, Industrial and Engineering Chemistry 35(12): 4566-4578.

18 Mauricio M. Dantus and Karen A. High, (1999), Evaluation of waste minimization alternatives under uncertainty: a multiobjective optimization approach. Computers & Chemical Engineering, 23(10), 1493-1508.

Antonis Kokossis, Michael C. Georgiadis, Efstratios N. Pistikopoulos (Eds.)
PROCEEDINGS OF THE 33rd European Symposium on Computer Aided Process Engineering (ESCAPE33), June 18-21, 2023, Athens, Greece
 http://dx.doi.org/10.1016/B978-0-443-15274-0.50108-6

Architectural Design of Chemical and Biological Pathways through Parameter Sensitivity Oriented Mixed Integer Formulations

Emir Topac,[a] Emrullah Erturk,[a] Ozgun Deliismail,[b] Sahin Sarrafi,[b] Hasan Sildir,[a]

[a]*Department of Chemical Engineering, Gebze Technical University, Kocaeli, 41400, Türkiye*
[b]*SOCAR Turkey R&D and Innovation Co., Izmir, 35800, Türkiye*

hasansildir@gtu.edu.tr

Abstract

A mixed integer nonlinear programming (MINLP) formulation is developed for the automatized design of chemical and biological pathways. The proposed formulation is solved using rigorous algorithms and enables the selection of reactions whose rate constants have the highest sensitivity. Unlike the traditional parameter estimation approach, the training performance is introduced as constraint which is pre-defined to balance the architectural simplicity and fitting performance. Adjoint sensitivity expressions are additionally solved simultaneously in the formulation and the resulting parameter sensitivities are included in the objective. Binary variables are introduced in the linking constraints for the selection of a reaction subset from a more complex pathway, also tightening the search space for the optimization algorithm. The approach is implemented on a complex chemical/biological pathway which processes glucose as a primary feedstock to produce hydroxymethylfurfural (HMF). The reduced pathway prediction bounds are narrower ensuring a similar mean predictions, despite 25% reduction in the superstructure.

Keywords: Parameter identifiability; Mixed integer programming; Sensitivity; Uncertainty.

1. Introduction

Chemical and biological pathway models include a high number of rate expressions with tunable parameters to be estimated from the experimental data. The lack of spatial and temporal measurements might result in statistically ill-defined inverse problems with multiple solutions in parameter estimation (Mclean and Mcauley, 2012). Furthermore, the computational load and identifiability problems increase once the disturbances and measurement noises appear especially when the model architecture is large.

Usually, trial and error procedures or sequential methods are applied for the architectural synthesis and parameter identifiability tasks, which require significant manual effort in addition to decentralized computations on theoretical level. The most sensitive parameter selection based on fisher information matrix is compared to LASSO, which is least absolute shrinkage and selection operator, and applied to a biomechanical system in (Ramadan et al., 2018). Shapiro et *al*. developed a heuristic algorithm to eliminate some parameters and re-calculate the sensitivities until desired condition value is obtained (Shapiro et al., 2014).

A simultaneous approach using rigorous formulations to obtain the pathway architecture and corresponding parameters based on the parameter sensitivity values and their contribution to uncertainty is scarce and would have the potential to exploit available statistical knowledge from experimental data for a better model development. This study focuses on the development of a mixed-integer formulation to perform architectural synthesis and parameter estimation tasks based on sensitivities in addition to the model training performance.

2. Methodology

A typical reaction pathway is represented by ODEs and given by:

$$\frac{dy}{dt} = f(t, y, p) \tag{1}$$

where y is the vector of outputs which are typically concentrations; p is the vector of time-invariant parameters; t is the time; f is the vector of mathematical rate expressions.

The derivative of the terms in Eq. 1 results in forward sensitivity ODE for a particular parameter and output:

$$\frac{ds_{ij}}{dt} = \frac{df_j}{dy_j} s_{ij} + \frac{df_j}{dp_i} \tag{2}$$

where s_{ij} is the sensitivity of jth output, y_j, to i^{th} parameter, p_i. A simultaneous solution of Eq. 1 and Eq. 2 is required to address the specific considerations on parameter selection process to account for the model performance, additionally.

A modified mixed integer formulation is developed for the architectural design and estimation of its parameters who have the highest sensitivity, eliminating the parameters with relatively smaller impact on the outputs. The formulation after the ODEs are discretized based on orthogonal collocation enables the simultaneous solution based on rigorous optimization algorithms and is given by:

$$\begin{aligned}
& Min_{p,b} \quad -\sum s_{ij,k}^2 \\
& s.t. \\
& \frac{dy}{dt} = f(t, y, p) \\
& \frac{ds_{ij}}{dt} = \frac{df_j}{dy_j} s_{ij} + \frac{df_j}{dp_i} \quad i = 1,2, \dots, M \; j = 1,2, \dots, P \\
& s_{ij}(0) = 0 \quad i = 1,2, \dots, M \; j = 1,2, \dots, P \\
& y_j(0) = (0.25, 0, 0, 0, 0, 0) \quad j = 1,2, \dots, 6 \\
& \sum \lVert y_k - y_{m,k} \rVert \leq \varepsilon \\
& b_i p_{i,min} \leq p_i \leq b_i p_{i,max} \quad i = 1,2, \dots, P \\
& b_{min} \leq \sum b_i \leq b_{max} \quad b_i \in \{0,1\} \quad i = 1,2, \dots, P
\end{aligned} \tag{3}$$

where $s_{ij,k}$ is the sensitivity at sample time k; y_k is the model prediction at k; $y_{m,k}$ is the measurement at k; ε is the desired maximum training error; $p_{i,min}$ and $p_{i,max}$ are lower and upper bounds of ith parameter, p_i, respectively; b_i is the binary variable to account for the existence of p_i; b_{min} and b_{max} are lower and upper bounds of number of parameters desired in the ultimate architecture.

Solution of Eq. 3 requires an explicit formulation for sensitivity ODEs for the rigorous approach, which calls for an automatic or symbolic differentiation. There are various solver packages which deliver such expressions (Andersson et al., 2019), to avoid numerical approximation errors. On the other hand, sensitivity equations are mostly nonlinear and further mathematical reformulations and decompositions, in addition to dense discretization, might be essential for pathways with high complexity. In contrast to nonlinear optimization which applies to parameter estimation with fixed architectures, the proposed mixed integer formulation includes binary variables to account for the existence of particular parameter values which represent the reaction rate constant once they have a nonzero value. Linking constraints are also introduced to tighten the search space by fixing the parameter values while eliminating the corresponding reaction path from the network. The formulation is flexible and can address measurements and sensitivity considerations at desired sample times. Current formulation includes the selection of the parameters which deliver the maximum squared sensitivity at measurement points.

The parameter covariance matrix for a particular architecture (c_p) is subsequently calculated using bootstrapping method (Godo-Pla et al., 2019) for further analysis on the identifiability issues. The uncertainty in the outputs due to parameter variations is calculated using (Tellinghuisen, 2001):

$$c_y = S_k c_p S_k^T \tag{4}$$

where c_y is an under estimator of actual output covariance matrix; S_k is the sensitivity matrix at k.

The approach is implemented on a complex chemical/biological pathway and solved using BONMIN solver in PYOMO computation environment (Hart et al., 2011).

3. Results

The proposed approach is implemented on 5-Hydroxymethylfurfural (HMF) synthesis from glucose (Tang et al., 2017) which can be obtained through biological processes, providing a promising opportunity for sustainability considerations. The full network superstructure is shown in Fig. 1a, where several different reaction paths enable the production of a particular compound, which in turn contributes to identifiability of these paths. With a coupled and dependent set of ODEs, the unique estimation of the parameters becomes a challenging task and the proposed formulation is implemented to demonstrate the impact, unlike (Erturk et al., 2021) where sensitivity considerations are not included. Thus, in addition to six differential equations to represent the compounds, 72 additional sensitivity ODEs are introduced to account for 12 parameters in the full network. Once the simultaneous architectural synthesis and the parameter estimation are performed based on the selection of the most sensitive parameters, the resulting reduced network is presented in Fig. 1b.

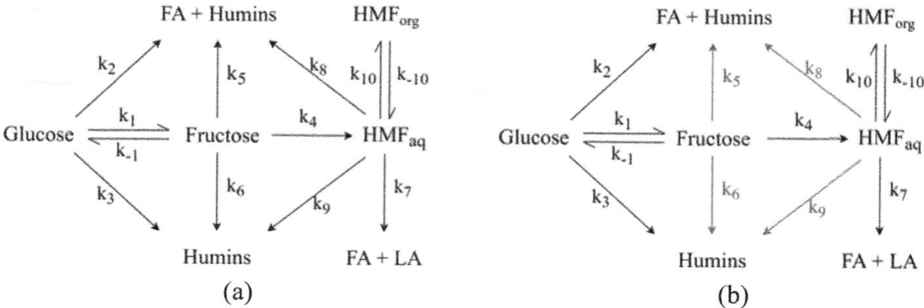

Figure 1. Reaction pathway
(a) Full network (b) Reduced network

The eliminated reaction paths from Fig. 1a are shown in Fig. 1b using red arrows. Next, the performance of the full network and the reduced network is compared in terms of prediction similarity and uncertainty. The uncertainty for the solutions is evaluated from the diagonal elements of c_y which is calculated from Eq. 4 after bootstrapping procedures are performed.

Fig. 2a and Fig. 2b include solutions of pathway ODEs for the full network and the reduced network, respectively, for Fructose and Formic Acid (FA) + Humins. With relatively smaller mean values, Levulinic Acid (LA) is excluded from figures. The mean predictions, which exclude the parameter covariance matrix and related propagation equations, show similar profile due to constraint in Eq. 3 to deliver a satisfactory training performance characterized by ε and shown in full lines. The uncertainty ranges are represented by dashed lines. In turn, a similar profile is obtained despite elimination of four reaction paths as shown in Fig. 1b.

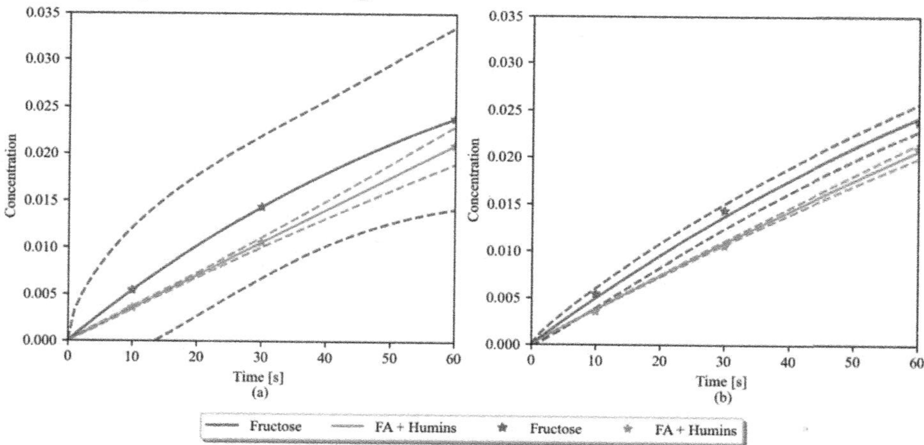

Figure 2. Mean solution and uncertainty interval for Fructose, FA+Humins
(a) Full network (b) Reduced network

As a result of numerical scale differences among pathway variables, the solution profiles for HMF$_{aq}$ and HMF$_{org}$, are presented in Fig. 3b, from which the latter is for the reduced network. Compared to the other components, the impact on HMF$_{org}$ uncertainty is less significant as the corresponding component related reactions have not changed (see Fig. 1a and Fig. 1b). However, some decrease in the uncertainty range is still observed due to a more robust parameter estimation which accounts for the identifiability.

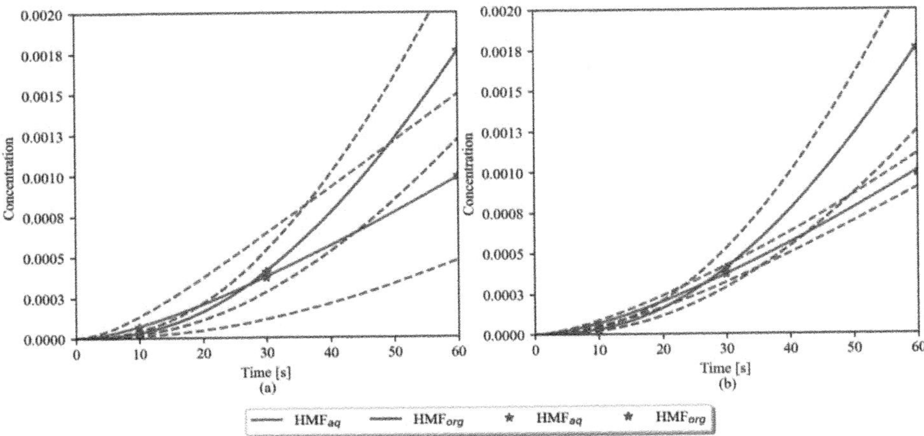

Figure 3. Mean solution and uncertainty interval for HMF_{aq}, HMF_{org} (a) Full network (b) Reduced network

Table 1 includes a comparison on the maximum and minimum dependent variable values based on uncertainty at final time. The prediction bounds for the reduced model is significantly more tightened compared to the full network. However, a more drastic reduction in uncertainty is favored by the elimination of the reaction paths associated with a particular component.

Table 1. Impact on output uncertainty

	Full Network [×100]		Reduced Network [×100]		Difference [%]
	Min	*Max*	*Min*	*Max*	
Glucose	19.15	21.16	19.96	20.32	82.09
Fructose	1.41	3.35	2.28	2.56	85.57
HMF_{aq}	0.05	0.15	0.09	0.11	80.58
HMF_{org}	0.12	0.23	0.13	0.23	5.45
FA+Humins	1.89	2.29	1.99	2.14	62.50
FA+LA	0.00	0.10	0.00	0.09	10.00

4. Discussion

Traditional parameter estimation problems for complex pathways usually process a fixed architecture and include the training error in the objective function under physically constrained rate expression parameters. In contrast to traditional approach, the proposed mixed-integer formulation performs simultaneous architectural design and parameter estimation by adding binary variables which consider the existence of a particular path and tighten the search space during the optimization through linking constraints. The model training performance is introduced as a constraint to balance the structural reduction and fitting performance, since elimination of some reaction paths from the superstructure hinders the fitting capability, in theory. Moreover, the simultaneous evaluation and consideration of sensitivity expressions enable the calculations to be performed by rigorous optimization algorithms, which can include advanced mathematical reformulations, decompositions, and approximations to account for various sensitivity requirements. Such a centralized processing of the complex problem delivers the best configuration and high identifiable parameters, especially when global or sophisticated solvers are utilized.

As a downside, the approach calls for more computational load compared to traditional methods, due to binary variables to represent the existence of parameters, evaluation of higher number of equations during optimization, additional nonlinearity terms, the rigorous solution algorithm delivers a more robust model architecture. Thus, subsequent tasks related to model update or control tasks become computationally more feasible.

A biochemical pathway is considered to demonstrate the impact of the approach. However, the problem formulation is flexible and can be further tailored to various applications. Current focus includes the development of sophisticated mathematical reformulations to address the computational problems when larger networks with more parameters are needed.

References

Andersson, J.A.E., Gillis, J., Horn, G., Rawlings, J.B., Diehl, M., 2019. CasADi: a software framework for nonlinear optimization and optimal control. Math. Program. Comput. 11, 1–36.

Erturk, E., Erdal, A., Sildir, H., 2021. Tam Sayili ve Surekli Optimizasyon Problemi ile Reaksiyon Ag Modellerinin Kucultulmesi. Konya Muhendis. Bilim. Derg. 9, 142–156.

Godo-Pla, L., Emiliano, P., Valero, F., Poch, M., Sin, G., Monclús, H., 2019. Predicting the oxidant demand in full-scale drinking water treatment using an artificial neural network: Uncertainty and sensitivity analysis. Process Saf. Environ. Prot. 125, 317–327.

Hart, W.E., Watson, J.P., Woodruff, D.L., 2011. Pyomo: Modeling and solving mathematical programs in Python. Math. Program. Comput. 3, 219–260. https://doi.org/10.1007/s12532-011-0026-8

Mclean, K.A.P., Mcauley, K.B., 2012. Mathematical modelling of chemical processes-obtaining the best model predictions and parameter estimates using identifiability and estimability procedures. Can. J. Chem. Eng. 90, 351–366. https://doi.org/10.1002/cjce.20660

Ramadan, A., Boss, C., Choi, J., Peter Reeves, N., Cholewicki, J., Popovich, J.M., Radcliffe, C.J., 2018. Selecting sensitive parameter subsets in dynamical models with application to biomechanical system identification. J. Biomech. Eng. 140.

Shapiro, B., Barton, M., Mavalvala, N., Mittleman, R., Youcef-Toumi, K., 2014. Selection of important parameters using uncertainty and sensitivity analysis. IEEE/ASME Trans. Mechatronics 20, 13–23.

Tang, J., Zhu, L., Fu, X., Dai, J., Guo, X., Hu, C., 2017. Insights into the kinetics and reaction network of aluminum chloride-catalyzed conversion of glucose in NaCl–H2O/THF biphasic system. ACS Catal. 7, 256–266.

Tellinghuisen, J., 2001. Statistical error propagation. J. Phys. Chem. A 105, 3917–3921. https://doi.org/10.1021/jp003484u

Antonis Kokossis, Michael C. Georgiadis, Efstratios N. Pistikopoulos (Eds.)
PROCEEDINGS OF THE 33rd European Symposium on Computer Aided Process Engineering (ESCAPE33), June 18-21, 2023, Athens, Greece
 http://dx.doi.org/10.1016/B978-0-443-15274-0.50109-8

Data-driven fouling modeling in crude oil heat exchanger

Pelin Dologlu[a], Alihan Tabak[a], A. Eren Vedin[a], Gizem Kayar[b]

[a]*SOCAR Turkey, Digital Transformation Department, Istanbul 34485, Turkey*
[b]*SOCAR STAR Oil Refinery, Process Department, Aliaga, Izmir 35800, Turkey*

Abstract

Unwanted materials in fluids deposit on heat transfer surfaces in time, which is called fouling, and it negatively affects heat exchangers regarding operation and maintenance costs. Therefore, fouling rate should be monitored to prevent unexpected costs and possible failures due to deposition limit of heat transfer area. Monitoring also allows process engineers to improve operating conditions to keep fouling under control for longer cleaning cycles and schedule cleanings effectively. In this work, to monitor fouling rate in the shell and tube crude oil preheat exchanger in the real refinery, the data-driven heat transfer rate prediction model is developed from processed and to-be-processed crude oil planning, on-line sensor and laboratory sample data. Results of various machine learning and deep learning techniques are compared, and as Multiple Linear Regression is found to be the best performing predictive model. Developed fouling rate model predictions are actively used for scheduling of crude oil preheat exchanger cleaning and improving operating conditions to reduce costs in the refinery.

Keywords: crude oil, pre-heat train exchangers, fouling, machine learning, multiple linear regression

1. Introduction

Petrochemical industries, especially oil refinery, include one of the most energy-required processes among other industries. Maintaining sustainability and reducing gas emissions while keeping the process profitable is a significant problem for refinery processes. This challenging problem can be solved with energy saving solutions (Mrayed et al., 2021). Since the crude distillation unit (CDU) is the front-end operation of the refinery, any intervention in CDU can reduce the energy consumption of the entire refinery by 24% (Yang et al., 2020). Crude oil preheat train (PHT) exchangers recover energy, which is necessary for CDU, by up to 70%. However, recovered energy via PHT exchangers decreases over time because of the deposition of unwanted materials on the heat transfer surface area, which is called fouling (Coletti et al., 2011). Therefore, the fouling rate should be monitored to prevent unexpected energy and maintenance costs, gas emissions and possible failures due to the deposition limit of the heat transfer area (Santamaria & Macchietto, 2019). Monitoring also allows process engineers to improve operating conditions to keep fouling under control for longer cleaning cycles and schedule cleanings effectively. Thus, monitoring of fouling in crude oil PHT exchangers has been studied by many researchers.

Fouling modeling and as a further step fouling prediction enable monitoring fouling in heat exchangers effectively. In literature, fouling is modeled in the first place as empirical and theoretical (Wang et al., 2019). Since, crude oil fouling contains many simultaneous and different mechanisms, empirical or theoretical models are not generalized and not enough to reflect the nonlinear nature of fouling properly (Sundar et al., 2020). Therefore, data-driven approach for fouling in heat exchanger is introduced to the literature in recent years. Neural network based methods for fouling monitoring and prediction of heat exchanger are preferred methods thanks to their natural capacity to exhibit nonlinear relationships easily (Davoudi & Vaferi, 2018). Also, Sundar et al. (Sundar et al., 2020) proposed a robust deep learning fouling prediction approach which learns heat exchanger flow and heat transfer physics. Furthermore, Hosseini et al. (Hosseini et al., 2022) estimated the fouling factor in a comprehensive study using five different algorithms namely Gaussian Process Regression, Decision Trees, Bagged Trees, Support Vector Regression and Linear Regression. Data-driven approaches are appropriate methods in Industry 4.0 where lots of data is generated and they have proved their success in heat exchangers to determine fouling in complex processes.

In this work, to monitor the fouling rate in the shell and tube crude oil preheat exchanger, the data-driven overall heat transfer rate prediction model is developed. Daily overall heat transfer rate predictions for the next 15 days are predicted from some of the processed and to-be-processed crude oil planning data, on-line sensor and laboratory sample data to calculate heat transfer rate based on the theoretical model. To the best of authors' knowledge, there are limited PHT exchanger fouling prediction models for long time horizon, that consider all processed and to-be-proses crude oil planning data, on-line sensor and laboratory sample data. In order to choose the best prediction model in real crude oil PHT exchangers, a comparison of different machine learning and deep learning techniques, which are XGBoost, Multiple Linear Regression, Random Forest, Extra Trees, Gradient Boosting, Decision Tree, LightGBM and Multilayer Perceptron are studied. The aim of this study is efficient scheduling of crude oil PHT exchanger cleaning and improving operating conditions by reducing energy loss to reduce costs in the refinery, while considering safety by preventing unexpected failures in normal operation. Moreover, proposed data-driven modeling approach can be generalized for other heat exchangers just by revising crude oil properties related inputs.

2. Materials and Methods

2.1. Crude Oil Preheat Train

PHT exchangers have a simple design, but they have an important role in terms of the economical aspect of refinery operations. Before crude oil is heated in the furnaces, crude oil passed through these PHT exchangers, which increases energy efficiency. In this work, the multi-pass shell and tube eight identical parallel heat exchangers in PHT, undergoing fouling deposition in both shell and tube side are modeled (see Figure 1). While crude oil flows from the shell-side, the overhead fluid of the atmospheric distillation column flows from the tube-side in this heat exchanger network. Crude oil is heated with overhead fluid while overhead fluid is cooled. If there is fouling in the heat exchanger, temperature and pressure differences are highly affected and energy efficiency decreases over time. To keep the process efficient and safe, the fouling level in the heat exchanger network in refinery should be conducted properly.

2.2. Fouling Model

Fouling can be defined as the deposition of unwanted materials on the heat transfer area, and it negatively affects heat transfer efficiency in operations. Cross section of one of heat exchanger from network and detailed fouling scheme of shell and tube heat exchanger can be seen in Figure 1. Fouling formation can be happen both inside and outside of tubes. Foulants in crude oil cause fouling inside tubes and crude oil flows at low temperature (T_c) which increases formation of fouling. Hot overhead fluid flows at shell side and T_h temperature, which causes fouling outside of tubes. Since one of the fouling indicators is the heat transfer rate, which can be calculated easily via a theoretical model, with the help of on-line sensors and laboratory data also known real PHT exchanger-specific coefficients. Therefore, in this work, the overall heat transfer rate is modeled to monitor fouling in the PHT exchanger network of real-time crude oil distillation unit operations. Energy efficiency of heat exchangers will be increased, and unexpected failures can be prevented by the help of this model. Also, maintenance of heat exchangers can be scheduled efficiently. Furthermore, data-driven modeling approaches can be generalized for other heat exchangers by revising crude oil related inputs.

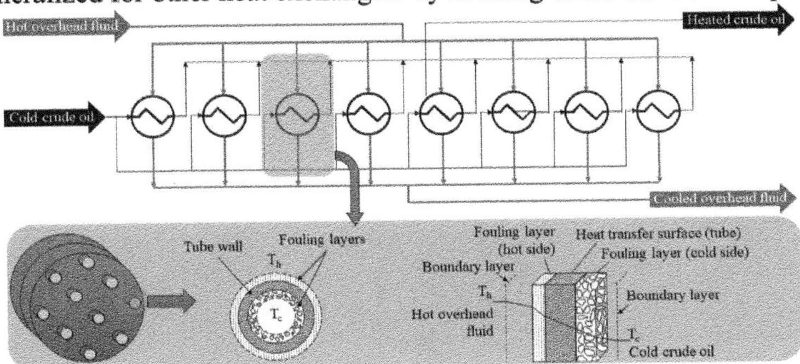

Figure 1. Crude oil PHT exchanger network and detailed fouling scheme of heat exchanger

2.3. Industrial Data Set Preparation For Modeling

30 different features related to PHT exchanger network and crude oil assays were used in this study. 16 of them belong to the online sensor data and the rest are laboratory sample data. In addition to online sensor and laboratory sensor data, unique crude oil properties are added to the dataset which is collected from crude oil assays. The planning department decides which crude oil will be used in the future and shares this information with the process units. In this study, to-be processed crude oil blend properties (viscosity, asphaltenes, naphthenes, total acid number, and API) are calculated as weighted averages of unique crude oil properties from crude oil assays according to the planning department's decision. Since different crude oil blend properties can aggregate effects on the process, the cumulative sums of these properties are also added to the dataset. Whole online sensor data, laboratory sample data and crude oil blend properties features have been fetched on a daily basis, and they are aggregated as one dataset. In addition, some new features to calculate the overall heat transfer rate were created by inputs that are already in the dataset. For instance, the shell-side delta temperature was calculated by the subtraction between the shell inlet temperature and the shell outlet temperature. Similarly, tube-side delta temperature was calculated by the subtraction between tube inlet temperature and tube outlet temperature. The delta pressure data for both shell and tube sides were calculated in this way, and they are added to the dataset because of possible indicators of fouling. Since the model predicts the overall heat transfer rate 15 days in advance, 15 days lagged data and the data which represents 1, 3, 7 and 15 days rolling means to see the trends of all inputs in the dataset, except laboratory sample data. Consequently, the final dataset has 947 sample points from real-time crude oil operations and 97 features with one target, which is the overall heat transfer rate.

2.4. Machine Learning and Deep Learning Algorithms

2.4.1. Multiple Linear Regression

Multiple Linear Regression (MLR) tries to find the linear relationship between a dependent variable (target) and one or more independent variables (predictors). The relationship between the target and predictors is represented as an equation that has an intercept and a coefficient for each predictor. The goal is to find the coefficients that minimize the difference between the predicted and actual values of the target. The obtained model can be used to predict the target value for a new set of predictor values. Linear Regression is a simple but powerful tool for modeling linear relationships and is widely used in various applications (Hosseini et al., 2022). The coefficients are estimated from the training data using normal equation as optimization algorithm in this work.

2.4.2. Decision Tree

Decision Tree is a non-parametric supervised learning algorithm. Decision Tree has the capability of predicting the target variable by learning simple decision rules inferred from the data features (Hosseini et al., 2022).

2.4.3. Ensemble Methods

Ensemble methods contain set of models, which are integrated to obtain final prediction. Many randomized decision trees on various sub-samples of dataset are used in Random Forest and Extra Trees to improve prediction accuracy. Also, techniques such as Gradient Boosting, XGBoost and LightGBM, that fit boosted decision trees by minimizing an error gradient (Hosseini et al., 2022), are used in this work.

2.4.4. Multilayer Perceptron

Multilayer Perceptron (MLP) is widely accepted technique as a promising tool for solving complex engineering problems in real industrial processes. The MLP can include any number of the layer with specified number of neurons in each layer, which transfers information between first and last layers (Davoudi & Vaferi, 2018). Since 1 hidden layer and 30 neurons give the best result, these parameters are used for MLP in this wok.

3. Results and Discussion

Correlation of features in the final dataset with the target -overall heat transfer rate- is observed, and the most correlated 8 features have been selected as inputs for modeling. Selected inputs are 15 days lagged data of shell-side delta temperature, 15 days lagged data of overall heat transfer rate, 15 days lagged data of viscosity, 1 day rolling mean of viscosity, 3 days rolling mean of viscosity, 7 days rolling mean of viscosity, 15 days rolling mean of viscosity and cumulative sum of asphaltene.

Different machine learning and deep learning techniques are compared their performances to find the best fouling prediction model in real-time operations. While MLR, Decision Tree, Random Forest, Extra Trees, Gradient Boosting, XGBoost and LightGBM were used as machine learning techniques, MLP was used as a deep learning technique. All algorithms are implemented in Python 3.8.

In this study, three different test periods were selected, and each test period contains 15 sample points. The reason why preferred 3 test periods is the business unit wanted to see the results of these periods whether models can catch the expected fouling results. Also, Mean Absolute Error (MAE) was preferred as the model evaluation metric for this work. After the results were tested in 3 different test periods, the best fouling prediction model for real-time operations is the MLR model. Figure 2 shows the comparison of the different model techniques' performances in terms of MAE for a test period.

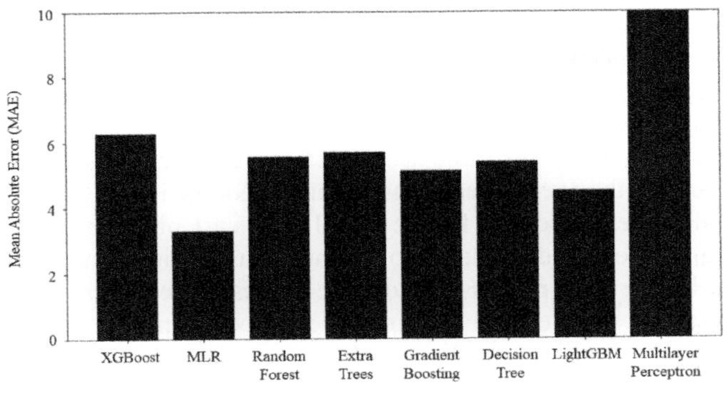

Figure 2. Different machine learning and deep learning techniques comparison

As it can be easily seen from Figure 2, the MLR model has the best performance with 3.29 MAE score. Multilayer Perceptron as a deep learning technique gives the highest MAE score. The reason why Multilayer Perceptron technique fails is deep learning techniques perform well on more sample point dataset than this study dataset. Besides, boosting techniques also failed because of the scarcity of data. MLR technique, which has the best results, is used for real-time PHT fouling modeling within 15 days horizon. One of the test set results for comparison of MLR fouling model predictions and actual overall heat transfer rate real-time operation of PHT exchanger values within 15 days horizon are shown in Figure 3. The results were deemed successful by the business unit. With the help of the fouling prediction model, the business unit can monitor fouling before reaching the critical limit, and can intervene in the process parameters by taking an action to slow down the fouling effect. Also, the business unit can schedule crude oil PHT exchanger cleaning efficiently and can improve operating conditions to reduce costs in the refinery. Furthermore, this modeling approaches can be generalized for other heat exchangers by revising crude oil related inputs.

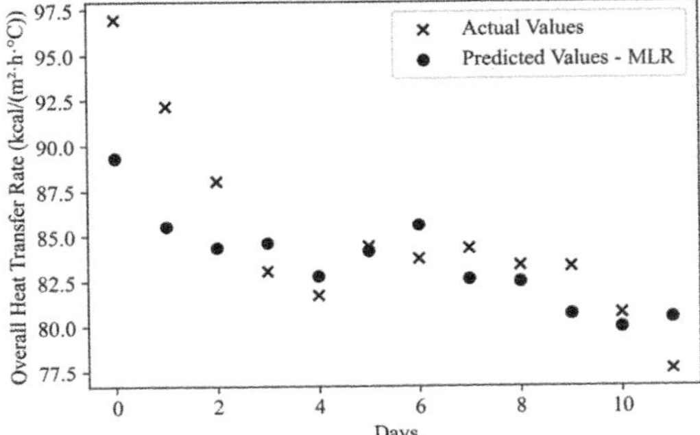

Figure 3. Fouling model predictions and actual values comparison of the test set

4. Conclusions

Oil refineries have the most energy-required processes, thus any energy saving solutions has a great impact on sustainability, gas emissions and operating costs. One of the energy saving solution is that energy efficiency increases while crude oil passed through PHT exchangers before crude oil is heated in the furnaces in CDU. However, fouling causes the deposition of a layer from unwanted materials in these heat exchangers, which decreases the heat transfer rate. If the fouling rate in PHT exchangers can be known before reaching the critical limit, the business unit can take action that slows down the fouling effect. Therefore, data-driven fouling rate prediction in the shell and tube crude oil PHT exchangers model is developed to real-time monitor fouling rate for a long time horizon. Since one of the fouling indicators is the heat transfer rate, which can be calculated easily via a theoretical model, the overall heat transfer rate is modeled to monitor fouling. Daily overall heat transfer rate predictions for the next 15 days are predicted from some of the processed and to-be-processed crude oil planning data, on-line sensor and laboratory sample data. Combining of all processed and to-be-proses crude oil planning data, on-line sensor and laboratory sample data are increased prediction accuracy and provides predicting capability to fouling modeling for long time horizon. In order to choose the best prediction model in real crude oil PHT exchangers, a comparison of different machine learning and deep learning techniques, which are XGBoost, Multiple Linear Regression, Random Forest, Extra Trees, Gradient Boosting, Decision Tree, LightGBM and Multilayer Perceptron are studied. Proposed data-driven modeling approach can be implemented on other heat exchangers easily just by revising crude oil related parameters. With the help of the fouling prediction model, the business unit can monitor fouling before reaching the critical limit, and can intervene in the process parameters by taking an action to slow down the fouling effect. Also, the business unit can schedule crude oil PHT exchanger cleaning efficiently and can improve operating conditions to reduce costs in the refinery.

References

Coletti, F., Macchietto, S., & Polley, G. T. (2011). Effects of fouling on performance of retrofitted heat exchanger networks: A thermo-hydraulic based analysis. *Computers & Chemical Engineering*, *35*(5), 907–917. https://doi.org/10.1016/J.COMPCHEMENG.2011.01.027

Davoudi, E., & Vaferi, B. (2018). Applying artificial neural networks for systematic estimation of degree of fouling in heat exchangers. *Chemical Engineering Research and Design*, *130*, 138–153. https://doi.org/10.1016/J.CHERD.2017.12.017

Hosseini, S., Khandakar, A., Chowdhury, M. E. H., Ayari, M. A., Rahman, T., Chowdhury, M. H., & Vaferi, B. (2022). Novel and robust machine learning approach for estimating the fouling factor in heat exchangers. *Energy Reports*, *8*, 8767–8776. https://doi.org/10.1016/J.EGYR.2022.06.123

Mrayed, S., Shams, M. Bin, Al-Khayyat, M., & Alnoaimi, N. (2021). Application of pinch analysis to improve the heat integration efficiency in a crude distillation unit. *Cleaner Engineering and Technology*, *4*, 100168. https://doi.org/10.1016/j.clet.2021.100168

Santamaria, F. L., & Macchietto, S. (2019). Integration of optimal cleaning scheduling and control of heat exchanger networks under fouling: MPCC solution. *Computers & Chemical Engineering*, *126*, 128–146. https://doi.org/10.1016/J.COMPCHEMENG.2019.04.012

Sundar, S., Rajagopal, M. C., Zhao, H., Kuntumalla, G., Meng, Y., Chang, H. C., Shao, C., Ferreira, P., Miljkovic, N., Sinha, S., & Salapaka, S. (2020). Fouling modeling and prediction approach for heat exchangers using deep learning. *International Journal of Heat and Mass Transfer*, *159*, 120112. https://doi.org/10.1016/J.IJHEATMASSTRANSFER.2020.120112

Wang, F. L., Tang, S. Z., He, Y. L., Kulacki, F. A., & Yu, Y. (2019). Heat transfer and fouling performance of finned tube heat exchangers: Experimentation via on line monitoring. *Fuel*, *236*, 949–959. https://doi.org/10.1016/J.FUEL.2018.09.081

Yang, K., Liu, S., He, C., Zhang, B., Chen, Q., & Pan, M. (2020). Improving energy saving of crude oil distillation units with optimal operations. *Journal of Cleaner Production*, *263*, 121340. https://doi.org/10.1016/J.JCLEPRO.2020.121340

Antonis Kokossis, Michael C. Georgiadis, Efstratios N. Pistikopoulos (Eds.)
PROCEEDINGS OF THE 33rd European Symposium on Computer Aided Process Engineering (ESCAPE33), June 18-21, 2023, Athens, Greece
 http://dx.doi.org/10.1016/B978-0-443-15274-0.50110-4

Explainable formation energy prediction for uncovering the relationship between the electronic structure and stability of the heterogeneous catalyst

Daeun Shin,[a, b] Dong Hyeon Mok,[c] Seoin Back,[c] Jonggeol Na[a, b]

[a]*Department of Chemical Engineering and Materials Science, Ewha Womans University, Seoul 03760, Republic of Korea*
[b]*Graduate Program in System Health Science and Engineering, Ewha Womans University, Seoul 03760, Republic of Korea*
[c]*Department of Chemical and Biomolecular Engineering, Institute of Emergent Materials, Sogang University, Seoul 04107, Republic of Korea*
jgna@ewha.ac.kr

Abstract

In recent studies, machine learning (ML) applications in the heterogeneous catalysis field for material properties prediction accelerate the catalyst discovery with desired properties. However, due to its high complexity, most ML models suffer from the black-box problem, which cannot provide a basis for prediction. Thus, reliable application and physical insight generation are challenging with conventional black-box models. Here, we developed an ML model that predicts formation energy (E_f) from the density of states (DOS). More importantly, by interpreting the model, we also confirmed the possibility of uncovering the relationship between the electronic structure of materials and their stability. Our model achieves successful performance demonstrating its superior capability of DOS featurization.
Keywords: Heterogeneous Catalyst, Stability, Artificial Intelligence, Convolutional Neural Network, Explainable AI (XAI)

1. Introduction

In the heterogeneous catalysis field, the discovery of new materials with desired properties is challenging because of its large parameter space (EsterhuizenJ.A., 2022). In this context, computational material discovery has grown continuously over the past decade thanks to the development of various crystal structure prediction algorithms. Previously, density functional theory (DFT) calculation was a major tool to search chemical space, however, due to its high computational cost, the demand for an efficient tool has been increasing. Recently, machine learning (ML) models, which generate predictions within sub-seconds, have emerged as a powerful approach to bypassing the formidable computational cost and navigating the large chemical space more efficiently (FungV., 2021). However, to maximize the prediction performance, ML models have become more complex with tremendous trainable parameters, which makes it nearly impossible to interpret the internal logic underlying the prediction (Doosun HongJaehoon 2022). Beyond mere prediction tools, in order for an ML model to play a conducive role in scientific advancement in the heterogeneous catalysis field, the model should be interpretable and able to provide meaningful knowledge and testable hypotheses. As an alternative, theory-infused models integrating physical interaction into the algorithm and thus, inherently interpretable were studied (WangSH., 2021). However, due to their restricted architecture and enforced simplicity, it is difficult to achieve high performance. Furthermore, since most of these models are system-specific, there is a limitation in extension to general application for various system. In order to satisfy both predictive performance and interpretability simultaneously, the implementation of the explainable AI (XAI) technique on the fully trained predictive model is in the spotlight. These studies are mainly conducted by utilizing descriptors related to material properties and obtaining the importance of each descriptor in prediction, for ease of interpretation and visualization (OmidvarNoushin 2021). Especially, since it is well-known that the electronic structure is closely related to physical properties of materials, there have been many efforts to utilize electronic features, such as *d*-band center, extracted from the density of state (DOS) (LinicHongliang 2012). Yet, the problem remains that it does not extend well to a wide range of data with only a few existing features identified by human experts. In addition, only constrictive interpretation is possible, which describes intricate physical interaction within the previously defined features. In this perspective, an approach for direct interpretation from the DOS signal itself, without human intervention in feature extraction, is required.

In this work, we developed a convolutional neural network (CNN) based ML model taking DOS directly as input for accurate formation energy prediction. Further, the possibility of translating the relationship between the electronic structure of materials and the crystal's thermodynamic stability via linking the XAI technique to the model is presented. Thanks to the convolution operation, which has strength in recognizing patterns of signal data, electronic features of DOS are self-explored inside the model with high flexibility. Particularly, our model, whose architecture is motivated by chemical background, is specialized to featurize DOS. After training, the SHapley Additive exPlanation (SHAP) (LundbergS.M., 2020) value of features discovered by the model was calculated to determine what factors on each input affected the prediction in which way. By visualizing how SHAP values show the pattern toward formation energy, common characteristics of stable materials can be confirmed.

2. Results

2.1. ML approach for accurate formation energy prediction

We developed a ML model with a chemically motivated architecture optimized to featurize partially projected DOS for accurate formation energy prediction. The model is built based on a convolutional neural network (CNN), which is well-established for learning high-level features from signal data with multiple channels (KiranyazSerkan 2019). Given that CNN takes input made up of separate channels and featurizes input preserving its spatial information, which indicates spin state according to energy here, it is advantageous to manage DOS input consisting of various projected states. Additionally, the convolution operation gathers input features h_k^{n-1} multiplying weight matrices w_{kj}^n, then aggregates them by sum to generate featurized output features represented as h_j^n (KimYoon 2014). Since each partial DOS represents a relative contribution of a particular state at total DOS, total DOS features can be obtained via aggregating features from partial DOS, which resembles convolution operation. Based on the similarity, the CNN-based featurizer generates abundant features of the total DOS from the input partial DOS.

Meanwhile, the sign of DOS is determined depending on its spin state, that is, positive and negative for up spin and down spin, respectively. Thus, total DOS is generally considered according to its spin state (YeoByung 2019). Inspired by the background, we introduced the CNN-based featurizer for each spin state, named spin

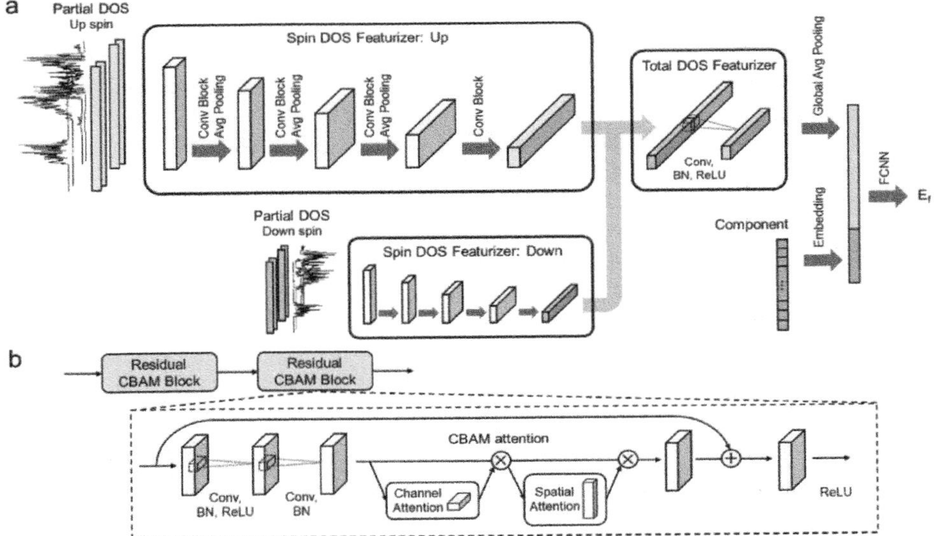

Figure 1. (a) Schematic representation of model architecture. Partial DOS in each spin state and component vector are fed into the model as different inputs. Three featurizers are used to generate total DOS feature maps. (b) Convolutional block architecture employing CBAM self-attention and residual connection. Each convolutional block is connected through average pooling.

DOS featurizer. As shown in Figure 1a, there are two spin DOS featurizers for each spin state respectively. Each featurizer takes orbital projected DOS as input according to the spin state and extracts features based on the convolution operation mentioned above. Note that, the featurizers are independent of each other, using separate weight matrices. The spin DOS featurizer is constructed with four convolutional blocks with two residual convolutional block attention module (CBAM) blocks (WooSanghyun 2018) for automatic and adaptive feature learning. Each block is designed with the aim of producing well-learned feature maps which abundantly contain meaningful features, and the detailed structure is represented in Figure 1b. The CBAM self-attention module performs dynamic feature selection that learns the importance of the features differently for each input, rather than imposing the same weight on all extracted features as general CNN does. In addition, the residual connection prevents the loss of these weighted features due to the depth of the network (DuXue 2020). After further learning through the subsequent convolutional blocks, total DOS feature maps of each spin state are obtained at the end of the featurizers.

The maps are concatenated in channel dimension and served as input to the following total DOS featurizer, which finally aggregates the integrated feature maps to form comprehensive feature maps covering all the spin and orbital states. The final feature maps generated through all the CNN-based featurizers described above are flattened and concatenated with component information followed by the embedding layer. Finally, it is taken as the input of fully connected neural network (FCNN) to predict formation energy. The key aspects of our model are optimal operation for aggregation of partially projected DOS and chemically inspired architecture considering DOS based on its spin state. This enables the model to maximize the number of high-quality features for formation energy prediction.

The model trained on DFT-calculated DOS from Materials Project (MP) database (JainAnubhav 2013), which includes 66,411 materials and their DOS data in total. Note that the training data consisted of only *s*, *p*, and *d*-block elements, but it is possible to include *f*-block elements in training/test data by integrating extra channels in the model. We generated input representations, which consist of up and down spin states of 9 orbitals from *s* to *d*. Each one comprises an individual input channel (total 18 channels), where StandardScaler of scikit-learn package was applied to remove the mean and to scale to unit variance. After preprocessing 18 channels, we added one extra channel representing a composition of materials, where one-hot encoding of all the elements in our dataset, 110, was used. Tensorflow 2.4.1 and Keras 2.4.0 are used as the backend for the model implementation. The model was trained for 300 epochs with an early stopping algorithm on an NVIDIA GPU with memory and CUDA 11.0. Additionally, we highlight that our model could be universally used for formation energy prediction tasks as it was trained with data consisting of diverse compositions of elements across more than 80.

2.2. Prediction Performance

Table 1. Prediction performance with 2e-ORR test set for different algorithms.

	DOS signal-based		**DOS feature-based***				
Model	Proposed	CNN	KRR	SVR	RF	XGBoost	DNN
MAE	**0.236**	0.273	0.271	0.280	0.356	0.307	0.274
RMSE	**0.372**	0.397	0.435	0.446	0.562	0.498	0.420

**d*-band center, Width, Skewness, Kurtosis, Number of fillings, Compositions

We then evaluated the prediction performance of our model and compare it with other ML models based on different input features and algorithms. All models were tested on the same dataset of electrochemical two-electron O_2 reduction reaction (2e-ORR) which contains DOS of 7165 binary alloys, and the overall performances are listed in Table 1. The DOS feature-based models (i.e. kernel ridge regressor (KRR) (SchlkopfBernhard 2018), support vector regressor (SVR) (AwadMariette 2015), random forest (RF) (BreimanLeo 2001), XGBoost (ChenTianqi 2016), deep neural network (DNN) with two fully connected layers (CourvilleIan 2016)), where 0–4th moment properties of 18 orbitals were used as inputs, underperform compared to the DOS signal-based models (i.e. proposed, CNN (SunDechun 2019)). This result reiterates that only a few existing features are insufficient to properly explain solid materials in the whole dataset. As such, since the excessive simplicity and low flexibility in fitting, these models are difficult to predict as accurately as models that turn all the information on DOS to account. Furthermore, the proposed model here outperforms conventional CNN, highlighting the effectiveness of chemically inspired architecture and CBAM managing input DOS effectively. As a result, our model achieved the highest performance in formation energy prediction with 0.237 eV/atoms of MAE, demonstrating its superior featurization ability.

2.3. XAI Interpretation

To unpack the black-box model and confirm the physical interaction between DOS and the formation energy, SHAP analysis was implemented. SHAP is a XAI technique introducing a game theoretic approach to explain the output of the ML model. It decomposes the predicted value into an additive sum of contributions of individual features on input. Due to its guaranteed mathematical background, SHAP values are one of the most-used approaches to understanding ML models. Prior to further analysis, we projected SHAP values obtained from 2e-ORR data into a three-dimensional space using uniform manifold approximation and projection (UMAP) (McInnesLeland 2018) to

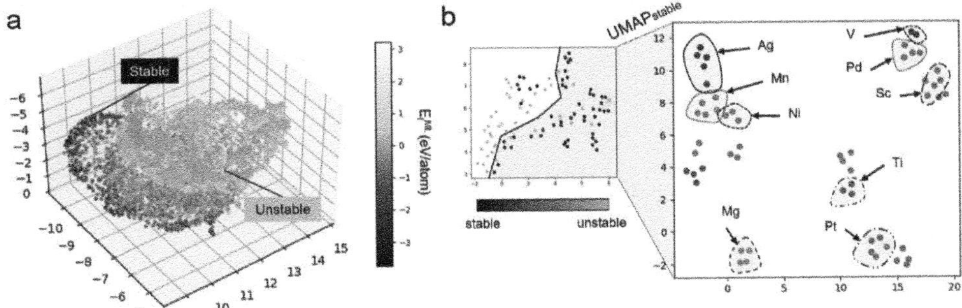

Figure 2. Visualization of SHAP value distribution for different datasets, generated by UMAP. a) Three-dimensional UMAP for 2eORR dataset. Darker colors indicate low E_f^{ML}. (b) Two-dimensional UMAP for the single-specie dataset. (left) According to the E_f^{ML}, less than 0.01 is categorized as stable and others as unstable. (right) Two-dimensional UMAP plotted again with only the data belonging to the stable region.

visualize whether the learned features show a tendency depending on material stability (Figure 2a). The unsupervised analysis displays the continuous change in feature importance on DOS in accordance with the formation energy. Particularly, the darker area, where the data with low formation energy is located close, demonstrates that stable materials share common features on DOS.

Inspired by this result, it is analyzed more specifically what common characteristics appear in SHAP values for materials that are predicted as stable. For ease of analysis, we preliminarily obtained DOS and their SHAP values from various crystal structures, i.e. simple cubic (SC), body-centered cubic (BCC), face-centered cubic (FCC), hexagonal close-packed (HCP), and rhombohedral crystal structure (RHO), for materials consisting of only a single specie, and projected the SHAP values into low dimension. Similar context to the previous result, it is confirmed that the materials predicted as stable and the materials predicted as unstable are distributed in separate regions on UMAP (Figure 2b left). Then, UMAP was performed again targeting only the data that belong to the darker area where 'stable' data are largely located (Figure 2b right). Surprisingly, the data are clustered by the specie, showing that data within the same element share similar features even if having different crystal structures. The similarities are not observed in the data with unstable crystal structures. These results demonstrate the possibility of XAI interpretation that deduces the common properties of stable materials and further uncovers the causal relationship between the electronic structure and the formation energy.

3. Conclusion and Future Works

Here we have developed a machine learning model which successfully predicts the formation energy from DOS with superior featurization ability. More importantly, we presented the potentiality of direct XAI interpretation of the DOS signal itself without any prior involvement of human experts in feature restriction. The result shown here is simply that features on DOS have commonality according to their formation energy. Thus, this observation will be developed in future works to reveal the 'which' factor of DOS is decisive in formation energy prediction, and finally, it will shed light on the causal relationship between the electronic structure of materials and the crystal's thermodynamic stability. Ultimately, it will contribute beyond the stability screening by the black-box model to knowledge and hypothesis generation in the field. (EsterhuizenJ.A., 2022)

References

Awad, Mariette and Khanna, Rahul},. Efficient Learning Machines: Theories, Concepts, and Applications for Engineers and System Designers. Apress, 2015.

Breiman, Leo. "Random Forests." Machine Learning, 2001. 5-32.

Chen, Tianqi and Guestrin, Carlos. "XGBoost: A Scalable Tree Boosting System." Association for Computing Machinery, 2016. 785-794.

Courville, Ian Goodfellow and Yoshua Bengio and Aaron. Deep Learning. MIT Press, 2016.

Doosun Hong, Jaehoon Oh, Kihoon Bang, Soonho Kwon, Se-Young Yun, Hyuck No Lee. "Interpretable Deep learning Model for Analyzing the Relationship between the Electronic Structure and Chemisorption Property." J. Phys. Chem. Lett., 2022. 8628-8634.

Du, Xue and Liao, Kuo and Shen, Xiaofeng. "Secondary Radar Signal Processing Based on Deep Residual Separable Neural Network." 2020 IEEE International Conference on Power, Intelligent Computing and Systems (ICPICS), 2020. 12-16.

Esterhuizen, J.A., Goldsmith, B.R., Linic, S. "Interpretable machine learning for knowledge generation in heterogeneous catalysis." Nat Catal 5, 2022. 175-184.

Fung, V., Hu, G., Ganesh, P. "Machine learned features from density of states for accurate adsorption energy prediction." Nat Commun 12, 2021. 88.

Jain, Anubhav and Ong, Shyue Ping and Hautier, Geoffroy and Chen, Wei and Richards, William Davidson and Dacek, Stephen and Cholia, Shreyas and Gunter, Dan and Skinner, David and Ceder, Gerbrand and others. "Commentary: The Materials Project: A materials genome approach to accelerating materials innovation." American Institute of PhysicsAIP, 2013. 011002.

Kim, Yoon. "Convolutional Neural Networks for Sentence Classification." Association for Computational Linguistics, 2014.

Kiranyaz, Serkan and Ince, Turker and Abdeljaber, Osama and Avci, Onur and Gabbouj, Moncef. "1-D Convolutional Neural Networks for Signal Processing Applications." ICASSP 2019 - 2019 IEEE International Conference on Acoustics, Speech and Signal Processing (ICASSP), 2019. 8360-8364.

Linic, Hongliang Xin and Adam Holewinski and Neil Schweitzer and Eranda Nikolla and Suljo. "Electronic structure engineering in heterogeneous catalysis: Identifying novel alloy catalysts based on rapid screening for materials with desired electronic properties." Topics in Catalysis, 2012. 376-390.

Lundberg, S.M., Erion, G., Chen, H. "From local explanations to global understanding with explainable AI for trees." Nat Mach Intell 2, 2020. 56-67.

McInnes, Leland and Healy, John and Melville, James. "UMAP: Uniform Manifold Approximation and Projection for Dimension Reduction." arXiv, 2018.

Omidvar, Noushin and Pillai, Hemanth S. and Wang, Shih-Han and Mou, Tianyou and Wang, Siwen and Athawale, Andy and Achenie, Luke E. K. and Xin, Hongliang. "Interpretable Machine Learning of Chemical Bonding at Solid Surfaces." J. Phys. Chem. Lett., 2021. 11476-11487.

Schlkopf, Bernhard and Smola, Alexander J. and Bach, Francis. Learning with Kernels: Support Vector Machines, Regularization, Optimization, and Beyond. The MIT Press, 2018.

Sun, Dechun and Chen, Yang and Liu, Jiaao and Li, Yu and Ma, Rui. "Digital Signal Modulation Recognition Algorithm Based on VGGNet Model." 2019 IEEE 5th International Conference on Computer and Communications (ICCC), 2019. 1575-1579.

Wang, SH., Pillai, H.S., Wang, S. "Infusing theory into deep learning for interpretable reactivity prediction." Nat Commun 12, 2021. 5288.

Woo, Sanghyun and Park, Jongchan and Lee, Joon-Young and Kweon, In So. "CBAM: Convolutional Block Attention Module." Springer International Publishing, 2018. 3-19.

Yeo, Byung Chul and Kim, Donghun and Kim, Chansoo and Han, Sang Soo. "Pattern Learning Electronic Density of States." Scientific Reports, 2019. 5879.

Antonis Kokossis, Michael C. Georgiadis, Efstratios N. Pistikopoulos (Eds.)
PROCEEDINGS OF THE 33rd European Symposium on Computer Aided Process Engineering (ESCAPE33), June 18-21, 2023, Athens, Greece
 http://dx.doi.org/10.1016/B978-0-443-15274-0.50111-6

Modular stochastic optimization for optimal rainwater harvesting system design

Qiao Yan Soh,[a] Edward O'Dwyer,[a] Salvador Acha,[a] Nilay Shah[z]

[a]Sargent Center for Process Systems Engineering, Department of Chemical Engineering, Imperial College London, London SW7 2AZ, UK
qiaoyan.soh13@imperial.ac.uk

Abstract

Rainwater Harvesting (RWH) systems can serve a dual functionality as a flood mitigation structure as well as providing local water availability. Optimisation-based design strategies must be transferrable enough to incorporate the influence of the local climate and case-specific catchment area characteristics into the design process, which can be a significant endeavour when required for every individual implementation. To increase the accessibility of optimisation methods in the appropriate sizing of RWH systems, this paper presents a modularised optimisation model, where tank components and dynamics are contained as individual blocks. These blocks can then be pieced together to produce a full system model, allowing optimisation models to be easily built for any combination and design of RWH system. This is implemented with a multi-tank RWH system, where an evaluation of the optimised system configuration showed a good balance between the dual objectives of providing improved flood mitigation and local water reuse, in comparison to an existing system derived through alternative sizing strategies.

Keywords: Multi-scale integration, Rainwater harvesting, Stochastic optimization

1. Introduction

Large-scale rainwater harvesting (RWH) systems have demonstrated potential as a strategy for simultaneously reducing flood risks by temporarily holding excess water, and minimising water stress by satisfying non-potable water demands with locally harvested water. As the performance of these systems has been demonstrated to be very sensitive to the local climate and catchment area (Semaan, 2020), appropriately sizing these structures has remained a key barrier to further implementations of RWH systems.

Optimisation models have been found to be effective and successful in deriving suitably sized RWH systems (Okoye, 2015; Nnaji, 2017), but with the range of possible system configurations, optimisation-based strategies are not the most scalable of sizing methods since the creation of optimisation models for each individual implementation of a RWH system can require significant time, effort, and expertise. Hence, it is common to find that design guidelines published by public water managers utilise simplified functions, for example of the expected average demand and catchment area to calculate an appropriate tank size. These are normally found to prioritise the flood mitigation objectives, especially in highly developed urban centres as these are likely to have a stable freshwater supply network to satisfy any local water demands. However, this would result in overly sized systems which can reduce the return on investment for the RWH system and incur further costs in the form of utilising the limited space resources available within a city (Matos, 2013).

Hence, to improve the accessibility of and encourage more widespread use of optimisation-model-based strategies for sizing urban RWH systems, this paper proposes a modularised optimisation model tool. This provides basic optimisation model blocks for a single tank, with a range of possible tank draining configurations which can be further customised through functional constraint blocks. This declarative format would allow individuals with expertise outside of implementing optimisation methods to utilise them as an effective system sizing strategy.

2. Methods

The modularised optimisation model is presented in the following sections and applied to a multi-tank RWH system. The tool is used to derive an optimal system size for a multi-tank RWH system, and the output system configuration is evaluated for its ability to achieve the dual objectives of flood mitigation and local water reuse. The computational requirements are measured as a function of the model solution time, as this could be a key factor for increased uptakes of optimisation-model based strategies.

2.1. Modular optimization model structure

The proposed modularised optimisation framework consists of function blocks which correspond to the dynamics of a given tank component and build up to a fully stochastic Mixed Integer Linear Program (MILP) model that looks to select an optimal tank height, area, and any applicable orifice heights for the range of rainfall scenarios given to the model.

The component function blocks correspond not only to typical tank dynamics such as the mass balance and flow equations, but also provide design specifications and behaviours including:

- Penalty functions for exceeding threshold overflows or wasted capacities
- Common optimisation objectives including system size reduction, and overflow minimisations.
- Freshwater use for satisfying unmet demand

To provide a turnkey functionality to this framework, further tank configuration blocks were defined which correspond specifically to the different possible combinations of ways a given tank may be drained. These blocks can be initialised immediately and solved when supplied with a desired input parameter set but can also be further supplemented with component function blocks to implement further tank behaviours and or penalty functions for multi-objective systems. A basic implementation workflow for a single tank would hence involve:

(1) Create a base optimisation model through a tank configuration block
(2) Add any additional system dynamics and constraints
(3) Define an objective function, including any penalties
(4) Populate the model with rainfall, demand, and design parameter options
(5) Solve the derived model.

Under this framework, multi-tank systems can be implemented and optimised sequentially through a series of much smaller MILPs, which typically would be much more tractable than a single, large optimisation problem.

2.2. Case study configuration

The framework was applied to a multi-tank RWH system aimed at simultaneously reducing local flooding risks and providing non-potable water to its service area. This is formed of four specialized tanks, as illustrated in Figure 1. The ‘separation’ tank serves as an initial receiver for all inflow and acts as a filter that directs flow between the ‘harvesting’ and ‘detention’ tanks. These tanks both store water, but separately for the

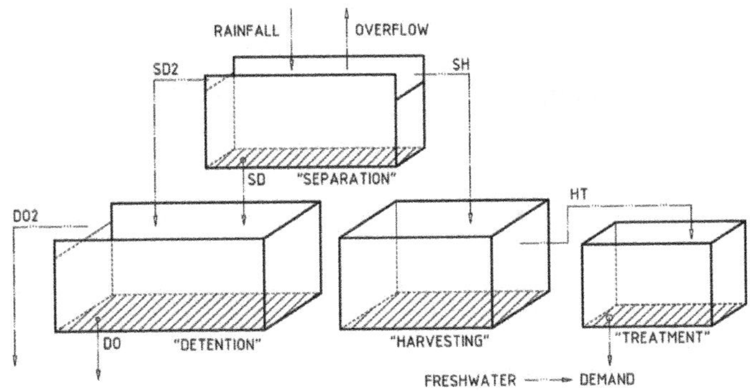

Figure 1: Multi-tank RWH system configuration and flow interactions

purposes of local reuse and temporarily holding excess to prevent downstream inundation respectively. Finally, a 'treatment' tank is used to hold treated, ready-to-use water.
Under the modularised framework, each individual tank is optimised individually with non-linear flow divisions within the system managed using an external flow division function. The flow of information between the individual tank optimisation models, as well as their corresponding model outputs, are highlighted in Figure 2.
The input data comes from a historical timeseries, which has been segregated into 24-hour rolling window samples. Scenarios are then selected randomly from the set of samples, with the set diversity ensured using a minimum threshold value for a variance measure, defined as the sum of the variance of the rainfall values in each timestep across the entire simulation horizon. When the minimum threshold is not reached, the set is reselected.

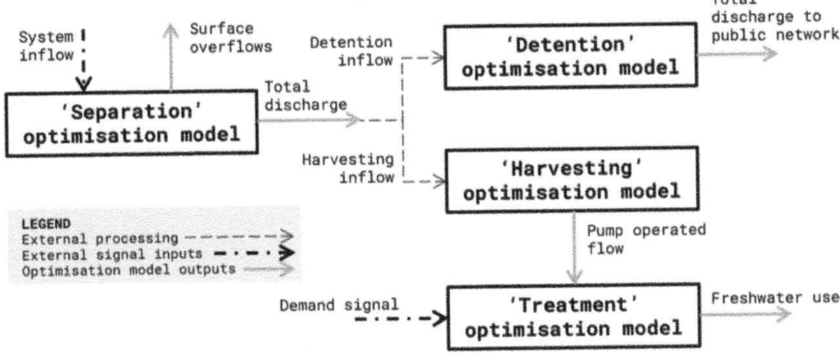

Figure 2: Information flow and interactions between optimisation models in a multi-tank RWH system

Table 1: Implemented objective and penalty functions for each tank optimisation model.

Tank	Objective	Penalty function
Separation	Minimise overflows	Wasted capacity
Detention	Minimise tank capacity	Overflow threshold
Harvesting	Maximise outflow to treatment tank	Overflow threshold
Treatment	Minimise freshwater top-up volumes	Overflow threshold

3. Results

The performance of a modularised optimisation model is dependent on the following model parameters:

- *System configuration parameters:* The set of possible tank areas, tank heights and orifice heights.
- *Solution time:* The total allowable time for the model to find a solution.
- *Allowable capacity:* The maximum volume of space the derived tank can occupy.
- *Penalty costs:* The penalty values that influence the objective function value, if applicable in the specific model.
- *Stochasticity:* The number of rainfall scenarios used.

The sets of possible parameter values provided for the optimization models are selected as a large possible design space for a system that is to serve a 1.57 ha residential estate. Lower vertical space allowances are used as the underground make-up of a city is typically carefully subdivided between multiple utilities and different layers are hence reserved for distinct, specific purposes. The objective and penalty functions for each of the four tank systems are shown in Table 1, where the overflow threshold penalty assigns a cost to the overflow volume in exceedance of the threshold volume defined for a given timestep, and the wasted capacity penalty function implements a three-tiered cost for any wasted tank volumes above 10% of the total tank capacity.

For effective adoption of a novel sizing strategy, the proposed method needs to be sufficiently easy to implement, whilst providing system configurations that provide adequate levels of flood mitigation and yield collection. Hence, the following subsections discusses the impact of stochasticity on the required solution time and derived solution quality for the modularised optimisation framework.

3.1. Stochasticity and solution time

The model's behaviour as the set of rainfall scenarios increases is characterised by the computational time required to derive an optimal solution. The set of rainfall scenarios is selected separately for each set size to ensure that each set meets the minimum threshold variance value. This was applied to each of the optimisation models in the multi-tank RWH system individually, such that the behaviour of each tank configuration model block can be characterised.

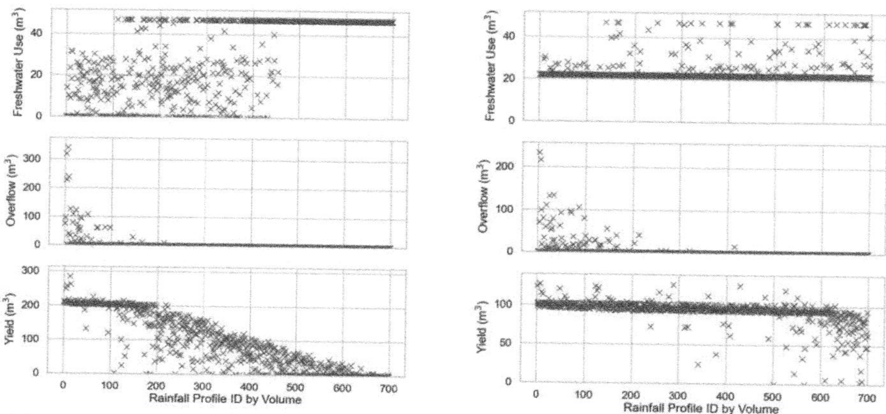

(a) Modularised system performance (b) Base system performance

Figure 3: Scatterplot of key performance indicator values, in order of decreasing input rainfall volume.

The number of scenarios and size of the derived optimisation model have minimal impact on the computational requirement of the individual tank optimisations. On the other hand, the 'Separation' tank, which has the highest number of openings and hence constraints and variables, is significantly more computationally intensive than the other tank design types and requires eight-times the amount of computational time to derive an optimal solution.

3.2. Performance of the derived configuration

The output optimal configuration for the full system remained the same despite increasing the number of scenarios used with the modular optimisation model. This optimised output is shown, in comparison to an existing configuration in Table 2. The total time required to derive a full configuration with 25 rainfall scenarios selected with a minimum variance value of 50 was 1.5 hours when implemented on an Intel Core i7-8650U with 16.0 GB ram.

The performance of the derived configuration is determined for its ability to reduce surface overflow risks and provide local water availability using a simulation model. A statistical measure of the key performance indicators is generated by running simulations for 700 days of the highest rainfall volumes extracted from an eight-year historical timeseries, and the KPI values, sorted by descending input rainfall volumes are demonstrated in Figure 3.

Statistical performance metrics are then extracted from these figures to provide quantitative indicators of the system performance. Freshwater use indicates the amount of freshwater required to ensure that local non-potable water demands are fully met. The scatterplots above demonstrate that the existing configuration is limited in its ability to provide water for irrigation purposes in a timely manner and hence has a minimum freshwater requirement for all simulated scenarios. On the other hand, the optimised system is observed to be able to provide for all the required water demand for 16.3% of the simulated scenarios and only fails to provide sufficient water for reuse when the inflow volumes fall below a total of 230 m^3.

Table 2: System configurations for an existing RWH system, and an optimised configuration for the same catchment area.

	Existing configuration		Optimised configuration	
Tank	*Height (m)*	*Area (m^2)*	*Height (m)*	*Area (m^2)*
Separation	1.23	5.02	2.5	80
Detention	2.00	264.52	0.5	10
Harvesting	2.40	38.20	2.5	80
Treatment	2.00	5.00	1.25	100
Total	**636.84 m^3**		**530 m^3**	
Orifice	*Height (m)*		*Height (m)*	
SH	0.52		1.5	
SD2	0.66		1.75	
DO2	1.8		0.4	

The optimised configuration has a higher peak overflow volume of 285m^3, but the proportion of scenarios which demonstrated overflow is reduced from 10.86% to 5.86%. Thus, with the optimised configuration, overflow events are less likely to occur, but this comes at the price of having higher overflow volumes if they do. The existing configuration is better at harvesting some amount of water for 99.7% of the simulated scenarios, while the optimal configuration only harvested water in 78.3% of the simulated scenarios. Hence, the optimal configuration is much better at providing water availability during larger rain events, whilst the existing configuration provides a much more consistent yield performance over the possible range of rainfall volumes.

4. Conclusions

The results demonstrated that the modularised optimisation model can provide a full RWH system configuration for a multi-tank system with low computational time limits of 1.5 minutes. Under this optimisation-based sizing strategy, there is an improved balance between the dual objectives of improving local reuse of harvested water and reducing flood risks, especially in comparison to an existing RWH system configuration which had not been derived using optimisation-based methods.

Thus, the modularised optimisation tool presented in this paper can encourage more implementations of multi-objective RWH systems by improving the accessibility of optimisation-based sizing strategies. Through the tank block framework, optimisation models can easily be built if the desired configuration of discharge openings is known. Further behavioural specifications can also be appended to the model if this is required, and a full optimisation model, specific to the design for the given catchment, can be derived much more quickly in comparison to a full, single optimisation model.

The performance of a stochastic optimisation model, however, remains highly influenced by the data quality used for its input scenarios. Therefore, the performance of the modularised tool continues to be dependent on the availability of high-resolution rainfall data. To mitigate this, there is currently work in progress to integrate such a sizing strategy with a rainfall design model to derive a robust RWH system configuration.

Acknowledgements

This research is supported by the Singapore Ministry of National Development and the National Research Foundation, Prime Minister's Office under the Land and Liveability National Innovation Challenge (L2 NIC) Research Programme (L2 NIC Award No. L2NICTDF1-2017-3). Any opinions, findings, and conclusions or recommendations expressed in this material are those of the author(s) and do not reflect the views of the Singapore Ministry of National Development and National Research Foundation, Prime Minister's Office, Singapore.

References

M. Semaan, S. D. Day, M. Garvin, N. Ramakrishnan, and A. Pearce, "Optimal sizing of rainwater harvesting systems for domestic water usages: A systematic literature review," *Resour. Conserv. Recycl. X*, vol. 6, no. November 2019, p. 100033, 2020.

C. O. Okoye, O. Solyali, and B. Akintug, "Optimal sizing of storage tanks in domestic rainwater harvesting systems: A linear programming approach," *Resour. Conserv. Recycl.*, vol. 104, pp. 131–140, 2015.

C. C. Nnaji, P. G. C. Emenike, and I. T. Tenebe, "An Optimization Approach for Assessing the Reliability of Rainwater Harvesting," *Water Resour. Manag.*, vol. 31, no. 6, pp. 2011–2024, 2017.

C. Matos, C. Santos, S. Pereira, I. Bentes, and M. Imteaz, "Rainwater storage tank sizing: Case study of a commercial building," *Int. J. Sustain. Built Environ.*, vol. 2, no. 2, pp. 109–118, 2013.

Q. Y. Soh, E. O'Dwyer, S. Acha, and N. Shah, "Stochastic optimal design for large-scale rainwater harvesting and detention systems," in *ECOS 2022 - the 35th international conference on efficiency, cost, optimization, simulation and environmental impact of energy systems*, 2022.

Antonis Kokossis, Michael C. Georgiadis, Efstratios N. Pistikopoulos (Eds.)
PROCEEDINGS OF THE 33rd European Symposium on Computer Aided Process Engineering (ESCAPE33), June 18-21, 2023, Athens, Greece
 http://dx.doi.org/10.1016/B978-0-443-15274-0.50112-8

Knowledge-based model and simulations to support decision making in wastewater treatment processes

Claire Valentin,[a] Frédéric Lagoutière,[b] Jean-Marc Choubert,[c] Françoise Couenne,[a] Christian Jallut,[a]

[a]Univ Lyon, Université Claude Bernard Lyon 1, CNRS, LAGEPP UMR 5007, 43 boulevard du 11 novembre 1918, 69100, Villeurbanne, France
[b] Univ Lyon, Université Claude Bernard Lyon 1, CNRS, Institut Camille Jordan, UMR 5208, 43 boulevard du 11 novembre 1918, 69622 Villeurbanne Cedex, France
[c]INRAE REVERSAAL, 69625 Villeurbanne, France
claire.valentin@univ-lyon1.fr

Abstract

In this paper, mass and momentum balances are used to model and simulate the most significant phenomena occurring in a continuous urban sludge clarifier. This model is designed to be the core of a digital twin in the future. This has two consequences: the choice of a one-dimensional model only and a numerical scheme for simulation that allows a reasonable runtime. We propose a different way of writing of the dynamic model of the clarifier. Instead of using volume fractions and velocity of solid particles as state variables, we use volume fractions and flux of solid particles. This approach, which is used for conservation law systems, gives more stable simulation results for hyperbolic systems.
This paper focuses on comparative discussions of these two different versions of the model. The numerical simulation scheme is presented. In addition, the simulation is compared with experimental data obtained in a full-scale wastewater treatment plant.

Keywords: Urban sludge continuous settling, Dynamic mass and momentum balances, nonlinear hyperbolic system, Rusanov numerical approximation

1. Introduction

The efficiency of wastewater treatment facilities is a worldwide major problem. Urban wastewater treatment plants are regulated by European directives, such as 91/271/EEC. The biological treatment unit of a treatment plant purifies wastewater before it is released into the natural environment by using a clarifier. The quality of the clarified water strongly depends on the instantaneous hydraulic loads arriving upstream, on the design of the equipment and on the operating conditions. The settling of the sludge in the clarifier enables the separation of more concentrated sludge that is pumped down and clarified water that is released up into the environment in an overflow. Thus, the quality of the plant effluent is highly dependent on the performance of the clarifier. In order to optimize its operation, modelling and simulation are the first steps to perform.

In the literature, the modelling of clarifiers is obtained through mass and momentum balance equations for solid particles. The latter is written as a static or dynamic partial differential equation (Chauchat (2013), Garrido (2003)) or replaced by a constitutive relation representing the velocity of solid particles (Li (2014)). In the case where the momentum balance is represented by a PDE, the global model is weakly hyperbolic and the position of the so-called sludge blanket corresponds to the front of a shock wave. This shock is related to a discontinuity in the solid particle volume fraction and a peak value in the flux.

In this paper, we present two choices of state variables for the sludge settling modelling in a clarifier (Valentin (2022)) and discuss the simulation results with respect to experimental data.

2. A 1-D schematic representation of the clarifier

Fig. 1 shows a 1-D schematic representation of the clarifier. It is connected to the wastewater treatment process at three points, one inlet and two outlets:

• one inlet where the sludge, consisting of fluid and particles, flows into the clarifier by gravity from the upstream biological aeration tank at the volume flow rate $Q_f(t)$ and with a particle concentration $C_f(t)$. The sludge feed is situated at a depth of $z = z_f$,

• one top outlet for clarified water at $z = 0$, with a volume flow rate $Q_e(t)$ and particle concentration $C_e(t)$,

• one bottom outlet for compressed sludge at $z = z_b$, with a volume flow rate $Q_u(t)$ and particle concentration $C_u(t)$. Some of the compressed sludge is recirculated back into the aeration tank at a volume flow rate $Q_{ur}(t)$ and some is extracted from the clarifier at a volume flow rate $Q_{ue}(t)$ such that $Q_u(t) = Q_{ur}(t) + Q_{ue}(t)$.

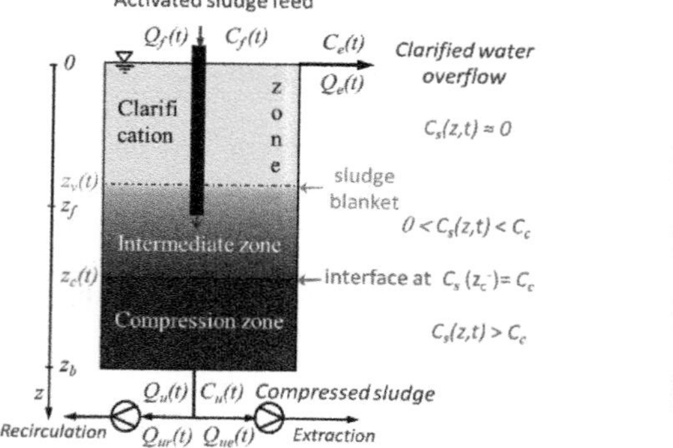

Figure 1: Schematic view of a clarifier

A	$1175\ m^2$
A_k	$0.02\ m/s$
ε_c	$4.1\ 10^{-2}$
n_r	2
n_s	11
ρ_l	$1000\ kg/m^3$
ρ_s	$1030\ kg/m^3$
σ_0	$0.5\ kg/ms^2$
z_b	$2.8\ m$
z_f	$1.8\ m$

Table 1: Model parameter values

The open-air settler content can be divided into two moving interfaces separating three zones:

• The upper interface is the sludge blanket, located at depth $z_v(t)$. It separates the clarification zone (which contains no or very few particles) from the intermediate zone,

• The lower interface is defined by the intermediate/compression threshold and is located at depth $z_c(t)$ where a change in the particles behaviour occurs as the particle concentration $C_s(z,t)$ exceeds the critical threshold C_c. Above this threshold, interparticle stress comes into effect. Under $z_c(t)$, the liquid phase flows through a porous network of concentrated particles, Toorman (1996). The model parameter values are given in Table 1.

3. A 1-D dynamic model of the clarifier

The knowledge-based model includes two dynamic mass and momentum balances of urban sludge added with physical constitutive equations under standard simplifying assumptions, Valentin (2022). As the liquid and solid phase densities ρ_l and ρ_s are constant, the two most natural state variables are the particle volume fraction, ε_s, and the particle flux, $f_s = \varepsilon_s v_s$ with v_s the particle velocity. The corresponding hyperbolic system of conservation laws is then defined, if $\varepsilon_s > 0$, by:

$$\partial_t \varepsilon_s + \partial_z f_s = \frac{f_{1s}}{\rho_s}\, \delta_f \tag{1}$$

$$\partial_t f_s + \partial_z \left(\frac{{f_s}^2}{\varepsilon_s}\right) = \varepsilon_s g \left(1 - \frac{\rho_l}{\rho_s}\right) - \frac{\partial_z \sigma_e(\varepsilon_s)}{\rho_s} + \frac{r(\varepsilon_s)(\varepsilon_s v_m - f_s)}{\rho_s \varepsilon_s (1 - \varepsilon_s)} + \frac{f_{2s}}{\rho_s}\, \delta_f \tag{2}$$

with $\sigma_e(\varepsilon_s)$ the interparticle stress between the particles, $r(\varepsilon_s)$ the liquid/solid drag force and v_m the average velocity of the liquid/solid mixture (also the total volume flux of the suspension). f_{1s} and f_{21s} are the source terms representing the sludge feed inlet in the particle mass and momentum balances equations respectively. They depend on Q_f and C_f. δ_f is a Dirac function that represents the location of the sludge feed at $z = z_f$. The two boundary conditions are $v_s(0, t) = -\frac{Q_e}{A}$ and $v_s(z_b, t) = \frac{Q_u}{A}$.

Various kind of constitutive equations that describe the compression and drag phenomena have been proposed in the literature, Li (2014). $r(\varepsilon_s) = \frac{\rho_l g}{A_k} {\varepsilon_s}^{2/(3-n_r)}$ is proposed in Chauchat (2013) and the following $\sigma_e(\varepsilon_s)$ expression, that depends on the particle volume fraction, in Garrido (2003). If $\varepsilon_s > \varepsilon_c$ (in the compression zone), $\sigma_e(\varepsilon_s) = \sigma_0 \frac{\varepsilon_s^{n_s} - \varepsilon_c^{n_s}}{\varepsilon_c^{n_s}}$ else (in the two other zones), $\sigma_e(\varepsilon_s) = 0$ (see Table 1). $\sigma_e(\varepsilon_s)$ is a continuous function on [0, z_b] but its disadvantage is that it is zero over the upper part of the spatial domain which makes the system weakly hyperbolic only.

Most of the papers in the literature present a model based on a dynamic particle mass balance with the state variable, ε_s, coupled with a constitutive equation that gives the so-called batch or hindered settling velocity (Garrido, 2003), (Li, 2014). This motivated us and others, Chauchat (2013), Valentin (2022) to use first the state variables (ε_s, v_s) according to the following hyperbolic system of two PDEs:

$$\partial_t \varepsilon_s + \partial_z (\varepsilon_s v_s) = \frac{f_{1s}}{\rho_s}\, \delta_f \tag{3}$$

$$\partial_t v_s + \partial_z \left(\frac{{v_s}^2}{2}\right) = g \left(1 - \frac{\rho_l}{\rho_s}\right) - \frac{\partial_z \sigma_e(\varepsilon_s)}{\rho_s \varepsilon_s} + \frac{r(\varepsilon_s)(v_m - v_s)}{\rho_s \varepsilon_s (1 - \varepsilon_s)} + \frac{f_{21s}}{\rho_s}\, \delta_f \tag{4}$$

This representation is based on the temporal and spatial derivatives of products such as $\varepsilon_s v_s$. As shock waves appear on the two state variables, such product derivatives may lead to mathematical problems. Moreover, the difficulty of numerically solving the model based on the state variables (ε_s, v_s) and the oscillations obtained on the spatial profiles of the velocity v_s led us to return to the more "natural" state variables that correspond to those used in the so-called conservative approach, the particle volume fraction, ε_s, and the particle flux, $f_s = \varepsilon_s v_s$ (equations (1) and (2)) and to include the interparticle stress in the flux.

4. Numerical discretization scheme

The simulations were carried out using explicit Euler time-discretization and an efficient numerical scheme adapted to hyperbolic and weakly hyperbolic nonlinear PDE systems: a finite volume method spatial-discretization with the Rusanov approximation of the fluxes (Godlewski (1996), LeVeque (2002)).

The state variable vector x and the flux vector $F_s(x)$ are defined if $\varepsilon_s > 0$ by $x = \begin{pmatrix} \varepsilon_s \\ f_s \end{pmatrix}$ and $F_s(x) = \begin{pmatrix} f_s \\ \frac{f_s{}^2}{\varepsilon_s} + \frac{\sigma_e(\varepsilon_s)}{\rho_s} \end{pmatrix}$. The state variables are considered as uniform in each volume i of the mesh and equal to the average values $\bar{x}_i^k$ at time step k.

For time step k and volume i, the flux at the input interface $i - \frac{1}{2}$ is approximated by:

$$F_{i-\frac{1}{2}}^k = \frac{1}{2}\left(F_s{}^k(\bar{x}_{i-1} + F_s{}^k(\bar{x}_i)\right) - \frac{\omega_s{}^k}{2}\left(\bar{x}_i^k - \bar{x}_{i-1}^k\right)$$

with $\omega_s{}^k = \max_i(\rho_i{}^k)$, $\rho_i{}^k$ the spectral radius of the volume i which depends on $\bar{x}_i^k$.

Then:

$$\bar{x}_i^{k+1} = \bar{x}_i^k + \frac{\Delta t}{\Delta z}\left(F_{i-\frac{1}{2}}(\bar{x}_i, t) - F_{i+\frac{1}{2}}(\bar{x}_i, t)\right) + \Delta t\left(S_1(\bar{x}_i) + S_2\right)$$

where we assume that $S_1(\bar{x}_i)$ is a good approximation of $\frac{1}{\Delta z}\int_{i-1/2}^{i+1/2} S_1(x)dz$

and that the variable time-step Δt respects the CFL (Courant-Friedrichs-Lewy) condition: $\Delta t = \gamma \frac{\Delta z}{\omega_s{}^k}$ with $0 < \gamma < 1$. As $\sigma_e(\varepsilon_s) = 0$ in the clarification and intermediate zones, the system is only weakly hyperbolic.

5. Simulation of a transient state experiment of continuous settling

The model simulation is compared to experimental data obtained in a full-scale settler operated under the following transient state scenario (Fig.3): the sludge feed rate was abruptly increased by a magnitude of two from a value corresponding to a stationary profile at $t =$ 1.40 am (660 *m3/h*) and abruptly decreased *8 hrs* later (370 *m3/h*). The flow rates and concentrations at the sludge inlet and outlet were measured on-line as well as the sludge blanket position.

Simulations based on this model were performed with various discretization parameters such as spatial mesh size, convergence condition (Courant-Friedrichs-Lewy). They are compared to the experimental data. Measured sludge feed flow rate, $Q_f(t)$, recirculation flow rate, $Q_{ur}(t)$ and extraction flow rate, $Q_{ue}(t)$ as well their mean values are given in Fig.3.
An N-node spatial mesh was used to run the simulations of the discretized model presented in section 4. with the constitutive equations and boundary conditions given in section 3.

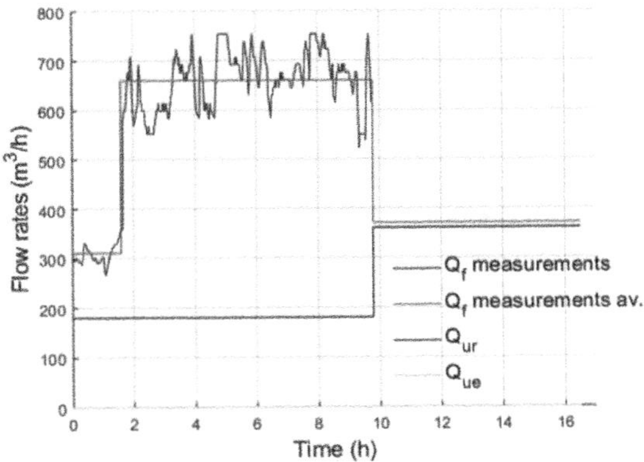

Figure 3: Applied flow rates

The calculated sludge blanket position $z_v(t)$ corresponds to the location where the maximum gradient of the solid concentration is reached. Fig 4.a shows that the simulated and measured sludge blankets positions are very close.

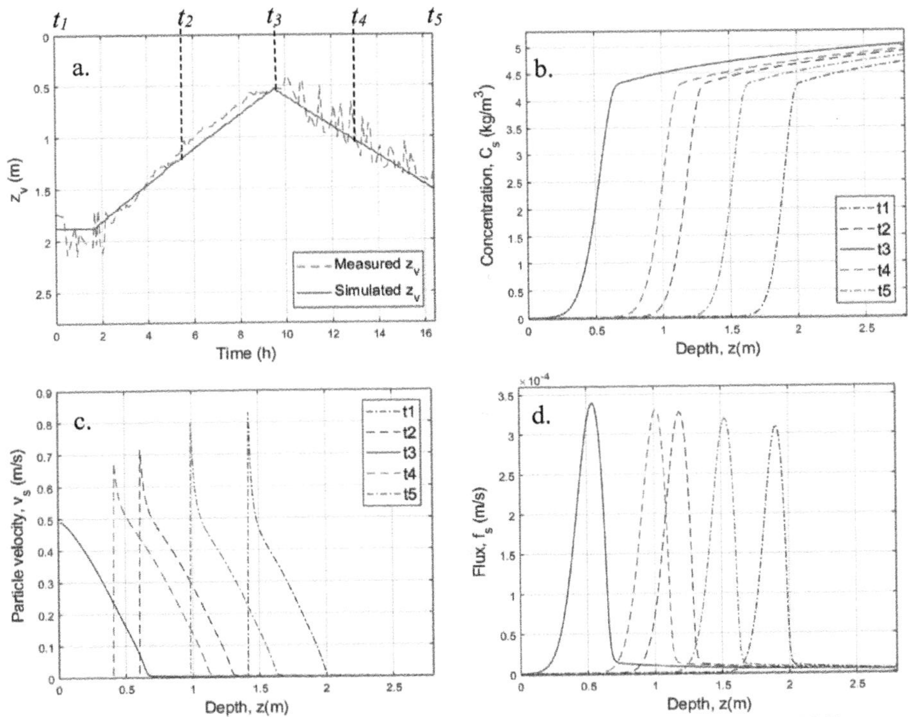

Figure 4: Simulated $z_v(t)$ and $C_S(z)$, $v_s(z)$, $f_s(z)$ spatial profiles (N=200)

Five spatial profiles at t_1, t_2, t_3, t_4 and t_5 are shown in Fig 4.b to Fig 4.d. Fig 4.b presents particle concentration spatial profiles. The particle concentration at the top of the clarifier (z=0) is close to zero. It increases sharply at the depth of the sludge blanket and then increases more moderately to the bottom as soon as the critical threshold is exceeded. The effect of the interparticle stress below the critical threshold is also well highlighted.

Spatial profiles $v_s(z)$ and $f_s(z)$ are shown in Fig. 4.c and Fig. 4.d. The shock wave is also clearly visible at the "peak" location. The particle velocity at the bottom of the settler $v_s(z_b)$ is very low compared to the velocity at the sludge blanket position, *5.3* 10^{-4} *m/s*, and is set by the Q_u flow rate.

Good results are obtained with this numerical scheme for a mesh size of up to 800 volumes and $\gamma = 0.99$ in the CFL condition. Choosing a lower γ value gives the same results but with a longer runtime due to a smaller Δt. The runtime of a simulation with 200 volumes on a workstation equipped with an Intel Xeon at 3.8GHz is 8 *min* 15 *s*. According to the nature of the numerical scheme, the runtime with N = 800 is 16 times longer that makes 2 *hrs* 12 *min*. Although the simulation gives good results with N = 800 with less numerical diffusion, as this model will be the core of a decision support system and then a digital twin, a compromise must be decided. A spatial discretization of N = 200 seems suitable.

6. Conclusions

As a conclusion, the well-suited conservative state variables are the volume fraction and the volume flux of the solid particles. The expression of the momentum balance includes the compression stress between the particles based on a nonlinear constitutive law taken from Garrido (2003). A numerical scheme with explicit Euler time-discretization and a finite volume method spatial-discretization with the Rusanov approximation of the fluxes works well. The simulation results are close to experimental results by using a set of well-chosen parameter values. It can be used for prediction and decision support for other scenarios of operation. The runtime with a spatial mesh of 100 volumes is still about 2 min 4 s on a workstation equipped with Intel Xeon at 3.8GHz, which may be a limitation of this approach.

Acknowledgements

The authors would like to thank the Auvergne-Rhone-Alpes region's PAI 2020 program, along with the C-StaRRE 4.0 project for their financial support, in particular for Hadi ABOU DAHER's internship (Grant number 22PACKR-CVALENTIN-6887).

References

J. Chauchat, S. Guillou, D. Pham van Bang and K. Dan Nguyen, 2013, Modeling sedimentation-consolidation in the framework of a one-dimensional two-phase flow model, Journal of Hydraulic Research, 51 (3), pp. 293-305.

P. Garrido, F. Concha, R. Burger, 2003, Settling velocities of particulate systems: 14. Unified model of sedimentation, centrifugation and filtration of flocculated suspensions. Int. J. Mineral rocessing, vol. 72, pp.57-74.

E. Godlewski, P.A. Raviart, 1996, Numerical Approximation of Hyperbolic Systems of Conservation Laws, Applied Mathematical Sciences (AMS, volume 118), Springer.

R. LeVeque, 2002, Finite volume methods for hyperbolic problems, Cambridge university press, vol. 31.

B. Li and M.K. Stenstrom, 2014, Research advances and challenges in one-dimensional modeling of secondary settling Tanks - A critical review, Water Research, vol. 65, pp. 40-63.

E.A. Toorman, 1996, Sedimentation and self-weight consolidation: general unifying theory. Geotechnique, vol. 46 (1), pp.103–113.

C. Valentin, N. Chassin, F. Couenne, J.M. Choubert and C. Jallut, 2022, 1-D Dynamic knowledge based model of urban sludge continuous-flow settling process. Comparison with experimental results, prepublication submitted to a journal, 9 pages (hal-03678231v1).

Antonis Kokossis, Michael C. Georgiadis, Efstratios N. Pistikopoulos (Eds.)
PROCEEDINGS OF THE 33rd European Symposium on Computer Aided Process Engineering (ESCAPE33), June 18-21, 2023, Athens, Greece
 http://dx.doi.org/10.1016/B978-0-443-15274-0.50113-X

Homotopic Method for binodal curves computation in ternary liquid-liquid separation

Nataliya Shcherbakova,[a] Kevin Roger,[a] Joseph Gergaud,[b] Olivier Cots[b]

[a]*Laboratoire de Génie Chimique, Université de Toulouse, CNRS, INP, UPS, Toulouse, France*

[b]*IRIT, Université de Toulouse, CNRS, INP, UPS, Toulouse, France*

Abstract

This paper analyses the intrinsic geometric structure of binodal curves of phase separation diagrams of ternary mixtures. A new computational method is presented, and the efficiency of the proposed approach is tested for the Flory-Huggins model identification of water – acetone – hexadecane mixture.

Keywords: liquid-liquid separation, bitangent plane, binodal curve, differential homotopy

1. Introduction

Controlling the phase separation in industrial processes requires detailed knowledge of thermodynamic diagrams of multi-component mixtures (Koningsveld et al., 2001). Many computational problems arise in this context: identification of model parameters from the experimental data, localization of phase separation envelopes, detection of multi-phase regions, etc. Despite the enormous progress in this domain in the past decades, there is still a real need for new fast and reliable algorithms, for instance, in industrial applications using bio-sourced mixtures or in polymer production.

This paper focuses on the analysis of liquid-liquid separation of ternary mixtures at constant temperature and pressure. The topological structure of phase separation diagrams is then defined by binodal and spinodal curves delimiting stable, metastable and unstable domains in the composition space. Knowing these domains is crucial for modeling the demixing process in a given mixture. The experimental detection of spinodals that delimits the material stability domain is a highly non-trivial task. Instead, the various techniques allows measuring the composition of coupled coexisting phases, which can be used to establish a reliable model in order to predict the spinodal location *a-posteori*.

The standard method of numerical computation of binodal curves uses the Newton-Raphson-like iterative algorithm to solve a system of algebraic equations over the mesh covering the state space of the diagram, while the classical least square method is commonly used for model identification. Often this approach fails to meet at least one of the quality criteria characterizing the "good" model: the binodal curve shape and the bundle of tie-lines indicating the coupled phases composition. Inspired by the ideas of D.J. Korteweg (Levelt Sengers, 2002) applied to the topological analysis of the Gibbs free energy surface, in this paper, we propose a new numerical method for computing binodal curves by a kind of differential homotopy method. The key idea is to consider binodals as the 2D projections of certain smooth curves in 4D configuration space formed by the pairs of points of the phase diagram. This geometrical point of view is presented in Section 2, where a set of ordinary differential equations (ODE) describing the 4D curve as an integral curve of a smooth vector filed in 4D space is derived. A conventional ODE solver can numerically compute this curve at desired precision.

In Section 3, the developed approach is tested on the water-acetone-hexadecane mixture, characterized by two non-miscibility binary gaps among its components. The Flory-Huggins model parameters, including the triple interaction term, are computed by solving a non-trivial minNLP problem associated with a criterion, which accounts for the intrinsic geometry of the binodal curve. The computations performed using Mathematica 9 software yield very promising results that meet both quality criteria cited above.

2. Binodal and spinodal curves: thermodynamics vs. geometry

2.1 Phase coexistence conditions in multi-component mixtures

The physicochemical properties of an N-component system can be described by the Gibbs free energy $G(P, T, n)$, where $n = (n_1, \dots, n_N)$, n_i being the number of moles of i-th component, P the pressure and T the temperature of the system. Being a homogeneous function of 1-st order with respect to n_i, G can be expressed in terms of chemical potentials $\mu_i = \frac{\partial G}{\partial n_i}$ as $G(P,T,n) = \sum_{i=1}^{N} n_i \mu_i(P,T,n)$. The infinitesimal changes in the state of the system obey the fundamental Gibbs equation

$$dG = -S\,dT + V\,dP + \sum_{i=1}^{N} \mu_i dn_i \tag{1}$$

where S and V are the entropy and the volume of the system.

All expressions above rest valid for each phase of a closed system maintained at thermodynamic equilibrium with two coexisting phases. Numbering the phases by a and b, the equilibrium condition reads $dG = dG^a + dG^b = 0$. Since in a closed system $n_i = n_i{}^a + n_i{}^b$ and $n_{tot} = \sum_{i=1}^{N} n_i$ are constant, it follows that $dn_i{}^a = -dn_i{}^b$, and hence Eq.(1) implies that

$$T^a = T^b, \quad P^a = P^b, \quad \mu_i{}^a(P^a, T^a, n^a) = \mu_i{}^b(P^b, T^b, n^b),\ i = 1, \dots, n \tag{2}$$

In this paper, it is assumed that both phases can be described by the same thermodynamic model, though in principle, two different expressions can model be used for chemical potentials of phases a and b. In addition, only the isobaric isothermal conditions will be considered, so the first two of Eqs. (2) become trivial.

The physicochemical properties of real mixtures are usually expressed in terms of molar, volume or mass fractions. In particular, the Flory-Huggins model used in Section 3 to describe the mixture of a polymer with water and a solvent, employs the volume fractions. Assuming the total volume of the mixture to be equal to the sum of partial volumes of components, and denoting by $x_i = v_i/V$ the volume fraction of i–th component and by $g(P,T,x) = G(P,T,n(x))/R\,V$ the free Gibbs energy of mixing per unit of volume, Eqs. (2) can be rewritten in the form

$$\frac{\partial g^a(x^a)}{\partial x_i^a} = \frac{\partial g^b(x^b)}{\partial x_i^b}, \qquad i = 1, \ldots, N-1,$$
$$g^a(x^a) - g^b(x^b) - \sum_{i=1}^{N-1} \frac{\partial g^a(x^a)}{\partial x_i{}^a}(x_i{}^a - x_i{}^b) = 0 \tag{3}$$

In the binary case (N=2) these conditions mean that the graph of function $g(x)$ admits a bitangent line, as it is shown in Fig. 1a.

2.2 Ternary case: binodal curves and bitangent planes of the Gibbs free energy surface

To simplify notations, from now on g'_i stays for $\frac{\partial g}{\partial x_i}$, the symbols $\nabla_x g$ and $D_x^2 g$ are used for the gradient and for the Hessian of function g, and (|) for the scalar product in $\mathbb{R}^2$.

In case of ternary mixtures Eqs.(3) describe the specific properties of the surface, referred to as surface W, defined by equation $z = g(x_1, x_2)$ in a 3D Cartesian space endowed with coordinates x_1, x_2, z, as it shown in Fig. 1b. The material stability condition implies that such a surface has a physical meaning only in the sub-domain of the volume fraction space where the function $g(x)$ is convex:

$$\Omega = \{x = (x_1, x_2): x_1, x_2 \in [0,1], x_1 + x_2 \leq 1, \mathrm{D}_x^2 g \text{ is weakly pos. definite}\}$$

The projection on the x-plane of the set of singular points of surface W, i.e. the points where $D_x^2 g$ is singular, defines the spinodal curve of the phase diagram. The first two of Eqs.(3) guarantee the existence of a pair of points P_1 and P_2 on the surface W with collinear normals, whereas the latter of Eqs.(3) says that these points belong to the same plane tangent to W. The projection of the segment P_1P_2 on x-plane is called a tie-line. The one-parametric family of conodal points P_1, P_2 define two directrices of a certain ruled surface in 3D space. Their projections on the x-plane correspond to the two branches of binodal curve on the phase diagram. These branches can meet each other at the critical point of the phase diagram. Since by definition the critical point also belongs to the spinodal, these are the only common point between the binodal and the spinodal curves, and the whole binodal curve lies on the same side with respect to the spinodal curve. Fig.1b. illustrates all these concepts for the mixture having one critical point.

2.3 Differential equations of binodal and spinodal curves

As it follows from the above analysis, the binodal curves are formed by the pairs of points. So it is natural to describe these curves in the 4D space $\boldsymbol{\Sigma} = \boldsymbol{\Omega} \times \boldsymbol{\Omega}$ defined as follows:

$$\boldsymbol{\Sigma} = \{\boldsymbol{q} = (\boldsymbol{q}_1, \boldsymbol{q}_2) \in \mathbb{R}^4 : \boldsymbol{q}_1 = \boldsymbol{x}^a \in \boldsymbol{\Omega},\ \boldsymbol{q}_2 = \boldsymbol{x}^b \in \boldsymbol{\Omega}\}$$

Consider now the three co-dimension one smooth sub-manifolds in Σ associated to the zero-levels of the functions

$$F_1(q) = g_1'(q_1) - g_1'(q_2), \qquad F_2(q) = g_2'(q_1) - g_2'(q_2),$$
$$F_3(q) = g(q_2) - g(q_1) + (\nabla_x g(q_1) | q_2 - q_1), \quad q \in \Sigma \tag{4}$$

These sub-manifolds can intersect each other forming a one-dimensional sub-manifold $\boldsymbol{B} = \{\boldsymbol{q} \in \boldsymbol{\Sigma}\text{: } \boldsymbol{F}_i(\boldsymbol{q}) = \boldsymbol{0},\ \boldsymbol{i} = \boldsymbol{1}, \boldsymbol{2}, \boldsymbol{3}\}$ whose orthogonal projections on two exemplars of $\boldsymbol{\Omega}$

define the two branches of the binodal curve. In what follows B will be referred to as the generalized binodal curve. This curve can be seen as the integral curve of the vector field $\boldsymbol{q} \longrightarrow \boldsymbol{V}(\boldsymbol{q}) \in \boldsymbol{T_q B}$. By definition,

$$\nabla_q F_i(q)V(q) = 0, \qquad i = 1,2,3 \tag{5}$$

Define a pair of functions $\boldsymbol{\Phi}_i$ and $\boldsymbol{\Psi}_i$ such that $(\boldsymbol{\Phi}_i, \boldsymbol{\Psi}_i)^T = \mathbf{D}_x^2 \boldsymbol{g}(\boldsymbol{q}_i)(\boldsymbol{q}_1 - \boldsymbol{q}_2)$, for $i=1,2$. Then Eqs. (5) are equivalent to the following conditions:

$$\Phi_1 V_1 + \Psi_1 V_2 = 0 \quad \text{and} \quad \mathrm{D}_x^2 g(q_1)(V_1, V_2) = \mathrm{D}_x^2 g(q_2)(V_3, V_4) \tag{6}$$

The straightforward computation shows that $\boldsymbol{V}(\boldsymbol{q})$ can be expressed as

$$V_1 = \Psi_1 \, det \, \mathrm{D}_x^2 g(q_2), \qquad V_2 = -\Phi_1 \, det \, \mathrm{D}_x^2 g(q_2)$$
$$V_3 = \Psi_2 \, det \, \mathrm{D}_x^2 g(q_1), \qquad V_4 = -\Phi_2 \, det \, \mathrm{D}_x^2 g(q_1). \tag{7}$$

Clearly, the critical points of surface W correspond to the singular points of $\boldsymbol{V}(\boldsymbol{q})$.

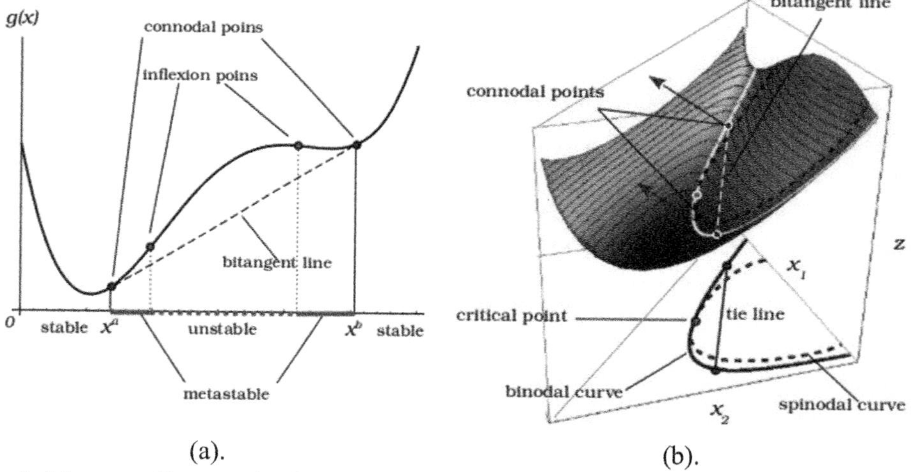

(a). (b).

Fig.1. bitangent lines to the free Gibbs energy surface W in binary (a) and ternary (b) cases, and their projections on the composition space.

Remark. The same argument can be used to define the spinodal curve as the integral curve of the vector field $\boldsymbol{\nabla}_x \boldsymbol{S}(\boldsymbol{x})^{\perp}$, where $\boldsymbol{S}(\boldsymbol{x}) = \mathbf{det}\,\mathbf{D}_x^2 \boldsymbol{g}(\boldsymbol{x})$.

The described construction is a geometrical interpretation of the differential homotopy method (Allgower and Georg, 2003) for computing smooth curves defined implicitly. Using Eq.(5) instead of Eqs.(3) it is possible to compute binodal curves by a conventional ODE solver without solving the set of algebraic Eqs.(3) or Eqs.(2) by a Newton-Raphson type iterative algorithm, as it is usually done. The only iterative step concerns the computation of the starting point of the generalized binodal (or spinodal) curve, which reduces to a 2D problem. An algorithm based on the same principles was described in detail in N. Shcherbakova at al. 2017 and O. Cots et al. 2021.

3. Case study: Flory-Huggins parameters of water-acetone-hexadecane

3.1 The Flory-Huggins model with ternary interaction term

The classical Flory-Huggins model defines the excess free energy of mixing per unit of volume according to the expression

$$g = \varphi_1 \ln \varphi_1 + \frac{1}{N_2}\varphi_2 \ln \varphi_2 + \frac{1}{N_3}\varphi_3 \ln \varphi_3 + \sum_{i,j=1,i<j}^{3} \chi_{ij}\varphi_i\varphi_j + \beta\, \varphi_1\varphi_2\varphi_3 \quad (8)$$

Here φ_i, $i=1,2,3$ are the volume fractions of water, the solvent and the polymer respectively, N_2 and N_3 are the number of segments in the molecules of the solvent and of the polymer referred to the water segment, χ_{ij} and β are the binary and ternary interaction coefficients. In this paper N_2, N_3, χ_{ij} and β are assumed to be constant. Notice that the cross term $\beta\, \varphi_1\varphi_2\varphi_3$ contains several physical effects. Indeed, it is the simplest possible correction of the Flory-Huggins model that accounts for composition-dependent binary interaction parameters as well as for the failure of the total volume conservation hypothesis when mixing the components. Since the present work aims to propose a generic robust method for binodal identification, it is not necessary to investigate the real physical meaning of the cross term here.

In further computations the volume fractions of acetone and of polymer are chosen as the independent variables so that $x_1 = \varphi_2$, $x_2 = \varphi_3$ and $\varphi_1 = 1 - x_1 - x_2$. Inserting Eq.(8) into Eqs.(4) yields three algebraic conditions to be verified along the binodal curve. Due to the particular form of Eq. (8), these conditions linearly depend on six scalar parameters: $\chi_{12}, \chi_{13}, \chi_{23}, \beta, r_1 = N_2^{-1}$, and $r_2 = N_3^{-1}$.

3.2 Case study: water-acetone-hexadecane

The further analysis is based on the experimental data obtained at $P = 1\ atm$ and T=23°C using Raman's spectroscopy to measure each species fraction for a given sample. 14 measurements of coupled compositions were acquired defining tie-lines $q^k \in \Sigma$ (black points in Fig. 2) and 17 non-coupled composition measurements $x_m \in \Omega$ (white circles in Fig. 2) defining the phase-separation envelope. The latter series contains a pair of points defining the miscibility gap x^a, x^b of the binary mixture acetone-hexadecane. Using Eqs.(3) for this binary mixture, χ_{23} and N_2 were computed in terms of x^a, x^b. Analogously, χ_{13} and N_3 were found using the miscibility gap limits of water-hexadecane mixture reported in the literature. Due to the significant variation of these data in different sources, the term χ_{13} in Eq. (8) was replaced by $\chi_{13} + \delta$ in the next step of the computation. In this second step the remaining parameters χ_{12}, β, δ were found by solving the following non-linear optimization problem:

$$\min_{\chi_{12},\beta,\delta} \sum_{k=1}^{14} \frac{F_1^{\,2}(q^k) + F_2^{\,2}(q^k)}{\left(1 - F_3^{\,2}(q^k)\right)^2} \quad (9)$$

The particular form of criterion Eq.(9) push the minNLP solver toward the solutions which guarantee that each point q_k corresponds to a pair of points on W having collinear normal vectors, whereas the denominator term penalize those pairs of points that do not belong the same bitangent line.

The described method was implemented using Mathematica 9 package, the result is shown Fig. 2. Grey lines correspond to the measured tie-lines, and the dashed lines denote the computed tie-lines. The thick black curve corresponds to the computed binodal curve, which satisfies both quality criteria discussed above. The right part of Fig.2 shows the acetone vertex zone in high resolution, showing an excellent correspondence between experimental and model identification results.

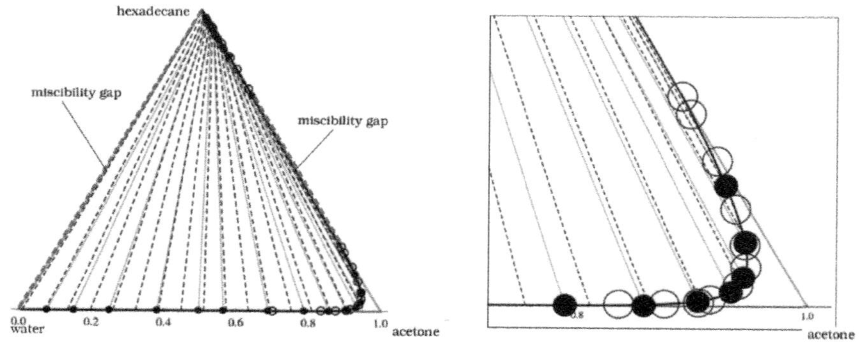

Fig.2: water-acetone-hexadecane mixture, experimental data (black and white points) vs. numerical computation of the binodal curve using Flory-Huggins model (black curve).

4. Conclusion

The geometric description of binodal curves provides a deeper insight into their internal structure. In particular, it allows reducing numerical computation of binodal and spinodal curves to standard ODEs integration, gaining in execution time and accuracy. The numerical method proposed in this paper can be adapted to any thermodynamic model. We plan to implement the resulting differential homotopy algorithm in the next version of the SMITH code (Cots et al, 2021).

References

R. Koningsveld, W.H.Stockmayer, E. Nies. Polymer Phase Diagrams. Oxford Univ. Press, 2001

J. Levelt Sengers, How fluids unmix. Royal Netherlands Acad. of Arts and Sc., Amsterdam, 2002

E. Allgower, K. Georg.Introduction to numerical continuation methods. Classica in Applied Mathematics, vol.45, SIAM, Philadelphia, USA, 2003

O. Cots, N. Shcherbakova, J. Gergaud. SMITH: differential homotopy and automatic differentiation for computing thermodynamic diagrams of complex mixtures. Computer Aided Chem. Eng., 50, pp. 1081-1086, 2021

N. Shcherbakova, I. Rodriguez, J. Abidskov, V. Gerbaud. A Novel Method for Detecting and Computing Univolatility Curves in Ternary Mixtures. Chem. Eng. Sci., 173, pp. 21-36, 2017

Mathematica 9. http://www.wolfram.com/mathematica/

Antonis Kokossis, Michael C. Georgiadis, Efstratios N. Pistikopoulos (Eds.)
PROCEEDINGS OF THE 33rd European Symposium on Computer Aided Process Engineering (ESCAPE33), June 18-21, 2023, Athens, Greece
 http://dx.doi.org/10.1016/B978-0-443-15274-0.50114-1

Modelling of iron oxide reduction with hydrogen

Emiliano Salucci,[a] Snigdha Ghosh,[b] Vincenzo Russo,[c] Henrik Grénman,[a] Henrik Saxén[a]

[a]Åbo Akademi University, Henrikinkatu 2, Turku 20500, Finland
[b]IIT (ISM) Dhanbad, Jharkhand, Dhanbad 826001, India
[c]University of Naples Federico II, Via Vicinale Cupa Cintia 26, Napoli 80126, Italy

Abstract

The steel industry represents one of the key sectors of our society and it is, therefore, essential to devise new strategies to drastically reduce its carbon dioxide emissions. From this point of view, a potentially beneficial solution would be to replace the carbon sources used in the reduction processes of minerals with hydrogen. This requires a deeper understanding of the kinetics of iron oxide reduction in order to provide crucial elements for selecting the best-operating conditions and designing new technologies. In this work, an mathematical model was developed, based on the assumptions of the shrinking core model, to describe the reaction kinetics and mass transport phenomena occurring during iron ore reduction. Experimental tests were conducted to provide detailed data to support this work. The model results qualitatively agree with what was found in the laboratory tests.

Keywords: Hydrogen-based DRI, green-ironmaking, iron oxides, kinetics-modelling.

1. Introduction

The aim of reducing greenhouse gas emissions is driving all manufacturing and industrial sectors to seek alternative solutions that are viable in the short term and offer excellent margins for improvement in the years ahead. With around 30% of global direct industrial emissions (2.6 Gt/year), steelmaking represents one of the largest CO_2 emitting sectors (Energy Agency 2020). The classical production process, carried out in a blast furnace (BF), involves the use of iron ores (i.e., iron oxides) which are reduced to metallic iron using carbon monoxide. This gas component is produced at partial oxidation of coal and reacts with the oxygen in the ore to produce iron and CO_2 (Hutny et al. 1991). As a result of the very high operating temperatures of the BF process, which locally exceed 2000°C, iron melts and slag forms in the lower part of the furnace. Developed 40 years ago, the Midrex process represents a possible alternative to conventional iron production. The reduction reaction in this process occurs at lower temperatures than in the BF and it usually operates on reformed natural gas. The gas-solid reactions produce direct reduced iron in solid state (DRI), and thus avoids melting (Atsushi et al. 2010). To achieve the complete decarbonization of this sector, it is necessary to redesign current production facilities and to replace all or most of the carbon source by hydrogen (Patisson et al. 2021). The kinetics of iron oxide reduction by hydrogen has been extensively studied and the literature shows that it is influenced by many factors and parameters. Several studies have focused on the thermodynamics of the reduction. At operating temperatures above 570°C, the reactive network consists of three reversible reactions in series, starting with the reduction of hematite (Fe_2O_3) and passing through the intermediates magnetite (Fe_3O_4) and wustite (FeO); at lower temperatures, wustite does not form due to its instability (Zhang et al. 2013). To investigate the kinetic dependence of the reduction on the main operating conditions, several works have examined in depth the effect of temperature (Bernasowski 2014) or the composition of the reducing gas (Wei et al. 2018) on the system. Many authors focused their efforts on the determination of the rate-limiting step of the heterogeneous process and on the importance of diffusion phenomena. For this reason, the study of morphology, geometry, and other properties of the solid is crucial. Several authors have investigated the effect of the size of the iron ore (e.g., particle diameter) (Bonalde et al. 2005; Hou et al. 2012), the initial porosity (Kazemi et al. 2017) or the gas flow rate to assess the resistance to gaseous diffusion around a pellet (Hou et al. 2012). The main mathematical models used to describe the process and evaluate the kinetic parameters are the Avrami Model (Piotrowski et al. 2005), often used in metallurgical work, the Shrinking Core Model (SCM) (Moon et al. 1998) or the Grain Model (Bonalde et al. 2005). As for the assessment of the rate-limiting step, different conclusions have been made, mainly proposing an intraparticle diffusive regime, a kinetic regime, or a variable regime, depending on the reduction stage considered (Lin et al. 2003). Given the variety of experimental conditions and modelling assumptions, a wide range of activation energies, are reported in the literature. The activation energy for individual intermediate reactions or for the whole process can vary between 18 kJ/mol and 246 kJ/mol (Pineau et al. 2006). To accurately describe the chemical kinetics and mass transfer phenomena of the reduction process, an extensive kinetic study is required. This work focuses on the development of mathematical models for the description of iron oxide reduction based on a large set of experiments, which will be useful in the future for the development and design of hydrogen-based DRI processes.

2. Materials and methods

2.1. Experimental Setup

To support the development of a mathematical model capable of describing the reduction process, several experimental tests have been conducted using an Autochem2910. This device, which is normally used for the characterization of catalysts (e.g., TPR, TPO etc.), makes it possible to study the reduction of iron ores in a small packed-bed reactor. The experimental apparatus consists of a U-tube glass, resistant to high temperatures, with branches of different diameters. Ore fines, loaded in the branch with the largest diameter, is held by a porous support placed above the reduction of the tube diameter. The reactive bed (diameter 8 mm, height approx. 1 mm) consists of hematite powder (purity 97%), with a particle diameter of less than 5 μm. The reduction of hematite was carried out using a gas mixture containing the reducing species (H_2 purity 99.9%) and an inert gas (Ar, purity 99.9%). The temperature was measured using a thermocouple placed inside the tube. The composition of the gas leaving the bed was analyzed using a thermal conductivity detector (TCD) and a mass spectrometer, from which the degree of iron oxide reduction can be estimated. The tests were conducted isothermally at a specific H_2/Ar ratio and total volumetric gas flow rate.

2.2. Mathematical Model

A mathematical model has been developed to describe the reactive and diffusive phenomena present in a bed of iron oxide particles through which a gas mixture containing a reducing gas species flow. The reactive network consists of several equilibrium reactions in series, the number of which varies depending on the operating temperature. Table 1 shows the reactions occurring at temperatures above 570°C:

Table 1. Reaction network, $T > 570°C$, and enthalpies (Zhang et al. 2013)

Reaction fronts (*j*)	Reduction reaction	ΔH^0 [kJ/mol]
1	$3\,\underbrace{Fe_2O_3}_{\text{Hematite}} + H_2 \Leftrightarrow 2\,\underbrace{Fe_3O_4}_{\text{Magnetite}} + H_2$	−11.72
2	$\underbrace{Fe_3O_4}_{\text{Magnetite}} + H_2 \Leftrightarrow 3\,\underbrace{FeO}_{\text{Wustite}} + H_2O$	+77.44
3	$\underbrace{FeO}_{\text{Wustite}} + H_2 \Leftrightarrow \underbrace{Fe}_{\text{Iron}} + H_2O$	+23.85

To describe the intraparticle reductive process, the mathematical model was developed based on the key assumptions of the SCM, where the reactions only take place on the surface separating two zones within the solid, an unreacted core, consisting of pure solid reactant, and a reacted outer layer, consisting of pure solid product. The reaction interface moves inwards as the core is consumed. By extending the SCM to the series reaction system, three different intraparticle reaction fronts can be described, as shown in Figure1.

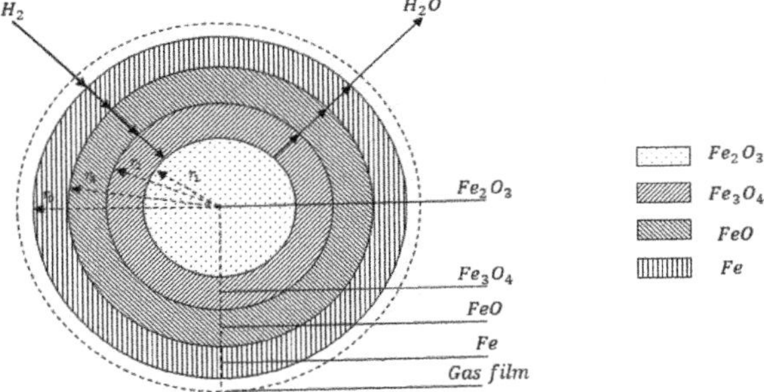

Figure 1. Three-interface SCM, reduction of hematite particle.

2.2.1. Mass Balances

The balance equations useful for describing the degrees of reduction of iron oxides over time (t) and along the axial coordinate of the reactor (z) were expressed as

$$\frac{\partial X_j(t,z)}{\partial t} = \frac{N_j(t,z)}{n_{OO} \cdot a_j} \tag{1}$$

where X_j is the degree of conversion of the j-th reaction front (Fe_2O_3 reduction, etc.), n_{OO} is the initial oxygen content in the system while a_j is the proportional oxygen content parameter depending on the type of iron oxide considered. N_j represents the reduction rate of the j-th reaction front defined under the assumptions of the SCM. N_j is given by

$$N_j(t,z) = \frac{P}{R_g T}\frac{3}{r_0}\frac{1}{A_j(t,z)\big|_f} \cdot \left(\frac{R_j(t,z)\big|_f}{W(t,z)\big|_f} - y_{eq,j}\right) \tag{2}$$

where P and T represent the operating pressure and temperature, R_g is the gas constant, r_0 is the particle radius of the ore, while $y_{eq,j}$ represents the fraction of H_2 at the equilibrium of the j-th reaction.

In the initial stages of reduction, with the formation of the intermediate species magnetite and wustite, a triple reaction front is established in the particles. Once the reduction of hematite has been completed, at any t and z ($X_1(t,z) = 1$), the equations useful for describing the entire reactive process at that specific location must change as the model no longer describes a triple but a double reaction front. Identically, this occurs when FeO is the only residual oxide, effectively establishing a single reaction front. The formulae of the coefficients (A_j, W, R_j), which define the reduction rate, are closely related to the number of available reaction fronts. The value of f defines the number of reaction fronts present at time t and the z-coordinate, according to

$$\begin{cases} \text{if } X_1(t,z) < 1 & f=3 \quad j=1,2,3 \\ \text{if } X_1(t,z) = 1 \cap X_2(t,z) < 1 & f=2 \quad j=2,3 \\ \text{if } X_1(t,z) = 1 \cap X_2(t,z) = 1 \cap X_3(t,z) < 1 & f=1 \quad j=3 \end{cases} \tag{3}$$

As an example, Eqs. (4-5) give the formulas for $R_1|_{f=3}$ and $W|_{f=3}$, which are functions of the terms representing kinetic (A_j), intra-particle diffusive (B_j) and external diffusive (F) resistance, respectively

$$\begin{aligned} R_1(t,z)|_3 &= (A_1A_2A_3)\cdot y_{H_2}(t) + [A_1A_2(B_3+F)]\cdot y_{eq3} \\ &+[A_1A_3B_2 + A_1A_3(B_3+F) + A_1B_2(B_3+F)]\cdot y_{eq2} \\ &+[A_3B_1(A_2+B_2) + A_3B_1(A_2+B_2) + B_1(A_2+B_2)(B_3+F) + A_2A_3B_2 \\ &\quad + A_2A_3(B_3+F) + A_2B_2(B_3+F)]\cdot y_{eq1} \end{aligned} \tag{4}$$

$$\begin{aligned} W(t,z)|_3 = \{&(A_1+B_1)\cdot[A_3(A_2+B_2+B_3+F) + (A_2+B_2)\cdot(B_3+F)] \\ &+ A_2[A_3(B_2+B_3+F) + B_2(B_3+F)]\}\cdot y_{eq1} \end{aligned} \tag{5}$$

The complex derivation of these equations is reported in the literature (Hara et al. 1974). Examples of typical formulations of the A_j and B_j terms are given in Eqs. (6-7) while, in the first instance, due to the small size of the particles, the external transport resistance is considered negligible (F=0).

$$A_1(t,z)|_3 = \frac{1}{(1-X_1(t,z))^{\frac{2}{3}}}\cdot\frac{1}{k_1\cdot\left(1+\frac{1}{K_1}\right)} \tag{6}$$

$$B_1(t,z)|_3 = \frac{(1-X_2(t,z))^{\frac{1}{3}} - (1-X_1(t,z))^{\frac{1}{3}}}{(1-X_1(t,z))^{\frac{1}{3}}\cdot(1-X_2(t,z))^{\frac{1}{3}}}\cdot\frac{r_0}{D_1} \tag{7}$$

where k_j and K_j represent the kinetic constant and the equilibrium constant of the j-th reaction front, while D_j is the intra-particle diffusion coefficient that depends on the morphological characteristics of the ferrous oxide. The kinetic (Chu et al. 2005), thermodynamic (Zhang et al. 2013), and diffusive (Takahashi et al. 1971) parameters used in the model were taken from the literature.

Under dynamic conditions, two mass balance equations were developed for the gas species involved in the reduction reaction

$$\frac{\partial c_i(t,z)}{\partial t} = \underbrace{-\frac{u}{h}\cdot\frac{\partial c_i(t,z)}{\partial z}}_{\text{Convection}} + \underbrace{\frac{D_B}{h^2}\cdot\frac{\partial^2 c_i(t,z)}{\partial z^2}}_{\text{Diffusion}} + \underbrace{\frac{1}{(1-\varepsilon_B)}\cdot\sum_{j|f}^{3}\nu_{i,j}\cdot N_j(t,z)}_{\text{Reaction}} \tag{8}$$

where c_i is the gas concentration in the reactor (H_2 or H_2O), u represents the velocity of the gas, h is the height of the reactor and D_B is the axial dispersion coefficient. In the term describing the chemical reaction, ν is the stoichiometric coefficient of the i-th gas species of the j-th reaction while ε_B is the void degree in the bed. Finally, the sum of the reductive contributions (N_j) must consider the number of active reaction fronts.

The boundary conditions of the system of partial differential equations were defined at the reactor inlet (z = 0) and at the outlet (z = 1):

$$\left.\frac{\partial c_i(t,z)}{\partial z}\right|_{z=0} = c_{\text{bulk},i} \qquad \left.\frac{\partial c_{H_2}(t,z)}{\partial z}\right|_{z=1} = 0 \tag{9}$$

2.2.2. Numerical Methods

The mathematical model was developed in MATLAB 2022 by solving the system of partial differential equations, setting 50 grid points for the axial dimensionless coordinates of the reactor. The numerical method used is the finite difference method, which is based on backward differentiation formulas.

3. Results

In this work, an intensive simulation study was conducted to assess the responsiveness of the developed mathematical model using kinetic parameters reported in the literature. As an example, Figure 2 shows the main results from a simulation conducted at 600 °C, total gas volume flow rate of 20 mL/min and H_2/Ar.

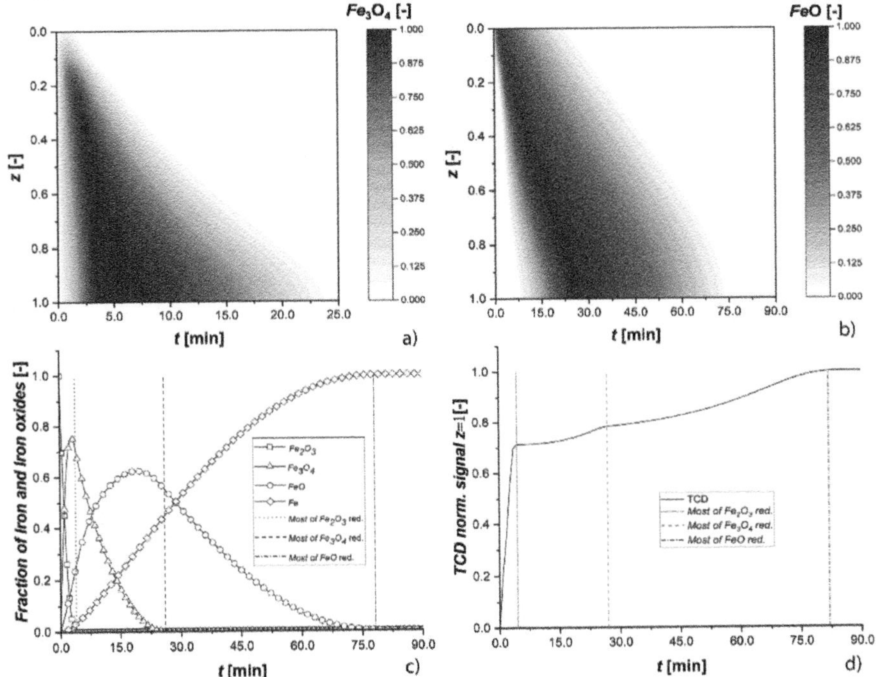

Figure 2. Axial profiles over time of the oxides Fe_3O_4 (a) and FeO (b). Simulated overall fractions of the oxides and iron over time (c). Simulated TCD signal from off-gas composition (d).

The contour diagrams in Figures 2a and 2b show the axial profiles of the Fe_3O_4 and FeO fractions in the particles over time, respectively, while Figure 2c shows the overall composition of the system. The reduction of Fe_2O_3 is seen to be completed extremely rapidly, promoting the formation of high amounts of Fe_3O_4 throughout the reactor after only a few minutes. Close to the entrance of the reactor (z=0), magnetite does not accumulate in large quantities due to its rapid reduction to wustite, but near the end of the reactor (z=1) its accumulation and depletion are considerably slower. The reasons lie in the high consumption of H_2 by the reduction in the overlying areas of the bed and the resistance to intra- and inter-particle transport. Due to hydrogen starvation, the amount of H_2 reaching the lower part of the reactor does not provide the driving force required to reduce the oxides. For this reason, about 25 minutes are required to complete the reduction of Fe_3O_4. The reaction of FeO to iron turns out to be the slowest: at the reactor inlet a complete reduction of the particles is observed within about 30 minutes, but close to the outlet the time required is more than doubled. This reveals the impact of the reactor bed height on the overall reaction time. Figure 2d shows a (normalized) TCD signal calculated based on the simulated off-gas composition (H_2, H_2O and Ar). This was found to show very many common features with the TCD signals measured in the experiments. Therefore, the objective of forthcoming work is to estimate kinetic parameters of the model by fitting it to the empirical TCD curves. However, the parameter estimation will not be undertaken until data from a more extensive set of reduction experiments will be available.

4. Conclusions

In this work, an in-depth modeling study was conducted to develop a model capable of describing the entire reductive process of a hematite bed exerted to hydrogen-containing gas. For this purpose, the sensitivity of the model to the variation of several structural parameters was assessed. The acquisition of accurate experimental data using the experimental system make it possible to estimate kinetic parameters of the individual reaction steps, which is the focus of forthcoming work.

Acknowledgement

We acknowledge support from Business Finland to the Towards Fossil-free Steel project.

References

M. Atsushi, H. Uemur, & T. Sakaguchi, 2010. MIDREX processes. Kobelco Technology Review, 29, 50–57.

M. Bernasowski, 2014. Theoretical study of the hydrogen influence on iron oxides reduction at the blast furnace process. Steel Research International, 85(4), 670–678.

A. Bonalde, A. Henriquez, & M. Manrique, 2005. Kinetic Analysis of the Iron Oxide Reduction Using Hydrogen-Carbon Monoxide Mixtures as Reducing Agent. ISIJ Int., 45(9), 1255–1260.

M. Chu, J. Yagi, & F. Shen. (2005). Modelling on blast furnace process and innovative ironmaking technologies. Northeastern University Press (Shenyang).

Energy Agency, I. 2020. Iron and Steel Technology Roadmap Towards more sustainable steelmaking Part of the Energy Technology Perspectives series.

Y. Hara, M. Tsuchiya, & S. Kondo. 1974. Intraparticle Temperature of Iron-Oxide Pellet during the Reduction. Tetsu to Hagane. 60, 1261–1270.

B. Hou, H. Zhang, H. Li, & Q. Zhu, 2012. Study on kinetics of iron oxide reduction by hydrogen. Chinese Journal of Chemical Engineering, 20(1), 10–17.

W. P. Hutny, G. K. Lee, & J. T. Price, 1991. Fundamentals of coal combustion during injection into blast furnace. In Prog. Energy Combust. Sci, 17.

M. Kazemi, M. S. Pour, & D. Sichen, 2017. Experimental and Modeling Study on Reduction of Hematite Pellets by Hydrogen Gas. Metallurgical and Materials Transactions B: Process Metallurgy and Materials Processing Science, 48(2), 1114–1122.

H. Y. Lin, Y. W. Chen, & C. Li, 2003. The mechanism of reduction of iron oxide by hydrogen. Thermochimica Acta, 400(1–2), 61–67.

I. J. Moon, C. H. Rhee, & D. J. Min, 1998. Reduction of hematite compacts by H2-CO gas mixtures. Steel Research, 69(8), 302–306.

F. Patisson, O. Mirgaux, & J. P. Birat, 2021. Hydrogen steelmaking. Part 1: Physical chemistry and process metallurgy. Materiaux et Techniques, 109(3–4).

A. Pineau, N. Kanari, & I. Gaballah, 2006. Kinetics of reduction of iron oxides by H2. Part I: Low temperature reduction of hematite. Thermochimica Acta, 447(1), 89–100.

K. Piotrowski, K. Mondal, H. Lorethova, L. Stonawski, T. Szymański, & T. Wiltowski, 2005. Effect of gas composition on the kinetics of iron oxide reduction in a hydrogen production process. International Journal of Hydrogen Energy, 30(15), 1543–1554.

R. Takahashi, J. Yagi & Y. Omori, 1971. Reduction Rate of Iron Oxide Pellets with Hydrogen. Science Reports of the Research Institutes, Tohoku University. Ser. A, Physics, Chemistry and Metallurgy, 23, 9–30.

Z. Wei, J. Zhang, B. Qin, Y. Dong, Y. Lu, Y. Li, W. Hao, & Y. Zhang, 2018. Reduction kinetics of hematite ore fines with H2 in a rotary drum reactor. Powder Technology, 332, 18–26.

W. Zhang, J. Zhang, Q. Li, Y. He, B. Tang, M. Li, Z. Zhang, & Z. Zou, 2013. Thermodynamic Analyses of Iron Oxides Redox Reactions. The 8th Pacific Rim International Congress on Advanced Materials and Processing, 777–789.

Antonis Kokossis, Michael C. Georgiadis, Efstratios N. Pistikopoulos (Eds.)
PROCEEDINGS OF THE 33rd European Symposium on Computer Aided Process Engineering (ESCAPE33), June 18-21, 2023, Athens, Greece
 http://dx.doi.org/10.1016/B978-0-443-15274-0.50115-3

Parameter estimation approach applied to microalgae-bacteria photobioreactor

Irina Bausa-Ortiz,[a,b] Raul Muñoz,[a,c] Smaranda P. Cristea,[a,b] Cesar de Prada [a,b]

[a]*Institute of Sustainable Processes, Universidad de Valladolid, Dr. Mergelina s/n, Valladolid 47011, Spain*
[b]*Dpt. of Systems Engineering and Automatic Control, School of Industrial Engineering, Universidad de Valladolid, Prado de la Magdalena 3-5, Valladolid 47011, Spain*
[c]*Dpt. of Chemical Engineering and Environmental Technology, School of Industrial Engineering, Universidad de Valladolid, Dr. Mergelina, s/n, Valladolid 47011, Spain*
irina.bausa@uva.es

Abstract

Mechanistic models are commonly used in wastewater treatment systems based on microalgae-bacteria consortia to study microalgae-bacteria interactions and for design and control purposes. Generally, in the proposed mathematical models, the parameters concerning the activity of bacteria and microalgae are obtained through calibration or parameter estimation. Microalgae-bacteria models frequently involve many nonlinear differential equations and parameters, which creates difficulties with the optimization for parameter estimation, resulting in non-optimum convergence in some cases. This work deals with formulating the optimization problem in the parameter estimation context. The goal is to propose an improved approach for parameter estimation when several outputs and parameters are involved in the optimization problem. The proposed approach solves a series of increasing-complexity optimization problems to estimate process parameters gradually, avoiding convergence problems.

Keywords: Modeling, Parameter Estimation, Optimization, Wastewater treatment.

1. Introduction

Microalgae-based technologies for wastewater treatment have received a renovated interest in the last years in the context of resource recovery and circular economy (Muñoz and Guieysse, 2006). The increasing number of microalgae-based applications has contributed to the development of new models for studying the main processes, factors, and variables affecting microalgae growth in different cultures media, including wastewater (Solimeno, Gómez-Serrano and Acién, 2019), (Casagli *et al.*, 2021). Generally, in microalgae-bacteria models, model parameters are obtained through calibration or parameter estimation (Solimeno, *et al.*, 2017), (Casagli *et al.*, 2021).

The main goal of this study is to propose an improved approach for parameter estimation when several outputs and parameters are involved in the optimization problem. The proposed approach solves a series of increasing-complexity optimization problems to estimate process parameters gradually, avoiding convergence problems. The proposed approach represents an alternative to dealing with large optimization problems in parameter estimation, which are common in these systems. Parameter estimation via dynamic optimization is realized in each optimization step to fit simulated outputs to experimental data. This methodology has been tested in a photobioreactor for wastewater treatment, in which processes concerning the activity of microalgae and bacteria, the influence of sunlight and nutrients concentration in the growth of microorganisms, chemical equilibrium reactions, and transfer of gases to the atmosphere are involved. Parameter estimation has been performed in a four-step optimization sequence. The first optimization step involves only two system outputs and five parameters. Finally, five model outputs were considered in the optimization problem: Total Suspended Solids (TSS) concentration, dissolved Total Organic Carbon (TOC) concentration, dissolved Inorganic Carbon (IC) concentration, dissolved ammonium concentration, and dissolved oxygen (DO) concentration. Simulation results show that model outputs matched with experimental data. The proposed optimization approach results in a better fit between experimental and simulated data and lower convergence time compared to the whole optimization problem.

2. Methodology for parameter estimation

The proposed approach solves a series of increasing-complexity optimization problems to estimate process parameters gradually, avoiding convergence problems. The idea is first to formulate one easier parameter estimation problem involving a subset of system outputs and parameters, replacing the other outputs variables with their experimental values, and then use these estimated values as a starting point for the next step of the optimization problem. The approach is oriented to increase the sub-set dimension until all the system outputs are being included in the optimization problem. The sub-sets selection is performed based on sensitivity analysis. Parameter estimation via dynamic optimization is realized in each optimization step to fit simulated outputs to experimental data.

This methodology has been tested in the photobioreactor for wastewater treatment used in the anoxic–aerobic algal–bacterial photobioreactor configuration with biomass recycling described in (Alcántara *et al.*, 2015). The anoxic reactor model and the settler model of this facility were described in (Bausa *et al.*, 2021, 2022). The data used in this study corresponded with 47 days of experimentation, these data were used for parameter estimation and model validation. In this work, the model BIO_ALGAE 2 (Solimeno, Gómez-Serrano and Acién, 2019) has been used to describe the photobioreactor. The modeling and simulation of the photobioreactor are coded in dynamic simulation software PROOSIS® (EA International, 2022).

2.1. Sensitivity analysis

Previous to parameter estimation, the sensitivity analysis was conducted to identify the parameters with the greatest impact on the model outputs. For this purpose, a subset of the most influential parameters on model outputs was analyzed.

Sensitivity analysis was carried out using PROOSIS®. The results of sensitivity analysis indicated that model outputs are especially sensitive to the maximum specific growth rate of microalgae (μ_{ALG}) and heterotrophic bacteria (μ_H); the decay rate of microalgae ($k_{death,ALG}$) and heterotrophic bacteria ($k_{death,H}$); and the gas-liquid mass transfer coefficients for ammonia ($K_{la,NH3}$), oxygen ($K_{la,O2}$), and carbon dioxide ($K_{la,CO2}$). These decision variables agree with decision variables used in other publications on the topic (Casagli *et al.*, 2021), (Solimeno *et al.*, 2017).

2.1.1. Sensitivity analysis as a guide to sub-sets selection

Results of sensitivity analysis are used here also as a guide to determine the best selection of groups of model outputs to consider. In addition, prior knowledge of system dynamic should be considered.

In the first step of the optimization problem, two model outputs were considered: TSS concentration and DO concentration. The TSS concentration is related to many of the processes of the model and its value depends on the concentration of all particulate components in the photobioreactor. Furthermore, DO concentration is involved in many of the processes concerning to microalgae and bacteria activity and mass transfer processes. Both model outputs depend mainly on the maximum specific growth rate of microalgae (μ_{ALG}), the decay rate of microalgae ($k_{death,ALG}$), and the mass transfer coefficient for oxygen ($K_{la,O2}$).

Inorganic carbon is included as model output in the second step of the optimization. Microalgae use carbon dioxide as a carbon source for growth. As expected, μ_{ALG} is the parameter with the greatest influence on IC concentration. Also, inorganic carbon is highly affected by the gas-liquid mass transfer coefficient for carbon dioxide ($K_{la,CO2}$).

In the third step, TOC concentration is also considered as model output. In microalgae-bacteria processes, heterotrophic bacteria oxidate the organic matter present in the wastewater, for that reason TOC concentration is mainly affected by the maximum specific growth rate (μ_H) and the decay rate of heterotrophic bacteria ($k_{death,H}$).

Finally, dissolved ammonium concentration is included as model output in the fourth step. Ammonium concentration is mainly affected by μ_{ALG}, $k_{death,ALG}$, and the mass transfer coefficient for ammonia ($K_{la,NH3}$). In order to reduce the size of the vector of parameters, mass transfer coefficients for ammonia and carbon dioxide were expressed as a function of the oxygen one (1) (Casagli *et al.*, 2021).

$$\frac{K_{la,j}}{K_{la,O2}} = \left(\frac{D_{Sj}}{D_{O2}}\right)^{0.5} \tag{1}$$

where $D_{Sj}[m^2s^{-1}]$ is the diffusivity coefficient for the gas j.

2.2. Parameter estimation

In this work, the fair function estimator (Huber, 2011) is used as a robust objective function J in the parameter estimation problem. Unlike Least Square Method, the fair function estimator reduces the effect of outliers, thus making it more robust in nature.

Dynamic optimization problem is represented by (2):

$$\min_{\hat{\boldsymbol{\theta}}} J(\hat{\theta}, \theta) = \sum_{i=1}^{n} c^2 \left[\frac{|\varepsilon_j|}{c} - log\left(1 + \frac{|\varepsilon_j|}{c}\right)\right] \tag{2}$$

Subject to restrictions (3) and (4)

$$\frac{d\mathbf{x}(t)}{dt} = \mathbf{f}(\mathbf{x}(t), \mathbf{u}(t), \boldsymbol{\theta}, t) \tag{3}$$

$$\hat{\mathbf{y}}(t) = \mathbf{g}(\mathbf{x}(t), \mathbf{u}(t), \boldsymbol{\theta}, t)$$

$$\underline{\boldsymbol{\theta}} \le \hat{\boldsymbol{\theta}} \le \overline{\boldsymbol{\theta}} \tag{4}$$

Where $\varepsilon_j = \left(y_j - \hat{y}_j\right)/\sigma_j$ represents the error between available process measurements (y) and their estimated values $(\hat{y})$ limited between user-defined minimum and maximum values, $c \in \mathbb{R}+$ is a user defined fitting parameter to tune the slope for large residues, and σ is the standard deviation of experimental data. The approach to solving a parameter estimation problem in terms of optimization considers that for each value of the vector of parameters $\boldsymbol{\theta}$ (decision variables) the model provides one prediction of the system output $\hat{y}(\boldsymbol{\theta})$ in each experiment. The same sequence of process inputs $\mathbf{u}(t)$ is applied to the model. For each sample time t, the prediction error $\varepsilon_j(t)$ is an indicator of model goodness. Fig. 1 represents the estimation procedure in sequential optimization context.

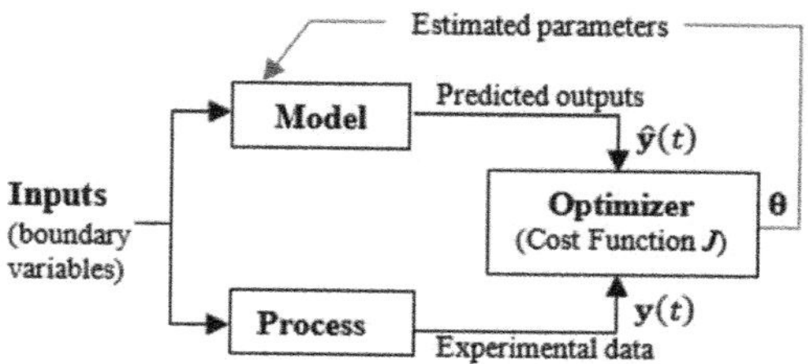

Figure 1. Estimator in sequential optimization

The dynamic optimization problem for the start-up optimization can be converted into a nonlinear programming (NLP) problem via control vector parameterization technique and a proper procedure for computing the cost function. In this work, the SNOPT nonlinear programming algorithm has been used in the PROOSIS® dynamic simulation environment to solve the optimization problem.

Parameter estimation results for each optimization step were provided in Table 1. Initial values for parameter estimation in Step 1 and decision variables ranges for optimization (Table 1) were defined in agreement with previous results reported in the literature.

Table 1. Values of estimated parameters in photobioreactor

Parameter	Value				Limits for Optimization
	Step 1	Step 2	Step 3	Step 4	
μ_{ALG} [d^{-1}]	1.627	0.990	1.062	1.062	0.4 – 2
$k_{death,ALG}$ [d^{-1}]	0.101	0.050	0.050	0.050	0.05 – 0.21
μ_H [d^{-1}]	1.656	1.000	1.210	1.211	1 – 6
$k_{death,H}$ [d^{-1}]	0.895	0.900	0.900	0.900	0.12 – 0.9
$K_{la,O2}$ [d^{-1}]	13.081	4.000	4.000	4.000	4 – 30

3. Results

In each step of the optimization problem, data from first 30 days were used for parameter estimation. Validation was performed using data from days 30 to 47. Figures 2 to 6 show simulation results of Step 4.

Fig. 2 present simulation results for the concentration of TSS. Fig. 2a) and 2b) represent data set used for parameter estimation and for validation, respectively. Although some discrepancies are observed in transient behavior prediction for TSS concentration, simulated average values are closely with experimental results.

Fig. 3 show simulation results for the dissolved oxygen concentration (S_O2). Fig. 3a) and 3b) represent data set used for parameter estimation and for validation, respectively.

Fig. 4 and Fig. 5 presents simulation results for the Inorganic Carbon concentration and Total Organic Carbon concentration, respectively. Fig. 4a) and 5a) represent data set used for parameter estimation. Fig. 4b) and 5b) represent data used for model validation.

Fig. 6 show simulation results for the dissolved ammonium concentration (S_O2). Fig. 6a) and 6b) represent data set used for parameter estimation and for validation, respectively.

The model was effective in reproducing dynamic behavior of different measured variables.

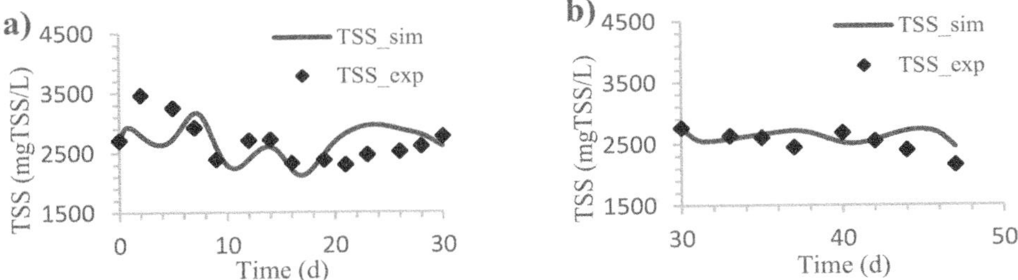

Fig. 2. Experimental and simulated Total Suspended Solids concentration

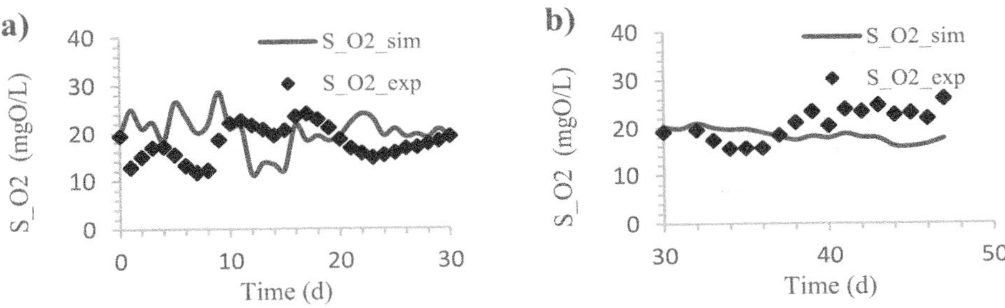

Fig. 3. Experimental and simulated dissolved oxygen concentration

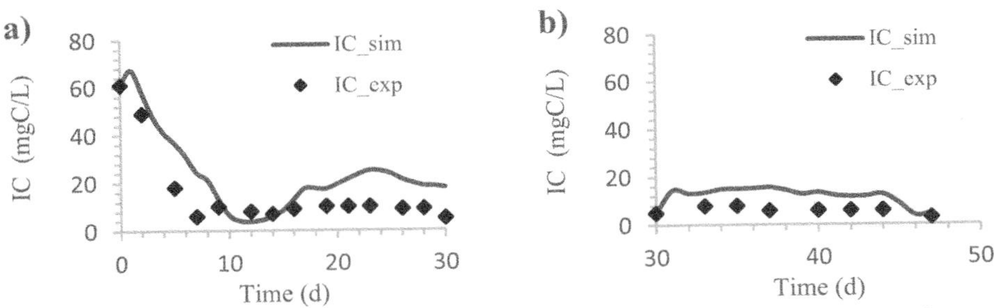

Fig. 4. Experimental and simulated dissolved Inorganic Carbon concentration

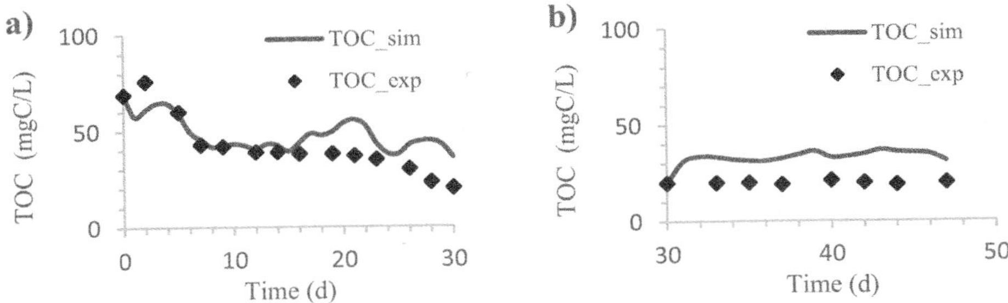

Fig. 5. Experimental and simulated dissolved Total Organic Carbon concentration

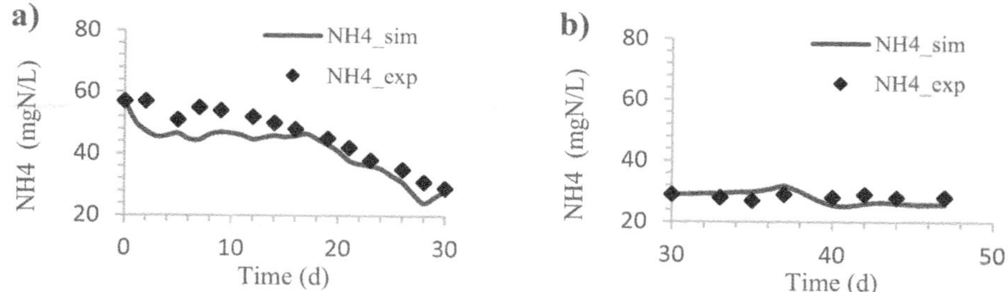

Fig. 6. Experimental and simulated dissolved ammonium concentration

4. Conclusions

Parameter estimation via dynamic optimization was realized in four optimization steps. The proposed optimization approach results in a better fit between the experimental and simulated data, and lower convergence time with respect to the optimization problem statement considering all the system outputs from the beginning.

Future work is oriented to design state-observers to estimate non-measurable variables.

5. Acknowledgments

This research was supported by Regional Government of Castilla y León and EU-FEDER program (CLU 2017-09, CL-EI-2021-07, UIC 233, UIC 315), by Regional Government of Castilla y León and the European Social Found (Order EDU/601/2020), and by the Project a-CIDiT (PID2021-123654OB-C31).

References

C. Alcántara *et al.*, 2015, 'Evaluation of wastewater treatment in a novel anoxic-aerobic algal-bacterial photobioreactor with biomass recycling through carbon and nitrogen mass balances', *Bioresource Technology*, 191, pp. 173–186. https://doi.org/10.1016/j.biortech.2015.04.125.

I. Bausa *et al.*, 2021, 'Modelo para la estimación de la concentración de biomasa en una instalación reactor anóxico-fotobiorreactor aerobio de algas y bacterias para el tratamiento de aguas residuales domésticas', https://doi.org/10.17979/spudc.9788497498043.427.

I. Bausa *et al.*, 2022, 'Modeling and simulation of anoxic-aerobic algal bacterial photobioreactor for nutrients removal', *COMPUTER-AIDED CHEMICAL ENGINEERING, 51*. Elsevier, pp. 151–156. https://doi.org/http://dx.doi.org/10.1016/B978-0-323-95879-0.50026-6.

F. Casagli *et al.*, 2021, 'ALBA: A comprehensive growth model to optimize algae-bacteria wastewater treatment in raceway ponds', *Water Research*, 190. https://doi.org/10.1016/j.watres.2020.116734.

EA International, 2022, *EcosimPro | PROOSIS - Modelling and Simulation Toolkits and Services*. Available at: https://www.ecosimpro.com/ (Accessed: 18 July 2022).

P. J. Huber, 2011, 'Robust Statistics', in M. Lovric (ed.) *International Encyclopedia of Statistical Science*. Springer, Berlin, https://doi.org/https://doi.org/10.1007/978-3-642-04898-2_594.

R. Muñoz and B. Guieysse, 2006, 'Algal-bacterial processes for the treatment of hazardous contaminants: A review', *Water Research*, 40(15), pp. 2799–2815. https://doi.org/10.1016/j.watres.2006.06.011.

A. Solimeno *et al.*, 2017, 'Integral microalgae-bacteria model (BIO_ALGAE): Application to wastewater high-rate algal ponds', *Science of the Total Environment*, 601–602, pp. 646–657. https://doi.org/10.1016/j.scitotenv.2017.05.215.

A. Solimeno, C. Gómez-Serrano, and F.G. Acién, 2019, 'BIO_ALGAE 2: improved model of microalgae and bacteria consortia for wastewater treatment', *Environmental Science and Pollution Research*, 26(25), pp. 25855–25868. https://doi.org/10.1007/s11356-019-05824-5.

Antonis Kokossis, Michael C. Georgiadis, Efstratios N. Pistikopoulos (Eds.)
PROCEEDINGS OF THE 33rd European Symposium on Computer Aided Process Engineering (ESCAPE33), June 18-21, 2023, Athens, Greece
 http://dx.doi.org/10.1016/B978-0-443-15274-0.50116-5

An Event-Based Continuous-Time MILP for Short-Term Crude Oil Scheduling

İrem Marttin,[a,b] Sena Kurban,[a] Gizem Kuşoğlu Kaya,[a] Özgür Kabak,[b] Y. İlker Topcu[b]

[a]*Turkish Petroleum Refinery, 41780, Körfez, Kocaeli, Turkey*
[b]*Industrial Engineering Department, Faculty of Management, Istanbul Technical University, 34367, Sarıyer, Istanbul, Turkey*
irem.marttin@tupras.com.tr

Abstract

In complex real refinery systems, it is essential to find an optimal plan in a short time to maintain operations' efficiency. An effective tool for industrial use in refinery crude oil scheduling is still lacking. The aim of this study is to develop an optimization tool to support efficient decision-making by ensuring the stability of operations and considering economic profit. An event-based continuous-time Mixed Integer Linear Program is developed to handle crude oil operations from vessels to crude distillation units as well as to satisfy product demands. This new formulation describes real-world operational applications of the TUPRAS Refinery.
Keywords: Crude oil operations, Short-term scheduling, Mixed-integer linear programming, Optimization.

1. Introduction

In the literature, several approaches have been proposed to handle the management of crude oil scheduling. The early approaches proposed by Shah (1996) and Lee et al. (2002) focused on discrete-time representation. However, recently, continuous-time representation has received great attention since it reduces complexity and problem size. Yadav and Shaik (2012) proposed a state-task-network approach using event-based continuous-time representation to manage crude oil operations. In this study, a similar approach is implemented in a complex petroleum refinery. Therefore, the discrete-time approach model, developed by Dologlu et al. (2022), has been modified as a continuous model approach to better reflect the refinery. Unlike the discrete model, many new features have been added in the continuous-time approach model, such as tank rest times, the amount of crude oil remaining in the pipe at the beginning and after the unloading, the unloading of different crude oils from the same vessel, and the decision of crude oil to be selected in case more than one crude oil arrives on the same day in the plan.

2. Model Formulation and Problem Definition

Similar to Dologlu et al. (2022), TUPRAS Refinery including one docking station, 12 tanks and 2 Crude Distillation Units (CDUs) whose structure is illustrated in Fig. 1 is studied in this paper. During a given planning horizon, a vessel arrives at the docking station and different types of crude oil are unloaded from vessel into tanks through pipes. Finally, the appropriate crude oil mixtures in the tanks are charged into the CDUs. Before charging operations, the crude oil mixtures in the tanks must meet the quality requirements. Moreover, about 50 types of crude oil reach and are processed at the refinery. Having too many types of crude oil can lead to operational difficulties for processing in CDUs due to their specific properties. In addition, since the tanks are not dedicated to a single operation and all the unloading, transfer and charging operations can be carried out in all tanks, the scheduling problem becomes more complex.

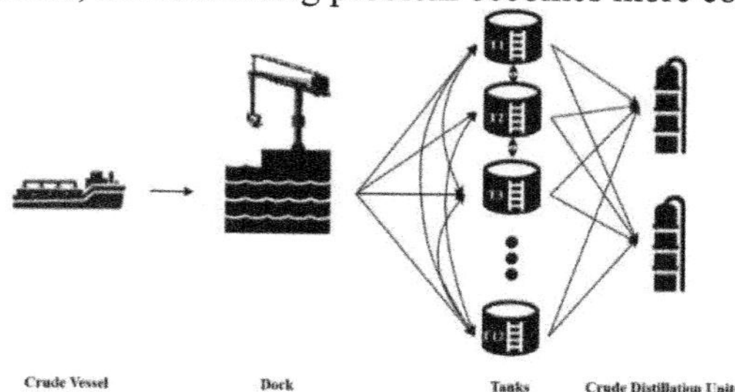

Figure 1. An overview of crude oil operations

The system addressed includes $i \in I$ crude oil tanks and $h \in H$ crude oil types. Four main operations of the crude oil system are described here as tasks under a set, and the four subsets are: $k(a)$, $ar(a)$ and $as(a)$ for unloading, transfer and charging tasks, respectively. Crude oils' descriptive properties such as American Petroleum Institute (API) and Sulfur are in the $p \in P$ set. Over the planning horizon, k vessels arrive at the system as well as operations take place along a chain of n events. The set of arriving crude oil types is K_h and T_k is the set of tanks to which vessel k can be unloaded. For simplicity, unloading operations are set to specific events. nk' and nk'' are the set of unloading events, which is for the main unloading and for the remaining part in the pipe, respectively. The following data is used for scheduling the operations: (1) scheduling time horizon, PH; (2) arrival time of vessels, $KT_{k,t}$ (hr); (3) amount of arriving crude oil vessels, KV_k (m^3); (4) demurrage DM_k (\$/hr) and lay time of vessels LT_k (hr); (5) crude oil properties $PR_{h,p}$; (6) Tank properties (min. & max. capacity of tanks, LC_{ih} and UC_{ih}; tank rest times after unloading and transfer, R; (7) Initial tank volume and compositions, $VI_{h,i}$ (m^3); (8) bounds on the flowrates of the streams for unloading ($FU_{ih,}$), transfer (FT_a), and charging (FC_{as}). Additional parameters are required for problem formulation. B_h is the set for crude oils that cannot mix with each other. Besides, since the rest time applies only to the tank that receives a transfer, A_{ps} and A_{ng} parameters are used to separate receiving and sending transfer tanks for a transfer operation, respectively. Moreover, there are two other vessel types; the first is the optional crudes that planners must choose one out of two or three different crude oil types. There may be more than one optional crude in a planning period, so they were handled in groups. GN_k is for the group numbers, O_t is the set of all optional crudes and O_l is the different types of crude in group l. Second is the vessels carrying two or three different crude oils and all of them should be unloaded sequentially. S_a is the set of these concurrent crudes. SF_k specifies the ones that are not the last one in the unloading sequence so the remaining pipe volume (VP) of them is not going to be planned separately since they will be unloaded one after another and by_k parameter takes value 1 for them. To introduce the limitations for charging operations, the related parameters are set as: The minimum charging hours for a charging task, MH. Also, the CDUs have minimum-maximum API and maximum Sulfur limits, LA_j, UA_j and US_j, respectively. The described problem is formulated in Mixed-Integer Linear Programming (MILP) with event-based continuous time representation with the assumption of perfect mixing in the tanks.
The optimum values of the following decision variables are determined to minimize the demurrage cost of vessels: (1) the start and end times of all operations, $s_{a,n}$ and $f_{a,n}$; (2) the

start and end times of operations in tanks, $ts_{n,i}$ and $tf_{n,i}$; (3) the start and end times of unloading and transfer operations, rs_n and rf_n; (4) the total volume of tanks, $vt_{i,n}$, the volume of components in the tanks, $v_{i,h,n}$; (5) the amount of crude oil that flows from one unit to another in unloading, $bk_{k,n,i}$, transfer, $bh_{ar,n,h}$, charging, $bs_{as,n,h,i}$; (6) delay time of vessel unloading, g_k; (7) total amount of flowing, $b_{a,n}$. Binary variables for operational decisions are also defined. $w_{a,n}$ shows if a task is assigned to an event. It also applies to all of the operations in specific, $wt_{k,n,i}$, $ws_{as,n,i}$, $wtb_{k,n,i}$ for unloading, charging and unloading of the remaining crude in the pipe, respectively. $w_{k,n}$ represents whether a vessel is unloaded and $wb_{k,n}$ is the same for unloading the crude in the pipe. The selection of optional crude oils is represented by q_k. $e_{n,i}$ shows whether the volume in the tank is above the minimum capacity, and $t_{n,i,h}$ shows whether a crude oil type exists in a tank.
The objective function is the minimization of the demurrage cost of vessels, Eq. (1). Tank components balance, total tank volume and capacity limits are given in Eqs. (2-4), respectively. The calculation of processed volumes at each event for transfer and charging tasks are handled in Eqs. (5a-b) and for unloading is handled in two different equations for the pipe part and the main part of the crude, Eqs. (6a-b). Eqs (7a-c) guarantee that if any flow occurs at any event, then the binaries take value 1 where M refers to a big number. In Eqs. (8a-10c), timing and sequencing rules are handled for each of the tasks. For continuous operation of CDUs, it is ensured that charging starts from the beginning and total operation time of each CDU should be equal to the time horizon *PH*, Eqs. (11a-b). Besides, since transfer operations are not preferred, they are eliminated in Eq. (12).

$$Cost = \sum_k g_k * DM_k \tag{1}$$

$$v_{i,h,n} = v_{i,h,n-1} + \sum_k bk_{k,n,i} + \sum_k wtb_{k,n,i} * VP + \sum_{ar \in A_{ps}} bh_{ar,n,h} - \sum_{ar \in A_{ng}} bh_{ar,n,h} - \sum_{as} bs_{as,n,h,i} \;\; \forall n \geq 2, n'_k, n''_k, h \in K_h, i \in T_k \tag{2}$$

$$vt_{i,n} = \sum_h v_{i,h,n} \;\; \forall n, i \tag{3}$$

$$LC_i \leq vt_{i,n} \leq UC_i \;\; \forall n, i \tag{4}$$

$$FT_{ar} * (f_{ar,n} - s_{ar,n}) = b_{ar,n} \;\; \forall ar, n \qquad FC_{as} * (f_{as,n} - s_{as,n}) = b_{as,n} \;\; \forall as, n \tag{5a,5b}$$

$$f_{k,n} - s_{k,n} = \sum_i wtb_{k,n,i} * \frac{VP}{FU_i} \;\; \forall k, n''_k \qquad f_{k,n} - s_{k,n} = \sum_{i \in T_k} \frac{bk_{k,n,i}}{FU_i} \;\; \forall k, n'_k \tag{6a,6b}$$

$$b_{ar,n} \leq M * w_{ar,n} \;\; \forall ar, n \qquad b_{as,n} \leq M * w_{as,n} \;\; \forall as, n \qquad b_{k,n} \leq M * w_{k,n} \;\; \forall k, n'_k, n''_k \tag{7a,7b,7c}$$

$$f_{k,n} = s_{k,n} \;\; \forall k, n \qquad f_{k,n} \geq s_{k,n} \;\; \forall k, n'_k, n''_k \qquad s_{k,n} \geq f_{k,n-1} \;\; \forall k, n \tag{8a,8b,8c}$$

$$f_{ar,n} \leq s_{ar,n} + 60 * 24 * w_{ar,n} \;\; \forall ar, n \qquad f_{ar,n} \geq s_{ar,n} \;\; \forall ar, n \qquad s_{ar,n} \geq f_{ar,n-1} \;\; \forall ar, n \tag{9a,9b,9c}$$

$$f_{as,n} \leq s_{as,n} + 60 * 24 * w_{as,n} \;\; \forall as, n \qquad f_{as,n} \geq s_{as,n} \;\; \forall as, n \qquad s_{as,n} = f_{as,n-1} \;\; \forall n > 1, as \tag{10a,10b,10c}$$

$$\sum_n b_{as,n} \geq FC_{as} * PH * 24 \;\; \forall as \qquad s_{as,n} = 0 \;\; \forall n = 1, as \tag{11a,11b}$$

$$w_{ar,n} = 0 \;\; \forall ar, n \tag{12}$$

Eq. (13) ensures that all the volume of vessel should be unloaded except the part in the pipe. The main unloading and pipe unloading for each of the vessels are assigned to an event in Eqs. (14a-b). Eqs. (15a-b) state the same condition for optional crudes for the one that is chosen. If there is any unloading in an event, then the binary related to the tank involving also takes the value 1 (Eqs. (16a-b)). Eq. (17) guarantees that one of the crude oils must be selected for each of the optional crude oil vessel groups. Unloading flow boundaries are in Eqs. (18a-b). The unloading can start only after the arrival of vessel and same for optional crudes (Eqs. (19a-b)). In Eqs. (20a-b), the unloading delay time subject to demurrage cost is calculated for both types of vessels. Eq. (21) ensures that unloading of remaining crude from the previous vessel should start after the next vessel's arriving. After the pipe crude, the main part of the next vessel should be unloaded immediately (Eq. (22)). Vessels are unloaded sequentially and the concurrent crudes in a vessel are unloaded repeatedly, Eqs. (23-24). Eqs. (25a-b) state that the crude in the pipe should be unloaded to the set of tanks T_k in which the next vessel can be unloaded. Lastly, Eqs. (26a-28b) are about sequencing unloading and transfer operations.

$$\sum_{i\in T_k} bk_{k,n,i} = (KV_k - (VP * (1 - by_k))) * w_{k,n} \ \forall k, n'_k \quad (13)$$

$$wb_{k,n} = 1 \ \forall k \notin O_t, n''_k \qquad wk_{k,n} = 1 \ \forall k \notin O_t, n'_k \quad (14a,14b)$$

$$wb_{k,n} = q_k \ \forall k \in O_t, n''_k \qquad wk_{k,n} = q_k \ \forall k \in O_t, n'_k \quad (15a,15b)$$

$$\sum_{i\in T_k} wt_{k,n,i} \le w_{k,n} * I \ \forall k, n'_k \qquad \sum_i wtb_{k,n,i} = wb_{k,n} \ \forall k, n''_k \quad (16a,16b)$$

$$\sum_{k\in O_l} q_k = 1 \ \forall l \quad (17)$$

$$bk_{k,n,i} \le M * wt_{k,n,i} \ \forall k, n'_k, i \in T_k \qquad bk_{k,n,i} \ge VP * wt_{k,n,i} \ \forall k, n'_k, i \in T_k \quad (18a,18b)$$

$$s_{k,n} \ge KT_k \ \forall k \notin O_t, n'_k \qquad s_{k,n} \ge KT_k * q_k \ \forall k \in O_t, n'_k \quad (19a,19b)$$

$$g_k \ge f_{k,n} - KT_k - LT_k \ \forall k \notin O_t, n'_k \qquad g_k + 24 * PH * (1 - q_k) \ge f_{k,n} - KT_k - LT_k \ \forall k \in O_t, n'_k \quad (20a,20b)$$

$$s_{k,n} + GN_k * 24 * PH * (1 - q_k) \ge KT_{k+1} - GN_{k+1} * 24 * PH * (1 - q_{k+1}) \ \forall k \in K_b, n''_k \quad (21)$$

$$f_{k,n} + GN_k * 24 * PH * (1 - q_k) \ge s_{k+1,n+1} - GN_{k+1} * 24 * PH * (1 - q_{k+1}) \ \forall k \in K_b, n''_k \quad (22)$$

$$s_{k,n} - GN_k * 24 * PH * (1 - q_k) \le rf_{n-1} \ \forall k, n'_k \quad (23)$$

$$s_{k+1,n+1} \le f_{k,n} \ \forall k \in S_a, n'_k \quad (24)$$

$$wtb_{k,n,i} = 0 \ \forall k \notin O_t, n''_k, i \notin T_k \qquad wtb_{k,n,i} \le (1 - q_{k+1}) \ \forall k \in O_t, n''_k, i \notin T_k \quad (25a,25b)$$

$$rs_n \le s_{ar,n} \ \forall ar, n \qquad rf_n \ge f_{ar,n} \ \forall ar, n \quad (26a,26b)$$

$$rs_n \le s_{k,n} \ \forall k, n \qquad rf_n \ge f_{k,n} \ \forall k, n \quad (27a,27b)$$

$$rs_n \le rf_n \ \forall n \qquad rs_n \ge rf_{n-1} \ \forall n \quad (28a,28b)$$

Eqs. (29a, 30b) relates tank and crude oil-based flows to event-based flow amounts. Tank times for operations should be associated with task times and are handled in Eqs. (31a-b) for charging; in Eqs. (32-36) for unloading; in Eqs. (37-38b) for transfer operations considering the tank rest times. Events in the tanks should be sequenced, Eqs. (39a, 39b). Eq. (40) ensures no simultaneous input-output at any event. Simultaneous unloading and transfer operations are also not allowed due to using the same line, Eq. (41).

$$\sum_{i\in T_k} bk_{k,n,i} = b_{k,n} \ \forall k, n'_k \qquad \sum_i wtb_{k,n,i} * VP = b_{k,n} \ \forall k, n''_k \quad (29a,29b)$$

$$\sum_h bh_{ar,n,h} = b_{ar,n} \ \forall ar, n \qquad \sum_i \sum_h bs_{as,n,h,i} = b_{as,n} \ \forall as, n \quad (30a,30b)$$

$$ts_{n,i} \le s_{as,n} + 24 * 60 * (1 - ws_{as,n,i}) \ \forall as, n, i \qquad tf_{n,i} \ge f_{as,n} - 24 * 60 * (1 - ws_{as,n,i}) \ \forall as, n, i \quad (31a,31b)$$

$$ts_{n,i} \le s_{k,n} + 24 * 60 * (1 - wt_{k,n,i}) \ \forall k, n'_k, i \in T_k \quad (32)$$

$$tf_{n,i} \ge f_{k,n} + R_k * (1 - SF_k) - 24 * PH * (1 - wt_{k,n,i}) \ \forall k, n'_k, i \in T_k \quad (33)$$

$$ts_{n,i} \le s_{k,n} + 24 * 60 * (1 - wtb_{k,n,i}) \ \forall k, n''_k, i \quad (34)$$

$$tf_{n,i} \ge f_{k,n} + R_k * (1 - wt_{k+1,n+1,i}) - 24 * PH * (1 - wtb_{k,n,i}) \ \forall k \notin O_t, n''_k, i \in T_k \quad (35)$$

$$tf_{n,i} \ge f_{k,n} + R_k * (1 - wt_{k+1,n+1,i}) - 24 * PH * (1 - wtb_{k,n,i}) - 24 * PH * (1 - q_{k+1}) \ \forall k \in O_t, n''_k, i \in T_k \quad (36)$$

$$ts_{n,i} \le s_{ar,n} + 24 * 60 * (1 - w_{ar,n}) \ \forall n, i, ar \quad (37)$$

$$tf_{n,i} \ge f_{ar,n} + R_{ar} - 24 * 60 * (1 - w_{ar,n}) \ \forall n, i, ar \in A_{ps} \qquad tf_{n,i} \ge f_{ar,n} - 24 * 60 * (1 - w_{ar,n}) \ \forall n, i, ar \in A_{ng} \quad (38a,38b)$$

$$ts_{n,i} \leq tf_{n,i} \; \forall n,i \qquad ts_{n,i} \geq tf_{n-1,i} \; \forall n,i \tag{39a,39b}$$

$$\sum_k wt_{k,n,i} + \sum_k wtb_{k,n,i} + \sum_{ar \in A_{ps}} w_{ar,n} + 0.5 * \sum_{ar \in A_{ng}} w_{ar,n} + 0.25 * \sum_{as} ws_{as,n,i} \leq 1 \quad \forall n'_k, n''_k, i \in T_k \tag{40}$$

$$w_{ar,n} = 0 \; \forall n'_k, n''_k, ar \tag{41}$$

If a tank's volume is above the minimum capacity, $e_{n,i}$ takes value 1 meaning that there is a significant amount of crude that is to be considered for mixing rules, Eq (42). Mixing constraint is handled in Eq. (43). Operational rules for charging are considered as follows: The binding of related variables is given in Eq. (44). Minimum hours of charging for a charging task is handled in Eq. (45). API and Sulfur limitations are given in Eqs. (46-47).

$$e_{n,i} * UC_i \geq vt_{i,n-1} - LC_i \qquad \forall n > 2, i \tag{42}$$

$$t_{n,i,h} + wt_{k,n,i} \leq (1 - e_{n,i}) + 1 \qquad \forall n'_k \geq 2, i, h \in B_h, k \in K_h \tag{43}$$

$$\sum_h bs_{as,n,h,i} \leq ws_{as,n,i} * M \; \forall as, n, i \tag{44}$$

$$b_{as,n} \geq w_{as,n} * FC_{as} * MH \; \forall n, as \tag{45}$$

$$LA_{as} * \sum_i \sum_h bs_{as,n,h,i} \leq \sum_i \sum_h bs_{as,n,h,i} * p_{h,\text{"API"}} \leq UA_{as} * \sum_i \sum_h bs_{as,n,h,i} \; \forall n, as \tag{46}$$

$$\sum_i \sum_h bs_{as,n,h,i} * p_{h,\text{"Sulfur"}} \leq US_{as} * \sum_i \sum_h bs_{as,n,h,i} \; \forall n, as \tag{47}$$

3. Results and Discussions

The developed continuous-time model is applied for 14-days planning horizon that contains 5 vessels shipped 7 different types of crude oil, 2 CDUs, 12 tanks (2 tanks are in maintenance), including 13 crude oil types. The volume of vessels varies between 40,000 and 175,000 m^3 and the total volume of 5 vessels is about 535,000 m^3. The MILP model results are shown in Fig. 2 for the first 10 days of the planning horizon.

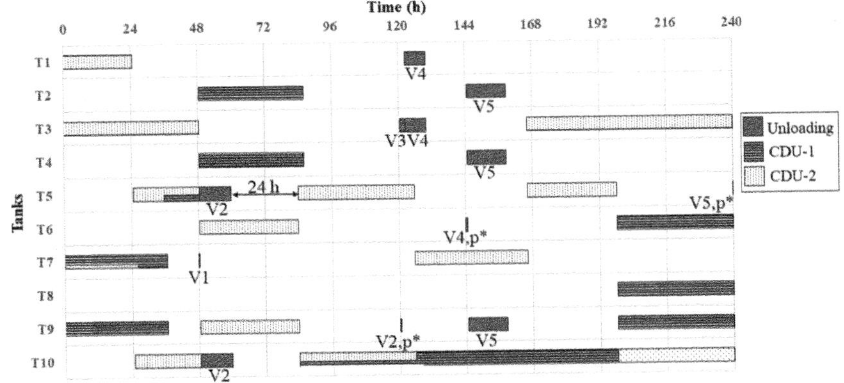

Figure 2. Gantt chart of optimal solution for first 10 days of planning period

Unlike the discrete model developed by Dologlu et al. (2022), in this model, a continuous charge flow is provided to the units. There is a 24-hour rest time in the tanks after unloading. For example, T5 was charged to CDU 24 hours after V2 unload to T5 was completed. The unloading of crude oil remaining in the pipe can be monitored. For example, V1 is the crude oil remaining in the pipe at the start of planning. Immediately after, V2 was unloaded. V3 and V4 are simultaneous incoming vessels and their unloading to the T3 tank appears to be sequentially scheduled. V2 vessel carries 2

different types of optional crude oil. The model made the decision of crude oil in line with the tank occupancy and contents, so as not leading to any demurrage cost.
The main motivation of this study is to compare the computational performance and behavior of models with different time approaches, such as discrete and continuous time, for the real refinery problem. To test the performance of discrete and continuous time models, four cases with different time horizons, 8-days, 14-days, 19-days, and 26-days were studied. All MILP models are executed in GAMS v 28.2.0 on an Intel(R) Xeon(R) Gold 6252 CPU 2.10 GHz machine with 12 GB of RAM using the CPLEX solver. In all cases, the objective function of minimizing the demurrage costs was used. The overall solutions of the four cases are presented in Table 1. When the results of the continuous time model are compared to the discrete time model, the number of constraints, variables, and binary variables is higher, so the model has a longer CPU time. This is because the constraints such as the rest time of the tanks and the evaluation of different unloading states, which were previously ignored in the discrete model, were added to the model to reflect the real refinery case, and the model became more complex. Thus, it is an expected result that the variables, constraints, and CPU time increase in the transition to the continuous model. In addition, as the planning period increased, the number of constraints and variables, and the CPU time increased accordingly for each model time approach.

Table 1. Comparison of discrete and continuous time model results

	The discrete-time approach model				The continuous-time approach model			
Scenario	1	2	3	4	5	6	7	8
Planning time horizon (day)	8	14	19	26	8	14	19	26
Number of constraints	781	1594	2382	3633	10977	16008	21101	23690
Number of variables	1440	2796	4046	5920	5178	7541	9960	11191
Number of binary variables	197	350	509	811	926	1306	1696	1888
CPU time (s)	0.28	0.50	2.28	5.98	10.03	149.36	551.80	1567.59

4. Conclusion

In this paper, an event-based continuous time formulation is proposed to address the real-life problem of scheduling of crude-oil operations. Compared to the discrete time formulation, the continuous time formulation provides better production schedules as it reflects the operations of refinery with the added constraints and the continuous time setup. The computation time of the continuous time model increases as more event points are needed to solve refinery-scheduling problems over longer time horizons. Future research will focus on case-specific operational constraints and solving the non-linear problem when the tank contents are uniformly charged to the CDUs as a homogeneous mixture as well as accelerating the model solving time.

References

N. Shah, 1996, Mathematical programming techniques for crude oil scheduling, Computers & Chemical Engineering, 20, 1227-1232.

P. Dologlu, S. Kurban, İ. Marttin, N. Ataman, G.K. Kaya, F. İşeri, Ö. Kabak, Y. I. Topcu, 2022, A new MILP formulation for scheduling of crude oil operations, In Computer Aided Chemical Engineering, 51, 991-996.

S. Yadav, M.A. Shaik, 2012, Short-term scheduling of refinery crude oil operations, Industrial & Engineering Chemistry Research, 51(27), 9287-9299.

Y. H. Lee, S. H. Kim, 2002, Production-distribution planning in supply chain considering capacity constraints, Computers & Industrial Engineering, 43(1-2), 169-190.

Antonis Kokossis, Michael C. Georgiadis, Efstratios N. Pistikopoulos (Eds.)
PROCEEDINGS OF THE 33rd European Symposium on Computer Aided Process Engineering (ESCAPE33), June 18-21, 2023, Athens, Greece
 http://dx.doi.org/10.1016/B978-0-443-15274-0.50117-7

Modeling aqueous multi-electrolyte systems for the simulation and optimization of concentrated brine processes

Wajeha Tauqir[a], Adaeze C. Maduako[a], Xi Yang[a], George M. Bollas[a]

[a]*Department of Chemical and Biomolecular Engineering, UTC Institute for Advanced Systems Engineering, University of Connecticut, 159 Discovery Dr, Storr, CT, 06269, USA. (george.bollas@uconn.edu)*

Abstract

An advancement of the refined e-NRTL model (Bollas et al., 2008) is used for the modeling of osmotic and activity coefficients associated with aqueous multi-electrolyte systems, with an emphasis on electrolyte systems relevant to concentrated brine solutions. Binary interaction parameters associated with the ions in each electrolyte system are estimated from available activity and osmotic coefficient experimental data. Electrolytes studied are those common in brine solutions (e.g., Na^+, K^+, Li^+ Mg^{2+}, Ca^2, Cl^-, SO_4^{2-}). Integration of the thermodynamic model in the software package IDAES (Institute for the Design of Advanced Energy Systems), as a thermodynamic property module are presented, and shown to predict various thermodynamic properties (such as activity and osmotic coefficient, and vapor pressure) at higher molalities and temperature ranges. Accurate prediction of solution thermodynamics in brines, then, targets the accurate modeling of brine water treatment systems.

Keywords: Concentrated brines, electrolyte thermodynamics, water treatment

1. Introduction

Brine solutions are encountered in many natural and industrial processes. Because of their detrimental effects, these highly concentrated salt solutions are rendered dangerous to be disposed of directly to the environment. High salinity in liquid streams is also the major cause of scaling in pipes and equipment. Thus, there is a need to model and develop energy-intensive desalination technologies to increase the reuse of saline waters. Exploration of novel and advanced brine management options is challenging because of the lack of steady state and dynamic simulators that can accurately capture the thermodynamics and kinetics that dominate separation efficiency and processing options. Uncertainty and the high cost associated with the deployment of these systems on a larger scale make modeling a necessity before expensive piloting. The challenge in designing brine wastewater treatment technologies arise because of the high concentration of ions, which when in mixtures are difficult to predict in terms of their impact on the macroscopic properties of the solution. Comprehensive and precise thermodynamic models need to be developed that can accurately predict various thermodynamic properties associated with each ion in the system, over a wide range of molality and temperature ranges.

Electrolyte mixtures are non-ideal even at very low temperature and pressure. To completely model the non-ideality of electrolyte solutions, long-range electrostatic interactions between ions and short-range ion dipole and non-electrostatic interactions should be taken into account. One of the most important thermodynamic properties, which not only quantifies the non-ideality of solutions involving electrolytes but also predicts other properties, is the mean ionic activity coefficient. Thus, it is fundamental to have models that can accurately predict activity and osmotic coefficients, which can, then, be generalized to excess Gibbs free energy models and predict other solution properties, like enthalpy of mixing and heat capacity. Various activity coefficient models have been used to predict the non-ideal behavior of solutions. The most common of these are the e-NRTL model and its many extensions. The e-NRTL model was first developed for single salt, single solvent solutions. The Pitzer-Debye-Hückel equation was used to model the long-range interactions, while short-range interactions were modeled based on non-random two-liquid, local composition theory. Later, Chen & Evans (1986) modified the model for multi electrolyte systems. The thermodynamics of multi-electrolyte solutions differ largely from single electrolyte systems. Additional terms need to be considered when modeling non-ideality in multi-electrolyte aqueous systems, to cater to increased ions interaction. Besides water treatment, accurate electrolyte models are needed to simulate processes like geothermal energy production, hydraulic fracturing, thermochemical hydrogen production, and pharmaceutical processes(Chen, 2006).

Major ions responsible for salinity and scaling in aqueous solutions include: Na^+, K^+, Li^+, Ca^{2+}, Sr^{2+}, Mg^{2+}, Ba^{2+}, Cl^-, SO_4^{2-}, HCO^{3-}, and CO_3^{2-}. Studies have employed extensions and refinements of the e-NRTL model to binary, ternary and quaternary systems involving these ions. (Tanveer and Chen, 2016)developed a thermodynamic model for the aqueous Ca^{2+}, Na^+, K^+, Cl^- quaternary system based on the symmetric e-NRTL model. Honarparvar et al. (2017) studied ternary and quaternary systems involving Ba^{2+}, Na^+, Cl^-, SO_4^{2-} using an e-NRTL-based model. The e-NRTL model was also studied for the system containing Sr^{2+}, Na^+, Cl^-, and SO_4^{2-} ions by Honarparvar et al. (2018). Bhattacharia et al. (2015) performed thermodynamic modeling of the $KCl + H_2O$ binary system, and the $KCl + NaCl + H_2O$ ternary system using the e-NRTL model. In this paper, the refined e-NRTL model developed by Bollas et al. (2008) for multi-electrolyte solutions and improved later by Yang et al. (2023) for mixed solvent solutions is applied on 11 aqueous binary systems, with electrolyte concentrations up to their solubility limit, at 298.15 K. The electrolytes studied include three 1-1 electrolytes; NaCl, KCl, LiCl, four 2-1 electrolytes; $MgCl_2$, $SrCl_2$, $CaCl_2$, $BaCl_2$; three 2:1 electrolytes; K_2SO_4, Li_2SO_4, and Na_2SO_4, and one 2-2 electrolyte, $MgSO_4$. Binary interaction parameters for each aqueous electrolyte were estimated from experimental data on activity and osmotic coefficients. The model predicts accurately the liquid phase non-idealities and through mean ionic activity and osmotic coefficients, providing sound thermodynamic basis for the modeling of process systems that involve brine solutions.

2. Thermodynamic Modeling Framework

The refined e-NRTL model used in this work is explained in detail in (Bollas et al. (2008) and Yang et al. (2023). It expresses the excess Gibbs free energy of the electrolyte solution as a summation of long-range electrostatic contributions, accounting for the ion-ion interactions, and short-range interactions, accounting for local, ion-ion, ion-molecule and molecule-molecule interactions. The model of Yang et al. (2023) describes short-range interactions with a classic NRTL model, but it uses a thermodynamically consistent conversion of the Debye-Hückel equation to model long-range, electrostatic interactions. The short-range model requires two binary interaction parameters τ_{ij} and τ_{ji}, where $\tau_{ij} \neq \tau_{ji}$, and the non-randomness parameter α. Classically, the value of α is fixed to 0.2. Binary interaction parameters refer to water-electrolyte and electrolyte-electrolyte pairs, in the case of binary systems. In the case of ternary systems, two electrolytes sharing a common ion in an aqueous system will require six binary interaction parameters; four corresponding to the individual electrolyte-water pairs and two for the electrolyte-electrolyte asymmetric pairs. Other parameters that are treated as adjustable in the model include the distance of closest approach of the original Debye-Hückel theory, and ionic hydration. Hydration is included to describe its effect on the local composition during the short-range equation formulation. To do this, the mole fractions in the e-NRTL model are replaced by the effective mole fractions of the hydrated solution. To predict the effect of water structure on its neighborhood, hydration numbers are allowed to have negative values. The model with constant hydration numbers can predict mean ionic activity and osmotic coefficients with accuracy at low concentrations. To consider the effect of decreased water activity at increased molality, the stepwise hydration equilibria theory was employed, which accounts for the decreased availability of water molecules as the solution ionic strength increases.

3. Methodology

3.1. Thermodynamic Modeling of Binary Systems

The refined e-NRTL model was first implemented to model 11 aqueous binary systems which included two anions (Cl^-, and $SO4^{2-}$) and seven cations (Na^+, K^+, Li^+, Ca^{2+}, Sr^{2+}, Mg^{2+}, and Ba^{2+}). The model was simultaneously fitted against experimental data from Robinson (1959) of activity and osmotic coefficient with their molality ranging up to the solubility limit of each solution studied, at 298.15 K and 1 atm. The objective function in the non-linear regression algorithm included both the activity and osmotic coefficients to improve model precision and fidelity. Pure component parameters including ionic radii, hydration indices, and effective molar volume were taken at the values estimated with the data of electrolyte solution data reported in Yang et al. (2023). The Average Absolute Relative Deviation (AARD) was calculated for each system.

Table 1: AARD of original e-NRTL, refined e-NRTL with constant hydration and refined e-NRTL model with stepwise hydration equilibria for binary systems relevant to brines.

Electrolyte	Original e-NRTL Model		Refined e-NRTL Constant Hydration Model		Refined e-NRTL Stepwise Hydration Model	
	AARD% $\gamma_{\pm}^{m}$	AARD% Φ	AARD% $\gamma_{\pm}^{m}$	AARD% Φ	AARD % $\gamma_{\pm}^{m}$	AARD % Φ
			1-1			
NaCl	1.551	2.088	0.229	0.149	0.047	0.034
KCl	0.808	1.788	0.078	0.071	0.034	0.021
LiCl	18.4931	9.448	4.740	2.109	1.355	0.454
			2-1			
$MgCl_2$	14.150	7.083	1.565	0.566	0.804	0.343
$SrCl_2$	6.598	3.722	0.841	0.559	0.281	0.216
$CaCl_2$	18.132	8.421	4.725	2.141	0.806	0.387
$BaCl_2$	1.618	1.157	0.177	0.092	0.328	0.316
			1-2			
K_2SO_4	0.842	1.598	0.53259	0.641	0.53285	0.564
Li_2SO_4	2.105	1.955	0.311	0.265	0.208	0.221
Na_2SO_4	2.871	3.281	0.419	0.477	0.171	0.209
			2-2			
$MgSO_4$	3.496	10.836	2.431	5.344	2.714	5.064

3.2. Thermodynamic Modeling of Ternary Systems

For the thermodynamic modeling of ternary systems, involving two electrolytes with a common ion in an aqueous solution, six binary interaction parameters are required, four corresponding to the binary systems involved and two corresponding to the electrolyte-electrolyte pair interactions. The binary interaction parameters corresponding to electrolyte-water pair are fixed to the values fitted by the binary systems model. The electrolyte-electrolyte interaction parameters are, of course, unknown to the model, as these data have not been used in the fitting of binary systems. Commonly, it is reported that these parameters should be of negligible contribution to the model, especially for dilute solutions. However, this is not the case in highly concentrated brines. The electrolyte-electrolyte relevant binary parameters are, thus, fitted against available experimental data of activity and osmotic coefficients for the ternary system. As a first step, the constant hydration model was deployed. The stepwise hydration model will need to identify the affinity of individual ions of the same charge sign to hydration, which is the topic of on-going work. The hydration indices and the corresponding ionic volumes approximated using the binary system model are fixed (to the values obtain from the study of binary systems) during the fitting of parameters of ternary systems.

4. Results

The AARD values of the original e-NRTL model, the refined e-NRTL model with constant hydration and the refined e-NRTL model with stepwise hydration equilibria for all the binary systems studied are provided in Table 1. The corrected Akaike Information Criterion (AIC_c) and the Bayesian Information Criterion (BIC) were calculated for both the refined e-NRTL model and the original e-NRTL model. Both information metrics assess the quality of fit of each model and the potential of overparameterization. Lower values of AIC_c and BIC in the case of the refined e-NRTL model (5313.74 and 7320.13, respectively) compared to those of the e-NRTL model (20080.80 and 21714.84, respectively) indicate a good balance of goodness of fit. The large F-ratio also signifies a statistically good model. To visualize the accuracy improvement with the refined e-NRTL model, as well as the accuracy improvement of the two hypotheses of the impact and mechanism of hydration, the performance of the models is illustrated for three binary systems shown in Figs. 1-3. These correspond to activity and osmotic coefficients of represented 1-1, 1-2, 2-1 binary electrolyte systems, compared against measurements.

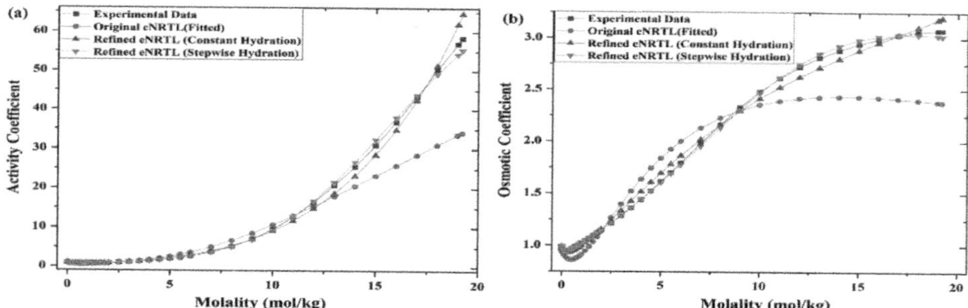

Figure 1: Comparison of the original e-NRTL, refined e-NRTL constant hydration, and refined eNRTL stepwise hydration model for the prediction of (a) activity coefficients of LiCl and (b) osmotic coefficient plot of LiCl..

Fig. 1 presents activity and osmotic coefficients of the LiCl + H_2O binary system. For these results, the binary parameters of the original e-NRTL model were fitted with the same algorithm used in all other studies, with an objective function that incorporates both activity and osmotic coefficient data. It can be clearly seen that at lower molalities, activity coefficient predictions from all three models are similar and quite accurate. At molalities higher than of 5 mol/kg the stepwise hydration model shows clear superiority.

To further demonstrate the performance of the refined e-NRTL with constant and stepwise hydration assumptions, the Na_2SO_4 + H_2O and $MgCl_2$ + H_2O binary systems are plotted in Figs. 2 and 3. For the case of Na_2SO_4 + H_2O, it can be observed that as the concentration increases beyond 1 mol/kg, the difference between experimental data and the original e-NRTL model becomes significant, especially in the prediction of osmotic coefficients. However, the results of both the constant and stepwise hydration equilibria assumptions with the refined e-NRTL model are accurate. Activity and osmotic coefficient plots for the $MgCl_2 + H_2O$ binary system predicted using all the three models are shown in Fig. 3. It can be observed from Fig. 3 that up to molality of 2 mol/kg activity and osmotic coefficients are predicted fairly well by all three models, but beyond that molality range, the stepwise hydration model fits the data best.

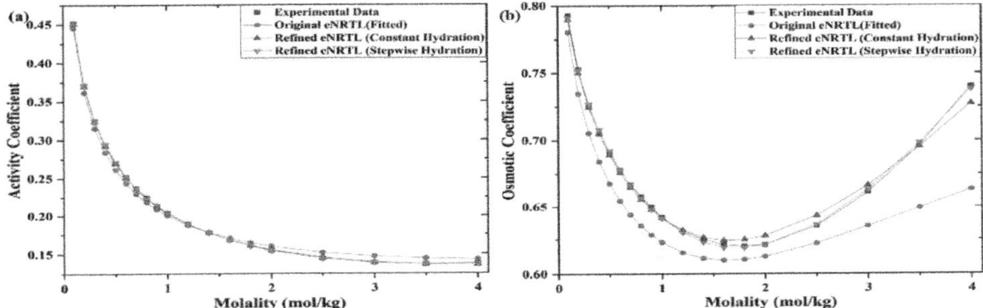

Figure 2: Comparison of the original e-NRTL, refined e-NRTL constant hydration, and refined e-NRTL stepwise hydration model for the prediction of (a) activity coefficient and (b) osmotic coefficient of Na_2SO_4..

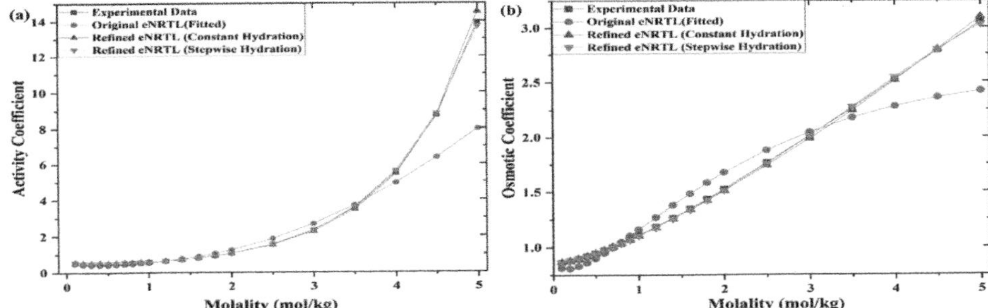

Figure 3: Comparison of the original e-NRTL, refined e-NRTL constant hydration, and refined eNRTL stepwise hydration model for the prediction of (a) activity coefficients and (b) osmotic coefficients of $MgCl_2$.

The ternary NaCl-KCl-H_2O system was studied as an example of ternary systems. Binary interaction parameters for the NaCl-H_2O and KCl-H_2O pairs remain the same for ternary system. The electrolyte-electrolyte binary interaction pairs were regressed against experimental osmotic coefficient data from Dinane et al. (2002). The binary interaction parameters for the NaCl-KCl aqueous system are summarized in Table 2. In Fig. 4, the corresponding predicted osmotic coefficient values are plotted against the experimental data. The agreement between model and experiment is very good, by just fitting the electrolyte-electrolyte binary interaction parameters.

Conclusion

The refined e-NRTL model was used to model the thermodynamics of brine solutions. 11 binary aqueous systems involving the following ion; Na^+, K^+, Li^+ Mg^{2+}, Ca^{2+}, Cl^-, SO42- were studied and one NaCl-KCl-H_2O ternary system was studied for validation of the extrapolation ability of the binary models to mixtures of electrolytes. The binary interaction parameters were fitted against experimental data of activity and osmotic coefficients of binary systems. The osmotic coefficients of the ternary system were predicted with fair accuracy. The simulation was performed on an Intel(R) Core(TM) i7-7700 processor with CPU of 3.60 GHz and 8.00 GB RAM. This thermodynamic model can be extended to predict other thermodynamic properties beyond the osmotic and activity coefficients presented here and at higher molalities. It is also being implemented in the user-friendly environment of IDEAS to develop open-source libraries of property models for highly concentrated aqueous and heterogeneous electrolyte solutions, embedded in block model libraries for processing and separations. These libraries are then intended to be utilized for process synthesis and intensification of brine treatment options.

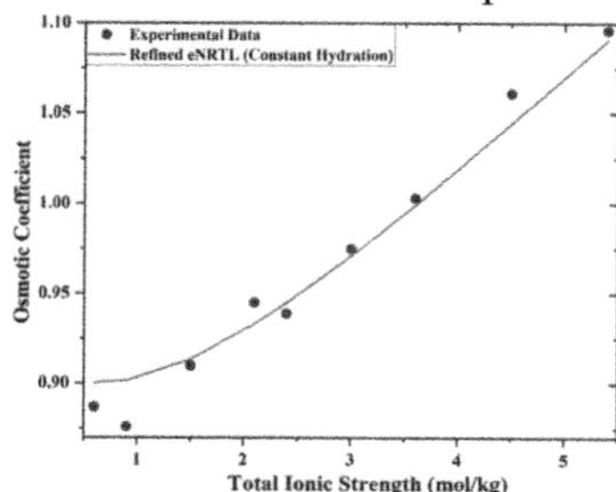

Figure 4: Prediction of osmotic coefficient of NaCl-KCl-H_2O ternary system.

Table 1: Binary interaction parameters for NaCl-KCl-H_2O ternary system

	Refined e-NRTL
$\tau_{H2O,NaCl}$	7.951
$\tau_{H2O,KCl}$	7.696
$\tau_{NaCl,H2O}$	-3.984
$\tau_{NaCl,KCl}$	1.3672
$\tau_{KCl,H2O}$	-3.830
$\tau_{KCl,NaCl}$	1.3672

Acknowledgments

Funding for this project was provided by the National Alliance for Water Innovation (NAWI) of the U.S. Department of Energy, Office of Energy Efficiency and Renewable Energy (EERE), Advanced Manufacturing Office, under Funding Opportunity Announcement Number DE-FOA-0001905.

References

Bhattacharia, S.K., Chen, C.-C., 2015. Fluid Phase Equilib 387, 169–77.

Bollas, G.M., Chen, C.C., Barton, P.I., 2008. AIChE Journal 54, 1608–24.

Chen, C.-C., 2006. Fluid Phase Equilib 241, 103–12.

Chen, C.C., Evans, L.B., 1986. AIChE Journal 32, 444–54.

Dinane, A., Guendouzi, M. el, Mounir, A., 2002. J Chem Thermodyn 34, 423–41.

Honarparvar, S., Saravi, S.H., Reible, D., Chen, C.C., 2018. Fluid Phase Equilib 470, 221–31.

Honarparvar, S., Saravi, S.H., Reible, D., Chen, C.C., 2017. Fluid Phase Equilib 447, 29–38.

Mehrpooya, M., Habibi, R., 2020. J Clean Prod 275, 123836.

Paukert Vankeuren, A.N., Hakala, J.A., Jarvis, K., Moore, J.E., 2017. Fluid. Environ Sci Technol 51, 9391–9402.

R. A. Robinson, R.H.S., 1959. Electrolyte Solutions: Second Revised Edition. Dover Publications.

Tanveer, S., Chen, C.C., 2016. Fluid Phase Equilib 409, 193–206.

Xi Yang, Paul. I. Barton, George. M. Bollas, 2023. In Preparation.

Antonis Kokossis, Michael C. Georgiadis, Efstratios N. Pistikopoulos (Eds.)
PROCEEDINGS OF THE 33rd European Symposium on Computer Aided Process Engineering (ESCAPE33), June 18-21, 2023, Athens, Greece
 http://dx.doi.org/10.1016/B978-0-443-15274-0.50118-9

Stable two-stage scenario generation via game-theoretic optimisation

Georgios L. Bounitsis, Lazaros G. Papageorgiou, Vassilis M. Charitopoulos

The Sargent Centre for Process Systems Engineering, Department of Chemical Engineering, University College London, Torrington Place, London WC1E 7JE, UK
v.charitopoulos@ucl.ac.uk

Abstract

Scenario generation methods constitute an important aspect towards efficient solution of Stochastic Programming (SP) problems and exploitation of big data. The ability of these methods to consistently provide scenario sets which guarantee stability on the solution of the stochastic programs is determinant of their performance. In this context, we present a modification of the existing Distribution and Moment Matching Problem (DMP) which is formulated as Mixed-Integer Linear Programming (MILP) model. The Nash bargaining approach is employed and the different statistical properties of the DMP are considered as players. Through this game-theoretic approach the impact of the user-defined parameters on the scenario generation procedure is investigated. Results from a capacity planning case study highlight the benefits of the proposed approach with respect to in-sample and out-of-sample stability.

Keywords: Scenario Generation; Stochastic Programming; Optimisation; Mixed-Integer Linear Programming (MILP); Nash equilibrium.

1. Motivation

Optimisation under uncertainty has emerged as a focal topic within the Process Systems Engineering (PSE) research agenda (Li & Grossmann, 2021; Charitopoulos et al., 2018). Existing mathematical frameworks can tackle optimisation problems under uncertainty for various degrees of risk aversion. In particular, risk-neutral Stochastic Programming (SP) takes advantage of scenario-based formulations to optimise the expected value of the problem over an assumed probability distribution. Recently, research interest has been directed towards the development of scenario tree generation (SG) or reduction approaches, which create a smaller and representative set of scenarios to efficiently solve computationally challenging stochastic programs. Among them, the optimisation-based SG techniques form an important category (Bertsimas & Mundru, 2022). Moment Matching Problem (MMP) is part of the latter and can generate a scenario tree by solving a statistical errors' minimisation problem (Høyland & Wallace, 2001). Recently, Bounitsis et al. (2022) proposed a Mixed-Integer Linear Programming (MILP) model for the Distribution and Moment Matching Problem (DMP), which was shown to overcome the so-called under-specification issues of the Nonlinear Programming (NLP) based counterparts (Calfa et al., 2014).
In this work, we present an extension to the work of Bounitsis et al. (2022) which aims to further enhance stability and performance of the SG framework, by mitigating the impact of the model's user defined parameters. A modification of the DMP MILP model is proposed, in which the Nash bargaining approach is used. The problem at hand is modelled as an MILP following a separable programming reformulation (Gjerdrum et al.

2001; Charitopoulos et al., 2020). Finally, the enhanced quality and stability of the optimisation-based SG methods are validated through bias, in-sample and out-of-sample tests (Kaut & Wallace, 2007).

The remainder of the article is organised as follows: in Section 2 a summary of the main theoretical aspects is conducted while the detailed methodology is outlined in Section 3. In Section 4, the proposed model is employed to evaluate its stability on a capacity planning case study. Finally, conclusions are drawn in Section 5.

2. Preliminaries and literature review

2.1. Stochastic Programming

In Stochastic Programming (SP) problems the uncertainty is modelled via a known discrete probability distribution. Two-stage stochastic programming (TSSP) is the most common version of stochastic programming problems, in which the uncertainty is revealed at one step. Optimisation problems under uncertainty can be reformulated as stochastic programs using discrete realisations for the uncertain parameters. These are also called scenarios *(k)* and are specified by their values for the uncertain parameters $(\xi_1, \ldots, \xi_k\,,\ k = \{1, \ldots, K\})$ and corresponding probabilities of occurrence $(p_1, \ldots, p_k\,,$ $k = \{1, \ldots, K\})$. TSSP optimises simultaneously the objective function of the first-stage decisions and the expected value of the second-stage costs, which is formulated as:

$$\mathbb{E}[Q(x,\xi)] = \sum_{k=1}^{K} p_k \cdot Q(x,\xi_k) \tag{1a}$$

Thus, the *deterministic equivalent problem* for the TSSP is given by Eq. (1b):

$$\begin{aligned} \min_{x, z_1, \ldots, z_k} \quad & c^{\mathsf{T}} x + \sum_{k=1}^{K} p_k \cdot q_k^{\mathsf{T}} \cdot z_k \\ s.t. \quad & Ax \leq b \\ & T_k x + W_k z_k \leq h_k, \quad k = 1, \ldots, K \\ & x, z_k \geq 0, \quad k = 1, \ldots, K \end{aligned} \tag{1b}$$

In this formulation, k is the index of scenarios, $x \in \mathbb{R}^n$ denotes the "here-and-now" decisions, $z \in \mathbb{R}^m$ is the set of the "wait-and-see" decisions and the vector $\xi = (q, T, W, h)$ contains the data (known or uncertain) of the second stage problem.

2.2. Scenario generation methods

In cases of either continuous distributions or a very large number of possible realisations of the uncertain parameters, scenario tree generation methods can be used to create a smaller in size set of scenarios for the uncertain parameters, with specific values and probabilities, which replaces the original uncertainty set for the solution of the TSSP.

2.2.1. Distribution and Moment Matching Problem

Moment Matching Problem (MMP) for scenario tree generation was introduced by Høyland and Wallace (2001) and aims at the minimisation of the errors regarding the considered statistical properties between the original uncertain set and the final reduced set. A typical version of MMP accounts for first four statistical moments (i.e., mean, variance, skewness, kurtosis) and covariance between the parameters. The optimisation-based NLP formulation has been extended to simultaneously match the stochastic distribution of the uncertain parameters by minimising the errors regarding the empirical cumulative distribution function (ECDF) (Calfa et al., 2014). This problem is referred to as Distribution Matching Problem (DMP) and remains NLP. However, it may lack of stability as it suffers from under-specification issues, in which case either a unique scenario is assigned to several nodes, or zero probabilities are assigned to some nodes. Recently, both MMP and DMP have been reformulated as MILP problems by employing binary variables to indicate the selection of the scenarios from a known original scenario set (Kaut, 2021; Bounitsis et al., 2022). Regarding DMP, Bounitsis et al. (2022) proposed a scenario generation MILP model using L^1 and L^∞ norms to minimise the errors regarding the statistical properties. The model was integrated into a framework with copula-based sampling and clustering techniques and its unified impact seems to lead to significant mitigation of the under-specification issues of the NLP model.

2.2.2. Stability assessment of scenario generation methods

Stability measures assess the performance of a scenario generation or reduction method by testing its capability to generate different trees which lead to consistent expected values. Considering multiple scenario trees, T_c, which lead to stochastic solutions $\bar{x}_c$ of the scenario-based problem and omitting the notation for the "wait-and-see" variables, the in-sample and out-of-sample stability are achieved, if Eqs. (2)-(3) are true respectively (Kaut & Wallace, 2007):

$$f(\bar{x}_c; T_c) \approx f(\bar{x}_{c'}; T_{c'}) \tag{2}$$

$$f(\bar{x}_c; \xi) \approx f(\bar{x}_{c'}; \xi) \tag{3}$$

Regarding Eq. (3) a large reference tree R can be used as an approximation of the true stochastic process ξ in order to estimate the out-of-sample stability. Finally, bias is also an important measure for the evaluation of the quality of certain stochastic solutions $\bar{x}$:

$$B(\bar{x}) = f(\bar{x}; R) - \min_{x \in X} f(x; R) \tag{4}$$

Bias captures the error that fixed here-and-now decisions, $\bar{x}$, impose to the so-called full-space problem (FS) which is solved using the reference tree R (Kaut & Wallace, 2007).

2.3. Nash game-theoretic approach

According to Nash bargaining approach a fair solution for all players of a game can be achieved taking into consideration their initial position. Given a game with players $t \in T$, their corresponding objective functions and status quo points prior to joining the game (or generally a lower profit requirement point), denoted as π_t^{SQ}, then a fair solution point, π_t, can be achieved. The bargaining solution fairly contributes the payoff to the players obeying the axioms of Pareto optimality, symmetry, linear invariance, and independence of irrelevant alternatives. For a typical profit maximisation problem, the fair solution maximises the Nash product which is given by (Harsanyi, 1977):

$$\Phi = \prod_{t \in T} \left(\pi_t - \pi_t^{SQ}\right)^{a_t} \tag{5}$$

where a_t is the negotiation power of each player t. Hence, a player t enters the game only if $\pi_t \geq \pi_t^{sq}$, i.e., if the profit entering the game is greater than its status quo point. However, the objective function of Eq. (5) is nonlinear and nonconvex leading to computationally challenging problems. An approximate linearisation of the Nash product based on a separable programming approach has been proposed by Gjerdrum et al. (2001). Initially, a convexification is conducted by employing a logarithmic transformation and then the concave Nash product is linearised via a piecewise linear function of g prespecified grid points. So, the linear approximation of the Nash product is:

$$\tilde{\Phi} = \sum_{t \in T} \sum_{g \in G} a_t \cdot \ln(\tilde{\pi}_{tg} - \pi_t^{SQ}) \cdot \lambda_{tg} \tag{6}$$

where by defining SOS2 variables λ_{tg} and variables for the profit of player t at grid point g, $\tilde{\pi}_{tg}$, then the profit of each player t can be estimated as:

$$\pi_t = \sum_{g \in G} \tilde{\pi}_{tg} \cdot \lambda_{tg} \quad \forall\, t \in T \tag{7}$$

$$\sum_{g \in G} \lambda_{tg} = 1 \quad \forall\, t \in T \tag{8}$$

3. Methodology and mathematical developments

In the recent study by Bounitsis et al. (2022), performance and stability assessment of the deterministic models were conducted for obtained stochastic solutions from scenario trees of "slightly different sizes" (King and Wallace, 2012). However, results in Bounitsis et al. (2022) indicated that even user-defined parameters of the formulation may affect stability of the deterministic models. In this work, this remark is investigated by conducting stability tests for varying values of user-defined weights of the objective function. Moreover, aiming to improve stability, a modification of the DMP MILP SG model is proposed. The Nash bargaining approach is employed in the terms of the objective function and a separable programming based linearisation of the objective function (Eqs. (6)-(8)) assists to preserve the MILP formulation of the problem. Overall, the modified model can replace DMP MILP in the SG methodology by Bounitsis et al. (2022), which includes the preliminary steps of: (i) statistical analysis of uncertain data, (ii) simulation of distributions, (iii) copula sampling of original scenarios, (iv) clustering. Proposed Nash reformulation of the DMP MILP model is presented for the L^{∞}-norm-based formulation. Let us denote as $i \in I$ the uncertain parameters and $m \in M$ the considered statistical moments. Moreover, we introduce an index $t \in T = \{SM, ECDF, COV\}$ for the terms of the objective function of the original model, which correspond to the players of the game theoretic approach. The model can be written as:

$$max \quad \sum_{t \in T, g \in G} ln(\pi_t^{MAX} - \tilde{\pi}_{tg}) \cdot \lambda_{tg} \tag{9}$$

$s.t.$ Eqs. (7)-(8) and

$$\pi_t = r_t^{MAX} \quad \forall\, t \in T \tag{10}$$

$$r_{SM}^{MAX} \geq W_{im}^{SM} \cdot (d_{im}^{+} + d_{im}^{-}) \qquad \forall i \in I, m \in M \tag{11}$$

$$r_{COV}^{MAX} \geq W_{ii'}^{COV} \cdot \left(c_{ii'}^{+} + c_{ii'}^{-}\right) \qquad \forall i \in I, i' \in I: i < i' \tag{12}$$

$$r_{ECDF}^{MAX} \geq W_i^{ECDF} \cdot e_i \qquad \forall i \in I \tag{13}$$

The rest necessary equations of the model and the definition of the deviations ($d_{im}^{+}, d_{im}^{-}, e_i, c_{ii'}^{+}, c_{ii'}^{-}$) are identical to the original DMP MILP model which can be found in Bounitsis et al. (2022). Eq. (9) is the objective function and aims at the maximisation of the Nash product, which for this case is expressed using parameters π_t^{MAX} as the higher allowable statistical error for each term t. These can be obtained by solving the original DMP MILP for minimisation of each term separately. Then, Eqs. (11) – (13) capture the maximum errors for each term, i.e., statistical moments (SM), covariance matrix (COV) and ECDF. The weights on the deviations are defined as $W_{im}^{SM} = \overline{w}_{im}^{SM} / |\widetilde{M}_{im}|$ and $W_{ii'}^{COV} = \overline{w}_{ii'}^{COV} / |\tilde{C}_{ii'}|$, where $\widetilde{M}_{im}, \tilde{C}_{ii'}$ are the estimated statistical properties from the data and $\overline{w}_{im}^{SM}, \overline{w}_{ii'}^{COV}$ are the user-defined weights. Finally, W_i^{ECDF} is also a user-defined weight for the error on ECDF.

4. Case Study

A capacity planning problem under uncertainty is studied. Detailed description, model and data can also be found in Bounitsis et al. (2022). Briefly, it is an MILP profit maximisation problem, which selects between 11 candidate processes on the production of 5 final products from 5 raw materials. Uncertainty is revealed in the production yields of processes and/or demands of final products. The bias and stability of DMP NLP, MILP and its proposed Nash-based reformulation is investigated for sets of varying values of $\overline{w}_{im}^{SM}, \overline{w}_{ii'}^{COV}, w_i^{ECDF}$. Typical values 1, 10, 50 and, 100 are considered and 37 sets of values are constructed allowing 1 or 2 of the attributes to vary concurrently. Regarding Nash reformulation, 50 grid points are employed for the piece-wise linear approximations. Instances of increasing complexity are examined and are presented in Table 1. MILP models uses copula-based original scenarios as input.

***Table 1:** Instances of scenario reduction in the case study*

Instances	Uncertain parameters	Original scenarios	Final scenarios
1	2	1,000	10
2	4	2,000	20
3	8	2,000	40

The executions are performed in a Dell workstation with Intel® Core™ i9-10900K CPU @ 3.70 GHz and 32.00 GB RAM. NLP model is solved using the BARON solver, while MILP models using the GUROBI 9.5 solver within GAMS 38.2 modelling system. Although calculation of lower bound constitutes a significant drawback of DMP MILP, the Nash-based reformulations can achieve near optimal solutions in short execution times. Towards a fair comparison, time limit of 1,200s and an optimality gap tolerance of 1% are set for the MILP models.

Firstly, the computational studies reflect the enhanced quality of stochastic solutions that the scenario sets by the considered MILP models provide compared to their NLP counterparts. In particular, NLP SG models suffer from under-specification issues and may generate scenario sets which lead to extreme bias values. Results among examined weights' sets are reported in Table 2 and are indicative of NLP models non-competitiveness. Moreover, Nash-based models lead to the lowest bias values.

***Table 2:** Average (%) bias values for the investigated instances.*

Instance	NLP L^1	NLP L^∞	MILP L^1	MILP L^∞	NASH L^1	NASH L^∞
1	1.327	1.383	0.047	0.034	**0.027**	0.028
2	2.364	2.834	0.048	0.049	0.034	**0.029**
3	2.761	3.049	0.058	**0.042**	**0.042**	**0.042**

Focusing on the MILP SG methods, the distribution of their expected values for the in-sample and out-of-sample stability (as they are introduced in Eqs. (2) - (3), respectively) are visualised for instance 2 in Fig. 1. While all SG models display similar results for in-sample stability, the proposed Nash reformulations seem to improve the performance of the methods for out-of-sample stability testing, which pertains also to the quality of the provided stochastic solutions. Enhanced stability of the proposed reformulations can be interpreted by the considerably tighter interquartile ranges. Moreover, expected values lie closer to TSSP full-space (FS) solution and indicate improved quality of the stochastic solutions. Ultimately, the results attest that the DMP NLP model is the least stable of the investigated approaches. DMP MILP's stability has been studied in the work by Bounitsis et al. (2022) but is also validated in this work for the case of varying weights of the formulation. Finally, the results demonstrate that combination of DMP MILP with the Nash bargaining approach may further enhance the stability of the SG method.

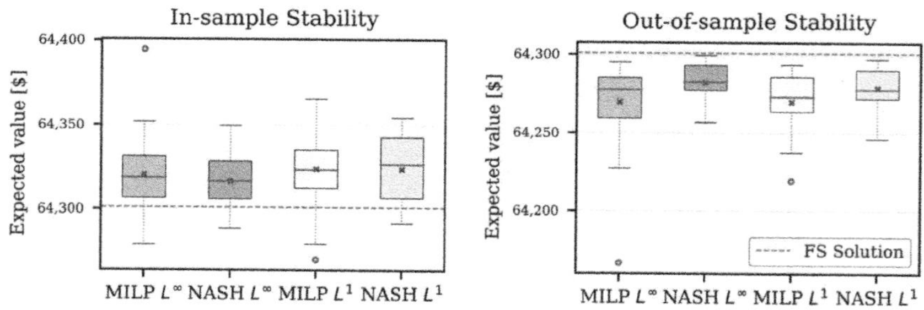

Figure 1: *In-sample & out-of-sample stability results for instance 2.*

5. Conclusions and future directions

Integration of Nash game-theoretic concepts with DMP MILP model can lead to a more stable version of the scenario generation method and further alleviate the under-specification drawbacks of the NLP models for the user-defined parameters. Future work within our research group focuses on: (i) the implementation of optimisation and decomposition techniques to improve the computational performance of the proposed model while maintaining stability, and (ii) the extension to multi-stage setting.

Acknowledgements: Financial support from the UK EPSRC under project EP/V034723/1, EP/V051008/1 & EP/V050168/1 is gratefully acknowledged.

References

D. Bertsimas, N. Mundru, 2022. Optimization-Based Scenario Reduction for Data-Driven Two-Stage Stochastic Optimization. Oper. Res.

G.L. Bounitsis, L.G. Papageorgiou, V.M. Charitopoulos., 2022. Data-driven scenario generation for two-stage stochastic programming. Chem. Eng. Res. Des. 187, 206–224

B.A. Calfa, A., Agarwal, I.E. Grossmann, J.M. Wassick, 2014. Data-driven multi-stage scenario tree generation via statistical property and distribution matching. Comput. Chem. Eng., 68, 7–23.

V.M. Charitopoulos, A. Aguirre, L.G. Papageorgiou, V. Dua, 2018. Uncertainty aware integration of planning, scheduling and multi-parametric control. Comput. Aid. Chem. Eng. 44, 1171-1176.

V.M. Charitopoulos, V. Dua, J.M. Pinto, L.G. Papageorgiou, 2020. A game-theoretic optimisation approach to fair customer allocation in oligopolies. Optim. Eng., 21(4), 1459–1486.

J. Gjerdrum, N. Shah, L.G. Papageorgiou, 2001. Transfer prices for multienterprise supply chain optimization. Ind. Eng. Chem. Res. 40, 1650–1660.

J. Harsanyi, 1977. Morality and the Theory of Rational Behavior. Soc Res. 44, 623-656.

K. Høyland, S. W. Wallace, 2001. Generating Scenario Trees for Multistage Decision Problems. Manag. Sci., 47(2), 295–307.

M. Kaut, 2021. Scenario generation by selection from historical data. Comput. Manag. Sci. 18, 411–429.

M. Kaut, S. W. Wallace, 2007. Evaluation of scenario-generation methods for stochastic programming. Pacific J. Optim., 3.

A. J. King, S. W. Wallace, 2012. Modeling with Stochastic Programming. Springer Series in Operations Research and Financial Engineering.

C. Li, I. E. Grossmann, 2021. A Review of Stochastic Programming Methods for Optimization of Process Systems Under Uncertainty. Front. Chem. Eng, 2, 622241.

Antonis Kokossis, Michael C. Georgiadis, Efstratios N. Pistikopoulos (Eds.)
PROCEEDINGS OF THE 33rd European Symposium on Computer Aided Process Engineering (ESCAPE33), June 18-21, 2023, Athens, Greece
 http://dx.doi.org/10.1016/B978-0-443-15274-0.50119-0

Global One-dimensional Root-finding Algorithm for Derivable and Continuous Functions

Andrea Galeazzi,[a,b] Kristiano Prifti,[a,b] Flavio Manenti[a,b,*]

[a]*Dipartimento di Chimica, Materiali e Ingegneria Chimica "Giulio Natta", Politecnico di Milano, Piazza Leonardo Da Vinci, 32, Milano, 20133, Italy*

[b]*Consorzio Interuniversitario Nazionale per la Scienza e Tecnologia dei Materiali, Via Giusti 9, Firenze, 50121, Italy*

[*]*flavio.manenti@polimi.it*

Abstract

Efficient and robust root finding is a core problem of chemical and process engineering solutions. Thermodynamics, nonlinear systems, finite difference methods for partial differential equations, and optimization are just some examples in which root finding is fundamental. In this work, an algorithm that can be integrated with any local root-finding method is proposed to solve the problem of finding robustly and efficiently all the roots of an objective function in a domain. The method requires the function to be continuous up until the second-order derivative and it is here applied for the one-dimensional case. With the definition of a proper mesh, the first derivative is investigated to find the sub-domains that contain the zeros of the objective function. The study of the derivative guarantees to find those zeros that cannot be identified using only Bolzano's existence criterion.

Keywords: One-dimension, Global Root-finding, Search Grid, Continuous and Derivable Function, Robust and Efficient Solution.

1. Introduction

This paper describes a method for solving the problem of computing all the simple roots of a given equation:

$$f(x) = 0 \tag{1}$$

in a predefined interval (a, b) where $f: [a, b] \subset \mathbb{R} \rightarrow \mathbb{R}$ is continuous and derivable up to the second order derivative, $f(x) \in C^2$. The method proposed may be applied on top of any local root-finding algorithms and the information required is the definition of a search grid and the computation of the function and its continuous first-order derivative.
Many different fields of science and technology make use of root-finding methods to solve various problems. One important example is the maximization or minimization of an objective function. In this case, a root-finding method can be applied to the first derivative of the objective function to find the optimum value. Another example of root-finding usage is in the numerical solution of non-analytical thermodynamic equations of state and the flashing problem (Bisotti et al., 2021; Quarteroni et al., 2007). Such problems are at the core of chemical process engineering solutions since the definition of the physical state of a mixture is the first step in fixing the degrees of freedom of a process simulation.
Being able to solve, efficiently and robustly, such thermodynamic problems brings an overall advantage in process modeling and a significant computational time reduction.

In this case, efficiency is defined in terms of the amount of functions evaluation needed to find a solution while higher robustness means a greater probability to find all the solutions in the investigated domain.
The method proposed in this work tries to tackle this challenge by first studying the function derivative to find with certainty all the domains in which the objective function has solutions and find them. This algorithm can be applied on top of any local root-finding solver, and it is based thoroughly on Bolzano's existence criterion (described in Section 2.1). The premise is that, given an appropriate search grid, or mesh, it guarantees to find all the solutions of the objective function inside a predefined domain.

2. Method

2.1. Basic Concepts

The method developed in this work is based on a very simple and well-known theorem which is Bolzano's existence criterion. It states that when f is a C^2 function in $[a, b] \subset \mathbb{R}$ and the following is true:

$$f(a) \cdot f(b) \leq 0 \tag{2}$$

at least a root is guaranteed to exist. However, the criterion does not give any information on the opposite case. If Equation (2) is not true it is impossible to state whether there is or not a solution between a and b since there could still be a zero. The application of this criterion to solve the root-finding problem is not new (Brent, 1971; Kearfott, 1987; Wood, 1991).
Thus, for this study, it has been decided to extend Bolzano's criterion using the function derivative since the key mathematical property is that between two adjacent roots (a, b) of the derivative of a $f(x) \in C^2$ function, where $f'(x) \geq 0 \vee f'(x) \leq 0 \ \forall x \in [a, b]$, f is strictly monotonic. Thus, if Equation (2) is true in the same interval $[a, b]$, a single root of f may only exist. Other authors have proposed different methods for solving this problem, for example by applying the concept of a Dominating Function of the derivative (Ying and Katz, 1989).
The definition of a search grid, or mesh, in this problem, is of utmost importance. Any possible failure of this whole method could be traced back mainly to this step. An appropriate mesh is defined such that it always contains no more than one single derivative root. To do so, a proper understanding of the function under study could be required. Otherwise, the most basic solution is to arbitrarily apply an extremely dense search grid. Some authors have addressed this problem by implementing an adaptive mesh refinement procedure (Gong et al., 2020; Razbani, 2015).
Thus, the case in which Equation (2) is not true that still has a solution can be ruled out if using an appropriate search grid in the first place since it can only happen when the roots of the derivative and the function are strictly coincident.
Should the search domain contain any equality or inequality constraints for x the problem must be split into multiple sub-problems and solved using the proposed method.

2.2. The Algorithm

The proposed algorithm can be briefly described in the following points:

1) Input: $f(x)$, $f'(x)$, global search domain, local f'search sub-domains (grid).
2) Evaluate f and f' at each sub-domain node.
3) Select only sub-domains where f and f' show a sign change between the nodes.
4) Start a local root-finding in the selected sub-domains for f';

5) Check for coincident zeros between f and f' at local derivative roots.
6) Generate new sub-domains, with adjacent non-coincident f' roots and selected f sub-domains from Step 2 as nodes.
7) Select only sub-domains where f shows a change of sign between the nodes.
8) Start a local root finding in the new sub-domains for f.
9) Collect all the zeros of f (coincident and non-coincident roots).

A flowchart of this method is shown in Figure 1.

A more in-depth description of the algorithm is reported below:

Step 1. The necessary inputs to the problem are defined: the objective function $f(x)$, its derivative $f'(x)$, the extrema of the global search domain $[a, b]$ where $f(x) \in C^2$, and the local search sub-domain nodes. The latter are crucial for the successful execution of this method. They may be defined arbitrarily but they must be generated in such a way as to always expect one and only one solution of f inside root-containing sub-domains. If this is accomplished then the robustness is guaranteed, otherwise, the search mesh should be refined accordingly. Having a good knowledge of the objective function to solve should result in a proper definition of the mesh.

Step 2. Each node of the local sub-domains is evaluated in $f(x)$ and $f'(x)$.

Step 3. Bolzano's existence criterion is applied to evaluate which sub-domain contains a root of the function and its derivative. Thus, Equation (2) is applied to select derivative subdomains with f roots and f sub-domains with sign change not bracketed by f' roots. This redundancy is mandatory to overcome any possible saddle point where the derivative does not change sign but a root still exists.

Step 4. Each f' bounded sub-domain thus selected is investigated using a local root-finding method since it must contain a root. In this Step, it could be advised to apply a bracketing method since it mathematically guarantees convergence and the downside of the slower convergence rate with respect to the open methods should be mitigated given the smaller local sub-domains. However, even an open method can be applied knowing that, in case of non-convergence or convergence outside the selected subdomain, it could be possible to fall back to a bracketing method and restart the root-finding.

Step 5. By evaluating the roots found in Step 4 in $f(x)$ it is possible to uncover any maxima or minima point that touch the axis at the extremum. They are coincident zeros of both $f(x)$ and $f'(x)$ and they are collected and saved for Step 9.

Step 6. The information gathered in the previous Step is now used to generate new sub-domains for the root-finding of the objective function with non-coincident roots as nodes. Each sub-domain is defined by nodes that are an adjacent pair of roots of f', thus not containing a coincident zero in between.

Step 7. Again, as in Step 3, Bolzano's existence criterion, see Equation (2), is applied to rule out all the subdomains in which a change of sign is not appearing.

Step 8. Inside each newly defined search sub-domain, a local root-finding method can be launched. Also, in this case, it is guaranteed to find a zero which may not be an extremum point thanks to Steps 5 and 6. This guarantee is given by the fact that, if there is a significant change in Step 7 and the derivative is passing through zero at the nodes of the subdomain there is one and only one root available on the objective function. In this Step, the choice of using an open method or a bracketing method is mainly arbitrary. It could be advisable to apply an open method, such as the Newton method if it is possible to generate an appropriate guess value (a trivial guess value could be positioned in the middle of the local sub-domain) for the function under investigation.

Step 9. All the roots of the objective function $f(x)$ found in Step 5 and Step 8 are then collected and the problem is solved.

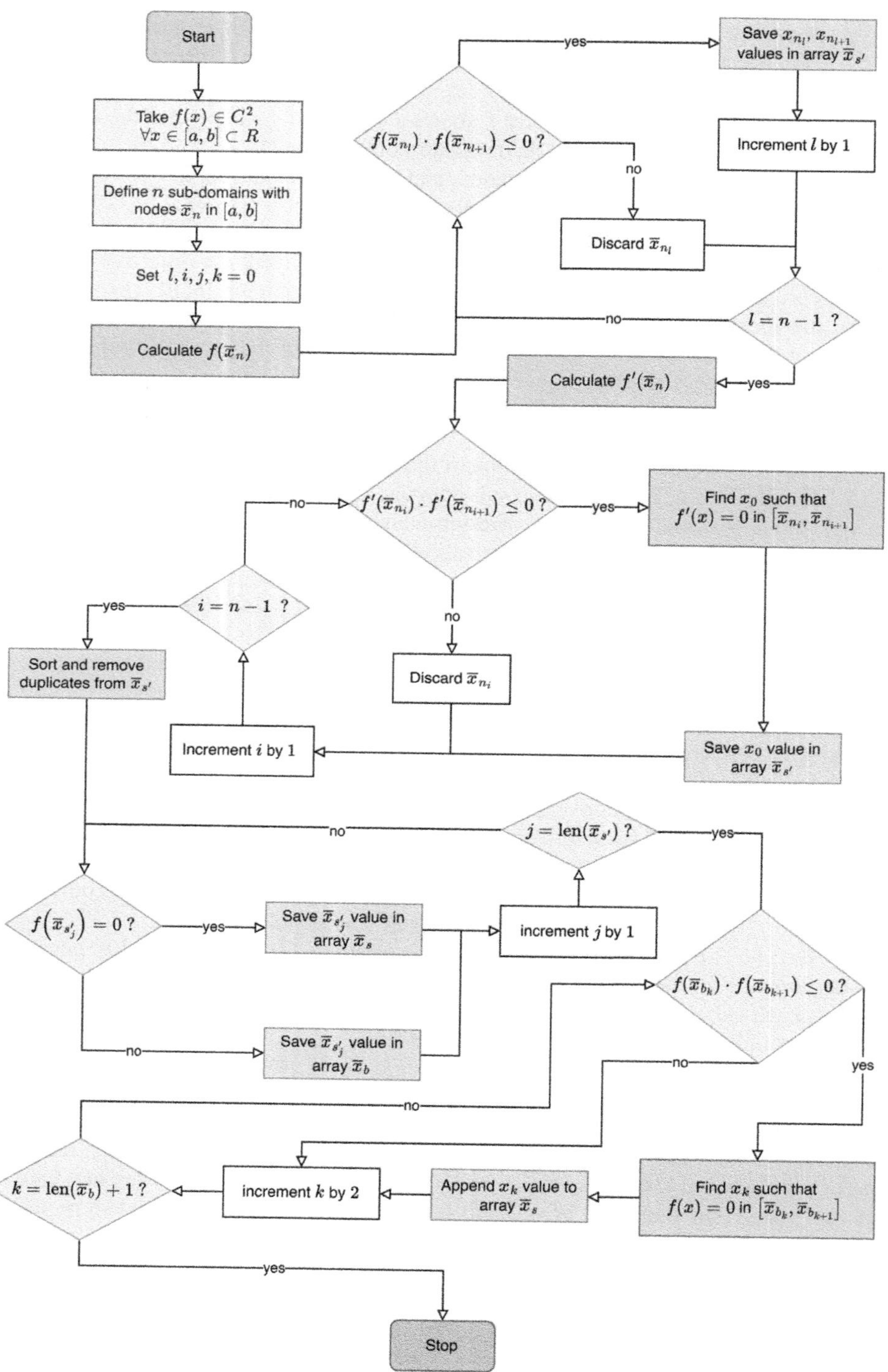

Figure 1: Flowchart of the global root finding algorithm.

3. Results & Discussion

3.1. Prototype Function Example

The method described in Section 2 has been applied to the following prototype function:

$$f(x) = \sin^2(x)\cos(x) \tag{3}$$

$$f'(x) = 2\cos(x)^2\sin(x) - \sin(x)^3 \tag{4}$$

chosen because it has an analytical solution and shows many coincident solutions to better showcase the method described above.

In Figure 2, the algorithm Steps presented in Section 3.1 are reported for the example function of Equations (4) and (5). In Figure 2.a, the function and its derivative are shown along with the initial derivative search grid. The investigated domain is $x \in [0,10]$ and the grid presented shows 30 equally distributed sub-domains. In Figure 2.b, Steps 2,3,4 are reported. The filled and dashed lines are the linear interpolation between each value of $f'(x)$ at each node. The former, indicate sub-domains with a sign change. In Figure 2.c, Steps 5,6,7,8 are shown. The updated search grid is created using the roots of $f'(x)$. Grayed areas are the excluded sub-domains given that there is a coincident solution, and the two roots are not adjacent. In Figure 2.d, Step 9 is reported. All the solutions inside $[0,10]$ are collected and the problem is solved.

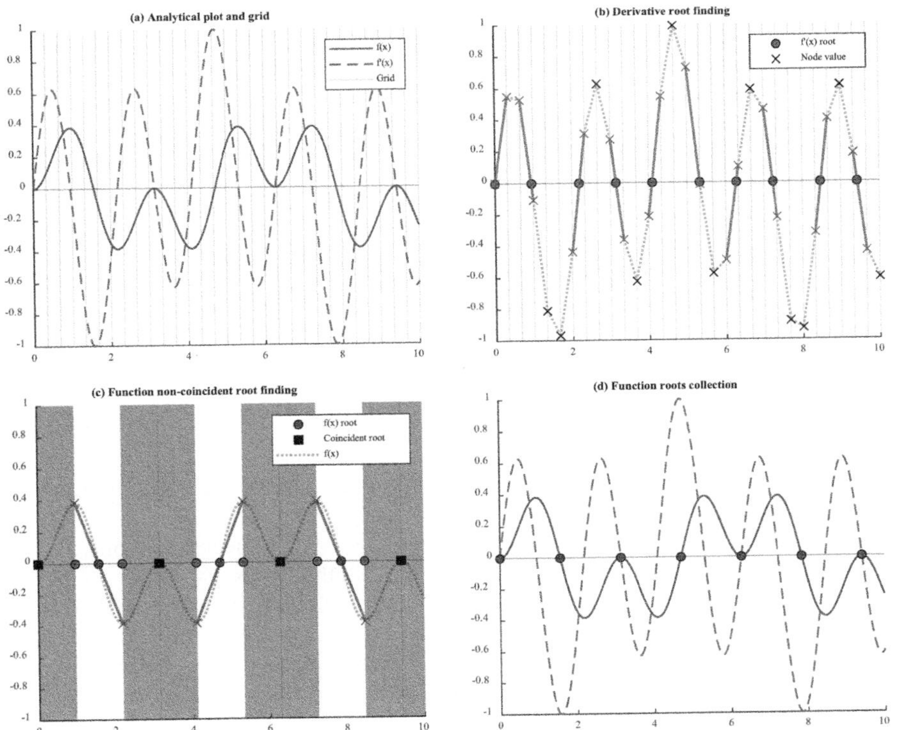

Figure 2: Condensed graphical representation of the algorithm Steps presented in Section 2.2 on the prototype function introduced in Section 3.1.

By using the proposed algorithm, it has been possible to solve the presented root finding problem by only using 12 local root findings (9 for the derivative and 3 for the objective function) and approximately 40 function evaluations (30 for the derivative and 10 for the objective function). If we consider a constant number of 10 iterations for every root-finding problem, we end up with a total of approximately 160 function evaluations.
In comparison, by using a more naive solution, which consists of directly launching 30 root-finding solvers (that do not guarantee finding a solution) the results would be that of using 300 function evaluations. Thus, in this case, using this method the computational time required is almost halved.

4. Conclusions

In conclusion, an algorithm has been proposed to solve reliably and efficiently the global root-finding problem of second-order continuous and derivable functions inside a domain. The method applied is based on an appropriate mesh and uses extensively Bolzano's existence criterion to define what subdomains to exclude from the root search. This method could benefit from a parallel computing development, possibly with a cloud approach (Galeazzi et al., 2021), to aid in the solution of chemical engineering problems, especially in thermodynamics.

References

F. Bisotti, A. Galeazzi, L. Galatioto, F. Masserdotti, A. Bigi, P. Gritti, F. Manenti, 2021, Implementing robust thermodynamic model for reliable bubble/dew problem solution in cryogenic distillation of air separation units. International Journal of Thermofluids, 10, 100083.

R.P. Brent, 1971, An algorithm with guaranteed convergence for finding a zero of a function, The Computer Journal, 14, 4, 422–425.

A. Galeazzi, R. Nasti, G.L. Bozzano, L. Verotta, S. Marzorati, F. Manenti, 2021, A Cloud Computing Application for the Supercritical Carbon Dioxide Extraction Using Coffee Grounds Silverskin, Computer Aided Chemical Engineering, 50, 31 European Symposium on Computer Aided Process Engineering, 1035–1040.

W. Gong, Y. Wang, Z. Cai, L. Wang, 2020, Finding Multiple Roots of Nonlinear Equation Systems via a Repulsion-Based Adaptive Differential Evolution. IEEE Transactions on Systems, Man, and Cybernetics: Systems, 50, 4,1499–1513.

R.B. Kearfott, 1987, Some tests of generalized bisection, ACM Trans. Math. Softw., 13, 3, 197–220.

A. Quarteroni, R. Sacco, F. Saleri, 2007, Rootfinding for Nonlinear Equations, Numerical Mathematics, Texts in Applied Mathematics, 247–284.

M.A. Razbani, 2015, Global root bracketing method with adaptive mesh refinement, Applied Mathematics and Computation, 268, 628–635.

G.R. Wood, 1991, Multidimensional bisection applied to global optimisation. Computers & Mathematics with Applications, 21, 6, 161–172.

X. Ying, I.N. Katz, 1989, A simple reliable solver for all the roots of a nonlinear function in a given domain, Computing, 41, 4, 317–333.

Antonis Kokossis, Michael C. Georgiadis, Efstratios N. Pistikopoulos (Eds.)
PROCEEDINGS OF THE 33rd European Symposium on Computer Aided Process Engineering (ESCAPE33), June 18-21, 2023, Athens, Greece
 http://dx.doi.org/10.1016/B978-0-443-15274-0.50120-7

A Demethanizer column Digital twin with non-conventional LSTM neural networks arrangement

Marta Mandis[a], Roberto Baratti[a], Jorge Chebeir[b], Stefania Tronci[a], José A. Romagnoli[b]

[a] *Dip. di Ingegneria Meccanica, Chimica e dei Materiali, Università degli Studi di Cagliari, Cagliari, 09123, Italy*

[b] *Department of Chemical Engineering, Louisiana State University, Baton Rouge, LA, 70809, USA*

Abstract

This work aims to develop a digital twin for a demethanizer column and provide a useful tool for monitoring and quality control of the NGL recovery process. For this purpose, a digital data-driven model was proposed to mimic real dynamics of a cold residue reflux (CRR) unit through the incorporation of physical knowledge. A non-conventional LSTM network arrangement was developed considering training test and validation data sets generated by the process simulator Aspen HYSYS®. This simulation model was built by considering realistic measurement noises to mimic the actual measures in a real plant. The obtained surrogate model was evaluated considering its ability to recreate the operation of the actual distillation column, estimating the temperature and composition transient profiles of the bottom column product and of every stage of the column. Overall, the model developed with the proposed LSTM network arrangement proves capable of successfully reconstructing the actual profiles of all the considered variables.

Keywords: Natural gas liquids recovery, LSTM Neural Networks, Digital twin, Distillation Column, Dynamic process simulation

1. Introduction

Natural gas is a hydrocarbon mixture constituted mostly of methane and a variable fraction of heavier hydrocarbons known as natural gas liquids (NGL). The latter constitute a source of additional profit as can be sold separately with a higher selling price and used as industrial feedstock materials. Additionally, with NGL separation it is possible to reduce CO_2 emissions related to natural gas combustion. The NGL recovery process under steady-state conditions has been extensively studied in the literature (Kidnay et al., 2011; Chebbi et al., 2010; Getu et al., 2013; Kherbeck & Chebbi, 2015). The associated dynamics of the NGL separation process have been the subject of recent studies and generally focused on the dynamics and control of product quality indices (Luyben, 2013; Chebeir et al., 2019; Mandis et al., 2022). For this, it is essential to have information on the plant composition targets, which can be obtained with the use of expensive composition analyzers. These analyzers also introduce long-time delays in the control loops due to measurements delays. To overcome these drawbacks, the replacement of measured concentrations with accurate real-time estimations represents one of the most widespread approaches. In this type of task, temporal dependencies play a key role to obtain adequate estimations and with the goal of addressing time-dependent problems Recurrent neural networks (RNNs) (Elman, 1990) were developed. One of the most useful RNNs when addressing long time dependency is the Long-Short Term Memory network (LSTM) (Hochreiter & Schmidhuber, 1997), designed to overcome the problem of the vanishing gradient. LSTM networks have been applied to different types of problems in chemical engineering including digital twin realization for plant monitoring and control (Qu et al., 2020; Zhu & Ji, 2022). A digital twin (Grieves, 2015) consists of a virtual model of whole plants or individual units which is interconnected with the actual equipment. The availability of reliable digital surrogate models is of great interest in the process industry. This tool exchanges information in real-time, and it can predict the dynamic evolution of a process over time. For this reason, a digital surrogate can provide a useful tool for performance monitoring, predictive asset maintenance, production optimization and advanced process control. This work aims to develop a data-driven digital model for a demethanizer column of a CRR unit, which is realized by using LSTM neural network with a non-conventional network architecture. The main goal is to obtain a surrogate model capable of approximating the real dynamics of the column with a much shorter computation time than the process simulator itself. In addition, it is intended to improve NGL recovery control strategies with real-time estimations of control variables.

2. Process Description

The dynamic operation of the CRR unit has been modelled in the process simulator Aspen HYSYS® considering realistic operating conditions (Chebeir et al., 2019). The raw gas is fed to the plant with a nominal flowrate of 4980 kmol/h, pressure and temperature of 5818 kPa and 35 °C, respectively, and a gas composition characterized by a low content of liquids (Chebbi et al., 2010). The main unit of the separation process, responsible for the removal of the light component of the raw gas, is the demethanizer column, given by a 30 stages distillation column with a reboiler. The raw gas entering the CRR unit is first cooled and then the liquid fraction is separated utilizing a flash tank. Three resulting feed streams enter the demethanizer column in the 2nd the 8th and the 26th trays. The column reflux is provided by compressing part of the demethanizer top product with a cryogenic compressor, ensuing a reflux stream of nearly pure methane. Finally, the remaining part of the column product is recompressed for commercialization while the NGL bottom product of the demethanizer column is sent to further separation in the fractionation train.

3. Methods

In this work, deep learning methods for the development of RNNs were used to build the surrogate data-driven model. The goals of the neural model were the achievement of accurate estimations for the prediction of the time evolution of temperature and compositions of all column trays and reboiler of the demethanizer column. The considered inputs were given by only easy-to-measure variables: pressure and temperature of all column feeds, reboiler duty, power of the cryogenic compressor and separator pressure and temperature. The selection of RNNs is based on the need to obtain an adaptive neural model capable to capture the existing temporal correlations between the involved variables. In particular, thanks to their ability to overcome the vanishing gradient problem, LSTM networks were used for digital twin realization.

3.1. Neural model architecture

In order to design a suitable model to represent the dynamics of the demethanizer column, the LSTM neural networks are arranged considering a particular architecture. The proposed network layout involves the utilization of a dedicated LSTM cell to predict the operations of each stage of the distillation column and the reboiler. Thus, each cell was demanded at modelling the dynamic separation of the corresponding tray, allowing the dynamics inherent in the separation process occurring in each column stage to be captured in the model. Furthermore, bidirectional connections have been applied by modifying the flow of information of a standard LSTM, interconnecting adjacent cells one to another to account for the interactions between adjacent stages and emulate the action of internal flows within the distillation column. The described model layout, whose structure is schematically represented in Figure 1a, is responsible to approximate the demethanizer operation at a certain instant of time. A schematic representation of a modified LSTM cell is depicted in Figure 1b. As can be seen, in the considered cell (referred to by the subscript s), the previous states for the cell are provided by combining the actual previous hidden states and internal cell states ($h_{t-1,s}$ and $c_{t-1,s}$) with the hidden states and internal cell states calculated by the adjacent LSTM cells (indicated with subscripts $s-1$ and $s+1$) in the current time.

For what concern the terminal cells of the network, mimicking the column top tray and the reboiler, the hidden states from the missing adjacent cell are replaced by the actual measurement of the corresponding variables of the reflux in a case and the liquid stream entering the reboiler in the other.

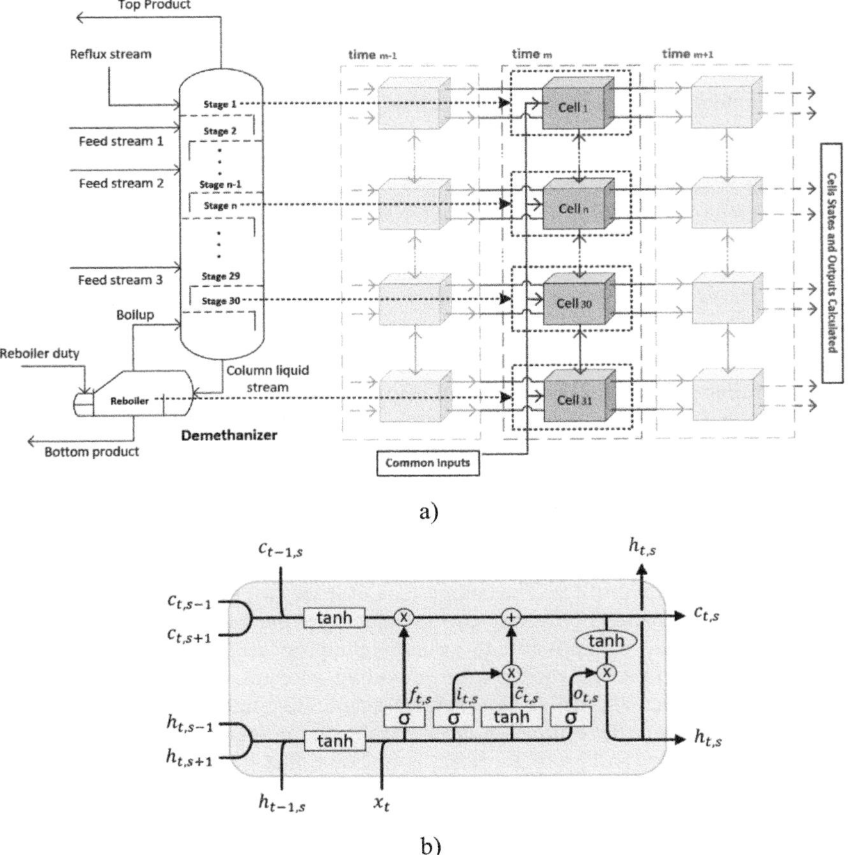

Figure 1: Schematic representation of: overall architecture a) and the modified LSTM cell b).

3.2. Datasets

All the data used in the work were generated using the simulated demethanizer column described in section 2 and collected in different datasets for the training, testing and validation campaigns. The datasets employed for training and testing of the neural models, considering a split ratio of 80:20, were two months of simulated plant operation.

Step changes of varying amplitude in the plant feed flow rate were applied with variation peaks of 10% of the plant feed nominal value. To validate the performance of the presented neural model, 2 days of plant operation were simulated by applying step variations of 5% in the plant feed flow rate nominal value.
All the datasets used in this work were registered by considering a sampling time of 20 seconds and without measurement delays, as the data were treated as historical plant data. Moreover, a measurement noise was applied to provide more realistic measured data. The values of the maximum allowed measurement noise applied are shown in Table 1.

Table 1: Values of maximum measurement noise considered for input and output data in train, test and validation datasets.

	Duty	**Temperatures**	**Pressures**	**Concentrations**
maximum measurement noise	1.3%*	0.1 [°C]	1.5%*	2%*

* *Referred to the maximum value of the considered variable for the given column stage.*

3.3. Neural model training

The LSTM neural networks described in section 3.1, have been implemented and trained in the Python programming environment by using the functions available in the open-source library PyTorch. The loss function considered in the optimization problem is given by the Mean Square Error (MSE), while the parameter updating was performed by the Adaptive Moment Estimation optimizer (ADAM) (Kingma & Ba, 2015). The network hyperparameters were obtained by multiobjective optimization using the NSGA-II algorithm available in the pymoo package (Blank & Deb, 2020).

4. Results

The performances of the developed neural model are evaluated by considering the column transient profiles predicted by the proposed neural model in the validation campaign for temperature, and key components composition. For sake of brevity, only the results obtained under the worst-case variation represented by the application of a 5% decrease in the feed plant nominal value for ethane composition are reported and shown in Figure 2. The ability of the model to estimate the remaining target variable is assessed by considering the Mean square error (MSE) between the actual and the predicted profiles in the same campaign. Figures 2a and 2b show respectively the actual and estimated ethane transient column profiles. The comparison of these two graphs gives a qualitative indication of the estimation performance of the model for the transient column profile of ethane composition. As it is possible to visualize the proposed model is able to predict the trend of ethane time evolution profile providing an estimation with no detectable deviations from the actual profile. To provide a quantitative measure of the accuracy of the obtained estimation, the estimation error of the column transient profiles for ethane composition is depicted in Figure 2c. Here, is it shown that the proposed neural model can accurately predict the ethane composition transient profiles for all the column trays. The estimation error increase where most of the ethane variations occur, indeed, it increases in the column terminal regions, where most of the separation takes place. The deviations of the predicted transient column profile are always within the maximum error reported in Table 1.
To visualize in detail the estimation obtained, Figure 2d shows the comparison between the estimated ethane composition profile and the measured profile in the tray where the higher estimation error is registered, the 29th column tray. As can be seen from the figure, the profile obtained with the neural model can correctly predict the evolution of the composition, with considerably lower variance than the considered measurement.

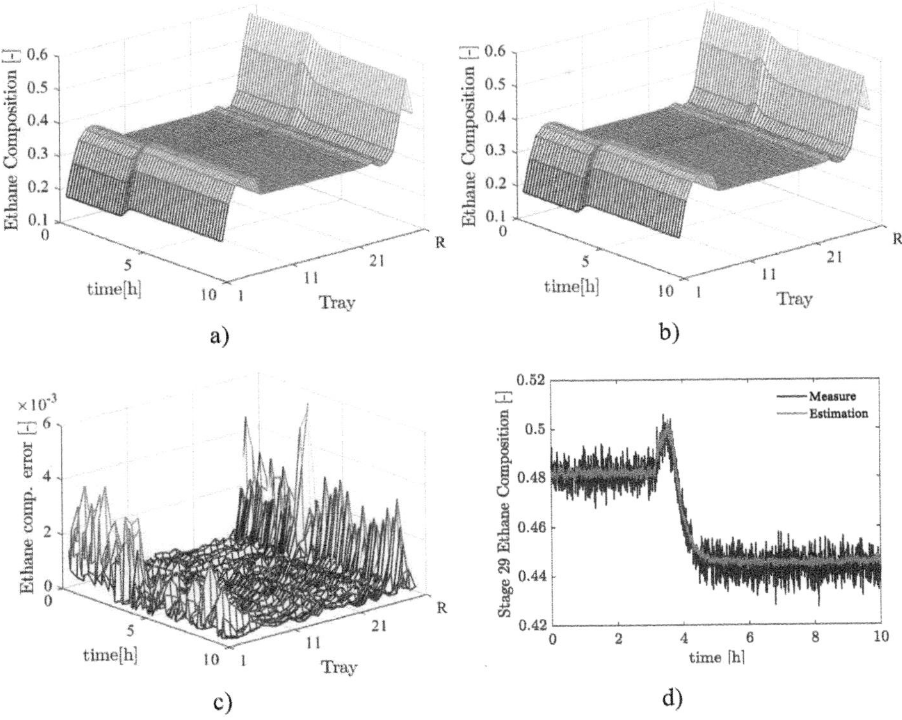

Figure 2: Ethane composition results obtained in the validation campaign under the decreasing variation of plant feed nominal value: actual column profile (a); predicted column profile (b); estimation error profile (c) and transient profiles of 29th tray (d).

Table 2: Mean square error obtained in the validation campaign under decreasing variation of 5% in the plant feed nominal value.

	Temp.	Methane	Ethane	Propane	Butanes	Pentanes	Hexane
MSE	0.0072	7.33e-7	4.31e-7	7.49e-8	1.22e-8	1.42e-9	2.27e-9

Analogue results are obtained for the temperature and the composition of the remaining components. Table 2 shows the MSE values obtained by considering the predictions of the transient composition profiles in all column trays and the actual profiles, for all the target variables in the validation campaign. As it can be noticed from the obtained MSE value, the use of the proposed model provides an accurate prediction of dynamic temperature profiles. Although it is not apparent from the MSE, it is reasonable to highlight that larger errors have been observed in the last three stages and in the reboiler, where most of the column temperature variation takes place. A possible action for reducing the mismatch is to increase the dataset to be used to train the network, simulating the system in different operating conditions (e.g., impose set-point changes). Concerning the prediction of the remaining components, also in these cases, the model is capable of accurately estimating the evolution over time with estimation errors resulting always lower than the assumed measurement errors, as can be inferred from the obtained MSE values, model's ability to filter the measurement noise is furthermore assessed.

5. Conclusions

In this work, a demethanizer column of a simulated CRR unit has been modelled by employing modified bidirectional LSTM neural networks exploring the possibility of its use as digital twins for the column. The results show that the developed neural model can approximate the column trend profiles of the target variables and has a good ability to filtrate measurement noise for the column composition profiles. It also proves to be able to predict the actual trend of the temperature transient profiles quite accurately for most of the column trays and to be a useful tool for temperature monitoring in the final stages of the column and the reboiler. As a result, the presented model may be a suitable candidate as a digital twin of the demethanizer column. On the other hand, the need for concentration measurements yields less attraction to be used as digital twins in a real plant.

References

Chebbi, R., Al-Amoodi, N.S., Abdel Jabbar, N.M., Husseini, G.A., Al Mazroui, K.A. (2010). Optimum ethane recovery in conventional turboexpander process. Chem. Eng. Res. Des. 88, 779-787.

Chebeir, J., Salas, S.D., Romagnoli, J.A. (2019). Operability assessment on alternative natural gas liquids recovery schemes. J. Nat. Gas Sci. Eng. 71, 102974.

Elman, J. L. (1990). Finding structure in time. Cognitive science, 14(2), 179-211.

Getu, M., Mahadzir, S., Long, N.V.D., Lee, M. (2013). Techno-economic analysis of potential natural gas liquid (NGL) recovery processes under variations of feed compositions. Chem. Eng. Res. Des. 91(7), 1272-1283.

Grieves, M. (2015). Digital Twin: Manufacturing Excellence through Virtual Factory Replication.

Hochreiter, S., & Schmidhuber, J. (1997). Long short-term memory. Neural computation, 9(8), 1735-1780.

Blank, J., Deb, K. (2020). pymoo: Multi-Objective Optimization in Python, in IEEE Access, vol. 8, pp. 89497-89509.

Kherbeck, L., Chebbi, R. (2015). Optimizing ethane recovery in turboexpander processes. J. Ind. Eng. Chem. 21, 292–297.

Kidnay, A.J., Parrish, W.R., McCartney, D.G. (2011). Fundamentals of Natural Gas Processing, 2nd Editio. ed, Fundamentals of Natural Gas Processing. CRC Press, Boca Raton, FL.

Kingma, D. P., & Ba, J. (2015). Adam: A Method for Stochastic Optimization. ICLR. 2015. arXiv preprint arXiv:1412.6980, 9.

Luyben, W.L. (2013). NGL demethanizer control. Ind. Eng. Chem. Res. 52, 11626–11638.

Mandis, M., Baratti, R., Jorge, A., Chebeir, J.A., Tronci, S., Romagnoli, J. A. (2022). Performance assessment of control strategies with application to NGL separation units. J Natural Gas Sci Eng vol. 106, 104763.

Qu, X., Song, Y., Liu, D., Cui, X., & Peng, Y. (2020). Lithium-ion battery performance degradation evaluation in dynamic operating conditions based on a digital twin model. Microelectronics Reliability, 114, 113857.

Tao, F., Zhang, L., Nee, A. Y. C., & Pickl, S. W. (2016). Editorial for the special issue on big data and cloud technology for manufacturing. The International Journal of Advanced Manufacturing Technology, 84(1-4), 1-3.

Zhang, S., Liu, C., Jiang, H., Wei, S., Dai, L., & Hu, Y. (2015). Feedforward sequential memory networks: A new structure to learn long-term dependency. arXiv preprint arXiv:1512.08301.

Zhu, X., & Ji, Y. (2022). A digital twin–driven method for online quality control in process industry. The International Journal of Advanced Manufacturing Technology, 119(5), 3045-3064.

Antonis Kokossis, Michael C. Georgiadis, Efstratios N. Pistikopoulos (Eds.)
PROCEEDINGS OF THE 33rd European Symposium on Computer Aided Process Engineering (ESCAPE33), June 18-21, 2023, Athens, Greece
 http://dx.doi.org/10.1016/B978-0-443-15274-0.50121-9

Soft modelling of spruce conversion into bio-oil through pyrolysis – Note I: steam explosion and LPMO-activated enzymatic saccharification

Matteo Gilardi,[a] Filippo Bisotti,[a,*] Olaf T. Berglihn,[a] Roman Tschentscher,[b] Vincent G.H. Eijsink,[c] Anikó Várnai,[c] Bernd Wittgens,[a]

[a]SINTEF Industry-Process Technology, Richard Birkelands vei 2B, Trondheim, 7034, Norway

[b]SINTEF Industry-Process Technology, Forskningsveien 1, Oslo N-0373, Norway

[c]Faculty of Chemistry, Biotechnology and Food Science, NMBU, Chr. Magnus Falsens vei 18, Aas N-1433, Norway

** filippo.bisotti@sintef.no (corresponding author)*

Abstract

Biorefinery converts biomass into energy and high-value bioproducts, such as bio-oil. The combined power and fuel generation is particularly attractive for energy transition and reduction of non-biogenic CO_2 emissions. Biomass conversion requires a series of complex treatments. The development of reliable models is essential for scaling up and designing. Soft-modelling approach provides a simplified description of the system without losing accuracy. This contribution presents a soft model for steam explosion (after impregnation with carbocation scavengers) and enzymatic saccharification (boosted with H_2O_2-driven oxidative depolymerization by LPMO enzymes). These are two key pre-treatments for the extraction of carbohydrate fraction prior to pyrolysis of lignin-rich solids to produce the bio-oil. The proposed saccharification model accurately predicts experimental data in the range of interest for steam explosion temperature and enzyme concentrations investigated. This approach can be considered general and is potentially applicable to every kind of biomass.

Keywords: biorefinery, steam explosion, enzymatic saccharification with LPMO, soft modelling in COCO, softwood cellulose and hemicellulose valorization.

1. Introduction and context

Softwood (such as Norwegian spruce) is a good feedstock for biorefinery. The main aim of the biorefinery is to convert the carbon captured by plants into valuable chemicals (fuels, for instance) or heat & power generation systems via several physical and (bio)chemical treatments (Cherubini, 2010). During downstream processing, biomass undergoes a series of pre-treatments to separate the cellulose/hemicellulose from the lignin fraction. Our model includes three main pre-treatment steps: 1) impregnation of spruce biomass with 2-naphthol, a carbocation scavenger that prevents lignin condensation during the subsequent steam explosion (Pielhop et al., 2016a); 2) steam explosion, which allows the breaking of biomass structure so that the hemicellulose and cellulose become accessible while partially solubilizing hemicellulose (Pielhop et al., 2016b); and 3) enzymatic saccharification, where the hemicellulose and cellulose fractions are depolymerized into their elemental sugars (i.e., pentose and hexose) by selective catalysts (i.e., enzymes) under mild conditions. We selected this process because it selectively solubilizes the carbohydrate fractions and at the same time yields a less condensed and more reactive lignin for further downstream processes (Pielhop et al., 2017; Hansen et al., 2022). The efficiency of the steam explosion and saccharification steps influences the productivity and the overall yield of the complete biorefinery process. Therefore, the present article focuses on modelling these steps. In our model, we take a closer look at the impact of 2-naphthol impregnation on steam explosion as well as LPMO-aided saccharification of Norwegian spruce. Finally, a sensitivity analysis was performed to detect the most influential parameters in designing and modelling softwood biorefineries based on such upstream processing.

2. Purpose and material

The present work investigates the direct scale-up and the process configuration of the upstream processes in a biorefinery described by (Hansen et al. 2022), via direct translation of the reported laboratory-scale biomass treatments. The selected process layout (Figure 1) is based on the experimental setup by (Hansen et al. 2022) with the following modifications: (1) after the steam explosion, the pH of the biomass slurry was set to 5.0 (pH optimum of the cellulases in the subsequent saccharification step) in a single step using NaOH, instead of using an acetic acid/acetate buffer solution and then a caustic solution, for scalability. (2) A filtration stage was added after the saccharification step to separate the lignin-rich saccharification residue (solids) from the sugar solution (liquid/aqueous phase). This step is expected to reduce the residual moisture content of the lignin-rich residue and thus, lead to energy savings in the drying step before the pyrolysis unit.

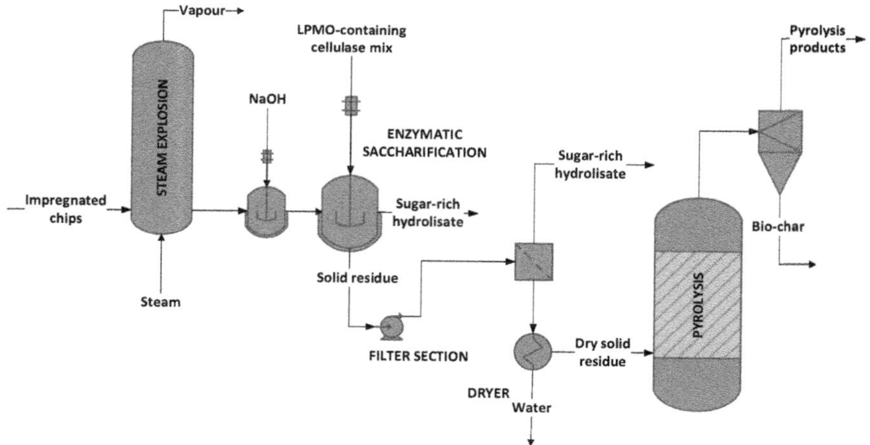

Figure 1 - Simplified layout of the investigated upstream processes of a softwood-based biorefinery. The picture does not include the impregnation step prior to the steam explosion.

3. Implementation

The steam explosion and enzymatic saccharification steps were modelled using COCO (a CAPE-open solution) simulation software (Figure 2). Due to the large complexities of bioprocesses, we adopted a soft-model approach instead of a first-principle one. The missing properties (i.e., enthalpy of formation, heat capacity, vapour pressure, and others) for some components, namely the cellulose, hemicellulose, and lignin constituents, have been estimated using Gani's methods (Constantinou and Gani, 1994). For the steam explosion step, we implemented a simplified kinetics scheme proposed by Lam (2009 and 2010), which has been suggested to be applicable for softwood in general. Due to the lack of measurements and data points collected during steam explosion, we do not have experimental data to validate the implemented model. For the enzymatic saccharification, on the other hand, the general expression of the soft model was taken from (Theuretzbacher et al., 2015), and it was validated and adjusted using the experimental saccharification data in (Hansen et al., 2022). The proposed model for saccharification conversion fits the experimental data (Figure 3), and the conversion model is a function of the steam explosion temperature (T_{SE} in [°C] and overall Cellic CTec2 enzyme concentration [E] expressed in [mg/$g_{biomass}$], as in (Hansen et al., 2022).

$$\gamma(t) = \gamma(0) + h \cdot [1 - \exp(-k_r \cdot t)] \quad (1)$$

Where γ is the conversion at the outlet of the saccharification reactor, $\gamma(0)$ the conversion at the inlet (called "instantaneous"; in Equation 2), h an adaptative parameter accounting for the conversion "jump" from the inlet and outlet conditions for ideal infinite residence time (see Equation 3), k_r the reaction constant (see Equation 4), and t the residence time.

$$\gamma(0) = 0.8607 \cdot [E] + 4.6 \quad (2)$$

$$h(T_{SE}, [E]) = (0.435 \cdot T_{SE} - 88.227) \cdot [E] + (0.15 \cdot T_{SE} - 24.275) \quad (3)$$

$$k_r(T_{SE}, [E]) = (1.3 \cdot 10^{-4} \cdot T_{SE} - 0.0355) \cdot [E] + (-4.17 \cdot 10^{-3} \cdot T_{SE} + 1.0937) \quad (4)$$

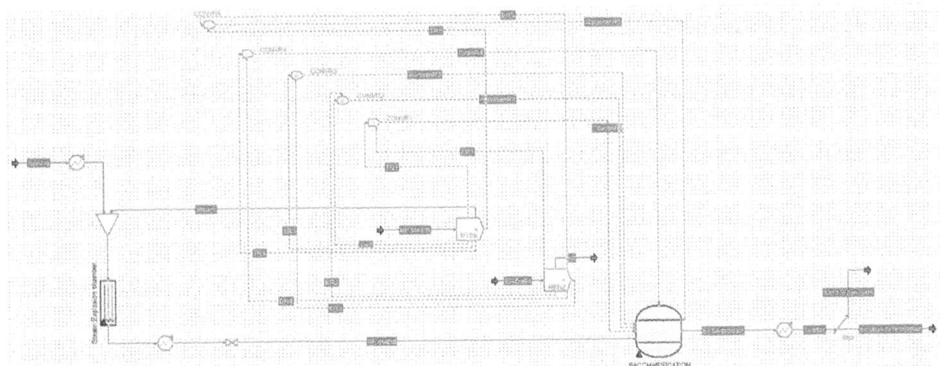

Figure 2 - Flowsheet for the steam explosion and enzymatic saccharification in COCO simulator. Yellow color indicates that the unit is handling a biphasic system of liquid and solid.

The saccharification curves in Figure 3 show that maximum glucose conversion is reached when the biomass undergoes a high-temperature steam explosion (220°C) and at the highest tested enzyme load (8 g/$kg_{biomass}$). The optimal residence time lies around 24 h, after which the saccharification yield reaches a plateau. Thus, longer saccharification times lead to a larger volume without substantial benefit to the process yield. As cellulose and hemicelluloses are depolymerized concomitantly by enzymes (Várnai et al., 2011), a similar yield trend is expected for the saccharification of the cellulose and hemicellulose fractions. Using the models here defined, the steam explosion chamber was simulated as a plug flow reactor and the saccharification reactor (which resembles a CSTR reactor with a large residence time) as a system at assigned conversion. The conversion was estimated according to the steam explosion operating temperature (i.e., steam) and the concentration of the LPMO-rich enzyme cocktail fed into the saccharification reactor.

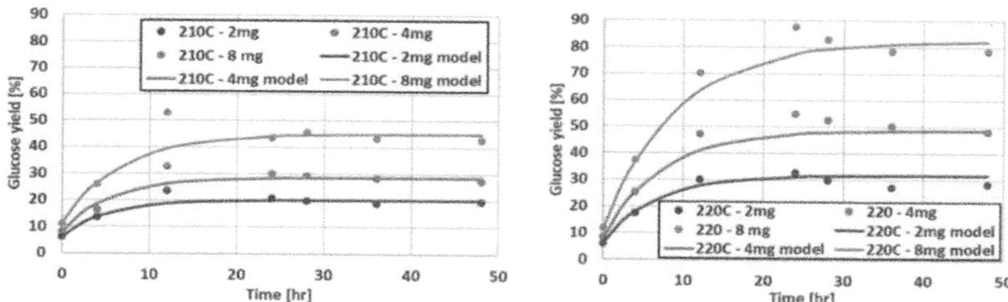

Figure 3 - Conversion model for enzymatic saccharification: experimental data (dots) and model (solid line) after the steam explosion (210°C–left and 220°C–right) using different enzyme doses expressed as $mg_{enzyme}/g_{biomass}$ (2–blue, 4–orange, 8–grey).

4. Sensitivity analysis and results

Additional parameters for the models were set as follows. The Norwegian spruce composition was adapted from (Wang et al., 2018). Hansen et al. report a residual humidity of 5 wt%. The steam explosion process is supposed to treat 100 kg/h of wet-impregnated spruce. The process requires steam at saturation pressure, corresponding to temperatures ranging from 210°C (18.5 bar) to 220°C (22.5 bar). The ratio between biomass and steam is 1:1 in mass. The steam explosion chamber volume has been set to 29 l (litres) and 58 l; with a fixed diameter of 0.3 m, hence the volume is directly proportional to the residence time. We performed a sensitivity analysis on the steam explosion temperature at two reactor volume levels to assess the impact of the operating conditions of this pre-treatment process on the overall upstream processing steps. Table 1 gathers the different case studies.

Table 1 - Case studies proposed in this work for the sensitivity analysis.

Case	Steam explosion temperature [°C]	Steam explosion reactor volume [l]
1	210	29
2	210	58
3	220	29
4	220	58

Table 2 gathers the increment or decrement in the mass yield of each product with respect to Case 1 (indicated in Table 1). Our main observations are that: (1) remarkably, the lignin fraction seems inert to both steam explosion and saccharification, thus the pyrolysis process is not affected; and (2) the composition of aqueous sugar solution depends on the steam explosion operating conditions. The latter one is partly because hemicellulose (arabinoxylan and galactoglucomannan) hydrolysis starts taking place already during steam explosion, depending on the severity. In Lam's model, hemicellulose hydrolysis is defined as a conversion to sugar (hexose and pentose) monomers and their corresponding degradation products HMF (hydroxymethyl-furfural) and furfural, respectively; the mass flow rate of the sugar monomer or secondary product being derived from the flowrate of its original corresponding polymer. The more severe the steam explosion conditions (i.e., high temperature and pressure) are, the larger the sugar yield from the cellulose and hemicellulose fractions is after saccharification. Both HMF and furfural are valuable for bio-oil production. We adopted Case 1 as a reference because it represents the operating condition with the lowest yield, but also the one associated with the lowest costs. Figure 4 depicts the products' mass composition at the steam explosion outlet and enzymatic saccharification reactors in the proposed case studies.

Table 2 - Predicted conversion of hemicellulose, increment (positive) or decrement (negative values) in pentose and hexose sugars mass (on biomass dry matter basis), at the outlet of the enzymatic saccharification reactor using Case 1 as the reference case.

Case	Arabinose	Galactose	Xylose	Mannose	HMF	Furfural
2	+7.61%	-94.28%	-95.66%	+14.61%	+12.99%	+7.65%
3	+6.74%	-44.47%	-56.53%	+15.04%	+7.31%	+4.52%
4	+8.72%	-98.65%	-99.20%	+19.90%	+13.67%	+7.92%

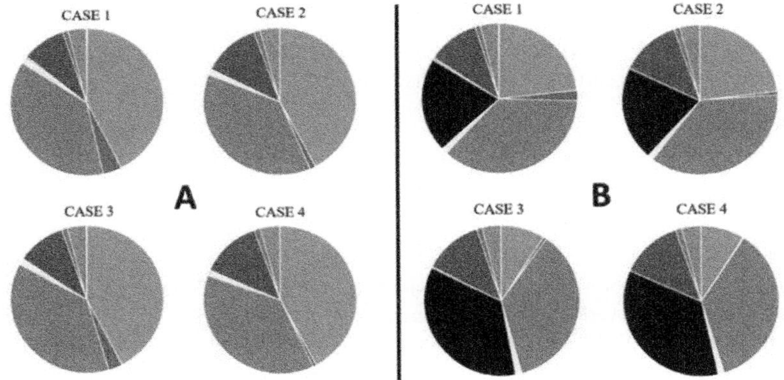

Figure 4 - Mass composition (dry weight basis) of the sugar solution obtained at the outlet of the steam explosion (A) and enzymatic saccharification (B) reactors in the four analyzed case studies (CASE). Legend: **lignin** (green), **glucan/cellulose** (grey), **glucose** (black), **mannan** (red), **mannose** (dark blue), **arabinose** (yellow), **HMF** (magenta), and **furfural** (light blue). Not mentioned species (arabinan, xylan, xylose, galactan, and galactose) have negligible content (< 0.1 wt%).

5. Conclusions and comments

The results show that the biomass pre-treatment model enables the scale-up of the impregnation–steam explosion–enzymatic saccharification process from laboratory to pilot scale while leading to the identification of the main parameters determining the selective and nearly complete removal of the polysaccharide fraction. At the projected scale, the steam explosion was the most determining step of the studied process setup since it had a direct influence on the subsequent enzymatic saccharification step. Correct tuning of the steam explosion operating conditions led to an optimal process yield. The hemicellulose and cellulose fractions were almost depleted after steam explosion and enzymatic saccharification. Hemicellulose (arabinoxylan and galactoglucomannan) was mainly degraded after steam explosion. The lignin fraction did not seem to undergo major chemical/physical transformation due to the impregnation pre-treatment. This emerges both from the experimental data and the model. Lignin remains in the insoluble fraction in the investigated process since its extraction and valorization require stronger operative conditions as in pyrolysis. The predicted outlet compositions from the steam explosion and enzymatic saccharification reactors at different steam explosion temperatures (210°C and 220°C) and reactor volumes (29 l and 58 l) point to the following conclusions: (1) A proper design of steam explosion and enzymatic saccharification reactors can lead to a very efficient decomposition of the hemicellulose and cellulose fractions of biomass, respectively, thus generating a lignin-rich solid residue. (2) Increasing steam explosion temperature and reactor volume is a good strategy to substantially improve the solubilization of the polysaccharide content. (3) Glucomannan removal during the steam explosion is significantly slower than arabinoxylan (or galactoglucomannan) removal as the greatest yield increase is observable for the mannose. (3) Larger residence times in the steam explosion chamber are associated with the formation of substantial amounts of secondary products (HMF and furfural). The temperature increment is not enough to fully deplete the intermediate products (galactose and xylose). Even though these are by-products generated from successive pentose and hexose sugars oxidation, they are valuable chemicals which can be blended in biodiesel. (4) High-temperature steam explosion benefits the glucan conversion occurring within enzymatic saccharification. If the temperature is shifted from 210°C to 220°C, the global final glucose yield rises from 50.8% up to 87.7%. On the contrary, steam explosion reactor volume does not impact the performance of downstream unit operations.

Acknowledgements

This work was supported by the Research Council of Norway through grants no. 257622 (Bio4Fuels) and 268002 (Enzymes4Fuels).

References

F. Cherubini, 2010, The biorefinery concept: Using biomass instead of oil for producing energy and chemicals, Energy Conversion and Management, 51(7), 1412-1421

L. Constantinou et R. Gani, 1994, New group contribution method for estimating properties of pure compounds, AIChE Journal, 40(10), 1697-1710

L.D. Hansen et al., 2022, 2-Naphthol impregnation prior to steam explosion promotes LPMO-assisted enzymatic saccharification of spruce and yields high-purity lignin. ACS Sustainable Chemistry & Engineering; 10(16), 5233-5242

P. Kumar et al., Lignocellulose biomass pyrolysis for bio-oil production: A review of biomass pre-treatment methods for production of drop-in fuels, Renewable and Sustainable Energy Reviews, 123, 109763

P.S.W. Lam, 2009, Kinetic modeling of pseudolignin formation in steam exploded woody biomass, Conference proceeding of 8th world congress of chemical engineering, Montreal, Canada, 23-27 August 2009

P.S.W. Lam, 2010, Chapter 5 – Modelling of polysaccharides depolymerization and pseudolignin formation of softwood during steam explosion, PhD Thesis

T. Pielhop et al., 2016a, Autohydrolysis pretreatment of softwood - enhancement by phenolic additives and the effects of other compounds. Green Chemistry, 18, 5239-5247

T. Pielhop et al., 2016b, Steam explosion pretreatment of softwood: the effect of the explosive decompression on enzymatic digestibility. Biotechnology for Biofuels, 9, 152

T. Pielhop et al., 2017, Pilot-scale steam explosion pretreatment with 2-naphthol to overcome high softwood recalcitrance. Biotechnology for Biofuels, 10(1), 130

F. Theuretzbacher et al., 2015, Steam explosion pretreatment of wheat straw to improve methane yields: Investigation of the degradation kinetics of structural compounds during anaerobic digestion, Bioresource Technology, 179, 299-305

A. Várnai, et al., 2011, Synergistic action of xylanase and mannanase improves the total hydrolysis of softwood. Bioresource Technology, 102(19), 9096-9104

Z. Wang et al., 2018, Chemical and structural factors influencing enzymatic saccharification of wood from aspen, birch and spruce, Biomass and Bioenergy, 109, 125-134

Antonis Kokossis, Michael C. Georgiadis, Efstratios N. Pistikopoulos (Eds.)
PROCEEDINGS OF THE 33rd European Symposium on Computer Aided Process Engineering (ESCAPE33), June 18-21, 2023, Athens, Greece
 http://dx.doi.org/10.1016/B978-0-443-15274-0.50122-0

Molecular Modelling and Optimization of Diesel Hydrotreating Processes

Xiaolin Bi,[a] Nan Zhang,[a] Robin Smith,[a]

[a]*Centre for Process Integration, Department of Chemical Engineering, University of Manchester, Oxford Road, Manchester, M13 9PL, UK*

Abstract

Diesel is an important transportation fuel produced in oil refineries, for which the quality should be strictly controlled to meet the diesel engine requirement and minimize associated air pollution issues. A key refining process to improve diesel quality is the diesel hydrotreating (DHT) process. The molecular type homologous series (MTHS) matrix is developed for the molecular characterization of petroleum fractions. The kinetic modelling and reaction modelling are built and improved to discuss the influence of operating conditions on the DHT processes. This work innovates to combine the catalyst deactivation model with the kinetic model to continuously monitor the catalyst deactivation during the DHT processes, which is essential to optimize the reaction operating conditions and maintain constant product yields and quality.

Keywords: Diesel Hydrotreating Processes, Optimization, Molecular Characterization, Catalyst Deactivation.

1. Introduction

Diesel hydrotreating (DHT) has become one of the essential fundamental processes in the petroleum-refining industry from environmental, technical, and economic points of view for over 60 years (Hoehn et al., 2017). The DHT process is mainly to obtain products with high quality and low content of pollution. In order to fulfill the applicable regulations of product quality, and remove the undesirable compounds, including but not solely limited to sulfur, nitrogen, olefins, and aromatics, the operating conditions of the reactor need to be improved. Establishing reaction modelling and simulation is a useful way to have deep research on phenomena occurring in the DHT reactor. Attributed to effective modelling and simulations, operating conditions, reactor design, catalyst formulation and process configuration would be optimized. In addition, effects of operating variables could be investigated and integration with other technologies could be implemented. (Mederos et al., 2012).

The loss of catalytic activity is a problem of great and continuing concern in the practice of industrial catalytic processes. During the industrial production process, with the extension of the operation time of the unit, the catalyst of hydrodesulfurization will experience a decline in activity. The degree of catalyst deactivation depends mainly on the operating conditions and the properties of the feed (Kohli et al., 2016). The loss of catalytic activity must be compensated by periodic increases in the reaction temperature in order to maintain the constant product yields and quality (Moghadassi et al., 2011). When the catalyst is deactivated or the catalyst activity drops below the economical operation of the unit, it is necessary to shut down the operation for catalyst regeneration or replacement with a new catalyst, which directly affects the economic benefits of the refinery. The annual cost to the industry for catalyst replacement and process shutdown is in the billions of dollars (Bartholomew et al., 2001). However, catalyst deactivation is inevitable. Therefore, it is significant to develop the catalyst deactivation model to investigate the catalyst deactivation regularity under ultra-deep desulfurization of diesel fuel, which shows practical guidance for refineries on predicting the catalyst usage periods and rationalising the production cycles of units.

In this research, the reaction network of a diesel hydrotreater is generated by incorporating most of the reaction types in this process, including hydrodesulfurization (HDS), hydrodenitrogenation (HDN), hydrodearomatization (HDA), and hydrocracking (HC). The current kinetic model of hydrodesulfurization has been proved by Vanrysselberghe et al. (1998) that expressed the reaction for dibenzothiophenes, which is the hardest sulfur component to remove. The three-phase hydrotreater model developed by Korsten and Hoffmann (1996) can be used in this research which contains a combination of algebraic and ordinary differential equations. The fourth-order Runge-Kutta method is used to numerically solve the first-order differential equations in GAMS.

2. Methodology

2.1. Molecular Type Homologous Series (MTHS) Method

The molecular type and homologous series method is used to characterize petroleum streams quickly and accurately at the molecular level. Pseudo-components could be generated based on their corresponding distillation profile and density using the pseudo-component approach, where each of them has specific boiling point, density and volume fraction. The molecular composition within each homologous series follows the gamma distribution shown in equations 1 and 2. The MTHS matrix is shown in Figure 1.

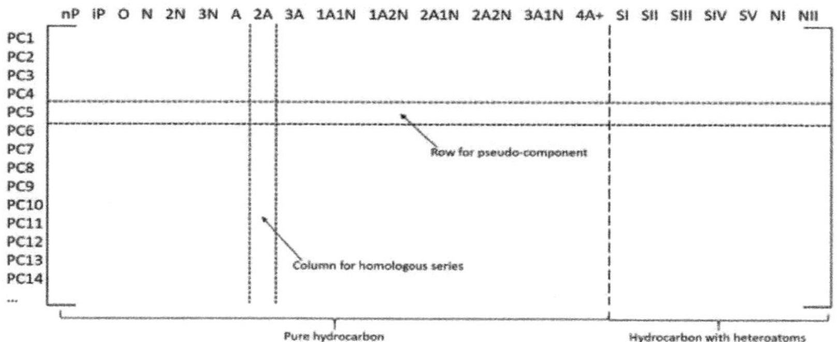

Figure 1 Pseudo-components based MTHS matrix

$$PDF(P) = \frac{(P-\eta)^{\alpha-1}\exp(-\frac{P-\eta}{\beta})}{\beta^{\alpha}\Gamma(\alpha)} \tag{1}$$

$$\Gamma(\alpha) = \int_0^{\infty} t^{\alpha-1}e^{-t}dt \tag{2}$$

where P is the property; α, β, η are three parameters of gamma distribution; PDF is the probability density function.

The objective function is to minimize the difference between the measured data and the predicted data:

$$OBJ = \sum_V (W_1 \cdot \frac{T_V^{msd} - T_V^{pred}}{T_V^{msd}})^2 + \sum_{PC} (W_2 \cdot \frac{BP_{PC}^{msd} - BP_{PC}^{pred}}{BP_{PC}^{msd}})^2 + \sum_P (W_3 \cdot \frac{P_P^{msd} - P_P^{pred}}{P_P^{msd}})^2 + \sum_{f \in PIONA} (W_4 \cdot \frac{C_f^{msd} - C_f^{pred}}{C_f^{msd}})^2 + \sum_{e \in (S,N,O)} (W_5 \cdot \frac{X_e^{msd} - X_e^{pred}}{X_e^{msd}})^2 \quad (3)$$

where, T is the boiling point of the distillation curve; BP_{PC} is the boiling point of the pseudo-components; P is bulk properties; C is the composition of PIONA; X is heteroatoms content; W is the weighting factors for each type of the properties; msd is the measured values; $pred$ is the predicted results based on the MTHS.

2.2. Kinetic Model of Hydrodesulfurization (HDS)

The kinetic model of HDS for the different sulfur compounds has been proposed by previous researchers. The Hougen-Waston rate equation is used to describe the hydrogenation and hydrogenolysis in the HDS process (Vanrysselberghe et al., 1998):

$$r_s = \frac{k_{s,\sigma} K_{H,\sigma} K_{s,\sigma} C_s C_{H_2}}{DEN_\sigma} + \frac{k_{s,\tau} K_{H,\tau} K_{s,\tau} C_s C_{H_2}}{DEN_\tau} \quad (4)$$

With

$$DEN_\sigma = (1 + \sum_i K_{i,\sigma} C_i + \sqrt{K_{H,\sigma} C_{H_2}})^3 \quad (5)$$

$$DEN_\tau = (1 + \sum_i K_{i,\tau} C_i + \sqrt{K_{H,\tau} C_{H_2}})^3 \quad (6)$$

where, r_s is the reaction rate of the sulfur compound; σ , τ are hydrogenolysis site and hydrogenation site; $k_{s,\sigma}$ and $k_{s,\tau}$ are the rate coefficients for the reactions of the sulfur compound on the two sites; $K_{i,\sigma}$ and $K_{i,\tau}$ are the adsorption coefficient of component i on the two sites; C_i is the liquid concentration of the component i.

2.3. Modelling of Three-Phase Diesel Hydrotreater

The three-phase catalytic reactor, the heart of the HDT unit, is the equipment that needs to be modelled. In HDT reactors, the gas phases (H_2) and liquid phases (hydrocarbons) are contacted with the solid phase (catalyst). The reactions occur between the dissolved gaseous reactants and the liquid phase reactants on the surface of the catalyst (Mederos and Ancheyta, 2007). The reactor model created by Korsten and Hoffmann (1996) is used based on the following assumptions:

a) The reactor operates under steady state conditions.
b) No radial concentration gradients.
c) Constant volumetric flowrates through the reactor.
d) Constant pressure through the reactor.
e) The catalyst particles are completely wetted.

2.3.1. Mass Balance

For component i of the gas phase:

$$\frac{u_G}{RT_G}\frac{\partial p_i^G}{\partial t} + k_i^G a_L \left(\frac{p_i^G}{H_i} - C_i^L\right) = 0 \quad (7)$$

$$z = 0, p_i^G = p_i^{G,0}$$

For gaseous components in the liquid phase:

$$u_L \frac{dC_i^L}{dz} + k_i^L a_L \left(\frac{p_i^G}{H_i} - C_i^L\right) + k_i^S a_S (C_i^L - C_i^S) = 0 \quad (8)$$

$$z = 0, C_i^L = C_i^{L,0}$$

At catalyst surface:

$$k_i^S a_S (C_i^L - C_i^S) = -\nu_i \rho_B \xi \eta r_i \quad (9)$$

Where, u_G is the superficial gas velocity; p_i^G is the partial pressure of gaseous i; Z is the length of the reactor; $k_i^G a_L$ and $k_i^S a_S$ are the gas−liquid liquid-solid mass-transfer coefficient; H_i is Henry's law coefficient; C_i^L and C_i^S are concentrations of component i in the liquid and on the surface of the catalyst; η is the effectiveness factor; r_i is the reaction rate.

2.3.2. Energy Balance

$$\frac{dT_R}{dz} = \frac{\xi n}{\rho_B u_L C_p} \sum_{i=1}^{n} r_i(-\Delta H_i) \tag{10}$$

where, C_p is the specific heat capacity of the feed; T_R is the reactor temperature.

2.4. Catalyst Deactivation Model

Dong and Gao (2009) analyzed the sulfur content of the product oil samples at different times of catalyst operation to calculate the corresponding hydrodesulfurization reaction rate constants. The relative activity of the catalyst is defined as the ratio of the reaction rate constant at different times of operation to the initial reaction rate constant. The catalyst deactivation models are established as follows:

$$k = A\,exp(-E_a/RT) \tag{11}$$

$$A = (1 - nt)A_0 \tag{12}$$

$$k = (1 - nt)k_0 \tag{13}$$

$$n = \rho^{61.305} T_{90\%}^{0.136} f_{Aro}^{0.111} \tag{14}$$

where, k is reaction rate constant; A is constant; E_a is activation energy; n is inactivation rate constant; f is total content of bicyclic and tricyclic aromatic hydrocarbons and colloid (ω%).

3. Case study

The bulk properties and composition information of the SRGO stream are given in Table 1.

Table 1. Bulk properties of SRGO stream

Density at 15°C (kg/m^3)	851.7	Distillation (D86)/ °C	
Sulfur content (m%)	1.51	IBP	220
Paraffin (m%)	44.4	10 vol%	279
Naphthene (m%)	21.9	30 vol%	295
Aromatics (m%)	17.3	50 vol%	310
Multi-ring aromatics (m%)	7.36	70 vol%	328
		90 vol%	359
		95 vol% 376	376
		FBP	379

The configuration of the reactor and its operating conditions are shown in the Table 2.

Table 2. Configuration of the reactor and operating conditions

Reactor configuration		Operating conditions	
Diameter (cm)	2.54	Temperature (K)	633
Length (cm)	45	Pressure (bar)	6.4
Density of bed (kg_{cat}/m^3)	780	H_2/oil	500
Catalyst characteristics		LHSV (hr^{-1})	1.5
Equivalent diameter (m)	1.30*10^{-3}	Hydrogen purity (%)	100
Density (kg/m^3)	1420		
Specific surface area (m^2/g)	200		

4. Results and Discussion

4.1. Feed Characterization

SRGO stream is divided into 25 pseudo-components, each having a 10K gap between 448K and 688K. Sulfur compounds are classified into 5 homologous series: SI (thiophene), SII (benzothiophene), SIII (dibenzothiophene), SIV (4-methyldibenzothiophene) and SV (4, 6-dimethyldibenzothiophene). The distribution of different types of sulfur compounds: thiophenes (TH) are commonly found at temperatures below 220°C, benzothiophenes (BT) make up the majority of the boiling range between 220°C and 257°C, and dibenzothiophenes (DBT) are substantially concentrated at temperatures above 330°C. Figure 2 and Table 3 show the comparison of distillation profiles and properties between the measured and predicted results. The biggest deviation of distillation profiles is at this point when the cumulative volume is 10%, and the biggest deviation of the properties is the content of multi-ring aromatics. These errors are still within acceptable limits.

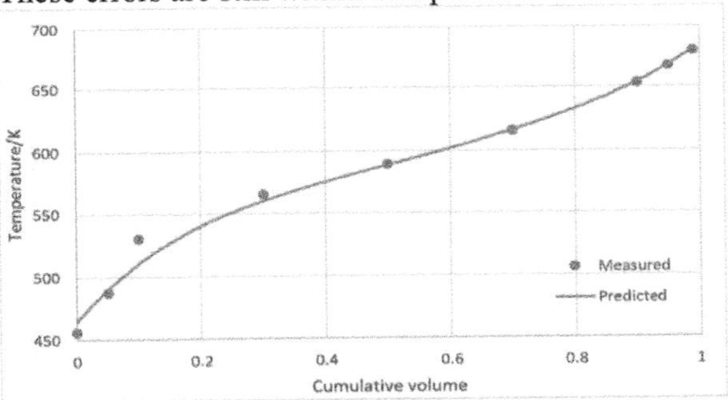

Figure 2. Comparison of distillation profiles between the measured and the predicted

Table 3. Comparison of the properties between the measured and predicted

Composition & Property	Measured	Predicted	Relative Error (%)
Density, kg/m^3	851.7	850.8	-0.11
Paraffin(m%)	44.4	43.13	-2.86
Naphthene (m%)	21.9	22.4	2.15
Aromatics (m%)	17.3	18.14	4.93
Multi-ring aromatics (m%)	7.36	7.8	5.83
Sulfur content (m%)	1.51	1.514	0.26

4.2. The Pilot Plant Diesel Hydrotreater

The comparison of compositions and properties of the hydrotreated product between the measured and the predicted at the operating condition of 633K, 6.4bar and 1.5 hr-1 are shown in Table 4. The relative error between predicted results and measured results for the nitrogen content is slightly larger, but it is still within an acceptable range. The errors between the predicted and measured for other content are lower than 5%.

Table 4. Composition and properties of the measured and predicted product

	Measured	Predicted	Error (%)
Paraffin (m%)	55.2	53.4	-3.26
Naphthene (m%)	31.8	31.2	-1.89
Aromatics (m%)	12.9	13.4	3.88
N (wtppm)	8.77	9.29	5.93
S (wtppm)	6.02	6.16	2.32

The rising trend of temperature in the axial position of the reactor is shown in Figure 3. It can be seen from the figure that the temperature at the inlet of the reactor rises significantly, and the temperature is in a steady increase after 5% axial of the reactor. This is because the sulfides, especially thiophene and benzothiophene, react rapidly at the inlet of the reactor. Therefore, the temperature also improves rapidly.

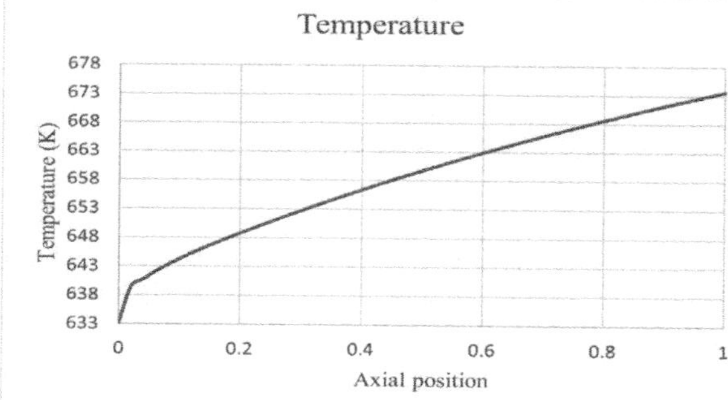

Figure 3. Temperature raise trend in the axial position of the reactor

5. Conclusion

In this research, the kinetic model and the mass balance of the three-phase diesel hydrotreater model are established based on the MTHS method. The feasibility of the model is confirmed in the case study by comparing the predicted results with the measured results, which is the basis for further work. Future work will include the optimization of the current modelling and the monitor continuously of the catalyst deactivation.

References

C. H. Bartholomew, 2001, Mechanisms of catalyst deactivation, Applied Catalysis A: General, 212(1-2), 17-60.

K. Dong, X. Gao, 2009, Deactivation model of hydrotreating catalyst for ultra low sulfur diesel fuel production, CHKI 40(7).

R. Hoehn, V. Thakkar, E. Yuh, 2017, Hydroprocessing for clean energy: design, operation, and optimization, John Wiley & Sons.

K. Kohli, R. Prajapati,S. K. Maity, M. Sau, M. O. Garg, 2016, Deactivation of hydrotreating catalyst by metals in resin and asphaltene parts of heavy oil and residues, Fuel, 175, 264-273.

H. Korsten, U. Hoffmann, 1996, Three-phase reactor model for hydrotreating in pilot trickle-bed reactors, AIChE J., 42, p. 1350.

F. S. Mederos, J. Ancheyta, 2007, Mathematical modeling and simulation of hydrotreating reactors: Cocurrent versus countercurrent operations, Applied Catalysis A: General, 332(1), 8-21.

F. S. Mederos, J. Ancheyta, I. Elizalde, 2012, Dynamic modeling and simulation of hydrotreating of gas oil obtained from heavy crude oil, Applied Catalysis A: General, 425, 13-27.

A. Moghadassi, N. Amini, O. Fadavi, M. Bahmani, 2011, Hydrocracking lumped kinetic model with catalyst deactivation in arak refinery hydrocracker unit.

V. Vanrysselberghe, R. Le Gall, G. F. Froment, 1998, Hydrodesulfurization of 4-methyldibenzothiophene and 4, 6-dimethyldibenzothiophene on a $CoMo/Al_2O_3$ catalyst: Reaction network and kinetics, Industrial & engineering chemistry research, 37(4), 1235-1242.

Antonis Kokossis, Michael C. Georgiadis, Efstratios N. Pistikopoulos (Eds.)
PROCEEDINGS OF THE 33rd European Symposium on Computer Aided Process Engineering (ESCAPE33), June 18-21, 2023, Athens, Greece
 http://dx.doi.org/10.1016/B978-0-443-15274-0.50123-2

Soft modelling of spruce conversion into bio-oil through pyrolysis – Note II: pyrolysis

Filippo Bisotti,[a,*] Matteo Gilardi,[a] Olaf T. Berglihn,[a] Roman Tschentscher,[b] Vincent G.H. Eijsink,[c] Anikó Várnai,[c] Bernd Wittgens,[a]

[a]*SINTEF Industry-Process Technology, Richard Birkelands vei 2B, Trondheim, 7034, Norway*

[b]*SINTEF Industry-Process Technology, Forskningsveien 1, Oslo N-0373, Norway*

[c]*Faculty of Chemistry, Biotechnology and Food Science, NMBU, Chr. Magnus Falsens vei 18, Aas N-1433, Norway*

* *filippo.bisotti@sintef.no (corresponding author)*

Abstract

Biorefinery converts biomass into valuable bioproducts. Cellulose and hemicellulose are degraded into simple sugars during enzymatic saccharification, while lignin remains unutilized. Lignin and residual cellulose can be subjected to pyrolysis, a key biorefinery step for improved valorization of biomass. During biomass pyrolysis thousands of compounds are formed, thus, a simplified approach must be adopted for the sake of modelling. Lumped kinetic models address a simplified description of the system by catching relevant features of the product mixture. This contribution presents a pyrolysis model where two previously validated kinetics are merged to get a more effective description of the product distribution. The model provides a quantitative characterization of the main product classes (aldehydes, carboxylic acids, phenols, furans, etc.) and can be used as a numerical tool for roughly scaling up pyrolysis reactors in a softwood-based biorefinery. This approach can be considered general and is applicable to all biomass types.

Keywords: biorefinery, pyrolysis, lumped kinetic, lignin valorization, bio-oil production.

1. Introduction

Biomass is composed of cellulose, hemicellulose, and lignin. Biochemical conversion of biomass, most commonly including steam explosion and enzymatic saccharification processes upstream, selectively depolymerizes and solubilizes the cellulose and hemicellulose fractions, leaving a lignin-rich solid residue. To increase the biorefinery yield, the lignin-rich residue can be subjected to thermochemical treatment to produce syngas and chemicals via gasification, heat and power via combustion, or bio-oil and biochar via pyrolysis (Kumar et al., 2020; Cherubini, 2010). The present work deals with the modelling of the pyrolysis of lignin-rich residue remaining after steam explosion and enzymatic saccharification of Norwegian spruce. An explicit, rigorous and descriptive model of the pyrolysis process, including thousands of species (considering also all intermediates) and reactions, would lead to a stiff and complex system, as exemplified by a few case studies where detailed kinetics have been adopted (e.g., Corbetta et al., 2015; Ranzi et al., 2014). Such models cannot be implemented in process simulators due to the lack of thermodynamic properties for most species present in the kinetics and due to the huge dimensions of nonlinear equations which challenge the numerical stability of the software itself (Manca et al., 2001). On the other hand, the lumped kinetics approach brings a substantial simplification of the kinetic schemes without losing accuracy (Ranzi et al., 2008). Lumped kinetics approach entails the simplification of the schemes to a limited number of chemical species (generally less than one hundred) and reactions (less than fifty), which can be handled by simulation software. In this work, we used the lumped kinetics over experimental pyrolysis data from Hansen (Hansen et al., 2022) and verified its capacity to predict product mixture composition.

2. Materials and methods

The adopted lumped kinetics describing the pyrolysis of lignin-rich residue is composed of two different kinetics models. The first part describing the conversion of the residual cellulose and hemicellulose fractions was adapted from the model proposed by (Ranzi et al. 2008). The second part describing the conversion of the lignin fraction, on the other hand, was adapted from the kinetics model proposed by Dussan (Dussan et al. 2018). The latter, the lignin-specific part of the model, is composed of four pseudo-lignin building blocks (LIG-1 to LIG-4) as in the model by Dussan (Dussan et al. 2008) while the stoichiometry of lignin pyrolysis reactions was modified. Overall, our lumped kinetics model contained 16 reactions and 49 species describing lignin pyrolysis. The kinetics model has been implemented in MATLAB 2022a code. The pyrolysis reactions occur in a fluidized bed reactor, where solid particles are circulated by the generated gas, and the gas itself guarantees continuous mixing inside the reactor. Thus, we modelled the fluidized bed as an isothermal plug flow reactor (PFR), with the following assumptions to simplify the calculations: the generated gas drags and entrains the small solid particles in a plug flow, while the gas and solid phases are vigorously mixed in the system, and this leads to almost isothermal conditions in the reactor volume. These simplifications provided us with a general simplified model that could be incorporated into a COCO simulation flowsheet, instead of using a descriptive and compute-intensive computational fluid dynamics (CFD) model. Furthermore, this work aims at the preliminary validation of the lumped kinetic model using experimental data from (Hansen et al. 2022).

3. Model equations and implementation

The pyrolysis reactor (fluidized bed) has been modelled as an isothermal PFR as follows. The material balance for each compound in the reactor is given in Equation (1):

$$\frac{d\dot{n}_k}{dV} = \sum_{j=1}^{NR} \nu_{kj} r_j \tag{1}$$

where $\dot{n}_k$ is the mole flow of a generic species and V is the volume of the reactor (domain of integration of the Ordinary Differential Equations system, ODE). The reaction rate (r_j) is defined as:

$$r_j = k_{R_j}(T) \cdot c_{reactant}^{\beta} \tag{2}$$

and the reaction order depends on the kinetics model adopted: first order (β=1) for the cellulose and hemicellulose reactants as in Ranzi (Ranzi et al. 2008) and second order (β=2) for the lignin pyrolysis as reported in Dussan (Dussan et al. 2018). The concentration of the reactant refers to the reactor volume. Reactions occur both in the solid and gas phases, thus the concentration (c) of the k-th specie should be defined as:

$$c_k = \frac{\dot{n}_k\,[\mathrm{kmol \cdot s^{-1}}]}{Q\,[\mathrm{m^3 \cdot s^{-1}}]} \quad (3)$$

We accounted for the variation of the volumetric flow (Q) which is the sum of both solid particles and gas species (in a plug flow the gas entrains the solid), as shown in (4):

$$Q = Q_{sol} + Q_{gas} \quad (4)$$

$$Q_{sol} = \frac{\sum \dot{m}_k^{sol}}{\sum \frac{\omega_k^{sol}(\dot{n}_{sol})}{\rho_k^{sol}}} \quad (5)$$

$$Q_{gas} = \frac{\sum \dot{m}_k^{gas}}{\rho_{gas}} \quad \text{with} \quad \rho_{gas} = \frac{P \cdot \sum x_i^{gas}(\dot{n}_{gas}) \cdot MW_i}{RT} \quad (6)$$

In the proposed expressions, $\dot{m}_k^{phase}$ is the mass flow of a generic species (k) in the corresponding phase (gas or solid), ω is the mass fraction, x is the molar one, and ρ the density of the phase. Thus, the gas and solid volumetric flows, calculated using Equations (5) and (6), respectively, are continuously updated with the composition of the respective phases from the material balance as the pyrolysis of biomass progresses. For the gas phase, we assumed ideal gas mixture behavior due to the high temperatures reached in pyrolysis ($T > 500°C$). The model requires additional inputs such as the reaction dimensions and the minimum fluidization velocity. The minimum fluidization velocity was calculated using Fogler's method (Fogler, 2020). For the calculation, we assumed that the lignin-rich biomass particles have a spherical shape with a diameter of 1 mm. It is noteworthy that a spherical shape is responsible for the largest pressure drop requirement to spark fluidization, thus, this is a conservative assumption. Typical dimensions of a pyrolysis reactor for industrial applications were adapted from (Zhou et al., 2019). The kinetic model described in this section has been assembled using MATLAB R2022a. The feed inlet composition is the outcome of previous calculations related to other unit operations for biomass pre-treatment. As feed to the pyrolysis, we selected the biomass that had been steam exploded at 220°C and saccharified with LPMO-cellulase solution with a dose of 8 $g_{enzymes}/kg_{biomass}$ (refer to Note I for details).

4. Results

We compared our model's results with the experimental observation reported by Hansen (Hansen et al., 2022). Then, we performed a sensitivity analysis on the pyrolysis temperature (in the range between 450 and 700°C). The pyrolysis process generates four main product streams: biochar (including unconverted biomass), syngas (including non-condensable components such as carbon monoxide, methane, and short aliphatic compounds), an aqueous phase, and bio-oil (comprising all other components). Table 1 gathers the results from the model under different scenarios. We considered our data obtained with 450°C and 500°C temperatures as relevant to contrast our results against the experimental observations (conducted at 500°C) in Hansen's work. Furthermore, we tested different pseudo-lignin building block combinations (LIG-1, LIG-2, LIG-3, or LIG-4 alone or the equimolar combination of all four building blocks; see Table 1) to characterize the biomass treated in the experiments since no details are available to fully describe the lignin fraction with the pseudo-lignin components approach by Dussan. There was a considerable mismatch between the experimental data and the model, with a tendency to overestimate the mass ratio of the bio-oil fraction (Table 1). There are several plausible explanations that could justify these deviations to some extent. First, the simplified model does not account for secondary (decomposition) reactions that generate lighter, non-condensable components in the gas phase, thus leading to an overestimation of the bio-oil and an underestimation of the gas fraction (Corbetta et al., 2015; Ranzi et al., 2014). Second, Dussan and Ranzi's models consider that a fraction of the gas and bio-oil constituents remain adsorbed to the solids even at high temperatures. Our model, on the other hand, assumes complete separation of the pyrolysis fractions, which, to some extent, could lead to underestimation of the biochar fraction and, at the same time, overestimation of the bio-oil and light gas fractions (compare the values to the ones in brackets in Table 1). Third, in the experimental setup by Hansen (Hansen et al. 2022), pyrolysis products are purged with cold nitrogen gas from a small fixed-bed system. The cold nitrogen may locally quench the solid bed, leading to a lower extent of pyrolysis and, consequently, to a higher amount of unconverted solid residue in the biochar fraction experimentally. Fourth, the mass ratio of the pyrolysis fractions substantially depends on the lignin model composition, i.e. the ratio of the four types of pseudo-lignin building blocks, as shown by the sensitivity analysis in Table 1. In the absence of available compositional data on the lignin-rich residue, the lignin model was approximated with the equimolar composition of the four pseudo-lignin building blocks based on the work by Dussan (Dussan et al. 2018), which may be suboptimal. The pseudo-lignin building blocks LIG-4 and, to some extent, LIG-2 are associated with a higher residual solid biochar fraction. Notably, (1) there are many other variables in the model that influence the pyrolysis fraction ratios, and (2) the differences between the outcomes with different building blocks are small compared to how far off we are from the experimental data. Moreover, these lignin building blocks (from LIG-1 to LIG-4 in Dussan) may not be optimal representatives of our system. Last but not least, the simplification of the process into a limited number of chemical species and reactions (i.e., the use of lumped kinetics) carries an additional source of error. The supporting material by Hansen (Hansen et al. 2022) reports a detailed product distribution and the classification of the thousands of reported species for the bio-oil fraction only. Estimating the distribution of especially low-boiling temperature species in the bio-oil and light gas fractions is not trivial. All in all, our results indicate that further refinement of the model is needed to reach the experimental outcome.

Table 1 - Comparison of the experimental mass distribution of pyrolysis product fractions between experimental data (Exp data) and model predictions with five pseudo-lignin building block combinations at 450°C and 500°C. Values before brackets assume that some species of the bio-oil and gas fractions remain entrapped in the solid bio-char fraction when recovering the pyrolysis fractions, while values in brackets assume complete separation of the pyrolysis fractions.

T = 500°C	**Exp data**	**Equal mix**	**Only LIG-1**	**Only LIG-2**	**Only LIG-3**	**Only LIG-4**
Bio-oil	0.35	0.708 (0.720)	0.800 (0.806)	0.710 (0.729)	0.761 (0.765)	0.675 (0.693)
Gas	0.20	0.081 (0.118)	0.133 (0.136)	0.078 (0.133)	0.082 (0.097)	0.044 (0.123)
Char	0.41	0.172 (0.124)	0.024 (0.015)	0.174 (0.100)	0.129 (0.110)	0.234 (0.137)
Water	0.04	0.039 (0.039)	0.044 (0.044)	0.039 (0.039)	0.030 (0.030)	0.048 (0.048)
T = 450°C						
Bio-oil	0.35	0.581 (0.598)	0.783 (0.793)	0.592 (0.629)	0.679 (0.685)	0.603 (0.620)
Gas	0.20	0.072 (0.108)	0.126 (0.132)	0.075 (0.145)	0.073 (0.089)	0.046 (0.114)
Char	0.41	0.310 (0.257)	0.051 (0.035)	0.293 (0.187)	0.221 (0.199)	0.299 (0.213)
Water	0.04	0.037 (0.037)	0.041 (0.041)	0.040 (0.040)	0.027 (0.027)	0.053 (0.053)

Next, we looked further into the impact of pyrolysis temperature on the conversion efficiency (described by the amount of unconverted biomass) and on the distribution of the types of chemicals being formed and consumed during pyrolysis (Figure 1). Pyrolysis was more complete at higher temperatures as indicated by a drop in the fraction of unconverted biomass remaining after pyrolysis (green bars in Figure 1), which can mainly be attributed to pyrolysis being endothermic and thus thermodynamically favored at higher temperatures. Increasing the process temperature facilitated char formation (brown bars in Figure 1), although overall char formation remained limited (mass fraction <0.08 w/w). Notably, the decrease in the proportion of unconverted biomass at higher temperatures was not comparable to the increase in char formation, resulting in an apparent decrease in the proportion of the solid "biochar" fraction that comprises unconverted biomass and char. In our model, lignin-rich biomass was primarily converted to lighter gaseous compounds and bio-oil species (listed in Figure 2). Water content and thus the amount of aqueous fraction after pyrolysis remained low (mass fraction <0.05 w/w), irrespective of the pyrolysis temperature. The differences between the product distributions above 600°C were negligible. Thus, we can conclude that the optimal temperature window is found between 600–650°C, with the lignin conversion being higher than 90% at such operating conditions.

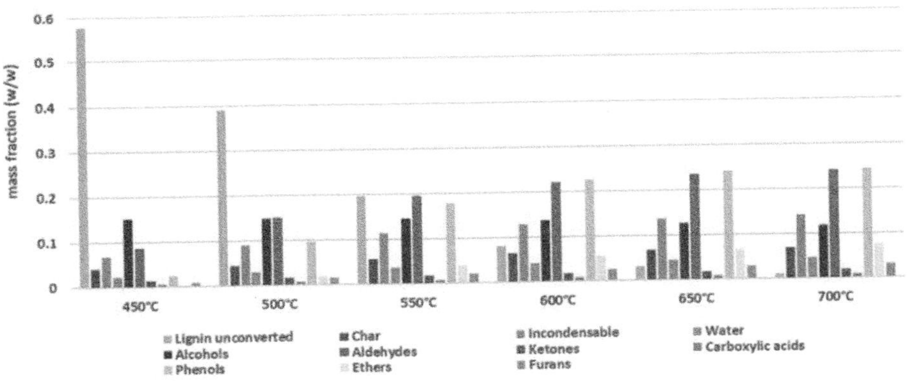

Figure 1 - Temperature effect on the distribution of pyrolysis products.

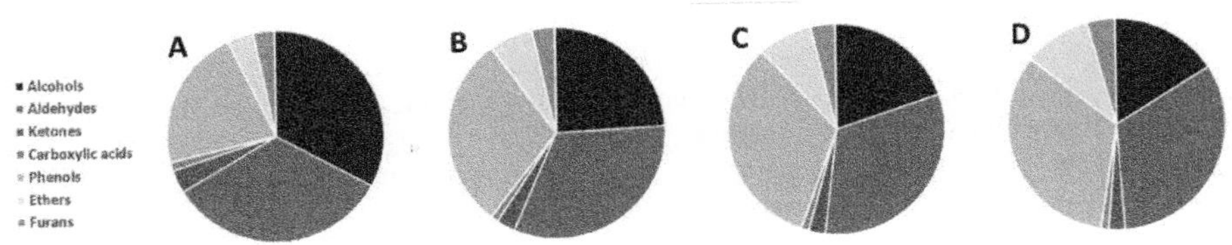

Figure 2 - The distribution of chemical compound types in the bio-oil generated at temperatures that are of interest for industrial pyrolysis: 500°C (A), 550°C (B), 600°C (C), and 700°C (D).

The relative proportion of bio-oil compounds formed at temperatures from 500 to 700°C is shown in Figure 2. We limited the upper temperature to 700°C because over that temperature lignin is almost depleted and secondary reactions result in excessive gas production, thus limiting the bio-oil production as described by Corbetta (Corbetta et al., 2015). In general, aldehydes, alcohols, and phenols, which are the main classes present in biofuels (Okolie et al., 2020) were the main constituents of the bio-oil (Figure 2). In particular, the alcohol fraction decreased when increasing process temperature, while the aldehyde and phenol contents followed opposite trends due to their relatively higher stability. Although phenols are valuable chemicals, their removal from bio-oil is preferred before fuel applications due to corrosiveness and low heating values (Fardhyanti et al., 2020). Furthermore, ethers and furans are appreciably formed above 450°C and make up about one-tenth of the bio-oil from the comparison between Figure 1 and Figure 2. Ketones and carboxylic acids are also present in the bio-oil mixtures in smaller amounts, and the pyrolysis temperature does not seem to affect substantially their formation. To fully evaluate the potential of bio-oil production in a biorefinery concept, the storage stability of the generated bio-oil will have to be considered in addition to composition.

5. Conclusions

The present work merged two lumped kinetics models describing pyrolysis of lignocellulosic biomass published by Ranzi for the cellulose and hemicellulose part, and Dussan for the lignin fraction to characterize the thermal conversion of lignin-rich biomass to biofuels. The model shows a qualitatively good agreement over a limited number of experimental data applied to scale up a spruce-based biorefinery combining biochemical and thermochemical conversion routes as in Hansen (Hansen et al., 2022). Although lumped kinetics itself, which simplifies the thermo- and bio-chemical processes into a comprehensive model, presents some limitations to the accuracy of the model, it allows flowsheet implementation and thereby easier and more efficient workflow during scale-up. Sensitivity analysis of the pyrolysis process allowed us to identify the optimal temperature window for operating a large-scale pyrolysis reactor, and, remarkably, it provided a rough estimate of the residual solid (biochar), bio-oil, and light gas fractions, as well as of the product distribution in the bio-oil.

Acknowledgements

This work was supported by the Research Council of Norway through grants no. 257622 (Bio4Fuels) and no. 268002 (Enzymes4Fuels).

References

F. Cherubini, 2010, The biorefinery concept: Using biomass instead of oil for producing energy and chemicals, Energy Conversion and Management, 51(7), 1412-1421

M. Corbetta et al., 2015, Multi-scale kinetic modelling and experimental investigation of syngas production from coal gasification in updraft gasifiers, Energy & Fuel, 29(6), 3972-3984

K. Dussan et al., 2019, A model of the chemical decomposition and pyrolysis kinetics of lignin, Proceedings of the Combustion Institute, 37, 2697-2704

D.S. Fardhyanti et al., 2020, The separation of phenolic compounds from bio-oil produced from pyrolysis of corncobs, AIP Conference Proceedings 2243, 020005

H.S. Fogler, 2020, Chapter 15 – Fluidized-bed reactors, in Elements of Chemical Reactions Engineering, fifth edition, Prentice Hall Editor

P. Kumar et al., 2020, Lignocellulose biomass pyrolysis for bio-oil production: A review of biomass pre-treatment methods for production of drop-in fuels, Renewable and Sustainable Energy Reviews, 123, 109763

L.D. Hansen et al., 2022, 2-Naphthol impregnation prior to steam explosion promotes LPMO-assisted enzymatic saccharification of spruce and yields high-purity lignin, ACS Sustainable Chemistry & Engineering, 10, 5233-5242

D. Manca et al., 2001, Numerical problems in the solution of oxidation and combustion models, Combustion Theory and Modelling, 5(2), 185-199

J.A. Okolie et al., 2021, Next-generation biofuels and platform biochemicals from lignocellulosic biomass, International Journal of Energy Research, 45, 14145-14169

E. Ranzi et al., 2014, Kinetic modeling of the thermal degradation and combustion of biomass, Chemical Engineering Science, 110, 2-12

E. Ranzi et al., 2008, Chemical kinetics of biomass pyrolysis, Energy & Fuel, 22, 4292-4300

G. Zhou et al., 2019, Simulation of coal pressurized pyrolysis process in an industrial-scale spout-fluid bed reactor, Advanced Powder Technology, 30, 3135-3145

Antonis Kokossis, Michael C. Georgiadis, Efstratios N. Pistikopoulos (Eds.)
PROCEEDINGS OF THE 33rd European Symposium on Computer Aided Process Engineering (ESCAPE33), June 18-21, 2023, Athens, Greece
 http://dx.doi.org/10.1016/B978-0-443-15274-0.50124-4

Numerical Modeling of Gas-liquid Hydrodynamics in a U-loop Bioreactor

Johan Le Nepvou De Carfort,[a] Tiago Pinto,[b] Ulrich Krühne,[a]

[a] *PROSYS, Department of Chemical and Biochemical Engineering, Technical University of Denmark, Søltofts Plads Building 228A, Kongens Lyngby, Denmark*
[b] *R&D Department, UNIBIO A/S, Roskilde, Denmark*

Abstract

A numerical model of a U-loop bioreactor is developed, with the aim of replicating the mass transfer characteristics of the real system. The bioreactor model includes a multiphase fluid dynamic model, coupled with a population balance model, to simulate the dispersed bubbles in the loop and estimate the gas-liquid interfacial surface area. The results from the numerical model are compared to experimental measurements of the mass transfer coefficient in the downgoing leg of the U-loop. The numerical model of the U-loop bioreactor is able to replicate the qualitative behavior of $K_L a$ as function of the operating parameters: liquid circulation, gas flow and bottom pressure, however the dependency on operational conditions is under-valued. Further validation of the individual models is required to develop more realistic and robust models, and will contribute to understanding the physical phenomena taking place in a U-loop. The numerical model also provides insights into the qualitative behavior of the multiphase flow in a U-loop bioreactor.
Keywords: Gas fermentation, Static Mixer, Computational Fluid Dynamics (CFD).

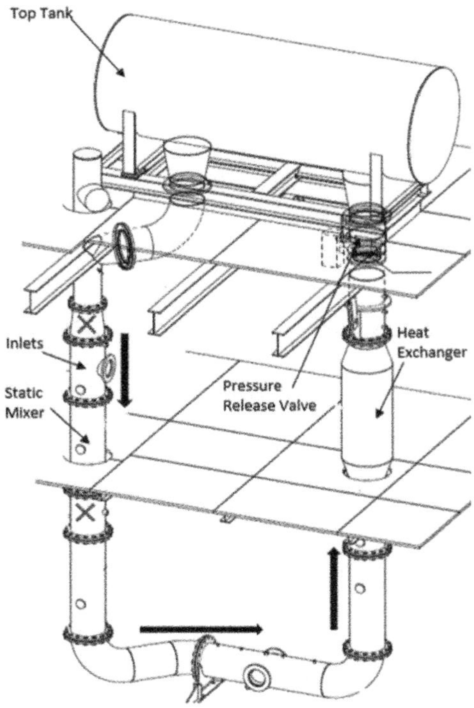

Figure 1 :Drawing of the 1725 L U-loop Bioreactor, UNIBIO A/S. ***Red cross****: location of dissolved oxygen measurements.* ***Black arrow****: Direction of the flow.*

1. Introduction

Fermentation has become an increasingly important chemical process in a wide range of applications, from pharmaceutical to bulk chemical and biofuel productions [1]. However, most fermentations are still run as batch processes and lack measurement-based control [2]. Other challenges with operating industrial scale fermentations relates to the inhomogeneous mixing conditions and mass transfer limitations, both related to the fluid dynamics of the system. This induces gradients in the fermenter, that lead to suboptimal conditions for the microorganisms and control issues, both having an impact on the process economic metrics. Thus, understanding and accurately modeling the fluid dynamic behavior of bioreactors is a key element for predicting the overall behavior of the fermentation process. The main challenge with modeling fluid behavior in large-scale equipment originates from the complex nature of turbulent flow, which limits the accuracy of deterministic models in macroscopic simulations at high Reynolds numbers. Computational fluid dynamics has become a useful tool for investigating hydrodynamic conditions in multiphase systems through modeling of the turbulence and fine discretization of the flow equations. Interesting case studies are the U-loop bioreactors currently used by UNIBIO A/S, a Danish company specialized in the production of Single Cell Proteins through continuous aerobic fermentation of *Methylococcus capsulatus* using methane gas as primary carbon source. An aerobic fermentation using a gaseous substrate, makes mass transfer between the phases of crucial importance to the process. The alternative bioreactor design, U-loop, greatly improves volumetric mass transfer, however, it is not yet fully understood which underlying phenomena are contributing to the enhanced mass transfer. This study therefore looks at developing and validate a numerical model of a U-loop bioreactor to better understand the mass transfer mechanism in such a system.

2. Materials & Methods

2.1. Description of the system

The design of the U-loop bioreactor consists of two distinguishable parts: the loop (Plug flow) section and the top tank (or degassing unit). Gases introduced at the start of the down-going leg are carried through the loop by the liquid and are finally released in the top tank. A sketch of the system is shown in Figure 1. A pump, located before the gas inlets, is in charge of circulating the liquid through the loop. The U-loop is equipped with static mixers, and a heat exchanger. The pressure in the loop can be adjusted using an expansion valve located after the heat exchanger. This study uses a production scale U-loop with an active volume of 1725 [L], a liquid height of 6 [m] and a loop pipe diameter of 0.4 [m]. The selected operating conditions are similar to the conditions during production (e.g. flowrates and pressure), to mimic industrial process conditions. To investigate the multiphase hydrodynamics, water is used instead of fermentation broth, and the gas phase only includes air and pure oxygen. This simplification is based on the assumption that the mass transfer is mainly driven by convective transport (not diffusion driven) and thus the global liquid phase mass transfer coefficient ($K_L a$) can be assumed equal for all compounds, and independent of the transfer direction. This simplification does consider the mass transfer of CO2 produced during fermentation. The model can later be coupled with biokinetic equations to predict the transfer rates and resulting concentration profiles for each individual gas.

2.2. Experimental Setup

The $K_L a$ in the U-loop is determined experimentally based on dissolved oxygen measurements. Due to the plug flow nature of the bioreactor, ideal mixing cannot be assumed for the entire system. Therefore, the component mass balance assuming ideal mixing (eq. 1), typically used to determine the $K_L a$ in stirred tanks, cannot be applied for the U-loop. However, $K_L a$ can be determined for a section of the loop based on two measurement points and the component mass balance over this loop section. Assuming steady state, and unidirectional mass transfer (i.e. from the gas phase to the liquid phase) the resulting mass balance (eq. 2) is an algebraic equation. Assuming perfect radial mixing, the dissolved oxygen measurements are used to determine the inlet and outlet oxygen concentrations, C_{in} and C_{out} respectively. The measurements are performed over a 1[m] vertical pipe section on the down-going leg of the loop. The dissolved oxygen is measured at the top and bottom of the section, and the pressure is only measured at the top. The equilibrium saturation concentration C^* of oxygen in water is generally given by Henry's Law (eq.3) and is adjusted for changing pressures.

$$\frac{dC}{dt} = K_L a\,(C^* - C) \qquad (eq.1)$$

$$0 = K_L a\,(C^* - C)V_L^{section} + F_{in}C_{in} - F_{out}C_{out} \qquad (eq.2)$$

$$C^* = HeP_{O_2} \qquad (eq.3)$$

He being the Henry Constant for oxygen in water (0.0013 $[mol\ kg^{-1}\ bar^{-1}]$ or 41.48 $[g\ m^{-3}\ bar^{-1}]$ at 25°C [3]), and $P_{O_2}$the partial pressure of oxygen (eq. 4).

$$P_{O_2} = y_{O_2}P \qquad (eq.4)$$

For this study, the ratio between air and pure oxygen is adjusted to achieve an oxygen inlet concentration of 30%. Since the O_2 liquid concentration C and saturation concentration C^*change over the pipe section, the driving force is estimated using the logarithmic mean (eq.5) of the driving force in (eq.6) and out (eq.7) of the pipe section.

$$(C^* - C)_m = \frac{(C^* - C)_{in} - (C^* - C)_{out}}{\ln\left(\frac{(C^* - C)_{in}}{(C^* - C)_{out}}\right)} \qquad (eq.5)$$

$$(C^* - C)_{in} = \left(HeP_{in}y_{O_2,in} - C_{in}\right) \qquad (eq.6)$$

$$(C^* - C)_{out} = \left(HeP_{out}y_{O_2,out} - C_{out}\right) \qquad (eq.7)$$

The inlet pressure corresponds to the measurement of the top pressure sensor, and the outlet pressure is estimated from the height difference (eq.8) (neglecting the pressure drop in the pipe). The inlet oxygen fraction in the gas phase $y_{O_2,in}$ is determined from the ratio of air and pure oxygen (eq.9) and the outlet oxygen fraction in the gas phase $y_{O_2,out}$ is determined from the mass balance (eq.10):

$$P_{out} = P_{in} + \rho_L g \Delta h \qquad (eq.8)$$

$$y_{O_2,in} = \frac{Q_{air}y_{air} + Q_{O_2}}{Q_{air} + Q_{O_2}} \qquad (eq.9)$$

$$y_{O_2,out} = \frac{Q_{air}y_{air} + Q_{O_2} - (C_{in} - C_{out})Q_{Uloop}}{Q_{air} + Q_{O_2} - (C_{in} - C_{out})Q_{Uloop}} \qquad (eq.10)$$

where Q_{Uloop} represents the circulation flow rate of the liquid phase in the loop, y_{air} the fraction of oxygen in air (21%), Q_{air} the inlet flowrate of air and $Q_{O_2}$the inlet flowrate of pure oxygen. From equations (eq. 2-10), $K_L a$ is determined based on the dissolved oxygen measurements in and out, as well as the pressure measurement at the section inlet. The latter measurement is performed at different operating conditions, by changing the liquid circulation rate, the gas inlet flow rate, and the operating pressure in the loop.

2.3. Computational Model

The aim of the computational model is to replicate the multiphase physics taking place in the bioreactor to predict the mass transfer rates at different operating conditions. The model will focus on the loop section of the bioreactor, as the aera of interest. Therefore, the fluid domain of the numerical model starts immediately after the pump, and ends immediately after the pressure reduction valve. For simplification, the heat exchanger, and the pipe bends on the horizontal part of the loop are neglected. The static mixers in the loop are modelled as a porous domain for which the model parameters are fitted to match the measurements. In this way, the pressure drop, fluid mixing and impact on the bubbly flow resulting from the static mixers can be recreated, without having to resolve the geometry of the static mixers. Modeling the exact geometry of the static mixers and the complete U-loop would require a fine discretization to resolve the flow, adding to the computational cost of the simulation. The geometry of the computational domain is shown in Figure 2. The fluid domain, as well as the temporal dimension are then discretized, to solve the model equations in a numerical way. Ideally, the discretization should not influence the results. Therefore, a discretization study is performed, where the results from different meshes and timesteps are assessed. Once the geometry has been defined, and discretized, the governing equations to be solved are selected. Based on the liquid phase Reynold's Number (eq.11) of around 4*10^5 in the pipe section, the flow regime is expected to be fully turbulent. The multiphase flow regime (e.g., Figure 3) depends on many factors (gas volume fraction, superficial gas velocity, pipe diameter and orientation, sparger type, etc. [5]) and is difficult to predict.

Figure 2: Fluid domain for the computational model

The bubble Reynolds number (eq.12) can provide some insight into the turbulence level at the scale of a single bubble. However, it requires knowing the bubble size and relative velocity.

$$Re = \frac{\rho_L u L}{\mu_L} \qquad (eq.11)$$

$$Re_{bubble} = \frac{\rho_L d_b |\overline{U_G} - \overline{U_L}|}{\mu_L} \qquad (eq.12)$$

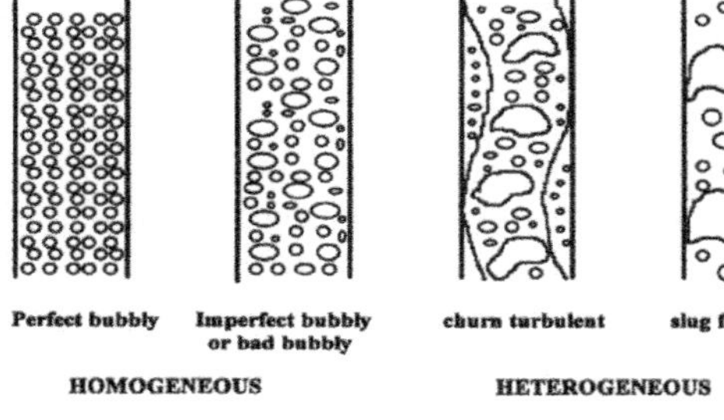

Figure 3: Gas-liquid flow regimes in a horizontal pipe (Adapted from [4])

For this study, the gas phase is assumed fully dispersed resulting in a perfect bubbly flow. The simulation parameters are shown in Table 1. The bubbles are assumed spherical, and the bubble size is modeled with a population balance model (PBM) describing the bubble coalescence and breakup. The bubble diameter is discretized into 5 bins: 0.86mm, 1.78mm, 2,7mm, 3.62mm and 4.54mm. The size fraction of each bin is determined from the population balance model. Finally, the mass transfer model is implemented. Based on the Film theory [11], the oxygen transfer rate can be expressed as shown in eq.13:

$$OTR = K_L a\,(C^* - C) \qquad (eq.13)$$

where OTR is the volumetric oxygen transfer rate $[g\,m^{-3}s^{-1}]$, C^* the saturation concentration of oxygen in water (or oxygen concentration at the surface of the bubble), and C the bulk oxygen concentration. The interfacial area a is determined from the PBM according to (eq.14) and (eq.15), d_p being the mean bubble Sauter diameter and r_β the volume fraction of air.

$$a = \frac{6\tilde{r_\beta}}{d_p} \qquad (eq.14)$$

$$\tilde{r_\beta} = \begin{cases} \max(r_\beta, r_{min}), & r_\beta < r_{max} \\ \max\left(\frac{1 - r_\beta}{1 - r_{max}} r_{max}\right), & r_\beta > r_{max} \end{cases} \qquad (eq.15)$$

Parameter	Value
Multiphase Model	Liquid phase: Continuous Gas phase: Fully dispersed + PBM (Ideal gas at 25°C)
Gravity	[0 -9.82 0] $[m/s^2]$
Reference pressure	1 [atm]
Coalescence Model	Prince and Blanch [7]
Breakup Model	Luo and Svendsen [8]
Liquid phase turbulence	Shear Stress Transport model [9]
Gas-liquid surface tension	0.072 $[N\ m^{-1}]$
Momentum Transfer Model	Schiller and Naumann [10]

Table 1: Simulation Parameters for the Numerical Study

r_{min} and r_{max} are model parameters. $\widetilde{r_\beta}$ is used as a modified volume fraction to avoid the overprediction of the interfacial area at high gas volume fractions, where the assumption of fully dispersed gas phase is no longer valid. The liquid mass transfer coefficient K_L (eq.16) is determined from the derivation of the Higbie's penetration theory for turbulent Newtonian flow [6]:

$$K_L = 0.301\left(\frac{D_\varphi}{\upsilon}\right)^{-0.5}(\varepsilon\upsilon)^{0.25} \qquad (eq.16)$$

where D_φ is the diffusion coefficient of oxygen in water, υ is the kinematic viscosity of water and ε is the turbulence eddy dissipation generated by the turbulence model.

2.4. Numerical Setup

The simulations are performed on ANSYS CFX 2019R2 on 24 hpc parallel cores, with three different conditions at the boundaries: the liquid phase inlet, the no-slip walls, and the pressure outlet. The relative pressure at the pressure outlet is adjusted to achieve the correct pressure profile in the loop (representing the pressure reduction valve). The gas phase is introduced into the system through a numerical source point with a bubble diameter of 2.7mm. All simulations are run in Transient mode for 200[s] simulation time, to ensure steady state is achieved (circulation time ≈14[s]). The fluid is initialized stationary. Due to the unsteady nature of turbulent transient simulations, the results are time averaged over the last 50 [s] simulation time.

3. Results and Discussion

3.1. Discretization Study

The simulation is performed for a range of generated meshes (from 15k to 140k mesh elements), each over a range of timesteps (0.001[s] to 0.1[s]). For each simulation, the gas holdup in the system is calculated and used to compare the results. In theory, the results should converge towards a single value as the discretization is made finer. The optimal discretization is the coarsest discretization that provides similar results as the finer discretization. The results from the discretization study are presented in Figure 4. For the simulations performed with the coarse meshes (<60k elements) the results are inconsistent and depend on the mesh and timestep. For the two finest meshes (>80k elements) the results seem to be independent on the discretization. One outlier is the simulation performed with a timestep of 0.1 [s] yielded a gas holdup above 30 %, whereas the other simulations resulted in gas holdups around 10%. Based on this analysis, the optimal discretization is 83k mesh with timestep of 0.05 [s].

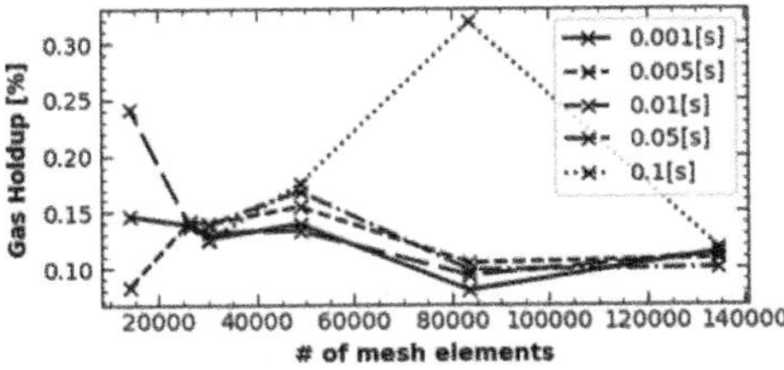

Figure 4: Gas Holdup as function of the number of mesh elements, for different

3.2. Model Validation

The measured and simulated K_La values are shown in Figure 5. The first run is used for calibration, where the length of the porous domain and r_{max} were used as fitting parameters. The numerical model is able to replicate an increase or decrease of the K_La at different operating conditions, however the dependency on operating conditions is under-valued. The discrepency between the simulated and measured K_La values can be explained by inaccuracies in the modeling, as well as inaccuracies in the experimental method. Regarding the numerical model, the choice of the models to be solved may be inadequate to provide realistic results. Here the choise of using a porous domain to simulate a static mixer may not be sufficient and requires further tuning, or resolving the exact geometry of the static mixer. The assumption of fully bubbly flow regime, which is the basis for the choice of multiphase model, may be inaccurate, which is discussed in the following section 3.3.. Lastly, the PBM may require additional bins and the bubble diameter range represented in the bins could be offset to the actual bubble size distribution in the loop. Regarding the experimental measurements, inaccuracy of the equipment (flowmeter, DO probes) due to calibration or malfunction is a potential source of error. Also, inaccurate assumptions used to derive K_La from the experimental data may contribute to the discrepency.

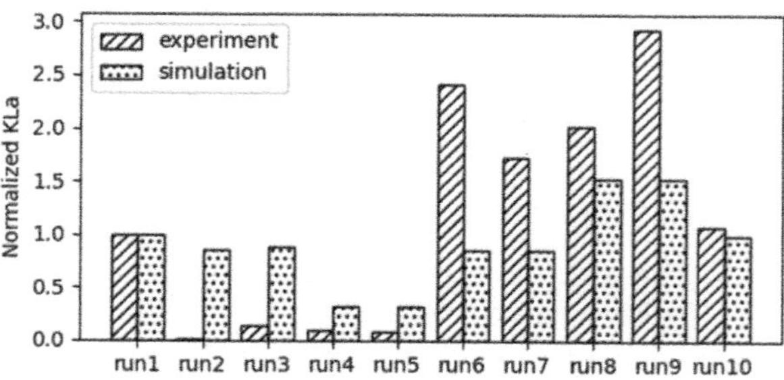

Figure 5: Measured and simulated K_La values, at different operating conditions. (Results are normalized due to confidentiality of the experimental results)

3.3. Discussion

It can be clearly noted that the gas is not distributed homogeneously in the loop. A gas fraction of 23% is reached in the down-going leg, whereas in the upgoing leg, the gas fraction is around 3%. In the horizontal section of the loop, the model predicts a large accumulation of gas in the top half of the pipe, indicating a phase split (gas volume fraction of 37%). This behavior can be expected in such a system, considering the buoyant forces, the bubbles will rise in water, with a velocity depending on the drag force [12]. This means the bubbles rise against the flow direction in the down-going leg, increasing the residence time of the bubbles, in the horizontal section the bubbles rise perpendicular to the flow direction, creating the phase split, and in the upgoing leg the bubbles rise along with the flow direction decreasing the residence time of the bubbles. This leads to an inhomogeneous distribution of K_La in the system, which is expected to be larger in the down-going leg, due to the phase split in the horizontal section that leads to a small interfacial surface area, and the low gas holdup in the upgoing leg. This can also be observed in the simulation results, where the K_La in the downgoing leg, the horizontal section and the upgoing leg is determined to 1692, 474 and 123 $[h^{-1}]$ respectively. The resulting gas volume fractions in some sections of the bioreactor indicated that the flow regime might be different from the assumed bubbly flow. The dispersed gas phase model requires that the bubbles are smaller than the mesh elements. If larger gas structures are expected, the dispersed model is not applicable, and the gas needs to be simulated as a continuous fluid where the gas-liquid interface is resolved. Here, it seems that in some regions of the bioreactor, the gas phase can be assumed fully dispersed, e.g. down- and up-going legs, whereas in other regions the interface needs to be resolved, e.g. in the horizontal loop section where a clear phase split occurs. The $\widetilde{r_\beta}$ parameter is used to correct the interfacial area, but the momentum transfer model considers only bubbles. Moreover, the validation of the multiphase model and the PBM is crucial for further studies to develop more accurate numerical models. This requires measurements of the multiphase behavior at a deeper level to directly validate each sub-model involved simulation.

4. Conclusion

The numerical model of the U-loop bioreactor is able to replicate the qualitative behavior of the measured $K_L a$, however the dependency on operational conditions is under-valued. As many sub-models are involved in the simulation of the bioreactor, it is difficult to evaluate the origin of the inaccuracy in predicting the measured $K_L a$. Therefore, further studies should include additional measurements of the multiphase flow (e.i. bubble size distribution, gas superficial velocity), to ensure a more rigorous validation of the sub-models. On the other hand, the numerical model provided insights into the qualitative behavior of the multiphase flow in a U-loop bioreactor, presenting some challenges related to inhomogeneous distribution of the mass transfer potential. The results can be used as a basis for upscaling and improving the design of the bioreactor. However, further studies including further validations are required for the model to be used for accurate prediction of the process, for process optimization and control purposes.

Reference

1. Capozzi, V., Fragasso, M., & Bimbo, F. (2021). Microbial resources, fermentation and reduction of negative externalities in food systems: Patterns toward sustainability and resilience. In *Fermentation* (Vol. 7, Issue 2).
2. Alarcon, C., & Shene, C. (2021). Fermentation 4.0, a case study on computer vision, soft sensor, connectivity, and control applied to the fermentation of a thraustochytrid. *Computers in Industry, 128.*
3. NIST Chemistry WebBook, Oxygen, Henry's Law data, Hensy's Law constant. URL: https://webbook.nist.gov/cgi/cbook.cgi?ID=C7782447&Mask=10#Solubility
4. Bouaifi, M., Hebrard, G., Bastoul, D., & Roustan, M. (2001). A comparative study of gas hold-up, bubble size, interfacial area and mass transfer coefficients in stirred gas-liquid reactors and bubble columns. *Chemical Engineering and Processing, 40*(2).
5. Hebrard, G., Bastoul, D., & Roustan, M. (1996). Influence of the gas sparger on the hydrodynamic behaviour of bubble columns. *Chemical Engineering Research and Design, 74*(3).
6. Kawase, Y., Halard, B., & Moo-Young, M. (1992). Liquid-Phase mass transfer coefficients in bioreactors. *Biotechnology and Bioengineering, 39*(11).
7. Prince, M.J. and Blanch, H.W. (1990), Bubble coalescence and break-up in air-sparged bubble columns. AIChE J., 36: 1485-1499.
8. Luo, H. and Svendsen, H.F. (1996), Theoretical model for drop and bubble breakup in turbulent dispersions. AIChE J., 42: 1225-1233.
9. Menter, F. R. (1994). Two-equation eddy-viscosity turbulence models for engineering applications. *AIAA Journal, 32*(8).
10. Schiller, L., & Naumann, Z. (1935). A drag coefficient correlation. In *Z.Ver.Deutsch.Ing* (Vol. 77, Issues 13–14).
11. Titchener-Hooker, N. (1995). Bioprocess Engineering Principles (by Pauline M. Doran). *TRENDS IN BIOTECHNOLOGY, 13*(1).
12. Kulkarni, A. A., & Joshi, J. B.(2005). Bubble formation and bubble rise velocity in gas-liquid systems: A review. In *Industrial and Engineering Chemistry Research* (Vol.44,Issue16).

Antonis Kokossis, Michael C. Georgiadis, Efstratios N. Pistikopoulos (Eds.)
PROCEEDINGS OF THE 33rd European Symposium on Computer Aided Process Engineering (ESCAPE33), June 18-21, 2023, Athens, Greece
 http://dx.doi.org/10.1016/B978-0-443-15274-0.50125-6

Effect of vaporisation models on the FCC riser modelling

Thabang W. Selalame, Raj Patel, Iqbal M. Mujtaba, Yakubu M. John

Department of Chemical Engineering, University of Bradford, Bradford, BD7 1DP, West Yorkshire, UK

Abstract

Traditional approach to FCC riser modelling usually involves the modelling of the gas oil feed vaporisation since the cracking reactions occur in the gas phase. The majority of these studies assume instantaneous vaporisation of feed which greatly simplifies the modelling effort. A limitation of this approach is that it greatly overestimates the rates of vaporisation occurring in the riser which may lead to erroneous predictions from the model. In response to this, various models have been proposed in the literature to better model the finite vaporisation rates in the riser. This paper investigates the effect of these various vaporisation models on the predictions from riser models. Specifically, the infinite conductivity and the finite conductivity modelling frameworks are investigated. We show for the first time in this paper that vaporisation models are largely insignificant to the predicted gas oil conversion profiles and gasoline yield for usual operation conditions of FCC risers. Additionally, the effect of vaporisation models on conversion and yield predictions only becomes important for riser nozzles with feed droplets exceeding 2mm in diameter. This is a key finding as it proves the validity of instantaneous vaporisation models for droplets smaller than 2mm.

Keywords: Vaporisation, FCC unit, Riser, Modelling, Simulation

1. Introduction

Process modelling and simulation have always been important tools used by the petroleum industry to improve safety, efficiency and profitability. Owing to climate change, more emphasis has been put on the accuracy of these models in order to comply with increasingly stringent environmental regulations.

The Fluid catalytic cracking (FCC) unit is one of the most important units in petroleum refinery (Fernandes, Verstraete et al. 2007). For this reason, the modelling of the FCC riser has received substantial commercial and academic attention since the introduction of the process over five decades ago. Significant progress has been made to model the complex hydrodynamics and the reactions occurring in the riser. However, an area that has not been fully explored is that of vaporisation of the feed. Several models have been proposed to study vaporisation of fuel droplets, although most of these studies focus on single droplets evaporating into an infinite and quiescent gaseous atmosphere (Finneran 2021). This is far from the conditions in the riser, where droplets vaporise in complex gas-solid-liquid convective environment. This study investigates the effect of different vaporisation models, that have been previously developed in the literature, on the FCC riser model predictions for feed conversion and product yield. Two droplet frameworks are investigated, infinite thermal conductivity (ITC) and finite thermal conductivity

(FTC) (Miller, Harstad et al. 1998). In the former, spatial temperature variation in the droplet are ignored, while the latter assumes a finite value for the liquid phase and considers the spatial temperature distribution in the droplet as it rises. A majority of FCC riser models in the literature have adopted either the ITC framework(Nayak, Joshi et al. 2005) or assumed instantaneous vaporisation of the feed at the point of injection (John, Mustafa et al. 2019). Very few papers have used FTC in their riser models. As such, the impact of FTC on the riser predictions remain unknown until now.

2. Model Description

The catalytic cracking of heavy fraction gas oil in the FCC riser is a heterogeneous gas-solid reaction system. The complex hydrodynamics of the fast fluidised circulating bed have been described previously. The feed to the riser is introduced as atomised spray droplets that first undergoes vaporisation in the initial stages of the riser before reaction can occur. Various models have been proposed to describe the vaporisation dynamics in the riser environment. In this paper, we compare the influence of these models on the conversion and yield profiles in the riser.
The vaporisation problem being investigated in this situation is that of discrete spherical mono-component droplets evaporating in a gas-droplet-solid convective environment. The fluid and far field properties such as temperature, concentration and pressure are varying with the height of the riser due to the cracking reactions. Firstly, the assumption of infinite thermal conductivity (ITC) and diffusivity for the droplet phase is adopted, so that the temperature and concentration is the same everywhere on the droplet. The evolution of the droplet temperature, mass and size with height of the riser bed is given by:

$$m_d C_{pd} \frac{dT_d}{dz} = \frac{\dot{q} - \dot{m} L_v}{v_d} \tag{1}$$

$$\frac{dm_d}{dz} = -\frac{\dot{m}}{v_d} \tag{2}$$

$$\frac{dd_d}{dz} = -\frac{2\dot{m}}{\pi \rho_d d_d^2 v_d} \tag{3}$$

Where m_d, d_d and T_d are the mass, diameter and temperature of the droplet, L_v is the latent heat of vaporisation, ρ_d is the liquid density and v_d is the velocity of the droplet. $\dot{q}$ and $\dot{m}$ are the rates of heat transfer and mass transfer at the droplet surface respectively. During the inert heating phase, $\dot{m} = 0$. Closures for $\dot{q}$ and $\dot{m}$ are provided via Nusselt and Sherwood number correlations. For convective environment (including the effect of Stefan flow)(Lupo 2017):

$$\dot{q} = \pi d_d^2\, Nu k_g \frac{T_d - T_\infty}{d_d} \tag{4}$$

$$\dot{m} = \pi d_d^2\, Sh \bar{\rho} D_{vap} \frac{Y_s - Y_\infty}{d_d} \tag{5}$$

$$Nu = \frac{Nu_0 \ln(1 + B_T)}{Le B_T} \tag{6}$$

$$Sh = \frac{Sh_0 \ln(1 + B_M)}{B_M (1 - Y_s)} \tag{7}$$

The subscript s and g represent the conditions at the droplet surface and the surrounding gas respectively. The heat and mass transfer is fully described by the dimensionless numbers: $B_M = (Y_s - Y_\infty)/(1 - Y_s)$, $B_T = (1 + B_M)^{\frac{1}{Le}} - 1$, $Le = \frac{k_g}{\bar{\rho} D_{vap} C_{pg}}$, $Nu = \frac{h d_d}{k_g}$, $Sh = \frac{k_c d_d}{D_{vap}}$. The boundary conditions are such that the far field properties(i.e Y_∞, T_∞) are evaluated in the ambient hydrocarbon

gas mixture and Y_s and T_d are at the surface of the droplet. The vapour properties are evaluated at a reference temperature and composition using the 1/3 rule (Csemány and Józsa 2021).
Subsequently, the gas and mixture properties are evaluated using the following mixing rules considering the vapour and ambient gas mixture:

$$C_{pg} = Y_{ref}C_{pv} + (1 - Y_{ref})C_{pm} \tag{8}$$

$$k_g = Y_{ref}k_v + (1 - Y_{ref})k_m \tag{9}$$

$$\mu_g = Y_{ref}\mu_v + (1 - Y_{ref})\mu_m \tag{10}$$

$$\rho_g = \left(\frac{Y_{ref}}{\rho_v} + \frac{1 - Y_{ref}}{\rho_m}\right)^{-1} \tag{11}$$

The subscripts "v" and "m" represent the vapour and ambient gas mixture respectively. C_p, k, μ and ρ properties are evaluated at T_{ref} and Y_{ref} for the vapour and evaluated at T_∞ and Y_∞ for the ambient gas mixture. The mixing laws in Equations (8)-(11) are based on ideal gas conditions (Csemány and Józsa 2021).
The subscript "0" in Equations (6) and (7) indicates that the correlations for the these transport numbers are evaluated in the limit of $B_M = B_T = 0$ (i.e negligible evaporation). The classical models for Nu_0 and Sh_0 for moderate Reynolds numbers are those provided by Ranz (1952). In this work, we denote this the Classsic model. The complexity with heat and mass transfer in the riser environment is that the flow field also consist of catalyst particle which enhance the heat transfer process. Several correlations have been suggested to account for this enhance heat transfer. In this work the models by Buchanan (1994) and Nayak consider, Joshi et al. (2005) are considered. A summary of these correlations are presented in Table 1.

Table 1: Nusselt number correlations for heat transfer to the droplets in the limit of negligible evaporation.

Model name	Reference	Model
Classic	Ranz (1952)	$Nu_0 = 2 + 0.552Re_d^{0.5}Pr^{0.33}$
Buchanan B*	Buchanan (1994)	$Nu_0 = \dfrac{\left(2 + 0.6Re_p^{\frac{1}{2}}\mathrm{Pr}^{\frac{1}{3}}\right)}{\left(1 + \frac{C_{pg}(T_g - T_d)}{L_v}\right)^n}$
Nayak**	Nayak, Joshi et al. (2005)	$Nu_0 = Nu_{0,classic} + \dfrac{\emptyset\varepsilon_s L_v Pr}{4C_p(T - T_d)}\left(Re_d - Re_p\dfrac{d_d}{d_p}\right)\left(1 - \dfrac{d_p}{d_d}\right)^2$

$Re_d = \frac{\rho_g|v_g - v_d|d_d}{\mu_g}$, $Re_p = \frac{\rho_s\varepsilon_s|v_g - v_d|d_d}{\mu_g}$, $Pr = \frac{c_{pg}\mu_g}{k_g}$,
Heat and mass transfer analogy is used to find Sh_0 from Nu_0. *During heat up $n = 0$ and $n = 0.7$ during boiling. **The Nayak is the classic model with an extra term that represents the effect of catalyst particles. $\emptyset$ is an adjustable parameter of the model.

3. Results

3.1. Droplet size and temperature

Figure 1 shows the evolution of droplet temperature and size along the height of the riser from the point of feed injection. The Buchanan A model is the limit of infinitely fast heat transfer from catalyst particles to the droplets (Buchanan 1994).

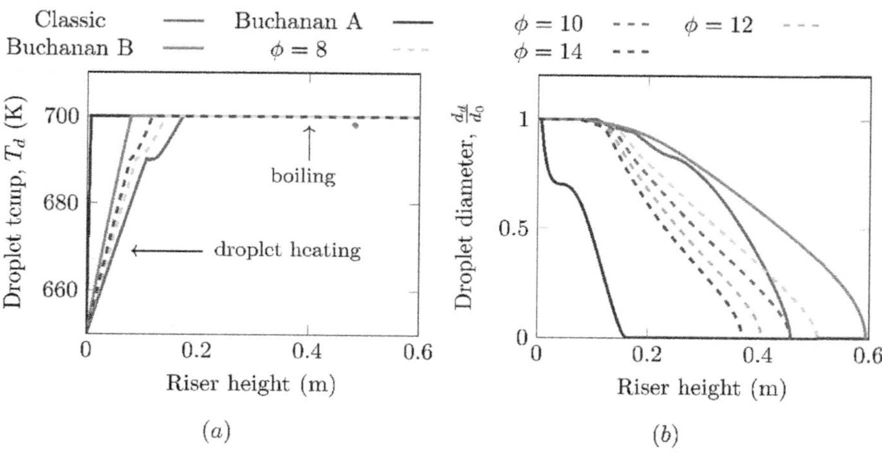

Figure 1:Simulation results for the steady state FCC riser (inputs: $\rho_d = 810$, $\varepsilon_{d0} = 0.012$, $d_d = 500\mu m$) comparing between different vaporisation models. (a) Evolution of droplet temperature from initial feed temperature to boiling, and (b) evolution of droplet size up the riser. Ø is the adjustable parameter in the Nayak heterogeneous model. The Buchanan models feature only inert heating and boiling, whereas Nayak and classic models have vaporisation beginning at $T_{vap} = 690K$.

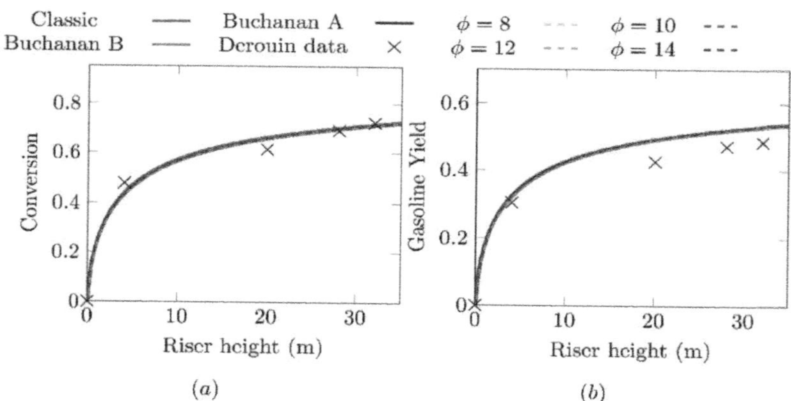

Figure 2: (a) Predictions of gas oil feed conversion, and (b) gasoline yiled along the riser for the various feed vaporisation models. (inputs: $\rho_d = 810$, $\varepsilon_{d0} = 0.012$, $d_d = 500\mu m$). Simulation results are compared against the plant data reported by Derouin, Nevicato et al. (1997).

Figure 1(a) shows that the Buchanan models have higher heat transfer rates in the heat up stages and therefore droplets reach boiling temperature faster. However, the factor $\left(1 + \frac{c_{pg}(T_g - T_d)}{L_v}\right)^{-0.7}$ in the Nusselt number for the Buchanan B model, which accounts for Leidenfrost effects, greatly reduces the heat transfer during boiling and hence results in longer droplet penetration as shown in Figure 1(b). It is shown in Figure 1(b) that the different vaporisation models predict different droplet penetration at the bottom of the riser. Additionally, all the models predict complete vaporisation within 0.6m from feed injection point, which is < 2% of the total height of the riser. However, Figure 2 shows that these differences do not translate to differences in conversion and yield profiles along the riser. This is a significant finding because it proves that for the ITC vaporisation framework; (a) hydrocarbon cracking reactions in the gas phase are the rate limiting step in the cracking process, and (b) vaporisation models are largely insignicant to the predictions of conversion and yield in the FCC riser, which gives validity to the commonly used assumption of instantaneous vaporisation.

3.2. Effect of modelling framework

In Figure 3, simulation for the ITC and the FTC modelling frameworks are compared. Figure 3(b) shows that all the FTC models predict slower vaporisation rates compared to their corresponding ITC counterparts in Figure 1(b). This is due to the spartial temperature variation in the droplet and heat transfer resistence caused by the finite thermal conductivity of the liquid phase. This resistence causes T_d to take longer to reach T_{bp}.

Beyond the initial vaporisation of the droplet phase (i.e in the range $690 \leq T_d \leq 700$), a complex coupling of gas-liquid-solid phases results, which determines the evolution of the gas oil conversion in Figure 3(c). In this temperature range, evaporation is slower since the heat transfer to the droplet is also used for droplet heat up. The added heat transfer resistence in the liquid leads to more energy being used up the droplet phase which leads to lower ambient (surrounding gas and catalyst) temperature and hence lower conversion in the FTC models. We find that exit gas temperatures of the Classic model between ITC and FTC differ by $\sim 15K$. An interesting find, is that the FTC models underpredict the conversion of the feed. This is a result of the aforementioned lower ambient temperatures in the FTC models. A fix for this, which has not been shown in this current work, is to use the effective thermal conductivity (ETC) modelling framework, which was developed to correct the liquid thermal conductivity of droplets vaporising in convective environments. In such systems, shear at the surfaces of the droplets causes internal circulations which enhaces the heat transfer in the droplet. A common problem in vaporisation modelling for the FCC riser is the difficult in validating the models since conditions are such that experimental determination of vaporisation rates is difficult.

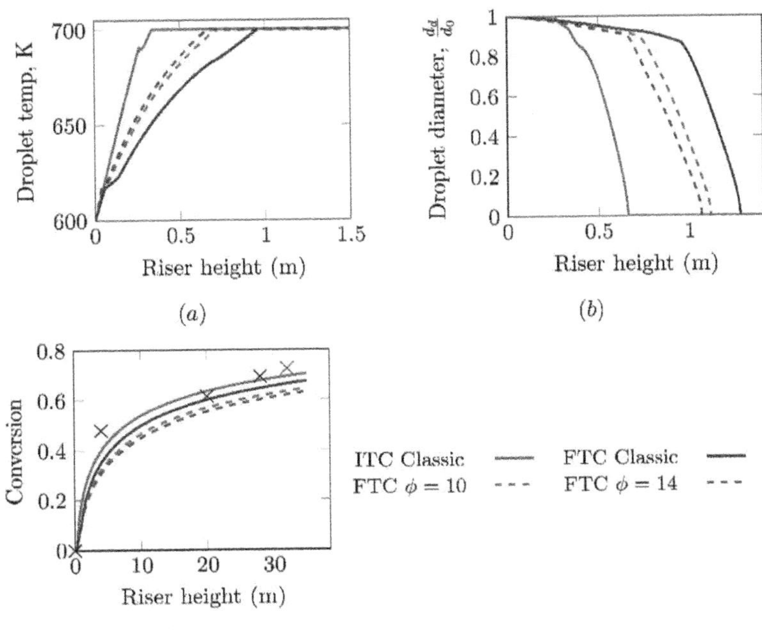

Figure 3:(a) Droplet temperature, T_d, (b) dimensionless droplet diameter, $\frac{d_d}{d_0}$, and (c) gas oil conversion for different models. All models use the same input values: ($d_0 = 500\mu m$, $T_{d0} = 600$, $T_{vap} = 690$, $T_{bp} = 700$, $\rho_d = 810$, $\varepsilon_{d0} = 0.012$. For the FTC model framework, initial surface droplet temperature, $T_{ds0} = 645$).

Using the conversion profile data, it is shown in Figure 3(c) that the ITC modelling framework best suits the data. It is also found in this work that this remains true for droplets $< 2mm$, beyond which the vaporisation starts slowing the cracking reactions.

4. Conclusions

The ITC and FTC vaporisation modelling frameworks were investigated in the context of the FCC riser modelling and simulation. Both modelling frameworks are able to predict the penetration of the droplet feed up the riser from the point of feed injection. It is revealed that the ITC framework predicts faster vaporisation compared to the FTC, and also shown that the FTC over-predicts the energy transferred to the droplets over their lifetime and therefore leading to under-prediction of the conversion of feed. For most application, the available conversion profile data suggests that the ITC framework is recommended. Importantly within the ITC framework, the various homogeneous and heterogeneous Nusselt and Sherwood correlations do not affect the predicted yield and conversion profiles (which is only valid for $d_{d0} < 2mm$). Beyond this droplet size range, the heterogeneous models should be used for better predictions of conversion.

5. References

Buchanan, J. S. (1994). "Analysis of Heating and Vaporization of Feed Droplets in Fluidized Catalytic Cracking Risers." Industrial & engineering chemistry research **33**(12): 3104-3111.

Csemány, D. and V. Józsa (2021). "A Two-Parameter Corresponding States Method for Calculating the Steady-State Evaporation Rate of C2–C9 n-Alkane Droplets in Air for Elevated Pressures and Temperatures." Flow, turbulence and combustion **107**(2): 283-305.

Derouin, C., D. Nevicato, M. Forissier, G. Wild and J.-R. Bernard (1997). "Hydrodynamics of Riser Units and Their Impact on FCC Operation." Industrial & engineering chemistry research **36**(11): 4504-4515.

Fernandes, J. L., J. J. Verstraete, C. I. C. Pinheiro, N. M. C. Oliveira and F. R. Ribeiro (2007). "Dynamic modelling of an industrial R2R FCC unit." Chemical Engineering Science **62**(4): 1184-1198.

Finneran, J. (2021). "On the evaluation of transport properties for droplet evaporation problems." International journal of heat and mass transfer **181**: 121858.

John, Y. M., M. A. Mustafa, R. Patel and I. M. Mujtaba (2019). "Parameter estimation of a six-lump kinetic model of an industrial fluid catalytic cracking unit." Fuel (Guildford) **235**: 1436-1454.

Lupo, G. (2017). "Detailed simulations of droplet evaporation."

Miller, R. S., K. Harstad and J. Bellan (1998). "Evaluation of equilibrium and non-equilibrium evaporation models for many-droplet gas-liquid flow simulations." International journal of multiphase flow **24**(6): 1025-1055.

Nayak, S. V., S. L. Joshi and V. V. Ranade (2005). "Modeling of vaporization and cracking of liquid oil injected in a gas–solid riser." Chemical engineering science **60**(22): 6049-6066.

Ranz, W. E. (1952). "Evaporation from drops, Parts I & II." Chem Eng Prog. **48**: 141-146.

Antonis Kokossis, Michael C. Georgiadis, Efstratios N. Pistikopoulos (Eds.)
PROCEEDINGS OF THE 33rd European Symposium on Computer Aided Process Engineering (ESCAPE33), June 18-21, 2023, Athens, Greece
 http://dx.doi.org/10.1016/B978-0-443-15274-0.50126-8

A joint model-based design of experiments approach for the identification of Kriging models in geological exploration

Philipp Deussen[a], Federico Galvanin[a,*]

[a]*Department of Chemical Engineering, University College London, Torrington Place, London WC1E 7JE, United Kingdom*

f.galvanin@ucl.ac.uk

Abstract

When exploring prospective mining locations, a central task is modelling rock attributes in the subsurface. The drilling needed to sample these locations is costly, so efficient sampling and reliable interpolation methods are needed. Kriging models (Gaussian Processes) are thus used, with the kernel and its parameters determined from data analysis of preliminary samples and expert judgement. New samples iteratively update the model, targeting exploitation (high ore grades) and exploration (minimising prediction variance).

The problem arises if the chosen kernel is incorrect or high uncertainty affects parameters. This paper thus suggests a joint model-based design of experiments (j-MBDoE) approach to target two objectives: maximising the distinguishability of candidate model predictions and reducing model uncertainty from parameter variance. Three different kernels in an Ordinary Kriging GP were used as candidate models. In-silico data was generated using one kernel and the optimal design strategy iteratively determined sampling locations to maximise model distinguishability with a constraint to ensure improved parameter estimates. Two models could be distinguished and the data approximated well with a limited number of drilling experiments while satisfactorily estimating kernel parameters.

Keywords: multiobjective optimization, joint model-based design of experiments, geostatistics, Gaussian processes, Kriging

1. Introduction

Geological exploration for metals is gaining relevance as industry seeks to secure supply chains for metals which are critical for industrial processes such as battery production. However, in the past decade, recovery rates and new discoveries have been low. The major bottleneck in mining value-dense metals (e.g. copper or nickel), is the identification and estimation of subsurface deposits. For resource and reserve models, Kriging models (Gaussian Processes) interpolate between sample data, which is obtained by drilling into the ground and taking sample cylinders of rock for visual and chemical analysis. Around targets of interest, closely spaced grids are drilled (infill drilling) for better estimates. Thus, optimal sampling can increase efficiency. Studies and industry practice show better performance of quantitative over qualitative design criteria (Jafrasteh and Suarez, 2020).

Model-based design of experiments (MBDoE) aims to design experiments for objectives such as obtaining the most informative data for parameter estimation (Franceschini and

Macchietto, 2008) or discrimination between candidate models (Schwaab et al., 2006). This paper proposes a joint experiment design approach (Galvanin et al., 2016) to locate samples to discriminate between three different candidate models for subsurface mineral concentrations while at the same time improving the precision of parameter estimates.

A metal concentration profile is generated in-silico by a function describing the spatial correlation of likely concentration grades and a preliminary model is fit based on ten initial samples. From there, the goal is to identify the valid model and its parameters more efficiently than methods such as closely spaced infill drilling. As a metric of success, model fit and parameter estimation statistics for candidate models may be compared.

2. Methodology

2.1. Fitting models using Gaussian Processes

A Gaussian Process (GP) interpolates between samples (Sahimi, 2011; Rossi & Deutsch, 2014). Its kernel is a correlation function $R(h)$, which relates the semivariance $\gamma(h)$ of likely unsampled concentrations at distance h and the overall concentration variance $\sigma_Z{}^2$.

$$\gamma(h) = \sigma_Z^2 - R(h) \tag{1}$$

Ordinary Kriging (OK) is a form of GP (Sahimi, 2011) giving the best unbiased linear estimate of the mean expected concentration at unsampled locations Z using the estimator $\hat{Z}$, and the variance of estimates (σ_{OK}^2) based on the expected square error to the mean.

$$\sigma_{OK}^2 = E[(Z - \hat{Z})^2] \tag{2}$$

The estimator is based on the samples Z_i and their relative importance weights w_i.

$$\hat{Z} = \sum_{i=1}^{N} w_i Z_i \tag{3}$$

The optimal weights w_i can be found by substituting (3) into (2):

$$\sigma_{OK}^2 = \sigma_Z^2 - 2\sum_{i=1}^{N} w_i R(Z, Z_i) + \sum_{i=1}^{N}\sum_{j=1}^{N} w_i w_j R(Z_i, Z_j) \tag{4}$$

Introducing constraint (5) ensures the unbiasedness of the calculated predictions, as they no longer depend on a previous estimate of population variance. Substituting (5) into (4):

$$\sum_{i=1}^{N} w_i = 1 \tag{5}$$

$$\sigma_{OK}^2 = 2\sum_{i=1}^{N} w_i\, \gamma(Z, Z_i) - \sum_{i=1}^{N}\sum_{j=1}^{N} w_i\, w_j\, \gamma(Z_i, Z_j) \tag{6}$$

Then, minimising the Kriging variance: $\partial\, \sigma_{BK}^2 \,/\, \partial\, w_i = 0$, results in:

$$\sum_{i=1}^{N} R(Z_i, Z_j)\, w_i = R(Z, Z_j) \tag{7}$$

This gives a solvable linear system of equations for all unsampled points. It comprises a matrix **W,** for weights of samples at unsampled locations (n_s x n_u), **P,** for correlations between these points (n_s x n_u) and **A,** for correlations between sampled points (n_s x n_s).

$$\mathbf{W} = \mathbf{A}^{-1}\mathbf{P} \tag{8}$$

Finally, the predictions from (3) and (6) are multiplied with **W** for all unsampled points.

2.2. Model-based design of experiments

The optimal experiment design comprises the objectives parameter estimation and model discrimination. Schwaab et al. (2006) propose a discrimination criterion, Ψ^{MD} (9), based on the difference in responses $\hat{Z}$, their uncertainty σ and the probabilities P of models m

and n. New samples reveal information in a Bayesian sense, since, once sampled, the GP must reflect sample values and the variance at the sampled point collapses to zero. So, instead of estimating the probabilities, P_m (11-12), using the χ^2 value, the total prediction variance, V (10) is used. For each coordinate (x,y) and set of kernel parameters ϑ:

$$\Psi^{MD}(x,y,\vartheta) = \sum_{m=1}^{M-1} \sum_{n=m+1}^{M} (P_m, P_n)\left[\hat{\mathbf{Z}}_m(x,y,\vartheta) - \hat{\mathbf{Z}}_n(x,y,\vartheta)\right]^{\mathrm{T}} \mathbf{V}_{m,n}^{-1}(x,y,\vartheta)\left[\hat{\mathbf{Z}}_m(x,y,\vartheta) - \hat{\mathbf{Z}}_n(x,y,\vartheta)\right] \quad (9)$$

$$V_{m,n}(x,y,\vartheta) = 2V(x,y) + V_m(x,y,\vartheta) + V_n(x,y,\vartheta) \quad (10)$$

$$\phi_m(\vartheta) = 1 \Big/ \sum_{x=1}^{X} \sum_{y=1}^{Y} \sum_{k=1}^{K} V_m(x,y,k,\vartheta) \quad (11)$$

$$P_m(\vartheta) = \frac{\phi_m(\vartheta)}{\sum_{m=1}^{M} \phi_m(\vartheta)} \quad (12)$$

To quantify parametric uncertainty (Franceschini and Macchietto, 2008), the kernel parameters $\boldsymbol{\vartheta}$ are perturbed from their initial estimates and the impact on the measured responses (Kriged concentration values), $\mathbf{Z}$, is stored in the sensitivity matrix $\mathbf{Q}$ (13).

$$\mathbf{Q}_{x,y}(\mathbf{Z},\boldsymbol{\vartheta}) = \begin{bmatrix} \frac{\partial Z_1}{\partial \vartheta_1} \\ \vdots \\ \frac{\partial Z_1}{\partial \vartheta_n} \end{bmatrix} \approx \begin{bmatrix} \frac{\hat{Z}_1 - \hat{Z}_1{}'}{\vartheta_1 - \vartheta_1{}'} \\ \vdots \\ \frac{\hat{Z}_1 - \hat{Z}_1{}'}{\vartheta_n - \vartheta_n{}'} \end{bmatrix} \quad (13)$$

However, in a GP, the information at a point depends on both the predicted mean and variance. A point with high variance is less informative than a point with low variance Thus, analogous to the local uncertainty in Ψ^{MD}, an uncertainty (14) (standard deviation) for each model response and parameter perturbation is defined and stored in $\mathbf{Q}^{\mathrm{Adj}}$ (15).

$$\sigma_{\mathrm{total}\,\frac{\partial \hat{Z}_n}{\partial \vartheta_n}}(x,y) = 2\,\sigma_{\exp}(x,y) + \sigma_{\hat{Z}_n,\vartheta_n}(x,y) + \sigma_{\hat{Z}_n{}',\vartheta_n{}'}(x,y) \quad (14)$$

$$\mathbf{Q}_{x,y}^{\mathrm{Adj.}}(\mathbf{y},\boldsymbol{\vartheta}) = \begin{bmatrix} \frac{1}{\sigma_{\mathrm{total},1}} \frac{\partial y_1}{\partial \vartheta_1} \\ \vdots \\ \frac{1}{\sigma_{\mathrm{total},n}} \frac{\partial y_1}{\partial \vartheta_n} \end{bmatrix} \approx \begin{bmatrix} \frac{1}{\left(2\sigma_{\exp} + \sigma_{y_1,\vartheta_1} + \sigma_{\hat{y}_1,\hat{\vartheta}_1}\right)} \left(\frac{y_1 - \hat{y}_1}{\vartheta_1 - \hat{\vartheta}_1}\right) \\ \vdots \\ \frac{1}{\left(2\sigma_{\exp} + \sigma_{y_n,\vartheta_n} + \sigma_{\hat{y}_n,\hat{\vartheta}_n}\right)} \left(\frac{y_1 - \hat{y}_1}{\vartheta_n - \hat{\vartheta}_n}\right) \end{bmatrix} \quad (15)$$

The Fisher information, $\mathbf{H}_{x,y}(\boldsymbol{\vartheta})$ (16), is the local information about parameters $\boldsymbol{\vartheta}$. Its trace gives a scalar criterion known as A-optimality (Franceschini and Macchietto, 2008).

$$\mathbf{H}_{x,y}(\boldsymbol{\vartheta}) = \mathbf{Q}_{x,y}^{\mathrm{Adj.}} \mathbf{Q}_{x,y}^{\mathrm{Adj.\,T}} \quad (16)$$

The joint design criterion, ϕ^{ID} (17), by Galvanin and co-workers (2016) combines Ψ^{MD} (9) and an A-optimal parameter precision criterion, Ψ^{PE} (18).

$$\phi^{ID} = arg\max\{\Psi^{MD}\}_{\phi \in D} \quad (17)$$

s.t.

$$\Psi^{PE} = \sum_{j=1}^{N_M} \left\|\mathbf{H}_j\right\| / N_M \leq \varepsilon \quad (18)$$

$$\varepsilon_{min} = \max(\Psi^{PE}) \quad \text{s.t} \quad \max(\Psi^{MD})$$
$$\varepsilon_{max} = \max(\Psi^{PE}) \quad \text{s.t} \quad \max(\Psi^{PE})$$

That is, ε_{min} ensures that the design criterion addresses model discrimination, while ε_{max} allows solutions that fully tilt towards parameter estimation. A slack variable, s_0, is added to avoid weakly efficient solutions (Mavrotas, 2009). The value of ε can be moved between the two extremes in order to explore the Pareto optimal set of suggested samples.

3. Case Study

3.1. Definition of candidate models

The candidate models are defined through the correlation function used in the kernel. In these kernels, the parameter s is the partial sill of semivariogram, which is the maximum semivariance (or minimum correlation) the probabilities of mineral concentration at two points in space will reach, minus the nugget effect. The nugget effect, n, is the difference in measured value from one point to a point just adjacent to it (on a smaller scale than the model can capture), r is the range (characteristic length at which the maximum variance between the probabilities of mineral concentration at two points in space is reached) and a is a numerical parameter. With the bounds $\delta_{(i,j)}(h)$ represented in a Boolean manner:

$$\delta_{(i,j)}(h) = \begin{cases} 1 \text{ if } h \in (i,j) \\ 0 \text{ otherwise} \end{cases}$$

Table 1 – Kernels used in formulation of candidate models

Type of Kernel (Source)	Kernel Expression
M1: Gaussian (Rossi & Deutsch, 2014)	$\gamma(h) = \left\{(s)\left(1 - e^{-3\frac{h^2}{r^2}}\right)\right\} \delta_{(0,r)}(h) + s\,\delta_{(r,\infty)}(h) + n\,\delta_{(0,\infty)}(h)$
M2: Exponential (Rossi & Deutsch, 2014)	$\gamma(h) = \left\{(s)\left(1 - e^{-3\frac{h}{r}}\right)\right\} \delta_{(0,r)}(h) + s\,\delta_{(r,\infty)}(h) + n\,\delta_{(0,\infty)}(h)$
M3: Spherical (Sahimi, 2011)	$\gamma(h) = \left\{(s)\left(\frac{3h}{2r} - \frac{1}{2}\left(\frac{h}{r}\right)^3\right)\right\} \delta_{(0,r)}(h) + s\,\delta_{(r,\infty)}(h) + n\,\delta_{(0,\infty)}(h)$

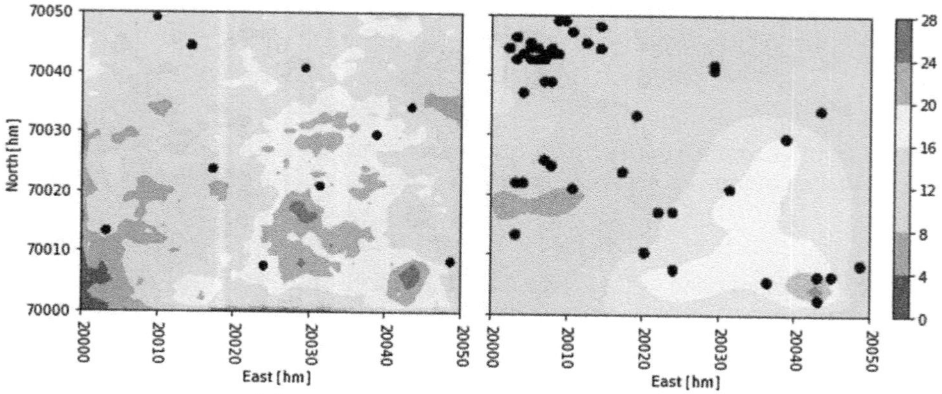

Figure 1 – Metal concentration (g/t) ground truth with initial samples marked in black (a) and (b) final model prediction of metal concentration with all samples marked in black.

3.2. Definition of Ground Truth and Design Space

The ground truth (Figure 1) was generated using the spherical kernel (M3 in Table 1) with a range $r = 40$ hm, sill of $s = 40$ $(g/t)^2$, and nugget effect of $n = 1$ $(g/t)^2$ over an area of 51E and 51N (5.1x5.1 km). The design variables are the coordinates of the new sample points (x,y). Ten preliminary samples were taken using Latin Hypercube Sampling.

3.3. Experimental Design

Fifteen optimal experiment designs were carried out based on the objectives in (17-18), adding 38 new sampling locations to the 10 preliminary samples. In each iteration, up to three samples were selected from the Pareto front (containing the optimal trade-offs between both objectives). One point was selected at either extreme (ε_{min} and ε_{max}) and a third in the middle, unless the maxima for both objectives coincided or were so close that the grid resolution permitted selecting only one location.

3.4. Model Identification Results

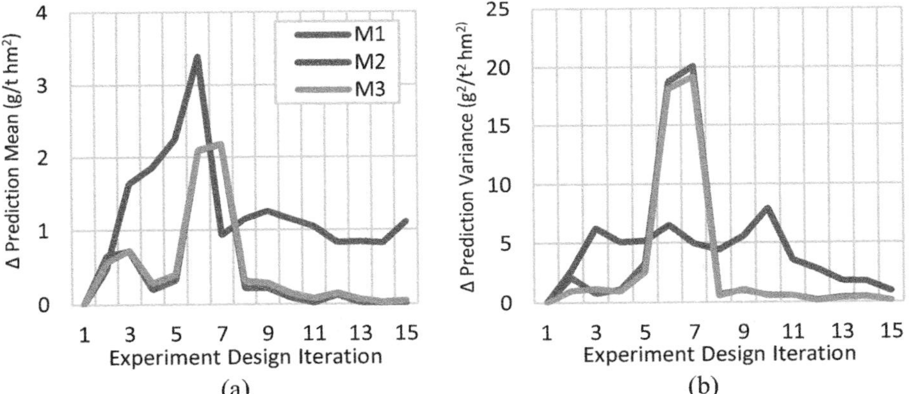

Figure 2 – Change in predicted mean (a) and variance (b) per hectare after 15 designs.

Each design iteration (1-15 in Figure 2) adds new samples that update **W** and with it the predicted mean and variance. The change in their magnitude shows how similar model predictions and new samples are, i.e. a model's predictive ability. As shown in Figure 2, M1 and M3 predict most new samples well and respond to them indistinguishably, because both kernels produce correlation structures that have similar shapes (and reflect the ground truth). The exception is a spike that results from suddenly improved parameter estimates (shown in Figure 3) that allow the model to capture a previously unknown characteristic of the data. M2 responds differently and shows constant large fluctuations, indicating a failure to represent the ground truth regardless of the design iteration.

The variance of parameter estimates is steadily reduced by new samples and starts converging after 10 design iterations as shown in Figure 3. Although the variances on the estimates for M1 are lower, they converge towards biased values. For M2, the variances are considerably higher, but they converge to the true value for the range and nugget effect. The estimates of the sill parameter, which is related to the maximum variability in the sample population, is systematically underestimated in all models. From Figure 1 (a) and (b) it becomes apparent that this is due to no samples hitting the small areas of tail concentration values. This is a known problem with Kriging interpolation, since tail values are likely to be less represented in the sample set (Rossi & Deutsch, 2014).

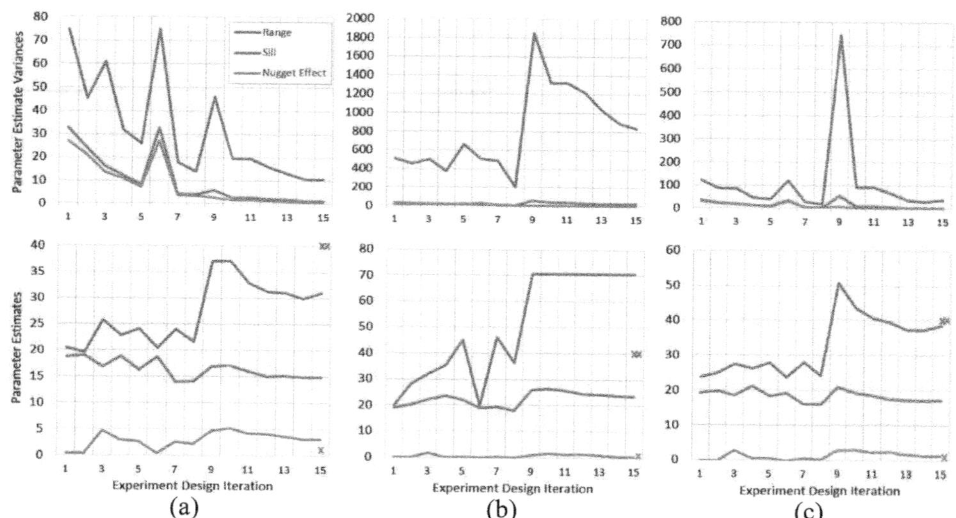

Figure 3 – Parameter estimates: variance (upper plots) and mean (lower plots) for M1 (a), M2 (b), M3 (c) following 15 sample designs. True parameter values are marked by the x.

4. Conclusions

A multiobjective MBDoE approach was applied to 3 Kriging models of an in-silico metal deposit, generated using one of the 3 models. The behaviour of model predictions vs. new samples allowed the exclusion of model M2 that did not correspond to the ground truth. Two models (M1 & M3) had similar predictions and were thus indistinguishable. Parameter estimates steadily improved with design iterations. Only the sill (variance-related) was underestimated in all models, as the tail ends of the concentration spectrum were insufficiently sampled, which is a common problem with Kriging models in geological modelling. The parameter estimates converged to true values in M3 but to biased values in M1. The approach succeeded at improving both design objectives, identifying two models that could have generated the data. Future work will aim to improve the estimation of the sill, e.g. by aiming for high-grade samples, and improve the discrimination capability for similar model outputs, e.g. through explorative designs.

References

Franceschini, G. and Macchietto, S., 2008, Model-based design of experiments for parameter precision: State of the art. Chem Eng Sc, 63, 4846-4872.

Galvanin, F., Enhong, C., Al-Rifai, N., Gavriilidis, A. and Dua, V. (2016) A joint model-based design approach for the identification of kinetic models in continuous flow laboratory systems. Comp Chem Eng, 95, 202-215.

Jafrasteh, B. and Suarez, A. (2021) Objective functions from Bayesian optimization to locate additional drillholes. *Computers and Geosciences*, 147.

Rossi, M. and Deutsch, C. (2014) Mineral Resource Estimation. 1st Ed., Springer.

Mavrotas, G. (2009) Effective implementation of the ε-constraint method in multi-objective mathematical programming problems. *Appl. Mathematics and Computation*, 213, pp 455-465.

Sahimi, M. (2011) Flow and Transport in Porous Media. 2nd Ed., Wiley-VCH Verlag.

Schwaab, M., Silva, F.M., Queipo, C.A., Barreto, Jr., A.G., Pinto, J.C. (2006) A new approach for sequential experimental design for model discrimination. *Chem.Eng.Sci.*, 61, 5791-5806.

Antonis Kokossis, Michael C. Georgiadis, Efstratios N. Pistikopoulos (Eds.)
PROCEEDINGS OF THE 33rd European Symposium on Computer Aided Process Engineering (ESCAPE33), June 18-21, 2023, Athens, Greece
 http://dx.doi.org/10.1016/B978-0-443-15274-0.50127-X

Reinforcement Learning for inventory management in multi-echelon supply chains

Guoquan Wu,[c] Miguel Ángel de Carvalho Servia,[b] Max Mowbray [a]

[a] *Department of Chemical Engineering, University of Manchester, Oxford Road, Manchester, M1 3AL, UK.*
[b] *Sargent Centre for Process Systems Engineering, Department of Chemical Engineering, Imperial College London, London, SW7 2AZ, United Kingdom*
[c] *Department of Chemical and Biomolecular Engineering, National University of Singapore, 117585, Singapore*

Abstract

Reinforcement Learning (RL) describes a set of model-free learning rules that enable identification of an approximately optimal feedback control policy function approximation for discrete, time uncertain process systems purely from data. This has led to interest in application of RL to supply chain management problems. However, many RL methods are dependent on estimating noisy first-order directions for policy improvement, and have been developed for video games and computer science applications. Here, we propose a deep RL method tailored for supply chain management. The algorithm deploys a derivative free approach to balance exploration and exploitation of a neural policy's parameter space, providing means to avoid low quality local optima. Additionally, the policy inherits ease in posing risk-sensitive formulations to learn a policy that respects constraints on the conditional value-at-risk. The performance of our algorithm is tested on a benchmark multi-echelon supply chain inventory management problem. The results demonstrate empirical improvements in performance over the first-order RL method, proximal policy optimization, and competitive performance with mathematical programming. Additionally, the risk-sensitive decisions generated can effectively handle low probability, high severity scenarios.

Keywords: Reinforcement Learning, multi-echelon supply chain management, inventory optimization,

1. Introduction

In recent years, many researchers have proposed methods to incorporate data-driven approaches into strategic and tactical supply chain, and scheduling, primarily for the associated benefits in accounting for uncertainty in decision-making. For example, in the context of long-term planning, Gao *et al.* (2019) proposed a two-stage distributionally robust optimization model, which takes advantage of industrial data to address the optimal design and operations of shale gas supply chains under uncertainty and distributional ambiguity. Distributionally robust optimization has also been exploited for tactical healthcare supply chain management as highlighted in Ash *et al.* (2022). The authors note how the hyperparameters of this approach (i.e. the size of the ambiguity set) could be used to bias the algorithm towards average cost performance or worst-case cost performance. This is a key observation, which underpins the flexibility of distributionally robust approaches. Others works have relaxed the assumption of distributional ambiguity and pursued stochastic optimization approaches. For example, Golari *et al.* (2017), explored the development of a multi-stage stochastic programming model for inventory control under uncertainty in a renewable energy supply. Robust and multi-parametric optimization has also been investigated (Charitopoulos *et al.*, 2019).

In the context of tactical supply chain management, traditional operations research methods are still posed with practical challenges. In these problems, we are interested in identifying recourse decisions as the system's uncertainty is realized. The challenges primarily relate to accounting for exogenous and endogenous uncertainty in closed loop decisions and the tractability of solving the associated mathematical program recursively. These challenges are compounded by complex network relationships present within the supply chain, and the size of the decision problem tends to grow combinatorially with the supply chain size.

More recently, research focus has been directed towards the development of Reinforcement Learning frameworks to overcome some of the challenges mentioned. For example, a recent work by Hubbs *et al.* (2020) presented a suite of benchmark problems for multi-echelon supply chain inventory management problems and studied the performance of various nominal mathematical programming implementations. Perez *et al.* (2021) extended this work to study divergent multi-echelon supply chains and included benchmark to a number of stochastic mathematical programming implementations. Both works demonstrated application of the RL algorithm, proximal policy optimization (PPO), which relies on estimating first-order directions for policy improvement, and make use of overparameterised policy function approximations. In this work, we attempt to improve the performance of RL and investigate the identification of risk-sensitive policies by making explicit connection between RL and derivative free optimization.

2. Reinforcement Learning and Derivative Free Optimization

2.1. Markov Decision Processes

Reinforcement Learning describes a set of algorithms, which generally aim to parameterize an approximately optimal policy function approximation that provides control decisions in a closed loop feedback control framework, known as the Markov Decision Process (MDP). Finite horizon MDPs are fully defined by a state space, $\boldsymbol{x} \in \mathbb{X} \subseteq \mathbb{R}^{n_x}$; a control space, $\boldsymbol{u} \in \mathbb{U} \subseteq \mathbb{R}^{n_u}$; a conditional probability density function, $P_{xx}: \mathbb{X} \times \mathbb{U} \times \mathbb{X} \rightarrow \mathbb{R}_+$, descriptive of the uncertain, discrete time evolution of the system, from the current state, $\boldsymbol{x}_t$, and control, $\boldsymbol{u}_t$, to a future state, $\boldsymbol{x}_{t+1}$; a stage cost function $\varphi: \mathbb{X} \times \mathbb{U} \times \mathbb{X} \rightarrow \Phi_{t+1} \in \mathbb{R}$, that ranks closed loop decisions at each control interaction; and a finite, discrete time horizon, such that $t \in \{0, \dots, T\}$. As decisions are identified within the context of state feedback, optimal decisions are provided by a policy, $\pi: \mathbb{X} \rightarrow \mathbb{U}$, which minimizes the expected sum of stage costs, $Z = \sum_{t=0}^{T} \Phi_{t+1}$ such that:

$$\pi^* = \underset{\pi \in \Pi}{\operatorname{argmin}}\, \mathbb{E}_\pi[Z] \tag{1}$$

where π defines a policy that belongs to the set of policies, Π, that satisfy input constraints. It is also worth briefly noting that this implies Z is a random variable described via some distribution, $Z \sim p_\pi(z)$, with values realized together with a given process trajectory, $\boldsymbol{\tau} = (\boldsymbol{x}_0, \boldsymbol{u}_0, \dots, \boldsymbol{x}_T)$. In this work, we assume that the problem of identifying a centralized re-order policy for a multi-echelon supply chain can be posed as an MDP.

2.2. Reinforcement Learning and derivative-free optimization

RL describes a set of policy learning methods that aim to, in general, identify a parametric policy function, $\pi(\boldsymbol{x};\, \theta)$, to approximate the optimal policy. This enables decision-making to generalize across continuous state spaces. The problem then reduces to identifying the policy parameters, such that:

$$\theta^* = \underset{\theta \in \Theta}{\operatorname{argmin}}\, \mathbb{E}_\pi[Z] \tag{2}$$

where $\theta \in \mathbb{R}^{n_\theta}$ are the policy parameters and Θ denotes the set of allowable policy parameters. The policy parameterization itself is identified through model-free parameter update rules. These are purely data-driven rules that estimate first-order directions to enable local gradient descent steps to improve the performance of policy parameters. However, importantly, although these directions may be estimated independently of gradient information from the dynamics themselves, they are often noisy and expensive to estimate. Additionally, the ultimate policy identified is known to be sensitive to policy initialization, indicating the optimization landscape is often characterized by multiple poor local minima.

Here, it is worth highlighting the strong connections between RL and derivative-free optimization (DFO). In DFO, one aims to identify an approximately optimal solution for a problem independently of gradient information, but rather purely via function evaluations. This is also property of DFO that enables it to optimize measures of objectives that are functions of random variables via, for example, a Monte Carlo method. Popular algorithms typically maintain a population of candidate solutions, evaluate the objective performance of each and then utilize information from the population to coordinate exploration and exploitation of the parameter space. This iterative process enables the DFO algorithm to identify better performing candidate solutions until a computational budget is exhausted or an exit tolerance satisfied. Examples include particle swarm optimization (PSO), evolutionary strategies (ES), and artificial bee colony (ABC), which can all be considered as heuristics inspired by phenomena observed within nature. It is well known that some algorithms are more explorative or exploitative than others. Hybridization of these algorithms has often been used to combine the respective strength of each algorithm to in one search process. Despite the potential application, these algorithms are known to perform poorly in highly dimensional spaces of $n_\theta \geq 1000$. This means we are required to construct a relatively low dimensional policy parameterization, given current RL practice is to massively overparameterize policy function approximations.

2.3. Risk-sensitive decision-making

As reflected in the discussion above, when deploying methods that consider uncertain variables in decision-making, typically one is posed with optimizing for some measure of the objective function. The choice of measure is typically guided by potential risk-tolerance of the operation. Optimizing in expectation is a popular choice, however, in some operations it may be desirable to consider what is happening in the tails of policy performance. For example, the conditional value-at-risk ($CVaR_\beta$) defines the expected performance of the policy observed with at least a probability level $\beta = (0{,}1)$ under the cumulative distribution function, $F_\pi(z) = \mathbb{P}(Z \leq z \mid \pi)$ associated with the policy:

$$CVaR_\beta = \mathbb{E}_\pi[Z | Z \geq F_\pi^{-1}(\beta)] \tag{3}$$

$F_\pi^{-1}(\beta)$ defines the inverse cumulative distribution function evaluated for probability β. Risk sensitive formulations that leverage the conditional value-at-risk are widely discussed elsewhere. However, in this work, we are interested in a formulation that incentivizes performance in expectation, but provides some notion of probabilistic robustness as follows:

$$\theta^* = \underset{\theta \in \Theta}{\operatorname{argmin}}\ \mathbb{E}_\pi[Z] \qquad s.t.\ CVaR_\beta \leq b \tag{4}$$

where $b \in \mathbb{R}$ is a bound that can be set by the implementation together with the probability level, β. This formulation is particularly appealing in supply chains because, for example, it enables one to place a bound on the expected costs observed with a given probability level under the policy.

2.4. Derivative-free optimization for robust reinforcement learning in multi-echelon supply chain inventory management

In this work, we investigate the application of derivative-free optimization for policy learning. Specifically, we hybridize standard in-house python implementations of ABC, PSO, ES and simulated annealing (SA) to more efficiently explore and exploit the parameter space.

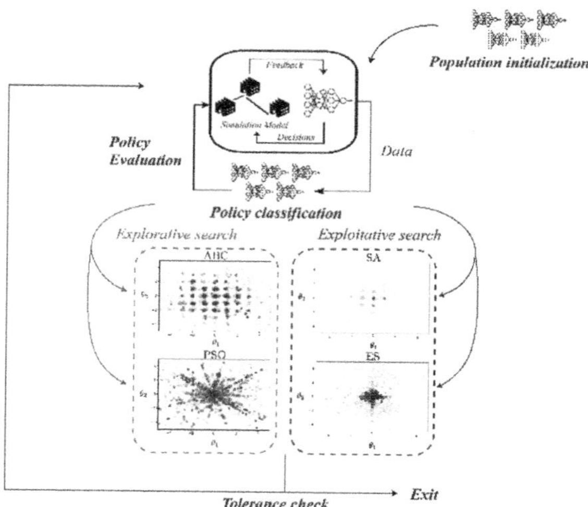

Figure 1: Figurative description of the proposed hybrid derivative free optimization algorithm. Poor performing candidate policies are passed to explorative search methods and high performing policies are passed to more exploitative methods.

Firstly a population of candidate policies is initialized. At each iteration of the algorithm, each of the candidate policies is evaluated within the supply chain inventory management problem and then the population is divided into disjoint sets according to policy performance. For example, the top 25% of policies comprise one subset, the policies ranked in the top 50-25% comprise an additional subset, and so on. The subsets of policies are then passed to one of the given DFO algorithms (ABC, ES, PSO and SA) depending on performance. The worst performing policies are passed to the most explorative algorithms, whereas the best are passed to the most exploitative. In this case, the bottom 50% are passed to ABC and PSO, with the top 50% passed to SA and and ES. This is demonstrated by Fig.1.

In case study, we seek to benchmark the performance of our algorithm in a multi-echelon supply chain inventory management problem with integer reorder decisions against the state-of-the-art RL method, proximal policy optimization (PPO), and mathematical programming implementations including offline nominal MILP (i.e. with no recourse decisions) and online shrinking horizon linear programming (with assumption that the integer constraints on decisions can be relaxed in the model and then the solution rounded appropriately).

3. Case Study

3.1. Operations Research Gym Benchmark

Here, we utilize the benchmarks and associated code provided by Hubbs *et al.* (2020). Specifically, we consider a supply chain network with $n_s = 4$ production stages. Stage 0 deals with exogenous consumer demand uncertainty described via a Poisson distribution with a rate of 20 units of inventory. Satisfaction of this demand is handled via a central reorder policy that places inventory order requests from stage n to $n+1$ $\forall n \in \{0, n_s - 1\}$. Stage n_s is assumed to have an infinite inventory to supply demand from stage $n_s - 1$ (and so in principle does not maintain a reorder policy itself). Production and transportation of the product is captured via a deterministic, constant delivery (lead) time.

The problem is defined as an MDP with discrete time dynamics described in Hubbs *et al.* (2020). The state is defined to preserve the Markov property and capture pipeline production and transportation. Specifically, the reorder history of the previous L_{max} time steps, where L_{max} is the maximum lead time associated with the delivery of product from all stages, are concatenated together with the current inventory level for stages with finite inventory. In this study $L_{max} = 10$. As a result, the state variables, $\boldsymbol{x} \in \mathbb{X} \subseteq \mathbb{R}^{(n_s-1)(L_{\max}-1)}$, are defined by the current inventory and the history of L_{max} controls. The controls, $\boldsymbol{u} \in \mathbb{U} \subseteq \mathbb{R}^{n_s-1}$, are the reorder quantities made at each stage in the supply chain, which are integer values and bounded between zero and an upper bound, which is stage dependent and defined as $\boldsymbol{u}_{ub} = [100,90,80]$, respectively. The aim of supply chain management is to maximize the expected net present value (NPV) of the supply chain after 30 discrete time steps. At each time step the cost function, φ, allocates the negative of discounted profit made by the total chain in the most recent time increment, which is comprised by total sales and various costs of holding inventory, procurement and unfulfilled demand at the respective stages.

3.2. Results

Here, we investigate two different formulations. The first formulation (F1) is provided by Eq. 2 and the second by Eq. 4 (F2). The optimization proceeds through unbiased Monte Carlo estimates of the objective measures. In the second formulation, we set the CVaR constraint to $b = -380$ and probability level $\beta = 0.9$.

It is clear from Fig. 2a) that the derivative free approach is able to outperform the policy gradient, PPO method, over the same training budget. Over 1000 Monte Carlo evaluations, the final policy of the hybrid approach observed expected NPV of \$454.0, whereas the offline MILP strategy presented in Hubbs *et al.* (2020) achieved \$378.5, and the shrinking horizon LP achieved \$485.4. In Fig 2b), the results of the approach under the risk sensitive formulation (F2) is presented. It is clear that hybrid algorithm makes progress to maximize expected NPV (Monte Carlo evaluation of the final policy achieves

$425) and satisfy the constraint on the CVaR (Monte Carlo evaluation of the final policy achieves $392.4). This provides probabilistic robustness in the policy performance. However, this robustness is at the cost of a greater required computational budget in policy optimization (over formulation F1), and reduced performance in expectation.

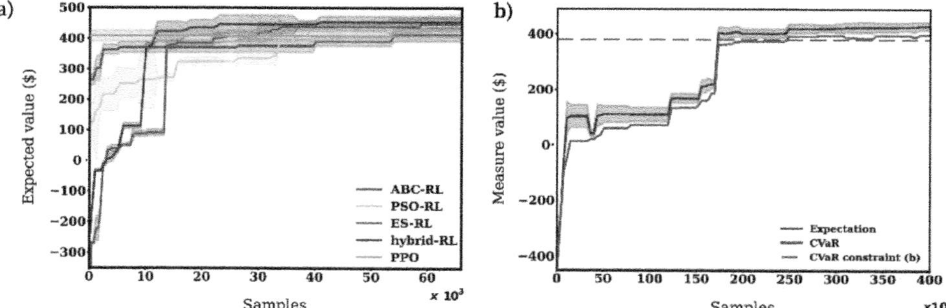

Figure 2: Policy optimization progress of a) the constituent stochastic search algorithms, together with the hybrid DFO algorithm (hybrid-RL) benchmarked against the final results of the state-of-the-art policy gradient method, PPO, evaluated over the same number of samples in F1; b) of the hybrid algorithm under F2. Presented from the perspective of NPV maximization.

4. Conclusions

In this work, we have demonstrated that derivative-free optimization methods are capable approaches to policy optimization tasks in RL in problems observed commonly in PSE and that compact, well-constructed policy function approximations can perform competitively with mathematical programming. RL with function approximation typically provides a nonlinear optimization problem and so derivative free approaches have value in e.g. escaping local minima. Additionally, they provide flexibility in optimizing risk-sensitive formulations via unbiased Monte Carlo estimators. However, it is likely that the limitations of derivative-free approaches in high dimensions will become substantial in industrial-scale supply chain networks. In future work, we will explore representation learning methods and exploit the structure of the supply chain network to reduce the effective dimensionality of these instances. Additionally, thorough comparison to mathematical programming formulations which consider uncertainty will be made.

References

[1] Gao, J., Ning, C. and You, F., 2019. Data-driven distributionally robust optimization of shale gas supply chains under uncertainty. AIChE Journal, 65(3), pp.947-963.

[2] Ash, C., Diallo, C., Venkatadri, U. and VanBerkel, P., 2022. Distributionally robust optimization of a Canadian healthcare supply chain to enhance resilience during the COVID-19 pandemic. Computers & Industrial Engineering, 168, p.108051.

[3] Golari, M., Fan, N. and Jin, T., 2017. Multistage stochastic optimization for production-inventory planning with intermittent renewable energy. Production and Operations Management, 26(3), pp.409-425.

[4] Charitopoulos, V.M., Papageorgiou, L.G. and Dua, V., 2019. Closed-loop integration of planning, scheduling and multi-parametric nonlinear control. Computers & Chemical Engineering, 122, pp.172-192.

[5] Hubbs, C.D., Perez, H.D., Sarwar, O., Sahinidis, N.V., Grossmann, I.E. and Wassick, J.M., 2020. Or-gym: A reinforcement learning library for operations research problems. arXiv preprint arXiv:2008.06319.

[6] Perez, H.D., Hubbs, C.D., Li, C. and Grossmann, I.E., 2021. Algorithmic approaches to inventory management optimization. Processes, 9(1), p.102.

Antonis Kokossis, Michael C. Georgiadis, Efstratios N. Pistikopoulos (Eds.)
PROCEEDINGS OF THE 33rd European Symposium on Computer Aided Process Engineering (ESCAPE33), June 18-21, 2023, Athens, Greece
 http://dx.doi.org/10.1016/B978-0-443-15274-0.50128-1

Monte Carlo Simulation of Photo Induced Atom-Transfer Radical Polymerization for Microscopic Properties

Rui Liu[a], Xiaowen Lin[a], Antonios Armaou[b, c, d], Xi Chen[a, e]

[a]*State Key Laboratory of Industrial Control Technology, College of Control Science and Engineering, Zhejiang University 310027, Hangzhou China.*
[b]*Chemical Engineering Department, University of Patras, Patras 26504, Greece.*
[c]*Chemical Engineering & Mechanical Engineering Departments, Pennsylvania State University, College Park, PA 16802 USA.*
[d]*College of Mechanical Engineering, Wenzhou University, 325035 Wenzhou China.*
armaou@psu.edu
[e]*National Center for International Research on Quality-targeted Process Optimization and Control, Zhejiang University 310027, Hangzhou China.*
xi_chen@zju.edu.cn

Abstract

Controlled radical polymerization (CRP) is a versatile way to control polymer microscopic structures through establishing a dynamic equilibrium between radical and dormant polymers. Atom transfer radical polymerization (ATRP), a typical case of CRP, has the ability to be conducted in mild conditions and produce uniformly distributed polymers. Moreover, the amount of metal addition can be greatly reduced in ATRP processes with the development of photochemical techniques. Photo induced ATRP (photoATRP) has a significant impact on sustainable production in industry. Modeling the microscopic quantities of polymers for photoATRP systems provides insights into the kinetic characteristics. In this work, a Monte Carlo (MC) approach, based on Gillespie's stochastic simulation algorithm, is developed for the photoATRP processes at microscopic resolution. The length dependency of the individual chains on the radical, dormant, and dead polymer information is captured. The polymerization rate of PhotoATRP is sensitive to light irradiation. The polymerization process model of photoATRP systems based on the MC method for temporal control over polymer growth is established. Kinetic characteristics are studied from the developed model under the condition of light source intermittency. The developed MC model is validated against the deterministic method of moments (MoM) and experimental results.

Keywords: Monte Carlo simulation; photochemical ATRP; microscopic properties; hybrid multiresolution model.

1. Introduction

In past decades, there has been a continuous evolution in controlled radical polymerization (CRP) for enabling to produce well-controlled polymers with narrow molecular weight distributions (MWD) (Pan et al., 2016). Atom transfer radical polymerization (ATRP) is a fundamental technique of CRP. The growth rate of radical polymers is controlled by manipulating activation and deactivation reactions in ATRP (Wang et al., 1995). Light irradiation as external stimuli maintains the stability of polymer growth through regenerating activators in ATRP systems (Ribelli et al., 2014). These photo induced ATRP (photoATRP) systems provide well-defined and uniformly distributed polymers under mild conditions (Lorandi et al., 2022). The temporal and spatial control over polymer growth in photoATRP can be efficiently achieved upon regulating the applied light irradiation (Dadashi-Silab et al., 2020).

Identifying the microscopic properties of polymers is of great significance and demand for photoATRP processes. Monte Carlo (MC) simulation is a powerful tool to accurately predict the microscopic structure of polymers and provides insights into the kinetic characteristics (Liu et al., 2021). In this work, a dynamic MC model is developed for photoATRP systems at microscopic resolution. Kinetic characteristics of temporal control with light switching are provided based on the developed MC model.

2. PhotoATRP systems

In ATRP systems, transition metal complexes and alkyl halides are used to establish the equilibrium between activation and deactivation reactions. Low oxidation state metal complexes as activators activate dormant polymers, and high oxidation state metal complexes as deactivators transform radical polymers into dormant polymers. The mechanistic scheme of the ATRP processes is presented in Figure 1(a). The oxidation-reduction reaction pair presents a trend in the direction of activation in ATRP processes due to the fact that radical polymers are constantly consumed in termination reactions. The concentration of activators decreases as the reaction progress. To maintain the stability of the equilibrium between activation and deactivation, and the concentrations between radical polymers and dormant polymers, a continuous supplement of activators to the reaction system is required. However, in photoATRP processes, activators can be regenerated through light irradiation, as indicated in Figure 1(b). The amount of metal catalyst loadings is decreased in photoATRP (Chen et al., 2016).

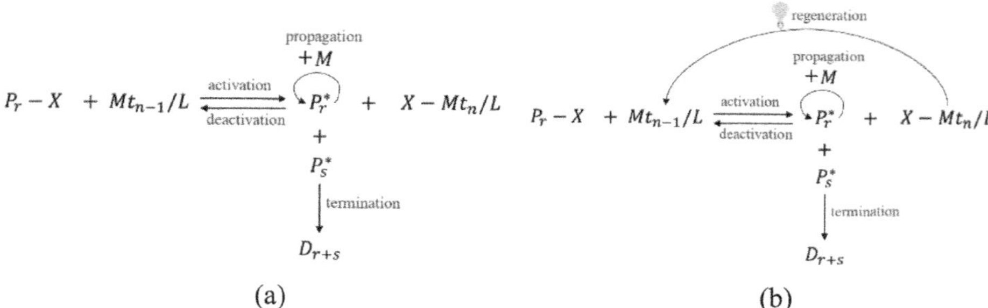

Figure 1 Mechanistic schemes for atom transfer radical polymerization (ATRP) processes. (a) Scheme of conventional ATRP. (b) Scheme of photoATRP.

Radical polymers are formed along with the regeneration of activators in photoATRP processes shown in Figure 1(b). Alkyl halides, ligands, and monomers can also lead to the formation of radicals in a photochemical way. A typical kinetic mechanism for photoATRP is listed in Table 1, where P_sX, D_s and P_s^* denote dormant polymers, dead polymers, and radical polymers with chain lengths of s, respectively; Mt_{n-1}/L and Mt_nX/L denote activators and deactivators, respectively; M denotes monomers, and L denotes ligands. Photo induced reactions 14 – 17 in Table 1 only take place under light irradiation.

Table 1. Kinetic mechanism for photoATRP

	description	chemical equation	rate constant
1	activation	$P_0X + Mt_{n-1}/L \rightarrow P_0^* + Mt_nX/L$	$k_{a0} = 2.00 \times 10^3\ mol^{-1}s^{-1}$
2	deactivation	$P_0^* + Mt_nX/L \rightarrow P_0X + Mt_{n-1}/L$	$k_{d0} = 5.00 \times 10^7\ mol^{-1}s^{-1}$
3	activation	$P_sX + Mt_{n-1}/L \rightarrow P_s^* + Mt_nX/L$	$k_a = 2.00 \times 10^2\ mol^{-1}s^{-1}$
4	deactivation	$P_s^* + Mt_nX/L \rightarrow P_sX + Mt_{n-1}/L$	$k_d = 2.80 \times 10^8\ mol^{-1}s^{-1}$
5	propagation	$P_0^* + M \rightarrow P_1^*$	$k_{p0} = 7.30 \times 10^2\ mol^{-1}s^{-1}$
6		$P_s^* + M \rightarrow P_{s+1}^*$	$k_p = 1.56 \times 10^4\ mol^{-1}s^{-1}$
7	termination by combination	$2P_0^* \rightarrow D_0$	$k_{t0} = 2.00 \times 10^9\ mol^{-1}s^{-1}$
8		$P_0^* + P_s^* \rightarrow D_s$	
9		$P_s^* + P_r^* \rightarrow D_{s+r}$	$k_t = 1.00 \times 10^8\ mol^{-1}s^{-1}$
10	catalytic radical termination	$P_0^* + Mt_{n-1}/L \rightarrow D_0 + Mt_{n-1}/L$	$k_{tx0} = 0.00\ mol^{-1}s^{-1}$
11		$P_s^* + Mt_{n-1}/L \rightarrow D_s + Mt_{n-1}/L$	$k_{tx} = 4.00 \times 10^3\ mol^{-1}s^{-1}$
12	transfer	$P_s^* + L \rightarrow P_0^* + D_s$	$k_{trL} = 2.80 \times 10^3\ mol^{-1}s^{-1}$
13		$P_s^* + P_0X \rightarrow P_0^* + D_s$	$k_{trX} = 2.30 \times 10^2\ mol^{-1}s^{-1}$
14	photo induced radical regeneration	$L + M \xrightarrow{hv} 2P_0^*$	$k_{rLM} = 1.50 \times 10^{-9}\ mol^{-1}s^{-1}$
15		$P_0X \xrightarrow{hv} 2P_0^*$	$k_{rX2} = 2.90 \times 10^{-9}\ s^{-1}$
16		$P_0X + L \xrightarrow{hv} 2P_0^*$	$k_{rL2} = 6.20 \times 10^{-6}\ mol^{-1}s^{-1}$
17		$Mt_nX/L + L \xrightarrow{hv} Mt_{n-1}/L + P_0^*$	$k_{rMt} = 1.00 \times 10^{-3}\ mol^{-1}s^{-1}$

3. Dynamic Monte Carlo simulation

In the dynamic MC simulation, the propensity of the μth reaction of the kinetic mechanism with M reactions is defined as:

$$R_\mu = k_\mu c_\mu, \forall \mu = 1,2, \ldots, M \tag{1}$$

where k_μ is the kinetic rate constant, and c_μ is the total concentration of the distinct molecular pairs that will react in the μth reaction. The probability $\mathbb{P}_\mu$ for reaction μ to take place is given by:

$$R_{SUM} = \sum_{v=1}^{M} R_v \tag{2}$$

$$\mathbb{P}_\mu = \frac{R_\mu}{R_{SUM}} \tag{3}$$

A random number rnd_1 with uniform distribution in the interval (0,1), i.e., $rnd_1 \in \mathcal{U}(0,1)$, is generated to select one reaction μ to occur in the current time interval based on the rule:

$$\sum_{\gamma=1}^{\mu-1} R_\gamma < rnd_1 \cdot R_{SUM} \leq \sum_{\gamma=1}^{\mu} R_\gamma \tag{4}$$

The dynamic time evolution of the kinetic system is described by determining the time interval between two consecutive reactions. To calculate the time interval τ, another random number $rnd_2 \in \mathcal{U}(0,1)$ is generated.

$$\tau = -\frac{1}{R_{SUM}}\ln(rnd_2) \tag{5}$$

The system states are updated after each reaction event, and accordingly, the simulation time is evaluated by the newly calculated τ. This process repeats until a preset reporting time or conversion is reached. A detailed discussion can be found in (Gillespie, 1977).

PhotoATRP processes are studied in this project through building dynamic MC models. In the implementation of the dynamic MC simulation for photoATRP systems, chain length information of radical polymers, dormant polymers, and dead polymers is stored in three individual arrays. The element value in the arrays denotes the chain length of one polymer. The arrays are updated following the selected reaction event. Specifically, if a propagation reaction is selected, an element in the array storing radical polymers is chosen at random, and the value of this element is increased by one. If an activation reaction is selected, one element in the dormant polymer array is chosen randomly and moved into the radical polymer array. If a termination by combination reaction is selected, two elements in the radical polymer array are chosen randomly and removed from the array, and the sum of the two element values is added to the dead polymer array. The instantaneous system states are recorded through tracking each radical, dormant and dead polymer chain.

4. Results and discussions

Simulations for photoATRP systems are conducted through the dynamic MC method and the deterministic MoM. The dynamic MC simulation is capable to provide the microscopic structures of the produced polymers. The deterministic MoM cannot be directly applied to calculate microscopic information for photoATRP systems as the moment equations for all the polymers with diverse chain lengths are difficult to solve due to stiffness. Furthermore, investigations of temporal control over polymerization rate are studied using intermittent light-on and light-off periods. All simulations were performed using Jupyter Notebook on the Anaconda distribution of Python language version 3.8 on a personal computer with RAM 32 GB and Intel i7-10700 CPU core @2.90GHz running Windows 10 OS.

4.1. Simulations of photoATRP for microscopic properties

A batch process where photoATRP takes place is simulated through developing a dynamic MC model and a deterministic method of moments (MoM) model. Tris 2-(dimethylamino)ethyl amine (Me_6TREN), ethyl 2-bromoisobutyrate (EBiB), Methyl acrylate (MA), CuBr/L and $CuBr_2$/L are used as ligands L, alkyl halides P_0X, monomers M, activators Mt_{n-1}/L, and deactivators Mt_nX/L, respectively. The initial concentrations of monomers, P_0^*, deactivators and ligands are 7.40 $mol \cdot L^{-1}$, 2.47×10^{-2} $mol \cdot L^{-1}$, 7.40×10^{-4} $mol \cdot L^{-1}$, and 3.70×10^{-3} $mol \cdot L^{-1}$, respectively. For brevity, the reaction conditions and kinetic rate constants can be found in (Ribelli et al., 2014).

As shown in Figure 2(a), the evolution of the average chain length calculated from the dynamic MC simulation is in good agreement with the results from the deterministic MoM and the experiment data. The evolution of concentrations for various reactants versus monomer conversion rate is shown in Figure 2(b). It can be observed that the concentrations of most of the reactants (monomer, activator, dormant polymer, dead polymer, deactivator, and ligand) from the developed dynamic MC model and the deterministic MoM are consistent with the literature (Ribelli et al., 2014). The concentrations of dead polymers (solid and dashed lines in purple) obtained from this work and the literature diverge during the period of 0% to 20% conversion rate. The concentration lines of dead polymers from this work are smoother. This divergence most likely results due to inaccuracy in the adaptive time-step algorithm of the old version *Predici*, the software that was adopted to do the simulation in the literature. The MWD result of all types of polymers at 80% monomer conversion rate is shown in Figure 2(c). Individual MWD results of radical polymers, dormant polymers, and dead polymers can be calculated by the developed dynamic MC model. It is challenging to obtain such microscopic distribution using the deterministic MoM, which is achieved through deriving and solving a large system of equations in the order of thousands.

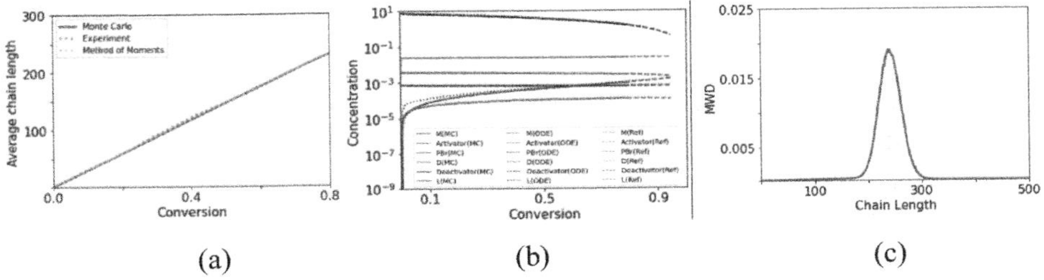

(a) (b) (c)

Figure 2 Simulation results for a photoATRP process. (a) Average chain length versus monomer conversion rate; the purple, yellow and red lines denote the results obtained from the MC simulation, MoM model and experiment, respectively. (b) Concentrations of reactants versus monomer conversion rate from different methods. (c) The molecular weight distribution (MWD) at 80% conversion obtained from the dynamic MC simulation, and the corresponding computational time is 9.79 hours. The MWD cannot be obtained from the deterministic MoM model.

4.2. Simulations on the temporal control of photoATRP

Simulations on the temporal control of the photoATRP processes are further conducted with intermittent on/off periods with $k_{ATRP} = 7.143 \times 10^{-5}$ ($k_{ATRP} = \frac{k_a}{k_d}$). In the implementation of the simulations, the kinetic rate values of photo induced reactions (reactions 14 – 17 in Table 1) are set to zero to simulate the system in dark.

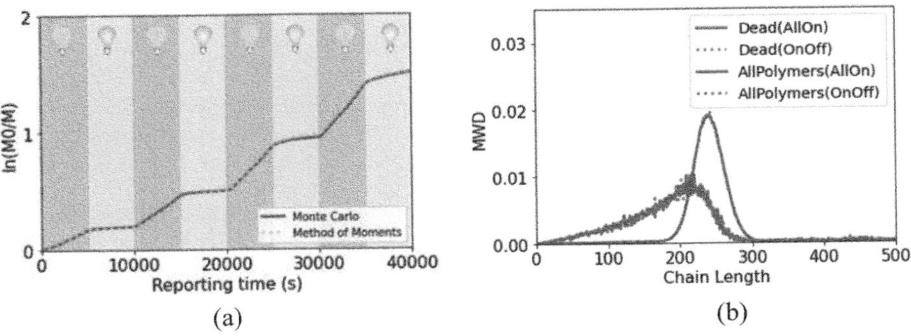

(a) (b)

Figure 3 Simulation results of temporal control. (a) temporal control of polymer growth obtained from the dynamic MC simulation and the deterministic method of moments model. (b) MWDs of different polymer types obtained from dynamic MC simulations under conditions of light all-time on and light on/off.

The evolution of the concentration of monomers using the MC and the MoM models is presented in Figure 3(a). We observe that switching the light off decreases the regeneration rate of radical polymers, leading to slower chain growth. The MWD results of dead polymers and all polymers are shown in Figure 3(b). The solid lines are the simulated results with all-time light irridiation, and the dotted lines are the results using intermittent 5000 second light-on and 5000 second light-off periods. As shown in Figure 3(b), reaction time can be affected by tuning the light on and off, and the MWDs are consistent under various light irridiation modes. It should be note that the MWDs of each type of polymers at any reporting time and conversion rates can be addressed using the developed MC model. The kinetic characteristics of polymerizations such as the polymerization rate, dynamic average chain length, dynamic concentrations of reactants, under diverse irradiation conditions can be investigated through using the MC simulations.

5. Conclusions

In this work, dynamic Monte Carlo models and deterministic Method of Moments models are developed to simulate the photoATRP processes. The developed dynamic MC models are capable to predict the microscopic structures of polymers and provide the kinetic characteristics of the simulated mechanisms on demand. Temporal control over polymer growth was further investigated based on the developed models. As advanced photoATRP systems develop, MC simulation has great application potential in providing a clear understanding of the details of various mechanisms, optimizing the qualities of products, and designing polymerization processes at a microscopic level.

Acknowledgments

The financial support of the "Pioneer" and "Leading Goose" R&D Program of Zhejiang (No. 2022C01176) is gratefully acknowledged by the first, second, and fourth authors. The financial support of the University of Patras, Medicus program (No. 81816) is gratefully acknowledged by the third author. The financial support from the China Scholarship Council is gratefully acknowledged by the first author. The authors would like to thank Prof. Krzysztof Matyjaszewski at Carnegie Mellon University and Prof. Dominik Konkolewicz at Miami University for offering experimental data, the simulated macroscopic information in their published paper, and analyzing the slight divergence in concentrations of dead polymers in Figure 2(b).

References

X. Pan, et al., 2016. Photomediated controlled radical polymerization. Prog. Polym. Sci. 62, 73-125.

J.Wang, et al., 1995. Controlled/"living" radical polymerization. atom transfer radical polymerization in the presence of transition-metal complexes. J. Am. Chem. Soc. 117(20), 5614–5615.

T. Ribelli, et al., 2014. How are radicals (re)generated in photochemical ATRP?. J. Am. Chem. Soc. 136(38), 13303–13312.

F. Lorandi, et al., 2022. Atom transfer radical polymerization: A mechanistic perspective. J. Am. Chem. Soc. 144(34), 15413–15430.

S. Dadashi-Silab, et al., 2020. Investigating temporal control in photoinduced atom transfer radical polymerization. Macromolecules. 53(13), 5280-5288.

R. Liu, et al., 2021. Adaptable parallel acceleration strategy for dynamic monte carlo simulations of polymerization with microscopic resolution. Ind. Eng. Chem. Res. 60(17), 6173–6187.

M. Chen, et al., 2016. Light-controlled radical polymerization: mechanisms, methods, and applications. Chem. Rev. 116(17), 10167-10211.

D. Gillespie., 1977. Exact stochastic simulation of coupled chemical reactions. J. Phys. Chem. 81(25), 2340–2361.

Antonis Kokossis, Michael C. Georgiadis, Efstratios N. Pistikopoulos (Eds.)
PROCEEDINGS OF THE 33rd European Symposium on Computer Aided Process Engineering (ESCAPE33), June 18-21, 2023, Athens, Greece
 http://dx.doi.org/10.1016/B978-0-443-15274-0.50129-3

Identification of a dynamic model of the N-vinylcaprolactam-co-glycidyl methacrylate microgel synthesis

Luise F. Kaven[a], Thomas Nevolianis[b], Nadja Wolter[c,d], Adel Mhamdi[a], Andrij Pich[c,d], Kai Leonhard, Alexander Mitsos[a,e]

[a]*Chair of Process Systems Engineering, RWTH Aachen University, Aachen, Germany*
[b]*Institut of Technical Thermodynamics, RWTH Aachen University, Aachen, Germany*
[c]*DWI – Leibniz Institute for Interactive Materials e.V., Aachen, Germany*
[d]*Functional and Interactive Polymers, Institute of Technical and Macromolecular Chemistry, RWTH Aachen University, Aachen Germany*
[e]*JARA-SOFT, Aachen, Germany*

amitsos@alumn.mit.edu

Abstract

Incorporating methacrylate into microgels allows for their application in catalysis, drug release systems, and enzyme immobilization. Detailed insight into the occurring reactions during the microgel synthesis enables the precise control of the microgel properties. In this contribution, we identify a kinetic model of the copolymerization of N-vinylcaprolactam-based microgels with functional glycidyl methacrylate monomers. We include propagation reaction rate constants calculated via quantum chemical methods. First, we estimate the five most sensitive propagation and initiation and termination rate constants using the enthalpy transfer rate from calorimetry experiments. For the parameter estimation, we use the quantum chemically calculated values as start values and the determined error scopes as bounds in the optimization. We compare our approach of incorporating computed parameter values to similar procedures. The results show that our estimation approach enables a sufficient accuracy of parameter values. Finally, we predict the distribution of functional groups within the microgel with the fully identified model.

Keywords: parameter estimation, quantum chemistry, synthesis modeling

1. Introduction

Microgels are crosslinked polymer networks with a reversible stimulus-specific swelling response to temperature, pH, or ionic strength. Microgels based on N-vinylcaprolactam (VCL) are temperature responsive and interesting due to their biocompatibility (Saunders 2009). Incorporating comonomers into pVCL microgels allows tuning their functionalities. Glycidyl methacrylate (GMA) is a promising comonomer thanks to its reactive epoxy group that is versatile in subsequent reactions. Hence, pVCL microgels functionalized with GMA experience many applications (Schäfer 2020, Li 2013, Zou 2019).

GMA is more reactive than VCL in the copolymerization leading to the heterogeneous composition of the resulting microgels. Thus, the authors of (Qiu 2005) performed a gradual feeding technique to produce more homogeneous microgels. Under batch conditions, (Häntzschel 2007) also showed via high-magnification SEM and TEM images that the microgels exhibit a core-shell structure with a GMA-rich core surrounded by a VCL-rich shell. Due to the complexity of the reaction network during microgel synthesis, understanding the relationship between the synthesis conditions and the final composition of the microgels is challenging. Thus, for designing tailored microgels, detailed reaction models provide insights and assist with developing microgels with defined monomer distributions.

In previous contributions, we have developed a mechanistic kinetic model for the aqueous microgel synthesis by precipitation polymerization (Janssen 2017, Jung 2019). To estimate the model parameters, we used a combination of quantum chemical calculations (Kröger 2017) and physical measurements (Jung 2019). We have applied the dynamic model to predict and guide the synthesis of microgels based on monomer N-isopropylacrylamide, comonomer vinyl ferrocene (VFc), and crosslinker N,N'-methylene bisacrylamide (BIS). In contrast to these monomer combinations, parameter values for the propagation rate constants and polymerization enthalpies of GMA/VCL and GMA/BIS systems in an aqueous solution remain unknown.

In our preceding work, we built a mechanistic model based on (Schneider 2020) for synthesizing VCL-based microgels functionalized with GMA. The model contained 14 unknown kinetic parameter values. For model identification, we followed a hybrid approach. First, using transition state theory and quantum chemical calculations, we computed 12 out of 14 parameters, namely the propagation rate constants and polymerization enthalpies of the GMA/VCL and GMA/BIS systems in an aqueous solution. We conducted experiments with four different initial compositions and measured the enthalpy transfer rate during the synthesis using a calorimetric device. Subsequently, we estimated the remaining parameter values of the five most sensitive propagation rates and the initiation and termination rates using the measured enthalpy transfer rate from the experiments. We used the quantum chemically calculated values as start values for the parameter estimation. We also added the computed error bounds from the quantum calculation as optimization constraints during the parameter estimation.

In this contribution, we further analyze the parameter estimation incorporating quantum chemically calculated parameter values. We estimate different combinations of unknown parameter values and conduct an identifiability analysis. Finally, we use the fully identified VCL/GMA microgel synthesis model to predict feeding strategies for a homogeneous microgel composition.

2. Materials and Methods

2.1. Microgel Synthesis and Reaction Calorimetry

We analyze the precipitation polymerization of VCL crosslinked with BIS and functionalized with GMA. As an initiator, we consider 2,2'-azobis(2-methylpropionamidine)dihydrochloride. Previously, we measured the enthalpy transfer

Table 1 Synthesis composition for experiments in the calorimeter.

Experiment	Molar ratio VCL:GMA	Molar amount of crosslinker BIS
1	90:10	2.0 mol%
2	90:10	2.6 mol%
3	93:7	2.0 mol%
4	95:5	2.6 mol%

rate during the reaction via reaction calorimetry (Nevolianis 2023). Details on the setup and compositions of the conducted experiments can be found in our previous work (Nevolianis 2023). In Table 1, we briefly outline the variations between the experiments used in the current work.

2.2. Quantum Chemical Calculations

In our previous work (Nevolianis 2023), we incorporated propagation rate constants of the VCL/GMA/BIS system that were calculated via quantum chemistry methods. There, we applied conventional transition state theory to determine propagation rate constants. Further, we calculate the enthalpies of reaction using the Gibbs-Helmholtz equation. Computational details can be found in (Nevolianis 2023).

2.3. Kinetic Model of Microgel Synthesis

We employ the reaction system of our kinetic synthesis model (Schneider 2020) and change the parameter values according to the VCL/GMA/BIS system. The main model assumptions are briefly summarized in the following. The dispersed system is modeled in a pseudo-homogeneous phase. The occurring reactions include initiation, propagation, termination, and crosslinking. The efficiency of the initiator decomposition is set to 50% (Jung 2020). The kinetics of the propagation reactions depend only on the radical at the terminal end of the chain, and propagation reactions make up for the enthalpy release. Termination only occurs due to disproportionation, and the kinetics of termination reactions do not depend on the terminal end of the active radical chains. The kinetics of the crosslinking reaction equals the kinetics of a propagation reaction with a BIS monomer.

2.4. Parameter Estimation and Identifiability Analysis

We use the five reaction enthalpy values involving GMA from the quantum chemical computations without further treatment in the synthesis model. We then analyze the sensitivity to identify the most sensitive propagation rate (kp) parameters. Subsequently, we perform parameter estimations to determine different combinations of parameter values. The initiation and termination rates are estimated for all sets as they remain unknown after the quantum chemical calculations. We split the conducted estimations into the following groups:

1) Estimate all k_ps with arbitrary start values and bounds
2) Estimate all k_ps using the quantum calculations as start values and their error bounds as constraints
3) Estimate the five most sensitive k_ps using the quantum calculations as start values and their error bounds as constraints
4) Insert all k_ps into the model and only estimate k_t and k_d

We perform the parameter estimation and the identifiability in Matlab, similar to (Jung 2019), using the presented copolymerization model and the calorimetric measurements. For part of the estimations, the calculated propagation rate constants act as start values, and the calculation uncertainties derived from Gaussian error propagation analysis of the quantum chemical method's uncertainty act as constraints in the optimization problem. For estimation (1), the start values for the propagation rates are 10, and the upper and lower bounds are 10^{-3} and 10^{5}, respectively.

3. Results and Discussion

3.1. Quantum Chemically Calculated and Estimated Kinetic Parameters

The calculated propagation rate constants for the VCL/GMA/BIS system in an ideally diluted aqueous solution at a temperature of 343 K are shown in the second column of Table 3. The applied bounds are presented in the third/fourth column of Table 3. Using the optimized parameter values from estimation (3), our model agrees to 97% on average to the measured enthalpy transfer rate from the experiments.

The results in Table 3 show that the homopolymerizations of GMA and VCL are approximately in the same order of magnitude when calculated via quantum chemistry. The most significant reaction occurs between VCL-radical and GMA. The reaction between GMA-radical and VCL plays a negligible role in the polymerization system. All reactions consuming GMA occur relatively fast. Thus, the experimental observation of a GMA-rich microgel core fits our calculated findings.

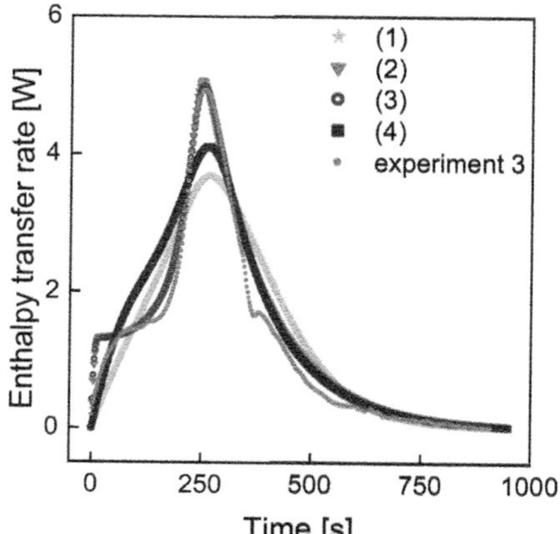

Figure 1 Enthalpy transfer rate for experiment 3 and the simulations with the estimated parameter values from estimation (1) to (4).

Table 1 Propagation rate constants [m3 · mol−1 · s−1] at 343 K calculated via quantum chemistry and optimized in the parameter estimations with corresponding upper and lower error bound from the quantum chemical calculations. * denotes estimated parameter values at the bounds.

Reactants	Calculated k_p	Lower error bound	Upper error bound	Optimized k_p (1)	Optimized k_p (2)	Optimized k_p (3)	Optimized k_p (4)
VCL-radical / VCL	86.8	10.4	727	28.7	23.8	116	
VCL-radical / GMA	516	61.6	4330	0.403	337	1300	
GMA-radical / GMA	65.5	7.8	549	10^{-3}*	18.2	64.9	
GMA-radical / VCL	4.6	0.5	38.5	1.59	0.74	4.6	
BIS-radical / VCL	34.3	4.1	287	733	9.76	287	
BIS-radical / BIS	8.1	0.967	67.9	10^{-3}*	67.9*		
GMA-radical / BIS	0.74	0.088	6.18	10^{-3}*	2.44		
BIS-radical / GMA	129	15.4	1080	4210	15.4*		
k_d [s^{-1}]		10^{-7}	10^{-3}	$6.15 \cdot 10^{-7}$	$3.48 \cdot 10^{-5}$	$8.22 \cdot 10^{-6}$	$8.97 \cdot 10^{-7}$
k_t		10^{-5}	10^{5}	10^{-5}*	244	1480	81.5

The enthalpy transfer rate simulated via the synthesis model containing the respective parameter values is shown in Figure 1, exemplary for Experiment 3. The parameter combinations from estimations (2) and (3) result in the best fit to the experimental data.

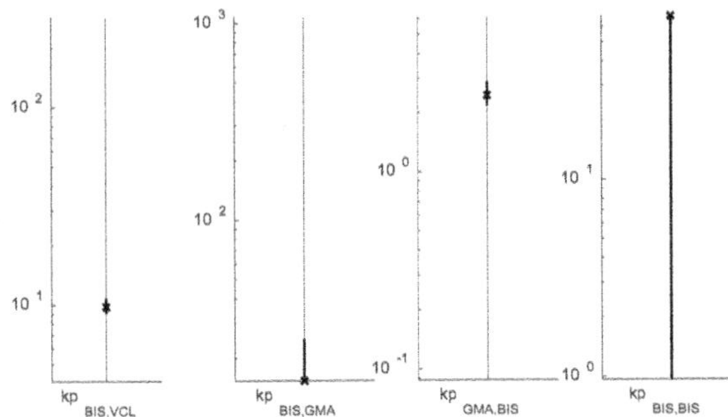

Figure 2 Identifiability analysis of selected parameters from estimation (2). The x indicates the optimal parameter value.

The estimation (1) allows a wide solution space with arbitrary start values and bounds gets stuck in a local minimum, indicating the need for deterministic global optimization. The simulation of estimation (4) only has two degrees of freedom and does not sufficiently follow the experimental trend. Altogether, estimation (3) is most suitable, as the fit to experimental data is eminent, and the computational effort is decreased due to reduced parameter values for estimation.

Figure 2 shows the identifiability result for the estimation set (2). For visualization, only the parameters with the most significant variability are shown. Here, all parameters appear to be sufficiently identifiable except $k_{p,BIS-BIS}$.

3.2. Process Simulations for Feeding Strategy

The identified model with parameters from estimation (3) allows for the assessment of feeding strategies. We simulate different feeding strategies of GMA to the batch reaction. We focus on two aspects, namely, the ratio of GMA initially in the reactor to the dosing feed and the dosing flow rate. The simulation includes the composition of Experiment 1 (see Table 1) and assumes that 7.5 ml of the total water volume is used for dosing. The ratio of GMA fed to GMA in the reactor varies from 0:1 (all GMA is initially in the reactor) to 1:1 and 1:0 (all GMA is fed to the reactor). Simultaneously, the dosing flow rate is set to 1.5, 3, and 6 mmol/s.

In Figure 3, the results of the simulation scenarios are shown. Here, the combined reaction rate of GMA is depicted over time. The combined reaction rate comprises all reactions where GMA is consumed. When assuming monodisperse particle growth, the time of consumption of GMA correlates to the incorporated position in the microgel. Thus, the combined reaction rate of GMA represents its composition in the microgel.

The progressions of Figure 3 indicate that adding all GMA at a dosing flow of 1.5 mmol/s is likely to achieve a homogeneous composition of GMA in the microgel.

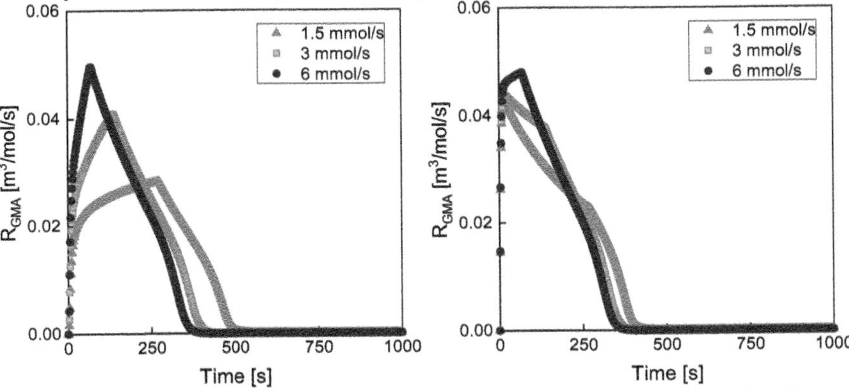

Figure 3 Left: Feeding strategy 1:0, all GMA is fed to the reactor, and the dosing flow rate varies. Right: Feeding strategy 1:1, half of the GMA is in the reactor at the start of the reaction, and half is fed at different dosing flow rates.

4. Conclusions

We studied the precipitation polymerization of VCL/GMA/BIS in an aqueous solution focusing on parameter estimation approaches. We showed that incorporating quantum chemically computed parameter values and their calculation error into the estimation problem improves the optimization result. Based on the identified synthesis model, we were able to derive an improved feeding strategy of GMA to the batch reaction to achieve a homogeneous GMA composition in the microgels by testing different scenarios. Our hybrid approach of quantum chemistry and parameter estimation from experimental data shows the potential of integrating theoretical derivations and semi-empirical knowledge. Further, we laid the ground to design and optimize the functionalization of VCL-based microgels with GMA in the future.

References

N. Häntzschel, F. Zhang, F. Eckert, A. Pich, M. A. Winnik, 2007, Poly(N-vinylcaprolactam-co-glycidyl methacrylate) aqueous microgels labeled with fluorescent LaF_3:Eu nanoparticles, Langmuir, 23, 21, 10793-10800

F. A. L. Janssen, M. Kather, L. C. Kröger, A. Mhamdi, K. Leonhard, A. Pich, A. Mitsos, 2017, Synthesis of poly(N-vinylcaprolactam)-based microgels by precipitation polymerization: process modeling and experimental validation, Ind. Eng. Chem. Res., 56, 14545-14556

F. Jung, F. A. L. Janssen, A. Ksiazkiewicz, A. Caspari, A. Mhamdi, A. Pich, A. Mitsos, 2019, Identifiability analysis and parameter estimation of microgel synthesis: a set-membership appoach, Ind. Eng. Chem. Res. 58, 13675-13685

F. Jung, A. Ksiazkiewicz, A. Mhamdi, A. Pich, A. Mitsos, 2020, Ind. Eng. Chem. Res., 59, 20437-20446

L. C. Kröger, W. A. Kopp, K. Leonhard, 2017, Prediction of chain propagation rate constants of polymerization reactions in aqueous NIPAM/BIS and VCL/BIS systems, Physical Chemistry B, 121, 2887-2895

P. Li, R. Xu, W. Wang, X. Li, Z. Xu, K. W. K. Yung, P. K. Chu, 2013, Thermosensitive poly(N-isopropylacrylamide-co-glycidyl methacrylate) microgels for controlled drug release, Colloids and Surfaces B: Biointerfaces, 101, 251-255

E. Mavroudakis, D. Cuccato, D. Moscatelli, 2015, On the use of quantum chemistry for the determination of propagation, copolymerization, and secondary reaction kinetics in free radical polymerization, Polymers, 7, 9, 1789-1819

T. Nevolianis, N. Wolter, L. Kaven, L. Krep, C. Huang, A. Mhamdi, A. Mitsos, A. Pich, K. Leonhard, 2023, Kinetic Modeling of a Poly(N-vinylcaprolactam-co-glycidyl methacrylate) Microgel Synthesis - A Hybrid In-Silico and Experimental Approach, Submitted to Ind. Eng. Chem. Res.

B. R. Saunders, N. Laajam, E. Daly, S. Teow, X. Hu and R. Stepto, Microgels: From responsive polymer colloids to biomaterials, Adv. Colloid Interface Sci., 2009, 147-148, 251–262

D. Schäfer, F. Fink, D. Kleinschmidt, K. Keisers, F. Thomas, A. Hoffmann, A. Pich, S. Herres-Pawlis, 2020, Enhanced catalytic activity of copper complexes in microgels for aerobic oxidation of benzyl alcohols, Chemical Communications, 56, 5601-5604

S. Schneider, F. Jung, O. Mergel, J. Kammertz, A. C. Nickel, T. Caumanns, A. Mhamdi, J. Mayer, A. Mitsos, F. A. Plamper, 2020, Model-based design and synthesis of ferrocene containing microgels, Polymer Chemistry, 11, 315-325

X. Qiu, S. A. Sukhishvili, 2005, Copolymerization of N-vinylcaprolactam and glycidyl methacrylate: Reactivity ratio and composition control, Polymer Chemistry, 44, 183-191

Z. Zou, E. Gau, I. El-Awaad, F. Jakob, A. Pich, U. Schwaneberg, 2019, Selective Functionalization of Microgels with Enzymes by Sortagging, Bioconjugate Chemistry, 30, 11, 2859-2869

Antonis Kokossis, Michael C. Georgiadis, Efstratios N. Pistikopoulos (Eds.)
PROCEEDINGS OF THE 33rd European Symposium on Computer Aided Process Engineering (ESCAPE33), June 18-21, 2023, Athens, Greece
 http://dx.doi.org/10.1016/B978-0-443-15274-0.50130-X

A Machine Learning-assisted Hybrid Model to Predict Ribbon Solid Fraction, Granule Size Distribution and Throughput in a Dry Granulation Process

Yan-Shu Huang,[a] David Sixon,[a] Phoebe Bailey,[a] Rexonni B. Lagare,[a] Marcial Gonzalez,[b,c] Zoltan K. Nagy,[a] Gintaras V. Reklaitis[a]

[a]*Davidson School of Chemical Engineering, Purdue University, West Lafayette, IN 47907, USA*
[b]*School of Mechanical Engineering, Purdue University, West Lafayette, IN 47907, USA*
[c]*Ray W. Herrick Laboratories, Purdue University, West Lafayette, IN 47907, USA*
huan1289@purdue.edu

Abstract

A quantitative model can play an essential role in controlling critical quality attributes of products and in designing the associated processes. One of the challenges in designing a dry granulation process is to find the optimal balance between improving powder flowability and sacrificing powder tabletability, both of which are highly affected by ribbon solid fraction and granule size distribution (GSD). This study is focused on developing a hybrid machine learning (ML)-assisted mechanistic model to predict ribbon solid fraction, GSD, and throughput for the purpose of implementing model predictive control of an integrated continuous dry granulation tableting process. It is found that the predictability of ribbon solid fraction and throughput are improved when modification is made to Johanson's model by incorporating relationships between roll compaction parameters and ribbon elastic recovery. Such relationships typically are either not considered or assumed to be a constant in the models reported in the literature. To describe the nature of the bimodal size distribution of roller compactor granules instead of only using traditional D_{10}, D_{50} and D_{90} values, the GSD is represented by a bimodal Weibull distribution with five fitting parameters. Furthermore, these five GSD parameters are predicted by ML models. The results indicate the ribbon solid fraction and screen size are the two most significant factors affecting GSD.

Keywords: dry granulation, roller compactor, machine learning, hybrid model

1. Introduction

The dry granulation process is an important route for producing a solid dosage form in the pharmaceutical industry. The roller compactor is the key unit operation in a dry granulation process. It includes two steps: (1) roll compaction in which powder blends are compressed between two counter-rotating rolls into a ribbon, and (2) milling in which these ribbons are crushed into granules. The benefits of dry granulation include improved blend uniformity and flowability by enlargement of particle size. Powder flowability plays a key role in determining the performance of the tablet manufacturing process and final drug product quality (Lagare et al., 2023). In addition, good powder flowability can reduce powder fouling and improve the on-line sensor performance, such as capacitance-based particulate flow rate sensor (Huang et al., 2022). However, particle size over-enlargement or over-compression of powders can compromise the tabletabiltiy (Herting & Kleinebudde, 2008). Finding the optimal balance between improving powder flowability and sacrificing powder tabletability, which are highly affected by ribbon solid fraction and GSD, becomes one of the challenges in designing a dry granulation process. Therefore, quantitative models to predict ribbon solid fraction and GSD are essential to optimally operate the roller compactor.

Mechanistic models such as Johanson's model(Johanson, 1965) are typically used to describe roll compaction and further predict ribbon solid fraction. However, one of the reasons for the unsatisfactory prediction accuracy of ribbon solid fraction is due to elastic recovery, which is either not considered or assumed to be a constant in the models reported in the literature (Keizer & Kleinebudde, 2020). Population balance models (PBM) can account for the milling step and can be used to predict GSD, but it is complicated to determine the breakage function in the PBM purely based on ribbon fracture physics. Given the unknown physical nature of ribbon elastic recovery and GSD, machine learning (ML) is a preferred alternative to developing a mechanistic model. Moreover, ML and mechanistic model components can be combined into a hybrid model to maintain high physical interpretability and feasibility. Therefore, the primary objective of this work is to develop and validate a hybrid model for the purpose of implementing model predictive control of an integrated continuous dry granulation tableting process.

2.Methodology

2.1. Parameter Estimation

Given n experiments and m output variables, the procedure to estimate model parameters can be formulated as a constrained optimization problem:

$$\min_{\hat{\theta}} J = \sum_{i=1}^{n} \sum_{j=1}^{m} w_j \left(y_{j,i} - \hat{y}_{j,i}\right)^2 \tag{1}$$

$$subject\ to\ \ f\left(u, \hat{\theta}, \hat{y}\right) = 0, \qquad \hat{\theta} \in \Omega_{\theta}\ , \qquad \hat{y} \in \Omega_y$$

where u and y are measurements of input and output variables. $\hat{\theta}$ and $\hat{y}$ are estimated model parameters and predicted output variables, bounded in compact sets Ω_θ and Ω_y, respectively. w_j is the weighting for residual of output variable y_j.

2.2. Machine Learning

Machine learning models are common alternatives when process outputs are difficult to predict by pure mechanistic models. The multiple linear regression (MLR) model and the neural network (NN) model are employed in this study. The NN models studied consist of one hidden layer, where the hyperbolic tangent function is applied.

2.3. Experimental Methods

The material used in this study was a blend of 90 % w/w microcrystalline cellulose Avicel PH102 and 10% w/w acetaminophen. In each experiment, the Alexanderwerks WP120 roller compactor was operated for three minutes to reach a steady state, at which point samples were collected. The ribbons of interest were collected after the powders were compressed into ribbons and then broken into smaller ribbons by the flake crusher. The granulator consists of a two-stage hammer mill with two different screen sizes (screen 1 and screen 2), which produces two granule samples (granule 1 and granule 2). An in-house flowrate sensor based on a Mettler Toledo ME 4001E weighing scale was located at the roller compactor exit to capture the throughput of granules.

In order to measure the ribbon thickness, a Fisherbrand Traceable digital caliper was used to measure thirty ribbons to provide a statistically significant sample size. Geopyc 1360 pycnometer was used to measure the ribbon envelop density (ρ_e). The ribbon solid fraction (γ_R) can be computed as:

$$\gamma_R = \frac{\rho_e}{\rho_t} \tag{2}$$

where ρ_t is the powder true density, which is measured by an Accupyc II 1340 pycnometer. The GSD of granule 1 and granule 2 samples were measured by the Canty SolidSizer, which measures the size and area of each particle. In this study, the circular equivalent diameter was used and the cumulative frequency of GSD was volume-based.

3. Results and Discussion

The input and output variables of the roller compactor are shown in Figure 1. Ribbon splitting is a phenomenon causing additional uncertainty in the model accuracy and can be avoided when the roll gap and roll pressure are low.

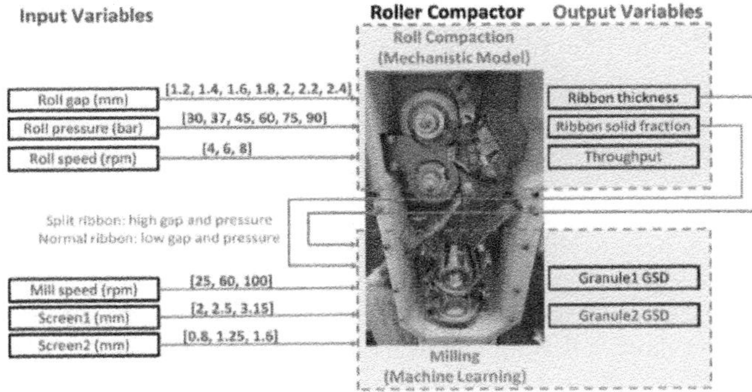

Figure 1. Roller compactor schematic.

3.1 Roll Compaction

When the materials transform from the slip condition to the non-slip condition, the stress gradients in slip region and nip region are equal. The critical angular roller position at which this occurs is known as the nip angle α and can be calculated by Johanson's model:

$$\frac{4\left(\frac{\pi}{2}-\alpha-\nu\right)\tan\delta_E}{\cot(A-\mu)-\cot(A+\mu)} - \frac{K\left(2\cos\alpha-1-\frac{S}{D_R}\right)\tan\alpha}{\cos\alpha} = 0 \tag{3}$$

$$\text{where } A = \frac{\alpha+\nu+\frac{\pi}{2}}{2}, \quad \nu = \frac{1}{2}\left[\pi - \sin^{-1}\left(\frac{\sin\phi_W}{\sin\delta_E}\right) - \phi_W\right], \quad \mu = \frac{\pi}{4} - \frac{\delta_E}{2} \tag{4}$$

Here, δ_E is the effective angle of internal friction and ϕ_W is wall friction angle, K is compressibility factor, S is roll gap, and D_R is roll diameter.

Given roll diameter D_R and roll width W, the peak pressure (P_{max}) applied on the powders at the minimum roll gap S is computed as follows:

$$P_{max} = \frac{2P_H A_{CS}}{W D_R F} \tag{5}$$

with the force factor, F, given by

$$F = \int_0^{\alpha} \left[\frac{\frac{S}{D_R}}{\left(1 - \frac{S}{D_R} - \cos\theta\right)\cos\theta}\right]^K \cos\theta \, d\theta \tag{6}$$

where P_H is hydraulic pressure (or roll pressure) and A_{cs} is area of the compact surface. Based on a compression power law, the ribbon solid fraction at the gap γ_G can be computed as follows:

$$\gamma_G = \gamma_0 (P_{max})^{\frac{1}{K}} \tag{7}$$

where γ_0 is the pre-consolidation solid fraction. However, γ_G is not easily measured because ribbon elastic relaxation makes ribbon density decrease when ribbons are released from the roll. Given the elastic recovery β, the ribbon solid faction γ_R is represented as follows:

$$\gamma_R = \frac{\gamma_G}{\beta} \tag{8}$$

Considering mass balance around the roll gap and roll speed N_R, the mass throughput can be calculated as follows:

$$\dot{M} = \pi D_R W S N_R \rho_t \gamma_G = \pi D_R W S N_R \rho_t \beta \gamma_R \tag{9}$$

In this work, 15 sets of training data and 4 sets of test data are used to validate and evaluate three roll compaction models. First, a data-driven MLR model is built as a benchmark. Secondly, roll compaction mechanistic models are highly sensitive to powder compressibility K and powder pre-consolidation solid fraction γ_0(Toson et al., 2019). Instead of using common regression approaches to estimate these two model parameters, this study estimates parameters in a constrained optimization framework as follows:

$$\min_{\phi_W, \delta_E, K, \gamma_0, \beta} J = \sum_{i=1}^{n} [W_\gamma (\gamma_{R,i} - \hat{\gamma}_{R,i})^2 + W_M (\dot{M}_i - \hat{\dot{M}}_i)^2] \tag{10}$$

Thirdly, elastic recovery is known to be a function of roll compaction parameters instead of a constant. Under the assumption that measured elastic recovery is the ratio of measured ribbon thickness to roll gap, the predicted elastic recovery β is formulated as:

$$\beta_{Model3} = 0.96 + 0.12\frac{S}{S_0} + 0.03\frac{P_H}{P_{H0}} + 0.01\frac{N_R}{N_{R0}} \tag{11}$$

where β_{Model3} has training error MAPE = 4.17 % and test error MAPE = 10.67%. To further improve the elastic recovery predictability, it is worth investigating ribbon splitting phenomenon and other nonlinear ML models in future studies. The prediction performances of the three roll compaction models are summarized in Table 1, which is calculated using the test sets. By incorporating the relationships between roll compaction parameters and ribbon elastic recovery, Model3 has the smallest mean absolute percentage error (MAPE) for both ribbon solid fraction and mass throughput, indicating the best model performance. The parity plot for Model3 is shown in Figure 2.

Table 1. Roll compaction model performance.

Model	MAPE(γ_R) [%]	MAPE($\dot{M}$) [%]
Model1 (MLR)	5.68	9.98
Model2 (Johanson's with constant β)	4.40	5.01
Model3 (Johanson's with $\beta = \beta_{model3}$)	2.86	4.61

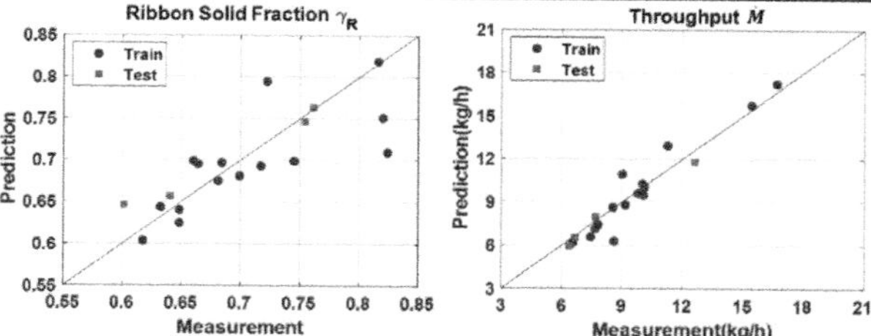

Figure 2. Performance of hybrid model of considering elastic recovery model (Model3).

3.2 Milling

The hammer milling step commonly produces granules with a bimodal size distribution, which is not adequately described by only using D_{10}, D_{50} and D_{90} values. Therefore, the entire cumulative size distribution $D_5, D_{10}, \ldots, D_{95}$ measured with the Canty SolidSizer is represented by a bimodal GSD, which can be characterized by a cumulative bimodal Weibull distribution $Q_3(x)$:

$$Q_3(x) = a\left(1 - e^{-\left(\frac{x}{p_1}\right)^{m_1}}\right) + (1-a)\left(1 - e^{-\left(\frac{x}{p_2}\right)^{m_2}}\right) \tag{12}$$

where a is the weighting of the two modes, p_1 and p_2 are the size parameters of the small mode and large mode, respectively, whereas $m_1$and m_2 represent the shape parameters of the associated modes. Utilizing these statistical model parameters provides a more interpretable approach to describing GSD and reduces the GSD parameter set from 19 to 5. The bimodal Weibull distribution parameters can be computed by solving an optimization problem:

$$\min_{a,\, p_1,\, p_2, m_1, m_2} \quad J = \sum_{i=5,10\ldots}^{95} \left(\frac{i}{100} - Q_3(D_i) \right)^2 \tag{13}$$

$subject\ to\ 0 \leq a \leq 1, \quad 0 < p_1 < p_2, \quad 1 < m_1,\ m_2$

To link the roll compaction and milling process, the ribbon solid fraction γ_R and ribbon thickness ($Rtck$) serve as inputs of the milling model. In addition, milling speed ($Mill$) and screen size(s) are also used to predict the GSD described by five bi-modal Weibull fitting parameters. The sensitivity analysis of both granule 1 and granule 2 are shown in Figure 3. The sensitivity analysis was determined by the MLR model coefficient, where inputs and outputs were both rescaled by dividing the minimum values. In terms of weighting a, ribbon solid fraction of granule 1 is dominated by ribbon solid fraction while that of granule 2 is more sensitive to screen size. For both granule 1 and granule 2, the size parameter p_1 is highly related to ribbon solid fraction. On the other hand, p_2 is dominated by screen size given that screen size determines the upper boundary of the particle size. Shape parameter m_1 is less sensitive to all process inputs compared to the shape parameter m_2, which is highly impacted by the solid ribbon fraction.

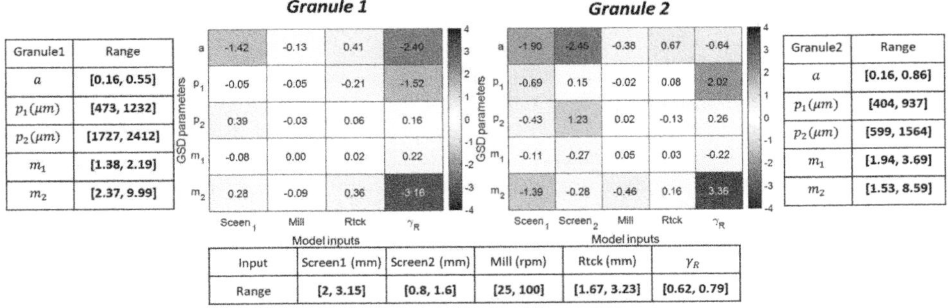

Granule1	Range
a	[0.16, 0.55]
$p_1(\mu m)$	[473, 1232]
$p_2(\mu m)$	[1727, 2412]
m_1	[1.38, 2.19]
m_2	[2.37, 9.99]

Granule2	Range
a	[0.16, 0.86]
$p_1(\mu m)$	[404, 937]
$p_2(\mu m)$	[599, 1564]
m_1	[1.94, 3.69]
m_2	[1.53, 8.59]

Input	Screen1 (mm)	Screen2 (mm)	Mill (rpm)	Rtck (mm)	γ_R
Range	[2, 3.15]	[0.8, 1.6]	[25, 100]	[1.67, 3.23]	[0.62, 0.79]

Figure 3. Sensitivity analysis of the milling process.

To evaluate the performance of the milling models, the MAPE of GSD is utilized and can be computed as:

$$MAPE = \sum_{i=1}^{n} \sum_{p=5,10,\ldots}^{95} \left| \frac{D_{p,i}^{pred} - D_{p,i}}{D_{p,i}} \right| \times 100\% \tag{14}$$

The evaluation of the milling models is summarized in Table 2, which is based on 20 training sets and 6 test sets. MLR and NN models are used to predict GSD. The NN models have two neurons in the hidden layer. While increasing the number of neurons can readily make the training error of the NN model smaller than that of the MLR model, the test error can become much worse due to overfitting. The NN model generally can handle nonlinearity better than the MLR model, but some constraints might be hard to enforce, e.g., predicted shape parameters m_1 and m_2 might be less than 1 in the test sets. Figure 4 demonstrates the predictability of the granule 2 GSD by using the MLR or NN models based on six test sets. It is worth noting that there exists a significant mismatch between measurement and NN predictions in Exp 6, but the NN model prediction seems more reasonable considering that screen 2 is 1.25 mm and Exp 6 has a smaller ribbon solid fraction compared to Exp 3, which should result in a smaller GSD. In summary, the hybrid model successfully predicts GSD, and further investigation of different ML models might be useful to enhance the model performance.

Table 2 Milling model performance.

Model	Granule1		Granule2	
	$MAPE_{train}$ (%)	$MAPE_{test}$ (%)	$MAPE_{train}$ (%)	$MAPE_{test}$ (%)
MLR	9.79	11.21	10.89	8.95
NN	9.66	12.20	13.03	8.37

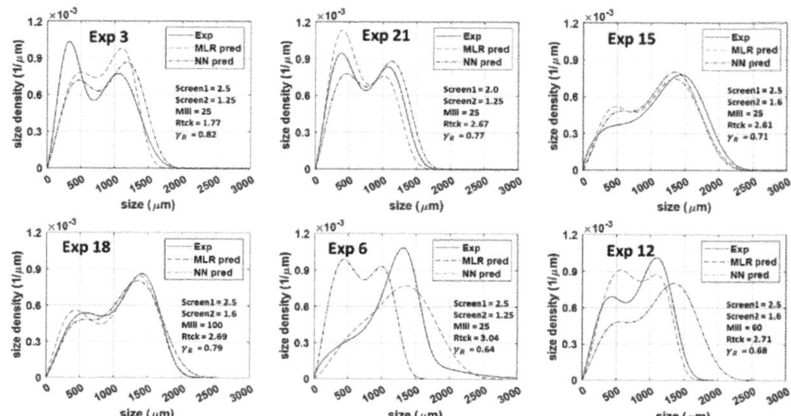

Figure 4. Model prefomance of granule2 based on test sets.

4. Conclusion

A hybrid model for the roller compactor is proposed that demonstrates satisfactory predictability of ribbon solid fraction, throughput, and GSD. To further improve the model performance, investigation on ribbon elastic recovery and splitting phenomenon and incorporation constraints into ML model could be important. Future work will include relating ribbon solid fraction and GSD to the tablet properties and implementing model predictive control of the integrated dry granulation tableting process.

Acknowledgement

This work was supported by the National Science Foundation under Grant No. 2140452 - CMMI-EPSRC: Right First Time Manufacture of Pharmaceuticals (RiFTMaP)

References

Herting, M. G., & Kleinebudde, P. (2008). Studies on the reduction of tensile strength of tablets after roll compaction/dry granulation. *European Journal of Pharmaceutics and Biopharmaceutics*, *70*(1), 372–379.

Huang, Y. -S., Medina-González, S., Straiton, B., Keller, J., Marashdeh, Q., Gonzalez, M., Nagy, Z., & Reklaitis, G. v. (2022). Real-Time Monitoring of Powder Mass Flowrates for Plant-Wide Control of a Continuous Direct Compaction Tablet Manufacturing Process. *Journal of Pharmaceutical Sciences*, *111*(1), 69–81.

Johanson, J. R. (1965). *A Rolling Theory for Granular Solids Theory and Method of Calculation.*

Keizer, H. L., & Kleinebudde, P. (2020). Elastic recovery in roll compaction simulation. *International Journal of Pharmaceutics*, *573*.

Lagare, R. B., Huang, Y.-S., Bush, C. O.-J., Young, K. L., Rosario, A. C. A., Gonzalez, M., Mort, P., Nagy, Z. K., & Reklaitis, G. v. (2023). Developing a Virtual Flowability Sensor for Monitoring a Pharmaceutical Dry Granulation Line. *Journal of Pharmaceutical Sciences.*

Toson, P., Lopes, D. G., Paus, R., Kumar, A., Geens, J., Stibale, S., Quodbach, J., Kleinebudde, P., Hsiao, W. K., & Khinast, J. (2019). Model-based approach to the design of pharmaceutical roller-compaction processes. *International Journal of Pharmaceutics: X*, *1*, 100005.

Antonis Kokossis, Michael C. Georgiadis, Efstratios N. Pistikopoulos (Eds.)
PROCEEDINGS OF THE 33rd European Symposium on Computer Aided Process Engineering (ESCAPE33), June 18-21, 2023, Athens, Greece
 http://dx.doi.org/10.1016/B978-0-443-15274-0.50131-1

A Data-driven Modeling Approach for Water Flow Dynamics in Soil

Zeyuan Song[a], Zheyu Jiang[a]*

[a]Oklahoma State University, 420 Engineering North, Stillwater, Oklahoma 74074 USA
Corresponding author: zheyu.jiang@okstate.edu

Abstract

Modeling and predicting soil moisture is essential for precision agriculture, smart irrigation, drought prevention, etc. Estimating root zone soil moisture from surface or near-surface soil moisture data is typically achieved by solving a hydrological model that describes water movement through soils. Advanced agro-hydrological models today use the Richards equation, a highly nonlinear, degenerate elliptic-parabolic partial differential equation that captures irrigation, precipitation, evapotranspiration, runoff, and drainage. State-of-the-art Richards equation solvers employ either a finite difference, finite element, or finite volume discretization framework in space. In this paper, we introduce a novel computational framework to solve generic n-dimensional Richards equation by introducing global random walk and deep neural network to a modified finite volume method (FVM). Furthermore, for n-dimensional Richards equation, we introduce multi-point flux approximation to the FVM framework. Through these innovations, our novel computational framework effectively utilizes the underlying physics behind the Richards equation, which enhances the speed and accuracy of the solution process. Through an illustrative case study, we demonstrate the efficiency and effectiveness of our computational framework and show that it correctly characterizes the physical relationships among soil moisture content, pressure head, and flux.

Keywords: Soil moisture, Richards equation, finite volume method, neural network, random walk

1. Introduction

Soil moisture is a key hydrological parameter that has significant importance to human society and environment. Accurate modeling and monitoring of soil moisture in crop fields, especially in the root zone (top 100 cm of soil), is essential for improving agricultural production and crop yield with the help of precision irrigation and farming tools. Recent studies also show that monitoring root zone soil moisture at suitable locations and adjusting irrigation schedules accordingly can reduce water use by 40-60% (Sadler et al., 2005). Improved irrigation infrastructures based on soil moisture knowledge could also prevent more than $30 billion/year in drought related agricultural losses in the US (Khand et al., 2018). This is especially important to US states like Oklahoma, where 41% of total water use goes directly to agricultural irrigation and almost 90% of the land area suffers from drought throughout the year (Droughts.gov, 2022). Estimating root zone soil moisture is typically achieved by solving a hydrological model that describes water movement through soils. Most of the advanced agro-hydrological models today incorporate the Richards equation (Richards, 1931), which captures irrigation, precipitation, evapotranspiration, runoff, and drainage dynamics of water in saturated and unsaturated porous medium such as soil:

$$\frac{\partial \theta(\varphi)}{\partial t} = \nabla \cdot [K(\theta)\nabla(\varphi + z)] - S \tag{1}$$

where φ stands for pressure head, θ denotes the soil moisture content, K is unsaturated hydraulic water conductivity, $t \in [0, T]$ denotes the time, z denotes the vertical depth, and S is the sink term associated with root water extraction, which we ignore without loss of generality. The flux is given by the Darcy's law: $q = -K(\theta)\nabla(\varphi + z)$. For unsaturated flow, both K and θ are highly nonlinear functions of φ. For instance, the widely used van Genuchten-Mualem correlations (van Genuchten, 1980) for θ(φ)and K(θ) are:

$$\theta(\varphi) = \begin{cases} \dfrac{\theta_s - \theta_r}{[1 + (\alpha|\varphi|)^n]^{\frac{n-1}{n}}} + \theta_r, & \varphi < 0 \\ \theta_s, & \varphi \geq 0 \end{cases}$$

$$K(\theta) = K_s\sqrt{\frac{\theta - \theta_r}{\theta_s - \theta_r}}\left\{1 - \left[1 - \left(\frac{\theta - \theta_r}{\theta_s - \theta_r}\right)^{\frac{n}{n-1}}\right]^{\frac{n-1}{n}}\right\}^2 \tag{2}$$

where α and n are van Genuchten parameters characterizing different soils, and θ_s, K_s and θ_r denote saturation soil moisture, residual water content, and saturated hydraulic conductivity, respectively. As a result, exact analytical solutions to Equation (1) often do not exist, and numerical solutions rely on discretization of spatial and temporal domains. In particular, the finite volume method (FVM) adopts an integral form of the Richards equation, which captures some valuable physical insights about water flow dynamics (Rathfelder and Abriola, 1994). In addition, FVM is a flexible framework that can be coupled with various other techniques such as the predictor-corrector method (Lai and Ogden, 2015), linearization scheme (Radu et al., 2015), and global random walk approach (Vamos, 2013). Recently, we developed a computationally efficient FVM-based framework for solving the Richards equation by integrating it with adaptive linearization scheme, global random walk method, and multi-layer neural networks (Song and Jiang, 2023). In an illustrative example, we showed that our novel data-driven framework not only generated fast and accurate solutions to the 1-D Richards equation, but also implicitly captured the underlying physical relationships among soil moisture content, pressure head, and flux. In this work, we generalize this computational framework to successfully solving 3-D Richards equation for the first time in the literature.

2. FVM Discretization

To apply FVM to Richards equation, we first integrate both sides of Equation (1) to obtain an integral form of the Richards equation over a higher-dimensional control V:

$$\int_V \frac{\partial \theta(\varphi)}{\partial t} dV = \int_V \nabla \cdot [K(\theta)\nabla(\varphi + z)]dV \tag{3}$$

Next, we can apply the divergence theorem to convert the volume integral on the RHS of Equation (3) into a surface integral over S_V by introducing the outward pointing unit normal vector denoted as **n**:

$$\int_V \frac{\partial \theta(\varphi)}{\partial t} dV = \int_{S_V} [K(\theta)\nabla(\varphi + z)] \cdot \mathbf{n}\, dS_V \tag{4}$$

By doing so, FVM is able to incorporate conservation laws, whereas other discretization methods cannot. Specifically, to obtain the discretized version of Equation (4) using FVM, we discretize the volume integral on the LHS of Equation (4) into a total of N_V small cells V_i with $i = 1, \cdots N_\omega$ whose volume is denoted as ϑ_i. Each V_i is associated with surfaces $\omega_{i,j}$ for $j = 1, \cdots N_{\omega_i}$ on the RHS of Equation (4) whose area is denoted as $\mathcal{A}(\omega_{i,j})$, and we use $[\cdot]_{\omega_i}$ and $\mathbf{n}_{\omega_i}$ respectively to denote the operator and outward pointing unit normal vector associated with $\omega_{i,j}$ upon discretization. Next, to discretize the time domain, we approximate the time derivative $\partial\theta(\varphi)$ as $\frac{\theta_i^{m+1,s+1} - \theta_i^m}{\Delta t}$, where $\theta_i^{m+1,s+1}$ represents the discretized θ in the ith small cell at the next time step $m + 1$ and iteration step $s + 1$, whereas θ_i^m is the converged θ value in the ith small cell at the current time step m. With this, the discretized version of Equation (4) using FVM becomes:

$$\frac{\theta_i^{m+1,s+1} - \theta_i^m}{\Delta t}\vartheta_i = \sum_{j=1}^{N_{\omega_i}} [K(\theta)\nabla(\varphi + z)]_{\omega_{i,j}} \cdot \mathbf{n}_{\omega_{i,j}} \mathcal{A}(\omega_{i,j}) \tag{5}$$

3. Data-Driven Global Random Walk Algorithm

In this section, we introduce a data-driven global random walk (DGRW) approach that seeks to implicitly encapsulate the physical knowledge describing particle movement and conservation within a control volume. Global random walk allows particles to move to neighboring cells simultaneously on any direction or stay in their current cells with a given probability. Thus, in 3-D case, there are a total of seven options for every particle. Let $\mathfrak{n}_{i,j,k}^{m,s}$ be the number of particles in the cell (i, j, k) at fixed-point iteration step s and time step m. Correspondingly, $\delta\mathfrak{n}_{i',j',k'}^{m,s}$, where $i' = i - 1, i, i + 1$, $j' = j - 1, j, j + 1$, and $k' = k - 1, k, k + 1$, denotes the number of particles moving from cell (i', j', k') to cell (i, j, k). Explicitly, we can write:

$$\mathfrak{n}_{i,j,k}^{m,s+1} = \delta\mathfrak{n}_{i+1,j,k}^{m,s} + \delta\mathfrak{n}_{i-1,j,k}^{m,s} + \delta\mathfrak{n}_{i,j+1,k}^{m,s} + \delta\mathfrak{n}_{i,j-1,k}^{m,s} + \delta\mathfrak{n}_{i,j,k+1}^{m,s} + \delta\mathfrak{n}_{i,j,k-1}^{m,s} + \delta\mathfrak{n}_{i,j,k}^{m,s} \tag{6}$$

As we may expect, having such physical knowledge is important for solving Richards equation, which governs the movement of water molecules in unsaturated and saturated soil systems. In fact, Suciu et al. (2021) adopted global random walk concepts and proposed a numerical framework for solving 1- and 2-D Richards equation. In their numerical framework, the authors assumed that the pressure head φ is proportional to the number of particles $\mathfrak{n}$ in a cell or control volume. With this assumption, soil moisture content in the cell of interest is simply proportional to the arithmetic mean of the number of particles. While this assumption is valid for diffusion equations (Vamoş et al., 2001), Richards equation is a highly nonlinear convection-diffusion equation, and the exact relationship between φ and $\mathfrak{n}$ remains unclear. In fact, we have shown that the numerical framework proposed by Suciu et al. (2021) failed to obtain an accurate solution for 1-D Richards equation (Song and Jiang, 2023).

Since the relationship between φ and $\mathfrak{n}$ may not be describable by any basic function, we decide to model the relationship using two multi-layer neural networks (MNNs). Although deep neural network with more layers and neurons could also be used, it is not required as MNN with less number of layers (e.g., 3) is sufficient for learning the relationship between φ and $\mathfrak{n}$ given enough neurons in the neural network (Hornik, 1991). In one of the MNNs, we approximate φ as a function of $\mathfrak{n}$, $\varphi_{i,j,k} = f(\mathfrak{n}_{i,j,k})$, at each fixed time step and iteration step. In the other MNN, we learn the inverse mapping

f^{-1} from pressure head information to the number of particles, $\mathfrak{n}_{i,j,k} = f^{-1}(\varphi_{i,j,k})$. During offline training, we first obtain reference solutions from the global random walk solvers developed by Sucui et al. (2021) (code available at GitHub repository: https://github.com/PMFlow/FlowBenchmark). We then add Gaussian noise to these reference solutions to 1) account for the possibly nonlinear relationship between φ and $\mathfrak{n}$, and 2) enhance its generalization performance and the quality of solution obtained (Song and Jiang, 2023). Once offline training is complete, we substitute f to Equation (5) and derive the following data-driven random walk formulation for the discretized Richards equation:

$$\begin{aligned}\mathfrak{n}_{i,j,k}^{m,s+1} = &\left[1-\left(r_{i+\frac{1}{2},j,k}^{m,s}+r_{i-\frac{1}{2},j,k}^{m,s}+r_{i,j+\frac{1}{2},k}^{m,s}+r_{i,j-\frac{1}{2},k}^{m,s}+r_{i,j,k+\frac{1}{2}}^{m,s}+r_{i,j,k-\frac{1}{2}}^{m,s}\right)\right]\mathfrak{n}_{i,j,k}^{m,s}\\ &+r_{i+\frac{1}{2},j,k}^{m,s}\mathfrak{n}_{i+1,j,k}^{m,s}+r_{i-\frac{1}{2},j,k}^{m,s}\mathfrak{n}_{i-1,j,k}^{m,s}+r_{i,j+\frac{1}{2},k}^{m,s}\mathfrak{n}_{i,j+1,k}^{m,s}+r_{i,j-\frac{1}{2},k}^{m,s}\mathfrak{n}_{i,j-1,k}^{m,s}\\ &+r_{i,j,k+\frac{1}{2}}^{m,s}\mathfrak{n}_{i,j,k+1}^{m,s}+r_{i,j,k-\frac{1}{2}}^{m,s}\mathfrak{n}_{i,j,k-1}^{m,s}\\ &+f^{-1}\left(\left(r_{i,j,k+\frac{1}{2}}^{m,s}-r_{i,j,k-\frac{1}{2}}^{m,s}\right)\Delta z-\frac{\theta^{m,s}\left(\mathfrak{n}_{i,j,k}^{m,s}\right)-\theta^{m-1}\left(\mathfrak{n}_{i,j,k}^{m-1,s}\right)}{L_{i,j,k}^{m,s}}\right)\end{aligned} \tag{6}$$

where $r_{i\pm\frac{1}{2},j,k}^{m,s} = \frac{K\left(\theta\left(\varphi_{i\pm\frac{1}{2},j,k}^{m,s}\right)\right)\Delta t}{(\Delta x)^2 L_{i\pm\frac{1}{2},j,k}^{m,s}}$, $r_{i,j\pm\frac{1}{2},k}^{m,s} = \frac{K\left(\theta\left(\varphi_{i,j\pm\frac{1}{2},k}^{m,s}\right)\right)\Delta t}{(\Delta y)^2 L_{i,j\pm\frac{1}{2},k}^{m,s}}$, and $r_{i,j,k\pm\frac{1}{2}}^{m,s} = \frac{K\left(\theta\left(\varphi_{i,j,k\pm\frac{1}{2}}^{m,s}\right)\right)\Delta t}{(\Delta z)^2 L_{i,j,k\pm\frac{1}{2}}^{m,s}}$. Here, we adopt an adaptive linearization scheme inspired by Mitra and Pop (2019) by adding the term $L_{i,j,k}^{m,s}\left(\varphi_{i,j,k}^{m,s+1}-\varphi_{i,j,k}^{m,s}\right)$ on the LHS of Equation (5).

In DGRW algorithm, Equation (6) passes through the inverse mapping f^{-1} learned from MNN to generate Richards equation solutions in terms of the particle distribution in each cell, $\mathfrak{n}_{i,j,k}$. To convert these solutions to physically meaningful solutions such as the pressure head, flux, and soil moisture content, we apply the trained mapping f^{-1} and Equation (2). We have proven that the DGRW algorithm is convergent, and the total error can be estimated using $\varepsilon = \max_{i}\left\{\frac{\left\|f\left(\mathfrak{n}_{i,j,k}^{m,s+1}\right)-f\left(\mathfrak{n}_{i,j,k}^{m,s}\right)\right\|_F}{\left\|f\left(\mathfrak{n}_{i,j,k}^{m,s+1}\right)\right\|_F}\right\} = \max_{i}\left\{\frac{\left\|\psi_{i,j,k}^{m,s+1}-\psi_{i,j,k}^{m,s}\right\|_F}{\left\|\psi_{i,j,k}^{m,s+1}\right\|_F}\right\}$, where $\|\cdot\|_F$ is the Frobenius norm.

4. An Illustrative Case Study

We modify the 2-D benchmark problem of Havercamp (1977) that describes groundwater reservoir recharge from a drainage trench by extending it to 3-D. The boundary conditions of the problem are set up as $\Omega = [0\text{m}, 2\text{m}]^3$, and $\Gamma_D = \Gamma_{D_1} \cup \Gamma_{D_2} \cup \Gamma_{D_3} \cup \Gamma_{D_4}$, in which:

$$\begin{aligned}\Gamma_{D_1} &= \{(x,y,z) \in \partial\Omega | x \in [0\text{m}, 1\text{m}] \wedge y \in \Omega \wedge z = 2\text{m}\},\\ \Gamma_{D_2} &= \{(x,y,z) \in \partial\Omega | x \in \Omega \wedge y \in [0\text{m}, 1\text{m}] \wedge z = 2\text{m}\},\\ \Gamma_{D_3} &= \{(x,y,z) \in \partial\Omega |\, x = 2\text{m} \wedge y \in \Omega \wedge z = [0\text{m}, 1\text{m}]\},\\ \Gamma_{D_4} &= \{(x,y,z) \in \partial\Omega |\, x \in \Omega \wedge y = 2\text{m} \wedge z = [0\text{m}, 1\text{m}]\}.\end{aligned}$$

The Dirichlet boundary condition on Γ_D results in the drainage process, whereas the Neumann boundary condition is applied on $\Gamma_N = \partial\Omega \setminus \Gamma_D$. The z-direction is point downward into the ground. The initial conditions describing hydrostatic equilibrium are modified as follows:

$$\varphi(\mathrm{x},\mathrm{y},\mathrm{z},\mathrm{t}) = \begin{cases} -2 + 2.2\dfrac{t}{\Delta t_D} \text{ on } \Gamma_{D_1} \text{ and } \Gamma_{D_2}, T \leq \Delta t_D, \\ 0.2 \text{ on } \Gamma_{D_1} \text{ and } \Gamma_{D_2}, T > \Delta t_D, \\ 1 - z \text{ on } \Gamma_{D_3} \text{ and } \Gamma_{D_4}, \end{cases}$$

$$\mathrm{q}(\mathrm{x},\mathrm{y},\mathrm{z},\mathrm{t}) = 0 \text{ on } \Gamma_N,$$

$$\varphi(\mathrm{x},\mathrm{y},\mathrm{z},0) = \begin{cases} 1 - \mathrm{z} \text{ on } \Omega \backslash \Gamma_D \\ -2 \text{ on } \Gamma_D \end{cases}.$$

We use the parameters listed in List (2016) to represent water flow dynamics in silt loam soil, in which $\theta_s = 0.396$, $\theta_r = 0.131$, $\alpha = 0.423$, $n = 2.06$, and $K_s = 0.0496$ in Equation (2). We choose $\Delta t_D = \frac{1}{16}$d, $\Delta t = \frac{1}{48}$d, and $T = \frac{3}{16}$d. A rectangular 3-D mesh with 9261 nodes is used, where $\Delta x = \Delta y = \Delta z = 0.1$. The convergence tolerance ε_0 is set to be 10^{-5}. Each MNN contains 3 layers and 10 neurons. Overall, 9261 noise-added reference solutions from global random walk solvers developed by Suciu et al. (2021) were obtained, with 70%, 15%, and 15% of them being used for training, validation, and testing, respectively. Bayesian regularization is used to train both MNNs.

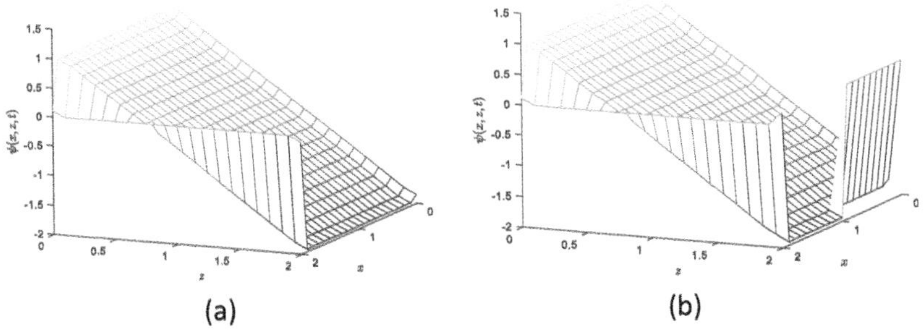

Figure 1. Pressure head obtained from (a) GRW algorithm, and (b) DGRW approach.

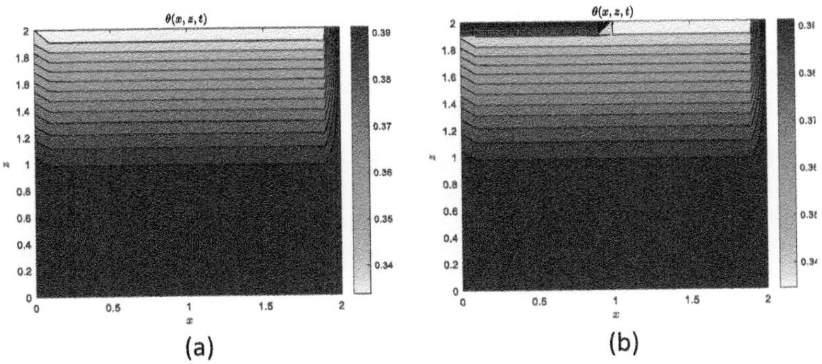

Figure 2. Pressure head obtained from (a) DGRW algorithm, and (b) GRW approach.

Figures 1 and 2 illustrate the pressure head and soil moisture results obtained using the GRW algorithm of Suciu et al. (2021) and our DGRW framework at $y = 1.7$m and $t = T = \frac{3}{16}$d. In general, our DGRW results show excellent agreement with the GRW results. Nevertheless, we observe that the GRW solutions did not produce jumps in pressure head and soil moisture content in the z-direction as x approaches 1m (Γ_{D_1}), whereas our

DGRW did predict these jumps. As the total simulated time is greater than Δt_D, the initial condition for pressure head should be 0.2m for $x = [0m, 1m]$ and -2m for $x = [1m, 2m]$ (Figure 1b). Similarly, jumps in soil moisture and flux in the z-direction are expected at $x = 1m$ as well. Therefore, by effectively characterizing the complex relationship between pressure head and the number of particles, our DGRW approach is capable of capturing the water flow dynamics embedded in the Richards equation.

5. Conclusion

In this work, we propose a novel data-driven DGRW framework to accurately solve 3-D Richards equation for the first time. DGRW synergistically integrates data-driven global random walk theory and FVM to encapsulate physical knowledge and insights describing water flow dynamics in soil systems. We compare the solutions of our DGRW framework and state-of-the-art GRW approach on a 3-D benchmark problem and successfully validate the accuracy and usefulness of our approach in capturing the underlying physics of water flow dynamics.

References

Drought.gov, 2022, Current U.S. Drought Monitor Conditions for Oklahoma, https://www.drought.gov/states/oklahoma, accessed March 2022.

R. Haverkamp, M. Vauclin, J. Touma, P.J. Wierenga, G. Vachaud, 1977, A comparison of numerical simulation models for one-dimensional infiltration, Soil Science Society of America, 41, 2, 285-294.

K. Hornik, 1991, Approximation capabilities of multilayer feedforward networks, Neural networks, 4, 2, 251-257.

K. Khand, S. Taghvaeian, A. Ajaz, 2018, Drought and Its Impact on Agricultural Water Resources in Oklahoma, https://extension.okstate.edu/fact-sheets/drought-and-its-impact-on-agricultural-water-resources-in-oklahoma.html, accessed March 2022.

W. Lai, F.L. Ogden, 2015, A mass-conservative finite volume predictor–corrector solution of the 1D Richards' equation, Journal of Hydrology, 523, 119-127.

F.A. Radu, J.M. Nordbotten, I.S. Pop, K. Kumar, 2015, A robust linearization scheme for finite volume based discretizations for simulation of two-phase flow in porous media, Journal of Computational and Applied Mathematics, 289, 134-141.

K. Rathfelder, L. M. Abriola, 1994, Mass conservative numerical solutions of the head-based richards equation, Water Resources Research, 30, 9, 2579-2586.

L.A. Richards, 1931, Capillary conduction of liquids through porous mediums, Physics, 1, 5, 318-333.

E. Sadler, R. Evans, K. Stone, and C. Camp, 2005, Opportunities for conservation with precision irrigation, Journal of Soil and Water Conservation, 60, 6, 371-378.

Z. Song, Z. Jiang, 2023, A data-driven random walk approach for solving water flow dynamics in soil systems, In: Proceedings Foundations of Computer-Aided Process Operations and Chemical Process Control Conference. San Antonio, TX.

N. Suciu, D. Illiano, A. Prechtel, F.A. Radu, 2021, Global random walk solvers for fully coupled flow and transport in saturated/unsaturated porous media, Advances in Water Resources, 152, 103935.

C. Vamos, N. Suciu, H. Vereecken, O. Nitzsche, H. Hardelauf, 2001, Global random walk simulations of diffusion, In Scientific Computing, Validated Numerics, Interval Methods, pp. 343-354.

M.T. van Genuchten, 1980, A closed-form equation for predicting the hydraulic conductivity of unsaturated soils, Soil Science Society of America Journal, 44, 5, 892-898.

Antonis Kokossis, Michael C. Georgiadis, Efstratios N. Pistikopoulos (Eds.)
PROCEEDINGS OF THE 33rd European Symposium on Computer Aided Process Engineering (ESCAPE33), June 18-21, 2023, Athens, Greece
 http://dx.doi.org/10.1016/B978-0-443-15274-0.50132-3

Numerical Simulation of No-Newtonian Fluids in Stirred-Tank with proximity impellers

Alvarado-Rodriguez C. E.[a*,b], Martinez-Herrera G.[a], Cortez-Gonzalez J.[c], Murrieta-Dueñas R[c].

[a]*Departamento de Ingeniería Química, División de Ciencias Naturales y Exactas, Universidad de Guanajuato, Noria Alta S/N, Guanajuato, 36050, México*

[b]*Consejo Nacional de Ciencia y Tecnología, Avenida Insurgentes Sur 1582, Crédito Constructor, Ciudad de México, 03940, México.*

[c]*Departamento de Ingeniería Química y Bioquímica, Tecnológico Nacional de Mexico/ Instituto Tecnológico Superior de Irapuato, 36821 Irapuato, Mexico,*

ce.alvarado@ugto.mx

Abstract

Many industrial applications involve stirred tanks with no-Newtonian rheology; hence the understanding of those mixing systems is required to improve final products. The present work aims to simulate the hydrodynamic no-Newtonian mixing system comprised of three impellers named: U-Type Anchor, Simple Hybrid, and Anchor with radial counterflow. As an industrial application, the water properties were considered for the Newtonian fluid, the ketchup and melted chocolate are used for shear-thinning and shear-thickening no-Newtonian fluids respectively. The performance is evaluated using power consumption, the homogeneity of concentration, velocity vector fields, and numerical mixing time. A modified version of the DualSPHysics code has been used to obtain the concentration fields and numerical mixing time. For the power consumption, the Herschel–Bulkley–Papanastasiou model was used to calculate the power for the different rheology. The numerical results suggest that the Simple Hybrid has the lowest power consumption and better performance.

Keywords: Stirred-Tank, Mixing times, CFD, Numerical simulation, SPH.

1. Introduction

Computational Fluid Dynamics (CFD) is a very useful tool to analyze in detail the hydrodynamics and mixing performance of stirred tanks in a variety of impellers. The CFD solves the fluid transport equations by computational methods. This can generate a huge amount of information that could not be obtained experimentally. It may also be used as an effective strategy to evaluate the performance of some processes. On mixing systems with a stirred tank, CFD allows obtaining axial and radial flow patterns, velocity profiles, mixing times, stagnation zones, or vortex formation in order to evaluate the performance of the impeller. In all processes, the agitation conditions vary with respect to the behavior of the different fluids, since these can present Newtonian or non-Newtonian behaviors. Recently, there is a great deal of interest in the performance of stirred tanks with no-Newtonian fluids using CFD. The existing literature on the no-Newtonian mixing system has been carried out by the eulerian-grid approach. Nevertheless, eulerian-grid numerical methods have certain drawbacks in dealing with deformable boundaries, moving interfaces, extremely large deformations, and especially grid regenerating (Abdolahzadeh et al., 2019). The present work aims to simulate the mixing in different rheology using four not reported close-clearance impellers, namely U-type anchor, anchor with radial counterflow, and simple hybrid. Prior to undertaking the simulation cases, a convergence analysis is carried out to establish adequate particle numbers for developing the simulation cases. Once the particle resolution is obtained by convergence analysis, the performance of impellers is evaluated using the macromixing parameters numerical mixing time and power consumption analysis. Subsequently, the velocity fields within the stirred tank are visualized for identifying pattern flows and possible stagnation zones.

2. The SPH numerical model

The SPH numerical model is based on the DualSPHysics model (Dominguez et al., 2022) shown in Eq. (1) to (6)

$$\frac{d\rho}{dt} = -\rho\nabla\cdot\mathbf{v}\,, \tag{1}$$

$$\frac{d\mathbf{v}}{dt} = -\frac{1}{\rho}\nabla P + \Gamma + g\,, \tag{2}$$

$$P = B\left[\left(\frac{\rho}{\rho_0}\right)^{\gamma} - 1\right], \tag{3}$$

where ρ is the density, t is time, **v** is the velocity vector, P is the pressure, Γ is the viscosity dissipative term, B is a reference pressure, $\gamma = 7$, ρ_0 is the reference density and g is the gravity acceleration.

The no-Newtonian fluids were simulated using the model reported in the DualSPHysics code based on the generalized Herschel–Bulkley–Papanastasiou (HBP) model (Papanastasiou, 1987; Mitsoulis, 2007). The HBP model can be used to simulate shear thinning or thickening materials in the absence of yield strength.

$$\tau^{ab} = \eta_{app}\gamma^{ab}\,, \tag{4}$$

$$\eta_{app} = \mu|\gamma|^{n_{HB}-1}\,, \tag{5}$$

$$\gamma = [\nabla\mathbf{v} + \nabla\mathbf{v}^T]\,, \tag{6}$$

where τ^{ab} is the viscosity stress tensor, η_{app} is the apparent viscosity, μ is the viscosity or consistency index, n is the power law index, $|\gamma|$ is the magnitude of the symmetric strain rate.

For this work the Eq. (7) was implemented in DualSPHysics follow the work reported by (Tartakovsky & Meakin, 2006) to include the mass transfer in the numerical simulations.

$$\frac{dC}{dt} = D\nabla^2 C - \nabla\cdot(C\mathbf{v}) + \mathbf{v}\cdot\nabla C\,, \tag{7}$$

where C is the concentration, and D is the diffusion coefficient.

The value of concentration is uploaded using the Verlet algorithm and the time step according to the Courant–Friedrichs–Lewy condition. The boundary conditions for Eq. (7) are implemented as Neumann conditions avoiding the mass transfer between the fluid particles to the boundary particles. With Eq. (7) is possible to calculate the mixing time considering advection-diffusion within a stirred tank to compare and evaluate the performance of the mixing between the impellers used in this work.

3. Geometry and impellers design

The diameter of tank is DT = 0.14 m, and the height of tank is HT = 0.245 m. Two elliptical capsules with 0.03 m of height are added to avoid stagnant zones on the bottom of the tank. In the Figure 1, it is shown the proposed unusual close-clearance impellers analyzed namely U-type anchor (1a), anchor with radial counterflow (1b) and simple hybrid (1c) defined by INOXPA as impellers for standard tanks. All impellers are used for a wide range of viscosities. The U-type anchor impeller has 4 scrapers that allow the removal of the material that adheres to the interior of the stirred tank. The recommended stirring rate works in a range of 3 and 120 rpm. The anchor with radial counterflow impeller is designed to facilitate radial flow. Its optimum working stirring rates range are from 3 to 120rpm. Finally, the self-designed modified simple hybrid impeller is intended to improve mixing times and fluid homogenization.

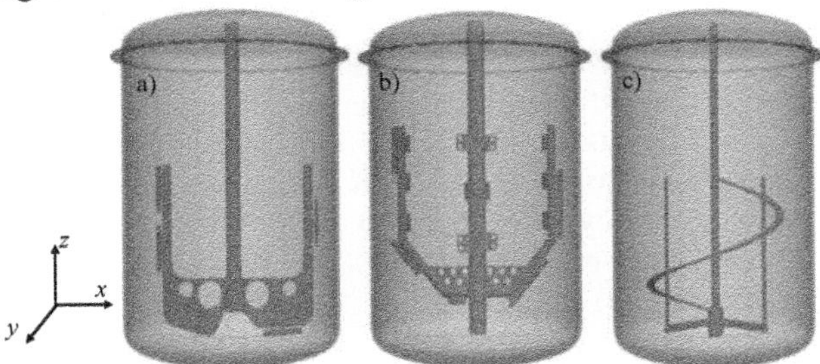

Figure 1. Design of the impellers described in section 3.1. a) U-type anchor, b) anchor with radial counterflow, c) simple hybrid.

4. Study Cases

The present work aims to simulate the hydrodynamic of no-Newtonian mixing systems using three impellers named: U-Type Anchor, Anchor with radial counterflow, and Simple Hybrid. As an industrial application, the water properties were considered for the Newtonian fluid, ketchup and melted chocolate are used for shear-thinning and shear-thickening no-Newtonian fluids, respectively. The performance is evaluated using power consumption, velocity vector fields, and numerical mixing time.

4.1. Convergence test.

Before proceeding to examine power consumption analysis, numerical mixing time, and velocity fields, it is important to execute a convergence test. The aim of the convergence test is to determine the adequate number of total particles required in the simulation cases. This test consists of increasing the number of particles in order to find the appropriate resolution of the simulation. In this work, the used particle numbers or particle resolution is defined as follows 75, 000, 125, 000, 250, 000, and 500, 000. In order to calculate the velocity profiles at different resolutions, it was selected the next coordinates $x = 0.036$ m, $y = 0.036$ m, and $z = 0.12$ m to evaluate the velocity at different times.

4.2. The mixing time

The mixing time test is one of the most common procedures for determining the performance of stirred tank system. A longer mixing time is mainly associated with an increase in power consumption due to the operation time. Consequently, the cost of energy is also related to the mixing time. Besides, prolonged mixing time also has an adverse effect on the quality of the product, for instance in crystal grains or damaged microorganisms (Abdolahzadeh et al., 2019). The initial condition of the numerical mixing time test is comprised of one solution concentration of $C_1 = 1\text{x}10^{-6}$ kg/m^3 that filled the superior fluid volume in the tank and another solution concentration of $C_2 = 3x10^{-6}$ kg/m^3 for the inferior fluid volume. The volume of each solution is the same and the initial condition of the mixing time system is shown in Fig. 2a. The criteria for fulfilling the mixing time test was that the final solution must have a concentration range between (0.49–0.51) kg/m^3 in 95 % of the fluid, according to reported literature (Delafosse et al., 2014). The diffusion coefficient of the solute was defined as $D = 7.5x10^{-7}$ m^2/s.

4.3. Power consumption

Power consumption displays the performance of the agitating process depending on the geometry of the impeller and the physical properties of the fluid. In the present work, power consumption (*Po*) is used as a parameter in order to evaluate the efficiency of the proposed new close-clearance impellers. The Power consumption was calculated using the measure tool post-processing in version 5 of the DualSPHysics code. For a shear-thinning fluid, the parameters of ketchup at 298.8 K were used. The pseudoplastic rheology parameters are defined as follows: a consistency index of K = 6.1 Pa.s, a power-law index of n = 0.4, and a density of ρ = 1,325 kg/m^3. For a shear-thickening fluid, the parameters of the melted chocolate at 319.15 K were chosen. The dilatant rheology is defined as follows; a consistency index of K = 0.57. Pa.s, a power-law index of n = 1.16, and a density of ρ = 1,600 kg/m^3, respectively. These parameters were taken from the work of (Joyner & Daubert, 2017).

5. Results

5.1. Convergence test

Figure 2 displays the velocity profiles as a function of time for different resolutions using the simple hybrid impeller. The maximum and minimum velocity values are developed in a range between 0.06 m/s and 0.22 m/s in a periodic way in time. In order to quantify adequate particle resolution, the root mean square error (RMSE) has been used to calculate the closeness between the velocity values for the highest particle resolution and the others lower particle resolutions.

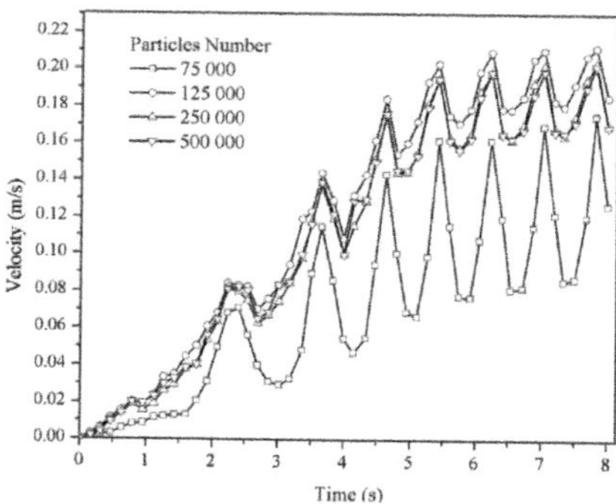

Figure 2. Design of the impellers described in section 3.1. a) U-type anchor, b) anchor with radial counterflow, c) simple Hybrid.

The obtained values of RMSE are 4.99x10^{-2}, 1.07x10^{-2}, and 3.14x10^{-3} for total particles of 75, 000, 125, 000 and 250, 000, respectively. In summary, the RMSE values from this convergence velocity test suggest that there is no significant difference between the resolutions with 250, 000 or 500, 000 particles. Because of that, it was decided to use 250, 000 for reducing the runtime simulation.

5.2. The mixing time

The mixing time, according to the restriction reported in section 4.2, for each impeller are 306, 804, and 95 seconds for the U-type anchor, anchor with radial counterflow, and simple hybrid, respectively. The simple hybrid impeller has less mixing time compared to the other two impellers based on an anchor geometry. These results can be explained by the flow dynamics within the stirred tank shown in Figure 3. The impellers with anchor geometry produce a mean flow parallel to the wall tank, reducing the movement in the *z*-direction, however, the helical geometry produces vortices that induce the movement in the *z*-direction improving the blend within the tank.

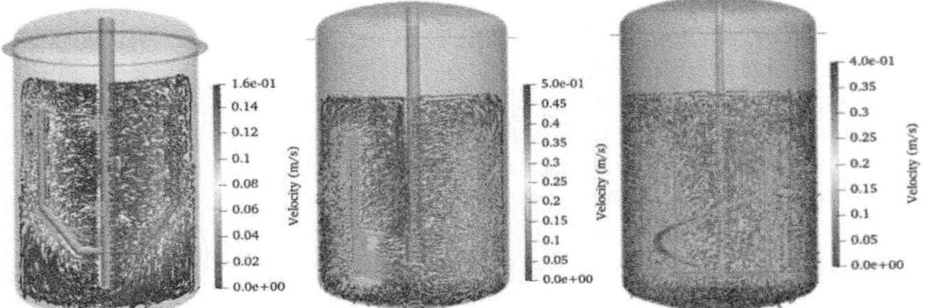

Figure 3. Velocity vectors within the tank. a) U-type anchor, b) anchor with radial counterflow, c) simple Hybrid.

5.3. Power consumption

In Figure 4 is shown the results obtained from the SPH simulations of the power consumption as a function of time for the Newtonian, shear-thickening, and pseudoplastic fluids at 75 rpm. The most striking result to emerge from Figure 4a is that the simple hybrid demonstrates the lowest power consumption among all impellers. In addition, the simple hybrid has also performed the lowest mixing time, as mentioned in section 5.2. In the opposite case, the anchor with radial counterflow impeller demonstrates the worst performance because it uses the highest amount of power energy consumption. It is suggested not to use Counter-current for this mixing system configuration in a Newtonian fluid. In Figure 4b are shown the results obtained from the power consumption of the ketchup as a pseudo-plastic fluid. What stands out in Fig 4b is the rapid increase of power consumption from 5 W to 45 W for the Simple Hybrid impeller in approximately 3 seconds.

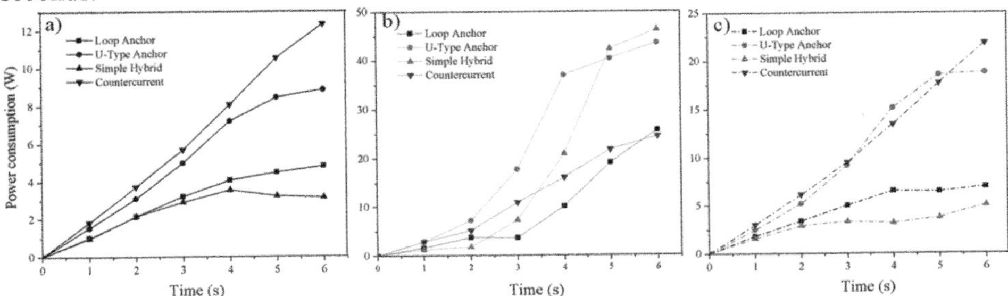

Fig. 4 Power consumption for each impeller and rheology studied.

Conversely, no difference greater than 20 W was observed using Counter-current. These results suggest that the Counter-current promotes the lowest power consumption in pseudo-plastic fluids. However, a mixing time test in a pseudo-plastic fluid may be recommended in order to gain additional insights into those impellers for future works. In summary, it has been clearly observed from Figure 4b that power consumption in pseudo-plastic rheology is substantially greater than the Newtonian case in all impellers. Finally, in Figure 4c are shown the results obtained from power consumption as a function of time for mixing melted chocolate solution, a shear-thickening fluid, for N = 75 rpm. First, what can be clearly seen in this figure is the similar trend of power input between shear-thickening and Newtonian fluids.

6. Conclusions

The aim of the present research was to assess the power consumption in a stirred tank using three studied impellers U-type anchor, anchor with radial counterflow, and simple hybrid for different rheology. The second aim of this study was to investigate the effects in the homogeneity evaluated by the numerical mixing time. The convergence test has determined that the adequate particle number for the proposed mixing configuration is 250, 000 approximately, this means a relationship between the initial distance particle and a diameter of the tank of 0.0162 approximately. On the other hand, the mixing time analysis reveals the importance of a helical blade in the impeller. Thus, the lack of a helical blade in the U-Type Anchor and anchor with radial counterflow impellers presents a poor performance. Thus, this study has contributed to understanding the performance of these unusual impellers. In relation to the study cases simulated, the numerical results suggest that a simple hybrid impeller has a better performance to homogenize and also demonstrates the lowest power consumption in Newtonian and shear-thickening fluids for a low stirring rate of 75 rpm. The investigation highlights the potential usefulness of the SPH method to evaluate the mixing system performance in a variety of rheology from Newtonian to no-Newtonian fluids.

References

T. C. Papanastasiou, (1987) Flows of materials with yield. J Rheol (N Y N Y) 31:385–404. https:// doi. org/ 10. 1122/1. 549926

E. Mitsoulis, (2007) Flows of viscoplastic materials: models and computations. Rheol Rev: 135–178

A. M. Tartakovsky, P. Meakin, (2006), Pore-scale modeling of immiscible and miscible fluid flows using smoothed particle hydrodynamics, Advances in Water Resources, 29 (10):1464-1478.

J. M. Domínguez, G. Fourtakas, C. Altomare, R. B. Canelas, A. Tafuni, O. García-Feal, I. Martínez-Estévez, A. Mokos, R. Vacondio, A. J. C. Crespo,B. D. Rogers, P. K. Stansby, M. Gómez-Gesteira, 2022, DualSPHysics: from fluid dynamics to multiphysics problems. Computational Particle Mechanics, 9(5), 867-895.

M. Abdolahzadeh, A. Tayebi, P. Omidvar, (2019), Mixing process of two phase non-newtonian fluids in 2d using smoothed particle hydrodynamics. Computers & Mathematics with Applications 78, 110–122.

INOXPA, . Electronic references. URL: https://docplayer.es/9359211-Agitadores-para-los-depositos-estandares.html.

H.S. Joyner, C. R. Daubert, (2017), Rheological Principles for Food Analysis. In: Nielsen, S.S. (eds) Food Analysis. Food Science Text Series. Springer, Cham. https://doi.org/10.1007/978-3-319-45776-5_29.

A. Delafosse, M.L. Collignon, S. Calvo, F. Delvigne, M. Crine, P. Thonart, D. Toye, (2014) Cfd-based compartment model for description of mixing in bioreactors. Chemical Engineering Science 106, 76–85.

Antonis Kokossis, Michael C. Georgiadis, Efstratios N. Pistikopoulos (Eds.)
PROCEEDINGS OF THE 33rd European Symposium on Computer Aided Process Engineering (ESCAPE33), June 18-21, 2023, Athens, Greece
 http://dx.doi.org/10.1016/B978-0-443-15274-0.50133-5

Hybrid Dynamic Surrogate Modelling for a Once-Through Steam Generator

Sindre Stenen Blakseth,[a,b] Leif Erik Andersson,[a] Rubén Mocholí Montañés,[a] Marit Jagtøyen Mazzetti,[a]

[a]*Department of Gas Technology, SINTEF Energy Research, Trondheim, Norway*
[b]*Department of Mathematical Sciences, Norwegian University of Science and Technology, Trondheim, Norway*

Abstract

Four surrogate modelling techniques are compared in the context of modelling once-through steam generators (OTSGs) for offshore combined cycle gas turbines (GTCCs): Linear and polynomial regression, Gaussian process regression and neural networks for regression. Both fully data-driven models and hybrid models based on residual modelling are explored. We find that speed-ups on the order of 10k are achievable while keeping root mean squared error at less than 1%. Our work demonstrates the feasibility of developing OTSG surrogate models suitable for real-time operational optimization in a digital twin context. This may accelerate the adoption of GTCCs in offshore industry and potentially contribute towards a 25% reduction in emissions from oil & gas platforms.

Keywords: Digital Twin, Surrogate Modelling, Residual Modelling, Gaussian Process Regression, Neural Networks

1. Introduction

Installation of combined cycle gas turbines (GTCCs) offshore can reduce emissions from oil & gas platforms by up to 25% [Mazzetti et al. (2014)]. A digital twin (DT) framework for GTCCs may accelerate the adoption of GTCCs by increasing their reliability and performance. To this end, accurate and trustworthy yet computationally efficient dynamic models for crucial GTCC components are needed. The once-through steam generator (OTSG) has been identified as a particularly important component because its response to transients in the gas turbine load largely governs the GTCC's overall behavior. SINTEF Energy Research has previously developed highly accurate Modelica models for OTSGs [Montañés et al. (2021)]. In this work, we explore four data-driven techniques for creating high-speed, DT-suitable surrogate models based on the Modelica model: Linear regression (LReg), polynomial regression (PReg), Gaussian process regression (GPR), and neural networks (NNs). We use these techniques to develop both purely data-driven models and hybrid models utilizing the residual modelling technique. The goal of this work is to evaluate the feasibility of developing DT-suitable OTSG surrogate models, e.g., for real-time optimization and process control based on model predictive control schemes. The work is intended to contribute towards accelerating the adoption of GTCCs, thereby reducing emissions from offshore energy production. Secondarily, it is of general interest to analyze the relative performance of the different techniques studied.

2. Once-Through Steam Generators

2.1 Background and Motivation

A waste heat recovery steam generator (HRSG) is a heat exchanger boiler system that recovers waste heat from a given heat source by producing steam from feedwater that is circulated through the heat exchanger. In the context of GTCCs, the heat source is the gas turbine exhaust, and the steam produced by the HRSG is passed through a steam turbine to generate electricity. Consequently, in comparison to standalone gas turbines, GTCCs offer significantly increased efficiency, and thus a corresponding reduction in CO_2 emissions for fixed power production.

In this work, we consider once-through steam generators (OTSGs), which are a particular type of HRSGs common in industrial applications. OTSG configurations are the preferred option in volume- and weight-constrained energy system environments, such as offshore oil and gas installations or floating production, storage and offloading systems. High-fidelity, physics-based, dynamic OTSG models facilitate simulation-based studies to better understand the inherent dynamics of the OTSG system, and to conduct control loop and control structure design studies. Traditionally, operation of OTSG systems has predominantly taken place under steady-state conditions. Therefore, it has been sufficient to consider transient conditions in the offline design phase. Faster models suitable for real-time operational optimization have consequently not received much attention. However, as intermittent renewable energy sources enter the offshore energy mix, computationally efficient models applicable to transient conditions are becoming increasingly important. This motivates the present study of surrogate models for OTSGs, i.e., low-fidelity OTSG models that are designed to capture the main characteristics of OTSG transient behavior while offering significant computational speed-up in comparison to traditional high-fidelity models.

2.2 High-Fidelity OTSG Modelling

Development of surrogate models generally requires data from which the main characteristics of the considered system can be extracted. In the present work, this data will stem from a previously published high-fidelity (hi-fi), dynamic OTSG model for offshore combined cycle applications [Montañés et al. (2021)]. The model is developed in the Modelica language and utilizes dynamic energy and mass balances to produce a set of differential algebraic equations describing the OTSG's transient behavior.

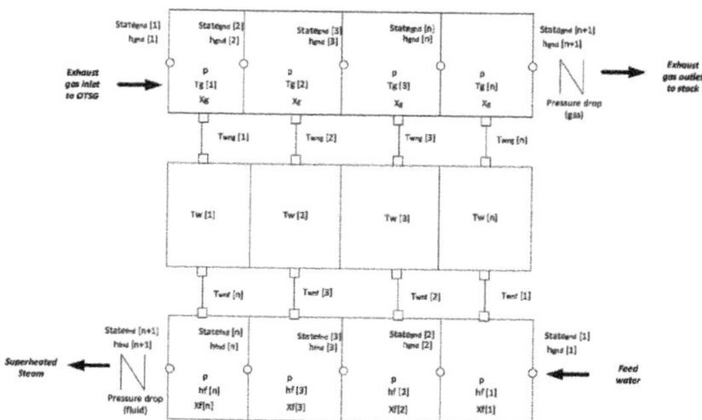

Figure 1: Discretization and main states of high-fidelity OTSG model [Montañés et al. (2021)].

Figure 1 (adapted from [Montañés et al. (2021)]) shows the OTSG model with discretization and main states. The model is based on 1-D lumped pressure flow pipes (for gas and water/steam sides) separated from each other by a wall model (representing the metal wall for heat transfer in a bundle of tubes). A total of n volumes and $n+1$ nodes are used to discretize the system in the direction of the flow. Overall, the model includes physical phenomena related to heat transfer, hydraulics, wall model thermal inertia, dynamic energy and mass balances on both the gas side and the water side [Montañés et al. (2021)]. We consider the OTSG operated under valve-throttling pressure control mode, i.e., with fixed pressure set point of the produced superheated steam [Nord and Montañés (2018), Zotică et al. (2022)]. Important boundary conditions for the model are gas turbine load (exhaust gas inlet mass flow and inlet specific enthalpy) and feed water mass flow rate (which is commonly a manipulated variable for OTSG operation).

3. Surrogate Modelling Techniques

In this section, we briefly describe the four surrogate modelling techniques considered in this work: linear and polynomial regression (Section 3.1), neural networks (Section 3.2) and Gaussian Process Regression (Section 3.3). We also describe how we generate the data needed to develop OTSG surrogate models using these techniques (Section 3.4) and how we evaluate the resulting models (Section 3.5). Common for all the surrogate models considered here is that their task is to predict the outlet temperature of the flue gas, as well as the outlet temperature and mass flow of the steam.

3.1 Linear and Polynomial Regression

Polynomial regression (PReg) amounts to approximating some target function $f(\cdot)$ using a polynomial constructed using N_i input data variables I_i, where $i = 1, \dots, N_i$. Mathematically, this can be expressed as finding coefficients $\alpha_{i,n}$ such that

$$\hat{f}_{\text{reg}} := a_0 + \sum_i \alpha_{i,1} I_i + \alpha_{i,2} I_i^2 + \cdots + \alpha_{i,N} I_i^d$$

is as close to the true target f as possible. We say that d is the degree of the interpolating polynomial. Linear regression (LReg) is simply the special case of $d = 1$. We used the Python package lmfit [Newville et al. (2014)] to implement our regression models.

3.2 Neural Networks for Regression

Owing to their universal function approximation properties and acclaimed empirical successes, neural networks (NNs) are well suited for surrogate modelling. Here, we consider fully connected feed-forward NNs (cf. [Nielsen (2015)] for an introduction). Such a network with n_l so-called *layers* can be expressed as

$$\hat{f}_{\text{NN}} := \varphi_{n_l}(\dots(\varphi_2(\varphi_1(I w_2 + b_1) w_2 + b_2)\dots) w_{n_l} + b_{n_l}),$$

where I is the NN's input vector, the matrices $w_1, \dots, w_{n_l}$ (weights) and vectors $b_1, \dots, b_{n_l}$ (biases) are tunable parameters, and $\varphi_1, \dots, \varphi_{n_l}$ are non-linear functions known as activation functions. Typically, stochastic gradient descent is used to tune the weights and biases such as to minimize the difference between $\hat{f}_{\text{NN}}$ and the target function f. We have implemented our NN models using the Python package pytorch [Paszke et al. (2019]. We use the LeakyReLU activation function with slope parameter 0.01, pytorch's MSELoss cost function, the Adam optimizer, and a learning rate of 1e-5.

3.3 Gaussian Process Regression

Gaussian Processes (GPs) are non-parametric, probabilistic kernel methods [Rasmussen and Williams (2006)] that aim to identify an unknown function $f: \mathcal{R}^{n_u} \rightarrow \mathcal{R}$ from data. It is assumed that the noisy observation of $f(\cdot)$ are given by

$$y = f(\boldsymbol{u}) + v,$$

where the noise v is Gaussian with zeros mean and variance σ_v^2, and $\boldsymbol{u}$ is the input, which is assumed to follow a multivariate Gaussian distribution. Smoothness properties of the underlying function f are enforced by the choice of mean and covariance function without relying on parametric assumptions [Snelson and Ghahramani (2006)]. A zero mean function and the automatic relevance (ARD) squared-exponential (SE) covariance function are chosen. GP depends on hyperparameters, which are usually unknown and need to be inferred from data. The marginal likelihood is used to estimate hyperparameters. The predictive distribution is the marginal of the normalized joint prior times the likelihood. The integral can be evaluated in closed form. The GP regression model was implemented in Python and the maximisation of the log marginal likelihood was solved with help of the SciPy package [Virtanen et al. (2020)].

3.4 Data Generation

Due to the unavailability of suitable operational data from a real GTCC, the high-fidelity OTSG model described in Section 2.2 will be used to create the reference data needed for developing and evaluating our surrogate models. As explained in Section 2.2, the gas turbine load is an essential boundary condition for the model and must therefore be prescribed. To cover a wide range of operating conditions, a randomized time series of one million data points at 1-second intervals, was generated. Every 100 seconds, a load change is occurring with 50% probability. The type of change (step or ramp), the new set point value and the duration of the transition (if it is a ramp) are randomized with the latter two drawn from uniform distributions on [40%, 100%] and [1s, 300s], respectively.

Using the generated gas turbine load sequence, the hi-fi model is used to generate two time series for each of the three outlet variables we aim to predict with our surrogate models. One time series is created using a coarse discretization in the hi-fi model (we call this the low resolution (LR) data), and the other is created using a fine discretization (denoted high resolution (HR) data). The HR data is used as ground truth, both during the model tuning process and the final evaluation. The LR data serves two purposes. Firstly, it can be used as additional input for the surrogate models. Secondly, it will allow us to explore the potential benefit of residual modelling. The concept of residual modelling is to model the residual $\varepsilon = HR - LR$ instead of the HR data directly [Willard et al. (2022)]. Predictions are then constructed as $\widehat{HR} = LR + \hat{\varepsilon}$, where $\hat{\varepsilon}$ is the model's approximation of ε. Both the HR and LR data are split into three subsets. The first 980k data points are included in the *training set* (used for tuning model parameters), the next 10k data points go in the *validation set* (used for tuning certain hyperparameters), and the final 10k data points constitute the *test set* (used for evaluating the models).

3.5 Model Evaluation

All our surrogate models depend on one or more so-called hyperparameters, i.e., parameters that define a model's structure but are not used directly to compute predictions. Examples include choice of input variables, polynomial degree (PReg), number of layers and neurons per layer (NN) and the number of data points to use for tuning (GPR). To facilitate a fair comparison between the different kinds of models, we conducted a grid search to identify good hyperparameter choices for all models. For each model instance (corresponding to a particular choice of hyperparameters), the normalized root mean squared error (NRMSE) of the models' predictions was computed with respect to the HR test data for each of the three target variables. The sum of the three NRMSEs was taken to represent the overall predictive accuracy of any particular model. For models using old HR data as input, it is important to consider that, when predicting more than one time step into the future, HR data from the previous time step will not be available. Then, the surrogate model's prediction from the previous time step must be used in its place. This may lead to divergent behavior, as is observed for some of our models (cf. Table 1).

4. Results

4.1 Predictive Accuracy

For each modelling technique and model input choice, Table 1 lists the lowest total NRMSE error observed in our grid search for both normal and residual models using the specified technique and input choice. The available input choices are 1) only inlet data (in), 2) inlet data and LR outlet data (in+LR), 3) same as 2) but also including old HR data (in+LR+old), and 4) same as 2), but also including the residual HR-LR (in+LR+res).

We observe that the most accurate model is a GPR residual model using in+LR+old input. However, several other models achieve comparable performance, and the performance different between the best PReg, NN and GPR models is generally insignificant from a practical perspective. Even the best LReg models are found to perform quite well, with a NRSME significantly smaller than 1% achievable for all three output variables. Moreover, while the use of LR data appears to be generally beneficial, we observe that only inlet data is sufficient to obtain NRSME values well below 1% for any given output variable. (Note that, in the table, we sum the NRMSE of each predicted variable.) Figure 2 shows the predictions of steam outlet mass flow made by the best model for each modelling technique (bold in Table 1). Based on Figure 2, it is difficult to identify any predominant failure mode for any of the different models.

	in		in+LR		in+LR+old		in+LR+res	
	Normal	Residual	Normal	Residual	Normal	Residual	Normal	Residual
LReg	0.0555	0.0154	0.0089	0.0089	**0.0088**	0.0514	19.4011	19.3999
PReg	0.0555	0.01525	0.0056	0.0056	***0.0031***	NAN	NAN	NAN
NN	0.0562	0.0107	0.0066	0.0061	0.0039	**0.0032**	0.0406	0.0107
GPR	*0.0369*	*0.0099*	*0.0045*	*0.0043*	0.0032	***0.0031***	*0.0036*	*0.0031*

Table 1: Normalized RMSE on the test set, summed over the three output variables, for given model type and input selection. Bold and italics are used for lowest error values in columns and rows, respectively. NAN indicates that the model diverged.

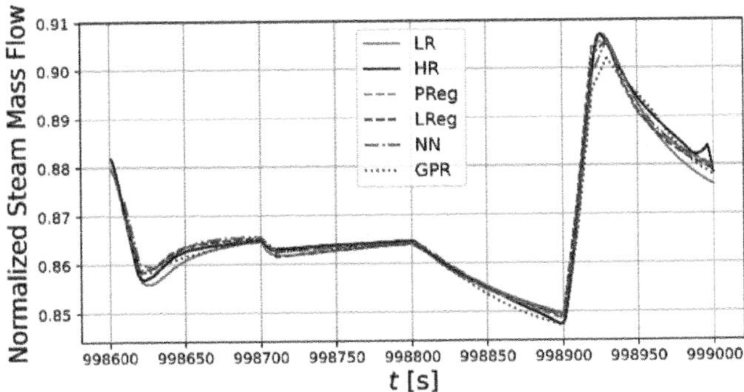

Figure 2: Predictions of steam outlet mass flow by best model within each category, shown along with the corresponding LR and HR data for a short interval within the test set.

4.2 Computational Efficiency

We use the number of CPU seconds (wall time) per simulated second of operations (CPUs/s), as measured on a standard, mid-range laptop, to quantify the computational efficiency of the various models. Using respectively the coarse and the fine spatial discretization, we measure 0.00289 CPUs/s and 0.1272 CPUs/s for the hi-fi-model. Thus, the reduced resolution yields a speed-up of roughly 44. In comparison, our linear regression model with only inlet data as input uses around 8e-6 CPUs/s, which corresponds to a speed-up of roughly 15k. The PReg model is roughly as fast as the LReg model, while the NN and GPR models are roughly 1–1.5 orders of magnitude slower than the LReg model. Consequently, for residual models and models using LR data as input, the LR hi-fi component always dominates the computational expense.

5. Discussion

An interesting takeaway from our numerical experiment is that blindly relying on more advanced techniques like neural networks and GPR to be better than simple techniques like regression is not advised. The relation between a model's theoretical representation capacity and its empirical predictive accuracy is not necessarily linear. Indeed, our results show that the relation is not even strictly increasing in general, as 1) our best PReg model outperforms our best NN model, and 2) increasing the number of neurons in the NN (and thereby its representation power) did not improve its accuracy on the *test* set.

Our results also show that the value of hybrid modelling depends on the quality of the physics-based model component, and how that component is integrated into the fully hybrid model. In our case, using LR predictions as input was generally observed to be useful, while residual modelling yielded mixed results. This illustrates the obvious, but easily overlooked fact that residual modelling is only beneficial if the relation between the input data and the residual is simpler than that between the input data and the data to be predicted. This criterion is not always met in practice, as evidenced by our results.

Finally, we conclude that it is feasible to construct OTSG surrogate models suitable for use in real-time optimization procedures within a digital twin framework. This motivates further work, which could include exploration of more advanced neural network-based techniques, such as Long Short-Term Memory networks. Additionally, we believe that looking into the robustness of the methods with respect to noisy data would be valuable from a practical perspective.

Acknowledgementss

This work was supported by DIGITAL TWIN, RCN project no. 318899.

References

Mazzetti, M. J., Nekså, P., Walnum, H. T., & Hemmingsen, A. K. T. (2014). Energy-efficiency technologies for reduction of offshore CO2 emissions. Oil and gas facilities, 3(01), 89–96.

Montañés, R. M., Skaugen, G., Hagen, B., & Rohde, D. (2021). Compact Steam Bottoming Cycles: Minimum Weight Design Optimization and Transient Response of Once-Through Steam Generators. Frontiers in Energy Research, 9, 687248.

Newville, M., Stensitzki, T., Allen, D. B., & Ingargiola, A. (2014). LMFIT: Non-Linear Least-Square Minimization and Curve-Fitting for Python (0.8.0). doi: 10.5281/zenodo.598352

Nielsen, M.A. (2015). Neural Networks and Deep Learning. Determination Press.

Nord, L. O., & Montañés, R. M. (2018). Compact steam bottoming cycles: Model validation with plant data and evaluation of control strategies for fast load changes. Applied Thermal Engineering, 142, 334–345.

Paszke, A. et al. (2019). PyTorch: An Imperative Style, High-Performance Deep Learning Library. In Advances in Neural Information Processing Systems, 32, 8024–8035.

Snelson, E. & Ghahramani, Z. (2005). Sparse Gaussian processes using pseudo-inputs. Advances in neural information processing systems, 18.

Virtanen, P. et al. (2020). SciPy 1.0: Fundamental Algorithms for Scientific Computing in Python. Nature Methods, 17(3), 261–272

Willard, J., Jia, X., Xu, S., Steinbach, M., & Kumar, V. (2022). Integrating scientific knowledge with machine learning for engineering and environmental systems. ACM Computing Surveys, 55(4), 1–37.

Williams, C. K., & Rasmussen, C. E. (2006). Gaussian processes for machine learning (Vol. 2, No. 3, p. 4). Cambridge, MA: MIT press.

Zotică, C., Montañés, R. M., Reyes-Lúa, A., & Skogestad, S. (2022). Control of steam bottoming cycles using nonlinear input and output transformations for feedforward disturbance rejection. IFAC-PapersOnLine, 55(7), 969–974.

Antonis Kokossis, Michael C. Georgiadis, Efstratios N. Pistikopoulos (Eds.)
PROCEEDINGS OF THE 33rd European Symposium on Computer Aided Process Engineering (ESCAPE33), June 18-21, 2023, Athens, Greece
 http://dx.doi.org/10.1016/B978-0-443-15274-0.50134-7

PINN-based Design of Experiment Concept for Process Model Parameter Identification

Aike Aline Tappe [a], Subiksha Selvarajan [a], Caroline Heiduk [b], Stephan Scholl [b] and René Schenkendorf [a,*]

[a]*Automation & Computer Sciences Department, Harz University of Applied Sciences, Friedrichstr. 57-59, 38855 Wernigerode, Germany; {atappe, sselvarajan}@hs-harz.de;*
[b]*TU Braunschweig, Institute for Chemical and Thermal Process Engineering, Langer Kamp 7, 38106 Braunschweig, Germany; {c.heiduk, s.scholl}@tu-braunschweig.de;*
[*]*Correspondence: rschenkendorf@hs-harz.de*

Abstract

In Process systems engineering, first-principles process models are standard. However, all model predictions, including any model-based design or control policy, suffer from imprecision. To increase the reliability of the process model, precise model parameters and model-based design of experiments (MBDoE) are essential. That is, experiments have to be designed for the specific purpose of model calibration. The same is true for data-driven engineering concepts. For example, in the case of physics-informed neural networks (PINNs), the first-principles model is incorporated into the structure of an artificial neural network, thus creating a hybrid framework in terms of model calibration. The PINN setting, in turn, results in different parameter sensitivities compared to the standard parameter identification approach, with profound consequences for the precision of parameter estimates. Also, due to the change in parameter sensitivities, traditional MBDoE concepts might fail. Alternatively, MBDoE strategies have to be adapted for PINN-based parameter identification and more precise parameter estimates. In particular, Monte Carlo simulations are combined with the PINN concept to choose appropriate experimental designs. As a case study, the proposed concept is demonstrated for a distributed-parameter system of an adsorption process model. The changes in parameter sensitivities and parameter uncertainties are critically discussed and, in terms of novelty, presented with the need for a mandatory MBDoE adaptation.

Keywords: design of experiments, data-driven engineering, physics-informed neural networks, sensitivities, uncertainty, partial differential equations, adsorption process

1. Introduction

Profit margins in the highly competitive and regulated pharmaceutical industry have continuously decreased in the past decades due to increased research and development costs (Behr et al., 2004). In the context of pharmaceutical manufacturing, process systems engineering can be used to design efficient and cost-effective production processes for drugs and other pharmaceutical products. This can involve optimizing the use of raw materials and energy, reducing waste and emissions, and ensuring that the final product meets the necessary quality and safety standards. In addition, process systems engineering can be used to model and simulate different production scenarios, helping to identify potential bottlenecks or inefficiencies in the manufacturing process. To gain an in-depth system understanding of the underlying physical processes, the temporal and spatial resolution of the underlying dynamic processes might be of relevance and interest. Thus, research activities on distributed parameter systems (DPS) have gained importance over the years. DPSs can be modeled using partial differential equations (PDEs), where, in particular, the model parameters determine the dynamic behavior of the system. However, the literature on parameter identification of PDEs is sparse (Zhang et al., 2017). Alternatively, recent advances in machine learning and artificial intelligence provide new approaches to modeling strategies and parameter identification problems alike. For example, if the process under study cannot be described with first-principles models, ML algorithms are very promising to compensate for that knowledge gap resulting in hybrid process models. However, there is a proven workflow for concepts such as DoE in relation to traditional modeling strategies, but little is known about how these concepts can be applied to hybrid models involving machine learning. Traditionally, the numerical solutions of PDE developed over the past decades typically include the discretization of system domains with mesh-based approaches, including finite differences, finite elements, and finite-volume methods. This strategy has several disadvantages: The individual mesh section is solved separately, and high-resolution systems are computationally expensive. ML technologies to solve these problems are welcome, and PINN methods have recently received a lot of attention (Raissi et al., 2019). In addition to solving PDE systems, the PINN framework enables a new perspective in discovering DPS, i.e., deriving the governing equations and identifying the corresponding model parameters directly. Still, to increase the reliability of the process model, precise model parameters and model-based design of experiments (MBDoE) are essential (Abt et al., 2018). In the case of PINNs, first-principles models are technically incorporated into the structure of artificial neural networks, thus creating a hybrid framework in terms of model calibration (Yazdani et al., 2019) as well. The PINN setting, in turn, results in different parameter sensitivities compared to the standard parameter identification approach (Selvarajan et al., 2022), with profound consequences for the precision of parameter estimates, rarely addressed in the current literature.

2. Methodology

2.1. Distributed parameter system

Following the notation in (Shukla et al., 2022), partial differential equation systems (PDEs) with the solution $u(\mathbf{x}, t)$, including model parameters p, defined in a domain Ω read as follows:

$$f\left(\mathbf{x}; \frac{\partial u}{\partial x_1}, \dots, \frac{\partial u}{\partial x_d}; \frac{\partial^2 u}{\partial x_1 \partial x_1}, \dots, \frac{\partial^2 u}{\partial x_1 \partial x_d}; \dots; p\right) = 0, \qquad \mathbf{x} = (x_1, \dots, x_d) \in \Omega, \quad (1)$$

with the boundary conditions:

$$B(u, \mathbf{x}) = 0 \text{ on } \partial\Omega\,. \quad (2)$$

In process system engineering, PDEs are used to describe the behavior of distributed parameter systems due to their ability to capture the spatial and temporal variation of the system's state variables. The heat equation and the wave equation, for example, are PDEs that describe how heat flows through a material or how waves propagate through a medium, respectively. However, distributed parameter systems can be challenging to analyze and control because the underlying PDEs often have complex solutions that cannot be obtained analytically. Instead, the solution needs to be approximated with numerical methods, which can be computationally intensive. However, several numerical methods, including finite difference methods, finite element methods, and spectral methods are standard in solving PDEs, but the choice of method depends on the specific form of the PDE and the desired accuracy of the solution.

2.2. Physics-informed neural networks

Physics-informed neural networks are beneficial for solving partial differential equations because they can handle complex, nonlinear equations that may be difficult to solve using classical numerical methods. These neural networks are trained using both the physical constraints of the problem and the data from the equation, allowing them to learn the underlying patterns and make accurate predictions. Additionally, physics-informed neural networks can be more efficient and scalable than classical numerical methods, particularly for high-dimensional or time-dependent problems. Technically, to solve a parametrized PDE system (Eq. (1)) through PINNs, a deep neural network $\hat{u}(\mathbf{x};\theta)$ with trainable meta-parameters θ is used. Boundary and initial constraints of the PDE system (Eq. (2)) and the PDE system itself (Eq. (1)) define the system identification problem and are represented in a separate term of the loss function, that governs the neural network training. Emphasis will be taken on the loss function composition in the following section.

2.3. System Identification

In PINNs, the loss function is typically composed of multiple terms that reflect different aspects of the problem. The specific terms used in the loss function will depend on the specific problem being tackled, but some common terms include: *Data loss*; this term measures the difference between the predicted output of the neural network and the known data for the problem. For example, if the goal is to solve a partial differential equation with known boundary conditions, the data loss term would measure the difference between the predicted solution and the known boundary conditions. *Physics loss*; this term measures the difference between the predictions of the neural network and the known physical laws and constraints that apply to the problem. For example, if the goal is to simulate the motion of a particle in a gravitational field, the physics loss would measure the difference between the predicted motion of the particle and the known laws of gravity. By combining these different loss terms into a single function, PINNs can learn to predict physical phenomena while also incorporating prior knowledge about the problem. This can help the model to generalize better and make more accurate predictions on unseen data. According to (Shukla et al., 2022), the resulting loss function reads as:

$$\mathcal{L}(\boldsymbol{\theta},p) = \mathcal{L}^{data}(\boldsymbol{\theta}) + \mathcal{L}^{pde}(\boldsymbol{\theta},p), \tag{3}$$

with the data loss term $\mathcal{L}^{data}$ defined for the measurements y^{data}:

$$\mathcal{L}^{data}(\boldsymbol{\theta}) = \sum_{i=1}^{m}\sum_{k=1}^{K} \|y^{data}(\rho_i,t_k) - h(\hat{u}(\rho_i,t_k,\boldsymbol{\theta}))\|_2^2, \tag{4}$$

with the measurement location $\rho_i \in (0,L)$ at K discrete time points and the model output $h(.)$. Moreover, the physics loss term $\mathcal{L}^{pde}$ relates to the governing equation as follows:

$$\begin{aligned}\mathcal{L}^{pde}(\boldsymbol{\theta},p) = &\frac{1}{|\mathcal{T}_f|}\sum_{\mathbf{x}\in\mathcal{T}_f}\left\| f\left(\mathbf{x};\frac{\partial\hat{u}}{\partial x_1},\dots,\frac{\partial\hat{u}}{\partial x_d};\frac{\partial^2\hat{u}}{\partial x_1\partial x_1},\dots,\frac{\partial^2\hat{u}}{\partial x_1\partial x_d};\dots;p\right)\right\|_2^2 \\ &+\frac{1}{|\mathcal{T}_b|}\sum_{\mathbf{x}\in\mathcal{T}_b}\|B(\hat{u},\mathbf{x})\|_2^2,\end{aligned} \tag{5}$$

Thus, the prediction of deep NN $\hat{u}$ has to fulfill the governing equation system and fit the experimental data. Using the PINN framework for parameter identification relates to toal least squares approach for a classical parameter identification problem, i.e., the combination of ordinary least squares (e.g., fitting the data) and input lest squares (e.g., fulfilling the equation system), see (Selvarajan et al., 2022) and references therein. However, this also implies that standard MBDoE concepts based on local parameter sensitivities. Fisher information matrix, and the Cramér-Rao inequality may fail (Selvarajan et al., 2022). By deliberately controlling the experimental setup (e.g., sensor positions, sampling rates, inlet profiles, etc.), a model-based design of experiments can provide more precise and accurate estimates of the model parameters. Practically, the basic prerequisite is that parameter sensitivity and uncertainty measures must be adapted to the underlying problem. In the case of PINN, the physical loss term (Eq. (5)) plays an important role.

3. Case study

The adsorption process separates and purifies chemicals and substances. It is used in the chemical, pharmaceutical, and food industries, as well as in environmental remediation and waste treatment, allowing the recovery of valuable resources. For better process understanding and process design, a reliable characterization of the kinetics is essential. In this study, we analyze the use of PINN-based kinetic parameter identification and its implications for the design of experimental concepts to ensure more precise parameter estimates with less experimental effort.

3.1. Mathematical model of the adsorption process

The following assumptions are used in the derivation of a so-called lumped kinetic model (LKM) of the adsorption process: (i) Radial gradients in column concentration were neglected. (ii) The column is isothermal and homogeneously packed. (iii) Constant volumetric flow rate v. (iv) Lumped contribution of internal and external mass transport resistances represented with a mass transfer coefficient k_m. (v) The axial dispersion coefficient D_{ax} is also considered constant. Thus, the governing partial differential equation system of the LKM reads as:

$$\frac{\partial C}{\partial t} = -v_z \frac{\partial C}{\partial z} + D_{ax} \frac{\partial^2 C}{\partial z^2} - \frac{1-\varepsilon}{\varepsilon} \frac{\partial q}{\partial t}, \tag{6}$$

$$\frac{\partial q}{\partial t} = k_m (q^* - q), \tag{7}$$

where t is the time and z is the space coordinate along the column axis; C and q are the concentrations in the liquid and solid phases, respectively, ε is the bed porosity, v_z is the average velocity of the fluid, k_m is the effective linear driving force mass transfer coefficient, and D_{ax} is the axial dispersion. Note that for the adsorption rate, a linear driving force model (Naidu and Mathews, 2021) is assumed. A detailed description of LKM is reported in the literature (Javeed et al., 2013). The boundary and initial conditions are given by:

$$\begin{gathered} t = 0 \text{ and } 0 \geq z \leq L \Rightarrow q = 0 \text{ and } C = 0 \\ t > 0 \text{ and } z = 0 \Rightarrow C = C_0 \\ t > 0 \text{ and } z = L \Rightarrow \partial C / \partial z = 0 \end{gathered} \tag{8}$$

Simulated data are generated based on the following parameter values: $t \in (0,1)$, $z \in (0,1)$, $C_0 = 1$, $D_{ax} = 0.4\text{x}10^{-4}$, $k_m = 10$, $\varepsilon = 0.8$ and $v_z = 0.1$ in a dimensionless form. In Figure 1a, the default experimental design is illustrated. Here, it is assumed that three sensor elements are available to measure the concentration C, where, for the default configuration, the concentration is selected at the inlet ($z = 0.0$), outlet ($z = 1.0$), and in the middle of the column ($z = 0.5$). Furthermore, for the default experimental design, an equal time sampling with $\Delta t = 0.1$ is postulated, as well as additive white measurement noise with a standard deviation of $\sigma = 0.01$.

3.2. PINN-based parameter identification problem and experimental design

Specifically, the following multi-layer perceptron (MLP) setting is applied: two hidden layers, an input and output layer with 16 nodes each, and a classic sigmoid activation function. The PINN-based parameter identification problem was implemented in Julia, using the packages *Lux, NeuralPDE, and OptimizationOptimJL.* Furthermore, since the sensor elements at the inlet ($z = 0$) and outlet ($z = 1$) can only measure constant concentration profiles under the given boundary constraints and the considered time horizon, the positions of the sensor elements are optimized to cover the entire dynamic, i.e., $C(z_i) = 0$ to $C(z_i) = 1$. The optimized sensor layout is shown in Figure 1b with $z_{opt} = [0.1,0.3,0.5]$. In addition to sensor location, the sampling rate of time recordings is essential as well. Here, the PINN-based MBDoE makes direct use of Eq. (1), i.e., the Pde system does not have to be solved numerically, but the local parameter sensitivities are derived based on the right-hand side of Eq. (1). The resulting parameter sensitivities of ε and v_z are given in Figure 2. As a high parameter sensitivity is crucial for more precise parameter estimates, the sampling interval is adapted accordingly. To this end (Figure 1b), 6 sample points are selected for higher parameter sensitivity regions for ε and 5 sample points for v_z, i.e., the overall number of data points is similar to the default experimental design.

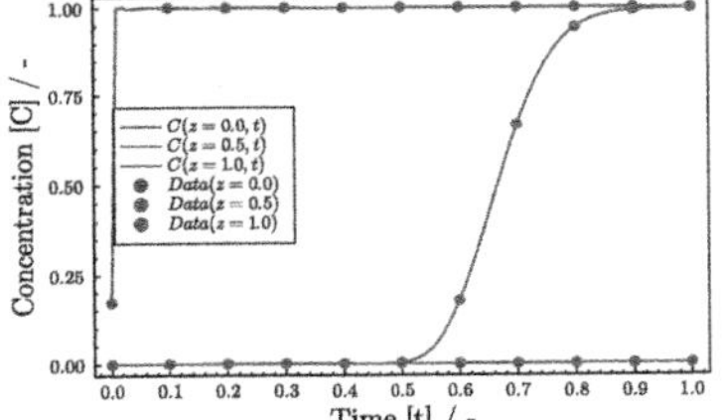

(a) Default experimental design

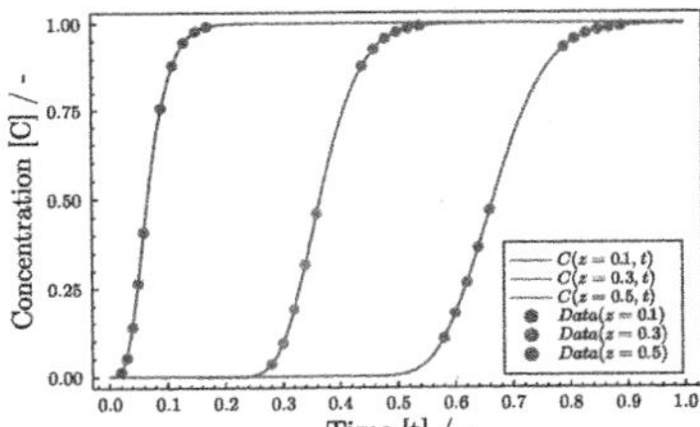

(b) Optimized experimental design

Figure 1: Dynamic profiles at three different sensor locations of the column/adsorption process. Time points of the actual measurement samples (simulated data) are highlighted.

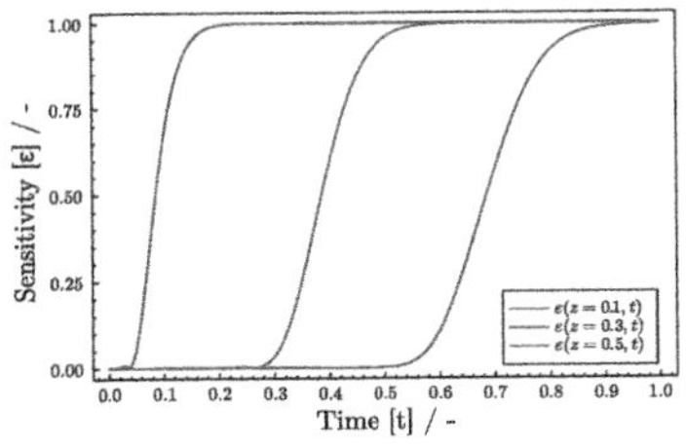

(a) Parameter sensitivity of ε

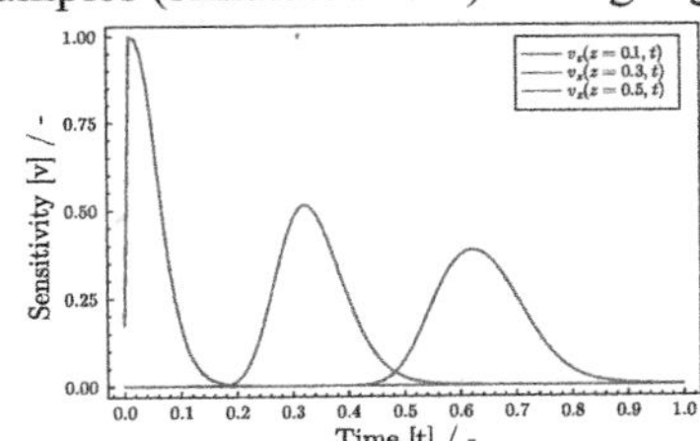

(b) Parameter sensitivity of v

Figure 2: Model parameter sensitivity plot; higher parameter sensitivity are considered for adapting the time sampling of the measurements.

3.3. Results

In Figure 3, the resulting parameter estimates are shown. A Monte Carlo simulation concept was used to repeat the PINN-based parameter identification step 200 times for the default and optimized experimental design. These parameter estimates suggest that the changed configuration in the sensor location and the sampling rate leads to a significant reduction in parameter uncertainties when the optimized experimental design configuration is applied. The results also reveal that the PINN-based parameter identification framework has a non-neglectable offset/bias in the parameter estimates, i.e., the true parameter estimates are rarely covered by the scatter plot results. The related mean values and standard deviations are listed in Table 1, indicating a positive effect of the optimized experimental design i.e., lower parameter uncertainties while using the same amount of experimental data.

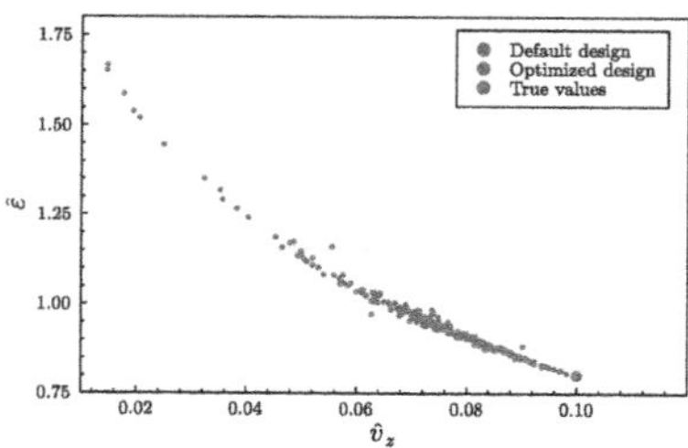

Figure 3: Scatter plot of the two estimated model parameters, including the result of the PINN-based parameter identification for the default and optimized experimental design.

Table 1: Uncertainties for the default and the optimized experimental design.

Parameter	True value	Mean & Std. (default)	Mean & Std. (optimized)
v_z	0.1	0.0718 ± 0.0164	0.0803 ± 0.0091
ε	0.8	0.9779 ± 0.1543	0.9101 ± 0.0570

4. Conclusions

As a result, the data collected using an optimal design is more informative and can be used to estimate the model parameters more precisely for the PINN-based parameter identification framework. This can lead to a better understanding of the process under study and can enable more accurate predictions of the behavior of the system. Additionally, using an optimal design can also help to reduce the number of experiments needed to obtain reliable parameter estimates, which can save time and resources. As demonstrated, while PINNs can be a powerful took for solving PDEs, they might be less optimal for precise parameter identification. Thus, further research on efficient uncertainty quantification and PINN-tailored design of experiment concepts is needed.

Acknowledgements

This research was funded by the German Research Foundation, grant number 444703025.

References

V. Abt, T. Barz, M. N. Cruz-Bournazou, C. Herewig, P. Kroll, J. Möller, R. Pörtner, R. Schenkendorf, 2018. Model-based tools for optimal experiments in bioprocess engineering. Current Opinion in Chemical Engineering 22, 244-252.

A. Behr, V. Brehme, C. Ewers, H. Grön, T. Kimmel, S. Küppers, I. Symietz, 2004. New developments in chemical engineering for the production of drug subsances, Engineering in Life Sciences 4(1), 15-24.

S. Javeed, S. Qamae, W. Ashraf, G. Warnecke, A. Seidel-Morgenstern, 2013. Analysis and numerical investigation of two dynamic models for liquid chromatography. Chemical Engineering Science 90, 17-31.

H. Naidu, A. P. Mathews, 2021. Linear driving force analysis of adsorption dynamics in stratified fixed-bed adsorbers. Separation and Purification Technology 257, 117955.

M. Raissi, P. Perdikaris, G. Karniadakis, 2019. Physics-informed neural networks: A deep learning framework for solving forward and inverse problems involving nonlinear partial differential equations. Journal of Computational Physics 378, 686-707.

S. Selvarajan, A. A. Tappe, C. Heiduk, S. Scholl, R. Schenkendorf, 2022. Process model inversion in the data-driven engineering context for improved parameter sensitivities. Processes 10(9), 1764.

K. Shukla, M. Xu, N. Trask, G. E. Karniadakis, 2022. Scalable algorithms for physics-informed neural and graph networks. Data-Centric Engineering 3.

A. Yazdani, L. Lu, M. Raissi, G. E. Karniadakis, 2019. Systems biology informed deep learning for inferring parameters and hidden dynamics.

X. Zhang, J. Cao, R. J. Carroll, 2017. Estimating varying coefficients for partial differential equation models. Biometrics 73(3), 949-959.

Antonis Kokossis, Michael C. Georgiadis, Efstratios N. Pistikopoulos (Eds.)
PROCEEDINGS OF THE 33rd European Symposium on Computer Aided Process Engineering (ESCAPE33), June 18-21, 2023, Athens, Greece
 http://dx.doi.org/10.1016/B978-0-443-15274-0.50135-9

Performance analysis of virtual platforms focused on multi-objective process optimization

Jaime David Ponce-Rocha,[a] César Ramírez-Márquez,[b] Ricardo Morales-Rodriguez,[c]

[a]Dirección de Ingeniería y construcción de plantas, CIATEQ, A.C. Centro de Tecnología Avanzada, Av. Del Retablo 150, Constituyentes Fovissste, Querétaro, Querétaro, 76150, México
[b]Departamento de Ingeniería Química, Universidad Michoacana de San Nicolás de Hidalgo, Ciudad Universitaria, Morelia, Michoacán, 58060, México
[c]Departamento de Ingeniería Química, División de Ciencias Naturales y Exactas, Universidad de Guanajuato, Noria Alta S/N, Noria Alta, Guanajuato, Guanajuato, 36050, México
jaime.ponce@ciateq.mx

Abstract

Several systematic methodologies for designing and optimizing conceptual processes have been reported by combining different computer-aided tools, mainly based on process simulators and programming environments that can help to improve any processes significantly. Thus, the objective of this work involves the performance analysis for virtual platforms focused on multi-objective process optimization, using a framework based on identical objective functions (economic, exergetic, and environmental targets), similar algorithms and hyperparameters (NSGA-II), for the separation and purification of the benzene, toluene, m-xylene mixture (BTX). Two computational interfaces were developed: Aspen Plus-Matlab (A) and Aspen Plus-Python (B), both using direct communication based on COM. Results were compared in terms of computational time, objective functions, and state variables values. Finally, both computational interfaces have a similar performance in all terms: interface B showed better computational time, but the use of the utopia-tracking approach let to conclude that interface A has better objective function values.

Keywords: Simulation-optimization, MOGA, process simulator, programming environments.

1. Introduction

The challenges associated with climate change, population growth, and resource limitations, require innovative solutions due to their inherent complexity and interdisciplinary nature (socio-techno-economic nexus), in this connection, chemical engineers can contribute through Process Systems Engineering (PSE) discipline, that offers techniques and tools for process synthesis, design, optimization and/or control problems at multi-scale levels, based on physical and chemical phenomena (Cameron et al., 2019). PSE techniques have a specific role in problem resolution, where two solution pathways for the synthesis and design of a product/process could be employed: 1) reality: experiment-based solution approach, and 2) virtual reality: typical computer-aided systems solution approach, which includes a) data and models, b) numerical solvers for simulation and/or optimization, and c) workflows: method or algorithm (Pistikopoulos et al., 2021). Several authors have reported methodologies based on virtual reality problem solutions focused on designing and optimizing chemical processes, combining different computer interfaces, linking process simulators and programming environments with several algorithms, mainly using direct communication through ActiveX-COM technology (Ponce-Rocha et al., 2022; Ruiz-Femenia et al., 2020; Sánchez-Ramírez et al., 2016). However, even though these methodologies have been applied in academic, research, and industrial fields, there is not a rigorous study to compare their performance in terms of objective function values and computational aspects. Thus, the objective of this work is to analyze the performance of two virtual platforms for multi-objective process optimization using identical objective functions (economic, exergetic, and environmental targets) with similar algorithms and hyperparameters (NSGA-II), for a separation and purification scheme, where the computational time, objective functions, and state variables values are used as performance criteria.

2. Methodology

2.1. Virtual platforms focused on multi-objective process optimization

Both virtual platforms proposed in this study were based on a systematic framework that includes four steps: 1) Optimization formulation: the multi-objective approach, state variables, and constraints definition, 2) Objective functions description: analysis and selection of objective functions associated with three targets: economy, exergy and environment (3E), 3) Optimal design: use of different computer-aided tools such as, Matlab and Phyton as programming environments, which allow the algorithm selection; both were combined with a process simulator (Aspen Plus), 4) Case study: separation and purification scheme.

2.2. Optimization formulation

The multi-objective optimization is represented as follows (1):

$$\underset{x}{Min}\ \underline{Z} = [\,Z_k(\underline{x})]\ k = 1, 2, \dots, k.\ \ s.t.\ \ \boldsymbol{h}(\underline{x}) = 0; \boldsymbol{g}(\underline{x}) \leq 0 \tag{1}$$

Where $\underline{Z}$ represents a vector with k objective functions (3E), $\underline{x}$ is the vector of state variables: number of stages (N_S), feed stages (F_S), and molar reflux ratios (R_R). $\boldsymbol{h}(\underline{x})$ and $\boldsymbol{g}(\underline{x})$ are the vectors to model equality and inequality constraints associated with the state variables ($\underline{x}$), mass purities (P), and mass recoveries (R).

2.3. Objective functions description

Generally, chemical processes have intrinsic characteristics of high energy consumption associated with the use of utilities such as cooling, heating, and electricity. In this sense, the combination of metrics based on 4E analysis (energy-exergy-economy-environment) has given an improvement in terms of process sustainability (Li et al., 2021; Ponce-Rocha et al., 2022). In this study, three of the four components of 4E analysis were selected to perform an optimization task: economic, exergetic, and environmental aspects.

2.3.1. Economy

The economic target was estimated using the Total Annual Cost (TAC), illustrated in Equation 2, which includes the equipment investment cost and payback period, and the sum of annualized utilities cost, referred to cooling, heating, and electrical services. To estimate the equipment cost the module factor approach introduced by Guthrie and modified by Ulrich was used, furthermore, the value of the chemical engineering plant cost index (CEPCI) was set at 801.30 (February-2022).

$$TAC = \frac{Investment\ cost}{Payback\ period} + Utilities \tag{2}$$

2.3.2. Exergy

The exergy target considers the exergy losses (σT_0), this metric is an adequate indicator to identify energy inefficiencies in the system, which considers the heating and cooling processes (Q), separation (Sep), and concentration changes ($Conc$) in all the equipment (j) and process streams (i) (see Equation 3).

$$\sigma T_0 = \Delta \dot{E}x_{In} - \Delta \dot{E}x_{Out}$$

$$\sigma T_0 = \sum_{j=1}^{n} \sum_{i=1}^{m} \left[\Delta \dot{E}x_{Q,j} + \Delta \dot{E}x_{Sep,i} + \Delta \dot{E}x_{Conc,i} \right] \tag{3}$$

2.3.3. Environment

The Life-Cycle Assessment (LCA) methodology enables the computation of environmental impact associated with a process, product, or activity. This methodology is normally applied in four steps (ISO 14040): 1) Goal and scope definition, our target focuses on minimizing the process environmental impact, therefore, a fixed production rate of the feed and products was selected, ("cradle-to-gate" analysis). Ecoindicator-99 (EI-99) is integrated as a life cycle assessment tool, based on 11 categories and 3 specific damage categories: human health, ecosystem quality, and resources. 2) Inventory analysis, this step provides the inputs and outputs of mass and energy, construction materials, and utilities associated with the process, and all the burdens must be expressed per unit of reference (Life Cycle Inventory-LCI). 3) Impact assessment, in this step the process data are translated into environmental information, Equation 4 represents the EI99 compute. Where β_b represents the total amount of chemical b released per unit of reference flow due to direct emissions, $\alpha_{b,k}$ is the damage caused in each category k per unit of chemical b released to the environment, ω_d is the weighting factor for damage in category d, and δ_d is the normalization factor for damage of category d. 4) Interpretation, the results are analyzed and a set of conclusions or recommendations for the system are formulated (Guillen-Gosalbez et al., 2007; Sánchez-Ramírez et al., 2016).

$$EI\text{-}99 = \sum_{b} \sum_{d} \sum_{k} \alpha_{b,k} \beta_b \delta_d \omega_d \tag{4}$$

2.4. Optimization tools

2.4.1. Muti-objective genetic algorithms

Both virtual optimization platforms were made considering the integration of a Multi-Objective Genetic Algorithm (MOGA) based on NSGA-II: Aspen Plus-Matlab (A) using the gamultiobj suite, and Aspen Plus-Python (B) using the PYMOO package. The hyperparameters for both algorithms were as follows: 50 generations, 200 individuals, 0.8 for mutation crossover, and the mutation function: feasible adaptable for gamultiobj, and polynomial for PYMOO.

2.4.2. Utopia-tracking approach

Once the optimal solutions set has been obtained by MOGA, the next challenge consists in selecting the optimal point, mainly when many objective functions exist, and the final solutions are non-dominated. The utopia-tracking approach was selected as an adequate tool to choose the best solution point through minimizing the normalized distances between the objective function values and the utopic point. Equation 5 shows the formulation for the utopia-tracking approach (Zavala and Flores-Tlacuahuac, 2012).

$$min\left\| \Phi - \Phi^{L,s} \right\|_p = \left(\sum_{i=1}^{n} |\Phi_i - \Phi_i^{L,s}|^p \right)^{\frac{1}{p}} \quad (5)$$

Where Φ is any point in the set of final solutions, $\Phi^{L,s}$ is the utopia point associated with the minimum value for each individual objective function, n represents the number of the objective functions (three for this study), and p represents the p-norm value (one for this study).

2.5. Case study

The case study considered the separation and purification of the benzene, toluene, and m-xylene (BTX) mixture. The NRTL-SRK thermodynamic model was used to calculate the phase equilibria (Cui et al., 2019). Figure 1 shows the separation and purification scheme, based on a direct distillation sequence (D_{101}, D_{102}). The feed stream (101) used in this study has an equimolar flow of 300 kmol/h at 50 °C and 5 bar. Streams 102, 104, and 105 correspond to the recovered and purified compounds; benzene, toluene, and m-xylene, respectively. Furthermore, the vectors to model equality and inequality constraints are defined below. Two assumptions were made for the objective functions computing: for the TAC estimation a payback period was fixed to five years, in addition, for the EI-99 a process lifetime was adjusted to ten years.

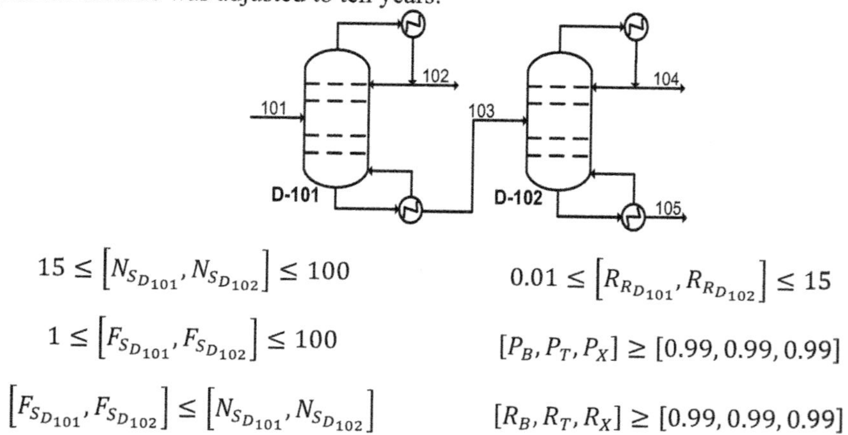

101
102
103
104
105
D-101
D-102
$15 \leq \left[N_{S_{D_{101}}}, N_{S_{D_{102}}}\right] \leq 100$
$0.01 \leq \left[R_{R_{D_{101}}}, R_{R_{D_{102}}}\right] \leq 15$
$1 \leq \left[F_{S_{D_{101}}}, F_{S_{D_{102}}}\right] \leq 100$
$[P_B, P_T, P_X] \geq [0.99, 0.99, 0.99]$
$\left[F_{S_{D_{101}}}, F_{S_{D_{102}}}\right] \leq \left[N_{S_{D_{101}}}, N_{S_{D_{102}}}\right]$
$[R_B, R_T, R_X] \geq [0.99, 0.99, 0.99]$

Figure 1. BTX separation and purification scheme and optimization constraints.

3. Results

Table 1 shows the employed optimal function values for both virtual platforms (VP) based on 3D Pareto front and employing the utopia-tracking approach as a tool to make optimal decisions. Besides, the computational times for each evaluation during the optimization step are shown, as well as the optimal state values. All calculations were made on laptop with an AMD Ryzen 7 5800 H processor @ 3.2 GHz, and 16 GB RAM.

To determine the best optimal results, normalized values of the objective functions were used, thus, the values of $\left\| \Phi - \Phi^{L,s} \right\|_p$ for the results of both virtual platforms were computed, getting a value of 0.45 (interface A), and 0.46 (interface B). For the selection of the optimal point, it was assumed that the objective functions have the same weight.

Table 1. Relevant indicators for the virtual platforms evaluated.

VP	Time (s)	TAC (USD/y)	σT_0 (kW)	EI-99 (mPt/y)	N_S D_{101}	F_S D_{101}	R_R D_{101}	N_S D_{102}	F_S D_{102}	R_R D_{102}
A	0.97	2.28E6	216.43	1.59E4	62	30	1.56	53	37	1.97
B	0.94	2.26E6	219.10	1.58E4	47	22	1.56	56	40	1.99
$\Phi^{L,S}$	-	**2.26E6**	**215.03**	**1.57E4**	-	-	-	-	-	-

Figure 2 shows the results, where the black dots correspond to interface A, and the blue dots correspond to interface B. Figure 2.a shows the 3D Pareto front for the final optimal solutions obtained in both virtual platforms, as well as a spider chart to show the relationship between state variables values from both platforms; the dots inside the red circles represent the optimal design for both interfaces and the values are shown in Table 1. Figure 2.b is illustrated to improve the visualization of the three-dimensional plot, thus the objective functions were plotted in 2D arrays, where the values were normalized to have a similar reference system.

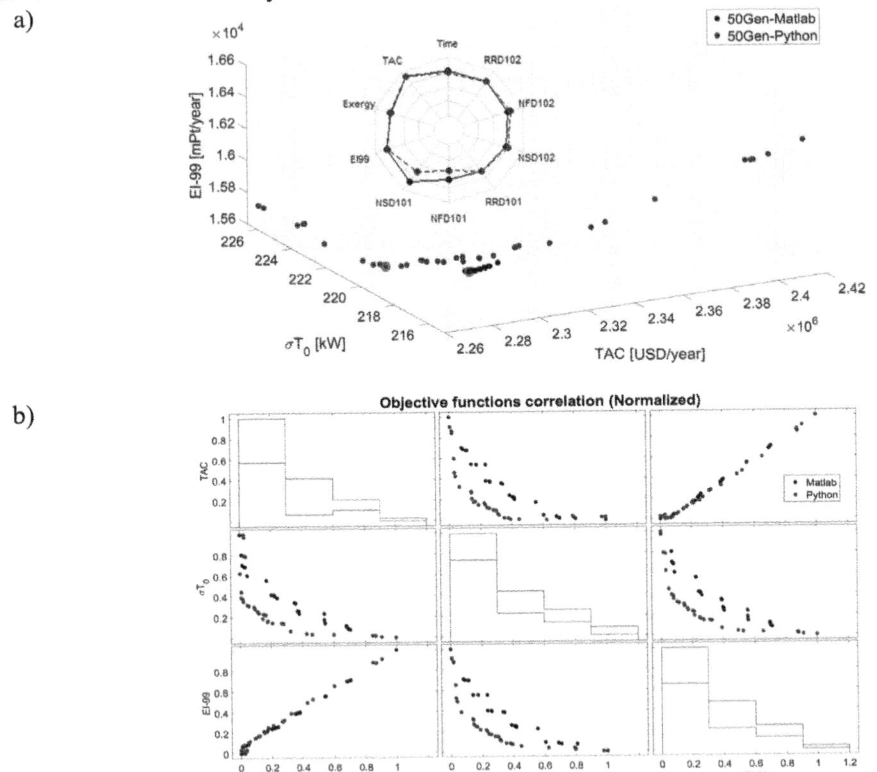

Figure 2. a) Comparative 3D Pareto plot. b) Objective functions normalized matrix.

The results showed that both computational interfaces have a similar performance in all terms, as well as a linear relation between economy-environment objective functions, and non-dominated solutions for economy-exergy, and economy-environment. In terms of computational time, (B) showed a better performance and, it might be possible to argue that Phyton is the best option because it has better performance in the TAC and the EI-99 objectives, but when the utopia-tracking approach is implemented, it is possible to observe that (A) had a lower value in terms of the three objectives normalized sum in the optimization problem, meaning that it is closer to the utopic value of the problem, which was generated through the minimum values obtained from both Pareto fronts. Based on the graphical results shown in Figure 2, it is likely to observe a close optimal solution for both algorithms, nevertheless, the limits and spread of both Pareto fronts are different for each solution and could be associated with the function coding of each step of the MOGA (fitness function selection, crossover, mutation, etcetera).

4. Conclusions

The proposed tools for selecting the best VP allowed concluded that, based on the decision-making criteria results, it was observed a better computational time for the virtual platform based on Aspen Plus-Python (B) (0.94 vs 0.97 seconds). Regarding the objective function, both VP have similar values, but the interface based on Aspen Plus-Matlab (A) obtained a better-normalized distance to the utopic point ($\Phi^{L,s}$) (0.45 vs 0.46). Although utopia-tracking offers a pathway to select the best solution in the optimal data sets from both Pareto fronts, a more detailed study should be carried out, which could tackle other case studies and focus on topics associated with the data set in the Pareto front and their performance indicators (e.g., distribution, convergence, etcetera).

Acknowledgments

Jaime David Ponce-Rocha acknowledges the financial support from "CIATEQ A.C. Advanced Technology Center", for the development of this project.

References

I.T. Cameron, S. Engell, C. Georgakis, N. Asprion, D. Bonvin, F. Gao, D.I. Gerogiorgis, I.E. Grossmann, S. Macchietto, H.A. Preisig, B.R. Young, 2019, Education in Process Systems Engineering: Why it matters more than ever and how it can be structured, Comput. and Chem. Eng., 126, 102-112.

C. Cui, X. Zhang, J. Sun, 2019, Design and optimization of energy-efficient liquid-only side-stream distillation configurations using a stochastic algorithm, Chemical Engineering Research and Design, 145, 48-52.

G. Guillen-Gosalbez, J.A. Caballero, L.J. Esteller, M. Gadalla, 2007, Application of life cycle assessment to the structural optimization of process flowsheets, Comput. Aided Chem. Eng., 24, 1163-1168.

Y. Li, T. Sun, Q. Ye, J. Li, Y. Xu, X. Jian, 2021, Economic and environmental assessment for purification of acetonitrile and isopropanol by reactive coupling extractive distillation, Separation and Purification Technology, 275, 119133.

E.N. Pistikopoulos, A. Barbosa-Povoa, J.H. Lee, R. Misener, A. Mitsos, G.V. Reklaitis, V. Venkatasubramanian, F. You, R. Gani, 2021, Process systems engineering – The generation next?, Comput. and Chem. Eng., 147, 107252.

J.D. Ponce-Rocha, M. Picón-Núñez, R. Morales-Rodriguez, 2022, A framework for optimal and flexible schemes design under uncertainty & sustainable aspects, Comput. Aided Chem. Eng., 51, 757–762.

R. Ruiz-Femenia, J. Javaloyes-Antón, R. Salcedo-Díaz, M.A.S.S. Ravagnani, J.A. Caballero, 2020, Integration of Chemical Process Simulators with Algebraic Modeling Languages, Comput. Aided Chem. Eng., 48, 1891-1896.

E. Sánchez-Ramírez, J.J. Quiroz-Ramírez, J.G. Segovia-Hernández, S. Hernández, J.M. Ponce-Ortega, 2016, Economic and environmental optimization of the biobutanol purification process, Clean Technology Environtal Policy, 18, 395-411.

Antonis Kokossis, Michael C. Georgiadis, Efstratios N. Pistikopoulos (Eds.)
PROCEEDINGS OF THE 33rd European Symposium on Computer Aided Process Engineering (ESCAPE33), June 18-21, 2023, Athens, Greece
 http://dx.doi.org/10.1016/B978-0-443-15274-0.50136-0

Mathematical Modeling of Anion Exchange Membrane Water Electrolyzer

Donggyun Lee[a] , Minsu Kim[a] , Il Moon[a *]

[a]Department of Chemical and Biomolecular Engineering, Yonsei University, Seoul 03722, Republic of Korea
email of the corresponding author: ilmoon@yonsei.ac.kr

Abstract

Rising energy demand, depletion of fossil fuels and global warming are driving research into eco-friendly alternative energy technologies. Currently, interest in hydrogen as alternative energy is increasing. Among various hydrogen production technologies, there are three major methods for producing hydrogen by water electrolysis: proton exchange membrane water electrolysis (PEMWE), alkaline water electrolysis (AWE), and anion exchange membrane water electrolysis (AEMWE). AEMWE has been attracting attention in recent years because it has the advantage of overcoming the limitations of both electrolysis techniques; Low current density, low pressure, and use of expensive noble metal catalysts.

High-efficiency hydrogen production and durability are determined by the design of the serpentine and the performance of the catalyst and exchange membrane. That is, the distribution of the temperature and pressure in the electrolyzer and the hydrogen production are affected by the design of the serpentine. Manufacturing and experimenting with various serpentine designs are costly and time-consuming. This can be solved by constructing the water electrolyzer system through computational fluid dynamics (CFD). In addition, it is possible to check the flow phenomena, distribution of pressure, temperature, and current density inside the system.

In this study, a mathematical model of the AEMWE system was established. The model for the electrochemical reaction was implemented through user defined functions (UDFs), and reliability was increased through validation with actual experimental data. This model will be used to analyze the efficiency and durability of the water electrolyzer in various serpentine designs.

Keywords: Hydrogen production, AEM water electrolysis, Mathematical modeling, Computational fluid dynamics

1. Introduction

Since traditional energy sources such as fossil fuels generate greenhouse gas (GHG) (e.g., methane, carbon dioxide), there is a need for alternative fuels. Hydrogen is considered an alternative energy source because it has a higher energy content than fossil fuels and can significantly reduce GHG. The hydrogen production process is divided into three categories: grey hydrogen produced through reforming of methane and water vapor, blue hydrogen produced by combining grey hydrogen with carbon capture and storage (CCS), and green hydrogen produced through water electrolysis. In particular, green hydrogen significantly reduces carbon dioxide emissions compared to other methods and is considered a more promising method because water is an infinite source.

AEM electrolysis is a method that combines the advantages of alkaline electrolysis and PEM electrolysis, and has various characteristics: First, hydrogen can be produced at low cost by using a relatively inexpensive non-platinum group metal catalyst. Second, the system is easy to handle due to the low operating temperature.

Here, we analyzed various indices for improving performance and durability through mathematical modeling of AEM electrolysis. Through our developed model, it is possible to identify the exact contribution of mass, momentum, and energy transport.

2. Mathematical model

Fig. 1 shows the schematic view of AEM electrolysis cell (AEMEC). AEMEC consists of plate, gas diffusion layer (GDL), catalyst layer (CL), and membrane. As a reactant, water passes through the flow channel and GDL in the plate, and the reaction occurs at the electrode. At the cathode, OH^- ions are generated through the reaction of Eq. (1) and move to the anode. At the anode, water and oxygen are produced through Eq. (2).

$$Cathode: 4H_2O + 4e^- \rightarrow 2H_2 + 4OH^- \quad (1)$$

$$Anode: 4OH^- \rightarrow 2H_2O + O_2 + 4e^- \quad (2)$$

The mixture model of ANSYS FLUENT was used to solve the two-phase flow, and the electrochemical model and source terms were applied as UDFs.

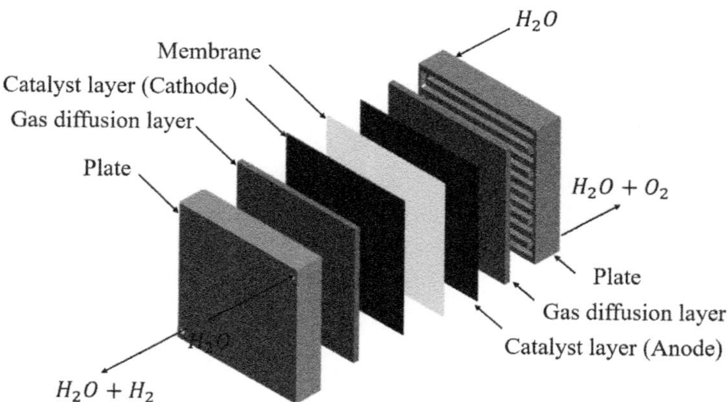

Fig. 1. Schematic view of AEM electrolysis cell.

2.1. Governing equations

The continuity equation for liquid-gas mixture is expressed as Eq. (3). $\overrightarrow{v_m}$ is the mass-averaged velocity, and ρ_m is the mixture density. α_k represents the volume fraction for the k phase.

$$\nabla \cdot (\rho_m \overrightarrow{v_m}) = S_m \tag{3}$$

$$\overrightarrow{v_m} = \frac{\sum_{k=1}^{n} \alpha_k \rho_k \overrightarrow{v_k}}{\rho_m} \tag{4}$$

$$\rho_m = \sum_{k=1}^{n} \alpha_k \rho_k \tag{5}$$

The momentum equation for a mixture can be obtained by summing the individual momentum equations for all phases. It is expressed as Eq. (6), where μ_m denotes the viscosity of the mixture

$$\nabla \cdot (\rho_m \overrightarrow{v_m} \overrightarrow{v_m}) = -\nabla p + \nabla \cdot \left(\mu_m \left(\nabla \overrightarrow{v_m} + \nabla \overrightarrow{v_m}^T\right)\right) + \nabla \cdot \left(\sum_{k=1}^{n} \alpha_k \rho_k \overrightarrow{v_{dr,k} v_{dr,k}}\right) + S_{mom} \tag{6}$$

$$\mu_m = \sum_{k=1}^{n} \alpha_k \mu_k \tag{7}$$

Drift velocity is related to relative velocity, which can be expressed as:

$$\vec{v}_{dr,k} = \overrightarrow{v_k} - \overrightarrow{v_m} \tag{8}$$

The energy equation for liquid-gas mixture is given by Eq. (9), where k_{eff} represents the effective conductivity. S_e includes volumetric heat sources.

$$\nabla \cdot \sum_{k=1}^{n} \left(\alpha_k \overrightarrow{v_k} (\rho_k E_k + p)\right) = \nabla \cdot \left(k_{eff} \nabla T\right) + S_e \tag{9}$$

The internal energy E_k for a compressible phase is given by Eq. (10), where h_k is the sensible enthalpy for k phase.

$$E_k = h_k - \frac{p}{\rho_k} + \frac{{v_k}^2}{2} \tag{10}$$

To solve the species conservation equation, the convection-diffusion equation for i th species predicts the local mass fraction of each species, Y_i. The species transport equation can be represented as follows:

$$\nabla \cdot (\rho \vec{v} Y_i) = -\nabla \cdot \vec{J_i} + S_i \tag{11}$$

$$\vec{J_i} = -\rho D_{i,m} \nabla Y_i \tag{12}$$

The source terms used in each governing equation are listed in Table 1.

Table 1
Source terms of governing equations.

Source Terms	Anode	Cathode
Mass: liquid phase	$S_m = \dot{N}_{H_2O,m} - \dot{N}_{H_2O,evap} + \frac{iM_{H_2O}}{2F}$	$S_m = -\dot{N}_{H_2O,m} - \dot{N}_{H_2O,evap} - \frac{iM_{H_2O}}{F}$
Mass: vapor phase	$S_m = \frac{iM_{O_2}}{4F} + \dot{N}_{H_2O,evap}$	$S_m = \frac{iM_{H_2}}{2F} + \dot{N}_{H_2O,evap}$
Momentum	$S_{mom} = -\frac{\mu}{k}\vec{v}$	
Energy	$S_e = (V_{cell} - 1.48V)i - \Delta H_{vap}\frac{\dot{N}_{H_2O,evap}}{M_{H_2O}}$	
Species: H_2	$S_{H_2} = 0$	$S_{H_2} = \frac{iM_{H_2}}{2F}$
Species: O_2	$S_{O_2} = \frac{iM_{O_2}}{4F}$	$S_{O_2} = 0$

2.2. Electrochemical model

In the electrochemical model, the total voltage is expressed as the sum of open circuit voltage (OCV), activation overpotential, and ohmic overpotential. OCV is the voltage when no current flows and is derived from the Nernst equation.

$$V = E_{OCV} + \eta_{act} + \eta_{Ohm} \tag{13}$$

$$E_{OCV} = 1.229 - 0.846 \times 10^{-3}(T - 298.15) + \frac{RT}{2F}(\ln p_{H_2} + 0.5 \ln p_{O_2}) \tag{14}$$

The activation overpotential is related to the electrochemical reaction at the electrode and is expressed by the Butler-Volmer equation (Eq. (15)).

$$\eta_{act} = \frac{RT_a}{2\alpha_a F}\operatorname{arcsinh}\left(\frac{i}{2i_{0,a}}\right) + \frac{RT_c}{2\alpha_c F}\operatorname{arcsinh}\left(\frac{i}{2i_{0,c}}\right) \tag{15}$$

$$i_0 = i_{0,ref}\, exp\left[-\frac{E_{act}}{R}\left(\frac{1}{T} - \frac{1}{T_{ref}}\right)\right] \tag{16}$$

Ohmic overpotential is a term related to resistance when current flows and is expressed by ohmic's law.

$$\eta_{Ohm} = R \times i \tag{17}$$

The relationship between current and voltage is expressed as Eq. (13), and the current according to the applied voltage is calculated through Newton's method through UDFs.

3. Results and discussion

Fig. 2 illustrates the validation with the experimental data. When the reference parameters were used in the electrochemical model, there was a difference from the experimental data. To solve this problem, the parameter values were estimated using a genetic algorithm and calibrated to the experimental data

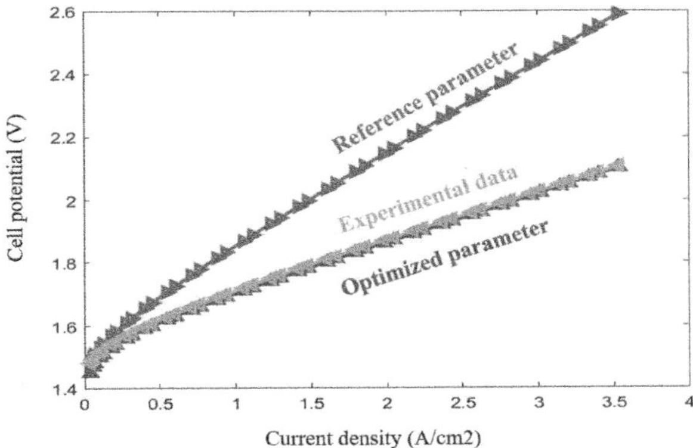

Fig. 2. Comparison between simulation results and experimental data.

Fig. 3 shows the gas volume fraction at the cathode using CFD simulation. In general, the gas volume fraction increases from inlet to outlet. Gas tends to accumulate at the tip of the cell.

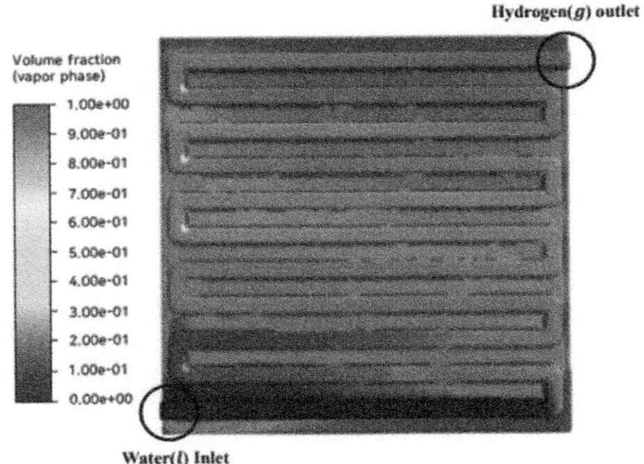

Fig. 3. Distribution of the gas volume fraction at the cathode.

In this study, 3-dimensional AEMEC model was constructed by combining a mixture model and an electrochemical model using UDFs. Durability and performance were analyzed by analyzing temperature and pressure distribution and hydrogen fraction in the cell. As future work, we plan to compare the effects of various serpentine designs, and find the optimal design achieving the highest performance and durability of cell system by optimizing serpentine configurations.

References

Hamish A.Miller, 2022, Green hydrogen from anion exchange membrane water electrolysis, Current Opinion in Electrochemistry, 36, 101122.

S. T. Revankar, 2010, Development of efficient flowsheet and transient modeling for nuclear heat coupled sulfur iodine cyclefor hydrogen production, Purdue Univ.

D. Xu, 2018, Earth-abundant oxygen electrocatalysts for alkaline anion exchange membrane water electrolysis: Effects of catalyst conductivity and comparison with performance in three electrode cell, ACS Catalysis, 9, 7.

I. Pushkareva, 2020, Comparative study of anion exchange membranes for low-cost water electrolysis, International Journal of Hydrogen Energy, 45, 26070.

H. A. Miller, 2020, Green hydrogen from anion exchange membrane water electrolysis: a review of recent developments in critical materials and operating conditions, Sustainable Energy & Fuels, 4, 2114.

Chaochao Cheng, 2021, Numerical investigation on the feasibility of metal foam as flow field in alkaline anion exchange membrane fuel cell, Applied Energy, 302, 117555.

Kui Jiao, 2014, Three-dimensional multiphase modeling of alkaline anion exchange membrane fuel cell, International Journal of Hydrogen Energy, 39, 5981.

Zhiwen Ma, 2021, A comprehensive modeling method for proton exchange membrane electrolyzer development, International Journal of Hydrogen Energy, 46, 17627.

Hao Deng, 2016, Modeling of hydrogen alkaline membrane fuel cell with interfacial effect and water management optimization, Renewable Energy, 91, 116.

Dang Saebea, 2019, Model based evaluation of alkaline anion exchange membrane fuel cells with water management, Chemical Engineering Journal, 374, 721.

D.S. Falcao, 2020, A review on PEM electrolyzer modelling: Guidelines for beginners, Journal of Cleaner Production, 261, 121184.

Haluk Görgün, 2006, Dynamic modelling of a proton exchange membrane (PEM) electrolyzer, International Journal of Hydrogen Energy, 31, 29.

Antonis Kokossis, Michael C. Georgiadis, Efstratios N. Pistikopoulos (Eds.)
PROCEEDINGS OF THE 33rd European Symposium on Computer Aided Process Engineering (ESCAPE33), June 18-21, 2023, Athens, Greece
 http://dx.doi.org/10.1016/B978-0-443-15274-0.50137-2

An Aided Optimization Method for Molecular Reconstruction of Crude Oil Based on Pre-optimized Dataset

Chen Zhang,[a,b] Tong Qiu,[a,b]

[a] *Institute of Process Systems Engineering, Department of Chemical Engineering, Tsinghua University, Beijing 100084, China*

[b] *Beijing Key Laboratory of Industrial Big Data System and Application, Tsinghua University, Beijing 100084, China*

qiutong@tsinghua.edu.cn

Abstract

As a method of converting easy-to-measure physical properties into detailed composition of feedstocks through distribution law and optimization calculation, molecular reconstruction method plays an important role in providing refineries with detailed composition data for molecular management. Introducing the knowledge of petroleum components, including the distribution pattern of homologues and the range of distribution parameters, can effectively enhance the effect of molecular reconstruction. We propose an aided optimization method to assist in the molecular reconstruction of crude oil. This method helps the optimization algorithm to obtain faster and more accurate results by using a pre-optimized dataset added to the initial population of the particle swarm optimization algorithm. The proposed aided optimization method was validated on 52 sets of crude oil data and achieved an average objective function reduction to 82.8% of the original and an average optimization time reduction to 87.5% of the original when using approximately 20% of the pre-optimized data points as aided data.

Keywords: Molecular reconstruction; Crude oil; Structure-oriented lumping; Pre-optimized dataset; Aided optimization method.

1. Introduction

The refining process is the separation, transformation and purification of oil molecules, in which the average molecular weight is gradually reduced, and various types of molecules are gathered together in an orderly manner. As the crude oil resources gradually become heavier, the government's environmental requirements for the production process become more and more stringent, and the description of the refining process becomes more and more detailed until it reaches the molecular level of description. Molecular management is a combined technical solution for petrochemical process system optimization based on the molecular level. Due to the limitations of analytical techniques and production costs, it is often difficult to perform an exhaustive experimental analysis of all oil products (Stratiev et al., 2019). Molecular reconstruction technology has gained wide research and application by using the physical properties that can be quickly measured to obtain detailed molecular compositions through optimization, which provide the basic data for molecular management (Wu and Zhang, 2010).

In recent years, besides the research on the methods and processes of molecular reconstruction itself (Bi and Qiu, 2019), a series of aided methods have been proposed to introduce information and knowledge in molecular reconstruction to improve the accuracy and computational speed. The aided methods mainly include (a) predicting the range of optimization parameters through historical data, to improve the speed and accuracy (Bi and Qiu, 2018); (b) based on the known distribution law of oil components, the virtual oil components are constructed to find similar oil components to assist in the optimization (Guan and Zhang, 2022); (c) achieving better prediction results by introducing neural network models in the optimization (Ma et al., 2022).

Considering that the distribution law of crude oil is not completely clear, an effective virtual oil fraction cannot be constructed. In this article, we propose an aided optimization method based on the results of molecular reconstruction of actual crude oil components. The similarity components of crude oil are found according to the similarity calculation method. By adding similar components to the initial particle population of the PSO algorithm, an enhancement to the optimization results is achieved. This aided optimization method not only demonstrates the enhancement effect of pre-optimized datasets on molecular reconstruction, but also shows the importance of introducing information and knowledge in molecular reconstruction.

2. Methods

2.1. Molecular construction based on SOL

The first step in the molecular reconstruction process is to represent molecules appropriately. We chose the structure-oriented lumping (SOL) method as the way to represent molecules and build a molecular library. Based on previous work, we use 21 structural increments to describe the composition of the molecule (Chen et al., 2018). The structure of each molecule can be described by the number of these 21 structural increments. As shown in Figure 1, a two-step approach is used to construct the molecular library by first constructing representative core structures, followed by adding different lengths of side chains to each core structure to complete all the molecules. For the physical properties of the constructed molecules, the group contribution (GC) method is used to calculate the relevant physical properties.

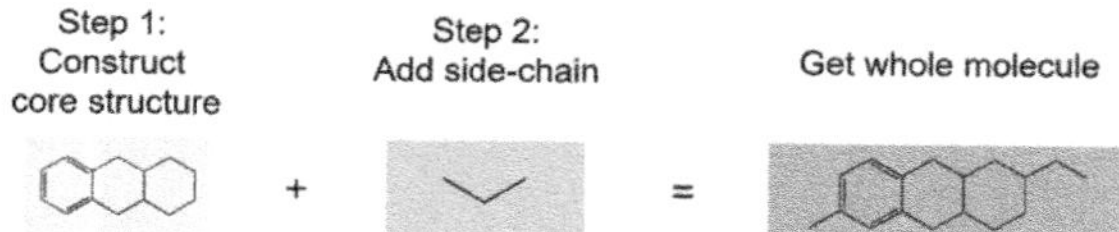

Figure 1. Molecular construction steps.

2.2. Calculation of mole fraction of molecules

After constructing the molecular library, it is necessary to describe how to calculate the mole fraction of each molecule. Considering the complexity of crude oil, a two-layer homologous series distribution is used to describe the distribution of the molecular library. As shown in Figure 2, the first layer of homologous series distribution is the homologous series distribution of the core structure. There are 19 types of core structures, and each type has 5 core structures. The exponential distribution is used to describe the probability density function in each core structure type, and the histogram distribution is used to describe the mole fraction of each core structure type. The second layer of homologous series distribution is the length distribution of the side chains. Side chains with a length of 1-50 carbon numbers are added to each core structure, and the gamma distribution is used to describe the carbon number distribution of the side chains.

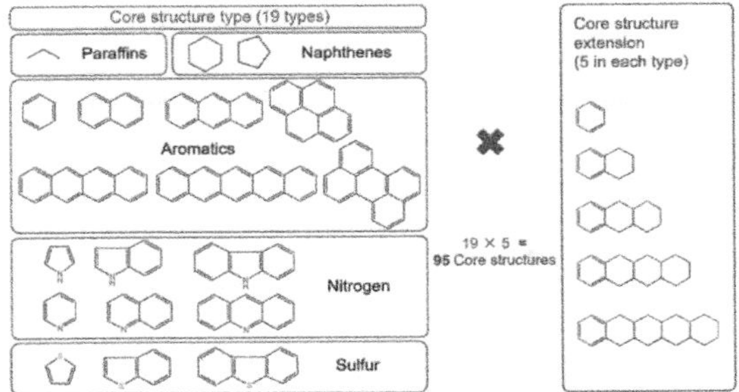

Figure 2. The core structure types and the homologous series distribution of the core structure.

According to the two-layer homologous series distribution and probability density function, the mole fractions of all molecules can be calculated. The physical properties of the reconstructed oil can be calculated based on the mole fractions of all molecules, the physical properties of all molecules and the mixing rule.

2.3. Optimization of molecular reconstruction of crude oil

The goal of the optimization of the molecular reconstruction is to make the difference between the calculated properties and the experimentally analyzed properties as small as possible, and the flow chart of the optimization is shown in Figure 3.

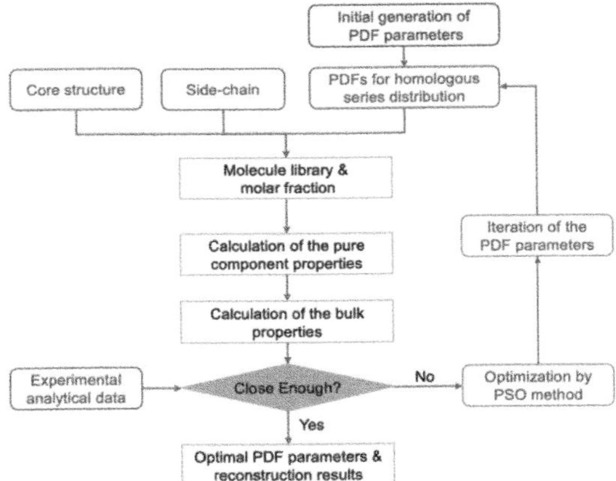

Figure 3. Flow chart of the optimization of molecular reconstruction.

The objective function is the weighted average of the absolute relative deviation values of physical properties, and the optimization variables are the parameters of the probability distribution, with a total of 227 optimization variables. The particle swarm optimization (PSO) is used as the optimization algorithm. The PSO can efficiently search for problems with large solution spaces and find candidate solutions without sufficient information. The particle swarm size is set to 200 based on the number of optimization variables, computational speed, and accuracy. The optimization process is calculated in MATLAB 2022a, and its own PSO algorithm is used for calculation.

2.4. Construction of pre-optimized dataset

Based on the molecular reconstruction proposed above, the molecular reconstruction optimization calculations are performed on the 52 sets of crude oil data we have, and the results (values of each optimization variable) are aggregated to obtain the pre-optimized dataset. The pre-optimized dataset will serve as the basic data used by the aided optimization method to assist other crude oils in molecular reconstruction, and it will also be used as a benchmark for comparing the results of the aided optimization method.

2.5. Using pre-optimized dataset to aid optimization

When new crude oil is available for molecular reconstruction, the process of using the pre-optimized dataset to aid in optimization can be divided into 3 steps.

Step 1, find similar data. Find data points in the pre-optimized dataset that are similar to the new crude oil. The Normalized Euclidean distance was used to measure the distance between the new crude oil data and data point in the pre-optimized dataset, which can effectively avoid the influence caused by different units of different properties and no pre-processing of data is required either. All 14 physical properties used in molecular reconstruction were used as the components of the distance calculation. The calculation formula of the Normalized Euclidean distance is shown in Equation 1, where u is the vector of a data point, v is the vector of another data point, and V is the variance of vectors u and v. According to the distance formula, the distance between the new crude oil data and all the data points in the pre-optimized dataset can be calculated, and the smaller the value, the more similar the two data set are.

$$distance = \sqrt{\sum_{i=1}^{n} \frac{(u_i - v_i)^2}{V_i}} \quad (1)$$

Step 2, introduce the pre-optimized data points. Based on the similarity data calculated in step 1, the pre-optimized data are sorted by similarity and the most similar pre-optimized data points are selected in preference. As shown in Figure 4, a certain number of pre-optimized data points are used as part of the initial particle population of the PSO algorithm in the solution of the optimization problem.

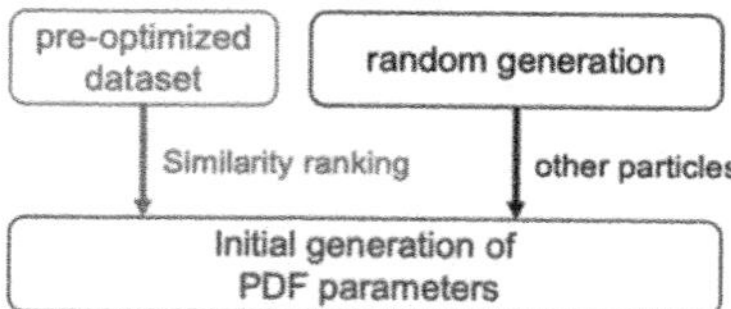

Figure 4. Initial particle and swarm construction process.

Step 3, perform the optimization calculation. After adding some of the pre-optimized data points to the initial particles, the remaining initial particles for the PSO algorithm are generated using a randomized approach as shown in Figure 4. Using these initial particles, the optimization computation is started following the process shown in Figure 3.

3. Results and discussion

The proposed aided optimization method is applied to the molecular reconstruction process of crude oil. Due to the limitation of the size of the available crude oil dataset (only 52 sets of crude oil with physical properties and calculation results data), we adopted the method of selecting one set of crude oil data as the test data one by one and the remaining 51 sets of data as the pre-optimized dataset for the aided optimization. The most similar 3/5/10/20 pre-optimized data points were selected as aided data points in the aided optimization calculation for each crude oil data, respectively. The optimization problem was solved in MATLAB 2022a, and the entire computation was performed on the Intel Xeon 4216 CPU @ 2.1GHz.

First, molecular reconstruction is performed on a set of crude oil data. The same values are used for all optimization parameters except the initial particle population. Calculations were performed for different number of aided data, and the results are shown in Table 1.

Table 1. The results obtained with the aided optimization method.

Number of Aided data	0	3	5	10	20
Objective Fun Value	100%	88.4%	82.5%	77.7%	97.4%
Optimization Time	100%	22.8%	19.8%	39.1%	34.4%
Property		Absolute relative error of the calculation result			
Carbon	0.27%	0.01%	0.01%	0.02%	0.11%
Hydrogen	1.70%	0.00%	0.00%	0.05%	0.60%
Sulfur	0.06%	0.00%	0.00%	0.00%	0.00%
Nitrogen	0.06%	0.00%	0.00%	0.00%	0.00%
Paraffins	0.05%	0.00%	0.00%	0.00%	0.00%
Naphthenes	0.06%	0.00%	0.00%	0.00%	0.00%
Aromatics	0.03%	0.00%	0.00%	0.00%	0.00%
5 vol% BP	5.65%	1.35%	0.21%	0.21%	5.65%
10 vol% BP	0.56%	0.13%	1.77%	3.86%	0.10%
30 vol% BP	4.54%	4.18%	0.40%	0.82%	0.31%
50 vol% BP	8.41%	0.17%	0.00%	7.40%	3.07%
70 vol% BP	0.01%	2.65%	3.49%	0.00%	2.31%
90 vol% BP	4.89%	7.17%	7.28%	4.32%	7.11%
95 vol% BP	7.55%	10.97%	10.11%	7.52%	10.11%

The objective function value and optimization time of the molecular reconstruction results without pre-optimized data are recorded as 100%, and the percentage share of the other results with pre-optimized data is calculated. It can be found that the inclusion of pre-optimized data points all reduces the objective function value and optimization time to some extent, respectively.

For the objective function, the value decreases gradually at first with the addition of pre-optimized data points and reaches the lowest level of 77.7% of the baseline value when 10 pre-optimized data are added (about 20% pre-optimized data). This indicates that the inclusion of pre-optimized data points effectively helps the whole population to find better results during the optimization. When 20 pre-optimized data were used (about 40% pre-optimized data), the objective function value increased to 97.4% of the baseline value. This situation indicates that only an appropriate proportion (around 20%) of all pre-optimized data points will work best, and the inclusion of more pre-optimized data points will not effectively help the population to find better results and generate positive effects due to the low similarity between them and the calculated data. By checking the similarity values, it was found that the pre-optimized data with distance values less than 4 had a great aided optimization effect, and the pre-optimized data with distance values greater than 4 had an insignificant or even counterproductive aided optimization effect.

For the optimization time, the addition of the pre-optimized data was also effective, which reached the lowest level of 19.8% of the baseline value when 5 pre-optimized data (about 10% pre-optimized data) were added. The addition of the pre-optimized data provides more information during the optimization and reduces the overall population disorder, allowing the algorithm to converge faster and more accurately. When 10 pre-optimized data are added, the optimization time rises to 39.1% of the baseline value, which is mainly due to the addition of more pre-optimized data, driving the algorithm to go for better results, which increases the number of iterations and makes the optimization time grow.

Then, all the data were calculated separately with the aided optimization method, and the results were recorded. The results are compared with those obtained without using the pre-optimized data, and the percentage share of the results is calculated. The average calculation results of all data are shown in Table 2 and Figure 5.

Table 2. The average calculation result of all data.

Number of Aided data	0	3	5	10	20
Objective Fun Value	100%	89.7%	85.5%	82.8%	89.1%
Optimization Time	100%	91.4%	85.1%	87.5%	101.3%

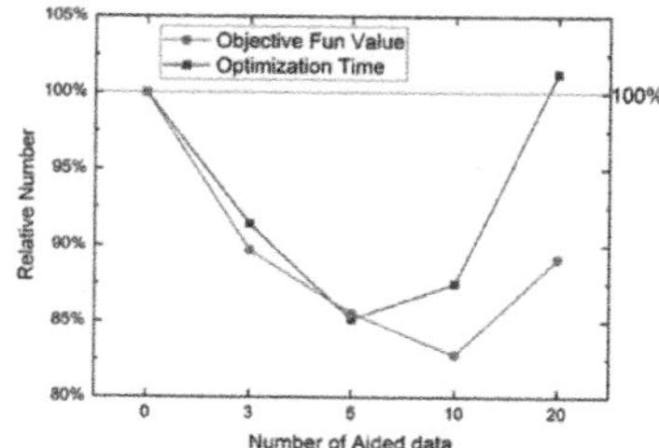

Figure 5. The trend of calculation results with the number of aided data.

According to Table 2 and Figure 5, when 10 pre-optimized data points are added, the average objective function value reaches the lowest level, reaching 82.8% of the baseline value. When 5 pre-optimized data points are added, the average optimization time reaches the lowest level, reaching 85.1% of the baseline value. Considering the objective function value and optimization time together, the best results can be achieved when 10 pre-optimized data (or about 20% pre-optimized data) are added in molecular reconstruction.

4. Conclusions

In conclusion, this novel work proposes an aided optimization method for molecular reconstruction of crude oil based on a pre-optimized dataset. With the pre-optimized dataset and similarity calculation, the results of historical molecular reconstruction can be introduced into the optimization of the molecular reconstruction as crude oil distribution knowledge, which plays an important role in enhancing the optimization results. By the characteristics of the PSO algorithm, the added pre-optimized data can effectively lead the whole population to a better solution, while the optimization time is simultaneously reduced due to the addition of more information and knowledge. The effectiveness of the proposed aided optimization method is verified on crude oil data and shows the potential for future application to real industrial scenarios to further improve the accuracy and speed of molecular reconstruction.

References

Bi, K., Qiu, T., 2019. Novel Naphtha Molecular Reconstruction Process Using a Self-Adaptive Cloud Model and Hybrid Genetic Algorithm–Particle Swarm Optimization Algorithm. Ind. Eng. Chem. Res. 58, 16753–16760.

Bi, K., Qiu, T., 2018. A high-performance molecular reconstruction method with parameter initialization based on PCA, in: Computer Aided Chemical Engineering. Elsevier, pp. 2005–2010.

Chen, J., Fang, Z., Qiu, T., 2018. Molecular reconstruction model based on structure oriented lumping and group contribution methods. Chinese Journal of Chemical Engineering 26, 1677–1683.

Guan, D., Zhang, L., 2022. Initial guess estimation and fast solving of petroleum complex molecular reconstruction model. AIChE Journal 68.

Ma, F., Zheng, X., Han, C., Wang, J., Sun, W., 2022. Molecular Reconstruction of Naphtha based on Physical Information Neural Network. IFAC-PapersOnLine 55, 186–191.

Stratiev, D., Shishkova, I., Tankov, I., Pavlova, A., 2019. Challenges in characterization of residual oils. A review. Journal of Petroleum Science and Engineering 178, 227–250.

Wu, Y., Zhang, N., 2010. Molecular Characterization of Gasoline and Diesel Streams. Ind. Eng. Chem. Res. 49, 12773–12782.

Antonis Kokossis, Michael C. Georgiadis, Efstratios N. Pistikopoulos (Eds.)
PROCEEDINGS OF THE 33rd European Symposium on Computer Aided Process Engineering (ESCAPE33), June 18-21, 2023, Athens, Greece
 http://dx.doi.org/10.1016/B978-0-443-15274-0.50138-4

A Global Optimization Algorithm for the Solution of Mixed-Integer Quadratic Adjustable Robust Optimization Problems under Endogenous Uncertainty

Byungjun Lee[a], Styliani Avraamidou[a*]

[a]*Department of Chemical and Biological Engineering, University of Wisconsin-Madison, 1415 Engineering Dr, Madison, WI 53706, USA*

Abstract

This work proposes an algorithm for the solution of the two-stage mixed-integer quadratic adjustable robust optimization (ARO) problem under endogenous uncertainty. This class of ARO problems is often used in Chemical Engineering application such as scheduling and control problems, but there are only a handful of approaches attempting to solve them. For our developed approach, the two-stage ARO problem is firstly reformulated as a tri-level mixed-integer quadratic programming problem, and the exact and global solution is obtained using multi-parametric programming. The proposed algorithm was verified and compared with other approaches such as the affine decision rules, the column-and-constraint generation algorithm, and full enumeration of extreme points. We showed that our algorithm has the ability to solve ARO problems that, to our knowledge, cannot be solved by any other algorithm. To assess the efficiency and the performance of the proposed algorithm, a computational study was performed.

Keywords: Adjustable robust optimization (ARO), Endogenous uncertainty, Multi-parametric programming, Tri-level programming

1. Introduction

Adjustable robust optimization (ARO) is an extension of robust optimization (RO), which includes recourse decisions in the second stage (Ben-Tal et al., 2009). In Chemical Engineering, the ARO problem formulation can be applied to solve a vast array of multi-stage decision making problems under uncertainty, such as batch scheduling problems (Shi and You, 2016) and explicit model predictive control (Tejeda-Iglesias et al., 2019). Although several methods have been proposed for the solution of this class of problems, including affine decision rules (Ben-Tal et al., 2004) and column-and-constraint generation algorithm (Zeng and Zhao, 2013), most solution methods are limited to the solution of continuous linear ARO problems, and solving the mixed-integer quadratic ARO problems still remains a challenge.

In this paper, we apply a multi-parametric programming approach to solve the two-stage mixed-integer quadratic ARO problems under endogenous uncertainty, as an extension of the previous work on the linear ARO problems (Avraamidou and Pistikopoulos, 2020). Section 2 presents the general formulation of the ARO problems and presents our solution algorithm. Then the proposed algorithm is verified with several numerical examples, and its advantages over the previously known algorithms are discussed in section 3. Lastly, section 4 presents the relation between the computational time and the total number of variables.

2. ARO through multi-parametric programming

The two-stage mixed-integer ARO problem with a convex quadratic objective function can be expressed as (1). The first-stage variable (x) is called "here-and-now" decision, which is determined prior to the realization of uncertainty (u). The second-stage variable (y) is called "wait-and-see" decision, which is determined after the uncertainty is revealed.

$$\begin{aligned} &\min_{x}(Q_x x + c_x)^T x + \max_{u\in\Omega_u}\min_{y}\left(Q_y y + c_y\right)^T y \\ &\text{s.t. } A_x x + A_y y + A_u u \le b \\ &\quad x \in X := \mathbb{R}^{p_c} \times \{0,1\}^{p_b},\ y \in Y := \mathbb{R}^{q_c} \times \{0,1\}^{q_b},\ u \in U := \mathbb{R}^{r_o}. \end{aligned} \tag{1}$$

In this work, uncertainty is assumed as a function of "here-and-now" decisions. This type of uncertainty is referred to as endogenous uncertainty, where "here-and-now" decisions constrain the feasible region of uncertainty (Lappas and Gounaris, 2018). At the initial step, problem (1) is reformulated as a tri-level mixed-integer quadratic programming (T-MIQP) problem (2).

$$\begin{aligned} &\min_{x} z^{(1)} = (Q_x x + c_x)^T x + \left(Q_y y + c_y\right)^T y \quad (1^{\text{st}}\text{ level}) \\ &\text{s.t. } \min_{u\in\Omega_u} z^{(2)} = -\left(Q_y y + c_y\right)^T y \quad (2^{\text{nd}}\text{ level}) \\ &\quad \text{s.t. } \min_{y} z^{(3)} = \left(Q_y y + c_y\right)^T y \quad (3^{\text{rd}}\text{ level}) \\ &\quad\quad \text{s.t. } A_x x + A_y y + A_u u \le b \\ &\quad\quad\quad x \in X := \mathbb{R}^{p_c} \times \{0,1\}^{p_b},\ y \in Y := \mathbb{R}^{q_c} \times \{0,1\}^{q_b},\ u \in U := \mathbb{R}^{r_o}. \end{aligned} \tag{2}$$

The tri-level problem (2) is then solved using multi-parametric approach based on the generalized piecewise affine decision rules (Avraamidou and Pistikopoulos, 2019). First, the 3rd level MIQP problem is solved to obtain the parametric solution $y(x,u)$ for each critical region. Similarly, resulting set of 2nd level MIQP problems are solved to find the parametric solutions $u(x)$. Then resulting set of single-level MIQP problems are solved for x, and the exact and global solution is selected after applying comparison procedure. The comparison procedure removes unfeasible critical regions and finds the optimum solution that minimizes the objective function $z^{(1)}$ among the remaining critical regions. The proposed algorithm is also illustrated in Figure 1.

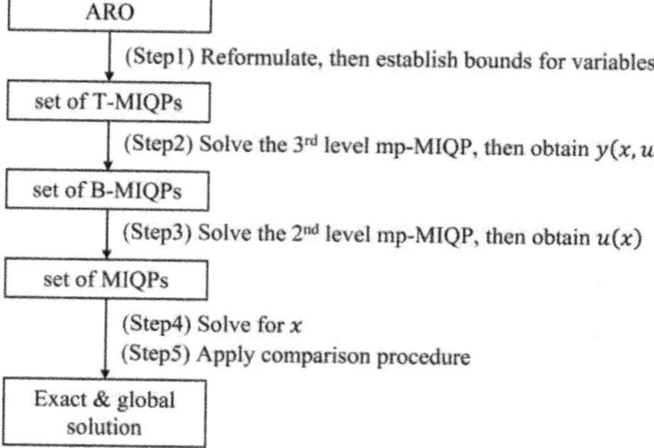

Figure 1. Algorithm to solve two-stage quadratic MI-ARO problem.

3. Numerical Examples

The first numerical example (3) has a convex, quadratic objective function in the first stage, which is a function of "here-and-now" decisions.

$$\begin{aligned}
&\min_x 2x_1^2 + x_1x_2 + 3x_2^2 + \max_{u\in\Omega_u} \min_y 12y_1 + 18y_2 \\
&\text{s.t.} \quad x_1 + x_2 \leq 10 \\
&\qquad y_1 \geq u_1 - x_1 \\
&\qquad y_2 \geq u_2 - x_2 \\
&\qquad x_1,\, x_2 \geq 0,\;\; y_1,\, y_2 \geq 0 \\
&\qquad \Omega_u = \{u\colon\; 12.1 \leq u_1 \leq 26.3,\;\; 13.2 \leq u_1 \leq 30.4\}
\end{aligned} \tag{3}$$

Problem (4) represents the tri-level formulation that is recast from the original two-stage ARO formulation (3).

$$\begin{aligned}
&\min_x z^{(1)} = 2x_1^2 + x_1x_2 + 3x_2^2 + 12y_1 + 18y_2 \\
&\text{s.t.} \quad \min_{u\in\Omega_u} z^{(2)} = -(12y_1 + 18y_2) \\
&\qquad \text{s.t.} \quad \min_y z^{(3)} = 12y_1 + 18y_2 \\
&\qquad\qquad \text{s.t.} \quad x_1 + x_2 \leq 10 \\
&\qquad\qquad\qquad y_1 \geq u_1 - x_1 \\
&\qquad\qquad\qquad y_2 \geq u_2 - x_2 \\
&\qquad\qquad\qquad x_1,\, x_2 \geq 0,\;\; y_1,\, y_2 \geq 0 \\
&\qquad\qquad\qquad \Omega_u = \{u\colon\; 12.1 \leq u_1 \leq 26.3,\;\; 13.2 \leq u_1 \leq 30.4\}
\end{aligned} \tag{4}$$

The proposed approach was verified by comparing it with both the affine decision rules and column-and-constraint generation algorithm. The objective function value and the corresponding optimum solution for each approach are summarized in Table 1.

Table 1. Objective function value and the corresponding optimum solution through multiparametric programming (MP), affine decision rules (AD), and column-and-constraint generation (CC).

Algorithm	z^*	x_1^*	x_2^*	y_1^*	y_2^*	u_1^*	u_2^*
MP	825.23	2.35	2.61	23.95	27.79	26.30	30.40
AD	825.23	2.35	2.61	23.95	27.79	26.30	30.40
CC	825.23	2.35	2.61	23.95	27.79	26.30	30.40

The second numerical example (5) has a convex, quadratic objective function in the second stage, which is a function of "wait-and-see" decisions.

$$\begin{aligned}
&\min_x 3x_1 + 2x_2 + 1500x_3 + 6000x_4 \\
&\qquad + \max_{u\in\Omega_u} \max_y (6y_1^2 + 5y_2^2 + 9y_3^2 + 4y_1y_2 + 2y_1y_3 + 6y_2y_3) \\
&\text{s.t.} \;\; x_1 + 2x_2 + 3x_3 + 4x_4 \leq 30 \\
&\qquad y_1 \geq u_1 - x_1 - x_2 \\
&\qquad y_2 \geq u_2 - 8x_3
\end{aligned}$$

$$y_3 \geq u_3 - 15x_4$$
$$x_1, x_2 \geq 0,\ x_3, x_4 \in \{0, 1\},\ y_1, y_2, y_3 \geq 0$$
$$\Omega_u = \{u\colon\ 10 \leq u_1, u_2, u_3 \leq 20\} \tag{5}$$

In this case, the affine decision rules are not applicable due to the nonlinear terms of uncertainties in the second stage, which are known to be computationally intractable in general. For the column-and-constraint generation algorithm, the sub-problem is non-convex, after using KKT conditions to transform the bi-level sub-problem into a single-level problem. Therefore, the multi-parametric approach has an advantage over the other two algorithms when the ARO problem includes the quadratic terms of "wait-and-see" decisions.
After the tri-level reformulation of the problem (5), POP toolbox (Avraamidou and Pistikopoulos, 2019) was used to obtain the third-level parametric solutions $y(x, u)$ and second-level parametric solutions $u(x)$. The decision rules for each critical region in the 3rd level and 2nd level are summarized in the Table 2.

Table 2. Decision rules and critical region definition for the 3rd level and 2nd level.

Level	Critical Region	Decision Rules	Additional Critical Region Definition
3	CR_1	$y_1 = -x_1 - x_2 + u_1,$ $y_2 = -8x_3 + u_3$ $y_3 = -15x_4 + u_4$	$-50 \leq x_1 + x_2 - u_1 \leq 0,\ -50 \leq 15x_4 - u_3 \leq 0,$ $6x_1 + 6x_2 + 16x_3 + 15x_4 - 6u_1 - 2u_2 - u_3 \leq 0,$ $2x_1 + 2x_2 + 40x_3 + 45x_4 - 2u_1 - 5u_2 - 3u_3 \leq 0,$ $x_1 + x_2 + 24x_3 + 135x_4 - u_1 - 3u_2 - 9u_3 \leq 0$
	CR_2	$y_1 = -x_1 - x_2 + u_1,$ $y_2 = -8x_3 + u_3$ $y_3 = 0$	$-50 \leq x_1 + x_2 - u_1 \leq 0,\ 15x_4 - u_3 \leq 0,$ $3x_1 + 3x_2 + 8x_3 - 3u_1 - u_2 \leq 0,$ $2x_1 + 2x_2 + 40x_3 - 2u_1 - 5u_2 \leq 0,$ $x_1 + x_2 + 24x_3 - u_1 - 3u_2 \leq 0$
	CR_3	$y_1 = 0,$ $y_2 = -8x_3 + u_3$ $y_3 = -15x_4 + u_4$	$x_1 + x_2 - u_1 \leq 0,\ -50 \leq 15x_4 - u_3 \leq 0,$ $16x_3 + 15x_4 - 2u_2 - u_3 \leq 0,$ $40x_3 + 45x_4 - 5u_2 - 3u_3 \leq 0,$ $8x_3 + 45x_4 - u_2 - 3u_3 \leq 0$
	CR_4	$y_1 = 0,$ $y_2 = -8x_3 + u_3$ $y_3 = 0$	$x_1 + x_2 - u_1 \leq 0,\ 15x_4 - u_3 \leq 0,$ $8x_3 - u_2 \leq 0$
	Common	–	$x_1 + 2x_2 + 3x_3 + 4x_4 \leq 30,\ -8x_3 + u_2 \leq 52,$ $x_1, x_2 \geq 0,\ x_3, x_4 \in \{0, 1\},$ $10 \leq u_1, u_2, u_3 \leq 20$
2	$CR_{1,1}$	$u_1 = 20,$ $u_2 = 20,$ $u_3 = 20$	$x_1 + x_2 \leq 20,\ 3x_4 \leq 4$ $6x_1 + 6x_2 + 16x_3 + 15x_4 \leq 180,$ $2x_1 + 2x_2 + 40x_3 + 45x_4 \leq 200,$ $x_1 + x_2 + 24x_3 + 135x_4 \leq 260$
	$CR_{2,1}$	$u_1 = 20,$ $u_2 = 20,$ $u_3 = 0$	$x_1 + x_2 \leq 20,$ $3x_1 + 3x_2 + 8x_3 \leq 80,$ $x_1 + x_2 + 20x_3 \leq 70$
	$CR_{3,1}$	$u_1 = 0,$ $u_2 = 20,$ $u_3 = 20$	$x_1 + x_2 \geq 0,\ 3x_4 \leq 4$ $4x_3 + 5x_4 \leq 16,$ $8x_3 + 45x_4 \leq 80$
	$CR_{4,1}$	$u_1 = 0,$ $u_2 = 20,$ $u_3 = 0$	$x_1 + x_2 \geq 0,$ $2x_3 \leq 5$
	Common	–	$x_1 + 2x_2 + 3x_3 + 4x_4 \leq 30,$ $x_1, x_2 \geq 0,\ x_3, x_4 \in \{0, 1\}$

Lastly, CPLEX was used to solve the resulting single-level mixed-integer quadratic programming problem. Table 3 summarizes the 1st level objective function values for all the critical regions, and only the first critical regions CR_1 and $CR_{1,1}$ were selected from the comparison procedure. Therefore, the optimum solution is $x = [13, 7, 1, 0]^T$ and $y = [0, 12, 20]^T$ with objective function value 7313. This value exactly matches with the optimum objective function value obtained from the full enumeration of extreme points.

Table 3. Objective function values and optimum solutions in the 1st level.

CR_i	CR_j	z^*	x_1	x_2	x_3	x_4	y_1	y_2	y_3	u_1	u_2	u_3
1	1	7313	13	7	1	0	0	12	20	20	20	20
2	1	2050	10	10	0	0	0	20	0	20	20	0
3	1	7260	0	0	1	0	0	12	20	0	20	20
4	1	2000	0	0	0	0	0	20	0	0	20	0

4. Computational Implementation

To evaluate the performance of the proposed algorithm, computational time of randomly generated mixed-integer ARO problems was measured. The ratio of the number of variables for continuous x, binary x, continuous y, binary y was set to be $1:1:1:1$, and the number of constraints is set to be equal to the total number of variables. Note that the solution time was measured using Intel Core i7 at 80GHz and 16GB RAM, MATLAB R2022b, and IBM ILOG CPLEX Studio 12.10.0.

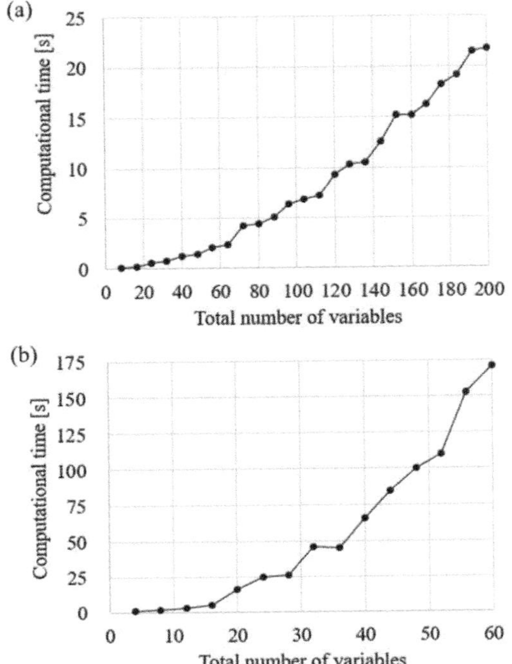

Figure 2. Measured computational time versus total number of variables for the (a) MILP-MILP structure and (b) MIQP-MIQP structure of the ARO problem.

Figure 2 represents the relation between the average computational time over five experiments and the total number of variables for the MILP-MILP structure and MIQP-MIQP structure of the ARO problem. Figure 2 also shows the exponentially increasing trend in the computational time as the total number of variables increases. In addition, it takes about 175s to solve the MIQP-MIQP structure of the ARO problem with 60 variables, about 70 times longer computational time compared to the MILP-MILP structure.

5. Conclusion

In this work, we developed an algorithm for the solution of the two-stage mixed-integer quadratic ARO problem under endogenous uncertainty. Initially, the ARO problem is reformulated into the tri-level mixed-integer quadratic programming problem, and it is solved multi-parametrically using the general piecewise affine decision rules. By solving the problem from the lowest level (3rd level) to the highest level (1st level), the exact and global solution is obtained after applying the comparison procedure. For the ARO problem with a quadratic objective function in the first stage, the proposed algorithm was verified by comparing its global solution with the global solutions from affine decision rules and column-and-constraint generation algorithm. In addition, we showed that these two well-known algorithms cannot solve the ARO problem with a quadratic objective function in the second stage, while our algorithm is able to solve this type of problem. Lastly, we analyzed the performance of the proposed algorithm by measuring the computational time for the randomly generated ARO problems with MILP-MILP structure and MIQP-MIQP structure.

References

S. Avraamidou, E. N. Pistikopoulos, 2019, Multi-parametric global optimization approach for tri-level mixed-integer linear optimization problems, Journal of Global Optimization, 74(3), 443-465.

S. Avraamidou, E. N. Pistikopoulos, 2020, Adjustable robust optimization through multi-parametric programming, Optimization Letters, 14(4), 873-887.

A. Ben-Tal, L. El Ghaoui, A. Nemirovski, 2009, Robust optimization, Princeton university press.

A. Ben-Tal, A. Goryashko, E. Guslitzer, A. Nemirovski, 2004, Adjustable robust solutions of uncertain linear programs, Mathematical programming, 99(2), 351-376.

N. H. Lappas, C. E. Gounaris, 2018, Robust optimization for decision-making under endogenous uncertainty, Computers & Chemical Engineering, 111, 252-266.

H. Shi, F. You, 2016, Adjustable robust optimization for scheduling of batch processes under uncertainty, Computer Aided Chemical Engineering, Vol. 38, 547-552.

M. Tejeda-Iglesias, N. H. Lappas, C. E. Gounaris, L. Ricardez-Sandoval, 2019, Explicit model predictive controller under uncertainty: An adjustable robust optimization approach, Journal of Process Control, 84, 115-132.

B. Zeng, L. Zhao, 2013, Solving two-stage robust optimization problems using a column-and-constraint generation method, Operations Research Letters, 41(5), 457-461.

Antonis Kokossis, Michael C. Georgiadis, Efstratios N. Pistikopoulos (Eds.)
PROCEEDINGS OF THE 33rd European Symposium on Computer Aided Process Engineering (ESCAPE33), June 18-21, 2023, Athens, Greece
 http://dx.doi.org/10.1016/B978-0-443-15274-0.50139-6

Cross-domain Fault Diagnosis for Chemical Processes through Dynamic Adversarial Adaptation Network

Ruoshi Qin,[a] Jinsong Zhao,[a,b]

[a]*State Key Laboratory of Chemical Engineering, Department of Chemical Engineering, Tsinghua University, Beijing 100084, China*
[b]*Beijing Key Laboratory of Industrial Big Data System and Application, Tsinghua University, Beijing 100084, China*

Abstract

During the past decade, deep learning models have shown great potential in chemical process fault diagnosis. However, a major assumption accepted by default in these previous studies is that the training and testing data possess the same feature distribution, which is generally invalid in industrial applications. The ability of data-driven models to generalize on various operation conditions are limited due to their tendency to overfit the training dataset. Motivated by the development of transfer learning, domain adaptation is utilized in this paper to settle the problem of knowledge transfer under the distribution discrepancy between the training and test datasets. The proposed model Dynamic Adversarial Adaptation Network (DAAN) can dynamically and quantitatively evaluate the importance of these two distributions and achieve accurate classification performance. Empirical evaluations based on Tennessee Eastman process validate the applicability and practicability of DAAN, which achieves the state-of-the-art transfer results in terms of diverse operating conditions and fault types.

Keywords: Domain adaptation; Fault diagnosis; Transfer learning; Tennessee Eastman process

1. Introduction

Fault diagnosis plays a significant role in chemical processes to avoid severe accidents. In recent years, significant advances have been made towards fault diagnosis with the assistance of deep learning. However, the excellent performance of current intelligent fault diagnosis methods based on deep learning is mainly attributed to the availability of large amounts of labeled failure data which is laborious and heavily dependent on experience. The entire equipment usually runs in a normal state and accessible fault samples are scarce as well (Qin et al., 2022). Another limitation on the practical application of these fault diagnosis methods is the continual changes in operating conditions. The distribution discrepancy between source training data and target test data makes a high requirement for the transferability of the diagnosis model (Wu et al., 2020).

In industrial production, the variation of ambient temperature and raw material components contribute to diverse working modes. When the fault samples in the test dataset are lack of labels and the training and testing data are subject to different distributions, the performance of most intelligent diagnostic models deteriorates considerably (Qin et al., 2022). Transfer learning puts forward an emerging frame to address the above problems. The concept of domain in transfer learning represents the corresponding data space and the data distribution. In this study, the source domain refers to the data space and distribution of the labeled training data, while the attribute of the unlabeled test data is known as the target domain. The challenge of transferring the learned knowledge from the source domain to the target domain is defined as the domain shift phenomenon. This problem makes the source domain classifier lose its generalization ability in the target domain. Domain adaptation is a preferred solution to the domain shift issue, aiming to develop a cross-domain classifier which can extract domain-invariant features to bridge the distribution discrepancy (Li et al., 2019). Fig.1 illustrates the principle of domain shift and domain adaptation.

The latest studies reveal that deep neural networks can learn transferable features sufficiently with unsupervised domain adaptation algorithms, such as maximum mean discrepancy (MMD), multiple kernel MMD and joint MMD (Long et al., 2017). Some innovative works raise deep transfer learning methods in the field of computer vision, including Domain Adaptive Neural Network (DANN) (Ganin et al., 2017) and Domain Adaptive Ensemble Learning (DAEL) (Zhou et al., 2021), but it is still rare to discover a powerful cross-domain fault diagnosis method for chemical processes.

In this paper, a novel adversarial domain adaptation transfer learning model called Dynamic Adversarial Adaptation Network (DAAN) is proposed for fault diagnosis in industrial chemical processes. The diagnostic framework is composed of four modules, a shared feature extractor for feature learning, a label classifier for category prediction, a global domain discriminator for marginal distribution alignment and local subdomain discriminators for conditional distribution alignment. Deep residual shrinkage network is designed as the backbone of the feature extractor to achieve multiscale characteristics from time-series data. The adversarial learning mode with an adaptive weighting strategy can dynamically and quantitatively evaluate the relative importance of marginal and conditional distribution. The alignment of the source and target feature space is finally realized through the overall learning proceed.

The rest of the paper is organized as follows. In Section 2, the module structures and the diagnosis framework of DAAN are introduced in detail. Section 3 elaborates the experiments on Tennessee Eastman process and compares the results among different popular transfer learning models. The conclusion and outlook of the proposed model are summarized in Section 4.

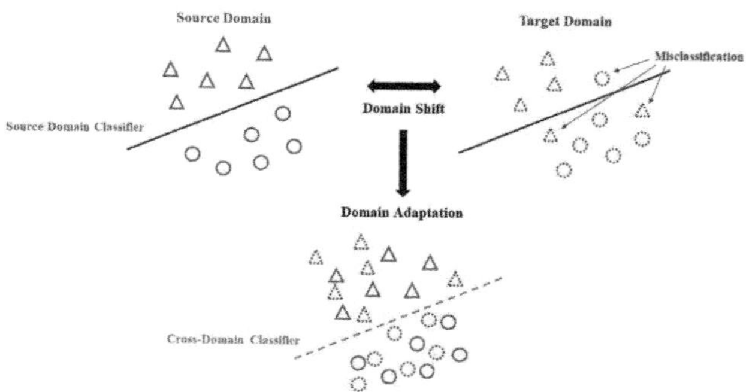

Fig.1. Illustration of domain shift and domain adaptation.

2. Dynamic Adversarial Adaptation Network

The proposed model Dynamic Adversarial Adaptation Network is displayed in Fig.2. Its skeleton consists of a feature extractor G_f, a label classifier G_y, a global domain discriminator G_d and a local subdomain discriminator G_d^c. In DAAN, G_f extracts the key features of data followed by G_d and G_d^c achieving the adaptation of marginal and conditional distributions respectively. The dynamic adversarial factor ω performs a quantitative evaluation of these two distributions. Under the action of G_y, the DAAN model is trained efficiently with the Gradient Reversal Layer (GRL) (Yu et al., 2019).

2.1. Feature Extractor

The raw data from different domains are directly fed into a shared feature extractor for certain generic characteristics. Deep residual shrinkage network is employed as the extractor which can obtain powerful feature learning capability and eliminate the noise-related information. The feature extractor includes multiple residual learning blocks and each of them is composed of convolution operation, batch normalization, activation function, and short connection. The structure of residual shrinkage building unit is a variant of classical ResNet, and a range of units are stacked to form the deep network. Sigmoid and rectified linear unit (ReLU) as non-linear transform tools are adopted to promote model convergence and alleviate the problem of gradient vanishing.

2.2. Domain Adaptation Network

2.2.1. Label Classifier

The label classifier G_y focuses on the discrimination of the input sample label from the source domain. With the effort of the supervised information and the extracted deep features, the classifier is trained based on the cross-entropy loss formulated as:

$$L_y = -\frac{1}{n_s}\sum_{x_i\in\mathcal{D}_s}\sum_{c=1}^{C} P_{x_i\to c}\log G_y\left(G_f(\boldsymbol{x}_i)\right) \tag{1}$$

where C is the number of classes, $P_{x_i\to c}$ is the probability of $\boldsymbol{x}_i$ belonging to class c.

2.2.2. Global Domain Discriminator

The global domain discriminator G_d takes the responsibility of align the marginal distributions between the source and target domains. Its loss objective function can be calculated in the formula:

$$L_g = \frac{1}{n_s+n_t}\sum_{x_i\in(\mathcal{D}_s\cup\mathcal{D}_t)} L_d\left(G_d\left(G_f(\boldsymbol{x}_i)\right), d_i\right) \tag{2}$$

where L_d is the domain discriminator loss and d_i is the original label of the input sample.

2.2.3. Local Subdomain Discriminator

The local subdomain discriminator G_d^c is designed to align the conditional distributions between the source and target domains. Its representation ability of merging multimode structure in two distributions makes the discriminator carry out more fine-grained domain adaptation. The local subdomain discriminator can be regarded class-wise to match the multi-domain data associated with the same class, and the global domain discriminator is able to be split into multi-class local subdomain discriminators. The output of the label predictor $G_y(\boldsymbol{x}_i)$ indicates the matching degree of each data point $\boldsymbol{x}_i$ to be categorized further. Therefore, the loss function of G_d^c can be described as:

$$L_l = \frac{1}{n_s+n_t}\sum_{c=1}^{C}\sum_{x_i\in(\mathcal{D}_s\cup\mathcal{D}_t)} L_d^c\left(G_d^c\left(\hat{y}_i^c G_f(\boldsymbol{x}_i)\right), d_i\right) \tag{3}$$

where L_d^c represents the cross-entropy loss of the subdomain discriminator associated with class c. $\hat{y}_i^c$ signifies the predicted probability distribution over the class c of the input sample $\boldsymbol{x}_i$.

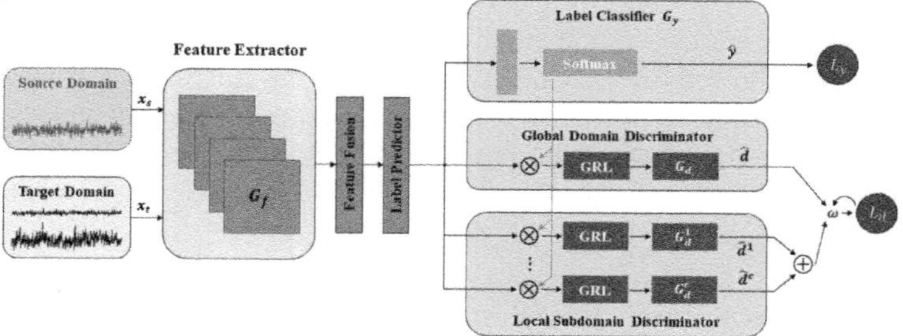

Fig.2. Architecture of Dynamic Adversarial Adaptation Network.

2.3. Dynamic Adversarial Training

Domain adversarial adaptation borrows the principle of Generative Adversarial Networks (GAN) to assist the model to learn transferable features. The adversarial learning procedure is a two-player game accompanied by the domain discriminator G_d and the feature extractor G_f. The former aims to distinguish the source domain from the target domain, while the latter keeps making puzzles for the former by extracting domain-invariant features. During this course, the parameters of G_f are optimized by maximizing the loss of G_d, and the ones of G_d are trained by minimizing the loss. With the loss of the label classifier G_y minimized, the loss function can be formalized as:

$$L(\theta_f, \theta_y, \theta_d) = \frac{1}{n_s} \sum_{\boldsymbol{x}_i \in \mathcal{D}_s} L_y\left(G_y\left(G_f(\boldsymbol{x}_i)\right), y_i\right) - \frac{\lambda}{n_s + n_t} \sum_{\boldsymbol{x}_i \in (\mathcal{D}_s \cup \mathcal{D}_t)} L_d\left(G_d\left(G_f(\boldsymbol{x}_i)\right), d_i\right) \quad (4)$$

where λ is a trade-off parameter. As the training converges, the parameters $\hat{\theta}_f$, $\hat{\theta}_y$, $\hat{\theta}_d$ come to the optimized state.

$$(\hat{\theta}_f, \hat{\theta}_y) = \arg\min_{\theta_f, \theta_y} L(\theta_f, \theta_y, \theta_d), (\hat{\theta}_d) = \arg\max_{\theta_d} L(\theta_f, \theta_y, \theta_d) \quad (5)$$

The dynamic adversarial factor ω is creatively built to evaluate the relative importance of the marginal and conditional distributions. In DAAN, the global $\mathcal{A}$-distance of the global domain discriminator and the local $\mathcal{A}$ -distance of the local subdomain discriminator can be denoted as:

$$d_{\mathcal{A},g}(\mathcal{D}_s, \mathcal{D}_t) = 2\left(1 - 2(L_g)\right), d_{\mathcal{A},l}(\mathcal{D}_s^c, \mathcal{D}_t^c) = 2(1 - 2(L_l^c)) \quad (6)$$

where $\mathcal{D}_s^c$ and $\mathcal{D}_t^c$ represent samples from class c. Eventually, the dynamic adversarial factor ω is estimated as:

$$\hat{\omega} = \frac{d_{\mathcal{A},g}(\mathcal{D}_s, \mathcal{D}_t)}{d_{\mathcal{A},g}(\mathcal{D}_s, \mathcal{D}_t) + \frac{1}{C}\sum_{c=1}^{C} d_{\mathcal{A},l}(\mathcal{D}_s^c, \mathcal{D}_t^c)} \quad (7)$$

ω starts to be initialized as 1 in the first epoch. The pseudo labels of the target domain can be obtained after each iteration, so the local distance for class c can be computed with cross entropy. Therefore, the learning objective of DAAN is concluded as follows:

$$L(\theta_f, \theta_y, \theta_d, \theta_d^c|_{c=1}^C) = L_y - \lambda\left((1 - \omega)L_g + \omega L_l\right) \quad (8)$$

2.4. DAAN-based fault diagnosis framework

The framework of the proposed DAAN-based fault diagnosis method is shown in Fig.3. The dual-stage procedures include offline training and online application. Historical data collected from the chemical process are utilized for model design. Benefitting from the transferred knowledge, the diagnosis network monitors the online process and outputs the fault classification result.

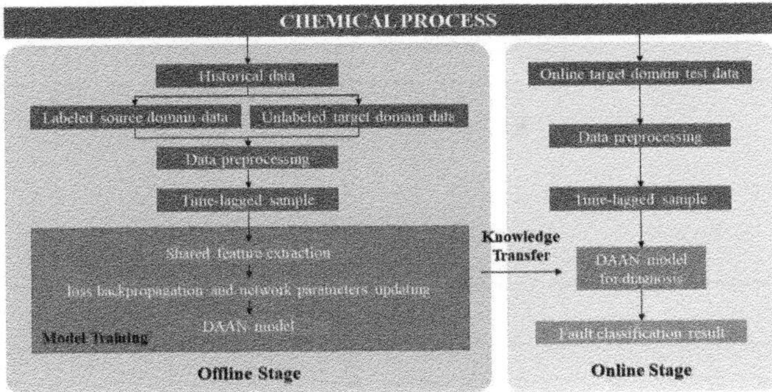

Fig.3. DAAN-based fault diagnosis framework.

3. Case Studies

In this section, the benchmark Tennessee Eastman process (TEP) is applied to evaluate the performance of the proposed model. TEP is widely used in chemical process monitoring tasks. The unit operations of TEP contain a reactor, a condenser, a recycle compressor, a vapor-liquid separator and a stripper. In this paper, the experiments are based on the revised version (Bathelt et al., 2015) which can be downloaded at http://depts.washington.edu/control/LARRY/TE/download.html. 19 fault types and 5 steady-state operating modes are introduced in this research to explore the multimode fault diagnosis performance. The fault types and working conditions are selected in the same way as the previously published work, and the datasets are prepared followed by its experience (Wu et al., 2020).

When the labeled fault data in the source mode and only one-batch unlabeled fault data in the target mode are involved in the training dataset, the unsupervised anomaly diagnosis tasks are presented to verify the model effectiveness. Different common transfer learning models are adopted in comparison with the proposed DAAN model, such as traditional deep transfer networks composed of stacked auto-encoder (DTN), domain adaptation networks along with CNN (DAN-CNN) and joint adaptation networks along with CNN (JAN-CNN). The fault identification performances are exhibited in Table 1. The average accuracy rates are calculated and summarized in Table 2. The classification results show the efficiency and robustness of DAAN.

Table.1. Unsupervised fault diagnosis performance comparison.

(a) DTN

	$\mathcal{D}_t(0)$	$\mathcal{D}_t(1)$	$\mathcal{D}_t(2)$	$\mathcal{D}_t(4)$	$\mathcal{D}_t(5)$
$\mathcal{D}_s(0)$	-	91.4%	92.0%	82.3%	91.2%
$\mathcal{D}_s(1)$	92.9%	-	93.6%	85.3%	89.4%
$\mathcal{D}_s(2)$	92.8%	92.5%	-	84.1%	93.1%
$\mathcal{D}_s(4)$	80.6%	83.5%	82.9%	-	79.2%
$\mathcal{D}_s(5)$	89.4%	90.7%	92.9%	77.8%	-

(b) DAN-CNN

	$\mathcal{D}_t(0)$	$\mathcal{D}_t(1)$	$\mathcal{D}_t(2)$	$\mathcal{D}_t(4)$	$\mathcal{D}_t(5)$
$\mathcal{D}_s(0)$	-	93.6%	93.1%	91.6%	91.8%

$\mathcal{D}_s(1)$	94.9%	-	95.4%	91.5%	95.6%
$\mathcal{D}_s(2)$	93.2%	93.3%	-	89.0%	**97.4%**
$\mathcal{D}_s(4)$	87.1%	84.4%	84.7%	-	82.2%
$\mathcal{D}_s(5)$	90.2%	92.6%	**97.4%**	86.3%	-

(c) JAN-CNN

	$\mathcal{D}_t(0)$	$\mathcal{D}_t(1)$	$\mathcal{D}_t(2)$	$\mathcal{D}_t(4)$	$\mathcal{D}_t(5)$
$\mathcal{D}_s(0)$	-	96.0%	96.4%	91.1%	**97.6%**
$\mathcal{D}_s(1)$	95.6%	-	**95.9%**	90.8%	96.4%
$\mathcal{D}_s(2)$	93.5%	95.6%	-	89.6%	97.3%
$\mathcal{D}_s(4)$	**88.8%**	89.7%	88.2%	-	90.6%
$\mathcal{D}_s(5)$	93.9%	95.8%	97.3%	86.9%	-

(d) DAAN

	$\mathcal{D}_t(0)$	$\mathcal{D}_t(1)$	$\mathcal{D}_t(2)$	$\mathcal{D}_t(4)$	$\mathcal{D}_t(5)$
$\mathcal{D}_s(0)$	-	**97.2%**	**96.8%**	**92.5%**	97.3%
$\mathcal{D}_s(1)$	**96.6%**	-	**95.9%**	**93.9%**	**96.7%**
$\mathcal{D}_s(2)$	**95.5%**	**96.1%**	-	**92.0%**	97.2%
$\mathcal{D}_s(4)$	87.2%	**91.7%**	**90.4%**	-	**91.1%**
$\mathcal{D}_s(5)$	**94.7%**	**96.0%**	97.3%	**88.4%**	-

Table.2. Average accuracy rate comparison of diagnostic models.

	DTN	DAN-CNN	JAN-CNN	DAAN
Average accuracy rate	87.9%	91.3%	93.4%	**94.2%**

4. Conclusions

In the present work, Dynamic Adversarial Adaptation Network is designed for fault diagnosis application in the chemical process. This adversarial domain adaptation model mainly consists of three components, label classifier, global domain discriminator and local subdomain discriminator with a powerful feature extractor attached to the front. Comprehensive results demonstrate that the model can obtain superior diagnostic performance compared with other transfer learning models. Further work is being conducted to improve this model, such as further alleviating the need for labeled data in the source domain and tapping its potential for industrial process tasks.

References

R. Qin, J. Zhao, 2022, High-Efficiency Generative Adversarial Network Model for Chemical Process Fault Diagnosis, *IFAC-PapersOnLine*, 55, 7, 732-737.

H. Wu, J. Zhao, 2020, Fault detection and diagnosis based on transfer learning for multimode chemical processes, *Computers and Chemical Engineering*, 135, 106731.

R. Qin, J. Zhao, 2022, Adaptive multiscale convolutional neural network model for chemical process fault diagnosis, *Chinese Journal of Chemical Engineering*, 50, 398-411.

X. Li, W. Zhang, Q. Ding, 2019, Cross-Domain Fault Diagnosis of Rolling Element Bearings Using Deep Generative Neural Networks, *IEEE Transactions on Industrial Electronics*, 66, 7, 5525-5534.

M. Long, H. Zhu, J. Wang, M.I. Jordan, 2017, Deep transfer learning with joint adaptation networks, *ICML'17: Proceedings of the 34th International Conference on Machine Learning*, 70, 2208-2217.

Y. Ganin, E. Ustinova, H. Ajakan, P. Germain, H. Larochelle, F. Laviolette, M. Marchand, V. Lempitsky, 2017, Domain-Adversarial Training of Neural Networks, *Domain Adaptation in Computer Vision Applications*, 189–209.

K. Zhou, Y. Yang, Y. Qiao, T. Xiang, 2021, Domain Adaptive Ensemble Learning, *IEEE Transactions on Image Processing*, 30, 8008-8018.

C. Yu, J. Wang, Y. Chen, M. Huang, 2019, Transfer Learning with Dynamic Adversarial Adaptation Network, *2019 IEEE International Conference on Data Mining (ICDM)*, 19303062.

A. Bathelt, N.L. Ricker, M. Jelali, 2015, Revision of the Tennessee Eastman Process Model, *IFAC-PapersOnLine*, 48, 8, 309-314.

Antonis Kokossis, Michael C. Georgiadis, Efstratios N. Pistikopoulos (Eds.)
PROCEEDINGS OF THE 33rd European Symposium on Computer Aided Process Engineering (ESCAPE33), June 18-21, 2023, Athens, Greece
 http://dx.doi.org/10.1016/B978-0-443-15274-0.50140-2

Design and performance analysis of energy efficient hydrogen liquefaction process

Ahmad Naquash[a], Moonyong Lee[a, *]

[a] *School of Chemical Engineering, Yeungnam University, Gyeongsan 38541, Republic of Korea*
**mynlee@yu.ac.kr*

Abstract

Hydrogen (H_2) has gained vital importance in reducing greenhouse gas emissions. Due to its low energy density, the long-time storage and long-distance transportation of H2 is a challenging issue. Liquefaction is a practical approach to enhance energy density; however, it is an energy and cost-intensive technique. In addition, variation in H_2 composition (ortho H_2/para H_2) with respect to temperature is exothermic, adding heat load to the refrigeration cycle. Most conceptual studies do not incorporate this heat load in specific energy consumption (SEC). This study presents two case scenarios, i.e., with heat load (adiabatic) and without heat load (isothermal), to produce liquid H_2 (LH_2) through stepwise cooling with an ortho to para conversion. The proposed adiabatic and isothermal processes show SEC of 12.2 and 6.9 kWh/kg, respectively. The adiabatic process requires a large quantity of refrigerant flowrates, i.e., 46 kg/s, compared to the isothermal process (26 kg/s). In both cases, 100% saturated LH_2 is produced with 99.7 % para conversion. This study can help achieve a sustainable green energy economy by enhancing the overall H_2 value chain.

Keywords: Hydrogen liquefaction, Clean energy, Ortho-para conversion, Adiabatic process, Isothermal process.

1. Introduction

The greenhouse gas emissions have now become a prime focus of industrialists and environmentalists owing to their negative impact on the atmosphere. These emissions can be reduced by introducing and promoting clean energy. Hydrogen (H_2) is extensively considered a clean and multi-purpose energy carrier that can be used in every sector of life, such as transportation, electricity generation, and industry [1]. H_2 importance cannot be emphasized more than the global H_2 demand is increased by approx. threefold since 1975 [2]. Due to this increasing demand and clean characteristics, H_2 may take the place of conventional fuels in the near future. For that purpose, intercontinental or intercountry transportation of H_2 is vital for the global energy mix. Alongside transportation, long-time H_2 storage is also of primary importance, especially in the case of excess H_2 or excess electricity as both H_2 and electricity are interconvertible.

Commercially, H_2 is liquefied in a three-stage process: pre-cooling (up to –193 °C), cooling (up to –243 °C), and liquefaction (up to –253 °C) [3]. Liquid nitrogen (N_2) is used as a refrigerant in the pre-cooling phase, while a combination of JT valves or expanders and H_2 itself is used in the subsequent stages. In recent times, mixed refrigerants (MR) have been suggested as an alternative to pure liquid N_2. Numerous studies have suggested different compositions of low-boiling hydrocarbons and N_2 as constituents of MR [4].

Similarly, helium (He) and neon (Ne) are also considered alongside H_2 in the cooling and liquefaction stages. The main objective behind trying different refrigerants and process configurations is to reduce the specific energy consumption (SEC) and improve the liquefaction process's thermodynamic (exergy) efficiency [5]. Industrial processes have SEC in the range of 12–15 kWh/kg_{LH2} and efficiency as low as 20–30% [6]. In an ideal case, the SEC for hydrogen liquefaction is ~2.7 kWh/kg_{LH2} [7]. This large gap between the ideal and the real process motivates the researchers to design conceptual processes wherein different MR compositions, and process configurations have been proposed. The sole objective of such processes is to bring down the SEC value as close to the ideal case as possible.

The additional challenge associated with the LH_2 process is ortho to the para conversion of H_2 isomers. At 25 °C, H_2 constitutes 75% ortho and 25% para having opposite rotational movement. The energy level of the ortho state is higher than the para-state, which means that by decreasing temperature, ortho H_2 tends to convert to para H_2 to maintain equilibrium by releasing energy. However, this ortho-to-para conversion (OPC) is slow due to the low rate of reaction, i.e., 0.0114 h^{-1}, and exothermic in nature, which releases a large amount of energy high enough to evaporate LH_2 [8]. Recently, Riaz et al. [9] have discussed the significance of OPC reaction and its heat of reaction on the energy consumption for the production of LH_2. In another study, Son et al. evaluated different OPC reactor configurations to mimic the actual process configurations [10]. In commercial plants, catalysts are embedded inside heat exchangers. In this way, the heat of the reaction is dissipated inside the heat exchanger. Similarly, liquid N_2 and liquid H_2 are also used as fluid media for isothermal operation.

The OPC reaction heat is critical to the SEC of the process. It is pertinent to discuss the issues and solutions adopted by the researchers from a process systems viewpoint. The literature survey shows that many integral studies have employed isothermal OPC reactors ignoring the reaction heat. Only a few studies, such as Riaz et al. [9] and Son et al. [10], have discussed the importance of this OPC reaction heat. In the present study, two process configurations are proposed considering the heat of the reaction and process simplicity. The proposed process employs two refrigeration cycles and two reactor configurations: isothermal and adiabatic. Both these configurations are evaluated based on energy and exergy perspectives to evaluate their maximum thermodynamic potential.

2. H_2 liquefaction process

H_2 liquefaction is an energy-intensive process mainly because of low boiling point of H_2 and exothermic OPC reaction. The process description and simulation details are provided in following subsections.

2.1. Process description

The proposed process is divided into two cases based on the reactor configurations.
Case I: Isothermal process
Case II: Adiabatic process

The H_2 feed stream (n-GH_2) is fed to a cryogenic heat exchanger (CHX-1) for pre-cooling. In CHX-1, the temperature is decreased from 25°C to −193.2°C (see Figure 1). After that, H_2 is fed to the equilibrium reactor (R-1) to convert 25% p-H_2 to 43% p-H_2. In case I, stream H4 is fed to CHX-2, where the temperature is further reduced to −252°C. In case II, stream H4 temperature increases to −183.2°C due to exothermic OPC reaction (see Figure 2). For further cooling stream, H4 is sent back to the first heat exchanger, CHX-1, where the temperature is reduced to −193.2°C. Then stream H5 is fed to CHX-2 to cool it till −252°C. The outlet stream is fed to another equilibrium reactor (R-2) to convert

43% p-H_2 into 99.7% p-H_2. The outlet stream H8 is at the temperature of −252°C due to heat being recovered through energy stream (Q) with a heat flow of -394.6 kW. Finally, H_2 is liquefied at 21 bar. The pressure of stream H8 is reduced to the LH_2 storage pressure of 1.3 bar by passing it through the expander K-5. At the K-5 outlet, 100% saturated liquid H_2 is obtained.

Case I is shown in Figure 1. In case I, the precooling MR (PMR) consists of C_1, C_2, C_3, n-C_4, i-C_4, n-C_5, and N_2. The PMR is first compressed and cooled in a series of compressors and coolers. Then, the PMR is cooled further in CHX-1 and expanded in expander. The outlet stream of expander is used to reduce temperature of hot streams.

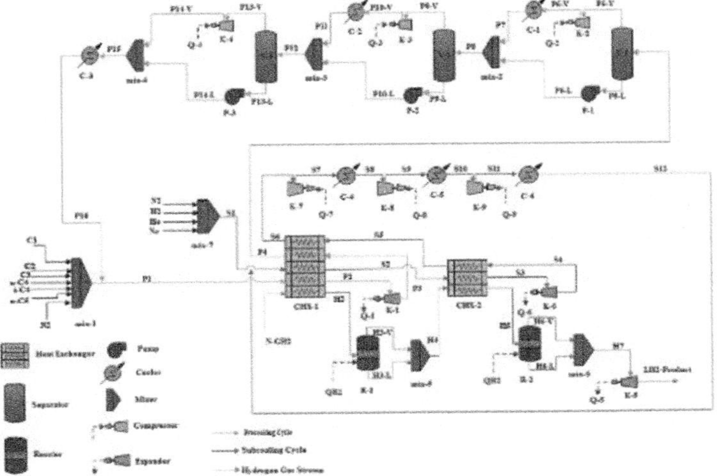

Figure 1 process flow diagram of case I

Case II is shown in Figure 2. In case II, the liquefaction MR (LMR) consists of H_2, He, and Ne. The LMR is first compressed and cooled in a series of compressors and coolers. Then, the LMR is cooled further in CHX-1 and -2 and expanded in expander. The outlet stream of expander is used to reduce temperature of hot streams in CHX-1 and -2.

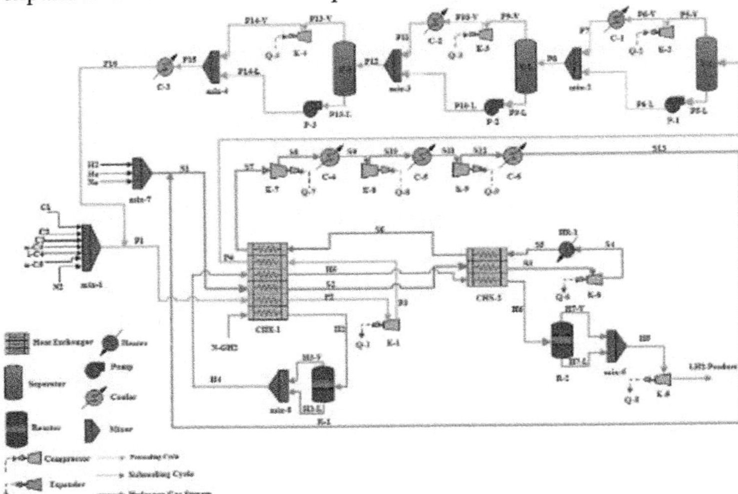

Figure 2 process flow diagram of case II

2.2. Process simulation

The proposed process was simulated in Aspen HYSYS v11 using the Peng–Robinson thermodynamic fluid package [11] for refrigerants and the Modified Benedict Webb Rubin equation of state for H_2. The feed conditions and design parameters were selected based on the base case, as shown in Table 1. The efficiency of the compressor and expander was adjusted to 75% and 80%, respectively.

Table 1 Design parameters for the H_2 liquefaction process [4]

Design parameters	Values
Feed temperature	25 °C
Feed pressure	21 bar
Feed flowrate	1 kg/s
Feed composition (mole fraction)	[oH_2/pH_2] = [0.75/0.25]
Pressure-drop across aftercoolers	0 bar
Pressure-drop across CHXs	0 bar
Temperature aftercooler	22 °C
H_2 liquefaction rate	100 %
Liquid H_2 pressure	1.3 bar

2.3. Energy analysis

The energy analysis of this study is conducted in terms of SEC. The design variables (Table 2), which include refrigerant mass flow rates, suction pressure, and discharge pressure of refrigeration cycles, are varied to observe their effect on the SEC.

2.4. Exergy analysis

Exergy analysis is an excellent tool for analyzing process inefficiencies based on the second law of thermodynamics. Exergy is of mainly two types: physical and chemical, which is calculated as:

$$ex_{ph,i} = (h_i - h_o) - T_o(s_i - s_o) \tag{1}$$

$$ex_{ch,i} = \sum x_i e_i^{CH} + RT_o \sum x_i \ln x_i \tag{2}$$

Where $ex_{ph,i}$ and $ex_{ch,i}$ is denoted as physical and chemical exergy, x_i is denoted as the mole fraction of ith component in a stream and e_i^{CH} is referred to as standard chemical exergy of ith component taken from [12]. The physical exergy values are taken from Aspen Hysys® stream properties, whereas chemical exergy values are calculated using equation 2. The exergy analysis is performed by calculating the exergy destruction of each piece of equipment. Exergy destruction of equipment explains the deviation of equipment performance from an ideal scenario. The higher the exergy destruction, the lower the process performance. The formulas used to calculate the exergy destruction of equipment can be found elsewhere [13].

3. Results and discussion: Process analysis

3.1. Energy analysis

In this study, the design variables, including refrigerant flow rates and suction/discharge pressures, were analyzed with respect to SEC. Table 2 lists the design variables, constraints, and the SECs of the base case and the proposed cases.

Table 2 Design variables and SEC of the base case, case I, and case II

PMR cycle	Base case	Case I	Case II
Suction / Discharge pressure (bar)	1.5 / 13.0	1.1 / 18.0	1.1 / 11.0
Refrigerants flowrate (kg/s)	84.7	19.3	31.4
CMR cycle			
Suction / Discharge pressure (bar)	2.9 / 35.6	-	-

Refrigerants flowrate (kg/s)	33.0	-	-
LMR cycle			
Suction / Discharge pressure (bar)	3.7 / 25.0	1.4 / 26.0	1.3 / 15.5
Refrigerants flowrate (kg/s)	1.9	7.1	16.4
SEC (kWh/kg$_{LH2}$)	**6.45**	**6.96**	**12.19**

According to Table 2, case II has the highest energy consumption of 12.19 kWh/kg$_{LH2}$ among all cases. Compared to the base case, SEC has increased by 8% and 53% in proposed cases I and II, respectively. Since the proposed cases have only two cycles (PMR and LMR), CMR will not be discussed. In addition, the suction pressure is reduced in PMR and LMR to withstand the MITA value within the exact range of 1-2 °C. Also, it is observed that the refrigeration load on the PMR cycle increases when suction pressure decreases. As a result, SEC increases. However, high suction pressure cannot achieve the desired cooling temperature. Therefore, the pre-cooling temperature was adjusted from -153.2 °C to -193.2 °C, reducing the LMR cycle's energy load. Moreover, the PMR cycle of the base case shows the highest MR flow rate (84.7 kg/s), with C_1, C_2, and C_4 contributing the most. On the other hand, the overall refrigerants flow rate is 78% and 60% less in the proposed case I and II, respectively.

Further, in proposed case I, PMR cycle, 19.26 kg/s of MR is utilized, whereas 31.382 kg/s is consumed in proposed case II. Moreover, all PMR refrigerants' flow rates are reduced in proposed cases, except C_3, N_2, and n-C_5, whose mass flow rate is increased. This is because only low boiling refrigerants are not energy efficient. So, an optimal blend of high and low boiling refrigerants is required to reduce energy consumption. In the LMR cycle, a new refrigerant was added, Ne, besides H_2 and He, to reach the required H_2 liquefaction temperature (-252.3°C). Here, a total mass flow rate of 7.06 kg/s and 16.4 kg/s is used in the proposed cases. Also, compared to the base case, a large amount of H_2 and He utilized 64.5% and 84% more in both proposed cases I and II, respectively.

3.2. Exergy analysis

Table 3 presents the exergy destruction values of each piece of equipment of the base case and the proposed cases.

Table 3 Equipment-wise exergy destruction of the base case, case I, and case II

Equipment	Case I (kW)	Case II (kW)
Compressors	4158.32	7542.97
Pumps	0.99	0.1
Coolers	4174.19	12083.33
Separators	195.36	-294.5
Reactors	-693.61	479.42
Expanders	20087.02	5582.97
CHXs	4824.24	16170.68
Total	**32746.5**	**41564.97**

The results of the proposed cases cannot be compared with the base case because of different process configurations. Therefore, the only case I and II are discussed in this section. In case-I, the expanders share 61% (20087.02 kW) of exergy destruction out of the total equipment destruction because their low-temperature operation promotes high exergy destruction. The CHXs show 4824.24 kW of exergy destruction. On the other hand, a high exergy destruction source in case-II caused by CHXs with 38.9 % (16170.68 kW). The main reason for exergy destruction in heat exchangers is the huge temperature difference between inlet and outlet streams. For example, in both cases, feed enters CHX-1 at 25°C, and its outlet temperature is -193.2°C (pre-cooling). Similarly, other streams also have a temperature gap of more than 50°C between in and out streams. Also, compressors share a significant amount of exergy destruction. Since OPC is an exothermic reaction, it contributes to exergy destruction as well. However, the energy gain from the OPC reaction is recycled, so it is calculated as inlet energy for reactors. Furthermore, the exergy destruction of reactors and phase separators in case-II shows a negative value which is thermodynamically not possible. Here the negative sign may appear due to the impact of other equipment. Since it is a cascaded process, all equipment is related to each other. Overall, case II has higher exergy destruction than case I.

4. Conclusions

In this study, the proposed case I is energy efficient as compared case II. The case II is energy intensive because of additional heat load of OPC reaction on refrigeration cycles. However, the case I is energy intensive as compared to the base case. Similarly, the exergy destruction analysis results conclude that the case I is 21.2% efficient than the case II. In case II, the CHXs are the major source of exergy destruction.

References

[1] IEA. GLOBAL TRENDS AND OUTLOOK FOR HYDROGEN. 2017.

[2] International Energy Agency (IEA). The Future of Hydrogen - Seizing today's opportunities. 2019. https://doi.org/10.1787/1e0514c4-en.

[3] Naquash A, Riaz A, Lee H, Qyyum MA, Lee S, Lam SS, et al. Hydrofluoroolefin-based mixed refrigerant for enhanced performance of hydrogen liquefaction process. Int J Hydrogen Energy 2022.

[4] Qyyum MA, Riaz A, Naquash A, Haider J, Qadeer K, Nawaz A, et al. 100% saturated liquid hydrogen production: Mixed-refrigerant cascaded process with two-stage ortho-to-para hydrogen conversion. Energy Convers Manag 2021;246:114659. https://doi.org/10.1016/j.enconman.2021.114659.

[5] Bi Y, Yin L, He T, Ju Y. Optimization and analysis of a novel hydrogen liquefaction process for circulating hydrogen refrigeration. Int J Hydrogen Energy 2021. https://doi.org/10.1016/J.IJHYDENE.2021.10.012.

[6] Krasae-in S, Stang JH, Neksa P. Development of large-scale hydrogen liquefaction processes from 1898 to 2009. Int J Hydrogen Energy 2010;35:4524–33. https://doi.org/10.1016/j.ijhydene.2010.02.109.

[7] Aasadnia M, Mehrpooya M. Large-scale liquid hydrogen production methods and approaches : A review. Appl Energy 2018;212:57–83. https://doi.org/10.1016/j.apenergy.2017.12.033.

[8] Sherif SA, Goswami DY, Stefanakos EK, Steinfeld A. Handbook of Hydrogen Energy. Taylor & Francis; 2014.

[9] Riaz A, Qyyum MA, Hussain A, Lee M. Significance of ortho-para hydrogen conversion in the performance of hydrogen liquefaction process. Int J Hydrogen Energy 2022. https://doi.org/10.1016/j.ijhydene.2022.09.022.

[10] Son H, Yu T, Hwang J, Lim Y. Simulation methodology for hydrogen liquefaction process design considering hydrogen characteristics. Int J Hydrogen Energy 2022;47:25662–78. https://doi.org/10.1016/J.IJHYDENE.2022.05.293.

[11] Peng D, Robinson DB. A New Two-Constant Equation of State. Ind Eng Chem Fundam 1976;15:59–64. https://doi.org/10.1021/i160057a011.

[12] Szargut J. Chemical exergies of the elements. Appl Energy 1989;32:269–86. https://doi.org/10.1016/0306-2619(89)90016-0.

[13] Naquash A, Qyyum MA, Islam M, Sial NR, Min S, Lee S, et al. Performance enhancement of hydrogen liquefaction process via absorption refrigeration and organic Rankine cycle-assisted liquid air energy system. Energy Convers Manag 2022;254:115200. https://doi.org/10.1016/j.enconman.2021.115200.